Electromagnetic Field Theory

About the Authors

Shankar Prasad Ghosh received his BE (Hons) in Electrical Engineering from National Institute of Technology, Durgapur and Master of Electrical Engineering degree from Jadavpur University, with specialisation in High Voltage Engineering.

Prof. Ghosh joined College of Engineering & Management, Kolaghat as Lecturer in 2002. He is presently working as Assistant Professor in the Department of Electrical Engineering at College of Engineering & Management, Kolaghat. He has authored and co-authored several books, all of which have been published by Tata McGraw-Hill. He has also published several papers in national and international conferences. He is currently pursuing his PhD. His areas of interest include electromagnetics, high voltage applications, and power systems among others.

Lipika Datta received her BE (Hons) in Electrical Engineering from National Institute of Technology, Durgapur, and ME in Computer Science and Engineering from West Bengal University of Technology with specialisation in Embedded Systems. She joined College of Engineering & Management, Kolaghat as Lecturer in 2006.

Prof. Datta is presently working as Assistant Professor in the Department of Computer Science and Engineering at College of Engineering & Management, Kolaghat. She is also pursuing her PhD at Calcutta University. Her areas of interest include embedded systems, operating systems, and computer architecture, among others.

Electromagnetic Field Theory

Shankar Prasad Ghosh
Assistant Professor
Department of Electrical Engineering
College of Engineering & Management, Kolaghat
Purba Medinipur, West Bengal

Lipika Datta
Assistant Professor
Department of Computer Science & Engineering and Information Technology
College of Engineering & Management, Kolaghat,
Purba Medinipur, West Bengal

Tata McGraw Hill Education Private Limited
NEW DELHI

McGraw-Hill Offices

New Delhi New York St Louis San Francisco Auckland Bogotá Caracas
Kuala Lumpur Lisbon London Madrid Mexico City Milan Montreal
San Juan Santiago Singapore Sydney Tokyo Toronto

Tata McGraw-Hill

Published by Tata McGraw Hill Education Private Limited,
7 West Patel Nagar, New Delhi 110 008.

Electromagnetic Field Theory

ISBN (13): 978-1-25-902951-6
ISBN (10): 1-25-902951-4

Vice President and Managing Director —McGraw-Hill Education: *Ajay Shukla*

Head—Higher Education Publishing and Marketing: *Vibha Mahajan*
Publishing Manager—SEM & Tech Ed.: *Shalini Jha*
Editorial Executive: *Koyel Ghosh*
Copy Editor: *Preyoshi Kundu*
Sr. Production Manager: *Satinder S Baveja*
Asst. Manager Production: *Anjali Razdan*

Marketing Manager—Higher Ed.: *Vijay Sarathi*
Sr. Product Specialist—SEM & Tech VOC.: *Tina Jajoriya*
Graphic Designer—Cover: *Meenu Raghav*

General Manager—Production: *Rajender P Ghansela*
Production Manager: *Reji Kumar*

Typeset at The Composers, 260, C.A. Apt., Paschim Vihar, New Delhi 110 063, and printed at
Pushp Print Services, B-39/12 A, Gali No.-1, Arjun Mohalla, Moujpur, Delhi – 110 053

Cover Printer: SDR Printers

RAXBCRCHDXDQY

The McGraw·Hill Companies

CONTENTS

PREFACE

Electromagnetic Field Theory is the study of characteristics of electric, magnetic and combined fields. The history of electromagnetism dates back over several thousand years, and in the twenty-first century, its applications are far reaching. Electromagnetism manifests as both electric fields and magnetic fields. In fact, all the forces involved in interactions between atoms can be explained by electromagnetic force, together with how these particles carry momentum by their movement. Use of any electric or magnetic or electromagnetic device involves the presence of electromagnetism. Its increased usefulness in science and engineering has made it gain an important position in various areas of technology or physical research.

Electromagnetic Field Theory is designed for undergraduate and postgraduate students studying electromagnetic field theory. The highlight of the book is its lucid and easy-to-understand language, with in-depth coverage of all the important topics sequentially which are supported by numerous illustrations. The book begins with a discussion on experimental laws and gradually synthesises them in the form of Maxwell's equations in an adoptive approach. The latter part of the book takes the axiomatic way of presentation, starting with Maxwell's equations, identifying each with the appropriate experimental law, and then specialising the general equations to static and time-varying situations for analysis.

Being one of the extreme events of the human intellect, electromagnetic theory is, therefore, the foundation of the technologies of electrical and computer engineering. The study of electromagnetism and electromagnetic field theory thus becomes imperative for all branches of engineering dealing with electricity and electronics and related applications.

Salient Features of the Book

- Simple and lucid language
- In-depth discussion on vector algebra and coordinate systems to build strong fundamentals for the course
- Comprehensive coverage of topics like electric and magnetic fields, wave propagation, wave guides and antenna
- Exhaustive treatment of electrostatics and electromagnetic waves and their applications
- Complexities of subject overcome through easy explanation, illustrative examples and simplified derivations
- Excellent pedagogy with several hundreds of solved examples, review questions, exercises and multiple-choice-questions.

Solved Examples: 300
Exercises: 157

Multiple Choice Questions: 189
Review Questions: 130

Illustrative examples are interspersed throughout the book at appropriate locations. Most questions have been selected carefully from different university question papers and competitive examinations. With so many years of teaching experience, we have found that such illustrations permit a level of understanding otherwise unattainable. As an aid to both, the instructor and the student, multiple-choice questions, review questions and the exercise problems provided at the end of each chapter progress from easy to hard levels.

Structure of the Book

The book is organized in seven chapters with the first chapter beginning with a discussion on vector which is the most basic requirement in the study of electromagnetism. In the subsequent chapters, the effects of non-varying electric charges, uniformly varying electric charges and varying electric charges with accelerating or decelerating velocity, have been discussed in detail. The concluding chapters cover some of the practical applications, such as transmission lines, waveguides.

Chapter 1 gives an introduction to vector analysis with an emphasis on scalar and vector fields, properties of vectors, vector algebra, coordinate systems, general curvilinear coordinates, differential elements, vector calculus, Gauss' divergence theorem, Stokes' theorem and classifications of vector fields. Since the book has been written with a vector approach of the fundamental quantities, and the knowledge of vector is indispensable for the study of electromagnetism, this chapter serves as a foundation for the study of electromagnetic theory.

Chapter 2 elaborates on the basic physics behind the interaction of static electric charges with the nature and associated phenomena. It explains electric fields and discusses electric charge, Coulomb's law, superposition of charges and electric fields, electric flux displacement and flux density, electric potential, potential gradient, equipotential surfaces, electric dipoles, Poisson's and Laplace's equations, uniqueness theorem, capacitance, method of images, electric boundary conditions and Dirac–Delta representations among others. **Chapter 3** provides the basis for understanding the physics behind the interaction of static magnetic charges, in the form of either magnets or moving electric charges, with the nature and associated phenomena. This chapter is divided into two parts. *Part I* deals with behavior of different electrical materials in an electric field and *Part II* deals with magnetostatic field. The Biot–Savart law and Ampere's law are discussed in this part. Chapters 2 and 3 together cover the concept of electrostatics to the core.

The following two chapters enlighten the concepts of electromagnetic fields and wave propagation. The functions of different practical electromagnetic devices and their accessories can be understood after completing these chapters. Both these chapters also open up the scope of different research activities related to electromagnetic wave phenomena. **Chapter 4** discusses electromagnetic fields with special emphasis on time-varying fields. The topics taken up are Faraday's law of induction, induced emf, and inconsistencies in Ampere's law, Maxwell's equations and potentials for electromagnetic fields. **Chapter 5** is on electromagnetic wave propagation, reflection, refraction and polarisation of electromagnetic waves. It also covers Helmholtz equation, properties of electromagnetic waves, standing electromagnetic waves, intrinsic impedance, propagation of uniform plane waves through different media, Poynting theorem and Poynting vector.

Chapter 6 covers transmission lines, their types, modes, parameters, equations, characteristic and input impedance, and the Smith chart. The concluding chapter, **Chapter 7**, explains waveguides and their properties and types—rectangular, circular-power losses and attenuation, cavity resonator and resonant cavities, dielectric slab waveguides, transmission line analogy, and finally, the applications of waveguides. Both these chapters describe two different applications of electromagnetic field theory—Transmission lines and waveguides, which provide the base for understanding the functionaries of many modern electromagnetic devices.

In addition, each chapter contains a summary and a list of important formulae for quick review. A large number of solved examples, short-answer-type question-answers, exercise problems, objective-type questions and review questions taken from different university question papers and other competitive examinations have also been included. These features make it an indispensable source of study of electromagnetic theory for students and practicing engineers alike. The book will help them improve their problem solving capability and also will guide them prepare for different competitive examinations.

Web Supplements

The following additional information is available at *http://www.mhhe.com/ghosh/eft1*

- Chapter on Antennas and Wave Propagation
- Additional Exercises

Acknowledgements

We are grateful to Prof. S N Bhadra, Professor, Department of Electrical Engineering, College of Engineering and Management, Kolaghat and Ex-Professor, IIT Kharagpur, Prof. C. K. Roy, Professor, Department of Electrical Engineering, College of Engineering and Management, Kolaghat and Ex-Professor, Jadavpur University and Prof. A. K. Chakraborty, Department of Electrical Engineering, NIT Agartala for their constant inspiration and encouragement in the field of academics.

We would also like to express our gratitude to the following reviewers for reviewing the book and providing constructive suggestions for the improvement of the book.

Amit Kant Pandit	*Mata Vaishno Devi University, Jammu and Kashmir*
Anirban Neogi	*Bengal Institute of Technology, West Bengal*
Shishir K Balita	*Silicon Institute of Technology, Odisha*
P G Polgawande	*Shri Shivaji Vidya Prasarak Sanstha's Bapusaheb Shivaji Rao Deore College of Engineering, Maharashtra*
M. Karthikeyan	*Paavai Engineering College, Tamilnadu*
C Murali	*Ramachandra College of Engineering, Andhra Pradesh*
Preeta Sharan	*Vemana Institute of Technology, Karnataka*
Amit Kant Pandit	*University College of Engineering, Rajasthan Technical University, Rajasthan*
Sanghamitra Chatterjee	*Camellia Institute of Technology, Kolkata*
Abhijit Lahiri	*Calcutta Institute of Engineering and Management, Kolkata*
K T Mathew	*Cochin University of Science and Technology, Kerala*
S C Raghavendra	*People's Education Society School of Engineering, Karnataka*

We are also thankful to the editorial and production staff of Tata McGraw Hill Education Private Limited for taking interest in publishing this edition. Last but not the least, we acknowledge the support offered by our family members, our daughter- Adrita, in particular, without whom this work would not have been successful.

S P Ghosh
L Datta

Feedback

Criticism and suggestions for improvement shall be gratefully acknowledged. Readers may contact the authors at *ghosh_shankar@rediffmail.com* and *lipika_13@rediffmail.com.*

Publisher's Note

Don't forget to write to us! We are always open to new ideas (the best ideas come from you!). You may send your comments to *tmh.elefeedback@gmail.com* (don't forget to mention the title and author name in the subject line). Piracy-related issues may be reported as well!

VISUAL WALKTHROUGH

Learning Objectives

This chapter deals with the following topics:

- *Sources of electrostatics*
- *Basic laws of electrostatics*
- *To acquire knowledge of fundamental quantities of electrostatics*
- *Boundary conditions in electrostatics*
- *Concepts of capacitance*

Learning Objectives offer an overview of chapter ideas. Each chapter opens with a list of objectives that will be discussed and explained throughout the chapter.

Introduction at the beginning of each chapter lays a foundation for the topics that have been explained in detail in the succeeding pages.

4.1 INTRODUCTION

In the previous chapters, we have studied different concepts of electrostatic and magnetostatic fields. In general, electrostatic fields are produced by stationary charges and magnetostatic fields are produced by motion of electric charges with uniform velocity (i.e. steady currents). However, if the current is time-varying, the field produced is also time-varying and is known as electromagnetic fields or waves.

In this chapter, we will discuss the concepts of electromagnetic fields and contribution of Maxwell to the laws of electromagnetism.

4.2 FARADAY'S LAW OF INDUCTION FOR TIME-VARYING FIELDS

English physicist Michael Faraday and American scientist Joseph Henry independently and simultaneously, in 1831, observed experimentally that any change in the magnetic environment of a coil of wire will cause a voltage (emf) to be induced in the coil. If the circuit is a closed one, this emf will cause flow of current. This phenomenon is known as *electromagnetic induction*. The results of Faraday and Henry's experiment led to two laws:

1. Neumann's Law: When a magnetic field linked with a coil or circuit is changed in any manner, the emf induced in the circuit is proportional to the rate of change of the flux-linkage with the circuit.

2. Lenz's Law: The direction of the induced emf is such that it will oppose the change of flux producing it.

Example 4.1 An infinite straight wire carries a current I is placed to the left of a rectangular loop of wire with width w and length l, as shown in Fig. 4.2.

(a) Determine the magnetic flux through the rectangular loop due to the current I.

(b) Suppose that the current is a function of time with $I(t) = a + bt$, where a and b are positive constants. What is the induced emf in the loop and the direction of the induced current?

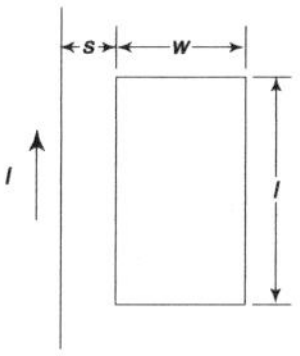

Fig. 4.2 *Rectangular loop near a wire*

Solution

(a) Using Ampere's law,

$$\oint \vec{B} \cdot d\vec{S} = \mu_0 I_{\text{enc}}$$

the magnetic field due to a current-carrying wire at a distance r away is

$$B = \frac{\mu_0 I}{2\pi r}$$

Illustrated Examples have been carefully selected from various university question papers and competitive examinations. These examples will help students achieve a level of understanding which is not possible through theory alone.

Notes offer additional information on the topics discussed in the book.

NOTE

Vector Representation of Surface

Differential surface area is defined as a vector whose magnitude corresponds to the area of the surface and whose direction is perpendicular to the surface.

$$\therefore \qquad d\vec{S} = dS\hat{a}_n$$

For closed surface, the outward normal direction is taken as positive direction. For an open surface, the normal created by positive periphery according to the right-hand cork screw rule is taken to be positive.

If the surface is not a plane, then the surface is subdivided into smaller elements that are considered to be plane and a vector surface is considered for each of these elemental surfaces. Vector addition of these elemental vector surfaces gives the total surface.

Summary

- Faraday's law states that the emf induced in a closed circuit is proportional to the rate of change of the magnetic flux-linkage and the direction of the current flow in the closed circuit is such that it opposes the change of the flux.

$$\mathcal{E} = -\frac{d\phi}{dt}$$

- Different forms of Faraday's law are:

Differential Form: $\nabla \times \vec{E} = -\frac{\partial \vec{B}}{\partial t}$

Integral Form: $\oint_C \vec{E} \cdot d\vec{l} = -\frac{d}{dt}\int_S \vec{B} \cdot d\vec{S}$

- Induced emf for different cases are as follows:

Stationary loop in time varying magnetic field (transformer emf)

$$\mathcal{E}_S = -\int_S \frac{\partial \vec{B}}{\partial t} \cdot d\vec{S}$$

Moving loop in static magnetic field (motional emf)

$$\mathcal{E}_m = \oint_C \vec{E}_m \cdot d\vec{l} = \oint_C (\vec{v} \times \vec{B}) \cdot d\vec{l}$$

Summary at the end of each chapter act as an overview of all the topics discussed in that particular chapter.

List of **Important Formulae** after Summary will help students review all the important formulae in a short period of time!

Important Formulae

Faraday's law	$\mathcal{E} = -\frac{d\phi}{dt}$
Integral form of Faraday's law	$\oint_C \vec{E} \cdot d\vec{l} = -\frac{d}{dt}\int_S \vec{B} \cdot d\vec{S}$
Differential form of Faraday's law	$\nabla \times \vec{E} = -\frac{\partial \vec{B}}{\partial t}$
Transformer EMF	$\mathcal{E}_S = -\int_S \frac{\partial \vec{B}}{\partial t} \cdot d\vec{S}$
Motional EMF	$\mathcal{E}_m = \oint_C (\vec{v} \times \vec{B}) \cdot d\vec{l}$
EMF induced in Faraday disc generator	$\mathcal{E}_m = \frac{1}{2}\omega B a^2$
Modified Ampere's circuital law	$\nabla \times \vec{H} = \vec{J} + \frac{\partial \vec{D}}{\partial t}$
Inhomogeneous wave equation	$\nabla^2 V - \mu\varepsilon \frac{\partial^2 V}{\partial t^2} = -\frac{\rho}{\varepsilon}$ and $\nabla^2 \vec{A} - \mu\varepsilon \frac{\partial^2 \vec{A}}{\partial t^2} = -\mu\vec{J}$

Exercises

[NOTE: * marked problems are important university problems]

- *Easy*

1. A conductor 1 cm in length is parallel to the z-axis and rotates at radius of 25 cm at 1200 r.p.m. Find the induced voltage, if the radial field is given by $\vec{B} = 0.5\vec{a}_r\ T$. [−157.08 mV]
2. A square conducting loop with sides 25 cm long is located in a magnetic field of 1 A/m varying at a frequency of 5 MHz. The field is perpendicular to the plane of the loop. What voltage will be read on a voltmeter connected in series with one side of the loop? [2.54 V]
3. A square loop of wire 25 cm has a voltmeter (of infinite impedance) connected in series with one side. Determine the voltage indicated by the meter when the loop is placed in an alternating field, the maximum intensity of which is 1 A/m. The plane of the loop is perpendicular to the magnetic field, the frequency is 10 MHz. [4.93 V]

Chapter-end **Exercises** have been divided into three levels as easy, medium and hard which will help students assess their level of understanding

Review Questions include theoretical questions to allow students to confirm their mastery of topics and concepts

Review Questions

[NOTE: * marked questions are important university questions.]

1. State and explain Faraday's law of electromagnetic induction.

*2. From the fundamental principle, establish the relation (i) $\nabla \times \vec{E} = -\frac{\partial \vec{B}}{\partial t}$ and (ii) $\nabla \times \vec{H} = \vec{J} + \frac{\partial \vec{D}}{\partial t}$.

3. Show that the electric field $\vec{E}$ induced by a time-varying magnetic field $\vec{B}$ is given by the expression $\nabla \times \vec{E} = -\frac{\partial \vec{B}}{\partial t}$.

Multiple Choice Questions

1. Maxwell's equations in differential form from Ampere's law are obtained from
   (a) M. M. F. area (b) Electric potential area
   (c) Magnetic flux volume (d) Electric current area.
2. Maxwell's equations are not completely symmetrical because
   (a) isolated magnetic charges do not exist.
   (b) it is difficult to get curl of a vector in spherical coordinates.
   (c) $\nabla \cdot \mathbf{D}$ is always zero.
   (d) $\nabla \times \mathbf{H}$ does not exist in free space.
3. A circular disc of 1 m radius rotates at 150 radians per second anticlockwise on ϕ-r plane with flux density $\mathbf{B} = 10^{-3}\boldsymbol{a}_z$ tesla. The voltage induced between two stationary brushes connected at the centre and at the circumference will be
   (a) 0.15 V (b) 0.075 V (c) 1.0 V (d) 0.85 V
4. A conductor 4 m long lies along the y-axis with a current 10.0 A in the a_y direction, the force on the conductor if the field in the region be $\boldsymbol{B} = 0.05\boldsymbol{a}_x$ T is
   (a) $-2.0\boldsymbol{a}_z$ N (b) $2.0\boldsymbol{a}_z$ N (c) $2.0\boldsymbol{a}_y$ N (d) $-2.0\boldsymbol{a}_x$ N

Set of **Multiple Choice Questions** at the end of each chapter serves as an exercise for students for quick understanding and analysis

1

VECTOR ANALYSIS

Learning Objectives

This chapter deals with the following topics:

- *Vector algebra and calculus*
- *Different laws of vector*

1.1 INTRODUCTION

Electromagnetics is the branch of physics or electrical engineering in which the electric and magnetic phenomena are studied. The basic knowledge for analysing the performance of any electrical network is the knowledge of circuit theory. However, the approach of circuit theory is a simplified approximation of a more exact field theory. Field theory is more difficult than circuit theory because of the larger number of variables involved. For most electromagnetic field problems, there are three space variables and thus, the solutions become complex.

This can be overcome by the use of vector analysis. Thus, knowledge of vector analysis is an essential prerequisite to the study of electromagnetic field theory. The use of vector analysis in the study of electromagnetic field theory results in less time for solutions.

1.2 SCALAR AND VECTOR QUANTITIES

A quantity that has only magnitude is said to be a *scalar* quantity. Examples of scalar quantities are time, mass, distance, temperature, work, electric potential, etc. Scalar quantities are represented by italic letters, e.g., A, B, a, b, and F.

A quantity that has both magnitude and direction is called a *vector* quantity. Examples of vector quantities are force, velocity, displacement, electric field intensity, etc. Vector quantities are represented by a letter with an arrow on the top, such as $\vec{A}$ and $\vec{B}$ or with a bold letter, such as **F**, **a**, **B**.

1.3 FIELDS

If at each point of a region, there is a value of some physical function, the region is called a *field*.

Fields may be classified as scalar fields and vector fields.

1.3.1 Scalar Fields

If the value of the physical function at each point is a scalar quantity, then the field is known as a *scalar field.*

Some examples of scalar fields are: temperature distribution in a building, sound intensity in a theatre, height of the surface of the earth above sea level and electric potential in a region.

A scalar field independent of time is called a *stationary* or *steady-state scalar field.*

1.3.2 Vector Fields

If the value of the physical function at each point is a vector quantity, then the field is known as a *vector field.*

Some examples of vector fields are: gravitational force on a body in space, wind velocity in the atmosphere and the force on a charge body placed in an electric field.

A time-independent vector is called a *stationary vector field.*

1.4 PROPERTIES OF VECTORS

We will consider the following essential properties that enable us to represent physical quantities as vectors:

1. Vectors can exist at any point in space.
2. Vectors have both the direction and the magnitude.
3. Any two vectors that have the same direction and magnitude are equal no matter where they are located in space; this is called *vector equality.*
4. *Unit vector*: A vector $\vec{A}$ has both magnitude and direction. The magnitude of $\vec{A}$ is a scalar written as A or $|\vec{A}|$. A *unit vector* $\hat{a}_A$ along $\vec{A}$ is defined as a vector whose magnitude is unity and its direction is along $\vec{A}$.

   In general, any vector can be represented by its magnitude and its direction as follows.

   $$\vec{A} = A\hat{a}_A = |\vec{A}|\,\hat{a}_A \tag{1.1}$$

   where A or $|\vec{A}|$ represents the magnitude of the vector and $\hat{a}_A$ direction of the vector $\vec{A}$. This $\hat{a}_A$ is called a unit vector.

   From Eq. (1.1), the unit vector is given as

   $$\hat{a}_A = \frac{\vec{A}}{A} = \frac{\vec{A}}{|\vec{A}|} \tag{1.2}$$

   Note that, $|\hat{a}_A| = 1$

   In Cartesian coordinates, the unit vectors along the three axes, e.g., x-axis, y-axis and z-axis, are represented as $(\hat{a}_x, \hat{a}_y, \hat{a}_z)$ or as $(\hat{i}, \hat{j}, \hat{k})$.
5. *Component vectors:* Any vector $\vec{A}$ in Cartesian coordinates may be represented as (A_x, A_y, A_z) or

   $$\vec{A} = A_x\hat{a}_x + A_y\hat{a}_y + A_z\hat{a}_z \tag{1.3}$$

   where A_x, A_y, A_z are called the *component vectors* in x, y and z directions, respectively.

   Using the Pythagorean Theorem, the magnitude of $\vec{A}$ is given as

   $$A = |\vec{A}| = \sqrt{A_x^2 + A_y^2 + A_z^2}$$

The unit vector along $\vec{A}$ is given as

$$\hat{a}_A = \frac{\vec{A}}{A} = \frac{\vec{A}}{|\vec{A}|} = \frac{A_x\hat{a}_x + A_y\hat{a}_y + A_z\hat{a}_z}{\sqrt{A_x^2 + A_y^2 + A_z^2}} \quad (1.4)$$

6. *Vector decomposition:* Choosing a coordinate system with an origin and axes, we can decompose any vector into component vectors along each coordinate axis. In Fig. 1.1, we choose Cartesian coordinates. A vector at P can be decomposed into the vector sum

$$\vec{A} = \vec{A}_x + \vec{A}_y + \vec{A}_z$$

where, $\vec{A}_x$ is the x-component vector pointing in the positive or negative x-direction, and $\vec{A}_y$ is the y-component vector pointing in the positive or negative y-direction and $\vec{A}_z$ is the z-component vector pointing in the positive or negative z-direction (Fig. 1.1).

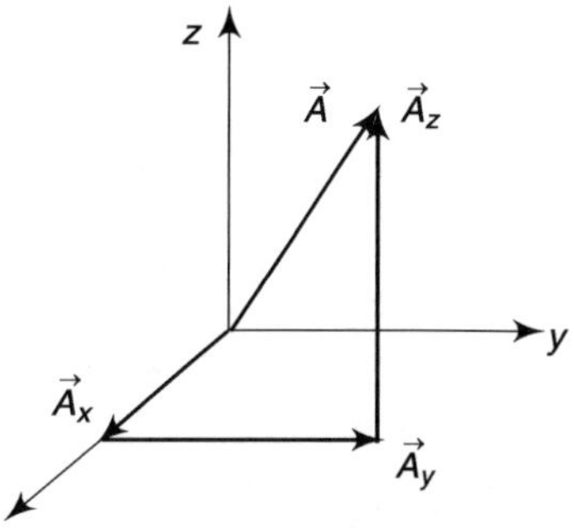

Fig. 1.1 *Vector decomposition*

7. *Direction angles and direction cosines of a vector:* The direction cosines of a vector are merely the cosines of the angles that the vector makes with the x, y, and z axes, respectively. We label these angles α (angle with the x-axis), β (angle with the y-axis), and γ (angle with the z-axis). Given a vector (A_x, A_y, A_z) in three-space, the *direction cosines* of this vector are given as

$$l = \cos\alpha = \frac{A_x}{\sqrt{A_x^2 + A_y^2 + A_z^2}} \quad (1.5a)$$

$$m = \cos\beta = \frac{A_y}{\sqrt{A_x^2 + A_y^2 + A_z^2}} \quad (1.5b)$$

$$n = \cos\gamma = \frac{A_z}{\sqrt{A_x^2 + A_y^2 + A_z^2}} \quad (1.5c)$$

Here, the *direction angles* α, β, γ are the angles that the vector makes with the positive x-, y- and z-axes, respectively. In formulas, it is usually the direction cosines that occur, rather than the direction angles. We have

$$\cos^2\alpha + \cos^2\beta + \cos^2\gamma = 1 \quad (1.5d)$$

Direction cosines define the orientation of a vector in three dimensions.

1.5 VECTOR ALGEBRA

We consider the addition, subtraction and multiplication of vectors.

1.5.1 Vector Addition and Subtraction

Let $\vec{A}$ and $\vec{B}$ be two vectors given as

$$\vec{A} = A_x\hat{a}_x + A_y\hat{a}_y + A_z\hat{a}_z$$

$$\vec{B} = B_x\hat{a}_x + B_y\hat{a}_y + B_z\hat{a}_z$$

We define a new vector, $\vec{C} = \vec{A} + \vec{B}$, the *vector addition* of $\vec{A}$ and $\vec{B}$ and is given as

$$\begin{aligned}\vec{C} &= (\vec{A} + \vec{B}) = (A_x\hat{a}_x + A_y\hat{a}_y + A_z\hat{a}_z) + (B_x\hat{a}_x + B_y\hat{a}_y + B_z\hat{a}_z)\\ &= (A_x + B_x)\hat{a}_x + (A_y + B_y)\hat{a}_y + (A_z + B_z)\hat{a}_z\end{aligned}$$

Similarly, we define a new vector, $\vec{D} = \vec{A} - \vec{B}$, the *vector subtraction* of $\vec{A}$ and $\vec{B}$ and is given as

$$\begin{aligned}\vec{D} &= (\vec{A} - \vec{B}) = (A_x\hat{a}_x + A_y\hat{a}_y + A_z\hat{a}_z) - (B_x\hat{a}_x + B_y\hat{a}_y + B_z\hat{a}_z)\\ &= (A_x - B_x)\hat{a}_x + (A_y - B_y)\hat{a}_y + (A_z - B_z)\hat{a}_z\end{aligned}$$

Physically, vector subtraction $(\vec{A} - \vec{B})$ is the addition of vector $\vec{A}$ and vector $\vec{B}$ after reversing the direction of vector $\vec{B}$.

Graphically, vector addition and subtraction are obtained by the triangle or parallelogram rules as explained below.

Triangular Rule of Vector Addition An arrow is drawn that represents the vector $\vec{A}$. The tail of the arrow that represents the vector $\vec{B}$ is placed at the tip of the arrow for $\vec{A}$ as shown in Fig. 1.2 (*a*). The arrow that starts at the tail of $\vec{A}$ and goes to the tip of $\vec{B}$ is defined to be the *vector addition*, $\vec{C} = \vec{A} + \vec{B}$. This is the *triangle rule of vector addition.*

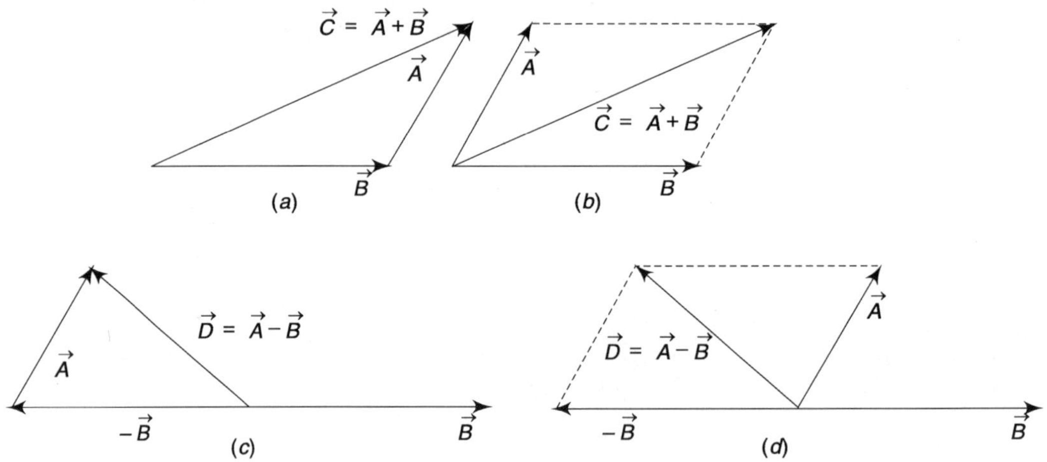

Fig. 1.2 *(a) Vector addition triangle rule, (b) vector addition parallelogram rule, (c) vector subtraction triangle rule, and (d) vector subtraction parallelogram rule*

Parallelogram Rule of Vector Addition The vectors $\vec{A}$ and $\vec{B}$ can be drawn with their tails at the same point. The two vectors form the sides of a parallelogram. The diagonal of the parallelogram corresponds to the vector $\vec{C} = \vec{A} + \vec{B}$, as shown in Fig. 1.2 (*b*). This is the *parallelogram rule of vector addition.*

Vector addition and subtraction satisfies the following properties:

1. *Commutivity:* The order of adding vectors does not matter.

$$\vec{A} + \vec{B} = \vec{B} + \vec{A}$$

2. *Associativity:* When adding three vectors, it does not matter which two we start with

$$\vec{A} + (\vec{B} + \vec{C}) = (\vec{A} + \vec{B}) + \vec{C}$$

3. *Identity element for vector addition:* There is a unique vector, 0, that acts as an identity element for vector addition.
   This means that for all vectors $\vec{A}$,

$$\vec{A} + \vec{0} = \vec{0} + \vec{A} = \vec{A}$$

4. *Inverse element for vector addition:* For every vector $\vec{A}$, there is a unique inverse vector

$$(-1)\vec{A} \equiv -\vec{A}$$

such that $$\vec{A} + (-\vec{A}) = 0$$

   This means that the vector $-\vec{A}$ has the same magnitude as $\vec{A}$, i.e., $|\vec{A}| = |-\vec{A}| = A$; but they point in opposite directions.

5. *Distributive law for vector addition:* Vector addition satisfies a distributive law for multiplication by a number.
   Let c be a real number. Then

$$c(\vec{A} + \vec{B}) = c\vec{A} + c\vec{B}$$

1.5.2 Vector Multiplication or Product

When two vectors $\vec{A}$ and $\vec{B}$ are multiplied, the result may be a scalar or a vector depending on how they are multiplied. There are two types of vector multiplication:

1. Scalar or Dot Product, and
2. Vector or Cross Product.

1.5.3 Scalar or Dot Product ($\vec{A} \cdot \vec{B}$)

Definition The scalar or dot product of two vectors $\vec{A}$ and $\vec{B}$, written as $\vec{A} \cdot \vec{B}$, is defined as (Fig. 1.3)

$$\vec{A} \cdot \vec{B} = AB \cos \theta_{AB} \tag{1.6}$$

where θ_{AB} is the smaller angle between $\vec{A}$ and $\vec{B}$,

$A = |\vec{A}|$ and $B = |\vec{B}|$ represent the magnitude of $\vec{A}$ and $\vec{B}$, respectively.

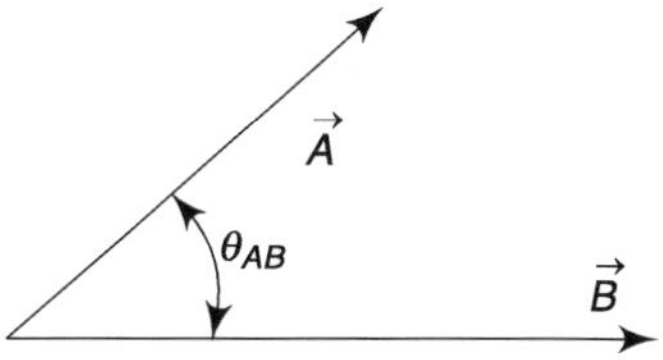

Fig. 1.3 *Dot product of two vectors*

The dot product can be positive, zero, or negative, depending on the value of cos θ_{AB}. The result of $\vec{A} \cdot \vec{B}$ is always a scalar quantity.

Dot product is geometrically defined as the product of magnitude of $\vec{B}$ and the projection of $\vec{A}$ onto $\vec{B}$ or vice versa. This is illustrated in Fig. 1.4 (*a*) and (*b*).

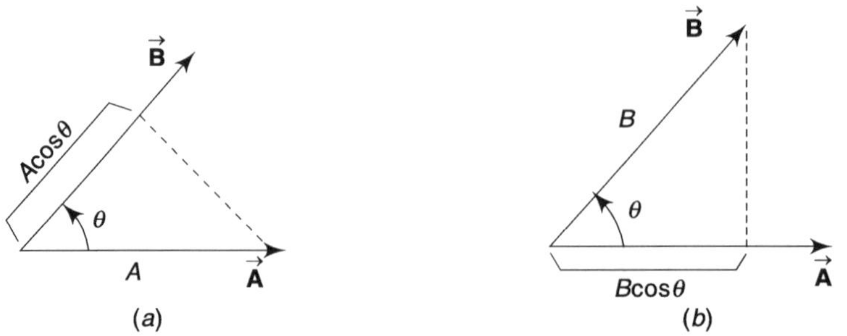

Fig. 1.4 *Projection of vectors and the dot product*

Properties of Dot Product

1. The first property is that the dot product is *commutative*

$$\vec{A}\cdot\vec{B}=\vec{B}\cdot\vec{A}$$

2. The second property involves the dot product between a vector $c\vec{A}$ which is a scalar and a vector $\vec{B}$

$$c\vec{A}\cdot\vec{B}=c(\vec{A}\cdot\vec{B})$$

3. The third property involves the dot product between the sum of two vectors $\vec{A}$ and $\vec{B}$ with a vector $\vec{C}$

$$(\vec{A}+\vec{B})\cdot\vec{C}=\vec{A}\cdot\vec{C}+\vec{B}\cdot\vec{C}$$

This shows that the dot product is *distributive*.

4. Since the dot product is commutative, similar relations are given, e.g.,

$$\vec{A}\cdot c\vec{B}=c(\vec{A}\cdot\vec{B})$$
$$\vec{C}\cdot(\vec{A}+\vec{B})=\vec{C}\cdot\vec{A}+\vec{C}\cdot\vec{B}$$

Vector Decomposition and Dot Product We now develop an algebraic expression for the dot product in terms of components. We choose a Cartesian coordinate system with the two vectors having component vector as

$$\vec{A}=A_x\hat{a}_x+A_y\hat{a}_y+A_z\hat{a}_z$$
$$\vec{B}=B_x\hat{a}_x+B_y\hat{a}_y+B_z\hat{a}_z$$

$$\begin{aligned}\therefore\quad \vec{A}\cdot\vec{B}&=(A_x\hat{a}_x+A_y\hat{a}_y+A_z\hat{a}_z)\cdot(B_x\hat{a}_x+B_y\hat{a}_y+B_z\hat{a}_z)\\&=A_xB_x(\hat{a}_x\cdot\hat{a}_x)+A_yB_y(\hat{a}_y\cdot\hat{a}_y)+A_zB_z(\hat{a}_z\cdot\hat{a}_z)+A_xB_y(\hat{a}_x\cdot\hat{a}_y)+A_xB_z(\hat{a}_x\cdot\hat{a}_z)\\&\quad+A_yB_x(\hat{a}_y\cdot\hat{a}_x)+A_yB_z(\hat{a}_y\cdot\hat{a}_z)+A_zB_x(\hat{a}_z\cdot\hat{a}_x)+A_zB_y(\hat{a}_z\cdot\hat{a}_y)\end{aligned}$$

Now, $\hat{a}_x\cdot\hat{a}_x=\hat{a}_y\cdot\hat{a}_y=\hat{a}_z\cdot\hat{a}_z=1\cos(0°)=1$

and $\hat{a}_x\cdot\hat{a}_y=\hat{a}_y\cdot\hat{a}_z=\hat{a}_z\cdot\hat{a}_x=\hat{a}_y\cdot\hat{a}_x=\hat{a}_z\cdot\hat{a}_y=\hat{a}_x\cdot\hat{a}_z=1\cos(90°)=0$

Hence, we get

$$\boxed{\vec{A}\cdot\vec{B}=(A_xB_x+A_yB_y+A_zB_z)} \quad \textbf{(1.7)}$$

Application of Dot Product

1. The dot product is to find the work done by a force $\vec{F}$ for a displacement of $\vec{D}$, given as

$$W = \vec{F} \cdot \vec{D} = FD \cos \theta$$

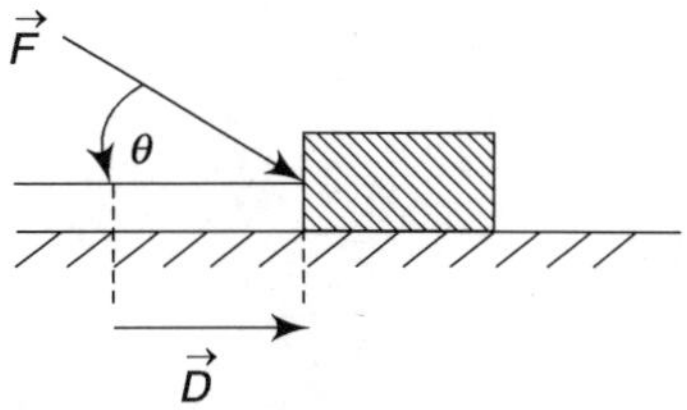

Fig. 1.5 *Work done by a force*

2. It is used to find out the line integral of a vector over a path, e.g., to calculate the electric potential between two points in an electric field.
3. It is used to find out the surface integral over a surface, e.g., to calculate the total charge enclosed by a surface placed in an electric field.

Example 1.1 Given the two vectors

$$\vec{A} = -7\hat{a}_x + 12\hat{a}_y + 3\hat{a}_z \text{ and } \vec{B} = 4\hat{a}_x - 2\hat{a}_y + 16\hat{a}_z$$

Find the dot product and the angle between the two vectors.

Solution The dot product between the two vectors is given as

$$\vec{A} \cdot \vec{B} = (A_x B_x + A_y B_y + A_z B_z) = (-7) \times 4 + 12 \times (-2) + 3 \times 16 = -4$$

Since the dot product is negative, it is expected that the angle between the two vectors will be greater than 90°.

Here, $|\vec{A}| = \sqrt{(-7)^2 + 12^2 + 3^2} = \sqrt{202}$

$|\vec{B}| = \sqrt{4^2 + (-2)^2 + 16^2} = \sqrt{276}$

Hence, the angle between the two vectors is given as,

$$\theta = \cos^{-1}\left(\frac{\vec{A} \cdot \vec{B}}{|\vec{A}||\vec{B}|}\right) = \cos^{-1}\left(\frac{-4}{\sqrt{202 \times 276}}\right) = 90.97°$$

1.5.4 Vector or Cross Product ($\vec{A} \times \vec{B}$)

Definition The cross product of two vectors $\vec{A}$ and $\vec{B}$, written as $(\vec{A} \times \vec{B})$, is defined as

$$\vec{A} \times \vec{B} = AB \sin \theta_{AB} \hat{a}_n \quad \textbf{(1.8)}$$

where $\hat{a}_n$ is the unit vector normal to the plane containing $\vec{A}$ and $\vec{B}$

The vector multiplication is called cross product due to the cross sign. It is also termed as vector product because the result is a vector.

The direction of the cross product is obtained from a common rule, called *right-hand rule.*

Right-hand Rule for Direction of Cross Product The direction of $\hat{a}_n$ is taken as the direction of right thumb when the fingers of the right hand rotate from $\vec{A}$ to $\vec{B}$ [Fig. 1.6 (*a*)]. Alternatively, the direction of $\hat{a}_n$ is taken as that of the advance of right-handed screw as $\vec{A}$ is turned into $\vec{B}$ [Fig. 1.6 (*b*)].

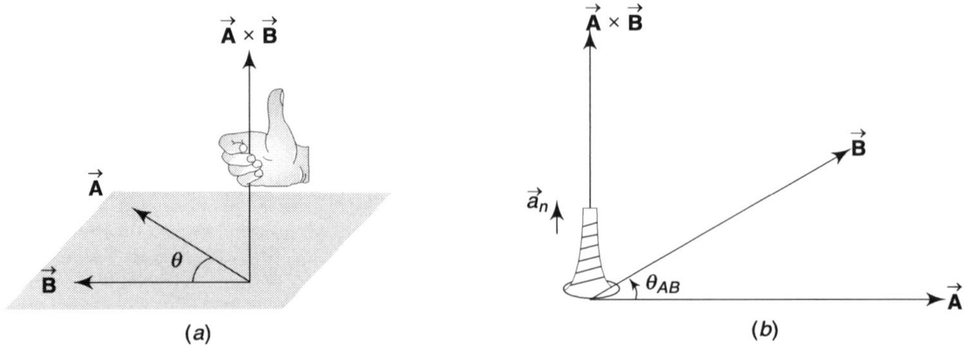

Fig. 1.6 *(a) Right-hand rule, and (b) Right-hand cork-screw rule*

We can give a geometric interpretation to the magnitude of the cross product by writing the definition as

$$|\vec{A} \times \vec{B}| = A(B \sin \theta)$$

The vectors $\vec{A}$ and $\vec{B}$ form a parallelogram. The area of the parallelogram equals the height times the base, which is the magnitude of the cross product. In Fig. 1.7, two different representations of the height and base of a parallelogram are illustrated. As depicted in Fig. 1.7 (*a*), the term $B \sin \theta$ is the projection of the vector $\vec{B}$ in the direction perpendicular to the vector $\vec{A}$.

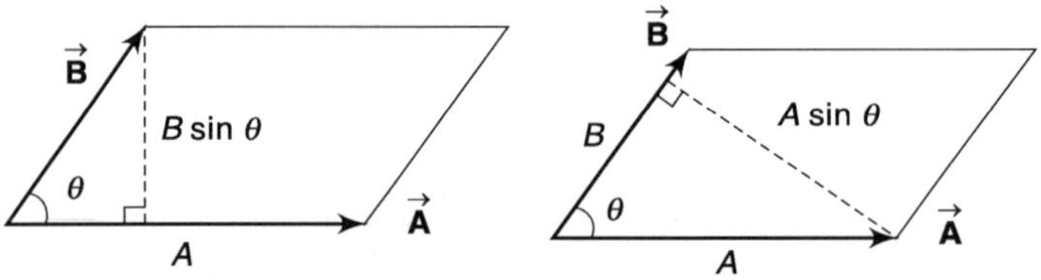

Fig. 1.7 *Projection of vectors and the cross product*

We could also write the magnitude of the cross product as

$$|\vec{A} \times \vec{B}| = (A \sin \theta)B$$

Now the term $A \sin \theta$ is the projection of the vector $\vec{A}$ in the direction perpendicular to the vector $\vec{B}$ as shown in Fig. 1.7(b).

Properties of Cross Product

1. The cross product is anti-commutative since changing the order of the *vector's* cross product changes the direction of the cross product vector by the right-hand rule:

$$\vec{A} \times \vec{B} = -\vec{B} \times \vec{A}$$

2. The cross product between a vector $c\vec{A}$ where c is a scalar and a vector $\vec{B}$ is

$$c\vec{A} \times \vec{B} = c(\vec{A} \times \vec{B})$$

Similarly, $$\vec{A} \times c\vec{B} = c(\vec{A} \times \vec{B})$$

3. The cross product is not associative.

$$\vec{A} \times (\vec{B} \times \vec{C}) \neq (\vec{A} \times \vec{B}) \times \vec{C}$$

4. The cross product between the sum of two vectors $\vec{A}$ and $\vec{B}$ with a vector $\vec{C}$ is,

$$(\vec{A}+\vec{B})\times\vec{C} = \vec{A}\times\vec{C} + \vec{B}\times\vec{C}$$

Similarly,
$$\vec{A}\times(\vec{B}+\vec{C}) = \vec{A}\times\vec{B} + \vec{A}\times\vec{C}$$

This shows that the cross product is distributive.

Vector Decomposition and Cross Product We now develop an algebraic expression for the cross product in terms of components. We choose a Cartesian coordinate system with the two vectors having component vector as

$$\vec{A} = A_x\hat{a}_x + A_y\hat{a}_y + A_z\hat{a}_z$$
$$\vec{B} = B_x\hat{a}_x + B_y\hat{a}_y + B_z\hat{a}_z$$

$$\therefore \quad \vec{A}\times\vec{B} = (A_x\hat{a}_x + A_y\hat{a}_y + A_z\hat{a}_z)\times(B_x\hat{a}_x + B_y\hat{a}_y + B_z\hat{a}_z)$$
$$= A_xB_x(\hat{a}_x\times\hat{a}_x) + A_yB_y(\hat{a}_y\times\hat{a}_y) + A_zB_z(\hat{a}_z\times\hat{a}_z) + A_xB_y(\hat{a}_x\times\hat{a}_y) + A_xB_z(\hat{a}_x\times\hat{a}_z)$$
$$+ A_yB_x(\hat{a}_y\times\hat{a}_x) + A_yB_z(\hat{a}_y\times\hat{a}_z) + A_zB_x(\hat{a}_z\times\hat{a}_x) + A_zB_y(\hat{a}_z\times\hat{a}_y)$$

Now, for right-handed coordinate system

$$\hat{a}_x\times\hat{a}_x = \hat{a}_y\times\hat{a}_y = \hat{a}_z\times\hat{a}_z = 0$$

and, $\hat{a}_x\times\hat{a}_y = \hat{a}_z = -\hat{a}_y\times\hat{a}_x \quad \hat{a}_y\times\hat{a}_z = \hat{a}_x = -\hat{a}_z\times\hat{a}_y \quad \hat{a}_z\times\hat{a}_x = \hat{a}_y = -\hat{a}_x\times\hat{a}_z$

This is illustrated in Fig. 1.8 (*a*) and (*b*).

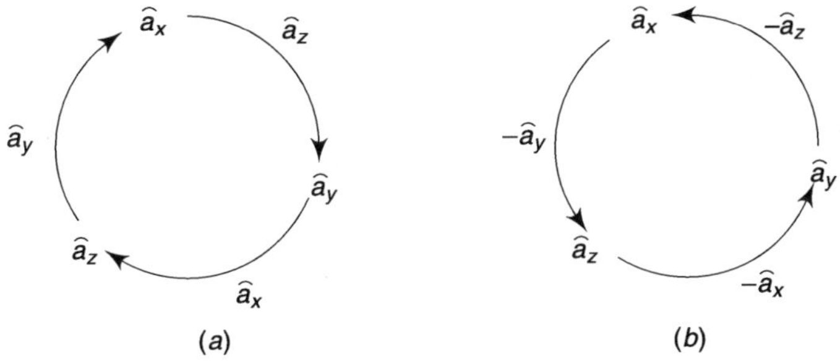

Fig. 1.8 *(a) Moving clockwise leads to positive results, and (b) Moving counterclockwise leads to negative results*

$$\therefore \quad \vec{A}\times\vec{B} = (A_yB_z - A_zB_y)\hat{a}_x + (A_zB_x - A_xB_z)\hat{a}_y + (A_xB_y - A_yB_x)\hat{a}_z$$
$$= \begin{vmatrix} \hat{a}_x & \hat{a}_y & \hat{a}_z \\ A_x & A_y & A_z \\ B_x & B_y & B_z \end{vmatrix}$$

Hence, we get

$$\boxed{\vec{A}\times\vec{B} = \begin{vmatrix} \hat{a}_x & \hat{a}_y & \hat{a}_z \\ A_x & A_y & A_z \\ B_x & B_y & B_z \end{vmatrix}} \qquad \textbf{(1.9)}$$

Example 1.2 Given the two vectors

$$\vec{A} = 8\hat{a}_x + 3\hat{a}_y - 10\hat{a}_z \text{ and } \vec{B} = -15\hat{a}_x + 6\hat{a}_y + 17\hat{a}_z$$

Find the cross product between the two vectors and the unit vector normal to the plane containing the two vectors $\vec{A}$ and $\vec{B}$.

Solution The cross product between the two vectors is given as

$$\vec{A}\times\vec{B} = \begin{vmatrix} \hat{a}_x & \hat{a}_y & \hat{a}_z \\ A_x & A_y & A_z \\ B_x & B_y & B_z \end{vmatrix} = (A_yB_z - A_zB_y)\hat{a}_x + (A_zB_x - A_xB_z)\hat{a}_y + (A_xB_y - A_yB_x)\hat{a}_z$$

$$= [3\times 17 - (-10)\times 6]\hat{a}_x + [(-10)\times(-15) - 8\times 17]\hat{a}_y + [8\times 6 - 3\times(-15)]\hat{a}_z$$

$$= 111\hat{a}_x + 14\hat{a}_y + 93\hat{a}_z$$

The unit vector normal to the plane containing the vectors $\vec{A}$ and $\vec{B}$ is given as

$$\hat{a}_n = \frac{\vec{A}\times\vec{B}}{|\vec{A}||\vec{B}|} = \frac{111\hat{a}_x + 14\hat{a}_y + 93\hat{a}_z}{\sqrt{8^2+3^2+(-10)^2}\sqrt{(-15)^2+6^2+17^2}} = \frac{111\hat{a}_x + 14\hat{a}_y + 93\hat{a}_z}{308.46}$$

$$= 0.36\hat{a}_x + 0.04\hat{a}_y + 0.30\hat{a}_z$$

Example 1.3 Two vectors are represented by $\vec{A} = 2\hat{i} + 2\hat{j}$, $\vec{B} = 3\hat{i} + 4\hat{j} - 2\hat{k}$. Find the dot and cross-products and the angle between the vectors. Show that $\vec{A}\times\vec{B}$ is at right angle to $\vec{A}$.

Solution Here, $\vec{A} = 2\hat{i} + 2\hat{j}$, $\quad \vec{B} = 3\hat{i} + 4\hat{j} - 2\hat{k}$

$\therefore$
$$\vec{A}\cdot\vec{B} = 2\times 3 + 2\times 4 + 0\times(-2) = 14$$

$\therefore$
$$\vec{A}\times\vec{B} = \begin{vmatrix} \hat{a}_x & \hat{a}_y & \hat{a}_z \\ A_x & A_y & A_z \\ B_x & B_y & B_z \end{vmatrix} = \begin{vmatrix} \hat{a}_x & \hat{a}_y & \hat{a}_z \\ 2 & 2 & 0 \\ 3 & 4 & -2 \end{vmatrix} = -4\hat{a}_x + 4\hat{a}_y + 2\hat{a}_z$$

Now,
$$A = \sqrt{2^2 + 2^2 + 0} = 2.83; \qquad B = \sqrt{3^2 + 4^2 + (-2)^2} = 5.39$$

$\because$
$$\vec{A}\cdot\vec{B} = AB\cos\theta$$

$\Rightarrow$
$$\theta = \cos^{-1}\left(\frac{\vec{A}\cdot\vec{B}}{AB}\right) = \cos^{-1}\left(\frac{14}{2.83\times 5.39}\right) = 23.2^\circ$$

For $\vec{A}\times\vec{B}$ to be at right angle to $\vec{A}$, $(\vec{A}\times\vec{B})\cdot\vec{A}$ should be zero.

$$(\vec{A}\times\vec{B})\cdot\vec{A} = 2\times(-4) + 2\times 4 + 0\times 2 = 0$$ **(Proved)**

Applications of Cross Product

1. To find the torque about a point *P* which can be described mathematically by the cross product of a vector from *P* to where the force acts, and the force vector.
2. To find the force experienced by a current carrying conductor placed in a magnetic field.

1.5.5 Triple Products

Multiplication of three vectors $\vec{A}$, $\vec{B}$ and $\vec{C}$ is called *vector triple product*. The product of three vectors is classified into two categories:

1. Scalar triple product, and
2. Vector triple product.

Scalar Triple Product For the three vectors $\vec{A}$, $\vec{B}$ and $\vec{C}$, scalar triple product is defined as

$$\vec{A}\cdot(\vec{B}\times\vec{C}) = \vec{B}\cdot(\vec{C}\times\vec{A}) = \vec{C}\cdot(\vec{A}\times\vec{B})$$

Since the result is a scalar quantity, this is known as *scalar triple product*.

If the components of three vectors $\vec{A}$, $\vec{B}$ and $\vec{C}$ are given as $\vec{A}=(A_x, A_y, A_z)$, $\vec{B}=(B_x, B_y, B_z)$, $\vec{C}=(C_x, C_y, C_z)$, respectively, then the scalar triple product is obtained by the determinant of a 3×3 matrix given as

$$\boxed{\vec{A}\cdot(\vec{B}\times\vec{C}) = \begin{vmatrix} A_x & A_y & A_z \\ B_x & B_y & B_z \\ C_x & C_y & C_z \end{vmatrix}}$$

Similarly, $\vec{B}\cdot(\vec{C}\times\vec{A}) = \begin{vmatrix} B_x & B_y & B_z \\ C_x & C_y & C_z \\ A_x & A_y & A_z \end{vmatrix}$

and $\vec{C}\cdot(\vec{A}\times\vec{B}) = \begin{vmatrix} C_x & C_y & C_z \\ A_x & A_y & A_z \\ B_x & B_y & B_z \end{vmatrix}$

We know from a determinant theorem that if any two columns or rows of a determinant are interchanged, its value remains the same, but the sign changes. Hence, we can write

$$\vec{B}\cdot(\vec{C}\times\vec{A}) = \begin{vmatrix} B_x & B_y & B_z \\ C_x & C_y & C_z \\ A_x & A_y & A_z \end{vmatrix} = -\begin{vmatrix} B_x & B_y & B_z \\ A_x & A_y & A_z \\ C_x & C_y & C_z \end{vmatrix} = (-)(-)\begin{vmatrix} A_x & A_y & A_z \\ B_x & B_y & B_z \\ C_x & C_y & C_z \end{vmatrix} = \begin{vmatrix} A_x & A_y & A_z \\ B_x & B_y & B_z \\ C_x & C_y & C_z \end{vmatrix}$$

Similarly,

$$\vec{C}\cdot(\vec{A}\times\vec{B}) = \begin{vmatrix} C_x & C_y & C_z \\ A_x & A_y & A_z \\ B_x & B_y & B_z \end{vmatrix} = -\begin{vmatrix} B_x & B_y & B_z \\ A_x & A_y & A_z \\ C_x & C_y & C_z \end{vmatrix} = (-)(-)\begin{vmatrix} A_x & A_y & A_z \\ B_x & B_y & B_z \\ C_x & C_y & C_z \end{vmatrix} = \begin{vmatrix} A_x & A_y & A_z \\ B_x & B_y & B_z \\ C_x & C_y & C_z \end{vmatrix}$$

Therefore, we can write that

$$\boxed{\vec{A}\cdot(\vec{B}\times\vec{C}) = \vec{B}\cdot(\vec{C}\times\vec{A}) = \vec{C}\cdot(\vec{A}\times\vec{B})} \quad \textbf{(1.10)}$$

Example 1.4 If the components of three vectors $\vec{A}$, $\vec{B}$ and $\vec{C}$ are given as,

$$\vec{A} = (A_x, A_y, A_z),$$
$$\vec{B} = (B_x, B_y, B_z),$$
$$\vec{C} = (C_x, C_y, C_z)$$

respectively, then show that $\vec{A}\cdot(\vec{B}\times\vec{C})$ is the volume of a parallelepiped of sides $\vec{A}$, $\vec{B}$ and $\vec{C}$.

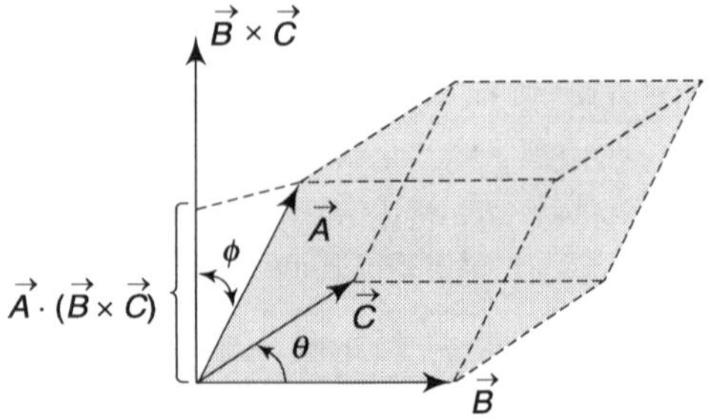

Fig. 1.9 *Scalar triple product*

Solution Here,

$$\vec{B}\times\vec{C} = BC\sin\theta\,\hat{a}_n = \text{Area of the rectangle with side } B \text{ and } C \text{ and perpendicular to the plane containing the vectors } \vec{B} \text{ and } \vec{C}$$
$$= P\hat{a}_n = \vec{P}$$

$\therefore$ $\vec{A}\cdot(\vec{B}\times\vec{C}) = \vec{A}\cdot\vec{P} = AP\cos\phi$

= Product of the component of A normal to BC-plane and the area of the parallelogram with sides B and C

= Volume of the parallelepiped of sides A, B and C

Vector Triple Product For the three vectors $\vec{A}$, $\vec{B}$ and $\vec{C}$, vector triple product is defined as

$$\vec{A}\times(\vec{B}\times\vec{C}) = \vec{B}(\vec{A}\cdot\vec{C}) - \vec{C}(\vec{A}\cdot\vec{B}) \tag{1.11}$$

Since the result is a vector quantity, this is known as *vector triple product*. It may be noted that $\vec{A}\times(\vec{B}\times\vec{C})$ is in the plane containing $\vec{B}$ and $\vec{C}$ and is perpendicular to $\vec{A}$.

Associative law does not hold good for vector triple product, i.e.,

$$\vec{A}\times(\vec{B}\times\vec{C}) \neq (\vec{A}\times\vec{B})\times\vec{C}$$

Rather,

$$(\vec{A}\times\vec{B})\times\vec{C} = -\vec{C}\times(\vec{A}\times\vec{B}) = -[\vec{A}(\vec{C}\cdot\vec{B}) - \vec{B}(\vec{C}\cdot\vec{A})] = \vec{B}(\vec{C}\cdot\vec{A}) - \vec{A}(\vec{C}\cdot\vec{B})$$

The concept of vector triple product is used in deriving wave equations from Maxwell's equations.

1.6 COORDINATE SYSTEMS

Coordinate system is defined as a system to describe uniquely the spatial variation of a quantity at all points in space.

All coordinate systems can be broadly classified into two categories:

1. *Orthogonal coordinate systems*: Three coordinate axes are perpendicular to each other.
2. *Non-orthogonal coordinate systems*: Coordinate axes are not perpendicular to each other.

From another point of view, the coordinate systems are of two types:

1. *Right-handed coordinate systems:* These systems follow the right-hand cork-screw rule. This means that if one rotates from the first coordinate axis towards the second coordinate axis, a right-hand screw will advance in the positive direction of the third coordinate axis.

2. *Left-handed coordinate systems:* These systems follow the opposite movement of a right-handed screw.

Here, we will discuss the most useful three right-handed orthogonal coordinate systems; namely,

1. Cartesian or rectangular coordinates,
2. Circular or cylindrical coordinates, and
3. Spherical coordinates.

1.6.1 Cartesian or Rectangular Coordinates (*x*, *y*, *z*)

A point P in Cartesian coordinates is represented as $P(x, y, z)$.

The ranges of coordinate variables are

$$\begin{aligned} -\infty < x < \infty \\ -\infty < y < \infty \\ -\infty < z < \infty \end{aligned} \tag{1.12}$$

From Fig. 1.10 (*b*), it is understood that any point in rectangular coordinates is the intersection of three planes (*i*) constant x-plane (*ii*) constant y-plane and (*iii*) constant z-plane, which are mutually perpendicular.

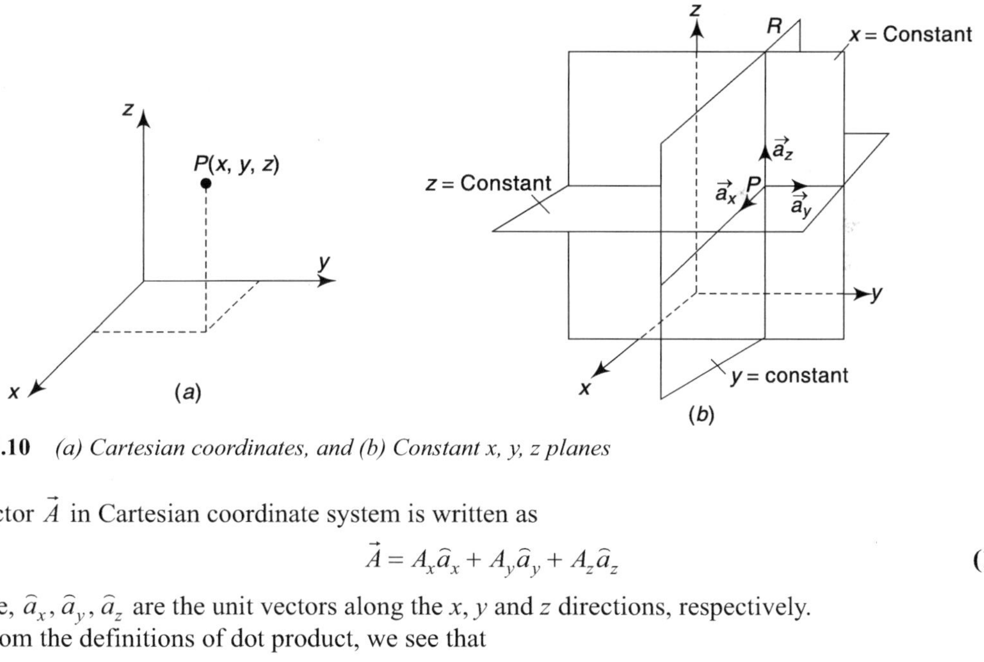

Fig. 1.10 *(a) Cartesian coordinates, and (b) Constant x, y, z planes*

A vector $\vec{A}$ in Cartesian coordinate system is written as

$$\vec{A} = A_x\hat{a}_x + A_y\hat{a}_y + A_z\hat{a}_z \tag{1.13}$$

where, $\hat{a}_x, \hat{a}_y, \hat{a}_z$ are the unit vectors along the x, y and z directions, respectively.

From the definitions of dot product, we see that

$$\begin{aligned} \hat{a}_x \cdot \hat{a}_x = \hat{a}_y \cdot \hat{a}_y = \hat{a}_z \cdot \hat{a}_z = 1 \\ \hat{a}_x \cdot \hat{a}_y = \hat{a}_y \cdot \hat{a}_z = \hat{a}_z \cdot \hat{a}_x = 0 \end{aligned} \tag{1.14}$$

From the definitions of cross product, we see that

$$\begin{aligned} \hat{a}_x \times \hat{a}_x = \hat{a}_y \times \hat{a}_y = \hat{a}_z \times \hat{a}_z = 0 \\ \hat{a}_x \times \hat{a}_y = \hat{a}_z; \quad \hat{a}_y \times \hat{a}_z = \hat{a}_x; \quad \hat{a}_z \times \hat{a}_x = \hat{a}_y \end{aligned} \tag{1.15}$$

1.6.2 Cylindrical or Circular Coordinates (r, ϕ, z)

A point P in cylindrical coordinates is represented as $P(r, \phi, z)$.

Here

r = radius of the cylinder passing through P = radial distance from the z-axis
ϕ = angle measured from the x-axis in the xy-plane, known as *azimuthal angle*
z = same as in Cartesian coordinates

The ranges of coordinate variables are

$$\begin{aligned} 0 &\le r < \infty \\ 0 &\le \phi < 2\pi \\ -\infty &< z < \infty \end{aligned} \quad \textbf{(1.16)}$$

From Fig. 1.11 (*b*), it is understood that any point in cylindrical coordinates is an intersection of three planes viz. (*i*) constant 'r' plane (a circular cylinder) (*ii*) constant ϕ plane (semi-infinite plane with its edge along the z-axis (*iii*) constant z-plane (parallel to xy-plane).

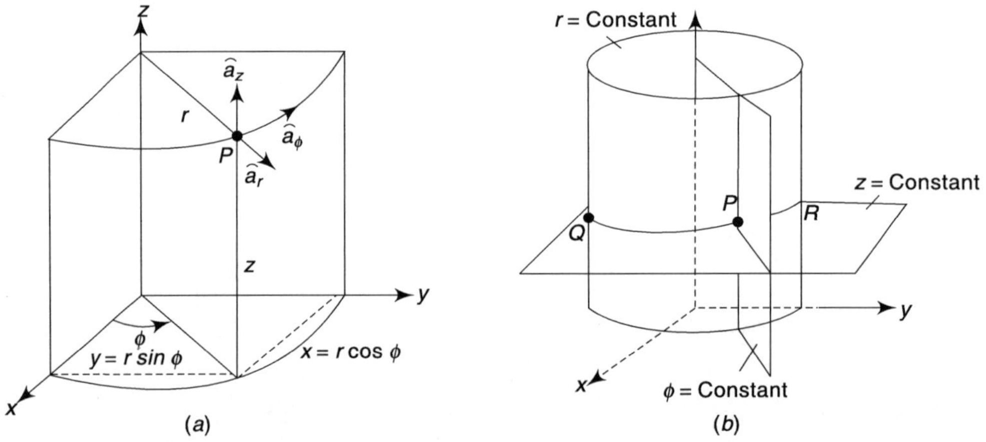

Fig. 1.11 *(a) Cylindrical coordinates, and (b) Constant r, ϕ, z planes*

A vector $\vec{A}$ in cylindrical coordinate system is written as

$$\vec{A} = A_r\hat{a}_r + A_\phi\hat{a}_\phi + A_z\hat{a}_z \quad \textbf{(1.17)}$$

where $\hat{a}_r$, $\hat{a}_\phi$, $\hat{a}_z$ are the unit vectors along the r, ϕ and z directions, respectively.

From the definitions of dot product, we see that

$$\begin{aligned} \hat{a}_r \cdot \hat{a}_r = \hat{a}_\phi \cdot \hat{a}_\phi = \hat{a}_z \cdot \hat{a}_z = 1 \\ \hat{a}_r \cdot \hat{a}_\phi = \hat{a}_\phi \cdot \hat{a}_z = \hat{a}_z \cdot \hat{a}_r = 0 \end{aligned} \quad \textbf{(1.18)}$$

From the definitions of cross product, we see that

$$\begin{aligned} &\hat{a}_r \times \hat{a}_r = \hat{a}_\phi \times \hat{a}_\phi = \hat{a}_z \times \hat{a}_z = 0 \\ &\hat{a}_r \times \hat{a}_\phi = \hat{a}_z; \quad \hat{a}_\phi \times \hat{a}_z = \hat{a}_r; \quad \hat{a}_z \times \hat{a}_r = \hat{a}_\phi \end{aligned} \quad \textbf{(1.19)}$$

Relations between Cartesian (x, y, z) and Cylindrical (r, ϕ, z) Coordinates The relationship between Cartesian (x, y, z) and cylindrical (r, ϕ, z) coordinates are obtained from Fig. 1.11(a) and are written as

$$r = \sqrt{x^2 + y^2} \qquad \phi = \tan^{-1}\left(\frac{y}{x}\right), \qquad z = z \tag{1.20}$$

and

$$x = r\cos\phi \qquad y = r\sin\phi \qquad z = z \tag{1.21}$$

The relationships between the unit vectors are obtained from Fig. 1.12 and are given as

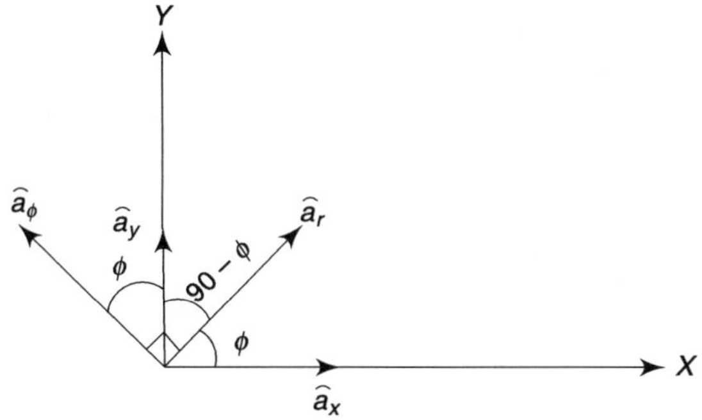

Fig. 1.12 *Unit vector transformation between Cartesian and cylindrical coordinates*

$$\begin{aligned} \hat{a}_x &= \cos\phi\,\hat{a}_r - \sin\phi\,\hat{a}_\phi \\ \hat{a}_y &= \sin\phi\,\hat{a}_r + \cos\phi\,\hat{a}_\phi \\ \hat{a}_z &= \hat{a}_z \end{aligned} \tag{1.22}$$

and

$$\begin{aligned} \hat{a}_r &= \cos\phi\,\hat{a}_x + \sin\phi\,\hat{a}_y \\ \hat{a}_\phi &= -\sin\phi\,\hat{a}_x + \cos\phi\,\hat{a}_y \\ \hat{a}_z &= \hat{a}_z \end{aligned} \tag{1.23}$$

The relationships between the component vectors (A_x, A_y, A_z) and (A_r, A_ϕ, A_z) are obtained by using Eqs. (1.22) and (1.23) and then rearranging the terms. This is given as

$$\begin{aligned} \vec{A} &= (A_x\cos\phi + A_y\sin\phi)\hat{a}_r + (-A_x\sin\phi + A_y\cos\phi)\hat{a}_\phi + A_z\hat{a}_z \\ &= (A_r\cos\phi - A_\phi\sin\phi)\hat{a}_x + (A_r\sin\phi + A_\phi\cos\phi)\hat{a}_y + A_z\hat{a}_z \end{aligned} \tag{1.24}$$

Thus, the relationships between the component vectors can be written in matrix forms as

$$\begin{bmatrix} A_r \\ A_\phi \\ A_z \end{bmatrix} = \begin{bmatrix} \cos\phi & \sin\phi & 0 \\ -\sin\phi & \cos\phi & 0 \\ 0 & 0 & 1 \end{bmatrix} \begin{bmatrix} A_x \\ A_y \\ A_z \end{bmatrix} \quad \text{and} \quad \begin{bmatrix} A_x \\ A_y \\ A_z \end{bmatrix} = \begin{bmatrix} \cos\phi & -\sin\phi & 0 \\ \sin\phi & \cos\phi & 0 \\ 0 & 0 & 1 \end{bmatrix} \begin{bmatrix} A_r \\ A_\phi \\ A_z \end{bmatrix} \tag{1.25}$$

1.6.3 Spherical or Polar Coordinates (ρ, θ, ϕ)

A point P in spherical coordinates is represented as $P(\rho, \theta, \phi)$.

Here,

ρ = distance of the point from the origin
= radius of a sphere centered at the origin and passing through the point P,
θ = angle between the z-axis and the position vector P, known as *colatitudes,* and
ϕ = angle measured from the x-axis in the xy-plane, known as *azimuthal angle* (same as in cylindrical coordinates).

The ranges of coordinate variables are

$$\begin{aligned} &0 \le \rho < \infty \\ &0 \le \theta < \pi \\ &0 \le \phi < 2\pi \end{aligned} \tag{1.26}$$

From Fig. 1.13 (*b*), it is understood that any point in spherical coordinates is an intersection of three planes, viz. (*i*) constant 'ρ' plane (a sphere with its centre at the origin), (*ii*) constant θ-plane (circular cone with z-axis as its axis and the origin at its vertex), and (*iii*) constant ϕ-plane (semi-infinite plane as in cylindrical coordinates).

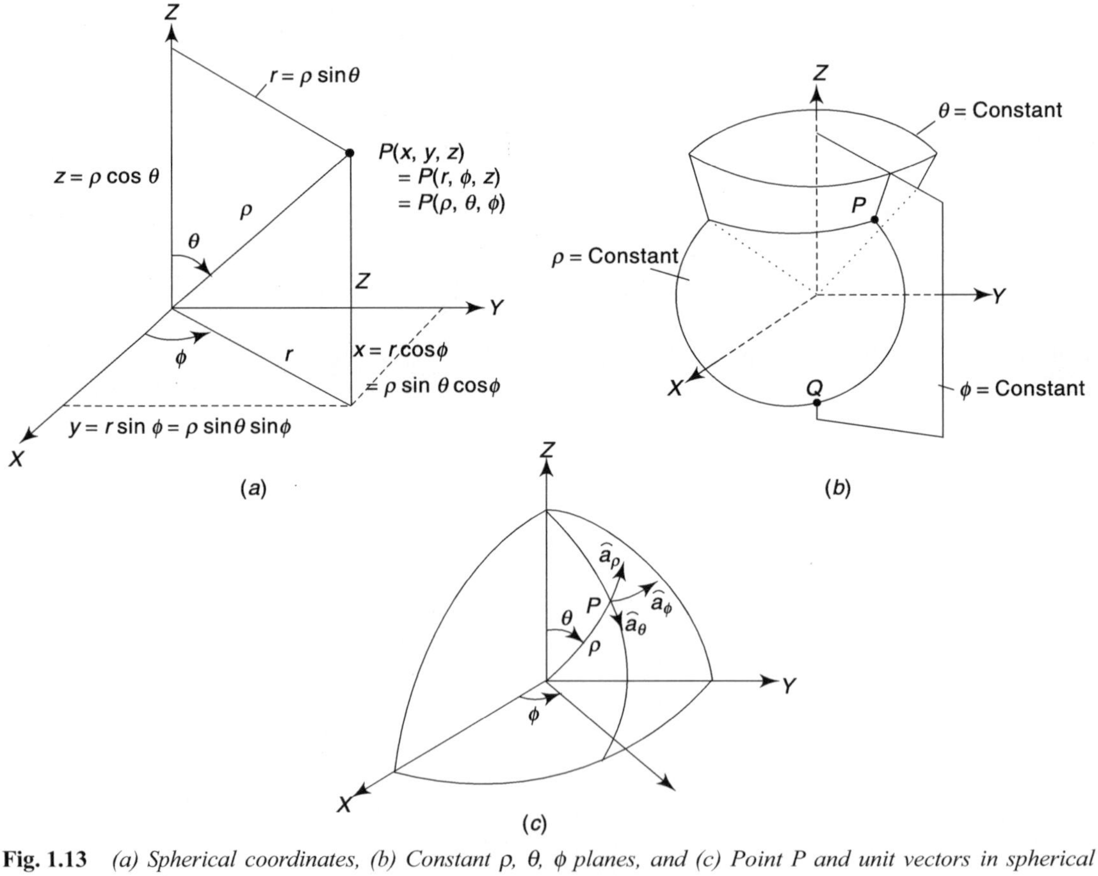

Fig. 1.13 *(a) Spherical coordinates, (b) Constant ρ, θ, ϕ planes, and (c) Point P and unit vectors in spherical coordinates*

A vector $\vec{A}$ in spherical coordinate system is written as

$$\vec{A} = A_\rho \hat{a}_\rho + A_\theta \hat{a}_\theta + A_\phi \hat{a}_\phi \tag{1.27}$$

where $\hat{a}_\rho, \hat{a}_\theta, \hat{a}_\phi$ are the unit vectors along the ρ, θ and ϕ directions, respectively.

From the definitions of dot product, we see that

$$\begin{aligned} \hat{a}_\rho \cdot \hat{a}_\rho = \hat{a}_\theta \cdot \hat{a}_\theta = \hat{a}_\phi \cdot \hat{a}_\phi = 1 \\ \hat{a}_\rho \cdot \hat{a}_\theta = \hat{a}_\theta \cdot \hat{a}_\phi = \hat{a}_\phi \cdot \hat{a}_\rho = 0 \end{aligned} \tag{1.28}$$

From the definitions of cross product, we see that

$$\begin{aligned} &\hat{a}_\rho \times \hat{a}_\rho = \hat{a}_\theta \times \hat{a}_\theta = \hat{a}_\phi \times \hat{a}_\phi = 0 \\ &\hat{a}_\rho \times \hat{a}_\theta = \hat{a}_\phi; \quad \hat{a}_\theta \times \hat{a}_\phi = \hat{a}_\rho; \quad \hat{a}_\phi \times \hat{a}_\rho = \hat{a}_\theta \end{aligned} \tag{1.29}$$

Relations between Cartesian (x, y, z) and Spherical (ρ, θ, ϕ) Coordinates The relationships between Cartesian (x, y, z) and spherical (ρ, θ, ϕ) coordinates can also be obtained from Fig. 1.13 (*a*) and can be written as

$$\boxed{\rho = \sqrt{x^2 + y^2 + z^2} \qquad \theta = \tan^{-1}\left(\frac{\sqrt{x^2+y^2}}{z}\right), \qquad \phi = \tan^{-1}\left(\frac{y}{x}\right)} \tag{1.30}$$

and

$$\boxed{x = \rho \sin\theta \cos\phi \qquad y = \rho \sin\theta \sin\phi \qquad z = \rho \cos\theta} \tag{1.31}$$

The relationships between the unit vectors are obtained from Fig. 1.14 and are given as

$$\begin{aligned} \hat{a}_x &= \sin\theta \cos\phi \,\hat{a}_\rho + \cos\theta \cos\phi \hat{a}_\theta - \sin\phi \hat{a}_\phi \\ \hat{a}_y &= \sin\theta \sin\phi \,\hat{a}_\rho + \cos\theta \sin\phi \hat{a}_\theta + \cos\phi \hat{a}_\phi \\ \hat{a}_z &= \cos\theta \,\hat{a}_\rho - \sin\theta \hat{a}_\theta \end{aligned} \tag{1.32}$$

and

$$\begin{aligned} \hat{a}_\rho &= \sin\theta \cos\phi \hat{a}_x + \sin\theta \sin\phi \hat{a}_y + \cos\theta \hat{a}_z \\ \hat{a}_\theta &= \cos\theta \cos\phi \hat{a}_x + \cos\theta \sin\phi \hat{a}_y - \sin\theta \hat{a}_z \\ \hat{a}_\phi &= -\sin\phi \hat{a}_x + \cos\phi \hat{a}_y \end{aligned} \tag{1.33}$$

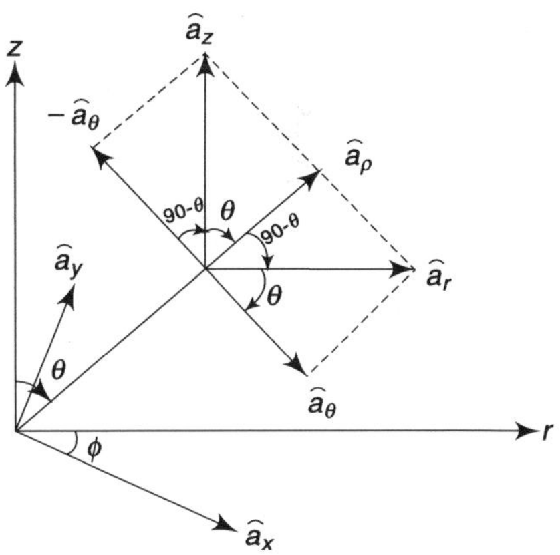

Fig. 1.14 *Unit vector transformation for Cartesian and spherical coordinates*

The relationships between the component vectors (A_x, A_y, A_z) and $(A_\rho, A_\theta, A_\phi)$ can be obtained by using Eqs. (1.32) and (1.33) and then rearranging the terms. This is written in matrix form as

$$\boxed{\begin{bmatrix} A_\rho \\ A_\theta \\ A_\phi \end{bmatrix} = \begin{bmatrix} \sin\theta\cos\phi & \sin\theta\sin\phi & \cos\theta \\ -\cos\theta\cos\phi & \cos\theta\sin\phi & -\sin\theta \\ -\sin\phi & \cos\phi & 0 \end{bmatrix} \begin{bmatrix} A_x \\ A_y \\ A_z \end{bmatrix}} \tag{1.34}$$

and

$$\begin{bmatrix} A_x \\ A_y \\ A_z \end{bmatrix} = \begin{bmatrix} \sin\theta\cos\phi & \cos\theta\cos\phi & -\sin\phi \\ \sin\theta\sin\phi & \cos\theta\sin\phi & \cos\phi \\ \cos\theta & -\sin\theta & 0 \end{bmatrix} \begin{bmatrix} A_\rho \\ A_\theta \\ A_\phi \end{bmatrix} \quad (1.35)$$

Relations between Cylindrical (r, ϕ, z) and Spherical (ρ, θ, ϕ) Coordinates The relationships between cylindrical (r, ϕ, z) and spherical (ρ, θ, ϕ) coordinates are obtained from Fig. 1.13 (a) and are written as

$$\rho = \sqrt{r^2 + z^2} \qquad \theta = \tan^{-1}\left(\frac{r}{z}\right), \qquad \phi = \phi \quad (1.36)$$

and

$$r = \rho\sin\theta \qquad \phi = \phi \qquad z = \rho\cos\theta \quad (1.37)$$

The relationships between the unit vectors are obtained from Fig. 1.15 and are given as

$$\begin{aligned} \hat{a}_\rho &= \sin\theta\,\hat{a}_r + \cos\theta\,\hat{a}_z \\ \hat{a}_\theta &= \cos\theta\cos\phi\,\hat{a}_r - \sin\theta\,\hat{a}_z \\ \hat{a}_\phi &= \hat{a}_\phi \end{aligned} \quad (1.38)$$

and

$$\begin{aligned} \hat{a}_r &= \sin\theta\,\hat{a}_\rho + \cos\theta\cos\phi\,\hat{a}_\theta \\ \hat{a}_\phi &= \hat{a}_\phi \\ \hat{a}_z &= \cos\theta\,\hat{a}_\rho - \sin\theta\,\hat{a}_\theta \end{aligned} \quad (1.39)$$

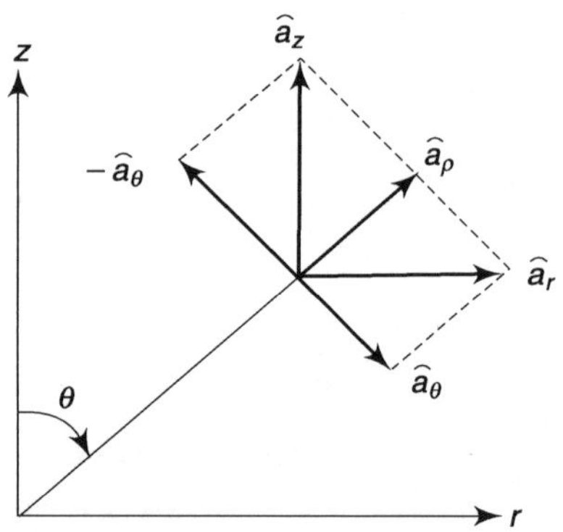

Fig. 1.15 *Unit vector transformation for cylindrical and spherical coordinates*

The relationships between the component vectors (A_ρ, A_θ, A_ϕ and (A_r, A_ϕ, A_z) can be obtained by using Eqs. (1.38) and (1.39) and then rearranging the terms. This is written in matrix form as

$$\begin{bmatrix} A_\rho \\ A_\theta \\ A_\phi \end{bmatrix} = \begin{bmatrix} \sin\theta & 0 & \cos\theta \\ \cos\theta & 0 & -\sin\theta \\ 0 & 1 & 0 \end{bmatrix} \begin{bmatrix} A_r \\ A_\phi \\ A_z \end{bmatrix} \quad (1.40)$$

and

$$\begin{bmatrix} A_r \\ A_\phi \\ A_z \end{bmatrix} = \begin{bmatrix} \sin\theta & \cos\theta & 0 \\ 0 & 0 & 1 \\ \cos\theta & -\sin\theta & 0 \end{bmatrix} \begin{bmatrix} A_\rho \\ A_\theta \\ A_\phi \end{bmatrix} \quad (1.41)$$

***Example 1.5** Convert the point $P(1, 3, 5)$ from Cartesian to cylindrical and spherical coordinates.

Solution At point P: $x = 1, y = 3, z = 5$

Conversion from Cartesian to cylindrical coordinates:

Hence,

$$r = \sqrt{x^2 + y^2} = \sqrt{1^2 + 3^2} = \sqrt{10} = 3.162$$

$$\phi = \tan^{-1}\left(\frac{y}{x}\right) = \tan^{-1}\left(\frac{3}{1}\right) = 71.565°$$

$$z = 5$$

Conversion from Cartesian to spherical coordinates:

Here,

$$\rho = \sqrt{x^2 + y^2 + z^2} = \sqrt{1^2 + 3^2 + 5^2} = \sqrt{35} = 5.916$$

$$\theta = \tan^{-1}\left(\frac{\sqrt{x^2 + y^2}}{z}\right) = \tan^{-1}\left(\frac{\sqrt{1^2 + 3^2}}{5}\right) = 32.311°$$

$$\phi = \tan^{-1}\left(\frac{y}{x}\right) = \tan^{-1}\left(\frac{3}{1}\right) = 71.565°$$

Therefore, the point P is written as

$$\boxed{P(1, 3, 5) = P(3.162, 71.565°, 5) = P(5.916, 32.311°, 71.565°)}$$

***Example 1.6** Transform the vector

$$\vec{A} = \frac{\sqrt{x^2 + y^2}}{\sqrt{x^2 + y^2 + z^2}}\hat{a}_x - \frac{yz}{\sqrt{x^2 + y^2 + z^2}}\hat{a}_z$$

to cylindrical and spherical coordinates.

Solution For $\vec{A}$, the components are given as

$$A_x = \frac{\sqrt{x^2 + y^2}}{\sqrt{x^2 + y^2 + z^2}} \qquad A_y = 0 \qquad A_z = -\frac{yz}{\sqrt{x^2 + y^2 + z^2}}$$

Transformation from Cartesian to cylindrical coordinates:

From Eq. (1.25),

$$\begin{bmatrix} A_r \\ A_\phi \\ A_z \end{bmatrix} = \begin{bmatrix} \cos\phi & \sin\phi & 0 \\ -\sin\phi & \cos\phi & 0 \\ 0 & 0 & 1 \end{bmatrix}\begin{bmatrix} A_x \\ A_y \\ A_z \end{bmatrix} = \begin{bmatrix} \cos\phi & \sin\phi & 0 \\ -\sin\phi & \cos\phi & 0 \\ 0 & 0 & 1 \end{bmatrix}\begin{bmatrix} \dfrac{\sqrt{x^2 + y^2}}{\sqrt{x^2 + y^2 + z^2}} \\ 0 \\ -\dfrac{yz}{\sqrt{x^2 + y^2 + z^2}} \end{bmatrix}$$

$$\therefore \qquad A_r = \frac{\sqrt{x^2+y^2}}{\sqrt{x^2+y^2+z^2}}\cos\phi$$

$$A_\phi = -\frac{\sqrt{x^2+y^2}}{\sqrt{x^2+y^2+z^2}}\sin\phi$$

$$A_z = -\frac{yz}{\sqrt{x^2+y^2+z^2}}$$

But, $x = r\cos\phi \quad y = r\sin\phi \qquad \therefore \sqrt{x^2+y^2} = r$

Substituting these relations, we get

$$\therefore \qquad A_r = \frac{r\cos\phi}{\sqrt{r^2+z^2}}$$

$$A_\phi = -\frac{r\sin\phi}{\sqrt{r^2+z^2}}$$

$$A_z = -\frac{rz\sin\phi}{\sqrt{r^2+z^2}}$$

Hence, the vector is expressed in cylindrical coordinates as

$$\vec{A} = \frac{r\cos\phi}{\sqrt{r^2+z^2}}\hat{a}_r - \frac{r\sin\phi}{\sqrt{r^2+z^2}}\hat{a}_\phi - \frac{rz\sin\phi}{\sqrt{r^2+z^2}}\hat{a}_z = \frac{r}{\sqrt{r^2+z^2}}(\cos\phi\,\hat{a}_r - \sin\phi\,\hat{a}_\phi - z\sin\phi\,\hat{a}_z)$$

$$\boxed{\vec{A} = \frac{r}{\sqrt{r^2+z^2}}(\cos\phi\,\hat{a}_r - \sin\phi\,\hat{a}_\phi - z\sin\phi\,\hat{a}_z)}$$

Transformation from Cartesian to spherical coordinates:

From Eq. (1.34),

$$\begin{bmatrix} A_\rho \\ A_\theta \\ A_\phi \end{bmatrix} = \begin{bmatrix} \sin\theta\cos\phi & \sin\theta\sin\phi & \cos\theta \\ \cos\theta\cos\phi & \cos\theta\sin\phi & -\sin\theta \\ -\sin\phi & \cos\phi & 0 \end{bmatrix} \begin{bmatrix} A_x \\ A_y \\ A_z \end{bmatrix}$$

$$= \begin{bmatrix} \sin\theta\cos\phi & \sin\theta\sin\phi & \cos\theta \\ \cos\theta\cos\phi & \cos\theta\sin\phi & -\sin\theta \\ -\sin\phi & \cos\phi & 0 \end{bmatrix} \begin{bmatrix} \dfrac{\sqrt{x^2+y^2}}{\sqrt{x^2+y^2+z^2}} \\ 0 \\ -\dfrac{yz}{\sqrt{x^2+y^2+z^2}} \end{bmatrix}$$

$$\therefore \quad A_\rho = \frac{\sqrt{x^2+y^2}}{\sqrt{x^2+y^2+z^2}}\sin\theta\cos\phi - \frac{yz}{\sqrt{x^2+y^2+z^2}}\cos\theta$$

$$A_\theta = \frac{\sqrt{x^2+y^2}}{\sqrt{x^2+y^2+z^2}}\cos\theta\cos\phi + \frac{yz}{\sqrt{x^2+y^2+z^2}}\sin\theta$$

$$A_\phi = -\frac{\sqrt{x^2+y^2}}{\sqrt{x^2+y^2+z^2}}\sin\phi$$

But, $x = \rho\sin\theta\cos\phi \qquad y = \rho\sin\theta\sin\phi \qquad z = \rho\cos\theta$

$$\therefore \quad \sqrt{x^2+y^2} = \rho\sin\theta \quad \text{and} \quad \sqrt{x^2+y^2+z^2} = \rho$$

Substituting these relations, we get

$$\therefore \quad A_\rho = \frac{\rho\sin^2\theta\cos\phi}{\rho} - \frac{\rho\sin\theta\sin\phi\rho\cos^2\theta}{\rho}$$

$$= (\sin^2\theta\cos\phi - \rho\sin\theta\cos^2\theta\sin\phi)$$

$$A_\theta = \frac{\rho\sin\theta\cos\theta\cos\phi}{\rho} + \frac{\rho\sin^2\theta\sin\phi\rho\cos\theta}{\rho}$$

$$= (\sin\theta\cos\theta\cos\phi + \rho\sin^2\theta\cos\theta\sin\phi)$$

$$A_\phi = -\frac{\rho\sin\theta\sin\phi}{\rho} = -\sin\theta\sin\phi$$

Hence, the vector is expressed in spherical coordinates as

$$\vec{A} = (\sin^2\theta\cos\phi - \rho\sin\theta\cos^2\theta\sin\phi)\hat{a}_\rho$$
$$+ \left(\sin\theta\cos\theta\cos\phi + \rho\sin^2\theta\cos\theta\sin\phi\right)\hat{a}_\theta - \sin\theta\sin\phi\hat{a}_\phi$$
$$= \sin\theta(\sin\theta\cos\phi - \rho\cos^2\theta\sin\phi)\hat{a}_\rho$$
$$+ \sin\theta\cos\theta(\cos\phi + \rho\sin\theta\sin\phi)\hat{a}_\theta - \sin\theta\sin\phi\hat{a}_\phi$$

$$\boxed{\vec{A} = \sin\theta\,(\sin\theta\cos\phi - \rho\cos^2\theta\sin\phi)\hat{a}_\rho + \sin\theta\cos\theta\,(\cos\phi + \rho\sin\theta\sin\phi)\hat{a}_\theta - \sin\theta\sin\phi\hat{a}_\phi}$$

Example 1.7 Express the following vectors in Cartesian coordinates:

(a) $\vec{A} = rz\sin\phi\,\hat{a}_r + 3r\cos\phi\,\hat{a}_\phi + r\cos\phi\sin\phi\,\hat{a}_z$

(b) $\vec{B} = \rho^2\hat{a}_\rho + \sin\theta\,\hat{a}_\phi$

Solution **(a)** $\vec{A} = rz\sin\phi\,\hat{a}_r + 3r\cos\phi\hat{a}_\phi + r\cos\phi\sin\phi\,\hat{a}_z$

Here, $A_r = rz\sin\phi$, $A_\phi = 3r\cos\phi$, $A_z = r\cos\phi\sin\phi$

$$\begin{bmatrix} A_x \\ A_y \\ A_z \end{bmatrix} = \begin{bmatrix} \cos\phi & -\sin\phi & 0 \\ \sin\phi & \cos\phi & 0 \\ 0 & 0 & 1 \end{bmatrix}\begin{bmatrix} A_r \\ A_\phi \\ A_z \end{bmatrix} = \begin{bmatrix} \cos\phi & -\sin\phi & 0 \\ \sin\phi & \cos\phi & 0 \\ 0 & 0 & 1 \end{bmatrix}\begin{bmatrix} rz\sin\phi \\ 3r\cos\phi \\ r\cos\phi\sin\phi \end{bmatrix}$$

$A_x = rz \sin\phi \cos\phi - 3r \sin\phi \cos\phi$
$A_y = rz \sin^2\phi + 3r \cos^2\phi$
$A_z = r \cos\phi \sin\phi$

But $r = \sqrt{x^2 + y^2}$, $\cos\phi = \dfrac{x}{\sqrt{x^2 + y^2}}$, $\sin\phi = \dfrac{y}{\sqrt{x^2 + y^2}}$

Substituting these values

$$A_x = \sqrt{x^2+y^2}\, z\frac{xy}{x^2+y^2} - 3\sqrt{x^2+y^2}\,\frac{xy}{x^2+y^2} = \frac{xyz}{\sqrt{x^2+y^2}} - \frac{3xy}{\sqrt{x^2+y^2}}$$

$$A_y = \sqrt{x^2+y^2}\, z\frac{y^2}{x^2+y^2} + 3\sqrt{x^2+y^2}\,\frac{x^2}{x^2+y^2} = \frac{y^2 z}{\sqrt{x^2+y^2}} + \frac{3x^2}{\sqrt{x^2+y^2}}$$

$$A_z = \sqrt{x^2+y^2}\,\frac{xy}{x^2+y^2} = \frac{xy}{\sqrt{x^2+y^2}}$$

Hence, the vector in Cartesian coordinates is written as

$$\vec{A} = \frac{1}{\sqrt{x^2+y^2}}[(xyz - 3xy)\hat{a}_x + (y^2 z + 3x^2)\hat{a}_y + xy\hat{a}_z]$$

(b) $\vec{B} = \rho^2 \hat{a}_\rho + \sin\theta\, \hat{a}_\phi$

Here, $B_\rho = \rho^2$, $B_\theta = 0$, $B_\phi = \sin\theta$

$$\begin{bmatrix} B_x \\ B_y \\ B_z \end{bmatrix} = \begin{bmatrix} \sin\theta\cos\phi & \cos\theta\cos\phi & -\sin\phi \\ \sin\theta\sin\phi & \cos\theta\sin\phi & \cos\phi \\ \cos\theta & -\sin\theta & 0 \end{bmatrix} \begin{bmatrix} \rho^2 \\ 0 \\ \sin\theta \end{bmatrix}$$

$$B_x = \rho^2 \sin\theta\cos\phi - \sin\theta\sin\phi$$
$$B_y = \rho^2 \sin\theta\sin\phi + \sin\theta\cos\phi$$
$$B_z = \rho^2 \cos\theta$$

But

$$\rho = \sqrt{x^2+y^2+z^2},\quad \sin\theta = \frac{\sqrt{x^2+y^2}}{\sqrt{x^2+y^2+z^2}},\quad \cos\theta = \frac{z}{\sqrt{x^2+y^2+z^2}},$$

$$\sin\phi = \frac{y}{\sqrt{x^2+y^2}},\quad \cos\phi = \frac{x}{\sqrt{x^2+y^2}}$$

Substituting these values

$$B_x = (x^2+y^2+z^2)\frac{\sqrt{x^2+y^2}}{\sqrt{x^2+y^2+z^2}}\frac{x}{\sqrt{x^2+y^2}} - \frac{\sqrt{x^2+y^2}}{\sqrt{x^2+y^2+z^2}}\frac{y}{\sqrt{x^2+y^2}}$$

$$= x\sqrt{x^2+y^2+z^2} - \frac{y}{\sqrt{x^2+y^2+z^2}}$$

$$= \frac{1}{\sqrt{x^2+y^2+z^2}}[x(x^2+y^2+z^2) - y]$$

$$B_y = (x^2+y^2+z^2)\frac{\sqrt{x^2+y^2}}{\sqrt{x^2+y^2+z^2}}\frac{y}{\sqrt{x^2+y^2}} + \frac{\sqrt{x^2+y^2}}{\sqrt{x^2+y^2+z^2}}\frac{x}{\sqrt{x^2+y^2}}$$

$$= y\sqrt{x^2+y^2+z^2} + \frac{x}{\sqrt{x^2+y^2+z^2}}$$

$$= \frac{1}{\sqrt{x^2+y^2+z^2}}[y(x^2+y^2+z^2)+x]$$

$$B_z = (x^2+y^2+z^2)\frac{z}{\sqrt{x^2+y^2+z^2}} = z\sqrt{x^2+y^2+z^2}$$

Hence, the vector in Cartesian coordinates is written as

$$\vec{B} = \frac{1}{\sqrt{x^2+y^2+z^2}}[\{x(x^2+y^2+z^2)-y\}\hat{a}_x + \{y(x^2+y^2+z^2)+x\}\hat{a}_y + z(x^2+y^2+z^2)\hat{a}_z]$$

***Example 1.8** Express the field $\vec{E} = 2xyz\hat{a}_x - 3(x+y+z)\hat{a}_z$, in cylindrical coordinates, and calculate $|\vec{E}|$ at the point $P\,(r=2, \phi=60°, z=3)$.

Solution In the cylindrical system,

$$\begin{bmatrix} E_r \\ E_\phi \\ E_z \end{bmatrix} = \begin{bmatrix} \cos\phi & \sin\phi & 0 \\ -\sin\phi & \cos\phi & 0 \\ 0 & 0 & 1 \end{bmatrix}\begin{bmatrix} E_x \\ E_y \\ E_z \end{bmatrix} = \begin{bmatrix} \cos\phi & \sin\phi & 0 \\ -\sin\phi & \cos\phi & 0 \\ 0 & 0 & 1 \end{bmatrix}\begin{bmatrix} 2xyz \\ 0 \\ -3(x+y+z) \end{bmatrix}$$

$$E_r = 2\,xyz\cos\phi$$
$$E_\phi = -2\,xyz\sin\phi$$
$$E_z = -3(x+y+z)$$

But, $x = r\cos\phi$, $y = r\sin\phi$, $z = z$

Substituting these values

$$E_r = 2r\cos\phi r\sin\phi z\cos\phi = 2r^2 z\sin\phi\cos^2\phi$$
$$E_\phi = -2r\cos\phi r\sin\phi z\sin\phi = -2r^2 z\sin^2\phi\cos\phi$$
$$E_z = -3(r\cos\phi + r\sin\phi + z)$$

Hence, the vector in Cartesian coordinates is written as

$$\vec{E} = r^2 z\sin 2\phi\cos\phi\hat{a}_r - r^2 z\sin 2\phi\sin\phi\hat{a}_\phi - 3(r\cos\phi + r\sin\phi + z)\hat{a}_z$$

At $P\,(r=2, \phi=60°, z=3)$, the vector is given as

$$E_r = 2\times 2^2\times 3\sin 60°\cos^2 60° = 24\frac{\sqrt{3}}{2}\times\frac{1}{4} = 3\sqrt{3} = 5.196$$

$$E_\phi = -2\times 2^2\times 3\sin^2 60°\cos 60° = -24\times\frac{3}{4}\times\frac{1}{2} = -9$$

$$E_z = -3(2\cos 60° + 2\sin 60° + 3) = -3\left(2\times\frac{1}{2} + 2\times\frac{\sqrt{3}}{2} + 3\right) = -3(4+\sqrt{3}) = -17.196$$

Hence, at P, $\vec{E} = 5.196\hat{a}_r - 9\hat{a}_\phi - 17.196\hat{a}_z$

$$\therefore \qquad |\vec{E}| = \sqrt{(5.196)^2 - 9^2 - (17.196)^2} = 20.092$$

Example 1.9 Given the vector field in 'mixed' coordinate variables as

$$\vec{V} = \frac{x\cos\phi}{r}\hat{a}_x + \frac{2yz}{r^2}\hat{a}_y + \left(1 - \frac{x^2}{r^2}\right)\hat{a}_z$$

Convert the vector completely in spherical coordinates.

Solution Here, $V_x = \dfrac{x\cos\phi}{r}$; $\quad V_y = \dfrac{2yz}{r^2}$; $\quad V_z = \left(1 - \dfrac{x^2}{r^2}\right)$

But
$$x = r\cos\phi, \quad y = r\sin\phi, \quad z = z$$

and
$$r = \sqrt{x^2 + y^2}, \qquad \cos\phi = \frac{x}{\sqrt{x^2 + y^2}}, \qquad \sin\phi = \frac{y}{\sqrt{x^2 + y^2}}$$

$$\therefore \qquad V_x = \frac{x^2}{x^2 + y^2}; \qquad V_y = \frac{2yz}{x^2 + y^2}; \qquad V_z = 1 - \frac{x^2}{x^2 + y^2} = \frac{y^2}{x^2 + y^2}$$

$$\therefore \qquad \vec{V} = \left(\frac{x^2}{x^2 + y^2}\right)\hat{a}_x + \left(\frac{2yz}{x^2 + y^2}\right)\hat{a}_y + \left(\frac{y^2}{x^2 + y^2}\right)\hat{a}_z$$

This is the vector in purely Cartesian coordinates.

Now, by relations of Eq. (1.34), we get

$$\begin{bmatrix} V_\rho \\ V_\theta \\ V_\phi \end{bmatrix} = \begin{bmatrix} \sin\theta\cos\phi & \sin\theta\sin\phi & \cos\theta \\ -\cos\theta\cos\phi & \cos\theta\sin\phi & -\sin\theta \\ -\sin\phi & \cos\phi & 0 \end{bmatrix} \begin{bmatrix} V_x \\ V_y \\ V_z \end{bmatrix}$$

$$= \begin{bmatrix} \sin\theta\cos\phi & \sin\theta\sin\phi & \cos\theta \\ -\cos\theta\cos\phi & \cos\theta\sin\phi & -\sin\theta \\ -\sin\phi & \cos\phi & 0 \end{bmatrix} \begin{bmatrix} \left(\dfrac{x^2}{x^2 + y^2}\right) \\ \left(\dfrac{2yz}{x^2 + y^2}\right) \\ \left(\dfrac{y^2}{x^2 + y^2}\right) \end{bmatrix}$$

$$\therefore \qquad V_\rho = \left(\frac{x^2}{x^2 + y^2}\right)\sin\theta\cos\phi + \left(\frac{2yz}{x^2 + y^2}\right)\sin\theta\sin\phi + \left(\frac{y^2}{x^2 + y^2}\right)\cos\theta$$

$$V_\theta = \left(\frac{-x^2}{x^2 + y^2}\right)\cos\theta\cos\phi + \left(\frac{2yz}{x^2 + y^2}\right)\cos\theta\sin\phi - \left(\frac{y^2}{x^2 + y^2}\right)\sin\theta$$

$$V_\phi = \left(\frac{-x^2}{x^2 + y^2}\right)\sin\phi + \left(\frac{2yz}{x^2 + y^2}\right)\cos\phi$$

However, $x = \rho \sin\theta \cos\phi$; $y = \rho \sin\theta \sin\phi$; $z = \rho \cos\theta$

Therefore, the vector components in spherical coordinate system are given as

$$\therefore \quad V_\rho = \frac{1}{\rho^2 \sin^2\theta}(\rho^2 \sin^3\theta \cos^3\theta + 3\rho^2 \sin^2\theta \sin^2\phi \cos\theta)$$

$$= (\sin\theta \cos^3\theta + 3\sin^2\phi \cos\theta)$$

$$V_\theta = \frac{1}{\rho^2 \sin^2\theta}(-\rho^2 \sin^2\theta \cos\theta \cos^3\phi + 2\rho^2 \sin\theta \cos^2\theta \sin^2\phi - \rho^2 \sin^3\theta \sin^2\phi)$$

$$= \left(2\frac{\cos^2\theta \sin^2\phi}{\sin\theta} - \cos\theta \cos^3\phi - \sin\theta \sin^2\phi\right)$$

$$V_\varphi = \left(2\frac{\cos\theta \cos\phi}{\sin\theta} - \cos^2\phi \sin\phi\right)$$

1.7 GENERAL CURVILINEAR COORDINATES

Curvilinear Coordinate System *Curvilinear coordinates* are a coordinate system for the Euclidean space based on some transformation that converts the standard Cartesian coordinate system to a coordinate system, with the same number of coordinates in which the coordinate lines are curved.

The name *curvilinear coordinates*, given by the French mathematician Lame, derives from the fact that the coordinate surfaces of the curvilinear systems are curved.

Let

$f_1(x, y, z), f_2(x, y, z), f_3(x, y, z)$ be three independent, unambiguous and smooth functions,
x, y, z be three independent space variables in the Cartesian coordinate system, and
u_1, u_2, u_3 be three constant parameters.

We set the equations

$$u_1 = f_1(x, y, z) \qquad u_2 = f_2(x, y, z) \qquad u_3 = f_3(x, y, z)$$

By this equation, three surfaces are defined that can be labelled by these parameters as shown in Fig. 1.16.

The common intersection of the surfaces u_1 = constant 1, u_2 = constant 2, u_3 = constant 3 defines one point in the space to which a set of three unique numbers (u_1, u_2, u_3) can be assigned. These numbers are called ***curvilinear coordinates*** of that point as illustrated in Fig. 1.16.

Position Vector in Curvilinear System We have the set of equations

$$u_1 = f_1(x, y, z) \qquad u_2 = f_2(x, y, z) \qquad u_3 = f_3(x, y, z)$$

These equations can be solved and the solution can be written in the form

$$\begin{aligned} x &= f_1'(u_1, u_2, u_3) \\ y &= f_2'(u_1, u_2, u_3) \\ z &= f_1'(u_1, u_2, u_3) \end{aligned} \qquad \textbf{(1.42)}$$

This defines the position of the point A in the Cartesian system (x, y, z) using the curvilinear coordinates (u_1, u_2, u_3).

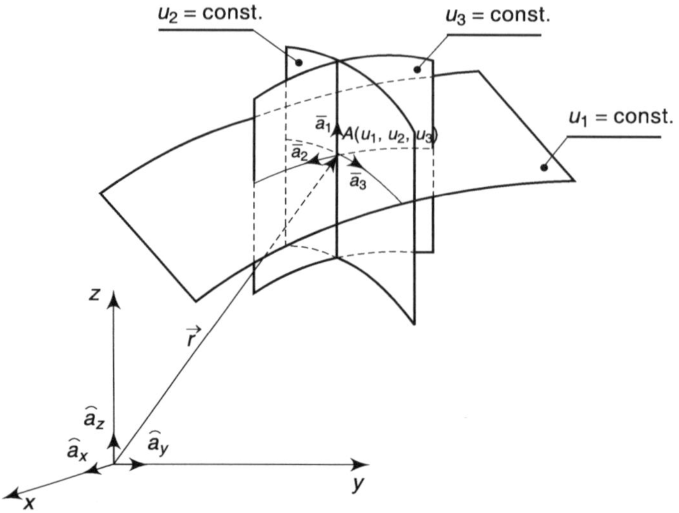

Fig. 1.16 *To the definition of the curvilinear coordinates of a point $A(u_1,u_2,u_3)$ in the space*

Here,

$$\vec{r} = x\hat{a}_x + y\hat{a}_y + z\hat{a}_z = x(u_1, u_2, u_3)\hat{a}_x + y(u_1, u_2, u_3)\hat{a}_y + z(u_1, u_2, u_3)\hat{a}_z \quad \textbf{(1.43)}$$

is the *position vector* of the point A and $\hat{a}_x, \hat{a}_y, \hat{a}_z$ are the unit vectors along the coordinate axes of the Cartesian system (see Fig. 1.16).

Base Vectors in Curvilinear System An elementary displacement of the point A in the space can be described by the differential formula

$$d\vec{r} = \frac{\partial \vec{r}}{\partial u_1} du_1 + \frac{\partial \vec{r}}{\partial u_2} du_2 + \frac{\partial \vec{r}}{\partial u_3} du_3 \quad \textbf{(1.44)}$$

or

$$d\vec{r} = du_1\hat{a}_1' + du_2\hat{a}_2' + du_3\hat{a}_3' \quad \textbf{(1.45)}$$

Here, $\hat{a}_1', \hat{a}_2', \hat{a}_3'$ are called the *base vectors* of the general curvilinear coordinate system at point $A(u_1, u_2, u_3)$. It is to be noted that the absolute values of $\hat{a}_1', \hat{a}_2', \hat{a}_3'$ are not equal to 1; they are generally not unit vectors. If $\hat{a}_1' \perp \hat{a}_2' \perp \hat{a}_3'$, we have the *orthogonal curvilinear coordinate system.*

Metric Coefficients The relation of Eq. (1.45) can be rewritten into the form

$$d\vec{r} = h_1(u_1, u_2, u_3)\, du_1\hat{a}_1 + h_2(u_1, u_2, u_3)\, du_2\hat{a}_2 + h_3(u_1, u_2, u_3)\, du_3\hat{a}_3$$

where we have put

$$\hat{a}_1' = h_1\hat{a}_1; \qquad \hat{a}_1' = h_2\hat{a}_2; \qquad \hat{a}_3' = h_3\hat{a}_3 \quad \textbf{(1.46)}$$

where $\hat{a}_1, \hat{a}_2, \hat{a}_3$ are now the units vectors of the same directions as vectors $\hat{a}_1', \hat{a}_2', \hat{a}_3'$. The functions h_1, h_2, h_3 are usually called the *metric coefficients*. The physical meaning of these coefficients can be understood when defining the length elements along the particular directions given by vectors $\hat{a}_1, \hat{a}_2, \hat{a}_3$ in the local curvilinear coordinate system at point $A(\vec{r})$ corresponding with the elementary displacements of $A(u_1, u_2, u_3)$ by du_1, du_2, du_3.

Differential Lengths in Curvilinear Coordinates The elementary displacement $d\vec{s}$ from point $A(\vec{r})$ to $A(\vec{r}+d\vec{r})$ can be described by

$$d\vec{s} = d\vec{r} = h_1 du_1 \hat{a}_1 + h_2 du_2 \hat{a}_2 + h_3 du_3 \hat{a}_3 \quad \textbf{(1.47a)}$$

$$\therefore \quad ds^2 = h_1^2 du_1^2 + h_2^2 du_2^2 + h_3^2 du_3^2 \quad \textbf{(1.47b)}$$

It is seen that the products h_1du_1, h_2du_2, h_3du_3 represent the lengths of projections of the elementary displacement $d\vec{s}$ onto the vectors $\hat{a}_1$, $\hat{a}_2$, $\hat{a}_3$, respectively. Consequently, the change of the curvilinear coordinate du_i can be transformed into corresponding displacement in space by multiplying it by h_i corresponding metric coefficient as shown in Fig. 1.17.

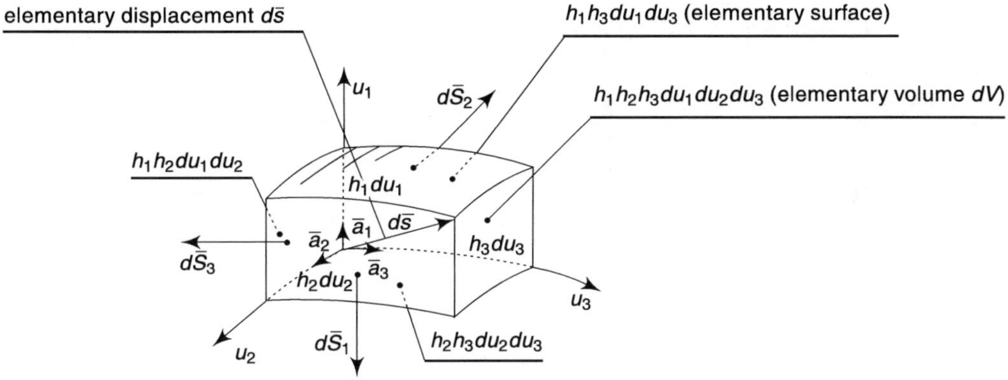

Fig. 1.17 *To the definition of the elementary displacement $d\vec{s}$, surface $d\vec{S}$ and volume dV, respectively*

Therefore, the metric coefficients can be represented in terms of the elementary lengths as

$$h_1 = \frac{ds_1}{du_1} \qquad h_2 = \frac{ds_2}{du_2} \qquad h_3 = \frac{ds_3}{du_3} \quad \textbf{(1.48)}$$

Differential Areas in Curvilinear Coordinates Elementary coordinate surface can be defined corresponding with the elementary changes of a couple of coordinates. It is explicitly defined by the relation

$$d\vec{S}_i = d\vec{s}_j \times d\vec{s}_k \quad \textbf{(1.49)}$$

Using expressions for elementary displacements $d\vec{s}_j$, $d\vec{s}_k$ from Eq. (1.47 *a*), we can write

$$d\vec{S} = h_2h_3du_2du_3\hat{a}_1 + h_1h_3du_1du_3\hat{a}_2 + h_1h_2du_1du_2\hat{a}_3 \quad \textbf{(1.50)}$$

In Eq. (1.50), the relations $\hat{a}_1 \times \hat{a}_2 = \hat{a}_3$, $\hat{a}_2 \times \hat{a}_3 = \hat{a}_1$, $\hat{a}_3 \times \hat{a}_1 = \hat{a}_2$, has been used. According to Eq. (1.50), a general elementary surface $d\vec{S}$ is composed of the three elementary surfaces $d\vec{S}_1$, $d\vec{S}_2$, $d\vec{S}_3$ oriented along the unit vectors $\hat{a}_1$, $\hat{a}_2$, $\hat{a}_3$, see Fig. 1.17.

Differential Volume in Curvilinear Coordinates Elementary volume element dV can be described by the relation

$$dV = d\vec{s}_i \cdot d\vec{S}_i \quad \textbf{(1.51)}$$

or using metric coefficients, we obtain

$$dV = h_1h_2h_3du_1du_2du_3 \quad \textbf{(1.52)}$$

1.8 DIFFERENTIAL ELEMENTS (LENGTHS, AREAS AND VOLUMES) IN DIFFERENT COORDINATE SYSTEMS

Now, we will consider the generalised curvilinear coordinates for the three different coordinate systems and determine the differential elements in these three coordinate systems.

To obtain the differential elements in length, area and volume, we consider the following figures:

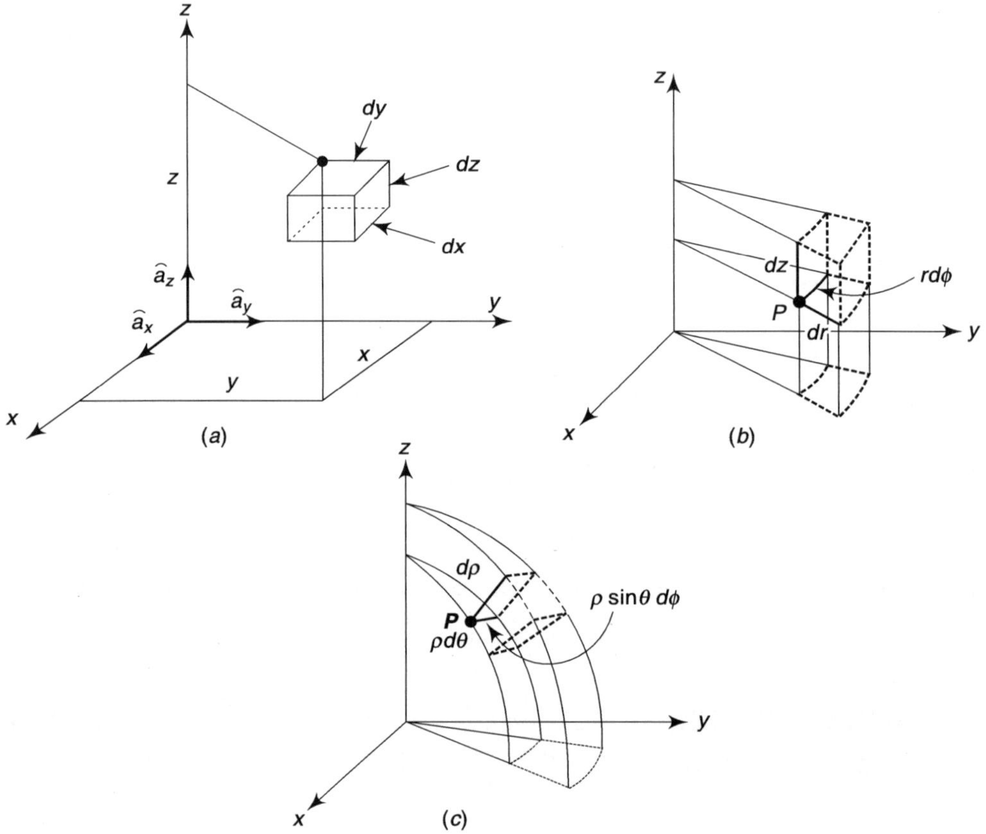

Fig. 1.18 *(a) Differential elements in Cartesian coordinates, (b) Differential elements in cylindrical coordinates, and (c) Differential elements in spherical coordinates*

Cartesian Coordinate System In this system as shown in Fig. 1.18 (*a*), the differential components of the general arc *ds* are identical with the differentials of the coordinates.

$\therefore \qquad u_1 = s_1 = x, \quad u_2 = s_2 = y, \quad u_3 = s_3 = z$

Hence, the metric coefficients are given as

$$h_1 = \frac{ds_1}{du_1} = 1 \qquad h_2 = \frac{ds_2}{du_2} = 1 \qquad h_3 = \frac{ds_3}{du_3} = 1$$

Differential displacement is given as

$$d\vec{s} = dx\hat{a}_x + dy\hat{a}_y + dz\hat{a}_z$$

Differential normal area is given as

$$d\vec{S} = dydz\hat{a}_x + dxdz\hat{a}_y + dxdy\hat{a}_z$$

Differential volume is given as

$$dV = dxdydz$$

NOTE

Vector Representation of Surface

Differential surface area is defined as a vector whose magnitude corresponds to the area of the surface and whose direction is perpendicular to the surface.

$$\therefore \qquad d\vec{S} = dS\hat{a}_n$$

For closed surface, the outward normal direction is taken as positive direction. For an open surface, the normal created by positive periphery according to the right-hand cork screw rule is taken to be positive.

If the surface is not a plane, then the surface is subdivided into smaller elements that are considered to be plane and a vector surface is considered for each of these elemental surfaces. Vector addition of these elemental vector surfaces gives the total surface.

Cylindrical Coordinate System In this system as shown in Fig. 1.18 (*b*), the differential components are given as

$$\therefore \qquad u_1 = r, \quad u_2 = \phi, \quad u_3 = z$$

The differential displacements are obtained from Fig. 1.18 (*b*) as

$$ds_1 = dr \quad ds_2 = rd_\phi \quad ds_3 = dz$$

Hence, the metric coefficients are given as

$$h_1 = \frac{ds_1}{du_1} = \frac{dr}{dr} = 1 \qquad h_2 = \frac{ds_2}{du_2} = \frac{rd\phi}{d\phi} = r \qquad h_3 = \frac{ds_3}{du_3} = \frac{dz}{dz} = 1$$

Differential displacement is given as

$$d\vec{s} = dr\hat{a}_r + rd\phi\hat{a}_\phi + dz\hat{a}_z$$

Differential normal area is given as

$$d\vec{S} = rd\phi dz\hat{a}_r + drdz\hat{a}_\phi + rdrd\phi\hat{a}_z$$

Differential volume is given as

$$dV = rdrd\phi dz$$

Spherical Coordinate System In this system as shown in Fig. 1.18 (*c*), the differential components are given as

$$\therefore \qquad u_1 = \rho, \quad u_2 = \theta, \quad u_3 = \phi$$

The differential displacements are obtained from Fig. 1.18 (*c*) as

$$ds_1 = d\rho \quad ds_2 = \rho d\theta \qquad ds_3 = \rho\sin\theta d\phi$$

Hence, the metric coefficients are given as

$$h_1 = \frac{ds_1}{du_1} = \frac{d\rho}{d\rho} = 1 \qquad h_2 = \frac{ds_2}{du_2} = \frac{\rho d\theta}{d\theta} = \rho \qquad h_3 = \frac{ds_3}{du_3} = \frac{\rho \sin\theta d\phi}{d\phi} = \rho \sin\theta$$

Differential displacement is given as

$$d\vec{s} = d\rho \hat{a}_\rho + \rho d\theta \hat{a}_\theta + \rho \sin\theta d\phi \hat{a}_\phi$$

Differential normal area is given as

$$d\vec{S} = \rho^2 \sin\theta d\theta d\phi \hat{a}_\rho + \rho \sin\theta d\rho d\phi \hat{a}_\theta + \rho d\rho d\theta \hat{a}_\phi$$

Differential volume is given as

$$dV = \rho^2 \sin\theta \, d\rho d\theta d\phi$$

These relations are given in Table 1.1.

Table 1.1 *Differential elements in different coordinate systems*

Differential Elements	Cartesian Coordinates	Cylindrical Coordinates	Spherical Coordinates
Length	$d\vec{s} = dx\hat{a}_x + dy\hat{a}_y + dz\hat{a}_z$	$d\vec{s} = dr\hat{a}_r + rd\phi\hat{a}_\phi + dz\hat{a}_z$	$d\vec{s} = d\rho\hat{a}_\rho + \rho d\theta\hat{a}_\theta + \rho\sin\theta d\phi\hat{a}_\phi$
Area	$d\vec{S} = dydz\hat{a}_x + dxdz\hat{a}_y + dxdy\hat{a}_z$	$d\vec{S} = rd\phi dz\hat{a}_r + drdz\hat{a}_\phi + rdrd\phi\hat{a}_z$	$d\vec{S} = \rho^2\sin\theta d\theta d\phi\hat{a}_\rho + \rho\sin\theta d\rho d\phi\hat{a}_\theta + \rho d\rho d\theta\hat{a}_\phi$
Volume	$dV = dxdydz$	$dV = rdrd\phi dz$	$dV = \rho^2 \sin\theta \, d\rho \, d\theta \, d\phi$

1.9 VECTOR CALCULUS

We now discuss the vector calculus, i.e., integrations and differentiations of vector.

1.9.1 Vector Integration or Vector Integrals

The integration of a vector may be obtained in three ways—line integral, surface integral and volume integral as discussed below.

Line Integral The line integral of a vector is the integral of the dot product of the vector and the differential length vector tangential to a specified path.

For vector $\vec{F}$ and a path l, the line integral is given by

$$\int_l \vec{F} \cdot d\vec{l} = \int_a^b |\vec{F}| \cos\theta dl \tag{1.53}$$

We consider a vector $\vec{F}$ and a specified path a–b. As the magnitude of the vector varies from point to point, the path is divided into a number of small line segments $d\vec{l}_1, d\vec{l}_2, \ldots$ with vector magnitudes $\vec{F}_1, \vec{F}_2, \ldots$ Thus, the total work done by the vector $\vec{F}$ from a to b is given as

$$W = \vec{F}_1 \cdot d\vec{l}_1 + \vec{F}_2 \cdot d\vec{l}_2 + \vec{F}_3 \cdot d\vec{l}_3 + \ldots = \sum_{i=1}^{n} \vec{F}_i \cdot d\vec{l}_i$$

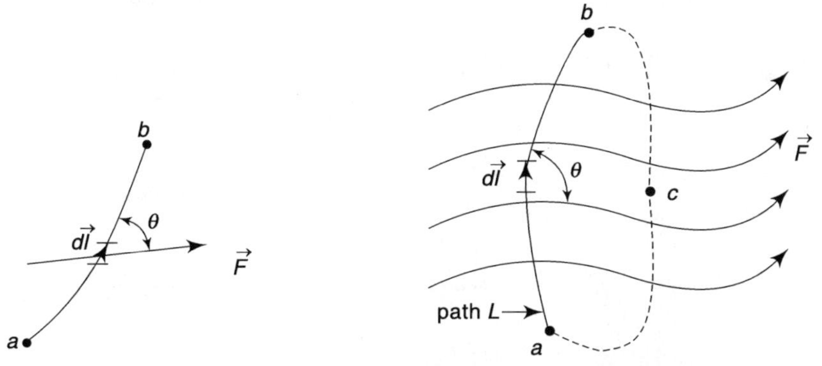

Fig. 1.19 *(a) Line integral of a vector, and (b) Path of integration of vector field $\vec{F}$*

If the lengths of the segments tend to zero, this work done can be written as line integral

$$W = \int_a^b \vec{F} \cdot d\vec{l}$$

If the path of integration is a closed curve, such as *abca,* the line integral becomes

$$W = \oint_l \vec{F} \cdot d\vec{l}$$

If the line-integration of a vector along a closed path is zero, i.e., $\oint_l \vec{F} \cdot d\vec{l} = 0$, then the vector is known as *conservative* or *lamellar vector*.

The concept of line integral is used to calculate the electric potential for a given electric field intensity or to calculate the total current enclosed by a closed path for a given magnetic field intensity.

***Example 1.10** Find the line integral of the vector $\vec{F} = (x^2 - y^2)\,\hat{a}_x + 2xy\hat{a}_y$ around a square of side *a* which has a corner at the origin, one side on the *x* axis and the other side on the *y* axis.

Solution

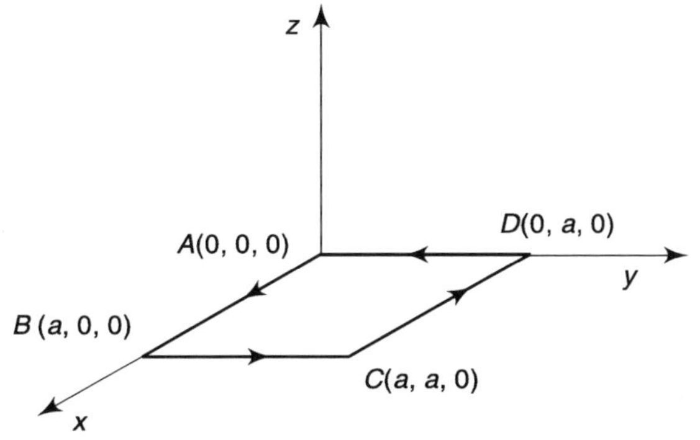

Fig. 1.20 *Line integral of a vector*

Here, $\oint_L \vec{F}\cdot d\vec{l} = \int_A^B \vec{F}\cdot d\vec{l} + \int_B^C \vec{F}\cdot d\vec{l} + \int_C^D \vec{F}\cdot d\vec{l} + \int_D^A \vec{F}\cdot d\vec{l}$

Now,

Along AB, $y = z = 0$, $dy = dz = 0$, $d\vec{l} = dx\hat{a}_x$

$$\therefore \qquad \int_A^B \vec{F}\cdot d\vec{l} = \int_{x=0}^{a} (x^2\hat{a}_x)\cdot(dx\hat{a}_x) = \int_{x=0}^{a} x^2 dx = \frac{a^3}{3}$$

Along BC: $x = a$, $z = 0$, $dx = dz = 0$, $d\vec{l} = dy\hat{a}_y$

$$\therefore \qquad \int_B^C \vec{F}\cdot d\vec{l} = \int_{y=0}^{a} [(a^2 - y^2)\hat{a}_x + 2ay\hat{a}_y]\cdot(dy\hat{a}_y) = \int_{y=0}^{a} 2ay dy = a^3$$

Along CD: $y = a$, $z = 0$, $dy = dz = 0$, $d\vec{l} = dx\hat{a}_x$

$$\therefore \qquad \int_C^D \vec{F}\cdot d\vec{l} = \int_{x=a}^{0} [(x^2 - a^2)\hat{a}_x + 2xa\hat{a}_y]\cdot(dx\hat{a}_x) = \int_{x=a}^{0} (x^2 - a^2)dx = \left[\frac{x^3}{3} - a^2 x\right]_a^0 = \frac{2}{3}a^3$$

Along DA: $x = z = 0$, $dx = dz = 0$, $d\vec{l} = dy\hat{a}_y$

$$\therefore \qquad \int_D^A \vec{F}\cdot d\vec{l} = \int_{y=a}^{0} 0\cdot(dy\hat{a}_y) = 0$$

Therefore, total line integral of the vector is given as

$$\oint_L \vec{F}\cdot d\vec{l} = \frac{a^3}{3} + a^3 + \frac{2}{3}a^3 = 2a^3$$

***Example 1.11** Compute the line integral of $\vec{F} = 6\hat{i} + yz^2\hat{j} + (3y + z)\hat{k}$ along triangular path shown in Fig. 1.21.

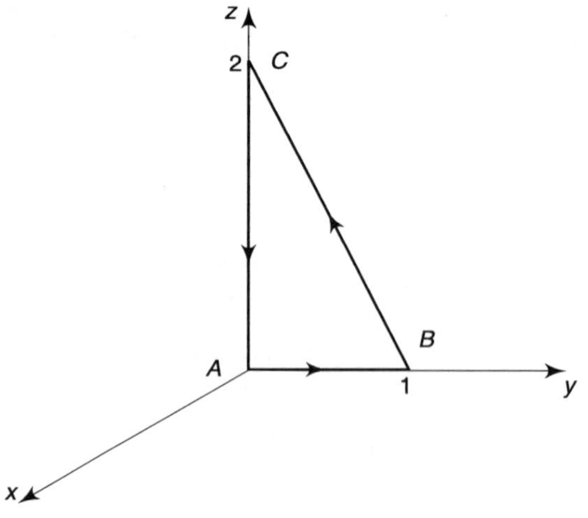

Fig. 1.21 *Triangular path of Example 1.11*

Solution Here, $\vec{F} = 6\hat{i} + yz^2\hat{j} + (3y + z)\hat{k}$

$$d\vec{l} = dx\hat{i} + dy\hat{j} + dz\hat{k}$$

$$\therefore \quad \vec{F} \cdot d\vec{l} = \{6\hat{i} + yz^2\hat{j} + (3y + z)\hat{k}\} \cdot (dx\hat{i} + dy\hat{j} + dz\hat{k}) = 6dx + yz^2dy + (3y + z)dz$$

The closed line integral is given as

$$\oint_L \vec{F} \cdot d\vec{l} = \int_A^B \vec{F} \cdot d\vec{l} + \int_B^C \vec{F} \cdot d\vec{l} + \int_C^A \vec{F} \cdot d\vec{l}$$

Along path AB, $dx = dz = 0$, $x = z = 0$; y varies from 0 to 1.

$$\therefore \quad \int_A^B \vec{F} \cdot d\vec{l} = \left[\int_0^1 yz^2 dy\right]_{z=0} = 0$$

Along path BC, $dx = 0$, $x = 0$; y varies from 1 to 0 and z varies from 0 to 2.
Also, for this path, the equation relating y and z is obtained as

$$\frac{y}{1} + \frac{z}{2} = 1 \quad \Rightarrow \quad z = 2(1 - y)$$

$$\therefore \quad \int_B^C \vec{F} \cdot d\vec{l} = \int_{y=1}^{0}\int_{z=0}^{2} yz^2dy + (3y + z)dz = \int_{y=1}^{0} y[2(1 - y)]^2\, dy + \int_{z=0}^{2}\left[3\left(\frac{2 - z}{2}\right) + z\right]dz$$

$$= \int_{y=1}^{0} 4(y^3 - 2y^2 + y)dy + \int_{z=0}^{2}\left(3 - \frac{z}{2}\right)dz$$

$$= 4\left[\frac{y^4}{4} - \frac{2y^3}{3} + \frac{y^2}{2}\right]_1^0 + \left[3z - \frac{z^2}{4}\right]_0^2$$

$$= 4\left[-\frac{1}{4} + \frac{2}{3} - \frac{1}{2}\right] + (6 - 1)$$

$$= \frac{14}{3}$$

Along path CA, $dx = dy = 0$, $x = y = 0$; z varies from 2 to 0.

$$\therefore \quad \int_C^A \vec{F} \cdot d\vec{l} = \left[\int_2^0 (3y + z)dz\right]_{y=0} = -2$$

By addition, we get the closed line integral as

$$\oint_L \vec{F} \cdot d\vec{l} = 0 + \frac{14}{3} - 2 = \frac{8}{3}$$

***Example 1.12** For a given vector, $\vec{F} = xy\hat{a}_x - 2x\hat{a}_y$, evaluate the line integral $\oint \vec{F} \cdot d\vec{l}$ over the path shown in Fig. 1.22.

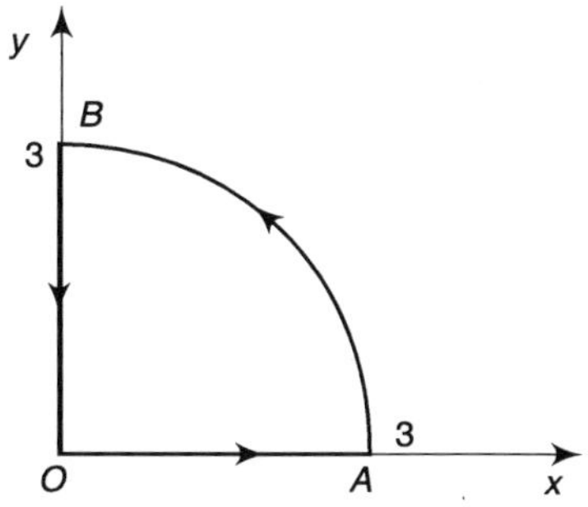

Fig. 1.22 *Path of Example 1.12*

Solution Here,

$$\vec{F} = xy\hat{a}_x - 2x\hat{a}_y, \; d\vec{l} = dx\hat{a}_x + dy\hat{a}_y + dz\hat{a}_z, \; d\vec{S} = dxdy\hat{a}_z$$

$$\therefore \qquad \vec{F}\cdot d\vec{l} = (xy\hat{a}_x - 2x\hat{a}_y)\cdot(dx\hat{a}_x + dy\hat{a}_y + dz\hat{a}_z) = (xydx - 2xdy)$$

Along the path OA, $y = 0$, x varies from 0 to 3.

$$\therefore \qquad \int_O^A \vec{F}\cdot d\vec{l} = 0$$

Along the path AB, x varies from 3 to 0 and y varies from 0 to 3. Also, along this path, the equation relating x and y is obtained as

$$x^2 + y^2 = 9$$

$$\therefore \qquad \int_A^B \vec{F}\cdot d\vec{l} = \int_{x=3}^{0}\int_{y=0}^{3} (xydx - 2xdy)$$

$$= \int_{x=3}^{0} x(\sqrt{9-x^2})dx - \int_{y=0}^{3} 2(\sqrt{9-y^2})dy$$

$$= -\frac{1}{3}(9-x^2)^{3/2}\Big|_3^0 - \left[y\sqrt{9-y^2} + 9\sin^{-1}\frac{y}{3}\right]_0^3$$

$$= -9\left(1+\frac{\pi}{2}\right)$$

Along the path BO, $x = 0$, y varies from 3 to 0.

$$\therefore \qquad \int_O^A \vec{F}\cdot d\vec{l} = 0$$

By addition, we get $\oint_L \vec{F}\cdot d\vec{l} = -9\left(1+\frac{\pi}{2}\right)$

$$\text{Also, } \nabla\times\vec{F} = \begin{vmatrix} \hat{a}_x & \hat{a}_y & \hat{a}_z \\ \frac{\partial}{\partial x} & \frac{\partial}{\partial y} & \frac{\partial}{\partial z} \\ xy & -2x & 0 \end{vmatrix} = -(x+2)\hat{a}_z$$

$$\therefore \qquad \int_S (\nabla\times\vec{F})\cdot d\vec{S} = \int_{x=0}^{3}\int_{y=0}^{3} [-(x+2)\hat{a}_z]\cdot(dxdy\hat{a}_z) = -\int_{x=0}^{3}\int_{y=0}^{3}(x+2)dxdy$$

$$= -\int_{y=0}^{3}\int_{x=0}^{\sqrt{9-y^2}} (x+2)dxdy = -\int_{y=0}^{3}\left(\frac{x^2}{2}+2x\right)_0^{\sqrt{9-y^2}} dy$$

$$= -\int_{y=0}^{3}\left(\frac{9-y^2}{2} + 2\sqrt{9-y^2}\right)dy = -\left[\frac{9}{2}y - \frac{y^3}{6} + y\sqrt{9-y^2} + 9\sin^{-1}\frac{y}{3}\right]_0^3$$

$$= -9\left(1+\frac{\pi}{2}\right)$$

Example 1.13 Calculate the circulation of $\vec{A} = r\cos\phi\hat{a}_r + z\sin\phi\hat{a}_z$ around the edge L of the wedge defined by $0 \le r \le 2$, $0 \le \phi \le 60°$, $z = 0$ as shown in Fig. 1.23.

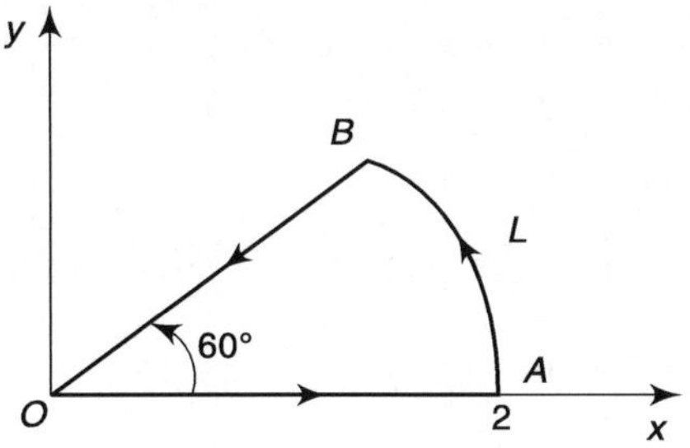

Fig. 1.23 *Path of Example 1.13*

Solution Here, $\vec{A} = r\cos\phi\hat{a}_r + z\sin\phi\hat{a}_z$

$$d\vec{l} = dr\hat{a}_r + rd\phi\hat{a}_\phi + dz\hat{a}_z$$

$$\therefore \quad \vec{A}\cdot d\vec{l} = (r\cos\phi\hat{a}_r + z\sin\phi\hat{a}_z)\cdot(dr\hat{a}_r + rd\phi\hat{a}_\phi + dz\hat{a}_z) = (r\cos\phi dr + z\sin\phi dz)$$

Since the path is on the *xy* plane, $dz = 0$.

$$\therefore \quad \vec{A}\cdot d\vec{l} = r\cos\phi dr$$

$$\oint_L \vec{A}\cdot d\vec{l} = \int_O^A \vec{A}\cdot d\vec{l} + \int_A^B \vec{A}\cdot d\vec{l} + \int_B^O \vec{A}\cdot d\vec{l}$$

Along the path OA, r varies from 0 to 2.

$$\therefore \quad \int_O^A \vec{A}\cdot d\vec{l} = \left[\int_0^2 r\cos\phi dr\right]_{\phi=0} = 2$$

Along the path AB, r is constant and so the integration is zero.

$$\therefore \quad \int_A^B \vec{A}\cdot d\vec{l} = 0$$

Along the path BO, r varies from 2 to 0.

$$\therefore \quad \int_B^O \vec{A}\cdot d\vec{l} = \left[\int_2^0 r\cos\phi dr\right]_{\phi=60^\circ} = \frac{1}{2}\times(-2) = -1$$

By addition, we get the closed line integral as, $\oint_L \vec{A}\cdot d\vec{l} = 2 + 0 - 1 = 1.$

Example 1.14 Using the concept of line integral, find the periphery of a circle of radius a.

Solution In cylindrical coordinates, the differential length is, $d\vec{l} = ad\phi\hat{a}_\phi$

By line integral, the periphery is obtained as

$$\int_{\phi=0}^{2\pi} \hat{a}_\phi\cdot d\vec{l} = \int_{\phi=0}^{2\pi} (\hat{a}_\phi)\cdot(ad\phi\hat{a}_\phi) = \int_{\phi=0}^{2\pi} ad\phi = 2\pi a$$

Surface Integral For a vector $\vec{F}$, continuous in a region containing a smooth surface S, the surface integral or the flux of $\vec{F}$ through S is defined as,

$$\psi = \int_S \vec{F} \cdot d\vec{S} = \int_S \vec{F} \cdot \hat{a}_n dS = \int_S |\vec{F}| \cos\theta dS \qquad \textbf{(1.54)}$$

where, $\hat{a}_n$ is the unit normal vector to the surface S.

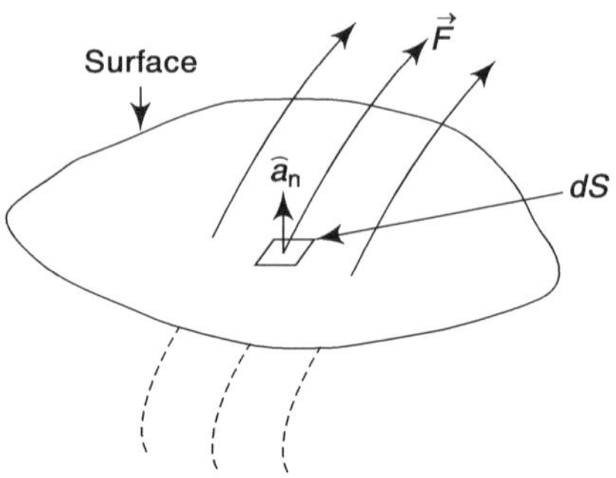

Fig. 1.24 *Surface integral*

We consider a surface S. We divide the surface into infinitesimal surfaces $d\vec{S}_1, d\vec{S}_2, \ldots$ which are treated as the vector quantities. Let $\vec{F}_1, \vec{F}_2, \ldots$ be the vector magnitudes at the elemental surfaces, respectively. Thus, the sum of the scalar products $\vec{F}_1 \cdot d\vec{S}_1$, $\vec{F}_2 \cdot d\vec{S}_2, \ldots$ is written as

$$\vec{F}_1 \cdot d\vec{S}_1 + \vec{F}_2 \cdot d\vec{S}_2 + \vec{F}_3 \cdot d\vec{S}_3 + \cdots = \sum_{i=1}^{n} \vec{F}_i \cdot d\vec{S}_i$$

If the elemental surfaces areas tend to zero, this can be written as surface-integral

$$\psi = \int_S \vec{F} \cdot d\vec{S}$$

If the surface is a closed surface, the surface integral is written as

$$\psi = \oint_S \vec{F} \cdot d\vec{S}$$

For a closed surface, the surface integral, $\oint_S \vec{F} \cdot d\vec{S}$ is referred to as the net outward flux of $\vec{F}$ from the surface. If the surface integral of a vector over a closed surface is zero, i.e., $\oint_S \vec{F} \cdot d\vec{S} = 0$, then the vector is known as *solenoidal vector*. The concept of surface integral is necessary to calculate the flux or current from the flux or current density over a surface.

Example 1.15 Using the concept of surface integral, find the surface area of a sphere of radius a.

Solution In spherical coordinates, the differential surface in the perpendicular direction is given as

$$d\vec{S} = a^2 \sin\theta d\theta d\phi \hat{a}_\rho$$

By surface integral, the surface area of the sphere is obtained as

$$S = \int_S \hat{a}_\rho \cdot d\vec{S} = \int_{\theta=0}^{\pi} \int_{\phi=0}^{2\pi} a^2 \sin\theta \, d\theta d\phi = a^2 \times 2\pi \times (-\cos\theta)_0^\pi = 4\pi a^2$$

***Example 1.16** Use the cylindrical coordinate system to find the area of a curved surface on the right circular cylinder having radius = 3 m and height = 6 m and $30° \le \phi \le 120°$.

Solution Here, the differential surface is given as

$$d\vec{S} = r d\phi dz \hat{a}_r$$

Taking the surface integral, the area of the curved surface is obtained as

$$S = \int_S \hat{a}_r \cdot d\vec{S} = \int_{\phi=\pi/6}^{2\pi/3} \int_{z=0}^{6} r d\phi dz \Big|_{r=3} = 3 \times 6 \times \left(\frac{2\pi}{3} - \frac{\pi}{6}\right) = 9\pi \quad \text{m}^2$$

***Example 1.17** Use the spherical coordinate system to find the area of the strip $\alpha \le \theta \le \beta$ on the spherical shell of radius 'a'. What results when $\alpha = 0$ and $\beta = \pi$?

Solution For a fixed radius of a, the elemental surface is

$$d\vec{S} = \rho^2 \sin\theta d\theta d\phi \hat{a}_\rho \Big|_{\rho=a} = a^2 \sin\theta d\theta d\phi \hat{a}_\rho$$

Hence, the area of the strip is given as

$$S = \int_S d\vec{S} \cdot \hat{a}_\rho = \int_{\theta=\alpha}^{\beta} \int_{\phi=0}^{2\pi} a^2 \sin\theta d\theta d\phi$$

$$= 2\pi \times a^2 \times (-\cos\theta)_\alpha^\beta$$

$$= 2\pi a^2 (\cos\alpha - \cos\beta)$$

For $\alpha = 0$ and $\beta = \pi$, the area is

$$S = 2\pi a^2 (\cos 0 - \cos\pi) = 4\pi a^2$$

This is the area of a sphere of radius a.

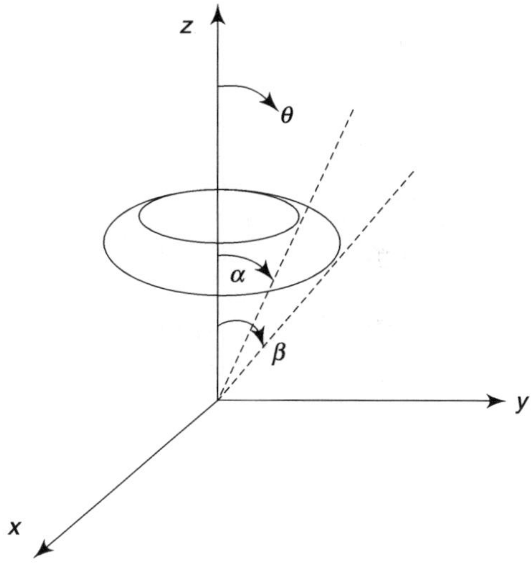

Fig. 1.25 *Arrangement of Example 1.17*

Example 1.18 Given $\rho_s = (x^2 + xy)$; calculate $\int \rho_s dS$ over the region $y \le x^2$, $0 < x < 1$.

Solution

$$\int \rho_s dS = \int_{x=0}^{1} \int_{y=0}^{x^2} (x^2 + xy) dx dy = \int_{x=0}^{1} \left(x^2 y + \frac{xy^2}{2}\right)_0^{x^2} dx = \int_{x=0}^{1} \left(x^4 + \frac{x^5}{2}\right)_0^{x^2} dx = \left(\frac{x^5}{5} + \frac{x^6}{12}\right)_0^1 = \frac{17}{60}$$

Example 1.19 Given $\vec{F} = \hat{a}_r \dfrac{k_1}{r} + \hat{a}_z k_2 z$; evaluate the scalar surface integral $\oint_s \vec{F} \cdot d\vec{S}$ over the surface of a closed cylinder about the z-axis specified by $z = \pm 3$ and $r = 2$.

Solution The cylinder has three surfaces as follows.

$$\oint_s \vec{F} \cdot d\vec{S} = \int_{\text{Circular 1}} \vec{F} \cdot d\vec{S} + \int_{\text{Circular 2}} \vec{F} \cdot d\vec{S} + \int_{\text{Curved}} \vec{F} \cdot d\vec{S}$$

For the upper circular surface

$$z = 3, \quad \hat{a}_n = \hat{a}_z$$

$$\int_{\text{Circular 1}} \vec{F} \cdot d\vec{S} = \int_{r=0}^{2} \int_{\phi=0}^{2\pi} \left(\hat{a}_r \frac{k_1}{r} + \hat{a}_z k_2 z\right) \cdot (r dr d\phi \hat{a}_z) \Big|_{z=3} = \int_{r=0}^{2} \int_{\phi=0}^{2\pi} 3k_2 r dr d\phi = 12\pi k_2$$

For the bottom circular surface

$$z = -3, \quad \hat{a}_n = -\hat{a}_z$$

$$\int_{\text{Circular 1}} \vec{F} \cdot d\vec{S} = \int_{r=0}^{2} \int_{\phi=0}^{2\pi} \left(\hat{a}_r \frac{k_1}{r} + \hat{a}_z k_2 z\right) \cdot (-r dr d\phi \hat{a}_z)\Bigg|_{z=-3} = 12\pi k_2$$

For the curved surface

$$r = 2, \quad \hat{a}_n = \hat{a}_r$$

$$\int_{\text{Circular 1}} \vec{F} \cdot d\vec{S} = \int_{\phi=0}^{2\pi} \int_{z=-3}^{3} \left(\hat{a}_r \frac{k_1}{r} + \hat{a}_z k_2 z\right) \cdot (r d\phi \hat{a}_r)\Bigg|_{r=2} = \int_{\phi=0}^{2\pi} \int_{z=-3}^{3} k_1 d\phi dz = 12\pi k_1$$

By addition, total surface area of the closed cylinder is given as

$$\oint_S \vec{F} \cdot d\vec{S} = 12\pi(k_1 + 2k_2)$$

Example 1.20 Evaluate $\oint_S \vec{r} \cdot \hat{a}_n dS$ where S is a closed surface.

Solution We know, $\vec{r} = x\hat{a}_x + y\hat{a}_y + z\hat{a}_z$

$$\therefore \quad \oint_S \vec{r} \cdot \hat{a}_n dS = \int_V \nabla \cdot \vec{r} dv = \int_V 3dv = 3V \quad \{\because \ \nabla \cdot \vec{r} = 3, \text{ see Example 36 (b)}\}$$

where V is the volume enclosed by the surface S.

Volume Integral The volume integral of a scalar quantity F over a volume V is written as

$$U = \int_V F dv \tag{1.55}$$

The concept of volume integral is necessary to calculate the charge or mass of an object, which are distributed in the volume.

Example 1.21 Using the concept of volume integral, find the volume of a sphere of radius a.

Solution In spherical coordinates, the differential volume is given as

$$dv = \rho^2 \sin\theta d\rho d\theta d\phi$$

where

$$0 \le \rho \le a$$
$$0 \le \theta \le \pi$$
$$0 \le \phi \le 2\pi$$

By volume integral, the volume of the sphere is obtained as

$$v = \int_v dv = \int_{\rho=0}^{a} \int_{\theta=0}^{\pi} \int_{\phi=0}^{2\pi} \rho^2 \sin\theta d\rho d\theta d\phi = 2\pi \times (-\cos\theta)_0^\pi \times \int_{\rho=0}^{a} \rho^2 d\rho = 4\pi \times \frac{a^3}{3} = \frac{4}{3}\pi a^3$$

Example 1.22 Obtain the expression for the volume of a sphere of radius a using the concept of volume integral.

Solution Here, the differential volume in spherical coordinates is given as

$$dv = \rho^2 \sin\theta d\rho d\theta d\phi$$

where

$$a \le r \le 0$$
$$0 \le \theta \le \pi$$
$$0 \le \phi \le 2\pi$$

The volume of the sphere is obtained by the volume integral as given.

$$V = \int_v \rho^2 \sin\theta d\rho d\theta d\phi = \int_{\phi=0}^{2\pi} \int_{\theta=0}^{\pi} \int_{\rho=0}^{a} \rho^2 \sin\theta d\rho d\theta d\phi = \frac{a^3}{3} \times 2\pi \times (-\cos\theta)_0^{\pi} = \frac{4}{3}\pi a^3$$

1.9.2 Vector Differentiations

In order to understand vector differentiation, we introduce an operator known as del operator or differential vector operator.

Differential Vector Operator (∇) or Del Operator The differential vector operator (∇) or Del or Nabla, in Cartesian coordinates, is defined as

$$\nabla = \frac{\partial}{\partial x}\hat{a}_x + \frac{\partial}{\partial y}\hat{a}_y + \frac{\partial}{\partial z}\hat{a}_z \tag{1.56}$$

This *Del is merely a vector operator but not a vector quantity*. When it operates on a scalar function, a vector is created. Since a vector, in general, is a function of the space and time both, the *del operator is a vector space function operator*. It is defined in terms of partial derivatives with respect to space.

In Eq. (1.56), *del* has been expressed in Cartesian coordinates. However, this operator can be expressed in general curvilinear coordinates as

$$\nabla = \frac{\partial}{\partial s_1}\hat{a}_1 + \frac{\partial}{\partial s_2}\hat{a}_2 + \frac{\partial}{\partial s_3}\hat{a}_3 \tag{1.57}$$

But, we know that, $ds_i = h_i du_i$; replacing this in Eq. (1.57), we get

$$\boxed{\nabla = \frac{1}{h_1}\frac{\partial}{\partial u_1}\hat{a}_1 + \frac{1}{h_2}\frac{\partial}{\partial u_2}\hat{a}_2 + \frac{1}{h_3}\frac{\partial}{\partial u_3}\hat{a}_3} \tag{1.58}$$

Substituting the values of h_i in different coordinate systems, we obtain the relation of *del* in three different coordinate systems as

$$\nabla = \frac{\partial}{\partial x}\hat{a}_x + \frac{\partial}{\partial y}\hat{a}_y + \frac{\partial}{\partial z}\hat{a}_z \quad \text{(Cartesian coordinates)} \tag{1.59a}$$

$$= \frac{\partial}{\partial r}\hat{a}_r + \frac{1}{r}\frac{\partial}{\partial \phi}\hat{a}_\phi + \frac{\partial}{\partial z}\hat{a}_z \quad \text{(Cylindrical coordinates)} \tag{1.59b}$$

$$= \frac{\partial}{\partial \rho}\hat{a}_\rho + \frac{1}{\rho}\frac{\partial}{\partial \theta}\hat{a}_\theta + \frac{1}{\rho\sin\theta}\frac{\partial}{\partial \phi}\hat{a}_\varphi \quad \text{(Spherical coordinates)} \tag{1.59c}$$

Corresponding to three different vector multiplications, there are three possible operations of ∇. These operations are:

1. Gradient of a scalar F, written as, ∇F;
2. Divergence of a vector $\vec{A}$, written as, $\nabla \cdot \vec{A}$;
3. Curl of a vector $\vec{A}$, written as, $\nabla \times \vec{A}$; and

Gradient of a Scalar

Definition The gradient of a scalar function is both the magnitude and the direction of the maximum space rate of change of that function.

Mathematical Expression of Gradient We consider a scalar function F. A mathematical expression for the gradient can be obtained by evaluating the difference in the field dF between the points P_1 and P_2.

Here, F_1 and F_2 are the contours on which F is constant.

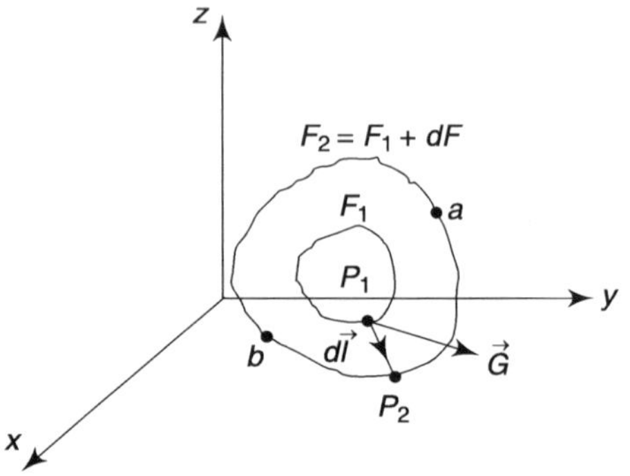

Fig. 1.26 *Gradient of a scalar quantity*

$$dF = \frac{\partial F}{\partial x}dx + \frac{\partial F}{\partial y}dy + \frac{\partial F}{\partial z}dz$$
$$= \left(\frac{\partial F}{\partial x}\hat{a}_x + \frac{\partial F}{\partial y}\hat{a}_y + \frac{\partial F}{\partial z}\hat{a}_z\right) \cdot (dx\hat{a}_x + dy\hat{a}_y + dz\hat{a}_z)$$
$$= \vec{G} \cdot d\vec{l}$$

where $\vec{G} = \left(\frac{\partial F}{\partial x}\hat{a}_x + \frac{\partial F}{\partial y}\hat{a}_y + \frac{\partial F}{\partial z}\hat{a}_z\right)$

and, $d\vec{l} = dx\hat{a}_x + dy\hat{a}_y + dz\hat{a}_z$ = differential length = differential displacement from P_1 to P_2

$\therefore$ $$dF = \vec{G} \cdot d\vec{l} = Gdl \cos\theta$$

$\therefore$ $$\frac{dF}{dl} = G\cos\theta; \quad \text{where } \theta \text{ is the angle between } \vec{G} \text{ and } d\vec{l}$$

For maximum $\left(\frac{dF}{dl}\right)$, $\theta = 0°$, i.e., when $d\vec{l}$ is in the direction of $\vec{G}$.

$\therefore$ $\left.\frac{dF}{dl}\right|_{\max} = \frac{dF}{dn} = G$; where, $\frac{dF}{dn}$ is the normal derivative.

Thus, by definition of gradient, we have the mathematical expression of gradient in Cartesian coordinates given as

$$\boxed{\text{Grad } F = \nabla F = \frac{\partial F}{\partial x}\hat{a}_x + \frac{\partial F}{\partial y}\hat{a}_y + \frac{\partial F}{\partial z}\hat{a}_z} \quad \textbf{(1.60)}$$

Physical Interpretation The gradient of a scalar quantity is the maximum space rate of change of the function.

For example, we consider a room in which the temperature is given by a scalar field T, so at any point (x, y, z) the temperature is $T(x, y, z)$ (assuming that the temperature does not change with time). Then, at any arbitrary point in the room, the gradient of T indicates the direction in which the temperature rises most rapidly. The magnitude of the gradient will determine how fast the temperature rises in that direction.

Gradient in General Curvilinear Coordinates Let $F(u_1, u_2, u_3)$ be a scalar function. The u_1 component of the gradient of F, by definition, is the maximum space rate of change of the function, i.e.,

$$(\text{Grad } F)_1 = \lim_{du_1 \to 0} \frac{F(G) - F(O)}{h_1 du_1} = \frac{1}{h_1}\frac{\partial F}{\partial u_1}$$

Similarly, considering directions 2 and 3 and so, the resultant expression of the gradient of F is given as

$$\boxed{\nabla F = \frac{1}{h_1}\frac{\partial F}{\partial u_1}\hat{a}_1 + \frac{1}{h_2}\frac{\partial F}{\partial u_2}\hat{a}_2 + \frac{1}{h_3}\frac{\partial F}{\partial u_3}\hat{a}_3} \quad \textbf{(1.61)}$$

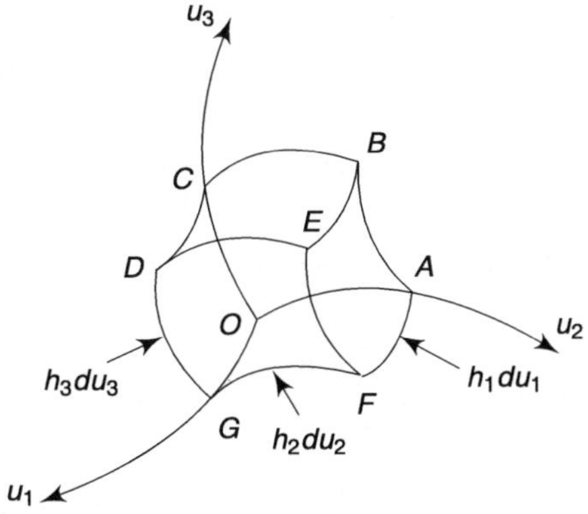

Fig. 1.27 *General Curvilinear coordinates*

Substituting the values of h_i from Section 1.8, we get the relation of gradient in three different coordinate systems as

$$\nabla F = \frac{\partial F}{\partial x}\hat{a}_x + \frac{\partial F}{\partial y}\hat{a}_y + \frac{\partial F}{\partial z}\hat{a}_z \quad \text{(Cartesian Coordinates)} \quad \textbf{(1.62}\textbf{\textit{a}}\textbf{)}$$

$$= \frac{\partial F}{\partial r}\hat{a}_r + \frac{1}{r}\frac{\partial F}{\partial \phi}\hat{a}_\phi + \frac{\partial F}{\partial z}\hat{a}_z \quad \text{(Cylindrical Coordinates)} \quad \textbf{(1.62}\textbf{\textit{b}}\textbf{)}$$

$$= \frac{\partial F}{\partial \rho}\hat{a}_\rho + \frac{1}{\rho}\frac{\partial F}{\partial \theta}\hat{a}_\theta + \frac{1}{\rho\sin\theta}\frac{\partial F}{\partial \phi}\hat{a}_\phi \quad \text{(Spherical Coordinates)} \quad \textbf{(1.62}\textbf{\textit{c}}\textbf{)}$$

Properties of Gradient

1. The magnitude of the gradient of a scalar function is the maximum rate of change of the function per unit distance.
2. The direction of the gradient of a scalar function is in the direction in which the function changes most rapidly.
3. The gradient of a scalar function at any point is always perpendicular to the surface that passes through the point and over which the function is constant (points *a* and *b* in Fig. 1.26).
4. The projection of the gradient of a scalar function (say, ∇S) in the direction of a unit vector $\hat{a}$, i.e., $\nabla S \cdot \hat{a}$ is known as the *directional derivative* of the function S along unit vector $\hat{a}$.

Example 1.23 Find the gradient of the following scalar fields:

(a) $F = x^2 y + e^z$
(b) $V = rz \sin\phi + z^2 \cos^2\phi + r^2$
(c) $S = \cos\theta \sin\phi \ln\rho + \rho^2 \phi$.

Solution **(a)** The gradient in Cartesian coordinates is given as

$$\nabla F = \frac{\partial F}{\partial x}\hat{a}_x + \frac{\partial F}{\partial y}\hat{a}_y + \frac{\partial F}{\partial z}\hat{a}_z = 2x\hat{a}_x + x^2\hat{a}_y + e^z\hat{a}_z$$

(b) The gradient in cylindrical coordinates is given as

$$\nabla V = \frac{\partial V}{\partial r}\hat{a}_r + \frac{1}{r}\frac{\partial V}{\partial \phi}\hat{a}_\phi + \frac{\partial V}{\partial z}\hat{a}_z$$

$$= (z \sin\phi + 2r)\hat{a}_r + \frac{1}{r}(rz\cos\phi - z^2\, 2\cos\phi\sin\phi)\hat{a}_\phi + (r\sin\phi + 2z\cos^2\phi)\hat{a}_z$$

$$= (z \sin\phi + 2r)\hat{a}_r + \left(z\cos\phi - \frac{z^2}{r}\sin 2\phi\right)\hat{a}_\phi + (r\sin\phi + 2z\cos^2\phi)\hat{a}_z$$

(c) The gradient in spherical coordinates is given as

$$\nabla S = \frac{\partial S}{\partial \rho}\hat{a}_\rho + \frac{1}{\rho}\frac{\partial S}{\partial \theta}\hat{a}_\theta + \frac{1}{\rho\sin\theta}\frac{\partial S}{\partial \phi}\hat{a}_\phi$$

$$= \left(\frac{\cos\theta\sin\phi}{\rho} + 2\rho\phi\right)\hat{a}_\rho + \frac{1}{\rho}\sin\theta\sin\phi\ln\rho\hat{a}_\theta + \frac{1}{\rho\sin\theta}(\cos\theta\cos\phi\ln\rho + \rho^2)\hat{a}_\phi$$

$$= \left(\frac{\cos\theta\sin\phi}{\rho} + 2\rho\phi\right)\hat{a}_\rho + \frac{\sin\theta\sin\phi}{\rho}\ln\rho\hat{a}_\theta + \left(\frac{\cot\theta}{\rho}\cos\phi\ln\rho + \rho\,\text{cosec}\,\theta\right)\hat{a}_\phi$$

Example 1.24 Find the gradient of the following scalar fields:

(a) $V = 4xz^2 + 3\,yz$
(b) $V = 2r(1 + z^2)\cos\phi$
(c) $V = \rho^2\cos\theta\cos\phi$

Solution

(a) $V = 4xz^2 + 3\,yz$

$$\nabla V = \frac{\partial}{\partial x}(4xz^2 + 3yz)\hat{a}_x + \frac{\partial}{\partial y}(4xz^2 + 3yz)\hat{a}_y + \frac{\partial}{\partial z}(4xz^2 + 3yz)\hat{a}_z = 4z^2\hat{a}_x + 3z\hat{a}_y + (8xz + 3y)\hat{a}_z$$

(b) $V = 2r(1 + z^2)\cos\phi$

$$\nabla V = \frac{\partial V}{\partial r}\hat{a}_r + \frac{1}{r}\frac{\partial V}{\partial \phi}\hat{a}_\phi + \frac{\partial V}{\partial z}\hat{a}_z = 2(1+z^2)\cos\phi\hat{a}_r - 2(1+z^2)\sin\phi\hat{a}_\phi + 4rz\cos\phi\hat{a}_z$$

(c) $V = \rho^2\cos\theta\cos\phi$

$$\therefore \quad \nabla V = \frac{\partial V}{\partial \rho}\hat{a}_\rho + \frac{1}{\rho}\frac{\partial V}{\partial \theta}\hat{a}_\theta + \frac{1}{\rho\sin\theta}\frac{\partial V}{\partial \phi}\hat{a}_\phi$$

$$= 2\rho\cos\theta\cos\phi\hat{a}_\rho - \rho\sin\theta\cos\phi\hat{a}_\theta + \rho\cot\theta\sin\phi\hat{a}_\phi$$

***Example 1.25** Find the rate at which the scalar function $V = r^2\sin 2\phi$ in cylindrical coordinates increases in the direction of the vector $\vec{A} = \hat{a}_r + \hat{a}_\phi$ at the point $\left(2, \frac{\pi}{4}, 0\right)$.

Or,

Find the gradient of the scalar function $V = r^2\sin 2\phi$ and the directional derivative of the function in the direction $(\hat{a}_r + \hat{a}_\phi)$ at the point $\left(2, \frac{\pi}{4}, 0\right)$.

Solution The gradient in cylindrical coordinates is given as

$$\nabla V = \frac{\partial V}{\partial r}\widehat{a}_r + \frac{1}{r}\frac{\partial V}{\partial \phi}\widehat{a}_\phi + \frac{\partial V}{\partial z}\widehat{a}_z = 2r \sin 2\phi \widehat{a}_r + 2r \cos 2\phi \widehat{a}_\phi = 2r(\sin 2\phi \widehat{a}_r + \cos 2\phi \widehat{a}_\phi)$$

The direction derivative is given as

$$\nabla V \cdot \widehat{a}_A = \nabla V \cdot \frac{\vec{A}}{|\vec{A}|} = (2r \sin 2\phi \widehat{a}_r + 2r \cos 2\phi \widehat{a}_\phi) \cdot \left(\frac{\widehat{a}_r + \widehat{a}_\phi}{\sqrt{2}}\right) = \sqrt{2}r \sin 2\phi + \sqrt{2}r \cos 2\phi$$

At $\left(2, \frac{\pi}{4}, 0\right)$, the directional derivative is given as

$$\nabla V \cdot \widehat{a}_A = \sqrt{2} \times 2 \sin \frac{\pi}{2} + \sqrt{2} \times 2 \cos \frac{\pi}{2} = 2\sqrt{2}$$

***Example 1.26** Find $\nabla\left(\frac{1}{r}\right)$, where $r = x\widehat{a}_x + y\widehat{a}_y + z\widehat{a}_z$

Or

Show that for a moving field point and a fixed source point, $\text{Grad}\left(\frac{1}{r}\right) = -\left(\frac{1}{r^2}\right)\widehat{a}_r$, where r is the distance between the source point and the field point.

Solution Here, $r = x\widehat{a}_x + y\widehat{a}_y + z\widehat{a}_z$

$$\therefore \quad r = \sqrt{x^2 + y^2 + z^2}$$

$$\therefore \quad \nabla\left(\frac{1}{r}\right) = \frac{\partial}{\partial x}\left(\frac{1}{\sqrt{x^2 + y^2 + z^2}}\right)\widehat{a}_x + \frac{\partial}{\partial y}\left(\frac{1}{\sqrt{x^2 + y^2 + z^2}}\right)\widehat{a}_y + \frac{\partial}{\partial z}\left(\frac{1}{\sqrt{x^2 + y^2 + z^2}}\right)\widehat{a}_z$$

$$= -\frac{1}{2}\frac{2x}{(x^2 + y^2 + z^2)^{3/2}}\widehat{a}_x - -\frac{1}{2}\frac{2y}{(x^2 + y^2 + z^2)^{3/2}}\widehat{a}_y - -\frac{1}{2}\frac{2z}{(x^2 + y^2 + z^2)^{3/2}}\widehat{a}_z$$

$$= -\frac{x\widehat{a}_x + y\widehat{a}_y + z\widehat{a}_z}{(x^2 + y^2 + z^2)^{3/2}}$$

$$= -\frac{\vec{r}}{r^3} \qquad \{\because \vec{r} = r\widehat{a}_r\}$$

$$= -\left(\frac{1}{r^2}\right)\widehat{a}_r$$

This problem can be solved easily in cylindrical coordinates as follows.

$$\nabla\left(\frac{1}{r}\right) = \frac{\partial}{\partial r}\left(\frac{1}{r}\right)\widehat{a}_r = -\left(\frac{1}{r^2}\right)\widehat{a}_r$$

$$\therefore \quad \boxed{\nabla\left(\frac{1}{r}\right) = -\frac{\vec{r}}{r^3} = -\left(\frac{1}{r^2}\right)\widehat{a}_r}$$

Divergence of a Vector

Definition Divergence of a vector at any point is defined as the limit of its surface integral per unit volume as the volume enclosed by the surface around the point shrinks to zero.

$$\therefore \qquad \text{div}\,\vec{F} = \nabla\cdot\vec{F} = \lim_{v\to 0}\left(\frac{\oint_S \vec{F}\cdot d\vec{S}}{v}\right) = \lim_{v\to 0}\left(\frac{\oint_S \vec{F}\cdot \hat{a}_n dS}{v}\right) \tag{1.63}$$

where v is the volume of an *arbitrarily* shaped region in space that includes the point, S is the surface of that volume, and the integral is a surface integral with $\hat{a}_n$ being the outward normal to that surface.

Figures 1.28 (*a*), (*b*) and (*c*) show three cases of positive, negative and zero divergence.

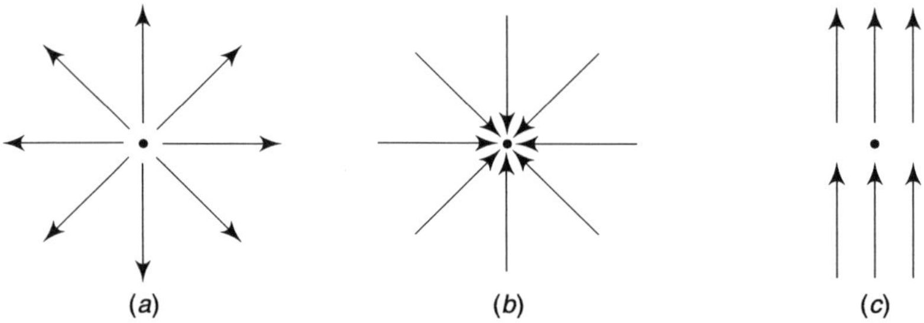

Fig. 1.28 *(a) Positive divergence, (b) Negative divergence, and (c) Zero divergence*

Mathematical Expression of Divergence We consider a hypothetical infinitesimal cubical box oriented along the coordinate axes around an infinitesimal region of space.

We consider a vector $\vec{V}$ at a point $P(x, y, z)$. Let, V_1, V_2, and V_3 be the components of $\vec{V}$ along the three coordinate axes.

In order to compute the surface integral, we see that there are six surfaces to this box, and the net content leaving the box is therefore, simply the sum of differences in the values of the vector field along the three sets of parallel surfaces of the box.

The component vectors are given as follows.

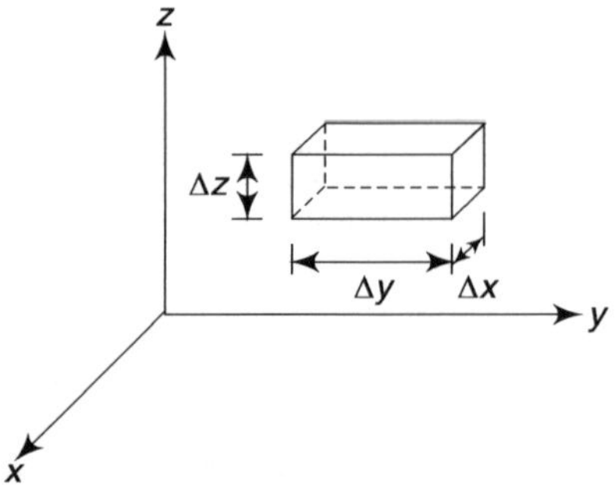

Fig. 1.29 *To derive expression for divergence in Cartesian coordinates*

Along x-direction: at front surface, $\left(V_1 + \frac{1}{2}\frac{\partial V_1}{\partial x}\Delta x\right)$

at back surface, $\left(V_1 - \frac{1}{2}\frac{\partial V_1}{\partial x}\Delta x\right)$

Along y-direction: at left surface, $\left(V_2 - \frac{1}{2}\frac{\partial V_2}{\partial y}\Delta y\right)$

at back surface, $\left(V_2 + \frac{1}{2}\frac{\partial V_2}{\partial y}\Delta y\right)$

Along z-direction: at top surface, $\left(V_3 + \frac{1}{2}\frac{\partial V_3}{\partial z}\Delta z\right)$

at bottom surface, $\left(V_3 - \frac{1}{2}\frac{\partial V_3}{\partial z}\Delta z\right)$

Therefore, the net outward flux of the vector is

$$\text{Along } x\text{-direction: } \left(V_1 + \frac{1}{2}\frac{\partial V_1}{\partial x}\Delta x\right)\Delta y\Delta z - \left(V_1 - \frac{1}{2}\frac{\partial V_1}{\partial x}\Delta x\right)\Delta y\Delta z = \frac{\partial V_1}{\partial x}\Delta x\Delta y\Delta z$$

Along y-direction: $\left(V_2+\frac{1}{2}\frac{\partial V_2}{\partial y}\Delta y\right)\Delta x\Delta z-\left(V_2-\frac{1}{2}\frac{\partial V_2}{\partial y}\Delta y\right)\Delta x\Delta z=\frac{\partial V_2}{\partial y}\Delta x\Delta y\Delta z$

Along z-direction: $\left(V_3+\frac{1}{2}\frac{\partial V_3}{\partial z}\Delta z\right)\Delta x\Delta y-\left(V_3-\frac{1}{2}\frac{\partial V_3}{\partial z}\Delta z\right)\Delta x\Delta y=\frac{\partial V_3}{\partial z}\Delta x\Delta y\Delta z$

The total net outward flow, considering all thee directions, is

$$\oint_S \vec{F}\cdot d\vec{S}=\left(\frac{\partial V_1}{\partial x}+\frac{\partial V_2}{\partial y}+\frac{\partial V_3}{\partial z}\right)\Delta x\Delta y\Delta z=\left(\frac{\partial V_1}{\partial x}+\frac{\partial V_2}{\partial y}+\frac{\partial V_3}{\partial z}\right)\Delta v$$

where $\Delta v=\Delta x\Delta y\Delta z$ is the infinitesimal volume of the cube.

Hence, the total net outward flow per unit volume is given as

$$\frac{\oint_S \vec{F}\cdot d\vec{S}}{\Delta v}=\left(\frac{\partial V_1}{\partial x}+\frac{\partial V_2}{\partial y}+\frac{\partial V_3}{\partial z}\right)$$

By definition, this is the divergence of the vector.

$$\therefore \quad \boxed{\operatorname{div}\vec{F}=\nabla\cdot\vec{F}=\left(\frac{\partial V_1}{\partial x}+\frac{\partial V_2}{\partial y}+\frac{\partial V_3}{\partial z}\right)} \tag{1.64}$$

Physical Interpretation The physical significance of the divergence of a vector field is the rate at which the density of a vector exits a given region of space.

In other words, divergence of a vector field at a given point is an operator that measures the magnitude of the source or sink, in terms of a signed scalar. More technically, the divergence represents the volume density of the outward flux of a vector field from an infinitesimal volume around a given point.

For example, we consider air as it is heated or cooled. The relevant vector field for this example is the velocity of the moving air at a point. If air is heated in a region, it will expand in all directions such that the velocity field points outward from that region. Therefore, the divergence of the velocity field in that region would have a positive value, as the region is a source. If the air cools and contracts, the divergence is negative and the region is called a sink.

Divergence in General Curvilinear Coordinates We consider the general curvilinear coordinates as shown in Fig. 1.29. We consider the vector

$$\vec{F}=F_1\hat{a}_1+F_2\hat{a}_2+F_3\hat{a}_3$$

In order to find a general expression for divergence of the vector, we need to calculate the closed surface integral of the vector per unit volume for the elemental volume as shown in Fig. 1.27(See page 41).

The contributions to the closed surface integral $\left(\oint_S \vec{F}\cdot d\vec{S}\right)$ through the six surfaces are given as follows.

1. Through surface $OABC$: Taken in the direction of the outward normal is

$$\int_{OABC}\vec{F}\cdot d\vec{S}=-F_1h_2h_3du_2du_3$$

2. Through surface *GFED*: $\int_{GFED} \vec{F}\cdot d\vec{S} = F_1 h_2 h_3 du_2 du_3 + \frac{\partial}{\partial u_1}(F_1 h_2 h_3) du_1 du_2 du_3$

3. Through surface *OCDG*: $\int_{OCDG} \vec{F}\cdot d\vec{S} = -\,F_2 h_1 h_3 du_1 du_3$

4. Through surface *ABEF*: $\int_{ABEF} \vec{F}\cdot d\vec{S} = F_2 h_1 h_3 du_1 du_3 + \frac{\partial}{\partial u_2}(F_2 h_1 h_3) du_1 du_2 du_3$

5. Through surface *OAFG*: $\int_{OAFG} \vec{F}\cdot d\vec{S} = -\,F_3 h_1 h_2 du_1 du_2$

6. Through surface *BCDE*: $\int_{BCDE} \vec{F}\cdot d\vec{S} = F_3 h_1 h_2 du_1 du_2 + \frac{\partial}{\partial u_3}(F_3 h_1 h_2) du_1 du_2 du_3$

Thus, the closed surface integral is given as

$$\oint_S \vec{F}\cdot d\vec{S} = \left[\frac{\partial}{\partial u_1}(F_1 h_2 h_3) + \frac{\partial}{\partial u_2}(F_2 h_1 h_3) + \frac{\partial}{\partial u_3}(F_3 h_1 h_2)\right] du_1 du_2 du_3$$

Thus, the surface integral per unit volume is obtained as

$$\frac{\oint_S \vec{F}\cdot d\vec{S}}{dv} = \frac{\oint_S \vec{F}\cdot d\vec{S}}{h_1 du_1 h_2 du_2 h_3 du_3} = \frac{1}{h_1 du_1 h_2 du_2 h_3 du_3}\left[\frac{\partial}{\partial u_1}(F_1 h_2 h_3) + \frac{\partial}{\partial u_2}(F_2 h_1 h_3) + \frac{\partial}{\partial u_3}(F_3 h_1 h_2)\right] du_1 du_2 du_3$$

$$= \frac{1}{h_1 h_2 h_3}\left[\frac{\partial}{\partial u_1}(F_1 h_2 h_3) + \frac{\partial}{\partial u_2}(F_2 h_1 h_3) + \frac{\partial}{\partial u_3}(F_3 h_1 h_2)\right]$$

By definition, this is the divergence of the vector.

$$\boxed{\nabla\cdot\vec{F} = \frac{1}{h_1 h_2 h_3}\left[\frac{\partial}{\partial u_1}(F_1 h_2 h_3) + \frac{\partial}{\partial u_2}(F_2 h_1 h_3) + \frac{\partial}{\partial u_3}(F_3 h_1 h_2)\right]} \qquad \textbf{(1.65)}$$

Substituting the values of h_i from Section 1.8, we get the relation of divergence in three different coordinate systems as

$$\nabla\cdot\vec{F} = \frac{\partial F_x}{\partial x} + \frac{\partial F_y}{\partial y} + \frac{\partial F_z}{\partial z} \qquad \text{(Cartesian coordinates)} \qquad \textbf{(1.66}\boldsymbol{a}\textbf{)}$$

$$= \frac{1}{r}\frac{\partial}{\partial r}(rF_r) + \frac{1}{r}\frac{\partial F_\phi}{\partial \phi} + \frac{\partial F_z}{\partial z} \qquad \text{(Cylindrical coordinates)} \qquad \textbf{(1.66}\boldsymbol{b}\textbf{)}$$

$$= \frac{1}{\rho^2}\frac{\partial}{\partial \rho}(\rho^2 F_\rho) + \frac{1}{\rho\sin\theta}\frac{\partial}{\partial\theta}(F_\theta\sin\theta) + \frac{1}{\rho\sin\theta}\frac{\partial F_\phi}{\partial\phi} \qquad \text{(Spherical coordinates)} \qquad \textbf{(1.66}\boldsymbol{c}\textbf{)}$$

Properties of Divergence

1. The result of the divergence of a vector field is a scalar.
2. Divergence of a scalar field has no meaning.
3. Divergence may be positive, negative or zero. A vector field with constant *zero divergence* is called *solenoidal*; in this case, no net flow can occur across any closed surface. For example, for

an incompressible fluid, if $\vec{V}$ denotes the quantity of the fluid at any point, then $\nabla \cdot \vec{V} = 0$, i.e., an incompressible fluid cannot diverge from, nor converge towards a point.

On the other hand, when the valve of a steam boiler is opened, there is a net outward flow of steam at each elemental volume. So, there exists a *positive divergence*. When an evacuated bulb is broken, there is a transient negative divergence in the space that was inside the bulb before breaking it.

Example 1.27 Find the divergence of the vector field, $\vec{F} = 2xy\hat{a}_x + z\hat{a}_y + yz^2\hat{a}_z$, at point (2, –1, 3).

Solution The divergence of the vector is given as

$$\nabla \cdot \vec{F} = \frac{\partial F_x}{\partial x} + \frac{\partial F_y}{\partial y} + \frac{\partial F_z}{\partial z} = \frac{\partial}{\partial x}(2xy) + \frac{\partial}{\partial y}(z) + \frac{\partial}{\partial z}(yz^2) = 2y + 2yz$$

At point, (2, –1, 3), the divergence if obtained as

$$\nabla \cdot \vec{F}\Big|_{(2,-1,3)} = 2y + 2yz = 2 \times (-1) + 2 \times (-1) \times 3 = -8$$

Example 1.28 If $\vec{A} = x^2 z\hat{i} - 2y^3 z^2 \hat{j} + xy^2 z\hat{k}$, find div $\vec{A}$ at the point (1, –1, 1).

Solution

$$\nabla \cdot \vec{A} = \frac{\partial}{\partial x}(x^2 z) + \frac{\partial}{\partial y}(-2y^3 z^2) + \frac{\partial}{\partial z}(xy^2 z) = 2xz - 6y^2 z^2 + xy^2$$

$$\therefore \qquad \nabla \cdot \vec{A}\Big|_{(1,-1,1)} = 2 - 6 + 1 = -3$$

Example 1.29 The electric field at a point P, expressed in cylindrical coordinate system is given by

$$\vec{E} = 16r^2 \sin\phi \hat{a}_r + 3r^2 \cos\phi \hat{a}_\phi$$

Find the value of divergence of the field, if the location of the point P is given by (1, 2, 3) m in Cartesian coordinate system.

Solution Here, $x = 1, \quad y = 2, \quad z = 3$

$$\therefore \qquad r = \sqrt{x^2 + y^2} = \sqrt{5}, \quad \phi = \tan^{-1}\left(\frac{y}{x}\right) = \tan^{-1}\left(\frac{2}{1}\right) = 63.43°, \quad z = 3$$

For the electric field, $\vec{E} = 16r^2 \sin\phi \hat{a}_r + 3r^2 \cos\phi \hat{a}_\phi$, the divergence in cylindrical coordinates is given as

$$\nabla \cdot \vec{E} = \frac{1}{r}\frac{\partial}{\partial r}(rE_r) + \frac{1}{r}\frac{\partial E_\phi}{\partial \phi} + \frac{\partial E_z}{\partial z} = \frac{1}{r}\frac{\partial}{\partial r}(16r^3 \sin\phi) + \frac{1}{r}\frac{\partial}{\partial \phi}(3r^2 \cos\phi) + \frac{\partial}{\partial z}(0)$$
$$= 48r \sin\phi - 3r \sin\phi$$
$$= 45r \sin\phi$$

Hence, the divergence of the field at point P is given as

$$\nabla \cdot \vec{E} = 45r \sin\phi = 45 \times \sqrt{5} \sin(63.43°) = 90$$

***Example 1.30** An electric field at point P, expressed in cylindrical coordinate system is given by

$$\vec{E} = 6r^2 \sin\phi \hat{a}_r + 2r^2 \cos\phi \hat{a}_\phi$$

Find the value of the divergence of the field if the location of the point P is given by (1, 1, 1) in Cartesian coordinate system.

Solution The divergence in cylindrical coordinate system is given as

$$\begin{aligned}\nabla\cdot\vec{E} &= \frac{1}{r}\frac{\partial}{\partial r}(rE_r) + \frac{1}{r}\frac{\partial E_\phi}{\partial \phi} + \frac{\partial E_z}{\partial z} = \frac{1}{r}\frac{\partial}{\partial r}(6r^3\sin\phi) + \frac{1}{r}\frac{\partial}{\partial \phi}(2r^2\cos\phi) + 0\\ &= 18r\sin\phi - 2r\sin\phi\\ &= 16r\sin\phi\end{aligned}$$

From the relation between the Cartesian and cylindrical coordinates, we have

$$r = \sqrt{x^2+y^2} \qquad \sin\phi = \frac{y}{\sqrt{x^2+y^2}}$$

$$\therefore \qquad \nabla\cdot\vec{E} = 16r\sin\phi = 16\times\sqrt{x^2+y^2}\times\frac{y}{\sqrt{x^2+y^2}} = 16y$$

Hence, the divergence at point (1, 1, 1) is

$$\therefore \qquad \nabla\cdot\vec{E} = 16y = 16\times 1 = 16$$

Example 1.31 Determine the divergence of the vector field given as

$$\vec{V} = \rho\cos\theta\hat{a}_\rho - \frac{1}{\rho}\sin\theta\hat{a}_\theta + 2\rho^2\sin\theta\hat{a}_\phi$$

Solution The divergence in spherical coordinate system is given as

$$\begin{aligned}\nabla\cdot\vec{V} &= \frac{1}{\rho^2}\frac{\partial}{\partial\rho}(\rho^2V_\rho) + \frac{1}{\rho\sin\theta}\frac{\partial}{\partial\theta}(V_\theta\sin\theta) + \frac{1}{\rho\sin\theta}\frac{\partial V_\phi}{\partial\phi}\\ &= \frac{1}{\rho^2}\frac{\partial}{\partial\rho}(\rho^3\cos\theta) + \frac{1}{\rho\sin\theta}\frac{\partial}{\partial\theta}\left(-\frac{1}{\rho}\sin^2\theta\right) + \frac{1}{\rho\sin\theta}\frac{\partial}{\partial\phi}(2\rho^2\sin\theta)\\ &= 3\cos\theta - \frac{1}{\rho^2\sin\theta}2\sin\theta\cos\theta + 0\\ &= \left(3 - \frac{2}{\rho^2}\right)\cos\theta\end{aligned}$$

Curl of a Vector

Definition The curl of a vector field, denoted as curl $\vec{F}$ or $\nabla\times\vec{F}$, is defined as the vector field having magnitude equal to the maximum circulation at each point and to be oriented perpendicularly to this plane of circulation for each point.

Mathematically, it is defined as the limit of the ratio of the integral of the cross product of the vector with outward drawn normal over a closed surface, to the volume enclosed by the surface, as the volume tends to zero.

$$\text{Curl } \vec{F} = \underset{v \to 0}{\text{Lim}} \left(\frac{\oint_S \vec{F} \times \hat{a}_n dS}{v} \right) = \underset{v \to 0}{\text{Lim}} \left(\frac{\oint_S \vec{F} \times d\vec{S}}{v} \right) \tag{1.67}$$

In other words, the component of curl of a vector in the direction of the unit vector $\hat{a}_n$ is the ratio of the line integral of the vector around a closed contour, to the area enclosed by the contour, as the area tends to zero.

$$\therefore \qquad \text{Curl } \vec{F} = \underset{\Delta S \to 0}{\text{Lim}} \left(\frac{\oint_l \vec{F} \cdot d\vec{l}}{\Delta S} \right) \hat{a}_n \tag{1.68}$$

where, the direction of the contour is obtained from right hand cork-screw rule.

Mathematical Expression of Curl In order to find an expression for the curl of a vector, we consider an elemental area in the *yz*-plane as shown in Fig. 1.30.

We define a vector $\vec{F}$ at the centre of the area $P(x, y, z)$.

The closed line integral of $\vec{F}$ around the path *abcd* is,

$$\oint_l \vec{F} \cdot d\vec{l} = \int_d^a \vec{F} \cdot d\vec{l} + \int_a^b \vec{F} \cdot d\vec{l} + \int_b^c \vec{F} \cdot d\vec{l} + \int_c^d \vec{F} \cdot d\vec{l}$$

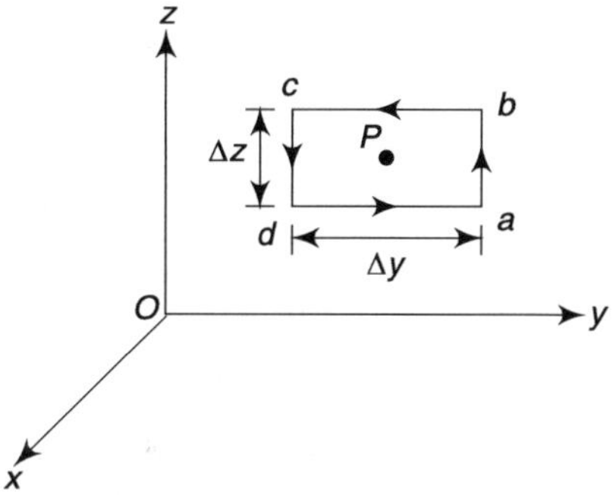

Fig. 1.30 *To derive expression for curl in Cartesian coordinates*

Now,

$$\int_d^a \vec{F} \cdot d\vec{l} = \left(F_y - \frac{\partial F_y}{\partial z} \frac{\Delta z}{2} \right) \Delta y$$

$$\int_a^b \vec{F} \cdot d\vec{l} = \left(F_z + \frac{\partial F_z}{\partial y} \frac{\Delta y}{2} \right) \Delta z$$

$$\int_b^c \vec{F} \cdot d\vec{l} = -\left(F_y + \frac{\partial F_y}{\partial z} \frac{\Delta z}{2} \right) \Delta y$$

$$\int_c^d \vec{F} \cdot d\vec{l} = -\left(F_z - \frac{\partial F_z}{\partial y} \frac{\Delta y}{2} \right) \Delta z$$

Summing up, we get

$$\oint_l \vec{F} \cdot d\vec{l} = \left(\frac{\partial F_z}{\partial y} - \frac{\partial F_y}{\partial z} \right) \Delta y \Delta z$$

Therefore, the *x*-component of the curl (since the area is considered in the *yz*-plane) of the vector is given as

$$\text{Curl}_x \vec{F} = \underset{\Delta S \to 0}{\text{Lim}} \frac{\oint_l \vec{F} \cdot d\vec{l}}{\Delta S} = \underset{\Delta S \to 0}{\text{Lim}} \left(\frac{\oint_l \vec{F} \cdot d\vec{l}}{\Delta y \Delta z} \right) = \left(\frac{\partial F_z}{\partial y} - \frac{\partial F_y}{\partial z} \right)$$

Similarly, considering the area in the *xy* and *xz* planes, we will get the other two components of the curl and can be written as

$$\text{Curl}_y \vec{F} = \left(\frac{\partial F_x}{\partial z} - \frac{\partial F_z}{\partial x} \right)$$

$$\text{Curl}_z \vec{F} = \left(\frac{\partial F_y}{\partial x} - \frac{\partial F_x}{\partial y} \right)$$

Thus, the curl of the vector, considering all three directions, is given as

$$\text{Curl } \vec{F} = \nabla \times \vec{F} = \left(\frac{\partial F_z}{\partial y} - \frac{\partial F_y}{\partial z} \right) \hat{a}_x + \left(\frac{\partial F_x}{\partial z} - \frac{\partial F_z}{\partial x} \right) \hat{a}_y + \left(\frac{\partial F_y}{\partial x} - \frac{\partial F_x}{\partial y} \right) \hat{a}_z \tag{1.69}$$

In matrix form, this can be written as

$$\nabla \times \vec{F} = \begin{vmatrix} \hat{a}_x & \hat{a}_y & \hat{a}_z \\ \frac{\partial}{\partial x} & \frac{\partial}{\partial y} & \frac{\partial}{\partial z} \\ F_x & F_y & F_z \end{vmatrix} \tag{1.70}$$

Physical Interpretation The physical significance of the curl of a vector at any point is that it provides a measure of the amount of rotation or angular momentum of the vector around the point.

We consider a stream on the surface of which floats a leaf, in the xy-plane.

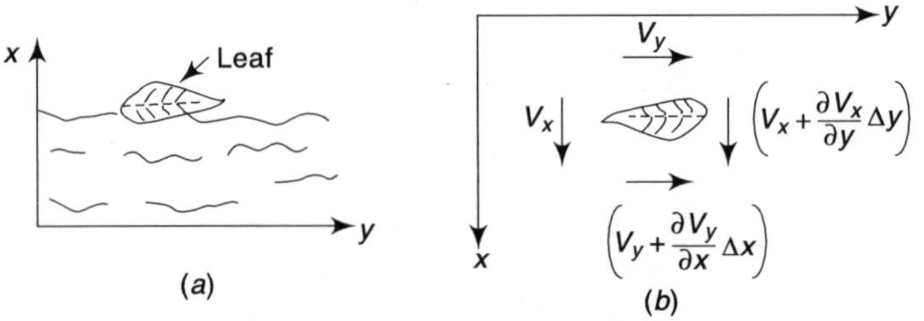

Fig. 1.31 *(a) Rotation of a floating leaf, and (b) Interpretation of curl*

If the velocity at the surface is only in y-direction and is uniform over the surface, there will be no circulation of the leaf.

But, if there are vertices or eddies in the stream, there will be rotational movement of the leaf.

The rate of rotation or angular velocity at any point is a measure of the curl of the velocity of the stream at that point.

In this case, the rotation is about the z-axis and the curl of velocity vector $\vec{V}$ in the z-direction is written as $(\nabla \times \vec{V})_z$. A positive value of $(\nabla \times \vec{V})_z$ implies a rotation from x to y, i.e., anticlockwise.

It is seen that

For positive value of $\frac{\partial V_y}{\partial x}$, rotation is anticlockwise.

For negative value of $\frac{\partial V_x}{\partial y}$, rotation is clockwise.

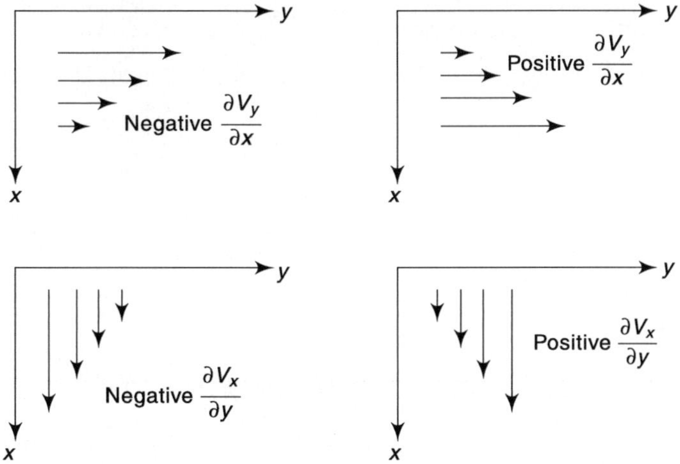

Fig. 1.32 *Interpretation of positive and negative velocity gradients*

The rate of rotation about the z-axis is therefore, proportional to the difference between these two quantities, i.e.,

$$(\nabla \times \vec{V})_z = \left(\frac{\partial V_y}{\partial x} - \frac{\partial V_x}{\partial y} \right)$$

Now, considering any point within the fluid, there may be rotations about the x and y axes, too. Thus,

$$(\nabla \times \vec{V})_x = \left(\frac{\partial V_z}{\partial y} - \frac{\partial V_y}{\partial z} \right)$$

$$(\nabla \times \vec{V})_y = \left(\frac{\partial V_x}{\partial z} - \frac{\partial V_z}{\partial x} \right)$$

A rotation about any axis can be expressed as the sum of the component rotations about the x, y and z axes. Since rotations have both magnitude and direction, the vector sum gives the resultant rotation as

$$\text{Curl } \vec{V} = \nabla \times \vec{V} = \left(\frac{\partial V_z}{\partial y} - \frac{\partial V_y}{\partial z} \right) \hat{a}_x + \left(\frac{\partial V_x}{\partial z} - \frac{\partial V_z}{\partial x} \right) \hat{a}_y$$

$$+ \left(\frac{\partial V_y}{\partial x} - \frac{\partial V_x}{\partial y} \right) \hat{a}_z = \begin{vmatrix} \hat{a}_x & \hat{a}_y & \hat{a}_z \\ \frac{\partial}{\partial x} & \frac{\partial}{\partial y} & \frac{\partial}{\partial z} \\ V_x & V_y & V_z \end{vmatrix}$$

Curl in General Curvilinear Coordinates We consider the general curvilinear coordinates as shown in Fig. 1.27 (see page 41). We consider the vector

$$\vec{F} = F_1 \hat{a}_1 + F_2 \hat{a}_2 + F_3 \hat{a}_3$$

In order to find a general expression for curl of the vector, we need to calculate the closed line integral of the vector per unit area for the elemental volume as shown in Fig. 1.27(See page 41).

The $\hat{a}_1$ component of curl of $\vec{F}$ is obtained by taking the line integral through the path *OABCO*. The contributions to the closed line integral $\left(\oint_l \vec{F}\cdot d\vec{l}\right)$ through the four paths are given as follows.

$$\oint_l \vec{F}\cdot d\vec{l} = \int_O^A \vec{F}\cdot d\vec{l} + \int_A^B \vec{F}\cdot d\vec{l} + \int_B^C \vec{F}\cdot d\vec{l} + \int_C^O \vec{F}\cdot d\vec{l}$$

$$= [F_2 h_2 du_2] + \left[F_3 h_3 du_3 + \frac{\partial}{\partial u_2}(F_3 h_3) du_3 du_2\right] - \left[F_2 h_2 du_2 + \frac{\partial}{\partial u_3}(F_2 h_2) du_2 du_3\right] - [F_3 h_3 du_3]$$

$$= \left[\frac{\partial}{\partial u_2}(F_3 h_3) - \frac{\partial}{\partial u_3}(F_2 h_2)\right] du_2 du_3$$

By definition of curl, this equals the $\hat{a}_1$ component of curl of $\vec{F}$, i.e., $(\nabla\times\vec{F})_1$ multiplied by the area of face *OABC*.

$$\therefore \qquad (\nabla\times\vec{F})_1\, h_2 h_3 du_2 du_3 = \left[\frac{\partial}{\partial u_2}(F_3 h_3) - \frac{\partial}{\partial u_3}(F_2 h_2)\right] du_2 du_3$$

$$\therefore \qquad (\nabla\times\vec{F})_1 = \frac{1}{h_2 h_3}\left[\frac{\partial}{\partial u_2}(F_3 h_3) - \frac{\partial}{\partial u_3}(F_2 h_2)\right]$$

By cyclic change of indices, we get the other two components of the curl and are written as

$$(\nabla\times\vec{F})_2 = \frac{1}{h_3 h_1}\left[\frac{\partial}{\partial u_3}(F_1 h_1) - \frac{\partial}{\partial u_1}(F_3 h_3)\right]$$

$$(\nabla\times\vec{F})_3 = \frac{1}{h_1 h_2}\left[\frac{\partial}{\partial u_1}(F_2 h_2) - \frac{\partial}{\partial u_2}(F_1 h_1)\right]$$

Hence, the expression for curl is given as

$$\nabla\times\vec{F} = \frac{1}{h_2 h_3}\left[\frac{\partial}{\partial u_2}(F_3 h_3) - \frac{\partial}{\partial u_3}(F_2 h_2)\right]\hat{a}_1 + \frac{1}{h_3 h_1}\left[\frac{\partial}{\partial u_3}(F_1 h_1) - \frac{\partial}{\partial u_1}(F_3 h_3)\right]\hat{a}_2$$
$$+ \frac{1}{h_1 h_2}\left[\frac{\partial}{\partial u_1}(F_2 h_2) - \frac{\partial}{\partial u_2}(F_1 h_1)\right]\hat{a}_3$$

or in matrix form

$$\boxed{\nabla\times\vec{F} = \left(\frac{1}{h_1 h_2 h_3}\right)\begin{vmatrix} h_1\hat{a}_1 & h_2\hat{a}_2 & h_3\hat{a}_3 \\ \frac{\partial}{\partial u_1} & \frac{\partial}{\partial u_2} & \frac{\partial}{\partial u_3} \\ F_1 h_1 & F_2 h_2 & F_3 h_3 \end{vmatrix}} \qquad \textbf{(1.71)}$$

Substituting the values of h_i from Section 1.8, we get the relation of curl in three different coordinate systems as

$$\nabla\times\vec{F} = \left(\frac{\partial F_z}{\partial y} - \frac{\partial F_y}{\partial z}\right)\hat{a}_x + \left(\frac{\partial F_x}{\partial z} - \frac{\partial F_z}{\partial x}\right)\hat{a}_y + \left(\frac{\partial F_y}{\partial x} - \frac{\partial F_x}{\partial y}\right)\hat{a}_z \quad \text{(Cartesian coordinates)} \qquad \textbf{(1.72a)}$$

$$\nabla \times \bar{A} = \left[\frac{1}{r}\frac{\partial A_z}{\partial \phi} - \frac{\partial A_\phi}{\partial z}\right]\hat{a}_r + \left[\frac{\partial A_r}{\partial z} - \frac{\partial A_z}{\partial r}\right]\hat{a}_\phi + \left(\frac{1}{r}\right)\left[\frac{\partial}{\partial r}(rA_\phi) - \frac{\partial A_r}{\partial \phi}\right]\hat{a}_z$$

(Cylindrical coordinates) **(1.72*b*)**

$$\nabla \times \vec{V} = \left(\frac{1}{\rho \sin\theta}\right)\left[\frac{\partial}{\partial \theta}(\sin\theta V_\phi) - \frac{\partial V_\theta}{\partial \phi}\right]\hat{a}_\rho + \frac{1}{\rho}\left[\frac{1}{\sin\theta}\frac{\partial V_\rho}{\partial \phi} - \frac{\partial}{\partial \rho}(\rho V_\phi)\right]\hat{a}_\theta + \frac{1}{\rho}\left[\frac{\partial}{\partial \rho}(\rho V_\theta) - \frac{\partial V_\rho}{\partial \theta}\right]\hat{a}_\phi$$

(Spherical coordinates) **(1.72*c*)**

Or in matrix form as

$$\nabla \times \vec{F} = \begin{vmatrix} \hat{a}_x & \hat{a}_y & \hat{a}_z \\ \frac{\partial}{\partial x} & \frac{\partial}{\partial y} & \frac{\partial}{\partial z} \\ F_x & F_y & F_z \end{vmatrix}$$

(Cartesian coordinates) **(1.73*a*)**

$$= \left(\frac{1}{r}\right)\begin{vmatrix} \hat{a}_r & r\hat{a}_\phi & \hat{a}_z \\ \frac{\partial}{\partial r} & \frac{\partial}{\partial \phi} & \frac{\partial}{\partial z} \\ F_r & rF_\phi & F_z \end{vmatrix}$$

(Cylindrical coordinates) **(1.73*b*)**

$$= \left(\frac{1}{\rho^2 \sin\theta}\right)\begin{vmatrix} \hat{a}_\rho & \rho\hat{a}_\theta & \rho\sin\theta\hat{a}_\phi \\ \frac{\partial}{\partial \rho} & \frac{\partial}{\partial \theta} & \frac{\partial}{\partial \phi} \\ F_\rho & \rho F_\theta & \rho\sin\theta F_\phi \end{vmatrix}$$

(Spherical coordinates) **(1.73*c*)**

Properties of Curl

1. The result of the curl of a vector field is another vector field.
2. Curl of a scalar field has no meaning.
3. If the value of curl of a vector field is zero, then the vector field is said to be *irrotational* or *conservative* field. Electrostatic field is one of such fields.

Example 1.32 Determine the curl of the following vector fields:

(a) $\vec{F} = x^2 y\hat{a}_x + y^2 z\hat{a}_y - 2xz\hat{a}_z$

(b) $\vec{A} = r^2 \sin\phi\hat{a}_r + r\cos^2\phi\hat{a}_\phi + z\tan\phi\hat{a}_z$

(c) $\vec{V} = \frac{\sin\phi}{\rho^2}\hat{a}_\rho - \frac{\cos\phi}{\rho^2}\hat{a}_\phi$

Solution

(a) $\vec{F} = x^2 y\hat{a}_x + y^2 z\hat{a}_y - 2xz\hat{a}_z$

The curl in Cartesian coordinate system is given as

$$\nabla \times \vec{F} = \begin{vmatrix} \hat{a}_x & \hat{a}_y & \hat{a}_z \\ \dfrac{\partial}{\partial x} & \dfrac{\partial}{\partial y} & \dfrac{\partial}{\partial z} \\ F_x & F_y & F_z \end{vmatrix} = \begin{vmatrix} \hat{a}_x & \hat{a}_y & \hat{a}_z \\ \dfrac{\partial}{\partial x} & \dfrac{\partial}{\partial y} & \dfrac{\partial}{\partial z} \\ x^2 y & y^2 z & -2xz \end{vmatrix} = -y^2\hat{a}_x + 2z\hat{a}_y - x^2\hat{a}_z$$

(b) $\vec{A} = r^2 \sin\phi \hat{a}_r + r\cos^2\phi \hat{a}_\phi + z\tan\phi \hat{a}_z$

The curl in cylindrical coordinate system is given as

$$\nabla \times \bar{A} = \left(\frac{1}{r}\right) \begin{vmatrix} \hat{a}_r & r\hat{a}_\phi & \hat{a}_z \\ \dfrac{\partial}{\partial r} & \dfrac{\partial}{\partial \phi} & \dfrac{\partial}{\partial z} \\ A_r & rA_\phi & A_z \end{vmatrix}$$

$$= \left[\frac{1}{r}\frac{\partial A_z}{\partial \phi} - \frac{\partial A_\phi}{\partial z}\right]\hat{a}_r + \left[\frac{\partial A_r}{\partial z} - \frac{\partial A_z}{\partial r}\right]\hat{a}_\phi + \left(\frac{1}{r}\right)\left[\frac{\partial}{\partial r}(rA_\phi) - \frac{\partial A_r}{\partial \phi}\right]\hat{a}_z$$

$$= \frac{1}{r}(z\sec^2\phi - 0)\hat{a}_r + (0-0)\hat{a}_\phi + \left(\frac{1}{r}\right)(2r\cos^2\phi - r^2\cos\phi)\hat{a}_z$$

$$= \frac{1}{r}\left[z\sec^2\phi \hat{a}_r + (2r\cos^2\phi - r^2\cos\phi)\hat{a}_z\right]$$

(c) $\vec{V} = \dfrac{\sin\phi}{\rho^2}\hat{a}_\rho - \dfrac{\cos\phi}{\rho^2}\hat{a}_\phi$

The curl in spherical coordinate system is given as

$$\nabla \times \vec{V} = \left(\frac{1}{\rho^2 \sin\theta}\right) \begin{vmatrix} \hat{a}_\rho & \rho\hat{a}_\theta & \rho\sin\theta\hat{a}_\phi \\ \dfrac{\partial}{\partial \rho} & \dfrac{\partial}{\partial \theta} & \dfrac{\partial}{\partial \phi} \\ V_\rho & \rho V_\theta & \rho\sin\theta V_\phi \end{vmatrix}$$

$$= \left(\frac{1}{\rho\sin\theta}\right)\left[\frac{\partial}{\partial\theta}(\sin\theta V_\phi) - \frac{\partial V_\theta}{\partial\phi}\right]\hat{a}_\rho + \frac{1}{\rho}\left[\frac{1}{\sin\theta}\frac{\partial V_\rho}{\partial\phi} - \frac{\partial}{\partial\rho}(\rho V_\phi)\right]\hat{a}_\theta$$

$$+ \frac{1}{\rho}\left[\frac{\partial}{\partial\rho}(\rho V_\theta) - \frac{\partial V_\rho}{\partial\theta}\right]\hat{a}_\phi$$

$$= \left(\frac{1}{\rho\sin\theta}\right)\left[\frac{\partial}{\partial\theta}\left(\sin\theta\frac{\cos\phi}{\rho^2}\right) - 0\right]\hat{a}_\rho + \frac{1}{\rho}\left[\frac{1}{\sin\theta}\frac{\partial}{\partial\phi}\left(\frac{\sin\phi}{\rho^2}\right) - \frac{\partial}{\partial\rho}\left(\rho\frac{\cos\phi}{\rho^2}\right)\right]\hat{a}_\theta$$

$$+ \frac{1}{\rho}\left[0 - \frac{\partial}{\partial\theta}\left(\frac{\sin\phi}{\rho^2}\right)\right]\hat{a}_\phi$$

$$= \left(\frac{1}{\rho\sin\theta}\right)\left(\frac{\cos\theta\cos\phi}{\rho^2}\right)\hat{a}_\rho + \frac{1}{\rho}\left(\frac{\cos\phi}{\rho^2\sin\theta} + \frac{\cos\phi}{\rho^2}\right)\hat{a}_\theta$$

$$\therefore \qquad \nabla \times \vec{V} = \left(\frac{1}{\rho^3} \cot \theta \cos \phi \right) \hat{a}_\rho + \frac{1}{\rho^3} \left(\frac{\cos \phi}{\sin \theta} + \cos \phi \right) \hat{a}_\theta$$

Example 1.33 Determine the divergence and curl of the following vector fields:

(a) $\vec{\phi} = yz\hat{a}_x + 4xy\hat{a}_y + y\hat{a}_z$

(b) $\vec{A} = r^2 z\hat{a}_r + r^3\hat{a}_\phi + 3rz^2\, \hat{a}_z$

(c) $\vec{F} = \frac{1}{\rho^2} \cos \theta \hat{a}_\rho + \rho \sin \theta \cos \phi \hat{a}_\theta + \cos \theta \hat{a}_\phi$

Solution

(a) $\vec{\phi} = yz\hat{a}_x + 4xy\hat{a}_y + y\hat{a}_z$

$$\therefore \qquad \nabla \cdot \vec{\phi} = \frac{\partial}{\partial x}(yz) + \frac{\partial}{\partial y}(4xy) + \frac{\partial}{\partial z}(y) = 0 + 4x + 0 = 4x$$

$$\therefore \qquad \nabla \times \vec{\phi} = \begin{vmatrix} \hat{a}_x & \hat{a}_y & \hat{a}_z \\ \frac{\partial}{\partial x} & \frac{\partial}{\partial y} & \frac{\partial}{\partial z} \\ yz & 4xy & y \end{vmatrix} = \hat{a}_x + y\hat{a}_y + (4y - z)\hat{a}_z'$$

(b) $\vec{A} = r^2 z\hat{a}_r + r^3\hat{a}_\phi + 3rz^2\, \hat{a}_z$

$$\therefore \quad \nabla \cdot \vec{A} = \frac{1}{r}\frac{\partial}{\partial r}(rA_r) + \frac{1}{r}\frac{\partial A_\phi}{\partial \phi} + \frac{\partial A_z}{\partial z} = \frac{1}{r}\frac{\partial}{\partial r}(r^3 z) + \frac{1}{r}\frac{\partial}{\partial \phi}(r^3) + \frac{\partial}{\partial z}(3rz^2) = 3rz + 0 + 6rz = 9rz$$

$$\therefore \qquad \nabla \times \vec{A} = \begin{vmatrix} \left(\frac{1}{r}\right)\hat{a}_r & \hat{a}_\phi & \left(\frac{1}{r}\right)\hat{a}_z \\ \frac{\partial}{\partial r} & \frac{\partial}{\partial \phi} & \frac{\partial}{\partial z} \\ F_r & rF_\phi & F_z \end{vmatrix} = \begin{vmatrix} \left(\frac{1}{r}\right)\hat{a}_r & \hat{a}_\phi & \left(\frac{1}{r}\right)\hat{a}_z \\ \frac{\partial}{\partial r} & \frac{\partial}{\partial \phi} & \frac{\partial}{\partial z} \\ r^2 z & r^4 & 3rz^2 \end{vmatrix} = 0 + (r^2 - 3z^2)\hat{a}_\phi + 4r^2\hat{a}_z$$

(c) $\vec{F} = \frac{1}{\rho^2} \cos \theta \hat{a}_\rho + \rho \sin \theta \cos \phi \hat{a}_\theta + \cos \theta \hat{a}_\phi$

$$\therefore \qquad \nabla \cdot \vec{F} = \frac{1}{\rho^2}\frac{\partial}{\partial \rho}(\rho^2 F_\rho) + \frac{1}{\rho \sin \theta}\frac{\partial}{\partial \theta}(F_\theta \sin \theta) + \frac{1}{\rho \sin \theta}\frac{\partial F_\phi}{\partial \phi}$$

$$= \frac{1}{\rho^2}\frac{\partial}{\partial \rho}(\cos \theta) + \frac{1}{\rho \sin \theta}\frac{\partial}{\partial \theta}(\rho \sin^2 \theta \cos \phi) + \frac{1}{\rho \sin \theta}\frac{\partial}{\partial \phi}(\cos \theta)$$

$$= 0 + \frac{2 \sin \theta \cos \theta}{\sin \theta} \cos \phi + 0$$

$$= 2 \cos \theta \cos \phi$$

$$\therefore \quad \nabla \times \vec{F} = \left(\frac{1}{\rho^2 \sin\theta}\right) \begin{vmatrix} \hat{a}_\rho & \rho\hat{a}_\theta & \rho\sin\theta\hat{a}_\phi \\ \dfrac{\partial}{\partial\rho} & \dfrac{\partial}{\partial\theta} & \dfrac{\partial}{\partial\phi} \\ \dfrac{1}{\rho^2}\cos\theta & \rho^2\sin\theta\cos\phi & \rho\sin\theta\cos\theta \end{vmatrix}$$

$$= \left(\frac{\cos 2\theta}{\rho\sin\theta} + \sin\phi\right)\hat{a}_\rho - \frac{1}{\rho}\cos\theta\hat{a}_\theta + \left(2\cos\phi + \frac{1}{\rho^3}\right)\sin\theta\hat{a}_\phi$$

Example 1.34 If $\vec{A} = xz^3\hat{i} - 2x^2yz\hat{j} + 2yz^4\hat{k}$, find Curl $\vec{A}$ at the point $(1, -1, 1)$.

Solution

$$\nabla \times \vec{A} = \begin{vmatrix} \hat{a}_x & \hat{a}_y & \hat{a}_z \\ \dfrac{\partial}{\partial x} & \dfrac{\partial}{\partial y} & \dfrac{\partial}{\partial z} \\ xz^3 & -2x^2yz & 2yz^4 \end{vmatrix} = (2z^4 + 2x^2y)\hat{a}_x + (3xz^2 - 0)\hat{a}_y + (-4xyz - 0)\hat{a}_z$$

$$= (2z^4 + 2x^2y)\hat{a}_x + 3xz^2\hat{a}_y - 4xyz\hat{a}_z$$

$$\therefore \quad \nabla \cdot \vec{A}\Big|_{(1,-1,1)} = (2-2)\hat{a}_x + 3\hat{a}_y + 4\hat{a}_z = 3\hat{a}_y + 4\hat{a}_z$$

***Example 1.35** **(a)** For a vector field $\vec{A}$, show explicitly that $\nabla \cdot \nabla \times \vec{A} = 0$, i.e., the divergence of the curl of any vector field is zero.
(b) For a scalar field V, show that $\nabla \times \nabla V = 0$, i.e., the curl of the gradient of any scalar field is zero.

Solution

(a) Let, $\vec{A} = A_x\hat{a}_x + A_y\hat{a}_y + A_z\hat{a}_z$

$$\therefore \quad \nabla \times \vec{A} = \begin{vmatrix} \hat{a}_x & \hat{a}_y & \hat{a}_z \\ \dfrac{\partial}{\partial x} & \dfrac{\partial}{\partial y} & \dfrac{\partial}{\partial z} \\ A_x & A_y & A_z \end{vmatrix} = \left(\frac{\partial A_z}{\partial y} - \frac{\partial A_y}{\partial z}\right)\hat{a}_x + \left(\frac{\partial A_x}{\partial z} - \frac{\partial A_z}{\partial x}\right)\hat{a}_y + \left(\frac{\partial A_y}{\partial x} - \frac{\partial A_x}{\partial y}\right)\hat{a}_z$$

$$= G_x\hat{a}_x + G_y\hat{a}_y + G_z\hat{a}_z \qquad \text{(Let)}$$

$$\therefore \quad \nabla \cdot \nabla \times \vec{A} = \frac{\partial G_x}{\partial x} + \frac{\partial G_y}{\partial y} + \frac{\partial G_z}{\partial z}$$

$$= \frac{\partial}{\partial x}\left(\frac{\partial A_z}{\partial y} - \frac{\partial A_y}{\partial z}\right) + \frac{\partial}{\partial y}\left(\frac{\partial A_x}{\partial z} - \frac{\partial A_z}{\partial x}\right) + \frac{\partial}{\partial z}\left(\frac{\partial A_y}{\partial x} - \frac{\partial A_x}{\partial y}\right)$$

$$= \left(\frac{\partial^2 A_z}{\partial x\partial y} - \frac{\partial^2 A_y}{\partial x\partial z}\right) + \left(\frac{\partial^2 A_x}{\partial y\partial z} - \frac{\partial^2 A_z}{\partial y\partial x}\right) + \left(\frac{\partial^2 A_y}{\partial z\partial x} - \frac{\partial^2 A_x}{\partial z\partial y}\right)$$

$$= 0 \left\{\because \frac{\partial^2 A_z}{\partial x\partial y} = \frac{\partial^2 A_z}{\partial y\partial x} \text{ and so on}\right\}$$

$$\boxed{\therefore \quad \nabla \cdot \nabla \times \vec{A} == 0}$$

(b) $\nabla V = \frac{\partial V}{\partial x} + \frac{\partial V}{\partial y} + \frac{\partial V}{\partial z}$

$$\therefore \quad \nabla \times \nabla V = \begin{vmatrix} \hat{a}_x & \hat{a}_y & \hat{a}_z \\ \frac{\partial}{\partial x} & \frac{\partial}{\partial y} & \frac{\partial}{\partial z} \\ \frac{\partial V}{\partial x} & \frac{\partial V}{\partial y} & \frac{\partial V}{\partial z} \end{vmatrix} = \left(\frac{\partial^2 V}{\partial y \partial z} - \frac{\partial^2 V}{\partial z \partial y}\right)\hat{a}_x + \left(\frac{\partial^2 V}{\partial z \partial x} - \frac{\partial^2 V}{\partial x \partial z}\right)\hat{a}_y + \left(\frac{\partial^2 V}{\partial x \partial y} - \frac{\partial^2 V}{\partial y \partial x}\right)\hat{a}_z$$

$$= 0$$

$$\boxed{\therefore \quad \nabla \times \nabla V = 0}$$

NOTE

*These two vector identities are known as **null identities**.*

Example 1.36

***(a)** Find the divergence and curl of the vector field: $\vec{E} = \left(\frac{x}{r}\right)\hat{a}_x$, where $r = x\hat{a}_x + y\hat{a}_y + z\hat{a}_z$.

***(b)** Prove that $\nabla \cdot \vec{r} = 3$ and $\nabla \times \vec{r} = 0$ where, $r = x\hat{a}_x + y\hat{a}_y + z\hat{a}_z$.

Solution

(a) Here, $r = x\hat{a}_x + y\hat{a}_y + z\hat{a}_z$

$$\therefore \quad r = \sqrt{x^2 + y^2 + z^2}$$

$$\vec{E} = \left(\frac{x}{r}\right)\hat{a}_x = \frac{x}{\sqrt{x^2 + y^2 + z^2}}\hat{a}_x$$

$$\therefore \quad \nabla \cdot \vec{E} = \frac{\partial}{\partial x}\left(\frac{x}{\sqrt{x^2 + y^2 + z^2}}\right) = \frac{\sqrt{x^2 + y^2 + z^2} - x\frac{1}{2}\frac{2x}{\sqrt{x^2 + y^2 + z^2}}}{x^2 + y^2 + z^2} = \frac{x^2 + y^2 + z^2 - x^2}{(x^2 + y^2 + z^2)^{3/2}}$$

$$= \frac{y^2 + z^2}{(x^2 + y^2 + z^2)^{3/2}} = \frac{y^2 + z^2}{r^3}$$

$$\therefore \quad \nabla \times \vec{E} = \begin{vmatrix} \hat{a}_x & \hat{a}_y & \hat{a}_z \\ \frac{\partial}{\partial x} & \frac{\partial}{\partial y} & \frac{\partial}{\partial z} \\ \frac{x}{\sqrt{x^2 + y^2 + z^2}} & 0 & 0 \end{vmatrix} = \frac{\partial}{\partial z}\left(\frac{x}{\sqrt{x^2 + y^2 + z^2}}\right)\hat{a}_y - \frac{\partial}{\partial y}\left(\frac{x}{\sqrt{x^2 + y^2 + z^2}}\right)\hat{a}_z$$

$$= -\frac{1}{2}\frac{x2z}{(x^2 + y^2 + z^2)^{3/2}}\hat{a}_y + \frac{1}{2}\frac{x2y}{(x^2 + y^2 + z^2)^{3/2}}\hat{a}_z$$

$$= -\frac{xz}{(x^2+y^2+z^2)^{3/2}}\hat{a}_y + \frac{xy}{(x^2+y^2+z^2)^{3/2}}\hat{a}_z$$

$$= -\frac{xz}{r^3}\hat{a}_y + \frac{xy}{r^3}\hat{a}_z$$

(b) Here, $r = x\hat{a}_x + y\hat{a}_y + z\hat{a}_z$

$$\therefore \quad \nabla \cdot \vec{r} = \frac{\partial}{\partial x}(x) + \frac{\partial}{\partial y}(y) + \frac{\partial}{\partial z}(z) = 1 + 1 + 1 = 3$$

$$\therefore \quad \nabla \times \vec{r} = \begin{vmatrix} \hat{a}_x & \hat{a}_y & \hat{a}_z \\ \frac{\partial}{\partial x} & \frac{\partial}{\partial y} & \frac{\partial}{\partial z} \\ x & y & z \end{vmatrix} = \left(\frac{\partial z}{\partial y} - \frac{\partial y}{\partial z}\right)\hat{a}_x + \left(\frac{\partial x}{\partial z} - \frac{\partial z}{\partial x}\right)\hat{a}_y + \left(\frac{\partial y}{\partial x} - \frac{\partial x}{\partial y}\right)\hat{a}_z = 0$$

$$\therefore \quad \boxed{\nabla \cdot \vec{r} = 3} \quad \text{and} \quad \boxed{\nabla \times \vec{r} = 0}$$

Example 1.37 If $\vec{v} = \vec{\omega} \times \vec{r}$, prove that $\vec{\omega} = \frac{1}{2}\text{Curl}\,\vec{v}$, where ω is a constant vector.

Solution Here, $\vec{r} = x\hat{a}_x + y\hat{a}_y + z\hat{a}_z$ and let, $\vec{\omega} = \omega_x\hat{a}_x + \omega_y\hat{a}_y + \omega_z\hat{a}_z$

$$\therefore \quad \text{Curl}\,\vec{v} = \nabla \times \vec{v} = \nabla \times (\vec{\omega} \times \vec{r}) = \nabla \times \begin{vmatrix} \hat{a}_x & \hat{a}_y & \hat{a}_z \\ \omega_x & \omega_y & \omega_z \\ x & y & z \end{vmatrix}$$

$$= \nabla \times [(\omega_y z - \omega_z y)\hat{a}_x + (\omega_z x - \omega_x z)\hat{a}_y + (\omega_x y - \omega_y x)\hat{a}_z]$$

$$= \begin{vmatrix} \hat{a}_x & \hat{a}_y & \hat{a}_z \\ \frac{\partial}{\partial x} & \frac{\partial}{\partial y} & \frac{\partial}{\partial z} \\ (\omega_y z - \omega_z y) & (\omega_z x - \omega_x z) & (\omega_x y - \omega_y x) \end{vmatrix}$$

$$= \left[\frac{\partial}{\partial y}(\omega_x y - \omega_y x) - \frac{\partial}{\partial z}(\omega_z x - \omega_x z)\right]\hat{a}_x + \left[\frac{\partial}{\partial z}(\omega_y z - \omega_z y) - \frac{\partial}{\partial x}(\omega_x y - \omega_y x)\right]\hat{a}_y$$

$$+ \left[\frac{\partial}{\partial x}(\omega_z x - \omega_x z) - \frac{\partial}{\partial y}(\omega_y z - \omega_z y)\right]\hat{a}_z$$

$$= 2\omega_x\hat{a}_x + 2\omega_y\hat{a}_y + 2\omega_z\hat{a}_z$$

$$= 2(\omega_x\hat{a}_x + \omega_y\hat{a}_y + \omega_z\hat{a}_z)$$

$$= 2\vec{\omega}$$

$$\boxed{\therefore \quad \vec{\omega} = \frac{1}{2}\text{Curl}\,\vec{v}}$$

Laplacian (∇^2) Operator The Laplacian operator (∇^2) can operate both on scalar as well as vector field.

Laplacian (∇^2) of a Scalar The Laplacian operator (∇^2) of a scalar field is the divergence of the gradient of the scalar field upon which the operator operates.

Practically, it is a single operator, which is the composite of gradient and divergence operators.

The Laplacian of a scalar field is also a scalar field.

The Laplacian of a scalar field F in Cartesian coordinate system is written as

$$\nabla^2 F = \nabla \cdot \nabla F = \left(\frac{\partial}{\partial x}\hat{a}_x + \frac{\partial}{\partial y}\hat{a}_y + \frac{\partial}{\partial z}\hat{a}_z\right)\cdot\left(\frac{\partial F}{\partial x}\hat{a}_x + \frac{\partial F}{\partial y}\hat{a}_y + \frac{\partial F}{\partial z}\hat{a}_z\right) = \frac{\partial^2 F}{\partial x^2} + \frac{\partial^2 F}{\partial y^2} + \frac{\partial^2 F}{\partial z^2} \quad \textbf{(1.74)}$$

Similarly, the expression of Laplacian of a scalar in other two coordinate systems can be obtained.

The gradient of the scalar field F in curvilinear coordinate system is given as

$$\nabla F = \frac{1}{h_1}\frac{\partial F}{\partial u_1}\hat{a}_1 + \frac{1}{h_2}\frac{\partial F}{\partial u_2}\hat{a}_2 + \frac{1}{h_3}\frac{\partial F}{\partial u_3}\hat{a}_3 = \vec{A}\,(\text{say}) = A_1\hat{a}_1 + A_2\hat{a}_2 + A_3\hat{a}_3$$

where, $A_1 = \dfrac{1}{h_1}\dfrac{\partial F}{\partial u_1}$; $\quad A_2 = \dfrac{1}{h_2}\dfrac{\partial F}{\partial u_2}$; $\quad A_3 = \dfrac{1}{h_3}\dfrac{\partial F}{\partial u_3}$

Divergence of this vector in curvilinear coordinate system is given as

$$\nabla \cdot \vec{A} = \nabla \cdot \nabla F = \nabla^2 F = \frac{1}{h_1 h_2 h_3}\left[\frac{\partial}{\partial u_1}(A_1 h_2 h_3) + \frac{\partial}{\partial u_2}(A_2 h_1 h_3) + \frac{\partial}{\partial u_3}(A_3 h_1 h_2)\right]$$

Substituting the values of A_1, A_2 and A_3, we have

$$\boxed{\nabla^2 F = \frac{1}{h_1 h_2 h_3}\left[\frac{\partial}{\partial u_1}\left(\frac{h_2 h_3}{h_1}\frac{\partial F}{\partial u_1}\right) + \frac{\partial}{\partial u_2}\left(\frac{h_1 h_3}{h_2}\frac{\partial F}{\partial u_2}\right) + \frac{\partial}{\partial u_3}\left(\frac{h_1 h_2}{h_3}\frac{\partial F}{\partial u_3}\right)\right]} \quad \textbf{(1.75)}$$

Substituting the values of h_i from Section 1.8, we get the relation of Laplacian of scalar field in three different coordinate systems as

$$\nabla^2 F = \frac{\partial^2 F}{\partial x^2} + \frac{\partial^2 F}{\partial y^2} + \frac{\partial^2 F}{\partial z^2} \quad \text{(Cartesian coordinates)} \quad \textbf{(1.76}\textbf{\textit{a}}\textbf{)}$$

$$= \frac{1}{r}\frac{\partial}{\partial r}\left(r\frac{\partial F}{\partial r}\right) + \frac{1}{r^2}\frac{\partial^2 F}{\partial \phi^2} + \frac{\partial^2 F}{\partial z^2} \quad \text{(Cylindrical coordinates)} \quad \textbf{(1.76}\textbf{\textit{b}}\textbf{)}$$

$$= \frac{1}{\rho^2}\frac{\partial}{\partial \rho}\left(\rho^2\frac{\partial F}{\partial \rho}\right) + \frac{1}{\rho^2 \sin\theta}\frac{\partial}{\partial \theta}\left(\sin\theta\frac{\partial F}{\partial \theta}\right) + \frac{1}{\rho^2 \sin^2\theta}\frac{\partial^2 F}{\partial \phi^2} \quad \text{(Spherical coordinates)} \quad \textbf{(1.76}\textbf{\textit{c}}\textbf{)}$$

NOTE

If the Laplacian of a scalar field V is zero in a region, i.e., $\nabla^2 V = 0$, then the scalar field V is said to be ***harmonic*** *(containing sine or cosine terms) in that region and the equation $\nabla^2 V = 0$ is known as* ***Laplace's equation****. We will learn about Laplace's equation in more detail in Chapter 2.*

Laplacian (∇^2) of Products of Scalar Fields If a scalar field is represented as the product of two other scalar functions, then the Laplacian of that scalar field in Cartesian coordinate system is written as

$$\nabla^2(uv) = (\nabla^2 u)v + u(\nabla^2 v) + 2(\nabla u)\cdot(\nabla v) \tag{1.77}$$

Proof: By definition of scalar Laplacian in Cartesian coordinate system

$$\nabla^2(uv) = \frac{\partial^2}{\partial x^2}(uv) + \frac{\partial^2}{\partial y^2}(uv) + \frac{\partial^2}{\partial z^2}(uv)$$

Now,

$$\frac{\partial^2}{\partial x^2}(uv) = \frac{\partial}{\partial x}\left[\frac{\partial}{\partial x}(uv)\right] = \frac{\partial}{\partial x}\left[\frac{\partial u}{\partial x}v + u\frac{\partial v}{\partial x}\right] = \frac{\partial^2 u}{\partial x^2}v + \frac{\partial u}{\partial x}\frac{\partial v}{\partial x} + \frac{\partial u}{\partial x}\frac{\partial v}{\partial x} + u\frac{\partial^2 v}{\partial x^2}$$
$$= \frac{\partial^2 u}{\partial x^2}v + 2\frac{\partial u}{\partial x}\frac{\partial v}{\partial x} + u\frac{\partial^2 v}{\partial x^2}$$

Similarly,

$$\frac{\partial^2}{\partial y^2}(uv) = \frac{\partial^2 u}{\partial y^2}v + 2\frac{\partial u}{\partial y}\frac{\partial v}{\partial y} + u\frac{\partial^2 v}{\partial y^2}$$

$$\frac{\partial^2}{\partial z^2}(uv) = \frac{\partial^2 u}{\partial z^2}v + 2\frac{\partial u}{\partial z}\frac{\partial v}{\partial z} + u\frac{\partial^2 v}{\partial z^2}$$

By summation, we get

$$\nabla^2(uv) = \frac{\partial^2 u}{\partial x^2}v + 2\frac{\partial u}{\partial x}\frac{\partial v}{\partial x} + u\frac{\partial^2 v}{\partial x^2} + \frac{\partial^2 u}{\partial y^2}v + 2\frac{\partial u}{\partial y}\frac{\partial v}{\partial y} + u\frac{\partial^2 v}{\partial y^2} + \frac{\partial^2 u}{\partial z^2}v + 2\frac{\partial u}{\partial z}\frac{\partial v}{\partial z} + u\frac{\partial^2 v}{\partial z^2}$$
$$= \left(\frac{\partial^2 u}{\partial x^2} + \frac{\partial^2 u}{\partial y^2} + \frac{\partial^2 u}{\partial z^2}\right)v + u\left(\frac{\partial^2 v}{\partial x^2} + \frac{\partial^2 v}{\partial y^2} + \frac{\partial^2 v}{\partial z^2}\right)$$
$$+ 2\left(\frac{\partial u}{\partial x}\hat{a}_x + \frac{\partial u}{\partial y}\hat{a}_y + \frac{\partial u}{\partial z}\hat{a}_z\right)\cdot\left(\frac{\partial v}{\partial x}\hat{a}_x + \frac{\partial v}{\partial y}\hat{a}_y + \frac{\partial v}{\partial z}\hat{a}_z\right)$$
$$= (\nabla^2 u)v + u(\nabla^2 v) + 2(\nabla u)\cdot(\nabla v)$$

Laplacian (∇^2) of a Vector The Laplacian of a vector is defined as the gradient of divergence of the vector minus the curl of curl of the vector; i.e.,

$$\boxed{\nabla^2 \vec{F} = \nabla(\nabla \cdot \vec{F}) - \nabla \times \nabla \times \vec{F}} \tag{1.78}$$

In terms of general curvilinear coordinate system, Laplacian of a vector is written as

$$\nabla^2 \vec{F} = \frac{1}{h_1 h_2 h_3}\left[\frac{\partial}{\partial u_1}\left(\frac{h_2 h_3}{h_1}\frac{\partial}{\partial u_1}\right) + \frac{\partial}{\partial u_2}\left(\frac{h_1 h_3}{h_2}\frac{\partial}{\partial u_2}\right) + \frac{\partial}{\partial u_3}\left(\frac{h_1 h_2}{h_3}\frac{\partial}{\partial u_3}\right)\right]\vec{F}$$
$$= \frac{1}{h_1 h_2 h_3}\left[\frac{\partial}{\partial u_1}\left(\frac{h_2 h_3}{h_1}\frac{\partial \vec{F}}{\partial u_1}\right) + \frac{\partial}{\partial u_2}\left(\frac{h_1 h_3}{h_2}\frac{\partial \vec{F}}{\partial u_2}\right) + \frac{\partial}{\partial u_3}\left(\frac{h_1 h_2}{h_3}\frac{\partial \vec{F}}{\partial u_3}\right)\right]$$
$$= \frac{1}{h_1 h_2 h_3}\left[\begin{aligned}&\frac{\partial}{\partial u_1}\left\{\frac{h_2 h_3}{h_1}\frac{\partial}{\partial u_1}(F_1\hat{a}_1 + F_2\hat{a}_2 + F_3\hat{a}_3)\right\} + \frac{\partial}{\partial u_2}\left\{\frac{h_1 h_3}{h_2}\frac{\partial}{\partial u_2}(F_1\hat{a}_1 + F_2\hat{a}_2 + F_3\hat{a}_3)\right\}\\ &+ \frac{\partial}{\partial u_3}\left\{\frac{h_1 h_2}{h_3}\frac{\partial}{\partial u_3}(F_1\hat{a}_1 + F_2\hat{a}_2 + F_3\hat{a}_3)\right\}\end{aligned}\right] \tag{1.79}$$

This is the general expression for vector Laplacian in curvilinear coordinate system.

Derivation of Vector Laplacian in Cartesian Coordinate System In Cartesian coordinate system

$$u_1 = x,\ u_2 = y,\ u_3 = z$$
$$h_1 = h_2 = h_3 = 1$$

From Eq. (1.76),

$$\nabla^2 \vec{F} = \left[\begin{array}{l} \dfrac{\partial}{\partial x}\left\{\dfrac{\partial}{\partial x}(F_x\hat{a}_x + F_y\hat{a}_y + F_z\hat{a}_z)\right\} + \dfrac{\partial}{\partial y}\left\{\dfrac{\partial}{\partial y}(F_x\hat{a}_x + F_y\hat{a}_y + F_z\hat{a}_z)\right\} \\ \qquad + \dfrac{\partial}{\partial z}\left\{\dfrac{\partial}{\partial z}(F_x\hat{a}_x + F_y\hat{a}_y + F_z\hat{a}_z)\right\} \end{array} \right]$$

Now

$$\frac{\partial}{\partial x}(F_x\hat{a}_x + F_y\hat{a}_y + F_z\hat{a}_z) = \left(\hat{a}_x\frac{\partial F_x}{\partial x} + F_x\frac{\partial \hat{a}_x}{\partial x}\right) + \left(\hat{a}_y\frac{\partial F_y}{\partial x} + F_y\frac{\partial \hat{a}_y}{\partial x}\right) + \left(\hat{a}_z\frac{\partial F_z}{\partial x} + F_z\frac{\partial \hat{a}_z}{\partial x}\right)$$

Since the unit vectors in Cartesian coordinate system are considered to be constant, we have

$$\frac{\partial \hat{a}_x}{\partial x} = 0 \qquad \frac{\partial \hat{a}_y}{\partial x} = 0 \qquad \frac{\partial \hat{a}_z}{\partial x}$$

$$\therefore \quad \frac{\partial}{\partial x}(F_x\hat{a}_x + F_y\hat{a}_y + F_z\hat{a}_z) = \left(\hat{a}_x\frac{\partial F_x}{\partial x} + \hat{a}_y\frac{\partial F_y}{\partial x} + \hat{a}_z\frac{\partial F_z}{\partial x}\right)$$

$$\therefore \quad \frac{\partial}{\partial x}\left\{\frac{\partial}{\partial x}(F_x\hat{a}_x + F_y\hat{a}_y + F_z\hat{a}_z)\right\} = \frac{\partial}{\partial x}\left(\hat{a}_x\frac{\partial F_x}{\partial x} + \hat{a}_y\frac{\partial F_y}{\partial x} + \hat{a}_z\frac{\partial F_z}{\partial x}\right)$$

$$= \hat{a}_x\frac{\partial^2 F_x}{\partial x^2} + \hat{a}_y\frac{\partial^2 F_y}{\partial x^2} + \hat{a}_z\frac{\partial^2 F_z}{\partial x^2} \tag{1.80a}$$

Similarly,

$$\frac{\partial}{\partial y}\left\{\frac{\partial}{\partial y}(F_x\hat{a}_x + F_y\hat{a}_y + F_z\hat{a}_z)\right\} = \hat{a}_x\frac{\partial^2 F_x}{\partial y^2} + \hat{a}_y\frac{\partial^2 F_y}{\partial y^2} + \hat{a}_z\frac{\partial^2 F_z}{\partial y^2} \tag{1.80b}$$

and,

$$\frac{\partial}{\partial z}\left\{\frac{\partial}{\partial z}(F_x\hat{a}_x + F_y\hat{a}_y + F_z\hat{a}_z)\right\} = \hat{a}_x\frac{\partial^2 F_x}{\partial z^2} + \hat{a}_y\frac{\partial^2 F_y}{\partial z^2} + \hat{a}_z\frac{\partial^2 F_z}{\partial z^2} \tag{1.80c}$$

By summation of Eqs. (1.80*a*) to (1.80*c*), we get

$$\nabla^2 \vec{F} = \left(\hat{a}_x\frac{\partial^2 F_x}{\partial x^2} + \hat{a}_y\frac{\partial^2 F_y}{\partial x^2} + \hat{a}_z\frac{\partial^2 F_z}{\partial x^2}\right) + \left(\hat{a}_x\frac{\partial^2 F_x}{\partial y^2} + \hat{a}_y\frac{\partial^2 F_y}{\partial y^2} + \hat{a}_z\frac{\partial^2 F_z}{\partial y^2}\right)$$
$$+ \left(\hat{a}_x\frac{\partial^2 F_x}{\partial z^2} + \hat{a}_y\frac{\partial^2 F_y}{\partial z^2} + \hat{a}_z\frac{\partial^2 F_z}{\partial z^2}\right)$$

$$= \left(\frac{\partial^2 F_x}{\partial x^2} + \frac{\partial^2 F_x}{\partial y^2} + \frac{\partial^2 F_x}{\partial z^2}\right)\hat{a}_x + \left(\frac{\partial^2 F_y}{\partial x^2} + \frac{\partial^2 F_y}{\partial y^2} + \frac{\partial^2 F_y}{\partial z^2}\right)\hat{a}_y + \left(\frac{\partial^2 F_z}{\partial x^2} + \frac{\partial^2 F_z}{\partial y^2} + \frac{\partial^2 F_z}{\partial z^2}\right)\hat{a}_z$$

$$= (\nabla^2 F_x)\hat{a}_x + (\nabla^2 F_y)\hat{a}_y + (\nabla^2 F_z)\hat{a}_z$$

$$\therefore \quad \nabla^2 \vec{F} = (\nabla^2 F_x)\hat{a}_x + (\nabla^2 F_y)\hat{a}_y + (\nabla^2 F_z)\hat{a}_z \tag{1.81}$$

Since the unit vectors in cylindrical and spherical coordinate systems are not constants, evaluation of vector Laplacian in these two coordinate systems become tedious. However, we can write the final expressions of vector Laplacian in these two coordinate systems as given.

Cylindrical coordinate system

$$\nabla^2 \vec{F} = \left(\nabla^2 F_r - \frac{2}{r^2}\frac{\partial F_\phi}{\partial \phi} - \frac{1}{r^2}F_r\right)\hat{a}_r + \left(\nabla^2 F_\phi + \frac{2}{r^2}\frac{\partial F_r}{\partial \phi} - \frac{1}{r^2}F_\phi\right)\hat{a}_\phi + \nabla^2 F_z \hat{a}_z \tag{1.82}$$

Spherical coordinate system

$$\begin{aligned} \nabla^2 \vec{F} = &\left[\nabla^2 F_\rho - \frac{2}{\rho^2}\left(F_\rho + \cot\theta F_\theta + \text{cosec}\,\theta \frac{\partial F_\phi}{\partial \phi} + \frac{\partial F_\theta}{\partial \theta}\right)\right]\hat{a}_\rho \\ &+ \left[\nabla^2 F_\theta - \frac{1}{\rho^2}\left(\text{cosec}^2\,\theta F_\theta - 2\frac{\partial F_\rho}{\partial \theta} + 2\cot\theta\,\text{cosec}\,\theta \frac{\partial F_\phi}{\partial \phi}\right)\right]\hat{a}_\theta \\ &+ \left[\nabla^2 F_\phi - \frac{1}{\rho^2}\left(\text{cosec}^2\,\theta F_\phi - 2\,\text{cosec}\,\theta \frac{\partial F_\rho}{\partial \phi} - 2\cot\theta\,\text{cosec}\,\theta \frac{\partial F_\theta}{\partial \phi}\right)\right]\hat{a}_\phi \end{aligned} \tag{1.83}$$

NOTE

The Laplacian of a vector field is zero if and only if the Laplacian of each of its components is independently zero.

Example 1.38 Find the Laplacian of the following scalar fields:

(a) $F = \sqrt{x^2 + y^2 + z^2}$

(b) $F = rz \sin\phi + r^2 + z^2 \cos^2\phi$

(c) $F = e^{-\rho} \sin\theta \cos\phi$

Solution

(a) $F = \sqrt{x^2 + y^2 + z^2}$

$$\nabla^2 F = \frac{\partial^2}{\partial x^2}(\sqrt{x^2 + y^2 + z^2}) + \frac{\partial^2}{\partial y^2}(\sqrt{x^2 + y^2 + z^2}) + \frac{\partial^2}{\partial z^2}(\sqrt{x^2 + y^2 + z^2})$$

Now, $\frac{\partial}{\partial x}(\sqrt{x^2+y^2+z^2}) = \frac{x}{\sqrt{x^2+y^2+z^2}}$

$$\therefore \qquad \frac{\partial^2}{\partial x^2}(\sqrt{x^2+y^2+z^2}) = \frac{\partial}{\partial x}\left(\frac{x}{\sqrt{x^2+y^2+z^2}}\right) = \frac{\sqrt{x^2+y^2+z^2} - x\frac{x}{\sqrt{x^2+y^2+z^2}}}{(x^2+y^2+z^2)}$$

$$= \frac{y^2+z^2}{(x^2+y^2+z^2)^{3/2}}$$

Similarly,

$$\frac{\partial^2}{\partial y^2}(\sqrt{x^2+y^2+z^2}) = \frac{x^2+z^2}{(x^2+y^2+z^2)^{3/2}}$$

$$\frac{\partial^2}{\partial z^2}(\sqrt{x^2+y^2+z^2}) = \frac{x^2+y^2}{(x^2+y^2+z^2)^{3/2}}$$

Hence, the Laplacian is given as

$$\nabla^2 F = \frac{y^2+z^2}{(x^2+y^2+z^2)^{3/2}} + \frac{x^2+z^2}{(x^2+y^2+z^2)^{3/2}} + \frac{x^2+y^2}{(x^2+y^2+z^2)^{3/2}} = \frac{2}{\sqrt{x^2+y^2+z^2}}$$

(b) $F = rz \sin\phi + r^2 + z^2 \cos^2\phi$

$$\nabla^2 F = \frac{1}{r}\frac{\partial}{\partial r}\left(r\frac{\partial F}{\partial r}\right) + \frac{1}{r^2}\frac{\partial^2 F}{\partial \phi^2} + \frac{\partial^2 F}{\partial z^2}$$

$$= \frac{1}{r}\frac{\partial}{\partial r}\left\{r\frac{\partial}{\partial r}(rz\sin\phi + r^2 + z^2\cos^2\phi)\right\} + \frac{1}{r^2}\frac{\partial^2}{\partial\phi^2}(rz\sin\phi + r^2 + z^2\cos^2\phi)$$

$$+ \frac{\partial^2}{\partial z^2}(rz\sin\phi + r^2 + z^2\cos^2\phi)$$

$$= \frac{1}{r}\frac{\partial}{\partial r}\{rz\sin\phi + 2r^2\} + \frac{1}{r^2}\frac{\partial}{\partial\phi}(rz\cos\phi - 2z^2\cos\phi\sin\phi) + \frac{\partial}{\partial z}(r\sin\phi + 2z\cos^2\phi)$$

$$= \frac{1}{r}(z\sin\phi + 4r) + \frac{1}{r^2}(-rz\sin\phi - 4z^2\cos 2\phi) + 2\cos^2\phi$$

$$= 4 + \frac{z}{r}\sin\phi - \frac{z}{r}\sin\phi - 4\frac{z^2}{r^2}\cos 2\phi + 2\cos^2\phi$$

$$= 4 + 2\cos^2\phi - 4\frac{z^2}{r^2}\cos 2\phi$$

(c) $F = e^{-\rho} \sin\theta \cos\phi$

$$\nabla^2 F = \frac{1}{\rho^2}\frac{\partial}{\partial\rho}\left(\rho^2 \frac{\partial F}{\partial \rho}\right) + \frac{1}{\rho^2 \sin\theta}\frac{\partial}{\partial\theta}\left(\sin\theta \frac{\partial F}{\partial\theta}\right) + \frac{1}{\rho^2 \sin^2\theta}\frac{\partial^2 F}{\partial\phi^2}$$

$$= \frac{1}{\rho^2}\frac{\partial}{\partial\rho}\{\rho^2(-1)e^{-\rho}\sin\theta\cos\phi\} + \frac{1}{\rho^2\sin\theta}\frac{\partial}{\partial\theta}\{\sin\theta\, e^{-\rho}\cos\theta\cos\phi\}$$

$$- \frac{1}{\rho^2\sin^2\theta}\frac{\partial}{\partial\phi}(e^{-\rho}\sin\theta\sin\phi)$$

$$= -\frac{1}{\rho^2}\frac{\partial}{\partial\rho}\{e^{-\rho}\rho^2\sin\theta\cos\phi\} + \frac{1}{\rho^2\sin\theta}e^{-\rho}\cos 2\theta\cos\phi - \frac{1}{\rho^2\sin^2\theta}(e^{-\rho}\sin\theta\cos\phi)$$

$$= -\frac{1}{\rho^2}\sin\theta\cos\phi(2\rho e^{-\rho} - e^{-\rho}\rho^2) + \frac{e^{-\rho}}{\rho^2}\frac{\cos\phi}{\sin\theta}(\cos 2\theta - 1)$$

$$= \sin\theta\cos\phi\left(e^{-\rho} - \frac{2}{\rho}e^{-\rho}\right) - \frac{e^{-\rho}}{\rho^2}\frac{\cos\phi}{\sin\theta}(1 - \cos 2\theta)$$

$$= \sin\theta\cos\phi\left(e^{-\rho} - \frac{2}{\rho}e^{-\rho}\right) - \frac{e^{-\rho}}{\rho^2}\frac{\cos\phi}{\sin\theta}2\sin^2\theta$$

$$= \sin\theta\cos\phi\left(e^{-\rho} - \frac{2}{\rho}e^{-\rho}\right) - \frac{e^{-\rho}}{\rho^2}2\sin\theta\cos\phi$$

$$= e^{-\rho}\sin\theta\cos\phi\left(1 - \frac{2}{\rho} - \frac{2}{\rho^2}\right)$$

Example 1.39 Find the Laplacian of the scalar fields:

(a) $S = x^2 y + xyz$
(b) $S = r^2 z(\cos\phi + \sin\phi)$
(c) $S = \cos\theta\sin\phi\ln\rho + \rho^2\phi$

Solution

(a) $S = x^2 y + xyz$

$$\nabla^2 S = \frac{\partial^2 S}{\partial x^2} + \frac{\partial^2 S}{\partial y^2} + \frac{\partial^2 S}{\partial z^2} = \frac{\partial}{\partial x}(2xy + yz) + \frac{\partial}{\partial y}(x^2 + xz) + \frac{\partial}{\partial z}(xy) = 2y$$

(b) $S = r^2 z(\cos\phi + \sin\phi)$

$$\nabla^2 S = \frac{1}{r}\frac{\partial}{\partial r}\left(r\frac{\partial S}{\partial r}\right) + \frac{1}{r^2}\frac{\partial^2 S}{\partial\phi^2} + \frac{\partial^2 S}{\partial z^2}$$

$$= \frac{1}{r}\frac{\partial}{\partial r}[2r^2 z(\cos\phi + \sin\phi)] + \frac{1}{r^2}\frac{\partial}{\partial\phi}[r^2 z(\cos\phi - \sin\phi)] + \frac{\partial}{\partial z}[r^2(\cos\phi + \sin\phi)]$$

$$= 4z(\cos\phi + \sin\phi) - z(\cos\phi + \sin\phi)$$

$$= 3z(\cos\phi + \sin\phi)$$

(c) $S = \cos\theta\sin\phi\ln\rho + \rho^2\phi$

$$\nabla^2 S = \frac{1}{\rho^2}\frac{\partial}{\partial\rho}\left(\rho^2\frac{\partial S}{\partial\rho}\right) + \frac{1}{\rho^2\sin\theta}\frac{\partial}{\partial\theta}\left(\sin\theta\frac{\partial S}{\partial\theta}\right) + \frac{1}{\rho^2\sin^2\theta}\frac{\partial^2 S}{\partial\phi^2}$$

$$\therefore \quad \nabla^2 S = \frac{1}{\rho^2}\frac{\partial}{\partial \rho}[\rho \cos\theta \sin\phi + 2\rho^3\phi] - \frac{1}{\rho^2 \sin\theta}\frac{\partial}{\partial\theta}[\sin^2\theta \sin\phi \ln\rho] + \frac{1}{\rho^2\sin^2\theta}\frac{\partial}{\partial\phi}[\cos\theta\cos\phi\ln\rho + \rho^2]$$

$$= \frac{1}{\rho^2}\left[\cos\theta\sin\phi + 6\rho^2\phi\right] - \frac{1}{\rho^2\sin\theta}[2\sin\theta\cos\theta\sin\phi\ln\rho] - \frac{1}{\rho^2\sin^2\theta}[\cos\theta\sin\phi\ln\rho]$$

$$= \left[\frac{\cos\theta\sin\phi}{\rho^2} + 6\phi\right] - \left[\frac{2\cos\theta\sin\phi\ln\rho}{\rho^2}\right] - \left[\frac{\cos\theta\sin\phi\ln\rho}{\rho^2\sin^2\theta}\right]$$

$$= \frac{\cos\theta\sin\phi}{\rho^2}(1 - 2\ln\rho - \operatorname{cosec}^2\theta\ln\rho) + 6\phi$$

***Example 1.40** Prove that $\nabla^2\left(\frac{1}{r}\right) = 0$, with usual meaning of r.

Solution

$$\nabla^2\left(\frac{1}{r}\right) = \frac{\partial^2}{\partial x^2}\left(\frac{1}{\sqrt{x^2+y^2+z^2}}\right) + \frac{\partial^2}{\partial y^2}\left(\frac{1}{\sqrt{x^2+y^2+z^2}}\right) + \frac{\partial^2}{\partial z^2}\left(\frac{1}{\sqrt{x^2+y^2+z^2}}\right)$$

Now, $$\frac{\partial}{\partial x}\left(\frac{1}{\sqrt{x^2+y^2+z^2}}\right) = -\frac{x}{(x^2+y^2+z^2)^{3/2}}$$

$$\therefore \quad \frac{\partial^2}{\partial x^2}\left(\frac{1}{\sqrt{x^2+y^2+z^2}}\right) = \frac{\partial}{\partial x}\left(-\frac{x}{(x^2+y^2+z^2)^{3/2}}\right)$$

$$= -\frac{(x^2+y^2+z^2)^{3/2} - x\frac{3}{2}2x(x^2+y^2+z^2)^{1/2}}{(x^2+y^2+z^2)^3}$$

$$= \frac{3x^2 - (x^2+y^2+z^2)}{(x^2+y^2+z^2)^{5/2}}$$

$$= \frac{2x^2 - y^2 - z^2}{(x^2+y^2+z^2)^{5/2}}$$

Similarly, $$\frac{\partial^2}{\partial y^2}\left(\frac{1}{\sqrt{x^2+y^2+z^2}}\right) = \frac{2y^2 - z^2 - x^2}{(x^2+y^2+z^2)^{5/2}}$$

and, $$\frac{\partial^2}{\partial z^2}\left(\frac{1}{\sqrt{x^2+y^2+z^2}}\right) = \frac{2z^2 - x^2 - y^2}{(x^2+y^2+z^2)^{5/2}}$$

By addition, we get

$$\nabla^2\left(\frac{1}{r}\right) = \frac{2x^2-y^2-z^2}{(x^2+y^2+z^2)^{5/2}} + \frac{2y^2-z^2-x^2}{(x^2+y^2+z^2)^{5/2}} + \frac{2z^2-x^2-y^2}{(x^2+y^2+z^2)^{5/2}} = 0$$

$$\boxed{\nabla^2\left(\frac{1}{r}\right) = 0}$$

Example 1.41 Given the vector field, $\vec{E} = (x^2 + y^2)\hat{a}_x + (x + y)\hat{a}_y$; Find $\nabla^2 \vec{E}$.

Solution

$$\nabla^2 \vec{E} = (\nabla^2 E_x)\hat{a}_x + (\nabla^2 E_y)\hat{a}_y + (\nabla^2 E_z)\hat{a}_z$$

where, $\nabla^2 = \left(\frac{\partial^2}{\partial x^2} + \frac{\partial^2}{\partial y^2} + \frac{\partial^2}{\partial z^2}\right)$

Now, $\nabla^2 E_y$ will not survive since E_y is linear. Also, $\nabla^2 E_z$ vanishes as E_z does not exist.

$$\frac{\partial E_x}{\partial x} = 2x; \qquad \frac{\partial^2 E_x}{\partial x^2} = 2$$

$$\frac{\partial E_x}{\partial y} = 2y; \qquad \frac{\partial^2 E_x}{\partial y^2} = 2$$

$$\frac{\partial E_x}{\partial z} = 0; \qquad \frac{\partial^2 E_x}{\partial z^2} = 0$$

$\therefore$

$$\nabla^2 \vec{E} = \nabla^2 E_x \hat{a}_x = (2 + 2)\hat{a}_x = 4\hat{a}_x$$

1.10 GAUSS' DIVERGENCE THEOREM

Statement This theorem states that the divergence of a vector field over a volume is equal to the surface integral of the normal component of the vector through the closed surface bounding the volume.

Mathematically,

$$\int_V \nabla \cdot \vec{F} dv = \oint_S \vec{F} \cdot d\vec{S} = \oint_S \vec{F} \cdot \hat{a}_n dS \qquad \textbf{(1.84)}$$

where V is the volume enclosed by the closed surface S.

Proof By definition of divergence in Cartesian coordinates,

$$\nabla \cdot \vec{F} = \frac{\partial F_x}{\partial x} + \frac{\partial F_y}{\partial y} + \frac{\partial F_z}{\partial z}$$

$\therefore$

$$\int_V \nabla \cdot \vec{F} dv = \int_V \left(\frac{\partial F_x}{\partial x} + \frac{\partial F_y}{\partial y} + \frac{\partial F_z}{\partial z}\right) dxdydz$$

where, $dv = dxdydz$

We consider an elemental volume as shown in Fig. 1.33.

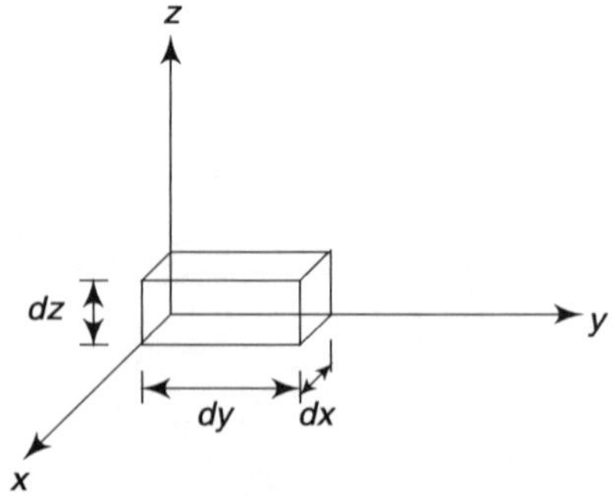

Fig. 1.33 *Elemental volume in Cartesian coordinates*

Let the rectangular volume have the dimensions dx, dy and dz along the x, y and z directions respectively.

Now, $\int_V \frac{\partial F_x}{\partial x} dxdydz = \int_S \left[\int \frac{\partial F_x}{\partial x} dx \right] dydz$

Here, $\int_{x_1}^{x_2} \frac{\partial F_x}{\partial x} dx = (F_{x2} - F_{x1})$

where, F_{x1} and F_{x2} are the x component of the vector on the back and front side of the element along x axis.

$$\therefore \qquad \int_V \frac{\partial F_x}{\partial x} dxdydz = \int_S (F_{x2} - F_{x1}) dydz = \int_S F_x dS_x$$

where, $dS_x = dydz$ is the x component of the surface area dS.

Thus, the above equation gives the surface integral of F_x with the x component of dS over the whole surface.

Similarly, considering $\frac{\partial F_y}{\partial y}$ and $\frac{\partial F_z}{\partial z}$ and summing up, we get

$$\therefore \qquad \int_V \nabla \cdot \vec{F} dv = \oint_S (F_x dS_x + F_y dS_y + F_z dS_z) = \oint_S \vec{F} \cdot d\vec{S}$$

where, $d\vec{S} = dS_x \hat{a}_x + dS_y \hat{a}_y + dS_z \hat{a}_z$ and $\vec{F} = F_x \hat{a}_x + F_y \hat{a}_y + F_z \hat{a}_z$.

Thus,

$$\int_V \nabla \cdot \vec{F} dv = \oint_S \vec{F} \cdot d\vec{S}$$

This is Gauss' divergence theorem.

NOTE

The divergence theorem is a mathematical statement of the physical fact that, in the absence of the creation or destruction of matter, the density within a region of space can change only by having it flow into or away from the region through its boundary. Intuitively, this theorem implies that the sum of all sources minus the sum of all sinks gives the net flow out of a region. *The divergence theorem is an important result for the mathematics of engineering, particularly in electrostatics and fluid mechanics.*

1.10.1 Green's Identities

These identities are corollaries of the divergence theorem and can be derived as follows.

We consider, $$\vec{A} = S_1 \nabla S_2 \qquad \textbf{(1.85)}$$

where S_1 and S_2 are scalar functions continuous together with their partial derivatives of first and second order.

$$\therefore \qquad \nabla \cdot \vec{A} = \nabla \cdot (S_1 \nabla S_2) = S_1 \nabla^2 S_2 + \nabla S_1 \cdot \nabla S_2 \qquad \text{\{by vector identity\}}$$

Applying divergence theorem

$$\int_V \nabla \cdot \vec{A} dv = \oint_S \vec{A} \cdot d\vec{S}$$

Substituting the value of $\nabla \cdot \vec{A}$ and $\vec{A}$, we get

$$\int_V (S_1 \nabla^2 S_2 + \nabla S_1 \cdot \nabla S_2) dv = \oint_S (S_1 \nabla S_2) \cdot d\vec{S} \tag{1.86}$$

This is the *first form of Green's identity*.

Now, interchanging the functions S_1 and S_2, we get

$$\int_V (S_2 \nabla^2 S_1 + \nabla S_2 \cdot \nabla S_1) dv = \oint_S (S_2 \nabla S_1) \cdot d\vec{S} \tag{1.87}$$

Subtracting Eq. (1.87) from Eq. (1.86), we get

$$\int_V (S_1 \nabla^2 S_2 - S_2 \nabla^2 S_1) dv = \oint_S (S_1 \nabla S_2 - S_2 \nabla S_1) \cdot d\vec{S} \tag{1.88}$$

This is the *second form of Green's identity.*

Using Green's identities, it can be proved that the specifications of both divergence and curl of a vector with boundary conditions are sufficient to make the function unique (known as *Uniqueness theorem*).

***Example 1.42** Given that $\vec{A} = 2xy\hat{a}_x + 3\hat{a}_y + yz^2\hat{a}_z$, evaluate $\oint_S \vec{A} \cdot d\vec{S}$, where S is the surface of the cube defined by $0 \le x \le 1, 0 \le y \le 1, 0 \le z \le 1$. Also, verify the result by using divergence theorem.

Solution We evaluate the surface integrals for the six surfaces as follows.

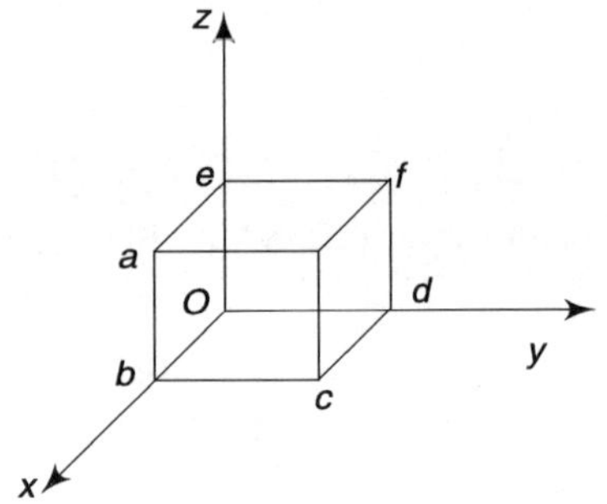

Fig. 1.34 *Surface integral of Example 1.42*

$$\oint_S \vec{A} \cdot d\vec{S} = \int_{x=0}^{1} \int_{y=0}^{1} A_z\Big|_{z=1} dxdy + \int_{x=0}^{1} \int_{y=0}^{1} (-A_z)\Big|_{z=0} dxdy + \int_{z=0}^{1} \int_{x=0}^{1} A_y\Big|_{y=1} dxdz$$

$$+ \int_{z=0}^{1} \int_{x=0}^{1} (-A_y)\Big|_{y=0} dxdz + \int_{y=0}^{1} \int_{z=0}^{1} A_x\Big|_{x=1} dydz + \int_{y=0}^{1} \int_{z=0}^{1} (-A_x)\Big|_{x=0} dydz$$

where $A_x = 2xy; \quad A_y = 3; \quad A_z = yz^2$

$$\therefore \quad \oint_S \vec{A}\cdot d\vec{S} = \int_{x=0}^{1}\int_{y=0}^{1}(y-0)dxdy + \int_{z=0}^{1}\int_{x=0}^{1}(3-3)dxdz + \int_{y=0}^{1}\int_{z=0}^{1}(2y-0)dydz$$

$$= \frac{1}{2} + 0 + 1$$

$$= \frac{3}{2}$$

Also, $\nabla\cdot\vec{A} = \frac{\partial}{\partial x}(2xy) + \frac{\partial}{\partial y}(3) + \frac{\partial}{\partial z}(yz^2) = 2y + 2yz = 2y(1+z)$

$$\therefore \quad \int_v \nabla\cdot\vec{A}dv = \int_{z=0}^{1}\int_{y=0}^{1}\int_{x=0}^{1} 2y(1+z)dxdydz = \int_{z=0}^{1}\int_{y=0}^{1} 2y(1+z)dydz = \int_{z=0}^{1}(1+z)dz = 1 + \frac{1}{2} = \frac{3}{2}$$

Since, $\int_v \nabla\cdot\vec{A}dv = \oint_S \vec{A}\cdot d\vec{S}$, divergence theorem is verified.

***Example 1.43** Given that $\vec{A} = \left(\frac{5r^3}{4}\right)\widehat{a}_r$ C/m^2 in cylindrical coordinates, evaluate both sides of divergence theorem for the volume enclosed by $r = 1$ m, $r = 2$ m, $z = 0$ and $z = 10$ m.

Solution Here, $d\vec{S} = rd\phi dz\widehat{a}_r$

$$\therefore \quad \oint_S \vec{A}\cdot d\vec{S} = \left[\int_{\phi=0}^{2\pi}\int_{z=0}^{10}\left(\frac{5r^3}{4}\widehat{a}_r\right)\cdot(rd\phi dz\widehat{a}_r)\right]_{r=2} - \left[\int_{\phi=0}^{2\pi}\int_{z=0}^{10}\left(\frac{5r^3}{4}\widehat{a}_r\right)\cdot(rd\phi dz\widehat{a}_r)\right]_{r=1}$$

$$= \left[\int_{\phi=0}^{2\pi}\int_{z=0}^{10}\frac{5r^4}{4}d\phi dz\right]_{r=2} - \left[\int_{\phi=0}^{2\pi}\int_{z=0}^{10}\frac{5r^4}{4}d\phi dz\right]_{r=1}$$

$$= \int_{\phi=0}^{2\pi}\int_{z=0}^{10}\frac{80}{4}d\phi dz - \int_{\phi=0}^{2\pi}\int_{z=0}^{10}\frac{5}{4}d\phi dz$$

$$= \int_{\phi=0}^{2\pi}\int_{z=0}^{10}\frac{75}{4}d\phi dz$$

$$= \frac{75}{4}\times 2\pi\int_{z=0}^{10}dz$$

$$= \frac{75}{4}\times 2\pi\times 10$$

$$= 375\pi$$

Also, $\nabla\cdot\vec{A} = \frac{1}{r}\frac{\partial}{\partial r}\left(r\frac{5r^3}{4}\right) = \frac{1}{r}\frac{\partial}{\partial r}\left(\frac{5r^4}{4}\right) = \frac{5}{4}\times\frac{1}{r}\times 4r^3 = 5r^2$

$$\therefore \quad \int_v (\nabla \cdot \vec{A})dv = \int_v 5r^2 r dr d\phi dz = \int_{r=1}^{2}\int_{\phi=0}^{2\pi}\int_{z=0}^{10} 5r^3 dr d\phi dz = 2\pi \times 10 \int_{r=1}^{2} 5r^3 dr = 100\pi\left[\frac{r^4}{4}\right]_1^2$$

$$= 375\pi$$

Since, $\int_v \nabla \cdot \vec{A} dv = \oint_S \vec{A} \cdot d\vec{S}$, divergence theorem is verified.

***Example 1.44** For the given vector $\vec{F} = (x^2 - y^2)\hat{a}_x + 2xy\hat{a}_y + (x^2 - xy)\hat{a}_z$, evaluate the surface integral $\oint_S \vec{F} \cdot d\vec{S}$ over the surface of the cube with centre at the origin and side length a.

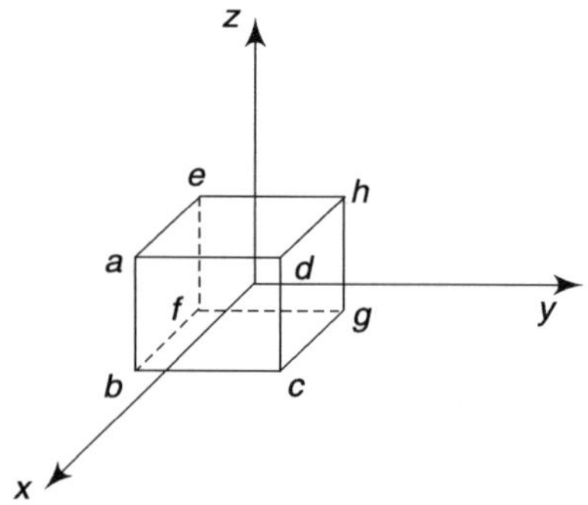

Fig. 1.35 *Surface integral of a vector*

Solution We calculate the surface integrals $\int_S \vec{F} \cdot d\vec{S}$ for all six surfaces and then add up to get the closed surface integral $\oint_S \vec{F} \cdot d\vec{S}$.

For surface *abcd*

$$x = \frac{a}{2}, \quad d\vec{S} = dydz\hat{a}_x$$

$$\int_{abcd} \vec{F} \cdot d\vec{S} = \int_S [(x^2 - y^2)\hat{a}_x + 2xy\hat{a}_y + (x^2 - xy)\hat{a}_z] \cdot (dydz\hat{a}_x) = \int_{y=-a/2}^{a/2}\int_{z=-a/2}^{a/2}\left(\frac{a^2}{4} - y^2\right)dydz$$

$$= \frac{a^2}{4}a \times a - a\frac{y^3}{3}\Bigg|_{-a/2}^{a/2} = \frac{a^4}{4} - \frac{a^4}{12} = \frac{a^4}{6}$$

For surface *efgh*

$$x = -\frac{a}{2}, \quad d\vec{S} = -dydz\hat{a}_x$$

$$\int_{efgh} \vec{F} \cdot d\vec{S} = \int_S [(x^2 - y^2)\hat{a}_x + 2xy\hat{a}_y + (x^2 - xy)\hat{a}_z] \cdot (-dydz\hat{a}_x) = -\int_{y=-a/2}^{a/2}\int_{z=-a/2}^{a/2}\left(\frac{a^2}{4} - y^2\right)dydz$$

$$= -\frac{a^4}{6}$$

For surface *cdhg*

$$y = \frac{a}{2}, \quad d\vec{S} = dxdz\widehat{a}_y$$

$$\int_{cdhg} \vec{F} \cdot d\vec{S} = \int_S [(x^2 - y^2)\widehat{a}_x + 2xy\widehat{a}_y + (x^2 - xy)\widehat{a}_z] \cdot (dxdz\widehat{a}_y) = \int_{x=-a/2}^{a/2} \int_{z=-a/2}^{a/2} 2xydxdz = 0$$

For surface *abfe*

$$y = -\frac{a}{2}, \quad d\vec{S} = -dxdz\widehat{a}_y$$

$$\int_{abfe} \vec{F} \cdot d\vec{S} = \int_S [(x^2 - y^2)\widehat{a}_x + 2xy\widehat{a}_y + (x^2 - xy)\widehat{a}_z] \cdot (-dxdz\widehat{a}_y) = -\int_{x=-a/2}^{a/2} \int_{z=-a/2}^{a/2} 2xydxdz = 0$$

For surface *adhe*

$$z = \frac{a}{2}, \quad d\vec{S} = dxdy\widehat{a}_z$$

$$\int_{adhe} \vec{F} \cdot d\vec{S} = \int_S [(x^2 - y^2)\widehat{a}_x + 2xy\widehat{a}_y + (x^2 - xy)\widehat{a}_z] \cdot (dxdy\widehat{a}_z) = \int_{x=-a/2}^{a/2} \int_{y=-a/2}^{a/2} (x^2 - xy)dxdy$$

$$= \frac{a^4}{12}$$

For surface *bcgf*

$$z = -\frac{a}{2}, \quad d\vec{S} = -dxdy\widehat{a}_z$$

$$\int_{bcgf} \vec{F} \cdot d\vec{S} = \int_S [(x^2 - y^2)\widehat{a}_x + 2xy\widehat{a}_y + (x^2 - xy)\widehat{a}_z] \cdot (-dxdy\widehat{a}_z) = -\int_{x=-a/2}^{a/2} \int_{y=-a/2}^{a/2} (x^2 - xy)dxdy$$

$$= -\frac{a^4}{12}$$

Addition of all six surface integrals gives

$$\oint_S \vec{F} \cdot d\vec{S} = 0$$

We can verify the result by using divergence theorem as follows:

$$\nabla \cdot \vec{F} = \frac{\partial F_x}{\partial x} + \frac{\partial F_y}{\partial y} + \frac{\partial F_z}{\partial z} = \frac{\partial}{\partial x}(x^2 - y^2) + \frac{\partial}{\partial y}(2xy) + \frac{\partial}{\partial z}(x^2 - xy)$$

$$= 2x + 2x + 0 = 4x$$

$$\therefore \quad \int_v \nabla \cdot \vec{F} dv = \int_{-a/2}^{a/2} \int_{-a/2}^{a/2} \int_{-a/2}^{a/2} 4xdxdydz = 2a^2[x^2]_{-a/2}^{a/2} = 0$$

$$\because \quad \int_v \nabla \cdot \vec{F} dv = \oint_S \vec{F} \cdot d\vec{S}, \text{ divergence theorem is verified.}$$

Example 1.45 Given that $\vec{A} = \frac{x^3}{3}\hat{a}_x$, evaluate both sides of the divergence theorem for the volume of a cube, 1 m on an edge, centered at the origin and with edges parallel to the axes.

Solution Here, $\vec{A} = \frac{x^3}{3}\hat{a}_x$

$$\therefore \qquad \vec{A}\cdot d\vec{S} = \left(\frac{x^3}{3}\hat{a}_x\right)\cdot(dydz\hat{a}_x) = \frac{1}{3}x^3\,dydz$$

$$\therefore \qquad \oint_S \vec{A}\cdot d\vec{S} = \left[\int_{y=-0.5}^{0.5}\int_{z=-0.5}^{0.5}\frac{1}{3}x^3\,dydz\right]_{x=0.5} - \left[\int_{y=-0.5}^{0.5}\int_{z=-0.5}^{0.5}\frac{1}{3}x^3\,dydz\right]_{x=-0.5}$$

$$= \frac{1}{24}\times 1\times 1 + \frac{1}{24}\times 1\times 1 = \frac{1}{12}$$

Also, $\nabla\cdot\vec{A} = \frac{\partial}{\partial x}\left(\frac{x^3}{3}\right) = x^2$

$$\therefore \qquad \int_v (\nabla\cdot\vec{A})dv = \int_{x=-0.5}^{0.5}\int_{y=-0.5}^{0.5}\int_{z=-0.5}^{0.5} x^2\,dxdydz = \left[\frac{x^3}{3}\right]_{-0.5}^{0.5}\times 1\times 1 = \frac{1}{12}$$

Hence, $\oint_S \vec{A}\cdot d\vec{S} = \int_v (\nabla\cdot\vec{A})dv$; thus, divergence theorem is verified.

***Example 1.46** Given that $\vec{F} = 10e^{-r}\hat{a}_r - 2z\hat{a}_z$, evaluate both sides of the divergence theorem for the volume enclosed by $r = 2$, $z = 0$ and $z = 5$.

Solution Here, $\vec{F} = 10e^{-r}\hat{a}_r - 2z\hat{a}_z$

$$d\vec{S} = rd\phi dz\hat{a}_r + drdz\hat{a}_\phi + rdrd\phi\hat{a}_z$$

$$\therefore \qquad \vec{F}\cdot d\vec{S} = (10e^{-r}\hat{a}_r - 2z\hat{a}_z)\cdot(rd\phi dz\hat{a}_r + drdz\hat{a}_\phi + rdrd\phi\hat{a}_z) = 10re^{-r}\,d\phi dz - 2rzdrd\phi$$

The cylinder has three surfaces as follows.

$$\oint_s \vec{F}\cdot d\vec{S} = \int_{\text{top}}\vec{F}\cdot d\vec{S} + \int_{\text{bottom}}\vec{F}\cdot d\vec{S} + \int_{\text{curved}}\vec{F}\cdot d\vec{S}$$

For the top surface, $z = 5$

$$\int_{\text{top}}\vec{F}\cdot d\vec{S} = \left.\int_{r=0}^{2}\int_{\phi=0}^{2\pi} -2rzdrd\phi\right|_{z=5} = -10\int_{r=0}^{2}\int_{\phi=0}^{2\pi} rdrd\phi = -10\times 2\pi\times\frac{2^2}{2} = -40\pi$$

For the bottom surface, $z = 0$

$$\int_{\text{bottom}}\vec{F}\cdot d\vec{S} = \left.\int_{r=0}^{2}\int_{\phi=0}^{2\pi} -2rzdrd\phi\right|_{z=0} = 0$$

For the curved surface, $r = 2$

$$\int_{\text{curved}}\vec{F}\cdot d\vec{S} = \left.\int_{\phi=0}^{2\pi}\int_{z=0}^{5} 10re^{-r}\,d\phi dz\right|_{r=2} = 20e^{-2}\times 2\pi\times 5 = 200\pi e^{-2}$$

By addition, total surface area of the closed cylinder is given as

$$\oint_s \vec{F}\cdot d\vec{S} = -40\pi + 0 + 200\pi e^{-2} = -40.63$$

Also, $\nabla\cdot\vec{F} = \frac{1}{r}\frac{\partial}{\partial r}(rF_r) + \frac{1}{r}\frac{\partial F_\phi}{\partial\phi} + \frac{\partial F_z}{\partial z} = \frac{1}{r}\frac{\partial}{\partial r}(10re^{-r}) + \frac{\partial}{\partial z}(-2z) = \frac{10e^{-r}}{r} - 10e^{-r} - 2$

$$\therefore \quad \int_v (\nabla\cdot\vec{F})dv = \int_{r=0}^{2}\int_{\phi=0}^{2\pi}\int_{z=0}^{5}\left(\frac{10e^{-r}}{r} - 10e^{-r} - 2\right) r dr d\phi dz$$

$$= 2\pi\times 5\times\int_{r=0}^{2}(10e^{-r} - 10re^{-r} - 2r)dr = -40.63$$

$\therefore \quad \oint_s \vec{F}\cdot d\vec{S} = \int_v (\nabla\cdot\vec{F})dv$ and hence, divergence theorem is verified.

***Example 1.47** Given the vector $\vec{D} = \frac{5\rho^2}{4}\hat{a}_\rho$ in spherical coordinates. Verify both sides of the divergence theorem for

(a) the volume enclosed by $\rho = 1$ and $\rho = 2$.

***(b)** the volume enclosed by $\rho = 2$ and $\theta = \frac{\pi}{4}$.

Solution Here, $\vec{D} = \frac{5\rho^2}{4}\hat{a}_\rho$

$$\therefore \quad d\vec{S} = \rho^2 \sin\theta d\theta d\phi \vec{a}_\rho$$

$$\therefore \quad \vec{D}\cdot d\vec{S} = \left(\frac{5\rho^2}{4}\hat{a}_\rho\right)\cdot(\rho^2\sin\theta d\theta d\phi\vec{a}_\rho) = \frac{5}{4}\rho^4\sin\theta d\theta d\phi$$

(a) Here, the surface integral is radially outward for the surface at $\rho = 2$ and radially inward at $\rho = 1$.

$$\therefore \quad \oint_S \vec{D}\cdot d\vec{S} = \left[\frac{5}{4}\int_{\theta=0}^{\pi}\int_{\phi=0}^{2\pi}\rho^4\sin\theta d\theta d\phi\right]_{\rho=2} - \left[\frac{5}{4}\int_{\theta=0}^{\pi}\int_{\phi=0}^{2\pi}\rho^4\sin\theta d\theta d\phi\right]_{\rho=1}$$

$$= 20\times 2\pi\times(-\cos\theta)_0^\pi - \frac{5}{4}\times 2\pi\times(-\cos\theta)_0^\pi$$

$$= 80\pi - 5\pi$$

$$= 75\pi$$

Also, $\nabla\cdot\vec{D} = \frac{1}{\rho^2}\frac{\partial}{\partial\rho}(\rho^2 D_\rho) = \frac{1}{\rho^2}\frac{\partial}{\partial\rho}\left(\frac{5}{4}\rho^4\right) = 5\rho$

$$\therefore \quad \int_v (\nabla\cdot\vec{D})dv = \int_{\rho=1}^{2}\int_{\theta=0}^{\pi}\int_{\phi=0}^{2\pi} 5\rho\rho^2\sin\theta d\rho d\theta d\phi = 5\left[\frac{\rho^4}{4}\right]_1^2\times 2\times 2\pi = 75\pi$$

$\therefore \quad \oint_S \vec{D}\cdot d\vec{S} = \int_v (\nabla\cdot\vec{D})dv$; and thus divergence theorem is verified.

(b) Here, the surface integral is non-vanishing only at $\rho = 2$.

$$\therefore \quad \oint_S \vec{D} \cdot d\vec{S} = \left[\frac{5}{4}\int_{\theta=0}^{\pi}\int_{\phi=0}^{2\pi} \rho^4 \sin\theta d\theta d\phi\right]_{\rho=2} = 20 \times 2\pi \times (-\cos\theta)_0^{\pi} = 80\pi$$

$$\text{Also, } \int_v (\nabla \cdot \vec{D}) dv = \int_{\rho=0}^{2}\int_{\theta=0}^{\pi/4}\int_{\phi=0}^{2\pi} 5\rho\rho^2 \sin\theta d\rho d\theta d\phi = 5\left[\frac{\rho^4}{4}\right]_0^2 \times 2 \times 2\pi = 80\pi$$

Thus, divergence theorem is verified.

***Example 1.48** Evaluate both sides of the divergence theorem.

$$\oint_S \vec{A} \cdot d\vec{S} = \int_v \nabla \cdot \vec{A} dv$$

For each of the following cases:

(a) $\vec{A} = xy^2 \hat{a}_x + y^3 \hat{a}_y + y^2 z\hat{a}_z$ and S is the surface of the cuboid defined by $0 < x < 1$, $0 < y < 1$, $0 < z < 1$.

(b) $\vec{A} = 2rz\hat{a}_r + 3z \sin\phi \hat{a}_\phi - 4r\cos\phi \hat{a}_z$ and S is the surface of the wedge $0 < r < 2$, $0 < \phi < 45°$, $0 < z < 5$.

Solution (a)

(a) Here, $\vec{A} = xy^2\hat{a}_x + y^3\hat{a}_y + y^2 z\hat{a}_z$

We evaluate the surface integrals for the six surfaces as follows.

$$\oint_S \vec{A} \cdot d\vec{S} = \int_{x=0}^{1}\int_{y=0}^{1} A_z\Big|_{z=1} dxdy + \int_{x=0}^{1}\int_{y=0}^{1} (-A_z)\Big|_{z=0} dxdy$$

$$+ \int_{z=0}^{1}\int_{x=0}^{1} A_y\Big|_{y=1} dxdz + \int_{z=0}^{1}\int_{x=0}^{1} (-A_y)\Big|_{y=0} dxdz$$

$$+ \int_{y=0}^{1}\int_{z=0}^{1} A_x\Big|_{x=1} dydz + \int_{y=0}^{1}\int_{z=0}^{1} (-A_x)\Big|_{x=0} dydz$$

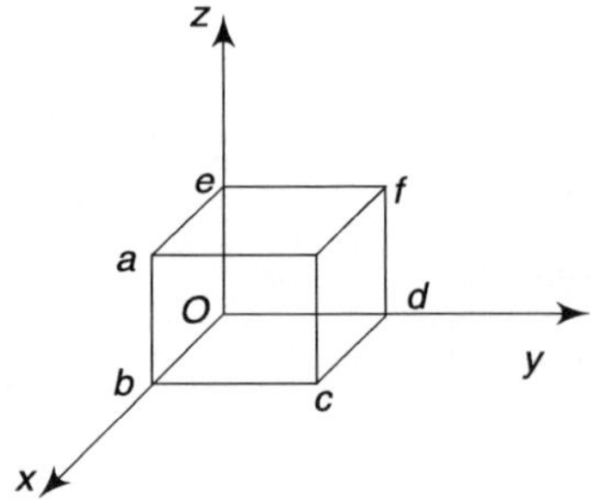

Fig. 1.36 *Surface integral of Example 1.48 (a)*

where $A_x = xy^2$; $A_y = y^3$; $A_z = y^2 z$

$$\therefore \quad \oint_S \vec{A} \cdot d\vec{S} = \int_{x=0}^{1}\int_{y=0}^{1} (y^2 - 0)dxdy + \int_{z=0}^{1}\int_{x=0}^{1} (1-0)dxdz + \int_{y=0}^{1}\int_{z=0}^{1} (y^2 - 0)dydz$$

$$= \frac{1}{3} + 1 + \frac{1}{3}$$

$$= \frac{5}{3}$$

$$\text{Also, } \nabla \cdot \vec{A} = \frac{\partial}{\partial x}(xy^2) + \frac{\partial}{\partial y}(y^3) + \frac{\partial}{\partial z}(y^2 z) = y^2 + 3y^2 + y^2 = 5y^2$$

$$\therefore \qquad \int_v \nabla \cdot \vec{A} dv = \int_{z=0}^{1} \int_{y=0}^{1} \int_{x=0}^{1} 5y^2 \, dxdydz = \int_{y=0}^{1} 5y^2 dy = \frac{5}{3}$$

Since, $\int_v \nabla \cdot \vec{A} dv = \oint_S \vec{A} \cdot d\vec{S}$, divergence theorem is verified.

(b) Here, $\vec{A} = 2rz\hat{a}_r + 3z \sin \phi \hat{a}_\phi - 4r \cos \phi \hat{a}_z$

We evaluate the surface integrals for the six surfaces as follows.

$$\oint_S \vec{A} \cdot d\vec{S} = \int_{z=0}^{5} \int_{\phi=0}^{45^\circ} A_r\Big|_{r=2} rd\phi dz + \int_{z=0}^{5} \int_{r=0}^{2} A_\phi\Big|_{\phi=45^\circ} drdz + \int_{z=0}^{5} \int_{r=0}^{2} (-A_\phi)\Big|_{\phi=0} drdz$$

$$+ \int_{\phi=0}^{45^\circ} \int_{r=0}^{2} A_z\Big|_{z=5} rdrd\phi + \int_{\phi=0}^{45^\circ} \int_{r=0}^{2} (-A_z)\Big|_{z=0} rdrd\phi$$

where $A_r = 2rz$; $\quad A_\phi = 3z \sin \phi$; $\quad A_z = -4r \cos \phi$

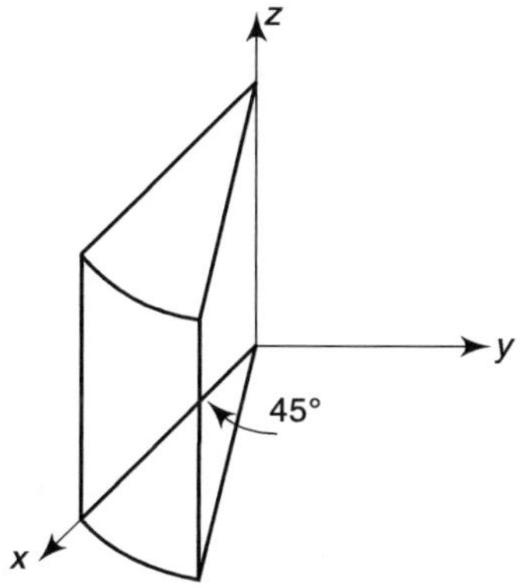

Fig. 1.37 *Surface integral of Example 1.48 (b)*

$$\therefore \qquad \oint_S \vec{A} \cdot d\vec{S} = \int_{z=0}^{5} \int_{\phi=0}^{45^\circ} 8zd\phi dz + \int_{z=0}^{5} \int_{r=0}^{2} \frac{3}{\sqrt{2}} zdrdz + 0 + \int_{\phi=0}^{45^\circ} \int_{r=0}^{2} -4rrdrd\phi + \int_{\phi=0}^{45^\circ} \int_{r=0}^{2} 4rrdrd\phi$$

$$= 25\pi + \frac{75}{\sqrt{2}}$$

Also, $$\nabla . \vec{A} = \frac{1}{r}\frac{\partial}{\partial r}(2r^2 z) + \frac{1}{r}\frac{\partial}{\partial \phi}(3z \sin \phi) + \frac{\partial}{\partial z}(-4r \cos \phi) = 4z + \frac{3z}{r} \cos \phi$$

$$\therefore \qquad \int_v \nabla \cdot \vec{A} dv = \int_{z=0}^{5} \int_{\phi=0}^{45^\circ} \int_{r=0}^{2} \left(4z + \frac{3z}{r} \cos \phi\right) rdrd\phi dz = 25\pi + \frac{75}{\sqrt{2}}$$

Since $\int_v \nabla \cdot \vec{A} dv = \oint_S \vec{A} \cdot d\vec{S}$, divergence theorem is verified.

Example 1.49 Find the total charge inside a cubical volume of 1 m on a side situated in the positive octant with three edges coincident with the x, y and z axis and one corner at the origin, if $\vec{D} = (x+3)\hat{a}_x$. Solve the problem using (a) divergence theorem, and (b) by integrating $\vec{D}$ over surface of the cube.

Solution Here, the flux density is given as, $\vec{D} = (x+3)\hat{a}_x$

We can find the total charge by two methods:

(a) by divergence theorem, or

(b) by evaluating the surface integrals of the flux density over surface of the cube.

(a) By divergence theorem

$$\nabla \cdot \vec{D} = \frac{\partial}{\partial x}(x+3) = 1$$

$$\therefore \qquad \int_v \nabla \cdot \vec{A} dv = \int_{z=0}^{1} \int_{y=0}^{1} \int_{x=0}^{1} 1 dxdydz = 1$$

(b) By evaluating the surface integrals of the flux density over the surface of the cube

Here, as the flux density is having only x-component, the surface integral reduces to

$$\oint_S \vec{D} \cdot d\vec{S} = \int_{y=0}^{1} \int_{z=0}^{1} D_x\big|_{x=1} dydz + \int_{y=0}^{1} \int_{z=0}^{1} (-D_x)\big|_{x=0} dydz = \int_{y=0}^{1} \int_{z=0}^{1} (4-3)dydz = \int_{y=0}^{1} \int_{z=0}^{1} dydz = 1$$

Thus, we see that in both the methods, the total charge inside the cube is found to be 1 Coulomb.

1.11 STOKES' THEOREM

Statement This theorem states that the line integral of a vector around a closed path is equal to the surface integral of the normal component of its curl over the surface bounded by the path.

Mathematically,

$$\oint_L \vec{F} \cdot d\vec{l} = \iint_S (\nabla \times \vec{F}) \cdot d\vec{S} = \iint_S (\nabla \times \vec{F}) \cdot \hat{a}_n dS \qquad \textbf{(1.89)}$$

where S is the surface enclosed by the path L. The positive direction of $d\vec{S}$ is related to the positive sense of defining L according to the right-hand rule.

Proof We consider an arbitrary surface S as shown in Fig. 1.38. We divide the surface into a large number of still smaller elements 1, 2, 3, ... etc. Taking the line integrals of all such small elements and summing up,

$$\oint_L \vec{F} \cdot d\vec{l} = \int_1 \vec{F} \cdot d\vec{l} + \int_2 \vec{F} \cdot d\vec{l} + \int_3 \vec{F} \cdot d\vec{l} \cdots$$

$$= \sum_k \int_k \vec{F} \cdot d\vec{l} = \sum_k \frac{\int_k \vec{F} \cdot d\vec{l}}{\Delta S_k} \Delta S_k$$

where ΔS_k is the area of the kth element.

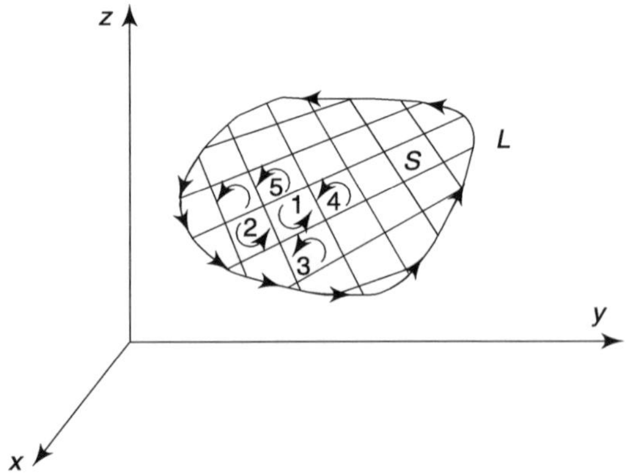

Fig. 1.38 *Arbitrary surface S in the yz plane bounded by closed path L*

(Due to cancelation of every interior path, the sum of the line integrals around k^{th} elements is the same as the line integral around the bounding path L, i.e., $\int_1 \vec{F}\cdot d\vec{l} + \int_2 \vec{F}\cdot d\vec{l} + \int_3 \vec{F}\cdot d\vec{l} \cdots = \oint_L \vec{F}\cdot d\vec{l}$)

As $\Delta S_k \to 0$, summation approaches integration, $\Sigma \to \int$, so that

$$\oint_L \vec{F}\cdot d\vec{l} = \operatorname*{Lim}_{\Delta S_k \to 0} \sum_k \left(\frac{\int_k \vec{F}\cdot d\vec{l}}{\Delta S_k} \right) \Delta S_k = \int_S \operatorname*{Lim}_{\Delta S_k \to 0} \left(\frac{\int_k \vec{F}\cdot d\vec{l}}{\Delta S_k} \right) \Delta S_k$$

Since, by definition of curl of a vector, the line integral divided by the surface area is the component of the curl normal to the surface, i.e., $\operatorname*{Lim}_{\Delta S_k \to 0} \left(\frac{\int_k \vec{F}\cdot d\vec{l}}{\Delta S_k} \right) = \nabla \times \vec{F}$, we can write

$$\oint_L \vec{F}\cdot d\vec{l} = \operatorname*{Lim}_{\Delta S_k \to 0} \sum_k \left(\frac{\int_k \vec{F}\cdot d\vec{l}}{\Delta S_k} \right) \Delta S_k = \int_S \operatorname*{Lim}_{\Delta S_k \to 0} \left(\frac{\int_k \vec{F}\cdot d\vec{l}}{\Delta S_k} \right) \Delta S_k = \int_S (\nabla \times \vec{F})\cdot d\vec{S}$$

$$\boxed{\therefore \ \oint_L \vec{F}\cdot d\vec{l} = \int_S (\nabla \times \vec{F})\cdot d\vec{S}}$$

NOTE

Stokes' theorem is a special case of Green's theorem in plane. To obtain Stokes' theorem from Green's theorem, we have to make two changes. First, the line integral in two dimensions (Green's theorem) is changed to a line integral in three dimensions (Stokes' theorem). Second, the double integral of curl $\vec{F}\cdot \hat{a}_k$ over a region R in the plane (Green's theorem) is changed to a surface integral of curl $\vec{F}\cdot \hat{a}_n$ over a surface S floating in space (Stokes' theorem).

***Example 1.50** Given that $\vec{F} = \hat{a}_x + y^2 z \hat{a}_y$. Verify Stokes' theorem for this vector field and the flat surface in the yz plane bounded by (0, 0, 0), (0, 1, 0), (0, 1, 1) and (0, 0, 1). Choose the contour in the clockwise direction.

Solution Here, $\vec{F}\cdot d\vec{l} = (\hat{a}_x + y^2 z \hat{a}_y)\cdot(dx\hat{a}_x + dy\hat{a}_y + dz\hat{a}_z) = (dx + y^2 z dy)$

$$\oint_L \vec{F}\cdot d\vec{l} = \int_{L_1} \vec{F}\cdot d\vec{l} + \int_{L_2} \vec{F}\cdot d\vec{l} + \int_{L_3} \vec{F}\cdot d\vec{l} + \int_{L_4} \vec{F}\cdot d\vec{l}$$

Along L_1: $dx = dy = 0$

$$\therefore \qquad \int_{L_1} \vec{F}\cdot d\vec{l} = \int_{z=0}^{1} (dx + y^2 z dy) = 0$$

Along L_2: $dx = 0$,

$$\therefore \qquad \int_{L_2} \vec{F}\cdot d\vec{l} = \int_{y=0}^{1} (y^2 z dy)_{z=1} = \left. \frac{y^3}{3} \right|_0^1 = \frac{1}{3}$$

Along L_3: $dx = dy = 0$

$$\therefore \quad \int_{L_3} \vec{F} \cdot d\vec{l} = 0$$

Along L_4: $dx = 0, z = 0$

$$\therefore \quad \int_{L_4} \vec{F} \cdot d\vec{l} = \int_{y=1}^{0} (dx + y^2 z dy) = 0$$

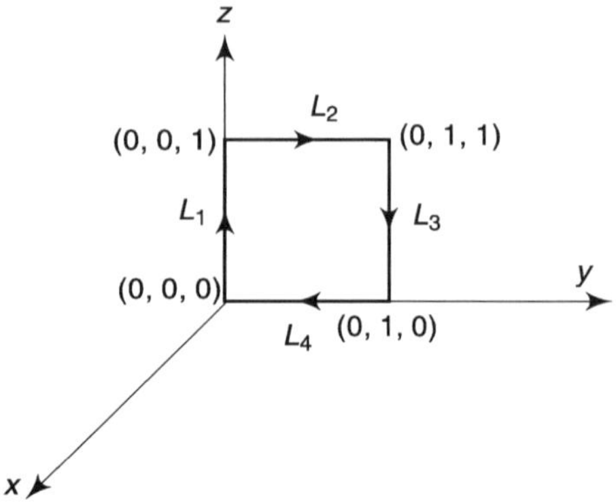

Fig. 1.39 *Line integral of Example 1.50*

Hence, the line integral is given as

$$\oint_L \vec{F} \cdot d\vec{l} = \int_{L_1} \vec{F} \cdot d\vec{l} + \int_{L_2} \vec{F} \cdot d\vec{l} + \int_{L_3} \vec{F} . d\vec{l} + \int_{L_4} \vec{F} \cdot d\vec{l}$$

$$= 0 + \frac{1}{3} + 0 + 0 = \frac{1}{3}$$

To evaluate surface integral, we have the elemental surface given as

$$d\vec{S} = dydz(-\hat{a}_x) = -dydz\hat{a}_x$$

Also, $\nabla \times \vec{F} = \begin{vmatrix} \hat{a}_x & \hat{a}_y & \hat{a}_z \\ \dfrac{\partial}{\partial x} & \dfrac{\partial}{\partial y} & \dfrac{\partial}{\partial z} \\ 1 & y^2 z & 0 \end{vmatrix} = -y^2 \hat{a}_x$

$$\therefore \quad \int_S (\nabla \times \vec{F}) \cdot d\vec{S} = \int_{z=0}^{1} \int_{y=0}^{1} (-y^2 \hat{a}_x) \cdot (-dydz\hat{a}_x) = \int_{z=0}^{1} \int_{y=0}^{1} y^2 dydz = \left. \frac{y^3}{3} \right|_0^1 = \frac{1}{3}$$

Since $\int_S (\nabla \times \vec{F}) \cdot d\vec{S} = \oint_L \vec{F} \cdot d\vec{l}$, Stokes' theorem is verified.

Example 1.51 Given $\vec{A} = r \sin \phi \hat{a}_r + r^2 \hat{a}_\phi$ in cylindrical coordinates. Verify the Stokes' theorem for the contour shown in Fig. 1.40.

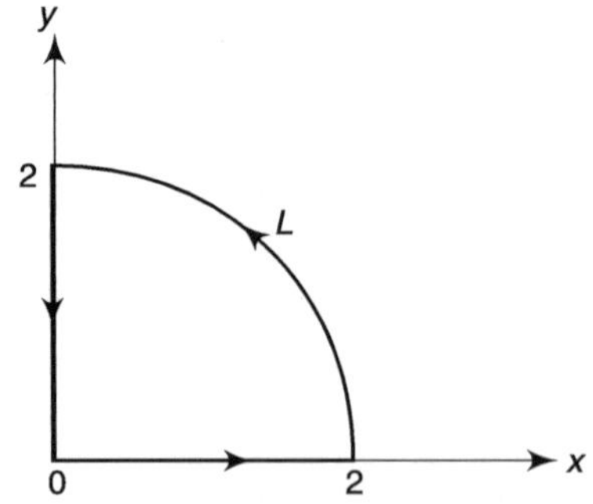

Fig. 1.40 *Contour of Example 1.51*

Solution Here,

$$\vec{A} \cdot d\vec{l} = (r \sin \phi \hat{a}_r + r^2 \hat{a}_\phi) \cdot (dr\hat{a}_r + rd\phi\hat{a}_\phi + dz\hat{a}_z)$$
$$= (r \sin \phi dr + r^3 d\phi)$$

$$\therefore \quad \oint_L \vec{A} \cdot d\vec{l} = \left[\int_{r=0}^{2} r \sin \phi dr\right]_{\phi=0} + \left[\int_{\phi=0}^{\pi/2} r^3 d\phi\right]_{r=2} + \left[\int_{r=2}^{0} r \sin \phi dr\right]_{\phi=\pi/2} = 0 + 4\pi - 2 = (4\pi - 2)$$

$$\text{Also, } \nabla \times \vec{A} = \begin{vmatrix} \frac{1}{r}\hat{a}_r & \hat{a}_\phi & \frac{1}{r}\hat{a}_z \\ \frac{\partial}{\partial r} & \frac{\partial}{\partial \phi} & \frac{\partial}{\partial z} \\ r \sin \phi & r^3 & 0 \end{vmatrix} = (3r - r \cos \phi)\hat{a}_z$$

$$\therefore \quad \int_S (\nabla \times \vec{A}) \cdot d\vec{S} = \int_{\phi=0}^{\pi/2} \int_{r=0}^{2} (3r - \cos \phi) r dr d\phi = \int_{\phi=0}^{\pi/2} (8 - 2 \cos \phi) d\phi = (8\phi - 2 \sin \phi)_0^{\pi/2} = (4\pi - 2)$$

Since $\int_S (\nabla \times \vec{F}) \cdot d\vec{S} = \oint_L \vec{F} \cdot d\vec{l}$, Stokes' theorem is verified.

***Example 1.52** Evaluate the line integral of vector function $\vec{F} = x\hat{i} + x^2 y\hat{j} + y^2 x\hat{k}$ around the square contour *ABCD* in the *xy*-plane as shown in Fig. 1.41. Also integrate $\nabla \times \vec{F}$ over the surface bounded by *ABCD* and verify that Stokes' theorem holds good.

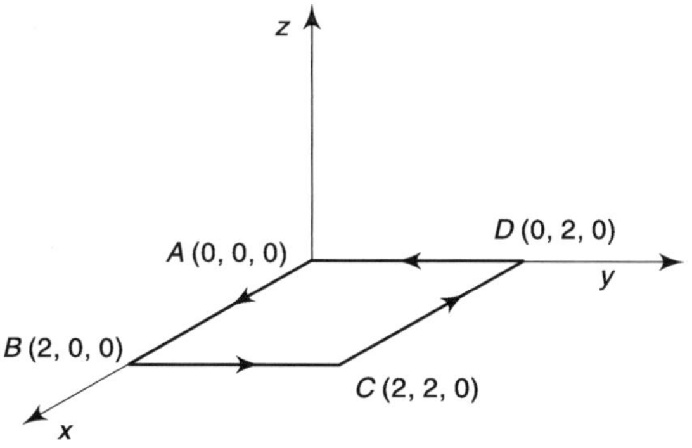

Fig. 1.41 *Contour of Example 1.52*

Solution Here,

$$\vec{F} = x\hat{i} + x^2 y\hat{j} + y^2 x\hat{k}$$

$$d\vec{l} = dx\hat{i} + dy\hat{j} + dz\hat{k}$$

$$\therefore \quad \vec{F} \cdot d\vec{l} = (x\hat{i} + x^2 y\hat{j} + y^2 x\hat{k}) \cdot (dx\hat{i} + dy\hat{j} + dz\hat{k}) = xdx + x^2 ydy + y^2 xdz$$

The closed line integral is given as

$$\oint_L \vec{F}\cdot d\vec{l} = \int_A^B \vec{F}\cdot d\vec{l} + \int_B^C \vec{F}\cdot d\vec{l} + \int_C^D \vec{F}\cdot d\vec{l} + \int_D^A \vec{F}\cdot d\vec{l}$$

Along the path AB, $dy = dz = 0$, $y = 0$, $z = 0$; x varies from 0 to 2

$$\therefore \qquad \int_A^B \vec{F}\cdot d\vec{l} = \int_0^2 x dx = \left[\frac{x^2}{2}\right]_0^2 = 2$$

Along the path BC, $dx = dz = 0$, $x = 2$, $z = 0$; y varies from 0 to 2

$$\therefore \qquad \int_B^C \vec{F}\cdot d\vec{l} = \int_0^2 x^2 y dy\Big|_{x=2} = 4\left[\frac{y^2}{2}\right]_0^2 = 8$$

Along the path CD, $dy = dz = 0$, $y = 2$, $z = 0$; x varies from 2 to 0

$$\therefore \qquad \int_B^C \vec{F}\cdot d\vec{l} = \int_2^0 x dx = -2$$

Along the path DA, $dx = dz = 0$, $x = 0$, $z = 0$; y varies from 2 to 0

$$\therefore \qquad \int_B^C \vec{F}\cdot d\vec{l} = \int_0^2 x^2 y dy\Big|_{x=0} = 0$$

By addition, we get the closed line integral as

$$\oint_L \vec{F}\cdot d\vec{l} = 2 + 8 - 2 + 0 = 8$$

Now, the curl of the vector is

$$\nabla\times\vec{F} = \begin{vmatrix} \hat{i} & \hat{j} & \hat{k} \\ \dfrac{\partial}{\partial x} & \dfrac{\partial}{\partial y} & \dfrac{\partial}{\partial z} \\ x & x^2 y & y^2 x \end{vmatrix} = 2xy\hat{i} - y^2\hat{j} + 2xy\hat{k}$$

$$\int_S (\nabla\times\vec{F})\cdot d\vec{S} = \int_{x=0}^{2}\int_{y=0}^{2} (2xy\hat{i} - y^2\hat{j} + 2xy\hat{k})\cdot(dxdy\hat{k}) = \int_{x=0}^{2}\int_{y=0}^{2} 2xydxdy = 2\left[\frac{x^2}{2}\right]_0^2 \times \left[\frac{y^2}{2}\right]_0^2 = 8$$

Thus, $\oint_L \vec{F}\cdot d\vec{l} = \int_S (\nabla\times\vec{F})\cdot d\vec{S}$; Stokes' theorem is verified.

1.12 CLASSIFICATIONS OF VECTOR FIELDS

A vector field is uniquely characterised by its divergence and curl. From these properties, vector fields are classified into four categories:

1. Solenoidal and Irrotational Vector Fields $(\nabla\cdot\vec{F} = 0, \nabla\times\vec{F} = 0)$
2. Non-solenoidal and Irrotational Vector Fields $(\nabla\cdot\vec{F} \neq 0, \nabla\times\vec{F} = 0)$
3. Solenoidal and Rotational Vector Fields $(\nabla\cdot\vec{F} = 0, \nabla\times\vec{F} \neq 0)$ and
4. Non-solenoidal and Rotational Vector Fields $(\nabla\cdot\vec{F} \neq 0, \nabla\times\vec{F} \neq 0)$.

1. *Solenoidal and Irrotational Vector Fields* **($\nabla \cdot \vec{F} = 0, \nabla \times \vec{F} = 0$)** For a given vector field $\vec{F}$, if its divergence and curl are both zero ($\nabla \cdot \vec{F} = 0, \nabla \times \vec{F} = 0$), the field is known as *solenoidal and irrotational vector fields*. Such a vector field has neither source nor sink of flux.

Examples of such vector fields are: linear motion of incompressible fluids, electrostatic fields in charge-free region, magnetic fields within a current-free region, gravitational fields in free space.

2. *Non-solenoidal and Irrotational Vector Fields* **($\nabla \cdot \vec{F} \neq 0, \nabla \times \vec{F} = 0$)** For a given vector field $\vec{F}$, if its divergence is not zero but curl is zero ($\nabla \cdot \vec{F} \neq 0, \nabla \times \vec{F} = 0$), then the field is known as *non-solenoidal and irrotational vector fields*.

Examples of such vector fields are: electrostatic fields in charged medium, gravitational force inside a mass, linear motion of compressible fluids.

3. *Solenoidal and Rotational Vector Fields* **($\nabla \cdot \vec{F} = 0, \nabla \times \vec{F} \neq 0$)** For a given vector field $\vec{F}$, if its divergence is zero but curl is non-zero ($\nabla \cdot \vec{F} = 0, \nabla \times \vec{F} \neq 0$), then the field is known as *solenoidal and rotational vector fields*.

Examples of such vector fields are: magnetic fields within a current carrying conductor, rotational motion of incompressible fluids.

4. *Non-solenoidal and Rotational Vector Fields* **($\nabla \cdot \vec{F} \neq 0, \nabla \times \vec{F} \neq 0$)** For a given vector field $\vec{F}$, if both the divergence and curl are non-zero, ($\nabla \cdot \vec{F} \neq 0, \nabla \times \vec{F} \neq 0$) then the field is known as *non-solenoidal and rotational vector fields*.

Examples of such vector fields: rotational motion of compressible fluids.

Example 1.53

(a) Determine the constant c such that the vector $\vec{F} = (x+ay)\hat{i} + (y+bz)\hat{j} + (x+cz)\hat{k}$ will be solenoidal.

(b) Find the value of constant Q to make $\vec{V} = (x+3y)\hat{i} + (y-2z)\hat{j} + (x+Qz)\hat{k}$, solenoidal.

(c) Find the constants a, b, c so that the vector

$$\vec{V} = (x+2y+az)\hat{i} + (bx-3y-z)\hat{j} + (4x+cy+2z)\hat{k} \text{ is irrotational.}$$

Solution

(a) $\vec{F} = (x+ay)\hat{i} + (y+bz)\hat{j} + (x+cz)\hat{k}$

The vector will be solenoidal if its divergence is zero.

$$\therefore \qquad \nabla \cdot \vec{F} = \frac{\partial}{\partial x}(x+ay) + \frac{\partial}{\partial y}(y+bz) + \frac{\partial}{\partial z}(x+cz) = 0$$

$$\Rightarrow \qquad 1 + 1 + c = 0$$

$$\therefore \qquad c = -2$$

(b) $\vec{V} = (x+3y)\hat{i} + (y-2z)\hat{j} + (x+Qz)\hat{k}$

The vector will be solenoidal if its divergence is zero.

$$\therefore \qquad \nabla \cdot \vec{V} = \frac{\partial}{\partial x}(x+3y) + \frac{\partial}{\partial y}(y-2z) + \frac{\partial}{\partial z}(x+Qz) = 0$$

$$\Rightarrow \qquad 1 + 1 + Q = 0$$

$$\therefore \qquad Q = -2$$

(c) $\vec{V} = (x + 2y + az)\hat{i} + (bx - 3y - z)\hat{j} + (4x + cy + 2z)\hat{k}$

The vector will be irrotational if its curl is zero.

$$\therefore \qquad \nabla \times \vec{V} = \begin{vmatrix} \hat{i} & \hat{j} & \hat{k} \\ \frac{\partial}{\partial x} & \frac{\partial}{\partial y} & \frac{\partial}{\partial z} \\ (x + 2y + az) & (bx - 3y - z) & (4x + cy + 2z) \end{vmatrix} = 0$$

$$\Rightarrow \qquad (c + 1)\hat{i} + (a - 4)\hat{j} + (b - 2)\hat{k} = 0$$

This vector will be zero if and only if its components are individually zero.

$\therefore \quad (c + 1) = 0 \quad \Rightarrow \quad c = -1$

$(a - 4) = 0 \quad \Rightarrow \quad a = 4$

and $\quad (b - 2) = 0 \quad \Rightarrow \quad b = 2$

Hence, $a = 4$, $b = 2$, $c = -1$.

***Example 1.54** If a scalar potential is given by the expression $\phi = xyz$, determine the potential gradient and also prove that the vector $\vec{F} = \text{grad } \phi = \text{grad } \phi$ is irrotational.

Solution Here, the potential is, $\phi = xyz$

The potential gradient is given as

$$\nabla \phi = \frac{\partial}{\partial x}(xyz)\hat{a}_x + \frac{\partial}{\partial y}(xyz)\hat{a}_y + \frac{\partial}{\partial z}(xyz)\hat{a}_z = (yz\hat{a}_x + xz\hat{a}_y + xy\hat{a}_z)$$

Now, we let, $\vec{F} = \nabla \phi = (yz\hat{a}_x + xz\hat{a}_y + xy\hat{a}_z)$

$$\therefore \qquad \nabla \times \vec{F} = \nabla \times \nabla \phi = \begin{vmatrix} \hat{a}_x & \hat{a}_y & \hat{a}_z \\ \frac{\partial}{\partial x} & \frac{\partial}{\partial y} & \frac{\partial}{\partial z} \\ yz & xz & xy \end{vmatrix} = (1 - 1)\hat{a}_x + (1 - 1)\hat{a}_y + (1 - 1)\hat{a}_z = 0$$

Hence, the vector $\vec{F} = \nabla \phi$ is irrotational.

1.13 HELMHOLTZ THEOREM

Statement A vector field is uniquely described within a region by its divergence and curl.

Explanation If the divergence and curl of any vector $\vec{F}$ are given as

$$\nabla \cdot \vec{F} = \rho_v \quad \text{and} \quad \nabla \times \vec{F} = \vec{\rho}_s \tag{1.90}$$

where ρ_v is the source density of $\vec{F}$, and

$\vec{\rho}_s$ is the circulation density of $\vec{F}$

both vanishing at infinity, then, according to Helmholtz theorem, we can write a vector field $\vec{F}$ as a sum of a component $\vec{F}_s$ whose divergence is zero $\nabla \cdot \vec{F}_s = 0$ (solenoidal) and a component $\vec{F}_i$ whose curl is zero $\nabla \times \vec{F}_i = 0$ (irrotational).

$\therefore \qquad \vec{F} = \vec{F}_s + \vec{F}_i$

Also, we know that the divergence of curl of a vector is zero and the curl of gradient of any scalar is zero. Using these two null identities, we can write $\vec{F}_s$ and $\vec{F}_i$ as follows.

$$\vec{F}_s = \nabla \times \vec{A} \quad \text{and} \quad \vec{F}_i = \nabla \phi \tag{1.91}$$

where $\vec{A}$ and φ are a vector and a scalar quantities respectively.
Thus, any vector can be represented by the Helmholtz theorem as

$$\boxed{\vec{F} = \nabla \phi + \nabla \times \vec{A}} \tag{1.92}$$

Proof We assume that there exist two vector fields $\vec{F}_1$ and $\vec{F}_2$ which show identical results for divergence and curl at least at one point, P (say) in volume v of interest. Besides, $\vec{F}_1$ and $\vec{F}_2$ both satisfy the boundary condition at the surface S bounded by volume; such that on the surface we can write

$$\vec{F}_1 \cdot \hat{a}_n = \vec{F}_2 \cdot \hat{a}_n$$

$$\Rightarrow \qquad \left(\vec{F}_1 - \vec{F}_2\right) \cdot \hat{a}_n = 0 \tag{1.93}$$

At point P, our assumption requires

$$\nabla \cdot \vec{F}_1 = \nabla \cdot \vec{F}_2$$

$$\Rightarrow \qquad \nabla \cdot (\vec{F}_1 - \vec{F}_2) = 0 \tag{1.94}$$

and

$$\nabla \times \vec{F}_1 = \nabla \times \vec{F}_2$$

$$\Rightarrow \qquad \nabla \times (\vec{F}_1 - \vec{F}_2) = 0 \tag{1.95}$$

From Eq. (1.94) and (1.95), we see that there exists a vector field $\vec{H} = (\vec{F}_1 - \vec{F}_2)$, which is solenoidal and irrotational $(\nabla \cdot \vec{H} = 0 \text{ and } \nabla \times \vec{H} = 0)$. Therefore, we can write

$$\vec{H} = \nabla \phi \tag{1.96}$$

where ϕ is a scalar field.

From Eq. (1.94) and (1.95),

$\nabla \cdot \vec{H} = 0 \quad \Rightarrow \quad \nabla \cdot \nabla \phi = 0 \quad \Rightarrow \quad \nabla^2 \phi = 0$

Now, by Green's first identity, we can write

$$\int_V (S_1 \nabla^2 S_2 + \nabla S_1 \cdot \nabla S_2) dv = \oint_S (S_1 \nabla S_2) \cdot d\vec{S}$$

If $S_1 = S_2 = \phi$ (say), then the identity becomes

$$\int_V (\phi \nabla^2 \phi + \nabla \phi \cdot \nabla \phi) dv = \oint_S (\phi \nabla \phi) \cdot d\vec{S}$$

$$\Rightarrow \qquad \int_V (\nabla \phi)^2 \, dv = \oint_S (\phi \nabla \phi) \cdot d\vec{S} \qquad \{\because \nabla^2 \phi = 0\} \tag{1.97}$$

By using Eq. (1.96), we have

$$\oint_S (\phi\nabla\phi)\cdot d\vec{S} = \oint_S \phi\vec{H}\cdot d\vec{S} = \oint_S \phi\vec{H}\cdot\hat{a}_n\,dS = 0 \qquad \{\because \text{by Eq. (1.95)}, \vec{H}\cdot\hat{a}_n = (\vec{F}_1 - \vec{F}_2)\cdot\hat{a}_n = 0\}$$

Hence, from Eq. (1.97), we have

$$\int_V (\nabla\phi)^2\,dv = 0$$

$$\Rightarrow \qquad \int_V |\vec{H}|^2\,dv = 0$$

$$\Rightarrow \qquad \vec{H} = 0$$

$$\Rightarrow \qquad (\vec{F}_1 - \vec{F}_2) = 0$$

$$\therefore \qquad \vec{F}_1 = \vec{F}_2$$

Hence, our initial assumption that $\vec{F}_1$ and $\vec{F}_2$ are two different vector fields showing identical values of divergence and curl is wrong.

Therefore, we conclude that no two vector fields can show identical divergence and curl everywhere. Hence, divergence and curl specifies a vector field uniquely.

1.13.1 Physical Interpretation of Helmholtz Theorem in Classical Electromagnetism

1. Helmholtz theorem is a profound mathematical theorem which provides the definite relationship between a vector field and mathematically defined source functions. The usual presentations of electromagnetic theory establish the principal sources of the electromagnetic field vectors, but do not give information whether all sources are included. Helmholtz theorem provides the basis for investigating the existence of other possible sources.
2. According to Helmholtz theorem, a general vector field can be written as the sum of a conservative field and a solenoidal field. Thus, we ought to be able to write electric and magnetic fields in this form. Second, a general vector field which is zero at infinity is completely specified once its divergence and its curl are given. Thus, we can guess that the laws of electromagnetism can be written as four field equations

$$\nabla\cdot\vec{E} = \text{something}$$
$$\nabla\times\vec{E} = \text{something}$$
$$\nabla\cdot\vec{H} = \text{something}$$
$$\nabla\times\vec{H} = \text{something}$$

and we can solve the field equations without even knowing the right-hand sides and any solutions we find will be unique.

In other words, there is only one possible steady electric and magnetic field which can be generated by a given set of stationary charges and steady currents.

If the right-hand sides of the above field equations are all zero, then the only physical solution is $\vec{E} = \vec{H} = 0$. This implies that steady electric and magnetic fields cannot generate themselves. Instead, they have to be generated by stationary charges and steady currents. So, if we come across a steady electric field, we know that if we trace the field-lines back, we shall eventually find a charge. Likewise, a steady magnetic field implies that there is a steady current flowing somewhere. All of these results follow from Helmholtz theorem prior to any investigation of electromagnetism.

3. Helmholtz theorem also provides a significant result pertaining to the meaning of the inverse square radial fields (such as, Newton's gravitational field and Coulomb's electrostatic fields). It can be established that the inverse square relation is determined by more elementary properties of the field and of the source function to which it relates.

Summary

- A quantity that has only magnitude is said to be a *scalar* quantity, such as time, mass, distance, temperature, work, electric potential, etc. A quantity that has both magnitude and direction is called a *vector* quantity, such as force, velocity, displacement, electric field intensity, etc.
- If the value of the physical function at each point is a scalar quantity, then the field is known as a *scalar field*, such as temperature distribution in a building. If the value of the physical function at each point is a vector quantity, then the field is known as a *vector field*, such as the gravitational force on a body in space.
- A unit vector $\hat{a}_A$ along $\vec{A}$ is defined as a vector whose magnitude is unity and its direction is along $\vec{A}$. In general, any vector can be represented as

$$\vec{A} = A\hat{a}_A = |\vec{A}|\,\hat{a}_A$$

where A or $|\vec{A}|$ represents the magnitude of the vector and $\hat{a}_A$ direction of the vector $\vec{A}$.

- Two vectors can be added together by the *triangle rule* or *parallelogram rule of vector addition.*
- The *dot product* of two vectors $\vec{A}$ and $\vec{B}$, written as, $\vec{A}\cdot\vec{B}$, is defined as

$$\vec{A}\cdot\vec{B} = AB\cos\theta_{AB}$$

where, θ_{AB} is the smaller angle between $\vec{A}$ and $\vec{B}$, and

$A = |\vec{A}|$ and $B = |\vec{B}|$ represent the magnitude of $\vec{A}$ and $\vec{B}$, respectively.

- The cross product of two vectors $\vec{A}$ and $\vec{B}$, written as $(\vec{A}\times\vec{B})$, is defined as

$$\vec{A}\times\vec{B} = AB\sin\theta_{AB}\,\hat{a}_n$$

where $\hat{a}_n$ is the unit vector normal to the plane containing $\vec{A}$ and $\vec{B}$. The direction of the cross product is obtained from a common rule, called the *right-hand rule.*

- Three orthogonal coordinate systems commonly used are Cartesian coordinates (x, y, z), cylindrical coordinates (r, ϕ, z) and spherical coordinates (ρ, θ, ϕ).
- The differential lengths in three coordinate systems are given respectively as

$$d\vec{l} = dx\hat{a}_x + dy\hat{a}_y + dz\hat{a}_z$$
$$d\vec{l} = dr\hat{a}_r + rd\phi\hat{a}_\phi + dz\hat{a}_z$$
$$d\vec{l} = d\rho\hat{a}_\rho + \rho d\theta\hat{a}_\theta + \rho\sin\theta d\phi\hat{a}_\phi$$

- The differential areas in three coordinate systems are given respectively as

$$d\vec{S} = dydz\widehat{a}_x + dxdz\widehat{a}_y + dxdy\widehat{a}_z$$
$$d\vec{S} = rd\phi dz\widehat{a}_r + drdz\widehat{a}_\phi + rdrd\phi\widehat{a}_z$$
$$d\vec{S} = \rho^2 \sin\theta d\theta d\phi\widehat{a}_\rho + \rho\sin\theta d\rho d\phi\widehat{a}_\theta + \rho d\rho d\theta\widehat{a}_\phi$$

- The differential volumes in three coordinate systems are given respectively as

$$dV = dxdydz$$
$$dV = rdrd\phi dz$$
$$dV = \rho^2 \sin\theta \, d\rho d\theta d\phi$$

- For vector $\vec{F}$ and a path l, the line integral is given by

$$\int_l \vec{F}\cdot d\vec{l} = \int_a^b |\vec{F}|\cos\theta dl$$

- If the path of integration is a closed curve, the line integral is the circulation of the vector around the path.
- If the line integration of a vector along a closed path is zero, i.e., $\oint_l \vec{F}\cdot d\vec{l} = 0$, then the vector is known as *conservative* or *lamellar*.
- For a vector $\vec{F}$, continuous in a region containing a smooth surface S, the surface integral or the flux of $\vec{F}$ through S is defined as

$$\psi = \int_S \vec{F}\cdot d\vec{S} = \int_S \vec{F}\cdot\widehat{a}_n dS = \int_S |\vec{F}|\cos\theta dS$$

where $\widehat{a}_n$ is the unit normal vector to the surface S.
- If the surface is a closed surface, the surface integral is the net outward flux of the vector.
- If the surface integral of a vector over a closed surface is zero, i.e., $\oint_S \vec{F}\cdot d\vec{S} = 0$, then the vector is known as *solenoidal vector*.
- The volume integral of a scalar quantity F over a volume V is written as

$$U = \int_V F dv$$

- The differential vector operator (∇) or Del or Nabla, defined in Cartesian coordinates as

$$\nabla = \frac{\partial}{\partial x}\widehat{a}_x + \frac{\partial}{\partial y}\widehat{a}_y + \frac{\partial}{\partial z}\widehat{a}_z$$

is merely *a vector operator, but not a vector quantity*. It is *a vector space function operator*, used for performing vector differentiations.
- The *gradient* of a scalar function is both the magnitude and the direction of the maximum space rate of change of that function.

- The gradient of a scalar quantity in three different coordinate systems is expressed respectively as

$$\nabla F = \frac{\partial F}{\partial x}\hat{a}_x + \frac{\partial F}{\partial y}\hat{a}_y + \frac{\partial F}{\partial z}\hat{a}_z \quad \text{(Cartesian coordinates)}$$

$$= \frac{\partial F}{\partial r}\hat{a}_r + \frac{1}{r}\frac{\partial F}{\partial \phi}\hat{a}_\phi + \frac{\partial F}{\partial z}\hat{a}_z \quad \text{(Cylindrical coordinates)}$$

$$= \frac{\partial F}{\partial \rho}\hat{a}_\rho + \frac{1}{\rho}\frac{\partial F}{\partial \theta}\hat{a}_\theta + \frac{1}{\rho \sin\theta}\frac{\partial F}{\partial \phi}\hat{a}_\phi \quad \text{(Spherical coordinates)}$$

- The *divergence* of a three-dimensional vector field at a point is a measure of how much the vector diverges or converges from that point.
- If the divergence of a vector is zero, then the vector is known as *solenoidal vector*.
- The divergence of a vector quantity in three different coordinate systems is expressed respectively as

$$\nabla \cdot \vec{F} = \frac{\partial F_x}{\partial x} + \frac{\partial F_y}{\partial y} + \frac{\partial F_z}{\partial z} \quad \text{(Cartesian coordinates)}$$

$$= \frac{1}{r}\frac{\partial}{\partial r}(rF_r) + \frac{1}{r}\frac{\partial F_\phi}{\partial \phi} + \frac{\partial F_z}{\partial z} \quad \text{(Cylindrical coordinates)}$$

$$= \frac{1}{\rho^2}\frac{\partial}{\partial \rho}(\rho^2 F_\rho) + \frac{1}{\rho \sin\theta}\frac{\partial}{\partial \theta}(F_\theta \sin\theta) + \frac{1}{\rho \sin\theta}\frac{\partial F_\phi}{\partial \phi} \quad \text{(Spherical coordinates)}$$

- The *curl* of a vector field, denoted curl $\vec{F}$ or $\nabla \times \vec{F}$, is defined as the vector field having magnitude equal to the maximum circulation at each point and to be oriented perpendicularly to this plane of circulation for each point.
- If the curl of a vector is zero, then the vector is known as *irrotational vector.*
- The curl of a vector quantity in three different coordinate systems is expressed respectively as

$$\nabla \times \vec{F} = \begin{vmatrix} \hat{a}_x & \hat{a}_y & \hat{a}_z \\ \frac{\partial}{\partial x} & \frac{\partial}{\partial y} & \frac{\partial}{\partial z} \\ F_x & F_y & F_z \end{vmatrix} \quad \text{(Cartesian coordinates)}$$

$$= \left(\frac{1}{r}\right)\begin{vmatrix} \hat{a}_r & r\hat{a}_\phi & \hat{a}_z \\ \frac{\partial}{\partial r} & \frac{\partial}{\partial \phi} & \frac{\partial}{\partial z} \\ F_r & rF_\phi & F_z \end{vmatrix} \quad \text{(Cylindrical coordinates)}$$

$$= \left(\frac{1}{\rho^2 \sin\theta}\right)\begin{vmatrix} \hat{a}_\rho & \rho\hat{a}_\theta & \rho\sin\theta\,\hat{a}_\phi \\ \frac{\partial}{\partial \rho} & \frac{\partial}{\partial \theta} & \frac{\partial}{\partial \phi} \\ F_\rho & \rho F_\theta & \rho\sin\theta\, F_\phi \end{vmatrix} \quad \text{(Spherical coordinates)}$$

- The *Laplacian operator* (∇^2) of a scalar field is the divergence of the gradient of the scalar field upon which the operator operates.

- The Laplacian operator (∇^2) of a scalar field in three different coordinate systems is expressed respectively as

$$\nabla^2 F = \frac{\partial^2 F}{\partial x^2} + \frac{\partial^2 F}{\partial y^2} + \frac{\partial^2 F}{\partial z^2} \quad \text{(Cartesian coordinates)}$$

$$= \frac{1}{r}\frac{\partial}{\partial r}\left(r\frac{\partial F}{\partial r}\right) + \frac{1}{r^2}\frac{\partial^2 F}{\partial \phi^2} + \frac{\partial^2 F}{\partial z^2} \quad \text{(Cylindrical coordinates)}$$

$$= \frac{1}{\rho^2}\frac{\partial}{\partial \rho}\left(\rho^2\frac{\partial F}{\partial \rho}\right) + \frac{1}{\rho^2 \sin\theta}\frac{\partial}{\partial \theta}\left(\sin\theta\frac{\partial F}{\partial \theta}\right) + \frac{1}{\rho^2 \sin^2\theta}\frac{\partial^2 F}{\partial \phi^2} \quad \text{(Spherical coordinates)}$$

- The *Laplacian of a vector* is defined as the gradient of divergence of the vector minus the curl of curl of the vector; i.e.,

$$\nabla^2 \vec{F} = \nabla(\nabla \cdot \vec{F}) - \nabla \times \nabla \times \vec{F}$$

- In the Cartesian coordinate system and only in the Cartesian coordinate system, vector Laplacian is written as

$$\nabla^2 \vec{F} = (\nabla^2 F_x)\hat{a}_x + (\nabla^2 F_y)\hat{a}_y + (\nabla^2 F_z)\hat{a}_z$$

- Gauss' *divergence theorem* is used to convert volume integral into surface integral and vice versa. According to this theorem, the divergence of a vector field over a volume is equal to the surface integral of the normal component of the vector through the closed surface bounding the volume.

$$\int_V \nabla \cdot \vec{F} dv = \oint_S \vec{F} \cdot d\vec{S} = \oint_S \vec{F} \cdot \hat{a}_n dS$$

where V is the volume enclosed by the closed surface S.

- According to Green's theorem, if $\vec{F}$ is a two-dimensional vector field, such that, $\vec{F} = (F_x\hat{a}_x + F_y\hat{a}_y)$, then

$$\oint_C (F_x dx + F_y dy) = \iint_R \left(\frac{\partial F_y}{\partial x} - \frac{\partial F_x}{\partial y}\right) dxdy$$

where C is a *positively oriented* boundary of the region R.

- Stokes' theorem is used to covert line integral into surface integral and vice versa. According to this theorem, the line integral of a vector around a closed path is equal to the surface integral of the normal component of its curl over the surface bounded by the path

$$\oint_L \vec{F} \cdot d\vec{l} = \iint_S (\nabla \times \vec{F}) \cdot d\vec{S} = \iint_S (\nabla \times \vec{F}) \cdot \hat{a}_n dS$$

where S is the surface enclosed by the path L. The positive direction of $d\vec{S}$ is related to the positive sense of defining L according to the right-hand rule.

- Helmholtz theorem states that a vector field is uniquely described within a region by its divergence and curl.

Important Formulae	
Dot product	$\vec{A}\cdot\vec{B} = AB\cos\theta_{AB}$
Cross product	$\vec{A}\times\vec{B} = AB\sin\theta_{AB}\hat{a}_n$
Differential lengths	$d\vec{l} = dx\hat{a}_x + dy\hat{a}_y + dz\hat{a}_z$, (Cartesian coordinate) $= dr\hat{a}_r + rd\phi\hat{a}_\phi + dz\hat{a}_z$, (Cylindrical coordinate) $= d\rho\hat{a}_\rho + \rho d\theta\hat{a}_\theta + \rho\sin\theta d\phi\hat{a}_\phi$, (Spherical coordinate)
Differential areas	$d\vec{S} = dvdz\hat{a}_x + dxdz\hat{a}_y + dxdy\hat{a}z$, (Cartesian coordinate) $= rd\phi dz\hat{a}_r + drdz\hat{a}_\phi + rdrd\phi\hat{a}_z$, (Cylindrical coordinate) $= \rho^2\sin\theta d\theta d\phi\hat{a}_\rho + \rho\sin\theta d\rho d\phi\hat{a}_\theta + \rho d\rho d\theta\hat{a}_\phi$ (Spherical coordinate)
Differential volume	$d\vec{S} = dxdydz$, (Cartesian coordinate) $= rdrd\phi dz$, (Cylindrical coordinate) $= \rho^2\sin\theta d\rho d\theta d\phi$, (Spherical coordinate)
Del operator in Cartesian coordinate	$\nabla = \frac{\partial}{\partial x}\hat{a}_z + \frac{\partial}{\partial y}\hat{a}_y + \frac{\partial}{\partial z}\hat{a}_z$
Gradient of a scalar	$\nabla F = \frac{\partial F}{\partial x}\hat{a}_x + \frac{\partial F}{\partial y}\hat{a}_y + \frac{\partial F}{\partial z}\hat{a}_z$ (Cartesian coordinates) $= \frac{\partial F}{\partial r}\hat{a}_r + \frac{1}{r}\frac{\partial F}{\partial \phi}\hat{a}_\phi + \frac{\partial F}{\partial z}\hat{a}_z$ (Cylindrical coordinates) $= \frac{\partial F}{\partial \rho}\hat{a}_\rho + \frac{1}{\rho}\frac{\partial F}{\partial \theta}\hat{a}_\theta + \frac{1}{\rho\sin\theta}\frac{\partial F}{\partial \phi}\hat{a}_\phi$ (Spherical coordinates)
Divergence of a vector	$\nabla\cdot F = \frac{\partial F_x}{\partial x} + \frac{\partial F_y}{\partial y} + \frac{\partial F_z}{\partial z}$ (Cartesian coordinates) $= \frac{1}{r}\frac{\partial}{\partial \gamma}(\gamma F_x) + \frac{1}{\gamma}\frac{\partial F_\phi}{\partial \phi} + \frac{\partial F_\theta}{\partial z}$ (Cylindrical coordinates) $= \frac{1}{\rho^2}\frac{\partial}{\partial \rho}(\rho^2 F_\theta) + \frac{1}{\rho\sin\theta}\frac{\partial}{\partial \theta}(F_z\sin\theta)$ $+ \frac{1}{\rho\sin\theta}\frac{\partial F_\phi}{\partial \phi}$ (Spherical coordinates)

Curl of a vector	$\nabla \times \vec{F} = \begin{vmatrix} \hat{a}_x & \hat{a}_y & \hat{a}_z \\ \frac{\partial}{\partial x} & \frac{\partial}{\partial y} & \frac{\partial}{\partial z} \\ F_x & F_y & F_z \end{vmatrix}$ (Cartesian coordinates) $= \left(\frac{1}{r}\right) \begin{vmatrix} \hat{a}_x & \hat{a}_y & \hat{a}_x \\ \frac{\partial}{\partial r} & \frac{\partial}{\partial \phi} & \frac{\partial}{\partial z} \\ F_r & F_\phi & F_z \end{vmatrix}$ (Cylindrical coordinates) $= \left(\frac{1}{\rho^2 \sin\theta}\right) \begin{vmatrix} \hat{a}_\theta & \rho\hat{a}_z & \rho\sin\theta\hat{a}_\phi \\ \frac{\partial}{\partial \rho} & \frac{\partial}{\partial \theta} & \frac{\partial}{\partial \phi} \\ F_\rho & \rho F_\theta & \rho\sin\theta F_\phi \end{vmatrix}$ (Spherical coordinates)
Laplacian of a scalar	$\nabla^2 F = \frac{\partial^2 F}{\partial x^2} + \frac{\partial^2 F}{\partial y^2} + \frac{\partial^2 F}{\partial z^2}$ (Cartesian coordinates) $= \frac{1}{r}\frac{\partial}{\partial r}\left(r\frac{\partial F}{\partial r}\right) + \frac{1}{r^2}\frac{\partial^2 F}{\partial \phi^2}\frac{\partial^2 F}{\partial z^2}$ (Cylindrical coordinates) $= \frac{1}{\rho^2}\frac{\partial}{\partial \rho}\left(\rho^2\frac{\partial F}{\partial \rho}\right) + \frac{1}{\rho^2 \sin\theta}\frac{\partial}{\partial \theta}\left(\sin\theta\frac{\partial F}{\partial \theta}\right)$ $+ \frac{1}{\rho^2 \sin^2\theta}\frac{\partial^2 F}{\partial \phi^2}$ (Spherical coordinates)
Laplacian of a vector in Cartesian coordinate	$\nabla^2 \vec{F} = (\nabla^2 F_x)\hat{a}_z + (\nabla^2 F_y)\hat{a}_y + (\nabla^2 F_z)\hat{a}_z$
Gauss' divergence theorem	$\int_V \nabla \cdot \vec{F} dv = \oint_S \vec{F} \cdot d\vec{S} = \oint_S \vec{F} \cdot \hat{a}_x dS$
Stokes' theorem	$\oint_L \vec{F} \cdot d\vec{l} = \iint_S (\nabla \times \vec{F}) d\vec{S} = \iint_S (\nabla \times \vec{F})\hat{a}_x dS$

Exercises

[Note: * marked problems are important university problems]

- ***Easy***

1. Given the vectors $\vec{A} = 5\hat{a}_x + 3\hat{a}_y + 6\hat{a}_z$ and $\vec{B} = 2\hat{a}_x + \hat{a}_y + 4\hat{a}_z$, find $\vec{A} \cdot \vec{B}$, $\vec{A} \times \vec{B}$ and the angle between $\vec{A}$ and $\vec{B}$. $[37, (6\hat{a}_x - 8\hat{a}_y - \hat{a}_z), 15.2°]$

2. Determine the gradient of the following scalar fields:

(a) $A = x^2 y + xyz$

(b) $S = r^2 z \cos 2\phi$

(c) $W = \dfrac{\sin\theta \sin\phi}{\rho^2}$

$$\left[\begin{array}{l}(a)\ \{y(2x+z)\hat{a}_x + x(x+z)\hat{a}_y + xy\hat{a}_z\};\\ (b)\ (2rz\cos 2\phi\hat{a}_r - 2rz\sin 2\phi\hat{a}_\phi + r^2\cos 2\phi\hat{a}_z);\\ (c)\left(-\dfrac{2\sin\theta\sin\phi}{\rho^3}\hat{a}_\rho + \dfrac{\cos\theta\sin\phi}{\rho^3}\hat{a}_\theta + \dfrac{\cos\phi}{\rho^3}\hat{a}_\phi\right)\end{array}\right]$$

3. Determine the divergence of the following vectors:

(a) $\vec{A} = \dfrac{\hat{a}_x}{\sqrt{x^2+y^2}}$

(b) $\vec{A} = r\sin\phi\hat{a}_r + 2r\cos\phi\hat{a}_\phi + 2z^2\hat{a}_z$

(c) $\vec{A} = \left(\dfrac{5}{\rho^2}\right)\sin\phi\hat{a}_\rho + \rho\cot\theta\hat{a}_\theta + \rho\sin\theta\cos\phi\hat{a}_\phi$

$$\left[(a) - \frac{x}{(x^2+y^{23/2})};\ (b)\ 4z;\ (c) - (1+\sin\theta)\right]$$

4. Determine the curl of the following vectors:

(a) $\vec{F} = (x^2 - z^2)\hat{a}_x + 2\hat{a}_y + 2xz\hat{a}_z$

(b) $\vec{F} = rz\sin\phi\hat{a}_r + 3rz^2\cos\phi\hat{a}_\phi$

(c) $\vec{F} = \dfrac{\sin\phi}{\rho^2}\hat{a}_\rho - \dfrac{\cos\phi}{\rho^2}\hat{a}_\phi$

$$\left[\begin{array}{l}(a) - 4z\hat{a}_y; (b) - 6rz\cos\phi\hat{a}_r + r\sin\phi\hat{a}_\phi + (6-1)z\cos\phi\hat{a}_z;\\ (c)\left(\dfrac{1}{\rho^3}\cot\theta\cos\phi\right)\hat{a}_\rho + \dfrac{1}{\rho^3}\left(\dfrac{\cos\phi}{\sin\theta} + \cos\phi\right)\hat{a}_\theta\end{array}\right]$$

5. Determine the Laplacian of the following scalar fields:

(a) $F = x^2 y + xyz$

(b) $F = r^2 z\cos 2\phi$

(c) $F = 10\,\rho\sin^2\theta\cos\phi$

(d) $F = (3x + 4y + 2z)(x - 2y + 4z)$

$$\left[(a)\ 2y;\ (b)\ 0;\ (c)\ \frac{10\cos\phi}{\rho}(1 + 2\cos 2\theta);\ (d)\ 6\right]$$

6. Show that the vector $\vec{F} = yz\hat{a}_x + xz\hat{a}_y + xy\hat{a}_z$ is both solenoidal and irrotational.

- ***Medium***

***7.** For a position vector $\vec{r}$, prove that $\nabla\left(\dfrac{1}{r}\right) = -\dfrac{\vec{r}}{r^3}$.

8. If (ρ, θ, ϕ) are spherical polar coordinates, show that,

$$\text{grad}(\cos\theta) \times \text{grad}\,\phi = \text{grad}\left(\frac{1}{\rho}\right) \quad \rho \neq 0$$

9. If a rigid body is rotating with an angular velocity $\vec{\omega}$, prove that $\vec{\omega} = \dfrac{1}{2}\text{curl}\,\vec{v}$, where $\vec{v}$ is the linear velocity. Give the physical meaning of the cross product of $\vec{\omega}$ and $\vec{r}$ where $\vec{r}$ is position vector.

*10. Given point P (–2, 6, 3) and vector $\vec{A} = y\hat{a}_x + (x+z)\hat{a}_y$. Express P and $\vec{A}$ in cylindrical and spherical coordinates. Evaluate $\vec{A}$ at P in the Cartesian, cylindrical and spherical systems.

$$\begin{bmatrix} P(-2,6,3) = P(6.32, 108.43°, 3) = P(7, 64.62°, 108.43°); \\ (6\hat{a}_x + \hat{a}_y), (-0.95\hat{a}_r - 6\hat{a}_\phi), (-0.86\hat{a}_\rho - 0.41\hat{a}_\theta - 6\hat{a}_\phi) \end{bmatrix}$$

11. Express the vector, $\vec{B} = \dfrac{10}{\rho}\hat{a}_\rho + \rho\cos\theta\hat{a}_\theta + \hat{a}_\phi$, in Cartesian and cylindrical coordinates. Find B (–3, 4, 0) and B (5, π/2, –2). $[(-2\hat{a}_x + \hat{a}_y); (2.407\hat{a}_r + \hat{a}_\phi + 1.167\hat{a}_z)]$

12. Express the vector, $3x\hat{a}_x - yz\hat{a}_y + x^2 z\hat{a}_z$ in cylindrical coordinates.

13. Express the vector, $2y\hat{a}_x - 2\hat{a}_y + x\hat{a}_z$ in spherical polar coordinates.

*14. Given that $\vec{A} = \dfrac{10x^3}{3}\hat{a}_x$, evaluate both sides of the divergence theorem for the volume of a cube, 2 m on an edge, centered at the origin and with edges parallel to the axes.

15. Given that $\vec{F} = 30e^{-r}\hat{a}_r - 4z\hat{a}_z$ in cylindrical coordinates, evaluate both sides of the divergence theorem for the volume enclosed by $r = 5$, $z = 0$ and $z = 10$.

*16. Given that $\vec{D} = \dfrac{10r^3}{4}\hat{a}_r$ in cylindrical coordinates, evaluate both sides of the divergence theorem for the volume enclosed by $r = 1$, $r = 2$, $z = 0$ and $z = 10$.

- ***Hard***

17. Prove that $\nabla^2 r^n = n(n+1)r^{n-2}$ and $\nabla^2\left(\dfrac{1}{r}\right) = 0$.

18. Given a vector field

$$\vec{D} = \rho\sin\phi\hat{a}_\rho - \frac{1}{\rho}\sin\theta\cos\phi\hat{a}_\theta + \rho^2\hat{a}_\phi$$

determine: (a) $\vec{D}$ at P (10, 150°, 330°).

(b) The component of $\vec{D}$ tangential to the spherical surface $\rho = 10$ at P.

(c) A unit vector at P perpendicular to $\vec{D}$ and tangential to the cone $\theta = 150°$.

$$\left[(a)\,(-5\hat{a}_\rho + 0.043\hat{a}_\theta + 100\hat{a}_\phi);\ (b)\,(0.043\hat{a}_\theta + 100\hat{a}_\phi);\ (c)\,(-0.999\hat{a}_\rho - 0.0499\hat{a}_\phi)\right]$$

19. Given the vector field in 'mixed' coordinate variables as

$$\vec{J} = \frac{3xz}{\rho^2}\hat{a}_x + \frac{3y\cos\theta}{\rho}\hat{a}_y + \left[2 - \frac{3y^2}{\rho^2} - \frac{3x^2}{\rho^2}\right]\hat{a}_z$$

Convert the vector completely in spherical coordinates.

20. For the vector $\vec{F} = (x+y)\hat{a}_x - x\hat{a}_y + x\hat{a}_z$, evaluate the line integral from point P_1 to point P_2 along the path L_1 and L_2 as shown in Fig. 1.42.

$$\left[-\frac{1}{2}, \frac{1}{2}\right]$$

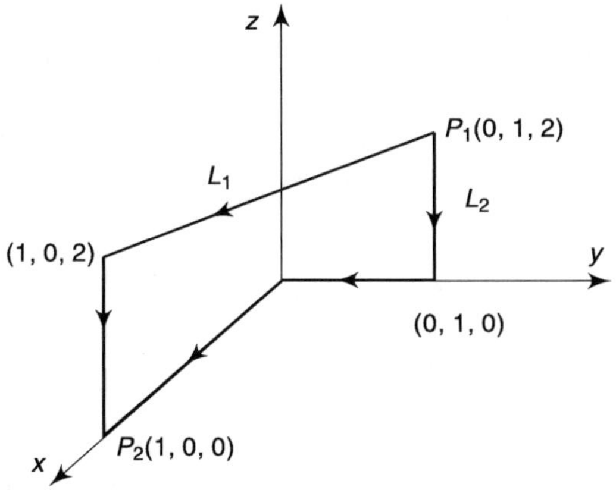

Fig. 1.42 *Contour of Example 21*

21. Given the vector $\vec{A} = \left(\dfrac{e^{-\rho}}{\rho}\right)\hat{a}_\theta$ in spherical coordinates. Verify both sides of Stokes' theorem for the curve bounded by the area shown in Fig. 1.43.

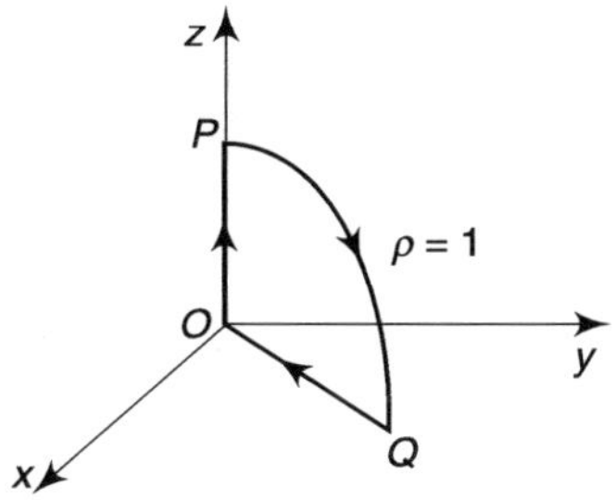

Fig. 1.43 *Arrangement of Problem 22*

Review Questions

1. What is the concept of field? Define scalar and vector fields by giving suitable example of each. What is the importance of unit vector?
2. **(a)** Discuss cross product and dot product in detail between two vectors.
   (b) Discuss the vector representation of a surface.
3. Define the divergence of a vector. Explain the physical significance of the term 'divergence of a vector field'.
4. Define the curl of a vector. Explain the physical significance of the term 'curl of a vector field'.
5. What is a 'gradient'? Give its physical interpretation.
6. Define generalised coordinate system. Find its gradient, divergence and curl equations.
7. Define and give examples for the following vector fields:
   (a) Solenoidal and irrotational
   (b) Non-solenoidal and irrotational
   (c) Solenoidal and rotational
   (d) Non-solenoidal and rotational
8. State and prove the Divergence theorem.

9. State and prove Stoke's theorem.
10. State and prove the Helmholtz theorem. What is the physical significance of this theorem?

Multiple Choice Questions

1. Choose the correct statement:
   (a) Divergence of curl of a vector is zero and curl of grad is zero.
   (b) Divergence of curl of a vector is zero and curl of grad is non-zero.
   (c) Divergence of curl of a vector is non-zero and curl of grad is non-zero.
   (d) Divergence of curl of a vector is non-zero and curl of grad is zero.
2. If the vectors $\vec{A}$ and $\vec{B}$ are conservative, then
   (a) $\vec{A} \times \vec{B}$ is solenoidal **(b)** $\vec{A} \times \vec{B}$ is conservative
   (c) $\vec{A} + \vec{B}$ is solenoidal **(d)** $\vec{A} - \vec{B}$ is solenoidal
3. The value of $\oint_C d\vec{l}$ along a circle of radius 2 units is
   (a) zero **(b)** 2π **(c)** 4π
4. Two vectors $\vec{A}$ and $\vec{B}$ are such that $\vec{A} + \vec{B} = n\vec{A}$ where n is a positive scalar. The angle between $\vec{A}$ and $\vec{B}$ is
   (a) $\frac{\pi}{2}$ **(b)** $\frac{3\pi}{4}$ **(c)** π **(d)** 2π
5. Which of the following equations is correct?
   (a) $\hat{a}_x \times \hat{a}_x = |\hat{a}_x|^2$ **(b)** $(\hat{a}_x \times \hat{a}_y) + (\hat{a}_y \times \hat{a}_x) = 0$
   (c) $\hat{a}_x \times (\hat{a}_y \times \hat{a}_z) = \hat{a}_x \times (\hat{a}_z \times \hat{a}_y)$ **(d)** $\hat{a}_r \cdot \hat{a}_\theta + \hat{a}_\theta \cdot \hat{a}_r = 0$
6. Match List-I (Term) with List-II (Type) and select the correct answer using the codes given below the lists:

List-I (Term)	List-II (Type)
A. curl $(\vec{F}) = 0$	1. Laplace equation
B. div $(\vec{F}) = 0$	2. Irrotational
C. div Grad $(\phi) = 0$	3. Solenoidal
D. div div $(\phi) = 0$	4. Not defined

Codes:

	A	B	C	D
(a)	2	3	1	4
(b)	4	1	3	2
(c)	2	1	3	4
(d)	4	3	1	2

7. Which one of the following is a meaningless combination?
   (a) grad div **(b)** div curl **(c)** curl grad **(d)** div curl
8. Which one of the following is zero?
   (a) grad div **(b)** div grad **(c)** curl grad **(d)** curl curl

Answers

1. (a) **2.** (a) **3.** (a) **4.** (c) **5.** (b) **6.** (a) **7.** (c) **8.** (c)

2

ELECTROSTATICS

Learning Objectives

This chapter deals with the following topics:

- *Sources of electrostatics*
- *Basic laws of electrostatics*
- *To acquire knowledge of fundamental quantities of electrostatics*
- *Boundary conditions in electrostatics*
- *Concepts of capacitance*

2.1 INTRODUCTION

The sources of electromagnetic fields are the presence of electric charges. An electrostatic field is considered to be a special case of electromagnetic field in which the sources are stationary. However, individual charges (e.g., electrons) are never stationary, having random velocities. The charges are referred to be stationary when any elemental macroscopic volume is considered and the net movement of charge through any face of the volume is zero.

In this chapter, we will learn the basics of electrostatics in detail.

2.2 ELECTRIC CHARGE

Electric charge (q) is a fundamental conserved property of some subatomic particles, which determines their electromagnetic interaction. An isolated electric charge creates an electric field around it and exerts force on all other charges within that field.

The SI unit of charge is Coulomb (C). The charge of an electron is 1.602×10^{-19}C. Thus, one Coulomb charge is defined as the charge possessed by $\left(\dfrac{1}{1.602\times 10^{-19}}\right)$ electrons.

$$\therefore \qquad 1 \text{ Coulomb charge} = \text{charge of } 6.24 \times 10^{18} \text{ electrons}$$

The total electric charge of an isolated system remains constant regardless of changes within the system itself. This is known as the law of *conservation of charge*. The law of *conservation of charge* states that charge can neither be created nor destroyed. A charge can however, be transferred from one body to another body.

2.3 COULOMB'S LAW

Statement This law states that the force between two point charges:

1. acts along the line joining the two charges.
2. is directly proportional to the product of the two charges.
3. is inversely proportional to the square of the distance between the charges.

Explanation We will consider two point charges Q_1 and Q_2 with separation distance R, as shown in Fig. 2.1 (*a*) and Fig. 2.1 (*b*).

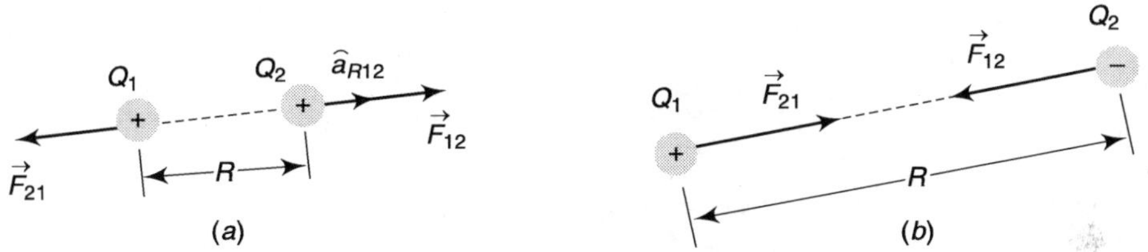

Fig. 2.1 *Coulomb interaction between two point charges (a) like charges, and (b) unlike charges*

The force exerted by Q_1 on Q_2 is

$$\vec{F}_{12}\ \alpha\ \frac{Q_1Q_2}{R^2}\hat{a}_{R12} \quad \Rightarrow \quad \vec{F}_{12} = k\frac{Q_1Q_2}{R^2}\hat{a}_{R12} \tag{2.1}$$

where k is the proportionality constant, which takes into account the effect of the medium in which the charges are placed and $\hat{a}_{R12}$ is a unit vector directed from Q_1 to Q_2.

In SI unit, charges expressed in Coulomb (C), the distance expressed in metre (m) and the force expressed in Newton (N), the proportionality constant is given as

$$k = \frac{1}{4\pi\varepsilon}$$

where ε = permittivity of the medium = $\varepsilon_0\varepsilon_r$

ε_0 = permittivity of free space = $\dfrac{1}{36\pi\times10^9} = 8.854\times10^{-12}$ F/m

ε_r = relative permittivity of the medium

Thus, Coulomb's law in SI unit becomes, from Eq. (2.1)

$$\boxed{\vec{F}_{12} = \frac{Q_1Q_2}{4\pi\varepsilon R^2}\hat{a}_{R12}} \tag{2.2}$$

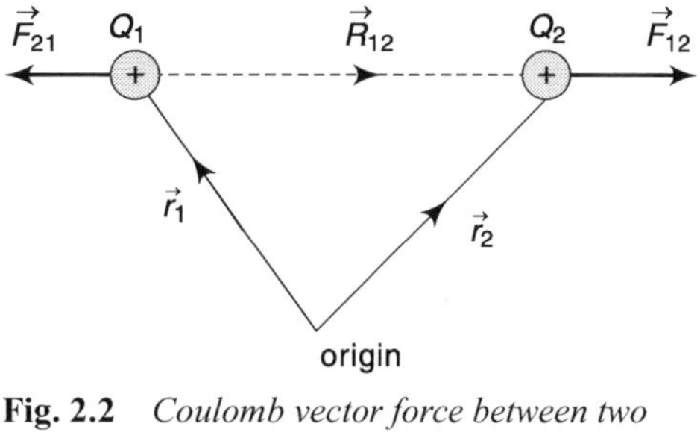

Fig. 2.2 *Coulomb vector force between two point charges*

Similarly, force exerted by Q_2 on Q_1 is

$$\vec{F}_{21}\ \alpha\ \frac{Q_1Q_2}{R^2}\hat{a}_{R21} \quad \Rightarrow \quad \vec{F}_{21} = k\frac{Q_1Q_2}{R^2}\hat{a}_{R21} = -\vec{F}_{12}$$

If two charges have the position vectors of $\vec{r}_1$ and $\vec{r}_2$, respectively, as shown in Fig. 2.2, then

Force on charge Q_2 due to charge Q_1 is

$$\vec{F}_{12} = \frac{Q_1 Q_2}{4\pi\varepsilon R^2}\hat{a}_{R12}$$

where $\vec{R}_{12} = (\vec{r}_2 - \vec{r}_1)$, $\quad R = |\vec{R}_{12}| \quad \therefore \hat{a}_{R12} = \frac{\vec{R}_{12}}{R}$

$$\therefore \quad \boxed{\vec{F}_{12} = \frac{Q_1 Q_2}{4\pi\varepsilon R^3}\vec{R}_{12} = \frac{Q_1 Q_2(\vec{r}_2 - \vec{r}_1)}{4\pi\varepsilon |\vec{r}_2 - \vec{r}_1|^3}} \qquad \textbf{(2.3)}$$

Similarly, force on charge Q_1 due to charge Q_2 is

$$\vec{F}_{21} = \frac{Q_1 Q_2}{4\pi\varepsilon R^3}\vec{R}_{21} = \frac{Q_1 Q_2(\vec{r}_1 - \vec{r}_2)}{4\pi\varepsilon |\vec{r}_1 - \vec{r}_2|^3} = -\vec{F}_{12}$$

Thus, we see that the force exerted by the charges on each other is equal in magnitude, but opposite in direction. It is also noticeable that the force between two like charges (charges of equal sign) is repulsive whereas the force between two unlike charges (charges of opposite sign) is attractive.

2.4 PRINCIPLE OF SUPERPOSITION OF CHARGES

If there is a number of charges Q_1, Q_2, …, Q_n placed at points with position vectors $\vec{r}_1, \vec{r}_2, \ldots, \vec{r}_n$, respectively, then the resultant force $\vec{F}$ on a charge Q located at point $\vec{r}$ is the vector sum of the forces exerted on Q by each of the charges Q_1, Q_2, …, Q_n.

$$\therefore \quad \vec{F} = \frac{QQ_1(\vec{r} - \vec{r}_1)}{4\pi\varepsilon |\vec{r} - \vec{r}_1|^3} + \frac{QQ_2(\vec{r} - \vec{r}_2)}{4\pi\varepsilon |\vec{r} - \vec{r}_2|^3} + \cdots + \frac{QQ_n(\vec{r} - \vec{r}_n)}{4\pi\varepsilon |\vec{r} - \vec{r}_n|^3}$$

$$\therefore \quad \boxed{\vec{F} = \frac{Q}{4\pi\varepsilon}\sum_{i=1}^{n}\frac{Q_i(\vec{r} - \vec{r}_i)}{|\vec{r} - \vec{r}_i|^3}} \qquad \textbf{(2.4)}$$

Example 2.1 Two point charges, $Q_1 = 50\mu C$ and $Q_2 = 10\mu C$ are located at (–1, 1, –3) m and (3, 1, 0) m, respectively. Find the force on Q_1.

Solution Here, $\vec{R}_{21} = (-1, 1, -3) - (3, 1, 0) = (-4\hat{i} - 3\hat{k})$

$$\therefore \quad \hat{a}_{21} = \frac{(-4\hat{i} - 3\hat{k})}{\sqrt{(-4)^2 + (-3)^2}} = \frac{-4\hat{i} - 3\hat{k}}{5}$$

Hence, the force on the charge Q_1 is

$$\vec{F}_{21} = \frac{Q_1 Q_2}{4\pi\varepsilon_0 R_{21}^2}\hat{a}_{21} = \frac{50 \times 10^{-6} \times 10 \times 10^{-6}}{4\pi \times \dfrac{10^{-9}}{36\pi} \times (5)^2} \times \left(\frac{-4\hat{i} - 3\hat{k}}{5}\right)$$

$$= 0.18(-0.8\hat{i} - 0.6\hat{k})N$$

The force has a magnitude of $0.18N$ and a direction given by the unit vector $(-0.8\hat{i} - 0.6\hat{k})$. In component form

$$\vec{F}_{21} = (-0.144\hat{i} - 0.108\hat{k})\text{N}$$

Example 2.2 Two small identical conducting spheres have charges of 2.0×10^{-9} C and -0.5×10^{-9} C, respectively. When they are placed 4 cm apart, what is the force between them? If they are brought into contact and then separated by 4 cm, what is the force between them?

Solution The force between the conducting spheres is given as

$$F = \frac{Q_1 Q_2}{4\pi\varepsilon r^2} = \frac{(2.0\times10^{-9})\times(-0.5\times10^{-9})}{4\pi\times8.854\times10^{-12}\times(4\times10^{-2})^2} = -0.5625\times19^{-5}\,N = 0.5625\times10^{-5}\,N\,\text{(attractive)}$$

When the two spheres are brought into contact and then separated, the charge of each sphere is

$$Q_1' = Q_2' = Q' = \frac{2.0\times10^{-9} - 0.5\times10^{-9}}{2} = 0.75\times10^{-9}\,\text{C}$$

In this case, the force between the conducting spheres is given as

$$F = \frac{Q_1 Q_2}{4\pi\varepsilon r^2} = \frac{Q^2}{4\pi\varepsilon r^2} = \frac{(0.75\times10^{-9})^2}{4\pi\times8.854\times10^{-12}\times(4\times10^{-2})^2} = 0.3164\times10^{-5}\,N\,\text{(repulsive)}$$

Example 2.3 Two particles each of mass 'm' and having a charge 'q' are suspended by string of length 'l' from a common point as shown in Fig. 2.3. Show that the angle 'θ' which each string makes with the vertical is obtained from $\dfrac{\tan^3\theta}{1+\tan^2\theta} = \dfrac{q^2}{16\pi\varepsilon_0 mgl^2}$.

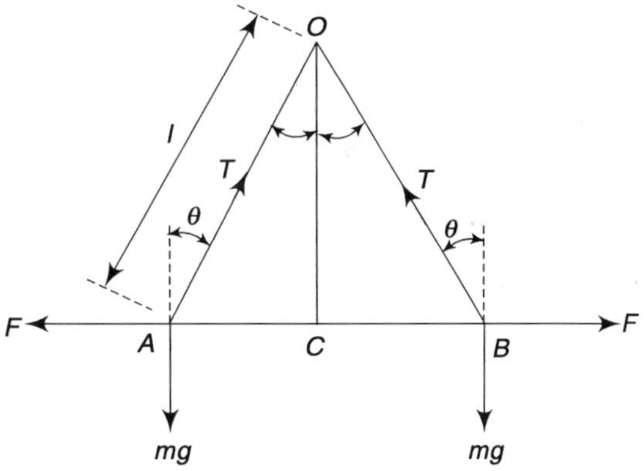

Fig. 2.3 *Two particles suspended by string from a common point*

Solution Force of repulsion, $F = \dfrac{q^2}{4\pi\varepsilon_0 (AB)^2}$

From ΔOAC,

$$\frac{AB/2}{l} = \sin\theta \quad \Rightarrow \quad AB = 2l\sin\theta$$

$$\therefore \qquad F = \frac{q^2}{4\pi\varepsilon_0(2l\sin\theta)^2} = \frac{q^2}{16\pi\varepsilon_0 l^2\sin^2\theta} \qquad \text{(i)}$$

Net force at point A $= (T\sin\theta - F)\hat{i} + (T\cos\theta - mg)\hat{j} = 0$

$$\therefore \qquad T\sin\theta = F$$

$$T\cos\theta = mg$$

$$\therefore \qquad \tan\theta = \frac{F}{mg}$$

Using Eq. (*i*), we get

$$\frac{q^2}{16\pi\varepsilon_0 l^2 \sin^2\theta mg} = \tan\theta$$

or
$$\frac{q^2}{16\pi\varepsilon_0 mgl^2} = \tan\theta\sin^2\theta = \frac{\sin^3\theta}{\cos\theta} = \frac{\sin^3\theta}{\cos^3\theta}\cos^2\theta = \frac{\tan^3\theta}{\sec^2\theta} = \frac{\tan^3\theta}{1+\tan^2\theta}$$

$\therefore$
$$\boxed{\frac{\tan^3\theta}{1+\tan^2\theta} = \frac{q^2}{16\pi\varepsilon_0 mgl^2}}$$

If the angle is very small, $\theta <<$, then $\theta \approx \theta$ and $\cos\theta \approx 1$ and hence we have

$$\boxed{\theta = \sqrt[3]{\frac{q^2}{16\pi\varepsilon_0 mgl^2}}}$$

Example 2.4 Three identical small spheres of mass m are suspended from a common point by threads of negligible masses and equal length l. A charge Q is divided equally among the spheres, and they come to equilibrium at the corners of a horizontal equilateral triangle whose sides are d. Show that

$$Q^2 = \frac{12\pi\varepsilon_0 mgd^3}{\left(l^2 - \frac{d^2}{3}\right)}$$

Solution The arrangement is shown in Fig. 2.4.

Let the charge on each sphere be q.

We will consider the net force acting on any one of the spheres; say, the sphere at point A.

The x component of the force of repulsion due to the spheres at points B and C is given as

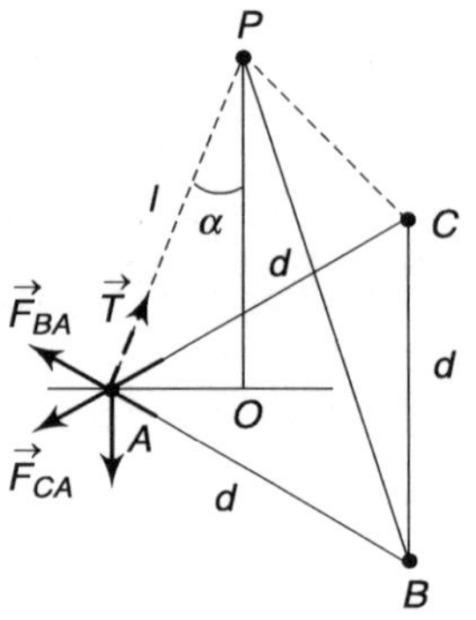

Fig. 2.4 *Three identical spheres suspended from a common point*

$$\vec{F} = \vec{F}_{BA}\cos 30^\circ + \vec{F}_{CA}\cos 30^\circ$$
$$= \frac{q^2}{4\pi\varepsilon_0 d^2}\cos 30^\circ \times 2(-\hat{i})$$
$$= \frac{\sqrt{3}q^2}{4\pi\varepsilon_0 d^2}(-\hat{i})$$

Net force on the sphere at point A is

$$F_{net} = (T\sin\alpha - F)\hat{i} + (T\cos\alpha - mg)\hat{j} = 0$$

$\therefore$
$$T\sin\alpha = F = \frac{\sqrt{3}q^2}{4\pi\varepsilon_0 d^2}$$
$$T\cos\alpha = mg$$

$\therefore$
$$\tan\alpha = \frac{\sqrt{3}q^2}{4\pi\varepsilon_0 mgd^2}$$

From Fig. 2.4, $\tan\alpha = \dfrac{AO}{OP} = \dfrac{d/\sqrt{3}}{\sqrt{AP^2 - AO^2}} = \dfrac{d/\sqrt{3}}{\sqrt{l^2 - (d/\sqrt{3})^2}} = \dfrac{d/\sqrt{3}}{\sqrt{l^2 - \dfrac{d^2}{3}}}$

$\therefore$
$$\frac{d/\sqrt{3}}{\sqrt{l^2 - \dfrac{d^2}{3}}} = \frac{\sqrt{3}q^2}{4\pi\varepsilon_0 mgd^2}$$

or
$$q^2 = \frac{4\pi\varepsilon_0 mgd^3}{3\sqrt{l^2 - \dfrac{d^2}{3}}}$$

Here, $q = \dfrac{Q}{3}$; putting this, we get,

$$\boxed{Q^2 = \frac{12\pi\varepsilon_0 mgd^3}{\sqrt{l^2 - \dfrac{d^2}{3}}}}$$

***Example 2.5** It is required to hold four equal point charges $+q$ each in equilibrium at the corners of a square. Find the point charge which will do this if placed at the centre of the square.

Solution Let the required charge at the centre be $-q'$. From Fig. 2.5, considering the forces acting on any one charge (say 1), due to all other charges, the resultant must be zero for equilibrium.

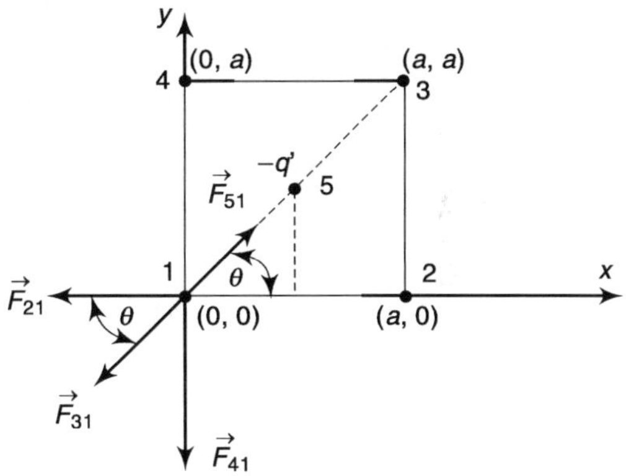

Fig. 2.5 *Arrangement of charges for Example 2.5*

Here,
$$\vec{F}_{21} = -\frac{q^2}{4\pi\varepsilon a^2}\hat{i}$$
$$\vec{F}_{41} = -\frac{q^2}{4\pi\varepsilon a^2}\hat{j}$$
$$F_{31} = \frac{q^2}{4\pi\varepsilon(\sqrt{2}a)^2} = \frac{q^2}{8\pi\varepsilon a^2}$$

$\therefore$
$$\vec{F}_{31} = -\frac{q^2}{8\pi\varepsilon a^2}(\cos\theta\,\hat{i} - \sin\theta\,\hat{j}) = -\frac{q^2}{8\sqrt{2}\pi\varepsilon a^2}\hat{i} - \frac{q^2}{8\sqrt{2}\pi\varepsilon a^2}\hat{j} \quad (\because \theta = 45°)$$

Similarly,
$$\vec{F}_{51} = \frac{qq'}{2\sqrt{2}\pi\varepsilon a^2}\hat{i} + \frac{qq'}{2\sqrt{2}\pi\varepsilon a^2}\hat{j}$$

For equilibrium,
$$\vec{F}_{21} + \vec{F}_{41} + \vec{F}_{31} + \vec{F}_{51} = 0$$

or
$$\frac{q}{8\sqrt{2}\pi\varepsilon a^2}(-2\sqrt{2}q\hat{i} - q\hat{i} + 4q'\hat{i} - 2\sqrt{2}q\hat{j} - q\hat{j} + 4q'\hat{j}) = 0$$

or
$$q' = \left(\frac{2\sqrt{2}+1}{4}\right)q = 0.956q$$

Thus, the required charge at the centre of the square is $-0.956q$.

***Example 2.6** It is required to hold three equal point charges $+q$ each in equilibrium at the corners of an equilateral triangle. Find the point charge which will do this if placed at the centre of the triangle.

Solution Let the required charge at the centre be $-q'$. From Fig. 2.6, considering the forces acting on any one charge (say 2), due to all other charges, the resultant must be zero for equilibrium.

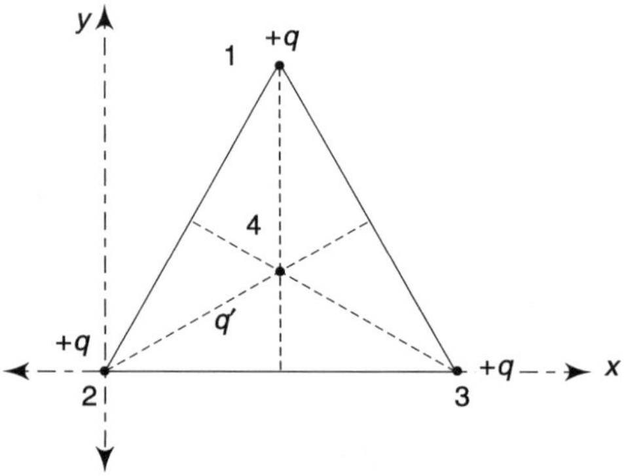

Fig. 2.6 *Arrangement of charges for Example 2.6*

Here,
$$\vec{F}_{12} = -\frac{q^2}{4\pi\varepsilon a^2}(\cos 60°\,\hat{i} + \sin 60°\,\hat{j})$$
$$= -\frac{q^2}{8\pi\varepsilon a^2}\hat{i} - \frac{\sqrt{3}q^2}{8\pi\varepsilon a^2}\hat{j}$$
$$\vec{F}_{32} = -\frac{q^2}{4\pi\varepsilon a^2}\hat{i}$$

Similarly,
$$\vec{F}_{42} = \frac{qq'}{4\pi\varepsilon(a/\sqrt{3})^2}(\cos 30°\,\hat{i} + \sin 30°\,\hat{j}) = \frac{3\sqrt{3}qq'}{8\pi\varepsilon a^2}\hat{i} + \frac{3qq'}{8\pi\varepsilon a^2}\hat{j}$$

For equilibrium,
$$\vec{F}_{12} + \vec{F}_{32} + \vec{F}_{42} = 0$$

or
$$-\frac{q^2}{8\pi\varepsilon a^2}\hat{i} - \frac{\sqrt{3}q^2}{8\pi\varepsilon a^2}\hat{j} - \frac{q^2}{4\pi\varepsilon a^2}\hat{i} + \frac{3\sqrt{3}qq'}{8\pi\varepsilon a^2}\hat{i} + \frac{3qq'}{8\pi\varepsilon a^2}\hat{j} = 0$$

or
$$\left(\frac{3\sqrt{3}qq'}{8\pi\varepsilon a^2} - \frac{q^2}{8\pi\varepsilon a^2} - \frac{q^2}{4\pi\varepsilon a^2}\right)\hat{i} + \left(\frac{3qq'}{8\pi\varepsilon a^2} - \frac{\sqrt{3}q^2}{8\pi\varepsilon a^2}\right)\hat{j} = 0$$

or
$$q' = \frac{q}{\sqrt{3}} = 0.577q$$

Thus, the required charge at the centre of the square is, $q' = -\dfrac{q}{\sqrt{3}} = -0.577q$.

***Example 2.7** Three point charges of 'q' are placed in air at the vertices of an equilateral triangle of side 'd'. Determine the magnitude and direction of the force on one charge due to other charges.

Solution We find out the force on charge at A due to the other two charges situated at points B and C.

Force on charge at A due to charge at B is

$$\vec{F}_{BA} = \frac{Q_1 Q_2}{4\pi\varepsilon r^2} = \frac{q^2}{4\pi\varepsilon d^2}$$

Force on charge at A due to charge at C is

$$\vec{F}_{CA} = \frac{q^2}{4\pi\varepsilon d^2}$$

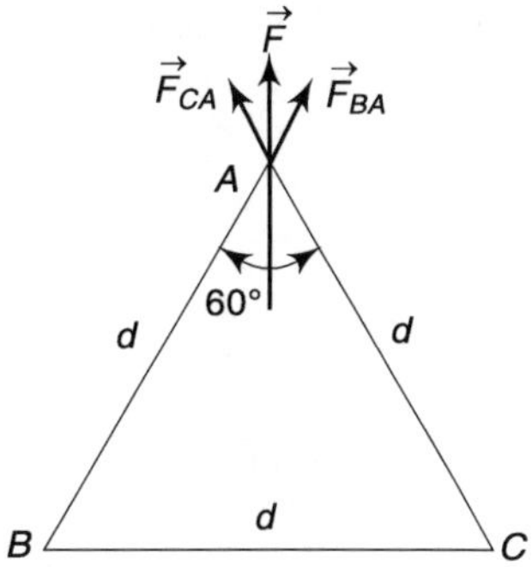

Fig. 2.7 *Arrangement of charges for Example 2.7*

From Fig. 2.7, it is observed that the horizontal components of the total force cancel out. Hence, the total force acts along the upward direction and its magnitude is given as

$$F = F_{BA}\cos 30° + F_{CA}\cos 30° = 2 \times \frac{q^2}{4\pi\varepsilon d^2} \times \cos 30°$$

$$= 2 \times \frac{q^2}{4\pi\varepsilon d^2} \times \frac{\sqrt{3}}{2} = \frac{\sqrt{3}q^2}{4\pi\varepsilon d^2}$$

$$\boxed{F = \frac{\sqrt{3}q^2}{4\pi\varepsilon d^2}}$$

Example 2.8 Four concentrated charges are located at the vertices of a plane rectangle as shown in Fig. 2.8. Find the magnitude and direction of the resultant force on Q_1.

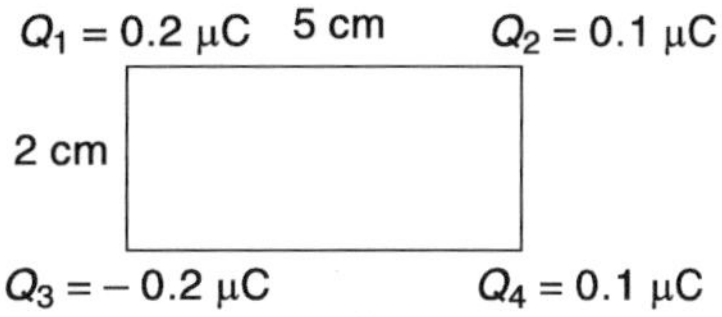

Fig. 2.8 *Arrangement of charges for Example 2.8*

Solution The resultant force on Q_1 charge is

F = Force of attraction between Q_3 and Q_1
+ Force of repulsion between Q_2 and Q_1
+ Force of repulsion between Q_4 and Q_1

Here,

$$F_{21} = \frac{0.2 \times 10^{-6} \times 0.1 \times 10^{-6}}{4\pi\varepsilon_0 \times 25 \times 10^{-4}} = 0.072\,N$$

along negative x axis.

$$F_{31} = \frac{0.2 \times 10^{-6} \times 0.2 \times 10^{-6}}{4\pi\varepsilon_0 \times 4 \times 10^{-4}} = 0.9\,N$$

along negative y axis.

$$F_{41} = \frac{0.2 \times 10^{-6} \times 0.1 \times 10^{-6}}{4\pi\varepsilon_0 \times 29 \times 10^{-4}} = 0.06\,N$$

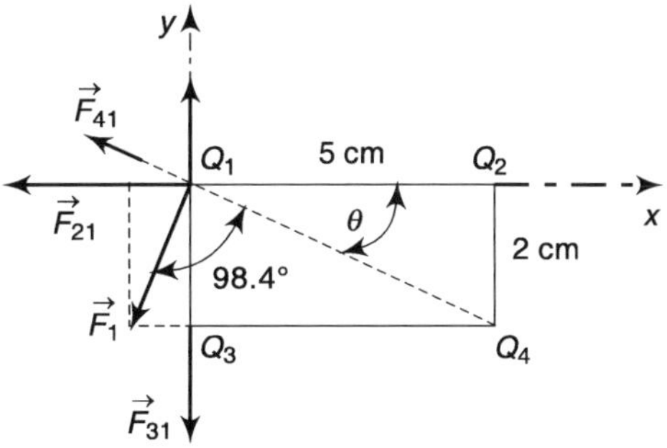

Fig. 2.9 *Forces for charge arrangement of Example 2.8*

with an angle θ to the negative x axis, where

$$\sin\theta = \frac{2}{\sqrt{29}} \quad \text{and} \quad \cos\theta = \frac{5}{\sqrt{29}}$$

Resolving the force F_{41} into two components, we get

$$F_{41x} = 0.06 \cos\theta = 0.06 \times \frac{5}{\sqrt{29}} = 0.058\,N$$

$$F_{41yx} = 0.06 \sin\theta = 0.06 \times \frac{2}{\sqrt{29}} = 0.023\,N$$

$\therefore$ Total force along the negative x axis is

$$F_x = 0.072 + 0.058 = 0.13\,N$$

$\therefore$ Total force along the negative y axis is

$$F_y = 0.9 - 0.023 = 0.877\,N$$

$\therefore$ Resultant force on the charge Q_1 is, $\vec{F}_1 = (-0.13\hat{i} - 0.877\hat{j}) = 0.89\angle -98.4°\,N$

2.5 ELECTRIC FIELD AND FIELD INTENSITY ($\vec{E}$)

The electrostatic force is a force that acts upon the objects at a distance, even when the objects are not in contact with one another. This implies that a charge creates a *field* at a distance, which in turn acts on the other charge.

Electric Field For an electric charge, there is a region in which it exerts a force on any other charge. This region where a particular charge exerts a force on any other charge located in that region, is called the electric field of that charge.

Electric Field Intensity $\vec{E}$ It is the force per unit charge when placed in the field.

$$\vec{E} = \lim_{Q \to 0} \frac{\vec{F}}{Q}$$

or simply

$$\boxed{\vec{E} = \frac{\vec{F}}{Q}} \tag{2.5}$$

It is seen that the field intensity is in the same direction as the force and is expressed in Newton per Coulomb (N/C) and volt per metre (V/m).

Thus, if a point charge Q_t is present at position vector $\vec{r}_t$, then the field intensity due to the charge Q at position vector $\vec{r}$ is

$$\vec{E} = \frac{Q_t}{4\pi\varepsilon R^2}\hat{a}_R = \frac{Q_t(\vec{r}_t - \vec{r})}{4\pi\varepsilon\,|\vec{r}_t - \vec{r}|^3}$$

Example 2.9 Show that the electric field intensity at a point due to a number of point charges is the vector sum of the electric field intensities due to individual point charges at that point.

Solution An electric charge q produces an electric field everywhere. The strength of the field created by a charge can be obtained by measuring the force on a positive test charge at some point in that field. The *electric field intensity* ($\vec{E}$) is defined as the force per unit charge when placed in that field.

If there is N number of source charges $q_1, q_2, \ldots, q_i$ acting on a test charge Q, then the net force on the test charge is

$$\vec{F} = \vec{F}_1 + \vec{F}_2 + \cdots \vec{F}_i + \cdots$$
$$= \frac{1}{4\pi\varepsilon}\left[\frac{q_1 Q}{r_1^2}\widehat{a}_{r1} + \frac{q_2 Q}{r_2^2}\widehat{a}_{r2} + \cdots + \frac{q_i Q}{r_i^2}\widehat{a}_{ri} + \cdots\right]$$
$$= Q\vec{E}$$

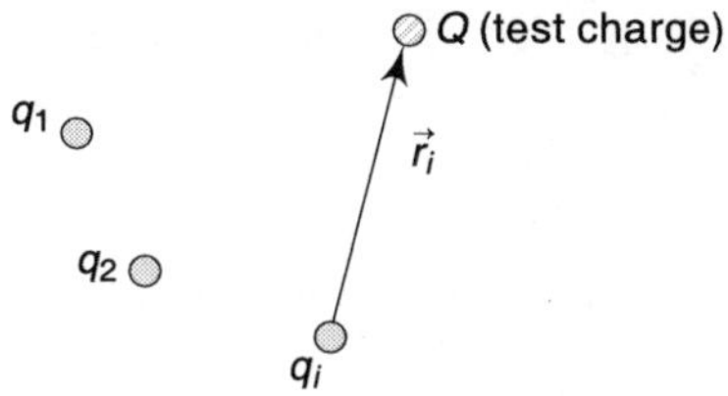

Fig. 2.10 *Assembly of point charges*

where

$$\boxed{\vec{E} = \frac{1}{4\pi\varepsilon}\sum_{i=1}^{N}\frac{q_i}{r_i^2}\widehat{a}_{ri}} \quad \textbf{(2.6)}$$

So, there is an electric field intensity around a charge q, given by

$$\boxed{\vec{E} = \frac{q}{4\pi\varepsilon r^2}\widehat{a}_r} \quad \textbf{(2.7)}$$

It is seen that the force between the charges is dependent on the test charge; but the electric field intensity is not. The test charge is only to detect the presence of the force.

Example 2.10 Point charges of 1mC and –2mC are located at (3, 2, –1) and (–1, –1, 4), respectively. Calculate the electric force on a 10nC charge located at (0, 3, 1) and the electric field intensity at that point.

Solution Here, $Q_1 = 10^{-3}$ C, $Q_2 = -2 \times 10^{-3}$ C, $Q = 10^{-9}$ C

Hence, the force on the charge is

$$\vec{F} = \sum_{i=1,2}\frac{QQ_i}{4\pi\varepsilon_0 R^2}\widehat{a}_R = \sum_{i=1,2}\frac{QQ_i(\vec{r}-\vec{r}_i)}{4\pi\varepsilon_0|\vec{r}-\vec{r}_i|^3}$$
$$= \frac{Q}{4\pi\varepsilon_0}\left[\frac{10^{-3}\times(-3\widehat{i}+\widehat{j}+2\widehat{k})}{\{(-3)^2+1^2+2^2\}^{3/2}} - \frac{2\times10^{-3}\times(\widehat{i}+4\widehat{j}-3\widehat{k})}{\{1^2+4^2+(-3)^2\}^{3/2}}\right]$$
$$= \frac{10\times10^{-9}\times10^{-3}\times36\pi}{4\pi\times10^{-9}}\left[\frac{(-3\widehat{i}+\widehat{j}+2\widehat{k})}{14\sqrt{14}} + \frac{(-2\widehat{i}-8\widehat{j}+6\widehat{k})}{26\sqrt{26}}\right]$$

$$\therefore \quad \vec{F} = (-6.507\widehat{i} - 3.817\widehat{j} + 7.506\widehat{k})\text{ mN}$$

Field intensity at that point

$$\vec{E} = \frac{\vec{F}}{Q} = (-6.507\widehat{i} - 3.817\widehat{j} + 7.506\widehat{k})\times\frac{10^{-3}}{10\times10^{-9}}$$
$$= (-650.7\widehat{i} - 381.7\widehat{j} + 750.6\widehat{k})\text{ kV/m}$$

***Example 2.11** Three positive charges of q, $2q$ and $3q$ are placed at the corners of an equilateral triangle as shown in Fig. 2.11. If the length of each side of the triangle is 'd', find the magnitude and direction of the electric field at the point bisecting the line joining q and $2q$.

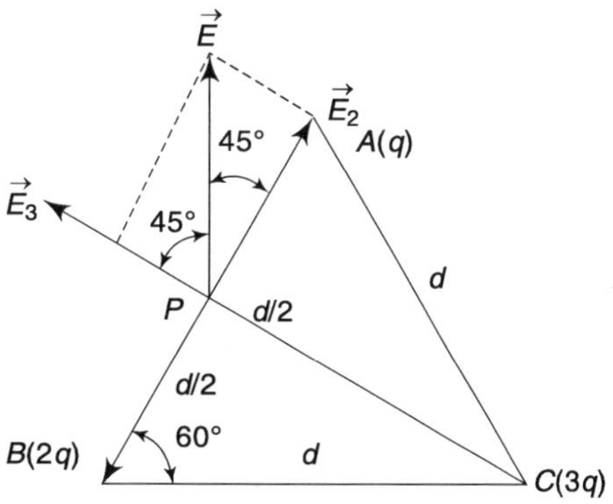

Fig. 2.11 *Arrangement of charges for Example 2.11*

Solution Electric field at point P due to the charge at point A is

$$E_1 = \frac{q}{4\pi\varepsilon r_1^2} = \frac{q}{4\pi\varepsilon(d/2)^2} = \frac{q}{\pi\varepsilon d^2}$$

Electric field at point P due to the charge at point B is

$$E_2 = \frac{2q}{4\pi\varepsilon(d/2)^2} = \frac{2q}{\pi\varepsilon d^2}$$

Electric field at point P due to the charge at point C is

$$E_3 = \frac{3q}{4\pi\varepsilon r_3^2} = \frac{3q}{4\pi\varepsilon(d\sin 60°)^2} = \frac{q}{\pi\varepsilon d^2}$$

Net resultant field at point P due to the charges at points A and B is

$$(E_2 - E_1) = \left(\frac{2q}{\pi\varepsilon d^2} - \frac{q}{\pi\varepsilon d^2}\right) = \frac{q}{\pi\varepsilon d^2}, \text{ along } PA \text{ direction}$$

Hence, the magnitude of the total field at point P is given as

$$E = \sqrt{\left(\frac{q}{\pi\varepsilon d^2}\right)^2 + \left(\frac{q}{\pi\varepsilon d^2}\right)^2} = \frac{\sqrt{2}q}{\pi\varepsilon d^2}$$

$$\boxed{E = \frac{\sqrt{2}q}{\pi\varepsilon d^2}}$$

***Example 2.12** Three charges q, $-q$ and q are situated at the vertices of an equilateral triangle of side 'a' as shown in Fig. 2.12. Find the field at the centroid of the triangle.

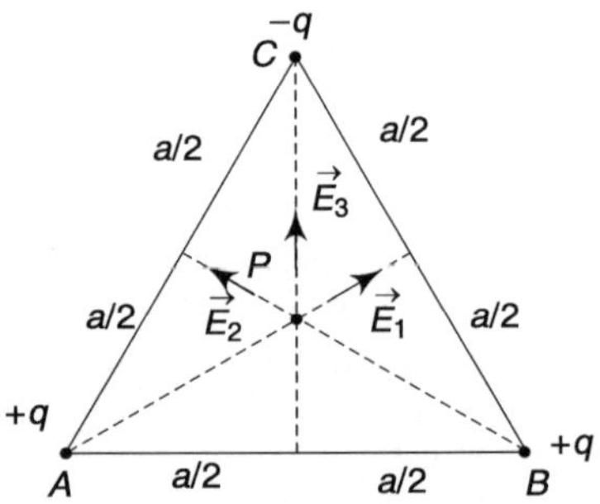

Fig. 2.12 *Arrangement of charges for Example 2.12*

Solution Field at point P due to the charge at point A is

$$\vec{E}_1 = \frac{q}{4\pi\varepsilon r_1^2} = \frac{q}{4\pi\varepsilon\left(\dfrac{a/2}{\sin 60°}\right)^2} = \frac{3}{4}\left(\frac{q}{\pi\varepsilon a^2}\right)$$

along AP direction

Field at point P due to the charge at point B is

$$\vec{E}_2 = \frac{q}{4\pi\varepsilon r_2^2} = \frac{q}{4\pi\varepsilon\left(\dfrac{a/2}{\sin 60°}\right)^2} = \frac{3}{4}\left(\frac{q}{\pi\varepsilon a^2}\right)$$

along BP direction

Net field at point P due to the charges at points A and B, in the direction perpendicular to AB is

$$E' = 2 \times \frac{3}{4}\left(\frac{q}{\pi\varepsilon a^2}\right)\cos 60° = \frac{3}{4}\left(\frac{q}{\pi\varepsilon a^2}\right)$$

Field at point P due to the charge at point C in the direction perpendicular to AB is

$$\vec{E}_3 = \frac{q}{4\pi\varepsilon r_3^2} = \frac{q}{4\pi\varepsilon\left(\dfrac{a/2}{\sin 60^\circ}\right)^2} = \frac{3}{4}\left(\frac{q}{\pi\varepsilon a^2}\right)$$

∴ Total field at the centroid is given as $E = (E' + E_3) = \frac{3}{4}\left(\frac{q}{\pi\varepsilon a^2}\right) + \frac{3}{4}\left(\frac{q}{\pi\varepsilon a^2}\right) = \frac{3}{2}\left(\frac{q}{\pi\varepsilon a^2}\right)$

$$\boxed{E = \frac{3}{2}\left(\frac{q}{\pi\varepsilon a^2}\right)}$$

Example 2.13 Three equal positive charges of 'q' each are located at three corners of a square of side 'l' as shown in Fig. 2.13. Determine the magnitude and direction of the electric field at the vacant corner of the square.

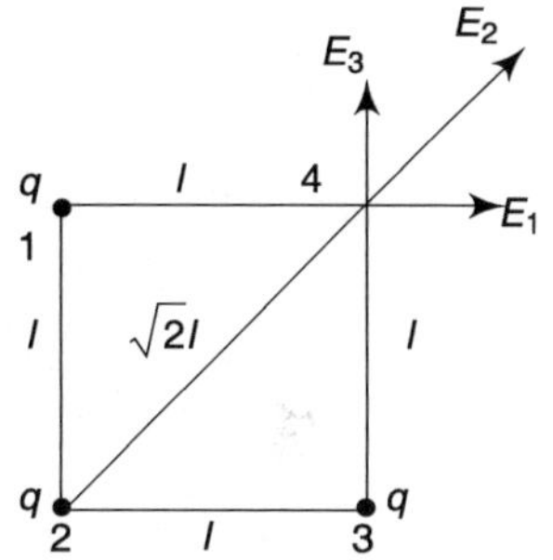

Fig. 2.13 *Field intensities for charge arrangement of Example 2.13*

Solution Field intensity at corner 4 due to charge 1 is

$$E_1 = \frac{q}{4\pi\varepsilon_0 l^2}; \text{ along the positive } x \text{ axis.}$$

Field intensity at corner 4 due to charge 3 is

$$E_3 = \frac{q}{4\pi\varepsilon_0 l^2}; \text{ along the positive } y \text{ axis.}$$

Resultant of these two field intensities is

$$E' = \sqrt{\left(\frac{q}{4\pi\varepsilon_0 l^2}\right)^2 + \left(\frac{q}{4\pi\varepsilon_0 l^2}\right)^2} = \sqrt{2}\frac{q}{4\pi\varepsilon_0 l^2}$$

with an angle 45° to the horizontal axis

Field intensity at corner 4 due to charge 2 is

$$E_2 = \frac{q}{4\pi\varepsilon_0(\sqrt{2}l)^2} = \frac{q}{8\pi\varepsilon_0 l^2}; \text{ with an angle 45° to the horizontal axis}$$

∴ Total field at corner 4 is given as

$$E = (E' + E_2) = \frac{\sqrt{2}q}{4\pi\varepsilon_0 l^2} + \frac{q}{8\pi\varepsilon_0 l^2} = \frac{(2\sqrt{2}+1)q}{8\pi\varepsilon_0 l^2} = 1.9142\left(\frac{q}{4\pi\varepsilon_0 l^2}\right);$$

with an angle 45° to the horizontal axis

$$\boxed{E = \frac{(2\sqrt{2}+1)q}{8\pi\varepsilon_0 l^2} = 1.9142\left(\frac{q}{4\pi\varepsilon_0 l^2}\right)}$$

***Example 2.14** Three sides of an equilateral triangle have uniform line charges 2μC/m, 1μC/m, and 1μC/m as shown in Fig. 2.14. Find the electric field at the centre of the triangle if each side is 50 m long.

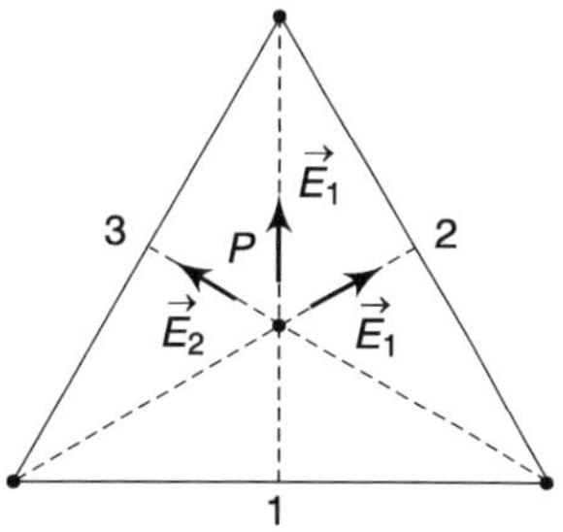

Fig. 2.14 *Arrangement of charges for Example 2.14*

Solution Here,

$$l = 50 \quad \text{cm}$$
$$\lambda_1 = 2 \quad \mu\text{C/m}$$
$$\lambda_2 = 1 \quad \mu\text{C/m}$$
$$\lambda_3 = 1 \quad \mu\text{C/m}$$

In this case, total field intensity at the centre is given as

$$\vec{E} = \vec{E}_1 + \vec{E}_2 + \vec{E}_3$$

where

$\vec{E}_1$ = Field at the centre due to side 1

$$= \frac{2\times 10^{-6} \times 50 \times 10^{-2}}{2\pi\varepsilon_0 r\sqrt{l^2 + 4r^2}}\widehat{a}_y \qquad \left[\text{Here, } r = \frac{50}{2}\tan 30° = \frac{25}{\sqrt{3}}\text{ cm}\right]$$

$$= \frac{2\times 10^{-6} \times 50 \times 10^{-2}}{2\pi\varepsilon_0 \frac{25}{\sqrt{3}}\sqrt{50^2 + 4\left(\frac{25}{\sqrt{3}}\right)^2}\times 10^{-4}}\widehat{a}_y$$

$$= \frac{6\times 10^{-6}}{\pi\varepsilon_0}\widehat{a}_y \quad \text{V/m}$$

$\vec{E}_2$ = Field at the centre due to side 2

$$= \frac{1\times 10^{-6} \times 50 \times 10^{-2}}{2\pi\varepsilon_0 r\sqrt{l^2 + 4r^2}}(-\widehat{a}_x \sin 60° - \widehat{a}_y \cos 60°) \qquad \left[\text{Here, } r = \frac{50}{2}\tan 30° = \frac{25}{\sqrt{3}}\text{ cm}\right]$$

$$= \frac{2\times 10^{-6} \times 50 \times 10^{-2}}{2\pi\varepsilon_0 \frac{25}{\sqrt{3}}\sqrt{50^2 + 4\left(\frac{25}{\sqrt{3}}\right)^2}\times 10^{-4}}(-\widehat{a}_x \sin 60° - \widehat{a}_y \cos 60°)$$

$$= \frac{3\times 10^{-6}}{\pi\varepsilon_0}(-\widehat{a}_x \sin 60° - \widehat{a}_y \cos 60°) \quad \text{V/m}$$

$\vec{E}_3$ = Field at the centre due to side 3

$$= \frac{1\times 10^{-6} \times 50 \times 10^{-2}}{2\pi\varepsilon_0 r\sqrt{l^2 + 4r^2}}(\widehat{a}_x \sin 60° - \widehat{a}_y \cos 60°) \qquad \left[\text{Here, } r = \frac{50}{2}\tan 30° = \frac{25}{\sqrt{3}}\text{ cm}\right]$$

$$= \frac{2\times 10^{-6} \times 50 \times 10^{-2}}{2\pi\varepsilon_0 \frac{25}{\sqrt{3}}\sqrt{50^2 + 4\left(\frac{25}{\sqrt{3}}\right)^2}\times 10^{-4}}(\widehat{a}_x \sin 60° - \widehat{a}_y \cos 60°)$$

$$= \frac{3\times 10^{-6}}{\pi\varepsilon_0}(\widehat{a}_x \sin 60° - \widehat{a}_y \cos 60°) \quad \text{V/m}$$

Hence, total field at the centroid is given as

$$\vec{E} = \vec{E}_1 + \vec{E}_2 + \vec{E}_3$$
$$= \frac{6\times10^{-6}}{\pi\varepsilon_0}\hat{a}_y + \frac{3\times10^{-6}}{\pi\varepsilon_0}(-\hat{a}_x \sin 60° - \hat{a}_y \cos 60°) + \frac{3\times10^{-6}}{\pi\varepsilon_0}(\hat{a}_x \sin 60° - \hat{a}_y \cos 60°)$$
$$= \frac{10^{-6}}{\pi\varepsilon_0}(6 - 6\cos 60°)\hat{a}_y$$
$$= \frac{10^{-6}\times 36\pi}{\pi\times10^{-9}}(6-3)\hat{a}_y$$
$$= 10^3 \times 36 \times 3\hat{a}_y$$
$$= 108 \text{ kV/m}$$

NOTE

If the charges are equal, $\vec{E}$ would be zero at the centre of the equilateral triangle.

2.6 DIFFERENT CHARGE DENSITIES

Due to a small number of charged particles the electric field can readily be computed using the superposition principle. But the computation becomes difficult if we have a very large number of charges distributed in some region in space. We will consider the system shown in Fig. 2.15.

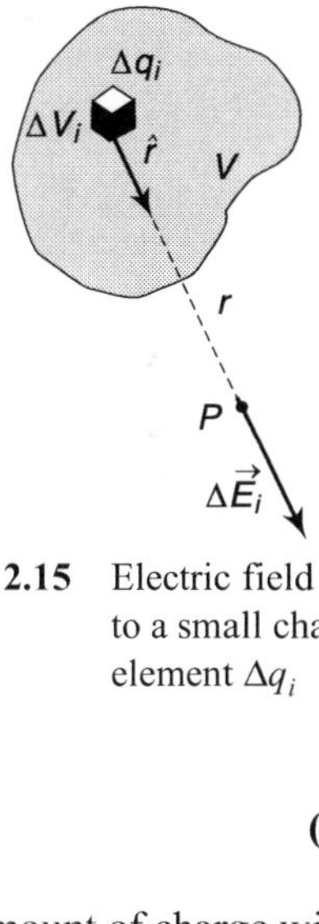

Fig. 2.15 Electric field due to a small charge element Δq_i

We define three types of charge densities as:

1. Volume charge density (ρ),
2. Surface charge density (σ), and
3. Line charge density (λ).

1. *Volume charge density ρ:* When the electric charge is continuously distributed throughout a region, then the volume charge density $[\rho(\vec{r})]$ at a point is defined as the charge Δq_i in a small volume element ΔV_i, divided by the volume with the limit of this ratio taken as the volume shrinks to zero around the point.

$$\rho(\vec{r}) = \lim_{\Delta V_i} \frac{\Delta q_i}{\Delta V_i} = \frac{dq}{dV} \text{ (C/m}^3\text{)} \quad \textbf{(2.8)}$$

The dimension of $\rho(\vec{r})$ is charge per unit volume (C/m^3) in SI units. The total amount of charge within the entire volume V is

$$Q = \sum_i \Delta q_i = \int_V \rho(\vec{r})dV \quad \textbf{(2.9)}$$

2. *Surface charge density* (σ)*:* When the charge is distributed over surface, the *surface charge density* is defined as the charge per unit area.

$$\sigma(\vec{r}) = \lim_{\Delta S\to 0} \frac{\Delta q_i}{\Delta S} = \frac{dq}{dS} \text{ (C/m}^2\text{)} \quad \textbf{(2.10)}$$

The dimension of σ is charge per unit area (C/m^2) in SI units. The total charge on the entire surface is

$$Q = \iint_S \sigma(\vec{r})dS \tag{2.11}$$

3. *Line charge density* (λ): When the charge is distributed over a line, the *line charge density* is defined as the charge per unit length.

$$\lambda(\vec{r}) = \lim_{\Delta l \to 0} \frac{\Delta q_i}{\Delta l} = \frac{dq}{dl} \text{ (C/m)} \tag{2.12}$$

where, the dimension of λ is charge per unit length (C/m^2).

The total charge is now an integral over the entire length

$$Q = \int_{\text{line}} \lambda(\vec{r})dl \tag{2.13}$$

2.7 ELECTRIC FIELDS DUE TO CONTINUOUS CHARGE DISTRIBUTION

The electric field at a point P due to each charge element dq is given by Coulomb's law as

$$d\vec{E} = \frac{1}{4\pi\varepsilon}\frac{dq}{r^2}\hat{a}_r$$

where r is the distance from/to and is the corresponding unit vector (see Fig. 2.10). Using the superposition principle, the total electric field $\vec{E}$ is the vector sum of all these infinitesimal contributions

$$\vec{E} = \frac{1}{4\pi\varepsilon}\int_V \frac{dq}{r^2}\hat{a}_r \tag{2.14}$$

This is an example of a *vector* integral that consists of three separate integrations, one for each component of the electric field.

2.7.1 Electric Field due to Line Charge Distribution

For line charge distribution with charge density λ (C/m), the total charge over a line is obtained as $Q = \int_{\text{line}} \lambda(\vec{r})dl$, so that the field intensity is given as

$$\vec{E} = \frac{1}{4\pi\varepsilon}\int_{\text{line}} \frac{\lambda(\vec{r})}{r^2}dl \tag{2.15}$$

Example 2.15 An infinite line extending along the z-axis carries a uniform line charge distribution of density λ (C/m) as shown in Fig. 2.16. Find the electric field intensity at any point $P(x, y, z)$.

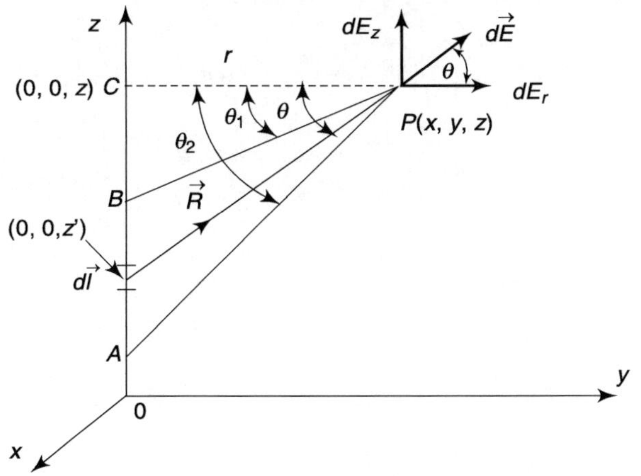

Fig. 2.16 *Determination of electric field due to line charge distribution*

Solution

We will consider an elemental length $d\vec{l}$ of the line at a distance z' from the origin. The field point is denoted by (x, y, z) and the source point by the primed coordinates (x', y', z')

Here, $dl = dz'$, charge in the element, $dQ = \lambda dl = \lambda dz'$

Also, $\vec{R} = (x, y, z) - (0, 0, z') = x\hat{a}_x + y\hat{a}_y + (z - z')\hat{a}_z = r\hat{a}_r + (z - z')\hat{a}_z$

$\therefore$
$$R^2 = |\vec{R}|^2 = x^2 + y^2 + (z - z')^2 = r^2 + (z - z')^2$$

$\therefore$
$$\frac{\vec{a}_R}{R^2} = \frac{\vec{R}}{|\vec{R}|^3} = \frac{r\hat{a}_r + (z - z')\hat{a}_z}{[r^2 + (z - z')^2]^{3/2}}$$

Substituting these values in Eq. (2.14), we get

$$\vec{E} = \frac{1}{4\pi\varepsilon}\int_{\text{line}} \frac{\lambda(\vec{R})}{R^2} dl = \frac{1}{4\pi\varepsilon}\int_{\text{line}} \frac{\lambda dl}{R^2}\hat{a}_R$$
$$= \frac{\lambda}{4\pi\varepsilon}\int \frac{r\hat{a}_r + (z - z')\hat{a}_z}{[r^2 + (z - z')^2]^{3/2}} dz'$$

Now,

$$R = r\sec\theta, \quad z' = z - r\tan\theta;$$

$\therefore$
$$(z - z') = r\tan\theta \qquad \text{and} \qquad dz' = -\,r\sec^2\theta\, d\theta$$

$\therefore$
$$\vec{E} = \frac{\lambda}{4\pi\varepsilon}\int \frac{r\hat{a}_r + (z - z')\hat{a}_z}{[r^2 + (z - z')^2]^{3/2}} dz'$$
$$= -\frac{\lambda}{4\pi\varepsilon}\int \frac{r\hat{a}_r + r\tan\theta\,\hat{a}_z}{[r^2 + r^2\tan^2\theta]^{3/2}} r\sec^2\theta\, d\theta$$
$$= -\frac{\lambda}{4\pi\varepsilon}\int \frac{r\hat{a}_r + r\tan\theta\,\hat{a}_z}{r^3\sec^3\theta} r\sec^2\theta\, d\theta$$

$$= -\frac{\lambda}{4\pi\varepsilon}\int \frac{\hat{a}_r + \dfrac{\sin\theta}{\cos\theta}\hat{a}_z}{r\sec\theta}d\theta$$

$$= -\frac{\lambda}{4\pi\varepsilon r}\int (\cos\theta\,\hat{a}_r + \sin\theta\,\hat{a}_z)d\theta$$

For a finite line charge, the field is given as

$$\vec{E} = -\frac{\lambda}{4\pi\varepsilon r}\int_{\theta_1}^{\theta_2}(\cos\theta\,\hat{a}_r + \sin\theta\,\hat{a}_z)d\theta$$

$$\therefore \qquad \vec{E} = \frac{\lambda}{4\pi\varepsilon r}[(\sin\theta_1 - \sin\theta_2)\hat{a}_r - (\cos\theta_1 - \cos\theta_2)\hat{a}_z]$$

For an infinite line charge distribution, $\theta_1 = 90°$; $\theta_2 = -90°$ and the z-components vanish, so that the field becomes

$$\boxed{\vec{E} = \frac{\lambda}{2\pi\varepsilon r}\hat{a}_r}$$

NOTE

If the line charge distribution is not along the z-axis, $\vec{r}$ is the perpendicular distance from the line to the field point and $\hat{a}_r$ is the unit vector along that distance directed from the source point to the field point.

Example 2.16 A line of length 'l' carries a uniform line charge 'λ' per unit length as shown in Fig. 2.17. Show that the electric field intensity in the median plane at a distance 'r' is

$$\vec{E} = \frac{\lambda l}{2\pi\varepsilon_0 r}\left[\frac{1}{(l^2 + 4r^2)^{1/2}}\right]\hat{a}_r$$

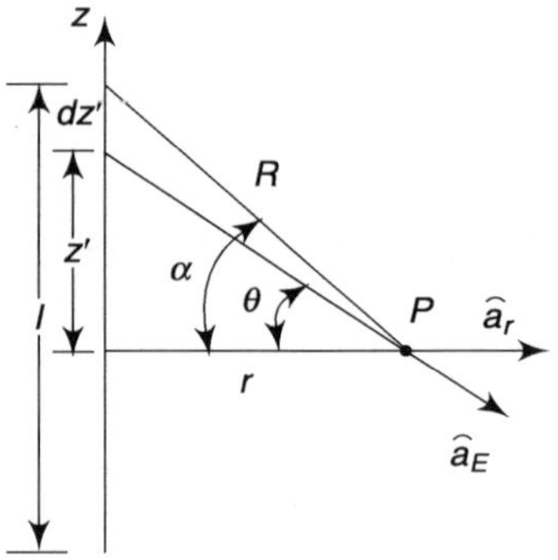

Fig. 2.17 *Field due to a charge carrying line charge*

Solution We will consider an elemental length dz' at a distance z'.

Charge on the elemental length is, $dQ = \lambda dz'$

$\therefore$ Field at point P due to this charge is,

$$d\vec{E} = \frac{dQ}{4\pi\varepsilon R_2}\hat{a}_E = \frac{\lambda dz'}{4\pi\varepsilon R^2}\hat{a}_E$$

From symmetry, it is observed that $\vec{E}$ has only a component along $\hat{a}_r$ direction.

$\therefore$ Total field at point P due to the line is

$$\vec{E} = \int_{-l/2}^{l/2}\frac{\lambda dz'}{4\pi\varepsilon R^2}\cos\theta\,\hat{a}_r = \int_{-l/2}^{l/2}\frac{\lambda r dz'}{4\pi\varepsilon(r^2 + z'^2)^{3/2}}\hat{a}_r$$

Here, $z' = r\tan\theta \;\Rightarrow\; dz' = r\sec^2\theta d\theta$

$$\begin{bmatrix} z' & -l/2 & l/2 \\ \theta & -\alpha & \alpha \end{bmatrix} \quad \text{where} \quad \alpha = \tan^{-1}\left(\frac{l}{2r}\right)$$

$$\therefore \qquad \vec{E} = \frac{\lambda}{4\pi\varepsilon}\int_{-\alpha}^{\alpha}\frac{r^2\sec^2\theta d\theta}{r^3\sec^3\theta}\hat{a}_r = \frac{\lambda}{4\pi\varepsilon r}\int_{-\alpha}^{\alpha}\cos\theta d\theta\hat{a}_r$$

$$= \frac{\lambda}{4\pi\varepsilon r}\sin\theta\Big|_{-\alpha}^{\alpha}\hat{a}_r$$

$$= \frac{\lambda}{4\pi\varepsilon r}2\sin\alpha\hat{a}_r$$

$$= \frac{\lambda}{2\pi\varepsilon r}\sin\alpha\hat{a}_r$$

Now, since $\alpha = \tan^{-1}\left(\frac{l}{2r}\right)$, $\therefore$ $\sin\alpha = \frac{l/2}{\sqrt{(l/2)^2 + r^2}} = \frac{l}{\sqrt{l^2 + 4r^2}}$

$$\therefore \qquad \boxed{\vec{E} = \frac{\lambda}{2\pi\varepsilon r}\frac{l}{\sqrt{l^2 + 4r^2}}\hat{a}_r}$$

NOTE

For infinite length line, $l \to \infty$, and the field intensity becomes

$$\vec{E} = \frac{\lambda l}{2\pi\varepsilon r}\frac{1}{l\sqrt{1 + 4(r/l)^2}}\hat{a}_r = \frac{\lambda l}{2\pi\varepsilon r}\hat{a}_r$$

Example 2.17 A non-conducting rod of length l with a uniform charge density λ and a total charge Q is lying along the x-axis as illustrated in Fig. 2.18. Compute the electric field at a point P, located at a distance y from the center of the rod along its perpendicular bisector.

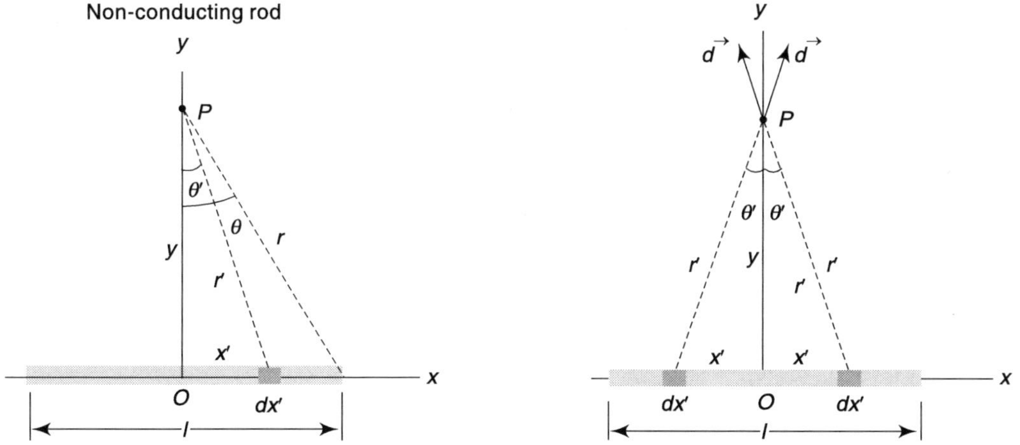

Fig. 2.18 *Symmetry argument showing that $E_x = 0$*

Solution We will follow a similar procedure as that outlined in earlier example. The contribution to the electric field from a small length element dx' carrying charge $dq = \lambda dx'$ is

$$dE = \frac{1}{4\pi\varepsilon_0}\frac{dq}{r'^2} = \frac{1}{4\pi\varepsilon_0}\frac{\lambda dx'}{x'^2 + y^2}$$

Using symmetry argument illustrated in Fig. 2.18, it is observed that the x-component of the electric field vanishes.

The y-component of dE is

$$dE_y = dE\cos\theta = \frac{1}{4\pi\varepsilon_0}\frac{\lambda dx'}{x'^2 + y^2}\frac{y}{\sqrt{x'^2 + y^2}}$$

$$= \frac{1}{4\pi\varepsilon_0}\frac{\lambda y dx'}{(x'^2 + y^2)^{3/2}}$$

By integrating over the entire length, the total electric field due to the rod is

$$E_y = \int dE_y = \frac{1}{4\pi\varepsilon_0}\int_{-l/2}^{l/2}\frac{\lambda y dx'}{(x'^2 + y^2)^{3/2}} = \frac{\lambda y}{4\pi\varepsilon_0}\int_{-l/2}^{l/2}\frac{dx'}{(x'^2 + y^2)^{3/2}}$$

Let, $x' = y\tan\theta'$ so that, $dx' = y\sec^2\theta' d\theta'$. The above integral becomes

$$E_y = \frac{\lambda y}{4\pi\varepsilon_0}\int_{-l/2}^{l/2}\frac{dx'}{(x'^2 + y^2)^{3/2}} = \frac{\lambda y}{4\pi\varepsilon_0}\int_{-\theta}^{\theta}\frac{y\sec^2\theta' d\theta'}{y^3(\tan^2\theta' + 1)^{3/2}} = \frac{\lambda}{4\pi\varepsilon_0}\frac{1}{y}\int_{-\theta}^{\theta}\frac{\sec^2\theta' d\theta'}{\sec^3\theta'}$$

$$= \frac{\lambda}{4\pi\varepsilon_0}\frac{1}{y}\int_{-\theta}^{\theta}\cos\theta' d\theta' = \frac{\lambda}{4\pi\varepsilon_0}\frac{1}{y}2\sin\theta = \frac{1}{4\pi\varepsilon_0}\frac{2\lambda}{y}\sin\theta$$

$$= \frac{1}{4\pi\varepsilon_0}\frac{2\lambda}{y}\frac{l/2}{\sqrt{y^2 + (l/2)^2}}$$

$$\boxed{E_y = \frac{1}{4\pi\varepsilon_0}\frac{2\lambda}{y}\frac{l/2}{\sqrt{y^2 + (l/2)^2}}}$$

NOTE

If $y \gg l$, the above expression reduces to the expression for a point charge given as

$$E_y \approx \frac{1}{4\pi\varepsilon_0}\frac{2\lambda}{y}\frac{l/2}{y} = \frac{1}{4\pi\varepsilon_0}\frac{\lambda l}{y^2}$$

$$= \frac{1}{4\pi\varepsilon_0}\frac{Q}{y^2}$$

On the other hand, if $l \gg y$, then we have

$$E_y \approx \frac{1}{4\pi\varepsilon_0}\frac{2\lambda}{y}$$

The characteristic behavior of E_y/E_0 (with $E_0 = Q/4\pi\varepsilon_0 l^2$) as a function of y/l is shown in Fig. 2.19.

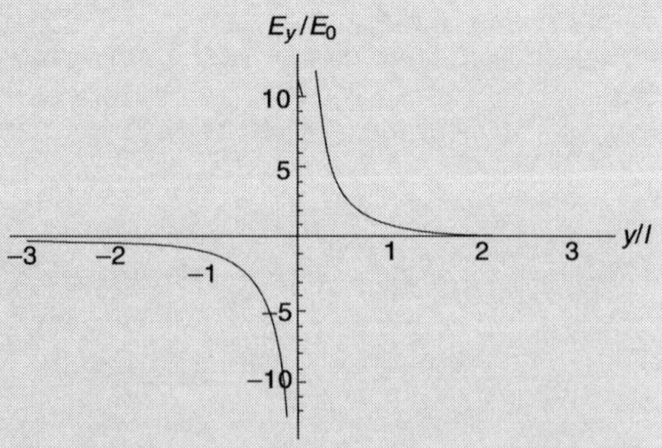

Fig. 2.19 *Electric field of a non-conducting rod as a function of y/l*

***Example 2.18** A line charge of length of $2m$ has a linear charge density of λ. Show that the electric field at a distance r perpendicular to the line charge from its middle point as shown in Fig. 2.20 is

$$E_r = \frac{\lambda}{2\pi\varepsilon_0 r\sqrt{\left(\frac{r}{m}\right)^2 + 1}}$$

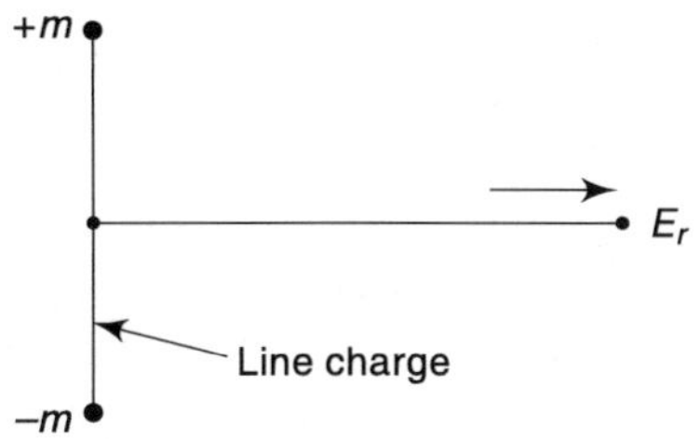

Fig. 2.20 *Arrangement of Example 2.17*

Solution We will consider an elemental length dz' of the line.

From the symmetry it is observed that all the z' components of the field will cancel each other. Also, there will be no ϕ component of the field.

Hence, the field will be only in the radial direction given as

$$d\vec{E} = \frac{\lambda dz'}{4\pi\varepsilon_0 R^2}\cos\alpha\,\hat{a}_r = \frac{\lambda dz'}{4\pi\varepsilon_0 (r^2 + z'^2)}\frac{r}{\sqrt{r^2 + z'^2}}\hat{a}_r$$

$$= \frac{\lambda}{4\pi\varepsilon_0}\frac{r}{(r^2 + z'^2)^{3/2}}dz'\hat{a}_r$$

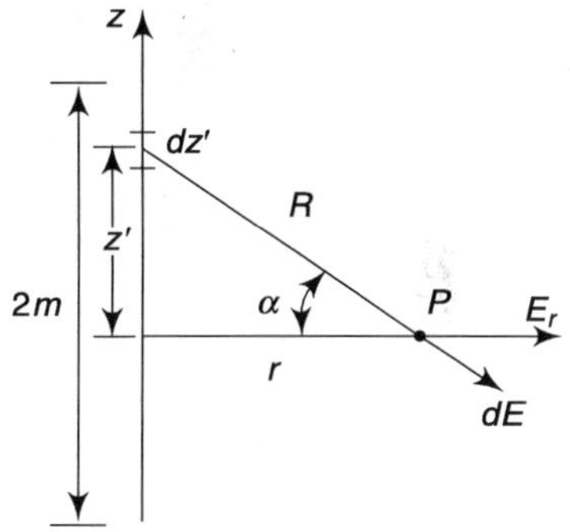

Fig. 2.21 *Field due to a charge carrying line*

So, the field due to the entire line is obtained as

$$\vec{E} = \int d\vec{E} = \frac{\lambda}{4\pi\varepsilon_0}\int_{-m}^{m}\frac{r}{(r^2 + z'^2)^{3/2}}dz'\hat{a}_r$$

$$= \frac{2\lambda}{4\pi\varepsilon_0}\int_0^m \frac{r}{(r^2 + z'^2)^{3/2}}dz'\hat{a}_r$$

$$= \frac{\lambda}{2\pi\varepsilon_0}\int_0^m \frac{r}{(r^2 + z'^2)^{3/2}}dz'\hat{a}_r$$

Let, $z' = r\tan\alpha \Rightarrow dz' = r\sec^2\alpha\, d\alpha$

$$\begin{bmatrix} z' & 0 & m \\ \alpha & 0 & \alpha_1 \end{bmatrix} \quad \text{where} \quad \alpha_1 = \tan^{-1}\left(\frac{m}{r}\right)$$

$\therefore$

$$\vec{E} = \frac{\lambda}{2\pi\varepsilon_0}\int_0^{\alpha_1}\frac{r r\sec^2\alpha d\alpha}{r^3\sec^3\alpha}\hat{a}_r$$

$$= \frac{\lambda}{2\pi\varepsilon_0 r}\int_0^{\alpha_1}\cos\alpha d\alpha\hat{a}_r$$

$$= \frac{\lambda}{2\pi\varepsilon_0 r}\sin\alpha_1\hat{a}_r$$

$$= \frac{\lambda}{2\pi\varepsilon_0 r}\frac{m}{\sqrt{r^2 + m^2}}\hat{a}_r$$

$$\left\{\because \tan\alpha_1 = \frac{m}{r} \quad \therefore \sin\alpha_1 = \frac{m}{\sqrt{r^2 + m^2}}\right\}$$

$$\therefore \qquad \boxed{\vec{E} = \frac{\lambda}{2\pi\varepsilon_0 r}\frac{1}{\sqrt{\left(\frac{r}{m}\right)^2 + 1}}\hat{a}_r}$$

***Example 2.19** A circular ring of radius 'R' carries a uniform charge distribution of line charge density 'λ' and is placed on the xy plane ($z = 0$ plane) with the axis the same as the z-axis as shown in Fig. 2.22.

(a) Show that

$$\vec{E}(0, 0, z) = \frac{\lambda R z}{2\varepsilon[R^2 + z^2]^{3/2}}\hat{a}_z$$

(b) What value of 'z' gives the maximum value of $\vec{E}$?

(c) If the total charge on the ring is Q, find $\vec{E}$ as $a \to 0$.

Solution (a) The field point is P (0, 0, z). The distance of the field point from the elemental length of the charge distribution is

$$\vec{r}' = -R\hat{a}_r + z\hat{a}_z$$

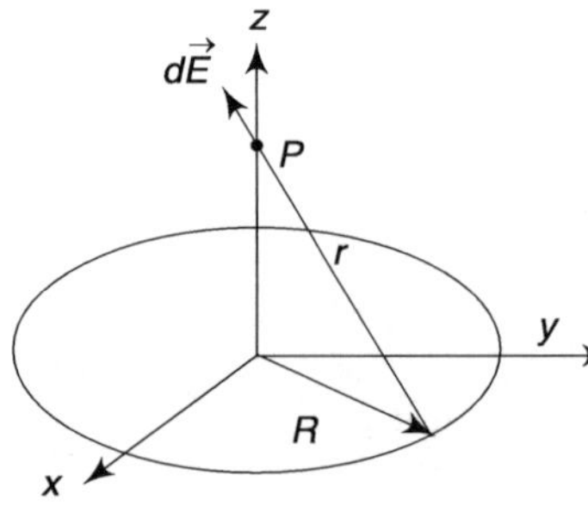

Fig. 2.22 *Field due to a ring of uniform charge distribution*

Hence, the electric field is given as

$$\vec{E} = \frac{\lambda}{4\pi\varepsilon_0}\int_0^{2\pi}\frac{R d\phi}{(R^2 + z^2)^{3/2}}(-R\hat{a}_r + z\hat{a}_z)$$

$$= \frac{\lambda}{4\pi\varepsilon_0}\frac{R}{(R^2 + z^2)^{3/2}}\left[-R\int_0^{2\pi}\hat{a}_r d\phi + z\int_0^{2\pi}\hat{a}_z d\phi\right]$$

$$= \frac{\lambda}{4\pi\varepsilon_0}\frac{R}{(R^2 + z^2)^{3/2}}$$

$$\left[-R\left\{\int_0^{2\pi}\hat{a}_x \cos\phi d\phi + \int_0^{2\pi}\hat{a}_y \sin\phi d\phi\right\} + 2\pi z\hat{a}_z\right]$$

$$= \frac{\lambda}{4\pi\varepsilon_0}\frac{R}{(R^2 + z^2)^{3/2}}[0 + 2\pi z\hat{a}_z]$$

$$= \frac{\lambda R z}{2\varepsilon_0(R^2 + z^2)^{3/2}}\hat{a}_z$$

$$\therefore \qquad \vec{E} = \frac{1}{4\pi\varepsilon_0}\frac{\lambda \times 2\pi R z}{(R^2 + z^2)^{3/2}}\hat{a}_z = \frac{1}{4\pi\varepsilon_0}\frac{Qz}{(R^2 + z^2)^{3/2}}$$

$$\boxed{\vec{E} = \frac{1}{4\pi\varepsilon_0}\frac{Qz}{(R^2 + z^2)^{3/2}}\hat{a}_z}$$

where $Q = \lambda \times 2\pi R$ is the total charge in the ring.

A plot of the electric field as function of z is shown in Fig. 2.23 (b).

Notice that the electric field at the centre of the ring vanishes. This is to be expected from symmetry arguments.

$$\therefore \qquad \boxed{\vec{E} = \frac{\lambda a z}{2\varepsilon_0(a^2 + z^2)^{3/2}}\hat{a}_z}$$

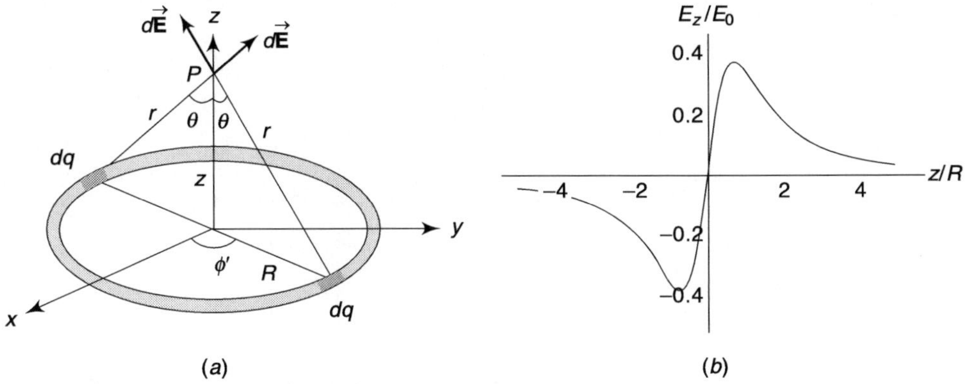

Fig. 2.23 *(a) Electric field directions, and (b) Electric field along the axis of symmetry of a non-conducting ring of radius R, with* $E_0 = \dfrac{Q}{4\pi\varepsilon_0 R^2}$

NOTE

At the centre of the loop, z = 0 and the electric field is also zero.

(b) For maximum value of the field intensity

$$\frac{d|\vec{E}|}{dz} = 0$$

$$\Rightarrow \quad \frac{d}{dz}\left[\frac{\lambda R z}{2\varepsilon_0 (R^2 + z^2)^{3/2}}\right] = 0$$

$$\Rightarrow \quad \frac{(R^2 + z^2)^{3/2} \times 1 - \frac{3}{2}(R^2 + z^2)^{1/2} \times 2z \times z}{(R^2 + z^2)^3} = 0$$

$$\Rightarrow \quad (R^2 + z^2)^{1/2}\,(R^2 + z^2 - 3z^2) = 0$$

$$\Rightarrow \quad R^2 - 2z^2 = 0$$

$$\Rightarrow \quad z = \pm\frac{R}{\sqrt{2}}$$

$$\therefore \quad \boxed{z = \pm\frac{R}{\sqrt{2}}}$$

(c) Since the charge is uniformly distributed, the line charge density is

$$\lambda = \frac{Q}{2\pi R}$$

Hence, the field is given as

$$\vec{E} = \frac{\lambda R z}{2\varepsilon (R^2 + z^2)^{3/2}}\,\hat{a}_z = \frac{\frac{Q}{2\pi R} R z}{2\varepsilon (R^2 + z^2)^{3/2}}\,\hat{a}_z = \frac{Qz}{4\pi\varepsilon (R^2 + z^2)^{3/2}}\,\hat{a}_z$$

As, $R \to 0$, the field is given as

$$\vec{E} = \frac{Q}{4\pi\varepsilon z^2}\hat{a}_z$$

2.7.2 Electric Field due to Surface Charge Distribution

For surface charge distribution with charge density σ (C/m^2), the total charge over a surface is obtained as $Q = \iint_S \sigma(\vec{r})dS$, so that the field intensity is given as

$$\vec{E} = \frac{1}{4\pi\varepsilon}\iint_S \frac{\sigma(\vec{r})}{r^2}dS \tag{2.16}$$

Example 2.20 Find the electric field intensity due to a uniformly charged infinite plane sheet with surface charge density of σ.

Solution We will consider an elemental surface dS as shown in Fig. 2.24.

The field at a point $P(0, 0, h)$ due to this elemental surface is

$$d\vec{E} = \frac{dQ}{4\pi\varepsilon R^2}\hat{a}_R$$

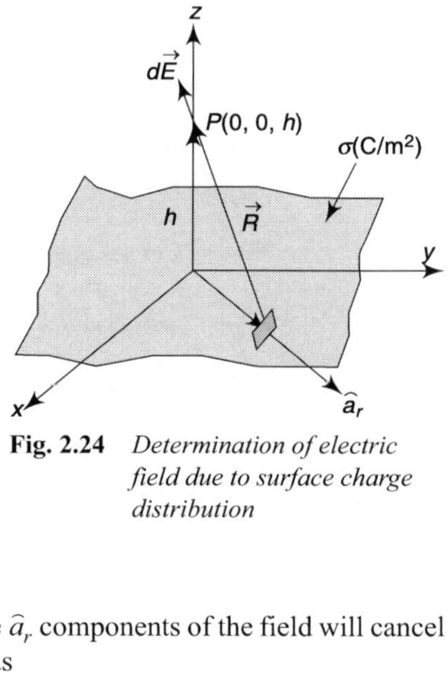

Fig. 2.24 *Determination of electric field due to surface charge distribution*

Here,

$$dQ = \sigma\, dS = \sigma r dr d\phi;\ \vec{R} = (-r\hat{a}_r + h\hat{a}_z)$$

$$\therefore R = |\vec{R}| = \sqrt{r^2 + h^2} \quad \therefore \hat{a}_R = \frac{\vec{R}}{R} = \frac{-r\hat{a}_r + h\hat{a}_z}{\sqrt{r^2 + h^2}}$$

Hence, the field is

$$d\vec{E} = \frac{dQ}{4\pi\varepsilon R^2}\hat{a}_R = \frac{\sigma\, r dr d\phi(-r\hat{a}_r + h\hat{a}_z)}{4\pi\varepsilon(r^2 + h^2)^{3/2}}$$

From the symmetry, it is understood that for an infinite plane, the $\hat{a}_r$ components of the field will cancel each other and the field will have only the z-component, given as

$$d\vec{E}_z = \frac{\sigma h\, r dr d\phi}{4\pi\varepsilon(r^2 + h^2)^{3/2}}\hat{a}_z$$

Thus, the total field is given as

$$\vec{E} = \int d\vec{E}_z = \frac{\sigma}{4\pi\varepsilon}\int_{\phi=0}^{2\pi}\int_{r=0}^{\infty}\frac{h\, r dr d\phi}{(r^2 + h^2)^{3/2}}\hat{a}_z = \frac{\sigma h}{4\pi\varepsilon}\times 2\pi\int_{r=0}^{\infty}\frac{r dr}{(r^2 + h^2)^{3/2}}\hat{a}_z$$

$$= \frac{\sigma h}{2\varepsilon}\int_{r=0}^{\infty}\frac{r dr}{(r^2 + h^2)^{3/2}}\hat{a}_z$$

Let $(r^2 + h^2) = p^2 \quad \therefore\ rdr = pdp$

$$\therefore \qquad \vec{E} = \frac{\sigma h}{2\varepsilon}\int_h^\infty \frac{pdp}{p^3}\hat{a}_z = \frac{\sigma h}{2\varepsilon}\int_h^\infty \frac{dp}{p^2}\hat{a}_z = \frac{\sigma h}{2\varepsilon}\left[-\frac{1}{p}\right]_h^\infty \hat{a}_z = \frac{\sigma}{2\varepsilon}\hat{a}_z$$

$$\therefore \qquad \vec{E} = \frac{\sigma}{2\varepsilon}\hat{a}_z$$

In general, for an infinite sheet of charge, the field is given as

$$\boxed{\vec{E} = \frac{\sigma}{2\varepsilon}\hat{a}_n}$$

where $\hat{a}_n$ is the unit vector normal to the plane.

NOTE

The field is independent of the distance of the field point from the plane. In case of a parallel plate capacitor, the electric field existing between the two plates of equal and opposite charges is given as $\vec{E} = \frac{\sigma}{2\varepsilon}\hat{a}_n + \frac{-\sigma}{2\varepsilon}(-\hat{a}_n) = \frac{\sigma}{\varepsilon}\hat{a}_n$.

Example 2.21 A circular disc of radius 'R' carries uniform charge distribution of surface charge density 'σ'. If the disc lies on the $z = 0$ plane, with its axis along the z-axis, (a) Find the electric field at a point P, along the z-axis that passes through the center of the disk perpendicular to its plane. Discuss the limit where $R >> z$. (b) From this, derive $\vec{E}$ field due to an infinite sheet of charge on the $z = 0$ plane.

Solution (a) By treating the disc as a set of concentric uniformly charged rings, the problem could be solved by using the result obtained in earlier. We will consider a ring of radius r' and thickness dr' as shown in Fig. 2.25.

By the symmetry argument, the electric field at P points in the $+z$-direction. Since the ring has a charge $dq = \sigma 2\pi r' dr'$, from Eq. (2.10–14), we see that the ring gives a contribution

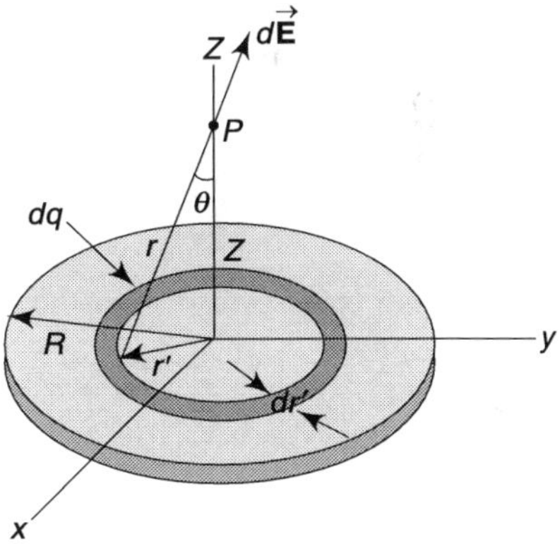

Fig. 2.25 *A uniformly charged disc of radius R*

$$dE_z = \frac{1}{4\pi\varepsilon_0}\frac{zdq}{(r'^2 + z^2)^{3/2}} = \frac{1}{4\pi\varepsilon_0}\frac{z2\pi\sigma r' dr'}{(r'^2 + z^2)^{3/2}}$$

Integrating from $r' = 0$ to $r' = R$, the total electric field at P becomes

$$E_z = \int dE_z = \frac{\sigma z}{2\varepsilon_0}\int_0^R \frac{r' dr'}{(r'^2 + z^2)^{3/2}}$$

$$= \frac{\sigma z}{4\varepsilon_0}\int_{z^2}^{R^2+z^2} \frac{du}{u^{3/2}} = \frac{\sigma z}{4\varepsilon_0}\left[\frac{u^{-1/2}}{(-1/2)}\right]_{z^2}^{R^2+z^2}$$

$$= -\frac{\sigma z}{2\varepsilon_0}\left[\frac{1}{\sqrt{R^2+z^2}} - \frac{1}{\sqrt{z^2}}\right] = \frac{\sigma}{2\varepsilon_0}\left[\frac{z}{|z|} - \frac{z}{\sqrt{R^2+z^2}}\right]$$

The above equation may be written as

$$E_z = \frac{\sigma}{2\varepsilon_0}\left[1 - \frac{z}{\sqrt{z^2+R^2}}\right], z > 0$$
$$= \frac{\sigma}{2\varepsilon_0}\left[-1 - \frac{z}{\sqrt{z^2+R^2}}\right], z < 0$$

The electric field $\frac{E_z}{E_0}\left(E_0 = \frac{\sigma}{2\varepsilon_0}\right)$ as a function of $\frac{z}{R}$ is shown in Fig. 2.26.

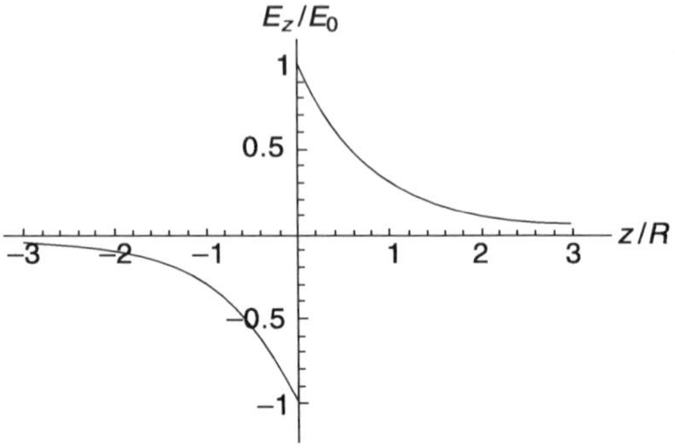

Fig. 2.26 *Electric field of a non-conducting plane of uniform charge density*

When $z >> R$, we will have the result similar to a point charge. By Taylor series expansion

$$1 - \frac{z}{\sqrt{z^2+R^2}} = 1 - \left(1 + \frac{R^2}{z^2}\right)^{1/2}$$
$$= 1 - \left(1 - \frac{1}{2}\frac{R^2}{z^2} + \cdots\right) \approx \frac{1}{2}\frac{R^2}{z^2}$$

$$\therefore \quad E_z = \frac{\sigma}{2\varepsilon_0}\frac{1}{2}\frac{R^2}{z^2} = \frac{1}{4\pi\varepsilon_0}\frac{\sigma\pi R^2}{z^2} = \frac{1}{4\pi\varepsilon_0}\frac{Q}{z^2}$$

This is the result for a point charge.

On the other hand, we may also consider the limit where $R >> z$. Physically, this means that the plane is very large, or the field point P is extremely close to the surface of the plane. The electric field in this limit becomes, in unit-vector notation

$$\vec{E} = \frac{\sigma}{2\varepsilon_0}\hat{k}, \quad z > 0$$
$$= -\frac{\sigma}{2\varepsilon_0}\hat{k}, \quad z < 0$$

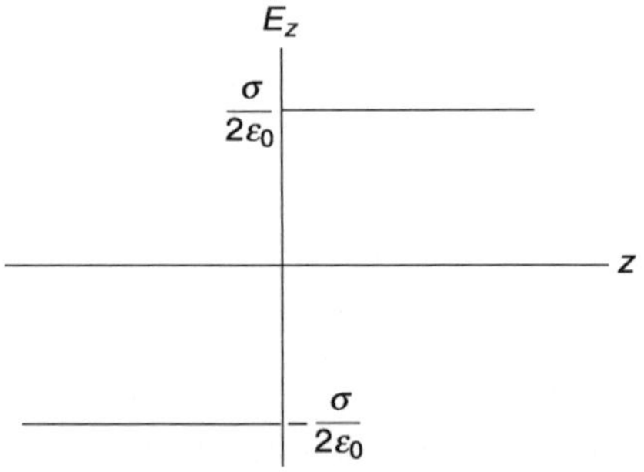

Fig. 2.27 *Electric field of an infinitely large non-conducting plane*

The plot of the electric field in this limit is shown in Fig. 2.27. When we cross the plane, there is a discontinuity in the electric field, given as

$$\Delta E_z = (E_{z+} - E_{z-}) = \frac{\sigma}{2\varepsilon_0} - \left(-\frac{\sigma}{2\varepsilon_0}\right) = \frac{\sigma}{\varepsilon_0}$$

which is obvious as has been explained in electric boundary conditions that if a given surface has a charge density σ, then the normal component of the electric field across that surface always exhibits a discontinuity with $\Delta E_n = \frac{\sigma}{\varepsilon_0}$.

(b) At $z = 0$, the field is given as

$$\therefore \quad \boxed{\vec{E}(0, 0, 0) = \frac{\sigma}{2\varepsilon}\hat{a}_z}$$

This is the field due to an infinite sheet of charge on the $z = 0$ plane.

Example 2.22 A thin annular disc of inner radius a and outer radius b carries a uniform surface charge density σ. Determine the electric field intensity at any point on the z axis when $z \geq 0$.

Solution The field point is P (0, 0, z) as shown in Fig. 2.28. The distance of the field point from the elemental length of the charge distribution is

$$\vec{R} = -r\hat{a}_r + z\hat{a}_z$$

Hence, the electric field is given as

$$\vec{E} = \frac{\sigma}{4\pi\varepsilon_0}\int_a^b\int_0^{2\pi}\frac{rdrd\phi}{(r^2+z^2)^{3/2}}(-r\hat{a}_r + z\hat{a}_z)$$

However, $\int_0^{2\pi}\hat{a}_r d\phi = 0$

$$\therefore \quad \vec{E} = \frac{\sigma}{4\pi\varepsilon_0}\int_a^b\int_0^{2\pi}\frac{rdrd\phi}{(r^2+z^2)^{3/2}}z\hat{a}_z$$

$$= \frac{\sigma}{4\pi\varepsilon_0}\times 2\pi z\int_a^b\frac{rdr}{(r^2+z^2)^{3/2}}\hat{a}_z$$

$$= \frac{\sigma z}{2\varepsilon_0}\left[\frac{1}{\sqrt{a^2+z^2}} - \frac{1}{\sqrt{b^2+z^2}}\right]\hat{a}_z$$

$$\boxed{\vec{E} = \frac{\sigma z}{2\varepsilon_0}\left[\frac{1}{\sqrt{a^2+z^2}} - \frac{1}{\sqrt{b^2+z^2}}\right]\hat{a}_z}$$

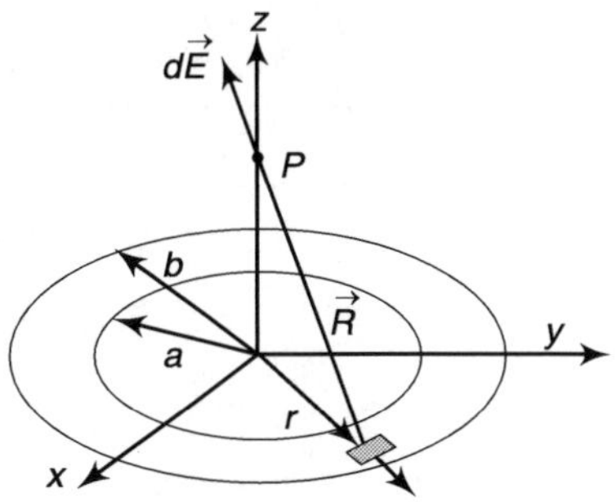

Fig. 2.28 *Field due to uniform charge distribution on an annular disc*

NOTE

(i) For an annular disc of very large outer radius, b → ∞, the field is given as

$$\vec{E} = \frac{\sigma z}{2\varepsilon_0}\left[\frac{1}{\sqrt{a^2+z^2}}\right]\hat{a}_z$$

(ii) For a solid finite disc of outer radius b, (a = 0) the field is given as

$$\vec{E} = \frac{\sigma z}{2\varepsilon_0}\left[\frac{1}{z} - \frac{1}{\sqrt{b^2+z^2}}\right]\hat{a}_z$$

(iii) For an infinite plane of charge, a → 0, b → ∞, and the field at any point is given as

$$\vec{E} = \frac{\sigma}{2\varepsilon_0}\hat{a}_z$$

2.7.3 Electric Field due to Volume Charge Distribution

For volume charge distribution with charge density ρ (C/m^3), the total charge over a surface is obtained as $Q = \int_V \rho(\vec{r})dV$, so that the field intensity is given as

$$\vec{E} = \frac{1}{4\pi\varepsilon}\int_V \frac{\rho(\vec{r})}{r^2}dV \quad \textbf{(2.17)}$$

Example 2.23 A sphere of radius a carries charge with a uniform volume density ρ (C/m^3). Determine the electric field intensity $\vec{E}$ at any distance r from the centre of the sphere.

Solution To find the field, we will consider an elemental volume dv within the sphere as shown in Fig. 2.29.

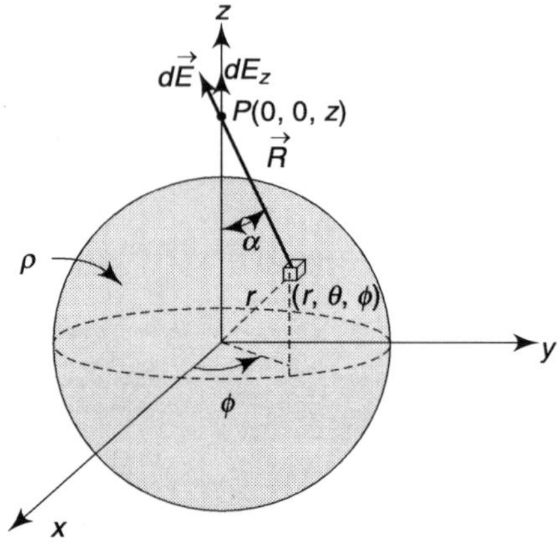

Fig. 2.29 *Field distribution due to volume charge distribution*

The field at a point $P(0, 0, h)$ due to this elemental volume is

$$d\vec{E} = \frac{dQ}{4\pi\varepsilon R^2}\hat{a}_R$$

Here, $\hat{a}_R = \cos\alpha\hat{a}_z + \sin\alpha\hat{a}_r$

Due to symmetry, field components E_x and E_y are zero.

$$\therefore \quad E_z = \vec{E}\cdot\hat{a}_z = \int dE\cos\alpha = \frac{\rho}{4\pi\varepsilon}\int\frac{dv\cos\alpha}{R^2} \quad \text{(i)}$$

Here, $dQ = \rho\, dv = \rho\, r^2\sin\theta dr d\theta d\phi$;

$R^2 = z^2 + r^2 - 2zr\cos\theta$

$r^2 = z^2 + R^2 - 2zR\cos\alpha$

$$\cos\alpha = \frac{z^2 + R^2 - r^2}{2zR}$$

$$\cos\theta = \frac{z^2 + r^2 - R^2}{2zr} \qquad \therefore \quad \sin\theta d\theta = \frac{R}{z}\frac{dR}{r}$$

Hence, the field is

$$E_z = \frac{\rho}{4\pi\varepsilon}\int\frac{dv\cos\alpha}{R^2} = \frac{\rho}{4\pi\varepsilon}\int_{\phi=0}^{2\pi}\int_{r=0}^{a}\int_{R=z-r}^{R=z+r} d\phi r^2\frac{RdR}{zr}dr\frac{z^2+R^2-r^2}{2zR}\frac{1}{R^2}$$

$$= \frac{\rho}{8\pi\varepsilon z^2}\times 2\pi\int_{r=0}^{a}\int_{R=z-r}^{R=z+r} r\left[1 + \frac{z^2-r^2}{R^2}\right]dRdr = \frac{\rho}{4\varepsilon z^2}\int_{r=0}^{a} r\left[R - \frac{z^2-r^2}{R}\right]_{z-r}^{z+r}dr$$

$$= \frac{\rho}{4\varepsilon z^2}\int_{r=0}^{a}4r^2dr = \frac{\rho}{4\varepsilon z^2}\frac{4}{3}a^3 = \frac{1}{4\pi\varepsilon}\frac{1}{z^2}\left(\frac{4}{3}\pi a^3\rho\right) = \frac{Q}{4\pi\varepsilon z^2}$$

$$\therefore \qquad E = \frac{Q}{4\pi\varepsilon z^2}\hat{a}_z$$

This result is obtained at point P (0, 0, z). From the symmetry of the charge distribution, it is clear that the field at any point P (r, θ, ϕ) is given as

$$\boxed{\vec{E} = \frac{Q}{4\pi\varepsilon r^2}\hat{a}_r}$$

NOTE

This result is identical to the field at the same point due to a point charge Q located at the origin or center of the spherical charge distribution.

2.8 ELECTRIC FLUX (DISPLACEMENT) (Ψ) AND FLUX DENSITY (DISPLACEMENT DENSITY) ($\vec{D}$)

2.8.1 Electric Flux (ψ)

Electric flux is the flux of the electric field. It is the total number of electric field lines passing through a given surface perpendicularly. It is expressed in Coulomb (C). It is the integral of the normal component of the vector $\vec{D}$ over a surface.

2.8.2 Electric Flux Density ($\vec{D}$)

Electric flux density is the total number of electric field lines per unit area passing through the area perpendicularly. It is expressed in Coulomb per square metre (C/m^2).

In case of a point charge, q, the electric displacement per unit area of a charge q at the centre of a sphere of radius r is

$$D = \frac{q}{4\pi r^2} \text{ (Coulomb/m}^2\text{)} \quad \textbf{(2.18)}$$

This is known as electric *displacement density* or *flux density*.

Now, for the same point charge, the electric field intensity at the centre of a sphere of radius r is

$$E = \frac{q}{4\pi \varepsilon r^2} \quad \textbf{(2.19)}$$

From Eqs. (2.18) and (2.19), we get

$$D = \varepsilon E \quad \textbf{(2.20)}$$

Here, flux density of a vector quantity ($\vec{D}$) and its direction is that of the normal to the surface element, which makes the displacement through the element of area a maximum. For a linear, isotropic medium, $\vec{D}$ is in the same direction as of $\vec{E}$.

In terms of flux density, $\vec{D}$, electric flux is defined as

$$\psi = \int_S \vec{D} \cdot d\vec{S} \quad \textbf{(2.21)}$$

In an SI unit, one line of electric flux emits from +1C charge and terminates on −1C charge. Hence, electric flux is expressed in Coulomb (C) and flux density in Coulomb per square metre (C/m^2).

Properties of Electric Flux

1. It is independent of the medium.
2. Its magnitude depends only upon the charge from which it is originated.
3. If a point charge in enclosed in an imaginary sphere of radius R, the electric flux must pass perpendicularly and uniformly through the surface of the sphere.
4. The electric flux density, i.e., flux per unit area, is then inversely proportional to R^2.

***Example 2.24** A point charge of 30nC is located at the origin while the plane $y = 2$ carries charge 20nC/m^2. Find the electric flux density $\vec{D}$ at point (0, 3, 4).

Solution The flux density component due to the point charge is given as

$$\vec{D}_1 = \varepsilon_0 \vec{E}_1 = \frac{Q}{4\pi R^2}\hat{a}_R = \frac{Q(\vec{r}-\vec{r}')}{4\pi|\vec{r}-\vec{r}'|^3} = \frac{30\times10^{-9}\{(0,3,4)-(0,0,0)\}}{4\pi|\{(0,3,4)-(0,0,0)\}|^3}$$

$$= \frac{30\times10^{-9}(3\hat{j}+4\hat{k})}{4\pi\times5^3} = (0.0573j + 0.0764\hat{k})\ \text{nC/m}^2$$

The flux density component due to the surface charge distribution is given as

$$\vec{D}_2 = \varepsilon_0\vec{E}_2 = \frac{\sigma}{2}\hat{a}_n = \frac{\sigma}{2}\hat{j} = \frac{20\times10^{-9}}{2}\hat{j} = 10\hat{j}\ \text{nC/m}^2$$

Hence, the total flux density due to the point charge and the surface charge distribution is

$$\vec{D} = \vec{D}_1 + \vec{D}_2 = (0.0573\hat{j} + 0.0764\hat{k}) + 10\hat{j} = (10.0573\hat{j} + 0.0764\hat{k})\ \text{nC/m}^2$$

2.9 ELECTRIC FIELD LINES (LINES OF FORCE) AND FLUX LINES

2.9.1 Electric Field Lines

These are the imaginary lines drawn in such a way that at every point, it has the direction of the electric field ($\vec{E}$). The number of lines per unit area is proportional to the magnitude of electric field strength (E).

2.9.2 Electric Flux Lines

These are the imaginary lines drawn in such a way that at every point, it has the direction of the electric flux density vector ($\vec{D}$). The number of flux lines per unit area is used to indicate the magnitude of the displacement density ($\vec{D}$).

In homogeneous, isotropic media, lines of force and lines of flux always have the same direction.

The field lines for a positive and a negative charge are shown in Figs. 2.30 (*a*) and (*b*). It is observed that the direction of field lines is radially outward for a positive charge and radially inward for a negative charge. For a pair of charges of equal magnitude but opposite sign (an electric dipole), the field lines are shown in Fig. 2.30 (*c*).

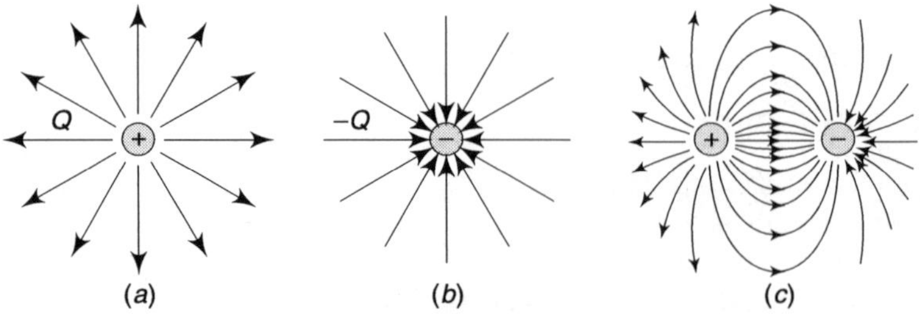

Fig. 2.30 *Field lines for (a) positive charge, (b) negative charge, and (c) an electric dipole*

Properties of Electric Field Lines

1. The direction of the electric field vector $\vec{E}$ at a point is tangential to the field lines.
2. Electric field lines never cross each other; otherwise, the field would be pointing in two different directions at the same point.
3. The field lines must begin on positive charges (or at infinity) and must terminate on negative charges (or at infinity).
4. Electric field lines are most dense around objects with the greatest amount of charge.
5. At locations where electric field lines meet the surface of an object, the lines are perpendicular to the surface.
6. The number of lines that originate from a positive charge or terminate on a negative charge must be proportional to the magnitude of the charge.
7. The number of lines per unit area through a surface perpendicular to the line is devised to be proportional to the magnitude of the electric field in a given region.

Example 2.25 Show that for electric lines of force

$$\frac{dx}{E_x} = \frac{dy}{E_y} = \frac{dz}{E_z}$$

Solution We will consider a test charge at point $P(x, y, z)$ in the field $\vec{E}$ as shown in Fig. 2.31. The force on test charge is directed along $\vec{E}$, to point $Q(x + \Delta x, y + \Delta y, z + \Delta z)$.

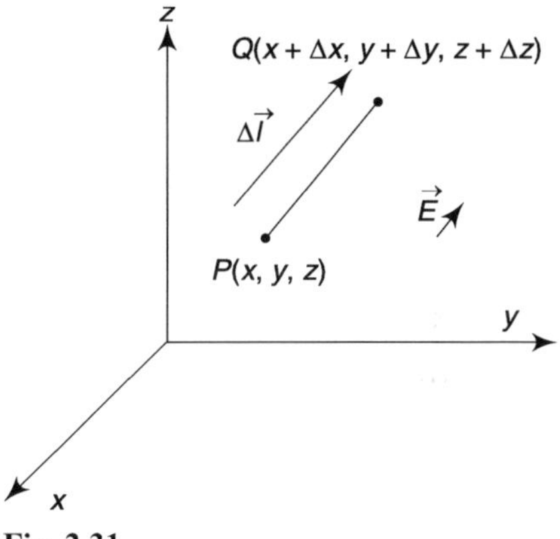

Fig. 2.31

The vector displacement of the test charge is

$$\Delta\vec{l} = (\Delta x\hat{i} + \Delta y\hat{j} + \Delta z\hat{k})$$

But, this is proportional to the field $\vec{E}$.

$$\therefore \qquad \vec{E} \propto \Delta\vec{l}$$

or $$(E_x\hat{i} + E_y\hat{j} + E_z\hat{k}) \propto (\Delta x\hat{i} + \Delta y\hat{j} + \Delta z\hat{k})$$

Now, two vectors are proportional if and only if their components are proportional by the same amount.

$$\therefore \qquad \frac{\Delta x}{E_x} = \frac{\Delta y}{E_y} = \frac{\Delta z}{E_z}$$

In the limiting case

$$\frac{dx}{E_x} = \frac{dy}{E_y} = \frac{dz}{E_z} \quad \text{in Cartesian coordinate}$$

$$\frac{dr}{E_r} = \frac{rd\phi}{E_\phi} = \frac{dz}{E_z} \quad \text{in cylindrical coordinate}$$

$$\frac{d\rho}{E_\rho} = \frac{\rho d\theta}{E_\theta} = \frac{\rho \sin\theta d\phi}{E_\phi} \quad \text{in spherical coordinate}$$

Example 2.26 Obtain the equation of flux line that passes through the point $P(-2, 7, 10)$ in the field $\vec{E}$ where

(a) $\vec{E} = 2(y-1)\hat{a}_x + 2x\hat{a}_y$ **(b)** $\vec{E} = e^y\hat{a}_x + (x+1)e^y\hat{a}_y$

Solution

(a) In two-dimensional system, the equation of flux lines is to be obtained from the equation

$$\frac{dy}{E_y} = \frac{dx}{E_x}$$

$$\Rightarrow \qquad \frac{dy}{dx} = \frac{E_y}{E_x} = \frac{2x}{2(y-1)} = \frac{x}{y-1}$$

$$\Rightarrow \qquad (y-1)dy = xdx$$

Integrating,

$$y^2 - 2y = x^2 + C$$

The constant C is evaluated from the coordinates of the point $(-2, 7, 10)$.

$$(7)^2 - 2 \times 7 = (-2)^2 + C \qquad \Rightarrow \quad C = 31$$

Hence, the equation of the flux lines is given as

$$y^2 - 2y = x^2 + 31 \qquad \Rightarrow \qquad (y-1)^2 - x^2 = 32$$

(b) In this case,

$$\frac{dy}{E_y} = \frac{dx}{E_x}$$

$$\Rightarrow \qquad \frac{dy}{dx} = \frac{E_y}{E_x} = \frac{(x+1)e^y}{e^y}$$

$$\Rightarrow \qquad dy = (x+1)dx$$

Integrating,

$$2y = x^2 + 2x + C$$

The constant C is evaluated from the coordinates of the point $(-2, 7, 10)$.

$$2 \times 7 = (-2)^2 + 2 \times (-2) + C \qquad \Rightarrow \quad C = 14$$

Hence, the equation of the flux lines is given as

$$2y = x^2 + 2x + 14 \qquad \Rightarrow \qquad 2y - (x+1)^2 = 13$$

2.10 GAUSS' LAW

Statement *Gauss' law,* also known as *Gauss' flux theorem,* states that the total electric displacement or electric flux through any closed surface surrounding charges is equal to the net positive charge enclosed by that surface.

Proof We consider a point charge Q located in a homogeneous isotropic medium of permittivity, ε. The electric field intensity at any point at a distance r from the charge will be

$$\vec{E} = \frac{Q}{4\pi\varepsilon r^2}\hat{a}_r \quad \textbf{(2.22)}$$

and the electric flux density is given as,

$$\vec{D} = \varepsilon\vec{E} = \frac{Q}{4\pi r^2}\hat{a}_r \quad \textbf{(2.23)}$$

Now, the electric flux through some elementary surface area dS as shown in Fig. 2.32 is

$$d\psi = DdS\cos\theta \quad \textbf{(2.24)}$$

where, θ is the angle between $\vec{D}$ and the normal to dS.

From Fig. 2.32, $dS\cos\theta$ is the projection of dS normal to the radius vector. By definition of a solid angle,

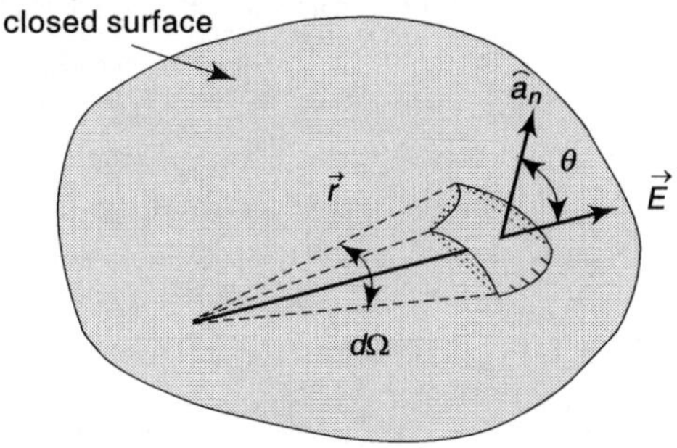

Fig. 2.32 *Determination of net electric flux through a closed surface*

$$dS\cos\theta = r^2 d\Omega \quad \textbf{(2.25)}$$

where, $d\Omega$ is the solid angle subtended at Q by the elementary surface of area dS.

Thus, total displacement or flux through the entire surface is

$$\psi = \oint_S d\psi = \oint_S DdS\cos\theta = \oint_S Dr^2 d\Omega = \frac{Q}{4\pi}\oint d\Omega \quad \text{\{using Eqs. (2.18), (2.22), (2.24) and (2.25)\}}$$

However, from the concept of calculus, the solid angle subtended by any closed surface is 4π steradian. Hence, total displacement or flux passing through the entire surface is,

$$\boxed{\psi = \oint_S \vec{D}\cdot d\vec{S} = Q = \int_v \rho dv} \quad \textbf{(2.26)}$$

This is the *integral form of Gauss' law.*

Applying divergence theorem in Eq. (2.26), we get

$$\oint_S \vec{D}\cdot d\vec{S} = \int_v \rho dv \;\Rightarrow\; \int_v (\nabla\cdot\vec{D})dv = \int_v \rho dv \;\Rightarrow\; \nabla\cdot\vec{D} = \rho$$

$$\boxed{\nabla\cdot\vec{D} = \rho} \quad \textbf{(2.27)}$$

This is the *differential form* or *point form of Gauss' law.*

NOTE

The surface to which Gauss' law is applied is known as Gaussian surface.

Steps useful when applying Gauss' law The following steps may be useful when applying Gauss' law:

1. First, the symmetry associated with the charge distribution is identified.
2. Then a *Gaussian surface* is identified and the direction of the electric field is determined.

3. The field space associated with the charge distribution is divided into different regions. For each region, the net charge enclosed by the Gaussian surface, Q_{enc} is calculated.
4. The electric flux ψ through the Gaussian surface for each region is calculated.
5. The magnitude of the electric field (E) is deduced by equating ψ with Q_{enc}/ε.

NOTE

Gaussian Surface—*a surface on which the magnitude of the electric field is constant over portions of the surface is known as Gaussian surface. This is a closed surface.*

2.10.1 Derivation of Gauss' Law from Coulomb's Law

Gauss' law can be derived from Coulomb's law, which states that the electric field due to a stationary point charge is:

$$\vec{E} = \frac{Q}{4\pi\varepsilon r^2}\hat{a}_r$$

where

$\hat{a}_r$ is the radial unit vector,
r is the radius, $|\vec{r}|$,
q is the charge of the particle, which is assumed to be located at the origin.

Using the expression from Coulomb's law, we get the total field at $\vec{r}$ by using an integral to sum the field at $\vec{r}$ due to the infinitesimal charge at each other point $\vec{r}'$ in space, to give

$$\vec{E} = \frac{1}{4\pi\varepsilon}\int_v \frac{\rho(\vec{r}')(\vec{r}-\vec{r}')}{|\vec{r}-\vec{r}|^3}dv$$

where ρ is the charge density. If we take the divergence of both sides of this equation with respect to $\vec{r}$, and use the known theorem

$$\nabla\cdot\left(\frac{\vec{r}'}{|\vec{r}'|^3}\right) = 4\pi\delta(r')$$

where $\delta(r')$ is the Dirac delta function, the result is

$$\nabla\cdot\vec{E} = \frac{1}{\varepsilon}\int_v \rho(\vec{r}')\delta(\vec{r}-\vec{r}')dv$$

Using the shifting property of the Dirac delta function, we get

$$\nabla\cdot\vec{E} = \frac{\rho}{\varepsilon}$$

which is the differential form of Gauss' law.

2.10.2 Derivation of Coulomb's Law from Gauss' Law

Gauss' law provides information only about the divergence of the electric field intensity, $\vec{E}$ and does not give any information about the curl $\vec{E}$. For this reason, Coulomb's law cannot be derived from Gauss' law alone. However, we assume that the electric field from a stationary point charge has spherical symmetry. With this assumption, which is exactly true if the charge is stationary, and approximately true if the charge is in motion, Coulomb's law can be proved from Gauss' law. We consider a spherical surface of radius r, centered at a point charge Q. Applying Gauss' law in integral form, we have

$$\oint_S \vec{E} \cdot d\vec{S} = \frac{Q}{\varepsilon}$$

By the assumption of spherical symmetry, the integrand is a constant and can be taken out of the integral as

$$4\pi r^2 \hat{a}_r \vec{E} = \frac{Q}{\varepsilon}$$

where $\hat{a}_r$ is a unit vector directed radially away from the charge. Again, by spherical symmetry, $\vec{E}$ is also in radially outward direction, and so we get

$$\vec{E} = \frac{Q}{4\pi\varepsilon r^2} \hat{a}_r$$

If another point charge q is placed on the surface, the force on that charge due to the charge Q is given as

$$\vec{F} = q\vec{E} = \frac{Qq}{4\pi\varepsilon r^2} \hat{a}_r$$

which is essentially equivalent to Coulomb's law.

Thus, the inverse-square law dependence of the electric field in Coulomb's law follows from Gauss' law.

NOTE

Coulomb's law applies only to stationary charges; whereas, Gauss' law holds for moving charges as well, and in this respect Gauss' law is more general than Coulomb's law.

2.11 APPLICATIONS OF GAUSS' LAW

Gauss' law provides an expedient tool for electric field computation. However, this law is applicable only to systems that possess certain symmetry; namely, systems with planar, cylindrical and spherical symmetry. The following table provides some examples of systems in which Gauss' law is applicable for electric field computation, with the corresponding Gaussian surfaces:

System	Symmetry	Gaussian Surface
Point charge	Spherical	Concentric sphere
Infinite rod	Cylindrical	Coaxial cylinder
Infinite plane	Planar	Gaussian "Pillbox"
Sphere, spherical shell	Spherical	Concentric sphere

2.11.1 Electric Field due to a Point Charge

Example 2.27 Determine the electric field at a distance r from a point charge Q. Use Gauss' law.

Solution We consider a point charge Q located at the origin. In order to determine the field $\vec{E}$ (or $\vec{D}$) at point P at a distance r from the charge, we follow the steps outlined in section 2.10.

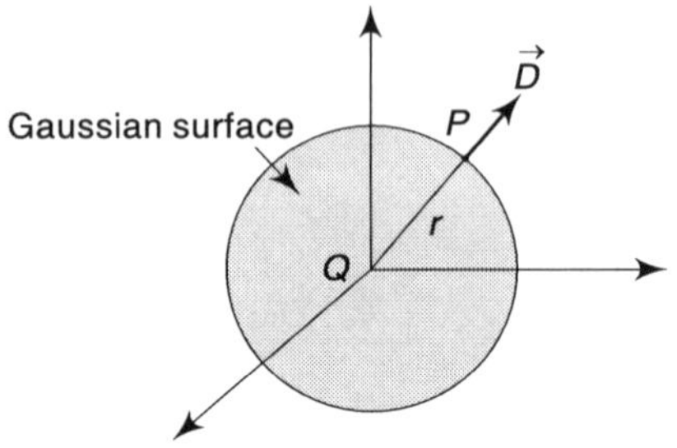

Fig. 2.33 *Gaussian surface of a point charge*

1. This problem possesses spherical symmetry.
2. We imagine a fictitious spherical surface of radius r, so that the point P is on the surface. So, the surface of the sphere with radius r is the Gaussian surface in this case.
3. The amount of charge enclosed by the Gaussian surface is the point charge $\boldsymbol{Q}$.
4. Since $\vec{E}$ is perpendicular to the Gaussian surface, i.e., $\vec{E} = E\hat{a}_r$ or, $\vec{D} = D\hat{a}_r$

   So, total flux through the Gaussian surface is

$$\oint_S \vec{D} \cdot d\vec{S} = D \int_{\phi=0}^{2\pi} \int_{\theta=0}^{\pi} r^2 \sin\theta d\theta d\phi = D4\pi r^2$$

5. Applying Gauss' law,

$$\Rightarrow \qquad \oint_S \vec{D} \cdot d\vec{S} = Q_{enc}$$

$$\Rightarrow \qquad D4\pi r^2 = Q$$

$$\therefore \qquad D = \frac{Q}{4\pi r^2}$$

In vector form

$$\boxed{\vec{D} = \frac{Q}{4\pi r^2}\hat{a}_r \quad \text{or,} \quad \vec{E} = \frac{Q}{4\pi\varepsilon r^2}\hat{a}_r} \qquad \textbf{(2.28)}$$

2.11.2 Electric Field due to Infinite Line Charge

***Example 2.28** Determine the electric field at a distance r from an infinite straight line carrying a uniform line charge distribution with line charge density λ. Use Gauss' law.

Solution We will consider an infinitely long wire of negligible radius with a uniform line charge density, λ. We calculate the field $\vec{E}$ at a distance r from the wire.

We shall solve the problem by following the steps outlined in Section 2.10.

1. This infinitely long wire possesses cylindrical symmetry.
2. The charge density is uniformly distributed throughout the length of the wire, and therefore, the electric field $\vec{E}$ must be radially outward from the axis of the wire [Fig. 2.34 (*a*)]. The magnitude of the electric field is constant on cylindrical surfaces of radius r. Therefore, a coaxial cylinder of any radius r is a Gaussian surface in this problem.
3. The amount of charge enclosed by the Gaussian surface (i.e., a cylinder of radius r and length l) [Fig. 2.34 (*b*)] is $Q_{enc} = \lambda l$.
4. The Gaussian surface consists of three parts, as indicated in Fig. 2.34 (*b*). Those are:
   (a) Left-end surface S_1
   (b) Right-end surface S_2
   (c) Curved surface S_3.

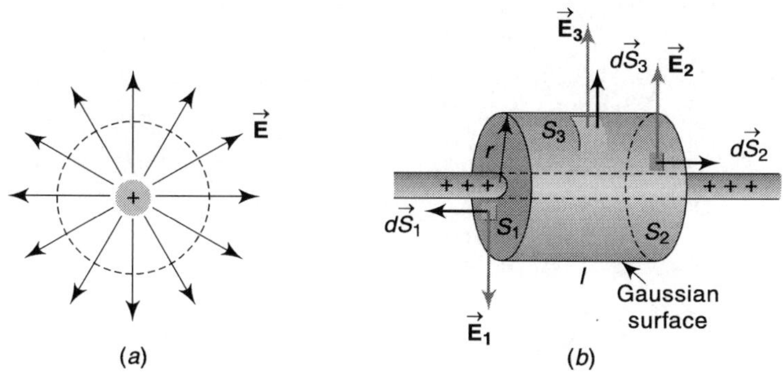

Fig. 2.34 *(a) Field lines for an infinite uniformly charged wire, and (b) Gaussian surface for a uniformly charged wire*

The flux through the Gaussian surface is

$$\psi = \oiint_S \vec{E}\cdot d\vec{S} = \oiint_{S_1} \vec{E}\cdot d\vec{S}_1 + \oiint_{S_2} \vec{E}\cdot d\vec{S}_2 + \oiint_{S_3} \vec{E}\cdot d\vec{S}_3 = 0 + 0 + E_3 S_3 = E(2\pi r l)$$

where $E_3 = E$ (say). As can be seen from the figure, no flux passes through the ends since the area vectors $d\vec{S}_1$ and $d\vec{S}_2$ are perpendicular to the electric field, which is directed radially outward.

5. Applying Gauss' law

$$E(2\pi r l) = \frac{\lambda l}{\varepsilon} \quad \Rightarrow \quad E = \frac{\lambda}{2\pi\varepsilon r} \quad \Rightarrow \quad D = \varepsilon E = \frac{\lambda}{2\pi r}$$

In vector form

$$\boxed{\vec{E} = \frac{\lambda}{2\pi\varepsilon r}\hat{a}_r \quad \text{or} \quad \vec{D} = \frac{\lambda}{2\pi r}\hat{a}_r} \qquad \textbf{(2.29)}$$

This is observed that the result is independent of the length l of the cylinder and only depends on the inverse of the distance r from the symmetry axis. The variation of E as a function of r is plotted in Fig. 2.35.

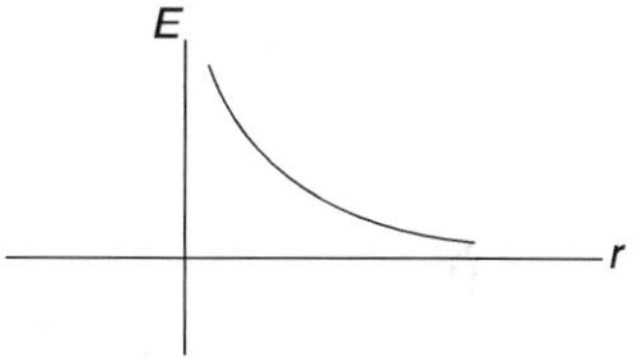

Fig. 2.35 *Electric field due to a uniformly charged rod as a function of r*

2.11.3 Electric Field due to an Infinite Plane Sheet of Charge

***Example 2.29** Determine the electric field intensity due to a uniformly charged infinite plane sheet with surface charge density σ.

Solution We will consider an infinitely large non-conducting plane in the *xy*-plane with uniform surface charge density σ. We want to determine the electric field everywhere in space. We will follow the steps as outlined in Section 2.9.

1. This infinitely large plane possesses a planar symmetry.
2. Since the charge is uniformly distributed on the surface, the electric field $\vec{E}$ is directed perpendicularly away from the plane $\vec{E} = E\hat{a}_z$. The magnitude of the electric field is constant on planes parallel to the non-conducting plane. Therefore, the Gaussian surface for this problem is a cylinder, which is often referred to as a 'pillbox' (Fig. 2.36).

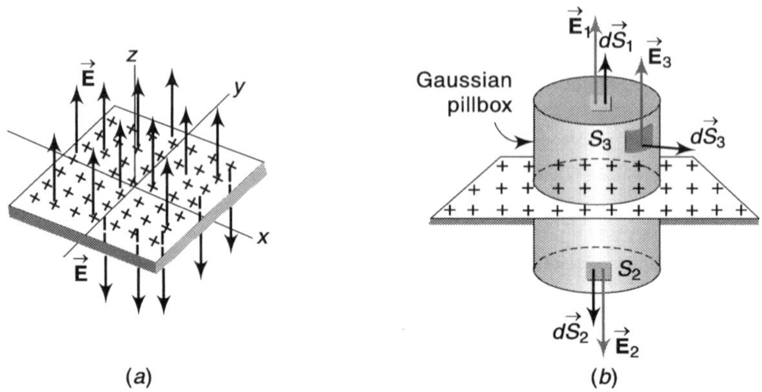

Fig. 2.36 *(a) Electric field for an infinite plane of charge, and (b) Gaussian surface for a large plane*

3. Since the charge is uniformly distributed over the surface, the charge enclosed by the Gaussian 'pillbox' is, $Q_{enc} = \sigma S$, where $S = S_1 = S_2$ is the area of the end-surfaces.
4. The Gaussian pillbox consists of three parts: two end-surfaces S_1 and S_2, and a curved surface S_3. The total flux through the Gaussian pillbox flux is

$$\psi = \oiint_S \vec{E} \cdot d\vec{S} = \oiint_{S_1} \vec{E} \cdot d\vec{S}_1 + \oiint_{S_2} \vec{E} \cdot d\vec{S}_2 + \oiint_{S_3} \vec{E} \cdot d\vec{S}_3 = E_1 S_1 + E_2 S_2 + 0 = (E_1 + E_2)S$$

Since the two ends are at the same distance from the plane, by symmetry, the magnitude of the electric field must be the same, $E_1 = E_2 = E$. Hence, the total flux can be rewritten as $\psi = 2ES$
5. By applying Gauss' law, we obtain

$$2ES = \frac{Q_{enc}}{\varepsilon} = \frac{\sigma S}{\varepsilon} \qquad E = \frac{\sigma}{2\varepsilon}$$

In vector form, the result can be written as

$$\boxed{\begin{aligned} \vec{E} &= \frac{\sigma}{2\varepsilon}\hat{a}_z, \quad z > 0 \\ &= -\frac{\sigma}{2\varepsilon}\hat{a}_z, \quad z < 0 \end{aligned}} \qquad \textbf{(2.30)}$$

Thus, we see that the electric field due to an infinite large non-conducting plane is uniform in space. The result is plotted in Fig. 2.37.

Note again the discontinuity in electric field as we cross the plane:

$$\Delta E_z = E_{z+} - E_{z-} = \frac{\sigma}{2\varepsilon} - \left(-\frac{\sigma}{2\varepsilon}\right) = \frac{\sigma}{\varepsilon}$$

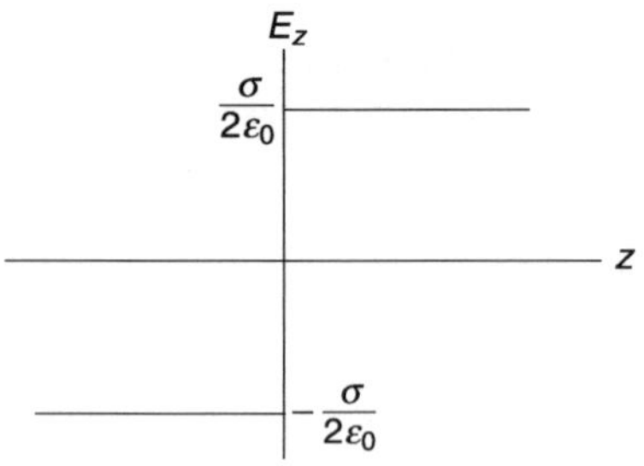

Fig. 2.37 *Electric field of an infinitely large non-conducting plane*

2.11.4 Electric Field due to a Uniformly Charge Sphere

***Example 2.30** Derive the expression for the electric field intensity at any point inside and outside of a sphere of radius 'a' due to a uniform spherical charge distribution of volume charge density of 'ρ'. Use Gauss' law.

Solution We consider a non-conducting solid sphere of radius a uniformly charged with volume charge density ρ (C/m^3). We want to determine the electric field everywhere inside and outside the sphere. We follow the steps outlined in Section 2.10.

1. This problem has a spherical symmetry.
2. In this case, the electric field $\vec{E}$ is radially symmetric and directed outward. The magnitude of the electric field is constant on spherical surfaces of radius r. Therefore, this spherical surface is the Gaussian surface.
3. The regions $r \le a$ and $r \ge a$ will be considered separately to find the charge enclosed and flux through the surface.

Case 1: $r \le a$
In this case, the Gaussian surface is a sphere of radius $r \le a$, as shown in Fig. 2.38 (a).

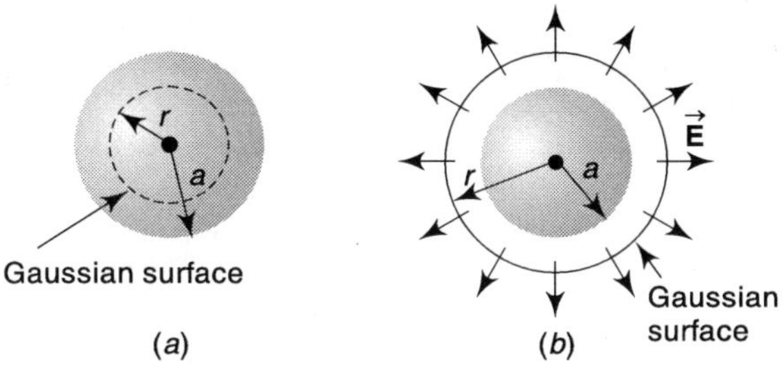

Fig. 2.38 *Gaussian surface for uniformly charged solid sphere, for (a) $r \le a$, and (b) $r > a$*

With uniform charge density, total charge enclosed is

$$Q_{\text{enc}} = \int_v \rho dv = \rho \int_{\phi=0}^{2\pi} \int_{\theta=0}^{\pi} \int_{r=0}^{r} r^2 \sin\theta dr d\theta d\phi = \rho\left(\frac{4}{3}\pi r^3\right)$$

The flux through the Gaussian surface is

$$\psi = \oint_S \vec{E}\cdot d\vec{S} = E\oint_S dS = E\int_{\phi=0}^{2\pi}\int_{\theta=0}^{\pi} r^2 \sin\theta d\theta d\phi = E(4\pi r^2)$$

Applying Gauss' law

$$\psi = \frac{Q_{\text{enc}}}{\varepsilon} \quad \text{or,} \quad E(4\pi r^2) = \frac{\rho}{\varepsilon}\left(\frac{4}{3}\pi r^3\right) \quad \text{or} \quad E = \frac{\rho r}{3\varepsilon}$$

In vector form

$$\boxed{\vec{E} = \frac{\rho r}{3\varepsilon}\hat{a}_r \qquad r \le a} \qquad \textbf{(2.31)}$$

Case 2: $r \ge a$
In this case, the Gaussian surface is a sphere of radius $r \ge a$, as shown in Fig. 2.38 (b). Since the radius of the Gaussian surface is greater than the radius of the sphere, all the charge is enclosed in the Gaussian surface, i.e.,

$$Q_{\text{enc}} = \int_v \rho dv = \rho \int_{\phi=0}^{2\pi} \int_{\theta=0}^{\pi} \int_{r=0}^{a} r^2 \sin\theta dr d\theta d\phi$$
$$= \rho\left(\frac{4}{3}\pi a^3\right)$$

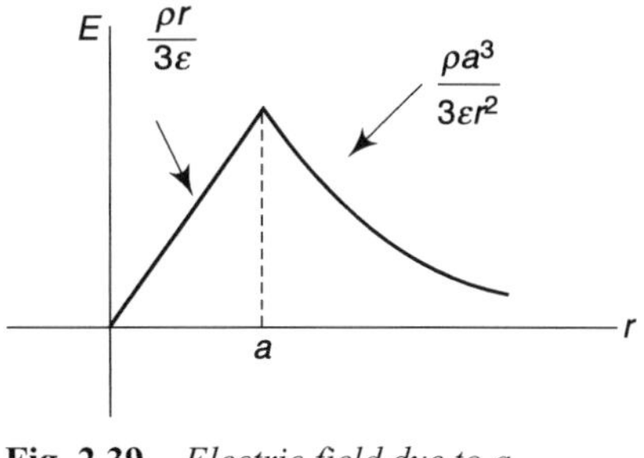

Fig. 2.39 *Electric field due to a uniformly charged sphere as a function of r*

Electric flux through the Gaussian surface is given by

$$\psi = \oint_S \vec{E} \cdot d\vec{S} = E \oint_S dS = E \int_{\phi=0}^{2\pi} \int_{\theta=0}^{\pi} r^2 \sin\theta d\theta d\phi = E(4\pi r^2)$$

Applying Gauss' law, we obtain

$$\psi = \frac{Q_{\text{enc}}}{\varepsilon} \quad \text{or} \quad E(4\pi r^2) = \frac{\rho}{\varepsilon}\left(\frac{4}{3}\pi a^3\right)$$

or

$$E = \frac{\rho a^3}{3\varepsilon r^2}$$

In vector form

$$\boxed{\vec{E} = \frac{\rho a^3}{3\varepsilon r^2}\hat{a}_r \qquad r \ge a} \qquad \textbf{(2.32)}$$

The results can be summarized as

$$\vec{E} = \frac{\rho r}{3\varepsilon}\hat{a}_r \qquad 0 < r \le a$$

$$= \frac{\rho a^3}{3\varepsilon r^2}\hat{a}_r \quad r \ge a$$

The field outside the sphere is the same as if all the charges were concentrated at the center of the sphere. The variation of E as a function of r is plotted in Fig. 2.39.

***Example 2.31** Using Gauss' law, determine the electric field intensity due to a uniformly charged infinite cylinder of radius 'a' having a line charge density of 'λ', at a distance 'r' from the axis of the cylinder.

Solution

1. The problem possesses cylindrical symmetry.

2. We consider a cylindrical Gaussian surface of radius r and coaxial with the charged cylinder. This is depicted in Fig. 2.40.

3. The flux and charge enclosed are found considering two cases:

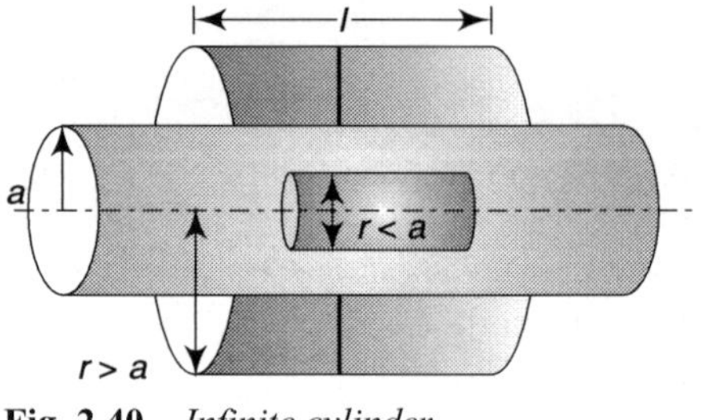

Fig. 2.40 *Infinite cylinder*

When $r > a$:

Applying Gauss' law for region outside the cylinder of length l

$$(2\pi r l) \times E = \frac{Q_{\text{enc}}}{\varepsilon} = \frac{\lambda l}{\varepsilon}$$

$$\therefore \qquad E = \frac{\lambda}{2\pi\varepsilon r}$$

In vector form $\vec{E} = \dfrac{\lambda}{2\pi\varepsilon r}\hat{a}_r$

When $r < a$:

In this case, we have to find out the charge enclosed by the Gaussian surface for a point inside the cylinder. This is obtained as follows.

Let ρ—the volume charge density

$\therefore$ Charge in the length l of the charged cylinder is

$$\lambda l = \pi a^2 l \rho \quad \Rightarrow \quad \rho = \frac{\lambda}{\pi a^2}$$

$\therefore$ Charge enclosed in the Gaussian cylinder is

$$Q_{\text{enc}} = \pi r^2 l \rho = \left(\frac{r^2}{a^2}\right) \lambda l$$

Applying Gauss' law in this region

$$(2\pi r l) \times E = \frac{Q_{\text{enc}}}{\varepsilon} = \left(\frac{r^2}{a^2}\right) \frac{\lambda l}{\varepsilon}$$

$\therefore$
$$E = \frac{\lambda r}{2\pi \varepsilon a^2}$$

In vector form,
$$\vec{E} = \frac{\lambda r}{2\pi \varepsilon a^2} \hat{a}_r$$

The results can be summarised as

$$\vec{E} = \frac{\lambda}{2\pi \varepsilon r} \hat{a}_r; \qquad r > a$$
$$= \frac{\lambda r}{2\pi \varepsilon a^2} \hat{a}_r; \qquad r < a$$

***Example 2.32** A thin spherical shell of radius a has a charge $+Q$ evenly distributed over its surface. Find the electric field both inside and outside the shell.

Solution

1. In this case, the charge distribution has spherical symmetry, with a surface charge density

$$\sigma = \frac{Q}{S} = \frac{Q}{4\pi a^2}$$

2. The electric field $\vec{E}$ must have radial symmetry and must be directed outward (Fig. 2.41). So, spherical surface is the Gaussian surface.
3. Flux and charge enclosed are determined by considering two regions $r \le a$ and $r \ge a$ separately.

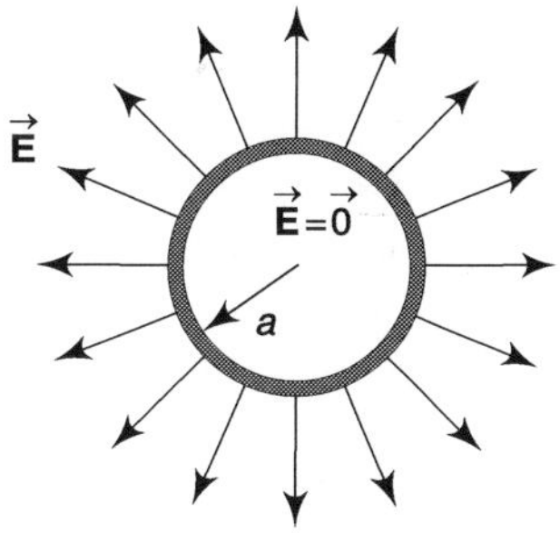

Fig. 2.41 *Electric field for uniform spherical shell of charge*

When $r \le a$:
In this case, the Gaussian surface is a sphere of radius $r \le a$, as shown in Fig. 2.42 (a).

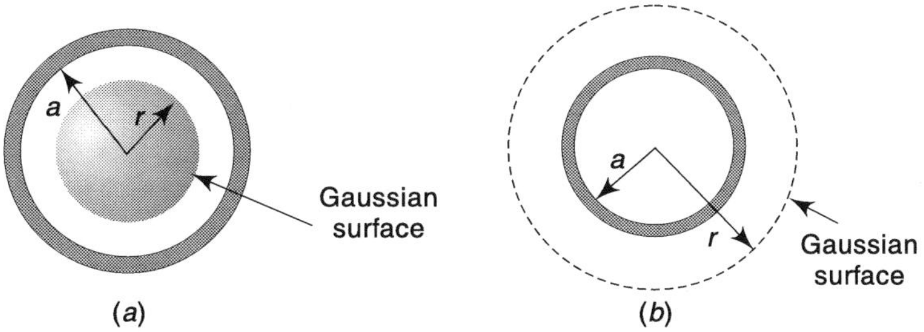

Fig. 2.42 *Gaussian surface for uniformly charged spherical shell for (a) $r \le a$, and (b) $r \ge a$*

Since all the charge is located on the surface of the shell, the charge enclosed by the Gaussian surface is zero, $Q_{enc} = 0$. Hence, by Gauss' law, we have

$$E = \frac{Q_{enc}}{\varepsilon} = \frac{0}{\varepsilon} = 0 \quad r < a$$

When $r \geq a$:

In this case, the Gaussian surface is a sphere of radius $r \geq a$, as shown in Fig. 2.42 (*b*). Since the radius of the Gaussian sphere is greater than the radius of the spherical shell, all the charge is enclosed, i.e., $Q_{enc} = Q$.

Total flux through the Gaussian surface is

$$\phi = \oint_S \vec{E} \cdot d\vec{S} = ES = E(4\pi r^2)$$

Applying Gauss' law

$$E(4\pi r^2) = \frac{Q}{\varepsilon}$$

$$\therefore \quad E = \frac{Q}{4\pi\varepsilon r^2} \quad r \geq a$$

The results can be summarized as

$$E = 0 \qquad r < a$$

$$= \frac{Q}{4\pi\varepsilon r^2} \qquad r \geq a$$

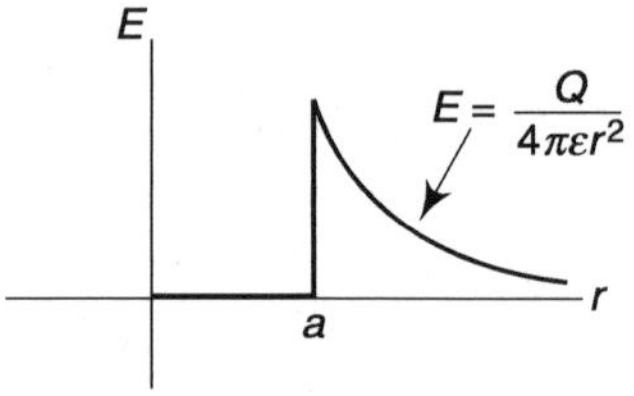

Fig. 2.42 (*c*) Electric field as a function of *r* due to a uniformly charged spherical shell

The variation of E as a function of r is plotted in Fig. 2.42 (*c*). We see a discontinuity of E as we cross the boundary at $r = a$. The change in the field intensity from the outer surface to the inner surface, is given by

$$\Delta E = E_+ - E_- = \frac{Q}{4\pi\varepsilon r^2} - 0 = \frac{\sigma}{\varepsilon_0}$$

***Example 2.33** A spherical volume charge density distribution is given by

$$\rho = \rho_0\left[1 - \frac{r^2}{a^2}\right]; \quad (r \leq a)$$

$$= 0; \qquad (r > a)$$

(a) Calculate the total charge Q.
(b) Find $\vec{E}$ everywhere for $0 < r \leq a$ and $r > a$.
(c) Show that the maximum value of $\vec{E}$ is at $r = 0.745a$.

Solution

(a) Total charge Q is obtained by finding the volume integral of charge density function ρ over the entire volume of sphere of radius $r = a$.

$$\therefore \quad Q = \iiint_{vol} \rho dv$$

Considering a thin spherical shell within the charged sphere as radius r and with thickness dr, the charge within the shell.

$$\therefore \qquad dQ = \rho \times 4\pi r^2 dr = \rho_0 \left[1 - \frac{r^2}{a^2}\right] \times 4\pi r^2 dr$$

$\therefore$ Total charge is

$$Q = \int_{r=0}^{a} dQ = \int_{r=0}^{a} \rho_0 \left[1 - \frac{r^2}{a^2}\right] 4\pi r^2 dr = 4\pi\rho_0 \int_{r=0}^{a} \left[r^2 - \frac{r^4}{a^2}\right] dr = 4\pi\rho_0 \left(\frac{r^3}{3} - \frac{r^5}{5a^2}\right)_0^a$$

$$= 4\pi\rho_0 \left(\frac{a^3}{3} - \frac{a^3}{5}\right)$$

$$= \frac{8}{15}\pi\rho_0 a^3$$

$$\therefore \qquad \boxed{Q = \frac{8}{15}\pi\rho_0 a^3}$$

(b) Field intensity at a distance r from the center of the sphere, outside ($r > a$) the sphere is

$$E = \frac{Q}{4\pi\varepsilon r^2} = \frac{2}{15}\frac{\rho_0 a^3}{\varepsilon r^2}$$

For field inside the sphere ($r \le a$), applying Gauss' law to the spherical surface

$$E \times (4\pi r^2) = \frac{Q}{\varepsilon} = \frac{1}{\varepsilon}\left[4\pi\rho_0 \int_0^r \left(1 - \frac{r^2}{a^2}\right) r^2 dr\right] = \frac{1}{\varepsilon} 4\pi\rho_0 \left(\frac{r^3}{3} - \frac{r^5}{5a^2}\right)$$

$$\Rightarrow \qquad E = \frac{\rho_0}{\varepsilon}\left(\frac{r}{3} - \frac{r^3}{5a^2}\right)$$

$$\boxed{\begin{aligned} \vec{E} &= \frac{\rho_0}{\varepsilon}\left(\frac{r}{3} - \frac{r^3}{5a^2}\right)\hat{a}_r; & r \le a \\ &= \frac{2}{15}\frac{\rho_0 a^3}{\varepsilon r^2}\hat{a}_r; & r > a \end{aligned}}$$

(c) For maximum value of E, $\frac{dE}{dr} = 0$.

$$\frac{d}{dr}\left[\frac{\rho_0}{\varepsilon}\left(\frac{r}{3} - \frac{r^3}{5a^2}\right)\right] = 0$$

$$\Rightarrow \qquad \frac{1}{3} - \frac{3r^2}{5a^2} = 0$$

$$\Rightarrow \qquad r^2 = \frac{5}{9}a^2$$

$$\Rightarrow \qquad r = 0.745a$$

Thus, the maximum value of $\vec{E}$ is at $r = 0.745a$.

Example 2.34 A long cylinder (charged) of radius 'a' has a volume charge density, $\rho = kr$, where 'k' is a constant and 'r' is the distance from the axis of the cylinder. Show that the electric field is given by

$$\vec{E} = \frac{kr^2}{3\varepsilon}\hat{a}_r; \quad r \le a$$
$$= \frac{ka^3}{3\varepsilon r}\hat{a}_r; \quad r \ge a$$

Solution

1. This problem possesses cylindrical symmetry.
2. The cylindrical surface of any radius 'r' is the Gaussian surface.
3. Flux and charge enclosed are determined considering two cases:

Inside the cylinder ($r \le a$)

We will consider a tubular cylinder of inner radius r and outer radius $(r + dr)$ located coaxially within the charged cylinder of radius a.

The charge contained in the tubing per unit length is

$$dQ = \rho 2\pi r dr \times 1 = kr2\pi r dr = 2\pi k r^2 dr$$

$\therefore$ Total charge per unit length contained in a cylinder of radius r is

$$Q = \int dQ = \int_0^r 2\pi k r^2 dr = \frac{2\pi k r^3}{3}$$

Applying Gauss' law to the Gaussian surface of radius r

$$2\pi r \times 1 \times E = \frac{Q}{\varepsilon} = \frac{1}{\varepsilon}\frac{2\pi k r^3}{3}$$

$$\Rightarrow \quad E = \frac{kr^2}{3\varepsilon}$$

$$\therefore \quad \vec{E} = \frac{kr^2}{3\varepsilon}\hat{a}_r$$

Outside the cylinder ($r \ge a$)

Here, total charge contained in the cylinder is

$$Q = \int dQ = \int_0^a 2\pi k r^2 dr = \frac{2\pi k a^3}{3}$$

Applying Gauss' law to the Gussian surface of radius r

$$2\pi r \times 1 \times E = \frac{Q}{\varepsilon} = \frac{1}{\varepsilon}\frac{2\pi k a^3}{3}$$

$$\Rightarrow \quad E = \frac{ka^3}{3\varepsilon r}$$

$$\therefore \quad \vec{E} = \frac{ka^3}{3\varepsilon r}\hat{a}_r$$

Thus, the electric field is given as

$$\boxed{\begin{aligned}\vec{E} &= \frac{kr^2}{3\varepsilon}\hat{a}_r; \quad r \le a \\ &= \frac{ka^3}{3\varepsilon r}\hat{a}_r; \quad r \ge a\end{aligned}}$$

2.12 ELECTRIC POTENTIAL (V)

2.12.1 Electric Potential (V)

The potential of a point is the work done to bring a unit positive charge from infinity to that point.

The unit of potential is Joule per Coulomb (J/C) or Volt. The dimensions of electric potential in terms of M, L, T and I are $[ML^2T^{-3}I^{-1}]$.

2.12.2 Potential Difference

The potential difference between two points is the change in the potential energy per unit charge as the charge tends to zero. This is depicted in Fig. 2.43.

$$V_{AB} = \lim_{Q \to 0}\left(\frac{W}{Q}\right) = -\int_A^B \vec{E} \cdot d\vec{l}$$

Explanation We will consider, a point charge Q, be moved from a point A to another point B in an electric field $\vec{E}$.

By Coulomb's law, the force on Q is, $\vec{F} = Q\vec{E}$

$\therefore$ Work done for a displacement of $d\vec{l}$ is, $dW = -\vec{F} \cdot d\vec{l} = -Q\vec{E} \cdot d\vec{l}$

Hence, the total work done in moving the charge Q from A to B, i.e., the potential energy required is

$$W = -Q\int_A^B \vec{E} \cdot d\vec{l}$$

The potential energy per unit charge $\left(\frac{W}{Q}\right)$, known as the *potential difference* between the two points A and B, denoted by V_{AB} is given as

$$\therefore \quad V_{AB} = \frac{W}{Q} = -\int_A^B \vec{E} \cdot d\vec{l} \qquad \textbf{(2.33)}$$

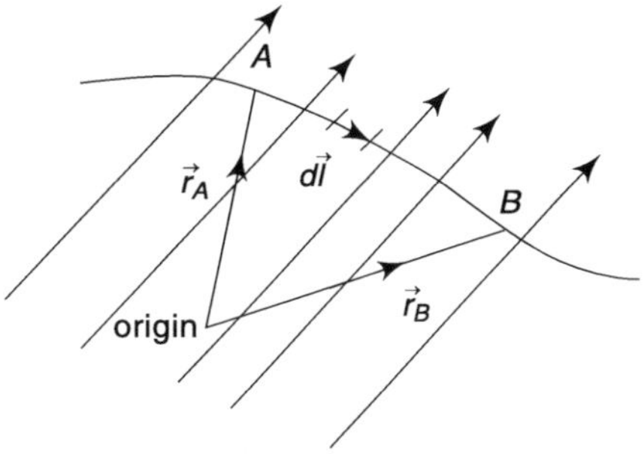

Fig. 2.43 *Potential difference due to a uniform electric field*

For example, if $\vec{E}$ is a field produced by a point charge Q at origin, i.e., $\vec{E} = \frac{Q}{4\pi\varepsilon r^2}\hat{a}_r$, then the potential difference between the points A and B is given as

$$V_{AB} = -\int_A^B \frac{Q}{4\pi\varepsilon r^2}\hat{a}_r \cdot dr\hat{a}_r = -\frac{Q}{4\pi\varepsilon}\int_A^B \frac{dr}{r^2} = \frac{Q}{4\pi\varepsilon}\left[\frac{1}{r_B} - \frac{1}{r_A}\right] = (V_B - V_A)$$

Thus, the potential difference between the points A and B may be considered to be the *potential* (or *absolute potential*) of B with respect to the *potential* (or *absolute potential*) of A.

In case of a point charge, the reference is taken to be at infinity with zero potential.

Hence, *potential* (or *absolute potential*) of a point is defined as the work done to bring a unit positive charge from infinity to that point.

$$\therefore \quad V = \frac{W}{Q} = -\int_\infty^R \frac{Q}{4\pi\varepsilon r^2}\hat{a}_r \cdot dr\hat{a}_r = -\frac{Q}{4\pi\varepsilon}\int_\infty^R \frac{dr}{r^2} = \frac{Q}{4\pi\varepsilon R}$$

$$\boxed{V = \frac{Q}{4\pi\varepsilon R}} \tag{2.34}$$

If the point charge Q is not located at the origin but at a point with position vector $\vec{r}$, then the potential of the point at a position vector $\vec{R}$ is

$$\boxed{V(\vec{R}) = \frac{Q}{4\pi\varepsilon\,|\vec{R} - \vec{r}\,|}} \tag{2.35}$$

Principle of Superposition of Potentials If there is a number of point charges $Q_1, Q_2, \ldots, Q_n$, located at position vectors $\vec{r}_1, \vec{r}_2, \ldots, \vec{r}_n$ respectively, then the potential at point $\vec{r}$ is

$$V(\vec{r}) = \frac{Q_1}{4\pi\varepsilon\,|\vec{r} - \vec{r}_1|} + \frac{Q_2}{4\pi\varepsilon\,|\vec{r} - \vec{r}_2|} + \cdots + \frac{Q_n}{4\pi\varepsilon\,|\vec{r} - \vec{r}_n|} = \frac{1}{4\pi\varepsilon}\sum_{i=1}^{n}\frac{Q_i}{|\vec{r} - \vec{r}_i|}$$

$$\boxed{V(\vec{r}) = \frac{1}{4\pi\varepsilon}\sum_{i=1}^{n}\frac{Q_i}{|\vec{r} - \vec{r}_i|}} \tag{2.36}$$

Potential due to Continuous Charge Distribution If the charge distribution is continuous, the potential at a point P can be found by summing over the contributions from individual differential elements of charge dq.

Consider the charge distribution shown in Fig. 2.44. Taking infinity as our reference point with zero potential, the electric potential at P due to dq is

$$dV = \frac{1}{4\pi\varepsilon}\frac{dq}{r}$$

Summing over contributions from all differential elements, we have

$$V = \frac{1}{4\pi\varepsilon}\int\frac{dq}{r}$$

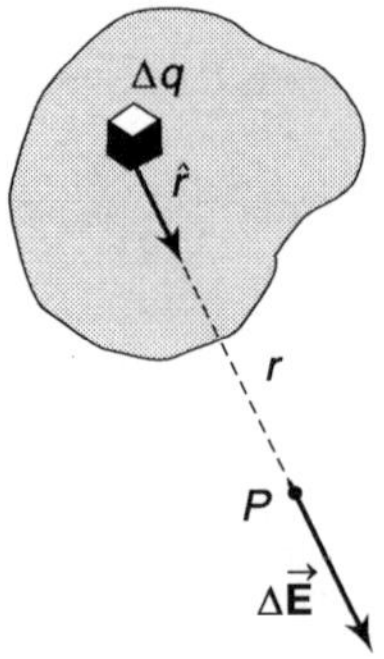

Fig. 2.44 *Continuous charge distribution*

For three different types of charge distribution, the potential at a point $\vec{r}$ is given below.

$$V(\vec{r}) = \frac{1}{4\pi\varepsilon}\int_l \frac{\lambda(\vec{r}\,')dl'}{|\vec{r} - \vec{r}\,'|}; \quad \text{for line charge distribution with density } \lambda\,(\text{C/m}) \tag{2.37}$$

$$= \frac{1}{4\pi\varepsilon}\int_S \frac{\sigma(\vec{r}\,')dS'}{|\vec{r} - \vec{r}\,'|}; \quad \text{for surface charge distribution with density } \sigma\,(\text{C/m}^2) \tag{2.38}$$

$$= \frac{1}{4\pi\varepsilon}\int_v \frac{\rho(\vec{r}\,')dv'}{|\vec{r} - \vec{r}\,'|}; \quad \text{for volume charge distribution with density } \rho\,(\text{C/m}^3) \tag{2.39}$$

Here, the primed coordinates refer to the source point and the unprimed coordinates refer to the field point (the point where the potential is to be calculated).

NOTE

$V = -\int \vec{E} \cdot d\vec{l}$; as $\vec{E}$ is in the radial direction, any contribution from a displacement in θ or ϕ direction is cancelled out by the dot product. Hence, $\vec{E} \cdot d\vec{l} = Edl \cos\theta = Edr$. Thus, the potential is independent of the path. For a closed path, $\oint_l \vec{E} \cdot d\vec{l} = 0$. Applying Stoke's theorem, $\oint_l \vec{E} \cdot d\vec{l} = \int_S (\nabla \times \vec{E}) \cdot d\vec{S} = 0$.

$\therefore$ $$\nabla \times \vec{E} = 0$$

*Thus, **the electrostatic field is conservative or irrotational**.*

***Example 2.35** Consider a uniformly charged ring of radius R and charge density λ. What is the electric potential at a distance z from the central axis?

Solution We shall consider a small differential element $d\vec{l} = Rd\phi'$ on the ring as shown in Fig. 2.45.

The charge carried by this element is

$$dq = \lambda dl = \lambda R d\phi'$$

Potential due to this element is

$$dV = \frac{1}{4\pi\varepsilon}\frac{dq}{r} = \frac{1}{4\pi\varepsilon}\frac{\lambda R d\phi'}{\sqrt{R^2 + z^2}}$$

Thus, the potential at point P due to the entire ring is

$$V = \int dV = \frac{1}{4\pi\varepsilon}\frac{\lambda R}{\sqrt{R^2+z^2}}\oint d\phi' = \frac{1}{4\pi\varepsilon}\frac{2\pi\lambda R}{\sqrt{R^2+z^2}}$$

$$= \frac{\lambda R}{2\varepsilon\sqrt{R^2+z^2}} = \frac{1}{4\pi\varepsilon}\frac{Q}{\sqrt{R^2+z^2}}$$

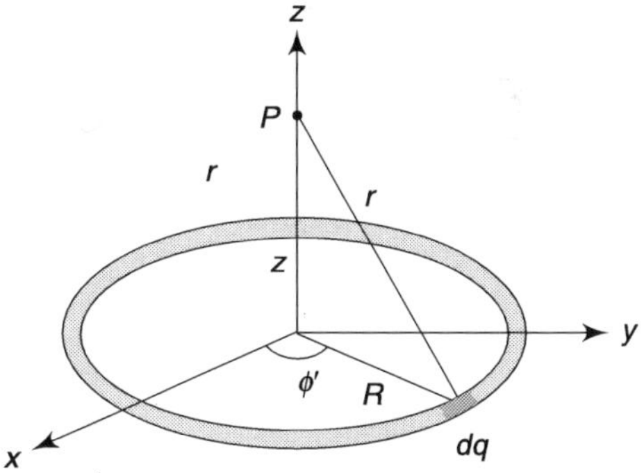

Fig. 2.45 *A non-conducting ring of radius R with uniform charge density λ*

$$\boxed{V = \frac{\lambda R}{2\varepsilon\sqrt{R^2+z^2}}}$$

where $Q = 2\pi\lambda R$ is the total charge on the ring.

Since the ring is uniformly charged, the electric field at point P will be along the axis, i.e., in the z direction. Hence, the field is given as

$$\vec{E} = E_z\hat{k} = -\frac{\partial V}{\partial z}\hat{k} = -\frac{\partial}{\partial z}\left(\frac{\lambda R}{2\varepsilon\sqrt{R^2+z^2}}\right)\hat{k} = \frac{\lambda R z}{2\varepsilon(R^2+z^2)^{3/2}}\hat{k} = \frac{1}{4\pi\varepsilon}\frac{Qz}{(R^2+z^2)^{3/2}}$$

$$\boxed{\vec{E} = \frac{\lambda R z}{2\varepsilon(R^2+z^2)^{3/2}}\hat{k}}$$

NOTE

In the limiting case with $z >> R$, the potential becomes, $V = \frac{Q}{4\pi\varepsilon z}$ and the field becomes $\vec{E} = \frac{Q}{4\pi\varepsilon z^2}\hat{k}$; which are the potential and field due to a point charge. Thus, if the distance of the field point is very large compared to the radius of the ring, the ring appears as a point charge.

***Example 2.36** Determine the electric potential at a distance 'r' form the centre of the sphere of Example 2.30.

Solution The results for the field intensity for Example 2.30 are as given below.

$$\vec{E} = \frac{\rho r}{3\varepsilon}\hat{a}_r; \qquad r < a$$

$$= \frac{\rho a^3}{3\varepsilon r^2}\hat{a}_r; \qquad r > a$$

The potential for the two cases is obtained as follows.

Outside the sphere ($r \geq a$)

Here,
$$E = \frac{\rho a^3}{3\varepsilon r^2}$$

Hence, the potential is, $V = -\int_{\infty}^{r} E dr = -\frac{\rho a^3}{3\varepsilon}\int_{\infty}^{r}\frac{dr}{r^2} = \frac{\rho a^3}{3\varepsilon r}$

$\therefore$
$$V = \frac{\rho a^3}{3\varepsilon r}$$

Inside the sphere ($r \leq a$)

Here,
$$E = \frac{\rho r}{3\varepsilon}$$

Hence, the potential is

$$V = V(r = a) - \int_a^r E dr = \frac{\rho a^2}{3\varepsilon} - \frac{\rho}{3\varepsilon}\int_a^r r dr = \frac{\rho a^2}{3\varepsilon} - \frac{\rho}{3\varepsilon}(r^2 - a^2) = \frac{\rho}{6\varepsilon}(3a^2 - r^2)$$

$\therefore$
$$V = \frac{\rho}{6\varepsilon}(3a^2 - r^2)$$

The results are summarised below.

$$V = \frac{\rho}{6\varepsilon}(3a^2 - r^2)_r; \qquad r < a$$

$$= \frac{\rho a^3}{3\varepsilon r}; \qquad r > a$$

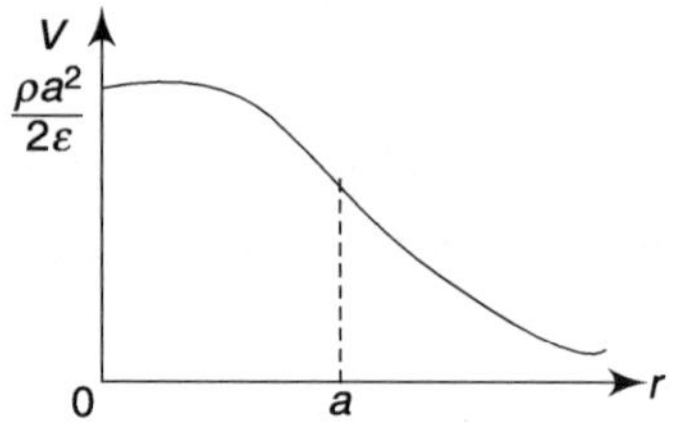

Fig. 2.46 *Variation of potential with the distance*

The variation of the potential with the distance is shown in Fig. 2.46.

***Example 2.37** A spherical conductor of radius 'R' contains a uniform surface charge density σ. Determine the field and potential due to the charge distribution.

Solution We will consider two cases:

Outside the sphere ($r \geq R$)

Applying Gauss' law to the Gaussian surface of radius r

$$4\pi r^2 E = \frac{Q}{\varepsilon} = \frac{4\pi R^2\sigma}{\varepsilon} \qquad \Rightarrow \qquad E = \frac{\sigma R^2}{\varepsilon r^2}$$

In vector form
$$\boxed{\vec{E} = \frac{\sigma R^2}{\varepsilon r^2}\hat{a}_r}$$

Hence, the potential is, $V = -\int_{\infty}^{r} E dr = -\frac{\sigma R^2}{\varepsilon}\int_{\infty}^{r}\frac{dr}{r^2} = \frac{\sigma R^2}{\varepsilon r}$

$\therefore$
$$\boxed{V = \frac{\sigma R^2}{\varepsilon r}}$$

Inside the sphere ($r \le R$)

Since there is no charge inside the sphere, $E = 0$. Therefore, the field intensity everywhere inside the spherical conductor is zero.

At $r = R$,
$$V = \frac{\sigma R}{\varepsilon}$$

Since E is zero inside the sphere, it requires no work to move a test charge inside and therefore, the potential is constant, being equal to the value at the surface of the sphere.

The results are summarised below.

$$\boxed{\begin{aligned}\vec{E} &= 0; & r < R\\ &= \frac{\sigma R^2}{\varepsilon r^2}\hat{a}_r; & r \ge R\end{aligned}} \quad \text{and} \quad \boxed{\begin{aligned}V &= \frac{\sigma R}{\varepsilon}; & r < R\\ &= \frac{\sigma R^2}{\varepsilon r}; & r \ge R\end{aligned}}$$

Example 2.38 Derive expression for $\vec{E}$ and V for a spherical volume of radius 'a' having a volume charge density:

(a) $\rho = \rho_0 \frac{a}{r}$; **(b)** $\rho = kr$, k is a constant;

(c) $\rho = \frac{k}{r^2}$, k is a constant; **(d)** $\rho = \rho_0 \frac{r}{a}$;

(e) $\rho = \rho_0 (r/a)^{3/2}$

where 'r' is the radial distance from the centre of the sphere.

Solution

(a) $\rho = \rho_0 \frac{a}{r}$

We will consider two cases:

Outside the sphere ($r \ge a$)

Applying Gauss' law to the Gaussian surface of radius r

$$4\pi r^2 E = \frac{Q}{\varepsilon} = \frac{1}{\varepsilon}\iiint_{\text{vol}} \rho \times 4\pi r^2 dr = \frac{1}{\varepsilon}\int_0^a \rho_0 \frac{a}{r} 4\pi r^2 dr = \frac{\rho_0 a^3}{2\varepsilon}$$

$\Rightarrow$
$$E = \frac{\rho_0 a^3}{2\varepsilon r^2}$$

Hence, the potential is, $V = -\int_{\infty}^{r} E dr = -\frac{\rho_0 a^3}{2\varepsilon}\int_{\infty}^{r}\frac{dr}{r^2} = \frac{\rho_0 a^3}{2\varepsilon r}$

Inside the sphere ($r \le a$)

Applying Gauss' law to the Gaussian surface of radius r

$$4\pi r^2 E = \frac{Q}{\varepsilon} = \frac{1}{\varepsilon}\iiint_{\text{vol}} \rho \times 4\pi r^2 dr = \frac{1}{\varepsilon}\int_0^r \rho_0 \frac{a}{r} 4\pi r^2 dr = \frac{\rho_0 a}{2\varepsilon} r^2$$

$\Rightarrow$
$$E = \frac{\rho_0 a}{2\varepsilon}$$

Hence, the potential is obtained as

$$V = \frac{\rho_0 a^3}{2\varepsilon a} - \int_a^r \frac{\rho_0 a}{2\varepsilon} dr = \frac{\rho_0 a^2}{2\varepsilon} - \frac{\rho_0 a}{2\varepsilon}(r-a) = \frac{\rho_0 a}{2\varepsilon}(2a - r)$$

The results are summarised below.

$$\boxed{\begin{aligned} \vec{E} &= \frac{\rho_0 a}{2\varepsilon}\hat{a}_r; & r < a \\ &= \frac{\rho_0 a^3}{2\varepsilon r^2}\hat{a}_r; & r \ge a \end{aligned}} \quad \text{and} \quad \boxed{\begin{aligned} V &= \frac{\rho_0 a}{2\varepsilon}(2a - r); & r < a \\ &= \frac{\rho_0 a^3}{2\varepsilon r}; & r \ge a \end{aligned}}$$

(b) $\rho = kr$, k is a constant

We will consider two cases:

Outside the sphere ($r \ge a$)

Applying Gauss' law to the Gaussian surface of radius r

$$4\pi r^2 E = \frac{Q}{\varepsilon} = \frac{1}{\varepsilon}\iiint_{\text{vol}} \rho \times 4\pi r^2 dr = \frac{1}{\varepsilon}\int_0^a kr 4\pi r^2 dr = 4\pi \frac{ka^4}{4\varepsilon}$$

$\Rightarrow$
$$E = \frac{ka^4}{4\varepsilon r^2}$$

Hence, the potential is, $V = -\int_\infty^r E dr = -\frac{ka^4}{4\varepsilon}\int_\infty^r \frac{dr}{r^2} = \frac{ka^4}{4\varepsilon r}$

Inside the sphere ($r \le a$)

Applying Gauss' law to the Gaussian surface of radius r

$$4\pi r^2 E = \frac{Q}{\varepsilon} = \frac{1}{\varepsilon}\iiint_{\text{vol}} \rho \times 4\pi r^2 dr = \frac{1}{\varepsilon}\int_0^r kr 4\pi r^2 dr = \pi k \frac{r^4}{\varepsilon}$$

$\Rightarrow$
$$E = \frac{kr^2}{4\varepsilon}$$

Hence, the potential is obtained as

$$V = \frac{ka^4}{4\varepsilon a} - \int_a^r \frac{kr^2}{4\varepsilon} dr = \frac{ka^3}{4\varepsilon} - \frac{k}{12\varepsilon}(r^3 - a^3) = \frac{k}{12\varepsilon}(4a^3 - r^3)$$

The results are summarised below.

$$\boxed{\begin{aligned} \vec{E} &= \frac{kr^2}{4\varepsilon}\hat{a}_r; & r < a \\ &= \frac{ka^4}{4\varepsilon r^2}\hat{a}_r; & r \ge a \end{aligned}} \quad \text{and} \quad \boxed{\begin{aligned} V &= \frac{k}{12\varepsilon}(4a^3 - r^3); & r < a \\ &= \frac{ka^4}{4\varepsilon r}; & r \ge a \end{aligned}}$$

(c) $\rho = \dfrac{k}{r^2}$, k is a constant

We will consider two cases:

Outside the sphere ($r \geq a$)

Applying Gauss' law to the Gaussian surface of radius r

$$4\pi r^2 E = \frac{Q}{\varepsilon} = \frac{1}{\varepsilon}\iiint_{\text{vol}} \rho \times 4\pi r^2 dr = \frac{1}{\varepsilon}\int_0^a \frac{k}{r^2} 4\pi r^2 dr = \frac{4\pi ka}{\varepsilon}$$

$$\Rightarrow \qquad E = \frac{ka}{\varepsilon r^2}$$

Hence, the potential is, $V = -\int_\infty^r E dr = -\dfrac{ka}{\varepsilon}\int_\infty^r \dfrac{dr}{r^2} = \dfrac{ka}{\varepsilon r}$

Inside the sphere ($r \leq a$)

Applying Gauss' law to the Gaussian surface of radius r

$$4\pi r^2 E = \frac{Q}{\varepsilon} = \frac{1}{\varepsilon}\iiint_{\text{vol}} \rho \times 4\pi r^2 dr = \frac{1}{\varepsilon}\int_0^r \frac{k}{r^2} 4\pi r^2 dr = \frac{4\pi kr}{\varepsilon}$$

$$\Rightarrow \qquad E = \frac{k}{\varepsilon r}$$

Hence, the potential is obtained as

$$V = \frac{ka}{\varepsilon a} - \int_a^r \frac{k}{\varepsilon r} dr = \frac{k}{\varepsilon} - \frac{k}{\varepsilon}\ln\left(\frac{r}{a}\right) = \frac{k}{\varepsilon}\left[1 + \ln\left(\frac{a}{r}\right)\right]$$

The results are summarised below.

$$\boxed{\begin{aligned} \vec{E} &= \frac{k}{\varepsilon r}\hat{a}_r; & r < a \\ &= \frac{ka}{\varepsilon r^2}\hat{a}_r; & r \geq a \end{aligned}} \quad \text{and} \quad \boxed{\begin{aligned} V &= \frac{k}{\varepsilon}\left[1 + \ln\left(\frac{a}{r}\right)\right]; & r < a \\ &= \frac{ka}{\varepsilon r}; & r \geq a \end{aligned}}$$

(d) $\rho = \rho_0 \dfrac{r}{a}$

We will consider two cases:

Outside the sphere ($r \geq a$)

Total charge enclosed by the sphere is

$$Q = \int_{r=0}^{a} dQ = \int_{r=0}^{a} \rho 4\pi r^2 dr = \frac{4\pi\rho_0}{a}\int_{r=0}^{a} r^3 dr = \pi\rho_0 a^3$$

Applying Gauss' law to the Gaussian surface of radius r

$$4\pi r^2 E = \frac{Q}{\varepsilon} = \frac{\pi\rho_0 a^3}{\varepsilon}$$

$$\Rightarrow \qquad E = \frac{\rho_0 a^3}{4\varepsilon r^2}$$

Hence, the potential is, $V = -\int_{\infty}^{r} E dr = -\dfrac{\rho_0 a^3}{4\varepsilon}\int_{\infty}^{r}\dfrac{dr}{r^2} = \dfrac{\rho_0 a^3}{4\varepsilon r}$

Inside the sphere ($r \le a$):

Charge enclosed by a spherical shell of radius r and thickness dr is

$$dQ_r = \rho 4\pi r^2 dr = \rho_0 \frac{r}{a} 4\pi r^2 dr = \frac{4\pi \rho_0}{a} r^3 dr$$

∴ Total charge enclosed

$$Q_r = \int_{r=0}^{r} dQ_r = \frac{4\pi \rho_0}{a}\int_{r=0}^{r} r^3 dr = \frac{\pi \rho_0 r^4}{a}$$

Applying Gauss' law to the Gaussian surface of radius r

$$4\pi r^2 E = \frac{Q}{\varepsilon} = \frac{1}{\varepsilon}\frac{\pi \rho_0 r^4}{a}$$

$$\Rightarrow \quad E = \frac{\rho_0 r^2}{4\varepsilon a}$$

Hence, the potential is obtained as

$$V = \frac{\rho_0 a^3}{4\varepsilon a} - \int_{a}^{r}\frac{\rho_0 r^2}{4\varepsilon a} dr = \frac{\rho_0 a^2}{4\varepsilon} - \frac{\rho_0}{4\varepsilon a}\left(\frac{r^3 - a^3}{3}\right) = \frac{\rho_0}{12\varepsilon a}(3a^3 - r^3 + a^3) = \frac{\rho_0}{12\varepsilon a}(4a^3 - r^3)$$

The results are summarised below.

$$\boxed{\begin{aligned}\vec{E} &= \frac{\rho_0 r^2}{4\varepsilon a}\hat{a}_r; \quad r < a \\ &= \frac{\rho_0 a^3}{4\varepsilon r^2}\hat{a}_r; \quad r \ge a\end{aligned}} \quad \text{and} \quad \boxed{\begin{aligned}V &= \frac{\rho_0}{12\varepsilon a}(4a^3 - r^3); \quad r < a \\ &= \frac{\rho_0 a^3}{4\varepsilon r}; \quad r \ge a\end{aligned}}$$

NOTE

Part (b) and (d) are same with $k = \dfrac{\rho_0}{a}$.

(e) $\rho = \rho_0 (r/a)^{3/2}$

We will consider two cases:

Outside the sphere ($r \ge a$)

Applying Gauss' law to the Gaussian surface of radius r

$$4\pi r^2 E = \frac{Q}{\varepsilon} = \frac{1}{\varepsilon}\iiint_{\text{vol}} \rho \times 4\pi r^2 dr = \frac{1}{\varepsilon}\int_0^a \rho_0 (r/a)^{3/2} 4\pi r^2 dr = \frac{4\pi\rho_0}{\varepsilon a^{3/2}}\frac{a^{9/2}}{9/2} = \frac{2}{9}\frac{4\pi\rho_0}{\varepsilon} a^3$$

$$\Rightarrow \quad E = \frac{2\rho_0}{9\varepsilon r^2} a^3$$

Hence, the potential is, $V = -\int_{\infty}^{r} E dr = -\dfrac{2\rho_0 a^3}{9\varepsilon}\int_{\infty}^{r}\dfrac{dr}{r^2} = \dfrac{2\rho_0 a^3}{9\varepsilon r}$

Inside the sphere ($r \le a$):

Applying Gauss' law to the Gaussian surface of radius r

$$4\pi r^2 E = \frac{Q}{\varepsilon} = \frac{1}{\varepsilon}\iiint_{\text{vol}} \rho \times 4\pi r^2 dr = \frac{1}{\varepsilon}\int_0^r \rho_0 (r/a)^{3/2}\, 4\pi r^2 dr = \frac{2}{9}\frac{4\pi\rho_0 r^{9/2}}{\varepsilon a^{3/2}}$$

$$\Rightarrow \qquad E = \frac{2\rho_0 r^{5/2}}{9\varepsilon a^{3/2}}$$

Hence, the potential is obtained as

$$V = \frac{2\rho_0 a^3}{9\varepsilon a} - \int_a^r \frac{2\rho_0 r^{5/2}}{9\varepsilon a^{3/2}} dr = \frac{2\rho_0 a^2}{9\varepsilon} - \frac{2\rho_0}{9\varepsilon a^{3/2}}\frac{2}{7}(r^{7/2} - a^{7/2}) = \frac{4\rho_0}{63\varepsilon a^{3/2}}\left(\frac{7}{2}a^{7/2} - r^{7/2} + a^{7/2}\right)$$

$$= \frac{4\rho_0}{63\varepsilon a^{3/2}}\left(\frac{9}{2}a^{7/2} - r^{7/2}\right)$$

The results are summarised below.

$$\vec{E} = \frac{2\rho_0 r^{5/2}}{9\varepsilon a^{3/2}}\hat{a}_r; \quad r < a$$
$$= \frac{2\rho_0}{9\varepsilon r^2}a^3\hat{a}_r; \quad r \ge a$$

and

$$V = \frac{4\rho_0}{63\varepsilon a^{3/2}}\left(\frac{9}{2}a^{7/2} - r^{7/2}\right); \quad r < a$$
$$= \frac{2\rho_0 a^3}{9\varepsilon r}; \quad r \ge a$$

2.13 PRINCIPLE OF SUPERPOSITION OF ELECTROSTATIC FIELDS

The principle of superposition states that the total resultant electric field at a point is the vector sum of the component fields.

We will consider a general system with the following charges:

1. a number of point charges $q_1, q_2, \ldots, q_n$;
2. a line charge distribution with density λ (C/m);
3. a surface charge distribution with density σ (C/m^2); and
4. a volume charge distribution with density σ (C/m^3).

The system is shown in Fig. 2.47.

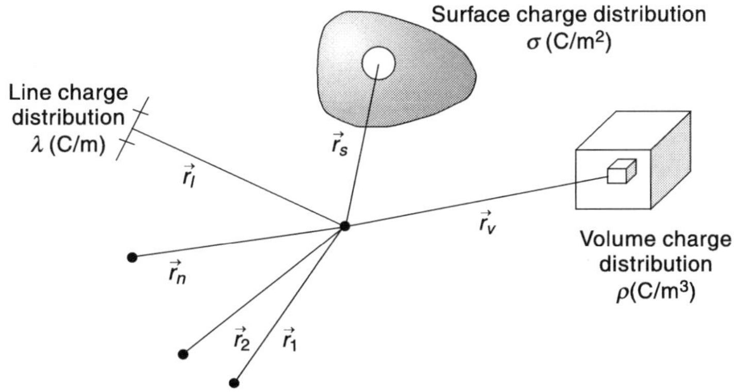

Fig. 2.47 *General system with different charge distributions*

Then, the resultant electric field at a point P is given by

$$\vec{E} = \vec{E}_{\text{due to point charges}} + \vec{E}_{\text{due to line charge distribution}} + \vec{E}_{\text{due to surface charge distribution}} + \vec{E}_{\text{due to volume charge distribution}}$$

$$= \frac{1}{4\pi\varepsilon}\left[\frac{q_1}{|\vec{r}_1|^2}\hat{a}_{r1} + \frac{q_2}{|\vec{r}_2|^2}\hat{a}_{r2} + \cdots + \frac{q_n}{|\vec{r}_n|^2}\hat{a}_{rn}\right] + \frac{1}{4\pi\varepsilon}\int_l \frac{\lambda dl}{|\vec{r}_l|^2}\hat{a}_{rl} + \frac{1}{4\pi\varepsilon}\int_S \frac{\sigma dS}{|\vec{r}_S|^2}\hat{a}_{rS} + \frac{1}{4\pi\varepsilon}\int_v \frac{\rho dv}{|\vec{r}_v|^2}\hat{a}_{rv}$$

$$\boxed{\vec{E} = \frac{1}{4\pi\varepsilon}\left[\sum_{i=1}^{n}\frac{q_i}{|\vec{r}_i|^2}\hat{a}_{ri} + \int_l \frac{\lambda dl}{|\vec{r}_l|^2}\hat{a}_{rl} + \int_S \frac{\sigma dS}{|\vec{r}_S|^2}\hat{a}_{rS} + \int_v \frac{\rho dv}{|\vec{r}_v|^2}\hat{a}_{rv}\right]} \quad \textbf{(2.40)}$$

The principle of superposition is also applicable for potential calculation. For the same system, the potential at the point P is given by

$$V = V_{\text{due to point charges}} + V_{\text{due to line charge distribution}} + V_{\text{due to surface charge distribution}} + V_{\text{due to volume charge distribution}}$$

$$= \frac{1}{4\pi\varepsilon}\left[\frac{q_1}{r_1} + \frac{q_2}{r_2} + \cdots + \frac{q_n}{r_n}\right] + \frac{1}{4\pi\varepsilon}\int_l \frac{\lambda dl}{r_l} + \frac{1}{4\pi\varepsilon}\int_S \frac{\sigma dS}{r_S} + \frac{1}{4\pi\varepsilon}\int_v \frac{\rho dv}{r_v}$$

$$\boxed{V = \frac{1}{4\pi\varepsilon}\left[\sum_{i=1}^{n}\frac{q_i}{r_i} + \int_l \frac{\lambda dl}{r_l} + \int_S \frac{\sigma dS}{r_S} + \int_v \frac{\rho dv}{r_v}\right]} \quad \textbf{(2.41)}$$

***Example 2.39** A point charge of -5π mCoulomb is located at (4, 0, 0) and a line charge of 3π mCoulomb/m is located along the y-axis. Find $\vec{D}$ at (4, 0, 3).

Solution Here, $\vec{D} = \vec{D}_Q + \vec{D}_L$

$$\vec{D}_Q = \frac{Q}{4\pi r^2}\hat{a}_r = \frac{Q}{4\pi}\frac{(\vec{r} - \vec{r}\,')}{|\vec{r} - \vec{r}\,'|^3} = \frac{-5\pi}{4\pi}\frac{(4, 0, 3) - (4, 0, 0)}{|(4, 0, 3) - (4, 0, 0)|^3} = -\frac{5}{4} \times \frac{3\hat{a}_z}{3^3} = -0.138\hat{a}_z \quad \text{mC/m}^2$$

Also,

$$\vec{D}_L = \frac{\lambda}{2\pi r}\hat{a}_r = \frac{\lambda}{2\pi}\frac{(\vec{r} - \vec{r}\,')}{|\vec{r} - \vec{r}\,'|^2} = \frac{3\pi}{2\pi}\frac{(4, 0, 3) - (0, 0, 0)}{|(4, 0, 3) - (4, 0, 0)|^2} = \frac{3}{2} \times \frac{4\hat{a}_x + 3\hat{a}_z}{25} = 0.24\hat{a}_x + 0.18\hat{a}_z \quad \text{mC/m}^2$$

Thus, total flux density is

$$\vec{D} = \vec{D}_Q + \vec{D}_L = -0.138\hat{a}_z + 0.24\hat{a}_x + 0.18\hat{a}_z = (240\hat{a}_x + 42\hat{a}_z) \quad \mu\text{C/m}^2$$

Example 2.40 **(a)** A line of length 'l' carries a charge λ per unit length. Show that the potential in the median plane is

$$V = \frac{\lambda}{4\pi\varepsilon}\ln\left(\frac{1 + \sin\alpha}{1 - \sin\alpha}\right) = \frac{\lambda}{4\pi\varepsilon}\ln\left(\frac{\sqrt{l^2 + 4r^2} + l}{\sqrt{l^2 + 4r^2} - l}\right); \text{ where, } \tan\alpha = \frac{l}{2r}$$

(b) A thin square loop carries a uniform charge of λ. Show that the potential at the centre of the loop is $V = \frac{2\lambda}{\pi\varepsilon}\ln(1 + \sqrt{2})$.

(c) A square that is 1 m on a side in air has a point charge $Q_1 = +1$ pC at the upper right corner, a point charge $Q_3 = -10$ pC at the lower right corner and a line distribution of charge density $\lambda = +10$ pC/m along the left edge. Find the potential at the point at the centre of the sphere.

Solution **(a)** We will consider an elemental length dz at a distance z as shown in Fig. 2.48.

Charge on the elemental length is, $dQ = \lambda dz$

$\therefore$ Potential at point P due to this charge is

$$dV = \frac{dQ}{4\pi\varepsilon R} = \frac{\lambda dz}{4\pi\varepsilon R}$$

$\therefore$ Total potential at point P due to the line is

$$V = \int dV = \int_{-l/2}^{l/2} \frac{\lambda dz}{4\pi\varepsilon R} = \frac{\lambda}{4\pi\varepsilon}\int_{-l/2}^{l/2} \frac{dz}{\sqrt{r^2+z^2}}$$

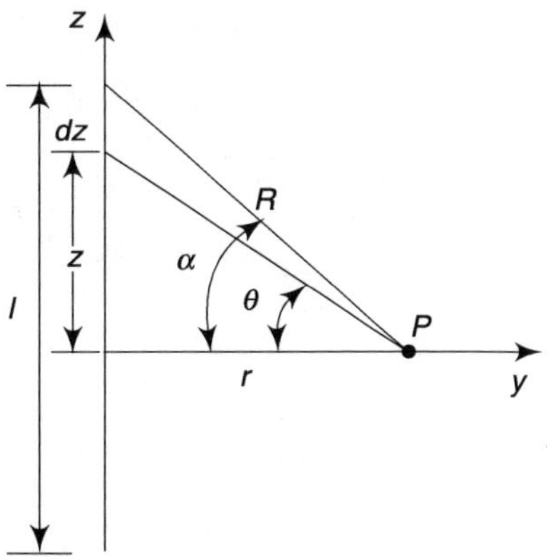

Fig. 2.48 *Potential due to a charge carrying line*

Let, $z = r\tan\theta \Rightarrow dz = r\sec^2\theta d\theta$

$$\begin{bmatrix} z & -l/2 & l/2 \\ \theta & -\alpha & \alpha \end{bmatrix} \quad \text{where} \quad \alpha = \tan^{-1}\left(\frac{l}{2r}\right)$$

$$\therefore \quad V = \frac{\lambda}{4\pi\varepsilon}\int_{-\alpha}^{\alpha}\frac{r\sec^2\theta d\theta}{r\sec^2\theta} = \frac{\lambda}{4\pi\varepsilon}\int_{-\alpha}^{\alpha}\sec\theta d\theta$$

$$= \frac{\lambda}{4\pi\varepsilon}\ln(\sec\theta + \tan\theta)\Big|_{-\alpha}^{\alpha}$$

$$= \frac{\lambda}{4\pi\varepsilon}\ln\left(\frac{1+\sin\alpha}{1-\sin\alpha}\right)$$

Now, since $\alpha = \tan^{-1}\left(\frac{l}{2r}\right)$, $\therefore \sin\alpha = \frac{l/2}{\sqrt{(l/2)^2+r^2}} = \frac{l}{\sqrt{l^2+4r^2}}$

$$V = \frac{\lambda}{4\pi\varepsilon}\ln\left(\frac{\sqrt{l^2+4r^2}+l}{\sqrt{l^2+4r^2}-l}\right)$$

$$\boxed{V = \frac{\lambda}{4\pi\varepsilon}\ln\left(\frac{1+\sin\alpha}{1-\sin\alpha}\right) = \frac{\lambda}{4\pi\varepsilon}\ln\left(\frac{\sqrt{l^2+4r^2}+l}{\sqrt{l^2+4r^2}-l}\right)}$$

(b) Here, total potential at the centre P is given as

V = Potential due to line AB + Potential due to line BC
+ Potential due to line CD + Potential due to line DA
$= V_{AB} + V_{BC} + V_{CD} + V_{DA}$ (Fig. 2.49)

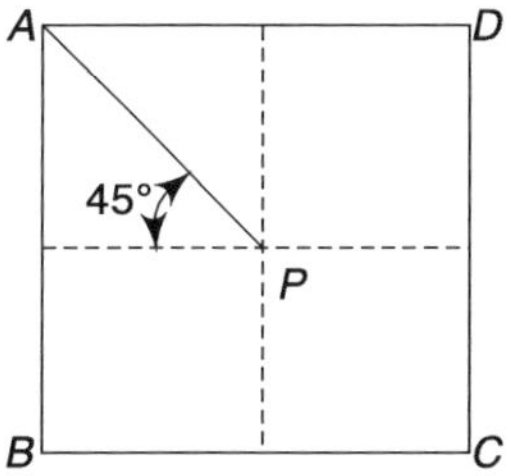

Fig. 2.49 *Potential due to square loop with line charge distribution*

Now from part (a), potential at point P due to line AB is

$$V_{AB} = \frac{\lambda}{4\pi\varepsilon}\ln\left(\frac{1+\sin 45°}{1-\sin 45°}\right)$$

$$= \frac{\lambda}{4\pi\varepsilon}\ln\left(\frac{1+\frac{1}{\sqrt{2}}}{1-\frac{1}{\sqrt{2}}}\right)$$

$$= \frac{\lambda}{4\pi\varepsilon}\ln\left(\frac{\sqrt{2}+1}{\sqrt{2}-1}\right) = V_{BC} = V_{CD} = V_{DA}$$

$$\therefore \quad V = 4\times\frac{\lambda}{4\pi\varepsilon}\ln\left(\frac{\sqrt{2}+1}{\sqrt{2}-1}\right) = \frac{\lambda}{\pi\varepsilon}\ln\left(\frac{\sqrt{2}+1}{\sqrt{2}-1}\right) = \frac{\lambda}{\pi\varepsilon}\ln\left[\frac{(\sqrt{2}+1)^2}{(\sqrt{2}-1)(\sqrt{2}+1)}\right] = \frac{2\lambda}{\pi\varepsilon}\ln(1+\sqrt{2})$$

$$\therefore \quad \boxed{V = \frac{2\lambda}{\pi\varepsilon}\ln(1+\sqrt{2})}$$

(c) The potential at the centre is given as

$$V = V_{Q_1} + V_{Q_2} + V_{\lambda}$$

$$= \frac{1\times 10^{-12}}{4\pi\varepsilon_0\sqrt{(0.5)^2+(0.5)^2}} + \frac{-10\times 10^{-12}}{4\pi\varepsilon_0\sqrt{(0.5)^2+(0.5)^2}} + \frac{\lambda}{4\pi\varepsilon_0}\ln\left(\frac{\sqrt{2}+1}{\sqrt{2}-1}\right)$$

$$= 0.012 - 0.12 + 0.159$$

$$= 0.044 \text{ Volt} = 44 \text{ mV} \quad \text{(Fig. 2.50)}$$

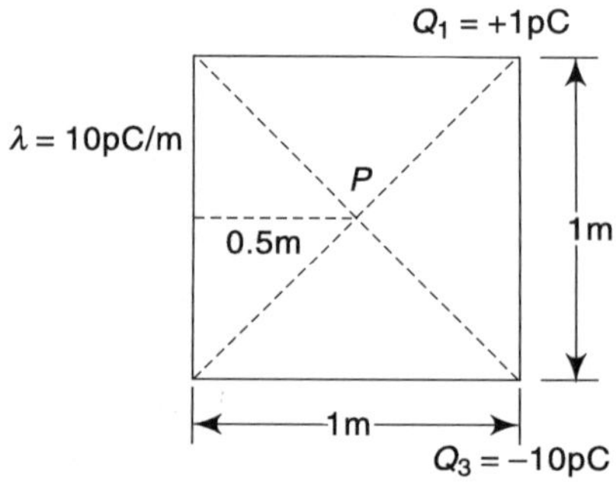

Fig. 2.50

2.14 POTENTIAL GRADIENT AND RELATION BETWEEN $\vec{E}$ AND V

2.14.1 Potential Gradient

We know that the line integral of electric field intensity $\vec{E}$ between any two points gives the potential difference between the points [see Eq. (2.31)]. For an elementary length ΔL, we can write the potential difference as

$$\Delta V = -\vec{E}\cdot\Delta L$$

Hence, an inverse relation must exist between the change of potential ΔV along the elementary length ΔL with the electric field $\vec{E}$ as $\Delta L \to 0$.

The rate of change of potential with respect to the distance is called the *potential gradient*.

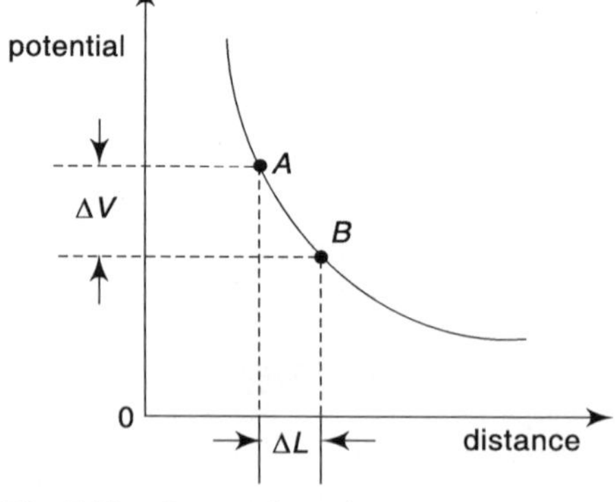

Fig. 2.51 *Potential gradient*

From Fig. 2.51, we can write that

$$\boxed{\text{Potential Gradient} = \frac{dV}{dL} = \lim_{\Delta L \to 0}\left(\frac{\Delta V}{\Delta L}\right)}$$

Relation between Potential V and Electric Field Intensity $\vec{E}$ We will consider two points in a surface separated by an infinitesimal distance $d\vec{l}$. The work done by an external electric field $\vec{E}$ in moving a unit positive charge from one point to the other is given as [see Eq. (2.31)]

$$dW = dV = -\vec{E} \cdot d\vec{l}$$

Since V is a function of the position coordinates, in Cartesian coordinates (x, y, z), the above can be written as

$$dV = -\vec{E} \cdot d\vec{l}$$

or
$$\frac{\partial V}{\partial x}dx + \frac{\partial V}{\partial y}dy + \frac{\partial V}{\partial z}dz = -\vec{E} \cdot d\vec{l}$$

or
$$\left(\frac{\partial V}{\partial x}\hat{i} + \frac{\partial V}{\partial y}\hat{j} + \frac{\partial V}{\partial z}\hat{k}\right) \cdot (dx\hat{i} + dy\hat{j} + dz\hat{k}) = -\vec{E} \cdot d\vec{l}$$

or
$$(\nabla V) \cdot d\vec{l} = -\vec{E} \cdot d\vec{l}$$

or
$$\vec{E} = -\nabla V$$

Thus, we have the relation that the *potential is the gradient of the electric field intensity.*

$$\boxed{\vec{E} = -\nabla V} \qquad \textbf{(2.42)}$$

From Eq. (2.42), we can also have two conclusions:

1. The magnitude of the electric field intensity is given by the maximum value of the rate of change of the potential, and
2. The maximum value is obtained when the direction of the displacement is opposite to the field direction.

Mathematically, we can think of $\vec{E}$ as the negative of the *gradient* of the electric potential V. Physically, the negative sign implies that if V increases as a positive charge and moves along some direction, say x, with $\frac{\partial V}{\partial x} > 0$, then there is a non-vanishing component of $\vec{E}$ in the opposite direction (i.e., $-E_x \neq 0$).

Example 2.41 Given the potential $V = \frac{10}{r^2}\sin\theta\cos\phi$, find the electric field intensity and flux density at $\left(4, \frac{\pi}{2}, 0\right)$.

Solution The electric field is given as

$$\vec{E} = -\nabla V = -\left[\frac{\partial V}{\partial r}\hat{a}_r + \frac{1}{r}\frac{\partial V}{\partial \theta}\hat{a}_\theta + \frac{1}{r}\frac{\partial V}{\partial \phi}\hat{a}_\phi\right]$$

$$= \frac{20}{r^3}\sin\theta\cos\phi\hat{a}_r - \frac{10}{r^3}\cos\theta\cos\phi\hat{a}_\theta + \frac{10}{r^3}\sin\phi\hat{a}_\phi$$

At $(4, \pi/2, 0)$, the field intensity is

$$\vec{E} = \frac{20}{4^3}\sin\left(\frac{\pi}{2}\right)\cos 0\hat{a}_r - \frac{10}{4^3}\cos\left(\frac{\pi}{2}\right)\cos 0\hat{a}_\theta + \frac{10}{4^3}\sin 0\hat{a}_\phi = \frac{5}{16}\hat{a}_r$$

Therefore, the flux density is given as

$$\vec{D} = \varepsilon_0\vec{E} = 8.854\times 10^{-12}\times\frac{5}{16}\hat{a}_r = 2.77\hat{a}_r \quad \text{pC/m}^2$$

2.15 EQUIPOTENTIAL SURFACES

Definition *Equipotential surface* is a surface with equal value of potential at every point on the surface. In other words, the locus of the points which have the same electric potential is known as *equipotential surface*. The surface obtained by joining the points with equal potential is known as *equipotential surface*.

The potential difference between any two points on an equipotential surface is zero.

Properties of Equipotential Surfaces The properties of equipotential surfaces are given here:

1. The electric field lines are perpendicular to the equipotential surfaces and are directed from higher to lower potentials.
2. The tangential component of the electric field along the equipotential surface is zero; otherwise non-vanishing work would be done to move a charge from one point on the surface to the other.
3. No work is required to move a particle along an equipotential surface.
4. The equipotential surfaces for a point charge or a sphere with uniform charge distribution are spheres concentric with the charge.
5. The equipotential surfaces for a line charge or a cylinder with uniform charge distribution are concentric cylinders centered on the axis of the charge distribution.
6. The equipotential surfaces for a flat surface with uniform charge distribution are planes parallel to the surface.

In Fig. 2.52, we illustrate some examples of equipotential surfaces.

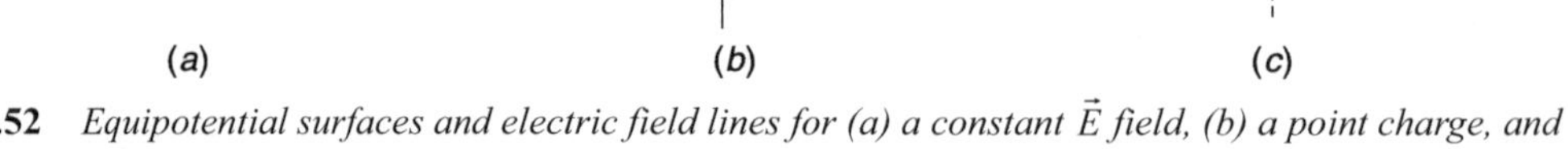

Fig. 2.52 *Equipotential surfaces and electric field lines for (a) a constant $\vec{E}$ field, (b) a point charge, and (c) an electric dipole*

Example 2.42 Prove that electric field lines are perpendicular to the equipotential surfaces.

Solution Since $\vec{E} = -\nabla V$, it can be shown that the direction of $\vec{E}$ is always perpendicular to the equipotential through the point. We give proof below (Fig. 2.53).

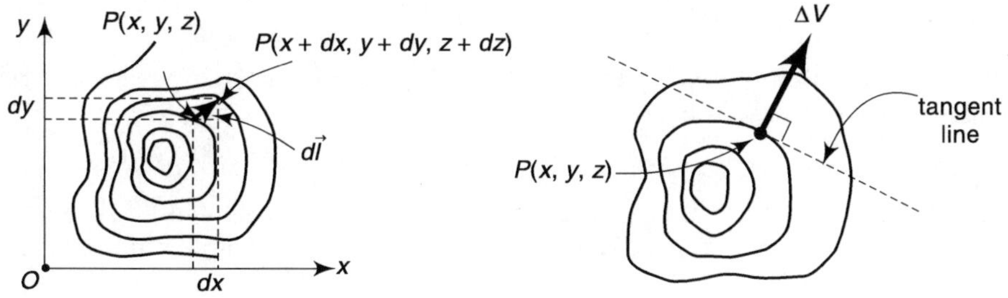

Fig. 2.53 *Change in V when moving from one equipotential curve to another*

Let the potential at a point $P(x, y, z)$ be $V(x, y, z)$. The difference in potential at a neighbouring point $P(x + dx, y + dy, z + dz)$ is given as

$$
\begin{aligned}
dV &= V(x + dx, y + dy, z + dz) - V(x, y, z) \\
&= \left[V(x, y, z) + \frac{\partial V}{\partial x} dx + \frac{\partial V}{\partial y} dy + \frac{\partial V}{\partial z} dz \right] - V(x, y, z) \\
&= \frac{\partial V}{\partial x} dx + \frac{\partial V}{\partial y} dy + \frac{\partial V}{\partial z} dz
\end{aligned}
$$

With the displacement vector given as $d\vec{l} = dx\hat{a}_x + dy\hat{a}_y + dz\hat{a}_z$, we can rewrite dV as

$$
\begin{aligned}
dV &= \frac{\partial V}{\partial x} dx + \frac{\partial V}{\partial y} dy + \frac{\partial V}{\partial z} dz = \left(\frac{\partial V}{\partial x} \hat{a}_x + \frac{\partial V}{\partial y} \hat{a}_y + \frac{\partial V}{\partial z} \hat{a}_z \right) \cdot (dx\hat{a}_x + dy\hat{a}_y + dz\hat{a}_z) \\
&= (\nabla V) \cdot d\vec{l} \\
&= -\vec{E} \cdot d\vec{l}
\end{aligned}
$$

If the displacement $d\vec{l}$ is along the tangent to the equipotential curve through $P(x, y, z)$, then $dV = 0$ because V is constant everywhere on the curve. This implies that $\vec{E}$ is perpendicular to V along the equipotential curve.

2.16 ELECTRIC DIPOLE

Two equal and opposite point charges separated by a distance, which is small compared to the distance at which the potential or the field is to be calculated, constitute an *electric dipole*.

Fig. 2.54 shows an electric dipole with $+q$ and $-q$ charges separated by a distance $\vec{d}$.

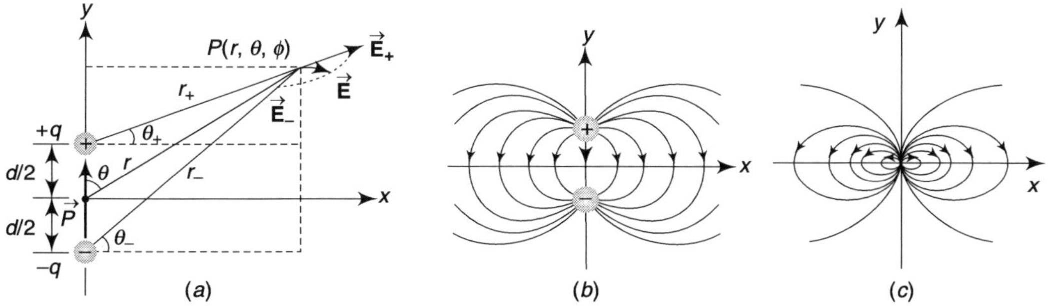

Fig. 2.54 *(a) Electric dipole, (b) Field lines for an electric dipole, and (c) Field lines for a pure electric dipole*

2.16.1 Dipole Moment

The product of the magnitude of one charge and the separation distance between the charges is called the *dipole moment* $(\vec{p})$. This is given as

$$\vec{p} = Q\vec{d} \tag{2.43}$$

This is a vector directed from the negative charge to the positive charge. Its unit is Coulomb-meter (C-m).

For an overall charge-neutral system having N charges, the electric dipole moment vector $(\vec{p})$ is defined as

$$\vec{p} = \sum_{i=1}^{N} Q_i \vec{r}_i \tag{2.44}$$

where $\vec{r}_i$ is the position vector of the charge Q_i.

Examples of dipoles include HCl, CO, H_2O and other polar molecules. In principle, any molecule in which the centres of the positive and negative charges do not coincide may be approximated as a dipole.

NOTE

An electric dipole is said to be a **pure dipole** *if $d \to 0$ and $Q \to \infty$ at the same rate so that the dipole moment $(\vec{p} = Q\vec{d})$ remains constant. The field lines for a pure dipole are shown in Fig. 2.54 (c).*

2.16.2 Electric Potential due to Dipole

We will consider the dipole shown in Fig. 2.54 (*a*). The potential at point $P(r, \theta, \phi)$ is given as

$$V = \frac{q}{4\pi\varepsilon}\left[\frac{1}{r_+} - \frac{1}{r_-}\right] = \frac{q}{4\pi\varepsilon}\left[\frac{r_- - r_+}{r_+ r_-}\right]$$

where r_+ and r_- are the distances between P and $+q$ and P and $-q$, respectively. If $r >> d$, then

$$(r_- - r_+) = d\cos\theta \quad \text{and} \quad r_- r_+ \approx r^2$$

$$\therefore \qquad V = \frac{q}{4\pi\varepsilon}\frac{d\cos\theta}{r^2} = \frac{qd\cos\theta}{4\pi\varepsilon r^2}$$

Since, $d\cos\theta = \vec{d}\cdot\hat{a}_r$, and dipole moment, $\vec{p} = Q\vec{d}$, we can write the potential as

$$\boxed{V = \frac{qd\cos\theta}{4\pi\varepsilon r^2} = \frac{q}{4\pi\varepsilon}\frac{\vec{d}\cdot\hat{a}_r}{r^2} = \frac{1}{4\pi\varepsilon}\frac{\vec{p}\cdot\hat{a}_r}{r^2} = \frac{1}{4\pi\varepsilon}\frac{\vec{p}\cdot\vec{r}}{|\vec{r}|^3}} \tag{2.45}$$

If the dipole centre is not at the origin, but at a position vector $\vec{r}'$, then the potential can be written as

$$\boxed{V(\vec{r}) = \frac{\vec{p}\cdot(\vec{r} - \vec{r}')}{4\pi\varepsilon|\vec{r} - \vec{r}'|^3}} \tag{2.46}$$

This is the potential at any point P due to an electric dipole.

NOTE

(i) *Equation (2.43) shows that the potential due to dipole varies inversely as the square of the distance $\left(V \alpha \frac{1}{r^2}\right)$, unlike as $\left(\frac{1}{r}\right)$ for a point charge.*

(ii) *For $\theta = 90°$, $V = 0$, i.e., the potential along the perpendicular bisector to dipole axis is zero and the perpendicular drawn at O will be an equipotential line of zero potential.*

2.16.3 Electric Field due to Dipole

The electric field due to the dipole with centre at the origin is given as

$$\vec{E} = -\nabla V = -\left(\frac{\partial V}{\partial r}\widehat{a}_r + \frac{1}{r}\frac{\partial V}{\partial \theta}\widehat{a}_\theta\right) \quad \text{(in spherical coordinates)}$$

$$= -\frac{\partial}{\partial r}\left(\frac{qd\cos\theta}{4\pi\varepsilon r^2}\right)\widehat{a}_r - \frac{1}{r}\frac{\partial}{\partial \theta}\left(\frac{qd\cos\theta}{4\pi\varepsilon r^2}\right)\widehat{a}_\theta$$

$$= \frac{qd\cos\theta}{2\pi\varepsilon r^3}\widehat{a}_r + \frac{qd\sin\theta}{4\pi\varepsilon r^3}\widehat{a}_\theta$$

$$\therefore \quad \boxed{\vec{E} = \frac{qd}{4\pi\varepsilon r^3}(2\cos\theta\widehat{a}_r + \sin\theta\widehat{a}_\theta) = \frac{p}{4\pi\varepsilon r^3}(2\cos\theta\widehat{a}_r + \sin\theta\widehat{a}_\theta)} \qquad \textbf{(2.47)}$$

The electric field due to an electric dipole may also be expressed as follows.

$$\vec{E} = -\nabla V = -\frac{1}{4\pi\varepsilon}\nabla\left(\frac{\vec{p}\cdot\vec{r}}{|\vec{r}|^3}\right) = -\frac{1}{4\pi\varepsilon}\left[\frac{1}{|\vec{r}|^3}\nabla(\vec{p}\cdot\vec{r}) + (\vec{p}\cdot\vec{r})\nabla\left(\frac{1}{|\vec{r}|^3}\right)\right] \quad \{\because \nabla(AB) = A\nabla B + B\nabla A\}$$

Since, $\nabla(\vec{p}\cdot\vec{r}) = \vec{p}$ and $\nabla\left(\frac{1}{|\vec{r}|^3}\right) = -\frac{3\vec{r}}{|\vec{r}|^5}$

$$\therefore \quad \vec{E} = \frac{1}{4\pi\varepsilon}\left[\frac{3(\vec{p}\cdot\vec{r})\vec{r}}{|\vec{r}|^5} - \frac{\vec{p}}{|\vec{r}|^3}\right]$$

or

$$\boxed{\vec{E} = \frac{1}{4\pi\varepsilon r^3}\left[\frac{3(\vec{p}\cdot\vec{r})\vec{r}}{r^2} - \vec{p}\right]} \qquad \textbf{(2.48)}$$

Thus, the field intensity due to a dipole varies inversely as the cube of the distance of the field point from the dipole $\left(E\alpha\frac{1}{r^3}\right)$, unlike as $\left(\frac{1}{r}\right)$ for a point charge.

NOTE

(iii) *For $\theta = 90°$, $\vec{E}_r = 0$ and $\vec{E}_\theta = \frac{p}{4\pi\varepsilon r^3}\widehat{a}_\theta$, i.e., the radial component of the field vanishes, but the angular component persists.*

(iv) *For $\theta = 0°$, $\vec{E}_r = \frac{p}{2\pi\varepsilon r^3}\widehat{a}_r$ and $\vec{E}_\theta = 0$, i.e., if the point is somewhere in alignment with the dipole axis, perpendicular component of the field will be zero.*

2.16.4 Electric Dipole in an External Electric Field

We will discuss the effects when an electric dipole is placed in an external field, say, $\vec{E} = E\hat{a}_x$. We will determine the torque experienced by the dipole and the potential energy of the dipole.

Torque on an Electric Dipole when placed in External Field From Fig. 2.54, we see that the unit vector which points in the direction of $\vec{P}$ is $(\cos\theta\hat{a}_x + \sin\theta\hat{a}_y)$. Thus, we have

$$\vec{p} = qd(\cos\theta\,\hat{a}_x + \sin\theta\,\hat{a}_y)$$

As seen from Fig. 2.55, since each charge experiences an equal but opposite force due to the field, the net force on the dipole is

$$\vec{F}_{\text{net}} = \vec{F}_+ + \vec{F}_- = 0$$

Even though the net force vanishes, the field exerts a torque on the dipole. The torque about the midpoint O of the dipole is

$$\begin{aligned}\vec{\tau} &= \vec{r}_+ \times \vec{F}_+ + \vec{r}_- \times \vec{F}_- \\ &= [(d/2)\cos\theta\,\hat{a}_x + (d/2)\sin\theta\,\hat{a}_y] \times (F_+\,\hat{a}_x) \\ &\quad + [-(d/2)\cos\theta\,\hat{a}_x - (d/2)\sin\theta\,\hat{a}_y] \times (-F_-\,\hat{a}_x) \\ &= (d/2)\sin\theta\,F_+(-\hat{a}_z) + (d/2)\sin\theta\,F_-(-\hat{a}_z) \\ &= Fd\sin\theta(-\hat{a}_z) \qquad \{\because F_+ = F_-\} \\ &= qEd\sin\theta(-\hat{a}_z) \\ &= pE\sin\theta(-\hat{a}_z) \qquad \{\because p = qd\}\end{aligned}$$

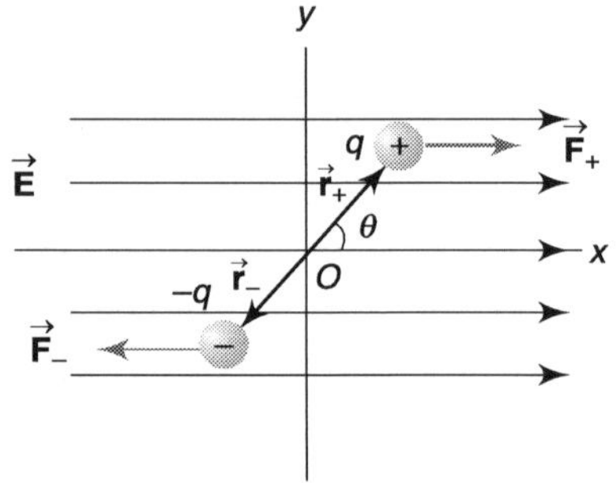

Fig. 2.55 *Electric dipole placed in a uniform field*

The direction of the torque is $(-\hat{a}_z)$, or into the page. The effect of the torque is to rotate the dipole *clockwise* so that the dipole moment becomes aligned with the electric field $\vec{E}$. The magnitude of the torque can be written as

$$\boxed{\tau = pE\sin\theta} \tag{2.49}$$

In vector form, the torque can be written as

$$\boxed{\vec{\tau} = \vec{p} \times \vec{E}} \tag{2.50}$$

Thus, we see that the cross product of the dipole moment with the electric field is equal to the torque.

Potential Energy of an Electric Dipole when placed in External Field The work done by the electric field to rotate the dipole by an angle $d\theta$ is

$$dW = -\tau d\theta = -pE\sin\theta\,d\theta$$

The negative sign indicates that the torque opposes any increase in θ. Therefore, the total amount of work done by the electric field to rotate the dipole from an angle θ_0 to θ is

$$W = \int_{\theta_0}^{\theta} (-pE\sin\theta)d\theta = pE(\cos\theta - \cos\theta_0)$$

The result shows that a positive work is done by the field when $\cos\theta > \cos\theta_0$. The change in potential energy of the dipole (ΔU) is the negative of the work done by the field, i.e.,

$$\Delta U = (U - U_0) = -W = -pE(\cos\theta - \cos\theta_0)$$

where, $U_0 = -pE\cos\theta_0$ is the potential energy at a reference point. We will choose our reference point to be $\theta_0 = 90°$ so that the potential energy is zero, i.e., $U_0 = 0$. Thus, in the presence of an external field, the electric dipole has a potential energy

$$\boxed{U = -pE\cos\theta = -\vec{p}\cdot\vec{E}} \tag{2.51}$$

NOTE

(i) *For $\theta = 0°$, $W = -pE$ and $\tau = 0$, i.e., potential energy is minimum. A system is at equilibrium when its potential energy is a minimum. This takes place when the dipole moment $\vec{p}$ is aligned parallel to $\vec{E}$, making potential energy a minimum, $U_{min} = -pE$.*

(ii) *For $\theta = 90°$, $W = 0$ and $\tau = pE$, i.e., potential energy is zero when the dipole is perpendicular to the electric field.*

(iii) *For $\theta = 180°$, $W = pE$ and $\tau = 0$, i.e., the potential energy is a maximum and, therefore, the system is unstable when $\vec{p}$ and $\vec{E}$ are anti-parallel.*

If the dipole is placed in a non-uniform field, there would be a net force on the dipole in addition to the torque, and the resulting motion would be a combination of linear acceleration and rotation. In Fig. 2.56, we consider the electric field $\vec{E}_+$ at $+q$ differs from the electric field $\vec{E}_-$ at $-q$.

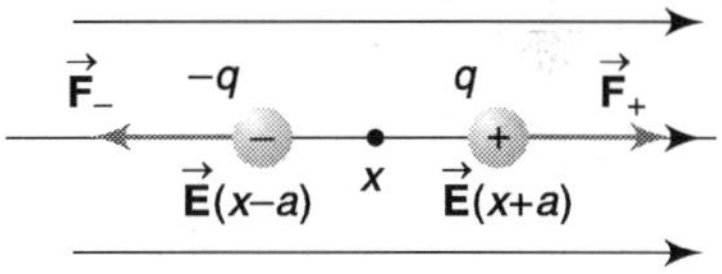

Fig. 2.56 *Force on a dipole*

Assuming the dipole to be very small, we expand the fields about x as

$$E_+(x+a) \approx E(x) + a\left(\frac{dE}{dx}\right), \quad E_-(x-a) \approx E(x) - a\left(\frac{dE}{dx}\right)$$

Hence, the force on the dipole is,

$$\vec{F}_e = q(\vec{E}_+ - \vec{E}) = 2qa\left(\frac{dE}{dx}\right)\hat{a}_x = p\left(\frac{dE}{dx}\right)\hat{a}_x \tag{2.52}$$

***Example 2.43** An electric dipole of dipole moment $100\hat{a}_z$ pC-m is located at the origin. Find the potential and field at point (0, 0, 10).

Solution The potential is given as

$$V = \frac{\vec{p}\cdot\vec{r}}{4\pi_0 r^3}$$

Here, $\vec{r} = (0, 0, 10) - (0, 0, 0) = 10\hat{a}_z$

$$V = \frac{\vec{p}\cdot\vec{r}}{4\pi\varepsilon_0 r^3} = \frac{(100\times 10^{-12}\hat{a}_z)\cdot(10\hat{a}_z)}{4\pi\times\dfrac{10^{-9}}{36\pi}\times 10^3} = 9\text{mV}$$

Field intensity is

$$\therefore \quad \vec{E} = \frac{1}{4\pi\varepsilon_0 r^3}\left[\frac{3(\vec{p}\cdot\vec{r})\vec{r}}{r^2} - \vec{p}\right] = \frac{1}{4\pi\times\dfrac{10^{-9}}{36\pi}\times 10^3}\left[\frac{3\times 1000\times 10^{-12}\times 10\hat{a}_z}{100} - 100\times 10^{-12}\,\hat{a}_z\right]$$

$$= 1.8\hat{a}_z \quad \text{mV/m}$$

***Example 2.44** Two dipole with dipole moments $-5\hat{a}_z$ nC-m and $9\hat{a}_z$ nC-m are located at points (0, 0, –2) and (0, 0, 3) respectively. Find the potential at the origin.

Solution The potential is given as

$$V = \sum_{k=1}^{2} \frac{\vec{p}_k \cdot \vec{r}_k}{4\pi\varepsilon_0 r_k^3} = \frac{1}{4\pi\varepsilon_0}\left[\frac{\vec{p}_1 \cdot \vec{r}_1}{r_1^3} + \frac{\vec{p}_2 \cdot \vec{r}_2}{r_2^3}\right]$$

Here,

$$\vec{p}_1 = -5\hat{a}_z, \quad \vec{p}_2 = 9\hat{a}_z,$$
$$\vec{r}_1 = (0, 0, 0) - (0, 0, -2) = 2\hat{a}_z; \quad r_1 = |\vec{r}_1| = 2$$
$$\vec{r}_2 = (0, 0, 0) - (0, 0, 3) = -3\hat{a}_z; \quad r_2 = |\vec{r}_2| = 3$$

$$\therefore \qquad V = \frac{1}{4\pi\varepsilon_0}\left[-\frac{10}{2^3} - \frac{27}{3^3}\right] = \frac{1}{4\pi \times \dfrac{10^{-9}}{36\pi}}\left[-\frac{10}{2^3} - \frac{27}{3^3}\right] = -20.25 \times 10^9 \text{ Volt}$$

***Example 2.45** Figure 2.57 shows two charges at points A and B in free space. Find the electric field at point P due to these charges. Is the result consistent with what may be expected if $d >> s$?

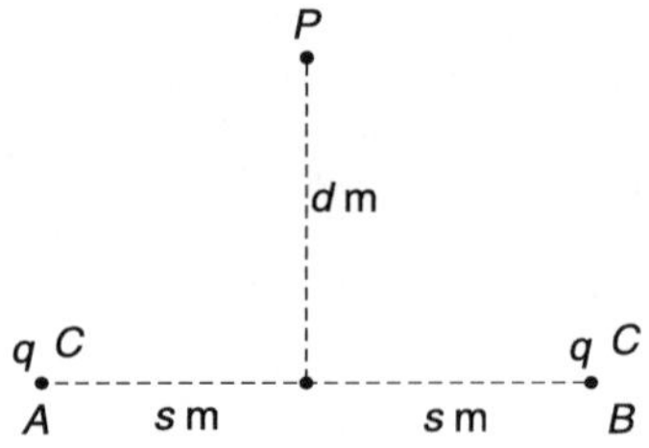

Fig. 2.57 *Arrangement of Example 2.45*

Solution Electric field at point P will be the resultant field due to the charges at point A and point B. This is obtained as follows.

$$\vec{E} = \frac{q}{4\pi\varepsilon(s^2 + d^2)} \times 2\sin\phi\,\hat{a}_z = \frac{q}{4\pi\varepsilon(s^2 + d^2)} \times 2 \times \frac{d}{\sqrt{s^2 + d^2}}\hat{a}_z$$
$$= \frac{qd}{2\pi\varepsilon(s^2 + d^2)^{3/2}}\hat{a}_z$$

When $s << d$, i.e., $\frac{s}{d} << 1$, the field reduces to

$$\vec{E} = \frac{qd}{2\pi\varepsilon d^3\left(\dfrac{s^2}{d^2} + 1\right)^{3/2}}\hat{a}_z = \frac{q}{2\pi\varepsilon d^2}\hat{a}_z$$

Example 2.46 An electron and a proton separated by a distance of 10^{-11} metre are symmetrically arranged along the z-axis with $z = 0$ as its bisecting plane. Determine the potential and field at point (3, 4, 12).

Solution The position vector is, $\vec{r} = 3\hat{a}_x + 4\hat{a}_y + 12\hat{a}_z$, $r = |\vec{r}| = 13$ m

The dipole moment is $\vec{p} = 1.6 \times 10^{-19} \times 10^{-11}\hat{a}_z = 1.6 \times 10^{-30}\hat{a}_z$

Thus, the potential at point (3, 4, 12) is given as

$$V = \frac{\vec{p} \cdot \vec{r}}{4\pi\varepsilon_0 r^3} = \frac{(1.6 \times 10^{-30}\hat{a}_z) \cdot (3\hat{a}_x + 4\hat{a}_y + 12\hat{a}_z)}{4\pi \times \dfrac{10^{-9}}{36\pi} \times 13^3} = \frac{1.6 \times 10^{-30} \times 12 \times 9}{10^{-9} \times 13^3} = 7.865 \times 10^{-23} \text{ Volt}$$

The electric field intensity at point (3, 4, 12) is given as

$$\vec{E} = \frac{1}{4\pi\varepsilon_0 r^3}\left[\frac{3(\vec{p}\cdot\vec{r})\vec{r}}{r^2} - \vec{p}\right]$$

$$= \frac{1}{4\pi \times \dfrac{10^{-9}}{36\pi} \times 13^3}\left[\frac{(3\times 1.6\times 10^{-30}\times 12)\times(3\hat{a}_x + 4\hat{a}_y + 12\hat{a}_z)}{13^2} - 1.6\times 10^{-30}\,\hat{a}_z\right]$$

$$= (4.189\hat{a}_x + 5.585\hat{a}_y + 10.2\hat{a}_z)\times 10^{-24} \quad \text{V/m}$$

Example 2.47 Figure 2.58 shows a linear quadrapole arrangement, with charges $+q, -2Q$ and $+Q$ disposed as indicated. Show that the potential due to this quadrapole at a large distance 'r' in comparison with the spacing '$2l$' is

$$V = \frac{Ql^2}{4\pi\varepsilon_0 r^3}(3\cos^2\theta - 1)$$

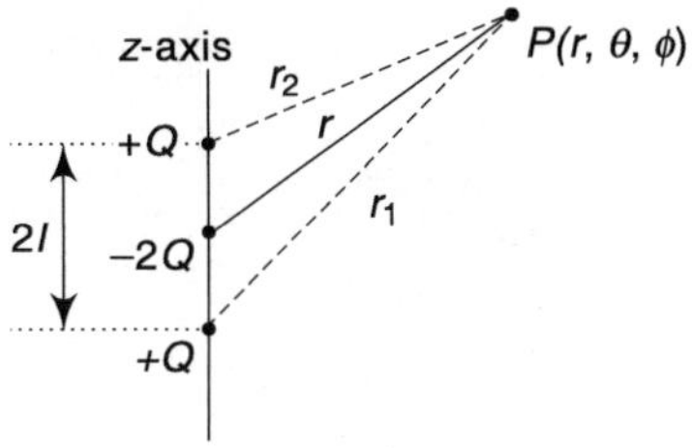

Fig. 2.58 *Linear quadrapole of Example 2.47*

Solution By the principle of superposition, the potential at point P is given as

$$V = \frac{1}{4\pi\varepsilon_0}\left[\frac{Q}{r_1} - \frac{2Q}{r} + \frac{Q}{r_2}\right] = \frac{Q}{4\pi\varepsilon_0 r}\left[\frac{r}{r_1} + \frac{r}{r_2} - 2\right] \qquad \text{(i)}$$

Now, using trigonometric relation for the triangle with sides r, r_1, and l, we have

$$r_1^2 = r^2 + l^2 + 2rl\cos\theta$$

$$\therefore \quad \left(\frac{r_1}{r}\right)^2 = 1 + \frac{l^2}{r^2} + \frac{2l}{r}\cos\theta$$

$$\therefore \quad \left(\frac{r_1}{r}\right) = \left[1 + \frac{l^2}{r^2} + \frac{2l}{r}\cos\theta\right]^{1/2}$$

$$\therefore \quad \left(\frac{r}{r_1}\right) = \left[1 + \frac{l^2}{r^2} + \frac{2l}{r}\cos\theta\right]^{-1/2}$$

$$= 1 - \frac{1}{2}\left(\frac{l^2}{r^2} + \frac{2l}{r}\cos\theta\right) + \frac{-\frac{1}{2}\left(-\frac{1}{2}-1\right)}{2}\left(\frac{l^2}{r^2} + \frac{2l}{r}\cos\theta\right)^2 + \cdots$$

$$= 1 - \frac{l}{r}\cos\theta + \frac{l^2}{r^2}\left(\frac{3\cos^2\theta - 1}{2}\right)$$

Similarly, we have

$$\left(\frac{r}{r_2}\right) = 1 + \frac{l}{r}\cos\theta + \frac{l^2}{r^2}\left(\frac{3\cos^2\theta - 1}{2}\right)$$

Thus, from (i), we get

$$V = \frac{Q}{4\pi\varepsilon_0 r}\left[\frac{r}{r_1} + \frac{r}{r_2} - 2\right]$$

$$= \frac{Q}{4\pi\varepsilon_0 r}\left[1 - \frac{l}{r}\cos\theta + \frac{l^2}{r^2}\left(\frac{3\cos^2\theta - 1}{2}\right) + 1 + \frac{l}{r}\cos\theta + \frac{l^2}{r^2}\left(\frac{3\cos^2\theta - 1}{2}\right) - 2\right]$$

$$= \frac{Q}{4\pi\varepsilon_0 r}\left[\frac{l^2}{r^2}(3\cos^2\theta - 1)\right]$$

$$\boxed{V = \frac{Ql^2}{4\pi\varepsilon_0 r^3}(3\cos^2\theta - 1)}$$

2.17 MULTIPOLE (OR FAR FIELD) EXPANSION OF ELECTRIC POTENTIAL

We will consider an arbitrary local charge distribution with volume charge density ρ in a volume V. It is assumed that the position coordinate of the field point is much larger than the position coordinate of the source point, i.e., ($\vec{r} >> \vec{r}'$). This is shown in Fig. 2.59.

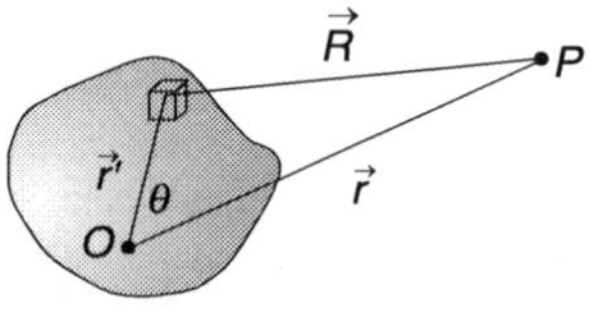

Fig. 2.59 *Multiple expansion of potential due to a volume charge distribution*

Then, the electric potential at point P with position coordinate $\vec{r}$ is given as,

$$V = \frac{1}{4\pi\varepsilon}\int_V \frac{\rho(\vec{r}')}{|\vec{R}|}dV = \frac{1}{4\pi\varepsilon}\int_V \frac{\rho(\vec{r}')}{|\vec{r} - \vec{r}'|}dV \quad \textbf{(2.53)}$$

Now, using the cosine law

$$\vec{R} = (\vec{r} - \vec{r}') = r^2 + (r')^2 - 2rr'\cos\theta = r^2\left[1 + \left(\frac{r'}{r}\right)^2 - 2\left(\frac{r'}{r}\right)\cos\theta\right] = r\sqrt{1+\varepsilon}$$

where $\varepsilon = \left(\frac{r'}{r}\right)\left(\frac{r'}{r} - 2\cos\theta\right)$, since ($\vec{r} >> \vec{r}'$), $\varepsilon << 1$

$$\therefore \quad \frac{1}{|\vec{R}|} = \frac{1}{r}(1+\varepsilon)^{-1/2} = \frac{1}{r}\left(1 - \frac{1}{2}\varepsilon + \frac{3}{8}\varepsilon^2 - \frac{5}{16}\varepsilon^3 + \ldots\right)$$

$$= \frac{1}{r}\left[1 - \frac{1}{2}\left(\frac{r'}{r}\right)\left(\frac{r'}{r} - 2\cos\theta\right) + \frac{3}{8}\left(\frac{r'}{r}\right)\left(\frac{r'}{r} - 2\cos\theta\right)^2 - \frac{5}{16}\left(\frac{r'}{r}\right)\left(\frac{r'}{r} - 2\cos\theta\right)^3 + \ldots\right]$$

$$= \frac{1}{r}\left[1 + \left(\frac{r'}{r}\right)(\cos\theta) + \left(\frac{r'}{r}\right)^2\left(\frac{3}{2}\cos^2\theta - \frac{1}{2}\right) + \left(\frac{r'}{r}\right)^3\left(\frac{5}{2}\cos^3\theta - \frac{3}{2}\cos\theta\right) + \ldots\right]$$

Here, the terms in parenthesis can be written in terms of Legendre polynomials as

$$\frac{1}{|\vec{R}|} = \frac{1}{r}\sum_{n=0}^{\infty}\left(\frac{r'}{r}\right)^n P_n\cos\theta \quad \textbf{(2.54)}$$

Putting this in Eq. (2.53), we get

$$V = \frac{1}{4\pi\varepsilon}\int_V \frac{\rho(\vec{r}')}{|\vec{r}-\vec{r}'|}dV = \frac{1}{4\pi\varepsilon}\int_V \frac{1}{r}\sum_{n=0}^{\infty}\left(\frac{r'}{r}\right)^n P_n \cos\theta\rho(\vec{r}')dV$$

$$= \frac{1}{4\pi\varepsilon}\sum_{n=0}^{\infty}\frac{1}{r^{n+1}}\int_V (r')^n P_n \cos\theta\rho(\vec{r}')dV \qquad \textbf{(2.55)}$$

$$\boxed{V = \frac{1}{4\pi\varepsilon}\sum_{n=0}^{\infty}\frac{1}{r^{n+1}}\int_V (r')^n P_n \cos\theta\rho(\vec{r}')dV} \qquad \textbf{(2.56)}$$

This can be written as

$$\boxed{V = \frac{1}{4\pi\varepsilon}\left[\frac{1}{r}\int_V \rho dV + \frac{1}{r^2}\int_V r'\cos\theta\rho dV + \frac{1}{r^3}\int_V (r')^2\left(\frac{3}{2}\cos^2\theta - \frac{1}{2}\right)\rho dV + \ldots\right]} \qquad \textbf{(2.57)}$$

This is the *multipole* or far field expansion of electric potential of a given charge distribution in terms of powers of $\frac{1}{r}$. The first term ($n = 0$) is called *monopole term*; the second term ($n = 1$) is called the *dipole term* and the third term ($n = 2$) is called the *quadrapole term*, the fourth term ($n = 3$) the *octopole*, and so on.

2.18 ENERGY STORED IN ELECTROSTATIC FIELDS

Electrostatic energy is the energy necessary to establish a given charge distribution in space.

Energy Stored in a Region with Discrete Charges

In order to determine the energy stored in an assembly of charges, we determine the amount of work done by an external source to assemble the charges.

We consider a region in space with no charge initially and zero field intensity. Point charges are brought from infinity to specific points in that space, one by one as shown in Fig. 2.60.

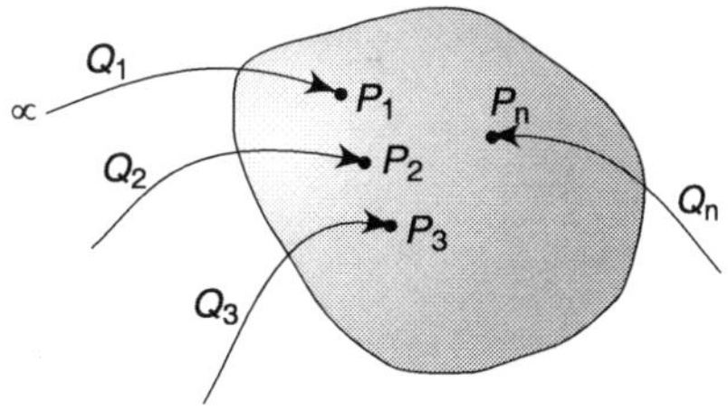

Fig. 2.60 *Assembly of charges*

Work done to bring the first charge Q_1 from infinity to point P_1 is

$$W_1 = 0 \qquad [\because \text{ initially charge-free region}]$$

and the field intensity due to charge Q_1 at a distance $\vec{r}$ is

$$\vec{E}_1 = \frac{Q_1}{4\pi\varepsilon r^2}\hat{a}_r$$

Work done to bring the second charge Q_2 from infinity to the point P_2 is

$$W_2 = Q_2 V_2 = Q_2\left[-\int_{\alpha}^{P_2}\vec{E}_1\cdot d\vec{l}\right] = Q_2\left[-\int_{\alpha}^{P_2}\frac{Q_1}{4\pi\varepsilon r^2}\hat{a}_r\cdot dr\hat{a}_r\right] = \frac{Q_1 Q_2}{4\pi\varepsilon r_{12}}$$

where r_{12} is the distance between the points P_1 and P_2.

Similarly, work done to bring the third charge Q_3 at point P_3 is found as follows.

$$V_3 = -\int_{\alpha}^{P_3} \vec{E}_1 \cdot d\vec{l} - \int_{\alpha}^{P_3} \vec{E}_2 \cdot d\vec{l} = \frac{Q_1}{4\pi\varepsilon r_{13}} + \frac{Q_2}{4\pi\varepsilon r_{23}}$$

$\therefore$
$$W_3 = Q_3 V_3 = \frac{Q_1 Q_3}{4\pi\varepsilon r_{13}} + \frac{Q_2 Q_3}{4\pi\varepsilon r_{23}}$$

Total energy in the system is

$$W = (W_1 + W_2 + W_3) = 0 + \frac{Q_1 Q_2}{4\pi\varepsilon r_{12}} + \frac{Q_1 Q_3}{4\pi\varepsilon r_{13}} + \frac{Q_2 Q_3}{4\pi\varepsilon r_{23}}$$

$$W = \frac{1}{4\pi\varepsilon}\left[\frac{Q_1 Q_2}{r_{12}} + \frac{Q_2 Q_3}{r_{23}} + \frac{Q_1 Q_3}{r_{13}}\right] \tag{2.58}$$

This is the energy obtained by arbitrarily bringing the charges in the sequence Q_1, Q_2 and Q_3. As energy depends on charges and potentials, the energy will be same if we reverse the sequence, i.e., we bring Q_3 first, then Q_2 and then Q_1.

In that case, total energy in the system is

$$W = (W_3 + W_2 + W_1) = 0 + Q_2 \frac{Q_3}{4\pi\varepsilon r_{32}} + Q_1 \frac{Q_3}{4\pi\varepsilon r_{31}} + Q_1 \frac{Q_2}{4\pi\varepsilon r_{21}}$$

$$W = \frac{1}{4\pi\varepsilon}\left[\frac{Q_1 Q_2}{r_{21}} + \frac{Q_2 Q_3}{r_{32}} + \frac{Q_1 Q_3}{r_{31}}\right] \tag{2.59}$$

Adding Eq. (2.58) and (2.59), we get

$$2W = Q_1\left[\frac{Q_3}{4\pi\varepsilon r_{31}} + \frac{Q_2}{4\pi\varepsilon r_{21}}\right] + Q_2\left[\frac{Q_1}{4\pi\varepsilon r_{12}} + \frac{Q_3}{4\pi\varepsilon r_{32}}\right] + Q_3\left[\frac{Q_1}{4\pi\varepsilon r_{13}} + \frac{Q_2}{4\pi\varepsilon r_{23}}\right]$$
$$= Q_1 V_1 + Q_2 V_2 + Q_3 V_3$$

(since the terms in the parentheses are the potentials at the corresponding nodes due to the other charges at the other nodes)

$\therefore$
$$W = \frac{1}{2}(Q_1 V_1 + Q_2 V_2 + Q_3 V_3)$$

Thus, for N point charges located at points P_1, P_2, …, P_N, the electrostatic energy stored in the charge system is given as

$$\boxed{W = \frac{1}{2}\sum_{i=1}^{N} Q_i V_i \quad \text{(Joule)}} \tag{2.60}$$

Energy Stored in a Region with Continuous Charge Distribution Instead of an assembly of point charges, if there were a volume of charge of density $\rho(\text{C/m}^3)$, then the electrostatic energy stored can be written as

$$W = \frac{1}{2}\int_v \rho dv V \qquad [dv \text{ – is an elemental volume and } V \text{ – is the potential}]$$

$$= \frac{1}{2}\int_v (\nabla \cdot \vec{D}) V dv \qquad (\text{using Gauss' law, } \nabla \cdot \vec{D} = \rho)$$

$$= \frac{1}{2}\int_v \left[\nabla \cdot (V\vec{D}) - \vec{D} \cdot (\nabla V)\right] dv \qquad \{\because \nabla \cdot (A\vec{B}) \equiv A(\nabla \cdot \vec{B}) + \vec{B} \cdot (\nabla A)\}$$

$$= \frac{1}{2}\int_S V\vec{D} \cdot d\vec{S} + \frac{1}{2}\int_v \vec{D} \cdot \vec{E} dv \qquad \{\because \vec{E} = -\nabla V\} \tag{2.61}$$

If we go on increasing the volume to trap all the charges, the surface area must shrink to keep the energy constant.

Now, $\vec{E}$ varies as $\frac{1}{r^2}$ and V varies as $\frac{1}{r}$ and surface area as r^2. Therefore, $V\vec{D}\alpha\frac{1}{r^3}$ and $S\alpha r^2$ and S $\alpha\, r^2$; So, the surface will reduce as $\frac{1}{r}$ and thus the surface integral will vanish. Hence, the energy stored in the charge system is

$$\boxed{W = \frac{1}{2}\int_v \vec{D} \cdot \vec{E} dv = \frac{1}{2}\int_v \varepsilon E^2 dv} \tag{2.62}$$

From this, we can define *electrostatic energy density* w (in Joule/m^3) as

$$\boxed{w = \frac{dW}{dv} = \frac{1}{2}\vec{D} \cdot \vec{E} = \frac{1}{2}\varepsilon E^2} \tag{2.63}$$

and the total energy stored in the system is

$$W = \int_v w dv \tag{2.64}$$

Example 2.48 Four point charges $Q_1 = 1\ nC$, $Q_2 = -2\ nC$, $Q_3 = 3\ nC$ and $Q_4 = -4\ nC$ are located at (0, 0, 0), (1, 0, 0), (0, 0, –1) and (0, 0, 1), respectively. Find the energy in the system.

Solution The energy in the system is given as

$$W = \frac{1}{2}(Q_1V_1 + Q_2V_2 + Q_3V_3 + Q_4V_4)$$

$$= \frac{Q_1}{2}\left[\frac{Q_2}{4\pi\varepsilon_0(1)} + \frac{Q_3}{4\pi\varepsilon_0(1)} + \frac{Q_4}{4\pi\varepsilon_0(1)}\right] + \frac{Q_2}{2}\left[\frac{Q_1}{4\pi\varepsilon_0(1)} + \frac{Q_3}{4\pi\varepsilon_0(\sqrt{2})} + \frac{Q_4}{4\pi\varepsilon_0(\sqrt{2})}\right]$$

$$+ \frac{Q_3}{2}\left[\frac{Q_1}{4\pi\varepsilon_0(1)} + \frac{Q_2}{4\pi\varepsilon_0(\sqrt{2})} + \frac{Q_4}{4\pi\varepsilon_0(2)}\right] + \frac{Q_4}{2}\left[\frac{Q_1}{4\pi\varepsilon_0(1)} + \frac{Q_2}{4\pi\varepsilon_0(\sqrt{2})} + \frac{Q_3}{4\pi\varepsilon_0(2)}\right]$$

$$= \frac{1}{4\pi\varepsilon_0}\left[Q_1Q_2 + Q_1Q_3 + Q_1Q_4 + \frac{Q_2Q_3}{\sqrt{2}} + \frac{Q_2Q_4}{\sqrt{2}} + \frac{Q_3Q_4}{2}\right]$$

$$= 9\left[-2 + 3 - 4 + \frac{-6}{\sqrt{2}} + \frac{8}{\sqrt{2}} - 6\right]$$

$$= -68.272 \quad \text{nJ}$$

Example 2.49 Calculate the energy of a sphere of charge of radius R in which the charge is uniformly distributed.

Solution The energy stored is the work done in bringing charges from infinity to the sphere. We imagine that the sphere is formed by assembly of various thin shells of charge.

We consider a small sphere of charge of radius r. Let ρ be the charge density.

$\therefore$ Total charge on the sphere $= \frac{4}{3}\pi r^3 \rho$

Suppose a small layer of charge dq in the form of thin shell of thickness dr is deposited on the sphere.

$$\therefore \qquad dq = \rho 4\pi r^2 dr$$

$\therefore$ Work done in bringing this charge from infinity is

$$dW = \text{potential at } r \times dq = \left(\frac{1}{4\pi\varepsilon_0} \frac{\frac{4}{3}\pi r^3 \rho}{r} \right) (\rho 4\pi r^2 dr) = \frac{4\pi\rho^2}{3\varepsilon_0} r^4 dr$$

$\therefore$ Total energy required to assemble charges so as to build up a sphere of radius R is

$$W = \frac{4\pi\rho^2}{3\varepsilon_0} \int_0^R r^4 dr = \frac{4\pi\rho^2}{15\varepsilon_0} R^5$$

Example 2.50 A charge distribution with spherical symmetry has the density

$$\rho = \rho_0, \quad 0 \le r \le R$$
$$= 0, \quad r > R$$

Determine the potential V everywhere and the energy stored in the region, $r < R$.

Solution From Examples 2.30 and 2.36, the potential and field inside the sphere ($r < R$) are given as

$$V = \frac{\rho_0}{6\varepsilon_0}(3a^2 - r^2) \quad \text{and} \quad \vec{E} = \frac{\rho_0 r}{3\varepsilon_0} \hat{a}_r$$

$\therefore$ Energy stored in the system is given as

$$W = \frac{1}{2}\varepsilon_0 \int_v E^2 dv = \frac{1}{2}\varepsilon_0 \int_v \left(\frac{\rho_0 r}{3\varepsilon_0} \right)^2 dv = \frac{1}{2} \frac{\rho_0^2}{9\varepsilon_0} \int_{r=0}^{R} \int_{\theta=0}^{\pi} \int_{\phi=0}^{2\pi} r^2 r^2 \sin\theta \, d\phi \, d\theta \, dr$$

$$= \frac{\rho^2}{18\varepsilon_0} \times 4\pi \times \left. \frac{r^5}{5} \right|_0^R = \frac{2\pi\rho_0^2}{45\varepsilon_0} R^5$$

$$\therefore \qquad \boxed{W = \frac{2\pi\rho_0^2}{45\varepsilon_0} R^5}$$

Example 2.51 A charge Q is placed on a spherical conductor of radius R. Calculate the electrostatic energy density at a distance r ($> R$) from the centre of the sphere. Hence, find the electrostatic energy of the system.

Solution Electric field at a distance r ($> R$) is, $E = \frac{Q}{4\pi\varepsilon_0 r^2}$

$\therefore$ Energy stored in the system is given as

$$W = \frac{1}{2}\varepsilon_0 \int_v E^2 dv = \frac{1}{2}\varepsilon_0 \int_v \left(\frac{Q}{4\pi\varepsilon_0 r^2} \right)^2 dv = \frac{Q^2}{32\pi^2\varepsilon_0} \int_{r=0}^{R} \int_{\theta=0}^{\pi} \int_{\phi=0}^{2\pi} \frac{1}{r^4} r^2 \sin\theta \, d\phi \, d\theta \, dr$$

$$= \frac{Q^2}{32\pi^2\varepsilon_0} \times 4\pi \int_R^\infty \frac{dr}{r^2} = \frac{Q^2}{8\pi\varepsilon_0} \left[-\frac{1}{r} \right]_R^\infty = \frac{Q^2}{8\pi\varepsilon_0 R}$$

$$\therefore \qquad \boxed{W = \frac{Q^2}{8\pi\varepsilon_0 R}}$$

2.19 POISSON'S AND LAPLACE'S EQUATIONS

In the earlier sections, we have determined the electric field $\vec{E}$ in a region using Coulomb's law or Gauss' law when the charge distribution is specified in the region or using the relation $\vec{E} = -\nabla V$ when the potential V is specified throughout the region. However, in practical cases, neither the charge distribution nor the potential distribution is specified, and the electrostatic conditions (charge and potential) are specified only at some boundaries. These types of problems are known as *electrostatic boundary value problems*. For these types of problems, the field $\vec{E}$ and the potential V are determined by using *Poisson's equation* or *Laplace's equation*.

For a linear, homogeneous material medium, Poisson's and Laplace's equations can easily be derived from Gauss' law.

$$\nabla \cdot \vec{D} = \rho$$

or

$$\nabla \cdot \varepsilon\vec{E} = \rho$$

or

$$\nabla \cdot \vec{E} = \frac{\rho}{\varepsilon}$$

or

$$\nabla \cdot (-\nabla V) = \frac{\rho}{\varepsilon} \quad (\because \vec{E} = -\nabla V)$$

or

$$\nabla^2 V = -\frac{\rho}{\varepsilon}$$

$$\boxed{\nabla^2 V = -\frac{\rho}{\varepsilon}} \qquad \textbf{(2.65)}$$

This equation is known as *Poisson's equation* which states that the potential distribution in a region depends on the local charge distribution.

In many boundary value problems, the charge distribution is involved on the surface of the conductors; for which the free volume charge density is zero, i.e., $\rho = 0$. In that case, Poisson's equation reduces to

$$\boxed{\nabla^2 V = 0} \qquad \textbf{(2.66)}$$

This equation is known as *Laplace's equation.*

Substituting the Laplacian operator ∇^2 as discussed in Chapter 1, the Laplace's equation in three different coordinate systems becomes

$$\frac{\partial^2 V}{\partial x^2} + \frac{\partial^2 V}{\partial y^2} + \frac{\partial^2 V}{\partial z^2} = 0 \quad \text{in Cartesian coordinates}$$

$$\frac{1}{r}\frac{\partial}{\partial r}\left(r\frac{\partial V}{\partial r}\right) + \frac{1}{r^2}\frac{\partial^2 V}{\partial \phi^2} + \frac{\partial^2 V}{\partial z^2} = 0 \quad \text{in cylindrical coordinates} \qquad \textbf{(2.67)}$$

$$\frac{1}{\rho^2}\frac{\partial}{\partial \rho}\left(\rho^2\frac{\partial V}{\partial \rho}\right) + \frac{1}{\rho^2 \sin\theta}\frac{\partial}{\partial \theta}\left(\sin\theta\frac{\partial V}{\partial \theta}\right) + \frac{1}{\rho^2 \sin^2\theta}\frac{\partial^2 V}{\partial \phi^2} = 0 \quad \text{in spherical coordinates}$$

Respective Poisson's equations can be written from Eq. (2.67) by replacing zero on the right hand side by $-\dfrac{\rho}{\varepsilon}$.

Example 2.52 Find whether the potential functions in a region of free space satisfy the Laplace's equation:

(a) $V = e^{-5x}\cos(13y)\sinh(12z)$, **(b)** $V = \dfrac{z\cos\phi}{r}$ **(c)** $V = \dfrac{30}{\rho^2}\cos\theta$

Solution

(a) $V = e^{-5x}\cos(13y)\sinh(12z)$

$$\therefore \quad \frac{\partial V}{\partial x} = -5e^{-5x}\cos(13y)\sinh(12z)$$

$$\therefore \quad \frac{\partial^2 V}{\partial x^2} = (-5)^2 e^{-5x}\cos(13y)\sinh(12z) = 25V \qquad \text{(i)}$$

$$\therefore \quad \frac{\partial V}{\partial y} = -13e^{-5x}\sin(13y)\sinh(12z)$$

$$\therefore \quad \frac{\partial^2 V}{\partial y^2} = (-13\times 13)e^{-5x}\cos(13y)\sinh(12z) = -169V \qquad \text{(ii)}$$

$$\therefore \quad \frac{\partial V}{\partial z} = 12e^{-5x}\cos(13y)\cosh(12z)$$

$$\therefore \quad \frac{\partial^2 V}{\partial y^2} = (12)^2 e^{-5x}\cos(13y)\sinh(12z) = 144V \qquad \text{(iii)}$$

Adding (i), (ii), and (iii), we get

$$\frac{\partial^2 V}{\partial x^2} + \frac{\partial^2 V}{\partial y^2} + \frac{\partial^2 V}{\partial z^2} = (25 - 169 + 144)V = 0$$

$$\therefore \quad \nabla^2 V = 0$$

Hence, this potential function satisfies Laplace's equation.

(b) $V = \dfrac{z\cos\phi}{r}$

$$\frac{\partial V}{\partial r} = -\frac{z\cos\phi}{r^2} \Rightarrow r\frac{\partial V}{\partial r} = -\frac{z\cos\phi}{r}$$

$$\therefore \quad \frac{\partial}{\partial r}\left(r\frac{\partial V}{\partial r}\right) = \frac{\partial}{\partial r}\left(-\frac{z\cos\phi}{r}\right) = \frac{z\cos\phi}{r^2}$$

$$\therefore \quad \frac{1}{r}\frac{\partial}{\partial r}\left(r\frac{\partial V}{\partial r}\right) = \frac{z\cos\phi}{r^3} = \frac{1}{r^2}V \qquad \text{(i)}$$

$$\frac{\partial V}{\partial \phi} = -\frac{z\sin\phi}{r}$$

$$\therefore \qquad \frac{\partial^2 V}{\partial \phi^2} = -\frac{z \cos \phi}{r}$$

$$\therefore \qquad \frac{1}{r^2}\frac{\partial^2 V}{\partial \phi^2} = -\frac{1}{r^2}\frac{z \cos \phi}{r} = -\frac{1}{r^2}V \qquad \textbf{(ii)}$$

$$\frac{\partial V}{\partial z} = \frac{\cos \phi}{r}$$

$$\therefore \qquad \frac{\partial^2 V}{\partial z^2} = 0 \qquad \textbf{(iii)}$$

Adding (i), (ii), and (iii), we get

$$\frac{1}{r}\frac{\partial}{\partial r}\left(r\frac{\partial V}{\partial r}\right) + \frac{1}{r^2}\frac{\partial^2 V}{\partial \phi^2} + \frac{\partial^2 V}{\partial z^2} = -\frac{1}{r^2}V + \frac{1}{r^2}V + 0 = 0$$

$$\therefore \qquad \nabla^2 V = 0$$

Hence, this potential function satisfies Laplace's equation.

(c) $V = \dfrac{30}{\rho^2}\cos\theta$

$$\frac{\partial V}{\partial \rho} = -\frac{2 \times 30}{\rho^3}\cos\theta \quad \Rightarrow \quad \rho^2 \frac{\partial V}{\partial \rho} = -\frac{60}{\rho}\cos\theta$$

$$\therefore \qquad \frac{\partial}{\partial \rho}\left(\rho^2 \frac{\partial V}{\partial \rho}\right) = \frac{60}{\rho^2}\cos\theta$$

$$\therefore \qquad \frac{1}{\rho^2}\frac{\partial}{\partial r}\left(\rho^2 \frac{\partial V}{\partial \rho}\right) = \frac{60}{\rho^4}\cos\theta = \frac{2}{\rho^2}V \qquad \textbf{(i)}$$

$$\frac{\partial V}{\partial \theta} = -\frac{30}{\rho^2}\sin\theta \quad \sin\theta\frac{\partial V}{\partial \theta} = -\frac{30}{\rho^2}\sin^2\theta$$

$$\therefore \qquad \frac{\partial}{\partial \theta}\left(\sin\theta\frac{\partial V}{\partial \theta}\right) = -\frac{30}{\rho^2} \times 2\sin\theta\cos\theta = -\frac{60}{\rho^2}\sin\theta\cos\theta$$

$$\therefore \qquad \frac{1}{\rho^2 \sin\theta}\frac{\partial}{\partial \theta}\left(\sin\theta\frac{\partial V}{\partial \theta}\right) = -\frac{60}{\rho^4}\cos\theta = -\frac{2}{\rho^2}V \qquad \textbf{(ii)}$$

$$\frac{\partial V}{\partial \phi} = 0$$

$$\therefore \qquad \frac{\partial^2 V}{\partial \phi^2} = 0 \qquad \textbf{(iii)}$$

Adding (i), (ii), and (iii), we get

$$\frac{1}{\rho^2}\frac{\partial}{\partial r}\left(\rho^2 \frac{\partial V}{\partial \rho}\right) + \frac{1}{\rho^2 \sin\theta}\frac{\partial}{\partial \theta}\left(\sin\theta\frac{\partial V}{\partial \theta}\right) + \frac{\partial^2 V}{\partial \phi^2} = \frac{2}{\rho^2}V - \frac{2}{\rho^2}V + 0 = 0$$

$$\therefore \qquad \nabla^2 V = 0$$

Hence, this potential function satisfies Laplace's equation.

2.20 UNIQUENESS THEOREM

Statement This theorem states that any solution of Laplace equation (or Poisson's equation) that satisfies the same boundary conditions must be the only solution, irrespective of the method of solution.

Proof The proof of this theorem follows a *proof by contradiction.*

Consider a volume v bounded by some surface S. Suppose we are given the charge density ρ throughout the volume, and the value of the scalar potential V on the surface. Suppose, for the sake of argument, that the solution of Laplace's (or Poisson's) equation is not unique. Let there be two potentials V_1 and V_2 and which satisfy the boundary conditions.

Thus,

$$\nabla^2 V_1 = -\frac{\rho}{\varepsilon} \quad \text{and} \quad \nabla^2 V_2 = -\frac{\rho}{\varepsilon} \text{ throughout } v, \text{ and}$$

$$V_1 = V_2 = V_S \text{ on } S.$$

$$\therefore \qquad \nabla^2 (V_1 - V_2) = 0 \tag{2.68}$$

and

$$(V_{1S} - V_{2S}) = 0 \tag{2.69}$$

According to vector identity for any scalar A and any vector $\vec{B}$

$$\nabla \cdot (A\vec{B}) \equiv A(\nabla \cdot \vec{B}) + \vec{B} \cdot (\nabla A)$$

Let, $A = (V_1 - V_2)$ and $\vec{B} = \nabla(V_1 - V_2)$

Thus,

$$\begin{aligned}\nabla \cdot [(V_1 - V_2)\nabla(V_1 - V_2)] &= (V_1 - V_2)[\nabla \cdot \nabla(V_1 - V_2)] + \nabla(V_1 - V_2) \cdot \nabla(V_1 - V_2) \\ &= (V_1 - V_2)\nabla^2(V_1 - V_2) + [\nabla(V_1 - V_2)]^2\end{aligned}$$

Integrating throughout the volume enclosed by the surface, we get

$$\int_v \nabla \cdot [(V_1 - V_2)\nabla(V_1 - V_2)]dv = \int_v (V_1 - V_2)\nabla^2(V_1 - V_2)dv + \int_v [\nabla(V_1 - V_2)]^2 dv \tag{2.70}$$

Applying divergence theorem to the left-hand side of Eq. (2.70), we get

$$\int_v \nabla \cdot [(V_1 - V_2)\nabla(V_1 - V_2)]dv = \int_S [(V_{1S} - V_{2S})\nabla(V_{1S} - V_{2S})] \cdot d\vec{S} = 0 \quad \{\text{using Eq. (2.69)}\}$$

Also, in the right-hand side of Eq. (2.70), $\nabla^2(V_1 - V_2)$ using Eq. (2.68).

Thus, Eq. (2.70) reduces to

$$\int_v [\nabla(V_1 - V_2)]^2 dv = 0$$

Here, the quantity $[\nabla(V_1 - V_2)]^2$ is always positive. The only way in which the volume integral of a positive definite quantity can be zero is if that quantity itself is zero throughout the volume. This is not necessarily the case for a non-positive definite quantity: we could have positive and negative contributions from various regions inside the volume which cancel one, hence, the integrand must be zero everywhere, so that the integral may be zero.

$$\therefore \qquad [\nabla(V_1 - V_2)]^2 = 0 \;\Rightarrow\; \nabla(V_1 - V_2) = 0 \;\Rightarrow\; (V_1 - V_2) = \text{constant } (k)$$

Now, applying the boundary condition that on the surface

$$(V_1 - V_2) = (V_{1S} - V_{2S}) = 0$$

$$\therefore \qquad V_1 = V_2$$

Hence, our initial assumption that V_1 and V_2 are two different solutions of Poisson's equation, satisfying the same boundary conditions, turns out to be incorrect.

The fact that the solutions to Poisson's equation are unique is very useful.

2.21 GENERAL PROCEDURE FOR SOLVING POISSON'S AND LAPLACE'S EQUATIONS

The following procedure may be followed for solving a boundary value problem, using Poisson's or Laplace's equation.

1. The equation is solved using either
   — direct integration when V is a function of one variable.
   — separation of variables if V is a function of more than one variable.
2. The unknown integration constants are found by applying the boundary conditions; so that the solution becomes unique.
3. Having obtained V, the field is obtained from the relation: $\vec{E} = -\nabla V$.
4. The charge induced on a conductor is obtained from the relation: $Q = \int_S \sigma dS$; where σ is the induced surface charge density, $\sigma = D_n = \varepsilon E_n$; where E_n is the normal component of the field.
5. The capacitance between two conductors is obtained from the relation: $C = \frac{Q}{V}$.

2.21.1 Solution of Laplace Equations in Cartesian Coordinates

Parallel Plate Electrode System

Example 2.53 The region between two conducting plates at $x = 0$ and $x = d$ is filled with perfect dielectric of uniform permittivity ε. If the plate at $x = d$ is maintained at a voltage V_0 and at $x = 0$ is grounded, find the potential distribution between the plates.

Solution Here, the space between the plates is filled with perfect dielectric material and there is no variation in the y and z direction. So, the problem is one-dimensional.

By Laplace's equation in Cartesian coordinates

$$\frac{\partial^2 V}{\partial x^2} + \frac{\partial^2 V}{\partial y^2} + \frac{\partial^2 V}{\partial z^2} = 0$$

As $\frac{\partial^2 V}{\partial y^2} = 0 = \frac{\partial^2 V}{\partial z^2}$, and V varies with x only the partial derivative becomes ordinary derivative.

$$\therefore \qquad \frac{d^2 V}{dx^2} = 0$$

Integrating twice, we get

$$V = Ax + B$$

Applying boundary conditions:

(1) At $x = 0$, $V = 0$,

$\therefore \quad B = 0$

(2) At $x = d$, $V = V_0$,

$\therefore \quad A = \dfrac{V_0}{d}$

Hence, the potential distribution between the parallel-plate electrode system is given as

$$\boxed{V = \left(\frac{V_0}{d}\right)x} \qquad \textbf{(2.71)}$$

Example 2.54 Two parallel planes of infinite extent in the x and y directions and separated by a distance d in the z-direction have a potential difference applied between them. The upper plane has a potential V_1 and the lower one has a potential V_0 ($V_1 > V_0$).

(a) By using Laplace's equation, find the potential distribution and electric field strength in the region between the planes.

(b) What will be the potential at z ($0 \le z \le d$) if the space between the planes is filled with electric charge of volume density $\rho = \rho_0 \dfrac{z}{d}$. Also, obtain the surface charge density on each plane.

Solution

(a) The planes being infinite, the potential V is a function of z only. Hence, Laplace's equation can be written as

$$\frac{d^2V}{dz^2} = 0$$

Integrating twice

$$V = C_1 z + C_2$$

where C_1 and C_2 are constants.

Applying boundary conditions, we get

$$\text{At } z = 0, V = V_0 \quad \Rightarrow \quad C_2 = V_0$$

$$\text{At } z = d, V = V_1 \quad \Rightarrow \quad C_1 = \frac{V_1 - V_0}{d}$$

Therefore, the potential is given as

$$V = \left(\frac{V_1 - V_0}{d}\right)z + V_0$$

Hence, the field intensity is given as

$$E = -\frac{dV}{dz} = \left(\frac{V_0 - V_1}{d}\right)$$

(b) If the volume charge density ρ is present, then by Poisson's equation

$$\frac{d^2V}{dz^2} = -\frac{\rho}{\varepsilon} = -\frac{\rho_0 z}{\varepsilon d}$$

Integrating twice

$$V = -\frac{\rho_0}{2\varepsilon d}\frac{z^3}{3} + C_1 z + C_2$$

where C_1 and C_2 are constants.

Applying boundary conditions, we get

$$\text{At } z = 0, V = V_0 \quad \Rightarrow \quad C_2 = V_0$$

$$\text{At } z = d, V = V_1 \quad \Rightarrow \quad V_1 = -\frac{\rho_0}{6\varepsilon d}d^3 + C_1 d + C_2 \quad \Rightarrow \quad C_1 = \frac{V_1 - V_0}{d} + \frac{\rho_0 d}{6\varepsilon}$$

Therefore, the potential is given as

$$V = -\frac{\rho_0}{6\varepsilon d}z^3 + \left(\frac{V_1 - V_0}{d} + \frac{\rho_0 d}{6\varepsilon}\right)z + V_0$$

Hence, the field intensity is given as

$$E = -\frac{dV}{dz} = \frac{\rho_0}{2\varepsilon d}z^2 - \left(\frac{V_1 - V_0}{d} + \frac{\rho_0 d}{6\varepsilon}\right)$$

The surface charge density on each plane is

$$\sigma|_{z=0} = \varepsilon E|_{z=0} = -\varepsilon\left(\frac{V_1 - V_0}{d} + \frac{\rho_0 d}{6\varepsilon}\right)$$

$$\sigma|_{z=d} = \varepsilon E|_{z=d} = \frac{\rho_0 d}{2} - \varepsilon\left(\frac{V_1 - V_0}{d} + \frac{\rho_0 d}{6\varepsilon}\right) = \frac{\rho_0 d}{3} - \varepsilon\left(\frac{V_1 - V_0}{d}\right)$$

2.21.2 Solution of Laplace Equations in Cylindrical Coordinates

Potential of Coaxial Cable

***Example 2.55** Using Laplace's equation, find the potential distribution within a coaxial cable of length 'L', having an inner conductor of radius 'a' and a outer conductor of radius 'b', if potential of 'V_0' is applied at the inner conductor with reference to outer conductor. Also, determine $\vec{E}$.

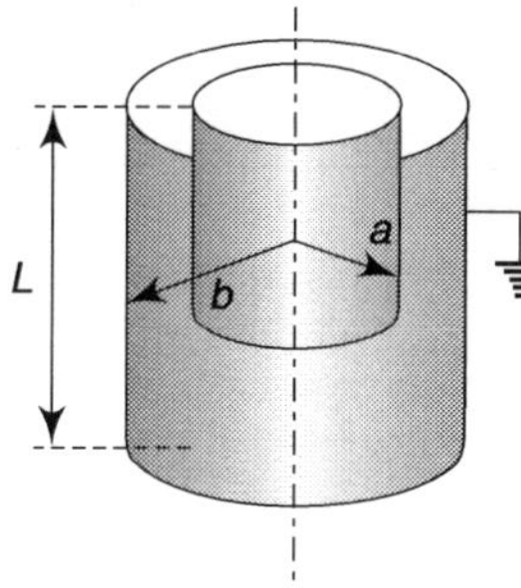

Fig. 2.61 *Coaxial cable*

Solution Since the medium between the concentric cables is filled with perfect dielectric, we use Laplace's equation

$$\nabla^2 V = 0$$

In cylindrical coordinates

$$\frac{1}{r}\frac{\partial}{\partial r}\left(r\frac{\partial V}{\partial r}\right) + \frac{1}{r^2}\frac{\partial^2 V}{\partial \phi^2} + \frac{\partial^2 V}{\partial z^2} = 0$$

Since the potential variation is only in the radial direction, the equation reduces to

$$\frac{1}{r}\frac{\partial}{\partial r}\left(r\frac{\partial V}{\partial r}\right) = 0$$

or
$$\frac{\partial}{\partial r}\left(r\frac{\partial V}{\partial r}\right)=0$$

Integrating
$$r\frac{\partial V}{\partial r}=A$$

or
$$\frac{\partial V}{\partial r}=\frac{A}{r}$$

Integrating
$$V = A\ln r + B$$

Applying boundary conditions:

(1) At $r = a$, $V = V_0$,
$\therefore\ V_0 = A\ln a + B$

(2) At $r = b$, $V = 0$,
$\therefore\ 0 = A\ln b + B$

Solving $A=\dfrac{V_0}{\ln(a/b)}$ and $B=-A\ln b=-\dfrac{V_0\ln b}{\ln(a/b)}$

Hence, the potential distribution for the co-axial cable is given as
$$V=\frac{V_0\ln r}{\ln(a/b)}-\frac{V_0\ln b}{\ln(a/b)}=\frac{V_0\ln(r/b)}{\ln(a/b)}$$

$$\boxed{V=\frac{V_0}{\ln(a/b)}\ln\left(\frac{r}{b}\right)} \qquad \textbf{(2.72a)}$$

Now, the field intensity is given as
$$\vec{E}=-\nabla V=-\left(\frac{\partial V}{\partial r}\hat{a}_r+\frac{1}{r}\frac{\partial V}{\partial \phi}\hat{a}_\phi+\frac{\partial V}{\partial z}\hat{a}_z\right)$$

Here, as the potential is a function of only r, the field intensity is given as
$$\vec{E}=-\nabla V=-\frac{\partial V}{\partial r}\hat{a}_r=-\frac{V_0}{\ln(a/b)}\frac{\partial}{\partial r}\left[\ln\left(\frac{r}{b}\right)\right]\hat{a}_r=-\frac{V_0}{\ln(a/b)}\times\frac{b}{r}\times\frac{1}{b}\hat{a}_r=\frac{V_0}{r\ln(b/a)}\hat{a}_r$$

$$\boxed{\vec{E}=\frac{V_0}{r\ln(b/a)}\hat{a}_r} \qquad \textbf{(2.72b)}$$

***Example 2.56** In cylindrical coordinates two (ϕ = constant) planes are insulated along z-axis. Neglect fringing and calculate the expression for V and $\vec{E}$ between the planes assuming that $V = 100$ Volt for $\phi = \alpha$ and $V = 0$ at $\phi = 0$ as shown in Fig. 2.62.

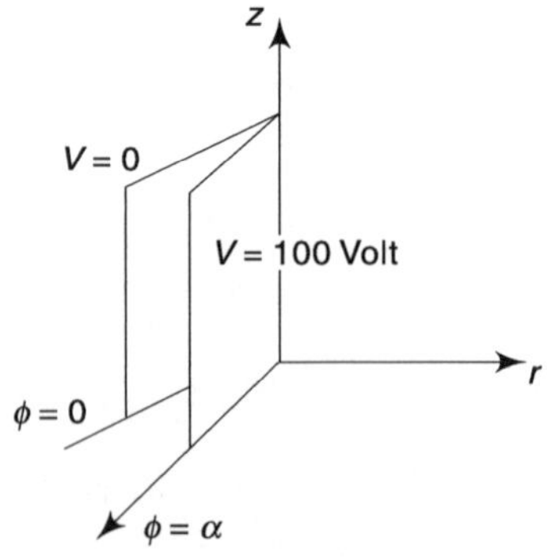

Fig. 2.62 *Arrangement of Example 2.56*

Solution Since the medium between the concentric cables is filled with perfect dielectric, we use Laplace's equation
$$\nabla^2 V = 0$$

In cylindrical coordinates

$$\frac{1}{r}\frac{\partial}{\partial r}\left(r\frac{\partial V}{\partial r}\right)+\frac{1}{r^2}\frac{\partial^2 V}{\partial \phi^2}+\frac{\partial^2 V}{\partial z^2}=0$$

Since the potential is constant with respect to r and z

$$\frac{1}{r^2}\frac{\partial^2 V}{\partial \phi^2}=0$$

or

$$\frac{\partial^2 V}{\partial \phi^2}=0$$

Integrating

$$\frac{\partial V}{\partial \phi}=A$$

Integrating

$$V = A\phi + B$$

Applying boundary conditions:

(1) At $\phi = \alpha$, $V = 100$,
$\therefore\ 100 = A\alpha + B$

(2) At $\phi = 0$, $V = 0$
$\therefore\ 0 = B$

Solving, $A = \dfrac{100}{\alpha}$ and $B = 0$

Hence, the potential distribution for the coaxial cable is given as

$$V = \frac{100}{\alpha}\phi \quad \text{Volt}$$

$$\boxed{V = \frac{100}{\alpha}\phi}$$

Now, the field intensity is given as

$$\vec{E} = -\nabla V = -\frac{1}{r}\frac{\partial V}{\partial \phi}\hat{a}_\phi$$

Here, as the potential is a function of only r, the field intensity is given as

$$\vec{E} = -\nabla V = -\frac{1}{r}\frac{\partial V}{\partial \phi}\hat{a}_\phi = -\frac{1}{r}\frac{\partial}{\partial \phi}\left[\frac{100}{\alpha}\phi\right]\hat{a}_\phi = -\frac{100}{\alpha}\left(\frac{1}{r}\right)\hat{a}_\phi$$

$$\boxed{\vec{E} = -\frac{100}{\alpha}\left(\frac{1}{r}\right)\hat{a}_\phi}$$

Example 2.57 Find the potential and electric field intensity for the region between two concentric right circular cylinders, where $V = 0$ at $r_a = 1$ mm and $V = 100$ V at $r_b = 20$ mm. Neglect fringing.

Solution Since, V varies with r only, Laplace's equation can be written as

$$\frac{1}{r}\frac{d}{dr}\left(r\frac{dV}{dr}\right)=0$$

Integrating twice

$$V = A\ln r + B$$

where A and B are constants.

Applying boundary conditions, we get

$$\begin{aligned}&\text{At } r = r_a = 0.001, V = 0 \quad \Rightarrow \quad 0 = A\ln(0.001) + B\\&\text{At } r = r_b = 0.02, V = 100 \quad \Rightarrow \quad 100 = A\ln(0.02) + B\end{aligned}$$

Solving, $A = 33.36$ and $B = 230.49$

$\therefore$
$$V = (33.36\ln r + 230.49)\text{ Volt}$$

Hence, the field intensity is given as

$$\vec{E} = -\nabla V = -\frac{\partial V}{\partial r}\hat{a}_r = -\frac{33.36}{r}\hat{a}_r\text{ V/m}$$

2.21.3 Solution of Laplace Equations in Spherical Coordinates

Potential of Spherical Shell Electrode System

***Example 2.58** Develop expressions for the potential difference and field intensity at any point between spherical shells in terms of the applied potential. Given: $V = V_0$; at $r = a$ and $V = 0$, at $r = b$ $(a < b)$. Hence, find $\vec{E}$.

Solution By Laplace's equation in spherical coordinates

$$\frac{1}{r^2}\frac{\partial}{\partial r}\left(r^2\frac{\partial V}{\partial r}\right)+\frac{1}{r^2\sin\theta}\frac{\partial}{\partial\theta}\left(\sin\theta\frac{\partial V}{\partial\theta}\right)+\frac{1}{r^2\sin^2\theta}\frac{\partial^2 V}{\partial\phi^2}=0$$

By symmetry of the problem, the field and, therefore, the potential function depends on the radial distance 'r' only.

$$\frac{1}{r^2}\frac{\partial}{\partial r}\left(r^2\frac{\partial V}{\partial r}\right)=0$$

or

$$\frac{\partial}{\partial r}\left(r^2\frac{\partial V}{\partial r}\right)=0$$

Integrating

$$\frac{\partial V}{\partial r}=\frac{A}{r^2}$$

Integrating once

$$V=-\frac{A}{r}+B$$

Applying boundary conditions:

(1) At $r = a$, $V = V_0$,

$$\therefore \quad V_0 = -\frac{A}{a} + B$$

(2) At $r = b$, $V = 0$,

$$\therefore \quad 0 = -\frac{A}{b} + B$$

Solving, $A = \dfrac{V_0 ab}{a-b}$ and $B = \dfrac{V_0 a}{a-b}$

Hence, the potential distribution for the spherical shell is given as

$$V = -\left(\frac{V_0 ab}{a-b}\right)\frac{1}{r} + \frac{V_0 a}{a-b} = \frac{V_0 ab}{a-b}\left(\frac{1}{r} - \frac{1}{b}\right) = \frac{V_0 a}{r}\left(\frac{b-r}{b-a}\right)$$

$$\boxed{V = \frac{V_0 a}{r}\left(\frac{b-r}{b-a}\right)} \qquad \textbf{(2.73a)}$$

Here, as the potential is a function of only r, the field intensity is given as

$$\vec{E} = -\nabla V = -\frac{\partial V}{\partial r}\hat{a}_r = \frac{V_0 ab}{b-a}\left[\frac{\partial}{\partial r}\left(\frac{1}{r} - \frac{1}{b}\right)\right]\hat{a}_r = \left(\frac{V_0 ab}{a-b}\right)\frac{1}{r^2}\hat{a}_r$$

$$\boxed{\vec{E} = \left(\frac{V_0 ab}{a-b}\right)\frac{1}{r^2}\hat{a}_r} \qquad \textbf{(2.73b)}$$

***Example 2.59** Region between the two coaxial cones is shown in Fig. 2.63. A potential V_1 exists at θ_1 and $V = 0$ at θ_2. The cone vertices are insulated at $r = 0$. Solve Laplace's equation to obtain potential at a cone at any angle θ.

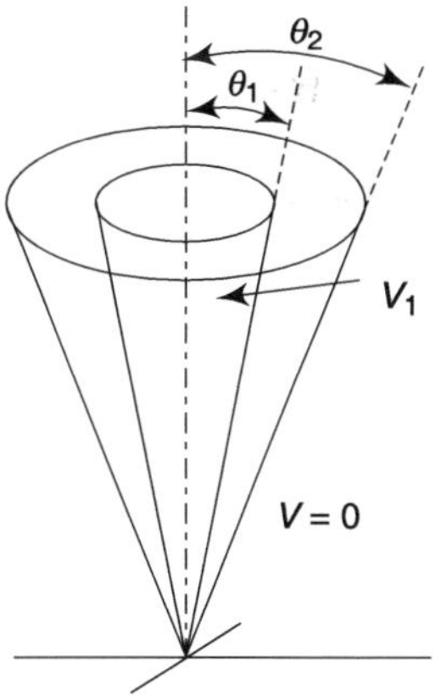

Fig. 2.63 *Coaxial cones of Example 2.59*

Solution Since, V varies with θ only, Laplace's equation can be written as

$$\frac{1}{r^2 \sin\theta}\frac{d}{d\theta}\left(\sin\theta \frac{dV}{d\theta}\right) = 0$$

Integrating

$$\sin\theta \frac{dV}{d\theta} = A \quad \Rightarrow \quad \frac{dV}{d\theta} = A \operatorname{cosec}\theta$$

Integrating once again

$$V = A \ln(\tan\tfrac{\theta}{2}) + B$$

where A and B are constants.

Applying boundary conditions, we get

$$\text{At } \theta = \theta_1, \quad V = V_1, \quad \Rightarrow \quad V_1 = A \ln\left(\tan\tfrac{\theta_1}{2}\right) + B$$

$$\text{At } \theta = \theta_2, \quad V = 0, \quad \Rightarrow \quad 0 = A \ln\left(\tan\tfrac{\theta_2}{2}\right) + B$$

Solving for A and B

$$A = \frac{V_1}{\ln\left(\tan\frac{\theta_1}{2}\right) - \ln\left(\tan\frac{\theta_2}{2}\right)}$$

$$B = -A\ln\left(\tan\frac{\theta_2}{2}\right) = -V_1\frac{\ln\left(\tan\frac{\theta_2}{2}\right)}{\ln\left(\tan\frac{\theta_1}{2}\right) - \ln\left(\tan\frac{\theta_2}{2}\right)}$$

Substituting the values of A and B and rearranging, we get

$$V = V_1\frac{\ln\left(\tan\frac{\theta}{2}\right) - \ln\left(\tan\frac{\theta_2}{2}\right)}{\ln\left(\tan\frac{\theta_1}{2}\right) - \ln\left(\tan\frac{\theta_2}{2}\right)}$$

Example 2.60 Two coaxial cones are insulated from each other at their vertices. Axes are along the z-axis. In spherical coordinates, the cones are presented by $\theta_1 = 10°$ and $\theta_2 = 45°$. The inner cone is at potential of 100 V and the potential of the outer cone is 0 V. Determine the potential for $\theta = 30°$.

Solution From Example 2.57, substituting the values, $V_1 = 100$, $\theta_1 = 10°$, $\theta_2 = 45°$, $\theta = 30°$ we get the potential given as

$$V = V_1\left\{\frac{\ln(\tan\theta/2) - \ln(\tan\theta_2/2)}{\ln(\tan\theta_1/2) - \ln(\tan\theta_2/2)}\right\} = 100\left\{\frac{\ln(\tan 15°) - \ln(\tan 22.5°)}{\ln(\tan 5°) - \ln(\tan 22.5°)}\right\} = 28.01\text{ V}$$

2.22 CAPACITOR AND CAPACITANCE

Capacitor: A capacitor is a device that stores electric charge and hence electrostatic energy. It consists of two conductors separated by an insulating medium.

Depending upon the shape and size of the conductors and insulating medium, capacitors are available in varying in shape and size, but the basic configuration is two conductors carrying equal but opposite charges (Fig. 2.64).

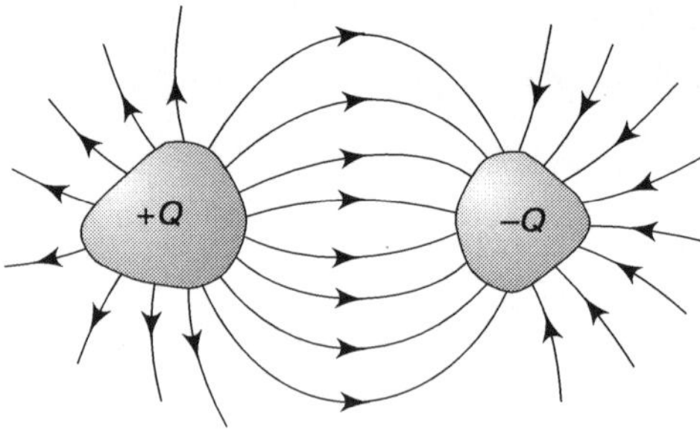

Fig. 2.64 *Basic configuration of a capacitor*

When the capacitor is *uncharged*, the charge on any one of the conductors is zero. During the charging process, a charge Q is moved from one conductor to the other, giving one conductor a charge $+Q$, and the other one a charge $-Q$. Thus, a potential difference ΔV is created, with the positively charged conductor at a higher potential than the negatively charged conductor. This must be remembered that whether charged or uncharged, the net charge on the capacitor as a whole is zero.

Experiments show that the amount of charge Q stored in a capacitor is linearly proportional to the electric potential difference between the conductors, ΔV. Thus, we may write

$$Q = C|\Delta V| \quad \Rightarrow \quad C = \frac{Q}{|\Delta V|}$$

where C is a positive proportionality constant called *capacitance*. Physically, capacitance is a measure of the capacity of storing electric charge for a given potential difference ΔV.

Capacitance: The capacitance of a capacitor is the ratio of the magnitude of charge on one conductor to the potential difference between the conductors.

$$C = \frac{Q}{V} = \frac{\varepsilon \oint_S \vec{E} \cdot d\vec{S}}{\int_l \vec{E} \cdot d\vec{l}} \qquad \textbf{(2.74)}$$

The SI unit of capacitance is the *farad* (F):

$$1\text{F} = 1 \text{ farad} = 1 \text{ coulomb/volt} = 1 \text{ C/V}$$

The dimension of capacitance in terms of M, L, T and I is L, T and I is $[M^{-1}L^{-2}T^4I^2]$.

Figure 2.65 (a) shows the symbol which is used to represent capacitors in circuits. For a polarised fixed capacitor which has a definite polarity, Fig. 2.65 (b) is sometimes used.

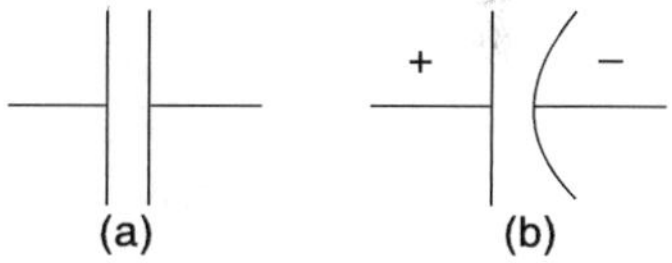

Fig. 2.65 *Symbol of capacitor*

Energy stored in Capacitors The energy stored in a capacitor is equal to the *work* done to charge it. Consider a capacitance C, holding a charge $+q$ on one plate and $-q$ on the other. Moving a small element of charge dq from one plate to the other against the potential difference $V = q/C$ requires the work dW:

$$dW = \frac{q}{C}dq$$

where W is the work in Joule
q is the charge in Coulomb
C is the capacitance in Farad

We can find the energy stored in a capacitance by integrating this equation. Starting with an uncharged capacitance ($q = 0$), the work done, in moving charge from one plate to the other until the plates have charge $+Q$ and $-Q$ respectively, is given as

$$W_{\text{charging}} = \int_0^Q \frac{q}{C}dq = \frac{1}{2}\frac{Q^2}{C} = \frac{1}{2}CV^2 = W_{\text{stored}}$$

$$\boxed{W_{\text{stored}} = \frac{1}{2}CV^2} \qquad \textbf{(2.75)}$$

where W is the energy in Joule
C is the capacitance in Farad
V is the voltage in Volt

General Procedure for Calculating Capacitance of Capacitor

1. A suitable coordinate system is chosen.
2. We assume that the two conductors carry charges $+Q$ and $-Q$.
3. The field intensity $\vec{E}$ is calculated using either Gauss' law or Coulomb's law.
4. The potential difference between the two conductors is calculated using the relation, $V = \int_l \vec{E} \cdot d\vec{l}$.
5. Finally, the value of the capacitance is obtained from the relation, $C = \dfrac{Q}{V}$.

We will now compute the capacitance in some systems with simple geometry.

2.22.1 Parallel Plate Capacitor

Example 2.61 Determine the capacitance of a parallel plate capacitor with plate separation d and area A.

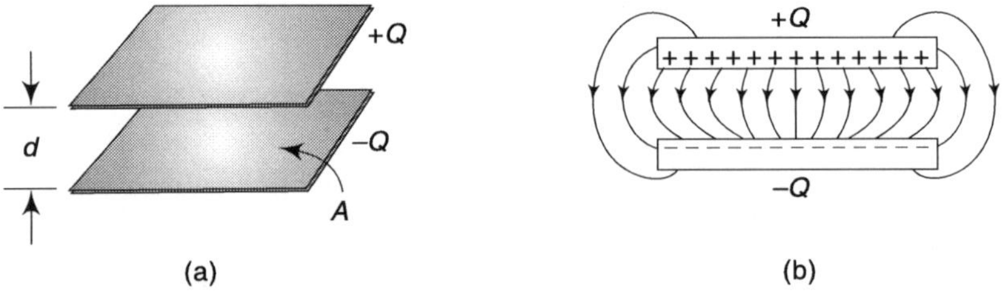

Fig. 2.66 *(a) Parallel plate capacitor, and (b) Electric field line for parallel plate capacitor*

Solution We shall consider two metallic plates of equal area A separated by a distance d, as shown in Fig. 2.66 (*a*). The charge on the top plate is $+Q$ while the charge on the bottom plate is $-Q$, both distributed uniformly on the plates.

Question: What do you mean by the *edge effects* and *fringing effects*?
Answer: In order to determine the capacitance C of parallel plate capacitor, the knowledge of the electric field between the plates is necessary. For an ideal capacitor, the plate separation d is very small compared with the dimensions of the plate. However, a real capacitor is finite in size. Thus, the electric field lines at the edge of the plates are not straight lines, and the field is not contained entirely between the plates. This is known as *edge effects*, and the non-uniform fields near the edge are called the *fringing fields*. In Fig. 2.41 (*b*), the field lines are drawn by taking into consideration these edge effects.

Neglecting the edge effects and fringing effects, assuming an ideal situation, where field lines between the plates are straight lines, the electric field is calculated using Gauss' law as

$$\oint_S \vec{E} \cdot d\vec{S} = \frac{Q_{\text{enc}}}{\varepsilon}$$

By choosing a Gaussian "pillbox" with surface area A to enclose the charge on the positive plate (see Fig. 2.66), the electric field in the region between the plates is given as

$$\vec{E} = \frac{\sigma}{\varepsilon}(-\hat{a}_z) = -\frac{Q}{\varepsilon A}\hat{a}_z$$

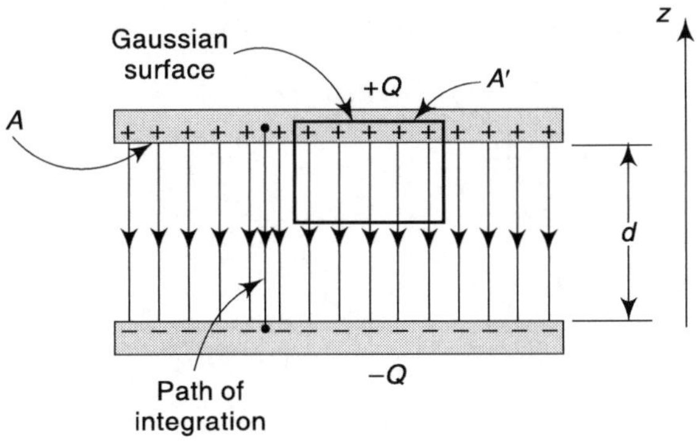

Fig. 2.67 *Gaussian surface for calculating the electric field between the plates*

The potential difference between the plates is

$$V = (V_+ - V_-) = -\int_-^+ \vec{E} \cdot d\vec{l} = -\int_0^d \left(-\frac{Q}{\varepsilon A}\hat{a}_z\right) \cdot dz\hat{a}_z = \frac{Qd}{\varepsilon A}$$

From the definition of capacitance, we have

$$\boxed{C = \frac{Q}{V} = \frac{\varepsilon A}{d}} \qquad \textbf{(2.76)}$$

NOTE:

The capacitance C depends only on the geometric factors A and d. The capacitance C is directly proportional to the area A since for a given potential difference V, a bigger plate can hold more charge. On the other hand, C is inversely proportional to d, the distance of separation, because the smaller the value of d, the smaller the potential difference V for a fixed Q.

Example 2.62 The permittivity of the dielectric material between the plates of a parallel plate capacitor varies uniformly from ε_1 at one plate to ε_2 at the other plate. Show that the capacitance is given by

$$C = \frac{A}{D}\frac{\varepsilon_2 - \varepsilon_1}{\ln(\varepsilon_2/\varepsilon_1)}$$

where A and D are the area of each plate and separation between the plates, respectively. Arrive at the value of C for $\varepsilon_1 = \varepsilon_2$.

Solution Let, at a distance x from the plate of dielectric constant ε_1, the permittivity be

$$\varepsilon = \varepsilon_1 + kx$$

Now, at $x = D$, the permittivity is ε_2

$$\therefore \qquad k = \frac{\varepsilon_2 - \varepsilon_1}{D}$$

$$\therefore \qquad \varepsilon = \varepsilon_1 + \left(\frac{\varepsilon_2 - \varepsilon_1}{D}\right)x$$

$\therefore$ Field intensity at a distance x from the plate of permittivity ε_1

$$E(x) = \frac{D}{\varepsilon} = \frac{\sigma}{\varepsilon}, \quad \sigma = \text{surface charge density on plate 1}$$

$$= \frac{\sigma}{\varepsilon_1 + \left(\frac{\varepsilon_2 - \varepsilon_1}{D}\right)x}$$

$\therefore$ Potential difference between the plates is

$$V = \int_0^D E(x)dx = \sigma \int_0^D \frac{1}{\varepsilon_1 + \left(\frac{\varepsilon_2 - \varepsilon_1}{D}\right)x} dx = \frac{\sigma D}{(\varepsilon_2 - \varepsilon_1)}\left[\ln\left(\frac{\varepsilon_2 - \varepsilon_1}{D}x + \varepsilon_1\right)\right]_{x=0}^{x=D}$$

$$= \frac{\sigma AD}{(\varepsilon_2 - \varepsilon_1)A} \ln\left(\frac{\varepsilon_2}{\varepsilon_1}\right)$$

$$= \frac{QD}{A(\varepsilon_2 - \varepsilon_1)} \ln\left(\frac{\varepsilon_2}{\varepsilon_1}\right) \quad \{\because Q = \sigma A\}$$

So, the capacitance of the parallel plate capacitor is

$$C = \frac{Q}{V} = \frac{A}{D} \frac{\varepsilon_2 - \varepsilon_1}{\ln\left(\frac{\varepsilon_2}{\varepsilon_1}\right)}$$

Example 2.63 The region between two conducting plates at $x = 0$ and $x = d$ is filled with perfect dielectric of non-uniform permittivity

$$\varepsilon = \frac{\varepsilon_0}{1 - \left(\frac{x}{2d}\right)}$$

If the plate at $x = d$ is maintained at a voltage V_0 and at $x = 0$ is grounded, find:

(a) The potential between the plates;
(b) The field intensity;
(c) Surface charge density, σ at $x = 0, d$;
(d) Capacitance per unit area for the plates.

Solution We know that, $\vec{E} = -\nabla V$ and $\rho = \nabla \cdot \vec{D} = \nabla \cdot (\varepsilon \vec{E}) = \varepsilon \nabla \cdot (-\nabla V) = -\nabla \cdot (\varepsilon \nabla V)$

Using the identity, for a scalar S and a vector $\vec{V}$,

$$\nabla \cdot (S\vec{V}) = S\nabla \cdot \vec{V} + \vec{V} \cdot \nabla S$$

Here,

$$\nabla \cdot (\varepsilon \vec{E}) = \varepsilon \nabla \cdot \vec{E} + \vec{E} \cdot \nabla \varepsilon$$

or

$$\varepsilon \nabla \cdot (-\nabla V) + (-\nabla V) \cdot \nabla \varepsilon = \rho$$

or

$$\varepsilon \nabla^2 V + \nabla \varepsilon \cdot \nabla V = -\rho$$

For a charge free region, $\rho = 0$, so that

$$\varepsilon \nabla^2 V + \nabla \varepsilon \cdot \nabla V = 0$$

Considering the variation of ε and V only with respect to x, the equation becomes

$$\varepsilon \frac{\partial^2 V}{\partial x^2} + \left(\frac{\partial \varepsilon}{\partial x}\right)\left(\frac{\partial V}{\partial x}\right) = 0$$

or

$$\frac{d}{dx}\left(\varepsilon \frac{dV}{dx}\right) = 0$$

Integrating, we get

$$\varepsilon \frac{\partial V}{\partial x} = A$$

or

$$\frac{\partial V}{\partial x} = \frac{A}{\varepsilon} = \frac{A}{\dfrac{\varepsilon_0}{1-\left(\dfrac{x}{2d}\right)}} = \frac{A}{\varepsilon_0}\left[1 - \frac{x}{2d}\right]$$

Integrating again, we get

$$V(x) = \frac{A}{\varepsilon_0}x - \frac{Ax^2}{4d\varepsilon_0} + B = \frac{A}{\varepsilon_0}\left(x - \frac{x^2}{4d}\right) + B$$

Applying the boundary conditions:

(1) At $x = 0$, $V = 0$,
$0 = 0 - 0 + B \Rightarrow B = 0$

(2) At $x = d$, $V = V_0$,

$$V_0 = \frac{A}{\varepsilon_0}\left(d - \frac{d^2}{4d}\right) = \frac{3}{4}\frac{Ad}{\varepsilon_0} \quad \Rightarrow \quad A = \frac{4}{3}\frac{\varepsilon_0 V_0}{d}$$

(a) So the potential is given as

$$V(x) = \frac{4}{3}\frac{V_0}{d}\left(x - \frac{x^2}{4d}\right)$$

(b) The field intensity is

$$\vec{E} = -\nabla V = -\frac{dV}{dx}\hat{a}_x = -\frac{4}{3}\frac{V_0}{d}\left(1 - \frac{x}{2d}\right)\hat{a}_x$$

(c) Flux density, $\vec{D} = \varepsilon \vec{E} = \varepsilon = \dfrac{\varepsilon_0}{1-\left(\dfrac{x}{2d}\right)} \times \left\{-\dfrac{4}{3}\dfrac{V_0}{d}\left(1 - \dfrac{x}{2d}\right)\hat{a}_x\right\} = -\dfrac{4}{3}\dfrac{\varepsilon_0 V_0}{d}\hat{a}_x$

So the surface charge densities are

$$|\sigma|_{x=0} = |\sigma|_{x=d} = \frac{4}{3}\frac{\varepsilon_0 V_0}{d}$$

(d) Charge on either plate per unit area is

$$\left|\frac{Q}{A}\right| = \frac{4}{3}\frac{\varepsilon_0 V_0}{d}$$

Thus, capacitance per unit area is given as

$$\frac{C}{A} = \frac{Q/A}{V_0} = \frac{4\varepsilon_0}{3d}$$

Example 2.64 The plates of a capacitor are squares, each of side length l, as shown in Fig. 2.68. The plates are inclined to each other at an angle α. The smallest distance between the plates is a. Calculate the capacitance when α is small.

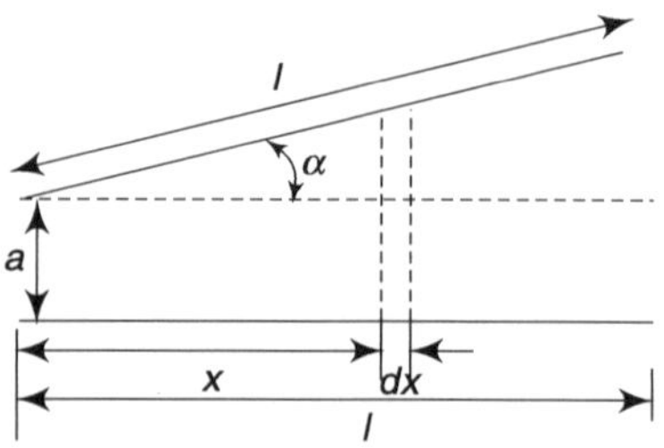

Fig. 2.68 *Capacitor of Example 2.64*

Solution We will consider an elemental strip of the capacitor of thickness dx at a distance x from the edge.

$\therefore$ Plate separation of the strip is

$$d = a + x\tan\alpha = (a + x\alpha) \quad \{\because \alpha <<\}$$

$\therefore$ Capacitance of this strip is, $dC = \dfrac{\varepsilon_0 l dx}{a + x\alpha}$

So, the capacitance of the complete capacitor is given as

$$C = \int dC = \int_0^l \frac{\varepsilon_0 l dx}{a + x\alpha}$$

Let, $(a + x\alpha) = p \quad \therefore \quad \alpha dx = dp$

x	0	l
p	a	$a+\alpha l$

$$\therefore \quad C = \int_0^l \frac{\varepsilon_0 l dx}{a + x\alpha} = \int_a^{a+\alpha l} \frac{\varepsilon_0 l dp}{\alpha p} = \frac{\varepsilon_0 l}{\alpha}\int_a^{a+\alpha l}\frac{dp}{p} = \frac{\varepsilon_0 l}{\alpha}\ln\left(\frac{a+\alpha l}{a}\right) = \frac{\varepsilon_0 l}{\alpha}\ln\left(1 + \frac{\alpha l}{a}\right)$$

$$\approx \frac{\varepsilon_0 l}{\alpha}\left[\left(\frac{\alpha l}{a}\right) - \frac{1}{2}\left(\frac{\alpha l}{a}\right)^2\right]$$

Neglecting the higher order terms as, α is small.

$$\therefore \quad \boxed{C = \frac{\varepsilon_0 l^2}{a}\left(1 - \frac{\alpha l}{2a}\right)}$$

***Example 2.65** A transmission line consists of a pair of long parallel conductors of radius r with a spacing D between centre in a medium of permittivity ε. Show that when D is large compared with r, the capacitance of the pair of conductors per unit length is given approximately by

$$C = \frac{\pi\varepsilon}{\ln(D/r)}$$

Solution We shall consider a line charge distribution of density λ between the two cylindrical conductors. If the charge on conductor A is positive, it will be negative on conductor B by electrostatic induction. (Fig. 2.69)

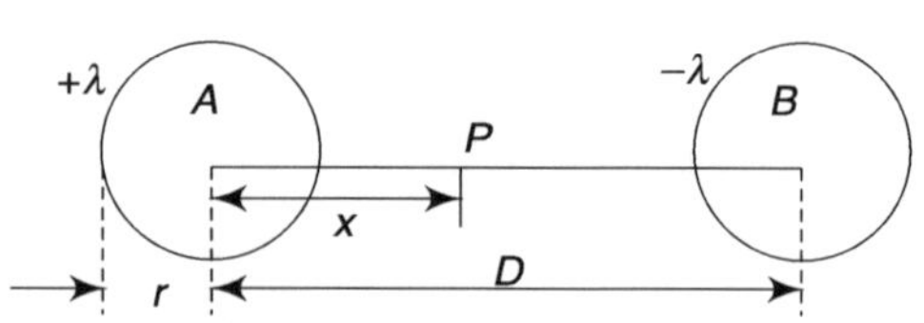

Fig. 2.69 *Capacitance between two wire transmission lines*

The electric field intensity at point P is due to both the charges and is written as

$$E = \frac{\lambda}{2\pi\varepsilon x} - \frac{\lambda}{2\pi\varepsilon(D-x)} = \frac{\lambda}{2\pi\varepsilon}\left[\frac{1}{x} - \frac{1}{D-x}\right]$$

The potential at P is given as

$$V = \int \vec{E} \cdot d\vec{l} = \frac{\lambda}{2\pi\varepsilon} \int_{r}^{D-r} \left[\frac{1}{x} - \frac{1}{D-x}\right] dx$$

$$= \frac{\lambda}{2\pi\varepsilon} [\ln x + \ln(D-x)]_r^{D-r}$$

$$= \frac{\lambda}{2\pi\varepsilon} [\ln(D-r) - \ln r + \ln(D-D+r) - \ln(D-r)]$$

$$= \frac{\lambda}{\pi\varepsilon} \ln\left(\frac{D-r}{r}\right)$$

Hence, the capacitance of the pair of conductors per unit length is given as

$$C = \frac{\lambda}{V} = \frac{\pi\varepsilon}{\ln\left(\frac{D-r}{r}\right)} = \frac{\pi\varepsilon}{\ln\left(\frac{D}{r}\right)} \quad \text{when} \quad D >> r$$

$\therefore$

$$\boxed{C = \frac{\pi\varepsilon}{\ln\left(\frac{D}{r}\right)}}$$

2.22.2 Cylindrical or Coaxial Capacitor

Example 2.66 A solid cylindrical conductor of radius a is surrounded by a coaxial cylindrical shell of inner radius b, as shown in Fig. 2.43. The length of both cylinders is L which is assumed to be much larger than $(b - a)$, the separation of the cylinders. Calculate the capacitance of the cylindrical capacitor.

Solution Since the length of both cylinders is L is assumed to be much larger than $(b - a)$, the separation of the cylinders, the edge effects can be neglected.

Let $+Q$ be the charge in the inner cylinder
$-Q$ be the charge in the outer cylindrical shell

We compute the electric field both inside and outside the capacitor.

Due to the cylindrical symmetry of the system, we choose the Gaussian surface to be a coaxial cylinder with length $l < L$ and radius r.

Region $r < a$:
The charge enclosed is zero, $Q_{enc} = 0$, since any net charge in a conductor must reside on its surface. Therefore, the field is also zero.

$\therefore$

$$\vec{E} = 0$$

Region $a < r < b$:
Using Gauss' law, we have

$$\oint_S \vec{E} \cdot d\vec{S} = \frac{Q_{enc}}{\varepsilon} \quad \text{or,} \quad E(2\pi rl) = \frac{1}{\varepsilon}\frac{Q}{L}l = \frac{\lambda l}{\varepsilon} \quad \text{or,} \quad E = \frac{\lambda}{2\pi\varepsilon r}$$

$\therefore$

$$\vec{E} = \frac{\lambda}{2\pi\varepsilon r}\hat{a}_r$$

where $\lambda = \frac{Q}{L}$ is the charge per unit length.

Region $r > b$:
The charge enclosed is zero, $Q_{enc} = \lambda l - \lambda l = 0$ since the Gaussian surface encloses equal but opposite charges from both conductors. Therefore, the field is also zero.

$$\therefore \qquad \vec{E} = 0$$

This is seen that the electric field is non-vanishing only in the region $a < r < b$.

The potential difference is given by

$$V = (V_+ - V_-) = -\int_l \vec{E} \cdot d\vec{l} = -\int_b^a \left(\frac{\lambda}{2\pi\varepsilon r} \hat{a}_r \right) \cdot dr\hat{a}_r = \frac{\lambda}{2\pi\varepsilon} \ln\left(\frac{b}{a}\right)$$

Thus, the capacitance of the coaxial capacitor is given as

$$\boxed{C = \frac{Q}{V} = \frac{\lambda L}{V} = \frac{2\pi\varepsilon L}{\ln(b/a)}} \qquad \textbf{(2.77)}$$

Once again, we see that the capacitance C depends only on the geometrical factors, L, a and b.

Example 2.67 A capacitance is made of two coaxial metallic cylinders of radii r_1 and r_2 $(r_1 < r_2)$ and length L $(L >> r_2)$. The region between r_1 and r_3 $(= \sqrt{r_1 r_2})$ is filled with a medium of dielectric constant K_1 and the remaining region is filled with a medium of dielectric constant K_2. Find the capacitance of the system (Fig. 2.70).

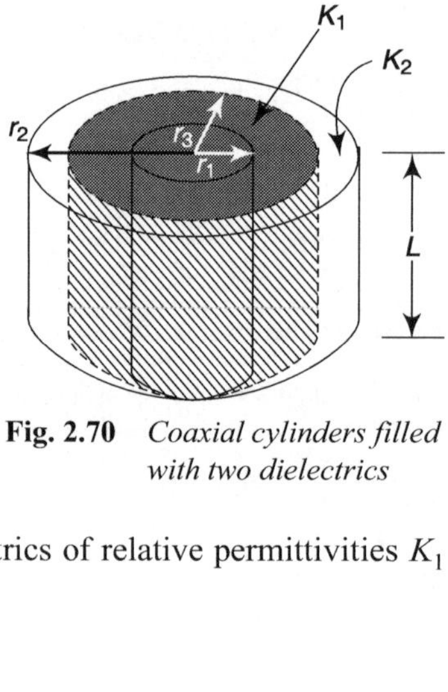

Fig. 2.70 *Coaxial cylinders filled with two dielectrics*

Solution Let, λ be the charge per unit length on the outer surface of the inner cylinder of radius r_1.

In the region, $r_1 < r < r_2$ applying Gauss's law

$$2\pi r L D = \lambda L \quad \Rightarrow \quad D = \frac{\lambda}{2\pi r}$$

Now, let, E_1 and E_2 be the electric field intensities in the two dielectrics of relative permittivities K_1 and K_2 respectively.

$$\therefore \qquad D = \varepsilon_0 K_1 E_1 = \varepsilon_0 K_2 E_2$$

$$\therefore \qquad E_1 = \frac{\lambda}{2\pi\varepsilon_0 K_1 r} \quad \text{and} \quad E_2 = \frac{\lambda}{2\pi\varepsilon_0 K_2 r}$$

Therefore, the potential difference between the cylinders is given as

$$V = \int_{r_1}^{r_3} E_1 dr + \int_{r_3}^{r_2} E_2 dr = \frac{\lambda}{2\pi\varepsilon_0 K_1} \int_{r_1}^{r_3} \frac{dr}{r} + \frac{\lambda}{2\pi\varepsilon_0 K_2} \int_{r_3}^{r_2} \frac{dr}{r} = \frac{\lambda}{2\pi\varepsilon_0 K_1} \ln\left(\frac{r_3}{r_1}\right) + \frac{\lambda}{2\pi\varepsilon_0 K_2} \ln\left(\frac{r_2}{r_3}\right)$$

$$= \frac{\lambda}{2\pi\varepsilon_0 K_1} \ln\left(\frac{\sqrt{r_1 r_2}}{r_1}\right) + \frac{\lambda}{2\pi\varepsilon_0 K_2} \ln\left(\frac{r_2}{\sqrt{r_1 r_2}}\right)$$

$$= \frac{\lambda}{2\pi\varepsilon_0 K_1} \ln\left(\sqrt{\frac{r_2}{r_1}}\right) + \frac{\lambda}{2\pi\varepsilon_0 K_2} \ln\left(\sqrt{\frac{r_2}{r_1}}\right)$$

$$= \frac{\lambda}{2\pi\varepsilon_0} \ln\left(\sqrt{\frac{r_2}{r_1}}\right)\left[\frac{1}{K_1} + \frac{1}{K_2}\right]$$

Hence, the capacitance of the system is given as

$$C = \frac{\lambda L}{V} = \frac{2\pi\varepsilon_0 L}{\ln\left(\sqrt{\frac{r_2}{r_1}}\right)\left[\frac{1}{K_1} + \frac{1}{K_2}\right]} = \frac{4\pi\varepsilon_0 L}{\ln\left(\frac{r_2}{r_1}\right)\left(\frac{K_1 + K_2}{K_1 K_2}\right)} = \frac{4\pi\varepsilon_0 K_1 K_2 L}{(K_1 + K_2)\ln\left(\frac{r_2}{r_1}\right)}$$

∴ $$\boxed{C = \frac{4\pi\varepsilon_0 K_1 K_2 L}{(K_1 + K_2)\ln\left(\frac{r_2}{r_1}\right)}}$$

Example 2.68 Find the capacitance per unit length between a cylindrical conductor of radius *a* and a ground plane parallel to the conductor axis and a distance *h* from it.

Solution This problem can be solved using the method of images. The grounded conducting plane is replaced by an image charge at a distance *h* inside the ground.

From Example 2.63, the potential between the actual charge and the image charge is

$$V_1 = \frac{\lambda}{\pi\varepsilon} \ln\left(\frac{2h - a}{a}\right)$$

So, the potential between the actual conductor and the ground plane is half this potential, i.e.,

$$V = \frac{1}{2} V_1 = \frac{\lambda}{2\pi\varepsilon} \ln\left(\frac{2h - a}{a}\right)$$

So, the capacitance per unit length between the conductor and the ground plane is given as

$$C = \frac{\lambda}{V} = \frac{2\pi\varepsilon}{\ln\left(\frac{2h - a}{a}\right)}$$

∴ $$\boxed{C = \frac{2\pi\varepsilon}{\ln\left(\frac{2h - a}{a}\right)}}$$

NOTE

If a << h, then, $C = \dfrac{2\pi\varepsilon}{\ln(2h/a)}$.

2.22.3 Spherical Capacitor

***Example 2.69** Find the capacitance of a spherical capacitor which consists of two concentric spherical shells of radii a and b ($a < b$), as shown in Fig. 2.71.

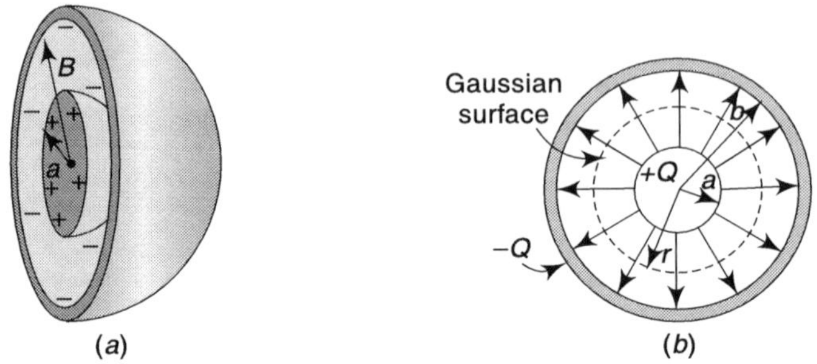

Fig. 2.71 *(a) spherical capacitor with two concentric spherical shells of radii a and b (b) Gaussian surface for calculating the electric field*

Solution Let the inner shell have a charge $+Q$ and the outer shell an equal but opposite charge $-Q$, both uniformly distributed over its surface. We want to calculate the capacitance of this spherical capacitor.

Similar to a cylindrical capacitor, the electric field is nonvanishing only in the region $a < r < b$. Using Gauss' law, we obtain

$$\oint_S \vec{E}\cdot d\vec{S} = \frac{Q_{\text{enc}}}{\varepsilon} \quad \text{or,} \quad E(4\pi r^2) = \frac{Q}{\varepsilon} \quad \text{or,} \quad E = \frac{Q}{4\pi\varepsilon r^2}$$

$$\therefore \qquad \vec{E} = \frac{Q}{4\pi\varepsilon r^2}\hat{a}_r$$

Therefore, the potential difference between the two conducting shells is

$$V = (V_+ - V_-) = -\int_l \vec{E}\cdot d\vec{l} = -\int_b^a \left(\frac{Q}{4\pi\varepsilon r^2}\hat{a}_r\right)\cdot dr\hat{a}_r = \frac{Q}{4\pi\varepsilon}\int_b^a \frac{dr}{r^2} = \frac{Q}{4\pi\varepsilon}\left(\frac{1}{a}-\frac{1}{b}\right) = \frac{Q}{4\pi\varepsilon}\left(\frac{b-a}{ab}\right)$$

Thus, the capacitance of the spherical capacitor is given as

$$\boxed{C = \frac{Q}{V} = \frac{4\pi\varepsilon}{\left(\frac{1}{a}-\frac{1}{b}\right)} = 4\pi\varepsilon\left(\frac{ab}{b-a}\right)} \qquad \textbf{(2.78)}$$

Again, the capacitance C depends only on the physical dimensions, a and b.

Capacitance of an Isolated Charged Sphere

***Example 2.70** Find the capacitance of a conducting sphere of radius R.

Solution For an isolated capacitor, the second conductor is assumed to be placed at infinity. From the result of a spherical capacitor, if the outer sphere is infinitely large, we can say the configuration as an *isolated sphere*.

Therefore, the capacitance of an isolated charged conducting sphere is obtained by putting the limit $b \to \infty$, in Eq. (2.77) as given below.

$$\lim_{b\to\infty} C = \lim_{b\to\infty} 4\pi\varepsilon\left(\frac{ab}{b-a}\right) = \lim_{b\to\infty} 4\pi\varepsilon\frac{a}{\left(1-\frac{a}{b}\right)} = 4\pi\varepsilon a$$

Thus, for a single isolated spherical conductor of radius R, the capacitance is given as

$$\boxed{C = 4\pi\varepsilon R} \qquad \textbf{(2.79)}$$

The above expression can also be obtained by noting that a conducting sphere of radius R with a charge Q uniformly distributed over its surface has the potential, $V = \frac{Q}{4\pi\varepsilon R}$, using infinity as the reference point having zero potential $V(\infty) = 0$; This gives

$$C = \frac{Q}{V} = \frac{Q}{Q/4\pi\varepsilon R} = 4\pi\varepsilon R$$

As expected, the capacitance of an isolated charged sphere only depends on its geometry, namely, the radius R.

Example 2.71 Derive an expression for the capacitance of a spherical capacitor consisting of two concentric spheres of radii a and b, the dielectric medium between the two spheres being air. Henceforth, show that the same expression can be written as $C = \frac{\varepsilon_0}{d}\sqrt{A_a A_b}$, where A_a and A_b are the surface areas of the two spheres with radii a and b respectively and d is their separation distance.

Solution From Example 2.69, the capacitance of the spherical capacitor is given as

$$C = 4\pi\varepsilon_0\left(\frac{ab}{b-a}\right)$$

Here, separation distance = d

Areas of the spheres are, $A_a = 4\pi a^2$ and $A_b = 4\pi b^2$

$$\therefore \qquad a = \sqrt{\frac{A_a}{4\pi}} \quad \text{and} \quad b = \sqrt{\frac{A_b}{4\pi}}$$

Therefore, the capacitance is written as

$$C = 4\pi\varepsilon_0\left(\frac{ab}{b-a}\right) = 4\pi\varepsilon_0\frac{1}{d}\sqrt{\frac{A_a}{4\pi}} \times \sqrt{\frac{A_b}{4\pi}} = \frac{\varepsilon_0}{d}\sqrt{A_a A_b}$$

$$\therefore \qquad \boxed{C = \frac{\varepsilon_0}{d}\sqrt{A_a A_b}}$$

Example 2.72 A concentric spherical conductor arrangement is shown in Fig. 2.72. If the capacitance of the arrangement is 0.1 nF, and a is 10 cm, find b.

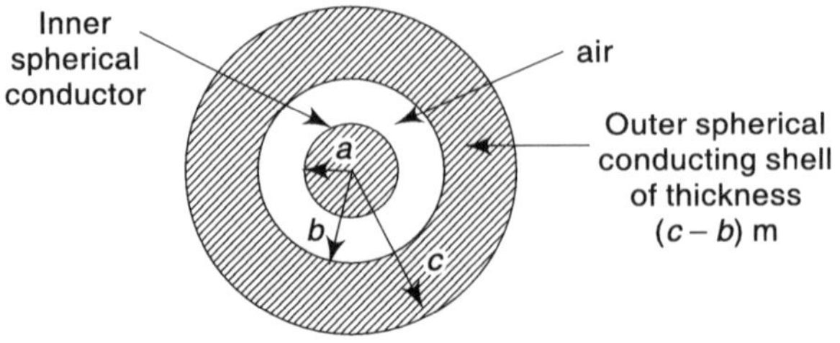

Fig. 2.72 *Spherical capacitor of Example 2.72*

Solution From Example 2.69, the capacitance of the spherical capacitor is given as

$$C = 4\pi\varepsilon_0\left(\frac{ab}{b-a}\right)$$

Here,

$$C = 0.1 \times 10^{-9}\ F,\ a = 10 \text{ cm}$$

$$\therefore \quad 0.1\times10^{-9} = 4\pi \times \frac{1}{36\pi\times10^9}\left(\frac{0.1b}{b-0.1}\right)$$

$$\Rightarrow \quad b = 11.25 \text{ cm}$$

2.23 METHOD OF IMAGES

The method of images, introduced by Lord Kelvin in 1848, is commonly used to determine V, $\vec{E}$, $\vec{D}$ and σ due to charges in the presence of conductors.

The image theory states that a given charge configuration above an infinite grounded perfect conducting plane may be replaced by the charge configuration itself, its image, and an equipotential surface in place of the conducting plane.

For image theory, two conditions must be satisfied:

1. The image charge(s) must be located in the conducting region.
2. The image charge(s) must be located such that on the conducting surface(s) the potential is zero or constant.

Definition of Images: The fictitious charges, placed in the region where the field is not required and producing the same field in the desired region as with the actual electrification of the surface, are defined as the electrical images.

Suppose we have a conducting surface in the vicinity of one or more point charges. The point charge will induce charges on the conducting surface.

Our purpose is to find the potential and field in the space outside the conductor not occupied by the charges. In this region, Laplace's equation is satisfied with suitable boundary conditions.

In the method of images, the actual electrification of the surface is replaced by one or more fictitious point charges in the region where the field or potential is not desired. The positions and the magnitudes of these fictitious charges are such that in the desired region, Laplace's equation is satisfied with the same conditions.

For example, the method of images suggests that for a point charge $+Q$ at a point (0, 0, d), an exactly equal and opposite charge $-Q$ is assumed to exist at point (0, 0, $-d$); replacing the conducting plane by an equipotential surface with zero potential. Typical examples of point, line and volume charge configurations are shown in Fig. 2.73 (a) and their images are shown in Fig. 2.73 (b).

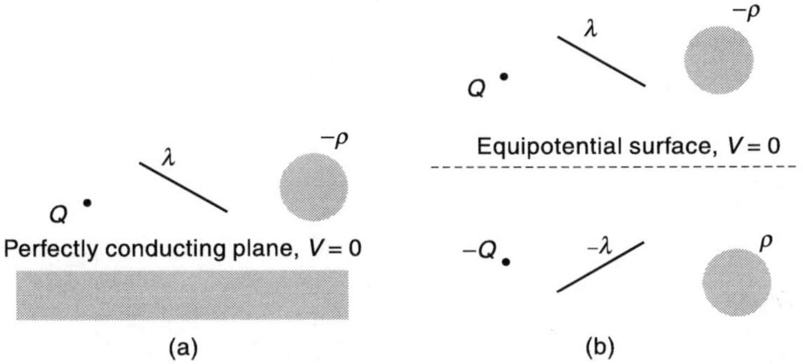

Fig. 2.73 *Image systems: (a) charge configurations in the presence of perfectly conducting plane, (b) image configurations*

By the method of images

$$\begin{pmatrix}\text{Field of an electrical charge}\\ \text{in front of a conducting plane}\end{pmatrix} = (\text{Field of the charge}) + \begin{pmatrix}\text{Field of its image}\\ \text{in the conducting plane}\end{pmatrix}$$

We will consider the following examples using method of images.

2.23.1 Applications of Method of Images

A Point Charge in front of an Infinite Grounded Plane

Example 2.73 An electrical system consists of a single charge q placed a distance d from an infinite earthed conducting plane. Determine

(a) Electric field and potential at any point $P(x, y, z)$;
(b) Electric field acting on the conductor surface;
(c) Surface charge induced;
(d) Force exerted by q on the conductor surface.

Also, draw the lines of force for the system.

Solution The charge and its image configurations are shown in Fig. 2.74.

(a) Potential and field at a point $P(x, y, z)$: The image charge $-q$ is placed at a distance d in the conducting region and the conducting grounded plane is replaced by an equipotential surface of zero potential.

For the charge q at (0, 0, d) and its image q' at (0, 0, $-d$), the potential at any point $P(x, y, z)$ is

$$V(x, y, z) = \frac{1}{4\pi\varepsilon}\frac{q}{\sqrt{x^2 + y^2 + (z-d)^2}} + \frac{1}{4\pi\varepsilon}\frac{q'}{\sqrt{x^2 + y^2 + (z+d)^2}}$$

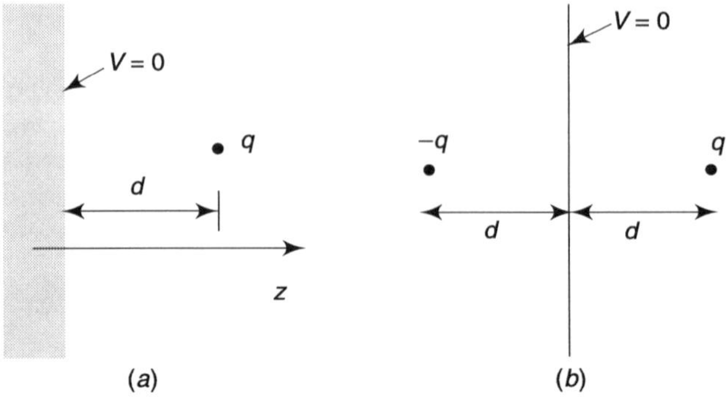

Fig. 2.74 *Point charge in front of conducting grounded plane (a) Charge configuration,and (b) Image configuration*

At $z = 0$, i.e., on the conducting plane, $V = 0$.

$$\therefore \qquad 0 = \frac{1}{4\pi\varepsilon}\frac{q}{\sqrt{x^2+y^2+d^2}} + \frac{1}{4\pi\varepsilon}\frac{q'}{\sqrt{x^2+y^2+d^2}}$$

$$\therefore \qquad q' = -q$$

Thus, we see that the magnitude of the image charge is $-q$.

Hence, the potential at any point $P(x, y, z)$ is

$$\boxed{V(x, y, z) = \frac{q}{4\pi\varepsilon}\left[\frac{1}{\sqrt{x^2+y^2+(z-d)^2}} - \frac{1}{\sqrt{x^2+y^2+(z+d)^2}}\right]}$$

The electric field at any point $P(x, y, z)$ can be found out by using the relation $\vec{E} = -\nabla V$ or may also be written as,

$$\vec{E} = \frac{1}{4\pi\varepsilon}\frac{q[x\hat{i}+y\hat{j}+(z-d)\hat{k}]}{[x^2+y^2+(z-d)^2]^{3/2}} + \frac{1}{4\pi\varepsilon}\frac{q'[x\hat{i}+y\hat{j}+(z+d)\hat{k}]}{[x^2+y^2+(z+d)^2]^{3/2}}$$

$$= \frac{q}{4\pi\varepsilon}\left\{\frac{[x\hat{i}+y\hat{j}+(z-d)\hat{k}]}{[x^2+y^2+(z-d)^2]^{3/2}} - \frac{[x\hat{i}+y\hat{j}+(z+d)\hat{k}]}{[x^2+y^2+(z+d)^2]^{3/2}}\right\} \quad (\because q' = -q)$$

(b) Electric field on the conductor surface: The electric field component E_z on the xy plane, i.e., on the conductor surface is

$$E_z(x, y, 0) = 2\times\frac{q}{4\pi\varepsilon}\left\{\frac{-d}{[x^2+y^2+d^2]^{3/2}}\right\} = \frac{-qd}{2\pi\varepsilon[x^2+y^2+d^2]^{3/2}}$$

The parallel field components E_x and E_y vanish as they should.

(c) Induced surface charge density and total induced surface charge: The induced surface charge density is given by, $\sigma = \varepsilon E_z$. The total induced surface charge is given by

$$q_{\text{induced}} = \iint \sigma(x, y)dxdy = \int_{-\infty}^{\infty}\int_{-\infty}^{\infty}\frac{-qddxdy}{2\pi[x^2+y^2+d^2]^{3/2}}$$

$$= -\frac{qd}{2\pi}\int_0^{2\pi}\int_0^{\infty}\frac{rdrd\phi}{(r^2+d^2)^{3/2}} \quad [r^2 = x^2+y^2,\ dxdy = rdrd\phi]$$

$$= -\frac{qd}{2\pi} \times 2\pi \int_d^\infty \frac{dp}{p^2} \qquad \begin{bmatrix} (r^2 + d^2) = p^2;\ rdr = pdp \\ r \quad 0 \quad \infty \\ p \quad d \quad \infty \end{bmatrix}$$

$$\therefore \qquad q_{\text{induced}} = -\frac{qd}{2\pi} \times 2\pi \left[-\frac{1}{d} + 0 \right] = -q$$

$$\therefore \qquad q_{\text{induced}} = -q$$

The field at the surface due to q is $\frac{1}{2}E_z$, since half the field at the surface is due to q and the other half is due to the image charge $-q$.

(d) Force experienced by the point charge: Thus, the force experienced by q is given by the integral of $\frac{1}{2}\sigma E_z$.

$$\therefore \qquad F = -\iint \frac{1}{2}\sigma E_z = -\iint \frac{1}{2}\varepsilon E_z^2 = -\frac{1}{2}\varepsilon \iint \left[\frac{-qd}{2\pi\varepsilon[x^2 + y^2 + d^2]^{3/2}} \right]^2 dxdy$$

$$= -\frac{q^2 d^2}{8\pi^2 \varepsilon} \int_{-\infty}^{\infty}\int_{-\infty}^{\infty} \frac{dxdy}{[x^2 + y^2 + d^2]^3} = -\frac{q^2 d^2}{8\pi^2 \varepsilon} \int_0^{2\pi}\int_0^{\infty} \frac{rdrd\phi}{[r^2 + d^2]^3} = -\frac{q^2 d^2}{4\pi\varepsilon} \int_0^{\infty} \frac{rdr}{[r^2 + d^2]^3}$$

$$= -\frac{q^2 d^2}{4\pi\varepsilon} \int_d^\infty \frac{dp}{p^5} \qquad \begin{bmatrix} (r^2 + d^2) = p^2;\ rdr = pdp \\ r \quad 0 \quad \infty \\ p \quad d \quad \infty \end{bmatrix}$$

$$= -\frac{q^2 d^2}{4\pi\varepsilon} \left[\frac{1}{-4d^4} \right] = -\frac{q^2}{16\pi\varepsilon d^2}$$

The negative sign indicates that the force is attractive.

$$\therefore \qquad \boxed{F = \frac{q^2}{16\pi\varepsilon d^2}}$$

Clearly, the force F is that between the charge $+q$ and its image $-q$ separated by a distance $2d$.

Lines of force: The lines of force originate from q and terminate normally on the conducting plane. In fact, the lines of force are the same as those for two point charges q and $-q$ separated by a distance $2d$. The lines on the left hand side have been shown by dotted lines as they do not exist in reality, as depicted in Fig. 2.75. Actually, the field in this region is zero, the conducting plane being grounded.

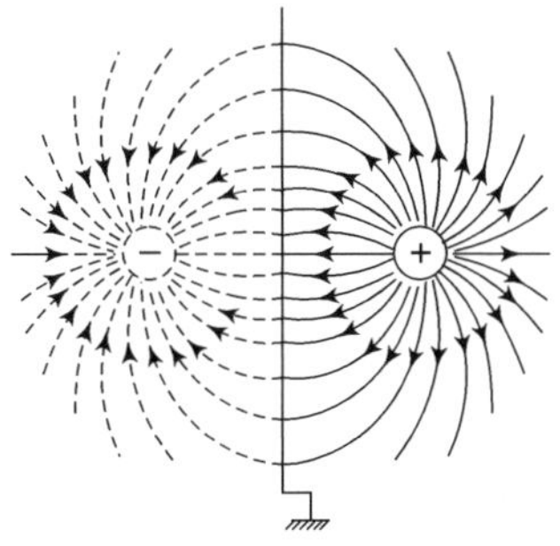

Fig. 2.75 *Lines of force between the point charge and its image*

A Charge in the Presence of a Grounded Conducting Sphere

***Example 2.74** A point charge 'Q' is situated at a distance 'd' from the centre of a grounded spherical conductor of radius 'a' (< 'd').

(a) Show that the image charge required for computing the field outside the spherical conductor is a point charge of value $-Q\frac{a}{d}$ lying at a distance $\frac{a^2}{d}$ from the centre of the conductor along the line joining the centre to the charge Q and on the side of Q.

(b) What is the induced charge on the surface of the conductor?

Solution We consider the point charge Q at a distance d from the centre of a grounded sphere of radius a, shown in Fig. 2.76.

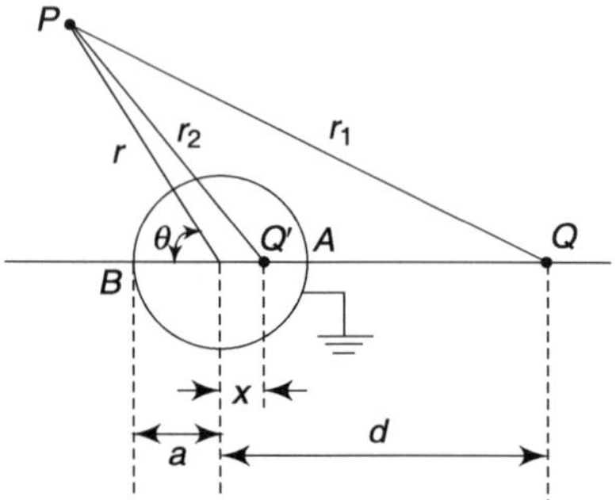

Fig. 2.76 *Point charge in front of conducting sphere*

(a) Image charge: Let Q' be the image charge located at a distance x from the centre of the sphere on the line joining the point charge Q and the centre of the sphere.

To find the values of x and Q', we shall consider two points on the sphere. If V_A and V_B are the potentials at these two points, then

Since the sphere is grounded, $V_A = 0$ and $V_B = 0$.

Now,

$$V_A = \frac{Q}{4\pi\varepsilon}\frac{1}{d-a} + \frac{Q'}{4\pi\varepsilon}\frac{1}{a-x} = 0$$

and

$$V_B = \frac{Q}{4\pi\varepsilon}\frac{1}{d+a} + \frac{Q'}{4\pi\varepsilon}\frac{1}{a+x} = 0$$

From the first equation, $Q' = -Q\frac{(a-x)}{(d-a)}$

Substituting this value in the second equation

$$\frac{Q}{(d+a)} = \frac{Q(a-x)}{(d-a)(a+x)}$$

or

$$(d-a)(a+x) = (d+a)(a-x)$$

or

$$2dx = 2a^2$$

∴

$$\boxed{x = \frac{a^2}{d}}$$

Putting this value, we get

$$Q' = -Q\frac{(a-a^2/2)}{(d-a)} = -Q\frac{(ad-a^2)}{d(d-a)}$$

∴

$$\boxed{Q' = -Q\frac{a}{d}}$$

(b) Potential and field at Point *P*: The potential at any point P is,

$$V = \frac{1}{4\pi\varepsilon}\left[\frac{Q}{r_1} + \frac{Q'}{r_2}\right] = \frac{1}{4\pi\varepsilon}\left[\frac{Q}{r_1} + \frac{(a/d)Q}{r_2}\right]$$

Here,

$$r_1 = \sqrt{d^2 + r^2 - 2dr\cos\theta}; \qquad r_2 = \sqrt{x^2 + r^2 - 2xr\cos\theta} = \sqrt{\left(\frac{a^2}{d}\right)^2 + r^2 - 2\frac{a^2}{d}r\cos\theta}$$

To find the charge density at a point on the sphere, we find the normal component of the field intensity E at $r = a$ as follows.

$$E_r = -\frac{\partial V}{\partial r}\bigg|_{r=a} = -\frac{\partial}{\partial r}\left[\frac{1}{4\pi\varepsilon}\left\{\frac{Q}{\sqrt{d^2 + r^2 - 2dr\cos\theta}} - \frac{aQ}{d}\frac{1}{\sqrt{\left(\frac{a^2}{d}\right)^2 + r^2 - 2\frac{a^2}{d}r\cos\theta}}\right\}\right]_{r=a}$$

$$= -\frac{Q}{4\pi\varepsilon}\frac{\partial}{\partial r}\left[\frac{1}{\sqrt{d^2 + r^2 - 2dr\cos\theta}}\right] + \frac{Q}{4\pi\varepsilon}\left(\frac{a}{d}\right)\frac{\partial}{\partial r}\left[\frac{1}{\sqrt{\left(\frac{a^2}{d}\right)^2 + r^2 - 2\frac{a^2}{d}r\cos\theta}}\right]_{r=a}$$

$$= \frac{Q}{4\pi\varepsilon}\left[\frac{r - d\cos\theta}{(d^2 + r^2 - 2dr\cos\theta)^{3/2}} - \left(\frac{a}{d}\right)\frac{r - \frac{a^2}{d}\cos\theta}{\left\{\left(\frac{a^2}{d}\right)^2 + r^2 - 2\frac{a^2}{d}r\cos\theta\right\}^{3/2}}\right]_{r=a}$$

$$= \frac{Q}{4\pi\varepsilon}\left[\frac{a - d\cos\theta}{(d^2 + a^2 - 2da\cos\theta)^{3/2}} - \left(\frac{a}{d}\right)\frac{\dfrac{ad - a^2\cos\theta}{d}}{\dfrac{(a^4 + a^2d^2 - 2a^3d\cos\theta)^{3/2}}{d^3}}\right]$$

$$= \frac{Q}{4\pi\varepsilon}\left[\frac{a - d\cos\theta}{(d^2 + a^2 - 2da\cos\theta)^{3/2}} - \frac{a^2d(d - a\cos\theta)}{a^3(a^2 + d^2 - 2ad\cos\theta)^{3/2}}\right]$$

$$= \frac{Q}{4\pi\varepsilon}\left[\frac{a - d\cos\theta}{(d^2 + a^2 - 2da\cos\theta)^{3/2}} - \left(\frac{d}{a}\right)\frac{(d - a\cos\theta)}{(d^2 + a^2 - 2da\cos\theta)^{3/2}}\right]$$

$$= \frac{Q}{4\pi\varepsilon a(d^2 + a^2 - 2da\cos\theta)^{3/2}}[a^2 - da\cos\theta - d^2 + da\cos\theta]$$

$\therefore$
$$E_r = \frac{Q(a^2 - d^2)}{4\pi\varepsilon a(d^2 + a^2 - 2da\cos\theta)^{3/2}}$$

$\therefore$
$$\boxed{\vec{E}_r = \frac{Q(a^2 - d^2)}{4\pi\varepsilon a(d^2 + a^2 - 2da\cos\theta)^{3/2}}\hat{a}_r}$$

Induced surface charge density and total induced surface charge The surface charge density of the induced charge is

$$\sigma = \varepsilon E_r = \frac{Q(a^2 - d^2)}{4\pi a(d^2 + a^2 - 2da\cos\theta)^{3/2}}$$

Since $d > a$, it is seen that σ is negative. The magnitude of the surface charge density is a maximum when the denominator is a minimum, i.e., when $\theta = 0°$. It decreases with increase in θ, and is a minimum when $\theta = 180°$. The variation of $|\sigma|$ with θ is shown in Fig. 2.77.

Therefore, the total charge induced on the surface of the sphere is given as

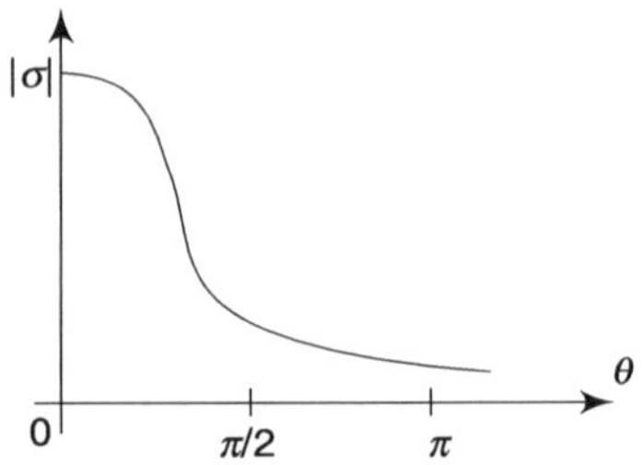

Fig. 2.77 Variation of $|\sigma|$ with θ

$$Q' = \iint_S \sigma dS = \int_{\phi=0}^{2\pi}\int_{\theta=0}^{\pi} \sigma a^2 \sin\theta\, d\theta\, d\phi$$

$$= \int_{\phi=0}^{2\pi}\int_{\theta=0}^{\pi} \frac{Q(a^2-d^2)}{4\pi a(d^2+a^2-2da\cos\theta)^{3/2}} a^2 \sin\theta\, d\theta\, d\phi$$

$$= \frac{Qa(a^2-d^2)}{4\pi} \times 2\pi \int_{\theta=0}^{\pi} \frac{\sin\theta\, d\theta}{(d^2+a^2-2da\cos\theta)^{3/2}}$$

$$= \frac{Qa(a^2-d^2)}{2} \int_{\theta=0}^{\pi} \frac{\sin\theta\, d\theta}{(d^2+a^2-2da\cos\theta)^{3/2}}$$

Let, $(d^2+a^2-2da\cos\theta) = p^2 \quad \therefore \sin\theta\, d\theta = \dfrac{pdp}{da}$

$$\therefore \qquad Q' = \frac{Qa(a^2-d^2)}{2da}\int_{d-a}^{d+a} \frac{pdp}{p^3} = \frac{Qa(a^2-d^2)}{2da}\int_{d-a}^{d+a} \frac{dp}{p^2}$$

$$= -\frac{Qa(a^2-d^2)}{2da}\left[\frac{1}{p}\right]_{d-a}^{d+a} = -\frac{Qa(a^2-d^2)}{2da}\left(\frac{1}{d+a} - \frac{1}{d-a}\right)$$

$$= -\frac{Qa(a^2-d^2)}{2da}\left(\frac{-2a}{d^2-a^2}\right)$$

$$= -\frac{Qa(a^2-d^2)}{2da}\left(\frac{2a}{a^2-d^2}\right)$$

$$= -Q\frac{a}{d}$$

$$\therefore \qquad \boxed{Q' = -Q\frac{a}{d}}$$

Force experienced by the point charge Since the induced charge on the sphere is replaced by the image charge, the force exerted on the charge Q is the force between the charge Q and its image Q'.

Thus,

$$F = \frac{QQ'}{4\pi\varepsilon(d-a^2/d)^2} = \frac{Q(-Q\,a/d)}{4\pi\varepsilon(d-a^2/2)^2} = -\frac{Q^2ad}{4\pi\varepsilon(d^2-a^2)^2}$$

Here also, the negative sign indicates that the force is attractive.

$$\therefore \qquad \boxed{F = \frac{Q^2ad}{4\pi\varepsilon(d^2-a^2)^2}}$$

If $d >> a$, the force becomes $F = \dfrac{Q^2a}{4\pi\varepsilon d^3}$.

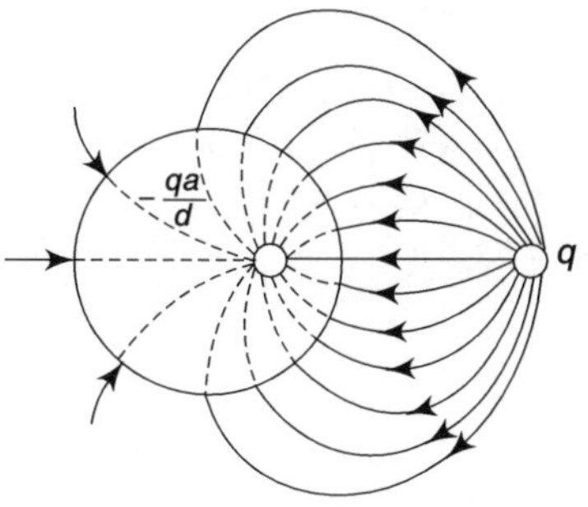

Fig. 2.78 *Lines of force between the point charge and its image*

Lines of force The lines of force originate from Q and terminate normally on the conducting sphere. In fact, the lines of force are between the two point charges Q and its image charge $-Q(a/d)$. This is shown in Fig. 2.78.

A Charge in the presence of a Conducting Sphere at Constant Potential

Example 2.75 What will be the image charge if the spherical conductor in Example 2.74 is not grounded, but has a potential 'V'?

Solution If the sphere is at a constant potential V, then the image system which will mimic such a sphere consists of two charges:

— one as already found Q' at a distance $x\left(=\frac{a^2}{d}\right)$ from the centre of the sphere, and

— another at the centre of the sphere, Q''.

Then, the total potential on the sphere is

$$V = (V_Q + V_{Q'} + V_{Q''})$$

But, from Example 2.74, we have, $(V_Q + V_{Q'}) = 0$

$$\therefore \qquad V_{Q''} = V$$

Also, since Q'' is a point charge at the centre of a sphere

$$V_{Q''} = \frac{Q''}{4\pi\varepsilon a} \quad \text{(on the sphere at } r = a)$$

$$\therefore \qquad Q'' = 4\pi\varepsilon a V$$

Thus, if the sphere is at a potential V, then the image system consists of

(i) Point charge $Q' = -Q\left(\frac{a}{d}\right)$ at a distance $x = \left(\frac{a^2}{d}\right)$ from the centre, and

(ii) Point charge $Q'' = 4\pi\varepsilon a V$ at the centre of the sphere.

The induced surface charge density in this case is

$$\sigma = \frac{Q(a^2 - d^2)}{4\pi a(d^2 + a^2 - 2da\cos\theta)^{3/2}} + \frac{Q''}{4\pi a^2} = \frac{Q(a^2 - d^2)}{4\pi a(d^2 + a^2 - 2da\cos\theta)^{3/2}} + \frac{\varepsilon V}{a}$$

The force experienced by the point charge is

$$F = \frac{QQ'}{4\pi\varepsilon(d - a^2/d)^2} + \frac{QQ''}{4\pi\varepsilon d^2} = \frac{Q(-Q\,a/d)}{4\pi\varepsilon(d - a^2/d)^2} + \frac{Q4\pi\varepsilon aV}{4\pi\varepsilon d^2} = -\frac{Q^2 ad}{4\pi\varepsilon(d^2 - a^2)^2} + \frac{QaV}{d^2}$$

This force may be attractive or repulsive depending upon which term is greater in the equation.

Example 2.76 A point charge 'Q' is located at point $(a, 0, b)$ between two semi-infinite conducting planes intersecting each other at right angles. Determine the potential at point $P(x, y, z)$ and the force on Q.

or

Two semi-infinite earthed conducting planes meet at right angles to each other. In the region between the planes, a point charge '+Q' is placed. The distance of the point charge from each plane is 'a'. Determine the force on the charge '+Q'.

Solution In this case, the number of image charges is

$$N = \left(\frac{360°}{\phi} - 1\right) = \left(\frac{360°}{90°} - 1\right) = 3$$

The image configuration is shown in Fig. 2.79.

The potential at $P(x, y, z)$ due to all four charges is

$$V = \frac{Q}{4\pi\varepsilon}\left[\frac{1}{r_1} - \frac{1}{r_2} + \frac{1}{r_3} - \frac{1}{r_4}\right]$$

where

$$r_1 = \sqrt{(x-a)^2 + y^2 + (z-b)^2}$$
$$r_2 = \sqrt{(x+a)^2 + y^2 + (z-b)^2}$$
$$r_3 = \sqrt{(x+a)^2 + y^2 + (z+b)^2}$$
$$r_4 = \sqrt{(x-a)^2 + y^2 + (z+b)^2}$$

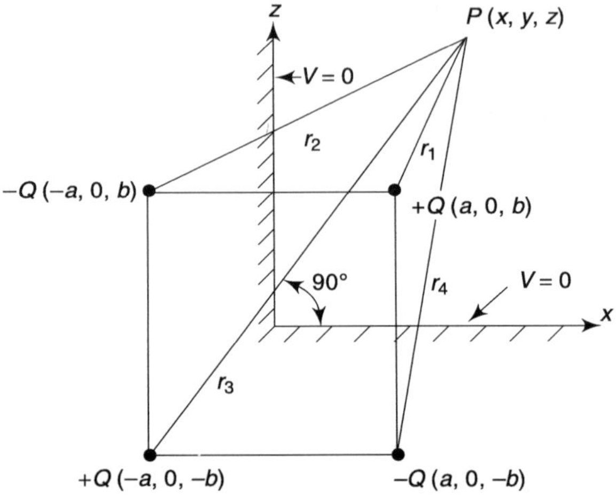

Fig. 2.79 *Image configuration for Example 2.76*

Total force on the +Q charge is

$$\vec{F} = (\vec{F}_1 + \vec{F}_2 + F_3)$$
$$= \frac{Q^2}{4\pi\varepsilon(2b)^2}(-\hat{a}_z) + \frac{Q^2}{4\pi\varepsilon(2a)^2}(-\hat{a}_x) + \frac{Q^2}{4\pi\varepsilon[(2a)^2 + (2b)^2]}(\cos\theta\,\hat{a}_x + \sin\theta\,\hat{a}_z)$$
$$= -\frac{Q^2}{16\pi\varepsilon b^2}\hat{a}_z - \frac{Q^2}{16\pi\varepsilon b^2}\hat{a}_x + \frac{Q^2}{16\pi\varepsilon(a^2+b^2)}\left\{\frac{a}{a^2+b^2}\hat{a}_x + \frac{b}{a^2+b^2}\hat{a}_z\right\}$$
$$= \frac{Q^2}{16\pi\varepsilon}\left[\left\{\frac{a}{(a^2+b^2)^{3/2}} - \frac{1}{a^2}\right\}\hat{a}_x + \left\{\frac{b}{(a^2+b^2)^{3/2}} - \frac{1}{b^2}\right\}\hat{a}_z\right]$$

If $a = b$, then the force becomes

$$\vec{F} = \frac{Q^2}{16\pi\varepsilon}\left[\left\{\frac{a}{(a^2+a^2)^{3/2}} - \frac{1}{a^2}\right\}\hat{a}_x + \left\{\frac{a}{(a^2+a^2)^{3/2}} - \frac{1}{a^2}\right\}\hat{a}_z\right]$$
$$= \frac{Q^2}{16\pi\varepsilon a^2}\left[\left(\frac{1}{2\sqrt{2}} - 1\right)\hat{a}_x + \left(\frac{1}{2\sqrt{2}} - 1\right)\hat{a}_z\right]$$
$$= \frac{(2\sqrt{2} - 1)Q^2}{32\pi\varepsilon a^2}$$

$$F = \frac{(2\sqrt{2}-1)Q^2}{32\pi\varepsilon a^2}$$

Example 2.77 An infinitely long line charge of uniform line charge density 'λ' is situated parallel to and at a distance 'x' from the grounded infinite plane conductor. Obtain the image charge and show that the induced surface charge on the conductor per unit length is '$-\lambda$'.

Solution The finite line charge λ is assumed at $x = 0$, $z = h$ and the image λ' is assumed at $x = 0$, $z = -h$, so that the two are parallel to y-axis.

The field at point $P(x, y, z)$ is

$$\vec{E} = (\vec{E}_\lambda + \vec{E}_{\lambda'}) = \frac{\lambda}{2\pi\varepsilon_0 r_1}\hat{a}_{r1} + \frac{\lambda'}{2\pi\varepsilon_0 r_2}\hat{a}_{r2}$$

Here,

$$\vec{r}_1 = (x, y, z) - (0, y, h) = (x, 0, z-h)$$
$$\vec{r}_2 = (x, y, z) - (0, y, -h) = (x, 0, z+h)$$

$\therefore$
$$\vec{E} = \frac{1}{2\pi\varepsilon_0}\left[\frac{\lambda\{x\hat{a}_x + (z-h)\hat{a}_z\}}{x^2 + (z-h)^2} + \frac{\lambda'\{x\hat{a}_x + (z+h)\hat{a}_z\}}{x^2 + (z+h)^2}\right]$$

Potential at P is

$$V = (V_\lambda + V_{\lambda'}) = -\frac{\lambda}{2\pi\varepsilon_0}\ln r_1 + \frac{-\lambda'}{2\pi\varepsilon_0}\ln r_2$$

At the conducting plane, $V = 0 \quad \Rightarrow \quad \lambda' = \lambda \quad$ (with $z = 0$)

$\therefore$
$$\vec{E} = \frac{\lambda}{2\pi\varepsilon_0}\left[\frac{\{x\hat{a}_x + (z-h)\hat{a}_z\}}{x^2 + (z-h)^2} - \frac{\{x\hat{a}_x + (z+h)\hat{a}_z\}}{x^2 + (z+h)^2}\right]$$

$\therefore$
$$V = -\frac{\lambda}{2\pi\varepsilon_0}\ln r_1 - \frac{-\lambda}{2\pi\varepsilon_0}\ln r_2 = \frac{-\lambda}{2\pi\varepsilon_0}\ln\left(\frac{r_1}{r_2}\right) = -\frac{\lambda}{2\pi\varepsilon_0}\ln\left[\frac{x^2 + (z-h)^2}{x^2 + (z+h)^2}\right]^{1/2}$$

The surface charge induced on the conducting plane is

$$\rho_s = D_n = \varepsilon_0 E_z|_{z=0} = -\frac{\lambda h}{\pi(x^2 + h^2)}$$

The induced charge per unit length on the conducting plane is

$$\rho_i = \int \rho_s dx = -\frac{\lambda h}{\pi}\int_{-\infty}^{\infty}\frac{dx}{x^2 + h^2} = -\frac{\lambda h}{\pi}\tan^{-1}\left(\frac{x}{h}\right)\Bigg|_{-\infty}^{\infty} \times \frac{1}{h} = -\frac{\lambda}{\pi}\left[\frac{\pi}{2} + \frac{\pi}{2}\right] = -\lambda$$

$\therefore$
$$\rho_i = -\lambda$$

Example 2.78 A pair of conducting planes meets at an angle of 60°. A point charge $+Q$ is located at a distance 'a' from both the planes. Find the electric field intensity induced at the foot of perpendicular.

Solution For $\phi = 60°$, the number of charges is

$$N = \left(\frac{360°}{60°} - 1\right) = 5$$

The image configuration is shown in Fig. 2.80.

We have to find the intensity at the point P.

Now,

$$AB = 2a, \quad \therefore BG = AB \sin 30° = 2a \times \frac{1}{2} = a$$

Also, $OB = (OG + BG) = (a + a) = 2a$

$$\therefore \quad AG = AB \cos 30° = 2a \times \frac{\sqrt{3}}{2} = \sqrt{3}a = OP$$

$$\therefore \; r_1 = PB = \sqrt{OB^2 + OP^2} = \sqrt{3a^2 + 4a^2} = \sqrt{7}a$$

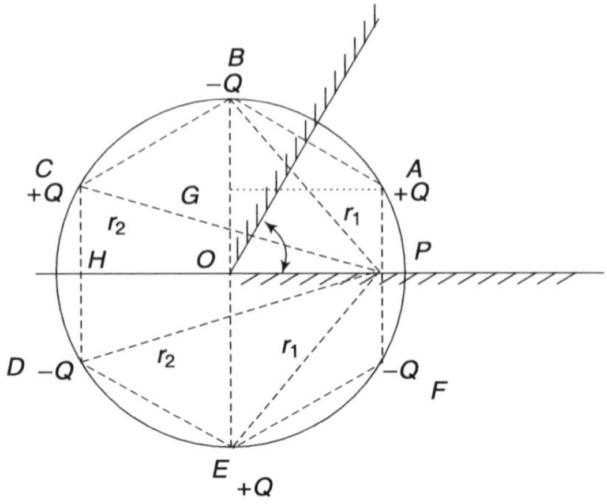

Fig. 2.80 *Image configuration of Example 2.78*

Also, $PH = 2PO = 2\sqrt{3}a$

$$\therefore \quad r_2 = \sqrt{PH^2 + CH^2} = \sqrt{12a^2 + a^2} = \sqrt{13}a$$

$\therefore$ Field intensity at P:

(i) due to charges at B and E is

$$\vec{E}_{BE} = 2 \times \frac{Q}{4\pi\varepsilon_0 r_1^2} \cos\theta_1 \hat{a}_n = \frac{Q}{2\pi\varepsilon_0 r_1^2}\frac{2a}{r_1}\hat{a}_n = \frac{Qa}{\pi\varepsilon_0 r_1^3}\hat{a}_n = \frac{Q}{\pi\varepsilon_0 a^2 7\sqrt{7}}\hat{a}_n \quad \{\because r_1 = \sqrt{7}a\}$$

[Since charges at B and C are equal and opposite, their horizontal components cancel each other and the resultant field intensity is in the vertical direction.]

(ii) due to charges at C and D is

$$\vec{E}_{CD} = -2 \times \frac{Q}{4\pi\varepsilon_0 r_2^2}\cos\theta_2 \hat{a}_n = -\frac{Q}{2\pi\varepsilon_0 r_2^2}\frac{a}{r_2}\hat{a}_n = -\frac{Qa}{2\pi\varepsilon_0 r_2^3}\hat{a}_n$$

$$= -\frac{Q}{2\pi\varepsilon_0 a^2 13\sqrt{13}}\hat{a}_n \quad \{\because r_2 = \sqrt{13}a\}$$

[Since charges at C and D are equal and opposite, their horizontal components of field intensity cancel each other, but vertical components give the resultant field intensity.]

(iii) due to charges at A and F is

$$\vec{E}_{AF} = -\frac{2Q}{4\pi\varepsilon_0 a^2}\hat{a}_n = -\frac{Q}{2\pi\varepsilon_0 a^2}\hat{a}_n$$

Hence, the total field intensity is given as

$$\vec{E} = (\vec{E}_{BE} + \vec{E}_{CD} + \vec{E}_{AF}) = \frac{Q}{\pi\varepsilon_0 a^2 7\sqrt{7}}\hat{a}_n - \frac{Q}{2\pi\varepsilon_0 a^2 13\sqrt{13}}\hat{a}_n - \frac{Q}{2\pi\varepsilon_0 a^2}\hat{a}_n$$

$$= -\frac{Q}{\pi\varepsilon_0 a^2}\left[\frac{1}{26\sqrt{13}} - \frac{1}{7\sqrt{7}} + \frac{1}{2}\right]\hat{a}_n$$

$$= -0.4566\frac{Q}{\pi\varepsilon_0 a^2}\hat{a}_n$$

Example 2.79 Two infinite intersecting planes are intersecting at right angle. A charge of 100nC is placed at (3, 4, 0). Find the electric potential and electric field intensity at (3, 5, 0).

Solution In this case, the number of image charges is

$$N = \left(\frac{360^\circ}{\phi} - 1\right) = \left(\frac{360^\circ}{90^\circ} - 1\right) = 3$$

The image configuration is shown in Fig. 2.81.

The potential at $P(x, y, z)$ due to all four charges is

$$V = \frac{Q}{4\pi\varepsilon_0}\left[\frac{1}{r_1} - \frac{1}{r_2} + \frac{1}{r_3} - \frac{1}{r_4}\right]$$

where

$$r_1 = \sqrt{(x-3)^2 + (y-4)^2 + z^2}$$
$$r_2 = \sqrt{(x+3)^2 + (y-4)^2 + z^2}$$
$$r_3 = \sqrt{(x+3)^2 + (y+4)^2 + z^2}$$
$$r_4 = \sqrt{(x-3)^2 + (y+4)^2 + z^2}$$

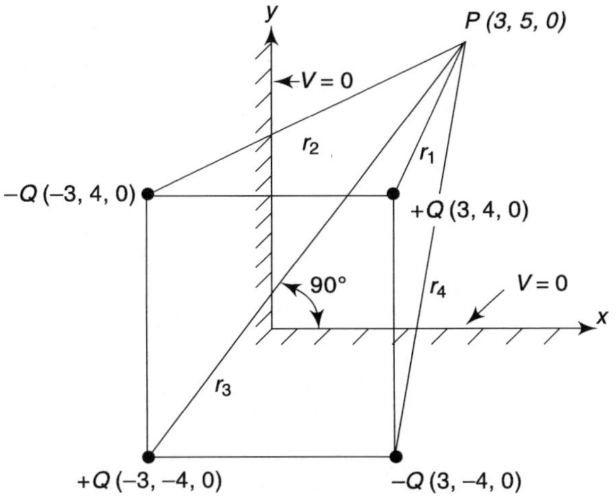

Fig. 2.81 *Image configuration for Example 2.79*

At P (3, 5, 0),

$$r_1 = 1, \quad r_2 = \sqrt{37}, \quad r_3 = \sqrt{17}, \quad r_4 = 9$$

So, the potential is given as

$$V = \frac{100 \times 10^{-9}}{4\pi \times \frac{10^{-9}}{36\pi}}\left[\frac{1}{1} - \frac{1}{\sqrt{37}} + \frac{1}{\sqrt{17}} - \frac{1}{9}\right] = 900\left[\frac{1}{1} - \frac{1}{\sqrt{37}} + \frac{1}{\sqrt{17}} - \frac{1}{9}\right] = 735.25 \text{ Volt}$$

Field intensity is given as

$$\vec{E} = -\nabla V = -\frac{\partial V}{\partial x}\hat{a}_x - \frac{\partial V}{\partial y}\hat{a}_y - \frac{\partial V}{\partial z}\hat{a}_z$$

However, $$\left.\frac{\partial V}{\partial x}\right|_{P(3,5,0)} = 900\left.\left[\frac{x-3}{r_1^3} + \frac{x+3}{r_2^3} - \frac{x+3}{r_3^3} + \frac{x-3}{r_4^3}\right]\right|_{x=3} = 19.8$$

$$\left.\frac{\partial V}{\partial y}\right|_{P(3,5,0)} = -891.36 \quad \text{and} \quad \left.\frac{\partial V}{\partial z}\right|_{P(3,5,0)} = 0$$

$\therefore$ $$\vec{E} = -19.8\hat{a}_x + 891.36\hat{a}_y \text{ V/m}$$

2.23.2 Method of Images at the Boundary between Dielectrics

In this case, as shown in Fig. 2.82, we no longer have the boundary condition that the potential is constant at the interface. We now must use the conditions that the normal components of the ***D***-fields and tangential components of the **E**-fields are continuous. Furthermore, we have the fields on both sides of the boundary.

We put the charge Q in A at a distance d from the interface and place an image charge Q' in A′ at the same distance from the interface. The potential from these two charges on the right side is

$$V(r, z) = \frac{1}{4\pi\varepsilon_2}\left[\frac{Q}{\sqrt{r^2+(z-d)^2}} + \frac{Q'}{\sqrt{r^2+(z+d)^2}}\right] \quad z > 0$$

To find the potential in the left side we put an effective charge Q'' in A (replacing Q).

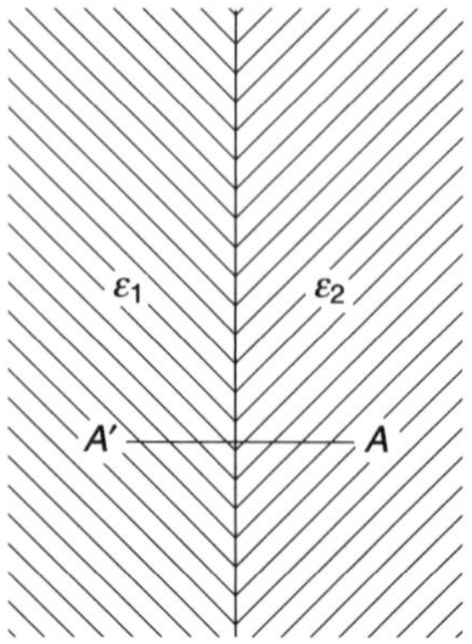

Fig. 2.82 *Method of images for dielectric-dielectric boundary*

$$\therefore \qquad V(r, z) = \frac{1}{4\pi\varepsilon_1}\frac{Q''}{\sqrt{r^2+(z-d)^2}}, \quad z < 0$$

From this we find that

$$\left.\frac{\partial V}{\partial z}\right|_{z+} = \frac{1}{4\pi\varepsilon_2}\frac{(Q-Q')d}{(r^2+d^2)^{3/2}}$$

$$\left.\frac{\partial V}{\partial z}\right|_{z-} = \frac{1}{4\pi\varepsilon_1}\frac{Q''d}{(r^2+d^2)^{3/2}}$$

$$\left.\frac{\partial V}{\partial r}\right|_{z+} = -\frac{1}{4\pi\varepsilon_2}\frac{(Q+Q')r}{(r^2+d^2)^{3/2}}$$

$$\left.\frac{\partial V}{\partial r}\right|_{z-} = -\frac{1}{4\pi\varepsilon_1}\frac{Q''r}{(r^2+d^2)^{3/2}}$$

To satisfy the boundary condition, we have

$$(Q - Q') = Q''$$

$$\frac{1}{\varepsilon_2}(Q+Q') = \frac{1}{\varepsilon_1}Q''$$

and

$$Q' = -\left(\frac{\varepsilon_1 - \varepsilon_2}{\varepsilon_1 + \varepsilon_2}\right)Q$$

$$Q'' = \left(\frac{2\varepsilon_1}{\varepsilon_1 + \varepsilon_2}\right)Q$$

From this, we could have obtained the results for a conducting plane if the metal were treated as a dielectric with infinite static dielectric function $\varepsilon_1 = \infty$; $Q' = -Q$; and $Q'' = 2Q$. Note that Q'' does not contribute to the potential inside the metal since the potential is divided by ε_1 which is infinite. The problem cannot be solved with a charge outside a dielectric sphere with a finite number of image charges.

2.24 ELECTRIC BOUNDARY CONDITIONS

If an electric field exists in a region consisting of two different media, then the conditions that the field must satisfy at the interface between the two media are called boundary conditions.

We consider three different interfaces:

1. Dielectric–Dielectric Boundary,
2. Conductor–Dielectric Boundary, and
3. Conductor–Free Space Boundary.

To determine the conditions, we use Maxwell's and Gauss' law

$$\oint_S \vec{D}\cdot d\vec{S} = Q \quad \text{and} \quad \oint_l \vec{E}\cdot d\vec{l} = 0$$

Also, the electric field intensity vector $(\vec{E})$ can be decomposed into two orthogonal components as

$$\vec{E} = \vec{E}_t + \vec{E}_n$$

where, $\vec{E}_t$ and $\vec{E}_n$ are the tangential and normal components of $\vec{E}$, respectively.

2.24.1 Dielectric–Dielectric Boundary Conditions

We consider two different media 1 and 2, characterised by the permittivities ε_1 and ε_2, respectively, shown in Fig. 2.83.

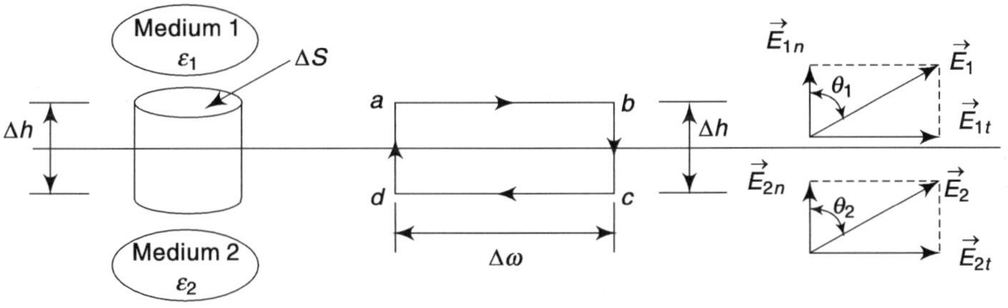

Fig. 2.83 *Dielectric-dielectric boundary conditions*

Applying Maxwell's equation for the closed path *abcda*

$$0 = E_{1t}\Delta\omega - E_{1n}\frac{\Delta h}{2} - E_{2n}\frac{\Delta h}{2} - E_{2t}\Delta\omega + E_{2n}\frac{\Delta h}{2} + E_{1n}\frac{\Delta h}{2}$$

Assuming the path to be very small with respect to the variation of $\vec{E}$, where $E_t = |\vec{E}_t|$ and $E_n = |\vec{E}_n|$, as $h <<$, we have

$$\boxed{E_{1t} = E_{2t}} \qquad \textbf{(2.80)}$$

Since, $\vec{D} = \varepsilon\vec{E} = \vec{D}_t + \vec{D}_n$, Eq. (2.80) can be written in terms of the flux density as

$$\boxed{\frac{D_{1t}}{\varepsilon_1} = \frac{D_{2t}}{\varepsilon_2}} \qquad \textbf{(2.81)}$$

Thus, *the tangential component of* $\vec{E}$ *is continuous at the boundary; but the tangential component of* $\vec{D}$ *is discontinuous.*

Now, applying Gauss' law to the pillbox (Gaussian surface), with $\Delta h \to 0$

$$D_{1n\Delta Q}\Delta S - D_{2n}\Delta S = \Delta Q = \sigma \Delta S$$

$$\boxed{(D_{1n} - D_{2n}) = \sigma} \tag{2.82}$$

where σ is the free surface charge density placed deliberately at the boundary. In general, no free charge is placed, so that, $\sigma = 0$. Hence, Eq. (2.81) can be written as

$$\boxed{D_{1n} = D_{2n}} \tag{2.83}$$

In terms of the field intensity, the boundary condition can be written as

$$\boxed{\varepsilon_1 E_{1n} = \varepsilon_2 E_{2n}} \tag{2.84}$$

Thus, *the normal component of* $\vec{D}$ *is continuous if there is no free charge at the interface, but normal component of* $\vec{E}$ *is always discontinuous at the boundary surface.*

Example 2.80 Show that for a charge-free medium, the electric boundary conditions can be expressed by the equation $\dfrac{\tan\theta_1}{\tan\theta_2} = \dfrac{\varepsilon_1}{\varepsilon_2}$, where the notations have their usual meanings.

Solution Let the fields make angle θ with the respective normal to the interface. Then we can combine the boundary conditions as

$$E_{1t} = E_{2t} \quad \text{or} \quad E_1 \sin\theta_1 = E_2 \sin\theta_2$$

and

$$\varepsilon_1 E_{1n} = \varepsilon_2 E_{2n} \quad \text{or} \quad \varepsilon_1 E_1 \cos\theta_1 = \varepsilon_2 E_2 \cos\theta_2$$

Combining

$$\boxed{\frac{\tan\theta_1}{\tan\theta_2} = \frac{\varepsilon_1}{\varepsilon_2}} \quad \text{or} \quad \boxed{\varepsilon_1 \cot\theta_1 = \varepsilon_2 \cot\theta_2} \tag{2.85}$$

This is the law of refraction of the electric field at a boundary free of charges.

2.24.2 Conductor–Dielectric Boundary Conditions

For a perfect conductor, the conductivity is infinite, $\sigma \to \infty$, or the resistivity is zero, $\rho \to 0$ and so, the field inside a conductor is zero, $E = 0$.

If some charges are introduced in the interior of such a conductor, the charges will move to the conductor surface and redistribute themselves in such a manner that the field inside the conductor is zero.

We consider the same procedure for finding the boundary conditions (Fig. 2.84).

Applying Maxwell's equation for the closed path *abcda*

$$0 = E_t \Delta\omega - E_n \frac{\Delta h}{2} - 0\frac{\Delta h}{2} - 0\Delta\omega + 0\frac{\Delta h}{2} + E_n \frac{\Delta h}{2}$$

$$\Rightarrow \qquad E_t = 0$$

$$\therefore \quad \boxed{D_t = 0 = E_t} \quad \textbf{(2.86)}$$

Thus, *the tangential component of* $\vec{E}$ *and* $\vec{D}$ *are both zero for a conductor–dielectric boundary.*

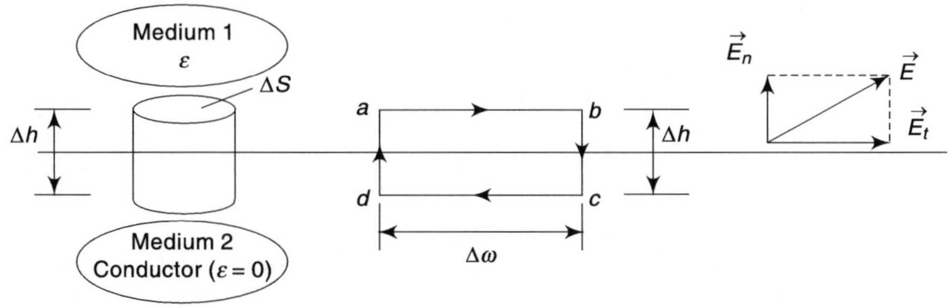

Fig. 2.84 *Conductor–dielectric boundary conditions*

Now, applying Gauss' law to the pillbox (Gaussian surface), with $\Delta h \to 0$,

$$D_n \Delta S - 0 \Delta S = \Delta Q = \sigma \Delta S \quad \Rightarrow \quad D_n = \sigma$$

$$\therefore \quad E_n = \frac{D_n}{\varepsilon} = \frac{\sigma}{\varepsilon}$$

$$\boxed{D_n = \sigma \quad \text{and} \quad E_n = \frac{\sigma}{\varepsilon}} \quad \textbf{(2.87)}$$

Thus, we can conclude that:

1. Inside a conductor, the electric field is always zero. This property is used for *electrostatic shielding*.
2. The electric field can only be external to the conductor and normal to its surface.
3. Since $\vec{E} = -\nabla V$ and $E = 0$ inside a conductor, there is no potential difference between any two points in the conductor, i.e., a conductor is an equipotential body.

2.24.3 Conductor-Free Space Boundary Conditions

These boundary conditions will be identical as those for a conductor–dielectric boundary except that ε will be replaced by ε_0, so that the boundary conditions for the tangential and normal components become

$$\boxed{D_t = 0 = E_t} \quad \text{and} \quad \boxed{D_n = \sigma \quad \text{and} \quad E_n = \frac{\sigma}{\varepsilon_0}} \quad \textbf{(2.88)}$$

***Example 2.81** A boundary exists at $z = 0$ between two dielectrics $\varepsilon_{r_1} = 2.5$ in the region $z < 0$, and $\varepsilon_{r_2} = 4$ in the region $z > 0$. The field in region of ε_{r1} is $\vec{E}_1 = -30\hat{i} + 50\hat{j} + 70\hat{k}$ V/m. Find the electric displacement vector in the second medium. Also, find the angle between electric field intensity in the second medium and the normal to the boundary surface.

Solution Since $z = 0$ is the boundary, $\hat{k}$ is the normal to the boundary plane, and the normal component of the field is

$$E_{1n} = \vec{E}_1 \cdot \hat{k} = 70$$

$\therefore$
$$\vec{E}_{1n} = 70\hat{k}$$

$\therefore$
$$\vec{E}_{1t} = \vec{E}_1 - \vec{E}_{1n} = -30\hat{i} + 50\hat{j}$$

By the boundary conditions for dielectric–dielectric interface, we get

$$\vec{E}_{2t} = \vec{E}_{1t} = -30\hat{i} + 50\hat{j}$$

and

$$\varepsilon_{r2}\vec{E}_{2n} = \varepsilon_{r1}\vec{E}_{1n} \quad \Rightarrow \quad \vec{E}_{2n} = \frac{\varepsilon_{r1}}{\varepsilon_{r2}}\vec{E}_{1n} = \frac{2.5}{4} \times 70\hat{k} = 43.75\hat{k}$$

$\therefore$
$$\vec{E}_2 = \vec{E}_{2n} + \vec{E}_{2t} = -30\hat{i} + 50\hat{j} + 43.75\hat{k}$$

Hence, the flux density is

$$\begin{aligned}\vec{D}_2 &= \varepsilon_2\vec{E}_2 = \varepsilon_0\varepsilon_{r2} \times (-30\hat{i} + 50\hat{j} + 43.75\hat{k}) \\ &= 8.854 \times 10^{-12} \times 4 \times (-30\hat{i} + 50\hat{j} + 43.75\hat{k}) \\ &= -1.061\hat{i} + 1.768\hat{j} + 1.547\hat{k} \text{ nC/m}^2\end{aligned}$$

The angle between electric field intensity in the second medium and the normal to the boundary surface is given as

$$\theta_2 = \tan^{-1}\left(\frac{\vec{E}_{2t}}{\vec{E}_{2n}}\right) = \tan^{-1}\left(\frac{\sqrt{30^2 + 50^2}}{43.75}\right) = 53.12°$$

Example 2.82 It is found that $\vec{E}_1 = 60\hat{i} + 20\hat{j} - 30\hat{k}$ mV/m at a particular point on the interface between air and a conducting surface. Find the flux density $\vec{D}$ and surface charge density ρ_s at that point.

Solution This is a dielectric–conductor interface. Here, the flux density is

$$\vec{D} = \varepsilon_0\vec{E}_1 = \frac{10^{-9}}{36\pi} \times (60\hat{i} + 20\hat{j} - 30\hat{k}) \times 10^{-3} = (0.531\hat{i} + 0.177\hat{j} - 0.265\hat{k}) \text{ pC/m}^2$$

The surface charge density is given as

$$\rho_s = \sigma = D_n = \sqrt{(0.531)^2 + (0.177)^2 + (-0.265)^2} \times (10^{-12}) = 0.619 \text{ pC/m}^2$$

***Example 2.83** Two extensive homogeneous isotropic dielectrics meet on plane $z = 0$ as shown in Fig. 2.85. For $z \geq 0$, $\varepsilon_{r_1} = 4$ and for $z \leq 0$, $\varepsilon_{r2} = 3$. A uniform electric field, $\vec{E}_1 = 5\hat{a}_x - 2\hat{a}_y + 3\hat{a}_z$ kV/m exists for $z \geq 0$. Find

(a) $\vec{E}_2$ for $z \leq 0$.
(b) The angles between electric field intensity and the normal to the boundary surface in both media.
(c) The energy densities in J/m^3 in both dielectrics.
(d) The energy within a cube of side 2 m centered at (3, 4, –5).

Solution

(a) Since, $\hat{a}_z$ is the normal to the boundary plane, the normal component is

$$E_{1n} = \vec{E}_1 \cdot \hat{a}_n = \vec{E}_1 \cdot \hat{a}_z = 3$$

$$\therefore \quad \vec{E}_{1n} = 3\hat{a}_z$$

$$\therefore \quad \vec{E}_{1t} = \vec{E}_1 - \vec{E}_{1n} = 5\hat{a}_x - 2\hat{a}_y$$

By boundary conditions

(1) $\vec{E}_{2t} = \vec{E}_{1t} = 5\hat{a}_x - 2\hat{a}_y$

and,

(2) $\varepsilon_{r2}\vec{E}_{2n} = \varepsilon_{r1}\vec{E}_{1n}$

$$\Rightarrow \vec{E}_{2n} = \frac{\varepsilon_{r1}}{\varepsilon_{r2}}\vec{E}_{1n} = \frac{4}{3}\vec{E}_{1n} = 4\hat{a}_z$$

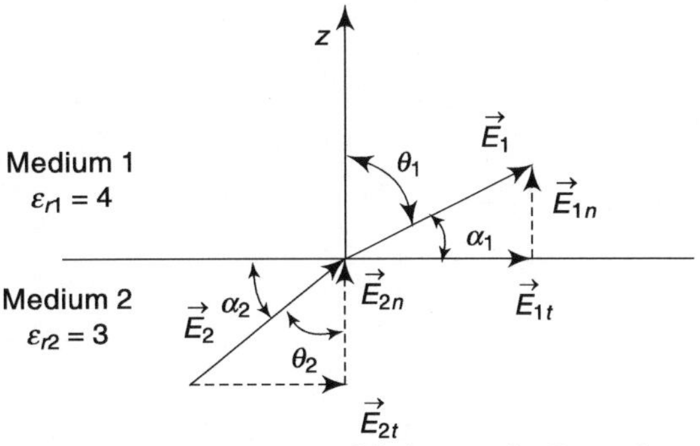

Fig. 2.85 *Arrangement of dielectrics for Example 2.83*

So, the field in second medium is given as

$$\vec{E}_2 = (\vec{E}_{2t} + \vec{E}_{2n}) = (5\hat{a}_x - 2\hat{a}_y + 4\hat{a}_z)\ \text{kV/m}$$

(b) Let, α_1 and α_2 be the angles $\vec{E}_1$ and $\vec{E}_2$ make with the interface while θ_1 and θ_2 are the angles they make with the normal to the interface.

$$\therefore \quad \alpha_1 = (90° - \theta_1) \quad \text{and} \quad \alpha_2 = (90° - \theta_2)$$

$$\therefore \quad \tan\theta_1 = \frac{E_{1t}}{E_{1n}} = \frac{\sqrt{5^2 + 2^2}}{3} = \frac{\sqrt{29}}{3} = 1.795$$

$$\therefore \quad \theta_1 = 60.9°$$

Similarly,

$$\therefore \quad \tan\theta_2 = \frac{E_{2t}}{E_{2n}} = \frac{\sqrt{5^2 + 2^2}}{4} = \frac{\sqrt{29}}{4} = 1.346$$

$$\therefore \quad \theta_2 = 53.4°$$

Hence, the angles between electric field intensity and the normal to the boundary surface in both media are given as

$$\theta_1 = 60.9° \quad \text{and} \quad \theta_2 = 53.4°$$

NOTE

The relation $\dfrac{\tan\theta_1}{\tan\theta_2} = \dfrac{\varepsilon_{r1}}{\varepsilon_{r2}}$ *is satisfied.*

(c) The energy densities are given as

$$W_{E1} = \frac{1}{2}\varepsilon_1|\vec{E}_1|^2 = \frac{1}{2} \times 4 \times \frac{10^{-9}}{36\pi}(25 + 4 + 9) \times 10^6 = 672\ \mu\text{J/m}^3$$

$$W_{E2} = \frac{1}{2}\varepsilon_2 |\vec{E}_2|^2 = \frac{1}{2} \times 3 \times \frac{10^{-9}}{36\pi}(25 + 4 + 16) \times 10^6 = 597\ \mu\text{J/m}^3$$

(d) At the centre (3, 4, –5) of the cube of side 2 m, $z = -5 < 0$; i.e., the cube is in region 2 with $2 \le x \le 4$, $3 \le y \le 5$, $-6 \le z \le -4$.

Hence, the energy within the cube is

$$W_E = \int w_{E_2} dv = \int_{x=2}^{4} \int_{y=3}^{5} \int_{z=-6}^{-4} w_{E_2} dz dy dx = w_{E_2} \times 2 \times 2 \times 2 = 597 \times 8\ \mu\text{J} = 4.776\ \text{mJ}$$

Example 2.84 A homogeneous dielectric ($\varepsilon_r = 2.5$) fills region 1 ($x \le 0$) while region 2 ($x \ge 0$) is free space.

(a) If $\vec{D}_1 = 12\hat{a}_x - 10\hat{a}_y + 4\hat{a}_z$ nC/m^2, find $\vec{D}_2$ and θ_2.

(b) If $E_2 = 12$ kV/m, and $\theta_2 = 60°$, find $\vec{E}_1$ and θ_1. Take θ_1 and θ_2 as the angles of $\vec{E}_1$ and $\vec{E}_2$ with the normal of the surface, respectively.

Solution

(a) Since, $\hat{a}_x$ is the normal to the boundary plane, the normal component is

$$D_{1n} = \vec{D}_1 \cdot \hat{a}_n = \vec{D}_1 \cdot \hat{a}_x = 12$$

$$\therefore \quad \vec{D}_{1n} = 12\hat{a}_x$$

$$\therefore \quad \vec{D}_{1t} = \vec{D}_1 - \vec{D}_{1n} = -10\hat{a}_y + 4\hat{a}_z$$

By boundary conditions

(1) $\vec{D}_{2t} = \dfrac{\varepsilon_{r2}}{\varepsilon_{r1}}\vec{D}_{1t} = \dfrac{1}{2.5}(-10\hat{a}_y + 4\hat{a}_z) = -4\hat{a}_y + 1.6\hat{a}_z$

and

(2) $\vec{D}_{2n} = \vec{D}_{1n} = 12\hat{a}_x$

So, the flux density in second medium is given as

$$\vec{D}_2 = (\vec{D}_{2t} + \vec{D}_{2n}) = (12\hat{a}_x - 4\hat{a}_y + 1.6\hat{a}_z)\ \text{nC/m}^2$$

$$\therefore \quad \tan\theta_2 = \frac{E_{2t}}{E_{2n}} = \frac{\sqrt{(-4)^2 + (1.6)^2}}{12} = 0.359$$

$$\therefore \quad \theta_2 = 19.75°$$

(b) Here, $\therefore\ \tan\theta_2 = \tan 60° = \dfrac{E_{2t}}{E_{2n}}$

$$\therefore \quad E_{2t} = \sqrt{3}E_{2n}$$

Also, $E_2 = \sqrt{E_{2t}^2 + E_{2n}^2} \Rightarrow 12 = \sqrt{4E_{2n}^3} \Rightarrow E_{2n} = 6$ V/m

$$\therefore \quad \vec{E}_{1t} = \vec{E}_{2t} = \sqrt{3} \times 6 = 10.39\ \text{V/m}$$

$$\therefore \quad E_{1n} = \frac{\varepsilon_{r2}}{\varepsilon_{r1}} E_{2n} = \frac{1}{2.5} \times 6 = 2.4 \text{ V/m}$$

$$\therefore \quad E_1 = \sqrt{E_{1t}^2 + E_{1n}^2} = \sqrt{(10.39)^2 + (2.4)^2} = 10.67$$

$$\therefore \quad \tan \theta_1 = \frac{E_{1t}}{E_{1n}} = \frac{10.39}{2.4} = 4.33$$

$$\therefore \quad \theta_1 = 77°$$

***Example 2.85** Region $y \leq 0$ consists of a perfect conductor while $y \geq 0$ is a dielectric medium ($\varepsilon_r = 2$). If there is a surface charge of 2nC/m^2 on the conductor, determine $\vec{E}$ and $\vec{D}$ at:

(a) A (3, –2, 2). **(b)** B (–4, 1, 5).

Solution

(a) Point A (3, –2, 2) is in the conductor as $y = -2 < 0$ at A.

$$\therefore \quad \vec{E} = \vec{D} = 0$$

(b) At point B (–4, 1, 5), $\rho_s = 2$ nC/m^2

$$\therefore \quad D_n = \rho_s = 2 \text{ nC/m}^2$$

$$\therefore \quad \vec{D} = 2\hat{a}_y \text{ nC/m}^2$$

$$\therefore \quad \vec{E} = \frac{\vec{D}}{\varepsilon_0 \varepsilon_r} = \frac{2 \times 10^{-9}}{\frac{10^{-9}}{36\pi} \times 2} = 36\pi\, \hat{a}_y = 113.1\hat{a}_y \text{ V/m}$$

2.25 DIRAC DELTA REPRESENTATION IN ELECTROSTATIC FIELDS

In field theory, most of the functions used are continuous and have continuous derivatives. However, there are certain functions, such as point charges, filamentary currents, current shells, etc., which have singularities at certain points. These functions require special treatment. One method is to treat these discrete sources as limiting cases of volume distributions for which one or more dimensions are allowed to become vanishingly small.

However, the most useful technique to treat these discrete sources is the use of *Dirac Delta function* (namely, *unit impulse function* in circuit theory).

The Dirac delta at the point $x = x_0$ is designated by $\delta(x - x_0)$ and at $x = 0$ it is designated by $\delta(x)$. It has the property that

$$\int_a^b \delta(x - x_0)dx = 1, \quad \textit{if } x_0 \text{ is in } (a, b)$$

$$= 0, \quad \textit{if } x_0 \text{ is not in } (a, b)$$

Thus, delta behaves as a very sharply peaked function of unit area. This function can be represented as a vanishingly thin Gaussian function of unit area as shown in Fig. 2.86.

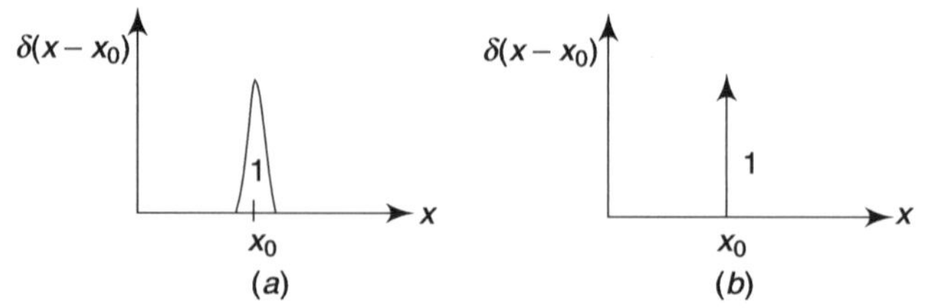

Fig. 2.86 *The Dirac delta: (a) Approximate form, and (b) Symbolic representation*

A special property of delta function may be stated as follows:

$$\int_a^b f(x)\delta(x-x_0)dx = f(x_0), \quad \text{if } x_0 \text{ is in } (a, b)$$
$$= 0, \qquad \text{if } x_0 \text{ is not in } (a, b)$$

This implies that the delta has the property of selecting the value of the function $f(x)$ at the point x_0; this property is known as *sampling property*.

NOTE

Technically, δ(x) is not a function at all, since its value is not finite at x = 0. In mathematics, it is known as a generalised function or distribution.

2.25.1 Three-Dimensional Dirac Delta Function

If $\vec{r}$ denotes the point (x, y, z) then the three-dimensional dirac delta function at the point $\vec{r} = \vec{r}_0$ is designated by $\delta(\vec{r} - \vec{r}_0)$ and is represented as

$$\int_V \delta(\vec{r}-\vec{r}_0)dV = 1 \quad \text{if } \vec{r}_0 \text{ is in } V$$
$$= 0 \quad \text{if } \vec{r}_0 \text{ is not in } V$$

This three-dimensional dirac delta also has the sampling property, which may be written as

$$\int_V f(\vec{r})\delta(\vec{r}-\vec{r}_0)dx = f(\vec{r}_0), \quad \text{if } \vec{r}_0 \text{ is in } V$$
$$= 0, \qquad \text{if } \vec{r}_0 \text{ is not in } \vec{r}_0$$

Three-dimensional dirac delta may be expressed in terms of one-dimensional deltas in different coordinates as follows.

$$\delta(\vec{r}-\vec{r}_0) = \delta(x-x_0)\delta(y-y_0)\delta(z-z_0) \quad \text{in Cartesian coordinates}$$
$$= \frac{\delta(r-r_0)\delta(\phi-\phi_0)\delta(z-z_0)}{r_0} \quad \text{in cylindrical coordinates}$$
$$= \frac{\delta(\rho-\rho_0)\delta(\theta-\theta_0)\delta(\phi-\phi_0)}{\rho_0^2 \sin\theta_0} \quad \text{in spherical coordinates}$$

2.25.2 Dirac Delta Representation of Point Charge and Related Equations of Electrostatics

If a point charge of q Coulomb is located within a volume V, then it can be written in terms of volume charge density ρ as follows.

$$\int_V \rho dV = q$$

If the charge q is located outside the volume V, then

$$\int_V \rho dV = 0$$

The above two equations can be combined if ρ is set equal to q multiplied by three-dimensional dirac delta. Thus, the charge density of a point charge q Coulomb located at $\vec{r} = \vec{r}_0$ may be written as,

$$\rho(\vec{r}) = q\delta(\vec{r} - \vec{r}_0)$$

In terms of one-dimensional deltas in Cartesian coordinates

$$\rho(\vec{r}) = q\delta(x - x_0)\delta(y - y_0)\delta(z - z_0)$$

Therefore, for a point charge located at origin, the volume charge density is written as

$$\rho(\vec{r}) = q\delta(x)\delta(y)\delta(z)$$

Similarly, line and surface distributions of charge may also be written in terms of Dirac delta.

For a line charge density of λ (C/m) along the z-axis, the equivalent volume charge density is given as

$$\rho(\vec{r}) = \lambda\delta(x)\delta(y)$$

For a surface charge density of σ (C/m^2) lying in the xy-plane, the equivalent volume charge density is given as

$$\rho(\vec{r}) = \sigma\delta(z)$$

Therefore, Gauss' law can be written in terms of Dirac delta function as

$$\nabla \cdot \vec{E} = \frac{q}{\varepsilon}\delta(r) = \frac{q}{\varepsilon}\delta(x)\delta(y)\delta(z) \qquad \text{for a point charge}$$

$$= \frac{\lambda}{\varepsilon}\delta(x)\delta(y) \qquad \text{for a line charge along } z\text{-axis}$$

$$= \frac{\sigma}{\varepsilon}\delta(z) \qquad \text{for a surface charge in the } xy\text{-plane}$$

Similarly, Poisson's equation can be written for a point charge q located at origin as

$$\nabla^2 V = -\frac{q}{\varepsilon}\delta(r)$$

Summary

- The quantitative expression for the affect of an electric charge and distance on electric force is given by *Coulomb's law*, which states that the force between two charges is

$$\vec{F}_{12} = \frac{Q_1 Q_2}{4\pi\varepsilon R^2}\hat{a}_{R12}$$

- If there is a number of charges $Q_1, Q_2, \ldots, Q_n$ placed at points with position vectors $\vec{r}_1, \vec{r}_2, \ldots, \vec{r}_n$, respectively, then the resultant force $\vec{F}$ on a charge Q located at point $\vec{r}$ is

$$\vec{F} = \frac{Q}{4\pi\varepsilon}\sum_{i=1}^{n}\frac{Q_i(\vec{r}-\vec{r}_i)}{|\vec{r}-\vec{r}_i|^3}$$

  This is known as principle of superposition of charges.
- The *electric field intensity* ($\vec{E}$) is defined as the force per unit charge when placed in an electric field. So, for a point charge, the field intensity is

$$\vec{E} = \frac{Q}{4\pi\varepsilon R^2}\hat{a}_R$$

- If there is a number of charges $q_1, q_2, \ldots, q_n$ placed at points with position vectors $\vec{r}_1, \vec{r}_2, \ldots, \vec{r}_n$, respectively, then the electric field intensity is

$$\vec{E} = \frac{1}{4\pi\varepsilon}\sum_{i=1}^{N}\frac{q_i}{r_i^2}\hat{a}_{ri}$$

  This is known as principle of superposition of field.
- The electric field intensity due to different continuous charge distribution is given as

$$\vec{E} = \frac{1}{4\pi\varepsilon}\int_{\text{line}}\frac{\lambda(\vec{r})}{r^2}dl \qquad \text{for line charge distribution}$$
$$= \frac{1}{4\pi\varepsilon}\iint_S\frac{\sigma(\vec{r})}{r^2}dS \qquad \text{for surface charge distribution}$$
$$= \frac{1}{4\pi\varepsilon}\int_V\frac{\rho(\vec{r})}{r^2}dV \qquad \text{for volume charge distribution}$$

- The electric flux density ($\vec{D}$) is defined as the total number of electric field lines per unit area passing through the area perpendicularly (in C/m^2). It is related to the field intensity as

$$\vec{D} = \varepsilon\vec{E}$$

  Hence, electric flux through a surface is given as

$$\psi = \int_S \vec{D}\cdot d\vec{S}$$

- Electric field lines are the imaginary lines drawn in such a way that at every point, it has the direction of the electric field ($\vec{E}$).
- Electric flux lines are the imaginary lines drawn in such a way that at every point, it has the direction of the electric flux density vector ($\vec{D}$).
- Gauss' law states that the total electric displacement or electric flux through any closed surface surrounding charges is equal to the net positive charge enclosed by that surface.
- Mathematically, it is expressed as

$$\psi = \oint_S \vec{D}\cdot d\vec{S} = Q = \int_v \rho dv, \qquad \text{integral form}$$
$$\nabla\cdot\vec{D} = \rho \qquad \text{differential form}$$

- The total work done in moving a unit positive charge from a point A to another point B is called the potential difference between the two points, given as

$$V_{AB} = \frac{W}{Q} = -\int_A^B \vec{E} \cdot d\vec{l}$$

This potential difference between the points A and B is also considered to be the *potential* (or absolute potential) of B with respect to the *potential* (or absolute potential) of A. In case of a point charge, the reference is taken to be at infinity with zero potential.

- *Potential* (or absolute potential) of a point is defined as the work done to bring a unit positive charge from infinity to that point. This is given as

$$V = \frac{Q}{4\pi\varepsilon R}$$

- If there is a number of point charges Q_1, Q_2, ..., Q_n, located at position vectors $\vec{r}_1, \vec{r}_2, \ldots, \vec{r}_n$ respectively, then the potential at point $\vec{r}$ is given as

$$V(\vec{r}) = \frac{1}{4\pi\varepsilon} \sum_{i=1}^{n} \frac{Q_i}{|\vec{r} - \vec{r}_i|}$$

This is known as *principle of superposition of potential.*

- The electric potential due to different continuous charge distribution is given as

$$V(\vec{r}) = \frac{1}{4\pi\varepsilon} \int_l \frac{\lambda(\vec{r}')dl'}{|\vec{r} - \vec{r}'|}; \quad \text{for line charge distribution with density } \lambda(\text{C/m})$$

$$= \frac{1}{4\pi\varepsilon} \int_S \frac{\sigma(\vec{r}')dS'}{|\vec{r} - \vec{r}'|}; \quad \text{for surface charge distribution with density } \sigma(\text{C/m}^2)$$

$$= \frac{1}{4\pi\varepsilon} \int_v \frac{\rho(\vec{r}')dv'}{|\vec{r} - \vec{r}'|}; \quad \text{for volume charge distribution with density } \rho(\text{C/m}^3)$$

- The rate of change of potential with respect to the distance is called the potential gradient. The relation between the potential and field intensity is written as

$$\vec{E} = -\nabla V$$

- The surface obtained by joining the points with equal potential is known as equipotential surface.
- Two equal and opposite point charges separated by a distance constitute an electric dipole.
- For an electric dipole with dipole moment $\vec{p}$ and centered at a position vector $\vec{r}'$, then the potential at a point $P(r, \theta, \phi)$ is given as

$$V(\vec{r}) = \frac{\vec{p} \cdot (\vec{r} - \vec{r}')}{4\pi\varepsilon |\vec{r} - \vec{r}'|^3}$$

- Similarly, for an electric dipole with dipole moment $\vec{p}$ and centered at origin, the field intensity at a point $P(r, \theta, \phi)$ is given as

$$\vec{E} = \frac{1}{4\pi\varepsilon r^3} \left[\frac{3(\vec{p} \cdot \vec{r})\vec{r}}{r^2} - \vec{p} \right]$$

- The torque on an electric dipole with dipole moment $\vec{p}$ when placed in external electric field is given as

$$\vec{\tau} = \vec{p} \times \vec{E}$$

- The potential energy of an electric dipole with dipole moment $\vec{p}$ when placed in external electric field is given as

$$U = -\vec{p} \cdot \vec{E}$$

- The electrostatic energy stored in an electric field is given as

$$W = \frac{1}{2}\sum_{i=1}^{N} Q_i V_i \qquad \text{for } N \text{ point charges}$$

$$= \frac{1}{2}\int_v \vec{D} \cdot \vec{E} dv \qquad \text{for continuous charge distribution}$$

- For a linear homogeneous material medium, *Poisson's equation* for electric potential is given as

$$\nabla^2 V = -\frac{\rho}{\varepsilon}$$

If the medium is charge-free (i.e., $\rho = 0$), Poisson's equation reduces to Laplace's equation, given as

$$\nabla^2 V = 0$$

- For electrostatic boundary value problems, the field $\vec{E}$ and the potential V are determined by solving Poisson's equation or Laplace's equation.
- Uniqueness theorem states that any solution of Laplace equation (or Poisson's equation) that satisfies the same boundary conditions must be the only solution, irrespective of the method of solution.
- A capacitor is a device that stores electric charge and hence electrostatic energy. The *capacitance* of a capacitor is the ratio of the magnitude of charge on one conductor to the potential difference between the conductors.

$$C = \frac{Q}{V} = \frac{\varepsilon \oint_S \vec{E} \cdot d\vec{S}}{\int_l \vec{E} \cdot d\vec{l}}$$

- The electrostatic energy stored in a capacitor is given as

$$W_{\text{stored}} = \frac{1}{2} C V^2$$

- Method of images is used for solving electrostatic boundary value problems involving and infinite conducting plane.
- The conditions that an electric field, existing in a region consisting of two different media, must satisfy at the interface between the two media are called electric boundary conditions. These are given as

For dielectric–dielectric interface:

$$E_{1t} = E_{2t}, \quad (D_{1n} - D_{2n}) = \sigma \quad \text{and} \quad D_{1n} = D_{2n} \text{ (when } \sigma = 0)$$

For dielectric–conductor interface:

$$D_t = 0 = E_t \quad \text{and} \quad D_n = \varepsilon E_n = \sigma$$

Important Formulae	
Coulomb's law	$\vec{F}_{12} = k\dfrac{Q_1Q_2}{R^2}\hat{a}_{R12} = \dfrac{Q_1Q_2(\vec{r}_2 - \vec{r}_1)}{4\pi\varepsilon\,\lvert\vec{r}_2 - \vec{r}_1\rvert^3}$
Principle of superposition of charges	$\vec{F} = \dfrac{Q}{4\pi\varepsilon}\sum_{i=1}^{x}\dfrac{Q_1(\vec{r} - \vec{r}_i)}{\lvert\vec{r} - \vec{r}_i\rvert^3}$
Electric field intensity	$\vec{E} = \dfrac{\vec{F}}{Q} = \dfrac{Q_t}{4\pi\varepsilon R^2}\hat{a}_R = \dfrac{Q_t(\vec{r}_t - \vec{r})}{4\pi\varepsilon\lvert\vec{r}_t - \vec{r}\rvert^3}$
Electric field intensity due to continuous charge distribution	$\vec{E} = \dfrac{1}{4\pi\varepsilon}\int_{\text{line}}\dfrac{\lambda(\vec{r})}{r^2}dl$ for line charge distribution $= \dfrac{1}{4\pi\varepsilon}\iint_s\dfrac{\sigma(\vec{r})}{r^2}dS$ for surface charge distribution $= \dfrac{1}{4\pi\varepsilon}\int_V\dfrac{\rho(\vec{r})}{r^2}dV$ for volume charge distribution
Electric flux density	$\vec{D} = \varepsilon\vec{E}$
Electric flux	$\psi = \oint_s \vec{D}\cdot d\vec{S}$
Gauss' law of electrostatics	$\psi = \oint_s \vec{D}\cdot d\vec{S} = Q = \int_v p dv,$ integral form $\nabla\cdot\vec{D} = \rho,$ differential form
Potential of a point due to point charge (s)	$V = \dfrac{Q}{4\pi\varepsilon R} = \dfrac{1}{4\pi\varepsilon}\sum_{i=1}^{n}\dfrac{Q_1}{\lvert\vec{r} - \vec{r}_i\rvert}$
Potential due to continuous charge distribution	$V(\vec{r}) = \dfrac{1}{4\pi\varepsilon}\int_l\dfrac{\lambda(\vec{r}')dl'}{\lvert\vec{r} - \vec{r}'\rvert};$ for line charge distribution with density λ(C/m) $= \dfrac{1}{4\pi\varepsilon}\int_S\dfrac{\sigma(\vec{r}')dS}{\lvert\vec{r} - \vec{r}'\rvert};$ for surface charge distribution with density $\sigma(\text{C/m}^2)$ $= \dfrac{1}{4\pi\varepsilon}\int_v\dfrac{\rho(\vec{r}')dv'}{\lvert\vec{r} - \vec{r}'\rvert};$ for volume charge distribution with density $\rho(\text{C/m}^3)$
Relation between potential and field	$\vec{E} = -\nabla V$
Potential due to electric dipole	$V(\vec{r}) = \dfrac{\vec{p}\cdot(\vec{r} - \vec{r}')}{4\pi\varepsilon\,\lvert\vec{r}' - \vec{r}'\rvert^3}$
Field due to electric dipole	$\vec{E} = \dfrac{1}{4\pi\varepsilon r^3}\left[\dfrac{3(\vec{p}\cdot\vec{r})\vec{r}}{r^2} - \vec{p}\right]$

Electrostatic energy store in an electric field	$W = \frac{1}{2}\sum_{i=1}^{N} Q_i V_i$ for N point charges $= \frac{1}{2}\int_v \vec{D}\cdot\vec{E}\,dv$ for continuous charge distribution
Poisson's equation	$\nabla^2 V = -\frac{\rho}{\varepsilon}$
Laplace's equation	$\nabla^2 V = 0$
Capacitance of a capacitor	$C = \frac{Q}{V} = \frac{\varepsilon\int_s \vec{E}\cdot d\vec{S}}{\int_l \vec{E}\cdot d\vec{l}}$
Energy stored in a capacitor	$W_{\text{stored}} = \frac{1}{2}CV^2$
Electric boundary conditions for dielectric–dielectric interface	$E_{1t} = E_{2t}$, $(D_{1n} - D_{2n}) = \sigma$ and $D_{1n} = D_{2n}$ (when $\sigma = 0$)
Electric boundary conditions for dielectric–conductor interface	$D_t = 0 = E_t$ and $D_n = \varepsilon E_n = \sigma$

Exercises

[NOTE: * marked problems are important university problems]

- ***Easy***

1. A thundercloud above earth sets up a vertical electric field of 40 V/m. A raindrop carrying a charge of 0.1 μC lies in this field. What is the electrostatic force exerted on this raindrop? [9×10^{15} N]

*2. Determine the electric field at a distance r from an infinite straight line carrying a uniform line charge distribution with line charge density λ.
$$\left[\vec{E} = \frac{\lambda}{2\pi\varepsilon r}\hat{a}_r\right]$$

*3. Derive an expression of electric field due to uniform charge distribution over an infinite plane with surface charge density σ.
$$\left[\vec{E} = \frac{\sigma}{2\varepsilon}\hat{a}_n\right]$$

*4. Determine the electric field due to spherical cloud of electrons giving the volume charge density of
$$\rho = -\rho_0; \quad 0 \le r \le a$$
$$= 0; \quad r > a$$
$$\left[\begin{aligned}\vec{E} &= -\frac{\rho_0 r}{3\varepsilon}\hat{a}_r; \quad 0 \le r \le a\\ &= \frac{-\rho_0 a^3}{3\varepsilon r^2}\hat{a}_r; \quad r \ge a\end{aligned}\right]$$

5. Deduce Laplace's equation in spherical coordinates and find whether the potential field $V = \frac{a}{r^3}\sin\theta$ volt in a region of free space satisfies it.

***6.** Two concentric coaxial cones have the same vertex at the origin and their common axis coincides with the positive z-axis. The cone vertices are insulated, and their semi-vertical angles are θ_1 and θ_2, respectively ($\theta_2 > \theta_1$). The outer cone is earthed and the inner cone has a potential V_1. Determine the potential in the region between the cones.

$$\left[V_1\left\{\frac{\ln(\tan\theta/2)-\ln(\tan\theta_2/2)}{\ln(\tan\theta_1/2)-\ln(\tan\theta_2/2)}\right\}\right]$$

7. Derive an expression for the capacitor of a spherical condenser consisting of two concentric spheres of radii a and b, the dielectric between the two spheres being air.

$$\left[C = 4\pi\varepsilon_0\left(\frac{ab}{a-b}\right)\right]$$

***8.** Show that the capacitance of an isolated sphere of radius R is $4\pi\varepsilon_0 R$.

***9.** A boundary exists at $z = 0$ between two dielectrics $\varepsilon_{r_1} = 2.5$ region $z < 0$, and $\varepsilon_{r_2} = 4$ region $z > 0$. The field in region of ε_{r_1} is $\vec{E}_1 = -30\hat{a}_x + 50\hat{a}_y + 70\hat{a}_z$ V/m. Find

(a) normal component of $\vec{E}_1$,
(b) tangential component of $\vec{E}_1$,
(c) the angle $\alpha_1 \le 90°$ between $\vec{E}_1$ and normal to the surface,
(d) normal component of $\vec{D}_2$,
(e) tangential component of $\vec{D}_2$,
(f) $\vec{D}_2$
(g) polarization in ε_{r_1} material,
(h) The angle α_2 between $\vec{E}_1$ and normal to the surface.

[70 V/m; 58.3 V/m; 39.8°; 1.55×10^{-9} C/m^2; 2.6×10^{-9} C/m^2;
$(-1.06\hat{a}_x + 1.77\hat{a}_y + 1.55\hat{a}_z)\times10^{-9}$ C/m^2;
$(-0.8\hat{a}_x + 1.33\hat{a}_y + 1.16\hat{a}_z)\times10^{-9}$ C/m^2; 53.1°]

- ***Medium***

***10.** Three point charges of 0.25 μC are placed in air at the vertices of an equilateral triangle of side 100 mm. Determine the magnitude and direction of the force on one charge due to other charges.
[97.43 mN]

11. Three point charges $q_1 = +10^{-6}$C, $q_2 = -10^{-6}$C, and $q_3 = 0.5\times10^{-6}$C are placed in air at the corners of an equilateral triangle of 50 cm side. Determine the magnitude and direction of the force on q_3.
[0.018 N]

***12.** Charges ($-e$) are placed at the vertices of an equilateral triangle of side a and a positive charge Q is placed at the centre of gravity of the triangle. What would be the value of Q if the force on any of the negative charges is zero?

$$\left[\frac{e}{\sqrt{3}}\right]$$

13. Three point electric charges are situated in a straight line 10 cm apart. They have charges 2.0 μC, -1.0 μC and 2.0 μC, with negative charge in the centre. Find the forces on each charge due to the other two.
[–1.8 N, –1.8 N, 0.9 N]

14. Three equal positive charges of 4×10^{-9} C each are located at three corners of a square of side 20 cm each. Determine the magnitude and direction of electric field at the vacant corner.
[1722.78 V/m; 45°]

15. Four concentrated charges are located at the vertices of a plane rectangle as shown. Find the magnitude and direction of the resultant force on Q_3.

[0.93N at angle + 84° to line joining Q_3Q_4]

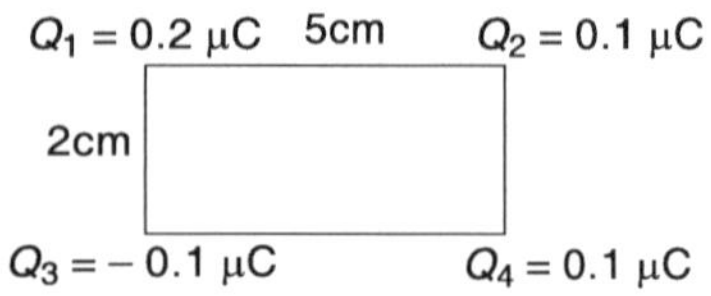

Fig. 2.87

16. Four equal charges are placed at the corners of a square of area 1 m^2. The force on each charge is 1 N. Determine the value of each charge. [7.62 μC]

17. Four point charges of $q_1 = +1 \times 10^{-8}$ C, $q_2 = -2 \times 10^{-8}$ C, $q_3 = +3 \times 10^{-8}$ C and $q_4 = +2 \times 10^{-8}$ C, respectively, are placed at the corners of a square of side 1 metre. If the medium is air, find the magnitude and direction of the field intensity at the point of intersection of the diagonals. Also, find the potential at that point. [509.19 Volt]

18. Two charges of similar sign and magnitude 1×10^{-12} Coulomb are located 1 metre apart. What is the potential at a point that is midway between the two charges and 50 cm from the line connecting the charges? What is the potential if the charges are of opposite sign?

[0.25 Volt, 0 Volt]

19. A spherical volume of radius R has a volume charge density given by $\rho = kr$ where r is the radial distance and k = constant. Find the expression for E and V in the region $0 \le r \le \infty$.

$$\left[\begin{aligned} \vec{E} &= \frac{kr^2}{4\varepsilon}\hat{a}_r; \quad r < R \\ &= \frac{kR^4}{4\varepsilon r^2}\hat{a}_r; \quad r \ge R \end{aligned}\right] \left[\begin{aligned} V &= \frac{k}{12\varepsilon}(4R^3 - r^3); \quad r < R \\ &= \frac{kR^4}{4\varepsilon r}; \quad r \ge R \end{aligned}\right]$$

20. The surface charge density of a charged thin circular disc of radius R is $\sigma(\vec{r}) = Ar$, where A is a constant and $\vec{r}$ is the position vector of a point of the thin disc from the centre of the disc. Calculate the electric field at any point on the axis of the disc.

$$\left[\vec{E} = \frac{Az}{2\varepsilon}\left\{\ln\left(\frac{R + \sqrt{R^2 + z^2}}{z}\right) - \frac{R}{\sqrt{R^2 + z^2}}\right\}\right]$$

21. A long charged cylinder of radius a has a volume density of charge $\rho = \gamma r$, where γ is a constant and r is the distance from the axis of the cylinder. Show that the electric field is given by

$$\begin{aligned} \vec{E}(r) &= \frac{\gamma r^2}{3\varepsilon}\hat{a}_r; \quad r < a \\ &= \frac{\gamma a^3}{3\varepsilon r}\hat{a}_r; \quad r > a \end{aligned}$$

***22.** A spherical volume charge distribution ρ is given by

$$\begin{aligned} \rho &= \rho_0\left(1 - \frac{r^2}{100}\right) \quad \text{for } r \le 10 \text{ mm} \\ &= 0 \quad \text{for } r > 10 \text{ mm} \end{aligned}$$

Show that the maximum value of electric field intensity E occurs at r = 7.45 mm. Obtain the value of E at r = 7.45 mm.

$$\left[1.656\frac{\rho_0}{\varepsilon_0}\right]$$

23. For a spherical charge distribution

$$\begin{aligned} \rho &= \rho_0(a^2 - r^2); \quad r < a \\ &= 0; \quad r > a \end{aligned}$$

(a) Find $\vec{E}$ and V for $r \geq a$
(b) Find $\vec{E}$ and V for $r \leq a$
(c) Find the total charge
(d) Show that $\vec{E}$ is maximum when $r = 0.145a$.

$$\left[(a)\frac{2\rho_0}{15\varepsilon_0 r^2}\hat{a}_r, \frac{2\rho_0}{15\varepsilon_0 r}; (b)\frac{\rho_0}{\varepsilon_0}\left(\frac{a^2 r}{3} - \frac{r^3}{5}\right)\hat{a}_r, \frac{\rho_0}{\varepsilon_0}\left(\frac{r^4}{20} - \frac{a^2 r^2}{6}\right) + \frac{2\rho_0}{15\varepsilon_0} + \frac{7\rho_0 a^4}{60\varepsilon_0}; (c)\frac{8\pi\rho_0}{15}\right]$$

24. A circular area of radius R has a point charge Q on its axis at a distance d from the centre of the area. Calculate the flux of the electric field due to Q through the circular area.

$$\left[\frac{Qd}{2\varepsilon_0}\left(\frac{1}{d} - \frac{1}{\sqrt{d^2 + R^2}}\right)\right]$$

25. Two infinite radial planes are inclined to each other at an angle α. There is an infinitesimal insulating gap at $r = 0$. Solve the one-dimensional Laplace's equation in cylindrical coordinate to obtain the potential V and field $\vec{E}$ as a function of ϕ. Boundary conditions are given as

$$V = 0 \quad \text{at} \quad \phi = 0 \quad \text{and} \quad V = V_0 \quad \text{at} \quad \phi = \alpha$$

$$\left[V = \frac{V_0}{\alpha}\phi; \quad \vec{E} = -\frac{1}{r}\frac{V_0}{\alpha}\hat{a}_\phi\right]$$

26. Two concentric conducting spherical shells of radii a and b ($b > a$), are charged to potentials V_1 and V_2, respectively. Determine the electric potential and field in the region between the shells. Also, determine the charge on the inner shell.

$$\left[V = \frac{ab}{b-a}\left(\frac{V_1 - V_2}{r}\right) + \frac{bV_2 - aV_1}{b-a}; \vec{E} = \frac{ab}{b-a}\left(\frac{V_1 - V_2}{r^2}\right)\hat{a}_r; \quad Q_1 = 4\pi\varepsilon ab\left(\frac{V_1 - V_2}{b-a}\right)\right]$$

27. The space between $x = 0$ and $x = d$ is filled with uniform charge of volume density ρ. The electric field at $x = d$ is zero. Calculate the potential difference between $x = 0$ and $x = d$.

$$\left[\frac{\rho d^2}{2\varepsilon_0}\right]$$

***28.** Prove that the capacitance between two lines is given by

$$C = \frac{\pi\varepsilon_0}{\ln\left(\frac{d}{r}\right)}$$

where d is the distance of separation between these lines and r is the radius of each line.

29. A point charge q is placed at a distance d from the centre of an earthed conducting sphere of radius a. Show, by solving Laplace's equation, that the potential at a point outside the sphere can be calculated by replacing the charges induced on the sphere by a point charge $-q(a/d)$ placed at a distance a^2/d from the centre of the sphere on the line joining the centre and the charge q.

- ***Hard***

30. Two point charges $-q$ and $+q/2$ are situated at the origin and at the point $(a, 0, 0)$ respectively. At what point along the axis does electric field vanish? Is this point a true minimum in the potential?

$$\left[\left(\frac{\sqrt{2}a}{\sqrt{2}-1}, 0, 0\right)\right]$$

31. An infinite charged sheet with surface charge density σ has a circular hole of radius a. The sheet is placed in the xy plane with its centre at origin. Using Coulomb's law or otherwise, find the potential V and the field $\vec{E}$ at any point at a distance z from the origin and long the positive z-direction.

$$\left[V = -\frac{\sigma}{2\varepsilon_0}(\sqrt{a^2+z^2}-a); \quad \vec{E} = -\frac{\sigma}{2\varepsilon_0}\left(\frac{z}{\sqrt{a^2+z^2}}\hat{a}_z\right)\right]$$

32. A line charge of length L has a uniform line charge distribution of λ. Determine an expression for the force exerted on a point charge q, located a distance r from the midpoint of the line charge distribution and perpendicular to the line charge.

$$\left[\vec{F} = \frac{q\lambda}{2\pi\varepsilon_0 r}\frac{L}{\sqrt{L^2+4r^2}}\hat{a}_r\right]$$

33. A charge density of σ (C/m^2) is uniformly spread over the area of a disc of radius a. Determine the force on a charge q that is placed a distance h from the centre of the disc and on a line that is perpendicular to the disc.

$$\left[\begin{aligned}\vec{F} &= \frac{q\sigma}{2\varepsilon_0}\left(1-\frac{h}{\sqrt{a^2+h^2}}\right)\hat{a}_z \quad h>0\\ &= -\frac{q\sigma}{2\varepsilon_0}\left(1+\frac{h}{\sqrt{a^2+h^2}}\right)\hat{a}_z \quad h<0\end{aligned}\right]$$

34. A point charge q is kept at a corner of a cube. Determine the flux of the electric field due to q through the three surfaces of the cube which do not meet at q.

$$\left[\frac{q}{8\varepsilon_0}\right]$$

35. A semi-infinite line extending from $-\infty$ to 0 along the z-axis carries a uniform line charge distribution of 10 nC/m. Find the field intensity at point P (0, 0, 4). If a charge of 5 μC is placed at point P, find the force acting on the charge. [$450\hat{a}_z$ V/m; $450\hat{a}_z$ μN]

36. Two water molecules each having a dipole moment 6.2×10^{-30} C-m point in the same direction and are inclined at an angle of 60° to the line joining their centres. Determine the potential energy due to their dipole–dipole interaction when their centres are 3.1×10^{-10} m apart. [18.1 meV]

37. Obtain by means of Laplace's equation, the potential distribution between two coaxial conducting cylinders of radii a and c with dielectric of constant ε_1 filling the region between a and b and a second dielectric of constant ε_2 filling the region between b and c. Given: $c > b > a$.

$$\left[\frac{\lambda}{2\pi\varepsilon_0}\left(\frac{1}{\varepsilon_1}\ln\frac{b}{a}+\frac{1}{\varepsilon_2}\ln\frac{c}{b}\right)\right]$$

38. The electrostatic potential in free space is given by $V = \alpha - \beta(x^2+y^2) = \gamma \ln\sqrt{x^2+y^2}$, where α, β and γ are constants. Find the charge density in the region. [$4\varepsilon_0\beta$]

39. A thin circular ring of radius R carries a uniform surface charge density σ. Calculate the electric potential and field at a point on the axis of the ring.

40. A single wire of radius r runs parallel to the ground at a height of h. Find an expression for its capacitance.

$$\left[C = \frac{2\pi\varepsilon_0}{\ln\left(\frac{h+\sqrt{h^2-r^2}}{r}\right)}\right]$$

***41.** Two co-axial conducting cylinders of radius 2 cm and 4 cm have a length of 1 m. The region between the cylinders contains a layer of dielectric from $r = c$ to $r = d$ with $\varepsilon_r = 4$. Find the capacitance if (i) $c = 2$ cm, $d = 3$ cm (ii) $d = 4$ cm and the volume of the dielectric is the same as in part (i). [(i)143 pF, (ii)177 pF]

***42.** A parallel plate capacitor has the region between the plates filled with a dielectric slab of relative permittivity ε_r. The plate dimensions are: width ω, length l and the plate separation is d. The capacitor is charged to a potential V, after which it is disconnected. The dielectric is now partially withdrawn in the l-dimension until only length x remains between the plates. What is the capacitance of this new system? What is now the potential across the capacitor?

$$\left[\frac{\varepsilon_0 \omega}{d}[l + (\varepsilon_r - 1)x]; \frac{\varepsilon_r l V}{l + (\varepsilon_r - 1)x}\right]$$

43. The permittivity of the dielectric material between the plates of a parallel plate capacitor varies uniformly from ε_1 at one plate to ε_2 at other plate. Show that the capacitance is given by

$$C = \frac{A}{d}\frac{\varepsilon_2 - \varepsilon_1}{\ln(\varepsilon_2/\varepsilon_1)}$$

where A and d are the area of each plate and separation between the plates respectively.

44. A parallel plate capacitor has rectangular plates of area A, but the plates are not exactly parallel. The separation at one edge is $(d - a)$ while at the other edge is $(d + a)$; $a << d$. Show that the capacitance is given approximately by

$$C = \frac{\varepsilon_0 A}{d}\left[1 + \frac{a^2}{3d^2}\right]$$

45. A positive point charge Q is placed at a height h from a flat conducting ground plane. Find the surface charge density ρ_s at a point on the ground plane, at a distance x along the plane measured from the point on the plane nearest to the charge.

$$\left[\frac{-Qh}{2\pi(h^2 + x^2)^{3/2}}\right]$$

46. Find the capacitance of an insulated conducting sphere of radius a when it is kept isolated. Determine the change in the capacitance when the sphere is placed at a distance b from an earthed conducting wall, where b is fairly large compared to a.

[$4\pi\varepsilon_0 a$; capacitance increases by $2\pi\varepsilon_0 a^2/b$]

47. A sheet charge of uniform density ρ_s extends in the entire xy-plane. Show that Gauss' law in differential form for the entire sheet charge is given by

$$\nabla \cdot \vec{E} = \frac{1}{\varepsilon_0}\rho_s \delta(z); \ \delta(z) \text{ is Dirac delta function}$$

Review Questions

[NOTE: * marked questions are important university questions.]

1. State and explain Coulomb's law in electrostatics. Express it mathematically for two point charges. Does the form depend on the system of units?

or

State and explain Coulomb's law and hence show that the electric field intensity is inversely proportional to the square of the distance between two point charges.

2. State Coulomb's law and explain the factors that influence the forces.

3. State and explain Coulomb's law for the vector force between two point charges in free space.

***4.** What is the superposition principle in electrostatics? What do you mean by the volume, surface and line charge densities? How can you express a point charge using the Dirac delta function?

5. Explain the superposition principle governing the forces between charges at rest.

6. (a) What do you mean by electrostatic potential and equipotential surfaces?
(b) Express the electrostatic potential in terms of the fundamental dimensions (MLTI).

***7.** (a) Define electric field $\vec{E}$ at a point and show that $\nabla \times \vec{E} = 0$.

or

Show that the electrostatic field is conservative.
(b) What are the dimensions of $\vec{E}$ in terms of the fundamental quantities *M*, *L*, *T* and *I*?
(c) Express the relationship between electric potential and field intensity.

or

What is 'potential' and 'potential gradient'? Prove that $\vec{E} = -\nabla V$, where, $\vec{E}$ is the electric field and V is the electric scalar potential.
(d) Show that the potential rise between two points is equal to the line integral of electric field intensity along the curved paths between the two points.
(e) State the properties of a static electric field.
(f) Show that the electric field intensity at a point due to a number of point charges is the vector sum of the electric filed intensities due to individual point charges at that point.

***8.** Discuss the term "electric field due to a static charge configuration".

***9.** Define the term "potential difference $V(A) - V(B)$ between points A and B in a static electric field". Give an energy interpretation to potential difference.

***10.** Define the term "potential difference $V(A) - V(B)$ between points A and B in a static electric field". Explain the concept of reference point and comment on its location.

***11.** Show that the work done in moving a charge from one point to another in an electrostatic field is independent of the path between the points.

***12.** Define the electric displacement vector $\vec{D}$ in the presence of dielectric. Obtain the expression for its divergence.

***13.** What are the equipotential surfaces for an infinite straight line of uniform charge density? Explain.

***14.** What are the equipotential surfaces for an infinite plane of uniform surface charge density? Explain.

***15.** (a) State and prove Gauss's theorem in electrostatics. What are the premises on which the theorem is based? Does the theorem differ from Coulomb's law?

or

Show that $div\,\vec{D} = \rho$, where $\vec{D}$ is the electric flux density and ρ is the volume charge density.
(b) Apply this theorem to calculate the electric field due to a (i) uniformly charged sphere and (ii) uniformly charged infinite cylinder.

16. State and prove Gauss' law in integral form, considering static charges in free space.

***17.** $\vec{E}$ is the electric field due to a point charge Q C at the origin in free space. Find $\int_S \vec{E} \cdot d\vec{a}$ where S is a spherical surface of radius R m and centre at origin.

18. (a) Show that the charge resides on the outer surfaces of a charged conductor.
(b) Prove that the electric field just outside a charged conductor is perpendicular to its surface. Also, determine the magnitude of this field.

19. If a total charge Q is uniformly distributed throughout the volume of a sphere of radius a, what would be the electric field at a distance r from the centre of the sphere?

20. State the boundary conditions on $\vec{E}$ and $\vec{D}$ at the interface of two different dielectric media, when there is a charge density on the interface.

or

Establish the boundary conditions for the electric field on the interface between two dielectric media of different relative permittivity.

21. What is an electric dipole? Calculate the electric field and potential in free space due to a dipole.

or

(a) Show that the electric field intensity due to a dipole varies inversely as the cube of the distance of the field point from the dipole.

(b) Show that the potential of a dipole at any point varies as the inverse square of the distance from centre of dipole to the point.

***22.** In electrostatics, what is meant by a physical dipole?

***23.** What is meant by a pure electric dipole?

24. For a physical dipole in the z-direction, located at the origin in free space, find the potential at a point $\left(r, \theta, \phi = \frac{\pi}{2}\right)$ (in spherical coordinates).

25. (a) Determine the potential energy of a dipole in an external electric field.

or

An electric dipole of moment $\vec{p}$ is placed in a uniform electric field $\vec{E}$. Show that its potential energy is $-\vec{p} \cdot \vec{E}$.

(b) Calculate also the torque on the dipole in a uniform electric field.

26. Show that the torque on a physical dipole $\vec{p}$ in a uniform electric field $\vec{E}$ is given by $\vec{p} \times \vec{E}$. Extend this result to a pure dipole.

27. Obtain Coulomb's law from Gauss' theorem.

28. (a) Derive Poisson's and Laplace's equations from fundamentals.

or

State and prove Poisson's equation in electrostatics. What form does it take when the charge density is zero? Illustrate the application of this equation to find the electric field and potential in two suitable cases of symmetric charge distribution.

(b) What are the importances of Poisson's and Laplace's equations in electrostatics?

29. Explain the Laplace's and Poisson's equations for steady electric field.

30. State and prove Uniqueness theorem in connection with Laplace's equation.

31. The solution to Laplace's equation in some region is uniquely determined if the value of the potential V is a specified function on all boundaries of the region. Prove it.

32. Obtain Green's integral identities and state their significance. Apply the first identity to show that the specifications of both divergence and curl of a vector with boundary conditions are sufficient to make the function unique.
What are Dirichlet and Neumann conditions?

33. What is an electrical image? State its usefulness in solving electrostatic problem.

or

Explain the method of electrical images in solving electrostatic problems.

34. What do you mean by the term 'energy density' in an electrostatic field? Show that in free space it is given by $\frac{1}{2}\varepsilon_0 E^2$, E being the electric field.

35. Show that the electromagnetic energy due to charged conductor in space is given by $\frac{1}{2}\int_v \vec{D}\cdot\vec{E}dv$ where fields $\vec{D}$ and $\vec{E}$ occupy whole of the space.

36. (a) Define capacitance of a conductor and explain its significance.

(b) Show that the electrostatic energy of a capacitor of capacitance C charged to a voltage V is $\frac{1}{2}CV^2$.

(c) Show that, when a capacitor of capacitance C is charged fully by connecting it across a battery of e.m.f. E, the energy expended by the battery is CE^2.

***37.** (i) Define capacitance. Express its units in two different ways.

(ii) As per the usual definition, show that a capacitance is always positive.

(iii) Sometimes, capacitance of a single conductor is referred to. What does this mean?

***38.** Define capacitance and explain why it is always a positive quantity.

39. Derive the expression for the energy stored in a capacitor.

40. (a) A conducting body is in the electric field of static charges. Explain why the net electric field at any point inside the conducting body will be zero.

(b) Use the result of (a) to show that

(i) the net volume charge density at any point inside the conductor is zero, and

(ii) the conductor is an equipotential body.

***41.** For a conducting body in the electric field of static charges, explain what will be the

(a) net electric field inside the conductor, and

(b) volume charge density at any point inside the conductor?

42. One medium is a dielectric with permittivity ε_1 and the other is a conductor. Find the angle θ_1 between the normal and a field line in medium 1 incident on the conductor (medium 2).

Multiple Choice Questions

1. Electric potential and electric field intensity inside a spherical shell are:

(a) Zero and constant respectively
(b) Both inversely proportional to radius
(c) Constant and zero respectively
(d) Zero and zero respectively

2. Poisson's and Laplace's equations govern the behaviour of electric scalar potential for:

(a) Charge free region
(b) A region of charge
(c) Charge free region and a region of charge, respectively
(d) Region of charge and charge free region, respectively

3. Electric field in a region containing space charges can be found using

(a) Laplace's equation
(b) Poisson's equation
(c) Coulomb's law
(d) Helmholtz equation

4. Capacitance of the earth of radius R is

(a) $2\pi\varepsilon_0 R$
(b) $4\pi\varepsilon_0 R$
(c) $\frac{4}{3}\pi\varepsilon_0 R^3$
(d) none

5. When an insulator is inserted between the plates of an air gap capacitor charged with constant charges, the electric field

(a) decreases
(b) increases
(c) remains the same
(d) becomes zero

6. On a perfect conductor surface
(a) The tangential component of **E** and normal component of **B** are zero
(b) The normal component of **D** is the surface charge density
(c) The tangential component of **H** is equal to the surface current density
(d) All of the above

7. Boundary conditions at the interface between perfect dielectrics are:
(a) Normal components of **E** and **B** and the tangential components of **H** and **D** are continuous.
(b) Tangential components of **E** and **H** and the normal components of **D** and **B** are continuous.
(c) Normal components of **E** and **H** and tangential components of **D** and **B** are continuous.
(d) Normal components of **B** and **H** and tangential components of **E** and **D** are continuous.

8. As one moves at right angle to a charged parallel-plate capacitor, the electric field E inside the capacitor
(a) Drops abruptly to zero
(b) Remains constant
(c) Approaches zero in a continuous and gradual way
(d) Approaches zero sinusoidally

9. An isolated metal sphere whose diameter is 10 cm has a potential of 8 kV. The energy density at the surface of the sphere is
(a) 0.011 J/m^2
(b) 0.11 J/m^2
(c) 1.1 J/m^2
(d) 11 J/m^2

10. An electron (mass = 9.1×10^{-31}kg, charge = 1.6×10^{-19} Coulomb) experience force equal to its weight in an electric field. The field strength is
(a) 5.6×10^{-11} N/C
(b) 5.6×10^{-10} N/C
(c) 5.6×10^{-12} N/C
(d) 5.6×10^{-13} N/C

11. Two metallic spheres, radii a and b, are connected by a thin wire. The separation is large compared to their dimensions. A charge Q is put into the system. The capacitance of the system is
(a) $4\pi\varepsilon_0 ab$
(b) $4\pi\varepsilon_0 \dfrac{a}{b}$
(c) $4\pi\varepsilon_0(a + b)$
(d) $4\pi\varepsilon_0 \log_e \dfrac{b}{a}$

12. Two concentric rings of radius 'a' and '$2a$' carrying equal and uniform charge densities revolve at the same angular speed 'ω' about their common axis. The ratio of flux densities due to the two rings at the centre will be
(a) 1 : 1
(b) 1 : 2
(c) 1 : 4
(d) 2 : 1

13. If $\vec{E}$ is the electric field intensity, $\nabla(\nabla \times \vec{E})$ is equal to
(a) $\vec{E}$
(b) $|\vec{E}|$
(c) null vector
(d) zero

14. An infinite number of concentric rings carry a charge Q each alternately positive and negative. The radii are 1, 2, 4, 8, … metres in geometric progression. The potential at the centre of the rings will be
(a) zero
(b) $\dfrac{Q}{12\pi\varepsilon_0}$
(c) $\dfrac{Q}{8\pi\varepsilon_0}$
(d) $\dfrac{Q}{6\pi\varepsilon_0}$

15. Two concentric spherical shells of radius 'R' and '$2R$' carry equal and opposite uniformly distributed charges over their surfaces. The electric field on the surface of the inner shell will be
(a) zero
(b) $\dfrac{Q}{4\pi\varepsilon_0 R^2}$
(c) $\dfrac{Q}{8\pi\varepsilon_0 R^2}$
(d) $\dfrac{Q}{16\pi\varepsilon_0 R^2}$

16. Two point charges Q and $-Q$ are located on two opposite corners of a square as shown in Fig. 2.88. If the potential at the corner A is taken as 1V, then the potential at B, the centre of the square will be

(a) zero (b) $\frac{1}{\sqrt{2}}$V

(c) 1 V (d) $\sqrt{2}$V

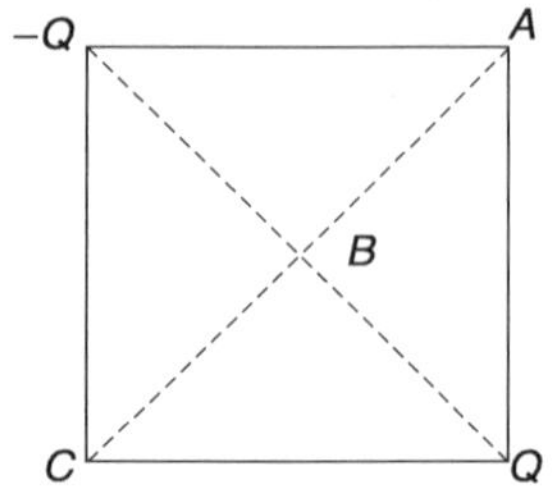

Fig. 2.88

17. The electric field across an interface is shown in figure. The surface charge density (in coulomb/m^2) on the surface is

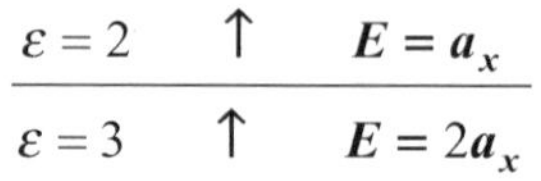

(a) $-4\varepsilon_0$ (b) $-3\varepsilon_0$

(c) $+3\varepsilon_0$ (d) $+4\varepsilon_0$

18. Match List I with List II and select the correct answer using the codes given below the lists:

List I (Expression)	List II (Law/theorem/equation)
A. $\nabla^2 V = -\frac{\rho_v}{\varepsilon}$	1. Gauss' Law
B. $\nabla^2 V = 0$	2. Divergence theorem
C. $\int_s \vec{D}\cdot d\vec{s} - Q$	3. Laplace's equation
D. $\int_v \nabla\cdot\vec{D}dv = \int_s \vec{D}\cdot d\vec{s}$	4. Poisson's equation
	5. Coulomb's law

Codes:

(a) A B C D
4 2 5 1

(b) A B C D
3 1 2 4

(c) A B C D
2 3 4 5

(d) A B C D
4 3 1 2

19. Two positive charges, Q coulombs each, are placed at points (0, 0, 0) and (2, 2, 0) while two negative charges, Q coulombs each in magnitude, are placed at points (0, 2, 0) and (2, 0, 0). The electric field intensity at the point (1, 1, 0) is

(a) zero (b) $\frac{Q}{8\pi\varepsilon_0}$ (c) $\frac{Q}{4\pi\varepsilon_0}$ (d) $\frac{Q}{16\pi\varepsilon_0}$

20. Two infinite parallel metal plates are charged with equal surface charge density of the same polarity. The electric field in the gap between the plates is

(a) The same as that produced by one plate
(b) Double of the field produced by one plate
(c) Dependent on coordinates of the field point
(d) Zero

21. Consider the following statements associated with a parallel plate capacitor:
    1. Capacitance is proportional to area of plates.
    2. Capacitance is inversely proportional to distance of separation of plates.
    3. The dielectric material is in a state of compression.

    Of these statements
    (a) 1, 2 and 3 are correct (b) 1 and 2 are correct
    (c) 1 and 3 are correct (d) 2 and 3 are correct

22. Two concentric metallic spheres of radii 'a' and 'b' carry charges $+Q$ and $-Q$ respectively as shown in the given figure. Potential at the centre 'P' will be:

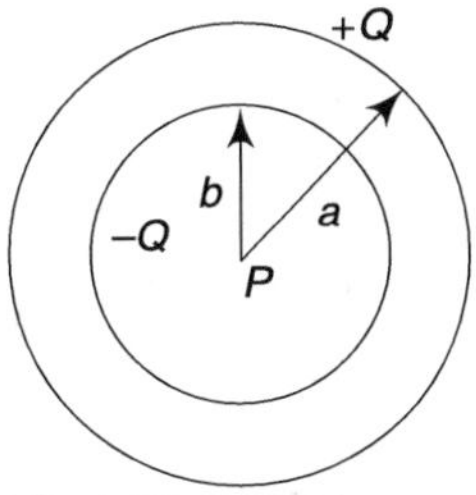

Fig. 2.89

(a) Zero (b) $\frac{Q}{4\pi\varepsilon_0 b}$

(c) $-\frac{Q}{4\pi\varepsilon_0 a}$ (d) $\frac{Q}{4\pi\varepsilon_0}\left(\frac{1}{a}-\frac{1}{b}\right)$

23. The electrostatic field on the surface of a conductor at a certain point is $0.3\hat{i} + 0.4\hat{j}$. If the normal to the surface of the conductor at that point makes an angle θ with respect to x-axis, the value of $\cos\theta$ will be
    (a) 0.8 (b) 0.75 (c) 0.6 (d) 0.5

24. A point charge $+Q$ is brought near a corner of two right angle conducting planes which are at zero potential as shown in Fig. 2.90. Which one of the following configurations describes the total effect of the charges for calculating the actual field in the first quadrant?

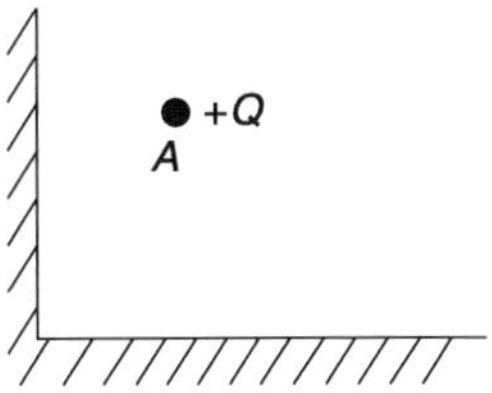

Fig. 2.90

(a)
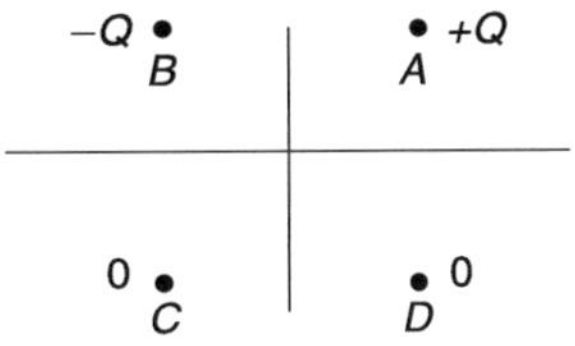

(b)
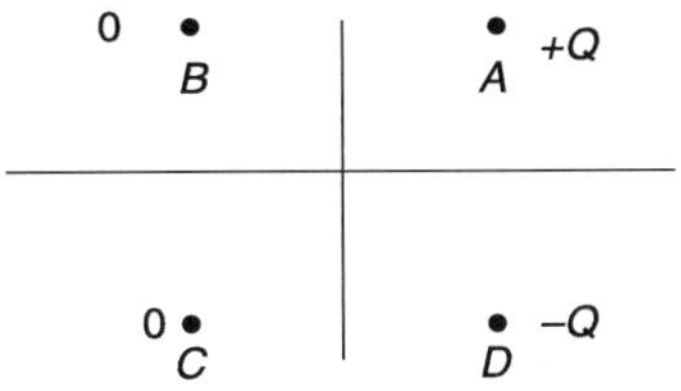

(c)
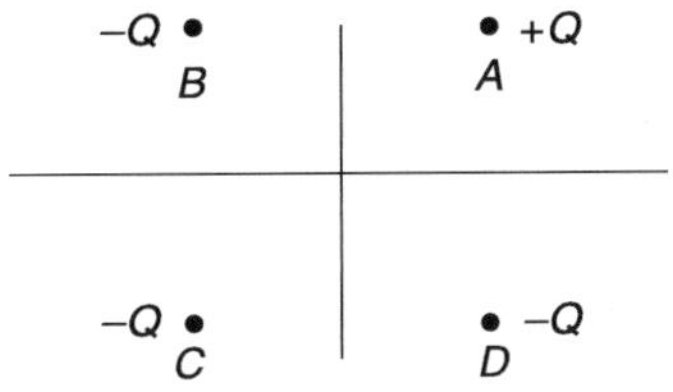

(d)
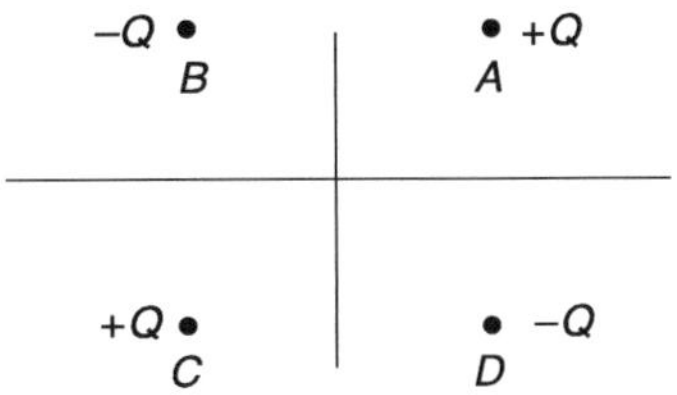

25. The relation between electric intensity E, voltage applied V and the distance d between the plates of a parallel plate condenser is
    (a) $E = \frac{V}{d}$ (b) $E = V \times d$ (c) $E = \frac{V}{(d)^2}$ (d) $E = V \times (d)^2$

26. A spherical conductor of radius a with charge q is placed concentrically inside an uncharged and unearthed spherical conducting shell of inner and outer radii r_1 and r_2 respectively. Taking the potential to be zero at infinity, the potential at any point P within the shell ($r_1 < r < r_2$) will be
    (a) $\frac{q}{4\pi\varepsilon_0 r}$ (b) $\frac{q}{4\pi\varepsilon_0 a}$ (c) $\frac{q}{4\pi\varepsilon_0 r_2}$ (d) $\frac{q}{4\pi\varepsilon_0 r_1}$

27. Inside a hollow conducting sphere
(a) electric field is zero.
(b) electric field is a non-zero constant.
(c) electric field changes with the magnitude of the charge given to the conductor.
(d) electric field changes with distance from the centre of the sphere.

28. In a uniform electric field, field lines and equipotentials
(a) are parallel to one another.
(b) intersect at 45°.
(c) intersect at 30°.
(d) are orthogonal.

29. Energy stored in a capacitor over a cycle, when excited by an a.c. source is
(a) the same as that due to a d.c. source of equivalent magnitude
(b) half of that due to a d.c. source of equivalent magnitude
(c) zero
(d) none of the above

30. Consider the following statements regarding field boundary conditions:
1. The tangential component of electric field is continuous across the boundary between two dielectrics.
2. The tangential component of electric field at a dielectric-conductor boundary is non-zero
3. The discontinuity in the normal component of the flux density at a dielectric–conductor boundary is equal to the surface charge density on the conductor.
4. The normal component of the flux density is continuous across the charge-free boundary between two dielectrics.

Of these statements
(a) 1, 2 and 3 are correct.
(b) 2, 3 and 4 are correct.
(c) 1, 2 and 4 are correct.
(d) 1, 3 and 4 are correct.

31. Match List I with List II and select the correct answer using the codes given below the lists: (Symbols have the usual meanings)

List I	List II
A. Poisson's equation	1. $\nabla^2\phi = 0$
B. Laplace's equation	2. $\nabla^2 E + K_0^2 E = 0$ where $K_0 = \omega\sqrt{\mu_0\varepsilon_0}$
C. Joule's equation	3. $\nabla^2\phi = -\rho/\varepsilon_0$
D. Helmholtz's equation	4. $\dfrac{d\rho}{dv} = U_j = \vec{E}\cdot\vec{J}$

Codes:

	A	B	C	D
(a)	2	1	4	3
(b)	3	4	1	2
(c)	3	1	4	2
(d)	2	4	1	3

32. By saying that the electrostatic field is conservative, we do *not* mean that
(a) It is the gradient of a scalar potential.
(b) Its circulation is identically zero.
(c) The work done in a closed path inside the field is zero.
(d) The potential difference between any two points is zero.

33. Point charges $Q_1 = 1$ nC and $Q_2 = 2$ nC are at a distance apart. Consider the following statements:
1. The force on Q_1 is repulsive.
2. The force on Q_2 is the same in magnitude as that on Q_1.
3. As the distance between them decreases, the force on Q_1 increases linearly.
4. The force on Q_2 is along the line joining them.
5. A point charge $Q_3 = -3$ nC located at the midpoint between Q_1 and Q_2 experiences no net force.

Which of the following statements are incorrect?
(a) 3 and 4 (b) 3 and 5 (c) 1, 3 and 5 (d) none

34. Plane $z = 10$ m carries charge 20 nC/m^2. The electric field intensity at the origin is:
(a) $-10\hat{a}_z$ V/m (b) $-18\pi\hat{a}_z$ V/m (c) $-72\pi\hat{a}_z$ V/m (d) $-360\pi\hat{a}_z$ V/m

35. Point charges 30 nC, −20 nC, and 10 nC are located at (−1, 0, 2), (0, 0, 0) and (1, 5, −1), respectively. The total flux leaving a cube of side 6 m centered at the origin is
(a) −20 nC (b) 10nC (c) 20 nC (d) 60 nC

36. The work done by the force $\vec{F} = 4\hat{a}_x - 3\hat{a}_y + 2\hat{a}_z$ N in giving a 1 nC charge a displacement of $10\hat{a}_x + 2\hat{a}_y - 7\hat{a}_z$ m is
(a) 103 nJ (b) 60 nJ (c) 64 nJ (d) 20 nJ

37. An electric potential field is produced by point charges 1 μC and 4 μC located at (−2, 1, 5) and (1, 3, −1), respectively. The energy stored in the field is:
(a) 2.57 mJ (b) 5.14 mJ (c) 10.28 mJ (d) none

38. The region $z \le 0$ is a dielectric of relative permittivity 2 while the region $z \ge 0$ is a dielectric of relative permittivity 5. If the electric field intensity in region $z \le 0$ is $\overline{E} = 10\hat{a}_z$ kV/m, the potential difference between (0, 0, -10^{-3}) m and (0, 0, 10^{-3}) m will be:
(a) zero (b) 70 V (c) 20 V (d) 14 V

Answers

1. (c),	**2.** (d),	**3.** (b),	**4.** (b),	**5.** (a),	**6.** (b),	**7.** (b)	**8.** (b),
9. (a),	**10.** (c),	**11.** (c),	**12.** (a),	**13.** (d),	**14.** (d),	**15.** (b),	**16.** (c),
17. (a),	**18.** (d),	**19.** (a),	**20.** (d),	**21.** (a)	**22.** (d),	**23.** (c),	**24.** (d),
25. (a),	**26.** (a),	**27.** (a),	**28.** (d)	**29.** (c),	**30.** (d),	**31.** (c),	**32.** (d),
33. (b),	**34.** (d),	**35.** (b)	**36.** (d),	**37.** (b),	**38.** (d)		

3

MAGNETOSTATICS

Learning Objectives

This chapter deals with the following topics:

- *Concepts of electric current*
- *Basic characteristics of conductors and dielectrics*
- *Basic laws of magnetostatics*
- *To acquire knowledge of fundamental quantities of magnetostatics*
- *Boundary conditions in magnetostatics*
- *Concepts of inductances*

3.1 INTRODUCTION

In the previous chapter, we have learned different aspects of electrostatic fields in free space or vacuum. In this chapter, we will discuss different phenomena of electric field in material space, like conductors and insulators.

In the first part of this chapter, we will discuss the properties of such electrical materials and the important relations governing the behaviour of steady electric current in those materials.

In the second part of this chapter, we will discuss the static magnetic fields, related laws and other terminologies in detail.

PART I

Behaviour of Different Electrical Materials in Electric Field

3.2 ELECTRIC CURRENT AND CURRENT DENSITY

3.2.1 Electric Current

Electric current is defined as the rate of flow of electric charges or electrons through a cross-sectional area.

If Q amount of charges flow through an area in time t, then the current is given as

$$I = \frac{Q}{t} \tag{3.1}$$

or in differential form

$$i = \frac{dq}{dt} \tag{3.2}$$

and the charge transferred between time t_0 and t is given by

$$q = \int_{t_0}^{t} i dt \tag{3.3}$$

NOTE

(i) *By convention, electric current flows in the direction opposite to the direction of the electrons. The current through a given area is the electric charge passing through the area per unit time.*

(ii) *As charge (Q) is expressed in Coulomb, the unit of electric current is Coulomb per second and it is given the name Ampere (A). Thus,*

1 A current = flow of 6.24×10^{18} electrons per second through an area

3.2.2 Electric Current Density

Current density ($\vec{J}$) at any point is a vector whose magnitude is the electric current per unit cross-sectional area and whose direction is normal to the cross-sectional area. Its unit is ampere per square metre (A/m^2).

$$\therefore \qquad \vec{J} = \frac{I}{A}\hat{n} \tag{3.4}$$

or in differential form $\vec{J} = \frac{dI}{dS_v}\hat{n}$ (3.5)

where dS_v is the projection of dS in a plane normal to that of the flow. This is depicted in Fig. 3.1.

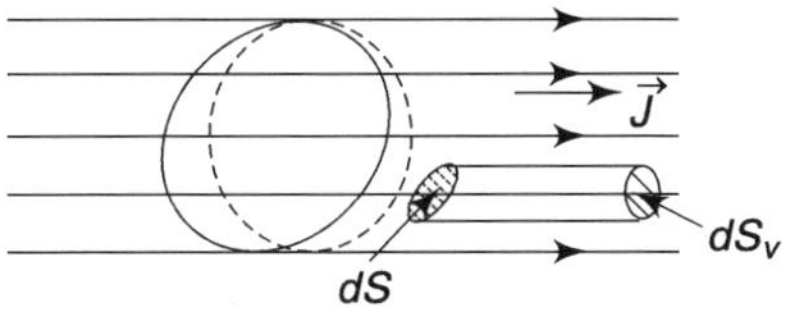

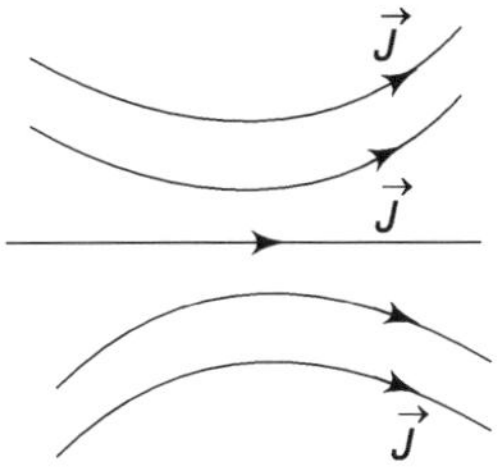

Fig. 3.1 *Current density distribution*

NOTE

(i) *For uniform cross section of the conductor, $\vec{J}$ is same in magnitude and direction everywhere.*

(ii) *For non-uniform cross section of the conductor, $\vec{J}$ is different in magnitude and direction at various points.*

Thus, total current flowing through a surface S is given as

$$I = \int_S \vec{J} \cdot d\vec{S} \tag{3.6}$$

Depending upon how I is produced, there are three different current densities:

1. Convection current density,
2. Conduction current density and
3. Displacement current density.

Example 3.1 In a cylindrical conductor to the region $0.01 \le r \le 0.02$, $0 < z < 1$ m and the current density is given by

$$\vec{J} = 10e^{-100r}\hat{a}_\phi \quad \text{A/m}^2$$

Find the total current crossing the extential of this region with $\phi =$ constant plane.

Solution Total current in the wire is given as

$$I = \int_S \vec{J}.d\vec{S} = \int_{r=0.01}^{0.02}\int_{z=0}^{1} [10e^{-100r}\hat{a}_\phi]\cdot[rdrdz\hat{a}_\phi]$$

$$= \int_{r=0.01}^{0.02}\int_{z=0}^{1} 10re^{-100r}drdz$$

$$= 10\int_{r=0.01}^{0.02} re^{-100r}dr$$

$$I = 10\left[\frac{re^{-100r}}{-100}\Bigg|_{0.01}^{0.02} - \int_{r=0.01}^{0.02}\frac{e^{-100r}}{-100}dr\right]$$

$$= 10\left[-\frac{1}{100}(0.02e^{-2} - 0.01e^{-1}) + \frac{e^{-100r}}{-100\times100}\Bigg|_{0.01}^{0.02}\right]$$

$$= 2\times10^{-3}e^{-1} - 310^{-3}e^{-2}$$

$$= 0.033 \text{ mA}$$

Example 3.2 Find the total current in a circular conductor of radius 4 mm if the current density varies according to $J = \frac{10^4}{r}$ A/m^2.

Solution Total current is given as

$$I = \int_S \vec{J}\cdot d\vec{S} = \int_{\phi=0}^{2\pi}\int_{r=0}^{0.004}\frac{10^4}{r}rdrd\phi = 2\pi\times10^4\int_{r=0}^{0.004}dr = 2\pi\times10^4\times0.004 = 80\pi \text{ A}$$

Example 3.3 In a certain region, the current density vector is given by,

$$\vec{J} = 3x\hat{a}_x + (y-3)\hat{a}_y + (2+z)\hat{a}_z \text{ A/m}^2$$

Find the total current flowing out of the surface of the box bounded by the five planes $x = 0$, $y = 0$, $z = 0$ and $(3x + z) = 3$.

Solution As shown in Fig. 3.2, we will consider the normal vector to be always pointing out of the box so that $\oint_S \vec{J} \cdot d\vec{S}$ gives the current flowing out of the surface.

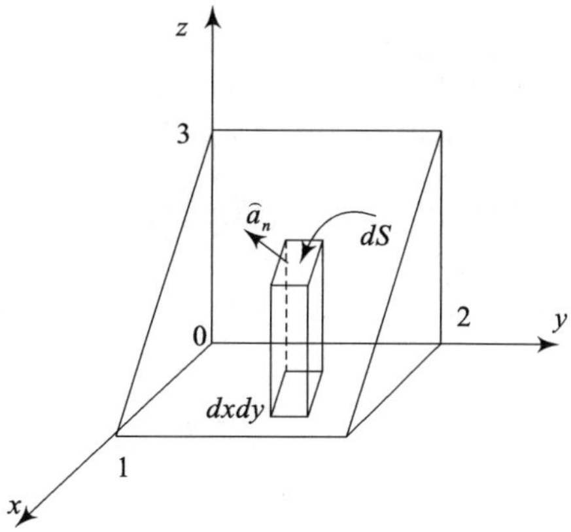

Fig. 3.2 *Closed surface of Example 3.3*

— **For the surface $x = 0$,**

$$\vec{J} = (y-3)\hat{a}_y + (2+z)\hat{a}_z$$
$$d\vec{S} = -dydz\hat{a}_z$$
$$\therefore \quad I = \int \vec{J} \cdot d\vec{S} = 0$$

— **For the surface $y = 0$,**

$$\vec{J} = 3x\hat{a}_x + -3\hat{a}_y + (2+z)\hat{a}_z$$
$$d\vec{S} = -dxdz\hat{a}_y$$

$$\therefore \quad I = \int \vec{J} \cdot d\vec{S} = \int_{x=0}^{1} \int_{z=0}^{3-3x} 3dzdx$$
$$= \int_{x=0}^{1} 3(3-3x)\,dx = \left[9x - \frac{9}{2}x^2\right]_0^1 = \frac{9}{2}\,A$$

— **For the surface $y = 2$,**

$$\vec{J} = 3x\hat{a}_x + \hat{a}_y + (2+z)\hat{a}_z$$
$$d\vec{S} = dxdz\hat{a}_y$$

$$\therefore \quad I = \int \vec{J} \cdot d\vec{S} = \int_{x=0}^{1} \int_{z=0}^{3-3x} -dzdx = \int_{x=0}^{1} (-3-3x)dx = \left[-3x + \frac{3}{2}x^2\right]_0^1 = -\frac{3}{2}\text{ A}$$

— **For the surface $z = 0$,**

$$\vec{J} = 3x\hat{a}_x + (y-3)\hat{a}_y + 2\hat{a}_z$$
$$d\vec{S} = -dxdy\hat{a}_z$$
$$\therefore \quad I = \int \vec{J} \cdot d\vec{S} = \int_{y=0}^{2} \int_{x=0}^{1} -2dxdy = -4\text{ A}$$

— **For the surface $(3x + z) = 3$,**

$$\vec{J} = 3x\hat{a}_x + (y-3)\hat{a}_y + (5-3x)\hat{a}_z$$
$$d\vec{S} = (3\hat{a}_x + 1\hat{a}_z)\,dxdy$$
$$\therefore \quad I = \int \vec{J} \cdot d\vec{S} = \int_{y=0}^{2} \int_{x=0}^{1} (9x + 5 - 3x)\,dxdy = \int_{x=0}^{1} (18x + 10 - 6x)\,dx = \int_{0}^{1} (12x+10)\,dx = 16\text{ A}$$

Adding the components of the currents, total current flowing out of the closed surface is

$$I = \left(0 + \frac{9}{2} - \frac{3}{2} - 4 + 16\right) = 15\text{ A}$$

Example 3.4 If $\vec{J} = \frac{1}{r^3}(2\cos\theta\,\hat{a}_r + \sin\theta\,\hat{a}_\theta)$ A/m^2, calculate the current passing through:

(a) A hemispherical shell of radius 20 cm.

(b) A spherical shell of radius 10 cm.

Solution Total current is given as, $I = \int \vec{J}\cdot d\vec{S}$

Here, $d\vec{S} = r^2 \sin\theta\, d\phi\, d\theta\,\hat{a}_r$

(a) Total current passing through a hemispherical shell of radius 20 cm is

$$I = \int_{\theta=0}^{\pi/2}\int_{\phi=0}^{2\pi} \frac{1}{r^3}(2\cos\theta\,\hat{a}_r + \sin\theta\,\hat{a}_\theta)\cdot(r^2\sin\theta\,d\phi\,d\theta\,\hat{a}_r)\Big|_{r=0.2}$$

$$= \int_{\theta=0}^{\pi/2}\int_{\varphi=0}^{2\pi} \frac{1}{r^3} 2\cos\theta r^2 \sin\theta\,d\phi\,d\theta\Big|_{r=0.2}$$

$$= \pi \times \frac{2}{r}\int_{\theta=0}^{\pi/2} \sin\theta d(\sin\theta)\Big|_{r=0.2}$$

$$= \frac{4\pi}{0.2}\left[\frac{\sin^2\theta}{2}\right]_0^{\pi/2} = 10\pi = 31.42 \text{ A}$$

(b) Total current passing through a spherical shell of radius 10 cm is

$$I = \int_{\theta=0}^{\pi}\int_{\phi=0}^{2\pi} \frac{1}{r^3}(2\cos\theta\,\hat{a}_r + \sin\theta\,\hat{a}_\theta)\cdot(r^2\sin\theta\,d\phi\,d\theta\,\hat{a}_r)\Big|_{r=0.1}$$

$$= \int_{\theta=0}^{\pi}\int_{\phi=0}^{2\pi} \frac{1}{r^3} 2\cos\theta r^2 \sin\theta\,d\phi\,d\theta\Big|_{r=0.1}$$

$$= 2\pi \times \frac{2}{r}\int_{\theta=0}^{\pi} \sin\theta d(\sin\theta)\Big|_{r=0.1}$$

$$= \frac{4\pi}{0.1}\left[\frac{\sin^2\theta}{2}\right]_0^{\pi}$$

$$= 0$$

Example 3.5 For the current density, $\vec{J} = 10z\sin^2\phi\,\hat{a}_r$ A/m^2, find the current through the cylindrical surface $r = 2$, $1 \le z \le 5$ m.

Solution Total current passing through the cylindrical surface is

$$I = \int \vec{J}\cdot d\vec{S} = \int_{z=1}^{5}\int_{\phi=0}^{2\pi}(10z\sin^2\phi\,\hat{a}_r)\cdot(rd\phi dz\hat{a}_r)\Big|_{r=2} = 10r\left[\frac{z^2}{2}\right]_1^5 \int_{\phi=0}^{2\pi}\sin^2\phi d\phi\Big|_{r=2}$$

$$= 10\times 2\times\frac{24}{2}\times\frac{2\pi}{2} = 240\pi = 754 \text{ A}$$

3.2.3 Convection Current

Definition The motion of charged particles in free space (vacuum) is said to constitute convection current.

Mathematical Equation We will consider a region with volume charge density ρ_v, in which the charges are moving under the influence of an electric field with an average velocity $\vec{U}$. This is shown below in Fig. 3.3.

$\therefore$ Current density, $J = \dfrac{dI}{dS_v}$

But, $dI = \dfrac{dQ}{dt} = \dfrac{\rho_v dv}{dt}$ [dv = volume of an infinitesimal cylinder = $dS_v dl$]

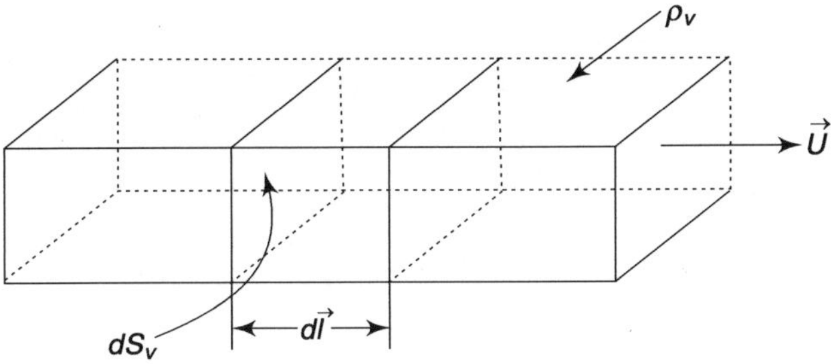

Fig. 3.3 *Flow of convection current*

$$\therefore \quad J = \frac{dI}{dS_v} = \frac{\rho_v dv}{dS_v dt} = \rho_v \frac{dS_v dl}{dS_v dt} = \rho_v \frac{dl}{dt} = \rho_v U \quad \left[U = \text{velocity of charge movement} = \frac{dl}{dt}\right]$$

In vector form,

$$\boxed{\vec{J} = \rho_v \vec{U}} \tag{3.7}$$

This is the convection current density.

Some Features of Convection Current

(i) Convection current does not require a conductor and does not obey Ohm's law.

(ii) Convection current is not electrostatically neutral.

(iii) Example of convection current is the motion of electrons from the cathode towards the anode in a vacuum tube.

(iv) If both positive as well as negative charges with charge densities ρ_{v+} and ρ_{v-} move with average velocities $\vec{U}_+$ and $\vec{U}_-$, respectively, then the positive charges will move in the direction of the electric field and the negative charges will move in the opposite direction and the total convection current density will be given as

$$\boxed{\vec{J} = \rho_{v+}\vec{U}_+ + \rho_{v-}\vec{U}_-} \tag{3.8}$$

3.2.4 Conduction Current and Ohm's Law

Definition The motion of the free electrons present in a conductor, by the influence of an electric field constitutes the *conduction current.*

Mathematical Equation When an electric field, $\vec{E}$ is applied, the force on an electron with charge, $-e$ is, $\vec{F} = -e\vec{E}$ As the electrons are not free in space, they will not be accelerated by the field; but they will suffer constant collisions with atomic lattice and drifts from one atom to another.

Let, m — mass of moving electron,

$\vec{U}$ — average drift velocity

By Newton's law,

$\frac{m\vec{U}}{\tau} = -e\vec{E}$ (τ is the average time interval between successive collisions)

$$\therefore \quad \vec{U} = -\frac{e\tau}{m}\vec{E} = -\mu\vec{E} \tag{3.9}$$

where, $\mu = \frac{e\tau}{m}$ is the mobility of electrons.

From Eq. (3.9), we see that, *drift velocity is directly proportional to the applied field.*

If there are N electrons per unit volume, the electron charge density is

$$\rho_v = -Ne$$

Thus, the conduction current density is

$$\vec{J} = \rho_v\vec{U} = Ne\mu\vec{E} = \frac{Ne^2\tau}{m}\vec{E} = \sigma\vec{E}$$

where, $\sigma = \frac{Ne^2\tau}{m}$ is the conductivity of the conductor.

$$\therefore \quad \boxed{\vec{J} = \sigma\vec{E}} \tag{3.10}$$

This is the *conduction current density.*

Some Features of Conduction Current

(i) From Eq. (3.10), it is seen that the current density is linearly dependent on the external electric field. This equation is known as the point form of *Ohm's law* which states that *the current density at any point in a conducting medium is directly proportional to the electric field.*

(ii) To maintain a steady current within a conductor, a continuous supply of electrons at one end and their removal at the other is necessary. So, conductor as a whole is *electrostatically neutral.*

Example 3.6 Find the current in a circular wire of radius 2 mm when the current density in the conductor is

$$\vec{J} = 30(1 - e^{-1000r})\hat{a}_z \text{ A/m}^2$$

Solution Total current in the wire is given as

$$I = \int_S \vec{J} \cdot d\vec{S} = \int_{\phi=0}^{2\pi}\int_{r=0}^{0.002} [30(1 - e^{-1000r})\hat{a}_z] \cdot [r dr d\phi \hat{a}_z]$$

$$= \int_{\phi=0}^{2\pi}\int_{r=0}^{0.002} 30(1-e^{-1000r})\, r dr d\phi$$

$$= 60\pi \int_{r=0}^{0.002} (1-e^{-1000r})\, r dr$$

$$I = 60\pi\left[\frac{r^2}{2}\Big|_0^{0.002}\right] - 60\pi\left[\frac{re^{-1000r}}{-1000}\Big|_0^{0.002} - \int_0^{0.002}\frac{e^{-1000r}}{-1000}dr\right]$$

$$= 60\pi\times 2\times 10^{-6} - 60\pi\left[-2\times 10^{-6}\times e^{-2} + \frac{e^{-1000r}}{-1000\times 1000}\Big|_0^{0.002}\right]$$

$$= 120\pi\times 10^{-6} + 120\pi\times 10^{-6}\times e^{-2} + 60\pi\times 10^{-6}\times e^{-2} - 60\pi\times 10^{-6}$$

$$= 60\pi\times 10^{-6}(1+3e^{-2})$$

$$= 265.03\ \mu\text{A}$$

***Example 3.7** The free charge density of copper is 1.81×10^{10} C/m^3. For a current density of 8×10^6 A/m^2, find the electric field intensity and the drift velocity. Take conductivity of copper as 5.8 $\times 10^7$ mho/m.

Solution Here, $\rho = 1.81\times 10^{10}\,\text{C/m}^3$, $J = 8\times 10^6\,\text{A/m}^2$, $\sigma = 5.8\times 10^7$

$\therefore$ Electric field intensity is

$$E = \frac{J}{\sigma} = \frac{8\times 10^6}{5.8\times 10^7} = 0.138\ \text{V/m}$$

$\therefore$ Drift velocity is

$$U = \frac{J}{\rho} = \frac{8\times 10^6}{1.81\times 10^{10}} = 4.42\times 10^{-4}\ \text{m/s}$$

***Example 3.8** The resistivity of seawater is about 25 Ω-cm. The charge carriers are chiefly Na^+ and Cl^- ions, and of each there are about 3×10^{20}/cm^3. If we fill a plastic tube 2 metre long with seawater and connect a 12-volt battery to the electrodes at each end, what is the resulting average drift velocity of the ions, in cm/s?

Solution The current in a conductor of cross-sectional A is related to the drift velocity U of the charge carriers as

$$I = JA = \rho UA = neUA$$

where n is the number of charge carriers per unit volume. Hence, Ohm's law can be written as

$$V = RI = \left(\frac{\rho l}{A}\right) JneUA = neU\rho l$$

$$\therefore \qquad U = \frac{V}{ne\rho l}$$

Substituting the values, we get

$$U = \frac{V}{nepl} = \frac{12}{(6\times10^{20})\,(1.6\times10^{-19})\,(25)\times(200)} = 2.5\times10^{-5}\ \text{cm/s}$$

3.2.5 Displacement Current

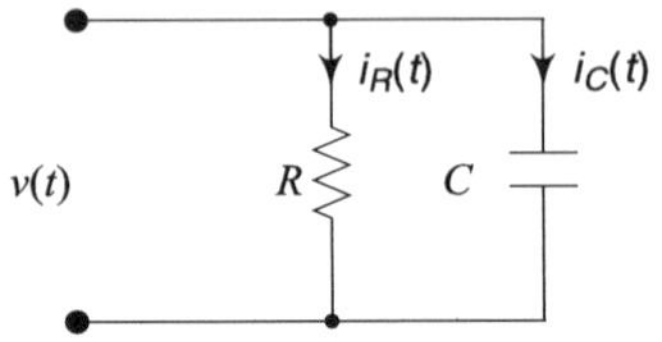

Fig. 3.4 *RC parallel circuit representing a lossy capacitor*

The concept of displacement current can be illustrated as in Fig. 3.4 by considering the currents in a simple parallel *RC* network (assume ideal circuit elements, for simplicity).

Here, $i_R(t)$ conduction current

$I_C(t)$ displacement current

From circuit theory,

$$i_R(t) = \frac{v(t)}{R} \qquad i_C(t) = C\frac{dv(t)}{dt}$$

In the resistor, the conduction current model is valid $(\vec{J}_R = \sigma_R \vec{E}_R)$. The ideal resistor electric field $(\vec{E}_R)$ and current density $(\vec{J}_R)$ are assumed to be uniform throughout the volume of the resistor. The conduction current model does not characterise the capacitor current. The ideal capacitor is characterised by large, closely-spaced plates separated by a perfect insulator ($\sigma_C = 0$) so that no charge actually passes throughout the dielectric $[\vec{J}_C(t) = \sigma_C \vec{E}_C(t)]$. The capacitor current measured in the connecting wires of the capacitor is caused by the charging and discharging the capacitor plates. Let $Q(t)$ be the total capacitor charge on the positive plate.

Hence, the capacitor current, also termed as the *displacement current* is given as

$$i_C(t) = i_d(t) = \frac{dQ(t)}{dt} = C\frac{dv(t)}{dt} = \frac{\varepsilon A}{d}\frac{dv(t)}{dt} = \varepsilon A\frac{d}{dt}\left[\frac{v(t)}{d}\right] = \varepsilon A\frac{dE(t)}{dt} = A\frac{d}{dt}[\varepsilon E(t)] = A\frac{dD(t)}{dt}$$

So, the *displacement current density* is given as

$$\vec{J}_d(t) = \frac{d\vec{D}(t)}{dt}$$

As $\vec{D}$ may vary with space, the displacement current density is written as

$$\boxed{\vec{J}_d(t) = \frac{\partial \vec{D}(t)}{\partial t}} \qquad \textbf{(3.11)}$$

Thus, displacement current for a closed surface is

$$I_d = \oint_S \vec{J}_d \cdot d\vec{S} = \oint_S \frac{\partial \vec{D}}{\partial t}\cdot d\vec{S}$$

Thus, the displacement current does not represent a current. It is only an *apparent current* representing the rate at which flow of charge takes place from electrode to electrode in the external circuit. Hence, the term '*displacement*' is justified.

***Example 3.9** Find the displacement current through a surface at a radius r ($a < r < b$) in a co-axial cylindrical capacitor of length l when a voltage $v = V_m \sin \omega t$ is applied; a and b being the radii of inner and outer cylinders respectively.

Solution From Example 2.53 of Chapter 2, we have

$$E = \frac{V}{r \ln (b/a)}$$

∴ Displacement current is

$$I_d = \varepsilon A \frac{dE}{dt} = \varepsilon 2\pi r l \frac{d}{dt}\left[\frac{V}{r \ln(b/a)}\right] = \frac{2\pi\varepsilon l}{\ln(b/a)}\frac{dV}{dt} = \frac{2\pi\varepsilon l}{\ln(b/a)}\frac{d}{dt}(V_m \sin \omega t)$$

$$= \frac{2\pi\varepsilon l}{\ln(b/a)} V_m \omega \cos \omega t$$

$$= C\omega V_m \cos \omega t$$

where, $C = \dfrac{2\pi\varepsilon l}{\ln (b/2)}$ is the capacitance of the co-axial cylindrical capacitor

∴ $$I_d = C\omega V_m \cos \omega t$$

Example 3.10 Find J_c and J_d for moist soil which has the conductivity $\sigma = 10^{-3}$ mho/m, $\varepsilon_r = 2.5$. Given: $E = 6 \times 10^{-6} \sin (9 \times 10^9 t)$ V/m.

Solution Conduction current density is

$$\vec{J}_C = \sigma \vec{E} = 10^{-3} \times 6 \times 10^{-6} \sin(9 \times 10^9 t) = 6 \times 10^{-9} \sin(9 \times 10^9 t) \text{ A/m}^2$$

Displacement current density is,

$$\vec{J}_d = \varepsilon \frac{\partial \vec{E}}{\partial t} = \varepsilon_0 \varepsilon_r \frac{\partial}{\partial t}[6 \times 10^{-6} \sin (9 \times 10^9 t)]$$

$$= 2.5 \times 8.854 \times 10^{-12} \times 6 \times 10^{-6} \times 10^9 \cos (9 \times 10^9 t)$$

$$= 1.19 \times 10^{-6} \cos (9 \times 10^9 t) \text{ A/m}^2$$

***Example 3.11** Show that the displacement through a capacitor is equal to the conduction current if the supply voltage is $v = V_m \sin \omega t$.

Solution Here the applied voltage is, $v = V_m \sin \omega t$
We consider a parallel plate capacitor of capacitance given as $C = \dfrac{\varepsilon A}{d}$

where, A is the area of the plate and d is the distance between the plates.
The conduction current is given as

$$I_d = C \frac{dv}{dt} = \frac{\varepsilon A}{d}\frac{d}{dt}[V_m \sin \omega t] = V_m \omega \frac{\varepsilon A}{d} \cos \omega t \qquad \textbf{(i)}$$

However, as the medium between the plates is dielectric, there is no conduction current in a capacitor. A displacement current may be considered to flow which is given as

$$I_d = \varepsilon A \frac{dE}{dt} = \varepsilon A \frac{d}{dt}\left[\frac{v}{d}\right] = \frac{\varepsilon A}{d}\frac{d}{dt}[V_m \sin \omega t] = V_m \omega \frac{\varepsilon A}{d} \cos \omega t \qquad \textbf{(ii)}$$

From (*i*) and (*ii*), it is seen that the displacement in a capacitor is equal to the conduction current.

***Example 3.12** **(a)** Show that the ratio of the amplitudes of conduction current density and displacement current density is $\left(\frac{\sigma}{\omega\varepsilon}\right)$ for the applied field $E = E_m \sin \omega t$. Assume $\mu = \mu_0$.

(b) What is the amplitude ratio if the applied field is $E = E_m e^{-t/\tau}$ where τ is real?

Solution

(a) Conduction current density is

$$\vec{J}_C = \sigma\vec{E} \qquad \therefore \quad J_C = |\vec{J}_C| = \sigma E \qquad \text{(i)}$$

Displacement current density is

$$\vec{J}_d = \varepsilon\frac{\partial\vec{E}}{\partial t}$$

$$\therefore \qquad J_d = |\vec{J}_d| = \varepsilon\frac{\partial E}{\partial t} = \varepsilon\frac{\partial}{\partial t}[E_m \sin \omega t] = \varepsilon E_m\omega \cos \omega t = j\omega\varepsilon E_m \sin \omega t = j\omega\varepsilon E \qquad \text{(ii)}$$

From (*i*) and (*ii*), the ratio of the amplitudes is

$$\boxed{\frac{J_C}{J_d} = \frac{\sigma}{\omega\varepsilon}}$$

(b) Conduction current density is

$$\vec{J}_C = \sigma\vec{E} \qquad \therefore \quad J_C = |\vec{J}_C| = \sigma E \qquad \text{(i)}$$

Displacement current density is

$$\vec{J}_d = \varepsilon\frac{\partial\vec{E}}{\partial t}$$

$$\therefore \qquad J_d = |\vec{J}_d| = \varepsilon\frac{\partial E}{\partial t} = \varepsilon\frac{\partial}{\partial t}[E_m e^{-t/\tau}] = \varepsilon\left(-\frac{1}{\tau}\right)E_m e^{-t/\tau} = -\left(\frac{\varepsilon}{\tau}\right)E \qquad \text{(ii)}$$

From (*i*) and (*ii*), the ratio of the amplitudes is

$$\boxed{\frac{J_C}{J_d} = -\frac{\sigma\tau}{\varepsilon}}$$

***Example 3.13** A copper wire carries a conduction current of 1A. Assuming that copper has about the same permittivity as free space, i.e., $\varepsilon = \varepsilon_0$, find:

(a) Displacement current in the wire at 100 MHz.

(b) Displacement current in the wire at 60 Hz.

For copper, $\sigma = 5.8 \times 10^7$ mho/m.

Solution Here, $I_C = 1A$; $\varepsilon = \varepsilon_0$; $\sigma = 5.8 \times 10^7$

From Problem 14 (a), we have

$$\frac{J_C}{J_d} = \frac{\sigma}{\omega\varepsilon} \quad \Rightarrow \quad J_d = J_C\frac{\omega\varepsilon}{\sigma} \quad \Rightarrow \quad I_d = I_C\frac{\omega\varepsilon}{\sigma} = I_C\frac{2\pi f\varepsilon}{\sigma}$$

Substituting the values

(a) Displacement current at $f = 100$ MHz is,

$$I_d = I_C \frac{2\pi f \varepsilon}{\sigma} = 1 \times \frac{2\pi \times 100 \times 10^6 \times 8.854 \times 10^{-12}}{5.8 \times 10^7} = 9.58 \times 10^{-8} \text{ A}$$

(b) Displacement current at $f = 60$ Hz is

$$I_d = I_C \frac{2\pi f \varepsilon}{\sigma} = 1 \times \frac{2\pi \times 60 \times 8.854 \times 10^{-12}}{5.8 \times 10^7} = 0.57 \times 10^{-16} \text{ A}$$

***Example 3.14** Charges are moving with a speed of v m/s along a line that has a charge density of λ C/m. Both v and λ are constants. Find the current through the line, and show that it is a steady current.

Solution We consider a strip of the line with thickness dl moved in time dt.

∴ Charge enclosed in the strip, $dq = \lambda dl = \lambda v dt$ $\left[\because v = \frac{dl}{dt}\right]$

The current is given as $I = \frac{dq}{dt} = \frac{\lambda v dt}{dt} = \lambda v$

Since both λ and v are constants, the current is a steady current.

***Example 3.15** Copper has 2 mobile electrons per atom, 0.09375×10^{26} atoms per kg, and a density of 9000 kg/m^3. The magnitude of the charge of an electron is 1.6×10^{-19} C.
(a) Find the mobile charge per unit length of a copper wire of radius 1 mm.
(b) If there is a steady current of 1 A through the wire, find the velocity of the mobile charges.

Solution
(a) Volume of the cylindrical strip of length 1 m and radius 1 mm is

$$v = \pi r^2 h = \pi \times (0.001)^2 \times 1 = \pi \times 10^{-6} \text{m}^3$$

∴ Mass of copper for this elemental volume is,

$$m = 900 \times \pi \times 10^{-6} = 9\pi \times 10^{-3} \text{ kg}$$

∴ Total atoms for this mass $= 9\pi \times 10^{-3} \times 0.09375 \times 10^{26}$
∴ Total electrons for these atoms $= 2 \times (9\pi \times 10^{-3} \times 0.09375 \times 10^{26})$
∴ Total charge per unit length,

$$\lambda = 2 \times (9\pi \times 10^{-3} \times 0.09375 \times 10^{26}) \times 1.6 \times 10^{-19} = 5.3014 \times 10^4 \text{ C/m}$$

(b) Here, $I = 1$A

From earlier problem, we got that $I = \frac{dq}{dt} = \frac{\lambda v dt}{dt} = \lambda v$

So, the velocity of the mobile charges is given as

∴ $$v = \frac{I}{\lambda} = \frac{1}{5.3014 \times 10^4} = 1.886 \times 10^{-5} \text{ m/s}$$

3.3 CLASSIFICATION OF ELECTRICAL MATERIALS

Depending upon the atomic structure, electrical materials are of three types:

1. Conductors
2. Insulators
3. Semiconductors

1. Conductors: Electrical materials, which have large number of free electrons or loosely bound valence-band electrons that can easily be knocked out of their orbit and constitute a large current, are known as *conductors*. Almost all metals and some liquids are good conductors.
2. Insulators: Electrical materials, where no free electrons are available and the valence-band electrons are tightly bound to the nucleus, are known as *insulators*. Examples of some insulators include glass, mica, plastics, etc.
3. Semiconductors: Electrical materials where there are no such free electrons present, but free electrons can easily be created by adding some impurities, are known as *semiconductors*. Examples of some semiconductors include germanium, silicon, etc. For example, germanium, a semiconductor, has approximately one trillion times (1×10^{12}) the conductivity of glass, an insulator, but has only about one thirty-millionth (3×10^{-8}) part of the conductivity of copper, a conductor.

3.4 BEHAVIOUR OF CONDUCTORS IN ELECTRIC FIELD

When an electric field $\vec{E}$ is applied to a conductor, the electrons will experience a force $(\vec{F} = -e\vec{E})$. An isolated conductor under a static electric field is shown in Fig. 3.5. As the electrons are not free in space, they will not be accelerated by the field; but they will suffer constant collisions with atomic lattice and drifts from one atom to another. This is called *drifting* of electrons. After some time, the electrons will attain a constant average velocity, called *drift velocity* $(\vec{U})$ which is directly proportional to the applied field $(\vec{U} = -\mu\vec{E};\ \mu$ is the mobility of electrons). The current associated with this drifting is known as the *drift current* or *conduction current* $(\vec{J} = \sigma\vec{E};\ \sigma$ is the conductivity of the conductor) as explained in Section 3.2.2.

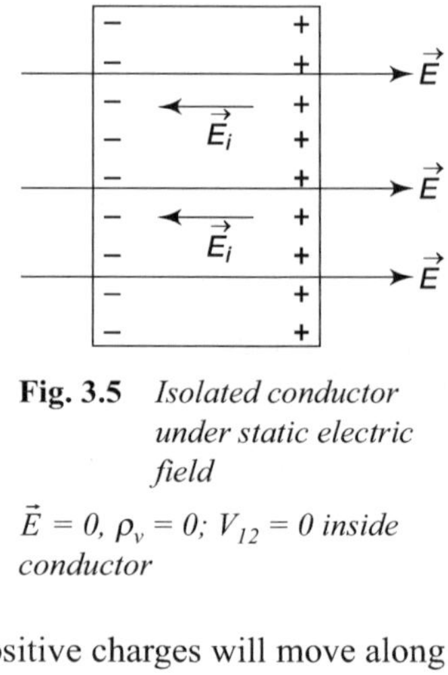

Fig. 3.5 *Isolated conductor under static electric field*

$\vec{E} = 0,\ \rho_v = 0;\ V_{12} = 0$ *inside conductor*

For a perfect conductor, the conductivity is infinite ($\sigma \to \infty$). As the conduction current is $(\vec{J} = \sigma\vec{E})$, to maintain a finite current density $(\vec{J})$, the field $\vec{E}$ must be zero for an isolated conductor. All positive charges will move along the direction of $\vec{E}$ and negative charges will move in the opposite direction. Thus, all charges will accumulate on the surface and these induced surface charges will set up an internal induced field, $\vec{E}_i$; which cancels the externally applied field $\vec{E}$. As field is zero inside a conductor, the potential inside a conductor is constant. Hence, a conductor is an equipotential body.

3.4.1 Properties of Conductors

From the discussion, we summarise the following properties of conductors:

1. The conductivity of a conductor in infinite.

2. Electric field inside a conductor is zero.
3. The charge density inside a conductor is zero.
4. The charge can exist on the surface of the conductor, giving rise to surface charge density.
5. The electric field at any point on the surface of a conductor is entirely perpendicular to the surface.
6. The conductor including its surface is an equipotential region.

3.4.2 Resistance of Conductor

Definition The ratio of the potential difference between the two ends of a conductor to the current flowing through the conductor is known as *resistance* of the conductor. This is given as

$$R = \frac{V}{I} = \frac{\rho l}{A}$$

Mathematical Equation We will consider a uniform conductor as shown in Fig. 3.6. Let

A uniform cross section,
L length of conductor,
V applied voltage

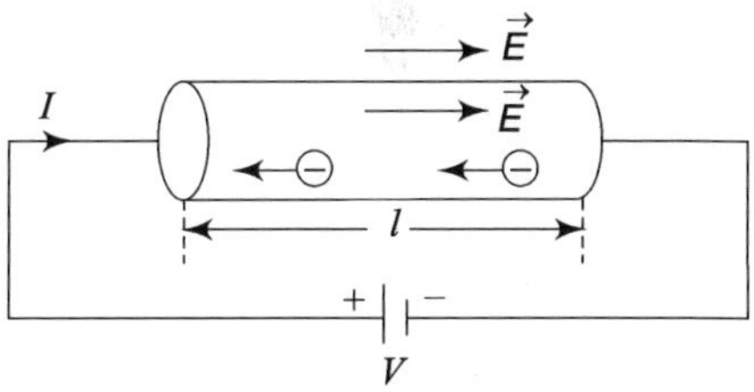

Fig. 3.6 *Conductor subjected to voltage V*

$\therefore$ Electric field $E = \frac{V}{l}$

$\therefore$ Current density $J = \frac{I}{A}$

By Ohm's law,

$$J = \sigma E$$

$$\Rightarrow \quad \frac{I}{A} = \sigma \frac{V}{l}$$

$$\Rightarrow \quad \frac{V}{I} = \frac{l}{\sigma A} = \frac{\rho l}{A}$$

$$\left(\rho = \frac{1}{\sigma} = \text{resistivity of the conductor}\right)$$

$$\therefore \quad \text{Resistance, } \boxed{R = \frac{\rho l}{A}} \qquad \textbf{(3.12)}$$

In general, resistance of a conductor with non-uniform cross section is given as

$$\boxed{R = \frac{V}{I} = \frac{\int \vec{E} \cdot d\vec{l}}{\int \vec{J} \cdot d\vec{S}} = \left(\frac{1}{\sigma}\right) \frac{\int \vec{E} \cdot d\vec{l}}{\int \vec{E} \cdot d\vec{S}}} \qquad \textbf{(3.13)}$$

$$(\because \ \vec{J} = \sigma \vec{E})$$

Example 3.16 A 3,000 km long cable consists of seven copper wires, each of diameter 0.73 mm, bundled together and surrounded by an insulating sheath. Calculate the resistance of the cable. Use $3 \times 10^{-6}\ \Omega - \text{cm}$ for the resistivity of the copper.

Solution The resistance R of a conductor is

$$R = \frac{\rho l}{A}$$

where ρ is the resistivity, l is the length of the conductor and A is its cross-sectional area. Since the cable consists of $N = 7$ copper wires, the total cross-sectional area is

$$A = N\pi r^2 = 7\frac{\pi d^2}{4} = 7\frac{\pi(0.073)^2}{4} = 0.0293 \text{ cm}^2$$

Thus, the resistance is

$$R = \frac{\rho l}{A} = \frac{(3\times 10^{-6})(3\times 10^{8})}{0.0293} = 3.1\times 10^{4}\,\Omega$$

Example 3.17 A truncated cone of altitude h and radii a and b for the right and the left ends, respectively is made of a material of resistivity ρ. Assuming that the current is distributed uniformly throughout the cross section of the cone, what is the resistance between the two ends?

Solution Consider a thin disk of radius r at a distance x from the left end. From Fig. 3.7(b), we have

$$\frac{b-r}{x} = \frac{b-a}{h}$$

$$\Rightarrow \qquad r = (a-b)\frac{x}{h} + b$$

Since resistance R is related to resistivity ρ by $R = \dfrac{\rho l}{A}$, where l is the length of the conductor and A is the cross-section, the contribution to the resistance from the disk having a thickness dy is

$$dR = \frac{\rho dx}{\pi r^2} = \frac{\rho dx}{\pi[b + {}^{(a-b)}x/h]^2}$$

By integration, we get

$$R = \int dR = \int_0^h \frac{\rho dx}{\pi[b + {}^{(a-b)}x/h]^2} = \frac{\rho h}{\pi ab}$$

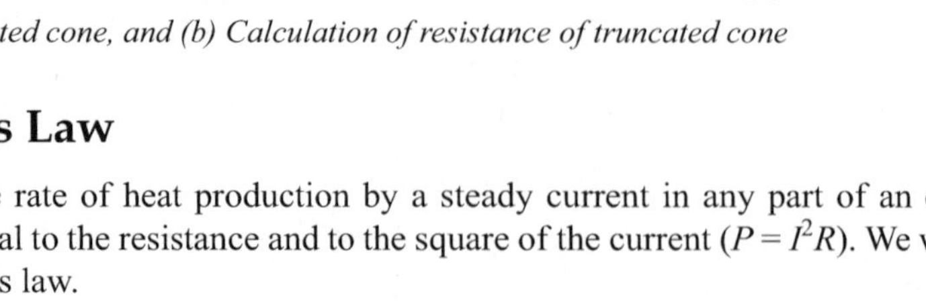

Fig. 3.7 *(a) Truncated cone, and (b) Calculation of resistance of truncated cone*

3.4.3 Joule's Law

Statement The rate of heat production by a steady current in any part of an electrical circuit is directly proportional to the resistance and to the square of the current ($P = I^2R$). We will derive the field equation for Joule's law.

Mathematical Equation

Let

$\vec{E}$ electric field intensity,
ρ_v volume charge density,
$\vec{U}$ average velocity of moving charges.

∴ Force experienced by the charge in a volume dv is

$$d\vec{F} = \rho_v dv\vec{E}$$

If the charges move a distance $d\vec{l}$ in time dt, such that $d\vec{l} = \vec{U}dt$, then the work done by the electric field is

$$dW = d\vec{F} \cdot d\vec{l} = \rho_v dv\vec{E} \cdot \vec{U}dt = \rho_v \vec{U} \cdot \vec{E}dvdt = \vec{J} \cdot \vec{E}dvdt$$

∴ Power (work done per unit time) supplied by the field is

$$dp = \frac{dW}{dt} = \vec{J} \cdot \vec{E}dv$$

∴ Power density (power per unit volume) is

$$\boxed{p = \vec{J} \cdot \vec{E}} \quad \textbf{(3.14)}$$

This is the *differential form* or *point form* of *Joule's law.*
Total power

$$P = \int_v pdv = \int_v \vec{J} \cdot \bar{E}dv$$

∴

$$\boxed{P = \int_v \vec{E} \cdot \vec{J}dv} \quad \textbf{(3.15)}$$

This is the *integral form* or *point form* of *Joule's law.*

For a linear conductor, $\vec{J} = \sigma\vec{E}$; thus, the power density is

$$p = \sigma\vec{E} \cdot \vec{E} = \sigma |E|^2$$

∴ Total power dissipation $P = \int_v \sigma E^2 dv$

For a conductor with uniform cross-section, $dv = dSdl$; so, total power dissipation is

$$P = \int_v EJdv = \int_l Edl \int_S JdS = VI$$

∴

$$\boxed{P = VI} \quad \textbf{(3.16)}$$

or,

$$P = \int_v \sigma E^2 dv = \sigma \int_v \left(\frac{V}{l}\right)^2 dv = \sigma \frac{V^2}{l^2} lA = \left(\frac{\sigma A}{l}\right) V^2 = \frac{V^2}{R}$$

$(A = \text{uniform cross section}, l = \textit{length})$

$$\boxed{P = \frac{V^2}{R}} \quad \textbf{(3.17)}$$

or,
$$P = \int_v EJdv = \int_v \frac{J}{\sigma} Jdv = \frac{1}{\sigma} J^2 lA = J^2 A^2 \frac{l}{\sigma A} = I^2 R$$

$\therefore$
$$\boxed{P = I^2 R} \quad \textbf{(3.18)}$$

This is another form of Joule's law which states that *the rate of heat dissipation varies as the square of the current in a linear conductor.*

Equations (3.16), (3.17) and (3.18) are the different forms of Joule's law for conductors with uniform cross sections.

Example 3.18 A parallel-palate capacitor with 10 square-cm plates and 0.4 cm plate separation contains a medium with permittivity $\varepsilon_r = 2$ and conductivity $\sigma = 4 \times 10^{-5}$ S/m. The potential difference between the plates is 200V. Determine the electric field intensity, the volume current density, power density, the power dissipation, the current and the resistance of the medium.

Solution Electric field intensity is

$$\vec{E} = -\frac{200}{0.004}\hat{a}_r = -50\hat{a}_r \text{ kV/m}$$

Current density

$$\vec{J} = \sigma\vec{E} = -4 \times 10^{-5} \times 50 \times 10^3 \hat{a}_r = -2\hat{a}_r \text{ A/m}^2$$

The current through the medium

$$I = \int_S \vec{J} \cdot d\vec{S} = 2 \times (0.1)^2 = 20 \text{ mA}$$

Power density is

$$p = \vec{J} \cdot \vec{E} = (-2\hat{a}_r) \cdot (-50 \times 10^3 \hat{a}_r) = 100 \text{ kW/m}^3$$

Total power dissipation in the medium is

$$P = \int_v pdv = 100 \times 10^3 \times (0.1)^2 \times 0.4 \times 10^{-2} = 4 \text{ W}$$

Hence, the resistance is

$$R = \frac{P}{I^2} = \frac{4}{(20 \times 10^{-3})^2} = 10 \text{ k}\Omega$$

3.4.4 Electromotive Force (EMF) and Kirchhoff's Voltage Law

We know, electric field intensity around any closed path vanishes; i.e., $\oint_l \vec{E} \cdot d\vec{l} = 0$.

Also, in a conducting medium, the current is, $I = \oint_s \vec{J} \cdot d\overline{S} = \oint_s \sigma\vec{E} \cdot d\overline{S}$

From these two equations, it follows that a purely electrostatic field cannot cause a current to circulate in a closed path (loop). In addition, there must also exist a source of energy to maintain the steady current in a closed loop.

The external source of energy may be non-electrical, such as

— a chemical reaction (battery),
— a mechanical drive (d.c. generator),
— a light-activated source (solar cell), or
— a temperature-sensitive device (thermocouple)

As these devices convert non-electrical energy into electrical energy; we consider them to be non-conservative elements, setting up a non-conservative field $\vec{E}_e$.

In general, the total electric field in a closed loop is

$$\vec{E} = \vec{E}_C + \vec{E}_e$$

where $\vec{E}_C$ is the static electric field due to charges

$\vec{E}_e$ is the field generated by other causes (emf-producing field)

Total power associated with the closed loop is

$$P = \int_v (\vec{E}_C + \vec{E}_e) \cdot \vec{J} dv = \oint_l I(\vec{E}_C + \vec{E}_e) \cdot d\vec{l}$$

(assuming that the steady current is uniformly distributed, such that $\vec{J} dv = I d\vec{l}$)

or,

$$P = \oint_l I\vec{E}_e \cdot d\vec{l} \qquad \left(\because \oint_l \vec{E}_C \cdot d\vec{l} = 0 \right)$$

By defining the *electromotive force* (*emf*) in the closed loop as

$$\boxed{\xi = \oint_l \vec{E}_e \cdot d\vec{l}} \tag{3.19}$$

we get

$$\boxed{P = \xi I} \tag{3.20}$$

Thus, power delivered to the loop is equal to the product of the emf and the current. Now, by Ohm's law

$$\vec{J} = \sigma \vec{E}$$

$$\Rightarrow \qquad \frac{\vec{J}}{\sigma} = \vec{E} = (\vec{E}_C + \vec{E}_e)$$

$$\Rightarrow \qquad \frac{\vec{I}}{A\sigma} = (\vec{E}_C + \vec{E}_e)$$

$$\Rightarrow \qquad \vec{I}\frac{R}{l} = (\vec{E}_C + \vec{E}_e) \qquad \left(\because R = \frac{l}{\sigma A} \right)$$

Taking line integral around a closed path,

$$\oint_l \vec{E}_C \cdot d\vec{l} + \oint_l \vec{E}_e \cdot d\vec{l} = I \oint_l \frac{R}{l} dl$$

$$\therefore \quad \boxed{\xi = IR_T} \tag{3.21}$$

where, R_T is the total resistance in the closed loop.
In general, for a closed circuit containing many resistances and emf sources,

$$\boxed{\sum \xi = I \sum R} \tag{3.22}$$

This is the *Kirchhoff's voltage law* (KVL).

Difference between Potential (V) and EMF (ξ)

1. Potential field, i.e., electric field generated by static charges, is conservative; but emf-producing field is non-conservative.

$$\therefore \quad \oint_l \vec{E}_C \cdot d\vec{l} = 0\text{; but, } \oint_l \vec{E}_e \cdot d\vec{l} \neq 0 = \xi \text{ (emf)}$$

2. Electric field produced by charges is not able to maintain a steady current; but emf-producing field can maintain a steady current.
3. Potential (V) is the negative of the line integral of the static field $\vec{E}_C$ while emf (ξ) is the line integral of $\vec{E}_e$. Thus, between two points a and b,

$$V_{ab} = (V_b - V_a) = -\int_a^b \vec{E}_C \cdot d\vec{l} \quad \text{and} \quad \xi_{ab} = \int_a^b \vec{E}_e \cdot d\vec{l}$$

4. Here, V_{ab} is independent of the path of integration between a and b, but ξ_{ab} is dependent on the path.

3.4.5 Kirchhoff's Current Law

The flux lines in a static electric field region begin and end on an electric charge and hence are discontinuous.
The tubes of steady current form closed circuit on themselves and are hence continuous. For this, steady current is said to be solenoidal, i.e., no sources or sinks are present.
Thus, as much current must flow into a volume as leaves it. In general, the integral of the normal component of the current density $\vec{J}$ over a closed surface S must be zero.

$$\therefore \quad \nabla \cdot \vec{J} = 0 \quad \Rightarrow \quad \boxed{\oint_S \vec{J} \cdot d\vec{S} = 0} \tag{3.23}$$

This relation for steady current applies to any volume. As illustration for the conductors shown in Fig. 3.8, $(I_1 - I_2 - I_3 + I_4 - I_5) = 0$

$$\therefore \quad \boxed{\sum I = 0} \tag{3.24}$$

This is the *Kirchhoff's current law* (KCL).

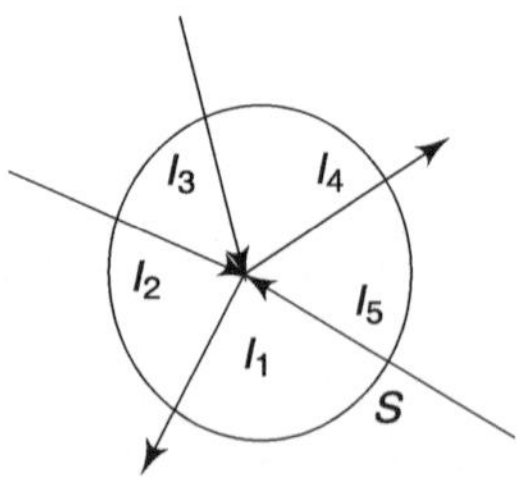

Fig. 3.8 *Currents in a closed surface*

3.4.6 Equation of Continuity and Relaxation Time

This equation is based on the *law of conservation of charge* which states that charges can neither be created nor be destroyed.

We consider a conductor carrying a surface current density $\vec{J}$, flowing perpendicular to the area $d\vec{S}$, having volume charge density ρ_v.

Thus, the total current coming out of the closed surface is

$$I = \oint_S \vec{J} \cdot d\vec{S} \qquad \textbf{(3.25)}$$

Now, charges cannot be created or destroyed. Since, the current is simply the motion of charge, the total current flowing out of some volume must be equal to the rate of decrease of charge within the volume.

$$\therefore \qquad I = -\frac{dQ}{dt} = -\frac{d}{dt}\int_v \rho_v dv$$

$$\therefore \qquad \oint_S \vec{J} \cdot d\vec{S} = -\int_v \frac{\partial \rho_v}{\partial t} dv \qquad \textbf{(3.26)}$$

Equation (3.26) is known as *the integral form of the equation of continuity.*

From Eq. (3.26), we get

$$\oint_S \vec{J} \cdot d\vec{S} = -\int_v \frac{\partial \rho_v}{\partial t} dv$$

$$\Rightarrow \qquad \oint_v \nabla \cdot \vec{J} dv = -\int_v \frac{\partial \rho_v}{\partial t} dv$$

$$\therefore \qquad \boxed{\nabla \cdot \vec{J} = -\frac{\partial \rho_v}{\partial t}} \qquad \textbf{(3.27)}$$

This is *the differential form* or *point form of equation of continuity* for time varying fields.

For steady current, ρ is constant, so that the equation of continuity for steady current becomes,

$$\nabla \cdot \vec{J} = 0 \qquad \textbf{(3.28)}$$

NOTE

Equation (3.28) shows that the net steady current through any closed surface is zero. This implies that for a steady current flowing through a conducting, the current density within the medium is solenoidal or continuous.

If the closed surface shrinks to a point, we have from Eq. (3.28) $\Sigma I = 0$, which is the statement of Kirchhoff's current law, stated as the algebraic sum of the currents at a point (called node) is zero.

Relaxation Time This is the time taken by a charge placed in the interior of a conductor to drop its value to 37% ($e^{-1} = 0.368 \approx 37\%$) of its initial value.

From Eq. (3.27) we have

$$-\frac{\partial \rho_v}{\partial t} = \nabla \cdot \vec{J} = \nabla \cdot (\sigma \vec{E}) = \sigma \nabla \cdot \vec{E} = \sigma \nabla \cdot \left(\frac{\vec{D}}{\varepsilon}\right) = \frac{\sigma}{\varepsilon} \nabla \cdot \vec{D} = \frac{\sigma}{\varepsilon} \rho_v$$

(Here, we have used, Ohm's law: $\vec{J} = \sigma \vec{E}$; Gauss' law: $\nabla \cdot \vec{D} = \rho_v$ and the constitutive relation: $\vec{D} = \varepsilon \vec{E}$)

$\therefore$
$$\frac{\partial \rho_v}{\rho_v} = -\frac{\sigma}{\varepsilon} \partial t$$

Integrating

$\int_{\rho_0}^{\rho_v} \frac{\partial \rho_v}{\rho_v} = -\int_0^t \frac{\sigma}{\varepsilon} \partial t$; where ρ_0 is the initial charge density at time $t = 0$.

$\therefore$
$$\boxed{\rho_v = \rho_0 e^{-(\sigma/\varepsilon)t} = \rho_0 e^{-t/\tau}} \qquad \textbf{(3.29)}$$

Equation (3.29) shows that whenever some charge is introduced at some interior point of a material, there is a decay of volume charge density ρ_v. This decay is associated with the movement of charge from the interior point to the surface of the material.

The time constant of the decay, is called the *relaxation time* or *rearrangement time*

$$\tau = \frac{\varepsilon}{\sigma} \qquad \textbf{(3.30)}$$

NOTE

(i) *The relaxation time is inversely proportional to the conductivity of the medium. This means that the value of τ is very small for good conductors and very large for a good dielectric. For example, for copper, $\sigma = 5.8 \times 10^7$ mho/m, and $\varepsilon_r = 1$, so that the value of relaxation time for copper is*

$$\tau = \frac{\varepsilon}{\sigma} = \frac{\varepsilon_0 \varepsilon_r}{\sigma} = \frac{8.854 \times 10^{-12} \times 1}{5.8 \times 10^7} = 1.53 \times 10^{-19} \text{ second}$$

(ii) *For good conductors, the relaxation time is so small that whenever any charge is placed inside the conductor, the charge will escape from the interior and will come to the conductor surface very quickly.*

(iii) *For a good dielectric, the relaxation time is so large that any charge placed inside it will almost remain within it.*

Example 3.19 Determine the relaxation time for silver having conductivity of $\sigma = 6.17 \times 10^7$ mho/m. If the charge density of 5 mC is placed inside a silver block, find the charge density in time $t = \tau$ and $t = 5\tau$.

Solution Relaxation time for silver is

$$\tau = \frac{\varepsilon}{\sigma} = \frac{\varepsilon_0}{\sigma} = \frac{8.854 \times 10^{-12}}{6.17 \times 10^7} \times 1.435 \times 10^{-19} \text{ second}$$

The initial charge inside the silver block, $\rho_0 = 5 \times 10^{-3}$ C

The charge decays as per the relation, $\rho_v = \rho_0 e^{-(\sigma/\varepsilon)t} = \rho_0 e^{-t/\tau}$

At $t = \tau$, the charge is

$$\rho_v = \rho_0 e^{-t/\tau} = 5\times 10^{-3}\exp\left(-\frac{\tau}{\tau}\right) = 5\times 10^{-3}\times 0.368 = 1.84 \text{ mC}$$

At $t = 5\tau$, the charge is

$$\rho_v = \rho_0 e^{-t/\tau} = 5\times 10^{-3}\exp\left(-\frac{5\tau}{\tau}\right) = 5\times 10^{-3}\times 0.368 = 0.034 \text{ mC}$$

***Example 3.20** Using $\nabla\cdot\vec{D} = \rho$, Ohm's law, and the equation of continuity, show that if at any instant a charge density ρ existed within a conductor, it would decrease to $\frac{1}{e}$ times this value in a time $\frac{\varepsilon}{\sigma}$ second. Calculate this time for a copper conductor.

Solution See Section 3.4.6.

$$\therefore \qquad \boxed{\rho_v = \rho_0 e^{-(\sigma/\varepsilon)t} = \rho_0 e^{-t/\tau}}$$

At

$$t = \frac{\varepsilon}{\sigma},\ \rho_v = \rho_0 e^{-1} = \frac{1}{e}\rho_0$$

This shows that if at any instant a charge density ρ existed within a conductor, it would decrease to $\frac{1}{e}$ times this value in a time $\frac{\varepsilon}{\sigma}$ second.

For copper, $\sigma = 5.8\times 10^7$ mho/m, and $\varepsilon_r = 1$, the value of this time is

$$\tau = \frac{\varepsilon}{\sigma} = \frac{\varepsilon_0\varepsilon_r}{\sigma} = \frac{8.854\times 10^{-12}\times 1}{5.8\times 10^7} = 1.53\times 10^{-19} \text{ second}$$

Example 3.21 Current density in a certain region is given by

$$\vec{J} = \frac{5}{r}\hat{a}_r + \frac{10}{(r^2+1)}\hat{a}_z \text{ A/m}^2$$

(a) Find the total current crossing the surface $z = 3$, $r < 6$, in the $\hat{a}_z$ direction.

(b) Find $\frac{\partial\rho_v}{\partial t}$.

(c) Calculate the total current crossing the closed surface bounded by $z = 0$, $z = 3$, $r = 1$ and $r = 6$.

(d) Is $\int_v \nabla\cdot\vec{J}dv = \oint_S \vec{J}\cdot d\vec{S}$?

Solution

(a) Total current is given as

$$I = \int_S \vec{J}\cdot dS\hat{a}_z = \int_{\phi=0}^{2\pi}\int_{r=0}^{6}\frac{10}{(r^2+1)} r\,dr\,d\phi = 10\pi \ln 37 = 113.44 \text{ A}$$

(b) By continuity equation

$$\nabla \cdot \vec{J} = -\frac{\partial \rho_v}{\partial t}$$

$$\therefore \quad \frac{\partial \rho_v}{\partial t} = -\nabla \cdot \vec{J} = -\left[\frac{1}{r}\frac{\partial}{\partial r}(rJ_r) + \frac{1}{r}\frac{\partial J_\phi}{\partial r} + \frac{\partial J_z}{\partial r}\right] = 0$$

(c) We find the currents at four surfaces:
At $z = 3$, $1 < r < 6$, $0 < \phi < 2\pi$ in $\hat{a}_z$ direction

$$I_1 = \int_S \vec{J} \cdot dS\hat{a}_z = \int_{\phi=0}^{2\pi}\int_{r=1}^{6} \frac{10}{(r^2+1)} r dr d\phi = 10\pi \ln 18.5 = 91.66 \text{ A} \quad \text{(outward)}$$

At $z = 0$, $1 < r < 6$, $0 < \phi < 2\pi$ in $\hat{a}_z$ direction

$$I_2 = \int_S \vec{J} \cdot dS(-\hat{a}_z) = -\int_{\phi=0}^{2\pi}\int_{r=1}^{6} \frac{10}{(r^2+1)} r dr d\phi = -10\pi \ln 18.5 = -91.66 \text{ A} \quad \text{(inward)}$$

At $r = 6$, $0 < z < 3$, $0 < \phi < 2\pi$ in $\hat{a}_r$ direction

$$I_3 = \int_S \vec{J} \cdot dS\hat{a}_r = \int_{\phi=0}^{2\pi}\int_{z=0}^{3} \frac{5}{r} r d\phi dz = 30\pi \text{ A} \quad \text{(outward)}$$

At $r = 1$, $0 < z < 3$, $0 < \phi < 2\pi$ in $\hat{a}_r$ direction

$$I_4 = \int_S \vec{J} \cdot dS(-\hat{a}_r) = -\int_{\phi=0}^{2\pi}\int_{z=0}^{3} \frac{5}{r} r d\phi dz = -30\pi \text{ A} \quad \text{(inward)}$$

Hence, total current crossing the closed surface is

$$I = \sum_{i=1}^{4} I_i = I_1 + I_2 + I_3 + I_4 = 0$$

(d) $\because \quad \nabla \cdot \vec{J} = 0 \qquad \therefore \quad \int_v (\nabla \cdot \vec{J})\, dv = 0$

$$\because \quad \int_v \nabla \cdot \vec{J} dv = \oint_S \vec{J} \cdot d\vec{S} = 0$$

3.4.7 Boundary Conditions for Conducting Medium

We consider a conductor–conductor boundary, as shown in Fig. 3.9, between two media of conductivities, σ_1 and σ_2 and permittivities, ε_1 and ε_2, respectively. We want to find out the boundary conditions that the current density will satisfy at the interface.
We will consider a coin-shaped surface, over the volume occupied by the surface

$$\therefore \quad \int_v (\nabla \cdot \vec{J}) dv = \oint_S \vec{J} \cdot d\vec{S} = 0$$

If the surface area of the coin is ΔS and thickness of the coin approaches zero, then

$$(J_{1n}\Delta S - J_{2n}\Delta S) = 0$$

$$\therefore \quad J_{1n} = J_{2n} \qquad \textbf{(3.31)}$$

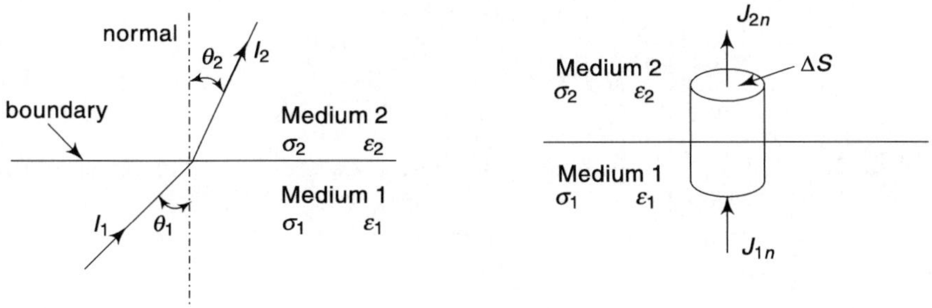

Fig. 3.9 *(a) Conductor to conductor boundary with change in direction of current, and (b) Conductor to conductor boundary*

Equation (3.31) implies that *the normal component of the electric current density $\vec{J}$ is continuous across the boundary.*

Further, since the electric field is conservative throughout the region, the tangential component of $\vec{E}$ is continuous.

$$\therefore \qquad E_{1t} = E_{2t}$$

$$\Rightarrow \qquad \frac{J_{1t}}{\sigma_1} = \frac{J_{2t}}{\sigma_2} \qquad \textbf{(3.32)}$$

(since by Ohm's law, $J = \sigma E$)

Equation (3.32) implies that *the tangential component of the electric current density $\vec{J}$ is discontinuous across the boundary. The ratio of the tangential components of the current densities at the interface is equal to the ratio of the conductivities.*

From Eqs. (3.31) and (3.32), we have

$$\frac{J_{1t}}{\sigma_1 J_{1n}} = \frac{J_{2t}}{\sigma_2 J_{2n}} \qquad \Rightarrow \qquad \frac{J_{1t}}{J_{1n}} \frac{J_{2n}}{J_{2t}} = \frac{\sigma_1}{\sigma_2}$$

Hence, from Fig. 3.9, we get

$$\boxed{\frac{\tan \theta_1}{\tan \theta_2} = \frac{\sigma_1}{\sigma_2}} \qquad \textbf{(3.33)}$$

If the boundary carries a surface charge with density, ρ_s (C/m^2), then by the boundary condition of normal component of electric displacement, D_n, we have

$$(D_{2n} - D_{1n}) = \rho_s$$

$$\Rightarrow \qquad (\varepsilon_2 E_{2n} - \varepsilon_1 E_{1n}) = \rho_s \qquad \textbf{(3.34a)}$$

Also, from Eq. (3.31), we get

$$J_{1n} = J_{2n} \quad \Rightarrow \quad \sigma_1 E_{1n} = \sigma_2 E_{2n} \qquad \textbf{(3.34b)}$$

Combining Eqs. (3.34*a*) and (3.34*b*), we have

$$\rho_s = \left(\varepsilon_2 \frac{\sigma_1}{\sigma_2} E_{1n} - \varepsilon_1 E_{1n} \right) = \left(\varepsilon_2 \frac{\sigma_1}{\sigma_2} - \varepsilon_1 \right) E_{1n} = \left(\frac{\varepsilon_2}{\sigma_2} - \frac{\varepsilon_1}{\sigma_1} \right) \sigma_1 E_{1n}$$

$$\therefore \qquad \boxed{\rho_s = \left(\frac{\varepsilon_2}{\sigma_2} - \frac{\varepsilon_1}{\sigma_1}\right) J_{1n}} \tag{3.35}$$

Example 3.22 Medium 1 ($z \geq 0$) has a dielectric constant of 2 and conductivity of 40 μS/m. Medium 2 ($z \leq 0$) has a dielectric constant of 5 and conductivity of 50 nS/m. If $\vec{J}_2$ has a magnitude of 2 A/m^2, and $\theta_2 = 60°$ with the normal to the interface, find $\vec{J}_1$ and θ_1. Also, find the surface charge density at the interface.

Solution Here, $J_{2n} = 2 \cos 60° = 1$ A/m^2 and $J_{2t} = 2 \sin 60° = 1.732$ A/m^2
From boundary condition of Eq. (3.34*b*)

$$J_{1n} = J_{2n} = 1 \text{ A/m}^2$$

From boundary condition of Eq. (3.32)

$$\frac{J_{1t}}{J_{2t}} = \frac{\sigma_1}{\sigma_2} = \frac{40 \times 10^{-6}}{50 \times 10^{-9}} = 800 \quad \Rightarrow \quad J_{1t} = 800 \times 1.732 = 1385.64 \text{ A/m}^2$$

$$\therefore \qquad J_1 = \sqrt{J_{1n}^2 + J_{1t}^2} = \sqrt{(1)^2 + (1385.64)^2} \approx 1385.64 \text{ A/m}^2$$

$$\therefore \qquad \theta_1 = \tan^{-1}\left(\frac{J_{1t}}{J_{1n}}\right) = \tan^{-1}\left(\frac{1385.64}{1}\right) = 89.96°$$

Surface charge density is obtained from Eq. (3.35), as

$$\rho_s = J_{1n}\left(\frac{\varepsilon_2}{\sigma_2} - \frac{\varepsilon_1}{\sigma_1}\right) = 1 \times \left(\frac{5}{50 \times 10^{-9}} - \frac{2}{40 \times 10^{-6}}\right) \times 8.854 \times 10^{-12} = 0.88 \text{ mC/m}^2$$

3.4.8 Laplace Equation for Conducting Media

For steady current,

$$\nabla \cdot \vec{J} = 0$$

$$\Rightarrow \qquad \nabla \cdot (\sigma \vec{E}) = 0 \qquad \text{(by Ohm's law, } \vec{J} = \sigma \vec{E})$$

$$\Rightarrow \qquad \sigma \nabla \cdot \vec{E} = 0$$

$$\Rightarrow \qquad \nabla \cdot (-\nabla V) = 0$$

$$\therefore \qquad \boxed{\nabla^2 V = 0} \tag{3.36}$$

Thus, the Laplace equation for electrostatic field is the same as that for steady current. This implies that the problems involving distributions of steady currents in conducting media can be handled in the same way as problems involving static field distributions in insulating media.

***Example 3.23**

(a) A uniformly thick metal plate in the shape of a quarter circle, has a constant DC voltage applied at the surfaces (planes) $V = 0$ and $V = V_1$. Compute the current density distribution in the plate.

(b) If the plate is 't' thick and the two radii are 'r_1' and 'r_2' ($r_1 < r_2$), what will be the maximum current density if the total current through the plate is 'I'?

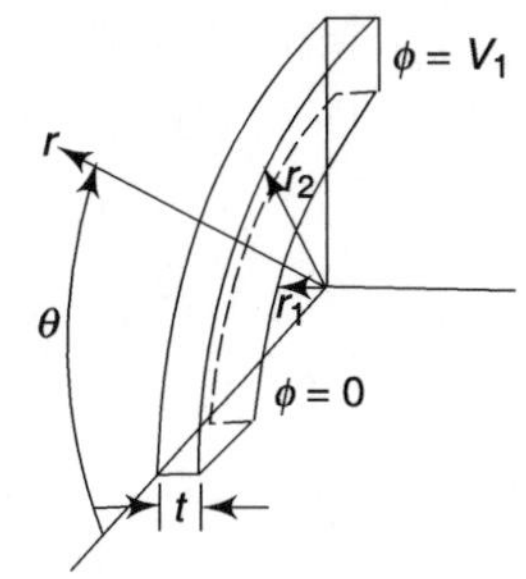

Fig. 3.10 *Metal plate of quarter circle shape*

Solution

This is depicted in Fig. 3.10.

(a) Since there is no charge inside the conductor, we apply Laplace's equation. By Laplace equation

$$\nabla^2 V = 0$$

In cylindrical coordinates

$$\frac{1}{r}\frac{\partial}{\partial r}\left(r\frac{\partial V}{\partial r}\right) + \frac{1}{r^2}\frac{\partial^2 V}{\partial \phi^2} + \frac{\partial^2 V}{\partial z^2} = 0$$

Considering the symmetry and the fact that V cannot change with r as the plate is a conductor, we have

$$\frac{1}{r^2}\frac{\partial^2 V}{\partial \phi^2} = 0$$

$$\Rightarrow \qquad \frac{\partial^2 V}{\partial \phi^2} = 0 \qquad (\because r \neq 0)$$

Integrating twice

$$V = k_1\phi + k_2$$

where k_1 and k_2 are the integration constants.

Applying the boundary conditions:

(1) At $\phi = 0$, $V = 0 \Rightarrow k_2 = 0$

(2) At $\phi = \dfrac{\pi}{2}$, $V = V_1 \Rightarrow k_1 = \dfrac{2V_1}{\pi}$

$$\therefore \qquad V = \frac{2V_1}{\pi}\phi$$

So, the field intensity is

$$\vec{E} = -\nabla V = -\frac{1}{r}\frac{\partial V}{\partial \phi}\hat{a}_\phi = -\frac{2V_1}{\pi r}\hat{a}_\phi$$

Hence, the current density distribution in the plate is given as

$$\boxed{\vec{J} = \sigma\vec{E} = -\frac{2V_1}{\pi r}\sigma\hat{a}_\phi}$$

(b) If the thickness of the plate is t, then total current is

$$I = \oint_S \vec{J}\cdot d\vec{S} = \int_{r_1}^{r_2} J t\,dr = \frac{2V_1}{\pi}\sigma t\int_{r_1}^{r_2}\frac{dr}{r} = \frac{2V_1\sigma t}{\pi}\ln\left(\frac{r_2}{r_1}\right)$$

$$\therefore \qquad \sigma = \frac{I\pi}{2tV_1\ln\left(\dfrac{r_2}{r_1}\right)}$$

Thus, the current density distribution in this case is

$$J = \frac{2V_1\sigma}{\pi r} = \frac{2V_1}{\pi r} \times \frac{I\pi}{2tV_1 \ln\left(\frac{r_2}{r_1}\right)} = \frac{I}{t\ln\left(\frac{r_2}{r_1}\right)}\frac{1}{r}$$

The current density will be maximum when r is minimum, i.e., $r = r_1$.

$\therefore$

$$\boxed{J_{max} = \frac{I}{t\ln\left(\frac{r_2}{r_1}\right)}\frac{1}{r_1}}$$

Example 3.24 Two infinitely conducting parallel plates, each of cross-sectional area A, are separated by a distance d. The potential difference between the plates is V_0. If the medium between the plates is homogeneous and has a finite conductivity σ, determine the resistance of the region between the plates. This is depicted in Fig. 3.11.

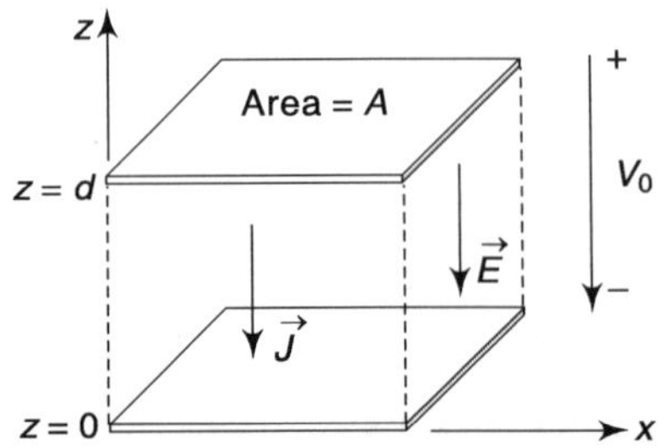

Fig. 3.11 *Parallel plates separated by conducting medium*

Solution We apply the Laplace equation for the region between the two infinitely conducting plates.

$$\nabla^2 V = 0$$

Assuming that the potential distribution to be a function of z only, we get

$$\frac{d^2V}{dz^2} = 0$$

Integrating twice, we get

$$V = k_1 z + k_2$$

where k_1 and k_2 are the constants of integration.
Applying the boundary conditions:
At $z = 0$, $V = 0 \Rightarrow k_2 = 0$

At $z = d$, $V = V_0 \Rightarrow k_1 = \frac{V_0}{d}$

Thus, the potential distribution between the plates is

$$V = \frac{V_0}{d}z$$

The electric field intensity is given as

$$\vec{E} = -\nabla V = -\frac{\partial V}{\partial z}\hat{a}_z = -\frac{V_0}{d}\hat{a}_z$$

The volume current density is

$$\vec{J} = \sigma\vec{E} = -\frac{\sigma V_0}{d}\hat{a}_z$$

The current through the surface normal to the current density, $\vec{J}$ is

$$I = \int_S \vec{J} \cdot d\vec{S} = \frac{\sigma A V_0}{d}$$

Thus, the resistance between the plates is

$$R = \frac{V_0}{I} = \frac{d}{\sigma A}$$

Example 3.25 Starting from the Maxwell's equation in differential form, obtain the Poisson's equation for the general situation in which the permittivity of the medium is not constant and is a function of position.

Solution From Maxwell's equation, $\nabla \cdot \vec{D} = \rho$

Also, $\nabla V = -\vec{E} = -\frac{\vec{D}}{\varepsilon}$

Taking divergence on both sides

$$\begin{aligned}
\nabla^2 V &= -\nabla \cdot \left(\frac{\vec{D}}{\varepsilon}\right) = -\left[\frac{1}{\varepsilon}\nabla \cdot \vec{D} + \vec{D}\nabla \cdot \left(\frac{1}{\varepsilon}\right)\right] \\
&= -\left[\frac{\rho}{\varepsilon} + \varepsilon\vec{E}\left(\frac{1}{\varepsilon^2}\right)\nabla \cdot \varepsilon\right] \\
&= -\left[\frac{\rho}{\varepsilon} + \frac{1}{\varepsilon}\vec{E}\nabla \cdot \varepsilon\right] \\
&= -\left[\frac{\rho}{\varepsilon} - \frac{1}{\varepsilon}\nabla V \nabla \cdot \varepsilon\right] \\
&= -\frac{1}{\varepsilon}[\rho - \nabla V \nabla \cdot \varepsilon]
\end{aligned}$$

This is the Poisson's equation for a medium whose permittivity is a function of position.

3.4.9 Principle of Duality

Two physical systems or phenomena are called dual if they are described by the mathematical equations of the same form.

There exists an analogy between $\vec{D}$ and $\vec{J}$ fields under static conditions. These two fields ($\vec{D}$ and $\vec{J}$) are described by the equations of the same mathematical form. This is listed in Table 3.1.

Table 3.1 *Duality between $\vec{D}$ and $\vec{J}$*

Sl. No.	$\vec{D}$-field	$\vec{J}$-field
1	$\nabla \cdot \vec{D} = 0$	$\nabla \cdot \vec{J} = 0$
2	$\nabla \times \vec{D} = 0$	$\nabla \times \vec{J} = 0$
3	$\vec{D} = \varepsilon \vec{E}$	$\vec{J} = \sigma \vec{E}$
4	$D_{n1} = D_{n2}$	$J_{n1} = J_{n2}$
5	$\frac{D_{t1}}{D_{t2}} = \frac{\varepsilon_1}{\varepsilon_2}$	$\frac{J_{t1}}{J_{t2}} = \frac{\sigma_1}{\sigma_2}$
	for dielectrics	for conductors

Example 3.26 Establish the relation $\frac{G}{\sigma} = \frac{C}{\varepsilon}$, where G and C are respectively the conductance and capacitance between the two electrodes; σ and ε are respectively the conductivity and permeability of the intervening medium.

Or

Deduce the relation $G = C\frac{\sigma}{\varepsilon}$ and show that the total electrical conductance between any configuration of conductors embedded in a conducting medium of σ may be obtained by replacing permittivity ε by σ in the expression for the capacitance of the configuration where they are embedded in a dielectric of permittivity ε.

Solution By definition of capacitance

$$C = \frac{Q}{V} = \frac{\int_S \vec{Q}_S \cdot d\vec{S}}{\int_1^2 \vec{E} \cdot d\vec{l}}$$

But, at the surface of conductors, by boundary condition, $\varepsilon E_n = Q_S$

$$\therefore \qquad C = \frac{\varepsilon \int_S \vec{E}_n \cdot d\vec{S}}{\int_1^2 \vec{E} \cdot d\vec{l}} \qquad \text{(i)}$$

For the current flow, the conductance is

$$\therefore \qquad G = \frac{I}{V} = \frac{\int_S \vec{J}_n \cdot d\vec{S}}{\int_1^2 \vec{E} \cdot d\vec{l}} = \frac{\sigma \int_S \vec{E}_n \cdot d\vec{S}}{\int_1^2 \vec{E} \cdot d\vec{l}} \qquad \text{(ii)}$$

Combining (*i*) and (*ii*), we get

$$G = C \times \frac{\sigma}{\varepsilon}$$

$$\therefore \qquad \boxed{\frac{G}{\sigma} = \frac{C}{\varepsilon}}$$

Thus, the conductance may be obtained at once by replacing ε by σ in the expression of the capacitance.

NOTE

Insulation resistance of the configuration is, $R = \frac{1}{G} = \frac{1}{C} \times \frac{\varepsilon}{\sigma}$.

Example 3.27 Use the principle of duality to find out the insulation resistance of:
(a) A parallel-plate capacitor of separation 'd' and area A;
(b) A coaxial cable of inner radius 'a' and outer radius 'b' filled with material of conductivity σ.
(c) A spherical shell consisting of a pair of concentric spheres with inner sphere of radius 'a' surrounded by a medium of conductivity 'σ' in contact with the inner surface of the outer sphere of radius 'b'.

Solution

(a) Here, the field intensity is, $E = \frac{V}{d}$

Total charge, $Q = DA = \varepsilon EA = \varepsilon \frac{V}{d} A$

$\therefore$ Capacitance is, $C = \frac{Q}{V} = \frac{\varepsilon A}{d}$

By principle of duality, the conductance is given by, $G = C\frac{\sigma}{\varepsilon} = \frac{\sigma A}{d}$
where, σ is the conductivity of the medium in mho/m.
$\therefore$ Insulation resistance is given as

$$\boxed{R = \frac{1}{G} = \frac{d}{\sigma A}}$$

(b) For coaxial cable, the field intensity at any radius r ($a < r < b$) is obtained by Gauss' law as follows.

$$2\pi r l E = \frac{\lambda l}{\varepsilon} \quad \Rightarrow \quad E = \frac{\lambda}{2\pi\varepsilon r}$$

So, the potential is given as $V = \int_a^b E dr = \frac{\lambda}{2\pi\varepsilon}\int_a^b \frac{dr}{r} = \frac{\lambda}{2\pi\varepsilon}\ln\left(\frac{b}{a}\right)$

So, the capacitance per unit length is $C = \frac{\lambda}{V} = \frac{2\pi\varepsilon}{\ln\left(\frac{b}{a}\right)}$

By principle of duality, the conductance is given by $G = C\frac{\sigma}{\varepsilon} = \frac{2\pi\sigma}{\ln\left(\frac{b}{a}\right)}$

$\therefore$ Insulation resistance is given as

$$R = \frac{1}{G} = \frac{\ln\left(\frac{b}{a}\right)}{2\pi\sigma}$$

(c) For spherical shell, the field intensity at any radius r ($a < r < b$) is obtained by Gauss' law as follows.

$$4\pi r^2 E = \frac{Q}{\varepsilon} \quad \Rightarrow \quad E = \frac{Q}{4\pi\varepsilon r^2}$$

So, the potential is given as $V = \int_a^b E dr = \frac{Q}{4\pi\varepsilon}\int_a^b \frac{dr}{r^2} = \frac{Q}{4\pi\varepsilon}\left(\frac{b-a}{ab}\right)$

So, the capacitance is $C = \frac{Q}{V} = \frac{4\pi\varepsilon ab}{(b-a)}$

By principle of duality, the conductance is given by $G = C\frac{\sigma}{\varepsilon} = \frac{4\pi\sigma ab}{(b-a)}$

∴ Insulation resistance is given as

$$\boxed{R = \frac{1}{G} = \frac{b-a}{4\pi\sigma ab} = \frac{1}{4\pi\sigma}\left(\frac{1}{1/a - 1/b}\right)}$$

3.5 BEHAVIOUR OF DIELECTRIC MATERIALS IN ELECTRIC FIELD

A *dielectric* or an *insulator* is a non-conducting material. In case of dielectrics, the valence band is completely filled and the energy gap between the conduction and valence band is very large so that free electrons are not normally available, except under the application of extremely high energy.

3.5.1 Properties of Dielectrics

The various properties of dielectric materials are as follows.

1. The dielectrics do not have any free charges, but all charges are bound and associated with the nearest atoms.
2. When an external electric field is applied, there is some displacement of the bound charges, thus creating small electric dipoles within the dielectric. This phenomenon is known as *polarisation.*
3. The electric field inside and outside a dielectric gets modified due to the presence of dipoles.
4. Dielectrics can store energy.

3.5.2 Dielectric Polarisation

The dielectric polarisation may be defined as a dynamical response of a system to an externally applied electric field. This is shown in Fig. 3.12.

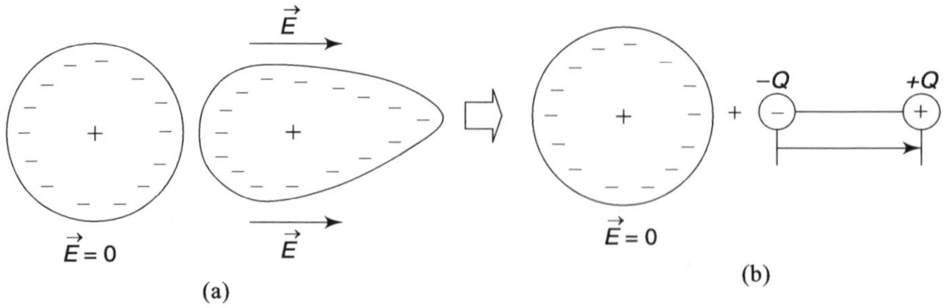

Fig. 3.12 *Polarisation of non-polar dielectric*

Macroscopic Polarisation To understand the macroscopic effect of an electric field on a dielectric, we consider an atom of the dielectric consisting of:

— a negative charge $-Q$ (electron cloud), and

— a positive charge $+Q$ (nucleus)

When an electric field $\vec{E}$ is applied:

— Positive charge is displaced from its equilibrium position in the direction of $\vec{E}$ by the force, $\vec{F}_+ = Q\vec{E}$.

— Negative charge is displaced to the opposite direction, by the force $\vec{F}_- = -Q\vec{E}$.

Thus, an electric dipole is created and the dielectric is said to be polarised.

Dipole moment is $\vec{p} = Q\vec{d}$

By superposition, the distorted charge distribution is equivalent to original distribution plus dipole.

In this way, on the presence of an electric field, all dipoles within a dielectric are aligned with the direction of the field.

The induced dipole field opposes the applied field. Thus, the entire dielectric is polarised.

If there are N dipoles in a volume Δv, the total dipole moment is

$$Q_1\vec{d}_1 + Q_2\vec{d}_2 + \ldots + Q_N\vec{d}_N = \sum_{i=1}^{N} Q_i\vec{d}_i$$

Microscopic Polarisation The type of polarisation on a microscopic scale is determined by the material. Dipoles, which are responsible for the polarisation, are of two types, as shown in Fig. 3.13:

Induced dipoles: those materials which exhibit polarisation only in the presence of an external field.

Permanent dipoles: those materials (ferroelectric) which exhibit permanent polarisation.

The connection between the macroscopic and microscopic polarisation is given through *polarisation vector* $\vec{P}$ as defined below.

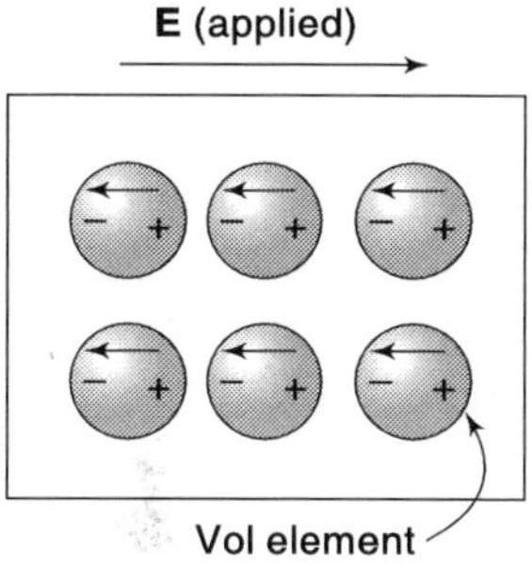

Fig. 3.13 *Polarisation of polar dielectric*

Polarisation Vector $(\vec{P})$ It is the dipole moment per unit volume of the dielectric,

$$\boxed{\vec{P} = \lim_{\Delta v \to 0}\left(\frac{\sum_{i=1}^{N} Q_i\vec{d}_i}{\Delta v}\right)} \quad \textbf{(3.37)}$$

Classification of Dielectric on the basis of Polarisation Based upon the effects of polarisation, dielectric materials are of two types:

Non-polar dielectric materials: These are the dielectric materials which have no free charges; all electrons are bound and associated with the nearest atoms. Hence, these materials do not have any dipole moment in the absence of any external electric field. An external electric field causes a small separation of the centres of the electron cloud and the positive ion core so that each infinitesimal element of volume behaves as an electric dipole. Example of such materials includes hydrogen, oxygen, nitrogen and other rare gases.

Polar dielectric materials: These are the dielectric materials in which the molecules or atoms possess a permanent dipole moment which is ordinarily randomly oriented, but which becomes more or less oriented by the application of an external electric field.

3.5.3 Calculation of Electric Field at an Exterior Point due to Polarised Dielectric

We consider a block of dielectric with polarisation $\vec{P}$ (dipole moment per unit volume). This is shown in Fig. 3.14.

We want to calculate the potential and field at an exterior point P.

By the relation of potential for a dipole

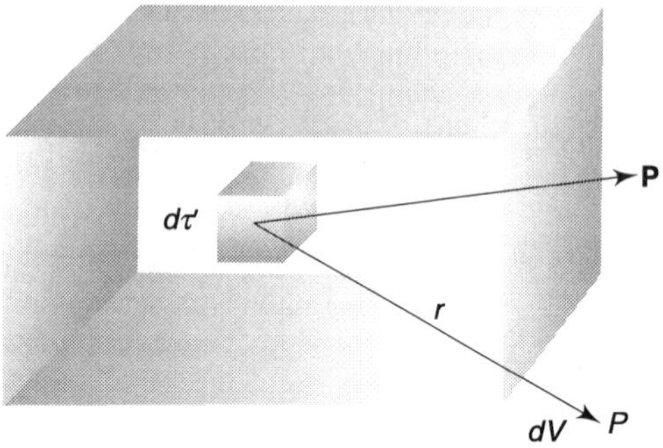

Fig. 3.14 *Block of dielectric with polarisation* ($\vec{P}$)

$$dV = \frac{1}{4\pi\varepsilon_0}\frac{\vec{P}\cdot\hat{a}_r}{r^2}d\tau' \tag{3.38}$$

for an elemental volume $d\tau'$ at (x', y', z') at the field point (x, y, z).

$$\therefore \qquad V = \frac{1}{4\pi\varepsilon_0}\int_v \frac{\vec{P}\cdot\hat{a}_r}{r^2}d\tau' \tag{3.39}$$

where $r = (x - x')\hat{i} + (y - y')\hat{j} + (z - z')\hat{k}$ $\quad \therefore \quad \frac{\hat{a}_r}{r^2} = \nabla'\left(\frac{1}{r}\right)$

From Eq. (3.39)

$$V = \frac{1}{4\pi\varepsilon_0}\int_v \vec{P}\cdot\nabla'\left(\frac{1}{r}\right)d\tau' \tag{3.40}$$

By the vector identity for a scalar S and a vector $\vec{V}$

$$\nabla\cdot(S\vec{V}) \equiv S(\nabla\cdot\vec{V}) + \vec{V}\cdot\nabla S$$

Putting $S = \left(\frac{1}{r}\right)$, $\vec{V} = \vec{P}$, we get

$$\nabla\cdot\left(\frac{\vec{P}}{r}\right) = \frac{1}{r}(\nabla\cdot\vec{P}) + \vec{P}\cdot\nabla\left(\frac{1}{r}\right) \quad \Rightarrow \quad \vec{P}\cdot\nabla\left(\frac{1}{r}\right) = \nabla\cdot\left(\frac{\vec{P}}{r}\right) - \frac{1}{r}(\nabla\cdot\vec{P})$$

From Eq. (3.40), we get

$$V = \frac{1}{4\pi\varepsilon_0}\int_v\left[\nabla'\cdot\left(\frac{\vec{P}}{r}\right) - \frac{1}{r}(\nabla'\cdot\vec{P})\right]d\tau'$$

$$= \frac{1}{4\pi\varepsilon_0}\int_S \frac{\vec{P}\cdot d\vec{S}}{r} - \frac{1}{4\pi\varepsilon_0}\int_v \frac{1}{r}(\nabla'\cdot\vec{P})d\tau' \tag{3.41}$$

(by divergence theorem)

Both of the terms in Eq. (3.41) have the form of potentials produced by the charge distribution; i.e.,

- a *surface charge density,* $\sigma_b = \vec{P} \cdot \hat{n}$
- a *volume charge density,* $\rho_b = -\nabla'(\vec{P})$

Thus,

$$V = \frac{1}{4\pi\varepsilon_0}\left[\int_S \frac{\sigma_b dS}{r} + \int_v \frac{\rho_b d\tau'}{r}\right] \tag{3.42}$$

Hence, the electric field is given as

$$\vec{E} = -\nabla V = \frac{1}{4\pi\varepsilon_0}\left[\int_S \frac{\sigma_b dS}{r^2}\hat{a}_r + \int_v \frac{\rho_b d\tau'}{r^2}\hat{a}_r\right] \tag{3.43}$$

due to the dielectric only.

To have the **macroscopic field**, we add the effects of the external charge distributions that are responsible for polarisation, so that

$$V = \frac{1}{4\pi\varepsilon_0}\left[\int_S \frac{(\sigma_f + \sigma_b)dS}{r} + \int_v \frac{(\rho_f + \rho_b)d\tau'}{r}\right]$$

and

$$\vec{E} = -\nabla V = \frac{1}{4\pi\varepsilon_0}\left[\int_S \frac{(\sigma_f + \sigma_b)dS}{r^2}\hat{a}_r + \int_v \frac{(\rho_f + \rho_b)d\tau'}{r^2}\hat{a}_r\right]$$

where σ_f and ρ_f are the free charge densities.

NOTE

Bound charges are those which cannot move within a dielectric; they are created due to displacement of charges during polarisation. Free charges are those which can move over macroscopic distance, such as electrons in a conductor.

- **The Bound Charge Densities, σ_b and ρ_b:**

The displacement of charges within the dielectric results in net volume and surface charge densities. This is explained below and also shown in Fig. 3.15.

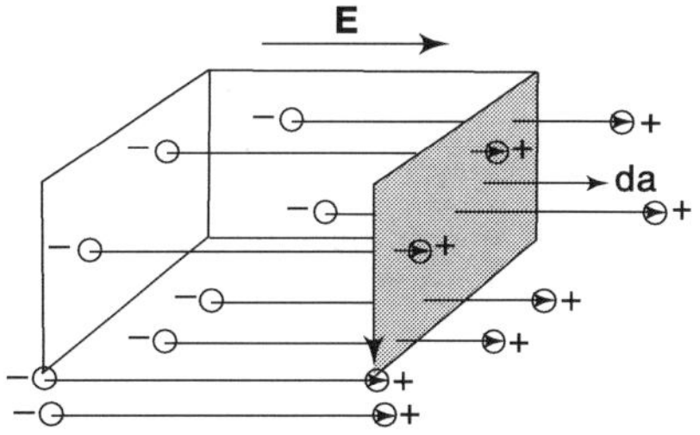

Fig. 3.15 *Illustration of bound charge densities*

Bound surface charge density σ_b: We consider a small volume inside the dielectric, where the electric field $\vec{E}$ is the resultant of an external field and the field due to the dipoles. The positive and negative charges are separated by an average distance l due to the influence of $\vec{E}$. Consider the element of surface dS and the charge which has crossed it. If we fix the origin in the negative charges we need consider only the movement of the positive charges. Then, the amount of charge dQ crossing dS is just the amount of positive charge within the volume $d\tau = \vec{l} \cdot d\vec{S}$

$$\therefore \qquad dQ = NQ\vec{l} \cdot d\vec{S} \qquad Q\vec{l} = \vec{p} \qquad NQ\vec{l} = \vec{P}$$

$$\therefore \qquad dQ = \vec{P} \cdot\cdot\, d\vec{S} \tag{3.44}$$

If dS is on the surface of the material, this charge accumulates there in a layer of thickness $\vec{l} \cdot \hat{n}$ (which is small, of molecular dimensions) and the charge can be treated as a surface layer with density

$$\boxed{\sigma_b = \frac{dQ}{dS} = \vec{P} \cdot \hat{n}} \tag{3.45}$$

Bound volume charge density ρ_b: The net charge flowing out of a volume τ across the elementary area dS of its surface is $\vec{P} .. d\vec{S}$ as found above in Eq. (3.44). Thus, the total charge flowing out of the surface bounding τ, is the integral of this over the surface, i.e.,

$$Q = \int_S \vec{P} \cdot d\vec{S}$$

and net charge remaining within is $-Q$.
If the density of this remaining charge is ρ_b, then

$$\int_v \rho_b d\tau = -Q = -\int_S \vec{P} \cdot d\vec{S} = -\int_v (\nabla \cdot \vec{P})\, d\tau$$

$$\therefore \qquad \boxed{\rho_b = -\nabla \cdot \vec{P}} \tag{3.46}$$

i.e., the bound charge density is numerically equal to minus the divergence of the polarisation vector. An important consequence of this is that if the polarisation is uniform within a region and its divergence is zero, then so is the bound volume charge density ρ_b.

Equations (3.45) and (3.46) show that where polarisation occurs, an equivalent volume charge density is formed throughout the dielectric while an equivalent surface charge density is formed over the surface of the dielectric.

Example 3.28 A dielectric cube of side L and centre at the origin has a polarisation vector given as, $\vec{P} = x\hat{i} + y\hat{j} + z\hat{k}$. Find the volume and surface bound charge densities and show explicitly that the total bound charge vanishes in this case.

Solution The bound surface charge density is, $\sigma_b = \vec{P} \cdot \hat{n}$. For each of the six sides of the cube, there exists a surface charge density. For the side located at $x = L/2$, the surface charge density is

$$\sigma_{b1} = \vec{P} \cdot \hat{i}\Big|_{x=L/2} = x\Big|_{x=L/2} = \frac{L}{2}$$

$\therefore$ The total bound surface charge is

$$Q_{bs} = \int_S \sigma dS = 6 \int_{-L/2}^{L/2} \int_{-L/2}^{L/2} \sigma dydz = \frac{6L}{2} L^2 = 3L^3$$

The bound volume charge density is

$$\rho_b = -\nabla \cdot \vec{P} = -(1+1+1) = -3$$

∴ The total bound volume charge is

$$Q_{bv} = \int_v \rho dv = -3\int dv = -3L^3$$

Hence, the total bound charge in the cube is

$$Q = Q_{bs} + Q_{bv} = 3L^3 - 3L^3 = 0$$

Thus, we see that the total bound charge vanishes in this case.

3.5.4 Electric Displacement, Susceptibility, Permittivity, Dielectric Constant and Dielectric Strength

Electric Displacement $(\vec{D})$ By Gauss' law in free space we know

$$\nabla \cdot \vec{E} = \frac{\rho_v}{\varepsilon_0}$$

where ρ_v is the total volume charge density = free volume charge density + bound volume charge density = $\rho_f + \rho_b$.

$$\therefore \quad \nabla \cdot \vec{E} = \frac{\rho_f + \rho_b}{\varepsilon_0} = \frac{\rho_f}{\varepsilon_0} + \frac{-\nabla \cdot \vec{P}}{\varepsilon_0}$$

$$\Rightarrow \quad \nabla \cdot \left(\vec{E} + \frac{\vec{P}}{\varepsilon_0}\right) = \frac{\rho_f}{\varepsilon_0}$$

$$\Rightarrow \quad \nabla \cdot (\varepsilon_0 \vec{E} + \vec{P}) = \rho_f$$

$$\therefore \quad \boxed{\nabla \cdot \vec{D} = \rho_f} \tag{3.47}$$

where $\vec{D}$ is a new vector field which has the same dimension as $\vec{P}$ and is known as *electric displacement*, given as

$$\boxed{\vec{D} = \varepsilon_0 \vec{E} + \vec{P}} \tag{3.48}$$

Equation (3.47) is *the modified Gauss' law*, which includes the effect of polarisation charges. In integral form it becomes

$$\int_S \vec{D} \cdot d\vec{S} = \int_v \rho_f dv \tag{3.49}$$

NOTE

(i) Equation (3.47) and (3.49) do not contain the permittivity ε and is thus independent of the medium.

(ii) Both div$\vec{D}$ and $\int_S \vec{D} \cdot d\vec{S}$ are unaffected by bound charges.

(iii) *Lines of $\vec{D}$ begin and end only on free charges.*

(iv) *Lines of $\vec{E}$ begin and end on either free or bound charges.*

(v) *In writing down expressions for the divergence of $\vec{E}$ and $\vec{P}$ we have implicitly assumed their existence. It should be noted that the space derivatives do not exist at a point charge or at the interface between two media. In such cases, the integral form of Gauss' law must be used.*

Electric Susceptibility (χ_e) The strength of polarisation $\vec{P}$ is directly proportional to the applied field $\vec{E}$.

$$\therefore \qquad \vec{P} \propto \vec{E} \quad \Rightarrow \quad \vec{P} = \text{constant} \times \vec{E} = \varepsilon_0 \chi_e \vec{E}$$

where the constant of proportionality is usually written as $\varepsilon_0\chi_e$ and χ_e is known as the *electric susceptibility* of the material.

Electric Susceptibility (χ_e): *The electric susceptibility χ_e of a dielectric material is a measure of how easily it is polarised in response to an electric field.*

It is roughly a measure of how susceptible (or sensitive) a dielectric is to an electric field.

Electric Permittivity (ε) and Dielectric Constant (ε_r) *Permittivity* is a quantity that describes how an electric field affects and/or is affected by an insulating medium.

The *dielectric constant* or *relative permittivity* is the ratio of the permittivity of a material to the permittivity of free space. It is a measure of the extent to which it concentrates electric flux lines. It is the ratio of permittivity of the dielectric to that of the free space.

From Eq. (3.48), substituting the value of polarisation in terms of applied field, we get

$$\vec{D} = \varepsilon_0 \vec{E} + \varepsilon_0 \chi_e \vec{E} = \varepsilon_0 (1 + \chi_e) \vec{E} = \varepsilon_0 \varepsilon_r \vec{E}$$

$$\vec{D} = \varepsilon \vec{E}$$

where
$$\boxed{\varepsilon = \varepsilon_0 \varepsilon_r}$$

and
$$\boxed{\varepsilon_r = (1 + \chi_e)}$$

Here, ε is the *permittivity* of the dielectric

ε_0 is the permittivity of free space $= \dfrac{1}{36\pi \times 10^9} = 8.854 \times 10^{-12}$ F/m, and

$\varepsilon_r = \dfrac{\varepsilon}{\varepsilon_0}$ is the *relative permittivity* or *dielectric constant* of the medium.

Dielectric Strength The maximum electric field that a dielectric material can withstand without breakdown is known as *dielectric strength* of the material.

If the applied field exceeds this value, the dielectric material starts conducting. This phenomenon of conduction in a dielectric material is termed as *dielectric breakdown*. The phenomenon depends on several factors like the nature of the materials, magnitude and time of application of the applied field, temperature, humidity, etc.

Example 3.29 Find the polarisation in a dielectric material with relative permittivity of 2.8 if the electric flux density is given as, $\vec{D} = 3 \times 10^{-7}$ C/m^2.

Solution The polarisation is given as

$$\vec{P} = \chi_e \varepsilon_0 \vec{E} = (\varepsilon_r - 1)\varepsilon_0 \vec{E} = (\varepsilon_r - 1)\varepsilon_0 \frac{\vec{D}}{\varepsilon_0 \varepsilon_r} \quad \{\because \varepsilon_r = (1 + \chi_e)\}$$

$$= \left(1 - \frac{1}{\varepsilon_r}\right)\vec{D}$$

$$= \left(1 - \frac{1}{2.8}\right) \times 3 \times 10^{-7}$$

$$= 1.93 \times 10^{-7} \text{ C/m}$$

Example 3.30 In a dielectric material, $E_x = 5$ V/m and $\vec{P} = \frac{1}{10\pi}(3\hat{i} - \hat{j} + 4\hat{k})$ nC/m^2. Calculate:

(a) χ_e, **(b)** $\vec{E}$ and **(c)** $\vec{D}$

Solution Here, $E_x = 5$ V/m and $\vec{P} = \frac{1}{10\pi}(3\hat{i} - \hat{j} + 4\hat{k})$ nC/m^2

(a) The polarisation is given as

$$\vec{P} = \chi_e \varepsilon_0 \vec{E}$$

$$\therefore \qquad \chi_e = \frac{\vec{P}}{\varepsilon_0 \vec{E}}$$

Considering only the x component

$$\chi_e = \frac{P}{\varepsilon_0 E} = \frac{3}{10\pi} \times 10^{-9} \times \frac{1}{\frac{10^{-9}}{36\pi} \times 5} = \frac{54}{25} = 2.16$$

(b) The term $\chi_e \varepsilon_0 = \frac{54}{25} \times \frac{10^{-9}}{36\pi} = \frac{3}{50\pi} \times 10^{-9}$

Hence, the electric field is

$$\vec{E} = \frac{\vec{P}}{\chi_e \varepsilon_0} = \frac{1}{10\pi}(3\hat{i} - \hat{j} + 4\hat{k}) \times 10^{-9} \times \frac{1}{\frac{3}{50\pi} \times 10^{-9}} = \left(5\hat{i} - \frac{5}{3}\hat{j} + \frac{20}{3}\hat{k}\right)$$

$$\therefore \qquad \vec{E} = (5\hat{i} - 1.67\hat{j} + 6.67\hat{k}) \text{ V/m}$$

(c) The electric flux density is given as

$$\vec{D} = \varepsilon_0 \vec{E} + \vec{P} = \frac{10^{-9}}{36\pi}(5\hat{i} - 1.67\hat{j} + 6.67\hat{k}) + \frac{10^{-9}}{10\pi}(3\hat{i} - \hat{j} + 4\hat{k})$$

$$= (139.7\hat{i} - 46.6\hat{j} + 186.3\hat{k}) \text{ pC/m}^2$$

3.5.5 Linear, Homogeneous, Isotropic Dielectrics

A dielectric material is said to be *linear* if $\vec{D}$ varies linearly with $\vec{E}$ and *non-linear* otherwise.

A dielectric material is said to be *homogeneous* if the permittivity (ε) or conductivity (σ) does not vary with space in a region and is thus, the same at all points in the region. The materials for which ε or σ is dependent on the space coordinates, is said to be *inhomogeneous*. The atmosphere is an example of an inhomogeneous medium.

A dielectric material is said to be *isotropic* if the electrical properties of the medium are independent of the direction, i.e., $\vec{D}$ and $\vec{E}$ are in the same direction. A material is *anisotropic* if $\vec{D}$, $\vec{E}$ and $\vec{P}$ are not parallel. For an anisotropic medium, ε or χ_e has nine components, collectively referred to as *tensor.* For an anisotropic medium, the relation between $\vec{D}$ and $\vec{E}$ is as given below.

$$\begin{bmatrix} D_x \\ D_y \\ D_z \end{bmatrix} = \begin{bmatrix} \varepsilon_{xx} & \varepsilon_{xy} & \varepsilon_{xz} \\ \varepsilon_{yx} & \varepsilon_{yy} & \varepsilon_{yz} \\ \varepsilon_{zx} & \varepsilon_{zy} & \varepsilon_{zz} \end{bmatrix} \begin{bmatrix} E_x \\ E_y \\ E_z \end{bmatrix} \quad \textbf{(3.50)}$$

3.6 BEHAVIOUR OF SEMICONDUCTOR MATERIALS IN ELECTRIC FIELD

If an isolated semiconductor is placed in an external electric field, the motion of free electrons will produce some electric field that will cancel the external applied field. Hence, under steady state equilibrium, the net electric field within an isolated semiconductor is zero.

Thus, from the electric point of view, the behaviour of a semiconductor is same as that of a conductor, as discussed in Section 3.4.

PART II

Magnetostatic Field

3.7 INTRODUCTION

In Chapter 2, we came to know that electric fields are created by the presence of static charges. However, if the charges move with a constant velocity, a *static magnetic field* (or *magnetostatic*) field is produced. In fact, a magnetostatic field is produced by a steady current which may be due to different sources, such as conduction currents as in a current carrying conductor, or magnetisation current as in permanent magnets.

In this part of the chapter, we have discussed the basic laws related with magnetostatics both in free space and in material space. We have also discussed the concepts of inductances and magnetic energy in this part of the chapter.

3.8 BIOT–SAVART LAW

Statement This law states that the magnetic field intensity $d\vec{H}$ produced at a point P due to a differential current element Idl is:

(i) directly proportional to the product of current I and differential length dl.

(ii) directly proportional to the sine of the angle between the element and the line joining P to the element, and

(iii) inversely proportional to the square of the distance r.

Explanation We consider an elemental current carrying conductor. Fig. 3.16 demmonstrates this.

Let,

dl elemental length of a current carrying conductor,

I steady current carried by the conductor,

r distance of the field point (P) from the elemental length,

$d\vec{H}$ magnetic field produced by the elemental length at the field point (P).

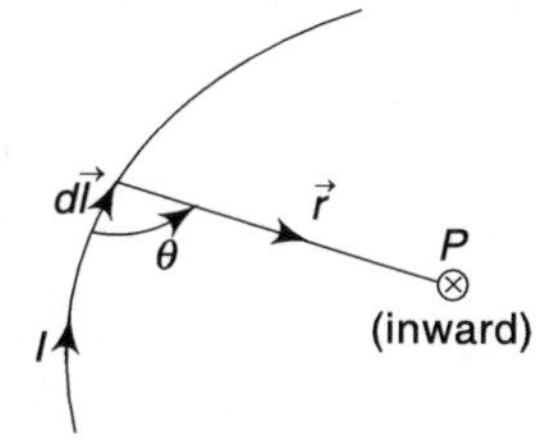

Fig. 3.16 *Magnetic field due to a steady current*

It is experimentally found that

$$\therefore \quad d\vec{H} \; \alpha \; Idl \quad d\vec{H} \; \alpha \; \sin\theta \quad \text{and} \quad d\vec{H} \alpha \frac{1}{r^2}$$

Combining

$$\therefore \quad d\vec{H} \; \alpha \frac{Idl \sin\theta}{r^2}$$

$$\therefore \quad d\vec{H} = k \frac{Idl \sin\theta}{r^2}$$

$$= \frac{Idl \sin\theta}{4\pi r^2} \quad \textbf{(3.51)}$$

where, k is the constant of proportionality. In SI unit, $k = \dfrac{1}{4\pi}$. Now, the product $Idl \sin\theta$ in vector form can be written as, $Id\vec{l} \times \hat{a}_r$, so that Eq. (3.51) becomes

$$\boxed{d\vec{H} = \frac{Id\vec{l} \times \hat{a}_r}{4\pi r^2} = \frac{Id\vec{l} \times \vec{r}}{4\pi r^3}} \quad \textbf{(3.52)}$$

where, $r = |\vec{r}|$ and $\vec{r} = r\hat{a}_r$. Hence, the total field produced by the conductor can be written as

$$\boxed{\vec{H} = \int_l \frac{Id\vec{l} \times \hat{a}_r}{4\pi r^2} = \int_l \frac{Id\vec{l} \times \vec{r}}{4\pi r^3}} \quad \textbf{(3.53)}$$

Equation (3.53) is the mathematical form of the Biot–Savart law. The direction of the field intensity is found by the *right-hand cork-screw rule* or *right-hand thumb rule* as illustrated in Fig. 3.17 (*a*) and Fig. 3.17 (*b*). As per *right-hand cork-screw rule*, we imagine a right-handed corkscrew being rotated along the wire in the direction of the current. The direction of rotation of the thumb gives the direction of the magnetic field. As per *right-hand thumb rule*, we imagine that we are holding the conductor in our right hand with the fingers curled around it. If the thumb points in the direction of the current, then the curled fingers show the direction of the magnetic field.

Similar to the different charge densities discussed in Chapter 2, we can have three types of current density distribution given as:

(i) line current density, I, given in Ampere,

(ii) surface current density, $\vec{K}$, given in Ampere/metre (A/m), and

(iii) volume current density, $\vec{J}$, given in Ampere/square metre (A/m^2)

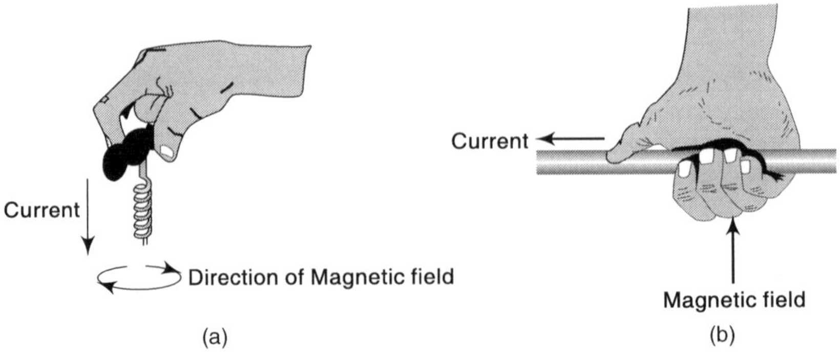

Fig. 3.17 *(a) Right-hand cork-screw rule (b) Right hand thumb rule*

These current densities are related to each other as

$$Id\vec{l} \equiv \vec{K}dS \equiv \vec{J}dv \tag{3.54}$$

In terms of these current densities, the Biot–Savart law can be written as

$$\boxed{\vec{H} = \int_l \frac{Id\vec{l} \times \hat{a}_r}{4\pi r^2} = \int_S \frac{\vec{K}dS \times \hat{a}_r}{4\pi r^2} = \int_v \frac{\vec{J}dv \times \hat{a}_r}{4\pi r^2}} \tag{3.55}$$

***Example 3.31** An infinitely long wire of negligible cross-section is carrying a current I, as shown in Fig. 3.18. Find the magnetic induction due to this current at a point, which is 'r' m away from the wire.

Or

(a) Derive an expression for the magnetic field due to a current in a straight wire of finite length.
(b) Find the magnetic field due to an infinitely long straight conductor carrying a steady current.

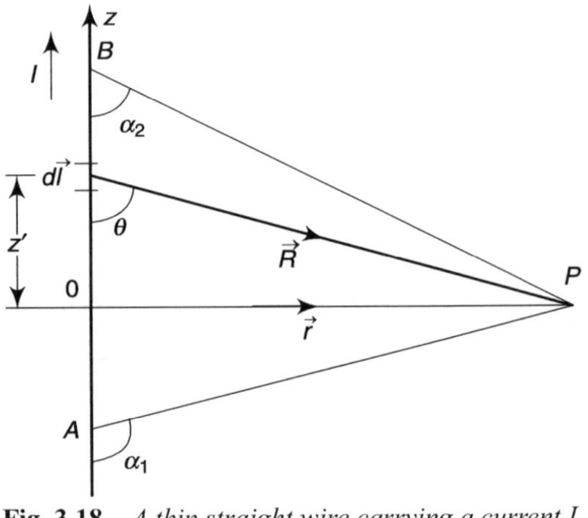

Fig. 3.18 *A thin straight wire carrying a current I*

Solution We consider a differential element $d\vec{l}$ carrying current I in the z-direction. The location of this source is represented by $dz'\hat{a}_z$.

The field point is located at some point $P(r, 0)$, given by the position vector, $r\hat{a}_r$.

The relative position vector which points from the source point to the field point is

$$\vec{R} = r\hat{a}_r - z'\hat{a}_z \quad \therefore \quad |\vec{R}| = \sqrt{r^2 + z'^2}$$

The product $d\vec{l} \times \vec{R} = (dz'\hat{a}_z) \times (r\hat{a}_r - z'\hat{a}_z) = rdz'\hat{a}_\phi$

Applying the Biot–Savart law, the magnetic field intensity due to this element is

$$d\vec{H} = \frac{Id\vec{l} \times \vec{R}}{4\pi R^3} = \frac{Irdz'}{4\pi(r^2 + z'^2)^{3/2}}\hat{a}_\phi$$

Hence, the total magnetic field intensity due to the entire wire is

$$\vec{H} = \int d\vec{H} = \int_A^B \frac{Irdz'}{4\pi(r^2 + z'^2)^{3/2}}\hat{a}_\phi$$

$$= \int_{\alpha_1}^{\alpha_2} \frac{-Irr\,\text{cosec}^2\theta\, d\theta}{4\pi r^3 \text{cosec}^3\theta}\hat{a}_\phi \qquad \{\text{Put, } z' = r\cot\theta,\ \therefore dz' = -r\,\text{cosec}^2\theta\, d\theta\}$$

$$\vec{H} = -\frac{I}{4\pi r}\int_{\alpha_1}^{\alpha_2} \sin\theta\, d\theta\hat{a}_\phi = -\frac{I}{4\pi r}(\cos\alpha_1 - \cos\alpha_2)\hat{a}_\phi = \frac{I}{4\pi r}(\cos\alpha_2 - \cos\alpha_1)\hat{a}_\phi$$

$$\therefore \qquad \boxed{\vec{H} = \frac{I}{4\pi r}(\cos\alpha_2 - \cos\alpha_1)\hat{a}_\phi}$$

We shall consider the following cases:

1. In the symmetric case where $\alpha_1 = (\pi - \alpha_2)$, the field point P is located along the perpendicular bisector. If the length of the wire is $2L$, then $\cos\alpha_2 = -\cos\alpha_1 = \frac{L}{\sqrt{L^2 + r^2}}$ and the magnetic field is given as

$$\vec{H} = \frac{I}{2\pi r}\frac{L}{\sqrt{L^2 + r^2}}\hat{a}_\phi$$

2. For a semi-infinite wire, $\alpha_1 = 90°$, $\alpha_2 = 0°$ and the magnetic field is given as

$$\vec{H} = \frac{I}{4\pi r}\hat{a}_\phi$$

3. For an infinite wire, $\alpha_1 = 180°$, $\alpha_2 = 0°$ and the magnetic field is given as

$$\vec{H} = \frac{I}{2\pi r}\hat{a}_\phi$$

In fact, in this case, the system possesses cylindrical symmetry, and the magnetic field lines are circular, as shown in Fig. 3.19. The direction of the magnetic field due to a long straight wire can be determined by the right-hand rule.

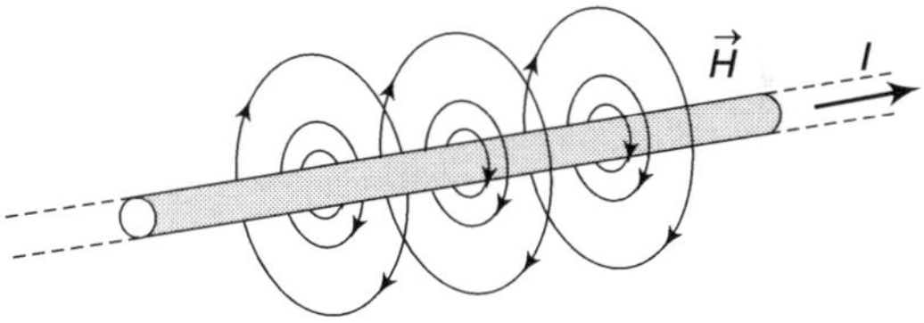

Fig. 3.19 *Magnetic field lines due to an infinite wire carrying current I*

***Example 3.32** A circular loop of radius r in the xy plane carries a steady current I, as shown in Fig. 3.20. What is the magnetic field at a point P on the axis of the loop, at a distance z from the centre?

Solution We consider a differential length on the wire, with the position vector, $\vec{dl} = r\,d\phi\hat{a}_\phi$.

The field point P is on the axis of the loop at a distance z from the centre; its position vector is given by $z\hat{a}_z$.

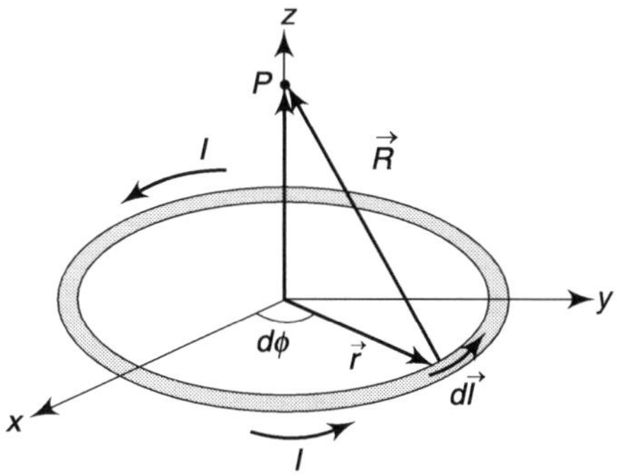

Fig. 3.20 *Magnetic field due to a circular loop carrying a steady current*

The relative position vector which points from the source point to the field point is

$$\vec{R} = -r\hat{a}_r + z\hat{a}_z$$

$$\therefore \qquad |\vec{R}| = \sqrt{r^2 + z^2}$$

The product

$$d\vec{l} \times \vec{R} = \begin{vmatrix} \hat{a}_r & \hat{a}_\phi & \hat{a}_z \\ 0 & rd\phi & 0 \\ -r & 0 & z \end{vmatrix} = rzd\phi\hat{a}_r + r^2 d\phi\hat{a}_z$$

Applying the Biot–Savart law, the magnetic field intensity due to this element is

$$d\vec{H} = \frac{Id\vec{l} \times \vec{R}}{4\pi R^3} = \frac{I}{4\pi(r^2 + z^2)^{3/2}}(rzd\phi\hat{a}_r + r^2 d\phi\hat{a}_z)$$

By symmetry, the contribution along the $\hat{a}_r$ add up to zero because the radial components produced by pairs of current elements 180° apart cancel each other.

$$\therefore \qquad H_r = 0$$

$$\therefore \qquad \vec{H} = \int dH_z\hat{a}_z = \int_0^{2\pi} \frac{Ir^2 d\phi}{4\pi(r^2 + z^2)^{3/2}}\hat{a}_z = \frac{Ir^2}{2(r^2 + z^2)^{3/2}}\hat{a}_z$$

$$\therefore \qquad \boxed{\vec{H} = \frac{Ir^2}{2(r^2 + z^2)^{3/2}}\hat{a}_z}$$

Example 3.33 A square loop of side $2a$ lies in the $z = 0$ plane and carries a current I in the anti-clockwise direction. Show that at the centre of the loop

$$\vec{H} = \frac{\sqrt{2}I}{\pi a}\hat{a}_z$$

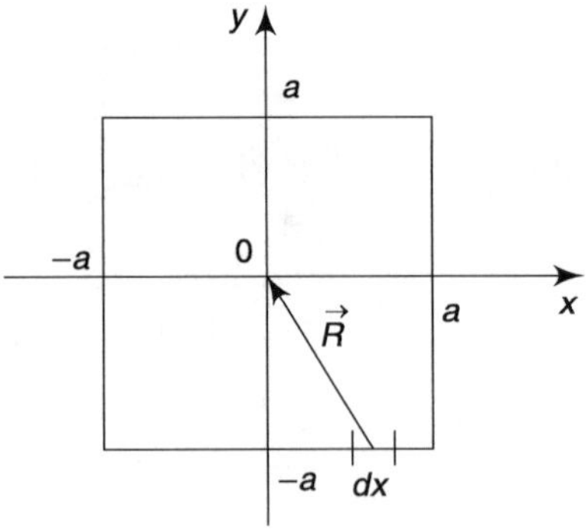

Fig. 3.21 *Square loop carrying current*

Solution Choosing a Cartesian coordinate system as shown in Fig. 3.21, by symmetry, each half-side contributes the same amount to $\vec{H}$ at the centre.

For half-side, $0 \le x \le a$, and $y = -a$, by the Biot-Savart law, we get

$$d\vec{H} = \frac{Id\vec{l} \times \vec{R}}{4\pi R^3} = \frac{(Idx\hat{a}_x) \times [-x\hat{a}_x + a\hat{a}_y]}{4\pi[x^2 + a^2]^{3/2}}$$

$$= \frac{Iadx\hat{a}_z}{4\pi[x^2 + a^2]^{3/2}}$$

So, the total field at the origin due to all the four sides is

$$\vec{H} = 8 \times \int d\vec{H} = 8\int_0^a \frac{Iadx\hat{a}_z}{4\pi[x^2 + a^2]^{3/2}}$$

$$= \frac{2}{\pi} Ia \int_0^a \frac{dx}{[x^2 + a^2]^{3/2}} \hat{a}_z$$

$$= \frac{2}{\pi} Ia \int_0^{\pi/4} \frac{a\sec^2\theta d\theta}{(a^2\tan^2\theta + a^2)^{3/2}} \hat{a}_z \qquad \left\{\begin{array}{ccc} x = a\tan\theta; & \therefore dx = a\sec^2\theta d\theta \\ x & 0 & a \\ \theta & 0 & \pi/4 \end{array}\right\}$$

$$= \frac{2I}{\pi a} \int_0^{\pi/4} \cos\theta \, d\theta \, \hat{a}_z$$

$$= \frac{2I}{\pi a} (\sin\theta)_0^{\pi/4} \hat{a}_z = \frac{2I}{\pi a} \times \frac{1}{\sqrt{2}} \hat{a}_z = \frac{\sqrt{2}I}{\pi a} \hat{a}_z$$

$$\therefore \qquad \boxed{\vec{H} = \frac{\sqrt{2}I}{\pi a} \hat{a}_z}$$

Example 3.34 Derive an expression for the magnetising force at any point on the axis of a single-turn coil carrying a steady current 'I':

(a) When the coil is in the form of a circle of radius 'r',
(b) When the coil is in the form of a rectangle of sides a and b,
(c) When it is square with sides of length '$2a$'.

Solution This is shown in Fig. 3.22.

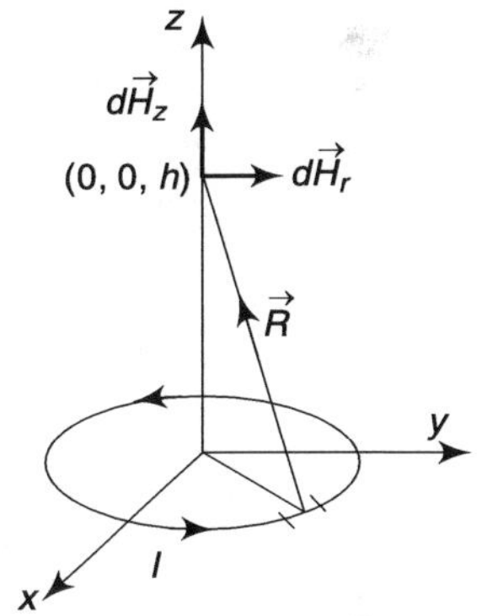

Fig. 3.22 *Circular current carrying loop*

(a) By the Biot–Savart Law,

$$d\vec{H} = \frac{Id\vec{l} \times \vec{R}}{4\pi R^3}$$

$$d\vec{l} = rd\phi\hat{a}_\phi;$$

$$\vec{R} = (0,0,h) - (x,y,0) = -x\hat{a}_x - y\hat{a}_y h\hat{a}_z = -r\hat{a}_r + h\hat{a}_z$$

$$\therefore \qquad d\vec{l} \times \vec{R} = \begin{vmatrix} \hat{a}_r & \hat{a}_\phi & \hat{a}_z \\ 0 & rd\phi & 0 \\ -r & 0 & h \end{vmatrix} = (rhd\phi\,\hat{a}_r + r^2 d\phi\,\hat{a}_z)$$

$$\therefore \qquad d\vec{H} = \frac{I}{4\pi(r^2 + h^2)^{3/2}} (rhd\phi\,\hat{a}_r + r^2 d\phi\,\hat{a}_z)$$

$$= dH_r\hat{a}_r + dH_z\hat{a}_z$$

By symmetry, the contributions along $\hat{a}_r$ add up to zero because the radial components produced by pairs of current elements 180° apart cancel out.

$$\therefore \qquad H_r = 0$$

Hence, the magnetic field is given as

$$\vec{H} = \int dH_z \hat{a}_z = \int_0^{2\pi} \frac{Ir^2 d\phi}{4\pi[r^2 + h^2]^{3/2}} \hat{a}_z = \frac{Ir^2}{2[r^2 + h^2]^{3/2}} \hat{a}_z$$

$$\boxed{\vec{H} = \frac{Ir^2}{2[r^2 + h^2]^{3/2}} \hat{a}_z}$$

NOTE

At the centre ($h = 0$), the field is, $\vec{H} = \frac{I}{2r}\hat{a}_z$

(b) Let P be a point at a height h above the plane of the loop.
The magnetic field due to the side AB is given as

$$dB_{AB} = \frac{\mu I}{4\pi r}(\cos \alpha_1 - \cos \alpha_2)$$

$$= \frac{\mu I}{4\pi r}[\cos \theta - \cos(\pi - \theta)]$$

$$= \frac{\mu I}{4\pi r} 2 \cos \theta$$

$$\therefore \quad dB_{AB} = \frac{\mu I}{2\pi r} \cos \theta$$

Similarly, the magnetic field due to the side CD is given as

$$dB_{CD} = \frac{\mu I}{2\pi r} \cos \theta$$

Since the flow of current in the two elements is in opposite direction, their cosine components will cancel each other and thus, only the axial components will add together.
Hence, the resultant field due to sides AB and CD is given as

$$dB_1 = (dB_{AB} + dB_{CD}) = \frac{\mu I}{2\pi r} \cos \theta \sin \alpha + \frac{\mu I}{2\pi r} \cos \theta \sin \alpha = \frac{\mu I}{\pi r} \cos \theta \sin \alpha$$

From Fig. 3.23, it is seen that $r = \sqrt{h^2 + \frac{b^2}{4}}$

$$\therefore \quad \cos \theta = \frac{a/2}{\sqrt{r^2 + \frac{a^2}{4}}} = \frac{a}{2\sqrt{h^2 + \frac{b^2}{4} + \frac{a^2}{4}}}$$

$$\therefore \quad \sin \alpha = \frac{b/2}{r} = \frac{b}{2\sqrt{h^2 + \frac{b^2}{4}}}$$

Putting these values, we get

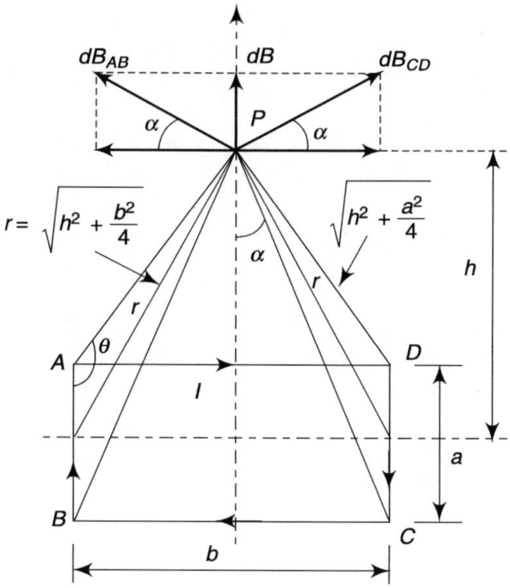

Fig. 3.23 *Rectangular current carrying loop*

$$dB_1 = \frac{\mu I}{\pi r}\cos\theta \sin\alpha = \frac{\mu I}{\pi r}\frac{a}{2\sqrt{h^2+\dfrac{b^2}{4}+\dfrac{a^2}{4}}}\frac{b}{2\sqrt{h^2+\dfrac{b^2}{4}}}$$

$$= \frac{\mu I ab}{4\pi\sqrt{h^2+\dfrac{b^2}{4}}\sqrt{h^2+\dfrac{b^2}{4}+\dfrac{a^2}{4}}\sqrt{h^2+\dfrac{b^2}{4}}}$$

$$= \frac{\mu I ab}{4\pi\left(h^2+\dfrac{b^2}{4}\right)\sqrt{h^2+\dfrac{b^2}{4}+\dfrac{a^2}{4}}}$$

Similarly, the magnetic field due to the other two sides BC and DA is given as

$$dB_2 = \frac{\mu I ab}{4\pi\left(h^2+\dfrac{a^2}{4}\right)\sqrt{h^2+\dfrac{b^2}{4}+\dfrac{a^2}{4}}}$$

Hence, total magnetic field due to the rectangular loop is given as

$$B = (dB_1 + dB_2) = \frac{\mu I ab}{4\pi\left(h^2+\dfrac{b^2}{4}\right)\sqrt{h^2+\dfrac{b^2}{4}+\dfrac{a^2}{4}}} + \frac{\mu I ab}{4\pi\left(h^2+\dfrac{a^2}{4}\right)\sqrt{h^2+\dfrac{b^2}{4}+\dfrac{a^2}{4}}}$$

$$= \frac{\mu I ab}{4\pi\sqrt{h^2+\dfrac{b^2}{4}+\dfrac{a^2}{4}}}\left[\frac{1}{h^2+\dfrac{a^2}{4}}+\frac{1}{h^2+\dfrac{b^2}{4}}\right]$$

$$B = \frac{\mu I ab}{4\pi\sqrt{h^2 + \frac{b^2}{4} + \frac{a^2}{4}}}\left(\frac{1}{h^2 + \frac{a^2}{4}} + \frac{1}{h^2 + \frac{b^2}{4}}\right)$$

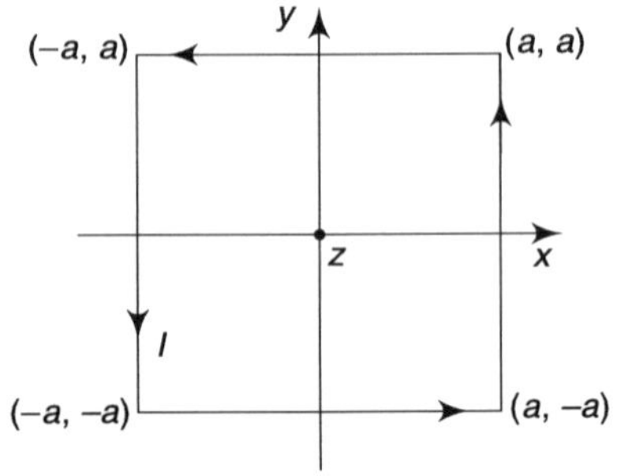

Fig. 3.24 *Square current carrying loop*

(c) Along the axis of the coil there will be only a z-component of magnetic field by symmetry. This is given in Fig. 3.24. In order to obtain the total field it is necessary to calculate only the z-component of the field generated by one side of the coil and then multiply by four. Consider the right-hand side.

Let $d\vec{l} = dy\hat{a}_y$ at (a, y)

The position of the element of length, $d\vec{l}$, is specified by the vector $\vec{r}$ where, $\vec{r} = a\hat{a}_x + y\hat{a}_y$.

The position of the point of observation along the z-axis is specified by the vector $\vec{R} = z\hat{a}_z$.

$$\therefore \quad (\vec{R} - \vec{r}) = -a\hat{a}_x - y\hat{a}_y + z\hat{a}_z$$

$$\therefore \quad d\vec{l} \times (\vec{R} - \vec{r}) = \begin{vmatrix} \hat{a}_x & \hat{a}_y & \hat{a}_z \\ 0 & dy & 0 \\ -a & -y & z \end{vmatrix} = zdy\hat{a}_x + ady\hat{a}_z$$

$$\therefore \quad |\vec{R} - \vec{r}| = \sqrt{a^2 + y^2 + z^2}$$

By the Biot–Savart law,

$$d\vec{H}_1 = \frac{Id\vec{l} \times \vec{R}}{4\pi R^3} = \frac{I}{4\pi(a^2 + y^2 + z^2)^{3/2}}(zdy\hat{a}_x + ady\hat{a}_z) = dH_x\hat{a}_x + dH_z\hat{a}_z$$

Since all the x components will add to zero, we have

$$d\vec{H}_1 = \frac{Id\vec{l} \times \vec{R}}{4\pi R^3} = \frac{Ia}{4\pi}\frac{dy}{(a^2 + y^2 + z^2)^{3/2}}\hat{a}_z$$

By integration

$$\vec{H}_1 = \int d\vec{H} = \frac{Ia}{4\pi}\int_{-a}^{a}\frac{dy}{(a^2 + y^2 + z^2)^{3/2}}\hat{a}_z = \frac{I}{4\pi}\frac{2a^2}{\pi(a^2 + b^2)\sqrt{2a^2 + z^2}}$$

As the coil has four sides, this must be multiplied by 4 to get the total magnetic field.

$$\therefore \quad \vec{H} = 4\vec{H}_1 = 4 \times \frac{I}{4\pi}\frac{2a^2}{(a^2 + z^2)\sqrt{2a^2 + z^2}}\hat{a}_z = \frac{2a^2 I}{\pi(a^2 + z^2)\sqrt{2a^2 + z^2}}\hat{a}_z$$

$$\therefore \quad \boxed{\vec{H} = \frac{2a^2 I}{\pi(a^2 + z^2)\sqrt{2a^2 + z^2}}\hat{a}_z}$$

Alternative Method: From the result of part (b), if $a = b = 2a$, then the result becomes

$$H = \frac{I2a2a}{4\pi\sqrt{h^2 + \frac{(2a)^2}{4} + \frac{(2a)^2}{4}}}\left(\frac{1}{h^2 + \frac{(2a)^2}{4}} + \frac{1}{h^2 + \frac{(2a)^2}{4}}\right) = \frac{2a^2 I}{\pi(a^2 + z^2)\sqrt{2a^2 + z^2}}$$

NOTE

At the centre (z = 0), $\vec{H} = \frac{\sqrt{2}I}{\pi a}\hat{a}_z = 0.9003\frac{I}{2a}\hat{a}_z$ as obtained in Example 3.33. This result can be compared with $\vec{H} = \frac{I}{2r}\hat{a}_z$ for a circular coil as obtained in Example 3.34(a).

***Example 3.35** A circular loop located on $x^2 + y^2 = 9, z = 0$ carries a direct current of 10 A along $\hat{a}_\phi$. Determine $\vec{H}$ at:

(i) (0, 0, 4) and **(ii)** (0, 0, −4)

Solution From Example 3.32, substituting the values

(i) $I = 10\text{A}, r = 3, h = 4,$

$$\vec{H}(0,0,4) = \frac{10 \times 3^2}{2[3^2 + 4^2]^{3/2}}\hat{a}_z = 0.36\hat{a}_z$$

(ii) $I = 10\text{A}, r = 3, h = -4,$

$$\vec{H}(0,0,-4) = \frac{10 \times 3^2}{2[3^2 + (-4)^2]^{3/2}}\hat{a}_z = 0.36\hat{a}_z$$

3.9 MAGNETIC FIELD INTENSITY ($\vec{H}$)

The magnetic field at a given point is specified by both a *direction* and a *magnitude* and is described by *magnetic field intensity,* $\vec{H}$.

The magnetic field intensity, $\vec{H}$, at any point is defined as the force experienced by a north pole of one Weber placed at that point. In other words, it is defined as the magnetomotive force per unit length produced by the steady current in a magnetic circuit.

Its unit is Newton per Weber (N/Wb) or Ampere per metre (A/m) or Ampere-turn per metre (AT/m).

3.10 MAGNETIC FLUX (ϕ) AND FLUX DENSITY ($\vec{B}$)

Magnetic flux is defined as the group of magnetic field lines emitted outward from the north pole of a magnet. It is measured in Weber and is denoted as ϕ.

Magnetic flux density ($\vec{B}$) is the amount of magnetic flux per unit area of a section, perpendicular to the direction of magnetic flux; i.e.,

$$B = \frac{\phi}{A} \tag{3.56}$$

Magnetic flux density ($\vec{B}$), also known as the *magnetic induction*, is a vector quantity. The unit of magnetic flux density is Weber per square metre or Tesla (T).

In general, for any arbitrarily shaped surface placed in a magnetic field with flux density vector $\vec{B}$, the total flux coming out of the surface is given as

$$\phi = \int_S \vec{B} \cdot d\vec{S} = \int_S \vec{B} \cdot \hat{a}_n dS \tag{3.57}$$

where, $\hat{a}_n$ is the unit vector normal to the surface.

The magnetic field intensity is related to the magnetic flux density as

$$\vec{B} = \mu \vec{H} \quad \textbf{(3.58)}$$

where, μ is a constant, called *permeability* of the medium. It is given as

$$\mu = \mu_0 \mu_r \quad \textbf{(3.59)}$$

where, μ_0 is the permeability of free space, known as *absolute permeability*, $= 4\pi \times 10^{-7}$ H/m
μ_r is the relative permeability.

NOTE

***Weber** is the unit of magnetic flux.*
***Tesla** is the unit of magnetic flux density.*
*The **Maxwell**, abbreviated as Mx, is the compound derived CGS unit of magnetic flux.*

1 Weber = 10^8 Maxwells = 10^8 magnetic field lines

3.11 GAUSS' LAW OF MAGNETOSTATIC INTERPRETATION OF DIVERGENCE OF MAGNETIC FIELD—MAXWELL'S EQUATIONS

The magnetic flux through any surface is the surface integral of the normal component of $\vec{B}$, i.e.,

$$\boxed{\phi = \int_S \vec{B} \cdot d\vec{S}} \quad \textbf{(3.60)}$$

As it is not possible to have an isolated magnetic pole, magnetic flux lines always close upon themselves.

Thus, the total magnetic flux through a closed surface must be zero.

$$\therefore \quad \oint_S \vec{B} \cdot d\vec{S} = 0 \quad \textbf{(3.61)}$$

This equation is known as *the law of conservation of magnetic flux* or *the integral form of Gauss' law of magnetostatic fields*.

Applying divergence theorem, we get

$$\oint_S \vec{B} \cdot d\vec{S} = \oint_v \nabla \cdot \vec{B} dv = 0$$

$$\therefore \quad \boxed{\nabla \cdot \vec{B} = 0} \quad \textbf{(3.62)}$$

This is known as the *differential form of Gauss' law of magnetostatic fields* which shows that magnetostatic fields have no sources or sinks and the magnetic field lines are continuous.

Example 3.36 The flux density at a point distance '*r*' from a long filamentary conductor is given by

$$\vec{B} = \frac{\mu_0 I}{2\pi r}\hat{a}_\phi$$

Find the flux crossing the portion of the conductor in the plane $\phi = \frac{\pi}{4}$ defined by $0.01 < r < 0.05$ m and $0 < z < 2$ m for a current of 2 A.

Solution The magnetic flux through any surface is given as

$$\phi = \int_S \vec{B}\cdot d\vec{S} = \int_{z=0}^{2}\int_{r=0.01}^{0.05}\left(\frac{\mu_0 I}{2\pi r}\hat{a}_\phi\right)\cdot(drdz\hat{a}_\phi) = \frac{\mu_0 I}{2\pi}\times 2\int_{r=0.01}^{0.05}\frac{dr}{r} = \frac{\mu_0 I}{\pi}\ln\left(\frac{0.05}{0.01}\right)$$

$$= \frac{\mu_0 I}{\pi}\ln 5 = 0.5123\mu_0 I \text{ Wb}$$

Example 3.37

(a) A radial field, $\vec{H} = \frac{2.39\times 10^6}{r}\cos\phi\,\hat{a}_r$ A/m, exists in free space. Find the magnetic flux ϕ crossing the surface defined by $-\pi/4 \le \phi \le \pi/4$, $0 \le z \le 1$ m.

(b) Compute the total magnetic flux ϕ crossing the $z = 0$ plane in cylindrical coordinates for $r \le 5\times 10^{-2}$ m if, $\vec{B} = \frac{0.2}{r}\sin^2\phi\,\hat{a}_r$ (T).

Solution

(a) $\vec{H} = \frac{2.39\times 10^6}{r}\cos\phi\,\hat{a}_r$

Magnetic flux crossing the given surface is given as

$$\phi = \int \vec{B}\cdot d\vec{S} = \int \mu_0\vec{H}\cdot d\vec{S} = \int_0^1\int_{-\pi/4}^{\pi/4}\left(\frac{2.39\times 10^6}{r}\mu_0\cos\phi\,\hat{a}_r\right)\cdot(rd\phi dz\hat{a}_r)$$

$$= \int_0^1\int_{-\pi/4}^{\pi/4}\frac{2.39\times 10^6}{r}\times 4\pi\times 10^{-7}\times r\cos\phi d\phi\, dz$$

$$= \int_0^1\int_{-\pi/4}^{\pi/4} 3\cos\phi\, d\phi\, dz$$

$$= 3[\sin\phi]_{-\pi/4}^{\pi/4} = 3\sqrt{2} = 4.24 \text{ Wb}$$

(b) $\vec{B} = \frac{0.2}{r}\sin^2\phi\,\hat{a}_r$

Magnetic flux crossing the given surface is given as

$$\phi = \int \vec{B}\cdot d\vec{S} = \int_0^{5\times 10^{-2}}\int_0^{2\pi}\left(\frac{0.2}{r}\sin^2\phi\,\hat{a}_z\right)\cdot(rdrd\phi\hat{a}_z)$$

$$= 0.2\int_0^{5\times 10^{-2}}\int_0^{2\pi}\sin^2\phi\, drd\phi = 0.2\times 5\times 10^{-2}\int_0^{2\pi}\sin^2 d\phi = 10^{-2}\int_0^{2\pi}\frac{1-\cos 2\phi}{2}d\phi$$

$$= 10^{-2}\left[\frac{\phi}{2} - \frac{\sin 2\phi}{4}\right]_0^{2\pi} = 10^{-2}[\pi - 0] = 3.14\times 10^{-2} \text{ Wb}$$

3.12 AMPERE'S CIRCUITAL LAW—INTERPRETATION OF CURL OF MAGNETIC FIELD

Statement This law states that the line integral of the magnetic field intensity ($\vec{H}$) around any closed path is equal to the direct current enclosed by the path.

$$\therefore \qquad \boxed{\oint_l \vec{H} \cdot d\vec{l} = I_{\text{enc}}} \qquad (3.63)$$

Proof We consider an infinite straight conductor carrying a steady current I along the z-direction as shown in Fig. 3.25. We consider a closed circular path of radius r enclosing the conductor.

By the Biot–Savart law, the magnetic field intensity at any point on the circle is

$$\vec{H} = \frac{I}{2\pi r} \hat{a}_\phi$$

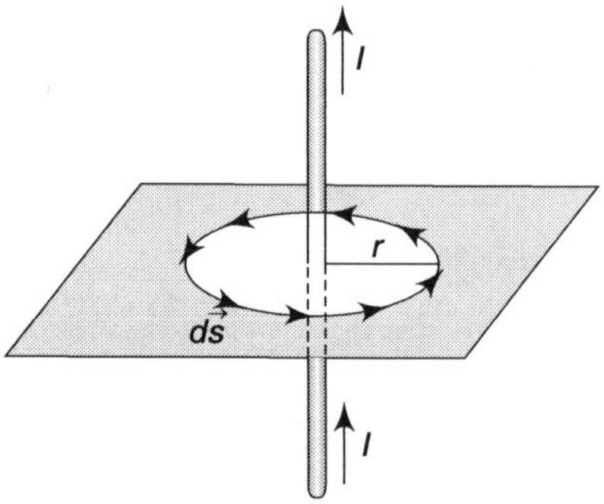

Fig. 3.25 *Amperian loop*

Now, the elemental length of the circle is, $d\vec{l} = r d\phi \, \hat{a}_\phi$

$$\therefore \qquad \vec{H} \cdot d\vec{l} = \left(\frac{I}{2\pi r} \hat{a}_\phi \right) \cdot (r d\phi \hat{a}_\phi) = \frac{I}{2\pi} d\phi$$

$$\therefore \qquad \oint_l \vec{H} \cdot d\vec{l} = \frac{I}{2\pi} \oint_l d\phi = \frac{I}{2\pi} \times 2\pi = I$$

$$\therefore \qquad \oint_l \vec{H} \cdot d\vec{l} = I$$

This proves the Ampere's circuital law.

From Eq. (3.61), applying Stoke's theorem, we get

$$\therefore \qquad \oint_S (\nabla \times \vec{H}) \cdot d\vec{S} = I = \oint_S \vec{J} \cdot d\vec{S}$$

Comparing the integrands, we get

$$\nabla \times \vec{H} = \vec{J} \qquad (3.64)$$

This is the *differential* or *point form of Ampere's circuital law.*

NOTE

The closed path on which Ampere's law is applied is known as ***Amperian path*** *or* ***Amperian loop****.*

Conditions for Ampere's Law For applying Ampere's law, two conditions must be satisfied:

1. At each point on the closed path, $\vec{H}$ is neither tangential nor normal to the path.
2. $\vec{H}$ has the same value at all points of the path where $\vec{H}$ is tangential.

NOTE

Ampere's law in magnetostatics is analogous to Gauss' law in electrostatics. The Biot–Savart law can be applied for a general current source; whereas Ampere's law can be applied only to the systems that possess certain symmetry. For example, in case of an infinite wire, the system possesses cylindrical symmetry and Ampere's law can be readily applied. However, when the length of the wire is finite, the Biot–Savart law must be used instead.

Biot–Savart Law	$\vec{H} = \int_l \frac{Id\vec{l} \times \hat{a}_r}{4\pi r^2} = \int_l \frac{Id\vec{l} \times \vec{r}}{4\pi r^3}$	general current source Ex: finite wire
Ampere' Law	$\oint_l \vec{H} \cdot d\vec{l} = I_{\text{enc}}$	current source that has certain symmetry Ex: infinite wire (cylindrical)

3.13 APPLICATIONS OF AMPERE'S LAW

Ampere's law can be applied to the following current configurations:

1. Infinitely long straight wires carrying a steady current I;
2. Infinitely large sheet of thickness b with a current density $\vec{J}$;
3. Infinite solenoid, and
4. Toroid.

We shall examine all four configurations in detail.

3.13.1 Infinitely Long Straight Wires Carrying a Steady Current I

***Example 3.38** An infinitely long cylindrical conductor of radius R carries a steady current I. Calculate the magnetic field at an interior and exterior point.

Or

A solid cylindrical conductor of a radius R has a uniform current density. Derive expression for $\vec{H}$ both inside and outside of the conductor. Plot the variation of $\vec{H}$, which is function of radial distance from the centre of the wire.

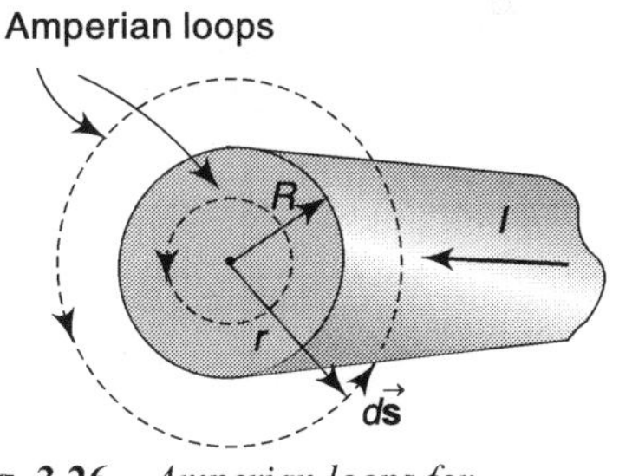

Fig. 3.26 *Amperian loops for calculating the magnetic field of a conducting wire of radius R*

Solution We will consider a long straight wire of radius R carrying a current I of uniform current density, as shown in Fig. 3.26. We want to find the magnetic field inside and outside the wire.

We will consider two cases:

Case (1) Outside the wire where, $r < R$

The Amperian loop completely encircles the current, i.e., $I_{\text{enc}} = I$. By Ampere's law, we get

$$\oint_l \vec{H} \cdot d\vec{l} = H \oint_l d\vec{l} = I$$

$$\Rightarrow \qquad H\int_0^{2\pi} r\,d\phi = I$$

$$\Rightarrow \qquad H(2\pi r) = I$$

$$\therefore \qquad \boxed{H = \frac{I}{2\pi r}}$$

Case (2): Inside the wire where, $r \geq R$

In this case, the amount of current encircled by the Amperian loop is proportional to the area enclosed, i.e.,

$$I_{\text{enc}} = \left(\frac{\pi r^2}{\pi R^2}\right)I = \left(\frac{r^2}{R^2}\right)I$$

By Ampere's law, we get

$$\oint_l \vec{H}\cdot d\vec{l} = I_{\text{enc}} \qquad = \left(\frac{r^2}{R^2}\right)I$$

$$\Rightarrow \qquad H(2\pi r) = \left(\frac{r^2}{R^2}\right)I$$

$$\therefore \qquad \boxed{H = \frac{Ir}{2\pi R^2}}$$

It is seen that the magnetic field is zero at the centre of the wire and increases linearly with r until $r = R$. Outside the wire, the field falls off as $1/r$. The variation of the field is shown in Fig. 3.27.

The results can be summarised in vector form as

$$\vec{H} = \frac{I}{2\pi r}\hat{a}_\phi; \quad r < R$$
$$= \frac{Ir}{2\pi R^2}\hat{a}_\phi; \quad r \geq R \qquad \textbf{(3.65)}$$

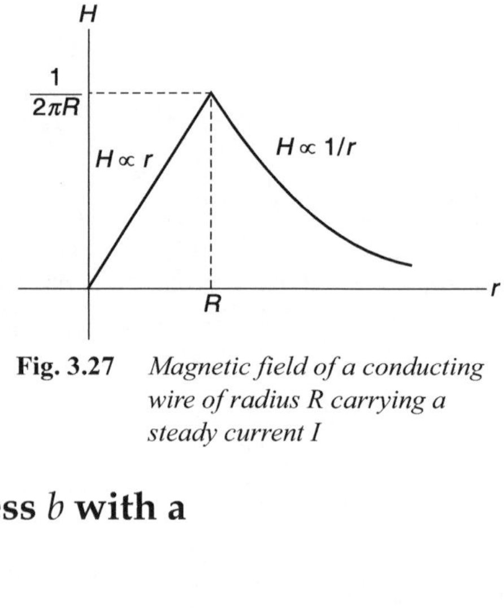

Fig. 3.27 *Magnetic field of a conducting wire of radius R carrying a steady current I*

3.13.2 Infinitely Large Sheet of Thickness b with a Current Density

***Example 3.39** Calculate the magnetic field intensity around a thin infinite current carrying conductor plane located in $z = 0$, having a surface current density K in y-direction, as shown in Fig. 3.28.

Solution The field can be thought of considering the sheet as current elements. The field will not vary with respect to y, $H_y = 0$. It will not vary with respect to x due to symmetry of the problem. The field will vary with respect to z-axis only.

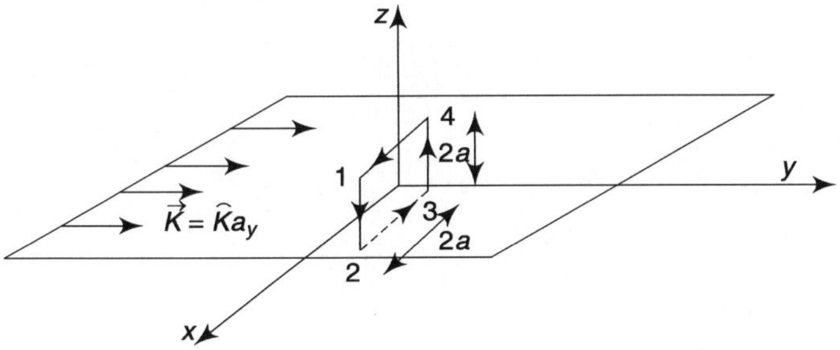

Fig. 3.28 *Current carrying conducting plane*

We will consider a path 1–2–3–4–1 with centre at P, each side as $2a$.
Considering the current elements symmetrically placed will cancel the z-components of the field; so the field exists only in x-direction.
By Ampere's circuital law

$$\oint \vec{H} \cdot d\vec{l} = I_{\text{enc}}$$

$$\Rightarrow \quad \left[\int_1^2 + \int_2^3 + \int_3^4 + \int_4^1\right] \vec{H} \cdot d\vec{l} = K2a$$

$$\Rightarrow \quad 0(2a) + H_x(2a) + 0(2a) + H_x(2a) = K2a$$

$$\Rightarrow \quad H_x = \frac{K}{2}$$

$$\therefore \quad \vec{H} = \frac{K}{2}\hat{a}_x$$

In general, the field can be written as

$\vec{H} = \frac{1}{2}\vec{K} \times \hat{a}_N;$ $\quad \hat{a}_N$ is the normal to the surface

$$\therefore \quad \boxed{\begin{aligned} \vec{H} &= \frac{1}{2}K\hat{a}_x & z > 0 \\ &= -\frac{1}{2}K\hat{a}_x & z < 0 \end{aligned}}$$

Example 3.40 Calculate the magnetic field intensity around an infinitely large sheet of thickness b lying in the xy plane with a uniform current density $\vec{J} = J_0\hat{a}_x$.

Solution The current sheet may be considered as a set of parallel wires carrying currents in the positive x direction. From Fig. 3.29(b), it is seen that the magnetic field at a point P above the plane is in the negative y-direction. The z-component vanishes after adding up the contributions from all wires. Similarly, we may show that the magnetic field at a point below the plane is in the positive y-direction.

The Amperian loops are shown in Fig. 3.29(c). Now we apply Ampere's law to find the magnetic field due to the current sheet.

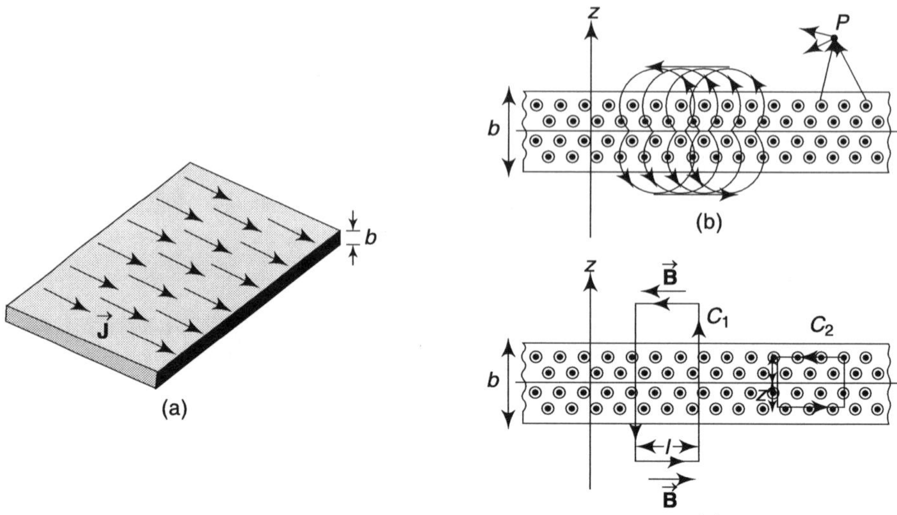

Fig. 3.29 *(a) An infinite sheet with current density $\vec{J} = J_0\hat{i}$, (b) Magnetic field of a current sheet, and (c) Amperian loops for the current sheets*

Field outside the Sheet For the field outside, we integrate along path C_1. The amount of current enclosed by C_1 is

$$I_{\text{enc}} = \int_S \vec{J} \cdot d\vec{S} = J_0 bl$$

Applying Ampere's law

$$\oint_l \vec{H} \cdot d\vec{l} = H(2l) = I_{\text{enc}} = J_0 bl$$

$\therefore$
$$H = \frac{J_0 b}{2}$$

It is seen that the magnetic field outside the sheet is constant, independent of the distance from the sheet.

Field inside the Sheet For the field inside the sheet, the amount of current enclosed by path C_2 is

$$I_{\text{enc}} = \int_S \vec{J} \cdot d\vec{S} = J_0(2|z|l)$$

Applying Ampere's law

$$\oint_l \vec{H} \cdot d\vec{l} = H(2l) = I_{\text{enc}} = J_0(2|z|l)$$

$\therefore$
$$H = J_0 |z|$$

At $z = 0$, the magnetic field vanishes, as required by symmetry. The results can be summarised in vector form as

$$\vec{H} = -\frac{J_0 b}{2}\hat{a}_y; \quad z > \frac{b}{2}$$
$$= -J_0 z\hat{a}_y; \quad -\frac{b}{2} < z < \frac{b}{2} \qquad (3.66)$$
$$= \frac{J_0 b}{2}\hat{a}_y; \quad z < -\frac{b}{2}$$

NOTE

In the limit where the sheet is infinitesimally thin, with b → 0, instead of volume current density $\vec{J} = J_0\hat{a}_x$, we have surface current $\vec{K} = K\hat{a}_x$, where $K = J_0 b$, in current/length (Ampere/metre). In this limit, the magnetic field becomes

$$\vec{H} = -\frac{K}{2}\hat{a}_y; \quad z > 0$$
$$= \frac{K}{2}\hat{a}_y; \quad z < 0 \qquad (3.67)$$

This result is the same as obtained in Example 3.39.

***Example 3.41** Planes $z = 0$ and $z = 4$ carry currents $\vec{K} = -10\hat{a}_x$ A/m and $\vec{K} = 10\hat{a}_x$ A/m respectively. Determine $\vec{H}$ at:
(i) (1, 1, 1) and **(ii)** (0, –3, 10)

Solution Let the parallel current sheets be as shown in Fig. 3.30.

$$\therefore \quad \vec{H} = \vec{H}_0 + \vec{H}_4$$

where $\vec{H}_0$ and $\vec{H}_4$ are the contributions due to the current sheets $z = 0$ and $z = 4$, respectively.

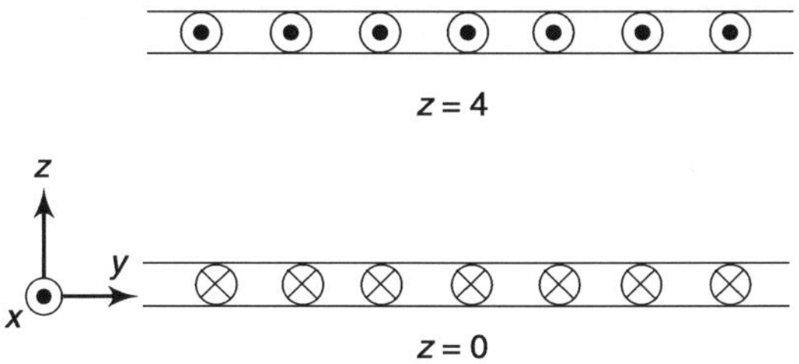

Fig. 3.30 *Two parallel current carrying planes*

From Example 3.39
(i) At (1, 1, 1), ($0 < z < 4$)

$$\vec{H}_0 = \frac{1}{2}\vec{K} \times \hat{a}_N = \frac{1}{2}(-10\hat{a}_x) \times \hat{a}_z = 5\hat{a}_y \text{ A/m}$$
$$\vec{H}_4 = \frac{1}{2}\vec{K} \times \hat{a}_N = \frac{1}{2}(10\hat{a}_x) \times (-\hat{a}_z) = 5\hat{a}_y \text{ A/m}$$
$$\therefore \quad \vec{H} = \vec{H}_0 + \vec{H}_4 = 5\hat{a}_y + 5\hat{a}_y = 10\hat{a}_y \text{ A/m}$$

(ii) At (0, –3, 10), ($z > 4$)

$$\vec{H}_0 = \frac{1}{2}\vec{K} \times \hat{a}_N = \frac{1}{2}(-10\hat{a}_x) \times \hat{a}_z = 5\hat{a}_y \text{ A/m}$$

$$\vec{H}_4 = \frac{1}{2}\vec{K} \times \hat{a}_N = \frac{1}{2}(10\hat{a}_x) \times (\hat{a}_z) = -5\hat{a}_y \text{ A/m}$$

$$\therefore \quad \vec{H} = \vec{H}_0 + \vec{H}_4 = 5\hat{a}_y - 5\hat{a}_y = 0$$

3.13.3 Solenoid

***Example 3.42** Find the field inside a solenoid of length L having N turns uniformly wound round a cylinder of radius a and carrying a current I.

Or

Derive a general expression for the field $\vec{H}$ at any point along the axis of a solenoid (uniform cylindrical coil wound on a non-magnetic frame). Sketch the variation of $\vec{H}$ from point to point along the axis.

Or

A solenoid of length L consists of N turns of wire carrying a current I. Show that the field at any point along the axis

$$\vec{H} = \frac{NI}{2L}(\cos\theta_2 - \cos\theta_1)\hat{a}_r$$

where, θ_1 and θ_2 are the angles subtended at the point by the end turns. Also, show that, if L is very large compared to the radius of the solenoid, then, at the centre of the solenoid, $\vec{H} = \frac{NI}{L}\hat{a}_r$.

Solution A *solenoid* is a loop of wire, wound around a metallic core in a helical form, which produces a magnetic field when an electric current passes through it. A solenoid is ideal if it is infinitely long with turns tightly packed.

Infinite Solenoid Figure 3.31 shows the magnetic field lines of a solenoid carrying a steady current I.

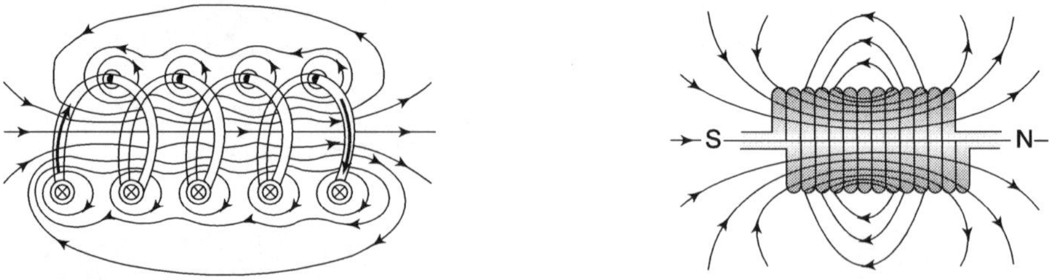

Fig. 3.31 *Magnetic field lines of a solenoid*

Some characteristics of infinite solenoid are:

(i) Its turns are closely spaced.

(ii) The resulting magnetic field inside the solenoid is fairly uniform (provided that the length of the solenoid is much greater than its diameter) and parallel to the axis.

(iii) The magnetic field outside the solenoid is zero.

The cross-sectional view of an ideal solenoid is shown in Fig. 3.32. To calculate $\vec{H}$, we consider a rectangular path (Amperian loop) of length l and width w and traverse the path in a counterclockwise manner. The line integral of $\vec{H}$ along this loop is

$$\oint_l \vec{H} \cdot d\vec{l} = \oint_1 \vec{H} \cdot d\vec{l} + \oint_2 \vec{H} \cdot d\vec{l} + \oint_3 \vec{H} \cdot d\vec{l} + \oint_4 \vec{H} \cdot d\vec{l}$$
$$= 0 + 0 + Hl + 0$$
$$= Hl$$

In the above equation, the contributions along sides 2 and 4 are zero because $\vec{H}$ is perpendicular to $d\vec{l}$. In addition, $\vec{H} = 0$ along side 1 because the magnetic field is non-zero only inside the solenoid.

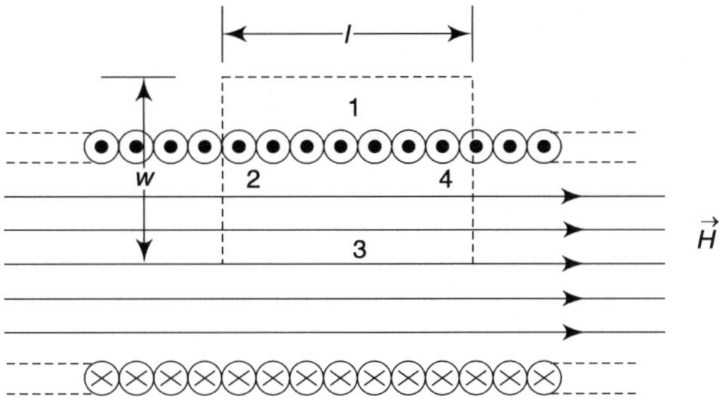

Fig. 3.32 *Amperian loop for calculating the magnetic field of an ideal solenoid*

Now, by Ampere's law, total current enclosed by the Amperian loop is, $I_{enc} = NI$ where, N is the total number of turns.
Applying Ampere's law

$$\oint_l \vec{H} \cdot d\vec{l} = I_{enc}$$

$\Rightarrow$ $$Hl = NI$$

$\therefore$ $$\boxed{H = \frac{NI}{l} = nI} \qquad \textbf{(3.68)}$$

where, $n = \dfrac{N}{l}$ = number of turns per unit length

Finite Solenoid To find the magnetic field due to a finite solenoid, the solenoid is assumed to be consisting of a large number of circular loops stacking together. Using the result obtained in Example 3.32, the magnetic field at a point P on the z axis may be calculated as follows.

We will consider a cross-section of tightly packed loops located at z' with a thickness dz', as shown in Fig. 3.33.

The amount of current flowing through is proportional to the thickness of the cross-section and is given by

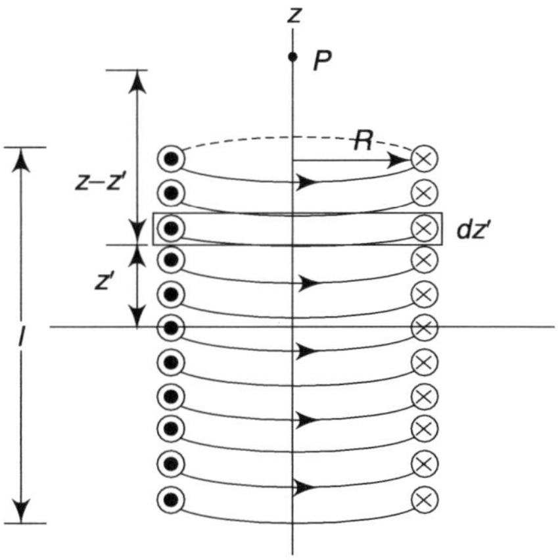

Fig. 3.33 *Finite solenoid*

$dI = I\left(\frac{N}{l}\right)dz = I(ndz')$; where, $n = \frac{N}{l}$ = number of turns per unit length

The contribution to the magnetic field at P due to this subset of loops is

$$dH = \frac{R^2}{2[(z-z')^2 + R^2]^{3/2}} dI = \frac{R^2}{2[(z-z')^2 + R^2]^{3/2}}(nIdz') \quad \textbf{(3.69)}$$

Integrating over the entire length of the solenoid, we get

$$H = \frac{nIR^2}{2}\int_{-l/2}^{l/2}\frac{dz'}{\left[(z-z')^2 + R^2\right]^{3/2}} = \frac{nIR^2}{2}\left.\frac{z'-z}{R^2\sqrt{(z-z')^2+R^2}}\right|_{-l/2}^{l/2}$$

$$= \frac{nI}{2}\left[\frac{\left(\frac{l}{2}\right)-z}{\sqrt{\left(z-\frac{l}{2}\right)^2 + R^2}} + \frac{\left(\frac{l}{2}\right)+z}{\sqrt{\left(z+\frac{l}{2}\right)^2 + R^2}}\right] \quad \textbf{(3.70)}$$

A plot of $\frac{H}{H_0}$, where $H_0 = nI = \frac{NI}{l}$ is the magnetic field of an infinite solenoid, as a function of $\frac{z}{R}$ is shown in Fig. 3.34 for $l = 10R$ and $l = 20R$. Note that the value of the magnetic field in the region $|z| < \frac{l}{2}$ is nearly uniform and approximately equal to H_0.

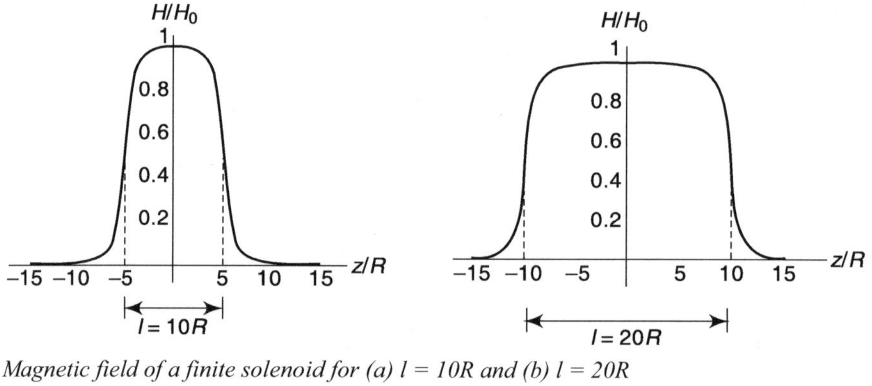

Fig. 3.34 *Magnetic field of a finite solenoid for (a) l = 10R and (b) l = 20R*

Alternative Way of Determination of Magnetic Field Intensity in Finite Solenoid

We consider the cross section of the solenoid as shown in Fig. 3.35 (*a*).

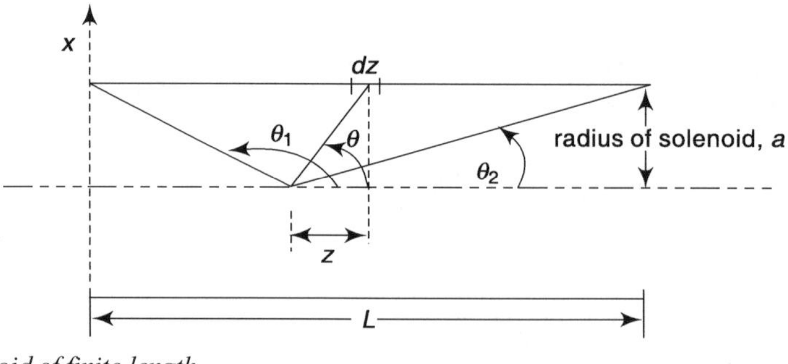

Fig. 3.35 (a) *Solenoid of finite length*

Since the solenoid consists of circular loops, using the result of Example 23 (a); the contribution to the magnetic field by an element of the solenoid of lenght dz is,

$$dH = \frac{Idla^2}{2[a^2+z^2]^{3/2}} = \frac{Ia^2Ndz}{2L(a^2+z^2)^{3/2}} \qquad \text{where,} \quad dl = \frac{N}{L}dz$$

From the figure, $\tan\theta = \frac{a}{z}$ $\quad \therefore\ dz = -a\ \text{cosec}^2\theta d\theta$

$$\therefore \qquad dz = -\frac{(a^2+z^2)}{a}\sin\theta d\theta$$

$$\therefore \qquad dH = -\frac{NI}{2L}\sin\theta d\theta$$

So, the field is given as,

$$H = \int dH = -\frac{NI}{2L}\int_{\theta_1}^{\theta_2}\sin\theta d\theta = \frac{NI}{2L}(\cos\theta_2 - \cos\theta_1)$$

$$\boxed{\therefore\ \vec{H} = \frac{NI}{2L}(\cos\theta_2 - \cos\theta_1)\hat{a}_z}$$

We consider three cases:

1. When P is at one end of the solenoid,

Here, $\qquad \theta_1 = 90°, \quad \cos\theta_2 = \frac{L}{\sqrt{L^2+a^2}}$

The magnetic field is given as, $\vec{H} = \frac{NI}{2L}\left(\frac{L}{\sqrt{L^2+a^2}} - 0\right)\hat{a}_z = \frac{NI}{2\sqrt{L^2+a^2}}\hat{a}_z$

$$\boxed{\vec{H} = \frac{NI}{2\sqrt{L^2+a^2}}\hat{a}_z}$$

NOTE

If L >> a, then $\vec{H} = \frac{NI}{2L}\hat{a}_z$

2. When P is at the centre of the solenoid,

Here, $\qquad \cos\theta_2 = \frac{L/2}{\sqrt{(L/2)^2+a^2}} = -\cos\theta_1$

The magnetic field is given as, $\vec{H} = \frac{NI}{2L}\left(\frac{L/2}{\sqrt{(L/2)^2+a^2}}\right)\hat{a}_z = \frac{NI}{\sqrt{L^2+4a^2}}\hat{a}_z$

$$\boxed{\vec{H} = \frac{NI}{\sqrt{L^2+4a^2}}\hat{a}_z}$$

NOTE

If L >> a, then $\vec{H} = \frac{NI}{L}\hat{a}_z$

3. When P is at the midway between one end and centre of the solenoid,
Here,

$$\cos\theta_2 = \frac{L/4}{\sqrt{(L/2)^2 + a^2}} = \frac{L}{\sqrt{L^2 + 16a^2}}$$

$$\cos(\pi - \theta_1) = \frac{3L/4}{\sqrt{(3L/4)^2 + a^2}} = \frac{3L}{\sqrt{9L^2 + 16a^2}}$$

$\therefore$

$$\cos\theta_1 = -\frac{3L}{\sqrt{9L^2 + 16a^2}}$$

The magnetic field is given as, $\vec{H} = \frac{NI}{2L}\left(\frac{L}{\sqrt{L^2 + 16a^2}} + \frac{3L}{\sqrt{9L^2 + 16a^2}}\right)\hat{a}_z$

NOTE

If L >> a, then $\cos\theta_1 = -\frac{3L}{\sqrt{9L^2 + 16a^2}} \approx -1$ *and the magnetic field is given as,*

$\vec{H} = \frac{NI}{2L}\left(1 + \frac{L}{\sqrt{L^2 + 16a^2}}\right)\hat{a}_z$.

The variation of the field from point to point is shown in Fig. 3.35 (*b*).

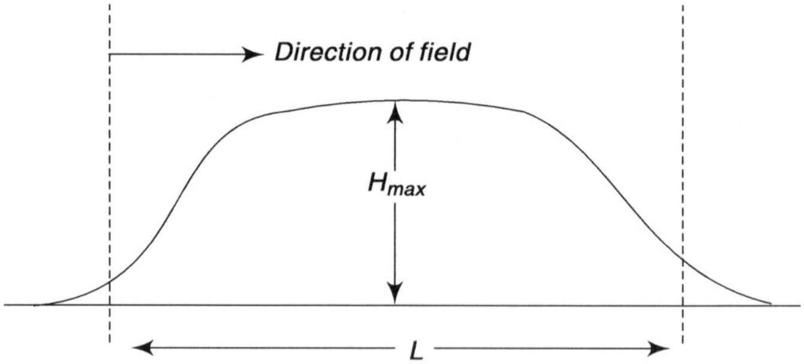

Fig. 3.35 (b) *Variation of magnetic field from point to point along the axis of a finite solenoid*

3.13.4 Toroid

***Example 3.43** A toroid of length L has N turns and carries a current I. Determine $\vec{H}$ inside and outside the toroid.

Solution A *toroid* consists of a circular ring-shaped magnetic core around which wire is coiled. A toroid may be considered as a solenoid wrapped around with its ends connected.

Thus, the magnetic field is completely confined inside the toroid and the field points in the azimuthal direction (clockwise due to the way the current flows, as shown in Fig. 3.36).

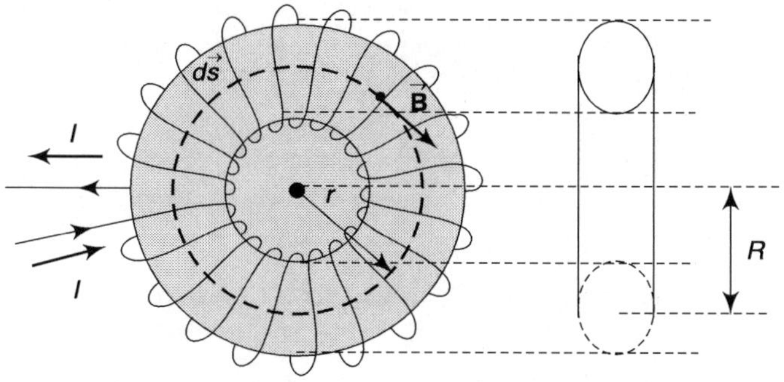

Fig. 3.36 *A toroid with N turns*

Since N wires cut the Amperian path, each carrying a current I, the net current enclosed by the Amperian path is

$$I_{\text{enc}} = NI$$

Applying Ampere's law to the Amperian path, we obtain

$$\oint_l \vec{H} \cdot d\vec{l} = \oint_l H dl = H \oint_l dl = H(2\pi r) = NI$$

$$\therefore \qquad \boxed{H = \frac{NI}{2\pi r}} \qquad \textbf{(3.71)}$$

where r is the distance measured from the centre of the toroid and is known as the mean radius. If the thickness of the toroid is much less than its mean radius, then $r \approx R$. Hence,

$$H = \frac{NI}{2\pi r} = \frac{NI}{2\pi R} = \frac{NI}{2L} \text{ inside the toroid}$$

where, L is the length of the wire.
Outside the toroid, net current enclosed = $(NI - NI) = 0$, and thus, $H = 0$.
To summarise the results

$$\begin{aligned} H &= \frac{NI}{2L}; \quad \text{inside the toroid} \\ &= 0 \quad ; \quad \text{outside the toroid} \end{aligned}$$

Example 3.44 Consider an infinitely long, cylindrical conductor of radius R carrying a current I with a non-uniform current density

$$J = \alpha r$$

where α is a constant. Find the magnetic field everywhere. This is shown in Fig. 3.37.

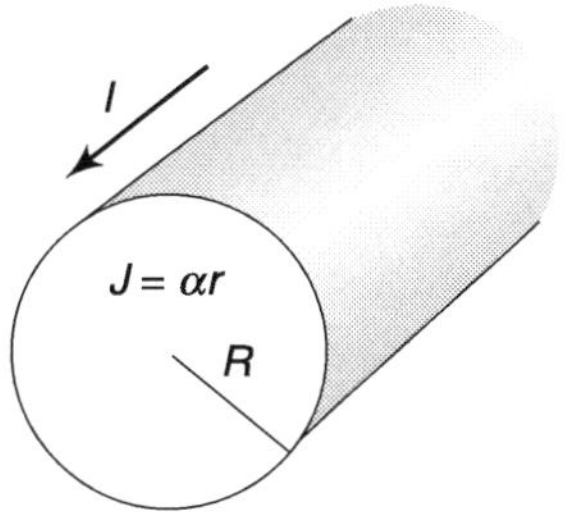

Fig. 3.37 *Non-uniform current density*

Solution By Ampere's law,

$$\oint \vec{B} \cdot d\vec{S} = \mu_0 I_{\text{enc}}$$

Here,

$$I_{\text{enc}} = \int \vec{J} \cdot d\vec{S} = \int (\alpha r)(2\pi r dr)$$

We will consider two cases:

(1) For $r < R$:

In this case

$$I_{\text{enc}} = \int (\alpha r)(2\pi r dr) = \int_0^r 2\pi\alpha r^2 dr = \frac{2}{3}\pi\alpha r^3$$

Applying Ampere's law, the magnetic field at any point inside the conductor is given as

$$B_i(2\pi r) = \frac{2}{3}\mu_0\pi\alpha r^3 \qquad \text{or,} \qquad B_i = \frac{\alpha\mu_0}{3}r^2$$

The direction of the magnetic field $\vec{B}_i$ is tangential to the Amperian loop that encloses the current.

(2) For $r > R$:

In this case

$$I_{\text{enc}} = \int (\alpha r)(2\pi r dr) = \int_0^R 2\pi\alpha r^2 dr = \frac{2}{3}\pi\alpha R^3$$

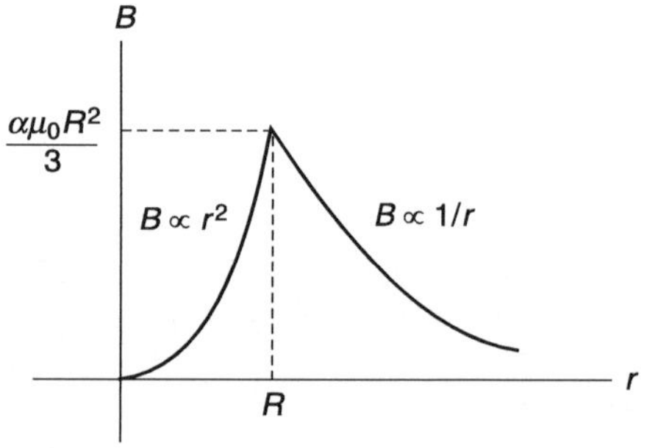

Fig. 3.38 *The magnetic field as a function of distance away from the conductor*

Applying Ampere's law, the magnetic field at any point outside the conductor is given as

$$B_o(2\pi r) = \frac{2}{3}\mu_0\pi\alpha R^3 \qquad \text{or,} \qquad B_o = \frac{\alpha\mu_0 R^3}{3r}$$

To summarise, the results are

$$B = \frac{\alpha\mu_0}{3}r^2 \qquad r < R$$
$$= \frac{\alpha\mu_0 R^3}{3r} \qquad r > R$$

A plot of B as a function of r is shown in Fig. 3.38.

Example 3.45 A coaxial cable has core of radius 'a' and sheath of radius 'b'. A current 'I' flows along the core, uniformly distributed across it, and returns along the sheath, uniformly distributed around it. Find the magnetic field intensity:

(i) Within the core ($r < a$).

(ii) Within the core-sheath space ($a \le r \le b$); and

(iii) Outside the sheath ($r > b$).

Solution **Within the core ($r < a$):**
Applying Ampere's law

$$\oint_l \vec{H} \cdot d\vec{l} = I_{\text{enc}} = \int \vec{J} \cdot d\vec{S}$$

Since, the current is distributed uniformly over the cross-section

$$\vec{J} = \frac{I}{\pi a^2}\hat{a}_z \qquad \text{and} \qquad d\vec{S} = rdrd\phi\hat{a}_z$$

$$\therefore \qquad I_{\text{enc}} = \int_0^r \int_0^{2\pi} \frac{I}{\pi a^2} rdrd\phi = \frac{I}{\pi a^2} \times 2\pi \times \frac{r^2}{2} = I\left(\frac{r^2}{a^2}\right)$$

$$\therefore \qquad \oint_l \vec{H} \cdot d\vec{l} = H_\phi \int_0^{2\pi} rd\phi = I\left(\frac{r^2}{a^2}\right) \quad \Rightarrow \quad H_\phi = \frac{Ir}{2\pi a^2}$$

$$\therefore \qquad \vec{H} = \frac{Ir}{2\pi a^2}\hat{a}_\phi$$

Within the core-sheath space ($a \le r \le b$):
Applying Ampere's law

$$\oint_l \vec{H} \cdot d\vec{l} = I_{\text{enc}} = \text{Total current} = I$$

$$\Rightarrow \qquad H_\phi \int_0^{2\pi} rd\phi = I \quad \Rightarrow \quad H_\phi = \frac{I}{2\pi r}$$

$$\therefore \qquad \vec{H} = \frac{I}{2\pi r}\hat{a}_\phi$$

Outside the sheath ($r > b$):
Applying Ampere's law

$$\oint_l \vec{H} \cdot d\vec{l} = (I_{\text{inner cond.}} + I_{\text{outer cond.}}) = (I - I) = 0$$

$$\Rightarrow \qquad H_\phi = 0$$

Note
If the thickness of the sheath is 't', then for the region, $b \le r \le (b + t)$,

$$I_{\text{enc}} = I + \int \vec{J} \cdot d\vec{S}$$

Here, $\vec{J}$ is the current density of the outer conductor and is along $-\hat{a}_z$.

$$\therefore \qquad \vec{J} = -\frac{I}{\pi[(b+t)^2 - t^2]}\hat{a}_z$$

$$\therefore \qquad I_{\text{enc}} = I - \frac{I}{\pi[(b+t)^2 - t^2]} \int_{\phi=0}^{2\pi} \int_{r=b}^{r} rdrd\phi = I\left\{1 - \frac{r^2 - b^2}{t^2 + 2bt}\right\}$$

By Ampere's law

$$H_\phi \int_0^{2\pi} r d\phi = I_{\text{enc}} \quad \Rightarrow \quad H_\phi = \frac{I}{2\pi r}\left\{1 - \frac{r^2 - b^2}{t^2 + 2bt}\right\}$$

$$\therefore \quad \vec{H} = \frac{I}{2\pi r}\left\{1 - \frac{r^2 - b^2}{t^2 + 2bt}\right\}\hat{a}_\phi$$

To summarise, the magnetic field is given as

$$\begin{aligned} \vec{H} &= \frac{Ir}{2\pi a^2}\hat{a}_\phi & 0 \le r \le a \\ &= \frac{I}{2\pi r}\hat{a}_\phi & a \le r \le b \\ &= \frac{I}{2\pi r}\left\{1 - \frac{r^2 - b^2}{t^2 + 2bt}\right\}\hat{a}_\phi & b \le r \le (b+t) \\ &= 0 & r \ge (b+t) \end{aligned}$$

Example 3.46 An infinitely long coaxial pair of circular conductors are located in free space and carry equal and opposite total static current I. The inner conductor is of radius a while the inner and outer radii of the outer conductor are b and c respectively. Sketch the variation of magnetic flux density over the range $(0, c)$ and show that the field inside the outer conductor $(b < r < c)$ is

$$B = \frac{\mu_0 I}{2\pi r}\left(\frac{c^2 - r^2}{c^2 - b^2}\right)$$

and zero for $r > c$.

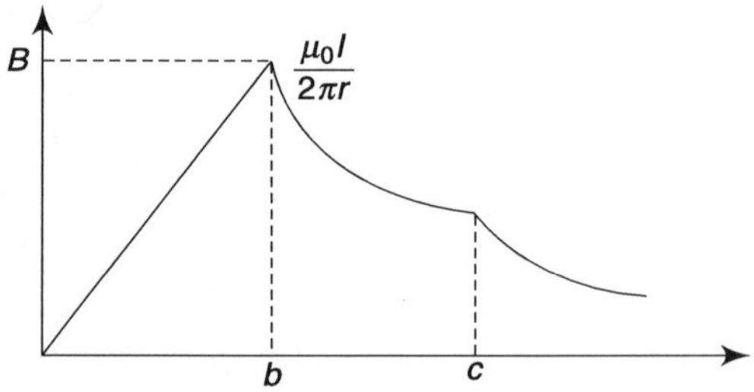

Fig. 3.39 *Variation of magnetic flux density within the sheath of coaxial cable*

Solution From Example 3.45, the thickness of the sheath in this problem can be written as $t = (c - b)$. Replacing this value, the magnetic flux density for the region $(b < r < c)$ is obtained as

$$B = \frac{\mu_0 I}{2\pi r}\left\{1 - \frac{r^2 - b^2}{t^2 + 2bt}\right\} = \frac{\mu_0 I}{2\pi r}\left[\frac{(c-b)^2 + 2b(c-b) - r^2 + b^2}{(c-b)^2 + 2b(c-b)}\right] = \frac{\mu_0 I}{2\pi r}\left(\frac{c^2 - r^2}{c^2 - b^2}\right)$$

The variation of the flux density is shown in Fig. 3.39.

Example 3.47 Using Ampere's circuital law in integral form, find $\vec{H}$ everywhere due to the current density in cylindrical co-ordinates:

$$\begin{aligned} \vec{J} &= 0, & 0 < r < a \\ &= J_0\hat{a}_z & a < r < b \\ &= 0 & b < r < \infty \end{aligned}$$

Solution $\vec{J} = 0, \quad 0 < r < a$

$= J_0 \hat{a}_z \quad a < r < b$

$= 0 \quad b < r < \infty$

$0 < r < a$:
By Ampere's law,

$$\oint \vec{H} \cdot d\vec{l} = \int \vec{J} \cdot d\vec{S} = 0 \qquad (\because \vec{J} = 0)$$

$$\Rightarrow \qquad H = 0$$

$a \le r < b$:
By Ampere's law,

$$\oint \vec{H} \cdot d\vec{l} = \int \vec{J} \cdot d\vec{S}$$

$$\Rightarrow \qquad H(2\pi r) = \int_{r=a}^{r} \int_{\phi=0}^{2\pi} (J_0 \hat{a}_z) \cdot (r dr d\phi \hat{a}_z) = \int_{r=a}^{r} \int_{\phi=0}^{2\pi} J_0 r dr d\phi = 2\pi J_0 \int_{r=a}^{r} r dr = J_0 \pi (r^2 - a^2)$$

$$\therefore \qquad \vec{H} = \frac{J_0 (r^2 - a^2)}{2r} \hat{a}_\phi$$

$b < r < \infty$:
By Ampere's law,

$$\oint \vec{H} \cdot d\vec{l} = \int \vec{J} \cdot d\vec{S} = 0 \qquad (\because \vec{J} = 0)$$

$$\Rightarrow \qquad H = 0$$

To summarise, the results are

$\vec{H} = 0$	$0 < r < a$
$= \dfrac{J_0 (r^2 - a^2)}{2r} \hat{a}_\phi$	$a \le r < b$
$= 0$	$b < r < \infty$

Example 3.48 Find the current distribution producing the following field distribution using Ampere's circuital law in differential form:

$$\vec{H} = J_0 r^2 \hat{a}_\phi, \qquad 0 < r < a$$

$$= J_0 \left(\frac{a^3}{r} \right) \hat{a}_\phi \qquad a < r < b$$

$$= 0 \qquad b < r < \infty$$

Solution By Ampere's law in differential form,

$$\nabla \times \vec{H} = \vec{J}$$

or,
$$\vec{J} = \left[\frac{1}{r} \frac{\partial H_z}{\partial \phi} - \frac{\partial H_\phi}{\partial z} \right] \hat{a}_r + \left[\frac{\partial H_r}{\partial z} - \frac{\partial H_z}{\partial r} \right] \hat{a}_\phi + \frac{1}{r} \left[\frac{\partial (r H_\phi)}{\partial r} - \frac{\partial H_r}{\partial \phi} \right] \hat{a}_z$$

Since, $\vec{H}$ is having only H_ϕ component which is a function of r only

$\therefore$
$$\vec{J} = \frac{1}{r}\frac{\partial}{\partial r}(rH_\phi)\hat{a}_z$$

$0 < r < a$:
The current distribution is given as

$$\vec{J} = \frac{1}{r}\frac{\partial}{\partial r}(rJ_0 r^2)\hat{a}_z = 3J_0 r\hat{a}_z$$

$a \le r < b$:
The current distribution is given as

$$\vec{J} = \frac{1}{r}\frac{\partial}{\partial r}\left(r\frac{J_0 a^3}{r}\right)\hat{a}_z = 0$$

$b < r < \infty$:
The current distribution is given as

$$\vec{J} = \frac{1}{r}\frac{\partial}{\partial r}(0)\hat{a}_z = 0$$

To summarise, the results are

$\vec{J} = 3J_0 r\hat{a}_z$	$0 < r < a$
$= 0$	$a \le r < b$
$= 0$	$b < r < \infty$

Example 3.49 Determine the current density function $\vec{J}$ associated with the magnetic field defined by:

(a) $\vec{H} = 3\hat{i} + 7\hat{j} + 2x\hat{k}$ A/m (Cartesian)
(b) $\vec{H} = 6r\hat{a}_r + 2r\hat{a}_\phi + 5\hat{a}_z$ A/m (cylindrical)
(c) $\vec{H} = 2\rho\hat{a}_\rho + 3\hat{a}_\theta + \cos\theta\,\hat{a}_\phi$ A/m (spherical)

Solution

(a) $\vec{H} = 3\hat{i} + 7\hat{j} + 2x\hat{k}$

By Ampere's law in Cartesian coordinates,

$$\vec{J} = \nabla \times \vec{H} = \begin{vmatrix} \hat{a}_x & \hat{a}_y & \hat{a}_z \\ \frac{\partial}{\partial x} & \frac{\partial}{\partial y} & \frac{\partial}{\partial z} \\ 3 & 7 & 2x \end{vmatrix} = -2\hat{a}_y \text{ A/m}^2$$

(b) By Ampere's law in cylindrical coordinates,

$$\vec{J} = \nabla \times \vec{H} = \begin{vmatrix} \frac{1}{r}\hat{a}_r & \hat{a}_\phi & \hat{a}_z \\ \frac{\partial}{\partial r} & \frac{\partial}{\partial \phi} & \frac{\partial}{\partial z} \\ H_r & rH_\phi & H_z \end{vmatrix}$$

$$= \left[\frac{1}{r}\frac{\partial H_z}{\partial \phi} - \frac{\partial H_\phi}{\partial z}\right]\hat{a}_r + \left[\frac{\partial H_r}{\partial z} - \frac{\partial H_z}{\partial r}\right]\hat{a}_\phi + \frac{1}{r}\left[\frac{\partial (rH_\phi)}{\partial r} - \frac{\partial H_r}{\partial \phi}\right]\hat{a}_z$$

$$= \left[\frac{1}{r}\frac{\partial}{\partial \phi}(5) - \frac{\partial}{\partial z}(2r)\right]\hat{a}_r + \left[\frac{\partial}{\partial z}(6r) - \frac{\partial}{\partial r}(5)\right]\hat{a}_\phi + \left(\frac{1}{r}\right)\left[\frac{\partial}{\partial r}(r2r) - \frac{\partial}{\partial \phi}(6r)\right]\hat{a}_z$$

$$= \left(\frac{1}{r}\right) \times 4r\hat{a}_z$$

$$= 4\hat{a}_z \text{ A/m}^2$$

(c) $\vec{H} = 2\rho\hat{a}_\rho + 3\hat{a}_\theta + \cos\theta\,\hat{a}_\phi$

By Ampere's law in spherical coordinates,

$$\vec{J} = \nabla \times \vec{H} = \frac{1}{\rho^2 \sin\theta}\begin{vmatrix} \hat{a}_\rho & \hat{a}_\theta & \hat{a}_\varphi \\ \dfrac{\partial}{\partial \rho} & \dfrac{\partial}{\partial \theta} & \dfrac{\partial}{\partial \phi} \\ H_\rho & \rho H_\theta & \rho \sin\theta\, H_\phi \end{vmatrix}$$

$$= \frac{1}{\rho\sin\theta}\left[\frac{\partial}{\partial \theta}(H_\phi \sin\theta) - \frac{\partial H_\theta}{\partial \phi}\right]\hat{a}_\rho + \left(\frac{1}{\rho}\right)\left[\frac{1}{\sin\theta}\frac{\partial H_\rho}{\partial \phi} - \frac{\partial}{\partial \rho}(\rho H_\phi)\right]\hat{a}_\theta$$

$$+ \frac{1}{\rho}\left[\frac{\partial}{\partial \rho}(\rho H_\theta) - \frac{\partial H_\rho}{\partial \theta}\right]\hat{a}_\phi$$

$$= \frac{1}{\rho\sin\theta}\left[\frac{\partial}{\partial \theta}(\cos\theta \sin\theta) - \frac{\partial}{\partial \phi}(3)\right]\hat{a}_\rho + \left(\frac{1}{\rho}\right)\left[\frac{1}{\sin\theta}\frac{\partial}{\partial \phi}(2\rho) - \frac{\partial}{\partial \rho}(\rho\cos\theta)\right]\hat{a}_\theta$$

$$+ \frac{1}{\rho}\left[\frac{\partial}{\partial \rho}(\rho 3) - \frac{\partial}{\partial \theta}(2\rho)\right]\hat{a}_\phi$$

$$= \frac{1}{\rho}\left(\frac{\cos 2\theta}{\sin\theta}\right)\hat{a}_\rho - \frac{1}{\rho}\cos\theta\,\hat{a}_\theta + \frac{3}{\rho}\hat{a}_\phi \text{ A/m}^2$$

Example 3.50

(a) A circular conductor of radius $r_0 = 1$ cm has an internal field,

$$\vec{H} = \frac{10^4}{r}\left(\frac{1}{a^2}\sin ar - \frac{r}{a}\cos ar\right)\hat{a}_\phi \text{ A/m}$$

where, $a = \dfrac{\pi}{2r_0}$. Find the total current in the conductor.

(b) A cylindrical conductor of radius 10^{-2} m has an internal magnetic field

$$\vec{H} = (4.77 \times 10^4)\left(\frac{r}{2} - \frac{r^2}{3 \times 10^{-2}}\right)\hat{a}_\phi \quad \text{A/m}$$

Find the total current in the conductor.

Solution

(a) $\vec{H} = \frac{10^4}{r}\left(\frac{1}{a^2}\sin ar - \frac{r}{a}\cos ar\right)\hat{a}_\phi = \frac{10^4}{r}\left[\frac{4r_0^2}{\pi^2}\sin\left(\frac{\pi r}{2r_0}\right) - \frac{2rr_0}{\pi}\cos\left(\frac{\pi r}{2r_0}\right)\right]\hat{a}_\phi$

$\left(\text{Putting } a = \frac{\pi}{2r_0}\right)$

By Ampere's law, the total current enclosed is,

$$I = \oint_l \vec{H}\cdot d\vec{l} = \int_0^{2\pi} \frac{10^4}{r}\left[\frac{4r_0^2}{\pi^2}\sin\left(\frac{\pi r}{2r_0}\right) - \frac{2rr_0}{\pi}\cos\left(\frac{\pi r}{2r_0}\right)\right]\hat{a}_\phi\cdot(rd\phi\hat{a}_\phi)\Bigg|_{r=r_0=0.01}$$

$$= \int_0^{2\pi} \frac{10^4}{r_0}\left[\frac{4r_0^2}{\pi^2}\sin\left(\frac{\pi r_0}{2r_0}\right) - \frac{2r_0r_0}{\pi}\cos\left(\frac{\pi r_0}{2r_0}\right)\right]r_0 d\phi$$

$$= 10^4 \int_0^{2\pi} \frac{4r_0^2}{\pi^2} d\phi$$

$$= 10^4 \times \frac{4r_0^2}{\pi^2} \times 2\pi = 10^4 \times \frac{8\times(0.01)^2}{\pi} = \frac{8}{\pi}\text{ A}$$

(b) $\vec{H} = \frac{10^4}{r}\left(\frac{1}{a^2}\sin ar - \frac{r}{a}\cos ar\right)\hat{a}_\phi$

By Ampere's law, the total current enclosed is,

$$I = \oint_l \vec{H}\cdot d\vec{l} = \int_0^{2\pi}(4.77\times10^4)\left(\frac{r}{2} - \frac{r^2}{3\times10^{-2}}\right)\hat{a}_\phi\cdot(rd\phi\hat{a}_\phi)\Bigg|_{r=10^{-2}}$$

$$= 4.77\times10^4\int_0^{2\pi}\left(\frac{r}{2} - \frac{r^2}{3\times10^{-2}}\right)rd\phi\Bigg|_{r=10^{-2}}$$

$$= 4.77\times10^4\left(\frac{10^{-2}}{2} - \frac{10^{-4}}{3\times10^{-2}}\right)\times10^{-2}\times2\pi$$

$$= 5\text{ A}$$

3.14 MAGNETIC POTENTIALS

Just like an electric potential, we can define a potential associated with magnetostatic field. In fact, the magnetic potentials are of two types:

1. Magnetic Scalar Potential, and
2. Magnetic Vector Potential

3.14.1 Magnetic Scalar Potential

The magnetic scalar potential is defined only in regions of space in the absence of currents.

We know from Ampere's law that $\nabla\times\vec{H} = \vec{J}$ for steady current. If the current density $\vec{J}$ is zero in some region of space, then we have

$$\nabla\times\vec{H} = 0$$

and so we can write the magnetic field $\vec{H}$ as the gradient of a scalar quantity as

$$\boxed{\vec{H} = -\nabla V_m} \quad \textbf{(3.72)}$$

where, V_m is called *the magnetic scalar potential.* It is expressed in Ampere.

Similar to the relation between the electric field intensity $\vec{E}$ and electric potential V, magnetic scalar potential can also be written as

$$V_{m,x,y} = -\int_y^x \vec{H} \cdot d\vec{l} \quad \textbf{(3.73)}$$

Since the divergence of magnetic flux density is zero,

$$\therefore \qquad \nabla \cdot \vec{B} = 0 \;\Rightarrow\; \nabla \cdot (\mu \vec{H}) = 0 \;\Rightarrow\; \nabla \cdot \vec{H} = 0 \qquad \nabla \cdot (-\nabla V_m) = 0$$

$$\therefore \qquad \boxed{\nabla^2 V_m = 0} \quad \textbf{(3.74)}$$

This shows that the magnetic scalar potential satisfies the Laplace equation in a source-free region ($\vec{J} = 0$).

Example 3.51 A coaxial cable has inner and outer radii a and b, respectively. The inner cable carries a current I. Find the magnetic scalar potential at a radius r inside the cable ($a < r < b$).

Solution From Section 3.13.1, the magnetic field intensity inside the cable at any radius r ($a < r < b$) is given as

$$\vec{H} = \frac{I}{2\pi r}\hat{a}_\phi$$

From the definition of magnetic vector potential, we have

$$\vec{H} = -\nabla V_m = -\left(\frac{\partial V_m}{\partial r}\hat{a}_r + \frac{1}{r}\frac{\partial V_m}{\partial \phi}\hat{a}_\phi + \frac{\partial V_m}{\partial z}\hat{a}_z\right) = -\frac{1}{r}\frac{\partial V_m}{\partial \phi}\hat{a}_\phi \quad \{\because V \text{ is a function of } r \text{ only}\}$$

Substituting the value of $\vec{H}$, we get

$$\frac{I}{2\pi r}\hat{a}_\phi = -\frac{1}{r}\frac{\partial V_m}{\partial \phi}\hat{a}_\phi$$

$$\Rightarrow \qquad \frac{\partial V_m}{\partial \phi} = -\frac{I}{2\pi}$$

Integrating,

$$V_m = -\frac{I}{2\pi}\phi + C$$

where, C is the integration constant.

Applying the boundary condition that $V_m = 0$ at $\phi = 0$ (i.e. zero current enclosed), we have $C = 0$.

$$\therefore \qquad V_m = -\frac{I}{2\pi}\phi$$

NOTE

If $\phi = 2\pi$ (anti-clockwise rotation), $V_m = -\frac{I}{2\pi} \times 2\pi = -I$; however, $\phi = 2\pi$ is the same point as $\phi = 0$ for which $V_m = 0$. This shows that V_m is not a single valued function.

For $\phi = 2n\pi$ (anti-clockwise rotation), $V_m = -\frac{I}{2\pi} \times 2n\pi = -nI$

For $\phi = -2n\pi$ (clockwise rotation), $V_m = -\frac{I}{2\pi} \times (-2n\pi) = nI$

So, there is total 2n number of values for V_m.

***Example 3.52** Prove that the magnetic scalar potential at (0, 0, z) due to a circular loop of radius 'a' is

$$V_m = \frac{I}{2}\left[1 - \frac{z}{(z^2 + a^2)^{1/2}}\right].$$

This is given in Fig. 3.40.

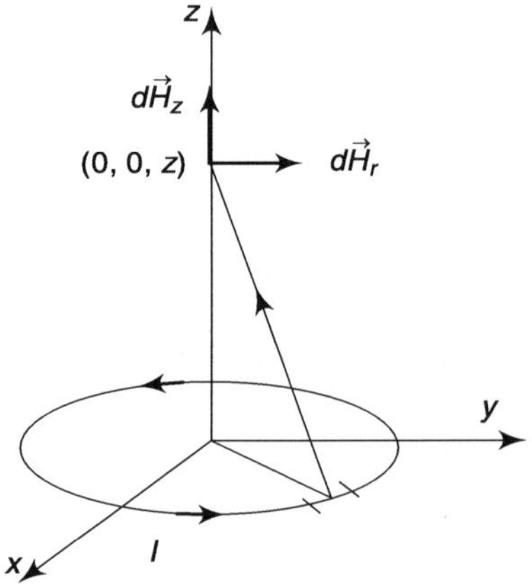

Fig. 3.40 *Circular current carrying loop*

Solution From Example 3.32, we have the magnetic field at (0, 0, z) due to the circular loop of radius a is given as

$$\vec{H} = \frac{Ia^2}{2[a^2 + z^2]^{3/2}}\hat{a}_z = -\nabla V_m$$

$$= -\left[\frac{\partial V_m}{\partial r}\hat{a}_r + \frac{1}{r}\frac{\partial V_m}{\partial \phi}\hat{a}_\phi + \frac{\partial V_m}{\partial z}\hat{a}_z\right]$$

Comparing both sides

$$\frac{\partial V_m}{\partial z} = -\frac{Ia^2}{2[a^2 + z^2]^{3/2}}$$

Integrating

$$V_m = -\int_{\infty}^{z} \frac{Ia^2}{2[a^2 + z^2]^{3/2}} dz$$

Let, $z = a \tan\theta \quad \therefore\ dz = a \sec^2 \theta d\theta$

$$\therefore \quad \tan\theta = \left(\frac{z}{a}\right) \quad \therefore \quad \sec^2\theta = \left(1 + \frac{z^2}{a^2}\right)$$

$$\therefore \quad \cos^2\theta = \frac{a^2}{a^2 + z^2}$$

$$\therefore \quad \sin^2\theta = (1 - \cos^2\theta) = \left(1 - \frac{a^2}{a^2 + z^2}\right) = \frac{z^2}{a^2 + z^2}$$

$$\therefore \quad \sin\theta = \frac{z}{\sqrt{a^2 + z^2}}$$

z	∞	z
θ	$\frac{\pi}{2}$	A

Here, $A = \tan^{-1}\left(\frac{z}{a}\right)$

$$\therefore \quad V_m = -\int_{\infty}^{z} \frac{Ia^2}{2[a^2 + z^2]^{3/2}} dz = -\int_{\pi/2}^{A} \frac{Ia^2 a \sec^2 \theta d\theta}{2a^3 \sec^3 \theta} = -\frac{I}{2} \int_{\pi/2}^{A} \cos \theta d\theta$$

$$= -\frac{I}{2}[\sin \theta]_{\pi/2}^{A} = -\frac{I}{2}(\sin \theta - 1) = \frac{I}{2}(1 - \sin \theta)$$

Putting the value $\sin \theta = \frac{z}{\sqrt{a^2 + z^2}}$, we get

$$\therefore \quad \boxed{V_m = \frac{I}{2}\left[1 - \frac{z}{\sqrt{a^2 + z^2}}\right]}$$

3.14.2 Magnetic Vector Potential

We know that the divergence of magnetic flux density is always zero everywhere ($\nabla \cdot \vec{B} = 0$). Hence, $\vec{B}$ can be expressed as the curl of some other vector function. We designate this vector as $\vec{A}$, which is known as *the magnetic vector potential.*

$$\therefore \quad \boxed{\vec{B} = \nabla \times \vec{A}} \tag{3.75}$$

Magnetic vector potential is expressed in Weber per metre (Wb/m) or in Newton per Ampere (*N/A*) or in Volt-second per metre (V-s/m); with its dimension as $MLI^{-1}T^{-2}$.

Now, by Ampere's law,

$$\nabla \times \vec{B} = \mu\vec{J} \quad \Rightarrow \quad \nabla \times (\nabla \times \vec{A}) = \mu\vec{J} \quad \Rightarrow \quad \nabla(\nabla \cdot \vec{A}) - \nabla^2 \vec{A} = \mu\vec{J}$$

If we let, $\nabla \cdot \vec{A} = 0$, which is called *Coulomb's gauge condition*, then we obtain

$$\boxed{\nabla^2 \vec{A} = -\mu\vec{J}} \tag{3.76}$$

This is similar to Poisson's equation of electrostatics, $\nabla^2 V = -\frac{\rho}{\varepsilon}$ whose solution is $V = \frac{1}{4\pi\varepsilon}\int_v \frac{\rho}{r} dv$. By comparison, we get the magnetic vector potential as

$$\vec{A} = \frac{\mu}{4\pi}\int_v \frac{\vec{J}}{r} dv \quad \text{for volume current density}$$

$$= \frac{\mu}{4\pi}\int_S \frac{\vec{K}}{r} dS \quad \text{for surface current density} \tag{3.77}$$

$$= \frac{\mu}{4\pi}\int_v \frac{Id\vec{l}}{r} \quad \text{for line current density}$$

The concept of magnetic vector potential is extremely useful for studying radiation in transmission lines, wave guides, antennas, etc.

NOTE

Magnetic scalar potential is defined only in regions where the current is zero ($\vec{J} = 0$); but magnetic vector potential is defined in regions with any finite value of current.

***Example 3.53** Find the vector magnetic potential and hence the magnetic flux density $\vec{B}$ due to an infinite wire carrying a current, at a point (i) inside, (ii) outside the wire.

Solution **(i) Inside the wire:**

Let a be the radius of the wire.

By symmetry, it is understood that only the z-component of the vector potential exists.

$$\nabla^2 A_z = -\mu J_z = -\frac{\mu I}{\pi a^2}$$

or

$$\frac{1}{r}\frac{\partial}{\partial r}\left(r\frac{\partial A_z}{\partial r}\right) = -\frac{\mu I}{\pi a^2}$$

Integrating

$$r\frac{\partial A_z}{\partial r} = -\frac{\mu I r^2}{2\pi a^2} + C_1$$

Since $r\frac{\partial A_z}{\partial r} = 0$ at $r = 0 \Rightarrow C_1 = 0$

Integrating again,

$$A_z = -\frac{\mu I r^2}{4\pi a^2} + C_2$$

Since $A_z = 0$ at $r = a, \Rightarrow C_2 = \frac{\mu I}{4\pi}$

$\therefore$

$$A_z = \frac{\mu I}{4\pi}\left[1 - \frac{r^2}{a^2}\right]$$

In vector form, the vector magnetic potential is given as

$$\vec{A} = \frac{\mu I}{4\pi}\left[1 - \frac{r^2}{a^2}\right]\hat{a}_z$$

Now, $\vec{B} = \nabla \times \vec{A}$

$\therefore$

$$B_r = (\text{curl } \vec{A})_r = \frac{1}{r}\frac{\partial A_z}{\partial \phi} - \frac{\partial A_\phi}{\partial z} = 0$$

$$B_\phi = (\text{curl } \vec{A})_\phi = \frac{\partial A_r}{\partial z} - \frac{\partial A_z}{\partial r} = \frac{\mu I r}{2\pi a^2}$$

$$B_z = (\text{curl } \vec{A})_z = \frac{1}{r}\frac{\partial (rA_\phi)}{\partial r} - \frac{1}{r}\frac{\partial A_r}{\partial \phi} = 0$$

Thus, the magnetic induction is

$$\vec{B} = \frac{\mu I r}{2\pi a^2}\hat{a}_\phi$$

(ii) Outside the wire:
Here,

$$\frac{1}{r}\frac{\partial}{\partial r}\left(r\frac{\partial A_z}{\partial r}\right) = 0$$

$$\Rightarrow \qquad \frac{\partial A_z}{\partial r} = \frac{C_1}{r} \qquad \textbf{(i)}$$

$$A_z = C_1 \ln r + C_2$$

At $r = a, A_z = 0 \quad \Rightarrow \quad C_2 = -C_1 \ln a$

$$\therefore \qquad A_z = C_1 \ln\left(\frac{r}{a}\right)$$

The constant C_1 is found from the boundary condition for $\frac{\partial A_z}{\partial r}$ at $r = a$.

Since $\vec{B} = \nabla \times \vec{A}$, $\therefore\ B_\phi = -\frac{\partial A_z}{\partial r}$

Now, B_ϕ must be continuous at $r = a$. From the result of (i), we get

$$-\frac{\partial A_z}{\partial r} = \frac{\mu I}{2\pi a} \qquad \textbf{(ii)}$$

From (i) and (ii), we get

$$\frac{C_1}{a} = -\frac{\mu I}{2\pi a} \quad \Rightarrow \quad C_1 = -\frac{\mu I}{2\pi}$$

$$\therefore \qquad A_z = -\frac{\mu I}{2\pi}\ln\left(\frac{r}{a}\right)$$

In vector form, the vector magnetic potential is given as

$$\vec{A} = -\frac{\mu I}{2\pi}\ln\left(\frac{r}{a}\right)\hat{a}_z$$

Proceeding in the same way as in (i), we get the magnetic induction as

$$\vec{B} = \frac{\mu I}{2\pi r}\hat{a}_\phi$$

To summarise the results

$$\vec{A} = \frac{\mu I}{4\pi}\left[1 - \frac{r^2}{a^2}\right]\hat{a}_z \qquad r < a$$

$$= -\frac{\mu I}{2\pi}\ln\left(\frac{r}{a}\right)\hat{a}_z \qquad r > a$$

$$\vec{B} = \frac{\mu I r}{2\pi a^2}\hat{a}_\phi \quad r < a$$

$$= \frac{\mu I}{2\pi r}\hat{a}_\phi \quad r > a$$

***Example 3.54** Using Ampere's circuital law in integral form, find $\vec{H}$ everywhere due to the current density in cylindrical co-ordinates:

$$\vec{J} = 0, \qquad 0 < r < a$$
$$= J_0\left(\frac{r}{a}\right)\hat{a}_z \qquad a < r < b$$
$$= 0 \qquad b < r < \infty$$

Also, find the vector magnetic potential.

Solution

$$\vec{J} = 0, \qquad 0 < r < a$$
$$= J_0\left(\frac{r}{a}\right)\hat{a}_z \qquad a < r < b$$
$$= 0 \qquad b < r < \infty$$

$0 < r < a$:

By Ampere's law,

$$\oint \vec{H}\cdot d\vec{l} = \int \vec{J}\cdot d\vec{S} = 0 \qquad (\because \vec{J} = 0)$$

$$\Rightarrow \qquad H = 0$$

$a \le r < b$:

By Ampere's law,

$$\oint \vec{H}\cdot d\vec{l} = \int \vec{J}\cdot d\vec{S}$$

$$\Rightarrow \quad H(2\pi r) = \int_{r=a}^{r}\int_{\phi=0}^{2\pi}\left(J_0\left(\frac{r}{a}\right)\hat{a}_z\right)\cdot(r\,dr\,d\phi\,\hat{a}_z) = \int_{r=a}^{r}\int_{\varphi=0}^{2\pi}\frac{J_0}{a}r^2\,dr\,d\phi = \frac{2\pi J_0}{a}\int_{r=a}^{r} r^2\,dr = \frac{J_0 2\pi}{3a}(r^3 - a^3)$$

$$\therefore \qquad \vec{H} = \frac{J_0}{3ar}(r^3 - a^3)\hat{a}_\phi$$

$b < r < \infty$:

By Ampere's law,

$$\oint \vec{H}\cdot d\vec{l} = \int \vec{J}\cdot d\vec{S} = 0 \qquad (\because \vec{J} = 0)$$

$$\Rightarrow \qquad H = 0$$

To summarise, the results are

$$\boxed{\begin{aligned} \vec{H} &= 0 && 0 < r < a \\ &= \frac{J_0}{3ar}(r^3 - a^3)\hat{a}_\phi && a \le r < b \\ &= 0 && b < r < \infty \end{aligned}}$$

The magnetic flux density is

$$\vec{B} = \mu_0 \vec{H} = \frac{\mu_0 J_0}{3ar}(r^3 - a^3)\hat{a}_\phi$$

Also, $\vec{B} = \nabla \times \vec{A} = \begin{vmatrix} \frac{1}{r}\hat{a}_r & \hat{a}_\varphi & \hat{a}_z \\ \frac{\partial}{\partial r} & \frac{\partial}{\partial \phi} & \frac{\partial}{\partial z} \\ A_r & A_\phi & A_z \end{vmatrix}$

As $\vec{B}$ has only $\hat{a}_\phi$ component, or since the current direction is in $\hat{a}_z$ direction, only A_z component of $\vec{A}$ will exist.

$\therefore$ $$A_r = 0 \qquad A_\phi = 0$$

$\therefore$ $$\nabla \times \vec{A} = -\frac{\partial A_z}{\partial r}\hat{a}_\phi$$

or, $$-\frac{\partial A_z}{\partial r} = \frac{\mu_0 J_0}{3ar}(r^3 - a^3)$$

Integrating,

$$A_z = \int_{r=a}^{r} -\frac{\mu_0 J_0}{3ar}(r^3 - a^3)dr = \int_{r=a}^{r} -\frac{\mu_0 J_0}{3a}\left(r^2 - \frac{a^3}{r}\right)dr$$

$$= -\frac{\mu_0 J_0}{3a}\left[\frac{r^3}{3} - a^3 \ln r\right]_a^r$$

$$= -\frac{\mu_0 J_0}{3a}\left[\left(\frac{r^3 - a^3}{3}\right) - a^3 \ln\left(\frac{r}{a}\right)\right]$$

$$A_z = \frac{\mu_0 J_0}{3a}\left[a^3 \ln\left(\frac{r}{a}\right) - \left(\frac{r^3 - a^3}{3}\right)\right]$$

$\therefore$ $$\boxed{\vec{A} = \frac{\mu_0 J_0}{3a}\left[a^3 \ln\left(\frac{r}{a}\right) - \left(\frac{r^3 - a^3}{3}\right)\right]\hat{a}_z}$$

***Example 3.55** Show that the vector potential due to moving charge q at a distance R is given by

$$\vec{A}(r) = \frac{\mu_0 q \vec{v}}{4\pi R},$$

$\vec{v}$ being velocity of charge.

Solution $\vec{A}(r) = \frac{\mu_0}{4\pi}\int \frac{Id\vec{l}}{R} = \frac{\mu_0}{4\pi R}\int Id\vec{l}$ ($\because$ Integration is w.r.t. $d\vec{l}$, R is constant)

$$= \frac{\mu_0}{4\pi R}\int \frac{dq}{dt}\vec{v}dt \qquad \left(\because I = \frac{dq}{dt},\quad \vec{v} = \frac{d\vec{l}}{dt}\right)$$

$$= \frac{\mu_0 \vec{v}}{4\pi R}\int dq$$

$$= \frac{\mu_0 q\vec{v}}{4\pi R}$$

$\therefore$

$$\boxed{\vec{A}(r) = \frac{\mu_0 q\vec{v}}{4\pi R}}$$

***Example 3.56** Using the concept of vector magnetic potential, find the magnetic flux density at a point due to a long straight filamentary conductor carrying a current I in the z-direction, as shown in Fig. 3.41.

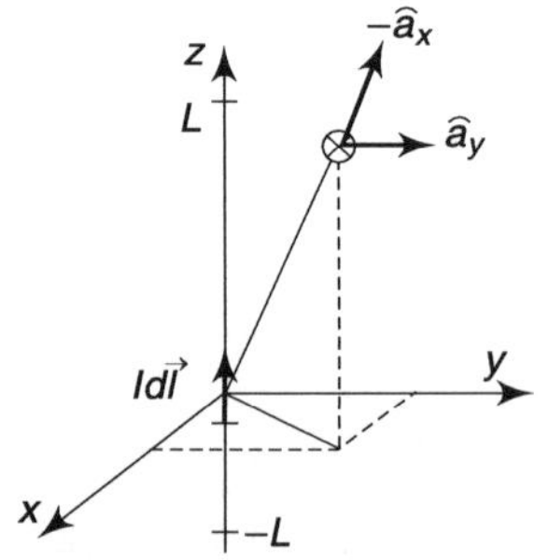

Fig. 3.41 *Long straight filamentary conductor*

Solution Vector magnetic potential

$$\vec{A}(r) = \frac{\mu_0}{4\pi}\int \frac{Id\vec{l}}{R} = \frac{\mu_0}{4\pi}\int_{-L}^{L} \frac{Idz}{R}\hat{a}_z$$

Current being in the z-direction, only A_z component of $\vec{A}$ will exist.

$\therefore$

$$A_z = \frac{\mu_0}{4\pi}\int_{-L}^{L}\frac{Idz}{R}$$

$$= 2\times\frac{\mu_0 I}{4\pi}\int_0^L \frac{dz}{\sqrt{x^2+y^2+z^2}}$$

$$= \frac{\mu_0 I}{2\pi}\int_0^L \frac{dz}{\sqrt{r^2+z^2}}$$

$$= \frac{\mu_0 I}{2\pi}\left[\ln(z+\sqrt{r^2+z^2})\right]_0^L$$

$$= \frac{\mu_0 I}{2\pi}\left[\ln(L+\sqrt{r^2+L^2}) - \ln r\right]$$

If $L >> r$, then

$$A_z = \frac{\mu_0 I}{2\pi}[\ln(L+\sqrt{r^2+L^2}) - \ln r] = \frac{\mu_0 I}{2\pi}[\ln 2L - \ln r] = \frac{\mu_0 I}{2\pi}\ln\left(\frac{2L}{r}\right)$$

$\therefore$

$$\boxed{\vec{A} = \frac{\mu_0 I}{2\pi}\ln\left(\frac{2L}{r}\right)\hat{a}_z}$$

Since only A_z exists, the magnetic flux density is given as

$$\vec{B} = \nabla\times\vec{A} = -\frac{\partial A_z}{\partial r}\hat{a}_\phi = -\frac{\mu_0 I}{2\pi}\frac{\partial}{\partial r}\left[\ln\left(\frac{2L}{r}\right)\right]\hat{a}_\phi = -\frac{\mu_0 I}{2\pi}\times\frac{r}{2L}\times\left(-\frac{2L}{r^2}\right)\hat{a}_\phi = \frac{\mu_0 I}{2\pi r}\hat{a}_\phi$$

$$\therefore \qquad \boxed{\vec{B} = \frac{\mu_0 I}{2\pi r}\hat{a}_\phi}$$

Example 3.57 Show that the vector magnetic potential at a distant point $P(x, y, z)$ due to a finite line current flowing through $-l \le z \le l$ along $\hat{a}_z$ is

$$\vec{A} = \frac{\mu_0 I l}{2\pi[x^2 + y^2 + z^2]^{1/2}}\hat{a}_z$$

Solution From Example 3.56, we have the vector magnetic potential given as

$$\vec{A}(r) = \frac{\mu_0 I}{4\pi}\int_{-l}^{l}\frac{d\vec{l}}{R} = 2\times\frac{\mu_0 I}{4\pi}\int_{0}^{l}\frac{d\vec{l}}{R} = \frac{\mu_0 I}{2\pi}\int_{0}^{l}\frac{dl}{\sqrt{x^2+y^2+z^2}}\hat{a}_z = \frac{\mu_0 I l}{2\pi[x^2+y^2+z^2]^{1/2}}\hat{a}_z$$

$$\therefore \qquad \boxed{\vec{A} = \frac{\mu_0 I l}{2\pi[x^2 + y^2 + z^2]^{1/2}}\hat{a}_z}$$

Example 3.58 A coaxial cable with the inner conductor of radius 'a' carries a current 'I' in $\hat{a}_z$ direction and the outer conductor of radius 'b' carries a current 'I' in the $-\hat{a}_z$ direction; between radius 'a' and 'b' there is no current. Find the magnetic vector potential at r, where $a < r < b$.

Solution At any point, $a < r < b$, $\vec{J} = 0$

$$\therefore \qquad \nabla^2\vec{A} = 0$$

In cylindrical coordinates,

$$\nabla^2\vec{A} \neq \nabla^2 A_r\hat{a}_r + \nabla^2 A_\phi\hat{a}_\phi + \nabla^2 A_z\hat{a}_z$$

Although in Cartesian coordinate systems, this type of equation holds.

In cylindrical coordinates, the z-component of the vector Laplacian is the scalar Laplacian of the z-component of $\vec{A}$.

i.e. $$\nabla^2 A\Big|_z = \nabla^2 A_z$$

The current has only the z-direction, so only A_z component will exist.

$$\therefore \qquad \nabla^2 A_z = 0$$

or, $$\frac{1}{r}\frac{\partial}{\partial r}\left(r\frac{\partial A_z}{\partial r}\right) = 0$$

Integrating twice, we get

$$A_z = C_1 \ln r + C_2$$

where, C_1 and C_2 are the integration constants.

Following the same procedure as in solved example 3.16, we have the results as

$$\boxed{\vec{A} = \frac{\mu_0 I}{2\pi}\ln\left(\frac{b}{r}\right)\hat{a}_z} \quad \text{and} \quad \boxed{\vec{H} = \frac{I}{2\pi r}\hat{a}_\phi}$$

***Example 3.59** An infinitely long conductor of radius a is placed such that its axis is along the z-axis. The vector magnetic potential, due to a direct current I_0 flowing along $\hat{a}_z$ in the conductor is given by:

$$\vec{A} = -\frac{I_0}{4\pi a^2}\mu_0(x^2 + y^2)\hat{a}_z \text{ Wb/m}$$

Find the corresponding $\vec{H}$. Also confirm the result using Ampere's law.

Solution The magnetic flux density is given as

$$\vec{B} = \nabla \times \vec{A} = \begin{vmatrix} \hat{a}_x & \hat{a}_y & \hat{a}_z \\ \frac{\partial}{\partial x} & \frac{\partial}{\partial y} & \frac{\partial}{\partial z} \\ 0 & 0 & -\frac{I_0}{4\pi a^2}\mu_0(x^2+y^2) \end{vmatrix} = -\frac{I_0}{2\pi a^2}\mu_0(y\hat{a}_x - x\hat{a}_y)$$

So, the magnetic field intensity is given as

$$\vec{H} = \frac{\vec{B}}{\mu_0} = -\frac{I_0}{2\pi a^2}(y\hat{a}_x - x\hat{a}_y)$$

We calculate the closed line integral of this field as follows

$$\begin{aligned}
\oint_L \vec{H}\cdot d\vec{l} &= -\frac{I_0}{2\pi a^2}\oint_L (y\hat{a}_x - x\hat{a}_y)\cdot(ad\phi\hat{a}_\phi) = -\frac{I_0}{2\pi a^2}\oint_L ad\phi(y\hat{a}_x - x\hat{a}_y)\cdot(\hat{a}_\phi) \\
&= -\frac{I_0}{2\pi a^2}\oint_L ad\phi(y\hat{a}_x - x\hat{a}_y)\cdot(-\sin\phi\hat{a}_x + \cos\phi\hat{a}_y) \\
&= -\frac{I_0}{2\pi a^2}\oint_L ad\phi(-y\sin\phi - x\cos\phi) \\
&= \frac{I_0}{2\pi a^2}\oint_L ad\phi(a\sin^2\phi + a\cos^2\phi) \qquad \{\because x = r\cos\phi \text{ and } y = r\sin\phi\} \\
&= \frac{I_0}{2\pi}\oint_L d\phi(\sin^2\phi + \cos^2\phi) \\
&= \frac{I_0}{2\pi}\oint_L d\phi = \frac{I_0}{2\pi}\times 2\pi = I_0
\end{aligned}$$

Since $\oint_L \vec{H}\cdot d\vec{l} = I_0$, Ampere's law is verified.

***Example 3.60** Given the magnetic vector potential, $\vec{A} = -\frac{\rho^2}{4}\hat{a}_z$ Wb/m, calculate the total flux crossing the surface $\phi = \pi/2$, $1 \le \rho \le 2$ m, $0 \le z \le 5$ m.

Solution The magnetic flux density is, $\vec{B} = \nabla\times\vec{A} = -\frac{\partial A_z}{\partial\rho}\hat{a}_\phi = \frac{\rho}{2}\hat{a}_\phi$

Differential surface is given as, $d\vec{S} = d\rho dz\hat{a}_\phi$

Hence, total flux crossing the given surface is given as

$$\phi = \int_S \vec{B} \cdot d\vec{S} = \int_{z=0}^{5} \int_{\rho=1}^{2} \frac{\rho}{2} \hat{a}_\phi \cdot d\rho dz \hat{a}_\phi = \frac{1}{2} \int_{z=0}^{5} \int_{\rho=1}^{2} \rho d\rho dz = \frac{1}{4}[\rho^2]_1^2 \times 5 = \frac{15}{4} \text{ Wb}$$

3.14.3 Derivation of Magnetic Vector Potential using Biot–Savart Law

By Biot–Savart law, the magnetic field intensity at a point $P(\vec{r})$ due to an element $d\vec{l}$ at a positional vector $\vec{r}'$ is

$$\vec{H} = \frac{1}{4\pi} \int_l \frac{I d\vec{l} \times \vec{R}}{R^3} \quad \textbf{(3.78)}$$

where, $\vec{R} = (\vec{r} - \vec{r}') = (x - x')\hat{i} + (y - y')\hat{j} + (z - z')\hat{k}$

Source point and field point are illustrated in Fig. 3.42. Hence,

$$\nabla\left(\frac{1}{R}\right) = -\frac{(x - x')\hat{i} + (y - y')\hat{j} + (z - z')\hat{k}}{[(x - x')^2 + (y - y')^2 + (z - z')^2]^{3/2}}$$

$$= -\frac{\vec{R}}{R^3} \quad \textbf{(3.79)}$$

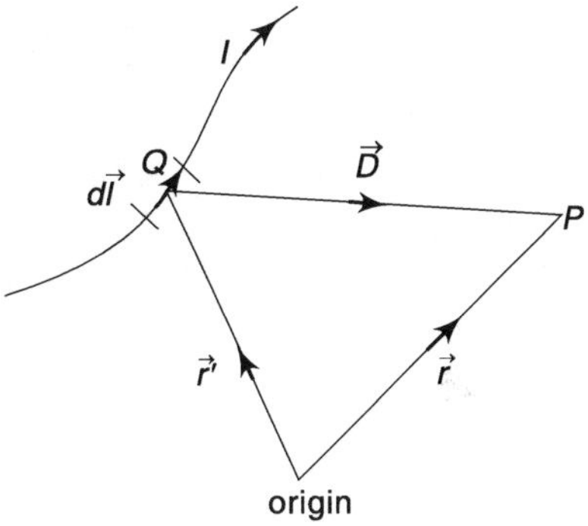

Fig. 3.42 *Illustration of source point (x′, y′, z′) and field point (x, y, z)*

From Eq. (3.78) and (3.79), we get

$$\vec{H} = -\frac{1}{4\pi} \int_l I d\vec{l} \times \nabla\left(\frac{1}{R}\right) = \frac{I}{4\pi} \int_l \nabla\left(\frac{1}{R}\right) \times d\vec{l} \quad \textbf{(3.80)}$$

We have the vector identity

$\nabla \times (S\vec{V}) = S\nabla \times \vec{V} + \nabla S \times \vec{V}$ where S is a scalar and $\vec{V}$ is a vector. Taking, $S = \left(\frac{1}{R}\right)$ and $\vec{V} = d\vec{l}$, we have

$$\nabla \times \left(\frac{d\vec{l}}{R}\right) = \frac{1}{R}(\nabla \times d\vec{l}) + \nabla\left(\frac{1}{R}\right) \times d\vec{l} = 0 + \nabla\left(\frac{1}{R}\right) \times d\vec{l}$$

{$\because \nabla$ operates w. r. t. (x, y, z) while $d\vec{l}$ is a function of (x', y', z'); $\therefore \nabla \times d\vec{l} = 0$}

$$\therefore \quad \nabla\left(\frac{1}{R}\right) \times d\vec{l} = \nabla \times \left(\frac{d\vec{l}}{R}\right) \quad \textbf{(3.81)}$$

From Eq. (3.80) and (3.81),

$$\vec{H} = \frac{I}{4\pi} \int_l \nabla \times \left(\frac{d\vec{l}}{R}\right) = \nabla \times \int_l \frac{I d\vec{l}}{4\pi R} \quad \textbf{(3.82)}$$

(Since integration and differentiation are w. r. t. two different sets of variable, we can interchange the order.)

$$\therefore \quad \vec{B} = \mu \vec{H} = \nabla \times \int_l \frac{\mu I d\vec{l}}{4\pi R} \quad \textbf{(3.83)}$$

By definition of magnetic vector potential as, $\vec{B} = \nabla \times \vec{A}$, we get

$$\boxed{\vec{A} = \int_l \frac{\mu I d\vec{l}}{4\pi R} = \int_S \frac{\mu \vec{K} dS}{4\pi R} = \int_v \frac{\mu \vec{J} dv}{4\pi R}}$$

3.14.4 Derivation of Magnetic Flux in Terms of Magnetic Vector Potential

We know that the magnetic flux coming out of a surface is given as

$$\phi = \int_S \vec{B} \cdot d\vec{S}$$

where, $\vec{B}$ is the magnetic flux density. Writing this in terms of magnetic vector potential as, $\vec{B} = \nabla \times \vec{A}$ and applying Stokes' theorem, we obtain

$$\phi = \int_S \vec{B} \cdot d\vec{S} = \int_S (\nabla \times \vec{A}) \cdot d\vec{S} = \oint_l \vec{A} \cdot d\vec{l}$$

$\therefore$
$$\boxed{\phi = \oint_l \vec{A} \cdot d\vec{l}} \quad \textbf{(3.84)}$$

Example 3.61 A current distribution gives rise to the vector magnetic potential $\vec{A} = x^2 y\hat{i} + y^2 x\hat{j} - 4xyz\hat{k}$ Wb/m. This is illustrated in Fig. 3.43. Calculate:

(a) $\vec{B}$ at (–1, 2, 5)

(b) The flux through the surface defined by $z = 1$, $0 \le x \le 1$, $-1 \le y \le 4$.

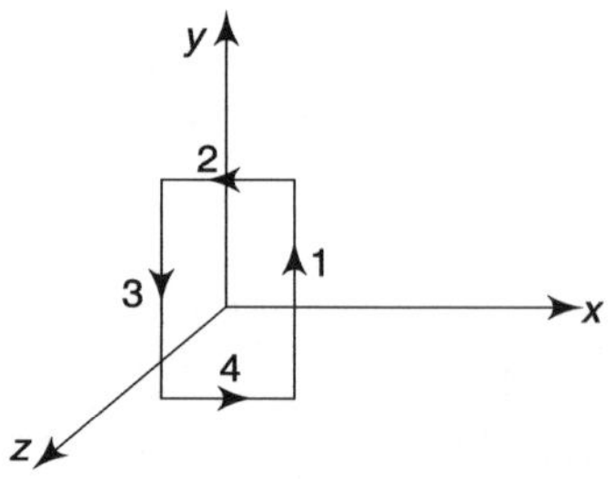

Fig. 3.43 *Arrangement of Example 3.61*

Solution

(a) $\vec{B} = \nabla \times \vec{A} = \begin{vmatrix} \hat{i} & \hat{j} & \hat{k} \\ \dfrac{\partial}{\partial x} & \dfrac{\partial}{\partial y} & \dfrac{\partial}{\partial z} \\ x^2 y & y^2 x & -4xyz \end{vmatrix} = -4xz\hat{i} + 4yz\hat{j} + (y^2 - x^2)\hat{k}$

$\therefore \quad \vec{B}_{(-1,2,5)} = -4 \times (-1) \times 5\hat{i} + 4 \times 2 \times 5\hat{j} + (4-1)\hat{k} = 20\hat{i} + 40\hat{j} + 3\hat{k}$ Wb/m^2

(b) Total flux is given as

$$\phi = \int_L \vec{A} \cdot d\vec{l} = \phi_1 + \phi_2 + \phi_3 + \phi_4$$

where L is the path bounding the surface S, ϕ_1, ϕ_2, ϕ_3, ϕ_4, are the fluxes along the segments of the path.

$$\phi_1 = \left[\int_{y=-1}^{4} (x^2 y\hat{i} + y^2 x\hat{j} - 4xyz\hat{k}) \cdot (dy\hat{j}) \right]_{x=1,\, z=1}$$

$$= \int_{y=-1}^{4} y^2 dy = \frac{y^3}{3}\bigg|_{-1}^{4} = \frac{65}{3}$$

$$\phi_3 = \left[\int_{y=4}^{-1} (x^2 y\hat{i} + y^2 x\hat{j} - 4xyz\hat{k}) \cdot (-dy\hat{j}) \right]_{x=0,\, z=1} = 0$$

$$\phi_2 = \left[\int_{x=1}^{0} (x^2 y\hat{i} + y^2 x\hat{j} - 4xyz\hat{k}) \cdot (-dx\hat{i}) \right]_{y=4,\, z=1} = -\int_{x=1}^{0} 4x^2 dx = -\frac{4x^3}{3}\Bigg|_1^0 = -\frac{4}{3}$$

$$\phi_4 = \left[\int_{x=0}^{1} (x^2 y\hat{i} + y^2 x\hat{j} - 4xyz\hat{k}) \cdot (dx\hat{i}) \right]_{y=-1,\, z=1} = -\int_{x=0}^{1} x^2 dx = -\frac{x^3}{3}\Bigg|_0^1 = -\frac{1}{3}$$

By summation, total flux through the surface is

$$\phi = \phi_1 + \phi_2 + \phi_3 + \phi_4 = \frac{65}{3} + 0 - \frac{4}{3} - \frac{1}{3} = 20 \text{ Wb}$$

3.15 FORCES DUE TO MAGNETIC FIELDS

There are three ways of experiencing force due to magnetic fields:

1. Force on a moving charged particle in a magnetic field,
2. Force on a current carrying conductor in a magnetic field, and
3. Force between two current carrying conductors (*Ampere's force law*).

3.15.1 Force on a Moving Charged Particle in a Magnetic Field

As discussed in Chapter 2, we know that the electric force on a stationary charge Q in an electric field $\vec{E}$, given by Coulomb's law is

$$\vec{F}_e = Q\vec{E} \tag{3.85}$$

Now if we allow this charge to move with a velocity $\vec{v}$ in the presence of a magnetic field, $\vec{B}$, then experimentally we have the following observations:

1. The magnitude of the magnetic force $\vec{F}_m$ exerted on the charged particle is proportional to both $\vec{v}$ and Q.
2. The magnitude and direction of $\vec{F}_m$ depends on both $\vec{v}$ and $\vec{B}$.
3. The magnetic force $\vec{F}_m$ vanishes when $\vec{v}$ is parallel to $\vec{B}$. However, when $\vec{v}$ makes an angle θ with $\vec{B}$, the direction of $\vec{F}_m$ is perpendicular to the plane formed by $\vec{v}$ and $\vec{B}$, and the magnitude of $\vec{F}_m$ is proportional to sin θ.
4. When the sign of the charge of the particle is switched from positive to negative (or vice versa), the direction of $\vec{F}_m$ also reverses.

The above observations can be summarized with the following equation

$$\vec{F}_m = Q\vec{v} \times \vec{B} \tag{3.86}$$

The magnitude of the force is given as,

$$F_m = |Q|\, vB \sin\theta \tag{3.87}$$

When the charged particle moves in the presence of both the electric as well as the magnetic field, the total force on the charge is

$$\vec{F} = \vec{F}_e + \vec{F}_m = Q\vec{E} + Q\vec{v} \times \vec{B} = Q(\vec{E} + \vec{v} \times \vec{B})$$

$$\therefore \qquad \boxed{\vec{F} = Q(\vec{E} + \vec{v} \times \vec{B})} \qquad \textbf{(3.88)}$$

This equation is known as *Lorentz force equation*, which relates the mechanical force to electrical force.

Comparison between Electric Force and Magnetic Force	
Electric Force ($\vec{F}_e = Q\vec{E}$)	**Magnetic Force ($\vec{F}_m = Q\vec{v} \times \vec{B}$)**
(i) It is in the same direction as the field $\vec{E}$.	(i) It is perpendicular to both $\vec{v}$ and $\vec{B}$.
(ii) It can perform work.	(ii) It cannot perform work.
(iii) It is independent of the velocity of charge.	(iii) It depends upon the velocity of charge.
(iv) It can produce change in kinetic energy.	(iv) It cannot produce change in kinetic energy.

3.15.2 Force on a Current Carrying Conductor in a Magnetic Field

We consider an element $d\vec{l}$ of a conductor carrying a current I. The direction of the vector $d\vec{l}$ is that of the current, so that $d\vec{l}$ is parallel to the velocity $\vec{v}$ of charge carriers inside the conductor.

∴ Number of charge carriers in the element $d\vec{l} = NAdl$

where, N is the number of charge carriers per unit volume;
A is the cross-sectional area of the conductor

Force on each of these charge carriers $= Q\vec{v} \times \vec{B}$

∴ Total force on all the charge carriers, i.e., the force acting on the conductor itself is

$$d\vec{F} = NAdlQ\vec{v} \times \vec{B} = NQvAd\vec{l} \times \vec{B} \qquad \{\because d\vec{l} \text{ is parallel to } \vec{v}\}$$
$$= Id\vec{l} \times \vec{B} \qquad \{\because I = NQvA\}$$

Hence, for a finite length of the conductor, the force exerted on it is

$$\boxed{\vec{F} = \int_l Id\vec{l} \times \vec{B}} \qquad \textbf{(3.89)}$$

Instead of line current, if we have surface current $\vec{K}$ or volume current $\vec{J}$, then the force equation becomes

$$\boxed{\vec{F} = \int_l Id\vec{l} \times \vec{B} = \int_S \vec{K}dS \times \vec{B} = \int_v \vec{J}dv \times \vec{B}} \qquad \textbf{(3.90)}$$

3.15.3 Force between Two Current Carrying Conductors (*Ampere's Force Law*)

We consider two loops C_1 and C_2 carrying currents I_1 and I_2, respectively as shown in Fig. 3.44.

Let

dl_1, dl_2 be the directed elements of lengths;

$R\widehat{R}_{12}$ be the directed distance between the elements;

$\widehat{R}_{12}$ be the unit vector drawn from 1 to 2.

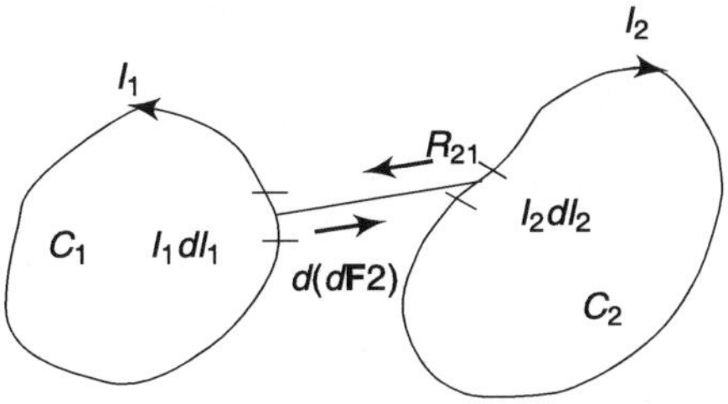

Fig. 3.44 *Force between two current carrying conductors*

According to the Biot–Savart law, both current carrying elements produce fields. Thus, the force on the element $I_2d\vec{l}_2$ due to the magnetic field $d\vec{B}_1$ produced by the element $I_1d\vec{l}_1$ is

$$d(d\vec{F}_2) = I_2d\vec{l}_2 \times d\vec{B}_1$$

But, from Biot–Savart law

$$d\vec{B}_1 = \frac{\mu I_1 d\vec{l}_1 \times \widehat{R}_{12}}{4\pi R^2}$$

Hence,

$$d(d\vec{F}_2) = \frac{\mu I_2 d\vec{l}_2 \times (I_1 d\vec{l}_1 \times \widehat{R}_{12})}{4\pi R^2}$$

Thus, the total force on loop C_2 due to loop C_1 is

$$\boxed{\vec{F}_2 = \frac{\mu I_1 I_2}{4\pi} \oint_{C_1} \oint_{C_2} \frac{d\vec{l}_2 \times (d\vec{l}_1 \times \widehat{R}_{12})}{R^2}} \tag{3.91}$$

This equation is essentially the law of force between two current elements; it is known as *Ampere's force law* and is analogous to Coulomb's law.

Similarly, force on loop C_1 due to loop C_2 can be found and it will be seen that $\vec{F}_1 = -\vec{F}_2$; thus $\vec{F}_1$ and $\vec{F}_2$ obey Newton's third law that action and reaction must be equal and opposite.

Example 3.62 A rectangular loop of length l and width w carries a steady current I_1. The loop is then placed near an infinitely long wire carrying a current I_2, as shown in Fig. 3.45. What is the magnetic force experienced by the loop due to the magnetic field of the wire?

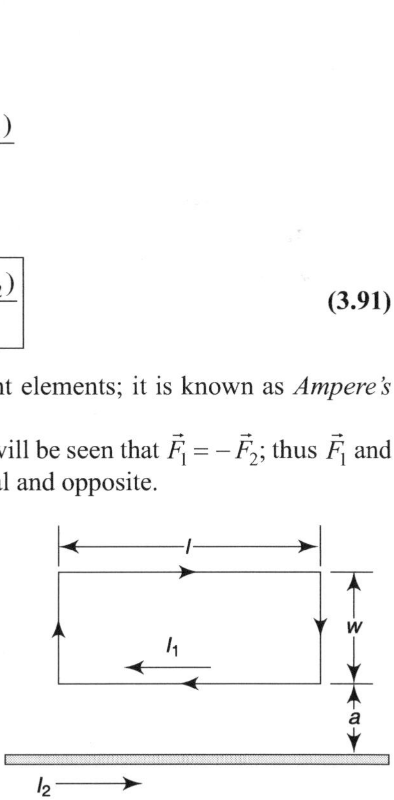

Fig. 3.45 *Magnetic force on a current loop*

Solution The forces are shown in Fig. 3.46.
The magnetic induction due to the infinitely long wire is

$$\vec{B}_2 = \frac{\mu I_2}{2\pi r}\widehat{a}_\phi$$

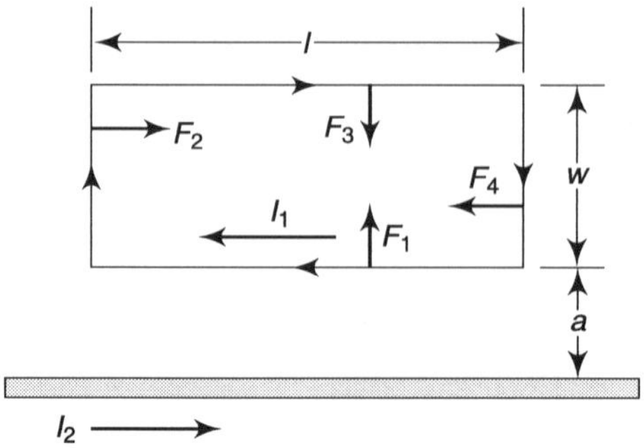

Fig. 3.46 *Magnetic forces on the loop and the wire*

The force on the loop is given as

$$\vec{F} = \vec{F}_1 + \vec{F}_2 + \vec{F}_3 + \vec{F}_4$$

Here,

$$\vec{F}_1 = I_1 \int d\vec{l}_1 \times \vec{B}_2 = I_1 \int_{z=l}^{0} dz\hat{a}_z \times \frac{\mu I_2}{2\pi r}\hat{a}_\phi \Bigg|_{r=a} = \frac{\mu I_1 I_2 l}{2\pi a}\hat{a}_r \quad \text{(repulsive)}$$

$$\vec{F}_3 = I_1 \int d\vec{l}_3 \times \vec{B}_2 = I_1 \int_{z=0}^{l} dz\hat{a}_z \times \frac{\mu I_2}{2\pi r}\hat{a}_\phi \Bigg|_{r=a+w} = -\frac{\mu I_1 I_2 l}{2\pi(a+w)}\hat{a}_r \quad \text{(attractive)}$$

$$\vec{F}_2 = I_1 \int d\vec{l}_2 \times \vec{B}_2 = I_1 \int_{r=0}^{w} dr\hat{a}_r \times \frac{\mu I_2}{2\pi r}\hat{a}_\phi = \frac{\mu I_1 I_2}{2\pi}\ln w\,\hat{a}_z \quad \text{(parallel)}$$

$$\vec{F}_4 = I_1 \int d\vec{l}_2 \times \vec{B}_2 = I_1 \int_{r=w}^{0} dr\hat{a}_r \times \frac{\mu I_2}{2\pi r}\hat{a}_\phi = -\frac{\mu I_1 I_2}{2\pi}\ln w\,\hat{a}_z \quad \text{(parallel)}$$

Thus, the total force on the loop is

$$\vec{F} = \vec{F}_1 + \vec{F}_2 + \vec{F}_3 + \vec{F}_4 = \frac{\mu I_1 I_2 l}{2\pi a}\hat{a}_r - \frac{\mu I_1 I_2 l}{2\pi(a+w)}\hat{a}_r = \frac{\mu I_1 I_2 l}{2\pi}\left[\frac{1}{a} - \frac{1}{a+w}\right]\hat{a}_r$$

***Example 3.63** Find the force between two straight, infinite, parallel wire carrying currents I_1 and I_2 separated by a distance d and placed in air.

Solution We consider two parallel wires separated by a distance d and carrying currents I_1 and I_2 in the x-direction, as shown in Fig. 3.47.

The magnetic force, $\vec{F}_{12}$, exerted on wire 1 by wire 2 may be computed as follows:

At an arbitrary point P on wire 1, the magnetic field due to current I_2 on wire 2 is

$$\vec{B}_2 = -\frac{\mu_0 I_2}{2\pi d}\hat{a}_y$$

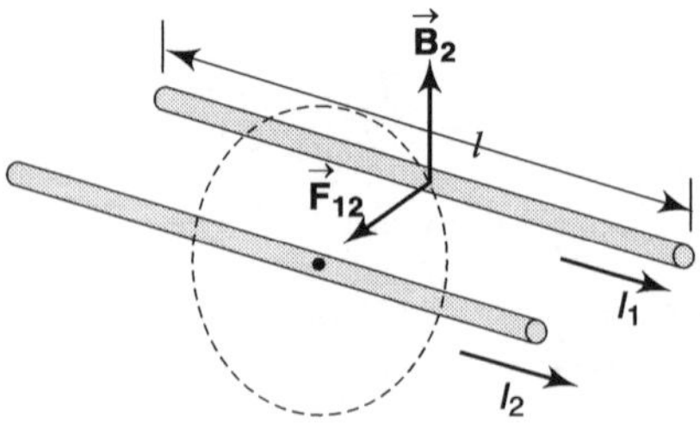

Fig. 3.47 *Force between two parallel wires*

This field which points in the direction perpendicular to wire 1, as depicted in the figure. Therefore the force on wire 2 due to wire 1 is

$$\vec{F}_{12} = I_1\vec{l} \times \vec{B}_2 = I_1(l\hat{a}_x) \times \left(-\frac{\mu_0 I_2}{2\pi d}\hat{a}_y\right) = -\frac{\mu_0 I_1 I_2}{2\pi d}\hat{a}_z$$

Clearly $\vec{F}_{12}$ points toward wire 2.

The same force will be acting on wire 1 due to wire 2.

In general, the magnitude of the force may be written as

$$\boxed{F = \frac{\mu_0 I_1 I_2}{2\pi d}}$$

The conclusion we can draw from this simple calculation is that the force between two parallel wires carrying currents:

— is attractive if the currents are in the same direction.

— is repulsive if the currents are in the opposite directions.

Example 3.64 Three very long parallel conductors are in free space. They lie in one plane spaced 50 cm. Each of the conductor carries a current of 100 amperes so that in the first and second the current has the same direction. What is the force acting a metre of the first, second (middle) and the third conductors?

Solution We use the results of the earlier problem to solve this problem.

Force on first conductor:

Force due to second conductor, $\vec{F}_{21} = \dfrac{\mu_0 I^2}{2\pi d}$ (attractive)

Force due to third conductor, $\vec{F}_{31} = \dfrac{\mu_0 I^2}{2\pi(d+d)} = \dfrac{\mu_0 I^2}{4\pi d}$ (repulsive)

Net force due these two conductors

$$F_1 = \frac{\mu_0 I^2}{2\pi d} - \frac{\mu_0 I^2}{4\pi d} = \frac{\mu_0 I^2}{4\pi d} = \frac{4\pi \times 10^{-7} \times (100)^2}{4 \times \pi \times 0.5} = 2\text{mN (attractive)}$$

Force on second conductor:

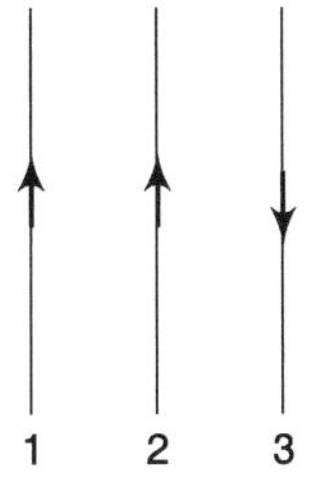

Force due to first conductor $\vec{F}_{12} = \dfrac{\mu_0 I^2}{2\pi d}$ (attractive)

Force due to third conductor $\vec{F}_{32} = \dfrac{\mu_0 I^2}{2\pi d}$ (repulsive)

Net force due these two conductors $F_1 = \dfrac{\mu_0 I^2}{2\pi d} - \dfrac{\mu_0 I^2}{2\pi d} = 0$

Force on third conductor:

Force due to first conductor $\vec{F}_{13} = \dfrac{\mu_0 I^2}{2\pi\,(d+d)} = \dfrac{\mu_0 I^2}{4\pi d}$ (repulsive)

Force due to second conductor $\vec{F}_{23} = \dfrac{\mu_0 I^2}{2\pi d}$ (repulsive)

Net force due to these two conductors

$$F_1 = \frac{\mu_0 I^2}{2\pi d} + \frac{\mu_0 I^2}{4\pi d} = \frac{3\mu_0 I^2}{4\pi d} = 3 \times \frac{4\pi \times 10^{-7} \times (100)^2}{4 \times \pi \times 0.5} = 6 \text{ mN (repulsive)}$$

Example 3.65 Determine the force between two parallel circular co-axial coils of radius '*R*', as shown in Fig. 3.48, which are a small distance '*d*' apart in free space and carry currents I_1 and I_2. Assume that each of the coils has a single turn.

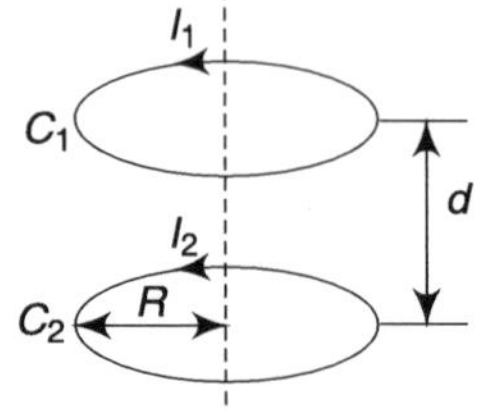

Fig. 3.48 *Parallel co-axial circular coils*

Solution As the distance between the coils is very small compared to the radius of the coils, the two coils can be treated as parallel current carrying wire. The force on one of the coils (say C_1) per unit length due to the other coil (say C_2) can be written as obtained in Example 3.63, as

$$F_0 = \frac{\mu_0 I_1 I_2}{2\pi d}$$

Therefore, the net force is given as $F = 2\pi R F_0 = \dfrac{\mu_0 I_1 I_2 R}{d}$

$$\therefore \qquad \boxed{F = \frac{\mu_0 I_1 I_2 R}{d}}$$

Example 3.66 Calculate the force between two parallel circular coaxial coils of nearly the same size and carrying current, being separated by a small distance in free space. For what distance between the coils is this force a maximum?

Solution This is shown in Fig. 3.49. We consider two coils C_1 and C_2 of radii R_1 and R_2 respectively, carrying currents I_1 and I_2, respectively. Let d be the separation distance between the centres of the coils. Force per unit length on either coil is

$$F_0 = \frac{\mu_0 I_1 I_2}{2\pi l}$$

where, l is the distance of the line ab.

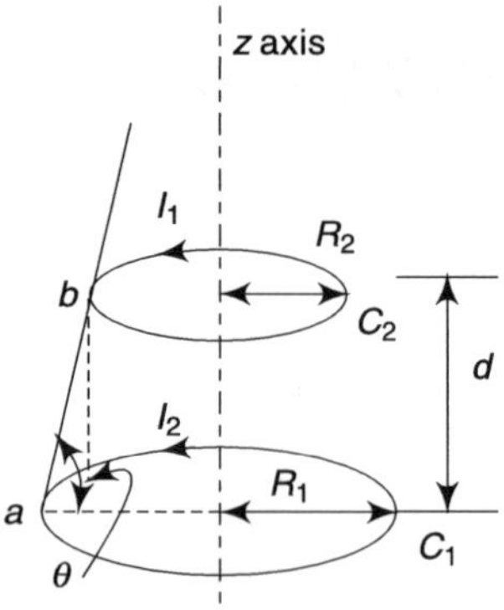

Fig. 3.49 *Parallel co-axial circular coils*

By symmetry, it is seen that the component of the force F_0 perpendicular to the z-axis will be cancelled when the entire coil is considered. The component of F_0 along the z-axis is

$$F_{0z} = F_0 \sin\theta = \frac{\mu_0 I_1 I_2}{2\pi l}\frac{d}{l} = \frac{\mu_0 I_1 I_2 d}{2\pi l^2}$$

For the entire coil, the length is $2\pi R_1 \approx 2\pi R_2$. Hence, the total force along the z-axis is

$$F = 2\pi R_2 F_{0z} = \frac{\mu_0 I_1 I_2 R_2 d}{l^2} = \mu_0 I_1 I_2 R_2 \frac{d}{d^2 + (R_1 - R_2)^2}$$

This force will be maximum when

$$\frac{dF}{dd} = 0$$

or, $$\frac{d}{dd}\left[\mu_0 I_1 I_2 R_2 \frac{d}{d^2 + (R_1 - R_2)^2}\right] = 0$$

or, $$\frac{d}{dd}\left[\frac{d}{d^2 + (R_1 - R_2)^2}\right] = 0$$

or, $$(R_1 - R_2)^2 + d^2 - 2d^2 = 0$$

or, $$d = (R_1 - R_2)$$

This implies that the force will be the maximum when the distance between the centres of the coils is equal to the difference between their radii.

3.16 MAGNETIC TORQUE

In Section 3.15, we have seen that a current carrying conductor placed in a magnetic field experiences a force that tends to move the conductor in a direction perpendicular to both the magnetic field and the conductor. However, if a current carrying coil is placed in a magnetic field, the magnetic force imparted may be a twisted force or moment which may rotate the conductor. In this section, we will find the magnetic torque and moment.

To determine the torque acting on a current loop in a magnetic field, we shall consider a rectangular current loop *abcd* carrying a current *I*, as shown in Fig. 3.50.

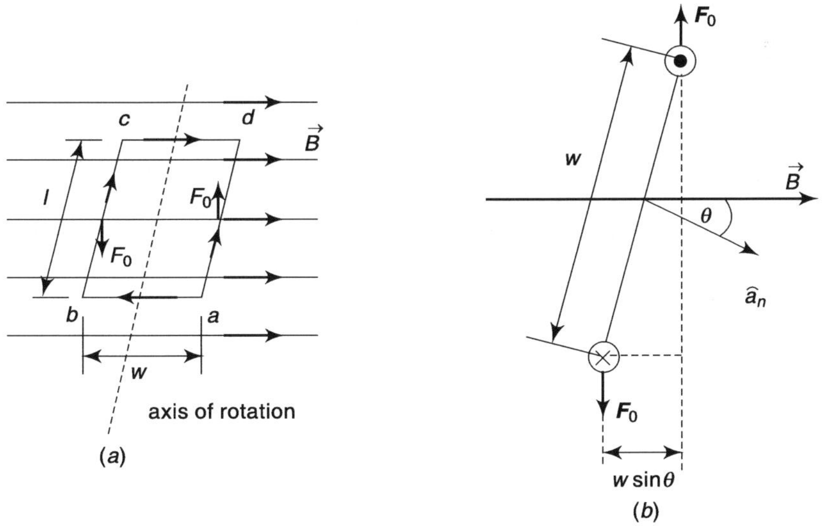

Fig. 3.50 *Rectangular loop in uniform magnetic field*

Let

l = length of the rectangular loop;
w = width of the rectangular loop;
$\vec{B}$ = uniform magnetic field

Total force acting on the loop is

$$\vec{F} = I\left[\int_a^b + \int_b^c + \int_c^d + \int_d^a d\vec{l} \times \vec{B}\right]$$

From Fig. 3.50, we notice that $d\vec{l}$ is parallel to $\vec{B}$ along sides *ab* and *cd* of the loop and no force is exerted on these sides. Thus, the total force acting on the loop is

$$\vec{F} = I\int_b^c d\vec{l} \times \vec{B} + I\int_d^a d\vec{l} \times \vec{B} = I\int_0^l dz\hat{a}_z \times \vec{B} + I\int_l^0 dz\hat{a}_z \times \vec{B} = \vec{F}_0 - \vec{F}_0 = 0$$

where, $|\vec{F}_0| = BIl$, as $\vec{B}$ is uniform. Thus, the force exerted on the loop is zero. However, the forces $\vec{F}_0$ and $-\vec{F}_0$ creates a torque.

If θ is the angle between $\vec{B}$ and the normal to the plane of the loop, then the torque is given by

$$|\vec{T}| = |\vec{F}_0|\, w \sin\theta \quad \text{or}$$

$$\boxed{T = BIlw \sin\theta = BIS \sin\theta} \qquad \textbf{(3.92)}$$

where, $S = lw$ is the area of the loop.

In vector form,

$$\boxed{\vec{T} = I\vec{S} \times \vec{B}} \qquad \textbf{(3.93)}$$

We define the *magnetic dipole moment* as

$$\boxed{\vec{m} = I\vec{S} = IS\hat{a}_n} \qquad \textbf{(3.94)}$$

where, $\hat{a}_n$ is the unit vector normal to the plane of the loop and its direction is determined by the right-hand rule.

Magnetic Dipole Moment **($\vec{m}$):** It is the product of current and area of the loop; its direction is normal to the plane of the loop; its unit is Am^2.

$$\vec{m} = I\vec{S} = IS\hat{a}_n$$

In terms of the magnetic dipole moment, the torque can be written as

$$\boxed{\vec{T} = \vec{m} \times \vec{B}} \qquad \textbf{(3.95)}$$

NOTE

(i) *This expression is applicable in determining torque on a planar loop of any arbitrary shape.*

(ii) *The torque is in the direction of the axis of rotation. It is directed so as to reduce the angle θ, so that $\vec{m}$ and $\vec{B}$ are in the same direction. In equilibrium, the loop is perpendicular to the magnetic field and thus, both torque and force on the loop are zero.*

3.17 MAGNETIC DIPOLE

A bar magnet or a small filamentary loop carrying a current is known as a *magnetic dipole*.

We will now find out the magnetic field produced by a magnetic dipole. We consider a circular loop of radius a carrying a current I as shown in Fig. 3.51 (a).

Magnetic vector potential at a point $P(r, \theta, \phi)$ is given as

$$\vec{A} = \frac{\mu I}{4\pi}\oint_l \frac{d\vec{l}}{R} \qquad \textbf{(3.96)}$$

Here, $d\vec{l} = ad\phi\hat{a}_\phi$.

At a given r and θ, $\vec{A}$ will be independent of ϕ. For the observation point P, we have

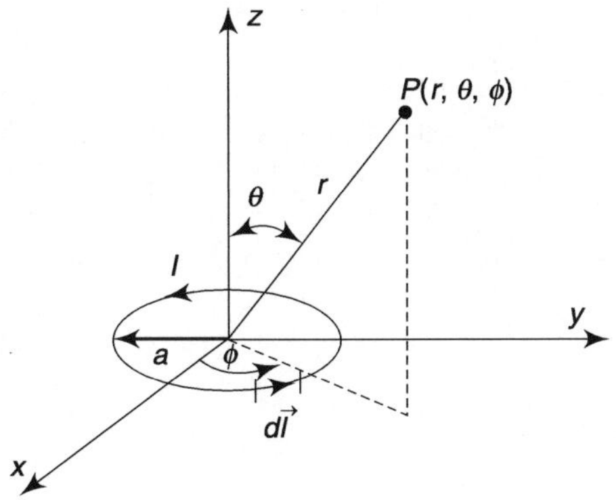

Fig. 3.51 *(a) Magnetic dipole*

$$\vec{r} = x\hat{a}_x + z\hat{a}_z \qquad \text{and} \qquad \vec{r}\,' = a\,\hat{a}_r = a(\cos\phi\,\hat{a}_x + \sin\phi\,\hat{a}_y)$$

$$\therefore \quad \vec{R} = (\vec{r} - \vec{r}\,') = (x\hat{a}_x + z\hat{a}_z) - a(\cos\phi\,\hat{a}_x + \sin\phi\,\hat{a}_y) = (x - a\cos\phi)\hat{a}_x - a\sin\phi\hat{a}_y + z\hat{a}_z$$

$$\therefore \quad R = |\vec{r} - \vec{r}\,'| = \sqrt{(x - a\cos\phi)^2 + (a\sin\phi)^2 + z^2} = [x^2 + z^2 - 2ax\cos\phi + a^2]^{1/2}$$
$$= [r^2 - 2ar\sin\theta\cos\phi + a^2]^{1/2}$$

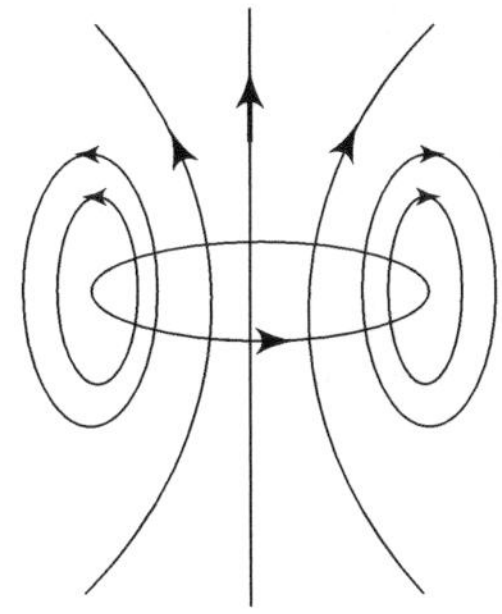

Fig. 3.51 *(b) Magnetic field lines for a filamentary conductor*

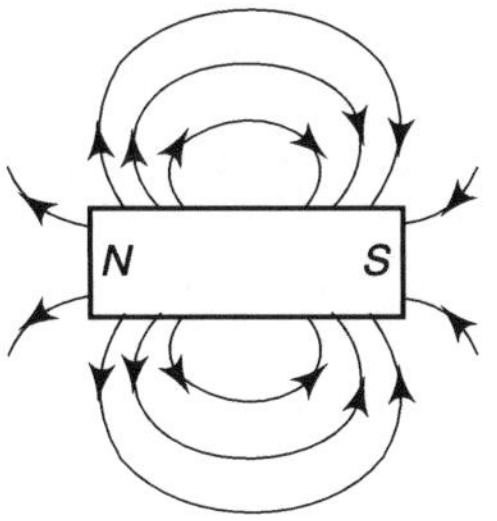

Fig. 3.51 *(c) Magnetic field lines for a bar magnet*

If the loop is small, i.e., the observation point is far away than the radius of the loop ($r >> a$), then we obtain

$$\therefore \quad \frac{1}{R} = [r^2 - 2ar\sin\theta\cos\phi + a^2]^{-1/2} \approx \frac{1}{r}\left(1 - \frac{2a\sin\theta\cos\phi}{r}\right)^{-1/2} \approx \frac{1}{r}\left(1 + \frac{1}{2}\frac{2a\sin\theta\cos\phi}{r}\right)$$
$$= \left(\frac{1 + (a/r)\sin\theta\cos\phi}{r}\right)$$

Substituting this in Eq. (3.97),

$$\vec{A} \approx \frac{\mu I}{4\pi}\int_{\phi=0}^{2\pi}\left(\frac{1 + (a/r)\sin\theta\cos\phi}{r}\right)ad\phi\hat{a}_\phi \approx \frac{\mu I(\pi a^2)\sin\theta}{4\pi r^2}\hat{a}_\phi$$

or
$$\boxed{\vec{A} \approx \frac{\mu I(\pi a^2)\sin\theta}{4\pi r^2}\hat{a}_\phi = \frac{\mu \vec{m} \times \hat{a}_r}{4\pi r^2}} \tag{3.97}$$

where $\vec{m} = I(\pi a^2)\hat{a}_z$ is the magnetic dipole moment defined as the product of the area of the loop and the magnitude of the circulation current flowing through the loop. The direction of the magnetic dipole moment of the current loop is perpendicular to the plane of the loop and is along the direction in which a right-handed screw would advance when moved along the direction of the current flow in the loop. From Eq. (3.100), the magnetic field produced by the magnetic dipole is obtained as

$$\vec{B} = \nabla \times \vec{A} = \frac{1}{r^2 \sin\theta}\begin{vmatrix} \hat{a}_r & r\hat{a}_\theta & r\sin\theta\hat{a}_\phi \\ \frac{\partial}{\partial r} & \frac{\partial}{\partial \theta} & \frac{\partial}{\partial \phi} \\ 0 & 0 & r\sin\theta\left[\frac{\mu I(\pi a^2)\sin\theta}{4\pi r^2}\right] \end{vmatrix}$$

$$= \left(\frac{1}{r\sin\theta}\right)\frac{\partial}{\partial\theta}\left(\sin\theta\frac{\mu I(\pi a^2)\sin\theta}{4\pi r^2}\right)\hat{a}_r - \left(\frac{1}{r}\right)\frac{\partial}{\partial r}\left(r\frac{\mu I(\pi a^2)\sin\theta}{4\pi r^2}\right)\hat{a}_\theta$$

$$= \frac{\mu I(\pi a^2)\cos\theta}{2\pi r^3}\hat{a}_r + \frac{\mu I(\pi a^2)\sin\theta}{4\pi r^3}\hat{a}_\theta$$

∴
$$\boxed{\vec{B} = \frac{\mu I(\pi a^2)}{4\pi r^3}(2\cos\theta\,\hat{a}_r + \sin\theta\,\hat{a}_\theta) = \frac{\mu m}{4\pi r^3}(2\cos\theta\,\hat{a}_r + \sin\theta\,\hat{a}_\theta)} \tag{3.98}$$

The magnetic field lines for a magnetic dipole are shown in Fig. 3.51 (*b*) and Fig. 3.51 (*c*).

Example 3.67 A small circular loop of radius 10 cm is centered at the origin and placed on the $z = 0$ plane. If the loop carries a current of 1 A along $\hat{a}_\phi$, calculate:

(a) The magnetic moment of the loop.
(b) The magnetic field intensity at (2, 2, 2).
(c) The magnetic flux density at (–6, 8, 10)

Solution **(a)** Magnetic moment is given as

$$\vec{m} = IS\hat{a}_n = 1 \times \pi r^2 \hat{a}_z = \pi \times (0.1)^2 \hat{a}_z = 0.01\pi = 0.03142\hat{a}_z \text{ A-m}^2$$

(b) Transforming the point (2, 2, 2) from the Cartesian coordinates into the spherical coordinates

$$r = \sqrt{2^2 + 2^2 + 2^2} = 2\sqrt{3}$$

$$\theta = \tan^{-1}\left(\frac{\sqrt{2^2 + 2^2}}{2}\right) = \tan^{-1}(\sqrt{2}) = 54.736°$$

∴
$$\cos\theta = \frac{1}{\sqrt{3}} \quad \text{and} \quad \sin\theta = \sqrt{\frac{2}{3}}$$

So, the magnetic field is given as

$$\vec{H} = \frac{\vec{B}}{\mu_0} = \frac{m}{4\pi r^3}(2\cos\theta\,\hat{a}_r + \sin\theta\,\hat{a}_\theta) = \frac{0.01\pi}{4\pi \times (2\sqrt{3})^3}\left(2 \times \frac{1}{\sqrt{3}}\hat{a}_r + \sqrt{\frac{2}{3}}\hat{a}_\theta\right)$$

$$= (69.44\hat{a}_r + 49.1\hat{a}_\theta)\ \mu\text{A/m}$$

(c) Transforming the point (–6, 8, 10) from the Cartesian coordinates into the spherical coordinates,

$$r = \sqrt{6^2 + 8^2 + 1^2} = 10\sqrt{2}$$

$$\theta = \tan^{-1}\left(\frac{\sqrt{6^2 + 8^2}}{10}\right) = \tan^{-1}(1) = 45°$$

$$\therefore \qquad \cos\theta = \sin\theta = \frac{1}{\sqrt{2}}$$

The magnetic flux density at point (–6, 8, 10) is given as

$$\vec{B} = \frac{m\mu_0}{4\pi r^3}(2\cos\theta\,\hat{a}_r + \sin\theta\,\hat{a}_\theta) = \frac{0.01\pi \times 4\pi \times 10^{-7}}{4\pi \times (10\sqrt{2})^3}\left(2 \times \frac{1}{\sqrt{2}}\hat{a}_r + \frac{1}{\sqrt{2}}\hat{a}_\theta\right)$$

$$= (1.571\hat{a}_r + 0.785\hat{a}_\theta)\quad \text{pWb/m}^2$$

3.18 MAGNETIC MATERIALS

The introduction of material media into the study of magnetism has very different consequences as compared to the introduction of material media into the study of electrostatics. When we dealt with dielectric materials in electrostatics, their effect was always to reduce $\vec{E}$ below what it would otherwise be, for a given amount of free electric charge. In contrast, when we deal with magnetic materials, their effect can be one of the following:

(i) To reduce $\vec{B}$ below what it would otherwise be, for the same amount of free electric current (*diamagnetic* materials);

(ii) To increase $\vec{B}$ a little above what it would otherwise be (*paramagnetic* materials);

(iii) To increase $\vec{B}$ a lot above what it would otherwise be (*ferromagnetic* materials).

We will discuss these effects in detail.

3.18.1 Magnetisation or Magnetic Polarisation

Magnetisation $(\vec{M})$ is defined as the amount of magnetic moment per unit volume. It is the property of some magnetic materials which describes magnetic field created by those materials themselves and the effects of some external magnetic field on those materials. It is expressed in Ampere per metre (A/m).

The origin of the magnetic moments that create the magnetization can be microscopic electric currents (bound current, I_b) due to:

— either the rotation of electrons around the positive nucleus;

— or the spin of the electrons.

Both of these electronic motions produce internal magnetic fields $(\vec{B}_i)$ that are similar to the magnetic field produced by a current loop. The equivalent current loop has the magnetic moment

$$\vec{m} = I_b S \hat{a}_n$$

where S is the area of the loop and I_b is the bound current.

Without an external magnetic field:

The sum of all $\vec{m}'s$ is zero due to random orientation.

When an external magnetic field $(\vec{B})$ is applied:

The magnetic moments of the electrons align themselves with $\vec{B}$ so that the net magnetic moment is not zero. Thus, the material exhibits some magnetisation.

If there are N atoms in a given volume Δv with magnetic moments $\vec{m}_i$, then *magnetisation* or *magnetic polarisation density* is defined as

$$\boxed{\vec{M} = \lim_{\Delta v \to 0} \left[\frac{\sum_{i=1}^{N} \vec{m}_i}{\Delta v} \right]} \tag{3.99}$$

3.18.2 Magnetisation in Maxwell's Equations

The behaviour of the magnetic fields $(\vec{B}, \vec{H})$, electric fields $(\vec{E}, \vec{D})$, charge density (ρ), current density $(\vec{J})$ is described by Maxwell's equations. The role of the magnetisation is described below, as also demonstrated in Fig. 3.52 (*a*), (*b*) and (*c*).

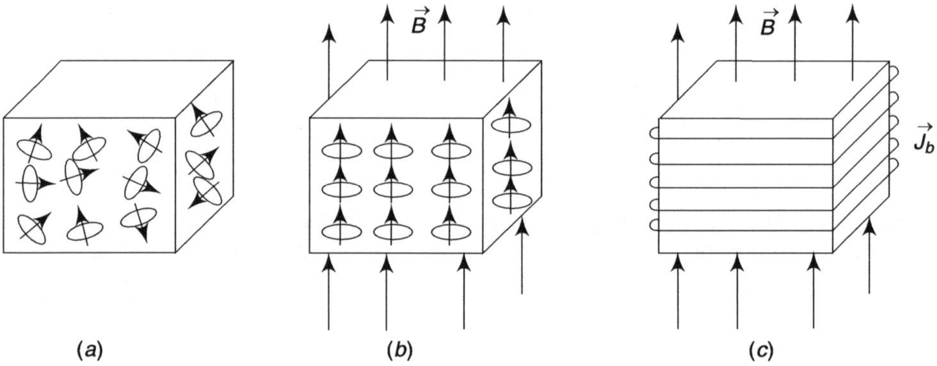

Fig. 3.52 *Magnetic dipoles (a) before applied $\vec{B}$ field, randomly oriented dipoles, (b) after $\vec{B}$ field is applied, aligned dipoles, and (c) aligned dipoles of (b) is equivalent to a bound surface current density $\vec{J}_b$*

Magnetisation Currents From the concept of magnetic dipole, the magnetic vector potential due to a magnetic moment $\vec{m}$ is

$$\vec{A} = \frac{\mu \vec{m} \times \vec{a}_r}{4\pi r^2}$$

For a differential volume dv', the magnetic moment is

$$d\vec{m} = \vec{M} dv' \quad \text{where } \vec{M} \text{ is the magnetisation}$$

Hence, vector magnetic potential due to this differential volume is

$$d\vec{A} = \frac{\mu \vec{M} \times \hat{a}_r}{4\pi r^2} dv' = \frac{\mu \vec{M} \times \vec{r}}{4\pi r^3} dv' = \frac{\mu \vec{M}}{4\pi} \times \nabla'\left(\frac{1}{r}\right) dv'$$

$$\left[\because \frac{\vec{R}}{R^3} = \nabla'\left(\frac{1}{R}\right)\right]$$

Summing over all differential volumes of the material, we obtain the vector magnetic potential as

$$\vec{A} = \int d\vec{A} = \frac{\mu}{4\pi}\int_{v'} \vec{M} \times \nabla'\left(\frac{1}{r}\right) dv' = \frac{\mu}{4\pi}\int_{v'} \frac{\nabla' \times \vec{M}}{r} dv' - \frac{\mu}{4\pi}\int_{v'} \nabla' \times \left(\frac{\vec{M}}{r}\right) dv'$$

$$\left[\because \vec{M} \times \nabla'\left(\frac{1}{r}\right) \equiv \frac{1}{r}\nabla' \times \vec{M} - \nabla' \times \left(\frac{\vec{M}}{r}\right)\right]$$

$$\therefore \quad \vec{A} = \frac{\mu}{4\pi}\int_{v'} \frac{\nabla' \times \vec{M}}{r} dv' + \frac{\mu}{4\pi}\oint_{S'} \left(\frac{\vec{M} \times \hat{a}_n}{r}\right) dS' \qquad \left\{\text{by vector identity, } \int (\nabla \times \vec{F})\, dv = -\oint_S \vec{F} \times d\vec{S}\right\}$$

$$\therefore \quad \vec{A} = \frac{\mu}{4\pi}\int_{v'} \frac{\vec{J}_b}{r} dv' + \frac{\mu}{4\pi}\oint_{S'} \frac{\vec{K}_b}{r} dS' \qquad \textbf{(3.100)}$$

where

$\vec{J}_b = \nabla \times \vec{M}$ = magnetisation volume current density or bound volume current density (A/m^2)

$\vec{K}_b = \vec{M} \times \hat{a}_n$ = magnetisation surface current density or bound surface current density (A/m)

Hence, total volume current density in Maxwell's equations is given by

$$\boxed{\vec{J} = \vec{J}_f + \nabla \times \vec{M} + \frac{\partial \vec{P}}{\partial t}} \qquad \textbf{(3.101)}$$

where, $\vec{J}_f$ is the free volume current density,

$\vec{J}_b$ is the bound volume current density,

$\vec{M}$ is the magnetic polarisation or magnetisation, and

$\vec{P}$ is the electric polarisation

Magnetic Susceptibility (χ_m) The *magnetic susceptibility* χ_m of a magnetic material is a measure of the degree of magnetisation of a material in response to an applied magnetic field. It is a dimensionless quantity. It is roughly a measure of how susceptible (or sensitive) a material is to a magnetic field.

Permeability (μ) *Permeability* (μ) is the degree of magnetisation of a material that responds linearly to an applied magnetic field.

In general, permeability is not a constant; it can vary with the position in the medium, the frequency of the applied field, humidity, temperature, and other parameters. In a nonlinear medium, the permeability can also depend on the strength of the magnetic field.

Relation between $\vec{B}$, $\vec{H}$ and $\vec{M}$ A magnetised material produces a secondary magnetic field ($\vec{B}_i$) as discussed in the preceding section, which may be thought to be generated by the bound current densities.

In free space, bound current densities are not present, i.e., $\vec{M} = 0$ and thus, we have

$$\nabla \times \vec{H} = \vec{J}_f \quad \Rightarrow \quad \vec{J}_f = \nabla \times \left(\frac{\vec{B}}{\mu_0} \right)$$

In material medium, $\vec{M} \neq 0$ and as a result $\vec{B}$ changes. We can write the total current density as the sum of free current density and bound current density, as

$$\vec{J} = \nabla \times \left(\frac{\vec{B}}{\mu_0} \right) = \vec{J}_f + \vec{J}_b = \nabla \times \vec{H} + \nabla \times \vec{M} = \nabla \times (\vec{H} + \vec{M})$$

$$\therefore \quad \boxed{\vec{B} = \mu_0(\vec{H} + \vec{M})} \qquad \textbf{(3.102)}$$

Equation (3.105) is so general that it is valid for any medium, linear or not. In a linear material, magnetisation is directly proportional to field intensity, so that

$$\vec{M} \infty \vec{H}$$

$$\boxed{\vec{M} = \chi_m \vec{H}} \qquad \textbf{(3.103)}$$

where, χ_m is the *magnetic susceptibility* of the medium.

Substituting Eq. (3.106) into Eq. (3.105), we get

$$\vec{B} = \mu_0(1 + \chi_m)\vec{H} = \mu_0\mu_r\vec{H} = \mu\vec{H}$$

$$\boxed{\vec{B} = \mu\vec{H}} \qquad \textbf{(3.104)}$$

where,

$\mu = \mu_0\mu_r$ is called the *permeability* of the medium, expressed in Henry per metre (H/m),

μ_0 is the permeability of free space, known as *absolute permeability*, $= 4\pi \times 10^{-7}$ H/m

$\mu_r = (1 + \chi_m) = \dfrac{\mu}{\mu_0}$ is the *relative permeability* of the medium, it is dimensionless.

NOTE

Compare this section with Section 3.5.

3.18.3 Classification of Magnetic Materials

Depending upon the values of the magnetic susceptibility (χ_m) or the relative permeability (μ_r), magnetic materials are broadly classified into three groups as

1. Paramagnetism,
2. Diamagnetism, and
3. Ferromagnetism.

The characteristics of these magnetic materials are given in Table 3.2.

Table 3.2 *Magnetic materials*

Type of Magnetism	Features	Ranges of μ_r and χ_m	Examples
1. Paramagnetism:	(i) The atoms or molecules have a permanent magnetic dipole moment. (ii) In the absence of any applied external magnetic field, the permanent magnetic dipoles in a paramagnetic material are randomly aligned and thus do not have any magnetisation ($\vec{M} = 0$) and thus, the average magnetic field ($\vec{B}_i$) is also zero. (iii) When placed in an external field ($\vec{H}$), the dipoles experience a torque that tends to align $\vec{m}$ with $\vec{H}$, thereby producing a net magnetisation ($\vec{M}$) parallel to $\vec{H}$. Since $\vec{B}_i$ is parallel to $\vec{H}$, it will tend to *enhance* the field. (iv) The magnetisation ($\vec{M}$) is not only in the same direction as ($\vec{H}$), but also linearly proportional to it. (v) Paramagnetism is temperature dependent.	$\mu_r > 1$, $\chi_m > 0$ (positive), although χ_m is usually of the order of 10^{-6} to 10^{-3}	Air, platinum, tungsten, potassium, aluminium, chromium, palladium, copper sulphate, manganese, etc.
2. Diamagnetism:	(i) The magnetic material does not have permanent magnetic dipoles. (ii) The presence of an external field $\vec{B}_0$ will induce magnetic dipole moments in the atoms or molecules. However, these induced magnetic dipoles are anti-parallel to $\vec{B}_0$, leading to a magnetisation $\vec{M}$ and average field anti-parallel to $\vec{B}_0$, and therefore, a *reduction* in the total magnetic field strength.	$\mu_r < 1$, $\chi_m < 0$ (negative), although χ_m is usually of the order of -10^{-5} to -10^{-9}	Copper ($\chi_m = -0.95 \times 10^{-5}$), gold ($\chi_m = -3.2 \times 10^{-5}$), silver ($\chi_m = -2.6 \times 10^{-5}$), lead, silicon, diamond, bismuth ($\chi_m = -16.6 \times 10^{-5}$), antimony, mercury, tin, zinc, alcohol, hydrogen, nitrogen, water, etc.
3. Ferromagnetism:	(i) Ferromagnetic materials are made up of small patches called *magnetic domains*. (ii) An externally applied field $\vec{B}_0$ will tend to line up those magnetic dipoles parallel to the external field. The strong interaction between neighbouring atomic dipole moments causes a *much stronger* alignment of the magnetic dipoles than in paramagnetic materials. (iii) The enhancement of the applied external field can be considerable, with the total magnetic field inside a ferromagnet 10^3 or 10^4 times greater than the applied field.		Iron, steel, nickel, cobalt

3.18.4 Hysteresis in Ferromagnetic Materials

The permeability μ of a ferromagnetic material is not a constant, since neither the total field nor the magnetisation $\vec{M}$ increases linearly with $\vec{H}$. Although the relation $\vec{B} = \mu_0(\vec{H} + \vec{M})$ is applicable for all types of magnetic materials, the relation between $\vec{B}$ and $\vec{H}$ for ferromagnetic materials is not unique, but is dependent on the previous magnetic history of the material. This phenomenon is known as *hysteresis*.

The variation of $\vec{B}$ as a function of the externally applied field $\vec{H}$ is shown in Fig. 3.53. The curve is known as a *hysteresis curve* or *magnetization curve* or *B–H curve*.

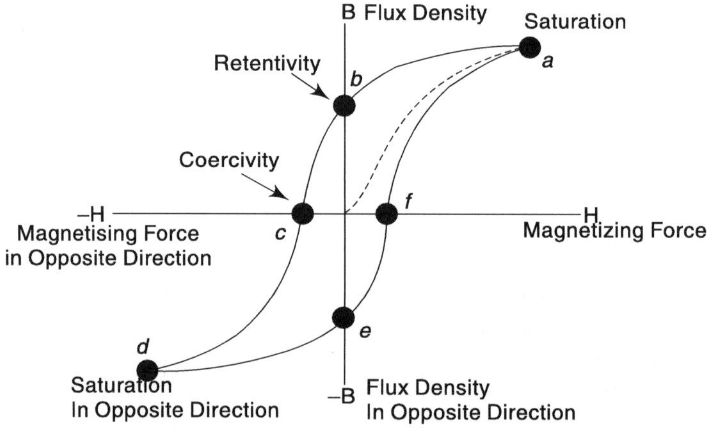

Fig. 3.53 *Hysteresis curve or magnetisation (B–H) curve*

Characteristics of Hysteresis Curve

Magnetic Saturation A ferromagnetic material that has never been previously magnetised or has been thoroughly demagnetised (virgin magnetic material) will follow the dashed line as $\vec{H}$ is increased. As the line demonstrates, the greater the amount of current applied, stronger is the magnetic field in the component $\vec{B}$. At point '*a*' almost all of the magnetic domains are aligned and an additional increase in the magnetising force will produce very little increase in magnetic flux. The material has reached the point of *magnetic saturation*.

Retentivity and Residual Magnetism When $\vec{H}$ is reduced to zero, the curve will move from point '*a*' to point '*b*'. At this point, it can be seen that some magnetic flux remains in the material, even though the magnetising force is zero. This is referred to as the point of *retentivity* on the curve and indicates the *remanence* or level of *residual magnetism* in the material.

Coercivity As the magnetising force is reversed, the curve moves to point '*c*', where the flux has been reduced to zero. This is called the point of *coercivity* on the curve. The reversed magnetising force has flipped enough of the domains, so that the net flux within the material is zero. The force required to remove the residual magnetism from the material is called the *coercive force* or *coercivity* of the material.

Hysteresis Loop and Hysteresis Loss As the magnetising force is increased in the negative direction, the material will again become magnetically saturated, but in the opposite direction (point '*d*').

Reducing $\vec{H}$ to zero brings the curve to point '*e*'. It will have a level of residual magnetism equal to that achieved in the other direction. Increasing $\vec{H}$ back in the positive direction will return $\vec{B}$ to zero. Notice that the curve did not return to the origin because some force is required to remove the residual magnetism. The curve will take a different path from point '*f*' back to the saturation point where it will complete a loop, called *hysteresis loop*. The area of a hysteresis loop gives the energy loss per unit volume during one complete cycle of periodic magnetisation of a ferromagnetic material. This is called *hysteresis loss*. This loss is in the form of heat.

From the hysteresis loop, a number of primary magnetic properties of a material can be determined as mentioned in Table 3.3.

Table 3.3 *Primary magnetic properties of magnetic material from hysteresis*

Property	Characteristics
Retentivity	It is the ability of a material to retain a certain amount of residual magnetic field when the magnetising force is removed after achieving saturation.
Residual Magnetism	It is the magnetic flux density that remains in a material when the magnetising force is zero. The residual magnetism and retentivity are the same when the material has been magnetised to the saturation point. However, the level of residual magnetism may be lower than the retentivity value when the magnetising force did not reach the saturation level.
Coercive Force	This is the amount of reverse magnetic field which must be applied to a magnetic material to make the magnetic flux return to zero.
Hysteresis Loss	This is the area of a hysteresis loop which gives the energy loss per unit volume during one complete cycle of periodic magnetisation.
Permeability	The slope of the curve at any point on the hysteresis loop gives the permeability of the material. The relative permeability is arrived at by taking the ratio of the material's permeability to the permeability in free space (air). $\mu_{\text{relative}} = \mu_{\text{material}}/\mu_{\text{air}}$ where, $\mu_{\text{air}} = 4\pi \times 10^{-7} = 1.256 \times 10^{-6}$ H/m

The shape of the hysteresis loop tells a great deal about the material being magnetised. The hysteresis curves of two different materials are shown in the graph in Fig. 3.54.

A material with a wider hysteresis loop has:

- Lower permeability
- Higher retentivity
- Higher coercivity
- Higher reluctance
- Higher residual magnetism
- Higher loss

A material with the narrower hysteresis loop has:

- Higher permeability
- Lower retentivity
- Lower coercivity
- Lower reluctance
- Lower residual magnetism
- Lower loss

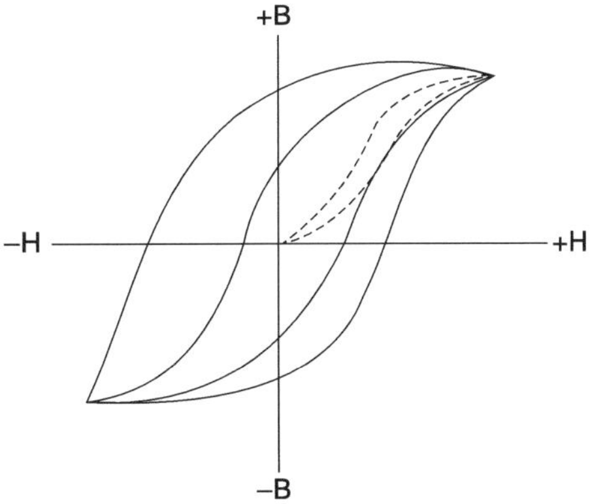

Fig. 3.54 *Different shapes of hysteresis curves*

3.19 MAGNETIC BOUNDARY CONDITIONS

Magnetic boundary conditions are the conditions that $\vec{B}$ or $\vec{H}$ (or $\vec{M}$) field must satisfy at the boundary between two different magnetic media. These are illustrated in Fig. 3.55.

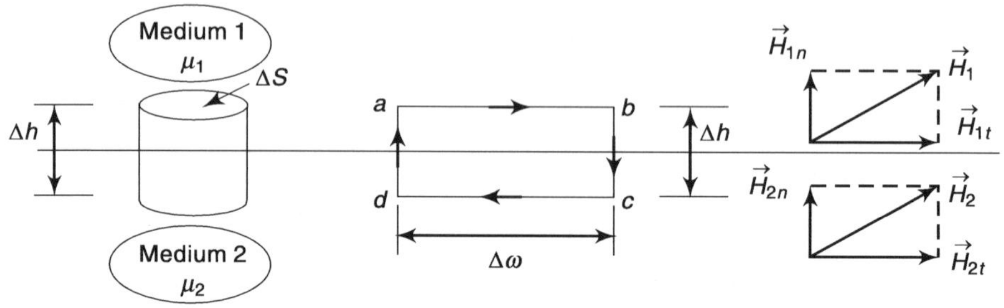

Fig. 3.55 *Magnetic boundary conditions*

To determine the conditions, we use Gauss' law of magnetostatics and Ampere's circuital law

$$\oint_S \vec{B} \cdot d\vec{S} = 0 \quad \text{and} \quad \oint_l \vec{H} \cdot d\vec{l} = I_{\text{enc}}$$

We will consider two different magnetic media 1 and 2, characterised by the permeabilities μ_1 and μ_2, respectively.

Applying Gauss' law to the pillbox (Gaussian surface), with $\Delta h \to 0$,

$$B_{1n}\Delta S - B_{2n}\Delta S = 0$$

$$\boxed{B_{1n} = B_{2n}} \tag{3.105}$$

In terms of the field intensity, the boundary condition can be written as

$$\boxed{\mu_1 H_{1n} = \mu_2 H_{2n}} \tag{3.106}$$

Thus, *the normal component of $\vec{B}$ is continuous, but normal component of $\vec{H}$ is discontinuous at the boundary surface.*

Now, applying Ampere's circuital law, assuming that the boundary carries a surface current $\vec{K}$ whose component normal to the plane of the closed path *abcda* is K (A/m)

$$K\Delta\omega = H_{1t}\Delta\omega - H_{1n}\frac{\Delta h}{2} - H_{2n}\frac{\Delta h}{2} - H_{2t}\Delta\omega + H_{2n}\frac{\Delta h}{2} + H_{1n}\frac{\Delta h}{2}$$

$$\therefore \qquad \boxed{(H_{1t} - H_{2t}) = K} \tag{3.107}$$

In terms of the flux density, we have

$$\boxed{\left(\frac{B_{1t}}{\mu_1} - \frac{B_{2t}}{\mu_2}\right) = K} \tag{3.108}$$

Thus, the tangential component of $\vec{H}$ is also discontinuous. The directions are specified by using the cross product as

$$\boxed{(\vec{H}_1 - \vec{H}_2) \times \hat{a}_{n21} = \vec{K}} \quad \text{or} \quad \boxed{(\vec{H}_1 - \vec{H}_2) = \vec{K} \times \hat{a}_{n12}} \tag{3.109}$$

where, $\hat{a}_{n21}$ is the unit vector normal to the boundary directed from medium 2 to medium 1.

If the media are not conductors, then the boundary is free of current, i.e., $K = 0$; then

$$H_{1t} = H_{2t} \quad \Rightarrow \quad \frac{B_{1t}}{\mu_1} = \frac{B_{2t}}{\mu_2} \quad \text{and} \quad B_{1n} = B_{2n} \tag{3.110}$$

If the fields make angle θ with the respective normal to the interface, then we can combine the boundary conditions as

$$\frac{B_1 \sin\theta_1}{\mu_1} = \frac{B_2 \sin\theta_2}{\mu_2} \quad \text{and} \quad B_1 \cos\theta_1 = B_2 \cos\theta_2$$

Combining,

$$\boxed{\frac{\tan\theta_1}{\tan\theta_2} = \frac{\mu_1}{\mu_2}} \quad \text{or} \quad \boxed{\mu_1 \cot\theta_1 = \mu_2 \cot\theta_2} \tag{3.111}$$

***Example 3.68** Two magnetic materials are separated by a surface $z = 0$; having permeabilities $\mu_1 = 4\mu_0$ H/m for region 1 where $z > 0$ and $\mu_2 = 7\mu_0$ H/m for region 2 where $z < 0$. There exists a surface current density $\vec{K}_s = 60\hat{i}$ A/m at the boundary $z = 0$. For field $\vec{B}_1 = (1\hat{i} - 2\hat{j} - 3\hat{k})$ mT in region 1, find the flux density $\vec{B}_2$ in region 2.

Solution Here, $z = 0$ is the boundary and hence, the normal component of flux density is

$$\vec{B}_{1n} = -3\hat{k}$$

∴ Tangential component is

$$\vec{B}_{1t} = \vec{B}_1 - \vec{B}_{1n} = (1\hat{i} - 2\hat{j})$$

Applying the boundary condition

$$\vec{B}_{2n} = \vec{B}_{1n} = -3\hat{k}$$

Applying the other boundary condition

$$\left(\frac{B_{1t}}{\mu_1} - \frac{B_{2t}}{\mu_2}\right) = K_s \text{ or in vector form, } \left(\frac{\vec{B}_{1t}}{\mu_1} - \frac{\vec{B}_{2t}}{\mu_2}\right) \times \hat{a}_n = \vec{K}_s$$

Here,

$$\left(\frac{1\hat{i} - 2\hat{j}}{4\mu_0} - \frac{\vec{B}_{2t}}{7\mu_0}\right) \times \hat{k} = 60\hat{i}$$

or,
$$[7\hat{i} - 14\hat{j} - 4(B_x\hat{i} + B_y\hat{j} + B_z\hat{k})] \times \hat{k} = 60 \times 28\mu_0\hat{i}$$

or,
$$-7\hat{j} - 14\hat{i} + 4B_x\hat{j} - 4B_y\hat{i} = 1680\mu_0\hat{i}$$

or,
$$(-14 - 4B_y)\hat{i} + (-7 + 4B_x)\hat{j} = 1680\mu_0\hat{i}$$

Equating the components

$$B_z = 0$$
$$B_x = \frac{7}{4} = 1.75$$
$$-14 - 4B_y = 1680\mu_0 = 1680 \times 4\pi \times 10^{-7} \quad \Rightarrow \quad B_y = -3.5$$

So, the magnetic flux density in region 2 is

$$\vec{B}_2 = (1.75\hat{i} - 3.5\hat{j} - 3\hat{k}) \text{ mT}$$

***Example 3.69** Consider an interface in *yz* plane. The region $x < 0$ is medium 1 with $\mu_{r1} = 4.5$ and magnetic field $\vec{H}_1 = 4\hat{i} + 3\hat{j} - 6\hat{k}$ A/m. The region $x > 0$ is medium 2 with $\mu_{r2} = 6$. Find $\vec{H}_2$ in medium 2 and angle made by $\vec{H}_2$ with normal to the interface.

Solution Here, $x = 0$ is the boundary and hence, the normal component of field intensity is

$$\vec{H}_{1n} = 4\hat{i}$$

∴ Tangential component is,

$$\vec{H}_{1t} = \vec{H}_1 - \vec{H}_{1n} = (3\hat{j} - 6\hat{k})$$

Applying the boundary condition

$$\frac{\vec{H}_{1n}}{\vec{H}_{2n}} = \frac{\mu_2}{\mu_1} \quad \Rightarrow \quad \vec{H}_{2n} = \left(\frac{\mu_1}{\mu_2}\right)\vec{H}_{1n} = \left(\frac{4.5}{6}\right) \times (4\hat{i}) = 3\hat{i}$$

Applying the other boundary condition

$$\vec{H}_{1t} = \vec{H}_{2t} = (3\hat{j} - 6\hat{k})$$

So, the magnetic field intensity in region 2 is

$$\vec{H}_2 = (3\hat{i} + 3\hat{j} - 6\hat{k}) \text{ A/m}$$

The angle made by $\vec{H}_2$ with the normal to the interface is obtained as follows.

$$\tan\theta_2 = \frac{H_{2t}}{H_{2n}} = \frac{\sqrt{3^2 + 6^2}}{3} = 2.236 \quad \Rightarrow \quad \theta_2 = 65.9°$$

Example 3.70 Given that $\vec{H}_1 = -2\hat{i} + 6\hat{j} + 4\hat{k}$ A/m in region $y - x - 2 \le 0$ where $\mu_1 = 5\mu_0$, calculate:

(a) $\vec{M}_1$ and $\vec{B}_1$.

(b) $\vec{H}_2$ and $\vec{B}_2$ in region $y - x - 2 \ge 0$ where $\mu_2 = 2\mu_0$.

Solution $y - x \le 2$ or, $y \le (x + 2)$ is in region 1. Let the surface of the plane be described by $f(x, y) = (y - x - 2)$; the unit vector normal to the plane is given as

$$\hat{a}_n = \frac{\nabla f}{|\nabla f|} = \frac{\hat{j} - \hat{i}}{\sqrt{2}}$$

(e) Magnetisation vector is

$$\vec{M}_1 = \chi_{m1}\vec{H}_1 = (\mu_{r1} - 1)\vec{H}_1 = (5-1)(-2 + 6\hat{j} + 4\hat{k}) = -8\hat{i} + 24\hat{j} + 16\hat{k} \text{ A/m}$$

$$\therefore \quad \vec{B}_1 = \mu_1\vec{H}_1 = \mu_0\mu_{r1}\vec{H}_1 = 4\pi \times 10^{-7} \times 5 \times (-2\hat{i} + 6\hat{j} + 4\hat{k})$$
$$= -12.57\hat{i} + 37.7\hat{j} + 25.13\hat{k} \quad \mu\text{Wb/m}^2$$

(f) In region 1, normal component of the field is

$$\vec{H}_{1n} = (\vec{H}_1 \cdot \hat{a}_n)\cdot \hat{a}_n = \left[(-2\hat{i} + 6\hat{j} + 4\hat{k})\cdot\left(\frac{\hat{j}-\hat{i}}{\sqrt{2}}\right)\right]\cdot\left(\frac{\hat{j}-\hat{i}}{\sqrt{2}}\right)$$
$$= \left(\frac{2+6}{2}\right)(\hat{j}-\hat{i})$$
$$= (-4\hat{i} + 4\hat{j})$$

Tangential component is

$$\vec{H}_{1t} = \vec{H}_1 - \vec{H}_{1n} = (-2\hat{i} + 6\hat{j} + 4\hat{k}) - (-4\hat{i} + 4\hat{j}) = (2\hat{i} + 2\hat{j} + 4\hat{k})$$

Applying boundary conditions

$$\vec{H}_{2t} = \vec{H}_{1t} = (2\hat{i} + 2\hat{j} + 4\hat{k}) \qquad \{\text{assuming } \vec{K} = 0\}$$

And $\vec{B}_{2n} = \vec{B}_{1n}$ or $\vec{H}_{2n} = \dfrac{\mu_1}{\mu_2}\vec{H}_{1n} = \dfrac{5}{2}(-4\hat{i} + 4\hat{j}) = (-10\hat{i} + 10\hat{j})$

$$\therefore \quad \vec{H}_2 = \vec{H}_{2t} + \vec{H}_{2n} = (2\hat{i} + 2\hat{j} + 4\hat{k}) + (-10\hat{i} + 10\hat{j}) = (-8\hat{i} + 12\hat{j} + 4\hat{k}) \text{ A/m}$$

$$\therefore \quad \vec{B}_2 = \mu_2\vec{H}_2 = \mu_0\mu_{r2}\vec{H}_1 = 4\pi \times 10^{-7} \times 2 \times (-8\hat{i} + 12\hat{j} + 4\hat{k})$$
$$= -20.11\hat{i} + 30.16\hat{j} + 10.05\hat{k} \ \mu\text{Wb/m}^2$$

3.20 SELF INDUCTANCE AND MUTUAL INDUCTANCE

Self Inductance (L) An electric current I flowing around a circuit produces a magnetic field ($\vec{B}$) and hence a magnetic flux $\left(\phi = \int_S \vec{B}\cdot d\vec{S}\right)$ through each turn of the circuit.

If the circuit has N identical turns, then the flux-linkage is defined as

$$\lambda = N\phi$$

Also, if the medium surrounding the circuit is linear, then $\lambda \,\alpha\, I$ or, $\lambda = LI$

where, L is a constant, known as *inductance* of the circuit.

$$\therefore \quad \boxed{L = \frac{\lambda}{I} = \frac{N\phi}{I}} \qquad \textbf{(3.112)}$$

Self-inductance (L): The ratio of the magnetic flux to the current is called the inductance, or more accurately self-inductance of the circuit.

The SI unit for inductance is Webber per Ampere (Wb/A) or Henry (in honour of Joseph Henry); 1H = 1Wb/A.

Inductor An *inductor* is a passive electrical device employed in electrical circuits for its property of inductance. Physically, the inductance L is a measure of an inductor's 'resistance' to the change of current; the larger the value of L, the lower the rate of change of current.

Example 3.71 Self-inductance of a solenoid:
Compute the self-inductance of a solenoid with turns N, length l, and radius R with a current I flowing through each turn, as shown in Fig. 3.56.

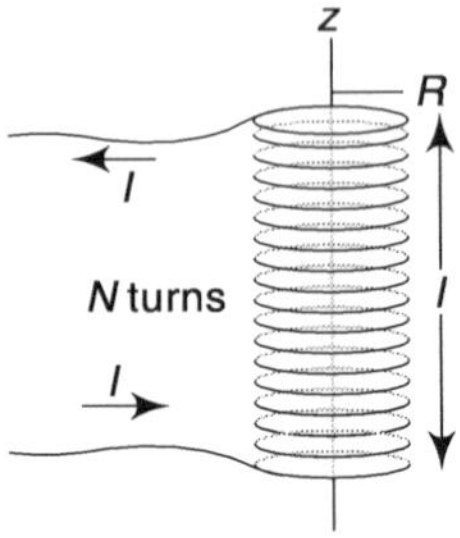

Fig. 3.56 *Solenoid*

Solution Ignoring edge effects and applying Ampere's law, the magnetic field inside a solenoid is given by

$$\vec{B} = \frac{\mu_0 NI}{l}\hat{k} = \mu_0 nI\hat{k}$$

where, $n = \frac{N}{l}$ is the number of turns per unit length. The magnetic flux through each turn is

$$\phi = BA = \mu_0 nI \cdot (\pi R^2) = \mu_0 nI\pi R^2$$

Thus, the self-inductance is

$$L = \frac{N\phi}{I} = \mu_0 n^2 \pi R^2 l$$

We see that L depends only on the geometrical factors (n, R and l) and is independent of the current I.

Coupled Inductor When the magnetic flux produced by an inductor links another inductor, these inductors are said to be *coupled.* For coupled inductors, there exists a *mutual inductance* that relates the current in one inductor to the flux linkage in the other inductor. Thus, there are three inductances defined for coupled inductors:

L_{11} — the self inductance of inductor 1
L_{22} — the self inductance of inductor 2
$L_{12} = L_{21}$ — the mutual inductance associated with both inductors.

Mutual Inductance (M)

> ***Mutual Inductance (M):*** Mutual inductance is the ability of one inductor to induce an e.m.f. across another inductor placed very close to it.

When two coils are placed very close to each other, the magnetic flux caused by current in one coil links with the other coil and induces some voltage in the second coil. This phenomenon is known as *mutual inductance.*

We consider two coils placed near each other, as shown in Fig. 3.57. The first coil has N_1 turns and carries a current I_1 which gives rise to a magnetic field $\vec{B}_1$. Since the two coils are close to each other, some of the magnetic field lines through

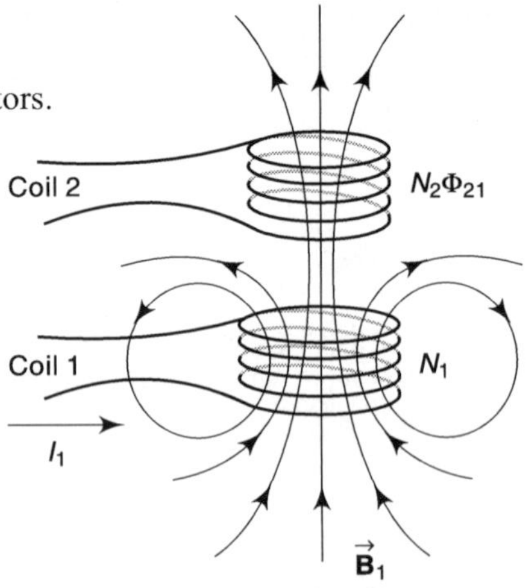

Fig. 3.57 *Changing current in coil 1 produces changing magnetic flux in coil 2*

coil 1 will also pass through coil 2. Let ϕ_{21} denote the magnetic flux through one turn of coil 2 due to I_1. Now, by varying I_1 with time, there will be an induced e.m.f. associated with the changing magnetic flux in the second coil:

$$\varepsilon_{21} = N_2 \frac{d\phi_{21}}{dt} = \frac{d}{dt} \iint_{\text{coil 2}} \vec{B}_1 \cdot d\vec{A}_2$$

The time rate of change of magnetic flux ϕ_{21} in coil 2 is proportional to the time rate of change of the current in coil 1 and thus the voltage can be written as

$$\varepsilon_{21} = N_2 \frac{d\phi_{21}}{dt} = N_2 \frac{d\phi_{21}}{dI_1} \times \frac{dI_1}{dt} = M_{21} \frac{dI_1}{dt}$$

where

$$\boxed{M_{21} = \frac{N_2 \phi_{21}}{I_1}} \qquad \textbf{(3.113)}$$

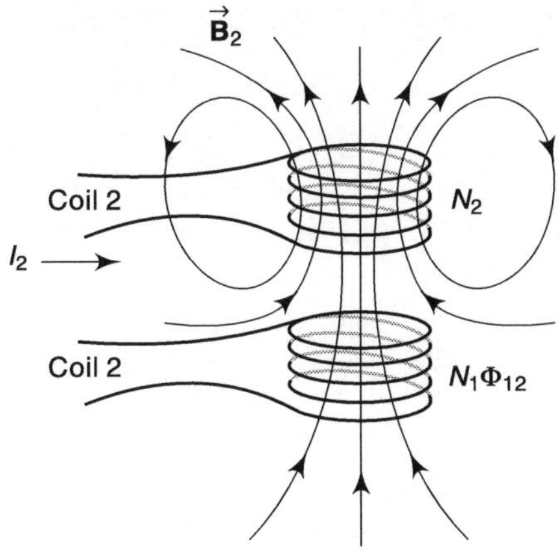

Fig. 3.58 *Changing current in coil 2 produces changing magnetic flux in coil 1*

is called the *mutual inductance.*

The mutual inductance M_{21} depends only on the geometrical properties of the two coils, such as the number of turns and the radii of the two coils.

In a similar manner, suppose instead there is a current I_2 in the second coil and it is varying with time (Fig. 3.58). Then the induced e.m.f. in coil 1 becomes

$$\varepsilon_{12} = N_1 \frac{d\phi_{12}}{dt} = \frac{d}{dt} \iint_{\text{coil 2}} \vec{B}_2 \cdot d\vec{A}_1$$

and a voltage is induced in coil 1.

This changing flux in coil 1 is proportional to the changing current in coil 2;

$$\varepsilon_{12} = N_1 \frac{d\phi_{12}}{dt} = N_1 \frac{d\phi_{12}}{dI_2} \times \frac{dI_2}{dt} = M_{12} \frac{dI_2}{dt}$$

where

$$\boxed{M_{12} = \frac{N_1 \phi_{12}}{I_2}} \qquad \textbf{(3.114)}$$

is another *mutual inductance.*

Using the *reciprocity theorem* which combines Ampere's law and the Biot-Savart law, it can be shown that the two mutual inductances are.

$$M_{12} \equiv M_{21} \equiv M \qquad \textbf{(3.115)}$$

Mutual Inductance between Two Coupled Inductors

Let,

L_1, L_2 — two inductors placed very close to each other.
$v_2(t)$ — open circuit voltage induced in L_2 by a current $i_1(t)$ in L_1
$v_1(t)$ — open circuit voltage induced in L_1 by a current $i_2(t)$ in L_2

So, when only $i_1(t)$ is flowing, the magnetic flux emerging from L_1 is given as

$$\phi_1 = \phi_{11}(\text{Linking with } L_1) + \phi_{12}(\text{Linking with } L_2)$$

$$\therefore \qquad v_1 = N_1 \frac{d\phi_1}{dt} = N_1 \frac{d\phi_1}{di_1}\frac{di_1}{dt} = L_1 \frac{di_1}{dt}$$

where $L_1 = N_1 \dfrac{d\phi_1}{di_1}$

and $v_2 = N_2 \dfrac{d\phi_{12}}{dt} = N_2 \dfrac{d\phi_{12}}{di_1}\dfrac{di_1}{dt} = M_{21}\dfrac{di_1}{dt}$

where $M_{21} = N_2 \dfrac{d\phi_{12}}{di_1}$ = Mutual Inductance of coil L_2 with respect to coil L_1

Now, when only $i_2(t)$ is flowing, the magnetic flux emerging from L_2 is given as

$$\phi_2 = \phi_{21} \text{ (Linking with } L_1) + \phi_{22} \text{ (Linking with } L_2)$$

$$\therefore \qquad v_2 = N_2 \frac{d\phi_2}{dt} = N_2 \frac{d\phi_2}{di_2}\frac{di_2}{dt} = L_2 \frac{di_2}{dt}$$

where, $L_2 = N_2 \dfrac{d\phi_2}{di_2}$

and $v_1 = N_1 \dfrac{d\phi_{21}}{dt} = N_1 \dfrac{d\phi_{21}}{di_2}\dfrac{di_2}{dt} = M_{12}\dfrac{di_2}{dt}$

where, $M_{12} = N_1 \dfrac{d\phi_{21}}{di_2}$ = Mutual Inductance of coil L_1 with respect to coil L_2

Example 3.72 A long solenoid with length l and a cross-sectional area A consists of N_1 turns of wire. An insulated coil of N_2 turns is wrapped around it, as shown in Fig. 3.59.

(a) Calculate the mutual inductance M, assuming that all the flux from the solenoid passes through the outer coil.

(b) Relate the mutual inductance M to the self-inductances L_1 and L_2 of the solenoid and the coil.

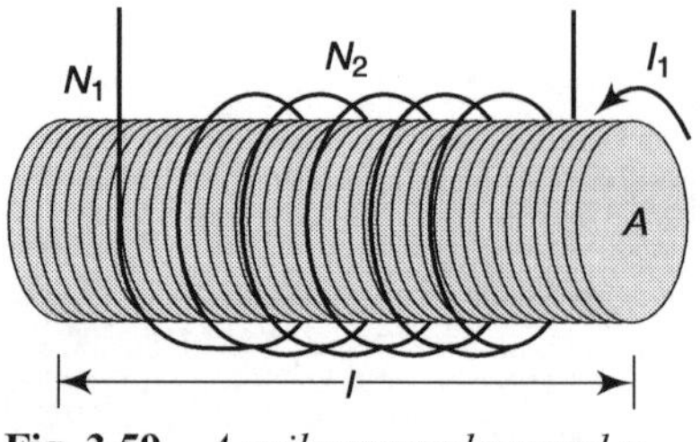

Fig. 3.59 *A coil wrapped around a solenoid*

Solutions:

(a) The magnetic flux through each turn of the outer coil due to the solenoid is,

$$\phi_{21} = BA = \frac{\mu_0 N_1 I_1}{l} A$$

where $B = \dfrac{\mu_0 N_1 I_1}{l}$ is the uniform magnetic field inside the solenoid. Hence, the mutual inductance between the solenoid and the coil is

$$M = \frac{N_2 \phi_{21}}{I_1} = \frac{\mu_0 N_1 N_2 A}{l}$$

(b) From Example 3.71, we see that the self-inductance of the solenoid with N_1 turns is given by

$$L_1 = \frac{N_1 \phi_{11}}{I_1} = \frac{\mu_0 N_1^2 A}{l}$$

where, ϕ_{11} is the magnetic flux through one turn of the solenoid due to the magnetic field produced by I_1.

Similarly, we have the self-inductance for the outer coil given as

$$L_2 = \frac{N_2\phi_{22}}{I} = \frac{\mu_0 N_2^2 A}{l}$$

Thus, in terms of L_1 and L_2, the mutual inductance can be expressed as

$$M = \frac{\mu_0 N_1 N_2 A}{l} = \sqrt{\frac{\mu_0 N_1^2 A}{l}\frac{\mu_0 N_2^2 A}{l}} = \sqrt{L_1 L_2}$$

NOTE

More generally, the mutual inductance is given as

$$M = k\sqrt{L_1 L_2} \qquad 0 \le k \le 1$$

where k is the coupling coefficient. In this example, we have k = 1 which means that all of the magnetic flux produced by the solenoid passes through the outer coil, and vice versa, in this idealisation.

***Example 3.73** Self-Inductance of Infinite Solenoid:

Find out the inductance of a long solenoid of radius '*r*' having '*N*' number of turns.

Or,

Calculate the self-inductance per unit length of an infinitely long solenoid.

Solution From Example 3.42, the magnetic field at the centre of the long solenoid is

$$H = \frac{NI}{l}$$

where, I is the current and l is the length of the solenoid.

If the solenoid is significantly long, compared to the diameter, H will be the same all over the cross-section. So, the total magnetic flux through the cross-section is

$$\phi = BA = \mu HA = \frac{\mu NIA}{l}$$

The flux linkage is the same through all the turns; as the solenoid is having N turns.

Flux linkage, $\lambda = N\phi = \dfrac{\mu IAN^2}{l}$

Thus, the inductance is given as $L = \dfrac{\lambda}{I} = \dfrac{\mu N^2 A}{l}$

$$\boxed{L = \frac{\mu N^2 A}{l}}$$

If the radius of the solenoid is r, then $A = \pi r^2$, then the inductance is given as $L = \dfrac{\pi\mu N^2 r^2}{l}$.

Example 3.74 Self-Inductance of Finite Solenoid:

Find the inductance of a solenoid of finite length 'l' with radius 'r' and 'N' number of turns.

Solution From Example 3.42, the magnetic field at an axial point P is given by

$$\vec{B} = \frac{\mu NI}{2l}(\cos\theta_2 - \cos\theta_1)\hat{a}_z$$

where, N is the number of turns, I is the current carrying, l is the length of the solenoid

Here, $\cos\theta_2 = \dfrac{l-z}{\sqrt{r^2+(l-z)^2}}$ and $\cos(\pi-\theta_1) = -\cos\theta_1 = \dfrac{z}{\sqrt{r^2+z^2}}$

$\therefore$
$$B = \frac{\mu NI}{2l}\left(\frac{l-z}{\sqrt{r^2+(l-z)^2}} + \frac{z}{\sqrt{r^2+z^2}}\right)$$

In a small length dz, the number of turns is $\left(\frac{N}{l}dz\right)$. The flux linking these turns is

$$d\phi = \frac{B(\pi r^2)N}{l}dz$$

Neglecting the variation of the magnetic flux over the cross-sectional area, total flux linking N turns is

$$\phi = \int d\phi = \int_{z=0}^{l} \frac{B(\pi r^2)N}{l}dz$$

$$= \frac{\mu N^2 I\pi r^2}{2l^2}\int_{z=0}^{l}\left(\frac{z}{\sqrt{r^2+z^2}} + \frac{l-z}{\sqrt{r^2+(l-z)^2}}\right)dz$$

$$= \frac{\mu N^2 I\pi r^2}{2l^2}\left[\sqrt{r^2+z^2} - \sqrt{r^2+(l-z)^2}\,\right]_0^l$$

$$= \frac{\mu N^2\pi r^2}{l^2}(\sqrt{r^2+l^2} - r)I$$

Hence, self-inductance of the solenoid is $L = \dfrac{\phi}{I} = \dfrac{\mu N^2\pi r^2}{l^2}(\sqrt{r^2+l^2} - r)$

$$\boxed{L = \frac{\mu N^2\pi r^2}{l^2}(\sqrt{r^2+l^2} - r)}$$

NOTE

If $l >> r$, $L = \dfrac{\mu N^2\pi r^2}{l}$ as obtained in Example 3.73.

***Example 3.75** Self-Inductance of Toroid of Circular Cross Section:

Obtain an expression for the self-inductance of a toroid of circular section with 'N' closely spaced turns.

Solution Let, r mean radius of the toroid,
N number of turns,
S radius of the coil.

From Example 3.43, we have the magnetic field

$$H = \frac{NI}{2\pi r}$$

$\therefore$ Total flux linkage per turn is $\phi = BA = \mu HA = \mu \frac{NI}{2\pi r} \pi S^2 = \frac{\mu NI}{2r} S^2$

Hence, the self-inductance of the toroid is $L = \frac{N\phi}{I} = \frac{\mu N^2 S^2}{2r}$

$$\boxed{L = \frac{\mu N^2 S^2}{2r}}$$

Example 3.76 Self-Inductance of Toroid of Rectangular Cross Section:
Calculate the self-inductance of a toroid which consists of N turns and has a rectangular cross section, with inner radius a, outer radius b and height h, as shown in Fig. 3.60 (a).

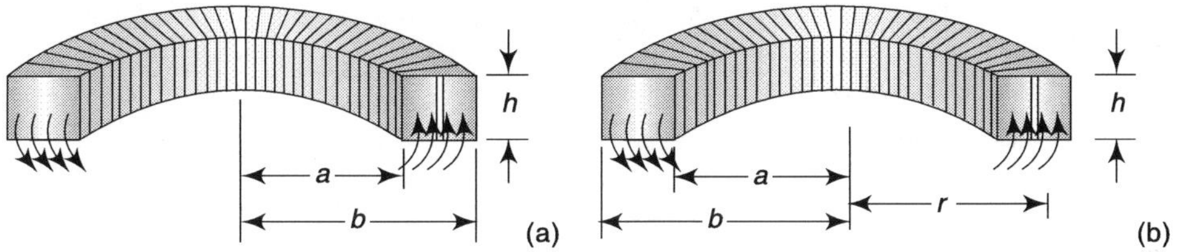

Fig. 3.60 *A toroid with N turns*

Solution According to Ampere's law, the magnetic field is given by

$$\oint \vec{B}.d\vec{l} = \oint Bdl = B\oint dl = B(2\pi r) = \mu_0 NI$$

or
$$B = \frac{\mu_0 NI}{2\pi r}$$

The magnetic flux through one turn of the toroid may be obtained by integrating over the rectangular cross section, with $dS = hdr$ as the differential area element [Fig. 3.60(b)].

$$\therefore \qquad \phi = \oint_S \vec{B} \cdot d\vec{S} = \int_a^b \left(\frac{\mu_0 NI}{2\pi r}\right) hdr = \left(\frac{\mu_0 NIh}{2\pi}\right)\int_a^b \frac{dr}{r} = \frac{\mu_0 NIh}{2\pi} \ln\left(\frac{b}{a}\right)$$

Hence, the self-inductance of the toroid is given as $L = \frac{N\phi}{I} = \frac{\mu_0 N^2 h}{2\pi} \ln\left(\frac{b}{a}\right)$

$$\boxed{L = \frac{\mu_0 N^2 h}{2\pi} \ln\left(\frac{b}{a}\right)}$$

NOTE

If a >> (b – a), the logarithmic term in the equation above may be expanded as

$$\ln\left(\frac{b}{a}\right) = \ln\left(1+\frac{b-a}{a}\right) \approx \frac{b-a}{a}$$

and the self-inductance becomes $L = \frac{\mu_0 N^2 h}{2\pi}\frac{b-a}{a} = \frac{\mu_0 N^2}{2\pi a}h(b-a) = \frac{\mu_0 N^2 A}{l}$

where, A is the cross-sectional area, and $l = 2\pi a$. We see that the self-inductance of the toroid in this limit has the same form as that of a solenoid.

Example 3.77 Self-Inductance of Coaxial Cable with Solid Inner Conductor:

Determine the self-inductance of a co-axial cable of inner radius 'a' and outer radius 'b' when the inner conductor is solid. A coaxial cable is shown in Fig. 3.61.

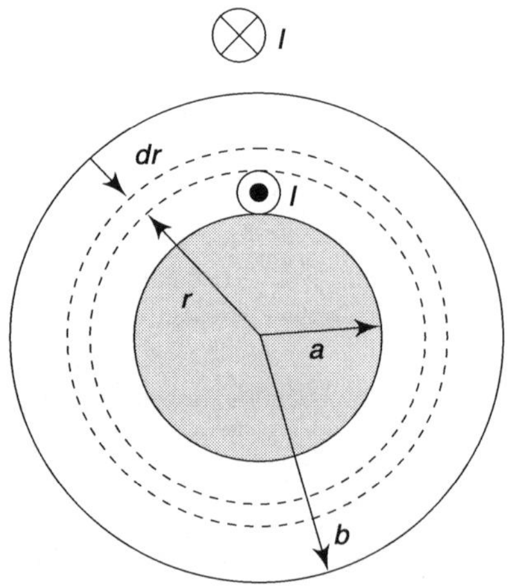

Fig. 3.61 *Coaxial cable*

Solution Here, we have to find two inductances:

1. Internal inductance, L_{in}, considering the flux linkages due to the inner conductor;
2. External inductance, L_{ext}, considering the flux linkages between the inner conductor and the outer conductors.

Now, magnetic energy stored in the inductor is

$$W_m = \frac{1}{2}LI^2 \qquad \Rightarrow \qquad L = \frac{2W_m}{I^2}$$

where, $W_m = \frac{1}{2}\int \vec{B}\cdot\vec{H}dv = \int \frac{B^2}{2\mu}dv$

$$\therefore \qquad L = \frac{2W_m}{I^2} = \frac{2}{I^2}\int\frac{B^2}{2\mu}dv$$

For $0 \le r \le a$:

By Ampere's law,

$$\vec{B}_1 = \frac{\mu I r}{2\pi a^2}\hat{a}_r \qquad \text{[See Example 3.35]}$$

$$\therefore \qquad L_{\text{in}} = \frac{2}{I^2}\int\frac{B_1^2}{2\mu}dv = \frac{2}{I^2}\int\frac{1}{2\mu}\left(\frac{\mu I r}{2\pi a^2}\right)^2 dv = \frac{2}{I^2}\int\frac{1}{2\mu}\left(\frac{\mu I r}{2\pi a^2}\right)^2 r dr d\phi dz$$

$$= \frac{\mu}{4\pi^2 a^4}\int_0^l dz\int_0^{2\pi} d\phi \int_0^a r^3 dr = \frac{\mu}{4\pi^2 a^4}\times l \times 2\pi \times \frac{a^4}{4} = \frac{\mu l}{8\pi}$$

$$\therefore \qquad \boxed{L_{\text{in}} = \frac{\mu l}{8\pi}}$$

For $a \le r \le b$:
By Ampere's law,

$$\vec{B}_2 = \frac{\mu I}{2\pi r}\hat{a}_r$$

$$\therefore \quad L_{ext} = \frac{2}{I^2}\int \frac{B_2^2}{2\mu}dv = \frac{2}{I^2}\int \frac{1}{2\mu}\left(\frac{\mu I}{2\pi r}\right)^2 dv = \frac{2}{I^2}\int \frac{1}{2\mu}\left(\frac{\mu I}{2\pi r}\right)^2 r dr d\phi dz$$

$$= \frac{\mu}{4\pi^2}\int_0^l dz \int_0^{2\pi} d\phi \int_a^b \frac{dr}{r} = \frac{\mu}{4\pi^2} \times l \times 2\pi \times \ln\left(\frac{b}{a}\right) = \frac{\mu l}{2\pi}\ln\left(\frac{b}{a}\right)$$

$$\therefore \quad \boxed{L_{ext} = \frac{\mu l}{2\pi}\ln\left(\frac{b}{a}\right)}$$

Hence, the total inductance of the coaxial cable is

$$L = (L_{in} + L_{ext}) = \frac{\mu l}{8\pi} + \frac{\mu l}{2\pi}\ln\left(\frac{b}{a}\right) = \frac{\mu l}{2\pi}\left[\frac{1}{4} + \ln\left(\frac{b}{a}\right)\right]$$

$$\boxed{L = \frac{\mu l}{2\pi}\left[\frac{1}{4} + \ln\left(\frac{b}{a}\right)\right]}$$

Example 3.78 Self-Inductance of Coaxial Cable with Cylindrical Conductors:

Determine the self-inductance of a coaxial cable with conducting cylinders of inner radius '*a*' and outer radius '*b*'. This is shown in Fig. 3.62.

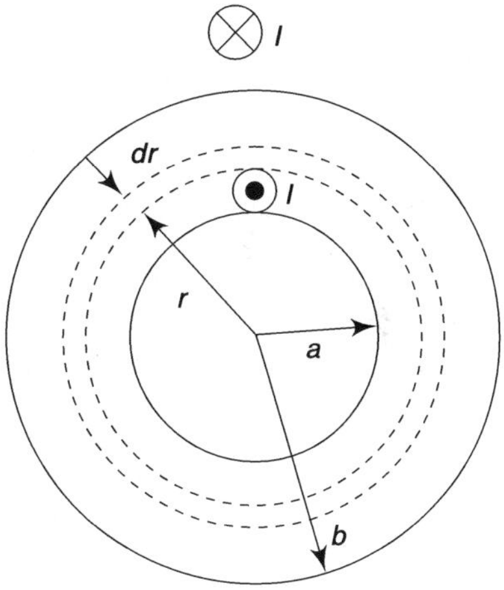

Fig. 3.62 *Coaxial cable*

Solution In this case, we will consider the region $a \le r \le b$ to find the self-inductance.

Let, l—length of the cable

We consider an element of the cable of thickness dr at a distance r from the centre.

Magnetic field at that point, $\vec{B} = \frac{\mu I}{2\pi r}\hat{a}_\phi$

Flux passing through the element is

$$d\phi = \vec{B}\cdot d\vec{S} = \left(\frac{\mu I}{2\pi r}\hat{a}_\phi\right)\cdot(drdz\hat{a}_\phi) = \frac{\mu I}{2\pi}dz\frac{dr}{r}$$

Hence, total flux linkage for the entire cable is

$$\phi = \int_S \vec{B}\cdot d\vec{S} = \frac{\mu I}{2\pi r}\int_0^l dz\int_a^b \frac{dr}{r} = \frac{\mu I l}{2\pi}\ln\left(\frac{b}{a}\right)$$

Hence, the self-inductance is given as $L = \frac{\phi}{I} = \frac{\mu l}{2\pi}\ln\left(\frac{b}{a}\right)$

$$\boxed{L = \frac{\phi}{I} = \frac{\mu l}{2\pi}\ln\left(\frac{b}{a}\right)}$$

***Example 3.79** Self-Inductance of Two-Wire Transmission Line: Determine the inductance per unit length of a two-wire transmission line with separation distance 'd'; each wire having radius 'a', as shown in Fig. 3.63.

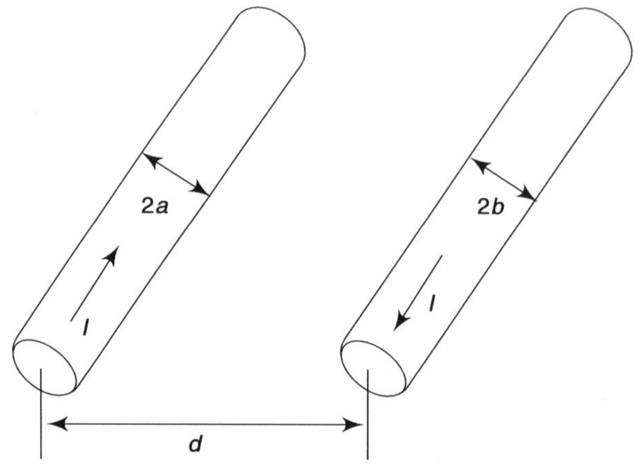

Fig. 3.63 *Two wire transmission line*

Solution We will consider two conductors A and B with radii a and b, respectively, and separation distance between them as d.

We assume that $d >> a$ or b, so that the field and flux through the conductors is nearly uniform.

We will use the results of Example 3.77.

Self-inductance of Conductor A:

Self-inductance per unit length due to internal flux is $L_{Ai} = \dfrac{\mu}{8\pi}$

Self-inductance per unit length due to external flux is $L_{Ae} = \dfrac{\mu}{2\pi}\ln\left(\dfrac{d}{a}\right)$

∴ Total self-inductance of conductor A is

$$L_A = (L_{Ai} + L_{Ae}) = \frac{\mu}{8\pi} + \frac{\mu}{2\pi}\ln\left(\frac{d}{a}\right) = \frac{\mu}{2\pi}\left[\frac{1}{4} + \ln\left(\frac{d}{a}\right)\right]$$

Self-inductance of Conductor B:

Self-inductance per unit length due to internal flux is $L_{Bi} = \dfrac{\mu}{8\pi}$

Self-inductance per unit length due to external flux is $L_{Be} = \dfrac{\mu}{2\pi}\ln\left(\dfrac{d}{b}\right)$

∴ Total self-inductance of conductor B is

$$L_B = (L_{Bi} + L_{Be}) = \frac{\mu}{8\pi} + \frac{\mu}{2\pi}\ln\left(\frac{d}{b}\right) = \frac{\mu}{2\pi}\left[\frac{1}{4} + \ln\left(\frac{d}{b}\right)\right]$$

Hence, self-inductance per unit length of the two wire transmission line is

$$L = (L_A + L_B) = \frac{\mu}{2\pi}\left[\frac{1}{4} + \ln\left(\frac{d}{a}\right)\right] + \frac{\mu}{2\pi}\left[\frac{1}{4} + \ln\left(\frac{d}{b}\right)\right] = \frac{\mu}{\pi}\left[\frac{1}{4} + \frac{1}{2}\ln\left(\frac{d^2}{ab}\right)\right] = \frac{\mu}{\pi}\left[\frac{1}{4} + \ln\left(\frac{d}{\sqrt{ab}}\right)\right]$$

$$L = \frac{\mu}{\pi}\left[\frac{1}{4} + \ln\left(\frac{d}{\sqrt{ab}}\right)\right]$$

If $a = b = r$, then the self-inductance per unit length is given as

$$L = \frac{\mu}{\pi}\left[\frac{1}{4} + \ln\left(\frac{d}{r}\right)\right]$$

Example 3.80 A long solenoid with length l and a cross-sectional area A consists of N_1 turns of wire. An insulated coil of N_2 turns is wrapped around it, as shown in Fig. 3.64.

(a) Calculate the mutual inductance M, assuming that all the flux from the solenoid passes through the outer coil.

(b) Relate the mutual inductance M to the self-inductances L_1 and L_2 of the solenoid and the coil.

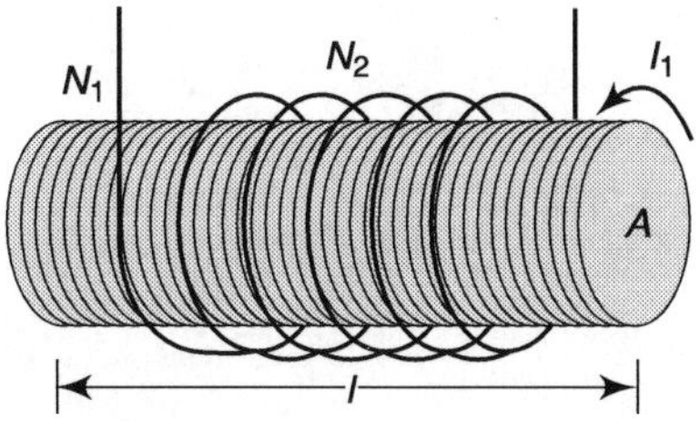

Fig. 3.64 *A coil wrapped around a solenoid*

Solution

(a) The magnetic flux through each turn of the outer coil due to the solenoid is

$$\phi_{21} = BA = \frac{\mu_0 N_1 I_1}{l} A$$

where $B = \frac{\mu_0 N_1 I_1}{l}$ is the uniform magnetic field inside the solenoid.

Hence, the mutual inductance between the solenoid and the coil is

$$M = \frac{N_2 \phi_{21}}{I_1} = \frac{\mu_0 N_1 N_2 A}{l}$$

$$M = \frac{\mu_0 N_1 N_2 A}{l}$$

(b) From Example 3.42, we see that the self-inductance of the solenoid with N_1 turns is given by,

$$L_1 = \frac{N_1 \phi_{11}}{I_1} = \frac{\mu_0 N_1^2 A}{l}$$

where ϕ_{11} is the magnetic flux through one turn of the solenoid due to the magnetic field produced by I_1.

Similarly, we have the self-inductance for the outer coil given as

$$L_2 = \frac{N_2 \phi_{22}}{I} = \frac{\mu_0 N_2^2 A}{l}$$

Thus, in terms of L_1 and L_2, the mutual inductance can be expressed as

$$M = \frac{\mu_0 N_1 N_2 A}{l} = \sqrt{\frac{\mu_0 N_1^2 A}{l}\frac{\mu_0 N_2^2 A}{l}} = \sqrt{L_1 L_2}$$

$$M = \sqrt{L_1 L_2}$$

NOTE

More generally, the mutual inductance is given as

$$M = k\sqrt{L_1 L_2} \qquad 0 \le k \le 1$$

where k is the coupling coefficient. In this example, we have k = 1 which means that all of the magnetic flux produced by the solenoid passes through the outer coil, and vice versa, in this idealisation.

Example 3.81 Two coils of N_1 and N_2 turns, respectively, are wound on a toroid of constant permeability μ. The various dimensions are as shown in Fig. 3.65. Determine the mutual inductance and self-inductance of the coils.

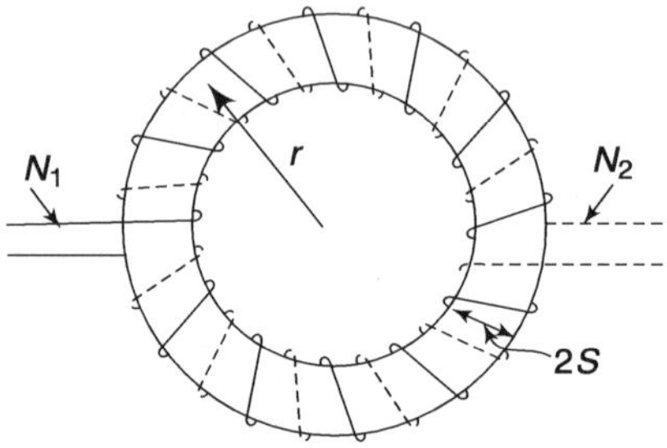

Fig. 3.65 *Two coils wound on a toroid*

Solution

Here, r—radius of the toroid,
S—radius of the winding,
N_1, N_2—number of turns of the coils,
I_1, I_2—currents in the coils

We consider I_1 current.

If $r >> S$, then the magnetic flux density is constant over the interior of the winding and is given as

$$B = \frac{\mu N_1 I_1}{2\pi r}$$

$\therefore$ Flux linking coil 1 due to current flowing in coil 1 is

$$\phi_{11} = N_1 \times B \times \pi S^2 = \frac{\mu N_1^2 I_1}{2\pi r}\pi S^2 = \frac{\mu N_1^2 I_1 A}{l}$$

where, $A = \mu S^2$ = area of winding cross-section and $l = 2\pi r$ = mean length of toroidal coil

Hence, self-inductance of coil 1 is $L_1 = \frac{\phi_{11}}{I_1} = \frac{\mu N_1^2 A}{l}$

Similarly, self-inductance of coil 2 is $L_2 = \frac{\phi_{22}}{I_2} = \frac{\mu N_2^2 A}{l}$

$$\therefore \qquad \boxed{L_1 = \frac{\mu N_1^2 A}{l} \quad \text{and} \quad L_2 = \frac{\mu N_2^2 A}{l}}$$

Now, flux linking coil 2 due to current flowing in coil 1 is

$$\phi_{21} = N_2 \times B \times \pi S^2 = \frac{\mu N_1 N_2 I_1}{2\pi r}\pi S^2 = \frac{\mu N_1 N_2 I_1 A}{l}$$

Similarly, flux linking coil 1 due to current flowing in coil 2 is

$$\phi_{12} = N_1 \times B \times \pi S^2 = \frac{\mu N_1 N_2 I_2}{2\pi r}\pi S^2 = \frac{\mu N_1 N_2 I_2 A}{l}$$

Hence, mutual inductance between the coils is $M = \frac{\phi_{21}}{I_1} = \frac{\phi_{12}}{I_2} = \frac{\mu N_1 N_2 A}{l}$

$$M = \frac{\mu N_1 N_2 A}{l}$$

NOTE

Just like in previous example, here also, $M = \sqrt{L_1 L_2}$ with unity coefficient of coupling.

Example 3.82 An infinite straight wire carrying current I is placed to the left of a rectangular loop of wire with width and length l, as shown in Fig. 3.66. Determine the mutual inductance of the system.

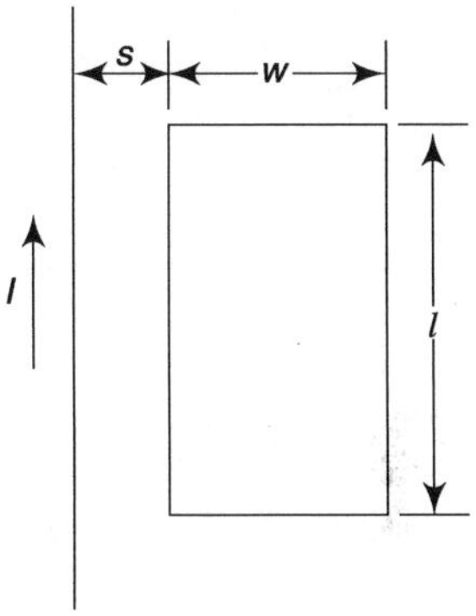

Fig. 3.66 *Rectangular loop placed near long straight current-carrying wire*

Solution To calculate the mutual inductance M, we first need to know the magnetic flux through the rectangular loop. Using Ampere's law, the magnetic field at a distance r away from the straight wire is given as

$$\vec{B} = \frac{\mu_0 I}{2\pi r}\hat{a}_\phi$$

The total magnetic flux ϕ through the loop can be obtained by summing over contributions from all differential area elements $dS = ldr$ as follows.

$$\phi = \int d\phi = \int \vec{B} \cdot d\vec{S} = \int_s^{s+w} \left(\frac{\mu_0 I}{2\pi r}\hat{a}_\phi\right) \cdot (ldr\hat{a}_\phi) = \frac{\mu_0 Il}{2\pi}\int_s^{s+w}\frac{dr}{r} = \frac{\mu_0 Il}{2\pi}\ln\left(\frac{s+w}{s}\right)$$

Hence, the mutual inductance is given as $M = \frac{\phi}{I} = \frac{\mu_0 l}{2\pi}\ln\left(\frac{s+w}{s}\right)$

$$M = \frac{\mu_0 l}{2\pi}\ln\left(\frac{s+w}{s}\right)$$

3.21 MAGNETIC ENERGY (ENERGY STORED IN MAGNETIC FIELDS)

Since an inductor in a circuit serves to oppose any change in the current through it, work must be done by an external source, such as a battery, in order to establish a current in the inductor. From the work–energy theorem, we conclude that energy can be stored in an inductor. The role played by an inductor in the magnetic case is analogous to that of a capacitor in the electric case.

The power or rate at which an external emf ε_{ext} works to overcome the self-induced emf ε_L and pass current I in the inductor is

$$P_L = \frac{dW_{ext}}{dt} = I\varepsilon_{ext}$$

If only the external emf and the inductor are present, then $\varepsilon_{ext} = \varepsilon_L$, which implies

$$P_L = \frac{dW_{ext}}{dt} = -I\varepsilon_L = -I\left(-L\frac{dI}{dt}\right) = IL\frac{dI}{dt}$$

If the current is increasing with $\frac{dI}{dt} > 0$, then $P > 0$, which means that the external source is doing positive work to transfer energy to the inductor. Thus, the internal energy of the inductor is increased. On the other hand, if the current is decreasing with $\frac{dI}{dt} < 0$, we then have $P < 0$. In this case, the external source takes energy away from the inductor, causing its internal energy to go down.

The total work done by the external source to increase the current form zero to I is then

$$W_{\text{ext}} = \int dW_{\text{ext}} = \int_0^I LIdI = \frac{1}{2}LI^2$$

This is equal to the *energy stored in the magnetic field* (W_m).

$$\boxed{W_m = \frac{1}{2}LI^2} \qquad \textbf{(3.116)}$$

NOTE

The above expression is analogous to the electric energy stored in a capacitor, $\left(W_e = \frac{1}{2}CV^2\right)$.

In order to evaluate the density of the stored energy in terms of the field quantities, we consider a long solenoid. The magnetic flux density within the solenoid is

$$B = \frac{\mu NI}{l}$$

where N is the number of turns, I is the current flowing and l is the length.

If the cross-sectional area of the coil is A, the flux through it is

$$\phi = \int_S \vec{B} \cdot d\vec{S} = BA = \frac{\mu ANI}{l}$$

$\therefore$ Total flux linkage $\lambda = N\phi = \frac{\mu AN^2 I}{l}$

So, the inductance of the solenoid is $L = \frac{\lambda}{I} = \frac{\mu AN^2}{l}$

Therefore, the magnetic energy stored in the solenoid is $W_m = \frac{1}{2}LI^2 = \frac{1}{2}\frac{\mu AN^2}{l}I^2$

Thus, the energy density (energy per unit volume) is

$$\frac{W_m}{V} = \frac{W_m}{Al} = \frac{1}{2}\frac{\mu N^2}{l^2}I^2 = \frac{1}{2}\mu\left(\frac{NI}{l}\right)^2 = \frac{1}{2}\mu H^2$$

$$\therefore \quad \boxed{\frac{dW_m}{dV} = \frac{1}{2}\mu H^2 = \frac{1}{2}\frac{B^2}{\mu} \quad (\text{Joule/m}^3)} \qquad \textbf{(3.117)}$$

Using this expression, magnetic energy can be expressed in different ways as follows.

(i) Using the relation that, $\vec{B} = \mu\vec{H}$, we have

$$\boxed{W_m = \frac{1}{2\mu}\int_{\text{vol}} \vec{B} \cdot \vec{B} dv = \frac{1}{2}\int_{\text{vol}} \vec{B} \cdot \vec{H} dv} \qquad \textbf{(3.118)}$$

(ii) $$W_m = \frac{1}{2\mu}\int_{\text{Vol}} \vec{B}\cdot\vec{B}dv = \frac{1}{2\mu}\int_{\text{Vol}} \vec{B}\cdot(\nabla\times\vec{A})dv = \frac{1}{2\mu}\left[\int_{\text{Vol}} \vec{A}\cdot(\nabla\times\vec{B})\,dv + \int_{\text{Vol}} \nabla\cdot(\vec{A}\times\vec{B})\,dv\right]$$

$$= \frac{1}{2\mu}\left[\int_{\text{Vol}} \vec{A}\cdot(\nabla\times\vec{B})\,dv + \int_S (\vec{A}\times\vec{B})\cdot d\vec{S}\right]$$

If V is chosen to include all space, then the surface S is at infinity, $\int_S (\vec{A}\times\vec{B})\cdot d\vec{S}$ will vanish.

$$W_m = \frac{1}{2\mu}\int_{\text{Vol}} \vec{A}\cdot(\nabla\times\vec{B})dv = \frac{1}{2\mu}\int_{\text{Vol}} \vec{A}\cdot(\nabla\times\vec{B})dv = \frac{1}{2\mu}\int_{\text{Vol containing } \vec{J}} \mu\vec{A}\cdot\vec{J}dv$$

$\{\because \nabla\times\vec{B} = \mu\vec{J}$ and $\nabla\times\vec{B} = 0$ for volume where $\vec{J} = 0\}$

$$\boxed{W_m = \frac{1}{2}\int_{\text{Vol containing } \vec{J}} \vec{A}\cdot\vec{J}dv} \quad \textbf{(3.119)}$$

(iii) $W_m = \frac{1}{2}\int_{\text{Vol}} \vec{A}\cdot\vec{J}dv = \frac{1}{2}\oint_l I(\vec{A}\cdot d\vec{l})$ $\{\because \vec{J}dv = Id\vec{l}\}$

$$\boxed{W_m = \frac{1}{2}I\oint_l \vec{A}\cdot d\vec{l} = \frac{1}{2}I\phi} \quad \textbf{(3.120)}$$

[by Eq. (3.84)]

***Example 3.83** Energy Stored in a Solenoid

A long solenoid with length l and a radius R consists of N turns of wire, A current I passes through the coil. Find the energy stored in the system.

Solution Here, the inductance of the solenoid is given by Example 3.22, as

$$L = \mu_0 n^2 \pi R^2 l$$

Hence, the magnetic energy stored in the system is

$$W_m = \frac{1}{2}LI^2 = \frac{1}{2}\mu_0 n^2 I^2 \pi R^2 l = \frac{1}{2\mu_0}(\mu_0 nI)^2(\pi R^2 l) = \frac{B^2}{2\mu_0}(\pi R^2 l)$$

since the magnetic field in a solenoid is, $B = \mu_0 nI$.

Here, the term $\pi R^2 l$ is the volume within the solenoid. Therefore, the magnetic energy density or energy per unit volume is given as

$$w_m = \frac{B^2}{2\mu_0}$$

Example 3.84 A long solenoid with length l and a radius R consists of N turns of wire. A current I passes through the coil. Find the energy stored in the system.

Solution From Example 3.73, we have self-inductance of a long solenoid given as

$$L = \frac{\pi\mu N^2 R^2}{l}$$

Hence, the magnetic energy stored in the solenoid is

$$W_m = \frac{1}{2}LI^2 = \frac{1}{2}\frac{\pi\mu N^2 R^2}{l}I^2 = \frac{1}{2}\frac{\pi\mu N^2 R^2 I^2}{l}$$

$$\boxed{W_m = \frac{1}{2}\frac{\pi\mu N^2 R^2 I^2}{l}}$$

The result can be expressed in terms of the magnetic field as

$$W_m = \frac{1}{2}\frac{\pi\mu N^2 R^2 I^2}{l} = \frac{1}{2\mu}\left(\frac{\mu NI}{l}\right)^2 (\pi R^2 l) = \frac{B^2}{2\mu}(\pi R^2 l)$$

Since $(\pi R^2 l)$ is the volume within the solenoid, and the magnetic field inside the solenoid is uniform, the term $\left(\frac{B^2}{2\mu}\right)$ may considered as the magnetic energy density.

Example 3.85 A toroid consists of N turns and has a rectangular cross section, with inner radius a, outer radius b and height h. Find the total magnetic energy stored in the toroid.

Solution From Example 3.75, the self-inductance of a toroid with rectangular cross-section is given as

$$L = \frac{\mu_0 N^2 h}{2\pi}\ln\left(\frac{b}{a}\right)$$

Hence, the magnetic energy stored in the toroid is

$$W_m = \frac{1}{2}LI^2 = \frac{\mu_0 N^2 h I^2}{4\pi}\ln\left(\frac{b}{a}\right)$$

Alternative method:

From Example 3.43, magnetic field of a toroid is $B = \frac{\mu NI}{2\pi r}$

$\therefore$ Magnetic energy density is

$$w_m = \frac{1}{2}\frac{B^2}{\mu} = \frac{1}{2\mu}\left(\frac{\mu NI}{2\pi r}\right)^2 = \frac{\mu N^2 I^2}{8\pi^2 r^2}$$

The total energy stored in the magnetic field can be found by integrating over the volume. We choose the differential volume element to be a cylinder with radius r, width dr and height h, so that $dv = 2\pi rhdr$.

Hence, the magnetic energy stored in the toroid is

$$W_m = \int w_m = \int_a^b \frac{\mu N^2 I^2}{8\pi^2 r^2} 2\pi rhdr = \frac{\mu N^2 h I^2}{4\pi}\int_a^b \frac{dr}{r} = \frac{\mu N^2 h I^2}{4\pi}\ln\left(\frac{b}{a}\right)$$

This is the same as obtained in the earlier method.

$$\therefore \quad \boxed{W_m = \frac{\mu N^2 h I^2}{4\pi}\ln\left(\frac{b}{a}\right)}$$

Example 3.86 A wire of nonmagnetic material with radius R and length l carries a current I which is uniformly distributed over its cross section. What is the magnetic energy inside the wire?

Solution Applying Ampere's law, the magnetic field at distance $r \le R$ can be obtained as

$$B(2\pi r) = \mu J(\pi r^2) = \mu\left(\frac{I}{\pi R^2}\right)(\pi r^2)$$

or

$$B = \frac{\mu I r}{2\pi R^2}$$

$\therefore$ Magnetic energy density is $w_m = \frac{1}{2}\frac{B^2}{\mu} = \frac{1}{2\mu}\left(\frac{\mu I r}{2\pi R^2}\right)^2 = \frac{\mu I^2 r^2}{8\pi^2 R^4}$

Hence, total magnetic energy stored inside the wire is

$$W_m = \int w_m = \int_0^R \frac{\mu I^2 r^2}{8\pi^2 R^4} 2\pi r l dr = \frac{\mu I^2 l}{4\pi R^4}\int_a^b r^3 dr = \frac{\mu I^2 l}{4\pi R^4}\left(\frac{R^4}{4}\right) = \frac{\mu I^2 l}{16\pi}$$

$$\boxed{W_m = \frac{\mu I^2 l}{16\pi}}$$

Summary

- The phenomenon of transferring electric charge from one point in a circuit to another is described by the term *electric current*. Electric current is defined as the rate of flow of electric charges or electrons through a cross-sectional area.

$$i = \frac{dq}{dt}$$

- *Current density vector*, $\vec{J}$ at any point is defined as the current through a unit normal area at that point.

$$\vec{J} = \frac{I}{A}\hat{n}$$

- Total current flowing through a surface S is given as

$$I = \int_S \vec{J} \cdot d\vec{S}$$

- Electric current is of three types:
  1. Convection current,
  2. Conduction current, and
  3. Displacement current
- The motion of charged particles in free space (vacuum) is said to constitute *convection current.*
- The motion of the free electrons present in a conductor, by the influence of an electric field constitutes the *conduction current*. The relation of conduction current given as

$$\vec{J} = \sigma \vec{E}$$

  is known as point form of *Ohm's law*.
- The current flowing in a capacitor is termed as *displacement current*, given as

$$i_d(t) = A\frac{dD(t)}{dt}$$

- From electrical point of view, materials can be classified as conductors ($\sigma >> 1$, $\varepsilon_r = 1$), dielectrics ($\sigma << 1$, $\varepsilon_r \geq 1$).
- The resistance of a conductor of uniform cross section is given as

$$R = \frac{\rho l}{A}$$

and in general, for non-uniform cross-section, it is given as

$$R = \frac{V}{I} = \frac{\int \vec{E} \cdot d\vec{l}}{\int \vec{J} \cdot d\vec{S}} = \left(\frac{1}{\sigma}\right) \frac{\int \vec{E} \cdot d\vec{l}}{\int \vec{E} \cdot d\vec{S}}$$

- The *Joule's law* states that the rate of heat production by a steady current in any part of an electrical circuit is directly proportional to the resistance and to the square of the current ($P = I^2R$. In differential and integral forms it is given as

$$p = \vec{J} \cdot \vec{E} \qquad \text{and} \qquad P = \int_v \vec{E} \cdot \vec{J} dv$$

- The *electromotive force* (EMF) in a closed loop is given as

$$\xi = \oint_l \vec{E}_e \cdot d\vec{l}$$

where $\vec{E}_e$ is the emf-producing field, i.e., the field generated by causes other than the static charges.
- The *Kirchhoff's current law* (KCL) in differential and integral forms is given as

$$\nabla \cdot \vec{J} = 0 \qquad \text{and} \qquad \oint_S \vec{J} \cdot d\vec{S} = 0$$

- The *continuity equation* of charges is given as

$$\nabla \cdot \vec{J} = -\frac{\partial \rho_v}{\partial t}$$

- *Relaxation time* $\left(\tau = \frac{\varepsilon}{\sigma}\right)$ of a material is the time taken by a charge placed in the interior of a conductor to drop its value to 37% ($e^{-1} = 0.368 \approx 37\%$) of its initial value.
- The boundary conditions for the current density for two different conducting media are given as

$$J_{1n} = J_{2n} \qquad \text{and} \qquad \frac{J_{1t}}{\sigma_1} = \frac{J_{2t}}{\sigma_2}$$

Combining these two conditions we get

$$\frac{\tan \theta_1}{\tan \theta_2} = \frac{\sigma_1}{\sigma_2}$$

where θ_1 and θ_2 are the angles with the normal in respective medium.

If the boundary carries a surface charge with density, ρ_s (C/m^2), then by the boundary conditions can be written as

$$\rho_s = \left(\frac{\varepsilon_2}{\sigma_2} - \frac{\varepsilon_1}{\sigma_1}\right) J_{1n}$$

- Laplace equation for conducting medium is given as, $\nabla^2 V = 0$.
- The *dielectric polarisation* may be defined as a dynamical response of a system to an externally applied electric field. Polarisation vector is defined as the dipole moment per unit volume of the dielectric, i.e.,

$$\vec{P} = \lim_{\Delta v \to 0} \left(\frac{\sum_{i=1}^{N} Q_i \vec{d}_i}{\Delta v} \right)$$

- The dielectric materials which have no free charges; all electrons that are bound and associated with the nearest atoms, are known as *non-polar dielectrics*. The dielectric materials, in which the molecules or atoms possess a permanent dipole moment which is ordinarily randomly oriented, but which becomes more or less oriented by the application of an external electric field, are known as *polar dielectrics*.
- The effect of macroscopic polarisation in a given volume of dielectric material is to induce some bound surface and volume charge densities in the dielectric, given as

$$\sigma_b = \vec{P} \cdot \hat{n} \qquad \text{and} \qquad \rho_b = -\nabla \cdot \vec{P}$$

- Electric displacement is given in terms of electric field and polarisation vector as

$$\vec{D} = \varepsilon_0 \vec{E} + \vec{P}$$

- The relation of polarisation vector with the electric field is given as

$$\vec{P} = \varepsilon_0 \chi_e \vec{E}$$

  Here, the quantity χ_e is known as the *electric susceptibility* of a dielectric material which gives a measure of how easily it polarises in response to an electric field.
- *Electric permittivity* (ε) is a physical quantity that describes how an electric field affects and is affected by a dielectric medium. It is determined by the ability of a material to polarise in response to the field, and thereby reduce the total electric field inside the material. Its relations are given as

$$\varepsilon = \varepsilon_0 \varepsilon_r \qquad \text{and} \qquad \varepsilon_r = (1 + \chi_e)$$

  The dielectric constant or relative permittivity (ε_r) is the ratio of the permittivity of a substance to the permittivity of free space.
- The maximum electric field that a dielectric can withstand without breakdown is known as *dielectric strength* of the material.
- A dielectric material is said to be homogeneous if the permittivity (ε) or conductivity (σ) does not vary with space in a region.
- A dielectric material is said to be *isotropic* if the electrical properties of the medium are independent of the direction, i.e. $\vec{D}$ and $\vec{E}$ are in the same direction.
- *Biot–Savart states* that the magnetic field intensity $d\vec{H}$ produced at a point P at a distance r from a differential current element Idl is:

$$d\vec{H} = \frac{Id\vec{l} \times \vec{r}}{4\pi r^3}$$

- Three current densities (line current density, I; surface current density $\vec{K}$ and volume current density, $\vec{J}$) are related to each other as

$$I d\vec{l} \equiv \vec{K} dS \equiv \vec{J} dv$$

- *The magnetic field intensity* $\vec{H}$ at any point is defined as the force experienced by a north pole of one Weber placed at that point. Its unit is Newton per Weber (N/Wb) or Ampere-turn per metre (AT/m).
- *Magnetic flux* is defined as the group of magnetic field lines emitted outward from the north pole of a magnet. It is measured in Weber and is denoted as ϕ.
- *Magnetic flux density* ($\vec{B}$) is the amount of magnetic flux per unit area of a section, perpendicular to the direction of magnetic flux; i.e.,

$$B = \frac{\phi}{A}$$

- It is a vector quantity (expressed in Weber per square metre) that specifies both the strength and direction of the magnetic field.
- The magnetic flux through a surface S is given as

$$\phi = \int_S \vec{B} \cdot d\vec{S}$$

- The magnetic field intensity is related to the magnetic flux density as

$$\vec{B} = \mu \vec{H}$$

where μ is a constant, called permeability of the medium. It is given as

$$\mu = \mu_0 \mu_r$$

where μ_0 is the permeability of free space, known as absolute permeability, $= 4\pi \times 10^{-7}$ H/m

μ_r is the relative permeability

- As it is not possible to have an isolated magnetic pole, the total magnetic flux through a closed surface must be zero. This is known as *Gauss' law of magnetostatics*. It is given as

$$\therefore \qquad \oint_S \vec{B} \cdot d\vec{S} = 0 \qquad \text{and} \qquad \nabla \cdot \vec{B} = 0$$

- *Ampere's circuital law* states that the line integral of the magnetic field intensity ($\vec{H}$) around any closed path is equal to the direct current enclosed by the path.

$$\oint_l \vec{H} \cdot d\vec{l} = I_{\text{enc}}$$

- The *magnetic scalar potential* (V_{m}) is defined as

$$\vec{H} = -\nabla V_m \qquad \text{if} \qquad \vec{J} = 0$$

- The *magnetic vector potential* ($\vec{A}$) is defined as

$$\vec{B} = \nabla \times \vec{A}$$

- The magnetic field due to a current distribution can be found using the concept of magnetic vector potential and using the relation as

$$\vec{A} = \frac{\mu}{4\pi}\int_v \frac{\vec{J}}{r}dv \quad \text{for volume current density}$$

$$= \frac{\mu}{4\pi}\int_S \frac{\vec{K}}{r}dS \quad \text{for surface current density}$$

$$= \frac{\mu}{4\pi}\int_v \frac{Id\vec{l}}{r} \quad \text{for line current density}$$

- The *Lorentz force equation* relates the force on a moving charged particle in the presence of a magnetic field and is given as

$$\vec{F} = Q(\vec{E} + \vec{v} \times \vec{B})$$

- The force in an element $d\vec{l}$ of a current carrying conductor carrying a current I, placed in a magnetic field is given as

$$d\vec{F} = Id\vec{l} \times \vec{B}$$

- The force between two current carrying conductors is given by *Ampere's force law* and is written as

$$\vec{F}_2 = \frac{\mu I_1 I_2}{4\pi}\oint_{C_1}\oint_{C_2} \frac{d\vec{l}_2 \times (d\vec{l}_1 \times \hat{R}_{12})}{R^2}$$

- The *magnetic dipole moment* is the product of current and area of the loop; its direction is normal to the plane of the loop; its unit is Am^2.

$$\vec{m} = I\vec{S} = IS\hat{a}_n$$

- The torque on a current carrying coil with magnetic moment $\vec{m}$ placed in a uniform magnetic field $\vec{B}$ is given as

$$\vec{T} = \vec{m} \times \vec{B} = I\vec{S} \times \vec{B} = IS\hat{a}_n \times \vec{B}$$

- A bar magnet or a small filamentary loop carrying a current is known as a *magnetic dipole.*
- *Magnetisation* $(\vec{M})$ is defined as the amount of magnetic moment per unit volume. It is expressed in Ampere per metre (A/m).
- The effect of magnetisation in a given volume of magnetic material is to induce some bound surface and volume current densities in the material, given as

  Magnetisation volume current density or bound volume current density $(A/m^2) = \vec{J}_b = \nabla \times \vec{M}$

  Magnetisation surface current density or bound surface current density $(A/m) = \vec{K}_b = \vec{M} \times \hat{a}_n$
- The *magnetic susceptibility* χ_m of a magnetic material is a measure of the degree of magnetisation of a material in response to an applied magnetic field. It relates the magnetic field to the magnetisation vector as

$$\vec{M} = \chi_m \vec{H}$$

- *Permeability* (μ) is the degree of magnetisation of a material that responds linearly to an applied magnetic field. It relates the magnetic field to the magnetic flux density as

$$\vec{B} = \mu\vec{H}$$

- For linear magnetic materials, the relation between different magnetic properties is given as,

$$\vec{B} = \mu\vec{H} = \mu_0\mu_r\vec{H} = \mu_0(1+\chi_m)\vec{H} = \mu_0(\vec{H}+\vec{M})$$

- Depending upon the values of the magnetic susceptibility (χ_m) or the relative permeability (μ_r), magnetic materials are broadly classified into three groups as
  1. Paramagnetic ($\mu_r \geq 1, \chi_m > 0$),
  2. Diamagnetic ($\mu_r \leq 1, \chi_m < 0$), and
  3. Ferromagnetic ($\mu_r >> 1, \chi_m >> 0$).
- The variation of $\vec{B}$ as a function of the externally applied field $\vec{H}$ is known as a *hysteresis curve* or *magnetisation curve* or *B-H curve.*
- The boundary conditions that the magnetic field must satisfy at the interface between two different magnetic media are

$$B_{1n} = B_{2n} \qquad \text{or} \qquad \mu_1 H_{1n} = \mu_2 H_{2n}$$

and

$$(\vec{H}_1 - \vec{H}_2) \times \hat{a}_{n12} = \vec{K} \qquad \text{or} \qquad H_{1t} = H_{2t} \text{ if } \vec{K} = 0$$

Combining these boundary conditions it can be written as

$$\frac{\tan \theta_1}{\tan \theta_2} = \frac{\mu_1}{\mu_2}$$

where θ_1 and θ_2 are the angles with the normal in respective medium.

- The ratio of the magnetic flux to the current is called the inductance, or more accurately *self-inductance* of the circuit.

$$L = \frac{\lambda}{I} = \frac{N\phi}{I}$$

- *Mutual inductance* is the ability of one inductor to induce an e.m.f. across another inductor placed very close to it.
- The relation between the self-inductances of two coils and their mutual inductance is written as

$$M = k\sqrt{L_1 L_2}$$

where k is the *coefficient of coupling.*

- Magnetic energy density (energy per unit volume) is given as

$$\frac{dW_m}{dV} = \frac{1}{2}\mu H^2 = \frac{1}{2}\frac{B^2}{\mu} \quad (\text{Joule/m}^3)$$

- Total magnetic energy stored can be written in different forms as

$$\begin{aligned} W_m &= \frac{1}{2}LI^2 \\ &= \frac{1}{2\mu}\int_{\text{vol}} \vec{B}\cdot\vec{B}dv = \frac{1}{2}\int_{\text{vol}} \vec{B}\cdot\vec{H}dv \\ &= \frac{1}{2}\int_{\text{Vol containing } \vec{J}} \vec{A}\cdot\vec{J}dv \\ &= \frac{1}{2}I\oint_l \vec{A}\cdot d\vec{l} = \frac{1}{2}I\phi \end{aligned}$$

Important Formulae	
Current	$I = \frac{Q}{t}$
Current density	$\vec{J} = \frac{I}{A}\hat{n}$
Total current flowing through a surface	$I = \int_S \vec{J} \cdot d\vec{S}$
Convection current density	$\vec{J} = \rho_s \vec{U} = \rho_{v+}\vec{U}_+ + \rho_{v-}\vec{U}_-$
Conduction current density (Ohm's law)	$\vec{J} = \sigma \vec{E}$
Displacement current density	$\vec{J}_d(t) = \frac{\partial \vec{D}(t)}{\partial t}$
Resistance	$R = \frac{V}{I} = \frac{\rho l}{A} = \frac{\int \vec{E} \cdot d\vec{l}}{\int \vec{J} \cdot d\vec{S}} = \left(\frac{1}{\sigma}\right)\frac{\int \vec{E} \cdot d\vec{l}}{\int \vec{E} \cdot d\vec{S}}$
Power dissipation (Joule's law)	$P = \int_v \vec{E} \cdot \vec{J} dv = VI = \frac{V^2}{R} = I^2 R$
Equation of continuity	$\nabla \cdot \vec{J} = -\frac{\partial \rho_v}{\partial t}$
Relaxation time	$\tau = \frac{\varepsilon}{\sigma}$
Biot–Savart law	$\vec{H} = \int_l \frac{I d\vec{l} \times \hat{a}_r}{4\pi r^2} = \int_S \frac{\vec{K} dS \times \hat{a}_r}{4\pi r^2} = \int_v \frac{\vec{J} dv \times \hat{a}_r}{4\pi r^2}$
Relation between magnetic field and flux density	$\vec{B} = \mu \vec{H}$
Magnetic field due to an infinite straight wire	$H = \frac{I}{2\pi r}\hat{a}_\phi$
Gauss' law of magnetostatics	$\int_S \vec{B} \cdot d\vec{S} = 0$ or $\nabla \cdot \vec{B} = 0$
Ampere's circuital law	$\oint_l \vec{H} \cdot d\vec{l} = I_{\text{enc}}$ or $\nabla \times \vec{H} = \vec{J}$
Magnetic field due to an infinite solenoid	$H = \frac{NI}{l}$
Magnetic scalar potential	$\vec{H} = -\nabla V_m$

Magnetic vector potential	$\vec{B} = \nabla \times \vec{A}$ and $\vec{A} = \frac{\mu}{4\pi}\int_v \frac{\vec{J}}{r} dv$ for volume current density $= \frac{\mu}{4\pi}\int_S \frac{\vec{K}}{r} dS$ for surface current density $= \frac{\mu}{4\pi}\int_v \frac{I d\vec{l}}{r}$ for line current density
Magnetic flux in terms of vector potential	$\phi = \int_l \vec{A} \cdot d\vec{l}$
Lorentz force equation	$\vec{F} = Q(\vec{E} + \vec{v} \times \vec{B})$
Force on a finite current element in magnetic field	$\vec{F} = \int_l I d\vec{l} \times \vec{B} = \int_S \vec{K} dS \times \vec{B} = \int_v \vec{J} dv \times \vec{B}$
Ampere's force law	$\vec{F} = \frac{\mu I_1 I_2}{4\pi} \oint_{C_1} \oint_{C_2} \frac{d\vec{l}_2 \times (d\vec{l}_1 \times \widehat{R}_{12})}{R^2}$
Magnetic torque	$\vec{T} = I\vec{S} \times \vec{B}$
Relation between $\vec{B}, \vec{H}$ and $\vec{M}$	$\vec{B} = \mu_0(\vec{H} + \vec{M})$
Magnetic boundary condition	$B_{1n} = B_{2n}$ and $(H_{1t} - H_{2t}) = K \infty (\overline{H}_1 - \overline{H}_2) \times \widehat{a}_{n21} = \overline{K}$
Self-inductance of a coil	$L = \frac{\lambda}{I} = \frac{N\phi}{I}$
Mutual inductance of two coils	$M = k\sqrt{L_1 L_2}$ $0 \le k \le 1$
Energy stored in magnetic field	$W_m = \frac{1}{2} LI^2 = \frac{1}{2} \int_{\text{vol}} \vec{B} \cdot \vec{H} dv$

Exercises

[Note: * marked problems are important university problems]

- ***Easy***

1. Find the current in a circular wire of radius 2 mm if the current density is given as $\vec{J} = 15(1 - e^{-1000r})\widehat{a}_z$ A/m^2. [0.133 mA]
2. Find the total current in a cylindrical conductor of radius 2 mm if the current density varies with the distance from the axis as $J = 10^3 e^{-400r}$ A/m^2. [7.51 mA]
3. Given the current density $\vec{J} = 10^3 \sin\theta\, \widehat{a}_r$ A/m^2 in spherical coordinates; find the total current crossing the spherical shell of radius 0.02 m. [3.95 A]
4. In a material for which $\sigma = 5.0$ S/m and $\varepsilon_r = 1$, the electric field intensity is $E = 250 \sin 10^{10}t$ (V/m). Find the conduction and displacement current densities and the frequency at which they have equal magnitudes. [1250 sin $10^{10}t$ (A/m^2); 22.1 cos $10^{10}t$ (A/m^2); 90 GHz]

5. Using Ampere's circuital law in integral form, find $\vec{H}$ everywhere due to the current density in cylindrical coordinates:

$$\begin{aligned}\vec{J} &= J_0 e^{-2r}\hat{a}_z \quad 0 < r < 0.5 \\ &= 0 \quad \text{elsewhere}\end{aligned}$$

$$\left[\begin{aligned}\vec{H} &= \frac{J_0}{4r}(1 - e^{-2r} - 2re^{-2r})\hat{a}_\phi \quad r < 0.5 \\ &= \frac{0.066}{r} J_0 \hat{a}_\phi \quad r \geq 0.5\end{aligned}\right]$$

6. Find the current distribution producing the following field distribution using Ampere's circuital law in differential form:

$$\begin{aligned}\vec{H} &= \frac{Ir}{2\pi a^2}\hat{a}_\phi, \quad r < a \\ &= \frac{I}{2\pi r}\hat{a}_\phi \quad r > a\end{aligned}$$

$$\left[\begin{aligned}\vec{J} &= \frac{I}{\pi a^2}\hat{a}_z \quad r < a \\ &= 0 \quad r > a\end{aligned}\right]$$

7. A thin cylindrical conductor of radius a and infinite length carries a current I. Find the magnetic field $\vec{H}$ at all points using Ampere's law.

$$\left[\begin{aligned}\vec{H} &= 0 \quad r < a \\ &= \frac{I}{2\pi r}\hat{a}_\phi \quad r > a\end{aligned}\right]$$

***8.** Determine the magnetic field $\vec{H}$ for a solid cylindrical conductor of radius a, where the current I is uniformly distributed over the cross section.

$$\left[\begin{aligned}\vec{H} &= \frac{Ir}{2\pi a^2}\hat{a}_\phi, \quad r < a \\ &= \frac{I}{2\pi r}\hat{a}_\phi \quad r > a\end{aligned}\right]$$

9. Determine the magnetic field $\vec{H}$ on the axis of a circular current loop of radius a, carrying a steady current I. What is the field at the centre of the loop?

10. Using Ampere's circuital law in integral form, find $\vec{H}$ everywhere due to the current density in cylindrical coordinates:

$$\begin{aligned}\vec{J} &= 4.5e^{-2r}, \quad 0 < r < 0.5 \text{ m} \\ &= 0 \quad \text{elsewhere}\end{aligned}$$

$$\left[\begin{aligned}\vec{H} &= \frac{1.125}{r}(1 - e^{-2r} - 2re^{-2r})\hat{a}_\phi, \quad r < 0.5 \text{ m} \\ &= \frac{0.297}{r}\hat{a}_\phi \quad r \geq 0.5 \text{ m}\end{aligned}\right]$$

***11.** A solid cylindrical conductor of radius R has a uniform current density. Derive an expression for $\vec{H}$ both inside and outside of the conductor. Plot the variation of $\vec{H}$ as a function of radial distance from the centre of the wire.

$$\left[\begin{aligned}\vec{H} &= \frac{Ir}{2\pi R^2}\hat{a}_\phi, \quad r < R \\ &= \frac{I}{2\pi R}\hat{a}_\phi \quad r > R\end{aligned}\right]$$

12. A 4 m long conductor lies along the y axis with a current of 10 A in $\hat{a}_y$ direction. Find the force on a conductor if the magnetic field in the region is given as $\vec{B} = 0.05\hat{a}_x$ Tesla. $[-2\hat{a}_z \text{ N}]$

13. A current sheet $\vec{K} = 9\hat{a}_y$ A/m is located at $z = 0$. The region 1 which is at $z < 0$ has $\mu_{r1} = 4$ and region 2 which is at $z > 0$ has $\mu_{r2} = 3$. Given $\vec{H}_2 = 14.5\hat{a}_x + 8\hat{a}_z$ A/m. Find $\vec{H}_1$.

$[\vec{H}_1 = 5.5\hat{a}_x + 6\hat{a}_z \text{ A/m}]$

- ***Medium***

***14.** A coaxial capacitor with inner radius a and outer radius b and length l has a dielectric of permittivity ε and an applied voltage $V_m \sin \omega t$. Determine the displacement current and compare with the conduction current.

$$\left[\frac{2\pi\varepsilon l}{\ln\left(\frac{b}{a}\right)}; I_C = I_D\right]$$

15. In cylindrical coordinates, $\vec{B} = \frac{2}{r}\hat{a}_\phi$ (T). Determine the magnetic flux crossing the plane surface defined by $0.5 \le r \le 2.5$ m and $0 \le z \le 2.0$ m. [6.44 Wb]

16. A radial field, $\vec{H} = \frac{2.39 \times 10^6}{r}\cos\phi\, \hat{a}_r$ A/m, exists in free space. Find the magnetic flux ϕ crossing the surface defined by $0 \le \phi \le \pi/4$, $0 \le z \le 1$ m. [2.12 Wb]

17. Find the flux passing the portion of the plane $\phi = \pi/4$ defined by $0.01 < r < 0.05$ m and $0 < z < 2$ m. A current filament of 2.5 A is along the z axis in the $\hat{a}_z$ direction, in free space. [1.61 μWb]

***18.** Using the vector potential concept find the magnetic intensity about a long straight wire carrying a current I.

$$\left[\frac{\mu_0 I}{2\pi r}\hat{a}_\phi\right]$$

19. In cylindrical coordinates $\vec{A} = 50r^2\hat{a}_z$ Wb/m is a vector magnetic potential, in a certain region of free space. Find $\vec{H}$, $\vec{B}$, $\vec{J}$ and using $\vec{J}$ find total current I crossing the surface $0 \le r \le 1$, $0 \le \phi \le 2\pi$ and $z = 0$.

$$\left[\vec{H} = -\frac{100}{\mu_0}r\hat{a}_\phi \text{ A/m}, \vec{B} = -100r\hat{a}_\phi \text{ Wb/m}^2, \vec{J} = -\frac{200}{\mu_0}\hat{a}_z \text{ A/m}^2, I = -500 \times 10^6 \text{ A}\right]$$

- ***Hard***

20. Find the leakage resistance between two transmission lines of finite radius a that are imbedded in a medium of conductivity σ and separated by a distance d.

$$\left[\frac{\ln(d/a)}{\pi\sigma}\right]$$

21. A long wire of radius r runs through a deep lake of conductivity σ at height h above the bottom which is a good conductor. Calculate the resistance per unit length between the wire and the lake bottom.

$$\left[\frac{1}{2\pi\sigma}\ln\left(\frac{2d}{r}\right)\right]$$

22. Find the insulation in a length l of a coaxial cable of inner conductor of radius a and outer conductor of radius b.

$$\left[\frac{1}{2\pi\sigma l}\ln\left(\frac{b}{a}\right)\right]$$

23. A curved rectangular bar forms a resistor. The curved sides are concentric circular arcs. If σ is the material conducting of the bar, l_0 is the length of the inner arc of radius r_0, $(r_0 + b)$ is the radius of the outer arc, and a is the width of the bar, calculate its electric resistance.

$$\left[\frac{l_0}{\sigma a r_0 \ln\left(1+\frac{b}{r_0}\right)}\right]$$

24. Show that the magnetic field intensity at any point $P(x, y, z)$ due to a current element $Idl\hat{a}_z$ located at origin may be expressed as

$$d\vec{H} = \frac{I\,dl}{4\pi}\frac{(-y\hat{a}_x - x\hat{a}_y)}{(x^2+y^2+z^2)^{3/2}}$$

25. A coaxial line carries the same current I up the inside conductor of radius R_1, the outer conductor of inner radius R_2 and outer radius R_3. Find the magnetic field at all distances r from the centre of the conductor.

$$\left[\begin{array}{ll} \vec{H} = \dfrac{Ir}{2\pi R_1^2}\hat{a}_\phi & 0 \le r \le R_1 \\ \quad = \dfrac{I}{2\pi r}\hat{a}_\phi & R_1 \le r \le R_2 \\ \quad = \dfrac{I}{2\pi r}\left(\dfrac{R_3^2 - r^2}{R_3^2 - R_2^2}\right)\hat{a}_\phi & R_2 \le r \le R_3 \\ \quad = 0 & r \ge R_3 \end{array}\right]$$

26. A coaxial conductor with an inner conductor of radius a and outer conductor of inner and outer radii b and c, respectively, carries current I in the inner conductor. Find the magnetic flux per unit length crossing a plane ϕ = constant between the conductors.

$$\left[\frac{\mu_0 I}{2\pi}\ln\frac{b}{a}\right]$$

***27.** Obtain an expression for magnetic vector potential in the region surrounding an infinitely long straight filamentary current I. Let a be the radius of the wire.

$$\left[\frac{\mu_0 I}{2\pi}\ln\left(\frac{a}{r}\right)\hat{a}_z\right]$$

28. An infinite current sheet lies in the plane $z = 0$ with $\vec{K} = K\hat{a}_y$. Obtain the magnetic vector potential everywhere.

$$\left[\vec{A} = -\frac{1}{2}\mu_0 K z\hat{a}_y \text{ Wb/m}\right]$$

29. An infinitely long solenoid of radius a having n_0 number of turns of wire per unit length carries the current I. Find the magnetic vector potential at a distance r $(> a)$ from the axis of the solenoid.

$$\left[\frac{\mu_0 n_0 I a^2}{2r}\right]$$

30. Find the vector magnetic potential in a plane bisecting a straight piece of thin wire of length $2L$ in free space and carrying steady current I. Therefrom find the magnetic flux density at a distance r from the wire.

$$\left[\begin{array}{ll} \vec{A} = \dfrac{\mu I}{4\pi}\left(1 - \dfrac{r^2}{d^2}\right)\hat{a}_z & r < a \\ \quad = -\dfrac{\mu I}{2\pi}\ln\left(\dfrac{r}{a}\right)\hat{a}_z & r > a \end{array}\right]\left[\begin{array}{ll} \vec{B} = \dfrac{\mu I r}{2\pi a^2}\hat{a}_\phi & r > a \\ \quad = \dfrac{\mu I}{2\pi r}\hat{a}_\phi & r > a \end{array}\right]$$

31. A wire carrying current I runs down the y axis to the origin, thence out to infinity along the positive x axis. Show that the magnetic field in the quadrant with $x, y > 0$ of the xy plane is given by

$$B_z = \frac{\mu_0 I}{4\pi}\left(\frac{1}{x} + \frac{1}{y} + \frac{x}{y\sqrt{x^2+y^2}} + \frac{y}{x\sqrt{x^2+y^2}}\right)$$

Review Questions

[Note: * marked questions are important university questions.]

1. Distinguish between conduction and displacement currents.

***2.** Explain how the current flowing through a capacitor differs from the normal conduction current.

3. Explain why conduction current is absent through the capacitor.

4. **(a)** Establish the continuity equation relating the charge density and current density at a point in a medium. Explain the significance of the equation. Explain the concept of displacement current and show its importance.

or

What is 'conduction' and 'displacement' current? Establish the relation $\vec{J} = \sigma\vec{E} + \frac{d\vec{D}}{dt}$ if both conduction current and displacement current are present. Here, σ—conductivity, $\vec{E}$—Electric field intensity, $\vec{D}$—Electric displacement vector, $\vec{J}$—current density vector and 'dt'—time derivative.

Express the principle of conservation of charge in differential form. What is the name of the differential equation? What is the form of the equation for steady current?

(b) Express the resistance R in terms of the fundamental quantities (M, L, T and I).

(c) Obtain Ohm's law between emf and current, starting from the current density $\vec{J}$ in the circuit and the electrical conductivity σ.

or

Starting from the relationship $\vec{J} = \sigma\vec{E}$, obtain Ohm's law $V = RI$.

(d) Show that in a conducting medium, the steady electric current density satisfies the equation $Div\,\vec{J} = 0$ and the potential V satisfies the Laplace's equation.

***5.** **(a)** Show that a charge placed anywhere in a conducting medium of conductivity σ and permittivity ε decays exponentially with a time constant ε/σ.

or

'Any initial charge density in a conductor dissipates in a characteristic time $\tau = \varepsilon/\sigma$, where ε is the permittivity and σ is the electrical conductivity of the material'. Establish this statement and discuss how τ determines the quality of a conductor.

(b) What do you mean by charge relaxation time? What is its typical value for a good conductor?

6. Show that Kirchhoff's current law is consistent with the principle of charge conservation and the voltage law is consistent with the principle of conservation of energy.

7. What is 'Principle of Duality'? Set up the analogy between $\vec{J}$ and $\vec{D}$.

8. Define polarisation. Explain how a dielectric acquires polarisation.

***9.** A certain volume of dielectric has a polarisation $\vec{P}$ C/m^2. Write the integral expression for the potential at any point due to this dielectric. Explain the different terms in this expression through a figure.

10. Explain the terms
    (a) linear dielectrics and
    (b) dielectric constant.

*11. State Biot-Savart law for the magnetic field $\vec{B}$ due to a steady line current element in free space. Hence, obtain the magnetic field due to a steady volume current configuration.

*12. Starting from Biot–Savart' law, obtain the expression for the magnetic field $\vec{B}$ due to a steady surface current in free space.

*13. State and explain Ampere's circuital law. Show how this law can be applied to find the magnetic field due to an indefinitely long straight conductor carrying a steady current.

14. How does Ampere's current law differ from Biot–Savart law?

15. Distinguish between magnetic vector potential, and magnetic scalar potential.

16. Show that magnetic scalar potential satisfies Laplace's equation in the absence of free currents in a linear material and for uniform magnetisation.

17. Explain the concept of vector magnetic potential. What is its unit? Explain why being potential, it is a vector quantity.

18. What is Lorentz force? A long straight conductor carries a current I. Determine the force per unit length of the conductor when it is placed in a uniform magnetic field B.

*19. Prove that the force on a closed filamentary circuit in a uniform magnetic field is zero.

20. What is Ampere's force law? Derive the expression.

21. Obtain the expression for the force experienced by two current carrying conductors. What is the direction of force when they are carrying current in similar direction and opposite direction?

*22. Explain the phenomena why a current carrying conductor is kept in magnetic field experience force.

23. Show that the torque acting on a loop of area $\vec{S}$ and carrying a current I, when placed in a uniform magnetic field $\vec{B}$ is given by $\vec{T} = I\vec{S} \times \vec{B}$. What is the potential energy of the current loop?

24. Show how a small current loop can be treated as a magnetic dipole. What is its dipole moment?

25. Derive an expression for the potential energy of a point magnetic dipole of moment M placed in a magnetic field B.

26. **(a)** A charged particle is moving in a magnetic field. Give the expression for the force acting on it. Does it gain energy from the field? Give reasons for your answer.
    (b) Show that a current element placed in a magnetic field experiences a force $\vec{F} = Id\vec{l} \times \vec{B}$.
    (c) Find the magnitude of the force per unit length acting on a conductor carrying a current, when it is placed in a magnetic field.
    (d) State and explain Ampere's Work law or, Ampere's Force law.

27. Explain the nature and behaviour of magnetic material. Define and explain the term magnetisation. Classify the different types of magnetic material? On what basis are they classified?

28. Derive the boundary conditions at the magnetic interfaces and show that $\dfrac{\tan \theta_1}{\tan \theta_2} = \dfrac{\mu_{r1}}{\mu_{r2}}$.

29. Establish the boundary conditions for the magnetic field on the interface between two dielectric media of different relative permeability.

    Or,

    State the boundary conditions satisfied by $\vec{B}$ and $\vec{H}$ at the interface of two magnetic media of different magnetic permeabilities assuming no free current.

*30. Explain Laplace's and Poisson's equations for steady magnetic field.

Multiple Choice Questions

1. Displacement current can flow through:
   (a) capacitor (b) inductor (c) resistor (d) none of the above
2. In general, magnetic lines of force are closed whereas electric lines of force are not closed, because:
   (a) Magnetic field is conservative and electric field is not.
   (b) Magnetic field is not conservative and electric field is conservative.
   (c) North and south poles exists together but positive and negative charges may exist separately.
   (d) None of the above
3. The direction of force on a conductor carrying current in the positive y-axis and placed in magnetic field directed in positive x-axis, will be:
   (a) Positive z-axis (b) Negative z-axis (c) Negative x-axis (d) Negative y-axis.
4. Which of the following statements is correct?
   (a) Magnetic field can always be expressed with the help of a scalar potential.
   (b) Magnetic field can always be expressed with the help of a vector potential.
   (c) Magnetic field can sometimes be expressed with the help of a scalar potential and sometimes only with the help of a vector potential.
   (d) Magnetic field can never be expressed with the help of a potential.
5. Given the conduction current density in a lossy dielectric as $J_c = 0.02 \sin 10^9 t$ (A/m^2). If the conductivity $\sigma = 10^3$ siemens/m and $\varepsilon_r = 6.5$, the displacement current density is
   (a) $1.15 \times 10^{-3} \cos 10^9 t$ (b) $1.15 \times 10^{-6} \cos 10^9 t$
   (c) $1.15 \times 10^{-3} \sin 10^9 t$ (d) $1.15 \times 10^{-6} \sin 10^9 t$
6. A long straight wire carries a current of 50 A. A proton (1.6×10^{-19} C) travelling at 10^7 m/s is 5.0 cm from the wire. The force acting on the proton if its velocity is directed towards the wire is
   (a) 3.2×10^{-6} N parallel to the wire. (b) 3.2×10^{-6} N perpendicular to the wire.
   (c) 0.0 (d) none of the above.
7. Two concentric rings of radius 'a' and '$2a$' carrying equal and uniform charge densities revolve at the same angular speed 'ω' about their common axis. The ratio of flux densities due to the two rings at the centre will be
   (a) 1 : 1 (b) 1 : 2 (c) 1 : 4 (d) 2 : 1
8. Two concentric square loops A and B carry equal currents in the same direction. The magnetic field at the centre due to the two loops A and B will be in the ratio
   (a) 1 : 1.414 (b) 1 : 1 (c) 1.414 : 1 (d) 2 : 1
9. The units of J_s and J_v, the surface and volume current densities are respectively
   (a) A/m^2, A/m^3 (b) A/m, A/m^2 (c) A/m, A/m^3 (d) A, A/m
10. If a vector field $\mathbf{A} = x\mathbf{a_x} + y\mathbf{a_y} + K\mathbf{a_z}$ represents a magnetic field, then the value of K must be
    (a) 1 (b) 2 (c) –1 (d) –2
11. The magnetic field intensity (in ampere/metre) at the centre of a circular coil of diameter 1 metre and carrying a current of 2 ampere is
    (a) 8 (b) 4 (c) 3 (d) 2
12. A straight conductor of length l moving with a velocity v in the presence of a magnetic field of flux density B directed at an angle θ with the direction of v experiences a force. Which of the following statement(s) are true for the magnitude of the force?
    1. It is independent of θ 2. It is proportional to l^2
    3. It is proportional to B 4. It is independent of v

Select the correct answer from the codes given below:

(a) 1, 2 and 3 **(b)** 4 alone **(c)** 3 alone **(d)** 2 and 4

13. Consider the following statements:

The force per unit length between two stationary parallel wire carrying (steady) currents:

1. is inversely proportional to the separation of wires.
2. is proportional to the magnitude of each current.
3. satisfies Newton's third law.

Of these statements

(a) 1 and 2 are correct **(b)** 2 and 3 are correct

(c) 1 and 3 are correct **(d)** 1, 2 and 3 are correct

14. A straight conductor of circular cross-section carries a current. Which of the following statements is true in this regard?

(a) No force acts on the conductor at any point.

(b) An axial force acts on the conductor tending to increase its length.

(c) A radial force acts towards the axis tending to reduce its cross-section.

(d) A radial force acts away from the axis tending to increase its cross-section.

15. The ratio of conduction current density to the displacement current density is (symbols have the usual meaning)

(a) $\frac{j\sigma}{\omega\varepsilon}$ **(b)** $\frac{\sigma}{j\omega\varepsilon}$ **(c)** $\frac{\sigma\omega}{j\varepsilon}$ **(d)** $\frac{\sigma\varepsilon}{j\omega}$

16. Consider the following statements associated with the basic properties of ideal conductors:

1. The resultant field inside is zero.
2. The net charge density in the interior is zero.
3. Any net charge resides on the surface.
4. The surface is always equipotential.
5. The field just outside is zero.

Of these statements

(a) 1, 2, 3 and 4 are correct **(b)** 3, 4 and 5 are correct

(c) 1, 2 and 3 are correct **(d)** 2 and 3 are correct

17. Conservation of charge implies that

(a) $\oiint \vec{J} \cdot d\vec{S} = -\frac{d}{dt}\iiint \rho dv$ **(b)** $\oiint \vec{J} \cdot d\vec{S} = 0$

(c) $\oiint \vec{J} \cdot d\vec{S} = \iiint \rho dv$ **(d)** $\oiint \vec{J} \cdot d\vec{S} = \rho$

18. The relaxation time (τ) in perfect dielectric is

(a) 0 **(b)** 1 **(c)** $1 < \tau < \alpha$ **(d)** α

19. Match List I with List II and select the correct answer using the codes given below the lists.

List I (Laws of magnetostatics)	List II (Equations with which the given laws are associated)
A. Ampere's Law	1. $\vec{F} = Q[\vec{E} + (\vec{v} \times \vec{B})]$
B. Biot–Savart Law	2. $\vec{B} = \frac{\mu_0 I}{4\pi}\int \frac{\mathbf{dl} \times \mathbf{R}}{R^2}$
C. Lorentz Force Law	3. $\nabla \times \vec{B} = \mu_0 \vec{J}$
D. Magnetic flux Law	4. $\nabla \cdot \vec{B} = 0$

Codes:

(a)	A	B	C	D
	3	2	1	4
(b)	A	B	C	D
	3	2	4	1
(c)	A	B	C	D
	2	3	4	1
(d)	A	B	C	D
	2	3	1	4

20. The region $z \leq 0$ is a perfect conductor. On its surface at the origin, the surface current density ***K*** is (5**i** – 6**j**) A/m. If the region $z > 0$ were free space, then the magnetic field intensity ***H*** in A/m, at the origin would be

(a) ***H*** = 0 **(b)** ***H*** = (5**i** – 6**j**) A/m
(c) ***H*** = (6**i** – 5**j**) A/m **(d)** ***H*** = –(6**i** + 5**j**) A/m

21. The magnitude of the magnetic flux density '*B*' at a distance '*R*' from an infinitely long straight current filament is

(a) $\frac{\mu_0 I}{2R}$ **(b)** $\frac{\mu_0 I}{2\pi R}$ **(c)** $\frac{\mu_0 I}{4\pi R}$ **(d)** $\frac{\mu_0 I}{8\pi R^2}$

22. If the vector $\vec{B} = x^2\hat{i} - xy\hat{j} - Cxz\hat{k}$ represents a magnetic field, the value of the constant *C* must be

(a) 0 **(b)** 1 **(c)** 2 **(d)** 3

23. A medium behaves like dielectric when the

(a) displacement current is just equal to the conduction current.
(b) displacement current is less than the conduction current.
(c) displacement current is much greater than the conduction current.
(d) displacement current is almost negligible.

24. If $\vec{A}$ and $\vec{J}$ are the vector potential and current density vectors associated with a coil, then $\vec{A} \cdot \vec{J} dv$ has the units of

(a) flux-linkage **(b)** power **(c)** energy **(d)** inductance

25. The energy stored in the magnetic field of a solenoid 30 cm long and 3 cm diameter wound with 1,000 turns of wire carrying a current of 10A is

(a) 0.015 Joule **(b)** 0.15 Joule **(c)** 0.5 Joule **(d)** 1.15 Joule

26. Kirchhoff's current law for direct currents is implicit in the expression

(a) $\nabla \cdot \vec{D} = \rho$ **(b)** $\int \vec{J} \cdot \hat{n} ds = 0$ **(c)** $\nabla \cdot \vec{B} = 0$ **(d)** $\nabla \cdot \vec{H} = \vec{J} + \frac{\partial \vec{D}}{\partial t}$

27. The relaxtion time of mica ($\sigma = 10^{-15}$ mho/m, $\varepsilon_r = 6$) is:

(a) 5×10^{-10} s **(b)** 10^{-6} s **(c)** 5 hour **(d)** 15 hour

28. Which one of the following is not characteristic of a static magnetic field?

(a) It is solenoidal **(b)** It is conservative
(c) It has no sinks or sources **(d)** Magnetic flux lines are always closed

29. Two identical coaxial circular coils carry the same current *I*, but in opposite directions. The magnitude of the magnetic field B at a point on the axis midway between the coils is

(a) Zero **(b)** The same as that produced by one coil
(c) Twice that produced by one coil **(d)** Half that produced by one coil

30. Two thin parallel wires carry currents along the same direction. The force experienced by one due to the other is

(a) Parallel to the lines **(b)** Perpendicular to the lines and attractive
(c) Perpendicular to the lines and repulsive **(d)** Zero

31. What is the unit of magnetic charge?
(a) Ampere-metre square **(b)** Coulomb
(c) Ampere **(d)** Ampere-metre

32. The concept of displacement current was a major contribution attributed to:
(a) Faraday **(b)** Lenz **(c)** Maxwell **(d)** Lorentz

33. The region $z < 0$ has $\mu_r = 6$ and the region $z > 0$ has $\mu_r = 4$. If the magnetic flux density in region $z > 0$ is $5\hat{a}_x + 8\hat{a}_z$ mWb/m^2, the magnetic field intensity in region $z < 0$ would be:
(a) $\dfrac{5\hat{a}_x + 8\hat{a}_z}{4\mu_0}$ mA/m **(b)** $\dfrac{5\hat{a}_x + 8\hat{a}_z}{6\mu_0}$ mA/m
(c) $\dfrac{5\hat{a}_x}{4\mu_0} + \dfrac{8\hat{a}_z}{6\mu_0}$ mA/m **(d)** $\dfrac{5\hat{a}_x}{6\mu_0} + \dfrac{8\hat{a}_z}{4\mu_0}$ mA/m

34. Plane defined by $z = 0$ carry surface current density $2\hat{a}_x$ A/m. The magnetic intensity 'H_y' in the two regions $-\infty < z < 0$ and $0 < z < \infty$ are respectively:
(a) $\hat{a}_y$ and $-\hat{a}_y$ **(b)** $-\hat{a}_y$ and $\hat{a}_y$
(c) $\hat{a}_x$ and $-\hat{a}_x$ **(d)** $-\hat{a}_x$ and $\hat{a}_x$

35. A solid cylindrical conductor of radius 'R' carrying a current 'I' has a uniform current density. The magnetic field intensity 'H' inside the conductor at the radial distance 'r' ($r < R$) is:
(a) zero **(b)** $\dfrac{I}{2\pi r}$ **(c)** $\dfrac{Ir}{2\pi R^2}$ **(d)** $\dfrac{IR^2}{2\pi r^3}$

Answers

1. (a)	**2.** (c)	**3.** (b)	**4.** (b)	**5.** (b)	**6.** (c)	**7.** (a)	**8.** (d)
9. (a)	**10.** (d)	**11.** (d)	**12.** (c)	**13.** (d)	**14.** (c)	**15.** (b)	**16.** (a)
17. (a)	**18.** (a)	**19.** (a)	**20.** (d)	**21.** (b)	**22.** (b)	**23.** (c)	**24.** (c)
25. (b)	**26.** (b)	**27.** (d)	**28.** (b)	**29.** (a)	**30.** (b)	**31.** (d)	**32.** (c)
33. (c)	**34.** (a)	**35.** (c)					

4 ELECTROMAGNETIC FIELDS

Learning Objectives

This chapter deals with the following topics:

- *Basic laws of electromagnetic induction*
- *To acquire knowledge of fundamental quantities for time-varying fields*

4.1 INTRODUCTION

In the previous chapters, we have studied different concepts of electrostatic and magnetostatic fields. In general, electrostatic fields are produced by stationary charges and magnetostatic fields are produced by motion of electric charges with uniform velocity (i.e. steady currents). However, if the current is time-varying, the field produced is also time-varying and is known as electromagnetic fields or waves.

In this chapter, we will discuss the concepts of electromagnetic fields and contribution of Maxwell to the laws of electromagnetism.

4.2 FARADAY'S LAW OF INDUCTION FOR TIME-VARYING FIELDS

The English physicist Michael Faraday and the American scientist Joseph Henry independently and simultaneously, in 1831, observed experimentally that any change in the magnetic environment of a coil of wire will cause a voltage (emf) to be induced in the coil. If the circuit is a closed one, this emf will cause flow of current. This phenomenon is known as *electromagnetic induction*. The results of Faraday and Henry's experiment led to two laws:

1. Neumann's Law: When a magnetic field linked with a coil or circuit is changed in any manner, the emf induced in the circuit is proportional to the rate of change of the flux-linkage with the circuit.

2. Lenz's Law: The direction of the induced emf is such that it will oppose the change of flux producing it.

These two laws together can be termed as *Faraday's law of electromagnetic induction*.

Statement of Faraday's Law The emf induced in a closed circuit is proportional to the rate of change of the magnetic flux-linkage and the direction of the current flow in the closed circuit is such that it opposes the change of the flux.

Explanation

Let $\vec{E}$—electric field,
$\vec{B}$—magnetic field,
ϕ—flux linking with the circuit,
V—induced emf in the circuit,
λ—total flux-linkage in a multi-turn coil,
C—the closed circuit binding an open surface S, placed in the magnetic field

By Faraday's law,
$$V = -\frac{d\phi}{dt} \tag{4.1}$$

In general, for a multi-turn coil, the emf induced is given as,
$$V = -\frac{d\lambda}{dt} \tag{4.2}$$

Now, the induced emf can be written in terms of electric field as
$$V = \oint_C \vec{E} \cdot d\vec{l} \tag{4.3}$$

and the total magnetic flux can be written in terms of magnetic field as
$$\lambda = \int_S \vec{B} \cdot d\vec{S} \tag{4.4}$$

From Eqs. (4.1) to (4.4), we get
$$\boxed{\oint_C \vec{E} \cdot d\vec{l} = -\frac{d}{dt}\int_S \vec{B} \cdot d\vec{S}} \tag{4.5}$$

This equation is termed as the *integral form of Faraday's law.*

From Eq. (4.5),
$$\oint_C \vec{E} \cdot d\vec{l} = -\frac{d}{dt}\int_S \vec{B} \cdot d\vec{S}$$

$$\Rightarrow \quad \int_S \nabla \times \vec{E} \cdot d\vec{S} = -\frac{d}{dt}\int_S \vec{B} \cdot d\vec{S} \quad \text{(applying Stoke's theorem)}$$

If the circuit is stationary then the time derivative can be moved inside the integral and it becomes a partial derivative. Hence,
$$\int_S \nabla \times \vec{E} \cdot d\vec{S} = -\int_S \frac{\partial \vec{B}}{\partial t} \cdot d\vec{S} \tag{4.6}$$

Thus, equating the integrands, we get
$$\boxed{\nabla \times \vec{E} = -\frac{\partial \vec{B}}{\partial t}}$$

This is the *differential form of Faraday's law.*

Proof of Faraday's Law We consider a closed circuit moving with a velocity $\vec{v}$ in a time varying magnetic field $\vec{B}$.

Force exerted on the circuit is, $\vec{F} = q(\vec{v} \times \vec{B})$

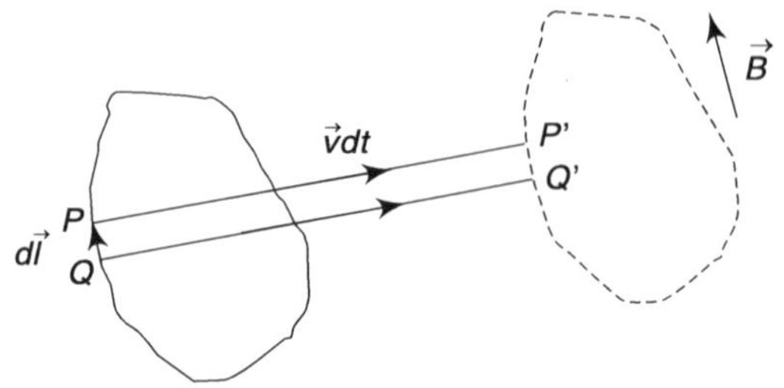

Fig. 4.1 *Closed circuit moving in a time varying field*

So, the intensity of electric field induced in the wire is,

$$\vec{E} = \frac{\vec{F}}{q} = (\vec{v} \times \vec{B})$$

So, induced emf is

$$V = \oint_l \vec{E} \cdot d\vec{l} = \oint_l \{(\vec{v} \times \vec{B})\} \cdot d\vec{l} \quad \textbf{(4.7)}$$

As the circuit moves, the element $d\vec{l}$ sweeps in time dt an area PP′QQ′ = $\vec{v}dt \times d\vec{l}$.

$\therefore$ Flux across this area = $\vec{B} \cdot (\vec{v}dt \times d\vec{l})$

$\therefore$ Flux over the entire band, $d\phi = \oint_S \vec{B} \cdot (\vec{v}dt \times d\vec{l})$

$$\therefore \quad \frac{d\phi}{dt} = \oint_S \vec{B} \cdot (\vec{v} \times d\vec{l}) = -\oint_S (\vec{v} \times \vec{B}) \cdot d\vec{l} \quad \textbf{(4.8)}$$

$$\{\because \vec{A} \cdot (\vec{B} \times \vec{C}) = \vec{C} \cdot (A \times \vec{B}) = -(\vec{B} \times \vec{A}) \cdot \vec{C}\}$$

Comparing Eq. (4.7) and (4.8), we get

$$\boxed{V = -\frac{d\phi}{dt}}$$

This is the induced emf given by Faraday's law.

NOTE

Weber is the unit of magnetic flux. Weber may be defined in terms of Faraday's law, as the amount of flux that induces an electromotive force of one volt when cut by a varying magnetic flux through a loop to the electric field around the loop per second.

Tesla *is the unit of magnetic flux density. It is defined as the magnetic flux density that produces a force of one Newton per metre in a conductor carrying a current of one Ampere at right angle to the magnetic field.*

The ***Maxwell****, abbreviated as Mx, is the compound derived CGS unit of magnetic flux. The unit was previously called a line. In a magnetic field of strength one Gauss (or Tesla), one Maxwell is the total flux across a surface of one square centimetre perpendicular to the field.*

1 Weber = 10^8 Maxwells = 10^8 magnetic field lines

Example 4.1 An infinite straight wire carries a current I is placed to the left of a rectangular loop of wire with width w and length l, as shown in Fig. 4.2.

(a) Determine the magnetic flux through the rectangular loop due to the current I.

(b) Suppose that the current is a function of time with $I(t) = a + bt$, where a and b are positive constants. What is the induced emf in the loop and the direction of the induced current?

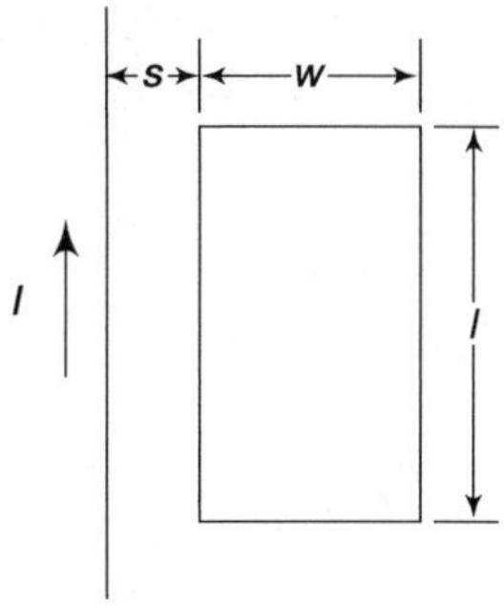

Fig. 4.2 *Rectangular loop near a wire*

Solution

(a) Using Ampere's law,

$$\oint \vec{B} \cdot d\vec{S} = \mu_0 I_{\text{enc}}$$

the magnetic field due to a current-carrying wire at a distance r away is

$$B = \frac{\mu_0 I}{2\pi r}$$

The total magnetic flux ϕ through the loop can be obtained by summing over contributions from all differential area elements $dS = ldr$

$$\phi = \int d\phi = \int \vec{B} \cdot d\vec{S} = \frac{\mu_0 Il}{2\pi} \int_{s}^{s+w} \frac{dr}{r} = \frac{\mu_0 Il}{2\pi} \ln\left(\frac{s+w}{s}\right)$$

NB: Here, the area vector has been chosen to point into the page so that $\phi > 0$

(b) According to Faraday's law, the induced emf is

$$V = -\frac{d\phi}{dt} = -\frac{d}{dt}\left[\frac{\mu_0 Il}{2\pi} \ln\left(\frac{s+w}{s}\right)\right] = -\frac{\mu_0 Il}{2\pi} \ln\left(\frac{s+w}{s}\right)\frac{dI}{dt} = -\frac{\mu_0 bl}{2\pi} \ln\left(\frac{s+w}{s}\right)$$

where $\frac{dI}{dt} = b$

The straight wire carrying a current I produces a magnetic flux into the page through the rectangular loop. By Lenz's law, the induced current in the loop must be flowing *counterclockwise* in order to produce a magnetic field out of the page to counteract the increase in inward flux.

Example 4.2 A rectangular loop of dimensions l and w moves with a constant velocity $\vec{v}$ away from an infinitely long straight wire carrying a current I in the plane of the loop, as shown in Fig. 4.3. Let the total resistance of the loop be R. What is the current in the loop at the instant the near side is a distance r from the wire?

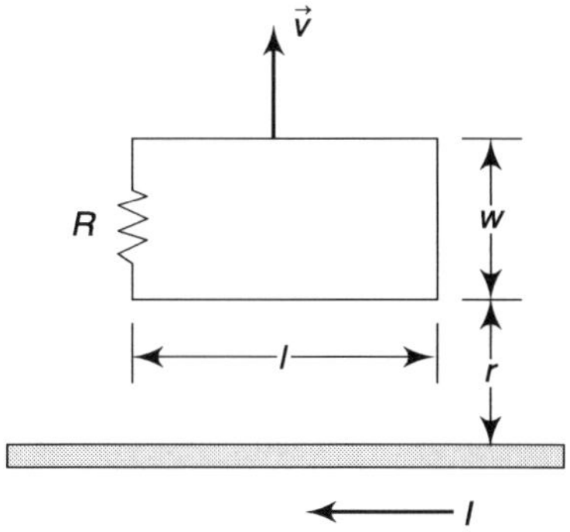

Fig. 4.3 *A rectangular loop moving away from a current-carrying wire*

Solution The magnetic field at a distance s from the straight wire is obtained using Ampere's law as

$$B = \frac{\mu_0 I}{2\pi s}$$

The magnetic flux through a differential area $dA = lds$ of the loop is

$$d\phi = \vec{B} \cdot d\vec{S} = \frac{\mu_0 I}{2\pi s} lds$$

where the area vector has been chosen to point into the page so that $\phi > 0$.

Integrating over the entire area of the loop, the total flux is

$$\phi = \frac{\mu_0 Il}{2\pi} \int_r^{r+w} \frac{ds}{s} = \frac{\mu_0 Il}{2\pi} \ln\left(\frac{r+w}{r}\right)$$

Differentiating with respect to t, we get the induced emf as

$$V = -\frac{d\phi}{dt} = -\frac{\mu_0 Il}{2\pi}\frac{d}{dt}\ln\left(\frac{r+w}{r}\right) = -\frac{\mu_0 Il}{2\pi}\left(\frac{1}{r+w} - \frac{1}{r}\right)\frac{dr}{dt} = \frac{\mu_0 Il}{2\pi}\frac{wv}{r(r+w)}$$

where $v = \dfrac{dr}{dt}$.

The induced current obtained as

$$I = \frac{|V|}{R} = \frac{\mu_0 Il}{2\pi R}\frac{wv}{r(r+w)}$$

Example 4.3 A rectangular loop of wire with dimensions a, b and lying in the xy-plane, is moving with a uniform velocity v in the direction of the positive y-axis in a static magnetic vector field pointing in the z-direction and varying sinusoidally in the y-direction as $B_z = B_0 \sin ky$. This is depicted in Fig. 4.4. Calculate the emf induced in the moving loop.

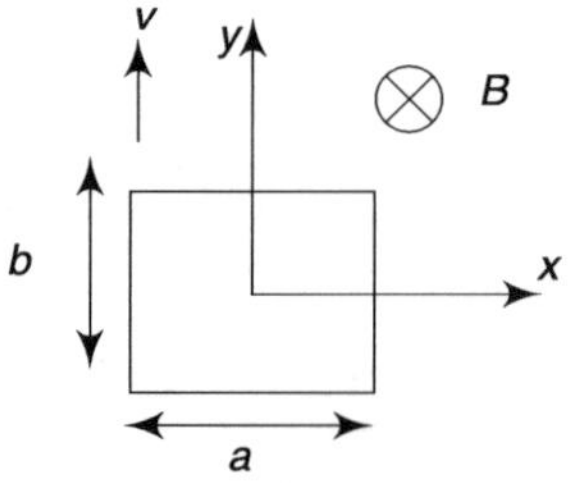

Fig. 4.4 *Rectangular loop moving in a magnetic field*

Solution We assume that at time $t = 0$, the front side of the loop is at $y = \dfrac{b}{2}$ and the rear side is at $y = -\dfrac{b}{2}$. At any time, t, the positions of the two sides are respectively, at

$$y_{\text{front}} = \left(vt + \frac{b}{2}\right) \text{ and } y_{\text{rear}} = \left(vt - \frac{b}{2}\right)$$

Total flux enclosed by the loop is

$$\phi = a\int_{y_{\text{rear}}}^{y_{\text{front}}} B_0 \sin ky dy = \frac{B_0 a}{k}[-\cos ky]_{y_{\text{rear}}}^{y_{\text{front}}}$$

$$= -\frac{B_0 a}{k}\left[\cos\left(vt - \frac{b}{2}\right) - \cos\left(vt + \frac{b}{2}\right)\right]$$

$$= 2\frac{B_0 a}{k}\sin\left(\frac{kb}{2}\right)\sin kvt$$

Hence, the emf induced is

$$V = -\frac{d\phi}{dt} = -\frac{d}{dt}\left[2\frac{B_0 a}{k}\sin\left(\frac{kb}{2}\right)\sin kvt\right] = -2B_0 av \sin\left(\frac{kb}{2}\right)\cos kvt$$

$\therefore$

$$\boxed{V = -2B_0 av \sin\left(\frac{kb}{2}\right)\cos kvt}$$

4.3 INDUCED EMF FOR TIME-VARYING FIELDS

According to Faraday's law, for a flux variation, there will be induced emf, given as $V = -\frac{d\phi}{dt}$. The variation of flux with time may be caused in three ways:

1. by having a stationary loop in a time-varying magnetic field;
2. by having a time-varying loop in a static magnetic field, and
3. by having a time-varying loop in a time-varying magnetic field.

1. Stationary loop in time varying magnetic field (transformer emf): In this case, the emf induced is

$$V_S = \oint_C \vec{E} \cdot d\vec{l} = -\int_S \frac{\partial \vec{B}}{\partial t} \cdot d\vec{S}$$

This emf induced by the time-varying current (producing time-varying magnetic field) in a stationary loop is called *transformer emf*. By Stoke's theorem,

$$\int_S \nabla \times \vec{E} \cdot d\vec{S} = -\int_S \frac{\partial \vec{B}}{\partial t} \cdot d\vec{S}$$

$$\therefore \qquad \boxed{\nabla \times \vec{E} = -\frac{\partial \vec{B}}{\partial t}}$$

Thus, $\nabla \times \vec{E} \neq 0$, i.e. time-varying electric field is not conservative. The work done in taking a charge about a closed path in a time-varying electric field is due to the energy from the time-varying magnetic field.

***Example 4.4** A circular wire loop of radius $a = 0.4$ m lies in the x–y plane with its axis along the z-axis. The vector magnetic field over the surface of the loop is, $\vec{H} = H_0 \cos \omega t \hat{a}_z$, where $H_0 =$ 200 μA/m and $f = 1$ MHz. Determine the emf induced in the loop.

Solution Since the loop is stationary, and the area is time-dependent, the transformer emf is induced. The emf induced is given as

$$\begin{aligned} V_S &= -\int_S \frac{\partial \vec{B}}{\partial t} \cdot d\vec{S} = -\mu_0 \int_S \frac{\partial \vec{H}}{\partial t} \cdot d\vec{S} = -\mu_0 \int_S \frac{\partial}{\partial t}(H_0 \cos \omega t \hat{a}_z) \cdot dS\hat{a}_z \\ &= -\mu_0 \int_S (-\omega H_0 \sin \omega t) dS \\ &= \mu_0 \omega H_0 \sin \omega t \int_S dS \\ &= \mu_0 \omega H_0 \sin \omega t (\pi a^2) \\ &= \omega \mu_0 \pi a^2 H_0 \sin \omega t \end{aligned}$$

Putting the values, we get the magnitude of the emf induced as

$$V_S = \omega \mu_0 \pi a^2 H_0 \sin \omega t = 2\pi \times 10^6 \times 4\pi \times 10^{-7} \times \pi (0.4)^2 (200 \times 10^{-6}) = 0.794 \text{ mV}$$

Example 4.5 A square conducting loop of sides $d = 25$ cm is located in a peak ac magnetic field of $H_m = 1$ A/m varying at a frequency $f = 5$ MHz. $\vec{H}$ is perpendicular to the plane of the loop. What voltage will be read on a voltmeter connected in series with one side of the loop?

Solution Since the loop is stationary, and the area is time-dependent, the transformer emf is induced. The emf induced is given as

$$V_S = -\int_S \frac{\partial \vec{B}}{\partial t} \cdot d\vec{S} = -\mu_0 \int_S \frac{\partial \vec{H}}{\partial t} \cdot d\vec{S} = -\mu_0 \int_S \frac{\partial}{\partial t}(H_m \cos \omega t \hat{a}_z) \cdot dS\hat{a}_z$$

$$= -\mu_0 \int_S (-\omega H_m \sin \omega t) dS$$

$$= \mu_0 \omega H_m \sin \omega t \int_S dS$$

$$= \mu_0 \omega H_m \sin \omega t \times d^2$$

$$= \omega \mu_0 \pi a^2 H_m \sin \omega t$$

Putting the values, we get the magnitude of the emf induced as

$$V_S = \mu_0 \omega H_m d^2 = 4\pi \times 10^{-7} \times 2\pi \times 5 \times 10^6 \times (0.25)^2 = 2.54\,\text{V}$$

Example 4.6 The circular loop conductor having a radius of 0.15 m is placed in the *xy* plane. This loop consists of a resistance of 20 Ω as shown in Fig. 4.5. If the magnetic flux density is

$$\vec{B} = 0.5 \sin 10^3 t\, \hat{a}_z\ \mathbf{T}$$

Find the current flowing through the loop.

Solution Here, since the loop is stationary and the magnetic field is time-varying, only the transformer emf is induced.

Transformer emf induced is

$$V_S = -\iint_S \frac{\partial \vec{B}}{\partial t} \cdot d\vec{S} = -\iint_S \frac{\partial}{\partial t}(0.5 \sin 10^3 t \hat{a}_z) \cdot (r dr\, d\phi\, \hat{a}_z)$$

$$= -0.5 \times 10^3 \cos 10^3 t \int_{r=0}^{0.15} \int_{\phi=0}^{2\pi} r dr\, d\phi$$

$$= -0.5 \times 2\pi \times 10^3 \cos 10^3 t \left[\frac{r^2}{2}\right]_0^{0.15}$$

$$= -10^3 \pi \cos 10^3 t \times 0.01125$$

$$= -35.34 \cos 10^3 t\ \text{V}$$

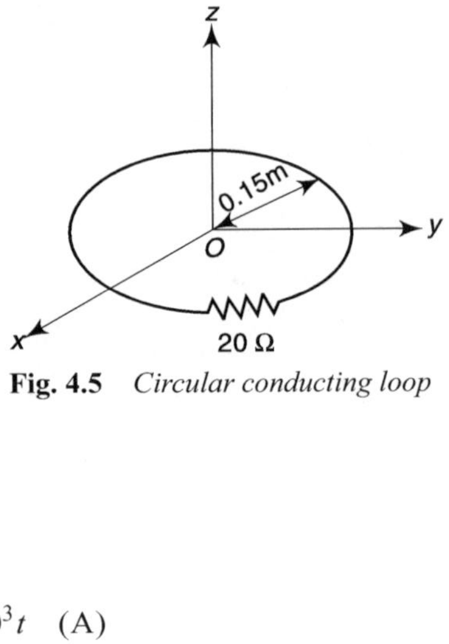

Fig. 4.5 *Circular conducting loop*

Hence, the current in the conductor is

$$I = \frac{V_S}{R} = \frac{-35.34 \cos 10^3 t}{20} = -1.767 \cos 10^3 t \quad (\text{A})$$

Example 4.7 An area of 0.65 m^2 in the plane $z = 0$ encloses a filamentary conductor. Find the induced voltage if

$$\vec{B} = 0.05 \cos 10^3 t \left(\frac{\hat{a}_y + \hat{a}_z}{\sqrt{2}}\right) \mathbf{Tesla}$$

Solution Since the filamentary conductor is fixed and placed in the *xy* plane, only transformer emf is induced.

Transformer emf induced is

$$V_S = -\iint_S \frac{\partial \vec{B}}{\partial t} \cdot d\vec{S} = -\iint_S \frac{\partial}{\partial t}\left[0.05 \cos 10^3 t \left(\frac{\hat{a}_y + \hat{a}_z}{\sqrt{2}}\right)\right] \cdot (dS\,\hat{a}_z)$$

$$= \frac{0.05 \times 10^3 \sin 10^3 t}{\sqrt{2}} \int_S dS$$

$$= \frac{0.05 \times 10^3 \sin 10^3 t}{\sqrt{2}} \times 0.65$$

$$= 22.98 \sin 10^3 t \quad \text{(volt)}$$

Example 4.8 A stationary 10-turn square coil of 1 metre side is situated with its lower left corner coincident with the origin and with sides x_1, y_1 along x-axis and y-axis. If the field $\vec{B}$ is normal to the plane of the coil and has its amplitude given by

$$B_0 = \sin\left(\frac{\pi x}{x_1}\right) \sin\left(\frac{\pi y}{y_1}\right) \quad \text{tesla,}$$

determine the rms value of the emf induced in the coil, if $\vec{B}$ varies harmonically at a frequency of 1 kHz.

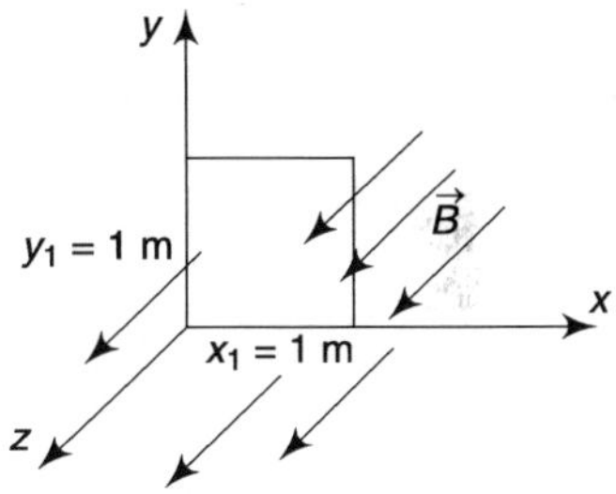

Fig. 4.6 *Square loop in a time-varying magnetic field*

Solution This is demonstrated in Fig. 4.6. Since $\vec{B}$ varies harmonically, it can be written as, $\vec{B} = B_0 \cos \omega t$.

Also, $x_1 = 1, y_1 = 1$

$\therefore$ $$B_0 = \sin(\pi x) \sin(\pi y) \quad \text{Tesla}$$

Here, only the transformer emf is induced given as

$$V_S = -\int_S \frac{\partial \vec{B}}{\partial t} \cdot d\vec{S} = -\int_S \frac{\partial \vec{B}}{\partial t} \cdot \hat{n} dS = -\int_S \frac{\partial (B_0 \cos \omega t)}{\partial t} dxdy = \left[\omega \int_S B_0 dxdy\right] \sin \omega t$$

Hence, the peak value of the induced emf is

$$V_{S_{\max}} = \omega \int_{y=0}^{1} \int_{x=0}^{1} \sin(\pi x) \sin(\pi y) dxdy = 2\pi f \left[-\frac{\cos \pi x}{\pi}\right]_0^1 \times \left[-\frac{\cos \pi y}{\pi}\right]_0^1 = 2\pi f \frac{2}{\pi} \times \frac{2}{\pi} = \frac{8f}{\pi}$$

$$\boxed{\mathcal{E}_{S_{\max}} = \frac{8f}{\pi}}$$

The rms value of the induced emf is

$$V_{S_{\text{rms}}} = \frac{8f}{\sqrt{2}\pi} = \frac{8 \times 1000}{\sqrt{2}\pi} = 18 \text{ kV}$$

2. Moving loop in static magnetic field (motional emf): When a charge is moving in the presence of a static magnetic field $\vec{B}$, the force on the charge is

$$\vec{F}_m = q\vec{v} \times \vec{B}$$

where, q is the amount of charge, $\vec{v}$ is the velocity of movement.

So, we define the motional electric field, $\vec{E}_m$ as

$$\vec{E}_m = \frac{\vec{F}_m}{q} = \vec{v} \times \vec{B} \quad \textbf{(4.9)}$$

Thus, if a conducting loop is moving with a velocity $\vec{v}$, then the emf induced in the loop is

$$\boxed{V_m = \oint_C \vec{E}_m \cdot d\vec{l} = \oint_C (\vec{v} \times \vec{B}) \cdot d\vec{l}} \quad \textbf{(4.10)}$$

This is called *motional emf* or *flux-cutting emf*, as found in electrical machines, motors, generator. By Stoke's theorem

$$\int_S (\nabla \times \vec{E}_m) \cdot d\vec{S} = \int_S \nabla \times (\vec{v} \times \vec{B}) \cdot d\vec{S}$$

$$\therefore \quad \boxed{\nabla \times \vec{E}_m = \nabla \times (\vec{v} \times \vec{B})}$$

*Example 4.9 Faraday Disc Generator

(Unipolar Generator or Homopolar Generator or Acyclic Generator or Disc Generator or Dynamo or Faraday Disc Generator) [Fig. 4.7(a) and (b)]

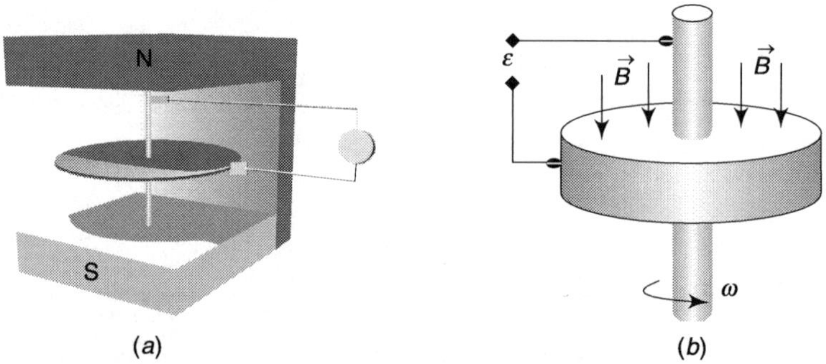

Fig. 4.7 *(a) Faraday disc generator (b) Faraday disc generator*

A Faraday disc generator is a homopolar DC electrical generator which consists of an electrically conductive disc made of magnetic material (flywheel) rotating in a uniform magnetic field with one electrical contact near the axis and the other near the periphery. This creates a potential difference between the two contact points, one in the centre of the disc the other on the outside of the disc. This generator is used for generating very high currents at low voltages in applications such as electrolysis, welding, etc.

Let ω—angular velocity of the disc,

$\vec{B}$—uniform magnetic field,

a—radius of the disc.

We consider an electron of charge e at any point on the disc at a distance r from the centre.

$\therefore$ Velocity of the electron is, $v = \Omega r$

Force on the electron is, $\vec{F} = e(\vec{v} \times \vec{B})$

Hence, the electric field acting on the electron at equilibrium is

$$\vec{E} = \frac{\vec{F}}{e} = \vec{v} \times \vec{B}$$

The magnitude of the field is $E = |\vec{E}| = vB \sin 90^\circ = \omega rB$

$$\therefore \qquad \boxed{E = |\vec{E}| = \omega rB}$$

The direction of this field is radially inward.

Thus, the emf between the centre and the peripheral rim of the disc is

$$V_m = \int_0^a \vec{E} \cdot d\vec{r} = \int_0^a \omega rB dr = \omega B\left[\frac{r^2}{2}\right]_0^a = \frac{1}{2}\omega Ba^2$$

$$\therefore \qquad \boxed{V_m = \frac{1}{2}\omega Ba^2}$$

This is the open circuit voltage between the two contact brushes.

***Example 4.10** A copper disc of 0.50 m diameter is rotated at a constant speed of 2,000 r.p.m. on a horizontal axis perpendicular to and through the centre of the disc, the axis lying in the magnetic meridian. Two brushes contact with the disc, one at the edge and the other at the centre. If the horizontal component of earth's field is 0.02 mWb/m^2 find the emf induced between the brushes.

Solution This is a Faraday disc problem. From Example 4.9, we get the emf induced as,

$$V_m = \frac{1}{2}\omega Ba^2 = \frac{1}{2} \times 2\pi fBa^2 = \pi\frac{N}{60}Ba^2 = \pi \times \frac{2000}{60} \times 0.02 \times 10^{-3} \times \left(\frac{0.5 \times 10^{-3}}{2}\right)^2$$
$$= 0.1308 \text{ mV}$$

***Example 4.11** A rod of length 'l' rotates about the z-axis with an angular velocity 'ω'. If $\vec{B} = B_0\hat{a}_z$, calculate the voltage induced on the conducting rod. This is shown in Fig. 4.8.

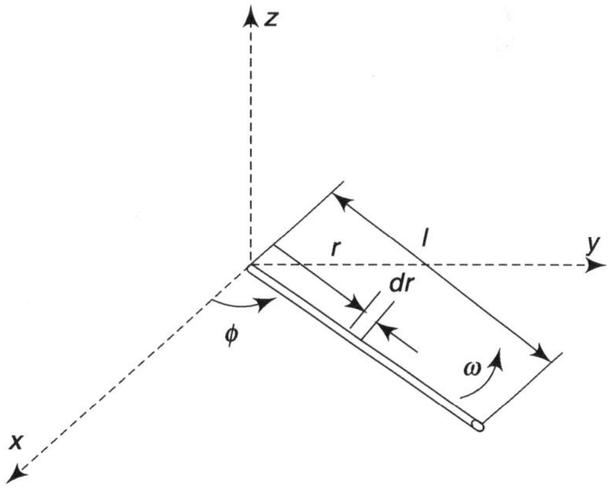

Fig. 4.8 *Metal rod rotating in stationary field*

Solution We consider an elemental section of the rod at any distance r. The velocity at that section is

$$\vec{v} = \omega r\hat{a}_\phi$$

$$\vec{v} \times \vec{B} = (\omega r\hat{a}_\phi) \times (B_0\hat{a}_z) = B_0\omega r\,\hat{a}_r$$

So, the induced motional emf is

$$V_m = \oint_C (\vec{v} \times \vec{B}) \cdot d\vec{l} = \oint_C (B_0\omega r\,\hat{a}_r) \cdot (dr\hat{a}_r)$$
$$= B_0\omega\int_0^l r dr = \frac{1}{2}B_0\omega l^2$$

Here, the induced field $\vec{v} \times \vec{B}$ is in the radial direction, the polarity of the moving end is positive with respect to the fixed end.

Example 4.12 A rectangular conducting loop with resistance of 0.2 Ω rotates at 500 r.p.m., as shown in Fig. 4.9. The vertical conductor at $r_1 = 0.03$ m is in the field $\vec{B}_1 = 0.25\hat{a}_r\ T$ and other conductor is at $r_2 = 0.05$ m and in the field $\vec{B}_2 = 0.8\hat{a}_r\ T$. Find the current flowing in the loop.

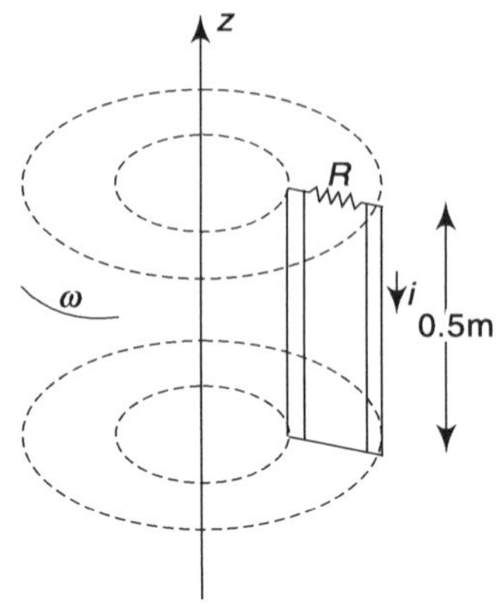

Fig. 4.9 *Rectangular conducting loop with resistance*

Solution Linear velocity of the inner conductor is,

$$\vec{v}_1 = \omega r_1 \hat{a}_\phi = 2\pi f r_1 = 2\pi \times \frac{500}{60} \times 0.03\hat{a}_\phi = 1.57\hat{a}_\phi \text{ m/s}$$

Linear velocity of the outer conductor is

$$\vec{v}_2 = \omega r_2 \hat{a}_\phi = 2\pi f r_2 = 2\pi \times \frac{500}{60} \times 0.05\hat{a}_\phi = 2.62\hat{a}_\phi \text{ m/s}$$

Since the magnetic field is not time-varying, only the motional emf will be induced.

Motional emf induced in the inner conductor is

$$V_{m1} = \oint_l (\vec{v}_1 \times \vec{B}_1) \cdot d\vec{l}_1 = \int_0^l (1.57\hat{a}_\phi \times 0.25\hat{a}_r) \cdot (dz\hat{a}_z)$$

$$= \int_{z=0}^{0.5} (-0.39\hat{a}_z) \cdot (dz\hat{a}_z)$$

$$= -0.39 \int_{z=0}^{0.5} dz = -0.196 \text{ V}$$

Motional emf induced in the outer conductor is

$$V_{m2} = \oint_l (\vec{v}_2 \times \vec{B}_2) \cdot d\vec{l}_2 = \int_0^l (2.62\hat{a}_\phi \times 0.8\hat{a}_r) \cdot (dz\hat{a}_z)$$

$$= \int_{z=0}^{0.5} (-2.09\hat{a}_z) \cdot (dz\hat{a}_z)$$

$$= -2.09 \int_{z=0}^{0.5} dz = -2.09 \times 0.5 = -1.047 \text{ V}$$

Hence, the current in the loop is given as

$$I = \frac{V_{m1} - V_{m2}}{R} = \frac{-0.196 - (-1.047)}{0.2} = \frac{0.85}{0.2} = 4.25 \text{ A}$$

3. Moving loop in time-varying magnetic field (general induction): Combining the above two results, for a moving conducting loop in a time-varying magnetic field, total emf induced is

$$V = \oint_C \vec{E} \cdot d\vec{l} = -\int_S \frac{\partial \vec{B}}{\partial t} \cdot d\vec{S} + \oint_C (\vec{v} \times \vec{B}) \cdot d\vec{l} \qquad \textbf{(4.11)}$$

$$V = -\int_S \frac{\partial \vec{B}}{\partial t} \cdot d\vec{S} \quad + \quad \oint_C (\vec{v} \times \vec{B}) \cdot d\vec{l}$$

$$\uparrow \qquad\qquad \uparrow$$

$$\begin{pmatrix}\text{transformer}\\ \text{emf}\end{pmatrix} \qquad \begin{pmatrix}\text{motional}\\ \text{emf}\end{pmatrix}$$

The field is given as

$$\nabla \times \vec{E} = -\frac{\partial \vec{B}}{\partial t} + \nabla \times (\vec{v} \times \vec{B}) \tag{4.12}$$

***Example 4.13** A conductor can slide over a static conductor in a time-varying magnetic field $\vec{B} = B_0 \cos \omega t(-\hat{a}_z)$ with a velocity $\vec{v} = v\hat{a}_x$, as shown in Fig. 4.10. The length of the sliding conductor is 'l'. Calculate the emf induced in the sliding conductor.

Solution In this case, the conductor is moving and the magnetic field is also time-varying. Therefore, both the transformer and motional emf's will be induced.

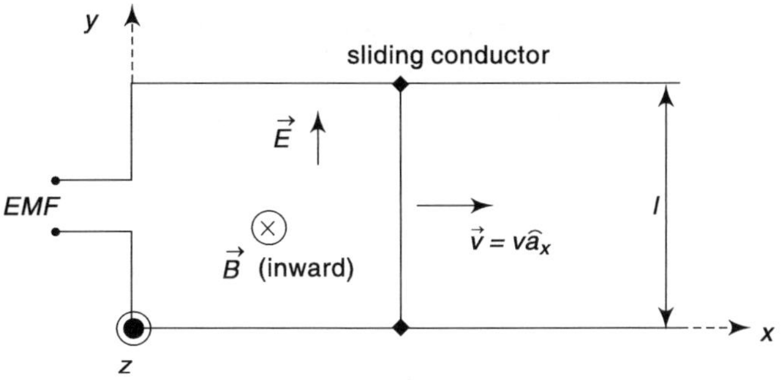

Fig. 4.10 *Sliding conductor in time-varying magnetic field*

Transformer emf is

$$V_S = -\int_S \frac{\partial \vec{B}}{\partial t} \cdot d\vec{S} = -\iint_S \frac{\partial}{\partial t}(-B_0 \cos \omega t\hat{a}_z) \cdot dxdy\hat{a}_z = -\int_0^x \int_0^l (\omega B_0 \sin \omega t)dxdy = -\omega x l B_0 \sin \omega t$$

If we let $x = 0$ at $t = 0$, then $x = vt$, and

$$V_S = -vlB_0 \omega t \sin \omega t$$

Motional emf is

$$V_m = \oint_C (\vec{v} \times \vec{B}) \cdot d\vec{l} = \oint_C \{(v\hat{a}_x) \times (-B_0 \cos \omega t d\hat{a}_z)\} \cdot dy\hat{a}_y$$

$$= \int_0^l vB_0 \cos \omega t dy$$

$$= vlB_0 \cos \omega t$$

Thus, the total induced emf is

$$V = V_S + V_m = B_0 lv(\cos \omega t - \omega t \sin \omega t)$$

***Example 4.14** **(a)** A conductor can slide over a static conductor in a static uniform magnetic field $\vec{B} = -B_0\hat{a}_z$ with a velocity $\vec{v} = v\hat{a}_x$, as shown in Fig. 4.11. The length of the sliding conductor is 'l'. Calculate the emf induced in the sliding conductor.

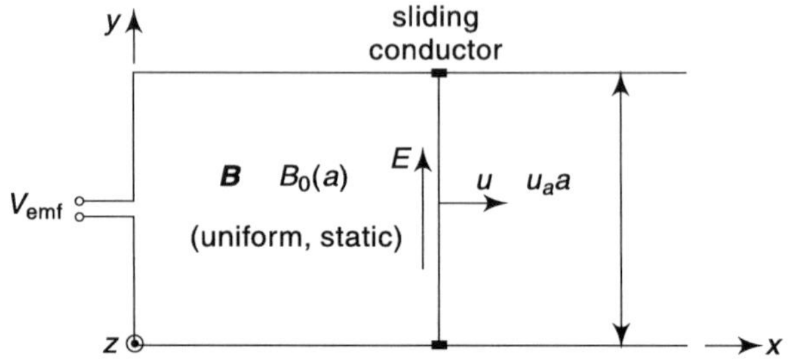

Fig. 4.11 *Sliding conductor in static uniform magnetic field*

(b) What will be the induced voltage, if the magnetic flux density is $\vec{B} = B_0 \cos \omega t(-\hat{a}_z)$ in part (a)?

Solution

(a) A particle of charge Q moving with a velocity $\vec{v}$ in a uniform $\vec{B}$ field experiences the force

$$\vec{F}_m = Q(\vec{v} \times \vec{B})$$

∴ motional field is

$$\vec{E}_m = \frac{\vec{F}_m}{Q} = (\vec{v} \times \vec{B}) = (v\hat{a}_x) \times (-B_0\hat{a}_z) = Bv\hat{a}_y$$

From Faraday's law, motional emf induced

$$V_m = \oint_l (\vec{v} \times \vec{B}) \cdot d\vec{l} = \int_0^l (Bv\hat{a}_y) \cdot (dy\hat{a}_y) = Bv\int_0^l dy = Blv$$

∴

$$\boxed{V_m = Blv}$$

(b) If the magnetic field is time-varying, both the transformer emf as well as the motional emf will be produced.

Transformer emf is

$$V_S = -\iint_S \frac{\partial \vec{B}}{\partial t} \cdot d\vec{S} = -\iint_S \frac{\partial}{\partial t}(-B_0 \cos \omega t\hat{a}_z) \cdot (dxdy\hat{a}_z) = -B_0\omega \int_0^x \int_0^l \sin \omega t dxdy = -B_0\omega lx \sin \omega t$$

If we let $x = 0$ at $t = 0$, then $x = vt$ and

$$V_S = -B_0 lv\omega t \sin \omega t$$

Motional emf is

$$V_m = \oint_l (\vec{v} \times \vec{B}) \cdot d\vec{l} = \int_0^l (v\hat{a}_x) \times (-B_0 \cos \omega t\hat{a}_z) \cdot (dy\hat{a}_y) = B_0 v \cos \omega t \int_0^l dy = B_0 lv \cos \omega t$$

Hence, the total emf induced in the conductor is

$$V = (V_S + V_m) = -B_0\omega lx \sin \omega t + B_0 lv \cos \omega t = B_0 lv[\cos \omega t - \omega t \sin \omega t]$$

***Example 4.15** A conducting rod of length l moves with a constant velocity $\vec{v}$ perpendicular to an infinitely long, straight wire carrying a current I, as shown in Fig. 4.12. What is the emf generated between the ends of the rod?

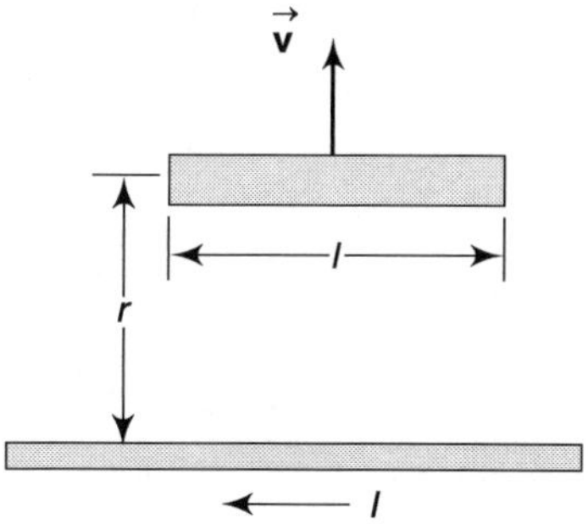

Fig. 4.12 *A bar moving away from a current-carrying wire*

Solution By Faraday's law, the motional emf is

$$|V_m| = Blv$$

where v is speed of the rod. However, the magnetic field due to the straight current-carrying wire at a distance r away is obtained using Ampere's law as

$$B = \frac{\mu_0 I}{2\pi r}$$

Thus, the emf between the ends of the rod is given by

$$|V_m| = \left(\frac{\mu_0 I}{2\pi r}\right) lv$$

Example 4.16 **(a)** An AC generator consisting of a single square loop of wire, with length of each side being 'a', is rotating in a steady magnetic field $\vec{B}$ at an angular velocity 'ω'. The axis of the loop rotation is perpendicular to the uniform field. Find the voltage induced in the loop.
(b) If the flux density varies harmonically with time as, $B = B_m \sin \omega t$, what will be the induced emf for (a)? This is demonstrated in Fig. 4.13.

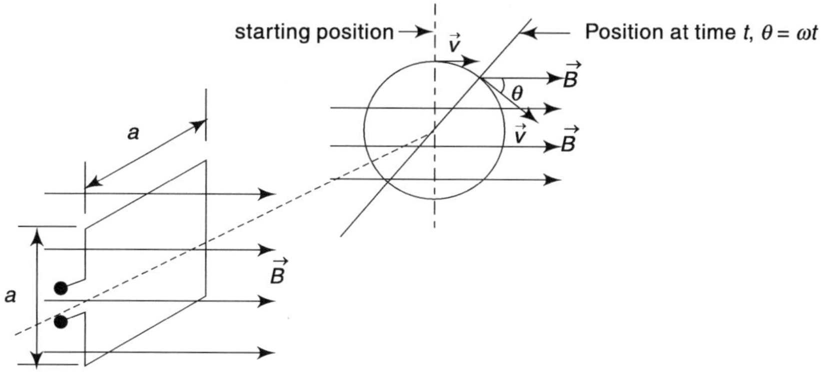

Fig. 4.13 *AC generator*

Solution

(a) Here, $\vec{v} \times \vec{B} = vB \sin \theta = vB \sin \omega t$

Since, only two sides of the loop cut the flux lines, $l = 2a$

$\therefore$ emf induced is

$$V_m = \oint (\vec{v} \times \vec{B}) \cdot d\vec{l} = vB \sin \omega t(2a) = \omega \frac{a}{2} B \sin \omega t(2a) = \omega a^2 B \sin \omega t = \omega AB \sin \omega t$$

where, $A = a^2$ = Area of the loop

$$\boxed{V_m = \omega a^2 B \sin \omega t = \omega AB \sin \omega t}$$

(b) In this case, in addition to the motional emf, transformer emf will be induced.

From (a), motional emf is

$$V_m = \omega a^2 (B_m \sin \omega t) \sin \omega t = \omega a^2 B_m \sin^2 \omega t$$

Transformer emf is $V_S = -\int \frac{\partial \vec{B}}{\partial t} \cdot d\vec{S}$

Here, $-\frac{\partial \vec{B}}{\partial t} = -\frac{\partial}{\partial t}(B_m \sin \omega t) = -B_m \omega \cos \omega t$

and, $\hat{a}_B \cdot \hat{a}_n = \cos \theta = \cos \omega t$

$$\therefore \qquad V_S = -\int \frac{\partial \vec{B}}{\partial t} \cdot d\vec{S} = -(B_m \omega \cos \omega t) \cos \omega t (a^2) = -\omega a^2 B_m \cos^2 \omega t$$

So, the total emf induced is

$$\begin{aligned} V &= (V_S + V_m) = \omega a^2 B_m \sin^2 \omega t - \omega a^2 B_m \cos^2 \omega t \\ &= \omega a^2 B_m (\sin^2 \omega t - \cos^2 \omega t) \\ &= -\omega a^2 B_m \cos 2\omega t \end{aligned}$$

$$\therefore \qquad \boxed{V = -\omega a^2 B_m \cos 2\omega t}$$

Example 4.17 A square coil of area of 100 cm^2 and 100 turns is rotated about an axis at right angles to a uniform magnetic field of 0.7 Wb/m^2 at a uniform speed of 1000 r.p.m. Evaluate the expression for the instantaneous value of induced e.m.f. Derive the formula used.

Solution Here, $A = 100 \text{ cm}^2$, $N = 100$, $B = 0.7 \text{ Wb/m}^2$, $\omega = \frac{2\pi \times 1000}{60} = \frac{100}{3}\pi$ rad/s

Using the formula derived in earlier problem, the expression for instantaneous value of the induced e.m.f. is given as

$$V_m = \omega AB \sin \omega t = \frac{100}{3}\pi \times 100 \times 10^{-4} \times 100 \times 0.7 \sin\left(\frac{100}{3}\pi t\right) = 73.3 \sin\left(\frac{100}{3}\pi t\right)$$

4.4 INCONSISTENCY IN AMPERE'S LAW FOR TIME-VARYING FIELDS

According to Ampere's circuital law in differential form we have

$$\nabla \times \vec{H} = \vec{J} \qquad \textbf{(4.13)}$$

Taking divergence on both sides

$$\nabla \cdot (\nabla \times \vec{H}) = \nabla \cdot \vec{J} = 0$$

(Since the divergence of curl of any vector is zero.)

However, according to the continuity equation of current, we have

$$\nabla \cdot \vec{J} = -\frac{\partial \rho}{\partial t} \neq 0$$

Therefore, we see that Ampere's law is not applicable for time-varying fields. In order to make it compatible for time varying fields, we modify the law.

4.4.1 Modified Ampere's Law for Time-Varying Fields

In order to have modified Ampere's law, we add a term to Eq. (4.13).

$$\nabla \times \vec{H} = \vec{J} + \vec{J}_d \tag{4.14}$$

where, $\vec{J}_d$ is to be determined and defined.

Taking divergence on both sides of Eq. (4.14),

$$\nabla \cdot (\nabla \times \vec{H}) = \nabla \cdot \vec{J} + \nabla \cdot \vec{J}_d = 0$$

$$\therefore \qquad \nabla \cdot \vec{J}_d = -\nabla \cdot \vec{J} = \frac{\partial \rho}{\partial t} = \frac{\partial}{\partial t}(\nabla \cdot \vec{D}) = \nabla \cdot \frac{\partial \vec{D}}{\partial t} \tag{4.15}$$

$$\therefore \qquad \boxed{\vec{J}_d = \frac{\partial \vec{D}}{\partial t}} \tag{4.16}$$

This $\vec{J}_d$ is known as *displacement current density* and $\vec{J}$ is the *conduction current density* ($\vec{J} = \sigma \vec{E}$).

NOTE

The concept of conduction and displacement currents have been discussed in Chapter 3, Section 3.2.

Now, substituting Eq. (4.15) into Eq. (4.14), we get,

$$\boxed{\nabla \times \vec{H} = \vec{J} + \frac{\partial \vec{D}}{\partial t}} \tag{4.17}$$

This is the *modified Ampere's law*.

4.5 MAXWELL'S EQUATIONS FOR TIME-VARYING FIELDS

Maxwell brought together the four basic laws governing electric and magnetic fields into one set of four equations, which completely describe the behaviour of any electromagnetic field. The form of these quantities is referred to as the *instantaneous form* (we can describe the fields at any point in time and space). The instantaneous form of Maxwell's equations may be used to analyse electromagnetic fields with any arbitrary time-variation. Maxwell's equations in differential and integral form are listed in Table 4.1.

Table 4.1 *Maxwell's Equations*

Sl. No.	Differential Form	Integral Form	Name
1	$\nabla \cdot \vec{D} = \rho_v$	$\oint_S \vec{D} \cdot d\vec{s} = \oint_v \rho_v dv = Q$	Gauss' law of Electrostatics
2	$\nabla \cdot \vec{B} = 0$	$\oint_S \vec{B} \cdot d\vec{s} = 0$	Gauss' law of Magnetostatic (non-existence of magnetic mono-pole)
3	$\nabla \times \vec{E} = -\frac{\partial \vec{B}}{\partial t}$	$\oint_l \vec{E} \cdot d\vec{l} = -\frac{\partial}{\partial t} \int_S \vec{B} \cdot d\vec{s}$	Faraday's law of electromagnetic induction
4	$\nabla \times \vec{H} = \vec{J} + \frac{\partial \vec{D}}{\partial t}$	$\oint_l \vec{H} \cdot d\vec{l} = \int_S \left(\vec{J} + \frac{\partial \vec{D}}{\partial t} \right) \cdot d\vec{s}$	Modified Ampere's circuital law

Word Statements

1. $\nabla \cdot \vec{D} = \rho_v$ or, $\oint_S \vec{D} \cdot d\vec{s} = \oint_v \rho_v dv = Q$

   The total electric displacement through any closed surface enclosing a volume is equal to the total charge within the volume. This is the Gauss' law for static electric fields.

2. $\nabla \cdot \vec{B} = 0$ or, $\oint_S \vec{B} \cdot d\vec{s} = 0$

   The net magnetic flux emerging through any closed surface is zero. In other words, the magnetic flux lines do not originate and end anywhere, but are continuous. This is the Gauss' law for static magnetic fields, which confirms the non-existence of magnetic monopole.

3. $\nabla \times \vec{E} = -\frac{\partial \vec{B}}{\partial t}$ or, $\oint_l \vec{E} \cdot d\vec{l} = -\frac{\partial}{\partial t} \int_S \vec{B} \cdot d\vec{s}$

   The electromotive force around a closed path is equal to the time derivative of the magnetic displacement through any surface bounded by the path. This is the Faraday's law for electromagnetic fields.

4. $\nabla \times \vec{H} = \vec{J} + \frac{\partial \vec{D}}{\partial t}$ or, $\oint_l \vec{H} \cdot d\vec{l} = \int_S \left(\vec{J} + \frac{\partial \vec{D}}{\partial t} \right) \cdot d\vec{s}$

   The magnetomotive force around a closed path is equal to the conduction current plus the time derivative of the electric displacement through any surface bounded by the path. This is the modified Ampere's law for time varying fields.

Physical Interpretation of Maxwell's Equations

1. $\nabla \cdot \vec{D} = \rho_v$ or, $\oint_S \vec{D} \cdot d\vec{s} = \oint_v \rho_v dv = Q$

   This implies that the divergence of electric flux density is the charge density. This is simple to understand, as ρ_v at any point is simply the charge per unit volume at that point and from the definition of divergence $\left(\nabla \cdot \vec{D} = \text{Lim} \frac{\oint_S \vec{D} \cdot d\vec{s}}{\Delta v} = \frac{\text{Total Electric Flux}}{\text{Volume}} = \rho_v \right)$, it is equal to $\nabla \cdot \vec{D}$.

   If a closed surface does not contain any charge, then obviously the total flux over the surface is zero.

This equation implies that the electric flux lines are not continuous; they originate from the positive charges and terminate on the negative charges.

2. $\nabla \cdot \vec{B} = 0$ or, $\oint_S \vec{B} \cdot d\vec{s} = 0$

This implies that the number of magnetic flux lines of force entering any region must be equal to the number of lines of force coming out of that region.

This is simple to understand because of the fact that there are no magnetic mono-poles on which the flux lines ends and thus, the magnetic flux lines are always continuous.

3. $\nabla \times \vec{E} = -\dfrac{\partial \vec{B}}{\partial t}$ or, $\oint_l \vec{E} \cdot d\vec{l} = -\dfrac{\partial}{\partial t}\int_S \vec{B} \cdot d\vec{s}$

By Stoke's theorem,

$$\oint_S (\nabla \times \vec{E}) \cdot d\vec{s} = \oint_l \vec{E} \cdot d\vec{l};$$

where, $d\vec{s} = |d\vec{s}|\,\hat{n}$; where, $\hat{n}$ is the unit vector normal to the direction of *ds*.

If the surface is reduced to an element of area *ds*, the LHS becomes $(\nabla \times \vec{E}) \cdot d\vec{s}$. Dividing through by $|d\vec{s}|$, the result is

$$(\nabla \times \vec{E}) \cdot \hat{n} = \frac{\oint_l \vec{E} \cdot d\vec{l}}{|d\vec{s}|}$$

If unit vector $\hat{n}$ is directed in such a way that $\oint_l \vec{E} \cdot d\vec{l}$ is maximum, then the direction of $(\nabla \times \vec{E})$ is given by $\hat{n}$ and its magnitude is $\dfrac{\oint_l \vec{E} \cdot d\vec{l}}{|d\vec{s}|}$. Thus, we conclude that:

- $(\nabla \times \vec{E})$ equals the emf per unit area in a direction of the area that results in a maximum emf around its edges.

This equation inter-relates the magnetic and electric fields and shows that a time-varying magnetic field can produce an electric field.

4. $\nabla \times \vec{H} = \vec{J} + \dfrac{\partial \vec{D}}{\partial t}$ or, $\oint_l \vec{H} \cdot d\vec{l} = \int_S \left(\vec{J} + \dfrac{\partial \vec{D}}{\partial t}\right) \cdot d\vec{s}$

Similar to the curl of electric field, this equation implies that $(\nabla \times \vec{H})$ equals the mmf per unit area in a direction of the area that results in a maximum mmf around its edge. This equation also signifies that conduction current as well as a changing electric field produces a magnetic field.

***Example 4.18** The electric field intensity of an electromagnetic wave in free space is given by

$$E_y = 0, \quad E_z = 0, \quad E_x = E_0 \cos \omega\left(t - \frac{z}{v}\right)$$

Determine the expression for the components of the magnetic field intensity $\vec{H}$. Also, find $\dfrac{E_x}{E_y}$.

Solution By Maxwell's equation

$$\nabla \times \vec{E} = -\frac{\partial \vec{B}}{\partial t} = -\mu_0 \frac{\partial \vec{H}}{\partial t}$$

or
$$\begin{vmatrix} \hat{i} & \hat{j} & \hat{k} \\ \dfrac{\partial}{\partial x} & \dfrac{\partial}{\partial y} & \dfrac{\partial}{\partial z} \\ E_x & E_y & E_z \end{vmatrix} = -\mu_0\left[\frac{\partial H_x}{\partial t}\hat{i} + \frac{\partial H_y}{\partial t}\hat{j} + \frac{\partial H_z}{\partial t}\hat{k}\right]$$

or
$$\begin{vmatrix} \hat{i} & \hat{j} & \hat{k} \\ \dfrac{\partial}{\partial x} & \dfrac{\partial}{\partial y} & \dfrac{\partial}{\partial z} \\ E_x & 0 & 0 \end{vmatrix} = -\mu_0\left[\frac{\partial H_x}{\partial t}\hat{i} + \frac{\partial H_y}{\partial t}\hat{j} + \frac{\partial H_z}{\partial t}\hat{k}\right]$$

or
$$\frac{\partial E_x}{\partial z}\hat{j} - \frac{\partial E_x}{\partial y}\hat{k} = -\mu_0\left[\frac{\partial H_x}{\partial t}\hat{i} + \frac{\partial H_y}{\partial t}\hat{j} + \frac{\partial H_z}{\partial t}\hat{k}\right]$$

Since, $E_x = E_0 \cos\omega\left(t - \frac{z}{v}\right)$; $\therefore \frac{\partial E_x}{\partial y} = 0$, and the above equation reduces to,

$$\frac{\partial E_x}{\partial z}\hat{j} = -\mu_0\left[\frac{\partial H_x}{\partial t}\hat{i} + \frac{\partial H_y}{\partial t}\hat{j} + \frac{\partial H_z}{\partial t}\hat{k}\right]$$

Comparing both sides

$H_x = 0$; $H_z = 0$ and,

$$\frac{\partial H_y}{\partial t} = -\frac{1}{\mu_0}\frac{\partial E_x}{\partial z} = \frac{1}{\mu_0}\frac{\partial}{\partial z}\left[E_0 \cos\omega\left(t - \frac{z}{v}\right)\right] = -\frac{E_0\omega}{\mu_0 v}\sin\omega\left(t - \frac{z}{v}\right)$$

$\therefore$
$$H_y = -\frac{E_0\omega}{\mu_0 v}\int \sin\omega\left(t - \frac{z}{v}\right)dt = \frac{E_0}{\mu_0 v}\cos\omega\left(t - \frac{z}{v}\right)$$

So, the magnetic field intensity is given as

$$H_y = \frac{E_0}{\mu_0 v}\cos\omega\left(t - \frac{z}{v}\right)$$

Now, $E_x = E_0 \cos\omega\left(t - \frac{z}{v}\right)$ and $H_y = \frac{E_0}{\mu_0 v}\cos\omega\left(t - \frac{z}{v}\right)$; so, the ratio,

$$\frac{E_x}{H_y} = \mu_0 v = \mu_0\frac{1}{\sqrt{\mu_0\varepsilon_0}} = \sqrt{\frac{\mu_0}{\varepsilon_0}} = \sqrt{\frac{4\pi\times 10^{-7}}{\frac{1}{36\pi}\times 10^{-9}}} = 120\pi = 377$$

NOTE

This is known as intrinsic impedance of free space as discussed in Chapter 5.

4.5.1 Time-Harmonic Fields (Sinusoidal Variations of Fields)

Time-harmonic fields are those fields that vary sinusoidally with time. They are easily expressed in phasors.

For any complex number, $z = (x + jy) = r\angle\theta = re^{j\theta}$, if $\theta = (\omega t + \phi)$ where, ϕ may be a function of time or space or a constant, then,

$$re^{j\theta} = re^{j\omega t}e^{j\phi}$$

$$\therefore \qquad \text{Re}(re^{j\theta}) = r\cos(\omega t + \phi) \quad \text{and} \quad \text{Im}(re^{j\theta}) = r\sin(\omega t + \phi)$$

Thus, for a sinusoidal current, $i(t) = I_0\cos(\Omega t + \phi)$, the complex form is $\text{Re}(I_0e^{j\omega t}e^{j\phi})$. Dropping the time factor $e^{j\omega t}$, the term $I_0e^{j\phi}$ is known as the phasor current, denoted by $\overline{I}_s$.

$$\therefore \qquad \overline{I}_s = I_0e^{j\phi} = I_0\angle\phi$$

$$\therefore \qquad i(t) = I_0\cos(\omega t + \phi) = \text{Re}(\overline{I}_s e^{j\omega t})$$

In general, a phasor can be a scalar or a vector. If a vector $\vec{A}(x, y, z, t)$ is a time-harmonic field, the phasor form of $\vec{A}$ is $\overline{A}(x, y, z)$; the relation being

$$\boxed{\vec{A} = \text{Re}[\overline{A}e^{j\omega t}]}$$

$$\therefore \qquad \overline{A}(x, y, z, t) = \text{Re}[\overline{A}(x, y, z)e^{j\omega t}]$$

$$\boxed{\frac{\partial \overline{A}}{\partial t} \rightarrow j\omega\overline{A}} \quad \text{and} \quad \boxed{\int \overline{A}dt \rightarrow \frac{\overline{A}}{j\omega}}$$

A general instantaneous vector electric field $[\vec{E}(x, y, z, t)]$ may be given by

$$\vec{E}(x, y, z, t) = E_x(x, y, z, t)\hat{a}_x + E_y(x, y, z, t)\hat{a}_y + E_z(x, y, z, t)\hat{a}_z$$

Each of the component scalars of the instantaneous vector electric field $[E_x, E_y, E_z]$ may be written in terms of the corresponding component phasors $[\overline{E}_x, \overline{E}_y, \overline{E}_z]$.

$$E_x(x, y, z, t) = E_{x0}(x, y, z)\cos(\omega t + \theta_x) = \text{Re}\left[E_{x0}(x, y, z)e^{j(\omega t + \theta_x)}\right] = \text{Re}\left[(E_{x0}(x, y, z)e^{j\theta_x})e^{j\omega t}\right]$$

$$E_y(x, y, z, t) = E_{y0}(x, y, z)\cos(\omega t + \theta_y) = \text{Re}\left[E_{y0}(x, y, z)e^{j(\omega t + \theta_y)}\right] = \text{Re}\left[(E_{yz0}(x, y, z)e^{j\theta_y})e^{j\omega t}\right]$$

$$E_z(x, y, z, t) = E_{z0}(x, y, z)\cos(\omega t + \theta_z) = \text{Re}\left[E_{z0}(x, y, z)e^{j(\omega t + \theta_z)}\right] = \text{Re}\left[(E_{z0}(x, y, z)e^{j\theta_z})e^{j\omega t}\right]$$

$$\therefore \qquad \vec{E}(x, y, z, t) = \text{Re}\{[\overline{E}_x(x, y, z, \omega)\hat{a}_x + \overline{E}_y(x, y, z, \omega)\hat{a}_y + \overline{E}_z(x, y, z, \omega)\hat{a}_z]e^{j\omega t}\}$$
$$= \text{Re}[\overline{E}(x, y, z, \omega)e^{j\omega t}]$$

This must be noted that $\overline{E}(x, y, z)$ is a vector phasor defined by three complex vector-components which are each defined by a magnitude and a phase.

4.5.2 Maxwell's Equations for Time-Harmonic Fields

To transform the instantaneous Maxwell's equations into time-harmonic forms, we replace all sources and field quantities by their phasor equivalents and replace all time-derivatives of quantities with $j\omega$ times the phasor equivalent. Thus, the Maxwell's equations for time-harmonic fields are given in Table 4.2.

Table 4.2 *Maxwell's Equations for Time-Harmonic Fields*

Sl. No.	Differential Form	Integral Form	Name
1	$\nabla . \bar{D} = \rho_v$	$\oint_S \bar{D} \cdot d\bar{s} = \oint_v \rho_v dv = Q$	Gauss' law of Electrostatic
2	$\nabla \cdot \bar{B} = 0$	$\oint_S \bar{B} \cdot d\bar{s} = 0$	Gauss' law of Magnetostatic (non-existence of magnetic mono-pole)
3	$\nabla \times \bar{E} = - j\omega\bar{B}$	$\oint_l \vec{E} \cdot d\vec{l} = - j\omega \int_S \bar{B} \cdot d\bar{s}$	Faraday's law of electromagnetic induction
4	$\nabla \times \bar{H} = \bar{J} + j\omega\bar{D}$	$\oint_l \bar{H} \cdot d\bar{l} = \int_S (\bar{J} + j\omega\bar{D}) \cdot d\bar{s}$	Modified Ampere's circuital law

Example 4.19 The instantaneous magnetic field is $\vec{H} = 2\cos(\omega t - 3y)\hat{a}_z$ A/m in a medium characterized by $\sigma = 0$, $\mu = 2\mu_0$, $\varepsilon = 5\varepsilon_0$. Calculate ω and $\vec{E}$. Assume a source-free region.

Solution Here, the phasor form of the magnetic field is obtained as follows.

$$\vec{H} = \text{Re}[\bar{H}e^{j\omega t}] = \text{Re}[2e^{-j3y}\hat{a}_z e^{j\omega t}] = 2\cos(\omega t - 3y)\hat{a}_z$$

$$\therefore \qquad \bar{H} = 2e^{-j3y}\hat{a}_z = \bar{H}_z\hat{a}_z$$

The phasor electric and magnetic fields are related by the time-harmonic Maxwell's equations in a source-free region ($\vec{J} = 0, \rho = 0$). $\bar{E}$ and $\bar{H}$ must satisfy all four equations:

1. $\nabla \times \bar{E} = - j\omega\mu\bar{H}$
2. $\nabla \times \bar{H} = j\omega\varepsilon\bar{E}$
3. $\nabla \cdot \bar{E} = 0$
4. $\nabla . \bar{H} = 0$

From Eq. (4), $\nabla \cdot \bar{H} = \dfrac{\partial \bar{H}_z}{\partial z} = 0$

From Eq. (2),

$$\begin{aligned}\bar{E} = \frac{1}{j\omega\varepsilon}\nabla \times \bar{H} &= \frac{1}{j\omega 5\varepsilon_0}\left(\frac{\partial \bar{H}_z}{\partial y}\hat{a}_x - \frac{\partial \bar{H}_z}{\partial x}\hat{a}_y\right) \\ &= \frac{1}{j\omega 5\varepsilon_0}\frac{\partial}{\partial y}(2e^{-j3y})\hat{a}_x = \frac{2}{j\omega 5\varepsilon_0}(-j3e^{-j3y})\hat{a}_x \\ &= \frac{-6/5}{\omega\varepsilon_0}e^{-j3y}\hat{a}_x \quad \text{(V/m)} \\ &= \bar{E}_x\hat{a}_x\end{aligned}$$

From Eq. (3), $\nabla \cdot \bar{E} = \dfrac{\partial \bar{E}_x}{\partial x} = 0$

From Eq. (1),

$$- j\omega\mu\bar{H} = \nabla \times \bar{E}$$

or, $$-j\omega 2\mu_0 2e^{-j3y}\hat{a}_z = \frac{\partial \bar{E}_x}{\partial z}\hat{a}_y - \frac{\partial \bar{E}_x}{\partial y}\hat{a}_z = -\frac{\partial \bar{E}_x}{\partial y}\hat{a}_z$$

or, $$-j\omega 4\mu_0 e^{-j3y}\hat{a}_z = -\frac{\partial}{\partial y}\left(\frac{-6/5}{\omega\varepsilon_0}e^{-j3y}\right)\hat{a}_z = \frac{6/5}{\omega\varepsilon_0}(-j3e^{-j3y})\hat{a}_z = -\frac{18}{5\omega\varepsilon_0}je^{-j3y}\hat{a}_z$$

Equating both sides, we get

$$\omega 4\mu_0 = \frac{18}{5\omega\varepsilon_0}$$

$$\Rightarrow \qquad \omega = \sqrt{\frac{9}{10\mu_0\varepsilon_0}} = \sqrt{\frac{9}{10}} \times \sqrt{\frac{1}{\mu_0\varepsilon_0}} = \sqrt{\frac{9}{10}} \times c = 0.95 \times 3 \times 10^8 = 2.846 \times 10^8 \text{ rad/s}$$

The frequency, $f = \dfrac{\omega}{2\pi} = \dfrac{2.846 \times 10^8}{2\pi} = 45.3 \text{ GHz}$

The phasor electric field is given as

$$\bar{E} = \frac{-6/5}{\omega\varepsilon_0}e^{-j3y} = \frac{-6/5}{2.846 \times 10^8 \times 8.854 \times 10^{-12}}e^{-j3y} = -476.2e^{-j3y} \quad \text{(V/m)}$$

Thus, the electric field is

$$\vec{E} = \text{Re}[\bar{E}e^{j\omega t}] = \text{Re}[-476.2e^{-j3y}e^{j\omega t}\hat{a}_x] = -476.2\cos(\omega t - 3y)\hat{a}_x \quad \text{(V/m)}$$

Example 4.20 **(a)** In free space, $\vec{D} = D_m \sin(\omega t + \beta z)\hat{a}_x$. Using Maxwell's equations, show that

$$\vec{B} = -\frac{\omega\mu_0 D_m}{\beta}\sin(\omega t + \beta z)\hat{a}_y$$

(b) In free space, $\vec{B} = B_m e^{j(\omega t + \beta z)}\hat{a}_y$. Using Maxwell's equations, show that

$$\vec{E} = -\frac{\omega B_m}{\beta}e^{j(\omega t + \beta z)}\hat{a}_x$$

Solution **(a)** By Maxwell's equation,

$$\nabla \times \vec{E} = -\frac{\partial \vec{B}}{\partial t} \text{ and } \vec{D} = \varepsilon_0\vec{E} \quad \text{or,} \quad \vec{E} = \frac{\vec{D}}{\varepsilon_0} \text{ for free space.}$$

$$\therefore \quad -\frac{\partial \vec{B}}{\partial t} = \nabla \times \vec{E} = \begin{vmatrix} \hat{a}_x & \hat{a}_y & \hat{a}_z \\ \dfrac{\partial}{\partial x} & \dfrac{\partial}{\partial y} & \dfrac{\partial}{\partial z} \\ \dfrac{D_m}{\varepsilon_0}\sin(\omega t + \beta z) & 0 & 0 \end{vmatrix} = \frac{D_m}{\varepsilon_0}\frac{\partial}{\partial z}[\sin(\omega t + \beta z)]\hat{a}_y = \frac{D_m\beta}{\varepsilon_0}\cos(\omega t + \beta z)\hat{a}_y$$

or, $$\vec{B} = -\frac{D_m\beta}{\varepsilon_0}\int \cos(\omega t + \beta z)\hat{a}_y dt = -\frac{D_m\beta}{\omega\varepsilon_0}\sin(\omega t + \beta z)\hat{a}_y$$

Also, for free space

$$\frac{\omega}{\beta} = v = c = \frac{1}{\sqrt{\mu_0\varepsilon_0}} \quad \Rightarrow \quad \frac{1}{\varepsilon_0} = \mu_0\left(\frac{\omega}{\beta}\right)^2$$

$$\therefore \quad \vec{B} = -\frac{D_m \beta}{\omega \varepsilon_0} \sin(\omega t + \beta z)\hat{a}_y = -\frac{D_m \beta}{\omega} \times \mu_0 \left(\frac{\omega}{\beta}\right)^2 \sin(\omega t + \beta z)\hat{a}_y = -\frac{\omega \mu_0 D_m}{\beta} \sin(\omega t + \beta z)\hat{a}_y$$

$$\therefore \quad \boxed{\vec{B} = -\frac{\omega \mu_0 D_m}{\beta} \sin(\omega t + \beta z)\hat{a}_y}$$

(b) By Maxwell's equation

$$\nabla \times \vec{E} = -\frac{\partial \vec{B}}{\partial t} = -\frac{\partial}{\partial t} B_m e^{j(\omega t + \beta z)} \hat{a}_y$$

or

$$\begin{vmatrix} \hat{a}_x & \hat{a}_y & \hat{a}_z \\ \frac{\partial}{\partial x} & \frac{\partial}{\partial y} & \frac{\partial}{\partial z} \\ E_x & E_y & E_z \end{vmatrix} = -B_m j\omega e^{j(\omega t + \beta z)} \hat{a}_y$$

Comparing both sides, we get

$$\left(\frac{\partial E_x}{\partial z} - \frac{\partial E_z}{\partial x}\right)\hat{a}_y = -B_m j\omega e^{j(\omega t + \beta z)} \hat{a}_y$$

or, $$\frac{\partial E_x}{\partial z} = -B_m j\omega e^{j(\omega t + \beta z)} \quad (\because E_z \text{ is not a function of } x)$$

or, $$E_x = \int -B_m j\omega e^{j(\omega t + \beta z)} dz = -B_m j\omega \frac{1}{j\beta} e^{j(\omega t + \beta z)} = -\frac{B_m \omega}{\beta} e^{j(\omega t + \beta z)}$$

$$\therefore \quad \boxed{\vec{E} = -\frac{\omega B_m}{\beta} e^{j(\omega t + \beta z)} \hat{a}_x}$$

Example 4.21 Starting from Maxwell's equations, establish Coulomb's law.

Solution We will consider a spherical surface of radius r, centered at a point charge Q. Applying Maxwell equation (Gauss' law) in integral form, we have

$$\oint_S \vec{E} \cdot d\vec{S} = \frac{Q}{\varepsilon}$$

By the assumption of spherical symmetry, the integrand is a constant and can be taken out of the integral as

$$4\pi r^2 \hat{a}_r . \vec{E} = \frac{Q}{\varepsilon}$$

where $\hat{a}_r$ is a unit vector pointing radially away from the charge. Again by spherical symmetry, $\vec{E}$ is also in radially outward direction, and so we get

$$\vec{E} = \frac{Q}{4\pi \varepsilon r^2} \hat{a}_r$$

If another point charge q is placed on the surface, the force on that charge due to the charge Q is given as

$$\vec{F} = q\vec{E} = \frac{Qq}{4\pi\varepsilon r^2}\hat{a}_r$$

This is essentially equivalent to Coulomb's law.

Example 4.22 Show that the equation continuity $\nabla \cdot \vec{J} + \frac{\partial \rho}{\partial t} = 0$ is contained in Maxwell's equations.

Solution By Maxwell's equation (modified Ampere's law),

$$\nabla \times \vec{H} = \vec{J} + \frac{\partial \vec{D}}{\partial t}$$

Taking divergence on both sides,

$$\nabla \cdot (\nabla \times \vec{H}) = \nabla \cdot \left(\vec{J} + \frac{\partial \vec{D}}{\partial t}\right) = 0$$

$\therefore$
$$\nabla \cdot \vec{J} + \nabla \cdot \frac{\partial \vec{D}}{\partial t} = 0$$

or,
$$\nabla \cdot \vec{J} + \frac{\partial}{\partial t}(\nabla \cdot \vec{D}) = 0$$

By Maxwell's equation, $\Delta \cdot \vec{D} = P$

$\therefore$
$$\boxed{\nabla \cdot \vec{J} + \frac{\partial \rho}{\partial t} = 0}$$

Example 4.23 Starting from Maxwell's equation

$$\nabla \times \vec{E} = -\frac{\partial \vec{B}}{\partial t} \quad \text{and} \quad \nabla \times \vec{H} = \vec{J} + \frac{\partial \vec{D}}{\partial t}$$

Show that $\nabla \cdot \vec{B} = 0$ and $\nabla \cdot \vec{D} = \rho$.

Solution $\nabla \times \vec{E} = -\frac{\partial \vec{B}}{\partial t}$

Taking divergence on both sides, we get

$$\nabla \cdot (\nabla \times \vec{E}) = 0 = -\nabla \cdot \left(\frac{\partial \vec{B}}{\partial t}\right)$$

$\therefore$
$$\nabla \cdot \left(\frac{\partial \vec{B}}{\partial t}\right) = 0 \quad \Rightarrow \quad \frac{\partial}{\partial t}(\nabla \cdot \vec{B}) = 0 \quad \Rightarrow \quad \nabla \cdot \vec{B} = \text{constant}$$

As an isolated magnetic monopole does not exist, we have

$$\boxed{\nabla \cdot \vec{B} = 0}$$

$$\nabla \times \vec{H} = \vec{J} + \frac{\partial \vec{D}}{\partial t}$$

Taking divergence on both sides, we get

$$\nabla \cdot (\nabla \times \vec{H}) = 0 = \nabla \cdot \left(\vec{J} + \frac{\partial \vec{D}}{\partial t}\right)$$

$$\therefore \qquad \nabla\cdot\left(\vec{J}+\frac{\partial\vec{D}}{\partial t}\right)=0 \;\Rightarrow\; \nabla\cdot\vec{J}=-\nabla\cdot\left(\frac{\partial\vec{D}}{\partial t}\right)$$

Again by continuity equation we know

$$\nabla\cdot\vec{J}+\frac{\partial\rho}{\partial t}=0$$

$$-\frac{\partial\rho}{\partial t}=-\nabla\cdot\left(\frac{\partial\vec{D}}{\partial t}\right)=-\frac{\partial}{\partial t}(\nabla\cdot\vec{D}) \;\Rightarrow\; \nabla\cdot\vec{D}=\rho$$

$$\boxed{\nabla\cdot\vec{D}=\rho}$$

4.6 TIME-VARYING POTENTIALS FOR EM FIELDS

Maxwell's equations define electromagnetic fields in terms of field quantities ($\vec{E}, \vec{H}, \vec{D}, \vec{B}$) and sources ($\vec{J}, \vec{D}$). However, complicated integrations have to be performed for solving Maxwell's equations directly for the electric and magnetic fields. The integration required to determine the fields can be simplified through the use of potentials (*magnetic vector potential* $\vec{A}$, *electric scalar potential* V). This is depicted in Fig. 4.14.

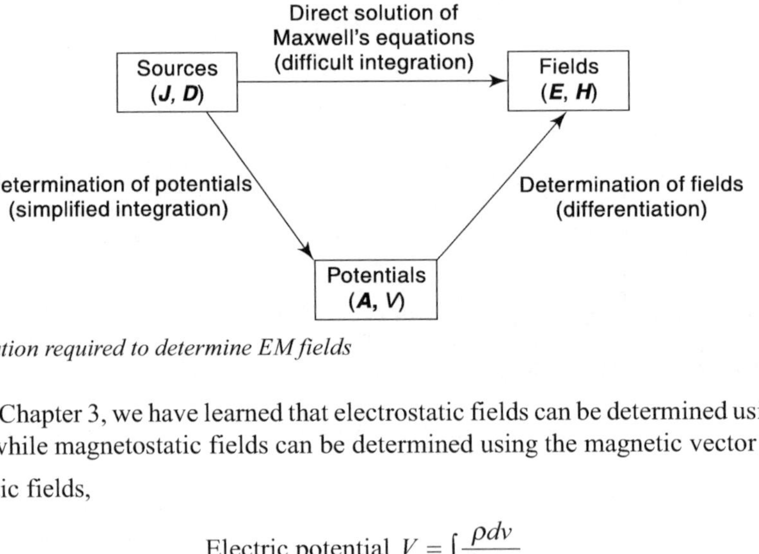

Fig. 4.14 *Integration required to determine EM fields*

In Chapter 2 and Chapter 3, we have learned that electrostatic fields can be determined using the electric scalar potential while magnetostatic fields can be determined using the magnetic vector potential.

We have, for static fields,

$$\text{Electric potential } V=\int_v \frac{\rho dv}{4\pi\varepsilon r}$$

$$\text{Magnetic vector potential } \vec{A}=\int_v \frac{\mu\vec{J}dv}{4\pi r}$$

We will find these expressions for time-varying fields.

We start with Gauss' law for magnetic field, which is same for static and dynamic fields.

$$\nabla\cdot\vec{B}=0 \;\Rightarrow\; \vec{B}=\nabla\times\vec{A} \tag{4.18}$$

By Faraday's law,

$$\nabla \times \vec{E} = -\frac{\partial \vec{B}}{\partial t} = -\frac{\partial}{\partial t}(\nabla \times \vec{A}) \quad \text{(by Eq. (4.16))}$$

$$\Rightarrow \qquad \nabla \times \left(\vec{E} + -\frac{\partial \vec{A}}{\partial t} \right) = 0 \qquad \textbf{(4.19)}$$

Since the curl of the gradient of a scalar field is zero, we can write from Eq. (4.19),

$$\left(\vec{E} + \frac{\partial \vec{A}}{\partial t} \right) = -\nabla V$$

$$\therefore \qquad \boxed{\vec{E} = -\nabla V - \frac{\partial \vec{A}}{\partial t}} \qquad \textbf{(4.20)}$$

Thus, we can find the fields $\vec{B}$ and $\vec{E}$ if the potentials $\vec{A}$ and V are known. The potentials at this point have been defined using only two of the four Maxwell's equations and are not completely described yet. We take divergence of Eq. (4.20) and apply Gauss' law, which is valid for time-varying fields.

$$\therefore \qquad \nabla \cdot E = \frac{\rho}{\varepsilon}$$

$$\Rightarrow \qquad \nabla \cdot \left(-\nabla V - \frac{\partial \vec{A}}{\partial t} \right) = \frac{\rho}{\varepsilon}$$

$$\Rightarrow \qquad -\nabla^2 V - \frac{\partial}{\partial t}(\nabla \cdot \vec{A}) = \frac{\rho}{\varepsilon}$$

$$\Rightarrow \qquad \nabla^2 V + \frac{\partial}{\partial t}(\nabla \cdot \vec{A}) = -\frac{\rho}{\varepsilon} \qquad \textbf{(4.21)}$$

Taking curl of Eq. (4.18) and applying modified Ampere's law, we get

$$\nabla \times \nabla \times \vec{A} = \nabla \times \vec{B} = \left(\mu \vec{J} + \mu \frac{\partial \vec{D}}{\partial t} \right) = \mu \vec{J} + \mu\varepsilon \frac{\partial \vec{E}}{\partial t} = \mu \vec{J} + \mu\varepsilon \frac{\partial}{\partial t}\left(-\nabla V - \frac{\partial \vec{A}}{\partial t} \right) \quad \text{[by Eq. (4.18)]}$$

$$\Rightarrow \qquad \nabla \times \nabla \times \vec{A} = \mu \vec{J} - \mu\varepsilon \nabla \left(\frac{\partial V}{\partial t} \right) - \mu\varepsilon \frac{\partial^2 \vec{A}}{\partial t^2}$$

$$\Rightarrow \qquad \nabla(\nabla . \vec{A}) - \nabla^2 \vec{A} = \mu \vec{J} - \mu\varepsilon \nabla \left(\frac{\partial V}{\partial t} \right) - \mu\varepsilon \frac{\partial^2 \vec{A}}{\partial t^2} \qquad \textbf{(4.22)}$$

To completely describe any vector, both the divergence (lamellar components) and the curl (solenoidal components) must be defined. So far we have defined the curl of $\vec{A}$, but not the divergence of $\vec{A}$. We may choose the divergence of $\vec{A}$ in such a way as to simplify the mathematics.

$$\text{Let,} \qquad \nabla \cdot \vec{A} = -\mu\varepsilon \frac{\partial V}{\partial t}$$

(This condition is known as Lorentz gauge condition or Lorentz condition for potentials.)

So, from Eq. (4.21) and (4.22)

$$\boxed{\nabla^2 V - \mu\varepsilon \frac{\partial^2 V}{\partial t^2} = -\frac{\rho}{\varepsilon}} \qquad \textbf{(4.23)}$$

and

$$\boxed{\nabla^2 \vec{A} - \mu\varepsilon \frac{\partial^2 \vec{A}}{\partial t^2} = -\mu \vec{J}} \quad \textbf{(4.24)}$$

Thus, $\vec{A}$ is connected with the vector $\vec{J}$ and V with the scalar ρ and both potentials satisfy the same form of equations. Such partial differential equations which relate the potentials to the sources have the basic form of *inhomogeneous wave equations* (fundamental equations defining wave behaviour).

For steady state, the time derivatives will be zero, so that the equations for potentials become

$$\nabla^2 V = -\frac{\rho}{\varepsilon}$$

and

$$\nabla^2 \vec{A} = -\mu \vec{J}$$

i.e., the potentials satisfy the *Poisson's equations* for time-varying conditions.

For time-harmonic fields, each partial derivative yields a $j\omega$ factor and the wave equations defining the potentials reduce to

$$\frac{\partial^2}{\partial t^2}[\,] = (j\omega)(j\omega)[\,] = -\omega^2[\,]$$

$$\nabla^2 V + \omega^2 \mu\varepsilon V = -\frac{\rho}{\varepsilon}$$

$$\nabla^2 \vec{A} + \omega^2 \mu\varepsilon \vec{A} = -\mu \vec{J}$$

4.6.1 Concept of Retarded Potentials

The concept of retarded potential describes the scalar or vector potential for electromagnetic fields of a time-varying current or charge distribution. The *retardation* between cause and effect is thereby essential; e.g., the signal takes a finite time, corresponding to the velocity of light, to propagate from the source point of the field to the point where an effect is produced or measured.

For static field, in which the charge density ρ, the current density $\vec{J}$, the electric field $\vec{E}$ and the potential V, and the magnetic field $\vec{B}$ and the potential $\vec{A}$ are all constant in time (i.e., they are functions of x, y, z; but not of t), then electric and magnetic potentials in free space is given by

$$V(x, y, z) = \frac{1}{4\pi\varepsilon_0} \int \frac{\rho(x', y', z')}{R} dv' \quad \textbf{(4.25a)}$$

and,

$$\vec{A}(x, y, z) = \frac{\mu_0}{4\pi} \int \frac{\vec{J}(x', y', z')}{R} dv' \quad \textbf{(4.25b)}$$

where, R is the distance between the source point (x', y', z') and the field point (x, y, z) and v' is a volume element at the point (x', y', z').

For time varying fields, in which, ρ, $\vec{J}$, $\vec{E}$, V, $\vec{B}$ and $\vec{A}$ are all time varying (i.e., they are functions of x, y, z and t), the potentials become

$$V(x, y, z, t) = \frac{1}{4\pi\varepsilon_0} \int \frac{\rho(x', y', z', t')}{R} dv' \quad \textbf{(4.26a)}$$

and

$$\vec{A}(x, y, z, t) = \frac{\mu_0}{4\pi}\int \frac{\vec{J}(x', y', z', t')}{R} dv' \tag{4.26b}$$

Thus, if $\rho(x', y', z', t')$ is the charge density at (x', y', z') at time t', then Eq. (4.26*a*) and (4.26*b*) give the potentials at (x, y, z) at some slightly later time t; the time difference being,

$$(t - t') = \frac{R}{c}, \qquad \text{where, } c \text{ is the speed of light} = 3 \times 10^8 \text{ m/s.}$$

So, if the charge density at (x', y', z') changes, the information about this change cannot reach the field point instantaneously; it takes a time (R/c) for the information to be transmitted.

$\therefore$

$$V(t) = \frac{1}{4\pi\varepsilon_0}\int \frac{\rho(t - R/c)}{R} dv' \tag{4.27a}$$

and

$\therefore$

$$\vec{A}(t) = \frac{\mu_0}{4\pi}\int \frac{\vec{J}(t - R/c)}{R} dv' \tag{4.27b}$$

The potentials so calculated are called *retarded potentials*.

Example 4.24 An infinite straight wire carries a current given by

$$I = 0 \qquad t \le 0$$
$$= I_0 \qquad t > 0$$

as shown in Fig. 4.15. Find the retarded potentials at a distance r from the axis of the wire.

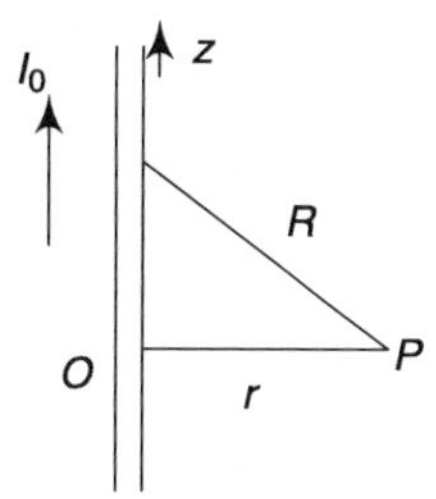

Fig. 4.15 *An infinite straight current-carrying wire*

Solution Since the wire is electrically neutral, the charge density, $\rho = 0$.

Hence, the scalar potential is zero, $\phi = 0$.

For simplicity, let the wire carries the current along the z-axis.

The retarded potential is given as

$$\vec{A}(\vec{r}, t) = \frac{\mu_0}{4\pi}\int_{-\infty}^{\infty} \frac{\vec{J}(t - R/c)}{R} dv' = \frac{\mu_0}{4\pi}\int_{-\infty}^{\infty} \frac{I(t - R/c)}{R} dz\hat{a}_z \qquad \{\because \vec{J}dv' = Id\vec{z}\}$$

It can be seen that a given current at a distance z from the origin (O) will contribute only if the distance from the observation point (P) is less than ct.

$\therefore$

$$I(t - R/c) = 0 \quad \text{when} \quad R^2 \ge c^2t^2 \quad \text{or,} \quad z^2 \ge (c^2t^2 - r^2)$$
$$= I_0 \quad \text{when} \quad R^2 < c^2t^2 \quad \text{or,} \quad z^2 < (c^2t^2 - r^2)$$

$\therefore$

$$\vec{A}(\vec{r}, t) = 0 \quad \text{when} \quad r \ge ct$$

and

$$\vec{A}(\vec{r}, t) = \frac{\mu_0 I_0}{4\pi}\int_{-\sqrt{c^2t^2 - r^2}}^{\sqrt{c^2t^2 - r^2}} \frac{dz}{\sqrt{z^2 + r^2}}\hat{a}_z = \frac{\mu_0 I_0}{2\pi}\int_{0}^{\sqrt{c^2t^2 - r^2}} \frac{dz}{\sqrt{z^2 + r^2}}\hat{a}_z$$

$$= \frac{\mu_0 I_0}{2\pi}\ln\left(\frac{\sqrt{c^2t^2 - r^2} + ct}{r}\right); \quad r < ct$$

The corresponding retarded electric field is given as

$$\vec{E}(\vec{r},t) = -\frac{\partial \vec{A}}{\partial t} = -\frac{\mu_0 I_0}{2\pi}\left(\frac{c}{\sqrt{c^2t^2 - r^2}}\right)\hat{a}_z$$

The corresponding retarded magnetic field is obtained as

$$\vec{B}(\vec{r},t) = \nabla \times \vec{A} = -\frac{\partial A_z}{\partial \phi}\hat{a}_\phi = \frac{\mu_0 I_0}{2\pi r}\left(\frac{ct}{\sqrt{c^2t^2 - r^2}}\right)\hat{a}_\phi$$

NB: As $t \to \infty$, from the results, we see that $\vec{E} = 0$ and $\vec{B} = \frac{\mu_0 I_0}{2\pi r}\hat{a}_\phi$, which are the results for static fields.

Summary

- Faraday's law states that the emf induced in a closed circuit is proportional to the rate of change of the magnetic flux-linkage and the direction of the current flow in the closed circuit is such that it opposes the change of the flux.

$$V = -\frac{d\phi}{dt}$$

- Different forms of Faraday's law are:

Differential Form: $$\nabla \times \vec{E} = -\frac{\partial \vec{B}}{\partial t}$$

Integral Form: $$\oint_C \vec{E} \cdot d\vec{l} = -\frac{d}{dt}\int_S \vec{B} \cdot d\vec{S}$$

- Induced emf for different cases are as follows:

Stationary loop in time varying magnetic field (transformer emf)

$$V_S = -\int_S \frac{\partial \vec{B}}{\partial t} \cdot d\vec{S}$$

Moving loop in static magnetic field (motional emf)

$$V_m = \oint_C \vec{E}_m \cdot d\vec{l} = \oint_C (\vec{v} \times \vec{B}) \cdot d\vec{l}$$

Moving loop in time varying magnetic field (general induction)

$$V = \underbrace{-\int_S \frac{\partial \vec{B}}{\partial t} \cdot d\vec{S}}_{\text{transformer emf}} + \underbrace{\oint_C (\vec{v} \times \vec{B}) \cdot d\vec{l}}_{\textit{motional emf}}$$

- The displacement current density is given as

$$\vec{J}_d = \frac{\partial \vec{D}}{\partial t}$$

- The modified Ampere's law for time-varying fields is given by considering displacement current density as

$$\nabla \times \vec{H} = \vec{J} + \frac{\partial \vec{D}}{\partial t}$$

- Maxwell's equations for time-varying fields are given as follows.

Differential Form	**Integral Form**
$\nabla \cdot \vec{D} = \rho_v$	$\oint_S \vec{D} \cdot d\vec{s} = \oint_v \rho_v dv = Q$
$\nabla \cdot \vec{B} = 0$	$\oint_S \vec{B} \cdot d\vec{s} = 0$
$\nabla \times \vec{E} = -\frac{\partial \vec{B}}{\partial t}$	$\oint_l \vec{E} \cdot d\vec{l} = -\frac{\partial}{\partial t} \int_S \vec{B} \cdot d\vec{s}$
$\nabla \times \vec{H} = \vec{J} + \frac{\partial \vec{D}}{\partial t}$	$\oint_l \vec{H} \cdot d\vec{l} = \int_S \left(\vec{J} + \frac{\partial \vec{D}}{\partial t} \right) \cdot d\vec{s}$

- Time-harmonic fields are those fields that vary sinusoidally with time. They are easily expressed in phasors.
- The relation between a time-harmonic field vector $\vec{A}(x, y, z, t)$ and its phasor form $\overline{A}(x, y, z)$ is given as

$$\vec{A} = \text{Re}[\overline{A} e^{j\omega t}]$$

- Maxwell's equations for time-harmonic fields are given as follows.

Differential Form	**Integral Form**
$\nabla \cdot \overline{D} = \rho_v$	$\oint_S \overline{D} \cdot d\overline{s} = \oint_v \rho_v dv = Q$
$\nabla \cdot \overline{B} = 0$	$\oint_S \overline{B} \cdot d\overline{s} = 0$
$\nabla \times \overline{E} = -j\omega \overline{B}$	$\oint_l \overline{E} \cdot d\overline{l} = -j\omega \int_S \overline{B} \cdot d\overline{s}$
$\nabla \times \overline{H} = \overline{J} + j\omega \overline{D}$	$\oint_l \overline{H} \cdot d\overline{l} = \int_S (\overline{J} + j\omega \overline{D}) \cdot d\overline{s}$

- Time-varying electric scalar potential $V(x, y, z, t)$ and magnetic vector potential $\vec{A}(x, y, z, t)$ satisfy the inhomogeneous wave equations $\left(\nabla^2 V - \mu\varepsilon \frac{\partial^2 V}{\partial t^2} = -\frac{\rho}{\varepsilon} \right)$ and $\left(\nabla^2 \vec{A} - \mu\varepsilon \frac{\partial^2 \vec{A}}{\partial t^2} = -\mu \vec{J} \right)$, provided Lorentz gauge condition $\left(\nabla \cdot \vec{A} = -\mu\varepsilon \frac{\partial V}{\partial t} \right)$ is assumed.

- The concept of retarded potential describes the scalar or vector potential for electromagnetic fields of a time-varying current or charge distribution. These retarded potentials are given as

$$V(t) = \frac{1}{4\pi\varepsilon_0}\int \frac{\rho(t - R/c)}{R} dv' \quad \text{and} \quad \vec{A}(t) = \frac{\mu_0}{4\pi}\int \frac{\vec{J}(t - R/c)}{R} dv'$$

where, R is the distance between the source point (x', y', z') and the field point (x, y, z) and c is the speed of light.

Important Formulae

Faraday's law	$V = -\frac{d\phi}{dt}$
Integral form of Faraday's law	$\oint_C \vec{E} \cdot d\vec{l} = -\frac{d}{dt}\int_S \vec{B} \cdot d\vec{S}$
Differential form of Faraday's law	$\nabla \times \vec{E} = -\frac{\partial \vec{B}}{\partial t}$
Transformer EMF	$V_S = -\int_S \frac{\partial \vec{B}}{\partial t} \cdot d\vec{S}$
Motional EMF	$V_m = \oint_C (\vec{v} \times \vec{B}) \cdot d\vec{l}$
EMF induced in Faraday disc generator	$V_m = \frac{1}{2}\omega B a^2$
Modified Ampere's circuital law	$\nabla \times \vec{H} = \vec{J} + \frac{\partial \vec{D}}{\partial t}$
Inhomogeneous wave equation	$\nabla^2 V - \mu\varepsilon \frac{\partial^2 V}{\partial t^2} = -\frac{\rho}{\varepsilon}$ and $\nabla^2 \vec{A} - \mu\varepsilon \frac{\partial^2 \vec{A}}{\partial t^2} = -\mu\vec{J}$

Exercises

[NOTE: * marked problems are important university problems]

- ***Easy***
  1. A conductor 1 cm in length is parallel to the z-axis and rotates at radius of 25 cm at 1200 r.p.m. Find the induced voltage, if the radial field is given by $\vec{B} = 0.5\vec{a}_r\ T$. [–157.08 mV]
  2. A square conducting loop with sides 25 cm long is located in a magnetic field of 1 A/m varying at a frequency of 5 MHz. The field is perpendicular to the plane of the loop. What voltage will be read on a voltmeter connected in series with one side of the loop? [2.54 V]
  3. A square loop of wire 25 cm has a voltmeter (of infinite impedance) connected in series with one side. Determine the voltage indicated by the meter when the loop is placed in an alternating field, the maximum intensity of which is 1 A/m. The plane of the loop is perpendicular to the magnetic field, the frequency is 10 MHz. [4.93 V]

4. A conducting cylinder of radius 7 cm and height 50 cm rotates at 600 r.p.m. in a radial magnetic field $\vec{B} = 0.10\hat{a}_r\, T$. Sliding contacts at the top and bottom are used to connect a voltmeter. Calculate the reading of the voltmeter. [–0.44 V]

***5.** A metal disc of radius a is rotated about its axis with a constant angular velocity ω in a uniform magnetic field $\vec{B}$ perpendicular to its plane. Calculate the potential difference between the rim and the centre of the disc. $\left[\frac{1}{2}\omega B a^2\right]$

***6.** A Faraday disc of 10 cm radius rotates at 500 r.p.m. in a uniform magnetic field of 250 mWb/m^2, the field being perpendicular to the plane of the disc. What is the potential difference between the rim and the axis assuming the diameter of the axis being very small? [0.651V]

- ***Medium***

7. A straight conductor of 0.2 m lies on the x-axis with one end at origin. The conductor is subjected to a magnetic flux density $\vec{B} = 0.04\hat{a}_y$ T and velocity $\vec{v} = 2.5 \sin 10^3 t\hat{a}_z$ m/s. Calculate the motional electric field intensity and emf induced in the conductor.
$[\vec{E} = -0.1 \sin 10^3 t\hat{a}_x$ (V/m); $-0.2 \sin 10^3 t\, V]$

8. A circular loop conductor lies in plane $z = 0$ and has a radius of 0.1 m and resistance of 5 Ω. Given: $\vec{B} = 0.2 \sin 10^3 t\hat{a}_z$ T. Determine the current in the loop. $\left[-1.26 \cos 10^3 t\hat{a}_x \text{ (A)}\right]$

9. Determine the emf developed about the path $r = 0.5$, $z = 0$ and at $t = 0$, if $\vec{B} = 0.1 \sin 377t\, \hat{a}_x$. [4.71 V]

- ***Hard***

10. A square loop of wire (side a) lies on a table near a very straight wire which carries a current I. as shown in the figure. Find the flux through the loop. If someone pulls the loop directly away from the wire at speed v, what emf is generated? $\left[\frac{\mu_0 a I}{2\pi}\ln\left(1+\frac{a}{d}\right), \frac{\mu_0 a^2 v I}{2\pi d(d+a)}\right]$

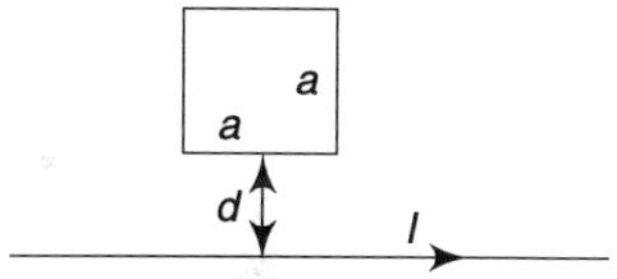

Fig. *Square loop wire and current carrying straight wire*

***11.** An electric vector $\vec{E}$ of an electromagnetic wave in free space is given by

$$E_x = E_z = 0 \quad \text{and} \quad Ey = Ae^{j\omega(t-z/c)}$$

Find the corresponding magnetic field components in free space.

$$\left[H_y = H_z = 0 \quad H_x = -\sqrt{\frac{\varepsilon_0}{\mu_0}}E_y\right]$$

Review Questions

[NOTE: * marked questions are important university questions.]

1. State and explain Faraday's law of electromagnetic induction.

***2.** From the fundamental principle, establish the relation (i) $\nabla \times \vec{E} = -\frac{\partial \vec{B}}{\partial t}$ and (ii) $\nabla \times \vec{H} = \vec{J} + \frac{\partial \vec{D}}{\partial t}$.

3. Show that the electric field $\vec{E}$ induced by a time-varying magnetic field $\vec{B}$ is given by the expression $\nabla \times \vec{E} = -\frac{\partial \vec{B}}{\partial t}$.

***4.** What are the limitations of Ampere's current law? How this law can be modified to time-varying field?

5. Explain the terms

(i) Motional emf

(ii) Static emf

***6.** Justify the statement 'Most electrical machines are working on electromagnetic principles rather than electrostatic principles'.

7. What do you mean by 'motional emf'? Find an expression for it.

8. (a) Write the Lorentz gauge condition, and hence derive the inhomogeneous wave equation for the scalar and vector potentials.

(b) What are retarded time and retarded potentials?

***9.** 'Ampere's law is bound to fail for non-steady currents'. Justify the statement. How did Maxwell remove this defect in Ampere's law?

Or

Explain how Maxwell modified Ampere's circuital law for steady currents.

Or

Show that Ampere's law for steady currents is not applicable for time-varying currents. Hence, explain the concept of displacement current and its intensity.

10. (a) Define scalar and vector potentials ϕ and $\vec{A}$ for an electromagnetic field, and show that under a gauge transformation of $\vec{A}$ and ϕ, the electromagnetic field equations are invariant.

(b) Show that under suitable conditions $\vec{A}$ and ϕ satisfy the inhomogeneous equations

$$\left(\nabla^2 - \frac{1}{c^2}\frac{\partial^2}{\partial t^2}\right)\vec{A} = -\mu_0\vec{J}; \qquad \left(\nabla^2 - \frac{1}{c^2}\frac{\partial^2}{\partial t^2}\right)\phi = -\frac{\rho}{\varepsilon_0}$$

where the symbols have their usual meanings.

11. (a) Write down the differential and integral forms of all Maxwell's equations for time-varying fields and discuss their physical significance. Rewrite them in an appropriate way for sinusoidal time variations.

Or

Write down Maxwell's electromagnetic field equations and explain the physical significance of each. Derive them for harmonically varying field.

(b) Write the integral or the large-scale forms of Maxwell's equations.

(c) Show how Maxwell's equations in free space imply local conservation of charge (continuity equation).

(d) Show that the Maxwell equations $\nabla \times \vec{E} = -\dfrac{\partial \vec{B}}{\partial t}$ and $\nabla \cdot \vec{B} = 0$ are compatible.

(e) State Maxwell's equation in differential form corresponding to Gauss' law for electric fields. Starting from the Maxwell's equation in differential form, obtain the Poisson's equation for the general situation in which the permittivity of the medium is not constant and is a function of position.

12. State Maxwell's equations in their general point form and derive their form for harmonically varying fields.

Multiple Choice Questions

1. Maxwell's equations in differential form from Ampere's law are obtained from
   (a) M. M. F. area (b) Electric potential area
   (c) Magnetic flux volume (d) Electric current area.
2. Maxwell's equations are not completely symmetrical because
   (a) isolated magnetic charges do not exist.
   (b) it is difficult to get curl of a vector in spherical coordinates.
   (c) $\nabla \cdot \mathbf{D}$ is always zero.
   (d) $\nabla \times \mathbf{H}$ does not exist in free space.
3. A circular disc of 1 m radius rotates at 150 radians per second anticlockwise on ϕ-r plane with flux density $\mathbf{B} = 10^{-3}\boldsymbol{a}_z$ tesla. The voltage induced between two stationary brushes connected at the centre and at the circumference will be
   (a) 0.15 V (b) 0.075 V (c) 1.0 V (d) 0.85 V
4. A conductor 4 m long lies along the y-axis with a current 10.0 A in the a_y direction, the force on the conductor if the field in the region be $\boldsymbol{B} = 0.05\boldsymbol{a}_x$ T is
   (a) $-2.0\boldsymbol{a}_z$ N (b) $2.0\boldsymbol{a}_z$ N (c) $2.0\boldsymbol{a}_y$ N (d) $-2.0\boldsymbol{a}_x$ N
5. A conducting rod of length L revolves about its mid-point at uniform angular speed ω in a uniform magnetic field B normal to its plane of revolution. The electric potential difference between the ends of the rod would be
   (a) zero (b) $B\omega^2 L/2$ (c) $-B\omega^2 L/2$ (d) $2B\omega^2 L$
6. Consider the following statements relating to circular disc rotating in a transverse magnetic field B Wb/m^2 as shown in the figure.

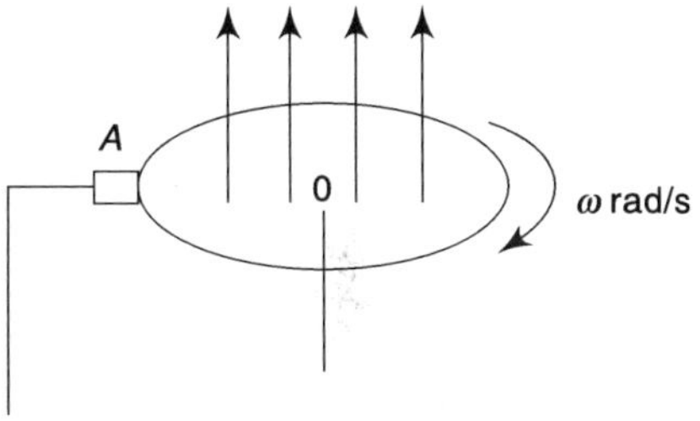

   The emf generated across outer rim A and the centre is proportional to
   1. angular velocity
   2. flux density
   3. square of the radius of the disc

   Of these statements
   (a) 1, 2 and 3 are correct. (b) 2 and 3 are correct.
   (c) 1 and 3 are correct. (d) 1 and 2 are correct.
7. A straight current-carrying conductor and two conducting loops A and B are shown in the figure. If the current in the straight wire is decreasing, then the induced currents in the two loops A and B will be

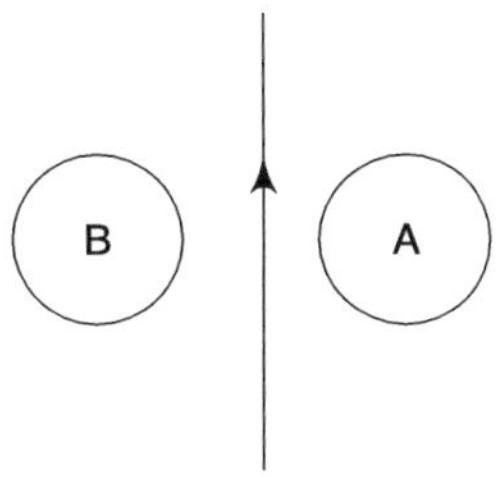

   (a) clockwise in both A and B.
   (b) anticlockwise in both A and B.
   (c) anticlockwise in A and clockwise in B.
   (d) clockwise in A and anticlockwise in B.
8. A circular loop is rotating about the y-axis as a diameter in a magnetic field $\vec{B} = B_0 \sin \omega t \hat{a}_x$ Wb/m^2. The induced emf in the loop is
   (a) due to transformer emf only
   (b) due to motional emf only
   (c) due to a combination of transformer and motional emf
   (d) zero.

9. Consider coils C_1, C_2, C_3, and C_4 (shown in the following figures) which are placed in time-varying field $\vec{E}(t)$ and electric field produced by the coils C_2', C_3', and C_4' carrying time-varying current $I(t)$ respectively:

1.

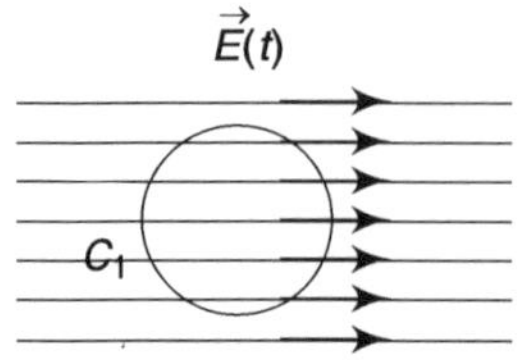

Time varying electric field $\vec{E}(t)$ parallel to the plane of the coil C_1

2.

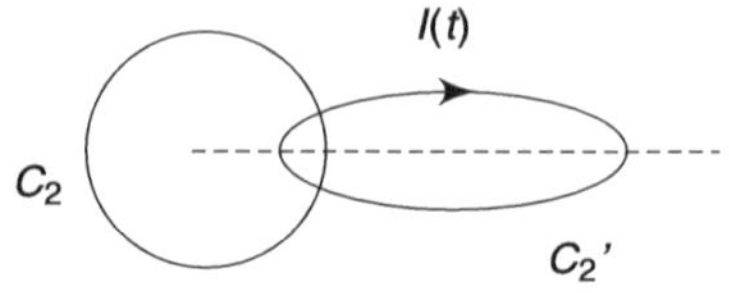

Coil planes are orthogonal

3.

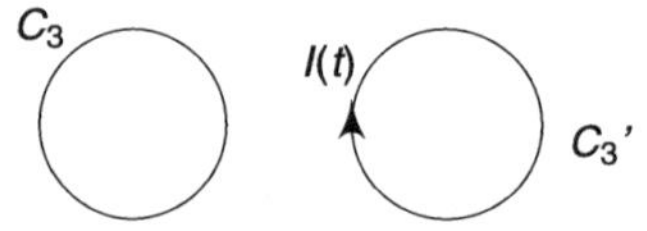

Co-planar coils

4.

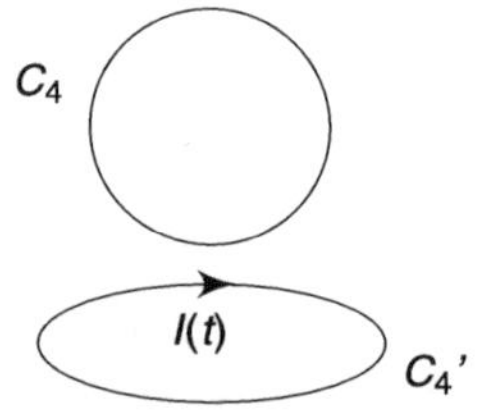

Coil planes are orthogonal

The electric field will induce an emf in the coils

(a) C_1 and C_2 (b) C_2 and C_3 (c) C_1 and C_3 (d) C_2 and C_4

10. Which of the following equations is/are not Maxwell's equation(s)?

1. $\nabla \cdot \vec{J} = -\dfrac{\partial \rho}{\partial t}$

2. $\nabla \cdot \vec{D} = \rho$

3. $\nabla \cdot \vec{E} = -\dfrac{\partial \vec{B}}{\partial t}$

4. $\oint \vec{H} \cdot d\vec{l} = \int_s \left(\sigma \vec{E} + \varepsilon \dfrac{\partial \vec{E}}{\partial t} \right) \cdot d\vec{s}$

Select the correct answer using the codes given below:

(a) 2 and 4 (b) 1 alone (c) 1 and 3 (d) 1 and 4

11. The laws of electromagnetic induction (Faraday's and Lenz's laws) are summarised in the following equations:

(a) $e = iR$

(b) $e = L\dfrac{di}{dt}$

(c) $e = -\dfrac{d\psi}{dt}$

(d) None of the above.

12. Which one of the following is not a Maxwell's equation for a static electromagnetic field in a linear homogeneous medium?

(a) $\nabla \cdot \vec{B} = 0$

(b) $\nabla^2 \vec{A} = \mu_0 \vec{J}$

(c) $\oint \vec{B} \cdot d\vec{l} = \mu_0 I$

(d) $\oint \vec{D} \cdot d\vec{S} = Q$

13. A loop is rotating about the y-axis in a magnetic field $\vec{B} = B_0 \sin \omega t \hat{a}_x$ Wb/m^2. The voltage induced in the loop is due to

(a) Motional emf

(b) Transformer emf

(c) A combination of motional and transformer emf

(d) None of these

14. Identify which of the following is not a Maxwell's equation for time-varying fields:

(a) $\nabla \cdot \vec{J} + \dfrac{\partial \rho_v}{\partial t} = 0$

(b) $\nabla \cdot \vec{D} = \rho_v$

(c) $\nabla \cdot \vec{E} = -\dfrac{\partial \vec{B}}{\partial t}$

(d) $\oint \vec{B} \cdot d\vec{S} = 0$

Answers

1. (a)	**2.** (a)	**3.** (b)	**4.** (a)	**5.** (a)	**6.** (a)	**7.** (d)	**8.** (c)
9. (c)	**10.** (c)	**11.** (c)	**12.** (b)	**13.** (c)	**14.** (c)		

5

PROPAGATION, REFLECTION, REFRACTION AND POLARISATION OF ELECTROMAGNETIC WAVES

Learning Objectives

This chapter deals with:

- *Basics of electromagnetic waves*
- *Concepts of propagation, polarisation, reflection and refraction of electromagnetic waves*

5.1 INTRODUCTION

In the previous chapter, we have studied Maxwell's equations where the existence of an electromagnetic wave has been mentioned. In fact, the existence of electromagnetic waves was first observed by Prof. Heinrich Hertz and thereafter has been established by several experiments and analysis.

In this chapter, we will discuss the different aspects of electromagnetic wave propagation, reflection, refraction and polarisation.

5.2 THREE-DIMENSIONAL WAVE EQUATIONS (HELMHOLTZ EQUATION)

Definition of Wave If a physical phenomenon that occurs at one place at a given time is reproduced at other places at later times, the time-delay being proportional to the space separation from the first location, then the group of phenomena constitutes a *wave*.

Maxwell's equations predict the propagation of electromagnetic energy away from time-varying sources (current and charge) in the form of waves.

Derivation of Wave Equation

Assumptions

1. We consider the medium to be a linear, homogeneous (i.e., quantities μ, ε, σ are constant throughout the medium) and isotropic (i.e., ε is a scalar constant so that $\vec{D}$ and $\vec{E}$ have everywhere the same direction).
2. We consider a source-free region of the medium.

By Maxwell's equations,

$$\nabla \times \vec{H} = \vec{J} + \frac{\partial \vec{D}}{\partial t} = \sigma \vec{E} + \varepsilon \frac{\partial \vec{E}}{\partial t} \tag{5.1}$$

Also, from Faraday's law,

$$\nabla \times \vec{E} = -\frac{\partial \vec{B}}{\partial t} = -\mu \frac{\partial \vec{H}}{\partial t}$$

Taking curl on both sides,

$$\nabla \times \nabla \times \vec{E} = -\mu \frac{\partial}{\partial t}(\nabla \times \vec{H})$$

$$\Rightarrow \quad \nabla(\nabla \cdot \vec{E}) - \nabla^2 \vec{E} = -\mu \frac{\partial}{\partial t}\left[\sigma \vec{E} + \varepsilon \frac{\partial \vec{E}}{\partial t}\right] \quad \text{(by Eq.(5.1))}$$

$$\Rightarrow \quad -\nabla^2 \vec{E} = -\mu\sigma \frac{\partial \vec{E}}{\partial t} - \mu\varepsilon \frac{\partial^2 \vec{E}}{\partial t^2} \quad \left(\because \ \nabla \cdot \vec{E} = \frac{1}{\varepsilon}\nabla \cdot \vec{D} = \frac{\rho}{\varepsilon} = 0; \text{as } \rho = 0\right)$$

Rearranging, we get,

$$\boxed{\nabla^2 \vec{E} - \mu\varepsilon \frac{\partial^2 \vec{E}}{\partial t^2} - \mu\sigma \frac{\partial \vec{E}}{\partial t} = 0} \tag{5.2}$$

This is the three-dimensional *vector wave equation* or *Helmholtz equation* in an absorbing medium or lossy dielectric medium.

Similarly, the equation in terms of magnetic field is obtained is follows.

From Eq. (5.1),

$$\nabla \times \vec{H} = \vec{J} + \frac{\partial \vec{D}}{\partial t} = \sigma \vec{E} + \varepsilon \frac{\partial \vec{E}}{\partial t}$$

Taking curl on both sides,

$$\nabla \times \nabla \times \vec{H} = \sigma \nabla \times \vec{E} + \varepsilon \frac{\partial}{\partial t}(\nabla \times \vec{E})$$

$$\Rightarrow \quad \nabla(\nabla \cdot \vec{H}) - \nabla^2 \vec{H} = -\mu\sigma \frac{\partial \vec{H}}{\partial t} + \varepsilon \frac{\partial}{\partial t}\left[-\mu \frac{\partial \vec{H}}{\partial t}\right] \quad \left(\because \ \nabla \times \vec{E} = -\frac{\partial \vec{B}}{\partial t} = -\mu \frac{\partial \vec{H}}{\partial t}\right)$$

$$\Rightarrow \quad -\nabla^2 \vec{H} = -\mu\sigma \frac{\partial \vec{H}}{\partial t} - \mu\varepsilon \frac{\partial^2 \vec{H}}{\partial t^2} \quad \left(\because \ \nabla \cdot \vec{H} = \frac{1}{\mu}\nabla \cdot \vec{B} = 0\right)$$

Rearranging, we get

$$\boxed{\nabla^2 \vec{H} - \mu\varepsilon \frac{\partial^2 \vec{H}}{\partial t^2} - \mu\sigma \frac{\partial \vec{H}}{\partial t} = 0} \tag{5.3}$$

This is the *Helmholtz equation* in terms of magnetic filed.

We will consider the modifications in the wave equations for different cases:

Wave Equation for Perfect Dielectric Medium In this case, the conductivity is zero; i.e. $\sigma = 0$. Hence the wave equations become

$$\boxed{\nabla^2 \vec{E} = \mu\varepsilon \frac{\partial^2 \vec{E}}{\partial t^2}} \quad \text{and} \quad \boxed{\nabla^2 \vec{H} = \mu\varepsilon \frac{\partial^2 \vec{H}}{\partial t^2}} \tag{5.4}$$

Wave Equation for Free Space In this case the wave equations take the forms as

$$\boxed{\nabla^2 \vec{E} = \mu_0\varepsilon_0 \frac{\partial^2 \vec{E}}{\partial t^2}} \quad \text{and} \quad \boxed{\nabla^2 \vec{H} = \mu_0\varepsilon_0 \frac{\partial^2 \vec{H}}{\partial t^2}} \tag{5.5}$$

Note that the quantity $\frac{1}{\sqrt{\mu\varepsilon}}$ has the dimension of velocity and in free space it becomes

$$\frac{1}{\sqrt{\mu_0\varepsilon_0}} = \frac{1}{\sqrt{4\pi \times 10^{-7} \times \dfrac{1}{36\pi \times 10^9}}} = 3 \times 10^8 \text{ m/s} = \text{velocity of light, } c$$

So, the electromagnetic wave propagates through free space with the velocity of light. In a homogeneous medium of permittivity ε and permeability μ, the wave propagates with a velocity, $v = \frac{1}{\sqrt{\mu\varepsilon}}$

Wave Equation for Time-Harmonic Fields For time-harmonic fields, the instantaneous (time-domain) vector $\vec{F}$ is related to the phasor (frequency-domain) vector $\boldsymbol{F}_s$ by

$$\vec{F} \Leftrightarrow \boldsymbol{F}_S \qquad \frac{\partial \vec{F}}{\partial t} \Leftrightarrow j\omega \boldsymbol{F}_s \qquad \frac{\partial^2 \vec{F}}{\partial t^2} \Leftrightarrow (j\omega)^2 \boldsymbol{F}_S$$

Using these relationships, the instantaneous vector wave equations are transformed into the phasor vector wave equations:

$$\nabla^2 \boldsymbol{E}_S = \mu\sigma(j\omega)\boldsymbol{E}_S + \mu\varepsilon(j\omega)^2 \boldsymbol{E}_S = j\omega\mu(\sigma + j\omega\varepsilon)\boldsymbol{E}_S$$

and

$$\nabla^2 \boldsymbol{H}_S = \mu\sigma(j\omega)\boldsymbol{H}_S + \mu\varepsilon(j\omega)^2 \boldsymbol{H}_S = j\omega\mu(\sigma + j\omega\varepsilon)\boldsymbol{H}_S$$

If we let, $j\omega\mu(\sigma + j\omega\varepsilon) = (j\omega\mu\sigma - \omega^2\mu\varepsilon) = \gamma^2$, then the wave equations become

$$\boxed{\nabla^2 \boldsymbol{E}_S - \gamma^2 \boldsymbol{E}_S = 0} \quad \text{and} \quad \boxed{\nabla^2 \boldsymbol{H}_S - \gamma^2 \boldsymbol{H}_S = 0} \tag{5.6}$$

Propagation constant (γ): The complex constant γ is defined as the *propagation constant*.

$$\gamma = \sqrt{j\omega\mu(\sigma + j\omega\varepsilon)} = (\alpha + j\beta) \tag{5.7}$$

Attenuation constant (α): The real part of the propagation constant is defined as the *attenuation constant* (α). The attenuation constant defines the rate at which the fields of the wave are attenuated as the wave propagates. Its unit is Neper per metre.

Phase constant (β): The imaginary part of the propagation constant is defined as the *phase constant* (β). The phase constant defines the rate at which the phase changes as the wave propagates. Its unit is radian per metre.

γ—propagation contstant (m^{-1})
α—attenuation constant (Neper/m)
β—phase constant (radian/m)

From Eq. (5.7),

$$\alpha^2 - \beta^2 = -\omega^2 \mu\varepsilon \quad \textbf{(5.8a)}$$

and

$$2\alpha\beta = \omega\mu\sigma \quad \textbf{(5.8b)}$$

$$\therefore \quad \alpha^2 + \beta^2 = \sqrt{(\alpha^2 - \beta^2)^2 + (2\alpha\beta)^2} = \omega\mu\sqrt{\omega^2\varepsilon^2 + \sigma^2} \quad \textbf{(5.8c)}$$

Solving Eq. (5.8*a*) and (5.8*c*), we get

$$\alpha = \sqrt{\frac{\omega\mu\sqrt{\omega^2\varepsilon^2+\sigma^2} - \omega^2\mu\varepsilon}{2}} = \omega\sqrt{\frac{\mu\sqrt{\varepsilon^2+\dfrac{\sigma^2}{\omega^2}} - \mu\varepsilon}{2}} = \omega\sqrt{\frac{\mu\varepsilon}{2}\left[\sqrt{1+\frac{\sigma^2}{\omega^2\varepsilon^2}} - 1\right]}$$

and

$$\beta = \sqrt{\frac{\omega\mu\sqrt{\omega^2\varepsilon^2+\sigma^2} + \omega^2\mu\varepsilon}{2}} = \omega\sqrt{\frac{\mu\sqrt{\varepsilon^2+\dfrac{\sigma^2}{\omega^2}} + \mu\varepsilon}{2}} = \omega\sqrt{\frac{\mu\varepsilon}{2}\left[\sqrt{1+\frac{\sigma^2}{\omega^2\varepsilon^2}} + 1\right]}$$

$$\boxed{\alpha = \omega\sqrt{\frac{\mu\varepsilon}{2}\left[\sqrt{1+\left(\frac{\sigma}{\omega\varepsilon}\right)^2} - 1\right]}} \quad \text{and} \quad \boxed{\beta = \omega\sqrt{\frac{\mu\varepsilon}{2}\left[\sqrt{1+\left(\frac{\sigma}{\omega\varepsilon}\right)^2} + 1\right]}} \quad \textbf{(5.9)}$$

Example 5.1 Prove that the one-dimensional wave equation for an electromagnetic wave propagating in the +*z*-direction is given by

$$\frac{\partial^2 E_x}{\partial z^2} = \frac{1}{v^2}\frac{\partial^2 E_x}{\partial t^2}$$

where *v* is the speed of propagation.

Solution We will consider a wave propagating in the +*z*-direction. The electric field component is $E_x(z, t)$.

Let, $z' = (z \pm vt)$ $\quad \therefore \quad \dfrac{\partial z'}{\partial z} = 1$ and $\dfrac{\partial z'}{\partial t} = \pm v$

Using the chain rule, the first two partial derivatives with respect to *z* are

$$\frac{\partial E_x(z')}{\partial z} = \frac{\partial E_x}{\partial z'}\frac{\partial z'}{\partial z} = \frac{\partial E_x}{\partial z'}$$

$$\frac{\partial^2 E_x}{\partial z^2} = \frac{\partial}{\partial z}\left(\frac{\partial E_x}{\partial z'}\right) = \frac{\partial^2 E_x}{\partial z'^2}\frac{\partial z'}{\partial z} = \frac{\partial^2 E_x}{\partial z'^2} \quad \textbf{(i)}$$

Similarly, the partial derivatives with respect to t are given by

$$\frac{\partial E_x}{\partial t} = \frac{\partial E_x}{\partial z'}\frac{\partial z'}{\partial t} = \pm v \frac{\partial E_x}{\partial z'}$$

$$\frac{\partial^2 E_x}{\partial t^2} = \frac{\partial}{\partial t}\left(\pm v \frac{\partial E_x}{\partial z'}\right) = \pm v \frac{\partial^2 E_x}{\partial z'^2}\frac{\partial z'}{\partial t} = v^2 \frac{\partial^2 E_x}{\partial z'^2} \qquad \textbf{(ii)}$$

Comparing (*i*) and (*ii*), we get

$$\boxed{\frac{\partial^2 E_x}{\partial z^2} = \frac{1}{v^2}\frac{\partial^2 E_x}{\partial t^2}} \qquad \textbf{(5.10)}$$

This is the *one-dimensional wave equation.*

NOTE

The wave equation is an example of a linear differential equation, which means that if $E_{x1}(z, t)$ and $E_{x2}(z, t)$ are the two solutions to the wave equation, then $[E_{x1}(z, t) \pm E_{x2}(z, t)]$ is also a solution. The implication is that electromagnetic waves obey the superposition principle.

Example 5.2 Verify that $E_y = f_1(x - v_0 t)$ is a solution of $\dfrac{\partial^2 E_y}{\partial x^2} = \mu\varepsilon \dfrac{\partial^2 E_y}{\partial t^2}$.

Solution Given: $E_y = f_1(x - v_0 t)$

$$\therefore \quad \frac{\partial E_y}{\partial x} = \frac{\partial f_1(x - v_0 t)}{\partial x}; \qquad \frac{\partial^2 E_y}{\partial x^2} = \frac{\partial^2 f_1(x - v_0 t)}{\partial x^2} \qquad \textbf{(i)}$$

$$\therefore \quad \frac{\partial E_y}{\partial t} = -v_0 \frac{\partial f_1(x - v_0 t)}{\partial t}; \qquad \frac{\partial^2 E_y}{\partial t^2} = v_0^2 \frac{\partial^2 f_1(x - v_0 t)}{\partial t^2} \qquad \textbf{(ii)}$$

From (*i*) and (*ii*), we get,

$$\frac{\partial^2 E_y}{\partial t^2} = v_0^2 \frac{\partial^2 E_y}{\partial x^2} = \frac{1}{\mu\varepsilon}\frac{\partial^2 E_y}{\partial x^2}$$

$$\therefore \quad \boxed{\frac{\partial^2 E_y}{\partial x^2} = \mu\varepsilon \frac{\partial^2 E_y}{\partial t^2}}$$

Example 5.3 Show that the function $F = e^{-\alpha z} \sin\left\{\frac{\omega}{v}(x - vt)\right\}$ satisfies the wave equation $\nabla^2 F = \left(\frac{1}{c^2}\right)\ddot{F}$ provided that the wave velocity is given by, $v = c\left(1 + \frac{\alpha^2 c^2}{\omega^2}\right)^{-1/2}$.

Solution Given, $F = e^{-\alpha z} \sin\left\{\frac{\omega}{v}(x - vt)\right\}$

$$\therefore \quad \nabla^2 F = \left(\frac{\partial^2 F}{\partial x^2} + \frac{\partial^2 F}{\partial y^2} + \frac{\partial^2 F}{\partial z^2}\right)$$

$$= -e^{-\alpha z}\left(\frac{\omega}{v}\right)^2 \sin\left\{\frac{\omega}{v}(x - vt)\right\} + 0 + (-\alpha)^2 e^{-\alpha z} \sin\left\{\frac{\omega}{v}(x - vt)\right\}$$

$$= \left[-\frac{\omega^2}{v^2} + \alpha^2\right] F$$

According to the given condition

$$\nabla^2 F = \left(\frac{1}{c^2}\right)\ddot{F} = \left(\frac{1}{c^2}\right)\frac{\partial^2 F}{\partial t^2}$$

or, $$\left[-\frac{\omega^2}{v^2} + \alpha^2\right] F = \left(\frac{1}{c^2}\right)(-\omega)^2 e^{-\alpha z} \sin\left\{\frac{\omega}{v}(x - vt)\right\} = -\frac{\omega^2}{c^2} F$$

or, $$\left(\alpha^2 - \frac{\omega^2}{v^2}\right) = -\frac{\omega^2}{c^2}$$

or, $$v^2 = \frac{\omega^2 c^2}{\omega^2 + \alpha^2 c^2} = \frac{c^2}{1 + \frac{\alpha^2 c^2}{\omega^2}} = c^2\left[1 + \frac{\alpha^2 c^2}{\omega^2}\right]^{-1}$$

$$\therefore \quad v = c\left[1 + \frac{\alpha^2 c^2}{\omega^2}\right]^{-1/2}$$

5.3 PROPERTIES OF ELECTROMAGNETIC WAVES

In a source-free region, the phasor vector wave equations are

$\nabla^2 \boldsymbol{E}_S - \gamma^2 \boldsymbol{E}_S = 0$ and $\nabla^2 \boldsymbol{H}_S - \gamma^2 \boldsymbol{H}_S = 0$

The operator in the above equations (∇^2) is the *vector Laplacian* operator. In rectangular coordinates, the vector Laplacian operator is given as below.

$$\vec{F} = F_x \hat{a}_x + F_y \hat{a}_y + F_z \hat{a}_z$$

$$\nabla^2 \vec{F} = (\nabla^2 F_x)\hat{a}_x + (\nabla^2 F_y)\hat{a}_y + (\nabla^2 F_z)\hat{a}_z$$

The phasor wave equations can then be written as

$$(\nabla^2 E_{xs})\hat{a}_x + (\nabla^2 E_{ys})\hat{a}_y + (\nabla^2 E_{zs})\hat{a}_z = \gamma^2 (E_{xs}\hat{a}_x + E_{ys}\hat{a}_y + E_{zs}\hat{a}_z)$$
$$(\nabla^2 H_{xs})\hat{a}_x + (\nabla^2 H_{ys})\hat{a}_y + (\nabla^2 H_{zs})\hat{a}_z = \gamma^2 (H_{xs}\hat{a}_x + H_{ys}\hat{a}_y + H_{zs}\hat{a}_z)$$

By equating the vector components on both sides of each phasor wave equation, individual wave equations for the phasor field components [(E_{xs}, E_{ys}, E_{zs}) and (H_{xs}, H_{ys}, H_{zs})] can be obtained.

$$\frac{\partial^2 E_{xs}}{\partial x^2} + \frac{\partial^2 E_{xs}}{\partial y^2} + \frac{\partial^2 E_{xs}}{\partial z^2} = \gamma^2 E_{xs}$$

$$\frac{\partial^2 E_{ys}}{\partial x^2} + \frac{\partial^2 E_{ys}}{\partial y^2} + \frac{\partial^2 E_{ys}}{\partial z^2} = \gamma^2 E_{ys}$$

$$\frac{\partial^2 E_{zs}}{\partial x^2} + \frac{\partial^2 E_{zs}}{\partial y^2} + \frac{\partial^2 E_{zs}}{\partial z^2} = \gamma^2 E_{zs}$$

$$\frac{\partial^2 H_{xs}}{\partial x^2} + \frac{\partial^2 H_{xs}}{\partial y^2} + \frac{\partial^2 H_{xs}}{\partial z^2} = \gamma^2 H_{xs}$$

$$\frac{\partial^2 H_{ys}}{\partial x^2} + \frac{\partial^2 H_{ys}}{\partial y^2} + \frac{\partial^2 H_{ys}}{\partial z^2} = \gamma^2 H_{ys}$$

$$\frac{\partial^2 H_{zs}}{\partial x^2} + \frac{\partial^2 H_{zs}}{\partial y^2} + \frac{\partial^2 H_{zs}}{\partial z^2} = \gamma^2 H_{zs}$$

The component fields of any time-harmonic electromagnetic wave (described in rectangular coordinates) must individually satisfy these six partial differential equations. In many cases, the electromagnetic wave will not contain all six components. An example of this is the *plane wave*, discussed in the next section.

5.3.1 Plane Waves

A wave is said to be a *plane wave*, if:

1. The electric field $\vec{E}$ and magnetic field $\vec{H}$ lie in a plane perpendicular to the direction of wave propagation.
2. The fields $\vec{E}$ and $\vec{H}$ are perpendicular to each other.

5.3.2 Uniform Plane Waves

A plane wave is said to be *uniform plane wave* if in addition to condition (1) and (2) above,

3. $\vec{E}$ and $\vec{H}$ are uniform in the plane perpendicular to the direction of propagation (i.e. $\vec{E}$ and $\vec{H}$ vary only in the direction of propagation).

Figure 5.1 shows the propagation of a uniform plane wave.

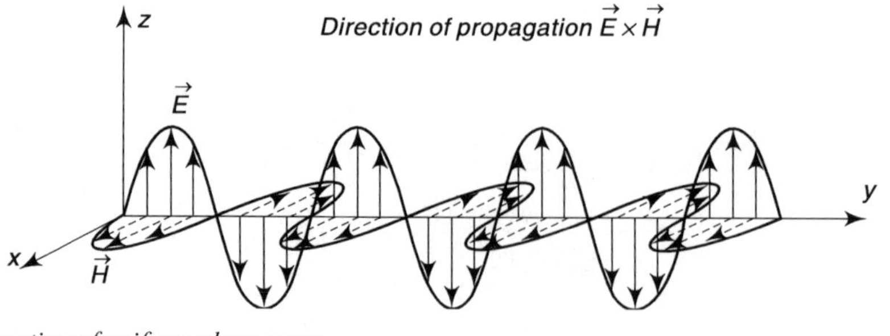

Fig. 5.1 *Propagation of uniform plane wave*

5.3.3 Transverse Electromagnetic (TEM) Wave

An electromagnetic wave which has no electric or magnetic field components in the direction of propagation (all components of $\boldsymbol{E}$ and $\boldsymbol{H}$ are perpendicular to the direction of propagation) is called a *transverse electromagnetic (TEM) wave*. All plane waves are TEM waves.

5.3.4 Solution of Wave Equation for Uniform Plane Wave

We consider the propagation of a uniform time-harmonic plane wave as shown in Fig. 5.2.

The uniform plane wave for this example has only a z-component of electric field and an x-component of magnetic field, which are both functions of only y. The *polarisation* of a plane wave is defined by the direction of the electric field (this example is a z-polarised plane wave). For this uniform plane wave, the component wave equations for the only two field components ($\boldsymbol{E}_{zs}$, $\boldsymbol{H}_{xs}$) can be simplified significantly given the field dependence on y only.

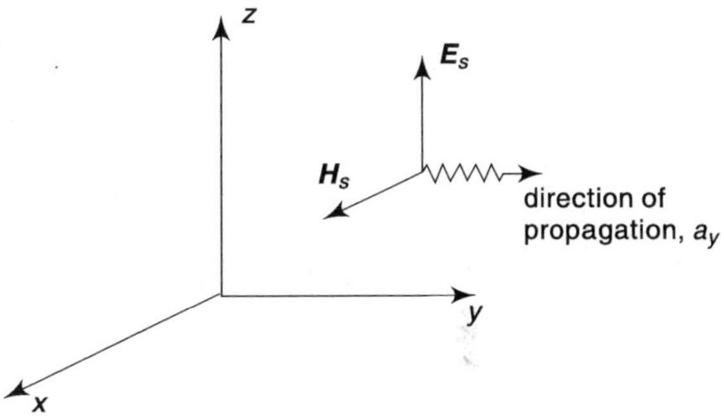

Fig. 5.2 *Uniform plane wave*

$$\boldsymbol{E}_S = E_{ZS}(y)\hat{a}_z$$
$$\boldsymbol{H}_S = H_{XS}(y)\hat{a}_x$$

$$\frac{\partial^2 E_{ZS}}{\partial x^2} + \frac{\partial^2 E_{ZS}}{\partial y^2} + \frac{\partial^2 E_{ZS}}{\partial z^2} = \gamma^2 E_{ZS}$$
$$\frac{\partial^2 H_{XS}}{\partial x^2} + \frac{\partial^2 H_{XS}}{\partial y^2} + \frac{\partial^2 H_{XS}}{\partial z^2} = \gamma^2 H_{XS}$$

The remaining single partial derivative in each component wave equation becomes a pure derivative since $\boldsymbol{E}_{zs}$ and $\boldsymbol{H}_{xs}$ are functions of y only.

$$\frac{d^2 E_{ZS}}{\partial y^2} - \gamma^2 E_{ZS} = 0 \tag{5.11a}$$

$$\frac{\partial^2 H_{XS}}{\partial y^2} - \gamma^2 H_{XS} = 0 \tag{5.11b}$$

The general solutions to the reduced wave equations are

$$\begin{aligned} E_{ZS}(y) &= E_1 e^{\gamma y} + E_2 e^{-\gamma y} \\ &= E_1 e^{(\alpha + j\beta)y} + E_2 e^{-(\alpha + j\beta)y} \\ &= E_1 e^{\alpha y} e^{j\beta y} + E_2 e^{-\alpha y} e^{-j\beta y} \end{aligned}$$

$$\begin{aligned} H_{XS}(y) &= H_1 e^{\gamma y} + H_2 e^{-\gamma y} \\ &= H_1 e^{(\alpha + j\beta)y} + H_2 e^{-(\alpha + j\beta)y} \\ &= H_1 e^{\alpha y} e^{j\beta y} + H_2 e^{-\alpha y} e^{-j\beta y} \end{aligned}$$

where (E_1, E_2) are constants (electric field amplitudes) and (H_1, H_2) are constants (magnetic field amplitudes).

The characteristics of the waves defined by the general field solutions above can be determined by investigating the corresponding instantaneous fields. We may focus on either the electric field or the magnetic field since they both have the same wave characteristics (they both satisfy the same differential equation).

$$\begin{aligned} E_z(y,t) &= \text{Re}\{E_{ZS}(y)e^{j\omega t}\} \\ &= \text{Re}\{(E_1 e^{\alpha y} e^{j\beta y} + E_2 e^{-\alpha y} e^{-j\beta y})e^{j\omega t}\} \\ &= \text{Re}\{E_1 e^{\alpha y} e^{j(\omega t + \beta y)} + E_2 e^{-\alpha y} e^{j(\omega t - \beta y)}\} \\ &= E_1 e^{\alpha y}\cos(\omega t + \beta y) + E_2 e^{-\alpha y}\cos(\omega t - \beta y) \end{aligned}$$

$$\therefore \quad E_z(y,t) = E_1 e^{\alpha y}\cos(\omega t + \beta y) \quad + \quad E_2 e^{-\alpha y}\cos(\omega t - \beta y)$$

$$\begin{pmatrix} Amplitude = E_1 e^{\alpha y} \\ Phase = \omega t + \beta y \\ \text{grows in } +\hat{a}_y \text{direction} \\ \text{decays in } -\hat{a}_y \text{direction} \end{pmatrix} \qquad \begin{pmatrix} Amplitude = E_2 e^{-\alpha y} \\ Phase = \omega t - \beta y \\ \text{decays in } +\hat{a}_y \text{direction} \\ \text{grows in } -\hat{a}_y \text{direction} \end{pmatrix}$$

$$\begin{matrix} \omega t + \beta y = \text{constant} \\ \beta y = \text{constant} - \omega t \\ (-\hat{a}_y \text{ travelling wave}) \end{matrix} \qquad \left.\begin{matrix} \omega t - \beta y = \text{constant} \\ \beta y = \omega t - \text{constant} \\ (+\hat{a}_y \text{ travelling wave}) \end{matrix}\right\} \Leftarrow \begin{pmatrix} \text{To locate a} \\ \text{point of} \\ \text{constant phase} \\ \text{on the wave} \end{pmatrix}$$

The velocity at which this point of constant phase moves is the *velocity of propagation* for the wave. Solving for the position variable y in the equations defining the point of constant phase gives

$$y = \pm\frac{1}{\beta}(\omega t - \text{constant}) \qquad (\pm\hat{a}_y \text{ travelling wave})$$

Given the y-coordinate of the constant phase point as a function of time, the vector velocity $\vec{v}$ at which the constant phase point moves is found by differentiating the position function with respect to time.

$$\vec{v} = \frac{dy}{dt}\hat{a}_y = \pm\frac{\omega}{\beta}\hat{a}_y \qquad (\pm\hat{a}_y \text{ travelling wave})$$

$$\therefore \quad \text{Velocity of wave propagation, } \boxed{v = \frac{\omega}{\beta}} \text{ (m/s)} \tag{5.12}$$

Now, $\omega = 2\pi f = \dfrac{2\pi}{T}$

Given a wave travelling at a velocity v, the wave travels one wavelength (λ) during one period (T).

$$\lambda = vT = \frac{v}{f} = \frac{\omega/\beta}{f} = \frac{2\pi}{\beta}$$

$$\therefore \qquad \boxed{\beta = \frac{2\pi}{\lambda}} \tag{5.13}$$

Note that for the field to be finite at infinity, it requires, $E_1 = 0$. Hence the electric field can be written as

$$E(y,t) = E_2 e^{-\alpha y}\cos(\omega t - \beta y)\hat{a}_z = E_0 e^{-\alpha y}\cos(\omega t - \beta y)\hat{a}_z$$

(say, $E_2 = E_0$)

Assuming a $+\boldsymbol{a}_y$ travelling uniform plane wave defined by an electric field of $\boldsymbol{E}_S = E_{ZS}\hat{a}_z = E_0 e^{-\gamma y}\hat{a}_z$ the corresponding magnetic field is found from the source-free Maxwell's equations.

$$\nabla \times \boldsymbol{E}_S = -j\omega\mu \boldsymbol{H}_S$$

$$\therefore \quad \boldsymbol{H}_S = -\frac{1}{j\omega\mu}\nabla \times \boldsymbol{E}_S = -\frac{1}{j\omega\mu}\left[\frac{\partial E_{ZS}}{\partial y}\hat{a}_x - \frac{\partial E_{ZS}}{\partial x}\hat{a}_y\right]$$

$$= -\frac{1}{j\omega\mu}\left[\frac{\partial}{\partial y}(E_0 e^{-\gamma y})\hat{a}_x\right]$$

$$= -\frac{1}{j\omega\mu}[-\gamma(E_0 e^{-\gamma y})\hat{a}_x]$$

$$= \frac{\gamma}{j\omega\mu}E_0 e^{-\gamma y}\hat{a}_x$$

$$= H_{XS}\hat{a}_x$$

Note that the direction of propagation for this wave is in the same direction as $\boldsymbol{E} \times \boldsymbol{H}$ $(\boldsymbol{a}_z \times \boldsymbol{a}_x = \boldsymbol{a}_y)$. This characteristic is true for all plane waves.

Example 5.4 The electric field in free space is given by,

$$\vec{E} = 50\cos(10^8 t + \beta x)\hat{a}_y \text{ V/m}$$

1. Find the direction of wave propagation.
2. Calculate β and the time it takes to travel a distance $\lambda/2$.
3. Sketch the wave at $t = 0$, $T/4$, $T/2$.

Solution:

1. From the positive sign in $(\omega t + \beta x)$, it is concluded that the wave is propagating in $-\hat{a}_x$ direction.
2. Here, $\beta = \dfrac{\omega}{c} = \dfrac{10^8}{3 \times 10^8} = \dfrac{1}{3} = 0.33$ rad/s

   As the wave is travelling at the speed of light, c

   $$\therefore \quad \frac{\lambda}{2} = ct_1 \quad \Rightarrow \quad t_1 = \frac{\lambda}{2c}$$

   But, $$\lambda = \frac{2\pi}{\beta} = 6\pi$$

   $$\therefore \quad t_1 = \frac{6\pi}{2c} = \frac{3\pi}{3 \times 10^8} = 31.42\,ns$$

3. At $t = 0$, $E_y = 50 \cos \beta x$

   At $t = T/4, E_y = 50\cos\left(\dfrac{\omega T}{4} + \beta x\right) = -50 \sin \beta x \quad \{\because \omega T = 2\pi\}$

   At $t = T/2, E_y = 50 \cos\left(\dfrac{\omega T}{2} + \beta x\right) = -50 \cos \beta x$

   The wave is plotted at $t = 0$, $T/4$, $T/2$ and shown in Fig. 5.3.

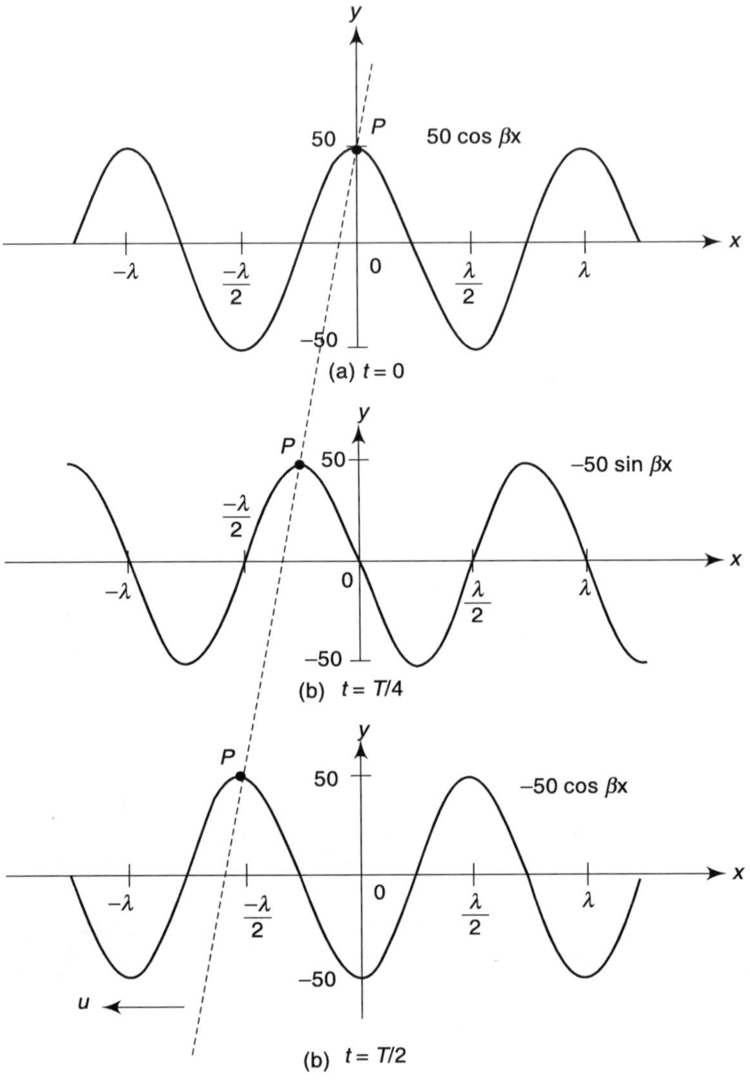

Fig. 5.3 *Wave of Example 5.4*

5.4 STANDING ELECTROMAGNETIC WAVES

We shall consider the situation where there are two sinusoidal plane electromagnetic waves, one travelling in the +x-direction, with the electric and magnetic fields given as

$$E_{1y}(x,t) = E_{10}\cos(\omega_1 t - \beta_1 x), \qquad B_{1z}(x,t) = B_{10}\cos(\omega_1 t - \beta_1 x)$$

and the other travelling in the –x-direction, with the electric and magnetic fields given as

$$E_{2y}(x,t) = -E_{20}\cos(\omega_2 t + \beta_2 x), \qquad B_{2z}(x,t) = B_{20}\cos(\omega_2 t + \beta_2 x)$$

For simplicity, we assume that these two electromagnetic waves have the same amplitudes i.e. $E_{10} = E_{20} = E_0$, $B_{10} = B_{20} = B_0$ and the same wavelengths, i.e., $\beta_1 = \beta_2 = \beta$, $\omega_1 = \omega_2 = \omega$. Using the superposition principle, the electric and the magnetic fields can be written as

$$E_y(x, t) = E_{1y}(x, t) + E_{2y}(x, t) = E_0[\cos(\omega t - \beta x) - \cos(\omega t + \beta x)]$$

and

$$B_z(x, t) = B_{1z}(x, t) + B_{2z}(x, t) = B_0[\cos(\omega t - \beta x) + \cos(\omega t + \beta x)]$$

Using the identities

$$\cos(\alpha \pm \beta) = \cos\alpha \cos\beta \mp \sin\alpha \sin\beta$$

The above expressions may be written as

$$E_y(x, t) = E_0[\cos\omega t \cos\beta x + \sin\omega t \sin\beta x - \cos\omega t \cos\beta x + \sin\omega t \sin\beta x] = 2E_0 \sin\omega t \sin\beta x \tag{5.14}$$

and

$$B_z(x, t) = B_0[\cos\omega t \cos\beta x + \sin\omega t \sin\beta x + \cos\omega t \cos\beta x - \sin\omega t \sin\beta x] = 2B_0 \cos\omega t \cos\beta x \tag{5.15}$$

This is observed that the total fields $E_y(x, t)$ and $B_z(x, t)$ still satisfy the wave equation $\frac{\partial^2 E_x}{\partial z^2} = \frac{1}{v^2}\frac{\partial^2 E_x}{\partial t^2}$, even though they no longer have the form of functions of $(\omega t \pm \beta x)$. The waves described by Eqs. (5.14) and (5.15) are *standing waves*, which do not propagate, but simply oscillate in space and time.

Spatial Dependence of Standing Wave The spatial dependence of the fields can be observed form Eqs. (5.14) and (5.15). From Eq. (5.14), we see that the total electric field is zero at all times if

$$\sin\beta x = 0$$

Or

$$x = \frac{n\pi}{\beta} = \frac{n\pi}{2\pi/\lambda} = \frac{n\lambda}{2}, \quad n = 0, 1, 2, \ldots$$

The planes that contain these points are called the *nodal planes* of the electric field.

On the other hand, when

$$\sin\beta x = \pm 1 \text{ or, } x = \left(n + \frac{1}{2}\right)\frac{\pi}{\beta} = \left(n + \frac{1}{2}\right)\frac{\pi}{2\pi/\lambda} = \left(\frac{n}{2} + \frac{1}{4}\right)\lambda, \quad n = 0, 1, 2, \ldots$$

The amplitude of the field is at its maximum $2E_0$. The planes that contain these points are called the *anti-nodal planes* of the electric field. It is to be noted that in between two nodal planes, there is an anti-nodal plane, and vice versa.

Similarly, for the magnetic field, the nodal points must contain points which satisfy the condition

$$\cos\beta x = 0$$

Or

$$x = \left(n + \frac{1}{2}\right)\frac{\pi}{\beta} = \left(n + \frac{1}{2}\right)\frac{\pi}{2\pi/\lambda} = \left(\frac{n}{2} + \frac{1}{4}\right)\lambda, \quad n = 0, 1, 2, \ldots$$

and the anti-nodal points for the magnetic field must contain the points which satisfy the condition,

$$\cos \beta x = \pm 1$$

Or

$$x = \frac{n\pi}{\beta} = \frac{n\pi}{2\pi/\lambda} = \frac{n\lambda}{2}, \quad n = 0, 1, 2, \ldots$$

Thus, we see that the nodal planes of $\vec{E}$ corresponds to the anti-nodal planes of $\vec{B}$ and vice-versa.

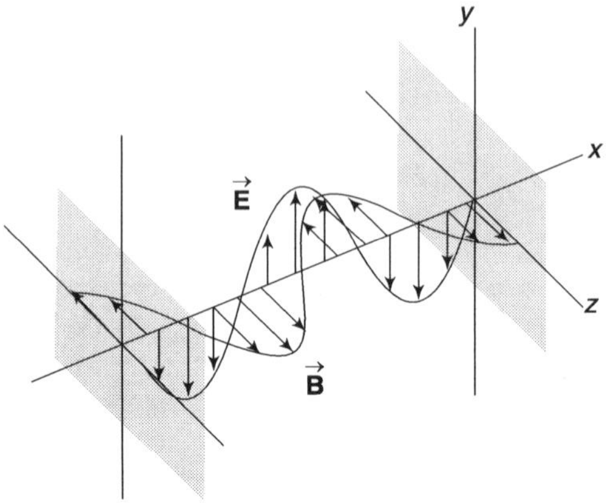

Fig. 5.4 *Formation of standing electromagnetic waves using two perfectly reflecting conductors*

Time Dependence of Standing Wave

Regarding the time-dependence, Eq. (5.15) shows that the electric field is zero everywhere when

$$\sin \omega t = 0$$

Or

$$t = \frac{n\pi}{\omega} = \frac{n\pi}{2\pi/T} = \frac{nT}{2}, \quad n = 0, 1, 2, \ldots$$

where, $T = 1/f = 2\pi/\omega$ is the period. However, this is precisely the maximum condition for the magnetic field.

NOTE

1. *Unlike the travelling electromagnetic wave in which the electric and the magnetic fields are always in phase, in standing waves, the two fields are 90° out of phase.*
2. *Standing waves can be formed by confining the electromagnetic waves within two perfectly reflecting conductors, as shown in Fig. 5.4.*

Example 5.5 Intensity of a Standing Wave

Compute the intensity of the standing electromagnetic wave given by

$$E_y(x, t) = 2E_0 \cos\omega t \cos\beta x, \qquad B_z(x, t) = 2B_0 \sin\omega t \sin\beta x$$

Solution The Poynting vector for the standing wave is

$$\vec{S} = \frac{\vec{E}\times\vec{H}}{\mu_0} = \frac{1}{\mu_0}(2E_0 \cos\omega t \cos\beta x\hat{j}) \times (2B_0 \sin\omega t \sin\beta x\hat{k})$$

$$= \frac{4E_0B_0}{\mu_0}(\sin\omega t \cos\omega t \sin\beta x \cos\beta x)\hat{i}$$

$$= \frac{E_0B_0}{\mu_0}(\sin 2\omega t \sin 2\beta x)\hat{i}$$

The time average of the Poynting vector is

$$\langle\vec{S}\rangle = \frac{E_0B_0}{\mu_0}\sin 2\beta x\langle\sin 2\omega t\rangle\hat{i} = 0$$

The result is to be expected since the standing wave does not propagate. Alternatively, we may say that the energy carried by the two waves travelling in the opposite directions to form the standing wave exactly cancel each other, with no net energy transfer.

5.5 PHASE VELOCITY AND GROUP VELOCITY OF ELECTROMAGNETIC WAVES

The velocity of a wave can be defined in many different ways, because of the presence of many different kinds of waves and different aspects or components of any given wave.
We can define the velocity of a wave in two different ways:

1. Phase velocity, and
2. Group velocity.

1. Phase Velocity

Definition The phase velocity of a wave is the rate at which the phase of the wave propagates in space. This is the speed at which the phase of any one frequency component of the wave travels.

Mathematical Equation We will consider the simple case of a pure travelling sinusoidal wave moving in the positive x direction with speed v, as illustrated in Fig. 5.5.

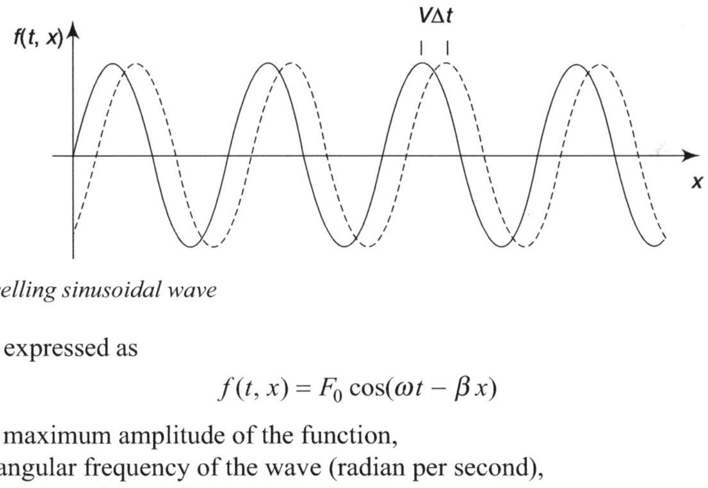

Fig. 5.5 *Pure travelling sinusoidal wave*

This wave can be expressed as

$$f(t, x) = F_0 \cos(\omega t - \beta x) \tag{5.16}$$

where, F_0 is the maximum amplitude of the function,
ω is the angular frequency of the wave (radian per second),
β is the wave number (radian per metre).

It may be noted that the function $f(t, x)$ is the fundamental solution of the one-dimensional wave equation given as

$$\frac{\partial^2 f}{\partial x^2} - \left(\frac{\beta}{\omega}\right)^2 \frac{\partial^2 f}{\partial t^2} = 0$$

From Eq. (5.16), for a constant phase point, we can write

$$(\omega t - \beta x) = \text{constant}$$

$$\therefore \quad \frac{dx}{dt} = \frac{\omega}{\beta}$$

Since ω is the number of radians of the wave that pass a given location per unit time, and $1/\beta$ is the spatial length of the wave per radian, it follows that $\frac{\omega}{\beta} = v$ is the speed at which the shape of the wave is moving, i.e., the speed at which any fixed phase of the cycle is displaced.

Consequently this is called the *phase velocity* of the wave, denoted by v_p:

In terms of the cyclical frequency and wavelength we have, $v_p = \lambda f$.

$$\therefore \quad \boxed{v_p = \frac{\omega}{\beta} = \lambda f = \frac{1}{\sqrt{\mu\varepsilon}}} \tag{5.17}$$

2. Group Velocity

Definition The velocity with which the overall shape of a wave amplitude, known as the modulation or envelope of the wave, propagates through a medium is known as the *group velocity* of the wave. Group velocity is the velocity with which the energy propagates; hence it is also known as *energy velocity*.

$$\therefore \quad \boxed{v_g = \frac{\Delta\omega}{\Delta\beta} = \frac{d\omega}{d\beta}} \tag{5.18}$$

Mathematical Equation We consider a plane wave propagating in the positive x-direction given as

$$f(t, x) = F_0 \cos(\omega t - \beta x)$$

If such a wave is modulated, there result a group of frequencies centered around the carrier frequency ω. If $(\omega + \Delta\omega)$ and $(\omega - \Delta\omega)$ are two such frequencies of the resulting group, the corresponding values of β are $(\beta + \Delta\beta)$ and $(\beta - \Delta\beta)$. Then the resulting wave consisting of the two superimposed waves is given as

$$f(x, t) = F_0 \cos[(\omega - \Delta\omega)t - (\beta - \Delta\beta)x] + F_0 \cos[(\omega + \Delta\omega)t - (\beta + \Delta\beta)x]$$

Using trigonometric identity, the two components of this signal can be expressed as

$$\cos[(\beta x - \omega t) \pm (\Delta\omega t - \Delta\beta x)] = \cos(\omega t - \beta x)\cos(\Delta\omega t - \Delta\beta x) \mp \sin(\omega t - \beta x)\sin(\Delta\omega t - \Delta\beta x)$$

On further simplification,

$$f(x, t) = 2F_0 \cos(\omega t - \beta x)\cos(\Delta\omega t - \Delta\beta x)$$

This can be somewhat loosely interpreted as a simple sinusoidal wave with the angular velocity ω, the wave number β, and the modulated amplitude $2F_0 \cos(\Delta\omega t - \Delta\beta x)$. In other words, the amplitude of the wave is itself a wave, and the phase velocity of this modulation wave is $v = \frac{\Delta\omega}{\Delta\beta}$. A typical plot of such a signal is shown in Fig. 5.6 for the case $\omega = 6$ rad/sec, $\beta = 6$ rad/metre, $\Delta\omega = 0.1$ rad/sec, $\Delta\beta = 0.3$ rad/metre.

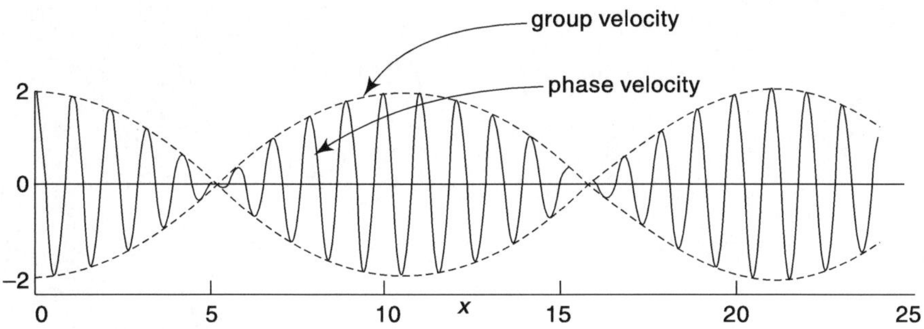

Fig. 5.6 *Phase and group velocities*

The phase velocity of the internal oscillations is $\frac{\omega}{\beta} = 1$ metre/sec, whereas the amplitude envelope wave (indicated by the dotted lines) has a phase velocity of $v = \frac{\Delta\omega}{\Delta\beta} = 0.33$ metre/sec. This is the phase velocity of the amplitude wave, but since each amplitude wave contains a group of internal waves, this speed is usually called the *group velocity*.

Relation between Phase Velocity and Group Velocity We know that the phase velocity is given as

$$v_p = \frac{\omega}{\beta} = \frac{1}{\sqrt{\mu\varepsilon}} = \frac{1}{\sqrt{\mu_0\varepsilon_0}}\frac{1}{\sqrt{\mu_r\varepsilon_r}} = \frac{c}{\sqrt{\mu_r\varepsilon_r}} = \frac{c}{\eta}$$

where, c is the speed of light and η is the refractive index of the medium.

Also, the group velocity is given as

$$v_g = \frac{\Delta\omega}{\Delta\beta} = \frac{d\omega}{d\beta}$$

$\therefore$ $$v_g = \frac{d}{d\beta}(\beta v_p) = v_p + \beta\frac{dv_p}{d\beta}$$

Now, $$\beta = \frac{2\pi}{\lambda}, \quad \therefore \quad \frac{d\beta}{d\lambda} = -\frac{2\pi}{\lambda^2} = -\frac{\beta}{\lambda}$$

$\therefore$ $$d\beta = -\beta\frac{d\lambda}{\lambda}$$

$\therefore$ $$v_g = v_p + \beta\frac{dv_p}{d\beta} = v_p - \lambda\frac{dv_p}{d\lambda}$$

$\therefore$ $$\boxed{v_g = v_p - \lambda\frac{dv_p}{d\lambda}} \tag{5.19}$$

In another way, $$v_p = \frac{\omega}{\beta} = \frac{c}{\eta}$$

$$\therefore \quad v_g = \frac{d\omega}{d\beta} = \frac{d}{d\beta}\left(\beta \frac{c}{\eta}\right) = \frac{c}{\eta} - \frac{c\beta}{\eta^2}\frac{d\eta}{d\beta} = v_p\left[1 - \frac{\beta}{\eta}\frac{d\eta}{d\beta}\right]$$

$$\therefore \quad \boxed{v_g = v_p\left[1 - \frac{\beta}{\eta}\frac{d\eta}{d\beta}\right]} \tag{5.20}$$

From these two relations between phase velocity and group velocity, we can draw the following conclusions:

1. For a *non-dispersive medium*, whose refractive index is constant $\left(\text{i.e.,}\ \frac{d\eta}{d\beta} = 0\right)$, independent of frequency (such as the vacuum), the phase velocity and group velocity are equal.
2. For a *dispersive medium*, whose refractive index is typically a function of wave number and therefore of frequency (such as air, water, glass, etc.), the phase velocity and group velocity are not same.

Dispersive media are of two types:

Normally dispersive media: where $\frac{dv_p}{d\lambda}$ is positive and group velocity is less than phase velocity.

Anomalously dispersive media: where $\frac{dv_p}{d\lambda}$ is negative and group velocity is more than phase velocity.

5.6 INTRINSIC IMPEDANCE

Definition The intrinsic impedance of the wave is defined as the ratio of the electric field and magnetic field phasors (complex amplitudes).

Mathematical Equation We consider an electric field given as

$$E_x = 0, \qquad E_y = 0, \qquad E_z = E_0 e^{-\gamma y}$$

By Maxwell's equation

$$\nabla \times \vec{E} = -\mu \frac{\partial \vec{H}}{\partial t} = -j\omega\mu\vec{H}$$

$$\Rightarrow \quad \begin{vmatrix} \hat{a}_x & \hat{a}_y & \hat{a}_z \\ \frac{\partial}{\partial x} & \frac{\partial}{\partial y} & \frac{\partial}{\partial z} \\ 0 & 0 & E_0 e^{-\gamma y} \end{vmatrix} = -j\omega\mu\vec{H}$$

$$\Rightarrow \quad \frac{\partial}{\partial y}[E_0 e^{-\gamma y}]\hat{a}_x = -j\omega\mu[H_x\hat{a}_x + H_y\hat{a}_y + H_z\hat{a}_z]$$

$$\Rightarrow \quad -\gamma E_0 e^{-\gamma y}\hat{a}_x = -j\omega\mu H_x\hat{a}_x$$

$$\Rightarrow \quad -\gamma E_z = -j\omega\mu H_x$$

$$\Rightarrow \quad \frac{E_z}{H_x} = \frac{j\omega\mu}{\gamma} = \frac{j\omega\mu}{\sqrt{j\omega\mu(\sigma + j\omega\varepsilon)}} = \sqrt{\frac{j\omega\mu}{\sigma + j\omega\varepsilon}}$$

Hence, the intrinsic impedance is given as

$$\boxed{\eta = \sqrt{\frac{j\omega\mu}{\sigma + j\omega\varepsilon}}} \tag{5.21}$$

5.7 PROPAGATION OF UNIFORM PLANE WAVES THROUGH DIFFERENT MEDIA

We shall consider the wave propagation along the y-direction, so that the electric field $\vec{E}$ has only z-component, E_z and the magnetic field has only x-component, H_x. Then the solution of wave equations gives,

$$E_z(y, t) = E_0 e^{-\alpha y} \cos(\omega t - \beta y)\hat{a}_z \text{ and, } H_x(y, t) = H_0 e^{-\alpha y} \cos(\omega t - \beta y)\hat{a}_x$$

where, $H_0 = \dfrac{E_0}{\eta}$, η = intrinsic impedance of the medium

We now consider the wave propagation through the following four media:

1. Wave Propagation through Imperfect (Lossy) Dielectric Medium
2. Wave Propagation through Perfect Dielectric Medium
3. Wave Propagation through Free Space
4. Wave Propagation through Conducting Medium (Good Conductors).

5.7.1 Wave Propagation through Imperfect (Lossy) Dielectric Medium

For lossy dielectric medium, we have the condition, $\dfrac{\sigma}{\omega\varepsilon} << 1$

∴ Attenuation constant is

$$\alpha = \omega\sqrt{\frac{\mu\varepsilon}{2}\left[\sqrt{1+\left(\frac{\sigma}{\omega\varepsilon}\right)^2} - 1\right]} = \omega\sqrt{\frac{\mu\varepsilon}{2}\left[1 + \frac{\sigma^2}{2\omega^2\varepsilon^2} - 1\right]} = \omega\sqrt{\frac{\mu\sigma^2}{4\omega^2\varepsilon^2}}$$

∴ $$\boxed{\alpha = \frac{\sigma}{2}\sqrt{\frac{\mu}{\varepsilon}}}$$

∴ Phase constant is

$$\beta = \omega\sqrt{\frac{\mu\varepsilon}{2}\left[\sqrt{1+\left(\frac{\sigma}{\omega\varepsilon}\right)^2} + 1\right]} = \omega\sqrt{\frac{\mu\varepsilon}{2}\left[1 + \frac{\sigma^2}{2\omega^2\varepsilon^2} + 1\right]} = \omega\sqrt{\mu\varepsilon}\left(1 + \frac{\sigma^2}{8\omega^2\varepsilon^2}\right)$$

∴ $$\boxed{\beta = \omega\sqrt{\mu\varepsilon}\left(1 + \frac{\sigma^2}{8\omega^2\varepsilon^2}\right)}$$

Here, $\omega\sqrt{\mu\varepsilon}$ is the phase shift for a perfect dielectric. The effect of a small amount of loss is to add the term $\omega\sqrt{\mu\varepsilon}\,\dfrac{\sigma^2}{8\omega^2\varepsilon^2}$ as a small correction factor.

$\therefore$ Velocity of wave propagation is

$$v = \frac{\omega}{\beta} = \frac{1}{\sqrt{\mu\varepsilon}\left(1 + \dfrac{\sigma^2}{8\omega^2\varepsilon^2}\right)} = \frac{1}{\sqrt{\mu\varepsilon}}\left[1 - \frac{\sigma^2}{8\omega^2\varepsilon^2}\right]$$

Here, $\dfrac{1}{\sqrt{\mu\varepsilon}}$ is the velocity of the wave propagation in a perfect dielectric (when $\sigma = 0$). The effect of a small amount of loss is to reduce slightly the velocity of wave propagation.

$\therefore$ Intrinsic impedance for the lossy dielectric is

$$\eta = \sqrt{\frac{j\omega\mu}{\sigma + j\omega\varepsilon}} = \sqrt{\frac{\mu}{\varepsilon}\left[\frac{1}{1 + \dfrac{\sigma}{j\omega\varepsilon}}\right]} == \sqrt{\frac{\mu}{\varepsilon}}\left[1 + j\frac{\sigma}{2\omega\varepsilon}\right] = |\eta| \angle\theta_\eta$$

where,
$$|\eta| = \frac{\sqrt{\mu/\varepsilon}}{\left[1 + \left(\dfrac{\sigma}{\omega\varepsilon}\right)^2\right]^{1/4}} \qquad \theta_\eta = \frac{1}{2}\tan^{-1}\left(\frac{\sigma}{\omega\varepsilon}\right)$$

Here too, the effect of loss is to add a small reactive component to the intrinsic impedance. Hence, if the electric field is given as $\vec{E} = E_0 e^{-\alpha y}\cos(\omega t - \beta y)\hat{a}_z$ the magnetic field will be

$$\vec{H} = H_0 e^{-\alpha y}\cos(\omega t - \beta y)\hat{a}_x = \frac{E_0}{|\eta|} e^{-\alpha y}\cos(\omega t - \beta y - \theta_\eta)\hat{a}_x$$

5.7.2 Wave Propagation through Perfect Dielectric Medium

In this case, $\sigma = 0$

$\therefore$ Attenuation constant, $\alpha = 0$

$\therefore$ Phase constant $\beta = \omega\sqrt{\mu\varepsilon}$

$\therefore$ Velocity of wave propagation $v = \dfrac{\omega}{\beta} = \dfrac{1}{\sqrt{\mu\varepsilon}}$

$\therefore$ Intrinsic impedance $\eta = \sqrt{\dfrac{\mu}{\varepsilon}}$

If the electric field is given as, $\vec{E} = E_0\cos(\omega t - \beta y)\hat{a}_z$ and, then the magnetic field will be

$$\vec{H} = H_0\cos(\omega t - \beta y)\hat{a}_x = \frac{E_0}{\eta}\cos(\omega t - \beta y)\hat{a}_x$$

Thus, we see that for a perfect dielectric medium, the wave propagates without any attenuation and the electric and magnetic fields are in phase with each other.

5.7.3 Wave Propagation through Free Space

In this case, $\sigma = 0, \quad \varepsilon = \varepsilon_0, \quad \mu = \mu_0$

$\therefore$ Attenuation constant, $\alpha = 0$

$\therefore$ Phase constant $\beta = \omega\sqrt{\mu_0\varepsilon_0} = \dfrac{\omega}{c}$

$\therefore$ Velocity of wave propagation $v = \dfrac{\omega}{\beta} = \dfrac{1}{\sqrt{\mu_0\varepsilon_0}} = c = 3 \times 10^8$ m/s, speed of light

This shows that the electromagnetic wave travels with speed of light in the free space. In other words, we can say that light is an electromagnetic wave.

$\therefore$ Intrinsic impedance in free space, $\eta_0 = \sqrt{\dfrac{\mu_0}{\varepsilon_0}} = \sqrt{\dfrac{4\pi \times 10^{-7}}{\dfrac{1}{36\pi \times 10^9}}} = 120\pi \approx 377\ \Omega$

If the electric field is given as, $\vec{E} = E_0 \cos(\omega t - \beta y)\hat{a}_z$ and, then the magnetic field will be

$$\vec{H} = H_0 \cos(\omega t - \beta y)\hat{a}_x = \frac{E_0}{\eta}\cos(\omega t - \beta y)\hat{a}_x$$

5.7.4 Wave Propagation through Conducting Medium (Good Conductors)

Here, the condition is, $\sigma >> \omega\varepsilon$, i.e., $\dfrac{\omega\varepsilon}{\sigma} << 1$

$\therefore$ Propagation constant,

$$\gamma = \sqrt{j\omega\mu(\sigma + j\omega\varepsilon)} = \sqrt{j\omega\mu\sigma\left(1 + \frac{j\omega\varepsilon}{\sigma}\right)} = \sqrt{j\omega\mu\sigma} = \sqrt{\omega\mu\sigma}\ \angle 45^\circ$$

$$= \sqrt{\omega\mu\sigma}\,(\cos 45^\circ + j\sin 45^\circ)$$

$$= \sqrt{\omega\mu\sigma}\left(\frac{1}{\sqrt{2}} + j\frac{1}{\sqrt{2}}\right)$$

$$= (\alpha + j\beta)$$

$\therefore$

$$\boxed{\alpha = \beta = \sqrt{\frac{\omega\mu\sigma}{2}}}$$

$\therefore$ Velocity of wave propagation, $v = \dfrac{\omega}{\beta} = \sqrt{\dfrac{2\omega}{\mu\sigma}}$

∴ Intrinsic impedance

$$\eta = \sqrt{\frac{j\omega\mu}{\sigma + j\omega\varepsilon}} = \sqrt{\frac{j\omega\mu}{\sigma\left(1 + j\dfrac{\omega\varepsilon}{\sigma}\right)}} = \sqrt{\frac{j\omega\mu}{\sigma}} = \sqrt{\frac{\omega\mu}{\sigma}}\angle 45^\circ$$

Thus, if the electric field is given as, $\vec{E} = E_0 e^{-\alpha y}\cos(\omega t - \beta y)\hat{a}_z$ and, then the magnetic field will be

$$\vec{H} = H_0 e^{-\alpha y}\cos(\omega t - \beta y)\hat{a}_x = \frac{E_0}{\sqrt{\dfrac{\omega\mu}{\sigma}}} e^{-\alpha y}\cos(\omega t - \beta y - 45^\circ)\hat{a}_x,$$

i.e., the magnetic field in a good conductor lags behind the electric field by 45°.

Skin Effect

Definition *Skin effect* is the tendency of an alternating electric current (AC) to distribute itself within a conductor in such a way that the current density is the largest near the surface of the conductor, while decreasing at greater depths. The electric current is practically confined at the 'skin' of the conductor, i.e. between the outer suface and a level called the *skin depth.*

Cause The skin effect is due to opposing eddy currents induced by the time varying magnetic field resulting from the alternating current.

Effect The skin effect causes the effective resistance of the conductor to increase with increase in frequency, thus reducing the effective cross-section of the conductor. Also, since power loss increases as resistance increases, power losses increase with an increase in frequency because of skin effect..

Skin Depth or Depth of Penetration (δ)

Definition It is defined as the depth in which the magnitude of the wave is attenuated to 37% (e^{-1}) of its original value. Variation inside a conductor is shown in Fig. 5.7.

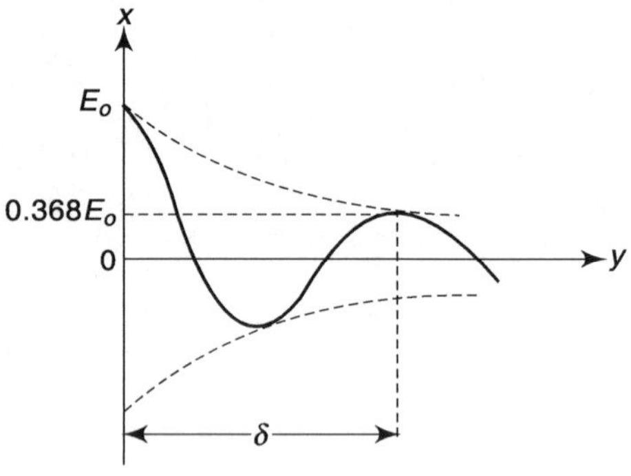

Fig. 5.7 *Variation of electric field inside a conductor (skin effect)*

Mathematical Equation From the field expression for a good conductor, the magnitude is, $E = E_0 e^{-\alpha y}$

At $y = \delta$, $\quad E = E_0 e^{-1} = E_0 e^{-\alpha\delta} \quad \Rightarrow \quad \delta = \dfrac{1}{\alpha}$

So, the general expression for the skin depth is

$$\boxed{\delta = \frac{1}{\omega\sqrt{\dfrac{\mu\varepsilon}{2}\left[\sqrt{1+\left(\dfrac{\sigma}{\omega\varepsilon}\right)^2} - 1\right]}}}$$

The phenomenon that the alternating fields and hence currents are confined within a small region of a conducting medium inside the surface is known as the *skin effect.*

For a good conductor, $\dfrac{\sigma}{\omega\varepsilon} >> 1$, the *skin-depth* is given by

$$\boxed{\delta = \frac{1}{\alpha} = \sqrt{\frac{2}{\omega\mu\sigma}}}$$

Hence, in a good conductor, the depth of penetration decreases with increasing frequency (ω). In terms of wavelength, λ, it is given as

$$\delta = \frac{1}{\alpha} = \frac{1}{\beta} = \frac{\lambda}{2\pi}$$

($\therefore \alpha = \beta$ for good conductors)

Hence, the wave gets attenuated even before one-cycle, about $\frac{1}{6}^{\text{th}}$ of a wavelength.

Surface Impedance (Z_s)

Definition The *surface impedance* is defined as the ratio of the tangential component of the electric field to the tangential component of the magnetic field. Also, according to the boundary condition, the tangential component of the magnetic field refers to surface current density. Hence, surface impedance is given as

$$Z_S = \frac{E_t}{H_t} = \frac{E_t}{K_S}$$

The *surface impedance* provides the boundary condition for fields outside the conductor and accounts for the dissipation and energy stored inside the conductor.

Mathematical Equation Now, the surface current density K_S represents the total conduction current per unit width flowing in the thin sheet. If we assume that the conductor is a flat plate with its surface at $x = 0$ plane, the current distribution in the x-direction will be given by

$$K = K_0 e^{-\gamma x}$$

where, K_0 is the current density at the surface.

If the thickness of the conductor is very much greater than the skin depth, so that there is no reflection from the back surface of the conductor, then the total conduction current per unit width of the conductor is given as

$$K_S = \int_0^\infty K dx = K_0 \int_0^\infty e^{-\gamma x} dx = -\frac{K_0}{\gamma}\left[e^{-\gamma x}\right]_0^\infty = \frac{K_0}{\gamma}$$

However, K_0 is the current density at the surface and is given as

$$K_0 = \sigma E_t$$

So, the surface impedance is given as

$$Z_S = \frac{E_t}{K_S} = \frac{\gamma}{\sigma}$$

For a good conductor, the propagation constant is given as, $\gamma = \sqrt{j\omega\mu\sigma}$

So, the surface impedance for a thick conductor is

$$Z_S = \frac{\gamma}{\sigma} = \frac{\sqrt{j\omega\mu\sigma}}{\sigma} = \sqrt{\frac{j\omega\mu}{\sigma}} = \eta$$

$$\therefore \qquad \boxed{Z_S = \sqrt{\frac{j\omega\mu}{\sigma}} = \eta}$$

Thus, we conclude that *for a good conductor whose thickness is much greater than the skin depth, the surface impedance is equal to the intrinsic impedance of the conductor.*

Skin Resistance (R_s)

Definition The real part of this impedance is known as *surface resistance* or *skin resistance*, R_S (Ω/m^2), and the imaginary part of this impedance is known as *surface reactance* or *skin reactance*, X_S (Siemens/ m^2).

Mathematical Equation From the results of intrinsic impedance or surface impedance for a good conductor, we can write

$$Z_S = \eta = \sqrt{\frac{j\omega\mu}{\sigma}} = \sqrt{\frac{\omega\mu}{\sigma}}\angle 45° = \left(\sqrt{\frac{\omega\mu}{2\sigma}} + j\sqrt{\frac{\omega\mu}{2\sigma}}\right)$$

$$\therefore \qquad R_S = \sqrt{\frac{\omega\mu}{2\sigma}} \quad \text{and} \quad X_S = \sqrt{\frac{\omega\mu}{2\sigma}}$$

Now,

$$R_S = \sqrt{\frac{\omega\mu}{2\sigma}} = \frac{1}{\sigma}\sqrt{\frac{\omega\mu\sigma}{2}} = \frac{1}{\sigma}\frac{1}{\delta}$$

$$\therefore \qquad \boxed{R_S = \frac{1}{\sigma\delta} = \sqrt{\frac{\omega\mu}{2\sigma}}}$$

Hence, we see that the surface resistance of a flat conductor at any frequency is equal to the DC resistance of thickness δ (skin depth) of the same conductor.

Complex Permittivity and Loss Tangent (tan θ)

For harmonically varying fields,

$$\nabla \times \boldsymbol{H}_s = (\sigma \boldsymbol{E}_s + j\omega\varepsilon \boldsymbol{E}_s) = (\sigma + j\omega\varepsilon)\boldsymbol{E}_s = j\omega\varepsilon\left[1 - j\frac{\sigma}{\omega\varepsilon}\right]\boldsymbol{E}_s = j\omega\varepsilon_c \boldsymbol{E}_s$$

where, $\varepsilon_c = \varepsilon\left[1 - j\dfrac{\sigma}{\omega\varepsilon}\right] = (\varepsilon' - j\varepsilon'') = \text{complex permittivity}$

The ratio, $\left|\dfrac{\varepsilon''}{\varepsilon'}\right| = \dfrac{\sigma}{\omega\varepsilon} = \dfrac{|\sigma \boldsymbol{E}_s|}{|\omega\varepsilon \boldsymbol{E}_s|} = \dfrac{|\boldsymbol{J}_{\text{conduction}}|}{|\boldsymbol{J}_{\text{displacement}}|} = \tan\theta$

The ratio of the imaginary part of the complex permittivity (ε'') to the real part of the complex permittivity (ε') is the ratio of the magnitude of the conduction current density to the magnitude of the displacement current density. This ratio is defined as the *loss tangent* or *loss angle* of the medium.

The loss tangent gives a measure of how lossy a medium is.

For a good (lossless or perfect) dielectric medium ($\sigma << \omega\varepsilon$), loss tangent is very small.

For a good conducting medium ($\sigma >> \omega\varepsilon$), loss tangent is very large.

For a lossy dielectric, loss tangent is of the order of unity.

NOTE

It is seen that $\theta = 2\theta_\eta$.

Example 5.6 In a medium, $\vec{E} = 16e^{-0.05x}\sin(2\times10^8 t - 2x)\hat{a}_z$ (V/m). Find (a) the propagation constant, (b) the wavelength, (c) the speed of the wave, and (d) the skin depth.

Solution Here, $\alpha = 0.05$, $\beta = 2$, $\omega = 2\times10^8$

1. Propagation constant, $\gamma = (\alpha + j\beta) = (0.05 + j2) = 2.000625\ \angle 88.568°$ per metre
2. Wavelength, $\lambda = \dfrac{2\pi}{\beta} = \dfrac{2\pi}{2} = 3.142$ m
3. Speed of the wave, $v = \dfrac{\omega}{\beta} = \dfrac{2\times10^8}{2} = 10^8$ m/s
4. Skin depth, $\delta = \dfrac{1}{\alpha} = \dfrac{1}{0.05} = 20$ m

***Example 5.7** The electric field intensity associated with a plane wave travelling in a perfect dielectric medium is given by

$$E_x(z,t) = 12\cos(2\pi\times10^7 t - 0.1\pi z)\ \text{V/m}$$

Find:

1. Velocity of propagation.
2. Intrinsic impedance.

Solution:

1. Velocity of propagation, $v = \dfrac{\omega}{\beta} = \dfrac{2\pi\times10^7}{0.1\pi} = 2\times10^8$ m/s
2. Now,

$$v = \sqrt{\frac{1}{\mu\varepsilon}} = \sqrt{\frac{1}{\mu_0\mu_r\varepsilon_0\varepsilon_r}} = \frac{c}{\sqrt{\mu_r\varepsilon_r}} = \frac{c}{\sqrt{\varepsilon_r}} \qquad \{\because \text{for perfect dielectric medium, } \mu_r = 1\}$$

$$\therefore \qquad \sqrt{\varepsilon_r} = \frac{c}{v} = \frac{3\times10^8}{2\times10^8} = \frac{3}{2}$$

$$\therefore \quad \text{Intrinsic impedance, } \eta = \sqrt{\frac{\mu}{\varepsilon}} = \sqrt{\frac{\mu_0\mu_r}{\varepsilon_0\varepsilon_r}} = 377\times\sqrt{\frac{1}{\varepsilon_r}} = 377\times\frac{2}{3} = 251.33\ \Omega$$

***Example 5.8** The electric field intensity of an electromagnetic wave in free space is given by

$$E_y = 0, \quad E_z = 0, \quad E_x = E_0\cos\omega\left(t - \frac{z}{v}\right)$$

Determine the expression for the components of the magnetic field intensity $\vec{H}$. Also, find $\dfrac{E_x}{H_y}$.

Solution: By Maxwell's equation,

$$\nabla\times\vec{E} = -\frac{\partial\vec{B}}{\partial t} = -\mu_0\frac{\partial\vec{H}}{\partial t}$$

or, $$\begin{vmatrix} \frac{\partial}{\partial x} & \frac{\partial}{\partial y} & \frac{\partial}{\partial z} \\ E_x & E_y & E_z \\ \hat{i} & \hat{j} & \hat{k} \end{vmatrix} = -\mu_0 \left[\frac{\partial H_x}{\partial t}\hat{i} + \frac{\partial H_y}{\partial t}\hat{j} + \frac{\partial H_z}{\partial t}\hat{k} \right]$$

or, $$\begin{vmatrix} \frac{\partial}{\partial x} & \frac{\partial}{\partial y} & \frac{\partial}{\partial z} \\ E_x & 0 & 0 \\ \hat{i} & \hat{j} & \hat{k} \end{vmatrix} = -\mu_0 \left[\frac{\partial H_x}{\partial t}\hat{i} + \frac{\partial H_y}{\partial t}\hat{j} + \frac{\partial H_z}{\partial t}\hat{k} \right]$$

or, $$\frac{\partial E_x}{\partial z}\hat{j} - \frac{\partial E_x}{\partial y}\hat{k} = -\mu_0 \left[\frac{\partial H_x}{\partial t}\hat{i} + \frac{\partial H_y}{\partial t}\hat{j} + \frac{\partial H_z}{\partial t}\hat{k} \right]$$

Comparing both sides, we get

$$H_x = 0, \quad H_z = 0$$

and

$$\frac{\partial H_y}{\partial t} = -\frac{1}{\mu_0}\frac{\partial E_x}{\partial z} = -\frac{1}{\mu_0}\frac{\partial}{\partial z}\left[E_0 \cos\omega\left(t - \frac{z}{v}\right)\right] = -\frac{E_0\omega}{\mu_0 v}\sin\omega\left(t - \frac{z}{v}\right)$$

$$\therefore \quad H_y = -\frac{E_0\omega}{\mu_0 v}\int \sin\omega\left(t - \frac{z}{v}\right)dt = \frac{E_0}{\mu_0 v}\cos\omega\left(t - \frac{z}{v}\right)$$

Thus, the components of the magnetic fields are:

$$H_x = 0, \quad H_z = 0 \quad \text{and} \quad H_y = \frac{E_0}{\mu_0 v}\cos\omega\left(t - \frac{z}{v}\right)$$

So, $$\frac{E_x}{H_y} = \mu_0 v = \mu_0 \times \frac{1}{\sqrt{\mu_0 \varepsilon_0}} = \sqrt{\frac{\mu_0}{\varepsilon_0}} = \sqrt{\frac{4\pi \times 10^{-7}}{\frac{1}{36\pi} \times 10^{-9}}} = 120\pi = 377\ \Omega$$

NB: This is known as the intrinsic impedance of free space.

Example 5.9 The electric field intensity associated with a plane wave travelling in a perfect dielectric medium having $\mu = \mu_0$ is given by

$$\vec{E} = 10\cos(6\pi \times 10^7 t - 0.4\pi z)\,\hat{i}\ \text{V/m}$$

Find the phase velocity, the permittivity of the medium and associated magnetic field vector $\vec{H}$. Velocity in free space = 3×10^8 m/s.

Solution Here, $\omega = 6\pi \times 10^7$ rad/s, $\beta = 0.4\pi$

$\therefore$ Phase velocity, $$v = \frac{\omega}{\beta} = \frac{6\pi \times 10^7}{0.4\pi} = 1.5 \times 10^8\ \text{m/s}$$

Now, $$v = \frac{1}{\sqrt{\mu\varepsilon}} = \frac{1}{\sqrt{\mu_0\varepsilon_0}} \times \frac{1}{\sqrt{\mu_r\varepsilon_r}} = \frac{c}{\sqrt{\varepsilon_r}} \quad (\because \mu_r = 1)$$

$\therefore$ $$\varepsilon_r = \left(\frac{c}{v}\right)^2 = \left(\frac{3\times 10^8}{1.5\times 10^8}\right)^2 = 4$$

By Maxwell's equation,

or $$\frac{\partial E_x}{\partial z}\hat{j} - \frac{\partial E_x}{\partial y}\hat{k} = -\mu_0\left[\frac{\partial H_x}{\partial t}\hat{i} + \frac{\partial H_y}{\partial t}\hat{j} + \frac{\partial H_z}{\partial t}\hat{k}\right]$$

Comparing both sides, we get

$$H_x = 0, H_z = 0$$

and

$$\nabla \times \vec{E} = -\frac{\partial \vec{B}}{\partial t} = -\mu_0\frac{\partial \vec{H}}{\partial t}$$

or, $$\begin{vmatrix} \frac{\partial}{\partial x} & \frac{\partial}{\partial y} & \frac{\partial}{\partial z} \\ E_x & E_y & E_z \\ \hat{i} & \hat{j} & \hat{k} \end{vmatrix} = -\mu_0\left[\frac{\partial H_x}{\partial t}\hat{i} + \frac{\partial H_y}{\partial t}\hat{j} + \frac{\partial H_z}{\partial t}\hat{k}\right]$$

or, $$\begin{vmatrix} \frac{\partial}{\partial x} & \frac{\partial}{\partial y} & \frac{\partial}{\partial z} \\ E_x & 0 & 0 \\ \hat{i} & \hat{j} & \hat{k} \end{vmatrix} = -\mu_0\left[\frac{\partial H_x}{\partial t}\hat{i} + \frac{\partial H_y}{\partial t}\hat{j} + \frac{\partial H_z}{\partial t}\hat{k}\right]$$

$$\begin{aligned}\frac{\partial H_y}{\partial t} &= -\frac{1}{\mu_0}\frac{\partial E_x}{\partial z} = -\frac{1}{\mu_0}\frac{\partial}{\partial z}\left[10\cos(6\pi\times 10^7 t - 0.4\pi z)\right] \\ &= -\frac{1}{\mu_0}[-0.4\pi\times 10\sin(6\pi\times 10^7 t - 0.4\pi z)] \\ &= \frac{4\pi}{\mu_0}\sin(6\pi\times 10^7 t - 0.4\pi z)\end{aligned}$$

$\therefore$ $$\begin{aligned} H_y &= \frac{4\pi}{\mu_0}\int \sin(6\pi\times 10^7 t - 0.4\pi z)dt \\ &= -\frac{4\pi}{\mu_0}\frac{1}{6\pi\times 10^7}\cos(6\pi\times 10^7 t - 0.4\pi z) \\ &= -\frac{1}{6\pi}\cos(6\pi\times 10^7 t - 0.4\pi z)\end{aligned}$$

Hence, the magnetic field is given as

$$\boxed{\vec{H} = -\frac{1}{6\pi}\cos(6\pi\times 10^7 t - 0.4\pi z)\hat{j}\ \text{A/m}}$$

Example 5.10 The x and y components of a circularly polarised electromagnetic wave in free space are

$$E_x = 2\sin(\omega t - \beta z)$$

$$E_y = 2\cos(\omega t - \beta z)$$

Find the expression for the displacement current density and draw a neat sketch showing the fields.

Solution Here, the magnitude of the electric field having circular polarisation and components as given is

$$|E| = \sqrt{E_x^2 + E_y^2} = \sqrt{2^2\cos^2(\omega t - \beta z) + 2^2\sin^2(\omega t - \beta z)} = 2$$

The displacement current density is

$$\vec{J}_d = \frac{\partial \vec{D}}{\partial t} = \frac{\partial}{\partial t}(\varepsilon \vec{E}) = j\omega\varepsilon \vec{E}$$

$$\therefore \qquad J_d = |\vec{J}_d| = |j\omega\varepsilon\vec{E}| = 2\omega\varepsilon E$$

$$\boxed{J_d = 2\omega\varepsilon E}$$

The electric field is plotted in the $z = 0$ plane for different values of ωt as shown in Fig. 5.8.

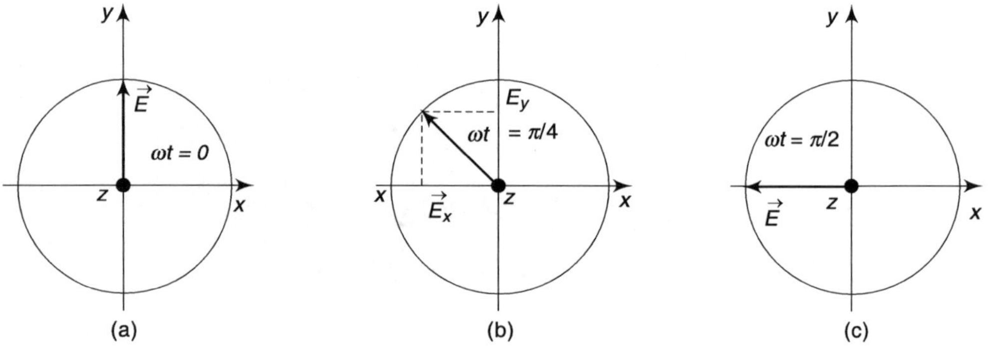

Fig. 5.8 *Electric field in z = 0 plane with three conditions: (a) $\omega t = 0$ (b) $\omega t = \pi/4$ (c) $\omega t = \pi/2$*

Example 5.11 Consider a monochromatic plane wave, where the electric field is given by

$$\vec{E} = E_0 e^{j(\omega t - kz)}\,\hat{i}$$

where E_0 is an arbitrary constant vector and other symbols have their usual meanings.

1. Show that the electric field vector lies in a direction perpendicular to the propagation.
2. Determine the corresponding magnetic field.
3. Calculate the wave impedance and show that this is equal to the intrinsic impedance of the medium.

Solution Here, $\vec{E} = E_0 e^{j(\omega t - kz)}\,\hat{i}$

By Maxwell's equation,

$$\nabla \times \vec{E} = -\frac{\partial \vec{B}}{\partial t} = -\mu_0 \frac{\partial \vec{H}}{\partial t}$$

or,
$$\begin{vmatrix} \frac{\partial}{\partial x} & \frac{\partial}{\partial y} & \frac{\partial}{\partial z} \\ E_x & 0 & 0 \\ \hat{i} & \hat{j} & \hat{k} \end{vmatrix} = -\mu_0 \left[\frac{\partial H_x}{\partial t}\hat{i} + \frac{\partial H_y}{\partial t}\hat{j} + \frac{\partial H_z}{\partial t}\hat{k} \right]$$

or,
$$\frac{\partial E_x}{\partial z}\hat{j} = -\mu_0 \left[\frac{\partial H_x}{\partial t}\hat{i} + \frac{\partial H_y}{\partial t}\hat{j} + \frac{\partial H_z}{\partial t}\hat{k} \right] \quad (\because \; E_x = f(t, z))$$

Comparing both sides, we get

$$H_x = 0, H_z = 0$$

and

$$\frac{\partial H_y}{\partial t} = -\frac{1}{\mu_0}\frac{\partial E_x}{\partial z} = -\frac{1}{\mu_0}\frac{\partial}{\partial z}[E_0 e^{j(\omega t - kz)}] = j\frac{E_0 k}{\mu_0} e^{j(\omega t - kz)}$$

$\therefore$
$$H_y = j\frac{E_0 k}{\mu_0}\int e^{j(\omega t - kz)} dt = j\frac{E_0 k}{\mu_0}\frac{1}{j\omega} e^{j(\omega t - kz)} = \frac{E_0 k}{\mu_0 \omega} e^{j(\omega t - kz)}$$

1. Here, the electric field propagates in the x direction and the magnetic field propagates in the y direction whereas the wave propagates in the z direction. Thus, we can say that the electric field vector lie in a direction perpendicular to the propagation.
   Since the electric field is directed across unit vector $\hat{i}$ and the magnetic field is directed across the unit vector $\hat{j}$, we conclude that the two fields are perpendicular to each other.
2. The corresponding magnetic field is given as

$$\boxed{\vec{H} = \frac{E_0 k}{\mu_0 \omega} e^{j(\omega t - kz)}\,\hat{j}}$$

3. The wave impedance is given as

$$\eta = \left|\frac{\vec{E}}{\vec{H}}\right| = \frac{\mu_0 \omega}{k} = \frac{\mu_0 2\pi f}{2\pi / \lambda} = \mu_0 f \lambda = \mu_0 c = \mu_0 \frac{1}{\sqrt{\mu_0 \varepsilon_0}} = \sqrt{\frac{\mu_0}{\varepsilon_0}} = 120\pi$$

   This is equal to the intrinsic impedance of the medium.

Example 5.12 In a region containing no charges and currents, the magnetic field is given by

$$\vec{H} = \hat{x} B_0 \sin \beta z \sin \omega t$$

where B_0, β and ω are constants. Using one of the Maxwell's curl equations at a time, find the two expressions for the associated electric field.

Solution By Maxwell's equation,

$$\nabla \times \vec{H} = \frac{\partial \vec{D}}{\partial t} = \varepsilon_0 \frac{\partial \vec{E}}{\partial t}$$

or,
$$\begin{vmatrix} \frac{\partial}{\partial x} & \frac{\partial}{\partial y} & \frac{\partial}{\partial z} \\ H_x & 0 & 0 \\ \hat{i} & \hat{j} & \hat{k} \end{vmatrix} = \varepsilon_0 \left[\frac{\partial E_x}{\partial t}\hat{i} + \frac{\partial E_y}{\partial t}\hat{j} + \frac{\partial E_z}{\partial t}\hat{k} \right]$$

or,
$$\frac{\partial H_x}{\partial z}\hat{j} - \frac{\partial H_x}{\partial y}\hat{k} = \varepsilon_0 \left[\frac{\partial E_x}{\partial t}\hat{i} + \frac{\partial E_y}{\partial t}\hat{j} + \frac{\partial E_z}{\partial t}\hat{k} \right]$$

or,
$$\frac{\partial}{\partial z}\left[B_0 \sin \beta z \sin \omega t\right]\hat{j} - \frac{\partial}{\partial y}\left[B_0 \sin \beta z \sin \omega t\right]\hat{k} = \varepsilon_0 \left[\frac{\partial E_x}{\partial t}\hat{i} + \frac{\partial E_y}{\partial t}\hat{j} + \frac{\partial E_z}{\partial t}\hat{k} \right]$$

Comparing both sides, we get

$$\frac{\partial E_y}{\partial t} = \frac{1}{\varepsilon_0}\frac{\partial}{\partial z}[B_0 \sin \beta z \sin \omega t] = \frac{B_0 \beta}{\varepsilon_0} \cos \beta z \sin \omega t \qquad \textbf{(i)}$$

and

$$\frac{\partial E_z}{\partial t} = \frac{1}{\varepsilon_0}\frac{\partial}{\partial y}[B_0 \sin \beta z \sin \omega t] = 0 \qquad \textbf{(ii)}$$

From (*i*),

$$E_y = \frac{B_0 \beta}{\varepsilon_0} \int \cos \beta z \sin \omega t dt = -\frac{B_0 \beta}{\omega \varepsilon_0} \cos \beta z \cos \omega t$$

From (*ii*),

$$E_z = \text{constant}$$

Thus, the two expressions for the associated electric field are given by,

$$\boxed{E_y = -\frac{B_0 \beta}{\omega \varepsilon_0} \cos \beta z \cos \omega t} \quad \text{and} \quad \boxed{E_z = \text{constant}}$$

Example 5.13

1. Determine α, β, γ, η, v and δ for a damp soil at 1 MHz. Given: $\varepsilon_r = 12$, $\sigma = 2 \times 10^{-2}$ mho/m, $\mu_r = 1$.
2. Determine the propagation constant γ for a material having $\mu_r = 1$, $\varepsilon_r = 8$ and $\sigma = 0.25$pS/m, if the wave frequency is 1.6 MHz.
3. Find the skin depth δ at a frequency of 1.6 MHz in aluminium where $\sigma = 38.2$MS/m and $\mu_r = 1$. Also, find γ and the wave velocity v.

Solution:

1. Here, $\dfrac{\sigma}{\omega \varepsilon} = \dfrac{2 \times 10^{-2}}{2\pi \times 10^6 \times 8.854 \times 10^{-12} \times 12} >> 1$

   Hence, the material is a good conductor.

$$\therefore \quad \alpha = \beta = \sqrt{\frac{\omega\mu\sigma}{2}} = \sqrt{\frac{2\pi \times 10^6 \times 4\pi \times 10^{-7} \times 2 \times 10^{-2}}{2}} = 0.281 \quad \text{per metre}$$

$$\therefore \quad \gamma = (\alpha + j\beta) = (0.281 + j0.281) = 0.3974 \angle 45° \quad \text{per metre}$$

$$\therefore \quad \eta = \sqrt{\frac{\omega\mu}{\sigma}} \angle 45° = \sqrt{\frac{2\pi \times 10^6 \times 4\pi \times 10^{-7}}{2 \times 10^{-2}}} \angle 45° = 19.869 \angle 45° \ \Omega$$

$$\therefore \quad v = \frac{\omega}{\beta} = \frac{2\pi \times 10^6}{0.281} = 22.36 \times 10^6 \text{ m/s}$$

$$\therefore \quad \delta = \frac{1}{\alpha} = \frac{1}{0.281} = 3.559 \text{ m}$$

2. In this case, $\dfrac{\sigma}{\omega\varepsilon} = \dfrac{0.25 \times 10^{-12}}{2\pi \times 1.6 \times 10^6 \times 8.854 \times 10^{-12} \times 8} << 1$

Hence, the material is a good dielectric.

$$\therefore \quad \alpha = \frac{\sigma}{2}\sqrt{\frac{\mu}{\varepsilon}} = \frac{0.25 \times 10^{-12}}{2} \times 377 \times \sqrt{\frac{1}{8}} = 16.666 \times 10^{-12} \approx 0$$

$$\therefore \quad \beta = \omega\sqrt{\mu\varepsilon}\left(1 + \frac{\sigma^2}{8\omega^2\varepsilon^2}\right) \approx \omega\sqrt{\mu\varepsilon}$$

$$= 2\pi \times 1.6 \times 10^6 \times \sqrt{4\pi \times 10^{-7} \times 8.854 \times 10^{-12} \times 8}$$

$$= 0.0948 \text{ rad/m}$$

$$\therefore \quad \gamma = (\alpha + j\beta) = j0.0948 = 0.0948 \angle 90° \quad \text{per metre}$$

3. Skin depth,

$$\delta = \sqrt{\frac{2}{\omega\mu\sigma}} = \sqrt{\frac{1}{\pi f \mu\sigma}} = \sqrt{\frac{1}{\pi 1.6 \times 10^6 \times 4\pi \times 10^{-7} \times 38.2 \times 10^6}} = 64.377 \ \mu\text{m}$$

$$\alpha = \beta = \frac{1}{\delta} = \frac{1}{64.377 \times 10^{-6}} = 15533.579 \quad \text{per metre}$$

$$\gamma = (\alpha + j\beta) = (15533.579 + j15533.579) = 21967.799 \angle 45° \quad \text{per metre}$$

$$\therefore \quad v = \frac{\omega}{\beta} = \frac{2\pi \times 1.6 \times 10^6}{15533.579} = 647.185 \text{ m/s}$$

***Example 5.14** A uniform plane wave in a lossy medium has a phase constant of 1.6 rad/s at 10^7 Hz and its magnitude is reduced by 60% for every 2 m travelled. Find the skin depth and speed of the wave.

Solution Here, $\beta = 1.6$ rad/s $\quad f = 10^7$ Hz

Since the reduction in magnitude for travel of $x = 2$ m is by 60%, we can find the attenuation constant as follows.

$$0.4E = Ee^{-\alpha \times 2} \quad \Rightarrow \quad -2\alpha = \ln(0.4) = -0.916 \quad \Rightarrow \quad \alpha = 0.458$$

Thus, the skin depth is, $\delta = \frac{1}{\alpha} = \frac{1}{0.458} = 2.183 \text{ m}$

Speed of the wave is, $v = \frac{\omega}{\beta} = \frac{2\pi f}{\beta} = \frac{2\pi \times 10^7}{1.6} = 3.927 \times 10^7 \text{ m/s}$

Example 5.15 Find the depth of penetration of a mega-cycle wave into copper which has conductivity $\sigma = 5.8 \times 10^7$ mho/m and a permeability approximately equal to that of free space.

Solution: Given: $\sigma = 5.8 \times 10^7$ mho/m, $f = 1$ MHz, $\mu = \mu_0 = 4\pi \times 10^{-7}$ H/m

$\therefore$ Depth of penetration is

$$\delta = \sqrt{\frac{2}{\omega\mu\sigma}} = \frac{1}{\sqrt{f\pi\mu\sigma}} = \frac{1}{\sqrt{1 \times 10^6 \times \pi \times 4\pi \times 10^{-7} \times 5.8 \times 10^7}} = 6.602 \times 10^{-5} \text{ m}$$

Example 5.16 A uniform plane wave is specified by $\vec{H} = 2e^{-j0.1z}\hat{i}$ A/m. If the velocity of the wave is 2×10^8 m/s and the relative permeability is 1.6, find the frequency, relative permittivity, wavelength and intrinsic impedance.

Solution Here, $\beta = 0.1$, $v = 2 \times 10^8$ m/s, $\mu_r = 1.6$

$\therefore$ Frequency, $f = \frac{\omega}{2\pi} = \frac{\beta v}{2\pi} = \frac{0.1 \times 2 \times 10^8}{2\pi} = 3.183 \text{ MHz}$

Now, $v = \frac{1}{\sqrt{\mu\varepsilon}} = \frac{c}{\sqrt{\mu_r \varepsilon_r}}$

$\therefore$ Relative permittivity $\varepsilon_r = \left(\frac{c}{v}\right)^2 \times \frac{1}{\mu_r} = \left(\frac{3 \times 10^8}{2 \times 10^8}\right)^2 \times \frac{1}{1.6} = 1.406$

$\therefore$ Wavelength $\lambda = \frac{2\pi}{\beta} = \frac{2\pi}{0.1} = 20\pi = 62.83 \text{ m}$

$\therefore$ Intrinsic impedance $\eta = \sqrt{\frac{\mu}{\varepsilon}} = \sqrt{\frac{\mu_0 \mu_r}{\varepsilon_0 \varepsilon_r}} = 120\pi \times \sqrt{\frac{1.6}{1.406}} = 402.123 \, \Omega$

***Example 5.17** An electric field vector $\vec{E}$ of an electromagnetic wave in free space is given by this expression $E_x = E_z = 0$,

$$E_y = Ae^{j\omega\left(t - \frac{z}{v}\right)}$$

Using Maxwell's equation for free space condition determine expressions for the components of the magnetic field vector $\vec{H}$.

Solution By Maxwell's equation,

$$\nabla \times \vec{E} = -\frac{\partial \vec{B}}{\partial t} = -\mu_0 \frac{\partial \vec{H}}{\partial t}$$

or, $$\begin{vmatrix} \frac{\partial}{\partial x} & \frac{\partial}{\partial y} & \frac{\partial}{\partial z} \\ E_x & E_y & E_z \\ \widehat{i} & \widehat{j} & \widehat{k} \end{vmatrix} = -\mu_0 \left[\frac{\partial H_x}{\partial t}\widehat{i} + \frac{\partial H_y}{\partial t}\widehat{j} + \frac{\partial H_z}{\partial t}\widehat{k} \right]$$

or, $$\begin{vmatrix} \frac{\partial}{\partial x} & \frac{\partial}{\partial y} & \frac{\partial}{\partial z} \\ 0 & E_y & 0 \\ \widehat{i} & \widehat{j} & \widehat{k} \end{vmatrix} = -\mu_0 \left[\frac{\partial H_x}{\partial t}\widehat{i} + \frac{\partial H_y}{\partial t}\widehat{j} + \frac{\partial H_z}{\partial t}\widehat{k} \right]$$

or, $$-\frac{\partial E_y}{\partial z}\widehat{i} + \frac{\partial E_y}{\partial x}\widehat{k} = -\mu_0 \left[\frac{\partial H_x}{\partial t}\widehat{i} + \frac{\partial H_y}{\partial t}\widehat{j} + \frac{\partial H_z}{\partial t}\widehat{k} \right]$$

or, $$-\frac{\partial E_y}{\partial z}\widehat{i} = -\mu_0 \left[\frac{\partial H_x}{\partial t}\widehat{i} + \frac{\partial H_y}{\partial t}\widehat{j} + \frac{\partial H_z}{\partial t}\widehat{k} \right] \qquad \{\because E_y = f(t, z)\}$$

Comparing both sides, we get

$$H_y = 0, H_z = 0$$

and

$$\frac{\partial H_x}{\partial t} = -\frac{1}{\mu_0}\frac{\partial E_y}{\partial z} = -\frac{1}{\mu_0}\frac{\partial}{\partial z}\left[Ae^{j\omega\left(t-\frac{z}{v}\right)} \right] = -\frac{1}{\mu_0}\left(\frac{-j\omega}{v}\right)Ae^{j\omega\left(t-\frac{z}{v}\right)} = \frac{Aj\omega}{\mu_0 v}e^{j\omega\left(t-\frac{z}{v}\right)}$$

$$\therefore \qquad H_x = \frac{Aj\omega}{\mu_0 v}\int e^{j\omega\left(t-\frac{z}{v}\right)}dt = \frac{Aj\omega}{\mu_0 v}\frac{1}{j\omega}e^{j\omega\left(t-\frac{z}{v}\right)} = \frac{A}{\mu_0 v}e^{j\omega\left(t-\frac{z}{v}\right)}$$

For free space, $v = \dfrac{1}{\sqrt{\mu_0 \varepsilon_0}}$

$$\therefore \qquad H_x = \frac{A}{\mu_0 \frac{1}{\sqrt{\mu_0 \varepsilon_0}}}e^{j\omega\left(t-\frac{z}{v}\right)} = A\sqrt{\frac{\varepsilon_0}{\mu_0}}e^{j\omega\left(t-\frac{z}{v}\right)} = \frac{A}{\eta}e^{j\omega\left(t-\frac{z}{v}\right)}$$

where, $\eta = \sqrt{\dfrac{\mu_0}{\varepsilon_0}}$ is the intrinsic impedance of free space.

Thus, the components of the magnetic fields are:

$$\boxed{H_x = 0, \quad H_z = 0} \quad \text{and} \quad \boxed{\therefore \quad H_x = \frac{A}{\eta}e^{j\omega\left(t-\frac{z}{v}\right)}}$$

***Example 5.18** The electric field intensity associated with a planar travelling wave in a dielectric medium is given by

$$E_x(z, t) = 10\cos(2\pi \times 10^7 t - 0.1\pi z)$$

Find the wavelength (λ), phase shift constant (β) and velocity of propagation (v).

Solution Phase shift constant, $\beta = 0.1\pi$

Wavelength, $\lambda = \dfrac{2\pi}{\beta} = 20$ m

Velocity of propagation, $v = \dfrac{\omega}{\beta} = \dfrac{2\pi \times 10^7}{0.1\pi} = 2 \times 10^8$ m/s

***Example 5.19** A lossy dielectric has an intrinsic impedance of $200\angle 30^\circ$ Ω at a particular frequency. If at that frequency, the plane wave propagating through the dielectric has the magnetic field component

$$\vec{H} = 10 \times e^{-\alpha x} \cos\left(\omega t - \frac{1}{2}x\right)\hat{a}_y \quad \text{A/m}$$

find $\vec{E}$ and α. Determine the skin depth and direction of the wave polarisation.

Solution

Here, $$H_0 = 10, \quad \beta = \frac{1}{2}, \quad \eta = 200\angle 30^\circ$$

∴ $$\frac{E_0}{H_0} = \eta = 200\angle 30^\circ \quad \Rightarrow \quad E_0 = 2000\angle 30^\circ$$

∴ $$E = \text{Re}[2000\angle 30^\circ e^{-\gamma x}] = 2000 e^{-\alpha x}\cos\left(\omega t - \frac{1}{2}x + \frac{\pi}{6}\right)$$

Since the wave is propagating in the $\hat{a}_x$ direction and the magnetic field is having only $\hat{a}_y$ component, the electric field will have only $-\hat{a}_z$ component [∴ direction of wave propagation is given by $\hat{a}_E \times \hat{a}_H$]. Hence, the electric field is given as

$$\vec{E} = -2000 e^{-\alpha x}\cos\left(\omega t - \frac{1}{2}x + \frac{\pi}{6}\right)\hat{a}_z \text{ V/m}$$

Now, we know that

$$\alpha = \omega\sqrt{\frac{\mu\varepsilon}{2}\left[\sqrt{1 + \left(\frac{\sigma}{\omega\varepsilon}\right)^2} - 1\right]} \quad \text{and} \quad \beta = \omega\sqrt{\frac{\mu\varepsilon}{2}\left[\sqrt{1 + \left(\frac{\sigma}{\omega\varepsilon}\right)^2} + 1\right]}$$

∴ $$\frac{\alpha}{\beta} = \frac{\sqrt{\left[\sqrt{1 + \left(\frac{\sigma}{\omega\varepsilon}\right)^2} - 1\right]}}{\sqrt{\left[\sqrt{1 + \left(\frac{\sigma}{\omega\varepsilon}\right)^2} + 1\right]}}$$

But $$\frac{\sigma}{\omega\varepsilon} = \tan\theta_\eta = \tan 60° = \sqrt{3}$$

$$\therefore \quad \frac{\alpha}{\beta} = \frac{\sqrt{\left[\sqrt{1+3}-1\right]}}{\sqrt{\left[\sqrt{1+3}+1\right]}} = \frac{1}{\sqrt{3}}$$

$$\therefore \quad \alpha = \beta \times \frac{1}{\sqrt{3}} = \frac{1}{2\sqrt{3}} = 0.2887 \text{ Np/m}$$

The skin depth is given as, $\delta = \dfrac{1}{\alpha} = 2\sqrt{3} = 3.464$ m

Since the wave has only z-component of the electric field, it is polarised along the z-direction.

Example 5.20 For sea water with $\sigma = 5$ mho/m and $\varepsilon_r = 80$, $\mu = \mu_0$, find the distance a radio signal can be transmitted at 25 kcps and 25 Mcps, if the range is taken to be the distance at which 90% of the wave amplitude is attenuated.

Solution The wave is attenuated by the factor $e^{-\alpha x}$. In this case, the wave is attenuated to 90%.

$$\therefore \quad e^{-\alpha x} = 0.1 \Rightarrow -\alpha x = \ln(0.1) \Rightarrow x = \frac{2.302}{\alpha}$$

Also, $$\alpha = \omega\sqrt{\frac{\mu\varepsilon}{2}\left(\sqrt{1+\frac{\sigma^2}{\omega^2\varepsilon^2}}-1\right)}$$

Substituting the given values, we have the following results.

For 25 kcps,

$$\alpha = 2\pi \times 25 \times 10^3 \sqrt{\frac{4\pi \times 10^{-7} \times 80 \times 8.854 \times 10^{-12}}{2}\left(\sqrt{1+\frac{5^2}{(2\pi \times 25 \times 10^3)^2(8.854 \times 10^{-12})^2}}-1\right)}$$

$$= 0.7025$$

$$\therefore \quad x = \frac{2.302}{0.7025} = 3.278 \text{ m}$$

For 25 Mcps,

$$\alpha = 2\pi \times 25 \times 10^6 \sqrt{\frac{4\pi \times 10^{-7} \times 80 \times 8.854 \times 10^{-12}}{2}\left(\sqrt{1+\frac{5^2}{(2\pi \times 25 \times 10^6)^2(8.854 \times 10^{-12})^2}}-1\right)}$$

$$= 21.969$$

$$\therefore \quad x = 21.969 = 0.105 \text{ m}$$

Example 5.21 For a uniform plane wave in fresh lake water $\sigma = 10^{-3}$ mho/m, $\varepsilon_r = 80$, $\mu = \mu_0$. Calculate α, β, η and λ for two frequencies 100 MHz and 10 kHz.

Solution **For 100 MHz frequency,**

$$\frac{\omega\varepsilon}{\sigma} = \frac{2\pi f\varepsilon}{\sigma} = \frac{2\pi \times 100 \times 10^6 \times 8.854 \times 10^{-12} \times 80}{10^{-3}} = 445$$

$$\therefore \qquad \frac{\omega\varepsilon}{\sigma} >> 1$$

So, the lake water acts as a good dielectric.

$$\therefore \quad \alpha = \frac{\sigma}{2}\sqrt{\frac{\mu}{\varepsilon}} = \frac{10^{-3}}{2}\sqrt{\frac{4\pi \times 10^{-7}}{8.854 \times 10^{-12} \times 80}} = 0.021 \text{ Neper/m}$$

$$\therefore \quad \beta = \omega\sqrt{\mu\varepsilon}\left(1 + \frac{\sigma^2}{8\omega^2\varepsilon^2}\right)$$

$$= 2\pi \times 100 \times 10^6 \sqrt{4\pi \times 10^{-7} \times 8.854 \times 10^{-12} \times 80}\left(1 + \frac{10^{-6}}{8(2\pi \times 100 \times 10^6)^2(8.854 \times 10^{-12} \times 80)^2}\right)$$

$$\approx 18.72 \text{ radian/m}$$

$$\therefore \quad \eta = \sqrt{\frac{\mu}{\varepsilon}} = \sqrt{\frac{\mu_0\mu_r}{\varepsilon_0\varepsilon_r}} = 120\pi \times \sqrt{\frac{1}{80}} = 42.13\ \Omega$$

$$\therefore \quad \lambda = \frac{2\pi}{\beta} = \frac{2\pi}{18.72} = 0.335 \text{ m}$$

For 10 kHz frequency,

$$\frac{\omega\varepsilon}{\sigma} = \frac{2\pi f\varepsilon}{\sigma} = \frac{2\pi \times 10 \times 10^3 \times 8.854 \times 10^{-12} \times 80}{10^{-3}} = 0.0445$$

$$\therefore \quad \frac{\omega\varepsilon}{\sigma} << 1$$

So, the lake water acts as a good conductor.

$$\therefore \quad \alpha = \sqrt{\frac{\omega\mu\sigma}{2}} = \sqrt{\frac{2\pi \times 10 \times 10^3 \times 4\pi \times 10^{-7} \times 10^{-3}}{2}} = 2\pi \times 10^{-3} \text{ Neper/m}$$

$$\therefore \quad \beta = \alpha = 2\pi \times 10^{-3} \text{ radian/m}$$

$$\therefore \quad \eta = \sqrt{\frac{\omega\mu}{\sigma}}\angle 45° = \sqrt{\frac{2\pi \times 10 \times 10^3 \times 4\pi \times 10^{-7}}{10^{-3}}}\angle 45° = 2\pi\sqrt{2} = 8.886\angle 45°\ \Omega$$

$$\therefore \quad \lambda = \frac{2\pi}{\beta} = \frac{2\pi}{2\pi \times 10^{-3}} = 1000 \text{ m}$$

Example 5.22

1. Earth is considered to be a good conductor when $\frac{\omega\varepsilon}{\sigma} << 1$. Determine the highest frequencies for which earth can be considered a good conductor if <<1 means less than 0.1. Assume the following constants:

$$\sigma = 5 \times 10^{-3} \text{mho/m}, \ \varepsilon = 10\varepsilon_0$$

Can α be assumed zero at these frequencies?

2. If the earth is considered to be a perfect dielectric for $\frac{\sigma}{\omega\varepsilon} < 0.1$; then at what frequencies may earth be considered a perfect dielectric? Assume: $\sigma = 5 \times 10^{-3}$mho/m and $\mu_r = 1$, $\varepsilon_r = 8$. Can α be assumed zero at these frequencies?

Solution

1. In this case, $\frac{\omega\varepsilon}{\sigma} < 0.1$

$$\therefore \quad \omega < \frac{0.1\sigma}{\varepsilon} < \frac{0.1 \times 5 \times 10^{-3}}{8.854 \times 10^{-12} \times 10}$$

$$\therefore \quad \omega < 5.647 \times 10^{6}$$

The highest frequency for which earth can be considered a good conductor is

$$f_c = \frac{\omega_c}{2\pi} = \frac{5.647 \times 10^6}{2\pi} = 0.899 \text{ MHz}$$

For a good conductor, $\alpha = \sqrt{\frac{\omega\mu\sigma}{2}} = \sqrt{\frac{5.647 \times 10^6 \times 4\pi \times 10^{-7} \times 5 \times 10^{-3}}{2}} = 0.133$ Neper/m

Hence, α cannot be zero.

2. In this case, $\frac{\sigma}{\omega\varepsilon} < 0.1$

$$\therefore \quad \omega > \frac{\sigma}{0.1\varepsilon} < \frac{5 \times 10^{-3}}{0.1 \times 8.854 \times 10^{-12} \times 8}$$

$$\therefore \quad \omega > 705.89 \times 10^{6}$$

The highest frequency for which earth can be considered a good conductor is

$$f_c = \frac{\omega_c}{2\pi} = \frac{705.89 \times 10^6}{2\pi} = 112.35 \text{ MHz}$$

For a good dielectric, $\therefore \quad \alpha = \frac{\sigma}{2}\sqrt{\frac{\mu}{\varepsilon}} = \frac{5 \times 10^{-3}}{2}\sqrt{\frac{4\pi \times 10^{-7}}{8.854 \times 10^{-12} \times 8}} = 0.333$ Neper/m

Hence, α is independent of frequency and whatever be the higher frequencies, α will be about 0.333 *Np/m*.

Example 5.23 For an aluminium wire having a diameter 2.6 mm, calculate the ratio of *ac* to *dc* resistance at (a) 10 MHz, (b) 2 GHz. This is shown in Fig. 5.9

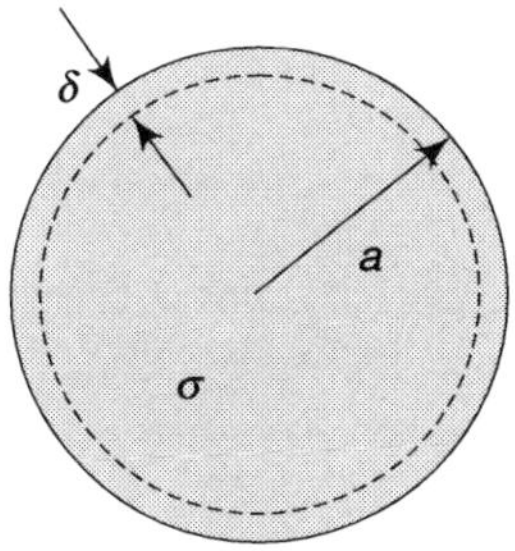

Fig. 5.9 *Wire of circular cross-section*

Solution Let l be the length of the wire
a be radius of the wire
For DC supply, the current will flow through the wire uniformly. So, the DC resistance is,

$$R_{DC} = \frac{\rho l}{A_{DC}} = \frac{\rho l}{\pi a^2}$$

When the supply is AC, the current flows only through the skin of the wire, so that the AC resistance is given as,

$$R_{AC} = \frac{\rho l}{A_{AC}} = \frac{\rho l}{\pi a^2 - \pi(a-\delta)^2} = \frac{\rho l}{\pi[a^2 - (a-\delta)^2]} = \frac{\rho l}{\pi[a^2 - a^2 + 2a\delta - \delta^2]}$$

$$= \frac{\rho l}{2\pi a \delta}$$

Therefore, the ratio of AC to DC resistance is,

$$\frac{R_{AC}}{R_{DC}} = \frac{a}{2\delta}$$

where, $\delta = \frac{1}{\alpha} = \sqrt{\frac{2}{\omega\mu\sigma}}$ is the skin depth.

$$\frac{R_{AC}}{R_{DC}} = \frac{a}{2\sqrt{\frac{2}{\omega\mu\sigma}}} = \frac{1}{2\sqrt{2}} a\sqrt{\omega\mu\sigma}$$

For aluminium, $\mu = \mu_0$, $\sigma = 3.5 \times 10^7$ S/m, so that,

$$\frac{R_{AC}}{R_{DC}} = \frac{1}{2\sqrt{2}} \frac{0.0026}{2} \sqrt{2\pi \times f \times 4\pi \times 10^{-7} \times 3.5 \times 10^7} = 0.00764\sqrt{f}$$

(a) For $f = 10$ MHz, $\frac{R_{AC}}{R_{DC}} = 0.00764\sqrt{10 \times 10^6} = 24.16$

(b) For $f = 2$ GHz, $\frac{R_{AC}}{R_{DC}} = 0.00764\sqrt{2 \times 10^9} = 341.67$

Example 5.24 Show that the ratio of very high-frequency resistance to DC resistance of a round conductor of radius r_0 and material with width of penetration δ can be written as

$$\frac{R_{(hf)}}{R_0} = \frac{r_0}{2\delta}$$

Solution Let l be the length of the wire
r_0 be radius of the wire

For DC supply, the current will flow through the wire uniformly. So, the DC resistance is,

$$R_{DC} = \frac{\rho l}{A_{DC}} = \frac{\rho l}{\pi r_0^2}$$

When the supply is AC of very high-frequency, the current flows only through the skin of the wire, so that the ac resistance is given as

$$R_{AC} = \frac{\rho l}{A_{AC}} = \frac{\rho l}{\pi r_0^2 - \pi(r_0-\delta)^2} = \frac{\rho l}{\pi\left[r_0^2 - (r_0-\delta)^2\right]} = \frac{\rho l}{\pi\left[r_0^2 - r_0^2 + 2r_0\delta - \delta^2\right]}$$

$$= \frac{\rho l}{2\pi r_0 \delta}$$

where, $\delta = \dfrac{1}{\alpha} = \sqrt{\dfrac{2}{\omega\mu\sigma}}$ is the skin depth.

Therefore, the ratio of very high-frequency resistance to DC resistance is

$$\boxed{\frac{R_{AC}}{R_{DC}} = \frac{r_0}{2\delta}}$$

5.8 POLARISATION

The *polarisation* of a uniform plane wave refers to the time-varying behaviour of the electric field at some point in space, i.e., the orientation of the electric field vector at a given instant of time in space.

5.8.1 Classification of Polarisation

A plane electromagnetic wave in which electric field vector vibrates harmonically along a fixed straight line perpendicular to the direction of wave propagation without changing its orientation is known as *plane polarised electromagnetic wave.* When two orthogonal plane polarised electromagnetic waves are superimposed, then the resultant vector rotates under certain conditions, giving rise to different polarisations like, *linear*, *circular* and *elliptical.*

Case (1): When two electric field vectors vibrate along the same axis: We will consider the propagation of two linearly polarised electromagnetic waves propagating along the same z-axis with electric field vectors vibrating along the x-axis.

The associated electric fields can be written as

$$E_1 = E_{x1} \cos(\omega t - \beta z + \phi_1)$$
$$E_2 = E_{x2} \cos(\omega t - \beta z + \phi_2)$$

where, E_{x1} and E_{x2} are the amplitudes of the waves, ω is the phase angle, β is the phase propagation constant and ϕ_1 and ϕ_2 are the phase angles of the two waves respectively.

Hence, the resultant wave can be written as

$$\begin{aligned} E &= E_1 + E_2 = E_{x1} \cos(\omega t - \beta z + \phi_1) + E_{x2} \cos(\omega t - \beta z + \phi_2) \\ &= E_{x1}[\cos(\omega t - \beta z)\cos\phi_1 - \sin(\omega t - \beta z)\sin\phi_1] + E_{x2}[\cos(\omega t - \beta z)\cos\phi_2 - \sin(\omega t - \beta z)\sin\phi_2] \\ &= (E_{x1}\cos\phi_1 + E_{x2}\cos\phi_2)\cos(\omega t - \beta z) - (E_{x1}\sin\phi_1 + E_{x2}\sin\phi_2)\sin(\omega t - \beta z) \\ &= E_x \cos\phi \cos(\omega t - \beta z) - E_x \sin\phi \sin(\omega t - \beta z) \qquad \left\{\begin{matrix}(E_{x1}\cos\phi_1 + E_{x2}\cos\phi_2) = E_x\cos\phi \\ (E_{x1}\sin\phi_1 + E_{x2}\sin\phi_2) = E_x\sin\phi\end{matrix}\right\} \end{aligned}$$

$\therefore \quad E = E_x \cos(\omega t - \beta z + \phi)$

This equation represents the resultant of the two waves and is a plane polarised wave vibrating along the same axis.

Case (2): When two electric field vectors vibrate perpendicular to each other: We will consider the superposition of two plane polarised electromagnetic waves propagating along the z-axis and vibrating along x and y axes respectively.

The associated electric fields can be written as

$$\vec{E}_x = E_{x0}e^{-\alpha z}\cos(\omega t - \beta z)\hat{a}_x$$
$$\vec{E}_y = E_{y0}e^{-\alpha z}\cos(\omega t - \beta z + \phi)\hat{a}_y$$

where, ω is the frequency of the waves, β is the propagation constant, α is the attenuation constant, E_{x0} and E_{y0} are the amplitudes of the x and y oriented electric fields respectively, and ϕ is the phase difference between the two fields.

In order to obtain the behaviour of the field with the variation of time at a given point in space, we shall consider the point $z = 0$. Then the two fields become,

$$\vec{E}_x = E_{x0}\cos\omega t\hat{a}_x \tag{5.22a}$$

$$\vec{E}_y = E_{y0}\cos(\omega t + \phi)\hat{a}_y \tag{5.22b}$$

At any instant of time, the total electric field is the vector sum of the two fields, given as

$$\vec{E} = \vec{E}_x + \vec{E}_y = E_{x0}\cos\omega t\hat{a}_x + E_{y0}\cos(\omega t + \phi)\hat{a}_y$$
$$= \sqrt{E_{x0}^2\cos^2\omega t + E_{y0}^2\cos^2(\omega t + \phi)}\ \angle\tan^{-1}\left(\frac{E_{y0}\cos(\omega t + \phi)}{E_{x0}\cos\omega t}\right)$$

Thus, both the magnitude and the direction of the resultant electric field change with time. The locus of the tip of the resultant field is obtained by eliminating the time parameter.

$$\cos\omega t = \frac{E_x}{E_{x0}} \quad\Rightarrow\quad \sin\omega t = \sqrt{1 - \frac{E_x^2}{E_{x0}^2}}$$

From Eq. (5.17*b*),

$$\frac{E_y}{E_{y0}} = \cos\omega t\cos\phi - \sin\omega t\sin\phi = \frac{E_x}{E_{x0}}\cos\phi - \sqrt{1 - \frac{E_x^2}{E_{x0}^2}}\sin\phi$$

$$\Rightarrow \quad \frac{E_x}{E_{x0}}\cos\phi - \frac{E_y}{E_{y0}} = \sqrt{1 - \frac{E_x^2}{E_{x0}^2}}\sin\phi$$

Squaring both sides,

$$\Rightarrow \quad \frac{E_x^2}{E_{x0}^2}\cos^2\phi + \frac{E_y^2}{E_{y0}^2} - 2\frac{E_xE_y}{E_{x0}E_{y0}}\cos\phi = \left(1 - \frac{E_x^2}{E_{x0}^2}\right)\sin^2\varphi$$

$$\Rightarrow \quad \frac{E_x^2}{E_{x0}^2}\cos^2\phi + \frac{E_x^2}{E_{x0}^2}\sin^2\phi + \frac{E_y^2}{E_{y0}^2} - 2\frac{E_xE_y}{E_{x0}E_{y0}}\cos\phi = \sin^2\phi$$

$$\therefore \quad \frac{E_x^2}{E_{x0}^2} - 2\frac{E_xE_y}{E_{x0}E_{y0}}\cos\phi + \frac{E_y^2}{E_{y0}^2} = \sin^2\phi \tag{5.23}$$

This is the general equation of an ellipse. In general, therefore, the tip of the field vector $\vec{E}$ draws an ellipse with the variation of time.

Three cases may appear, for which three different polarisations are represented.

1. Linear polarisation: A linearly polarised wave is one in which the tip of the electric field vector $\vec{E}$ traces a straight line as time varies.

Here, E_x and E_y may or may not have the same amplitude, but the phase difference between them is zero.

$$\therefore \qquad \phi = 0$$

From Eq. (5.23),

$$\frac{E_x^2}{E_{x0}^2} - 2\frac{E_x E_y}{E_{x0} E_{y0}} + \frac{E_y^2}{E_{y0}^2} = 0$$

$$\left(\frac{E_x}{E_{x0}} - \frac{E_y}{E_{y0}}\right) = 0$$

$$\therefore \qquad \boxed{\frac{E_x}{E_{x0}} = \frac{E_y}{E_{y0}} \quad \text{or} \quad E_y = \left(\frac{E_{y0}}{E_{x0}}\right) E_x} \qquad (5.24)$$

This represents a *linearly polarised wave*. The resultant polarisation vector oscillates along a line making an angle, $\theta = \tan^{-1}\left(\frac{E_{y0}}{E_{x0}}\right)$.

Depending upon the ratio of E_{y0} and E_{x0}, the slope of the line changes.

(a) If $E_{x0} = 0$, the line becomes vertical and the wave is said to be a *vertically polarised wave* and the wave is said to have *vertical polarisation*.

(b) If $E_{y0} = 0$, the line becomes horizontal and the wave is said to be a *horizontally polarised wave* and the wave is said to have *horizontal polarisation*.

(c) If $E_{x0} = E_{y0}$, the wave is said to be linearly polarised with polarisation angle of 45°.

A plane electromagnetic wave is said to be linearly polarised. The transverse electric field wave is accompanied by a magnetic field wave as shown in Fig. 5.10.

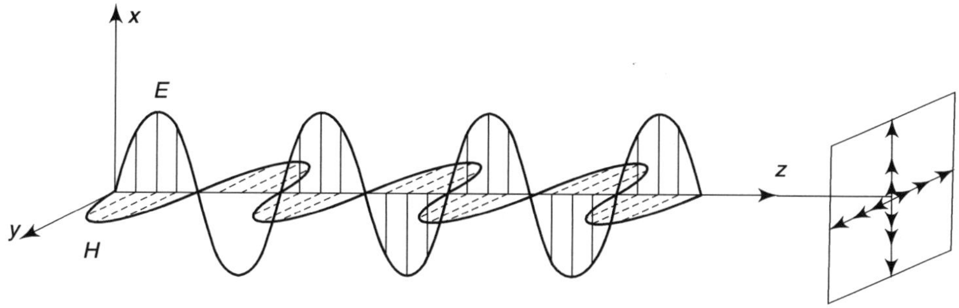

Fig. 5.10 *Linearly polarised wave*

2. Circular polarisation: A circularly polarised wave is one in which the tip of the electric field vector $\vec{E}$ traces a circle as time varies.

Here, the amplitudes of the two vectors are equal, i.e. $E_{x0} = E_{y0} = E_0$, but, their phases differ by $\pi/2$, i.e. $\phi = \pm\pi/2$.

From Eq. (5.23) $$\frac{E_x^2}{E_0^2} + \frac{E_y^2}{E_0^2} = 1$$

$$\therefore \quad \boxed{\frac{E_x^2}{E_0^2} + \frac{E_y^2}{E_0^2} = 1 \quad \text{or} \quad E_x^2 + E_y^2 = E_0^2} \qquad \textbf{(5.25)}$$

This is the equation of a circle and thus, the tip of the field vector traces out a circle at frequency ω; such a wave is said to be *circularly polarised*.

When $\vec{E}_x$ leads $\vec{E}_y$ by $\pi/2$, i.e. $\phi = -\pi/2$, rotation of the vector is counterclockwise, the wave is said to have *left- circular polarization* or *positive helicity*.

When $\vec{E}_x$ lags $\vec{E}_y$ by $\pi/2$, i.e. $\phi = \pi/2$, rotation of the vector is clockwise, the wave is said to have *right-circular polarisation* or *negative helicity*.

A circularly polarized wave consists of two perpendicular electromagnetic plane waves of equal amplitude and 90° difference in phase. The wave shown in Fig. 5.11 is right-circularly polarized.

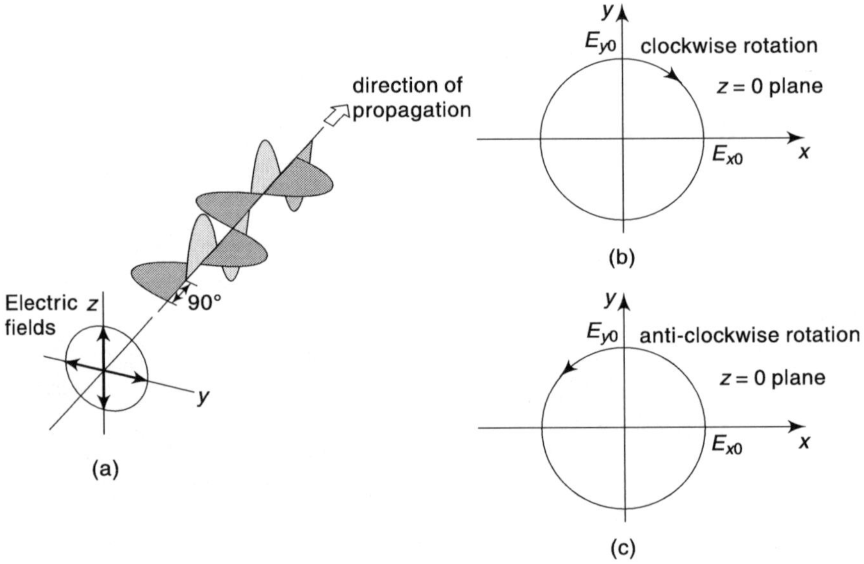

Fig. 5.11 *Circularly polarized wave (a) Circular polarisation, (b) Right circularly polarised wave, (c) Left circularly polarised wave*

3. Elliptical polarisation: An elliptically polarised wave is one in which the tip of the electric filed vector $\vec{E}$ traces an ellipse as time varies.

Here, the two fields neither have the same amplitude nor the phase difference is zero or $\pi/2$, i.e. $E_{x0} \neq E_{y0}$ and $\phi \neq 0$ or $\pm\pi/2$.

Then, from Eq. (5.23), we get an *elliptically polarised wave*. Similar to a circularly polarised wave, the sign of the phase angle ϕ determines the sense of rotation of the field vector.

If ϕ is positive, the wave is said to be *left-elliptically polarised.*

If ϕ is negative, the wave is said to be *right-elliptically polarised.*

Figure 5.12 shows an elliptically polarised wave.

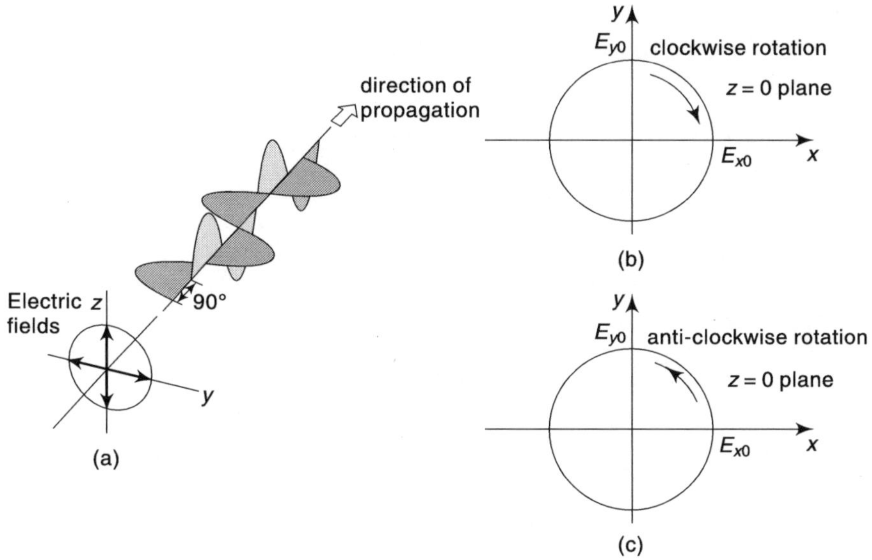

Fig. 5.12 *Elliptically polarised wave (a) Elliptical polarisation (b) Right elliptically polarised wave (c) Left elliptically polarised wave*

5.9 REFLECTION AND REFRACTION OF PLANE ELECTROMAGNETIC WAVES AT THE INTERFACE BETWEEN TWO DIELECTRICS

When a plane wave propagating in a homogeneous medium encounters an interface with a different medium, a portion of the wave is reflected from the interface while the remainder of the wave is transmitted. The proportion of reflection and transmission depends on the constitutive parameters of the media, i.e., ε, μ, σ.

We consider the reflection and refraction of a plane wave incident on a single boundary separating two different dielectric media. Two types of incidence may occur:

1. Normal Incidence, and
2. Oblique Incidence.

5.9.1 Normal Incidence

When a plane electromagnetic wave is incident normally at the interface between two dielectrics, part of the energy is transmitted and part of it is reflected.

Let E_i — Electric field strength of the incident wave striking the interface;
E_r — Electric field strength of the reflected wave leaving the interface;
E_t — Electric field strength of the transmitted wave propagated into the second dielectric;
H_i, H_r, H_t — corresponding magnetic field strengths;
ε_1, μ_1 — permittivity and permeability of the first dielectric;
ε_2, μ_2 — permittivity and permeability of the second dielectric

$\therefore \quad \eta_1 = \sqrt{\dfrac{\mu_1}{\varepsilon_1}}$ = intrinsic impedance of the first dielectric

$\therefore \quad \eta_2 = \sqrt{\dfrac{\mu_2}{\varepsilon_2}}$ = intrinsic impedance of the second dielectric

So,

$$E_i = \eta_1 H_i \tag{5.26a}$$

$$E_r = -\eta_1 H_r \tag{5.26b}$$

$$E_t = \eta_2 H_t \tag{5.26c}$$

According to the continuity of the tangential components of E and H,

$$(H_i + H_r) = H_t \tag{5.27a}$$

$$(E_i + E_r) = E_t \tag{5.27b}$$

From Eq. (5.27*a*) and Eq. (5.26*c*),

$$(H_i + H_r) = \frac{E_t}{\eta_2}$$

$$\Rightarrow \quad \frac{1}{n_1}(E_i - E_r) = \frac{1}{n_2}(E_i + E_r)$$

$$\Rightarrow \quad \frac{E_r}{E_i} = \frac{\eta_2 - \eta_1}{\eta_2 + \eta_1} \tag{5.28}$$

Also,

$$\frac{E_t}{E_i} = \frac{E_i + E_r}{E_i} = \left(1 + \frac{E_r}{E_i}\right) = \left(1 + \frac{\eta_2 - \eta_1}{\eta_2 + \eta_1}\right) = \frac{2n_2}{\eta_1 + \eta_2} \tag{5.29}$$

Similarly,

$$\frac{H_r}{H_i} = -\frac{E_r}{E_i} = \frac{\eta_1 - \eta_2}{\eta_1 + \eta_2} \tag{5.30}$$

$$\frac{H_t}{H_i} = \frac{\eta_1}{\eta_2}\frac{E_t}{E_i} = \frac{2n_1}{\eta_1 + \eta_2} \tag{5.31}$$

The *reflection coefficient* on *reflectance* is defined as the ratio of reflected wave to incident wave.

Similarly, *transmission co-efficient* or *transmittance* is defined as the ratio of transmitted wave to incident wave.

So, the reflectances and transmitances for electric and magnetic fields are given as:-

$$\text{Reflectance, } \Gamma = \frac{\eta_2 - \eta_1}{\eta_2 + \eta_1}; \quad \text{for electric field}$$

$$= \frac{\eta_1 - \eta_2}{\eta_1 + \eta_2}; \quad \text{for magnetic field}$$

$$\text{Transmittance, } \tau = \frac{2\eta_2}{\eta_1 + \eta_2}; \quad \text{for electric field}$$
$$= \frac{2\eta_1}{\eta_1 + \eta_2}; \quad \text{for magnetic field}$$

For non-magnetic dielectrics, $\mu_1 = \mu_2 = \mu_0$

Thus,

$$\frac{E_r}{E_i} = \frac{\sqrt{\frac{\mu_0}{\varepsilon_2}} - \sqrt{\frac{\mu_0}{\varepsilon_1}}}{\sqrt{\frac{\mu_0}{\varepsilon_2}} + \sqrt{\frac{\mu_0}{\varepsilon_1}}} = \frac{\sqrt{\varepsilon_1} - \sqrt{\varepsilon_2}}{\sqrt{\varepsilon_1} + \sqrt{\varepsilon_2}} \tag{5.32}$$

$$\frac{E_t}{E_i} = \frac{2\sqrt{\varepsilon_1}}{\sqrt{\varepsilon_1} + \sqrt{\varepsilon_2}} \tag{5.33}$$

$$\frac{H_r}{H_i} = \frac{\sqrt{\varepsilon_2} - \sqrt{\varepsilon_1}}{\sqrt{\varepsilon_2} + \sqrt{\varepsilon_1}} \tag{5.34}$$

and

$$\frac{H_t}{H_i} = \frac{2\sqrt{\varepsilon_2}}{\sqrt{\varepsilon_1} + \sqrt{\varepsilon_2}} \tag{5.35}$$

So, the reflectances and transmittances for non-magnetic dielectrics are

$$\text{Reflectance, } \Gamma = \frac{\sqrt{\varepsilon_1} - \sqrt{\varepsilon_2}}{\sqrt{\varepsilon_1} + \sqrt{\varepsilon_2}}; \quad \text{for electric field}$$
$$= \frac{\sqrt{\varepsilon_2} - \sqrt{\varepsilon_1}}{\sqrt{\varepsilon_2} + \sqrt{\varepsilon_1}}; \quad \text{for magnetic field}$$

$$\text{Transmittance, } \tau = \frac{2\sqrt{\varepsilon_1}}{\sqrt{\varepsilon_1} + \sqrt{\varepsilon_2}}; \quad \text{for electric field}$$
$$= \frac{2\sqrt{\varepsilon_2}}{\sqrt{\varepsilon_1} + \sqrt{\varepsilon_2}}; \quad \text{for magnetic field}$$

NOTE

(i) *Both Γ and τ are dimensionless and may be complex.*

(ii) $(1 + \Gamma) = \tau$

(iii) $0 \le |\Gamma| \le 1$

Example 5.25 An electromagnetic wave impinges on a metallic sheet. Compare the reflection coefficients for copper and iron if

$$\begin{aligned} \sigma &= 5.8\times 10^7 \text{ mho/m} && \text{for copper,}\\ &= 1\times 10^6 \text{ mho/m} && \text{for iron,}\\ \mu_r &= \mu_0 && \text{for copper,}\\ &= 1000\mu_0 && \text{for iron,}\\ \varepsilon_r &= 1 && \text{for copper and iron.} \end{aligned}$$

Take frequency to be 1 megacycle.

Solution Reflection coefficient, $\Gamma = \dfrac{\eta_2 - \eta_1}{\eta_2 + \eta_1}$

where, η_1 is the intrinsic impedance of air and η_2 is the intrinsic impedance of metal
For copper:

$$\eta_2 = \sqrt{\frac{j\omega\mu}{\sigma + j\omega\varepsilon}} = \sqrt{\frac{j\times 2\pi\times 10^6\times 4\pi\times 10^{-7}}{5.8\times 10^7 + j\times 2\pi\times 10^6\times 8.854\times 10^{-12}}}$$
$$= 3.68961\times 10^{-4}\angle 45^\circ$$
$$= (2.608951\times 10^{-4} + j2.60895\times 10^{-4})$$

Hence, reflection coefficient for copper is

$$\Gamma_{\text{copper}} = \frac{3.69\times 10^{-4}\angle 45^\circ - 377\angle 0^\circ}{3.69\times 10^{-4}\angle 45^\circ + 377\angle 0^\circ} \approx 1, \quad \text{almost perfect reflector}$$

For iron:

$$\eta_2 = \sqrt{\frac{j\omega\mu}{\sigma + j\omega\varepsilon}} = \sqrt{\frac{j\times 2\pi\times 10^6\times 4\pi\times 1000\times 10^{-7}}{1\times 10^6 + j\times 2\pi\times 10^6\times 8.854\times 10^{-12}}}$$
$$= 0.0889\angle 45^\circ$$

Hence, reflection coefficient for iron is

$$\Gamma_{\text{iron}} = \frac{0.0889\angle 45^\circ - 377\angle 0^\circ}{0.0889\angle 45^\circ + 377\angle 0^\circ} \approx 0.99, \quad \text{very good reflector}$$

***Example 5.26** $\vec{E}$ and $\vec{H}$ waves travelling in free space, are normally incident on the interface with a perfect dielectric with $\varepsilon_r = 3$. Compute the magnitudes of incident, reflected and transmitted $\vec{E}$ and $\vec{H}$ waves at the interface. Take $E_i = 1.5$ mV/m in medium 1.

Solution For medium 1 (free space), $\eta_1 = \sqrt{\dfrac{\mu_0}{\varepsilon_0}} = 120\pi = 377\ \Omega$

For medium 2 (perfect dielectric),

$$\eta_2 = \sqrt{\frac{\mu}{\varepsilon}} = \sqrt{\frac{\mu_0\mu_r}{\varepsilon_0\varepsilon_r}} = 120\pi\times\sqrt{\frac{\mu_r}{\varepsilon_r}} = 120\pi\times\sqrt{\frac{1}{3}} = 217.66\ \Omega$$

The transmission coefficient is given as

$$\tau = \frac{E_t}{E_i} = \frac{2\eta_2}{\eta_1 + \eta_2} = \frac{2 \times 217.66}{377 + 217.66} = 0.732$$

The reflection coefficient is given as

$$\Gamma_R = \frac{\eta_2 - \eta_1}{\eta_2 + \eta_1} = \frac{217.66 - 377}{217.66 + 377} = -0.268$$

Since the amplitude of the incident electric wave is $E_i = 1.5$ mV/m, we have

Amplitude of the incident magnetic field, $H_i = \frac{E_i}{\eta_1} = \frac{1.5 \times 10^{-3}}{377} = 3.98\ \mu\text{A/m}$

Amplitude of the transmitted electric field, $E_t = \tau \times E_i = 0.732 \times 1.5 = 1.098$ mV/m

Amplitude of the transmitted magnetic field, $H_t = \frac{E_t}{\eta_2} = \frac{1.098}{217.66} = 5.04\ \mu\text{A/m}$

Amplitude of the reflected electric field, $E_r = \Gamma_R \times E_i = -0268 \times 1.5 = -0.402$ mV/m

Amplitude of the reflected magnetic field, $H_r = \frac{E_r}{\eta_2} = \frac{-0.402}{217.66} = -1.85\ \mu\text{A/m}$

Example 5.27 Determine the normal incidence reflection coefficients for seawater, freshwater, and 'good' earth at frequencies of 60 Hz, 1 MHz and 1 GHz. Use $\varepsilon_r = 80$, $\sigma = 4$ mho/m for sea water, $\varepsilon_r = 80$, $\sigma = 5 \times 10^{-3}$ for fresh water; and $\varepsilon_r = 15$, $\sigma = 10 \times 10^{-3}$ for good earth.

Solution The reflection coefficient is given as, $\Gamma_R = \frac{\eta_2 - \eta_1}{\eta_2 + \eta_1}$

where, η_1 is the intrinsic impedance of air and η_2 is the intrinsic impedance of the other medium.

For air, $\eta_1 = \sqrt{\frac{\mu_0}{\varepsilon_0}} = 120\pi = 377\ \Omega$

For seawater,

$$\eta_2 = \sqrt{\frac{j\omega\mu}{\sigma + j\omega\varepsilon}} = \sqrt{\frac{j\omega\mu_0}{\sigma + j\omega\varepsilon_0\varepsilon_r}} = \sqrt{\frac{j2\pi \times 60 \times 4\pi \times 10^{-7}}{4 + j2\pi \times 60 \times 8.854 \times 10^{-12} \times 80}} = 0.0109\angle 45^\circ$$

For freshwater,

$$\eta_2 = \sqrt{\frac{j\omega\mu}{\sigma + j\omega\varepsilon}} = \sqrt{\frac{j\omega\mu_0}{\sigma + j\omega\varepsilon_0\varepsilon_r}}$$

$$= \sqrt{\frac{j2\pi \times 1 \times 10^6 \times 4\pi \times 10^{-7}}{5 \times 10^{-3} + j2\pi \times 1 \times 10^6 \times 8.854 \times 10^{-12} \times 80}} = 34.34\angle 24.16^\circ$$

For good water,

$$\eta_2 = \sqrt{\frac{j\omega\mu}{\sigma + j\omega\varepsilon}} = \sqrt{\frac{j\omega\mu_0}{\sigma + j\omega\varepsilon_0\varepsilon_r}}$$

$$= \sqrt{\frac{j2\pi \times 1 \times 10^9 \times 4\pi \times 10^{-7}}{10 \times 10^{-3} + j2\pi \times 1 \times 10^9 \times 8.854 \times 10^{-12} \times 15}} = 97.27\angle 0.34^\circ$$

So, the reflection coefficients are

$$\Gamma_{R(\text{sea water})} = \frac{0.0109\angle 45^\circ - 377}{0.0109\angle 45^\circ + 377} = 0.99\angle 180^\circ$$

$$\Gamma_{R(\text{fresh water})} = \frac{34.34\angle 24.16^\circ - 377}{34.34\angle 24.16^\circ + 377} = 0.847\angle 175.7^\circ$$

$$\Gamma_{R(\text{good earth})} = \frac{97.27\angle 0.343^\circ - 377}{97.27\angle 0.343^\circ + 377} = 0.742\angle 180^\circ$$

Example 5.28 In free-space ($z \le 0$), a plane wave with

$$\vec{H} = 10\cos(10^8 t - \beta z)\hat{a}_x \text{ mA/m}$$

is incident normally on a lossless medium ($\varepsilon = 2\varepsilon_0$, $\mu = 8\mu_0$) in region $z \ge 0$. Determine the reflected wave H_r, E_r and the transmitted wave H_t, E_t.

Solution For free space, the intrinsic impedance is, $\eta_1 = \sqrt{\dfrac{\mu_0}{\varepsilon_0}} = 120\pi$

For the lossless medium, the intrinsic impedance is, $\eta_2 = \sqrt{\dfrac{\mu_0\mu_r}{\varepsilon_0\varepsilon_r}} = \sqrt{\dfrac{\mu_0 \times 8}{\varepsilon_0 \times 2}} = 2\eta_1 = 240\pi$

Given that: $\vec{H}_i = 10\cos(10^8 t - \beta z)\hat{a}_x$ mA/m

For $z < 0$, $\beta_1 = \dfrac{\omega}{c} = \dfrac{10^8}{3 \times 10^8} = \dfrac{1}{3}$

For $z > 0$, $\beta_1 = \omega\sqrt{\mu\varepsilon} = \omega\sqrt{\mu_0\varepsilon_0}\sqrt{\mu_r\varepsilon_r} = \dfrac{\omega}{c} \times 4 = 4\beta_1 = \dfrac{4}{3}$

From the given form of the incident magnetic field, we may expect that

$$\vec{E}_i = E_{i0}\cos(10^8 t - \beta_1 z)\hat{a}_{Ei}$$

where,

$$\hat{a}_{Ei} = \hat{a}_{Hi} \times \hat{a}_{ki} = \hat{a}_x \times \hat{a}_z = -\hat{a}_y$$

and,

$$E_{i0} = \eta_1 H_{i0} = 120\pi \times 10 = 1200\pi$$

$\therefore$

$$\vec{E}_i = -1200\pi\cos\left(10^8 t - \frac{1}{3}z\right)\hat{a}_y \text{ mV/m}$$

Now,

$$\frac{E_r}{E_i} = \frac{\eta_2 - \eta_1}{\eta_2 + \eta_1} = \frac{2\eta_1 - \eta_1}{2\eta_1 + \eta_1} = \frac{1}{3}$$

$\therefore$

$$\vec{E}_r = -400\pi\cos\left(10^8 t + \frac{1}{3}z\right)\hat{a}_y \text{ mV/m}$$

$\therefore$

$$\vec{H}_r = \frac{\vec{E}_r}{120\pi} = -\frac{10}{3}\cos\left(10^8 t + \frac{1}{3}z\right)\hat{a}_x \text{ mA/m}$$

Similarly, $\dfrac{E_t}{E_i} = \dfrac{2\eta_2}{\eta_2 + \eta_1} = \dfrac{4}{3}$

$$\therefore \qquad \vec{E}_t = -1600\pi \cos\left(10^8 t - \frac{4}{3}z\right)\hat{a}_y \text{ mV/m}$$

$$\vec{H}_t = \frac{\vec{E}_t}{240\pi} = -\frac{20}{3}\cos\left(10^8 t - \frac{4}{3}z\right)\hat{a}_x \text{ mA/m}$$

5.9.2 Oblique Incidence

Any plane wave which is obliquely incident on a planar media interface can be represented by a linear combination of two special cases:

1. Perpendicular or horizontal polarisation, and
2. Parallel or vertical polarisation.

1. Perpendicular or Horizontal Polarisation When the electric field vector $\vec{E}$ is perpendicular to the plane of incidence, i.e., the electric vector is parallel to the boundary surface, it is called *perpendicular* or *horizontal polarization.*

Figure 5.13 illustrates a perpendicularly polarised wave.

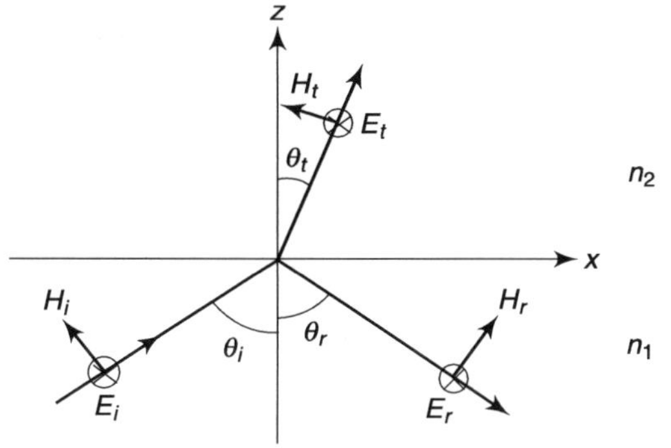

Fig. 5.13 *Perpendicular polarisation*

2. Parallel or Vertical Polarisation When the electric field vector $\vec{E}$ is parallel to the plane of incidence, i.e., the magnetic field is parallel to the boundary surface, it is called *parallel* or *vertical polarization.*

Figure 5.14 illustrates a parallel polarised wave.

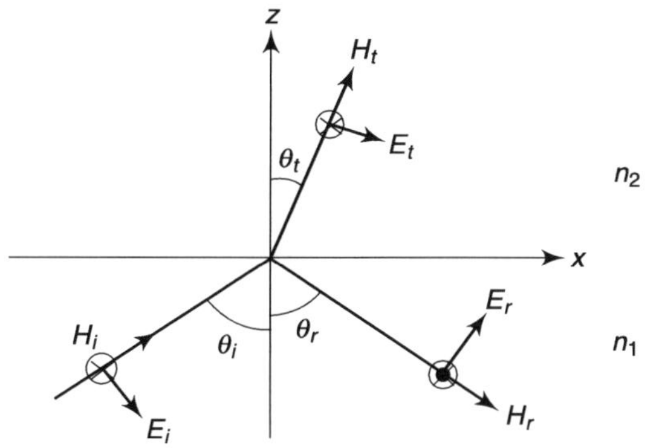

Fig. 5.14 *Parallel polarisation*

5.9.3 Reflection and Transmission Coefficients for Perpendicular and Parallel Polarisation

We consider Fig. 5.15.

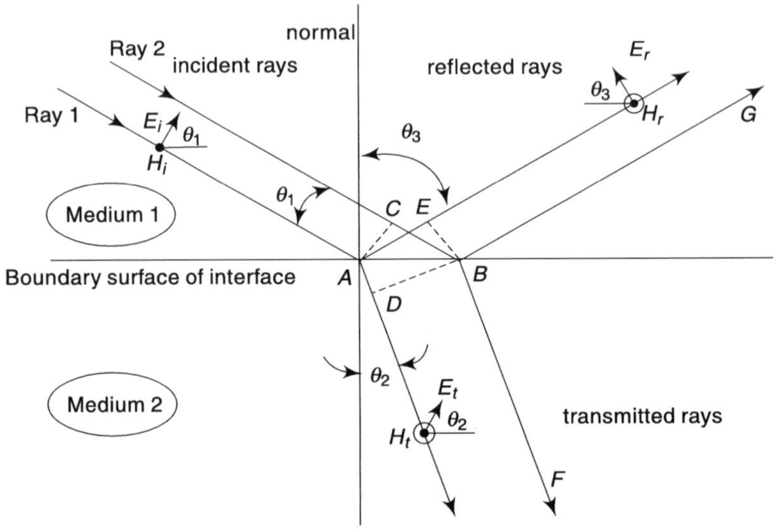

Fig. 5.15 *Reflection and transmission of plane waves*

For Fig. 5.15, the plane of the paper is the plane of incidence. The figure shows two rays of the EM wave:

Ray 1: reflected along *AE*, transmitted along *AD*
Ray 2: reflected along *BG*, transmitted along *BF*.

The directions:

AE and *BG* are parallel
AD and *BF* are parallel.

The line *AC*, which is perpendicular to the incident rays, represents the equi-phase surface in medium 1.

The line *DB*, which is perpendicular to the transmitted rays, represents the equi-phase surface in medium 2.

Ray 1 travels the distance *AD*,
Ray 2 travels the distance *CB*, and
Reflected Ray 1 travels the distance *AE*.

The time taken is the same for all three distances.

$$\therefore \qquad \frac{CB}{AD} = \frac{v_1 t}{v_2 t} = \frac{v_1}{v_2} = \frac{\sqrt{\mu_2 \varepsilon_2}}{\sqrt{\mu_1 \varepsilon_1}} = \frac{\sqrt{\varepsilon_2}}{\sqrt{\varepsilon_1}} = \sqrt{\frac{\varepsilon_2}{\varepsilon_1}} \qquad (\because \mu_1 = \mu_2)$$

$$\therefore \qquad \frac{AB \sin \theta_1}{AB \sin \theta_2} = \sqrt{\frac{\varepsilon_2}{\varepsilon_1}}$$

$$\therefore \quad \frac{\sin\theta_1}{\sin\theta_2} = \sqrt{\frac{\varepsilon_2}{\varepsilon_1}} \tag{5.36}$$

This equation is termed as *Snell's law*.

Also, $AE = CB \Rightarrow AB\sin\theta_3 = AB\sin\theta_1 \Rightarrow \theta_1 = \theta_3$

Perpendicular Polarisation For perpendicular polarisation, from the boundary condition,

$$(E_i + E_r) = E_t \tag{5.37a}$$

and

$$(H_i - H_r)\cos\theta_1 = H_t\cos\theta_2 \tag{5.37b}$$

From Eq. (5.26*a,b,c*),

$$\left(E_i\sqrt{\frac{\varepsilon_1}{\mu_1}} - E_r\sqrt{\frac{\varepsilon_1}{\mu_1}}\right)\cos\theta_1 = E_t\sqrt{\frac{\varepsilon_2}{\mu_2}}\cos\theta_2$$

$$\Rightarrow \quad (E_i\sqrt{\varepsilon_1} - E_r\sqrt{\varepsilon_1})\cos\theta_1 = E_t\sqrt{\varepsilon_2}\cos\theta_2 \qquad (\because \mu_1 = \mu_2)$$

$$\Rightarrow \quad (E_i - E_r)\sqrt{\frac{\varepsilon_1}{\varepsilon_2}}\cos\theta_1 = E_t\cos\theta_2$$

$$\Rightarrow \quad E_t = (E_i - E_r)\frac{\sin\theta_2\cos\theta_1}{\sin\theta_1\cos\theta_2} \quad \text{(by Eq. (5.36))} \tag{5.38}$$

From Eq. (5.37*a*) and Eq. (5.38), we get

$$(E_i + E_r) = (E_i - E_r)\frac{\sin\theta_2\cos\theta_1}{\sin\theta_1\cos\theta_2}$$

$$\Rightarrow \quad E_r(\sin\theta_2\cos\theta_1 + \cos\theta_2\sin\theta_1) = E_i(\sin\theta_2\cos\theta_1 - \cos\theta_2\sin\theta_1)$$

Hence the *reflection coefficient* is given as

$$\boxed{\text{Reflection Coefficient, } \Gamma = \frac{E_r}{E_i} = \frac{\sin(\theta_2 - \theta_1)}{\sin(\theta_2 + \theta_1)}}$$

It can be also written in the form

$$\boxed{\text{Reflection Coefficient, } \Gamma = \frac{E_r}{E_i} = \frac{\cos\theta_1 - \sqrt{\dfrac{\varepsilon_2}{\varepsilon_1} - \sin^2\theta_1}}{\cos\theta_1 + \sqrt{\dfrac{\varepsilon_2}{\varepsilon_1} - \sin^2\theta_1}}}$$

The transmission coefficient can be evaluated as follows.

$$\frac{E_t}{E_i} = \frac{E_i + E_r}{E_i} = \left(1 + \frac{E_r}{E_i}\right) = 1 + \frac{\sin(\theta_2 - \theta_1)}{\sin(\theta_2 + \theta_1)} = \frac{\sin(\theta_2 + \theta_1) + \sin(\theta_2 - \theta_1)}{\sin(\theta_2 + \theta_1)}$$

$\therefore$ $$\boxed{\text{Transmittance, } \tau = \frac{2 \sin \theta_2 \cos \theta_1}{\sin (\theta_2 + \theta_1)}}$$

Parallel Polarisation For parallel polarisation, from the boundary condition that the tangential component of vector $\vec{E}$ is continuous across the boundary,

$$(E_i - E_r) \cos \theta_1 = E_t \cos \theta_2 \qquad \textbf{(5.39a)}$$

and

$$(H_i + H_r) = H_t \qquad \textbf{(5.39b)}$$

From Eq. (5.39*b*),

$$\left(E_i \sqrt{\frac{\varepsilon_1}{\mu_1}} + E_r \sqrt{\frac{\varepsilon_1}{\mu_1}} \right) = E_t \sqrt{\frac{\varepsilon_2}{\mu_2}}$$

$\Rightarrow$ $$(E_i \sqrt{\varepsilon_1} + E_r \sqrt{\varepsilon_1}) = E_t \sqrt{\varepsilon_2} \qquad (\because \mu_1 = \mu_2)$$

$\Rightarrow$ $$(E_i + E_r) \sqrt{\varepsilon_1} = E_t \sqrt{\varepsilon_2}$$

$\Rightarrow$ $$E_t = (E_i + E_r) \sqrt{\frac{\varepsilon_1}{\varepsilon_2}} = (E_i + E_r) \frac{\sin \theta_2}{\sin \theta_1} \quad \text{(by Eq. (5.35))} \qquad \textbf{(5.40)}$$

From Eq. (5.39*a*) and Eq. (5.40), we get

$$(E_i - E_r) = (E_i + E_r) \frac{\sin \theta_2 \cos \theta_2}{\sin \theta_1 \cos \theta_1}$$

$\Rightarrow$ $$E_r (\sin \theta_2 \cos \theta_2 + \sin \theta_1 \cos \theta_1) = E_i (\sin \theta_1 \cos \theta_1 - \sin \theta_2 \cos \theta_2)$$

$\Rightarrow$ $$\frac{E_r}{E_i} = \frac{\sin \theta_1 \cos \theta_1 - \sin \theta_2 \cos \theta_2}{\sin \theta_2 \cos \theta_2 + \sin \theta_1 \cos \theta_1} = \frac{2 \sin \theta_1 \cos \theta_1 - 2 \sin \theta_2 \cos \theta_2}{2 \sin \theta_2 \cos \theta_2 + 2 \sin \theta_1 \cos \theta_1} = \frac{\sin 2\theta_1 - \sin 2\theta_2}{\sin 2\theta_2 + \sin 2\theta_1}$$
$$= \frac{\sin (\theta_1 - \theta_2) \cos (\theta_1 + \theta_2)}{\sin (\theta_1 + \theta_2) \cos (\theta_1 - \theta_2)} = \frac{\tan (\theta_1 - \theta_2)}{\tan (\theta_1 + \theta_2)}$$

Hence the *reflection coefficient* is given as

$$\boxed{\text{Reflection Coefficient, } \Gamma = \frac{E_r}{E_i} = \frac{\tan (\theta_1 - \theta_2)}{\tan (\theta_1 + \theta_2)}}$$

It can be also written in the form

$$\boxed{\text{Reflection Coefficient, } \Gamma = \frac{E_r}{E_i} = \frac{\dfrac{\varepsilon_2}{\varepsilon_1} \cos \theta_1 - \sqrt{\dfrac{\varepsilon_2}{\varepsilon_1} - \sin^2 \theta_1}}{\dfrac{\varepsilon_2}{\varepsilon_1} \cos \theta_1 + \sqrt{\dfrac{\varepsilon_2}{\varepsilon_1} - \sin^2 \theta_1}}}$$

The transmission coefficient can be evaluated as follows:

$$\frac{E_t}{E_i} = \frac{E_i + E_r}{E_i} = \left(1 + \frac{E_r}{E_i}\right) = 1 + \frac{\sin(\theta_1 - \theta_2)\cos(\theta_1 + \theta_2)}{\sin(\theta_1 + \theta_2)\cos(\theta_1 - \theta_2)}$$

$$= \frac{\sin(\theta_2 + \theta_1)\cos(\theta_1 - \theta_2) + \sin(\theta_1 - \theta_2)\cos(\theta_1 + \theta_2)}{\sin(\theta_1 + \theta_2)\cos(\theta_1 - \theta_2)}$$

$$= \frac{2\sin\theta_2\cos\theta_1}{\sin(\theta_1 + \theta_2)\cos(\theta_1 - \theta_2)}$$

$$\therefore \quad \boxed{\text{Transmittance, } \tau = \frac{2\sin\theta_2\cos\theta_1}{\sin(\theta_1 + \theta_2)\cos(\theta_1 - \theta_2)}}$$

NOTE

1. *The plane of incidence is the plane containing the incident wave and the normal to the interfacing surface.*
2. *The angle of incidence is defined as the angle between the direction of propagation and the normal to the boundary.*
3. *The terms 'horizontally' and 'vertically' polarized wave refer to the fact that the waves from horizontal and vertical antennas, respectively, would produce these particular orientations of electric and magnetic vectors in waves striking the surface of the earth.*

Example 5.29 Given two dielectric mediums: medium 1 is free space and medium 2 has $\varepsilon_2 = 4\varepsilon_0$ and $\mu = \mu_0$. Determine the reflection coefficients for oblique incidence $\theta_1 = 30°$ for

1. perpendicular polarisation;
2. parallel polarisation.

Solution:

1. For perpendicular polarisation, the reflection coefficient is

$$\Gamma_R = \frac{\cos\theta_1 - \sqrt{\frac{\varepsilon_2}{\varepsilon_1} - \sin^2\theta_1}}{\cos\theta_1 + \sqrt{\frac{\varepsilon_2}{\varepsilon_1} - \sin^2\theta_1}} = \frac{\cos 30° - \sqrt{\frac{4}{1} - \sin^2 30°}}{\cos 30° + \sqrt{\frac{4}{1} - \sin^2 30°}} = \frac{0.866 - \sqrt{4 - 0.25}}{0.866 + \sqrt{4 - 0.25}} = -0.382$$

2. For parallel polarisation, the reflection coefficient is

$$\Gamma_R = \frac{\frac{\varepsilon_2}{\varepsilon_1}\cos\theta_1 - \sqrt{\frac{\varepsilon_2}{\varepsilon_1} - \sin^2\theta_1}}{\frac{\varepsilon_2}{\varepsilon_1}\cos\theta_1 + \sqrt{\frac{\varepsilon_2}{\varepsilon_1} - \sin^2\theta_1}} = \frac{4\cos 30° - \sqrt{\frac{4}{1} - \sin^2 30°}}{4\cos 30° + \sqrt{\frac{4}{1} - \sin^2 30°}} = \frac{3.464 - \sqrt{4 - 0.25}}{3.464 + \sqrt{4 - 0.25}} = 0.283$$

5.9.4 Brewster Angle

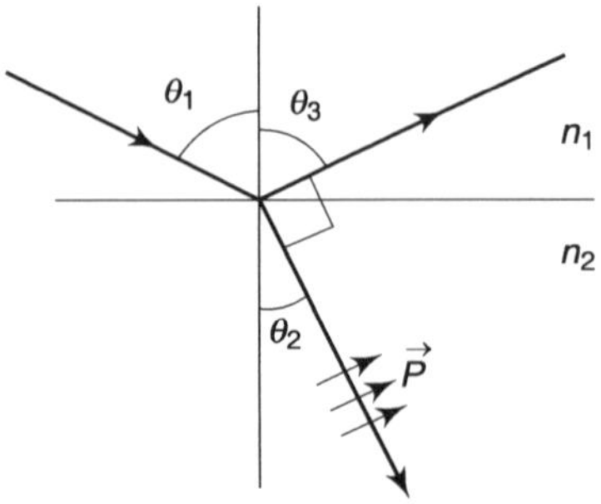

Fig. 5.16 *Brewster's angle*

Definition For a plane electromagnetic wave incident on a plane boundary between two dielectric media having different refractive indices, the angle of incidence at which transmittance from one medium to the other is unity, when the wave is linearly polarised with its electric vector parallel to the plane of incidence, is called *Brewster's angle*. This is demonstrated in Fig. 5.16.

For angle on incidence, θ_i (called θ_B) equal to *Brewster's angle* or *polarising angle*, the reflectance for parallel polarisation is zero and the reflected wave parallel polarisation only and is therefore, totally polarised.

Mathematical Equation We know, the reflectance, $\dfrac{E_r}{E_i} = \dfrac{\tan(\theta_1 - \theta_2)}{\tan(\theta_1 + \theta_2)}$

So, it is zero, if

$$(\theta_1 + \theta_2) = \frac{\pi}{2} \quad \text{i.e.} \quad (\theta_B + \theta_2) = \frac{\pi}{2}.$$

From Snell's law,

$$\eta_1 \sin\theta_B = \eta_2 \sin\left(\frac{\pi}{2} - \theta_B\right) = \eta_2 \cos\theta_B$$

$$\left[\begin{array}{l} \eta_1 = \sqrt{\dfrac{\mu_1\varepsilon_1}{\mu_0\varepsilon_0}}, \\ \eta_2 = \sqrt{\dfrac{\mu_2\varepsilon_2}{\mu_0\varepsilon_0}} \quad \text{are refractive indices} \end{array}\right]$$

$$\therefore \quad \boxed{\tan\theta_B = \frac{\eta_2}{\eta_1} = \sqrt{\frac{\varepsilon_2}{\varepsilon_1}}}$$

(since, $\mu_1 = \mu_2$)

At this angle, there is no reflected wave when the incident wave is parallel polarised.

NOTE

1. *Reflected wave is absent only when the incident wave is polarised in the plane of incidence. If the wave consisting of both parallel and perpendicular polarizations is incident on the boundary at Brewster's angle, the component with parallel polarisation is entirely transmitted into the second medium and the reflected wave consists only of the perpendicular polarisation.*
2. *Brewster's angle occurs for both $\eta_1 > \eta_2$ and $\eta_1 < \eta_2$.*

Brewster's Law The polarisation of an unpolarised wave upon reflection is stated in the form of *Brewster's law* as:

When unpolarised wave is incident on the surface of a dielectric like gas at Brewster angle (θ_B), the reflected wave is plane polarised with the plane of polarisation perpendicular to the plane of incidence, and the angle between the reflected and the refracted rays is 90°.

5.9.5 Total Internal Reflection

Let a wave be incident on the interface from the medium of higher refractive index, i.e., $\eta_1 > \eta_2$ or, $\varepsilon_1 > \varepsilon_2$.

When the angle of refraction θ_2 is $\frac{\pi}{2}$ (the largest possible value), the corresponding angle of incidence θ_1 is called the *critical angle* (θ_C).

From Snell's law,

$$\eta_1 \sin \theta_C = \eta_2 \sin \frac{\pi}{2} = \eta_2$$

$$\therefore \quad \boxed{\sin \theta_C = \frac{\eta_2}{\eta_1}}$$

For angles of incidence larger than the critical angle, we have from Snell's law,

$$\sin \theta_2 = \frac{\eta_1}{\eta_2} \sin \theta_1 > \frac{\eta_1}{\eta_2} \sin \theta_C$$

As, $\sin \theta_C = \frac{\eta_2}{\eta_1}$, we have, $\sin \theta_2 > 1$

Which shows that θ_2 cannot be a real angle. In fact, if $\sin\theta_2 > 1$, $\cos\theta_2$ will be imaginary, $\cos \theta_2 = \sqrt{1 - \sin^2 \theta_2} = j\delta$; where, δ is a real number.

So, the reflection coefficients become

$$\Gamma = \frac{E_r}{E_i} = \frac{\eta_2 \cos \theta_1 - j\eta_1\delta}{\eta_2 \cos \theta_1 + j\eta_1\delta}; \quad \text{for parallel polarisation}$$

$$= \frac{\eta_1 \cos \theta_1 - j\eta_2\delta}{\eta_1 \cos \theta_1 + j\eta_2\delta}; \quad \text{for parallel polarisation}$$

Thus, the coefficients for both polarisation is complex. The magnitude is

$$|\Gamma| = \left|\frac{E_r}{E_i}\right|^2 = \left(\frac{E_r}{E_i}\right) \cdot \left(\frac{E_r}{E_i}\right)^*$$ where, the asterisk (*) indicates complex conjugate.

$\therefore \quad |\Gamma| = 1$ for both polarisation; for all $\theta_1 \geq \theta_C$.

Thus, *total reflection occurs for all incident angles greater than or equal to the critical angle.*

NOTE

Brewster's angle: $\sin \theta_B = \frac{\eta_2}{\eta_1} \cos \theta_B$ ***Critical angle:*** $\sin \theta_C = \frac{\eta_2}{\eta_1}$

$$\frac{\sin \theta_B}{\sin \theta_C} = \cos \theta_B \leq 1$$

Thus, critical angle is normally greater than Brewster's angle.

Example 5.30 A wave is incident at an angle of 40° from air to teflon with relative permittivity $\varepsilon_r = 2.1$. Calculate the angle of transmission, Brewster's angle. What is the critical angle of the wave propagation from teflon to air?

Solution Since the media are non-magnetic, $\mu_1 = \mu_2$

The angle of transmission is given by the relation given in Eq. (5.36) as

$$\frac{\sin\theta_i}{\sin\theta_t} = \sqrt{\frac{\varepsilon_2}{\varepsilon_1}}$$

In this case,

$$\frac{\sin 40^\circ}{\sin\theta_t} = \sqrt{\frac{2.1}{1}} \quad \Rightarrow \quad \theta_t = 26.33^\circ$$

The Brewster's angle is given as

$$\tan\theta_B = \frac{\eta_2}{\eta_1} = \sqrt{\frac{\varepsilon_2}{\varepsilon_1}} = \sqrt{\frac{2.1}{1}} = \sqrt{2.1} \quad \Rightarrow \quad \theta_B = 55.39^\circ$$

The critical angle for wave propagation from teflon to air (i.e., from the medium of higher refractive index to the lower) is given as

$$\sin\theta_C = \frac{\eta_1}{\eta_2} = \sqrt{\frac{\varepsilon_1}{\varepsilon_2}} = \sqrt{\frac{1}{2.1}} \quad \Rightarrow \quad \theta_C = 43.63^\circ$$

Example 5.31 The dielectric constant of pure water is 80. Determine the Brewster angle for parallel polarisation and the corresponding angle of transmission.

If a plane wave of perpendicular polarisation impinges at this angle, find the reflection and transmission coefficients.

Solution For pure water, $\varepsilon_r = 80$

$$\text{Brewster's angle, } \theta_B = \tan^{-1}\left(\sqrt{\frac{\varepsilon_{r2}}{\varepsilon_{r1}}}\right) = \tan^{-1}(\sqrt{80}) = 83.62^\circ$$

Now, by Snell's law,

$$\frac{\sin\theta_1}{\sin\theta_2} = \sqrt{\frac{\varepsilon_{r2}}{\varepsilon_{r1}}} \quad \Rightarrow \quad \frac{\sin(83.62^\circ)}{\sin\theta_2} = \sqrt{80}$$

Hence, the transmission angle is given as

$$\theta_2 = \frac{\sin(83.62^\circ)}{\sqrt{80}} = 6.38^\circ$$

Reflection coefficient,

$$\Gamma = \frac{\cos\theta_1 - \sqrt{\left(\frac{\varepsilon_2}{\varepsilon_1}\right) - \sin^2\theta_1}}{\cos\theta_1 + \sqrt{\left(\frac{\varepsilon_2}{\varepsilon_1}\right) - \sin^2\theta_1}} = \frac{\cos(83.62^\circ) - \sqrt{(80) - \sin^2(83.62^\circ)}}{\cos(83.62^\circ) + \sqrt{(80) - \sin^2(83.62^\circ)}} = -0.9753$$

Transmission coefficient, $\tau = (1 + \Gamma) = (1 - 0.9753) = 0.0247$

Example 5.32 Find the critical angle of total internal reflection for glass ($\varepsilon_r = 4.0$), polyethylene ($\varepsilon_r = 2.25$) and polystyrene ($\varepsilon_r = 2.52$) to air surfaces.

Solution The critical angle of total internal reflection is given as

$$\theta_C = \sin^{-1}\left(\sqrt{\frac{1}{\varepsilon_r}}\right)$$

For glass, ($\varepsilon_r = 4$), $\theta_C = \sin^{-1}\left(\sqrt{\frac{1}{4}}\right) = 30°$

For polyethylene, ($\varepsilon_r = 2.25$), $\theta_C = \sin^{-1}\left(\sqrt{\frac{1}{2.25}}\right) = 41.8°$

For polystyrene, ($\varepsilon_r = 2.52$), $\theta_C = \sin^{-1}\left(\sqrt{\frac{1}{2.52}}\right) = 39.1°$

5.10 REFLECTION AND REFRACTION OF PLANE ELECTROMAGNETIC WAVES BY PERFECT CONDUCTOR

5.10.1 Normal Incidence

Let medium 1 be a perfect dielectric and medium 2 a perfect conductor, such that $\eta_2 = 0$. Let a wave be incident normally from medium 1 to medium 2.

The following points may be noted for conducting medium:

1. When a plane wave incidents normally upon the surface of the conducting medium, the wave is entirely reflected.
2. For field that varies with time, neither $\vec{E}$ nor $\vec{H}$ will exist within the conductor. So, no energy is transmitted through the conducting medium.
3. As there is no loss within a perfect conductor, there is no absorption of energy by the conducting medium.

∴ Transmission coefficient, $\tau = 0$, and

∴ Reflection coefficient, $\Gamma = -1$

Hence, the amplitudes of $\vec{E}$ and $\vec{H}$ of reflected wave are same as those of the incident wave, but they differ in the direction of flow.

Electric field of incident wave = $E_i e^{-j\beta x}$

Electric field of reflected wave = $E_r e^{-j\beta x}$

Boundary is the surface of the conductor given by, $x = 0$.

The boundary conditions to be applied are:

1. The tangential component of the electric field is continuous across the boundary.
2. Electric field $\vec{E}$ inside the conductor is zero; i.e. $(E_i + E_r) = 0 \Rightarrow E_r = -E_i$.

The amplitude of the reflected electric field is equal to that of the incident field, with phase reversal on reflection.

The resultant electric field strength at any point at a distance x from $x = 0$ plane is

$$E_T(x) = E_i e^{-j\beta x} + E_r e^{j\beta x} = E_i(e^{-j\beta x} - e^{j\beta x}) = -2jE_i \sin \beta x$$

$$\therefore \quad \bar{E}_T(x, t) = \text{Re}[-2jE_i \sin \beta x e^{j\omega t}] = 2E_i \sin \beta x \sin \omega t$$

This expression represents a standing wave. The magnitude of electric field varies sinusoidally with distance from the reflecting plane.

- $\bar{E}_T(x, t) = 0$ when $x = 0$ or multiples of $\lambda/2$.

$$\therefore \quad \beta x = \frac{2\pi}{\lambda} \times \frac{\lambda}{2} = \pi$$

- $\bar{E}_T(x, t)$ is maximum of $2E_i$, when x is odd multiples of $\lambda/4$.

$$\therefore \quad \beta x = \frac{2\pi}{\lambda} \times \frac{\lambda}{4} = \frac{\pi}{2}$$

As the electric field is reflected with the phase reversal, then the magnetic field is reflected without the phase reversal to satisfy the boundary conditions.

$$\therefore \quad H_r = H_i$$

$$\therefore \quad H_T(x) = H_i e^{-j\beta x} + H_r e^{j\beta x} = H_i(e^{-j\beta x} + e^{j\beta x}) = 2H_i \cos \beta x$$

$$\therefore \quad \bar{H}_T(x, t) = 2H_i \cos \beta x \cos \omega t$$

This also has a standing wave distribution.

- $\bar{H}_T(x, t) = 0$ when x is odd multiples of $\lambda/4$.
- $\bar{H}_T(x, t)$ is maximum, when x is multiples of $\lambda/2$.

Example 5.33 A 300 MHz uniform plane wave travelling in free space strikes a large block of copper ($\mu_r = 1$, $\varepsilon_r = 1$ and $\sigma = 5.8 \times 10^7$ S/m) normal to the surface. If the surface of copper lies in the yz plane and the wave is propagating in x-direction, write the complete time domain expressions for incident, reflected and transmitted waves in terms of the amplitude of the incident electric wave, E_i. This is illustrated in Fig. 5.17.

Solution: Since the wave is propagating in the x-direction, the electric field $\vec{E}$ will be in y-direction and the magnetic field $\vec{H}$ will be in z-direction.

Let the electric field in the incident wave be given as

$$\vec{E} = E_i \cos(\omega t - \beta z)\hat{a}_y$$

where, E_i is the amplitude.

Intrinsic impedance of medium 1 (free space) is

$$\eta_1 = 120\pi = 377\ \Omega$$

Intrinsic impedance of medium 2 (copper, which is a good conductor) is

$$\eta_2 = \sqrt{\frac{\omega\mu}{\sigma}} \angle 45° = \sqrt{\frac{2\pi \times 300 \times 10^6 \times 4\pi \times 10^{-7} \times 1}{5.8 \times 10^7}}$$

$$= 6.39 \times 10^{-3} \angle 45°$$

$$= (4.52 \times 10^{-3} + j4.52 \times 10^{-3})\ \Omega$$

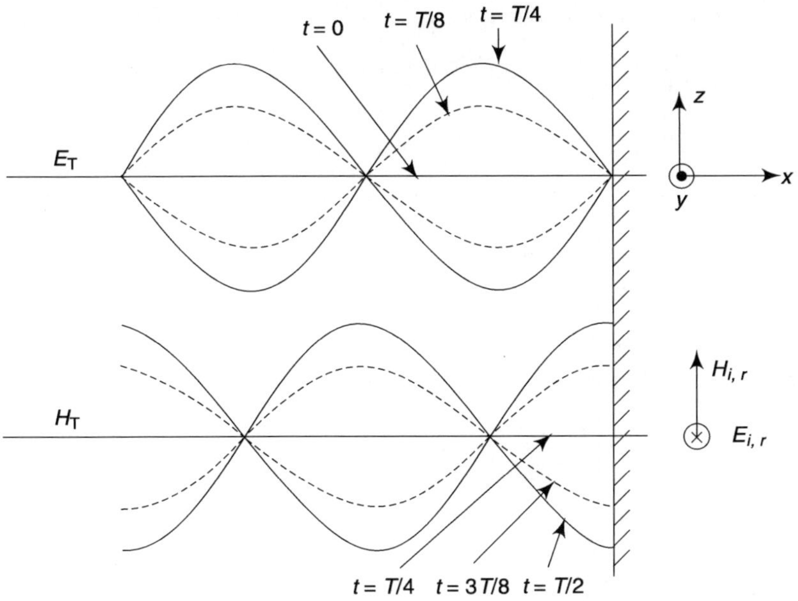

Fig. 5.17 *Standing wave distribution of E and H for normal incidence of EM wave at conducting surface*

Transmission coefficient is

$$\tau = \frac{E_t}{E_i} = \frac{2\eta_2}{\eta_1 + \eta_2} = \frac{2 \times 6.39 \times 10^{-3} \angle 45^\circ}{377 + (4.52 \times 10^{-3} + j4.52 \times 10^{-3})} = 3.37 \times 10^{-5} \angle 45^\circ$$

$\therefore$ Amplitude of the transmitted electric wave is

$$E_t = (3.37 \times 10^{-5} \angle 45^\circ) E_i$$

Reflection coefficient,

$$\Gamma_R = \frac{E_r}{E_i} = \frac{\eta_2 - \eta_1}{\eta_2 + \eta_1} = \frac{4.52 \times 10^{-3} + j4.52 \times 10^{-3} - 377}{(4.52 \times 10^{-3} + j4.52 \times 10^{-3}) + 377} = 1\angle 180^\circ$$

$\therefore$ Amplitude of the reflected electric wave is

$$E_r = (1\angle 180^\circ) E_i = -E_i$$

$\therefore$ Amplitude of the incident magnetic wave is

$$H_i = \frac{E_i}{\eta_1} = \frac{E_i}{377} = (2.65 \times 10^{-3}) E_i$$

$\therefore$ Amplitude of the transmitted magnetic wave is

$$H_t = \frac{E_t}{\eta_2} = \frac{(3.37 \times 10^{-5} \angle 45^\circ) E_i}{(6.39 \times 10^{-3} \angle 45^\circ)} = (5.27 \times 10^{-3}) E_i$$

$\therefore$ Amplitude of the reflected magnetic wave is

$$H_r = -\frac{E_r}{\eta_1} = -\frac{-E_i}{377} = (2.65 \times 10^{-3})E_i$$

Now, the electromagnetic wave in free space propagates with the velocity of light,

$$\therefore \qquad v = c = 3 \times 10^8$$

$$\beta = \frac{\omega}{v} = \frac{\omega}{c} = \frac{2\pi \times 300 \times 10^6}{3 \times 10^8} = 6.28 \text{ rad/m}$$

Hence, the complete time domain expressions for the waves are given below.

Incident waves:

$$\vec{E}_i = E_i \cos(2\pi \times 300 \times 10^6 t - 6.28x)\hat{a}_y \text{ V/m}$$
$$\vec{H}_i = (2.65 \times 10^{-3})E_i \cos(2\pi \times 300 \times 10^6 t - 6.28x)\hat{a}_x \text{ A/m}$$

Transmitted waves:

$$\vec{E}_t = (3.37 \times 10^{-5})E_i \cos(2\pi \times 300 \times 10^6 t - 6.28x + 45°)\hat{a}_y \text{ V/m}$$
$$\vec{H}_t = (5.27 \times 10^{-3})E_i \cos(2\pi \times 300 \times 10^6 t - 6.28x + 45°)\hat{a}_x \text{ A/m}$$

Reflected waves:

$$\vec{E}_r = -E_i \cos(2\pi \times 300 \times 10^6 t - 6.28x)\hat{a}_y \text{ V/m}$$
$$\vec{H}_r = (2.65 \times 10^{-3})E_i \cos(2\pi \times 300 \times 10^6 t - 6.28x + 45°)\hat{a}_x \text{ A/m}$$

Example 5.34 A 1MHz uniform plane wave travelling in free space strikes a large sheet of copper ($\mu_r = 1$, $\varepsilon_r = 1$ and $\sigma = 5.8 \times 10^7$ S/m) normal to the surface. If the surface of copper lies in the *yz* plane and the wave is propagating in *x*-direction, write the complete time domain expressions for incident, reflected and transmitted waves in terms of the amplitude of the incident electric wave, E_i.

Solution Since the wave is propagating in the *x*-direction, the electric field $\vec{E}$ will be in *y*-direction and the magnetic field $\vec{H}$ will be in *z*-direction.

Let the electric field in the incident wave be given as

$$\vec{E} = E_i \cos(\omega t - \beta z)\hat{a}_y$$

where, E_i is the amplitude.

Intrinsic impedance of medium 1 (free space) is

$$\eta_1 = 120\pi = 377\ \Omega$$

Intrinsic impedance of medium 2 (copper which is a good conductor) is

$$\eta_2 = \sqrt{\frac{\omega\mu}{\sigma}}\angle 45° = \sqrt{\frac{2\pi \times 300 \times 10^6 \times 4\pi \times 10^{-7} \times 1}{5.8 \times 10^7}} \approx 3.69 \times 10^{-4}\ \Omega$$

Transmission coefficient is

$$\tau = \frac{E_t}{E_i} = \frac{2\eta_2}{\eta_1+\eta_2} = \frac{2\times 3.69\times 10^{-4}\angle 45^\circ}{377+3.69\times 10^{-4}\angle 45^\circ} = 1.96\times 10^{-6}\angle 45^\circ$$

∴ Amplitude of the transmitted electric wave is

∴ $$E_t = (1.96\times 10^{-6}\angle 45^\circ)E_i$$

Reflection coefficient,

$$\Gamma_R = \frac{E_r}{E_i} = \frac{\eta_2-\eta_1}{\eta_2+\eta_1} = \frac{3.69\times 10^{-4}\angle 45^\circ - 377}{3.69\times 10^{-4}\angle 45^\circ + 377} = 1\angle 180^\circ = -1$$

∴ Amplitude of the reflected electric wave is

∴ $$E_r = (1\angle 180^\circ)E_i = -E_i$$

∴ Amplitude of the incident magnetic wave is

$$H_i = \frac{E_i}{\eta_1} = \frac{E_i}{377} = (2.65\times 10^{-3})E_i$$

∴ Amplitude of the transmitted magnetic wave is

$$H_t = \frac{E_t}{\eta_2} = \frac{(1.96\times 10^{-6}\angle 45^\circ)E_i}{(3.69\times 10^{-4}\angle 45^\circ)} = (4.95\times 10^{-3})E_i$$

∴ Amplitude of the reflected magnetic wave is

$$H_r = -\frac{E_r}{\eta_1} = -\frac{-E_i}{377} = (2.65\times 10^{-3})E_i$$

Now, the electromagnetic wave in free space propagates with the velocity of light

∴ $$v = c = 3\times 10^8$$

∴ $$\beta = \frac{\omega}{v} = \frac{\omega}{c} = \frac{2\pi\times 1\times 10^6}{3\times 10^8} = 0.02 \text{ rad/m}$$

Hence, the complete time domain expressions for the waves are given below.

Incident waves:

$$\vec{E}_i = E_i \cos(2\pi\times 10^6 t - 0.02x)\hat{a}_y \text{ V/m}$$
$$\vec{H}_i = (2.65\times 10^{-3})E_i \cos(2\pi\times 10^6 t - 0.02x)\hat{a}_x \text{ A/m}$$

Transmitted waves:

$$\vec{E}_t = (1.96\times 10^{-6})E_i \cos(2\pi\times 10^6 t - 0.02x + 45^\circ)\hat{a}_y \text{ V/m}$$
$$\vec{H}_t = (4.95\times 10^{-3})E_i \cos(2\pi\times 10^6 t - 0.02x + 45^\circ)\hat{a}_x \text{ A/m}$$

Reflected waves:

$$\vec{E}_r = -E_i \cos(2\pi\times 10^6 t - 0.02x)\hat{a}_y \text{ V/m}$$
$$\vec{H}_r = (2.65\times 10^{-3})E_i \cos(2\pi\times 10^6 t - 0.02x + 45^\circ)\hat{a}_x \text{ A/m}$$

5.10.2 Oblique Incidence

When a wave is incident obliquely at a conducting surface, two cases may arise:

(1) Horizontal Polarisation or Perpendicular Polarisation

In this case, the electric field is parallel to the boundary surface or perpendicular to the plane of incidence. The plane of incidence is the plane containing the incident ray and the normal to the surface.

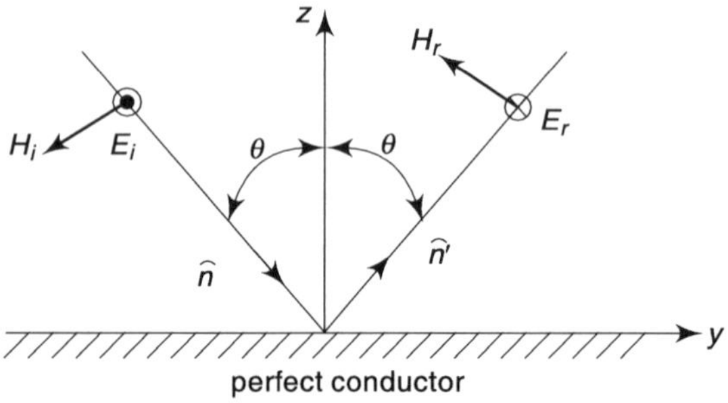

Fig. 5.18 *Horizontal polarisation for conducting surface*

With the coordinate system chosen as shown in Fig. 5.18,

$$\vec{E}_{\text{reflected}} = E_r e^{-j\beta \hat{n}' \cdot \vec{r}} = E_r e^{-j\beta(x \cos A + y \cos B + z \cos C)}$$

where, x, y, z are the components of the vector $\vec{r}$ and $\cos A$, $\cos B$, $\cos C$ are the components of the unit vector $\hat{n}$ along x, y and z axes; A, B, C are the angles that $\hat{n}$ makes with the positive x, y and z axes.

Here, $$A = \frac{\pi}{2}, \quad B = \left(\frac{\pi}{2} - \theta\right), \quad C = \theta$$

$\therefore$ $$\hat{n} \cdot \vec{r} = (y \sin \theta + z \cos \theta)$$

$\therefore$ $$\boxed{\vec{E}_{\text{reflected}} = E_r e^{-j\beta(y \sin \theta + z \cos \theta)}}$$

For incident wave, $$A = \frac{\pi}{2}, \quad B = \left(\frac{\pi}{2} - \theta\right), \quad C = (\pi - \theta)$$

$\therefore$ $$\hat{n} \cdot \vec{r} = (y \sin \theta - z \cos \theta)$$

$\therefore$ $$\boxed{\vec{E}_{\text{incident}} = E_i e^{-j\beta(y \sin \theta - z \cos \theta)}}$$

From the boundary conditions, $E_r = -E_i$

Thus, total electric field,

$$\vec{E} = E_i[e^{-j\beta(y \sin \theta - z \cos \theta)} - e^{-j\beta(y \sin \theta + z \cos \theta)}]$$

$\therefore$ $$\boxed{\vec{E} = 2jE_i e^{-j\beta y \sin \theta} \sin(\beta z \cos \theta) = 2jE_i e^{-j\beta_y y} \sin(\beta_z z)}$$

where, $\beta = \dfrac{2\pi}{\lambda}$; is the phase shift constant of the incident wave

$\beta_z = \beta \cos \theta$; is the phase shift constant in the z-direction

$\beta_y = \beta \cos \theta$; is the phase shift constant in the y-direction

The equation shows a standing wave distribution of electric field strength along the z-axis. The wavelength along the z-axis, $\lambda_z = \dfrac{2\pi}{\beta_z} = \dfrac{2\pi}{\beta \cos \theta} = \dfrac{\lambda}{\cos \theta}$ is greater than the wavelength of the incident wave (λ).

The velocity with which the standing wave travels along the y-direction is

$$v_y = \frac{\omega}{\beta_y} = \frac{\omega}{\beta \sin \theta} = \frac{v}{\sin \theta}$$

(2) Vertical Polarisation or Parallel Polarisation In this case, the magnetic field is parallel to the boundary surface and electric field is parallel to the plane of incidence.

E_i and E_r will have the directions as shown in Fig. 5.19, because the components parallel to the perfectly conducting boundary must be equal and opposite.

The magnetic field $\vec{H}$ will be reflected without phase reversal. The magnitudes of $\vec{E}$ and $\vec{H}$ are related by

$$\frac{E_i}{H_i} = \frac{E_r}{H_r} = \eta$$

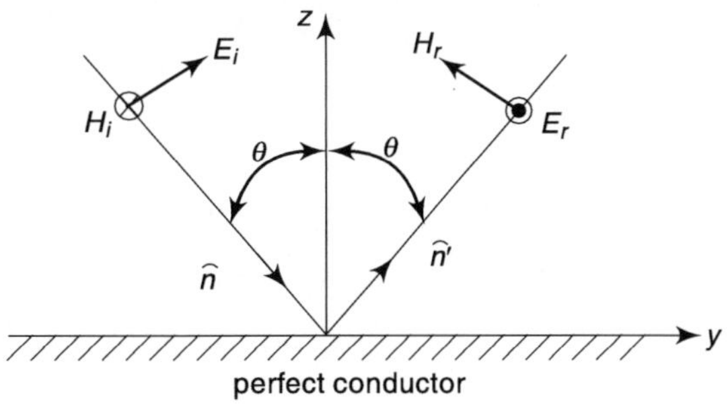

Fig. 5.19 *Vertical polarisation for conducting surface*

For incident wave,

$$\vec{H}_{\text{incident}} = H_i e^{-j\beta(y \sin\theta - z\cos\theta)}$$

For reflected wave,

$$\vec{H}_{\text{reflected}} = H_r e^{-j\beta(y \sin\theta + z\cos\theta)}$$

From the boundary conditions, $H_r = H_i$

∴ Total magnetic field is given as

$$\boxed{H = 2H_i e^{-j\beta_y y} \cos\beta_z z}$$

where, $\beta_z = \beta\cos\theta$ and $\beta_y = \beta\sin\theta$.

Thus, the magnetic field strength has a standing wave distribution in the z-direction with:

- H maximum at the conducting surface and multiples of $\lambda_z/2$
- $H = 0$ at odd multiples of $\frac{\lambda}{4}$ from the surface.

For electric field, we have to consider separately the components in the y and z directions.

For the incident wave,

$$E_i = \eta H_i \quad E_z = \eta\sin\theta H_i \quad E_y = \eta\cos\theta H_i$$

For the reflected wave,

$$E_r = \eta H_r \quad E_z = \eta\sin\theta H_r \quad E_y = -\eta\cos\theta H_r$$

Total z-components of the electric field strength

$$\boxed{E_z = 2\eta H_i e^{-j\beta_y y}\sin\theta\cos\beta_z z}$$

Total y-components of the electric field strength

$$\boxed{E_y = 2j\eta H_i e^{-j\beta_y y}\cos\theta\sin\beta_z z}$$

Both components have a standing wave distribution above the reflecting plane. However,

- E_z is maximum at the plane and multiples of $\lambda_z/2$ from the plane.
- E_y is minimum at the plane and multiples of $\lambda_z/2$ from the plane.

5.11 POYNTING THEOREM AND POYNTING VECTOR

5.11.1 Poynting Vector

It is defined as

$$\vec{S} = \vec{E} \times \vec{H}$$

where, $\vec{E}$ is the electric field and $\vec{H}$ is the magnetic field.

It represents the energy flux (in W/m^2) of an electromagnetic wave. It is named after its inventor, British physicist John Henry Poynting. This is illustrated in Fig. 5.20.

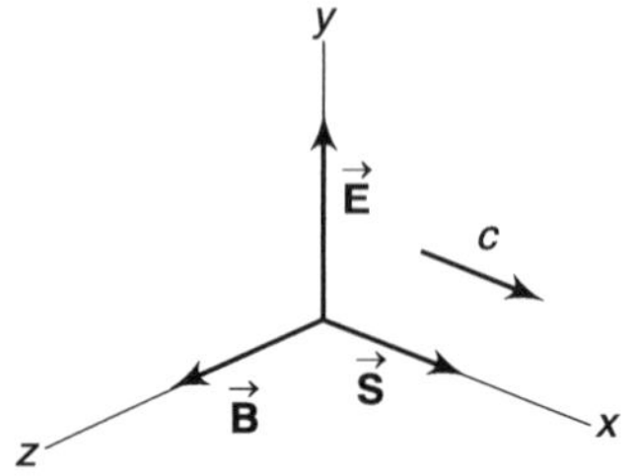

Fig. 5.20 *Poynting vector for a plane wave*

5.11.2 Poynting Theorem

Statement It states that the vector product, $\vec{S} = \vec{E} \times \vec{H}$ at any point is a measure of the rate of energy flow per unit area at that point. The direction of power flow is in the direction of the unit vector along the product $(\vec{E} \times \vec{H})$ and is perpendicular to both $\vec{E}$ and $\vec{H}$.

This theorem provides a statement about the conservation of energy of an electromagnetic wave.

Derivation By modified Ampere's circuital law in differential form,

$$\nabla \times \vec{H} = \vec{J} + \frac{\partial \vec{D}}{\partial t} = \vec{J} + \varepsilon \frac{\partial \vec{E}}{\partial t}$$

$$\Rightarrow \qquad \vec{J} = \nabla \times \vec{H} - \varepsilon \frac{\partial \vec{E}}{\partial t}$$

Multiplying both sides by $\vec{E}$, we get

$$\vec{E} \cdot \vec{J} = \vec{E} \cdot \nabla \times \vec{H} - \vec{E} \cdot \varepsilon \frac{\partial \vec{E}}{\partial t} \tag{5.41}$$

We use the vector identity

$$\nabla \cdot (\vec{E} \times \vec{H}) = \vec{H} \cdot (\nabla \times \vec{E}) - \vec{E} \cdot (\nabla \times \vec{H})$$

Thus, from Eq. (5.34),

$$\vec{E} \cdot \vec{J} = \vec{H} \cdot (\nabla \times \vec{E}) - \nabla \cdot (\vec{E} \times \vec{H}) - \vec{E} \cdot \varepsilon \frac{\partial \vec{E}}{\partial t}$$

$$= -\mu \vec{H} \cdot \frac{\partial \vec{H}}{\partial t} - \varepsilon \vec{E} \cdot \frac{\partial \vec{E}}{\partial t} - \nabla \cdot (\vec{E} \times \vec{H}) \tag{5.42}$$

$$\left(\because \ \nabla \times \vec{E} = -\frac{\partial \vec{B}}{\partial t} = -\mu \frac{\partial \vec{H}}{\partial t} \right)$$

Now, $\vec{H} \cdot \frac{\partial \vec{H}}{\partial t} = \frac{1}{2} \frac{\partial}{\partial t} (H^2)$ and $\vec{E} \cdot \frac{\partial \vec{E}}{\partial t} = \frac{1}{2} \frac{\partial}{\partial t} (E^2)$

So, from Eq. (5.42),

$$\vec{E}\cdot\vec{J} = -\frac{1}{2}\mu\frac{\partial}{\partial t}(H^2) - \frac{1}{2}\varepsilon\frac{\partial}{\partial t}(E^2) - \nabla\cdot(\vec{E}\times\vec{H})$$
$$= -\frac{\partial}{\partial t}\left(\frac{1}{2}\mu H^2 + \frac{1}{2}\varepsilon E^2\right) - \nabla\cdot(\vec{E}\times\vec{H})$$

Integrating over a volume,

$$\int_v \vec{E}\cdot\vec{J}dv = \int_v -\frac{\partial}{\partial t}\left(\frac{1}{2}\mu H^2 + \frac{1}{2}\varepsilon E^2\right)dv - \int_v \nabla\cdot(\vec{E}\times\vec{H})dv$$

$$\Rightarrow \qquad \int_v \vec{E}\cdot\vec{J}dv = -\frac{\partial}{\partial t}\int_v\left(\frac{1}{2}\mu H^2 + \frac{1}{2}\varepsilon E^2\right)dv - \oint_s(\vec{E}\times\vec{H})\cdot d\vec{s} \qquad (5.43)$$

where, s is the closed surface bounding the volume v.
On rearrangement, we get

$$\boxed{\int_v \vec{E}\cdot\vec{J}dv = -\frac{\partial}{\partial t}\int_v\left(\frac{1}{2}\mu H^2 + \frac{1}{2}\varepsilon E^2\right)dv - \oint_s(\vec{E}\times\vec{H})\cdot d\vec{s}}$$

In words,

$$\boxed{\begin{pmatrix}\text{Ohmic Power}\\ \text{Dissipated}\end{pmatrix} = \begin{pmatrix}\text{Rate of decrease in energy stored}\\ \text{in electric and magnetic field}\end{pmatrix} - \begin{pmatrix}\text{Total Power}\\ \text{leaving the volume}\end{pmatrix}}$$

This is the mathematical form of Poynting theorem.

Significance of Terms

1. The term $\int_v \vec{E}\cdot\vec{J}dv$ represents the rate at which the energy is being dissipated (i.e., the instantaneous power dissipated) within the total volume v. This is a generalisation of Joule's law.
2. The first term in the right hand side, $-\frac{\partial}{\partial t}\int_v\left(\frac{1}{2}\mu H^2 + \frac{1}{2}\varepsilon E^2\right)dv$ represents the time rate at which the stored electric field energy and the magnetic field energy is decreasing within the volume v.
3. The second term in the right-hand side, $\oint_s(\vec{E}\times\vec{H})\cdot d\vec{s}$ represents the rate at which energy is escaping or leaving the volume v through the closed surface s. This follows from the law of conservation of energy.

5.11.3 Poynting Theorem for Insulating Medium

In an insulating medium, $\vec{J} = 0$ and total energy density is

$$u = (u_E + u_M) = \frac{1}{2}\varepsilon E^2 \frac{1}{2}\mu H^2$$

So, from Eq. (5.33),

$$0 = -\frac{\partial}{\partial t}\int_v u dv - \oint_s \vec{S}\cdot d\vec{s} \qquad (\because\ \vec{S} = \vec{E}\times\vec{H})$$

$\Rightarrow$
$$\int_v \left(\nabla \cdot \vec{S} + \frac{\partial u}{\partial t}\right) dv = 0$$

$\therefore$
$$\boxed{\nabla \cdot \vec{S} + \frac{\partial u}{\partial t} = 0}$$

This is the differential form of Poynting theorem.

5.11.4 Poynting Theorem in Complex Form

Maxwell's equations in phasor form can be written as

$$\nabla \times \vec{H} = (\sigma + j\omega\varepsilon)\vec{E} \quad \text{and} \quad \nabla \times \vec{E} = -j\omega\mu\vec{H}$$

$\therefore$
$$\nabla \cdot (\vec{E} \times \vec{H}^*) = \vec{H}^* \cdot \nabla \times \vec{E} - \vec{E} \cdot \nabla \times \vec{H}^* = -j\omega\mu\vec{H} \cdot \vec{H}^* - (\sigma - j\omega\varepsilon)\vec{E} \cdot \vec{E}^* - \vec{E} \cdot \vec{J}^*$$

Integrating over the volume v surrounded by closed surface s,

$$\oint_s (\vec{E} \times \vec{H}^*) \cdot d\vec{s} = -j\omega \int_v (\mu\vec{H} \cdot \vec{H}^* - \varepsilon\vec{E} \cdot \vec{E}^*) dv - \int_v \sigma\vec{E} \cdot \vec{E}^* dv - \int_v \vec{E} \cdot \vec{J}^* dv \qquad \textbf{(5.44)}$$

Now,

$$\varepsilon\vec{E} \cdot \vec{E}^* = \text{Time-averaged stored electric energy} = U_E$$
$$\mu\vec{H} \cdot \vec{H}^* = \text{Time-averaged stored magnetic energy} = U_M$$
$$\vec{E} \times \vec{H} = \text{Complex Poynting Vector} = \text{Re}[\vec{S}] + \text{Im}[\vec{S}]$$

So, from Eq. (5.37),

$$\text{Re}\oint_s \vec{S} \cdot d\vec{s} + \text{Im}\oint_s \vec{S} \cdot d\vec{s} = -j\omega \int_v (U_M - U_E) dv - \sigma \int_v \vec{E} \cdot \vec{E}^* dv - \frac{1}{2}\text{Re}\int_v \vec{E} \cdot \vec{J}^* dv - \frac{1}{2}\text{Im}\int_v \vec{E} \cdot \vec{J}^* dv$$

This is the mathematical form of Poynting theorem in complex form. Separating the real and imaginary parts, we get

$$\text{Re}\oint_s \vec{S} \cdot d\vec{s} = -\sigma \int_v \vec{E} \cdot \vec{E}^* dv - \frac{1}{2}\text{Re}\int_v \vec{E} \cdot \vec{J}^* dv$$

$$\text{Im}\oint_s \vec{S} \cdot d\vec{s} = -j\omega \int_v (U_M - U_E) dv - \frac{1}{2}\text{Im}\int_v \vec{E} \cdot \vec{J}^* dv$$

The energy may dissipate or may circulate. The circulating energy is represented by the imaginary Poynting vector and the dissipative energy is represented by the real Poynting vector.

5.11.5 Average Power Calculation using Poynting Vector

For a time-harmonic field, the time-average Poynting vector is found by integrating the instantaneous Poynting vector over one period and dividing by the period.

$$p_{\text{ave}} = \frac{1}{T}\int_T (\vec{E} \times \vec{H}) dt$$

Writing the electric and magnetic fields in instantaneous form as

$$\vec{E} = |E| \cos(\omega t + \theta_E)\hat{a}_E \quad \vec{H} = |H| \cos(\omega t + \theta_H)\hat{a}_H$$

The instantaneous Poynting vector is given as,

$$\vec{S} = \vec{E} \times \vec{H} = |E||H| \cos(\omega t + \theta_E) \cos(\omega t + \theta_H)(\hat{a}_E \times \hat{a}_H)$$
$$= \frac{1}{2}|E||H|[\cos(2\omega t + \theta_E + \theta_H) + \cos(\theta_E - \theta_H)](\hat{a}_E \times \hat{a}_H)$$

(Using the trigonometric identity, $\cos x \cos y = \frac{1}{2}[\cos(x+y) + \cos(x-y)]$)

The time-average Poynting vector is then

$$p_{\text{ave}} = \frac{|E||H|}{2T}(\hat{a}_E \times \hat{a}_H)\left\{\int_T \cos(2\omega t + \theta_E + \theta_H)dt + \cos(\theta_E - \theta_H)\int_T dt\right\}$$

The time-average Poynting vector reduces to,

$$p_{\text{ave}} = \frac{|E||H|}{2T}(\hat{a}_E \times \hat{a}_H)\cos(\theta_E - \theta_H)T = \frac{1}{2}\text{Re}[(|E|\, e^{j\theta_E}\hat{a}_E) \times (|H|\, e^{j\theta_H}\hat{a}_H)]$$

$$\therefore \quad p_{\text{ave}} = \frac{1}{2}\text{Re}[\boldsymbol{E}_S \times \boldsymbol{H}_S^*] \tag{5.45}$$

By Eq. (5.45), the time-average Poynting vector is determined without integration.
The term in brackets in the above equation is defined as the phasor Poynting vector and is normally represented by $\boldsymbol{S}$.

$$\boldsymbol{S} = \boldsymbol{E}_S \times \boldsymbol{H}_S^*$$

$$\therefore \quad p_{\text{ave}} = \frac{1}{2}\text{Re}[\boldsymbol{E}_S \times \boldsymbol{H}_S^*] = \frac{1}{2}\text{Re}[\boldsymbol{S}]$$

All representations of the Poynting vector represent vector energy densities. Thus, to determine the total power passing through a surface, we must integrate the Poynting vector over that surface. The total time-average power passing through the surface s is

$$P_{\text{ave}} = \int_s p_{\text{ave}} \cdot ds = \frac{1}{2}\text{Re}\int_s [\boldsymbol{E}_S \times \boldsymbol{H}_S^*] \cdot ds = \frac{1}{2}\text{Re}\int_s \boldsymbol{S} \cdot ds$$

Average Power Density for Lossless (Perfect) Dielectric We consider an electric field propagating in the z-axis is given by the equation, $\vec{E}_x = E_0 \cos(\omega t - \beta z)$ and the associated magnetic field by the equation, $\vec{H}_y = \frac{E_0}{\eta}\cos(\omega t - \beta z)$, where, E_0 is the peak value of E_x at $t = 0$ and $z = 0$ and η is the intrinsic impedance of the dielectric.

The instantaneous Poynting vector is given as

$$\vec{P} = \vec{E} \times \vec{H}$$
$$P_z = E_x H_y = \frac{E_0^2}{\eta}\cos^2(\omega t - \beta z)$$

To find the average power, integrating over one cycle and dividing by the period T

$$p_{\text{ave}} = \frac{1}{T}\int_0^T \frac{E_0^2}{\eta}\cos^2(\omega t - \beta z)dt$$

$$= \frac{1}{T}\frac{E_0^2}{2\eta}\int_0^T [1 + \cos(2\omega t - 2\beta z)]dt$$

$$= \frac{1}{T}\frac{E_0^2}{2\eta}\left[t + \frac{1}{2\omega}\sin(2\omega t - 2\beta z)\right]_0^T$$

$$= \frac{1}{T}\frac{E_0^2}{2\eta}\left[T + \frac{1}{2\omega}\sin(4\omega - 2\beta z) - 0 - \frac{1}{2\omega}\sin(-2\beta z)\right] \qquad (\because \ \omega T = 2\pi)$$

$$= \frac{1}{2}\frac{E_0^2}{\eta} \quad \text{Watt/m}^2$$

So, the average power flowing through an area A normal to the z-axis is

$$\boxed{P_{\text{ave}} = \frac{1}{2}\frac{E_0^2}{\eta}A \quad \text{Watt}}$$

Average Power Density for Lossy Dielectric We consider an electric field propagating through a lossy dielectric in the z-axis is given by the equation, $\vec{E}_x = E_0 e^{-\alpha z}\cos(\omega t - \beta z)$ and the associated magnetic field by the equation, $\vec{H}_y = \frac{E_0}{\eta}e^{-\alpha z}\cos(\omega t - \beta z)$, where, E_0 is the peak value of E_x at $t = 0$ and $z = 0$ and η is the intrinsic impedance of the dielectric.

If $\eta = \eta_0 \angle\theta_\eta$, then the magnetic field can be written as

$$\vec{H}_y = \frac{E_0}{\eta_0}e^{-\alpha z}\cos(\omega t - \beta z - \theta_\eta)$$

Instantaneous Poynting vector is given as

$$\vec{P} = \vec{E} \times \vec{H}$$

$$P_z = E_x H_y = \frac{E_0^2}{\eta_0}e^{-2\alpha z}\cos(\omega t - \beta z)\cos(\omega t - \beta z - \theta_\eta)$$

$$= \frac{1}{2}\frac{E_0^2}{\eta_0}e^{-2\alpha z}[\cos(2\omega t - \beta z - \theta_\eta) + \cos(\theta_\eta)]$$

So, the time-average value of Poynting vector or average power is

$$p_{\text{ave}} = \frac{1}{T}\int_0^T P_z dt = \frac{1}{2}\frac{E_0^2}{\eta_0}e^{-2\alpha z}\int_0^T [\cos(2\omega t - \beta z - \theta_\eta) + \cos(\theta_\eta)]dt$$

$$\therefore \qquad \boxed{p_{\text{ave}} = \frac{1}{2}\frac{E_0^2}{\eta_0}e^{-2\alpha z}\cos\theta_\eta}$$

Average Power Density for Conducting Media For a perfect conducting medium, $\theta_\eta = 45°$. Therefore, the average power can be obtained from the earlier section by putting $\theta_\eta = 45°$.

$$\therefore \quad p_{\text{ave}} = \frac{1}{2\sqrt{2}}\frac{E_0^2}{\eta_0}e^{-2\alpha z} = \frac{1}{2\sqrt{2}}\frac{E_0^2}{\sqrt{\frac{\omega\mu}{\sigma}}}e^{-2\alpha z} = \frac{1}{4}\frac{E_0^2\sigma}{\sqrt{\frac{\omega\mu\sigma}{2}}}e^{-2\alpha z} = \frac{1}{4}E_0^2\sigma\delta\, e^{-2\alpha z}$$

where, $\delta = \sqrt{\frac{2}{\omega\mu\sigma}}$ is the skin depth.

$$\boxed{p_{\text{ave}} = \frac{1}{4}E_0^2\sigma\delta\, e^{-2\alpha z}}$$

5.12 ENERGY FLUX IN A PLANE ELECTROMAGNETIC WAVE

From average power calculation using Poynting vector,

$$\vec{p}_{\text{ave}} = \frac{1}{2}\text{Re}[\boldsymbol{E}_S \times \boldsymbol{H}_S^*]$$

Now, for a perfect dielectric medium,

$$\left|\frac{E_0}{H_0}\right| = \eta = \sqrt{\frac{\mu}{\varepsilon}} \text{ and if } \vec{E} = E_0 \cos(\omega t - \beta x)\hat{a}_E \quad \text{and} \quad \vec{H} = H_0 \cos(\omega t - \beta x)\hat{a}_H \text{, then}$$

$$\vec{p}_{\text{ave}} = \frac{1}{2}(E_0)\left(\frac{E_0}{\eta}\right)(\hat{a}_E \times \hat{a}_H) = \frac{1}{2}|E_0|^2\frac{1}{\sqrt{\frac{\mu}{\varepsilon}}}\hat{a}_n = \frac{1}{2}|E_0|^2\sqrt{\frac{\varepsilon}{\mu}}\hat{a}_n \tag{5.46}$$

where, $\hat{a}_n$ is the unit vector in the direction of wave propagation, normal to both $\vec{E}$ and $\vec{H}$.
Now, the electromagnetic energy density 'u' is given by the sum of the electric energy density $\left[u_E = \frac{1}{2}\vec{E}\cdot\vec{D}\right]$ and the magnetic energy density $\left[u_M = \frac{1}{2}\vec{B}\cdot\vec{H}\right]$.

$$\therefore \quad u = \frac{1}{2}\vec{E}\cdot\vec{D} + \frac{1}{2}\vec{B}\cdot\vec{H} = \frac{1}{2}\varepsilon E^2 + \frac{1}{2}\mu H^2 \quad (\because \ \vec{D} = \varepsilon\vec{E} \quad \vec{B} = \mu\vec{H})$$

where, $\vec{E} = E_0 \cos(\omega t - \beta x)\hat{a}_E$ and $\vec{H} = H_0 \cos(\omega t - \beta x)\hat{a}_H$

$$\therefore \quad u = \frac{1}{2}\varepsilon E_0^2 \cos^2(\omega t - \beta x) + \frac{1}{2}\mu H_0^2 \cos^2(\omega t - \beta x) = \frac{1}{2}[\varepsilon E_0^2 + \mu H_0^2]\cos^2(\omega t - \beta x) \tag{5.47}$$

So, the time-averaged energy density in the wave is

$$\langle U_d \rangle = \frac{1}{T}\int_0^T \frac{1}{2}[\varepsilon E_0^2 + \mu H_0^2]\cos^2(\omega t - \beta x)dt$$

$$= \frac{1}{4}[\varepsilon E_0^2 + \mu H_0^2] \quad \left(\because \ \langle \cos^2 \omega t \rangle = \frac{1}{2}\right)$$

$$= \frac{1}{4}\left[\varepsilon E_0^2 + \mu\frac{E_0^2}{\eta^2}\right] \quad \left(\because \ \eta = \sqrt{\frac{\mu}{\varepsilon}}\right)$$

$$= \frac{1}{4}\left[\varepsilon E_0^2 + \mu \frac{E_0^2}{\left(\sqrt{\frac{\mu}{\varepsilon}}\right)^2}\right]$$

$$= \frac{1}{2}\varepsilon |E_0|^2 \qquad \textbf{(5.48)}$$

From Eq. (5.46) and (5.48), we get

$$\vec{p}_{\text{ave}} = \frac{1}{2}\sqrt{\frac{\varepsilon}{\mu}}|E_0|^2 \hat{a}_n = \frac{1}{2}\varepsilon |E_0|^2 \sqrt{\frac{1}{\mu\varepsilon}}\hat{a}_n = \langle U_d \rangle \sqrt{\frac{1}{\mu\varepsilon}}\hat{a}_n$$

$$\therefore \qquad \boxed{\vec{p}_{\text{ave}} = \langle U_d \rangle \sqrt{\frac{1}{\mu\varepsilon}}\hat{a}_n = v\langle U_d \rangle \hat{a}_n} \qquad \textbf{(5.49)}$$

where, v is the velocity of the wave in the medium.

From this expression, we have the following conclusions:

1. The time averaged Poynting vector is the electromagnetic energy density multiplied by the wave velocity.
2. The time-averaged energy flow (Poynting vector) is in the direction of propagation of the wave and is equal to the phase velocity of the wave multiplied by the average energy density.
3. The ratio of Poynting vector to energy density is always less than the velocity of light.

$$\frac{\vec{p}_{\text{ave}}}{\langle U_d \rangle} = v = \frac{c}{\sqrt{\mu_r \varepsilon_r}} \le 3 \times 10^8 \text{ m/s}$$

4. For perfect dielectric, electric energy density is equal to magnetic energy density, as

$$u_E = \frac{1}{2}\varepsilon |E_0|^2 \quad \text{and} \quad u_M = \frac{1}{2}\mu |H_0|^2 = \frac{1}{2}\mu \left|\frac{E_0}{\sqrt{\mu/\varepsilon}}\right|^2 = \frac{1}{2}\varepsilon |E_0|^2$$

   So, time-averaged energy density is shared equally between electric and magnetic fields.

5. For a conductor, the time-averaged electric energy density is less than the time-averaged magnetic energy density, as

$$u_E = \frac{1}{2}\varepsilon |E_0|^2 \quad \text{and} \quad u_M = \frac{1}{2}\mu |H_0|^2 = \frac{1}{2}\mu \left|\frac{E_0}{\sqrt{\omega\mu/\sigma}}\right|^2 = \frac{1}{2}\frac{\sigma}{\omega}|E_0|^2$$

$$\therefore \qquad \frac{u_E}{u_M} = \left(\frac{\omega\varepsilon}{\sigma}\right) <<$$

Example 5.35 Calculate the power flow for a plane wave using Poynting theorem.

Solution Velocity of a uniform plane wave, $v = \frac{1}{\sqrt{\mu\varepsilon}}$

The total energy density due to electric and magnetic fields is given by $\frac{1}{2}(\varepsilon E^2 + \mu H^2)$

Rate of flow of energy per unit area,

$$P = \frac{1}{2}(\varepsilon E^2 + \mu H^2)v = \frac{1}{2}\left(\varepsilon\sqrt{\frac{\mu}{\varepsilon}}EH + \mu\sqrt{\frac{\varepsilon}{\mu}}EH\right)v \quad \left\{\because \frac{E}{H} = \sqrt{\frac{\mu}{\varepsilon}}\right\}$$

$$= \frac{1}{2}EH(2\sqrt{\varepsilon\mu})v$$

$$= vEH\sqrt{\varepsilon\mu}$$

$$= vEH\frac{1}{v}$$

$$= EH$$

Since the angle between $\vec{E}$ and $\vec{H}$ is 90°, the power flow can be written as

$$\boxed{\vec{P} = \vec{E} \times \vec{H}}$$

***Example 5.36** Show that the power flow along a concentric cable is the product of voltage and current. Use Poynting theorem.

Solution The magnetic field strength will be directed in circles about the axis, as shown in Fig. 5.21. By Ampere's law,

$$\oint_l \vec{H} \cdot d\vec{l} = I$$

or $$2\pi r H = I$$

or $$H = \frac{I}{2\pi r}$$

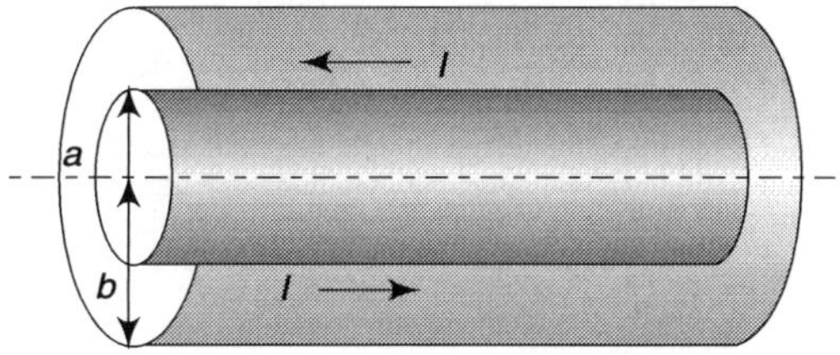

Fig. 5.21 *Concentric cable*

where, r is the radius of the circle being considered.
In vector form,

$$\vec{H} = \frac{I}{2\pi r}\hat{a}_\phi$$

The electric field will be radially outward. By Gauss' law,

$$E = \frac{q}{2\pi\varepsilon r}$$

∴ Voltage between the conductors is

$$V = \int_a^b E dr = \int_a^b \frac{q}{2\pi\varepsilon r}dr = \frac{q}{2\pi\varepsilon}\int_a^b \frac{dr}{r} = \frac{q}{2\pi\varepsilon}\ln\left(\frac{b}{a}\right)$$

∴ $$E = \frac{q}{2\pi\varepsilon r} = \frac{V}{r\ln\left(\frac{b}{a}\right)}$$

In vector form,

$$E = \frac{V}{r \ln\left(\frac{b}{a}\right)} \hat{a}_r$$

The Poynting vector is given as

$$\vec{P} = \vec{E} \times \vec{H}$$

Taking only the magnitudes,

$$P = E \times H = \frac{V}{r \ln\left(\frac{b}{a}\right)} \times \frac{I}{2\pi r} = \frac{VI}{2\pi \ln\left(\frac{b}{a}\right)} \frac{1}{r^2}$$

Total power flow along the cable is

$$W = \int_S (\vec{E} \times \vec{H}) \cdot d\vec{S} = \int_a^b \frac{VI}{2\pi \ln\left(\frac{b}{a}\right)} \frac{1}{r^2} \times 2\pi r dr = \frac{VI}{\ln\left(\frac{b}{a}\right)} \int_a^b \frac{dr}{r} = \frac{VI}{\ln\left(\frac{b}{a}\right)} \times n\left(\frac{b}{a}\right) = VI$$

***Example 5.37** A conductor of circular cross-section of radius 'a' is carrying a current 'I' that is uniformly distributed as shown in Fig. 5.22. Show that the surface integral of the Poynting vector over the surface of the conductor gives the total power dissipation in the conductor. Conductivity of the material is $\sigma\left(\frac{1}{\rho}\right)$.

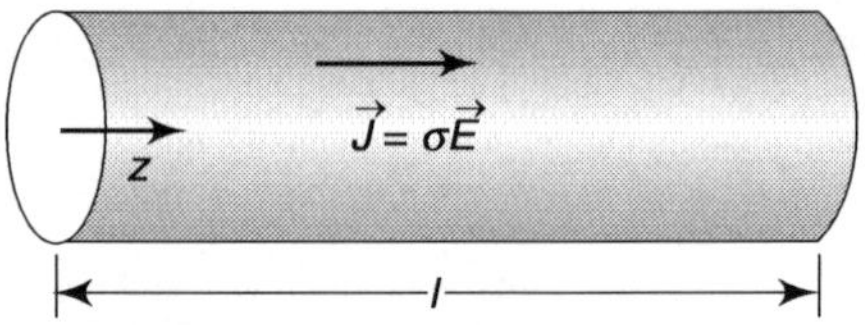

Fig. 5.22 *Circular conductor*

Solution From the relation, $\vec{J} = \sigma\vec{E} \Rightarrow \vec{E} = \frac{\vec{J}}{\sigma}$

So, the electric field will be parallel to the direction of current (here, z-direction).

$$\vec{J} = J_z \hat{a}_z$$

The magnetic field strength $\vec{H}$ will be in the circular (ϕ) direction.

$$\therefore \qquad \vec{H} = H_\phi \hat{a}_\phi$$

The Poynting vector is given as

$$\vec{P} = \vec{E} \times \vec{H} = \frac{\vec{J}}{\sigma} \times \vec{H} = \frac{J_z}{\sigma} \hat{a}_z \times H_\phi \hat{a}_\phi = -\frac{J_z H_\phi}{\sigma} \hat{a}_r$$

The Poynting vector is directed radially towards the axis of the conductor.
Now, magnetic field at any radius r, by Ampere's law is

$$H_\phi = \frac{I}{2\pi r} = \frac{J_z \pi r^2}{2\pi r} = \frac{J_z r}{2}$$

$$\therefore \qquad \vec{P} = -\frac{J_z^2 r}{2\sigma} \hat{a}_r$$

At the surface of the conductor, $r = a$.

$$\therefore \qquad \vec{P} = -\frac{J_z^2 a}{2\sigma}\hat{a}_r$$

So, the surface integral of this vector over the surface of the conductor is

$$W = -\oint \vec{P}\cdot d\vec{S} = -\int_S \left(-\frac{J_z^2 a}{2\sigma}\hat{a}_r\right)\cdot (d\vec{S}) = \int_0^l \left(\frac{J_z^2 a}{2\sigma}\hat{a}_r\right)\cdot (2\pi a\, dz\, \hat{a}_r) = \frac{J_z^2 \pi a^2}{\sigma} l$$

$$= \left(\frac{J_z}{\sigma} l\right)(J_z \pi a^2)$$

$$= (El)(I)$$

$$= VI$$

This is the total power dissipated in the conductor.

***Example 5.38** A long straight non-magnetic wire carries a steady current of I ampere. The resistance of the wire is R ohm/metre. Use the Poynting theorem to show that the energy flow into the wire is I^2R per metre.

Solution From Example 5.37, Poynting vector is

$$\vec{P} = -\frac{J_z^2 a}{2\sigma}\hat{a}_r = -\frac{(I/\pi a^2)^2 a}{2\sigma}\hat{a}_r = -\frac{I^2}{2\pi^2 \sigma a^3}\hat{a}_r$$

Using the Poynting theorem at the surface of the conductor of length l, energy flow into the wire is

$$W = -\oint \vec{P}\cdot d\vec{S} = -\left[-\frac{I^2}{2\pi^2 \sigma a^3}\hat{a}_r\right]\cdot (2\pi a l \hat{a}_r) = \frac{I^2 l}{\sigma \pi a^2} = I^2 \frac{l}{\sigma \pi a^2}$$

$$= I^2 \frac{\rho l}{\pi a^2} \qquad \left\{\because \sigma = \frac{1}{\rho}\right\}$$

$$= I^2 R$$

Example 5.39 An elliptically polarised wave in air has x and y components:

$$E_x = 4 \sin(\omega t - \beta z) \quad \text{Volt/m}$$

$$E_y = 8 \sin(\omega t - \beta z + 75°) \quad \text{Volt/m}$$

Find the Poynting vector. For air the intrinsic impedance is 367.7 ohm.

Solution By Maxwell's equation,

$$\nabla \times \vec{E} = -\frac{\partial \vec{B}}{\partial t} = -\mu_0 \frac{\partial \vec{H}}{\partial t}$$

or

$$\begin{vmatrix} \dfrac{\partial}{\partial x} & \dfrac{\partial}{\partial y} & \dfrac{\partial}{\partial z} \\ E_x & E_y & E_z \\ \hat{i} & \hat{j} & \hat{k} \end{vmatrix} = -\mu_0 \left[\frac{\partial H_x}{\partial t}\hat{i} + \frac{\partial H_y}{\partial t}\hat{j} + \frac{\partial H_z}{\partial t}\hat{k}\right]$$

or

$$\begin{vmatrix} \dfrac{\partial}{\partial x} & \dfrac{\partial}{\partial y} & \dfrac{\partial}{\partial z} \\ E_x & E_y & 0 \\ \hat{i} & \hat{j} & \hat{k} \end{vmatrix} = -\mu_0 \left[\frac{\partial H_x}{\partial t}\hat{i} + \frac{\partial H_y}{\partial t}\hat{j} + \frac{\partial H_z}{\partial t}\hat{k} \right]$$

or

$$-\frac{\partial E_y}{\partial z}\hat{i} + \frac{\partial E_x}{\partial z}\hat{j} = -\mu_0 \left[\frac{\partial H_x}{\partial t}\hat{i} + \frac{\partial H_y}{\partial t}\hat{j} + \frac{\partial H_z}{\partial t}\hat{k} \right]$$

Comparing both sides, we get

$$H_z = 0$$

and

$$\frac{\partial H_x}{\partial t} = \frac{1}{\mu_0}\frac{\partial E_y}{\partial z} = -\frac{1}{\mu_0}\frac{\partial}{\partial z}[8 \sin(\omega t - \beta z + 75°)] = -\frac{8\beta}{\mu_0}\cos(\omega t - \beta z + 75°)$$

$$H_x = -\frac{8\beta}{\mu_0}\int \cos(\omega t - \beta z + 75°)dt = -\frac{8\beta}{\omega\mu_0}\sin(\omega t - \beta z + 75°)$$

Similarly,

$$H_y = \frac{4\beta}{\omega\mu_0}\sin(\omega t - \beta z)$$

$\therefore$ Poynting vector,

$$\vec{P} = \vec{E} \times \vec{H} = \begin{vmatrix} \hat{i} & \hat{j} & \hat{k} \\ E_x & E_y & 0 \\ H_x & H_y & 0 \end{vmatrix} = (E_x H_y - E_y H_x)\hat{k}$$

$$= \left[4\sin(\omega t - \beta z)\frac{4\beta}{\omega\mu_0}\sin(\omega t - \beta z) + 8\sin(\omega t - \beta z + 75°)\frac{8\beta}{\omega\mu_0}\sin(\omega t - \beta z + 75°) \right]\hat{k}$$

$$= \frac{\beta}{\omega\mu_0}[16\sin^2 \sin(\omega t - \beta z) + 64\sin^2(\omega t - \beta z + 75°)]\hat{k}$$

$$= \sqrt{\frac{\varepsilon_0}{\mu_0}}[16\sin^2 \sin(\omega t - \beta z) + 64\sin^2(\omega t - \beta z + 75°)]\hat{k}$$

$$= \frac{1}{367.7}[16\sin^2 \sin(\omega t - \beta z) + 64\sin^2(\omega t - \beta z + 75°)]\hat{k}$$

$$\therefore \quad \vec{P} = 0.0435[\sin^2 \sin(\omega t - \beta z) + 4\sin^2(\omega t - \beta z + 75°)]\hat{k}$$

Example 5.40 The electric field of a uniform plane wave propagating in the positive z-direction is given by

$$\vec{E} = E_0 \cos(\omega t - \beta z)\hat{i} + E_0 \sin(\omega t - \beta z)\hat{j}$$

where, E_0 is a constant. Find (i) the corresponding magnetic field $\vec{H}$ and (ii) the Poynting vector. Evaluate the Poynting vector if $E_0 = 10$ V/m.

Solution (i) By Maxwell's equation,

$$\nabla \times \vec{E} = -\mu j\omega$$

or
$$\vec{H} = -\frac{1}{j\omega\mu}\begin{vmatrix} \frac{\partial}{\partial x} & \frac{\partial}{\partial y} & \frac{\partial}{\partial z} \\ E_x & E_y & 0 \\ \hat{i} & \hat{j} & \hat{k} \end{vmatrix} = -\frac{1}{j\omega\mu}\left[-\frac{\partial E_y}{\partial z}\hat{i} + \frac{\partial E_x}{\partial z}\hat{j}\right]$$

$$= -\frac{1}{j\omega\mu}[-\{-E_0\beta \cos(\omega t - \beta z)\}\hat{i} + \{E_0\beta \sin(\omega t - \beta z)\}\hat{j}]$$

$$= \frac{E_0\beta}{j\omega\mu}[\cos(\omega t - \beta z)\hat{i} + \sin(\omega t - \beta z)\hat{j}]$$

Hence, the magnetic field is given as

$$\boxed{\vec{H} = \frac{E_0\beta}{j\omega\mu}[\cos(\omega t - \beta z)\hat{i} + \sin(\omega t - \beta z)\hat{j}]}$$

(ii) The Poynting vector,

$$\vec{P} = \vec{E} \times \vec{H} = \begin{vmatrix} \hat{i} & \hat{j} & \hat{k} \\ E_x & E_y & 0 \\ H_x & H_y & 0 \end{vmatrix} = (E_x H_y - E_y H_x)\hat{k}$$

$$= \left[E_0 \cos(\omega t - \beta z)\frac{E_0\beta}{j\omega\mu}\sin(\omega t - \beta z) - E_0 \sin(\omega t - \beta z)\frac{E_0\beta}{j\omega\mu}\cos(\omega t - \beta z)\right]\hat{k}$$

$$= \frac{E_0^2\beta}{\omega\mu}\left\{\cos^2(\omega t - \beta z) + \sin^2(\omega t - \beta z)\right\}\hat{k} \qquad (\because\ j = 90°)$$

$$= \frac{E_0^2\beta}{\omega\mu}\hat{k}$$

$$= E_0^2\sqrt{\frac{\varepsilon}{\mu}}\,\hat{k}$$

$$= \frac{E_0^2}{\eta}\hat{k} \qquad \eta \text{ is the intrinsic impedance}$$

$\therefore$
$$\boxed{\vec{P} = \frac{E_0^2}{\eta}\hat{k}}$$

Example 5.41 If $\vec{E}(z,t) = 100\cos(\omega t - \beta z)\hat{a}_x$ V/m and $\vec{H}(z,t) = 2.65\cos(\omega t - \beta z)\hat{a}_y$ A/m, determine the time averaged Poynting vector at any position z.

Solution Time-average value of Poynting vector is given as,

$\therefore$
$$\boxed{P_{av} = \frac{1}{2}\frac{E_0^2}{\eta_0}e^{-2\alpha z}\cos\theta_\eta}$$

Here, $\alpha = 0, \quad \theta_\eta = 0°, \quad E_0 = 100, \quad \dfrac{E_0}{\eta_0} = 2.65$

Substituting these values, we get the time-average Poynting vector at any position as

$$P_{av} = \frac{1}{2} E_0 \frac{E_0}{\eta_0} e^{-2\alpha z} \cos\theta_\eta = \frac{1}{2} \times 100 \times 2.65 \times 1 \times \cos 0° = 132.5 \quad \text{Watt/m}^2$$

***Example 5.42** A plane electromagnetic wave having a frequency of 10 MHz has an average Poynting vector of 1 W/m^2. If the medium is lossless with relative permeability 2 and relative permittivity 3, find:

1. the velocity of propagation,
2. the wavelength,
3. the impedance of the medium, and
4. the r.m.s. electric field E.

Solution Given: f = 10 MHz, P = 1 W/m^2, $\mu_r = 2$, $\varepsilon_r = 3$,

1. Velocity of wave propagation is

$$v = \frac{1}{\sqrt{\mu\varepsilon}} = \frac{1}{\sqrt{\mu_0\mu_r\varepsilon_0\varepsilon_r}} = \frac{1}{\sqrt{\mu_0\varepsilon_0}}\frac{1}{\sqrt{\mu_r\varepsilon_r}} = \frac{c}{\sqrt{\mu_r\varepsilon_r}} = \frac{3\times10^8}{\sqrt{2\times3}} = 1.225\times10^8 \text{ m/s}$$

2. Wavelength is

$$\lambda = \frac{v}{f} = \frac{1.225\times10^8}{10\times10^6} = 12.25 \text{ m}$$

3. Impedance of the medium is

$$\eta = \sqrt{\frac{\mu}{\varepsilon}} = \sqrt{\frac{\mu_0\mu_r}{\varepsilon_0\varepsilon_r}} = \sqrt{\frac{\mu_0}{\varepsilon_0}}\sqrt{\frac{\mu_r}{\varepsilon_r}} = 120\,\pi \times \sqrt{\frac{2}{3}} = 307.81\,\Omega$$

4. RMS electric field is given as

$$P = EH = E\frac{E}{\eta} = \frac{E^2}{\eta} \quad \Rightarrow \quad E = \sqrt{P\eta} = \sqrt{1\times(120\pi)\times\sqrt{\frac{2}{3}}} = 17.54$$

This is the peak value of the field. Hence the RMS value is

$$E_{rms} = \frac{17.54}{\sqrt{2}} = 12.406 \text{ V/m}$$

***Example 5.43** A sinusoidal plane wave is transmitted through a medium whose breakdown strength is 30 kV/m (rms) and whose relative permittivity is 4. Determine the mean possible power flow density and the peak value of the associated magnetic field intensity.

Solution From section 5.11.5, the mean power density is

$$P_{av} = \frac{1}{2}\frac{E^2}{\eta} = \frac{E_{rms}^2}{\eta}$$

Here, $\eta = \sqrt{\frac{\mu}{\varepsilon}} = \sqrt{\frac{\mu_0 \mu_r}{\varepsilon_0 \varepsilon_r}} = \sqrt{\frac{\mu_0}{\varepsilon_0}}\sqrt{\frac{\mu_r}{\varepsilon_r}} = 120\,\pi \times \sqrt{\frac{1}{4}} = 60\,\pi\,\Omega$

$$\therefore \qquad P_{\text{av}} = \frac{(30 \times 10^3)^2}{60\pi} = 4774.65 \text{ kW/m}^2$$

Also, $H_{\text{rms}} = \frac{E_{\text{rms}}}{\eta} = \frac{P_{\text{av}}}{E_{\text{rms}}}$

$\therefore$ Peak value of the magnetic field is

$$H_0 = \sqrt{2} \times \frac{P_{\text{av}}}{E_{\text{rms}}} = \sqrt{2} \times \frac{4774.65 \times 10^3}{30 \times 10^3} = 225 \text{ A/m}$$

Example 5.44 In a non-magnetic medium,

$$\vec{E} = 4 \sin(2\pi \times 10^7 t - 0.8x)\hat{a}_z \text{ V/m}$$

Find:

(a) ε_r, η.

(b) The time-average power carried by the wave.

(c) The total power crossing 100 cm^2 of plane $2x + y = 5$.

Solution Since $\alpha = 0$ and $\beta \neq \frac{\omega}{c}$, the medium is not free space, but a lossless medium.

(a) Here, $\beta = 0.8$, $\omega = 2\pi \times 10^7$, $\mu = \mu_0$ (non-magnetic medium)

$$\therefore \qquad \beta = \omega\sqrt{\mu\varepsilon} = \omega\sqrt{\mu_0\mu_r\varepsilon_0\varepsilon_r} = \frac{\omega}{c}\sqrt{\varepsilon_r}$$

$$\therefore \qquad \varepsilon_r = \left(\frac{\beta c}{\omega}\right)^2 = \left(\frac{0.8 \times 3 \times 10^8}{2\pi \times 10^7}\right)^2 = \left(\frac{12}{\pi}\right)^2 = 14.59$$

$$\therefore \qquad \eta = \sqrt{\frac{\mu}{\varepsilon}} = \sqrt{\frac{\mu_0}{\varepsilon_0\varepsilon_r}} = \frac{120\pi}{12/\pi} = 10\pi^2 = 98.7\,\Omega$$

(b) Time-average power carried by the wave

$$P_{\text{av}} = \frac{1}{2}\frac{E_0^2}{\eta}\hat{i} = \frac{1}{2}\frac{(4/\sqrt{2})^2}{10\pi^2}\hat{i} = 81\hat{i} \quad \text{mW/m}^2$$

(c) On the plane $2x + y = 5$, $f = (2x + y - 5)$ $\quad \therefore\ \nabla f = (2\hat{i} + \hat{j})$

$\therefore$ unit vector normal to the plane is, $\hat{a}_n = \pm\frac{\nabla f}{|\nabla f|} = \pm\frac{(2\hat{i} + \hat{j})}{\sqrt{5}}$

Hence, the total power crossing the plane is

$$P_{av} = \int \vec{P}_{av} \cdot d\vec{S} = \vec{P}_{av} \cdot S\hat{a}_n = (81 \times 10^{-3}\hat{i}) \cdot (100 \times 10^{-4})\left[\frac{(2\hat{i} + \hat{j})}{\sqrt{5}}\right] = \frac{162 \times 10^{-5}}{\sqrt{5}} = 724.5\,\mu\text{W}$$

Example 5.45 Find the reflection coefficient and the transmission coefficient of an electric field wave travelling in air and incident normally on a boundary between air and a dielectric having permeability μ_0 and permittivity , $\varepsilon_r = 4$. Also, find the average power, P_i, P_r, P_t.

Solution For air, the intrinsic impedance is $\eta_1 = \sqrt{\dfrac{\mu_0}{\varepsilon_0}}$

For the dielectric medium, $\eta_2 = \sqrt{\dfrac{\mu_0}{\varepsilon_0 \varepsilon_r}} = \sqrt{\dfrac{\mu_0}{\varepsilon_0 \times 4}} = \dfrac{\eta_1}{2}$

$$\therefore \qquad \frac{E_r}{E_i} = \frac{\eta_2 - \eta_1}{\eta_2 + \eta_1} = \frac{\dfrac{\eta_1}{2} - \eta_1}{\dfrac{\eta_1}{2} + \eta_1} = -\frac{1}{3}$$

$$\therefore \qquad \frac{E_t}{E_i} = \frac{2\eta_2}{\eta_2 + \eta_1} = \frac{2\dfrac{\eta_1}{2}}{\dfrac{\eta_1}{2} + \eta_1} = \frac{2}{3}$$

Using the complex Poynting vector, we have

Incident Average Power, $P_i = \dfrac{1}{2}\text{Re}\{\vec{E}_i \times \vec{H}_i^*\} = \dfrac{E_i^2}{2\eta_1}$

Reflected Average Power, $P_r = \dfrac{1}{2}\text{Re}\{\vec{E}_r \times \vec{H}_r^*\} = -\dfrac{E_r^2}{2\eta_1} = -\dfrac{E_r^2}{18\eta_1} \qquad \left(\because \dfrac{E_r}{E_i} = -\dfrac{1}{3}\right)$

Transmitted Average Power, $P_t = \dfrac{1}{2}\text{Re}\{\vec{E}_t \times \vec{H}_i^*\} = \dfrac{E_t^2}{2\eta_2} = \dfrac{1}{2} \times \dfrac{2}{\eta_1} \times \dfrac{4}{9}E_i^2 = \dfrac{8}{18}E_i^2$

From the results, it is seen that one-ninth of the incident power is reflected and eight-ninth of it is transmitted into the second medium.

***Example 5.46** A plane wave is incident normally on a large sheet of copper. If the frequency and peak electric field of the incident wave is 100 MHz and 1 V/m respectively, find the power absorbed per unit area by the copper sheet.

Solution Given: $E_i = 1$ V/m, $f = 100 \times 10^6$ Hz
Assuming that the wave impinges from air, $\eta_1 = 377\ \Omega$

$$\eta_2 = \sqrt{\frac{j\omega\mu}{\sigma + j\omega\varepsilon}} = \sqrt{\frac{j \times 2\pi \times 100 \times 10^6 \times 4\pi \times 10^{-7}}{5.8 \times 10^7 + j \times 2\pi \times 100 \times 10^6 \times 8.854 \times 10^{-12}}}$$
$$= 3.68961 \times 10^{-3} \angle 45^\circ$$

Transmission coefficient is,

$$\frac{E_t}{E_i} = \frac{2\eta_2}{\eta_2 + \eta_1} = \frac{2 \times 3.68961 \times 10^{-3} \angle 45^\circ}{3.68961 \times 10^{-3} \angle 45^\circ + 377} = 1.96 \times 10^{-5} \angle 45^\circ$$

$$\therefore \qquad \frac{E_t}{1} = 1.96 \times 10^{-5} \angle 45^\circ \Rightarrow E_t = 1.96 \times 10^{-5} \angle 45^\circ \text{ V/m}$$

Depth of penetration,

$$\delta = \sqrt{\frac{2}{\omega\mu\sigma}} = \sqrt{\frac{2}{2\pi\times 100\times 10^6\times 4\pi\times 10^{-7}\times 5.8\times 10^7}} = 6.6\times 10^{-6}\text{ m}$$

∴ Power absorbed per unit area by the copper sheet is

$$p_{\text{ave}} = \frac{1}{4}E^2\sigma\delta = \frac{1}{4}(1.96\times 10^{-5})^2\times 5.8\times 10^7\times 6.6\times 10^{-6} = 3.67\times 10^{-8}\text{ W/m}^2$$

***Example 5.47** Assume a plane wave with $E = 1$ V/m and a frequency of 300×10^6 cps moving in free space, impinges on a thick copper located perpendicularly to the direction of propagation. Find:

1. E at the plane surface.
2. H at the same location.
3. δ, depth of penetration.
4. Conduction current density at the surface.
5. J, the conduction current density, at the depth of 10^{-2} mm.
6. Surface current density.
7. Surface impedance.
8. Power loss per square metre of surface area.

Assume $\sigma = 5.8 \times 10^7$mho/m. $\varepsilon = \varepsilon_0$, $\mu = \mu_0$.

Solution Given: $E_i = 1$ V/m, $f = 300 \times 10^6$ cps

1. Here, $\eta_1 = 377\ \Omega$

$$\eta_2 = \sqrt{\frac{j\omega\mu}{\sigma + j\omega\varepsilon}} = \sqrt{\frac{j\times 2\pi\times 300\times 10^6\times 4\pi\times 10^{-7}}{5.8\times 10^7 + j\times 2\pi\times 300\times 10^6\times 8.854\times 10^{-12}}}$$
$$= 6.39\times 10^{-3}\angle 45°$$

∴ Transmission coefficient is

$$\frac{E_t}{E_t} = \frac{2\eta_2}{\eta_2+\eta_1} = \frac{2\times 6.39\times 10^{-3}\angle 45°}{6.39\times 10^{-3}\angle 45° + 377} = 3.39\times 10^{-5}\angle 45°$$

$$\frac{E_t}{1} = 3.39\times 10^{-5}\angle 45° \Rightarrow E_t = 3.39\times 10^{-5}\angle 45°\text{ V/m}$$

2. $$\frac{H_t}{H_i} = \frac{2\eta_1}{\eta_2+\eta_1} = \frac{2\times 377}{6.39\times 10^{-3}\angle 45° + 377} = 2$$

$$H_t = 2H_i = 2\frac{E_i}{\eta_1} = \frac{2\times 1}{377} = 5.3\times 10^{-3}\text{ A/m}$$

3. Depth of penetration,

$$\delta = \sqrt{\frac{2}{\omega\mu\sigma}} = \sqrt{\frac{2}{2\pi\times 300\times 10^6\times 4\pi\times 10^{-7}\times 5.8\times 10^7}} = 3.81\times 10^{-6}\text{ m} = 3.81\ \mu\text{m}$$

4. Conduction current density at the surface is

$$J_0 = \sigma E_t = 5.8 \times 10^7 \times 3.39 \times 10^{-5} \angle 45° = 1966 \angle 45° \text{ A/m}^2$$

5. At a distance x below the surface, the conduction current density is given as

$$J = J_0 e^{-\gamma x}$$

where $\gamma = \sqrt{j\omega\mu\sigma} = \sqrt{2\pi \times 300 \times 10^6 \times 4\pi \times 10^{-7} \times 5.8 \times 10^7} \angle 45° = 3.7 \times 10^5 \angle 45°$

$$\text{Re}[\gamma] = 3.7 \times 10^5 \cos 45° = 2.62 \times 10^5$$

So, at a distance, $x = 0.01 \times 10^{-3}$ m, the conduction current density is

$$J = 1966 e^{-2.62 \times 10^5 \times 10^{-3}} \angle 45° = 144 \angle 45° \text{ A/m}^2$$

6. According to the boundary condition, the surface current density is equal to the tangential component of the magnetic field. From the result of (2), we have the surface current density given as

$$K = 5.3 \times 10^{-3} \text{ A/m}$$

7. Surface impedance is given as

$$Z_S = \sqrt{\frac{\omega\mu}{\sigma}} \angle 45° = \sqrt{\frac{2\pi \times 300 \times 10^6 \times 4\pi \times 10^{-7}}{5.8 \times 10^7}} \angle 45° = 6.39 \times 10^{-3} \angle 45° \ \Omega$$

8. Power loss per square metre of surface area is

$$p_{\text{ave}} = \frac{1}{4} E^2 \sigma \delta = \frac{1}{4} (3.39 \times 10^{-5})^2 \times 5.8 \times 10^7 \times 3.81 \times 10^{-6} = 6.35 \times 10^{-8} \text{ W/m}^2$$

Example 5.48 The electric field intensity in radiation field of an antenna located at the origin of a spherical coordinate system is given by

$$\vec{E} = E_0 \frac{\sin\theta \cos\theta}{r} \cos(\omega t - \beta r) \hat{a}_\theta$$

where E_0, ω and β are constants. Find:

1. the magnetic field associated with this electric field.
2. the Poynting vector, and
3. the total power radiated over a spherical surface of radius r centered at the origin.

Solution:

1. From Maxwell's equation,

$$\nabla \times \vec{E} = -\frac{\partial \vec{B}}{\partial t}$$

or
$$-\frac{\partial \vec{B}}{\partial t} = \begin{vmatrix} \frac{1}{r^2 \sin\theta}\hat{a}_r & \frac{1}{r\sin\theta}\hat{a}_\theta & \frac{1}{r}\hat{a}_\phi \\ \frac{\partial}{\partial r} & \frac{\partial}{\partial \theta} & \frac{\partial}{\partial \phi} \\ 0 & rE_0 \frac{\sin\theta}{r}\cos(\omega t - \beta r) & 0 \end{vmatrix}$$

or
$$-\frac{\partial \vec{B}}{\partial t} = \frac{\beta E_0}{r}\sin\theta\cos\theta\sin(\omega t - \beta r)\hat{a}_\phi$$

Integrating with respect to t,

$$\vec{B} = \frac{\beta E_0}{\omega r}\sin\theta\cos\theta\cos(\omega t - \beta r)\hat{a}_\phi$$

∴ Magnetic field is given as

$$\boxed{\vec{H} = \frac{\beta E_0}{\mu_0 \omega r}\sin\theta\cos\theta\cos(\omega t - \beta r)\hat{a}_\phi}$$

2. The Poynting vector is given as

$$\vec{P} = \vec{E}\times\vec{H} = \begin{vmatrix} \hat{a}_r & \hat{a}_\theta & \hat{a}_\phi \\ 0 & E_0\frac{\sin\theta\cos\theta}{r}\cos(\omega t - \beta r) & 0 \\ 0 & 0 & \frac{\beta E_0}{\mu_0\omega r}\sin\theta\cos\theta\cos(\omega t - \beta r) \end{vmatrix}$$

∴
$$\boxed{\vec{P} = \frac{\beta E_0^2}{\mu_0 \omega r^2}\sin^2\theta\cos^2\theta\cos^2(\omega t - \beta r)\hat{a}_r}$$

3. Total power radiated

$$W = \oint_S \vec{P}\cdot d\vec{S} = \int_{\theta=0}^{\pi}\int_{\phi=0}^{2\pi} \frac{\beta E_0^2 \sin^2\theta\cos^2\theta}{\mu_0\omega r^2}\cos^2(\omega t - \beta r) r^2 \sin\theta\, d\theta\, d\phi$$

$$= \frac{2\pi\beta E_0^2}{\mu_0\omega}\cos^2(\omega t - \beta r)\int_{\theta=0}^{\pi}\sin^2\theta\cos^2\theta\, d\theta$$

$$= \frac{8}{15}\frac{\pi\beta E_0^2}{\mu_0\omega}\cos^2(\omega t - \beta r)$$

Example 5.49 A uniform plane wave of frequency 'f' is normally incident from air onto a thick conducting sheet with conductivity σ, and $\varepsilon = \varepsilon_0$, $\mu = \mu_0$. Show that the proportion of power transmitted into the conductor (and then dissipated into heat) is given approximately by

$$\frac{P_{tr}}{P_{in}} = \frac{4R_s}{\eta} = \sqrt{\frac{8\omega\varepsilon_0}{\sigma}},\ R_s = \text{surface resistance}$$

Calculate this quantity for $f = 1\text{GHz}$ and copper, $\sigma = 5.8 \times 10^7$ Siemens/m.

Solution For a good conductor, $\sqrt{\dfrac{\omega\varepsilon_0}{\sigma}} << 1$

So, $$\frac{R_S}{\eta_0} = \frac{\sqrt{\dfrac{\omega\mu_0}{2\sigma}}}{\sqrt{\dfrac{\mu_0}{\varepsilon_0}}} = \sqrt{\frac{\omega\varepsilon_0}{2\sigma}} << 1$$

So, the complex characteristic impedance for a good conductor is

$$\eta = R_S(1 + j)$$

or $$\frac{\eta}{\eta_0} = (1 + j)\frac{R_S}{\eta_0} << 1$$

∴ Transmission coefficient,

$$\tau = \frac{2\eta}{\eta + \eta_0} \approx \frac{2\eta}{\eta_0}$$

∴ Reflection coefficient,

$$\Gamma = (\tau - 1) = \frac{2\eta}{\eta_0} - 1$$

So, the power transmission coefficient is

$$\frac{P_{tr}}{P_{in}} = 1 - |\Gamma|^2 = 1 - |\tau - 1|^2 = 1 - 1 - |\tau|^2 + 2\,\text{Re}\{\tau\} = 2\,\text{Re}\{\tau\} \qquad \text{(neglecting higher term as } \tau <<)$$

$$= 2\,\text{Re}\left[\frac{2\eta}{\eta_0}\right]$$

$$= \frac{4R_S}{\eta_0} \qquad (\because \eta = R_S(1 + j))$$

$$= 4\sqrt{\frac{\omega\varepsilon_0}{2\sigma}}$$

∴ $$\boxed{\frac{P_{tr}}{P_{in}} = \frac{4R_S}{\eta_0} = \sqrt{\frac{8\omega\varepsilon_0}{\sigma}}}$$

For copper at 1 GHz, the power transmission coefficient is

$$\frac{P_{tr}}{P_{in}} = \sqrt{\frac{8\omega\varepsilon_0}{\sigma}} = \sqrt{\frac{8 \times 2\pi \times 1 \times 10^9 \times 8.854 \times 10^{-12}}{5.8 \times 10^7}} = 8.76 \times 10^{-5} = 8.76 \times 10^{-3}\ \%$$

5.13 RADIATION PRESSURE

The electromagnetic wave transports not only energy but also momentum, and hence can exert a pressure, known as *radiation pressure* on a surface due to the absorption and reflection of the momentum by the surface.

Radiation Pressure for Perfectly Absorbing Body Radiation pressure is defined as the force per unit area exerted by electromagnetic radiation, and is given by,

$$P_{\text{rad}} = \langle U_d \rangle = \frac{|\langle \vec{S} \rangle|}{c} \qquad \text{[using Eq. (5.49)]}$$

where, $\langle U_d \rangle$ is the time-averaged electromagnetic energy density, $|\langle \vec{S} \rangle|$ is the magnitude of time-averaged Poynting vector and c is the speed of light.

We consider a plane electromagnetic wave propagating in free space and incident normally on the surface of a perfectly absorbing body.

Thus, the energy incident per second on a unit area of the surface is the energy contained in a cylinder of unity cross-sectional area and height c

$$\therefore \qquad W = \langle U_d \rangle c$$

where, $\langle U_d \rangle$ is the time-averaged electromagnetic energy density and c is the speed of light.
Also, the electromagnetic radiation consists of photons, each of which has the energy as given
Energy of a photon $= h\upsilon = mc^2$ (By Einstein's relation)
where h is Planck's constant and u is the frequency of the radiation.
Thus, the energy of the electromagnetic energy per second $W = \sum h\upsilon$

Therefore, the momentum of all the photons $= \sum mc = \sum \frac{h\upsilon}{c} = \frac{\langle U_d \rangle c}{c} = \langle U_d \rangle$

This is the rate of loss of momentum since the photons (or radiation) are completely absorbed by the surface. By Newton's second law of motion, the radiation exerts an equal and opposite force on the unit area of the surface; which is the radiation pressure.
Thus, radiation pressure is given as

$$\boxed{P_{\text{rad}} = \langle U_d \rangle = \frac{|\langle \vec{S} \rangle|}{c} = \frac{1}{2}\varepsilon\, |E_0|^2 = \frac{1}{4}\text{Re}[\vec{E} \cdot \vec{D}^* + \vec{B} \cdot \vec{H}^*]} \qquad \textbf{(5.50)}$$

Radiation Pressure for Perfectly Reflecting Body If the surface is a perfectly reflecting one, the velocity of the incident radiation changes from $+c$ to $-c$ and thus, the radiation pressure will be twice that of a perfectly absorbing surface. Thus, the radiation pressure for perfectly reflecting surface is given as

$$\boxed{P_{\text{rad}} = 2\langle U_d \rangle = 2\frac{|\langle \vec{S} \rangle|}{c} = \varepsilon\, |E_0|^2 = \frac{1}{2}\text{Re}[\vec{E} \cdot \vec{D}^* + \vec{B} \cdot \vec{H}^*]} \qquad \textbf{(5.51)}$$

Example 5.50 A plane electromagnetic wave with the $\vec{E}$ field amplitude 1 mV travelling in vacuum falls normally on a surface and is totally reflected. Calculate the radiation pressure exerted on the surface.

Solution The radiation pressure is given as

$$P_{\text{rad}} = 2\langle U_d \rangle = \varepsilon_0\, |E_0|^2 = 8.854 \times 10^{-12} \times (1 \times 10^{-3})^2 = 8.854 \times 10^{-6}\ \text{Pa}$$

Summary

- If a physical phenomenon that occurs at one place at a given time is reproduced at other places at later times, the time-delay being proportional to the space separation from the first location, then the group of phenomena constitutes a *wave*.
- Three dimensional wave equations (Helmholtz equations) in terms of electric and magnetic fields are given as

$$\nabla^2\vec{E} - \mu\varepsilon\frac{\partial^2\vec{E}}{\partial t^2} - \mu\sigma\frac{\partial\vec{E}}{\partial t} = 0 \quad \text{and} \quad \nabla^2\vec{H} - \mu\varepsilon\frac{\partial^2\vec{H}}{\partial t^2} - \mu\sigma\frac{\partial\vec{H}}{\partial t} = 0$$

- For perfect dielectric medium, the wave equations reduce to

$$\nabla^2\vec{E} = \mu\varepsilon\frac{\partial^2\vec{E}}{\partial t^2} \quad \text{and} \quad \nabla^2\vec{H} = \mu\varepsilon\frac{\partial^2\vec{H}}{\partial t^2}$$

- For free space, the wave equations reduce to

$$\nabla^2\vec{E} = \mu_0\varepsilon_0\frac{\partial^2\vec{E}}{\partial t^2} \quad \text{and} \quad \nabla^2\vec{H} = \mu_0\varepsilon_0\frac{\partial^2\vec{H}}{\partial t^2}$$

- For time-harmonic fields, the wave equations reduce to

$$\nabla^2\boldsymbol{E}_S - \gamma^2\boldsymbol{E}_S = 0 \quad \text{and} \quad \nabla^2\boldsymbol{H}_S - \gamma^2\boldsymbol{H}_S = 0$$

  where, γ is defined as the *propagation constant.*

$$\gamma = \sqrt{j\omega\mu(\sigma + j\omega\varepsilon)} = (\alpha + j\beta)$$

- The real part of the propagation constant (α) is defined as the *attenuation constant* (Neper/m). It is given as

$$\begin{aligned}
\alpha &= \omega\sqrt{\frac{\mu\varepsilon}{2}\left[\sqrt{1+\left(\frac{\sigma}{\omega\varepsilon}\right)^2} - 1\right]} && \text{in general} \\
&= \frac{\sigma}{2}\sqrt{\frac{\mu}{\varepsilon}} && \text{for lossy dielectric} \\
&= 0 && \text{for perfect dielectric} \\
&= 0 && \text{for free space} \\
&= \sqrt{\frac{\omega\mu\sigma}{2}} && \text{for good conductors}
\end{aligned}$$

- The imaginary part (β) is defined as the *phase constant* (Radian/m). It is given as

$$\begin{aligned}
\beta &= \omega\sqrt{\frac{\mu\varepsilon}{2}\left[\sqrt{1+\left(\frac{\sigma}{\omega\varepsilon}\right)^2} + 1\right]} && \text{in general} \\
&= \omega\sqrt{\mu\varepsilon}\left(1 + \frac{\sigma^2}{8\omega^2\varepsilon^2}\right) && \text{for lossy dielectric}
\end{aligned}$$

$$= \omega\sqrt{\mu\varepsilon} \quad \text{for perfect dielectric}$$

$$= \omega\sqrt{\mu_0\varepsilon_0} = \frac{\omega}{c} \quad \text{for free space}$$

$$= \sqrt{\frac{\omega\mu\sigma}{2}} \quad \text{for good conductors}$$

- A wave is said to be a *plane wave*, if:
  1. The electric field $\vec{E}$ and magnetic field $\vec{H}$ lie in a plane perpendicular to the direction of wave propagation.
  2. The fields $\vec{E}$ and $\vec{H}$ are perpendicular to each other.
- A plane wave is said to be *uniform plane wave* if
  1. The electric field $\vec{E}$ and magnetic field $\vec{H}$ lie in a plane perpendicular to the direction of wave propagation.
  2. The fields $\vec{E}$ and $\vec{H}$ are perpendicular to each other.
  3. $\vec{E}$ and $\vec{H}$ are uniform in the plane perpendicular to the direction of propagation (i.e., $\vec{E}$ and $\vec{H}$ vary only in the direction of propagation)
- *Standing waves* can be formed by confining the electromagnetic waves within two perfectly reflecting conductors. Unlike the travelling electromagnetic wave in which the electric and the magnetic fields are always in phase, in standing waves, the two fields are 90° out of phase.
- The *phase velocity* of a wave is the rate at which the phase of the wave propagates in space. This is given as

$$v_p = \frac{\omega}{\beta} = \lambda f = \frac{1}{\sqrt{\mu\varepsilon}}$$

- The velocity with which the overall shape of a wave amplitude, known as the modulation or envelope of the wave, propagates through a medium is known as the *group velocity* or *energy velocity* of the wave. This is given as

$$v_g = \frac{\Delta\omega}{\Delta\beta} = \frac{d\omega}{d\beta}$$

- The *intrinsic impedance* of the wave is defined as the ratio of the electric field and magnetic field phasors (complex amplitudes). It is given as

$$\eta = \left|\frac{\vec{E}}{\vec{H}}\right| = \sqrt{\frac{j\omega\mu}{\sigma + j\omega\varepsilon}} \quad \text{in general}$$

$$= \frac{\sqrt{\dfrac{\mu}{\varepsilon}}}{\left[1 + \left(\dfrac{\sigma}{\omega\varepsilon}\right)^2\right]^{1/4}} < \frac{1}{2}\tan^{-1}\left(\frac{\sigma}{\omega\varepsilon}\right) \quad \text{for lossy dielectric}$$

$$= \sqrt{\frac{\mu}{\varepsilon}} \quad \text{for perfect dielectric}$$

$$= \sqrt{\frac{\mu_0}{\varepsilon_0}} = 120\pi = 377\ \Omega \qquad \text{for free space}$$

$$= \sqrt{\frac{\omega\mu}{\sigma}} \angle 45^\circ \qquad \text{for good conductors}$$

- Velocity of electromagnetic wave propagation is given as

$$v = \frac{\omega}{\beta}$$

$$= \frac{1}{\sqrt{\mu\varepsilon}}\left[1 - \frac{\sigma^2}{8\omega^2\varepsilon^2}\right] \qquad \text{for lossy dielectric}$$

$$= \frac{1}{\sqrt{\mu\varepsilon}} \qquad \text{for perfect dielectric}$$

$$= \frac{1}{\sqrt{\mu_0\varepsilon_0}} = c = 3 \times 10^8 \text{ m/s, speed of light;} \qquad \text{for free space}$$

$$= \sqrt{\frac{2\omega}{\mu\sigma}} \qquad \text{for good conductors}$$

- The phenomenon that the alternating fields and hence currents are confined within a small region of a conducting medium inside the surface is known as the *skin effect* and the small distance from the surface of the conductor is known as *skin depth*. It is given as

$$\delta = \frac{1}{\omega\sqrt{\frac{\mu\varepsilon}{2}\left[\sqrt{1 + \left(\frac{\sigma}{\omega\varepsilon}\right)^2} - 1\right]}} \qquad \text{in general}$$

$$= \sqrt{\frac{2}{\omega\mu\sigma}} \qquad \text{for good conductors}$$

- The *surface impedance* of a conductor is defined as the ratio of the tangential component of the electric field to the tangential component of the magnetic field. The surface impedance for a thick conductor is

$$Z_S = \frac{\gamma}{\sigma} = \frac{\sqrt{j\omega\mu\sigma}}{\sigma} = \sqrt{\frac{j\omega\mu}{\sigma}} = \eta$$

- The real part of the intrinsic impedance is known as *surface resistance* or *skin resistance*, R_s (Ω/m^2). It is given as

$$R_S = \frac{1}{\sigma\delta} = \sqrt{\frac{\omega\mu}{2\sigma}}$$

- The ratio of the imaginary part of the complex permittivity (ε'') to the real part of the complex permittivity (ε') is the ratio of the magnitude of the conduction current density to the magnitude of the displacement current density. This ratio is defined as the *loss tangent* or *loss angle* of the medium.

$$\left|\frac{\varepsilon''}{\varepsilon'}\right| = \frac{\sigma}{\omega\varepsilon} = \frac{|\sigma \boldsymbol{E}_s|}{|\omega\varepsilon \boldsymbol{E}_s|} = \frac{|\boldsymbol{J}_{\text{conduction}}|}{|\boldsymbol{J}_{\text{displacement}}|} = \tan\theta$$

The loss tangent gives a measure of how lossy a medium is. For a good (lossless or perfect) dielectric medium ($\sigma << \omega\varepsilon$), loss tangent is very small. For a good conducting medium ($\sigma >> \omega\varepsilon$), loss tangent is very large. For lossy dielectric, loss tangent is of the order of unity.

- The *polarisation* of a uniform plane wave refers to the time-varying behaviour of the electric field at some point in space, i.e., the orientation of the electric field vector at a given instant of time in space.
- A plane electromagnetic wave in which electric field vector vibrates harmonically along a fixed straight line perpendicular to the direction of wave propagation without changing its orientation is known as *plane polarised electromagnetic wave*. When two orthogonal plane polarised electromagnetic waves are superimposed, then the resultant vector rotates under certain conditions, giving rise to different polarisations like, *linear*, *circular* and *elliptical*.
- If a plane electromagnetic wave is incident normally from medium 1 to medium 2, the reflection and transmission coefficients are given as

$$\text{Reflectance, } \Gamma = \frac{\eta_2 - \eta_1}{\eta_2 + \eta_1}; \quad \text{for electric field}$$

$$= \frac{\eta_1 - \eta_2}{\eta_1 + \eta_2}; \quad \text{for magnetic field}$$

$$\text{Transmittance, } \tau = \frac{2\eta_2}{\eta_1 + \eta_2}; \quad \text{for electric field}$$

$$= \frac{2\eta_1}{\eta_1 + \eta_2}; \quad \text{for magnetic field}$$

- The *plane of incidence* is the plane containing the incident wave and the normal to the interfacing surface.
- The *angle of incidence* is defined as the angle between the direction of propagation and the normal to the boundary.
- For a plane electromagnetic wave incident obliquely from medium 1 to medium 2, when the electric field vector $\vec{E}$ is perpendicular to the plane of incidence, i.e., the electric vector is parallel to the boundary surface, it is called *perpendicular* or *horizontal polarisation*. On the other hand, when the electric field vector $\vec{E}$ is parallel to the plane of incidence, i.e., the magnetic field is parallel to the boundary surface, it is called *parallel* or *vertical polarisation*.
- The reflection and transmission coefficients for perpendicular polarisation are given as

$$\text{Reflection coefficient, } \Gamma = \frac{E_r}{E_i} = \frac{\sin(\theta_2 - \theta_1)}{\sin(\theta_2 + \theta_1)} = \frac{\cos\theta_1 - \sqrt{\dfrac{\varepsilon_2}{\varepsilon_1} - \sin^2\theta_1}}{\cos\theta_1 + \sqrt{\dfrac{\varepsilon_2}{\varepsilon_1} - \sin^2\theta_1}}$$

$$\text{Transmittance, } \tau = \frac{2\sin\theta_2\cos\theta_1}{\sin(\theta_2 + \theta_1)}$$

- The reflection and transmission coefficients for perpendicular polarization are given as

$$\text{Reflection coefficient, } \Gamma = \frac{E_r}{E_i} = \frac{\tan(\theta_1 - \theta_2)}{\tan(\theta_1 + \theta_2)} = \frac{\frac{\varepsilon_2}{\varepsilon_1}\cos\theta_1 - \sqrt{\frac{\varepsilon_2}{\varepsilon_1} - \sin^2\theta_1}}{\frac{\varepsilon_2}{\varepsilon_1}\cos\theta_1 + \sqrt{\frac{\varepsilon_2}{\varepsilon_1} - \sin^2\theta_1}}$$

$$\text{Transmittance, } \tau = \frac{2\sin\theta_2\cos\theta_1}{\sin(\theta_1 + \theta_2)\cos(\theta_1 - \theta_2)}$$

- For a plane electromagnetic wave incident on a plane boundary between two dielectric media having a different refractive indices, the angle of incidence at which transmittance from one medium to the other is unity, when the wave is linearly polarised with its electric vector parallel to the plane of incidence, is called *Brewster's angle*. It is expressed as

$$\tan\theta_B = \frac{\eta_2}{\eta_1} = \sqrt{\frac{\varepsilon_2}{\varepsilon_1}}$$

  Total transmission or no reflection occurs when the angle of incidence is equal to the Brewster's angle.
- When the angle of refraction θ_2 is $\frac{\pi}{2}$ (the largest possible value), the corresponding angle of incidence θ_1 is called the *critical angle* (θ_C). It is expressed as

$$\sin\theta_C = \frac{\eta_2}{\eta_1}$$

  Total reflection occurs for all incident angles greater than or equal to the critical angle.
- If a plane wave is incident normally upon the surface of the conducting medium, the wave is entirely reflected. For field that varies with time, neither $\vec{E}$ nor $\vec{H}$ will exist within the conductor. Thus, in this case, where medium 1 is perfect dielectric and medium 2 is perfect conductor

  Transmission coefficient, $\tau = 0$, and
  Reflection coefficient, $\Gamma = -1$
  Hence, the amplitudes of $\vec{E}$ and $\vec{H}$ of reflected wave are same as those of the incident wave, but they differ in the direction of flow.
- The *Poynting vector* can be thought of as representing the energy flux (in W/m^2) of an electromagnetic field. It is given as

$$\vec{S} = \vec{E} \times \vec{H}$$

- The *Poynting theorem* states that the vector product, $\vec{S} = \vec{E} \times \vec{H}$ at any point is a measure of the rate of energy flow per unit area at that point. The direction of power flow is in the direction of the unit vector along the product $(\vec{E} \times \vec{H})$ and is perpendicular to both $\vec{E}$ and $\vec{H}$.
- *Radiation pressure* is defined as the force per unit area exerted by electromagnetic radiation, and is given by

$$P_{\text{rad}} = \langle U_d \rangle = \frac{|\langle \vec{S} \rangle|}{c}$$

  where, $\langle U_d \rangle$ is the time-averaged electromagnetic energy density, $|\langle \vec{S} \rangle|$ is the magnitude of the time-averaged Poynting vector and c is the speed of light.

Important Formulae

Three-Dimensional Wave Equation	$\nabla^2\vec{E} - \mu\varepsilon\frac{\partial^2\vec{E}}{\partial t^2} - \mu\sigma\frac{\partial\vec{E}}{\partial t} = 0$ and $\nabla^2\vec{H} - \mu\varepsilon\frac{\partial^2\vec{H}}{\partial t^2} - \mu\sigma\frac{\partial\vec{H}}{\partial t} = 0$
Propagation constant	$\lambda = \alpha + j\beta$
Attenuation constant	$\alpha = \omega\sqrt{\frac{\mu\varepsilon}{2}\left[\sqrt{1+\left(\frac{\sigma}{\omega\varepsilon}\right)^2} - 1\right]}$
Phase constant	$\beta = \omega\sqrt{\frac{\mu\varepsilon}{2}\left[\sqrt{1+\left(\frac{\sigma}{\omega\varepsilon}\right)^2} + 1\right]}$
Phase velocity	$v_p = \frac{\omega}{\beta} = \lambda f = \frac{1}{\sqrt{\mu\varepsilon}}$
Group velocity	$v_g = \frac{d\omega}{d\beta}$
Intrinsic impedance	$\eta = \sqrt{\frac{j\omega\mu}{\sigma + j\omega\varepsilon}}$
Skin depth	$\delta = \frac{1}{\omega\sqrt{\frac{\mu\varepsilon}{2}\left[\sqrt{1+\left(\frac{\sigma}{\omega\varepsilon}\right)^2} - 1\right]}}$
Surface impedance	$Z_S = \sqrt{\frac{j\omega\mu}{\sigma}} = \eta$
Skin resistance	$R_S = \frac{1}{\sigma\delta} = \sqrt{\frac{\omega\mu}{2\sigma}}$
Poynting vector	$\vec{S} = \vec{E} \times \vec{H}$

Exercises

[NOTE: * marked problems are important university problems]

• ***Easy***

***1** The electric field intensity associated with a plane wave travelling in a perfect dielectric medium is given by

$$E_x(z, t) = 10 \cos(2\pi \times 10^7 t - 0.1\pi z)$$

Calculate the velocity of propagation. Write down an expression for the magnetic field intensity associated with the wave if $\mu = \mu_0$. $\left[2 \times 10^8 \text{ m/s}, H_y(z, t) = \dfrac{10}{80\pi} \cos(2\pi \times 10^7 t - 0.1\pi z)\right]$

*2. In a lossless medium for which $\eta = 60\pi$, $\mu_r = 1$, and

$$\vec{H} = -0.1 \cos(\omega t - z)\hat{a}_x + 0.5 \sin(\omega t - z)\hat{a}_y \text{ A/m}$$

calculate ε_r, ω and $\vec{E}$.

$[4; 1.5 \times 10^8 \text{ rad/s}; \vec{E} = 94.25 \sin(1.5 \times 10^8 t - z)\hat{a}_x + 18.85 \cos(1.5 \times 10^8 t - z)\hat{a}_y \text{ V/m}]$

3. A uniform plane wave propagating in a medium has

$$\vec{E} = 2e^{-\alpha z} \sin(10^8 t - \beta z)\hat{a}_y \text{ V/m}$$

If the medium is characterised by $\varepsilon_r = 1$, $\mu_r = 20$ and $\sigma = 3$ mho/m, find α, β and $\vec{H}$.

$\left[61.4 \text{ Np/m}; 61.4 \text{ rad/m}; \vec{H} = -69.1e^{-61.4z} \sin\left(10^8 t - 61.42z - \dfrac{\pi}{4}\right)\hat{a}_x \text{ mA/m}\right]$

- ***Medium***

4. For the copper coaxial cable of inner conductor of radius, $a = 2$ mm, and outer conductor of inner radius $b = 6$ mm and thickness $t = 1$ mm, calculate the resistance of 2 m length of the cable at DC and at 100 MHz. [3.583 mΩ; 0.5484 Ω]
5. What is the polarisation of the electric field vector of a uniform plane wave travelling in the z-direction represented by
(a) $E = E_0(x + jy)e^{j\omega t}$ (b) $E = E_0(x + y)e^{j\omega t}$
Justify your answer. [circular, linear]
6. What values of A and β are required if the two fields $\vec{E} = 120\pi \cos(10^6 \pi t - \beta x)\hat{a}_y$ V/m and $\vec{H} = A \cos(10^6 \pi t - \beta x)\hat{a}_z$ A/m satisfy Maxwell's equations in linear, isotropic homogeneous medium where $\varepsilon_r = \mu_r = 4$ and $\alpha = 0$. [$A = 1$, $\beta = 0.042$]

- ***Hard***

7. In free space,

$$\vec{H} = 0.2 \cos(\omega t - \beta x)\hat{a}_z \text{ A/m}$$

Find the total power passing through (a) a square plate of side 10 cm on the plane $x = z = 1$; (b) a circular disc of radius 5 cm on the plane $x = 1$. [0; 59.22 mW]

8. Given a uniform plane wave in air as

$$\vec{E}_i = 40 \cos(\omega t - \beta z)\hat{a}_x + 30 \sin(\omega t - \beta z)\hat{a}_y \text{ V/m}$$

(a) Find $\vec{H}_i$.
(b) If the wave encounters a perfectly conducting plate normal to the z-axis at $z = 0$, find the reflected wave $\vec{E}_r$ and $\vec{H}_r$.
(c) What are the total $\vec{E}$ and $\vec{H}$ fields for $z \le 0$?
(d) Calculate the time-average Poynting vectors for $z \le 0$ and $z \ge 0$.

$$\left[\begin{aligned} &\vec{H}_i = -\frac{1}{4\pi} \sin(\omega t - \beta z)\hat{a}_x + \frac{1}{3\pi} \cos(\omega t - \beta z)\hat{a}_y \text{ mA/m;} \\ &\vec{E}_r = -40 \cos(\omega t + \beta z)\hat{a}_x - 30 \sin(\omega t + \beta z)\hat{a}_y \text{ V/m} \\ &\vec{H}_r = \frac{1}{3\pi} \cos(\omega t + \beta z)\hat{a}_y - \frac{1}{4\pi} \sin(\omega t + \beta z)\hat{a}_x \text{ mA/m} \end{aligned}\right]$$

*9. A uniform plane wave in air with

$$\vec{E} = 8 \cos (\omega t - 4x - 3z)\hat{a}_y \text{ V/m}$$

is incident on a dielectric slab ($z \geq 0$) with $\mu_r = 1.0$, $\varepsilon_r = 2.5$, $\sigma = 0$. Find:

(a) The polarisation of the wave
(b) The angle of incidence
(c) The reflected $\vec{E}$ field
(d) The transmitted $\vec{H}$ field.

[Perpendicular polarisation; 53.13°; $\vec{E}_r = -3.112 \cos (15 \times 10^8 t - 4x + 3z)\hat{a}_y$ V/m
$\vec{H}_t = (-17.69\hat{a}_x + 10.37\hat{a}_z) \cos (15 \times 10^8 t - 4x - 6.819z)$ mA/m]

10. In spherical coordinates, the spherical wave

$$\vec{E} = \frac{100}{r} \sin \theta \cos (\omega t - \beta r)\hat{a}_\theta \quad \text{(V/m)} \qquad \vec{H} = \frac{0.265}{r} \sin \theta \cos (\omega t - \beta r)\hat{a}_\phi \quad \text{(A/m)}$$

represents the electromagnetic field at large distances r from a certain dipole antenna in free space. Find the average power crossing the hemispherical shell $r = 1$ km, $0 \leq \theta \leq \pi/2$. [55.5 W]

11. The electric field intensity in radiation field of an antenna located at the origin of a spherical coordinate system is given by

$$\vec{E} = E_0 \frac{\sin \theta}{r} \cos (\omega t - \beta r)\hat{a}_\theta$$

where, E_0, ω and β are constants. Find:

(a) the magnetic field associated with this electric field.
(b) the Poynting vector.
(c) the total power radiated over a spherical surface of radius r centered at the origin.

Review Questions

[NOTE: * marked questions are important university questions.]

1. Derive the general solution of a wave equation.
2. Write short notes on the following:
   (a) Phase velocity and group velocity
   (b) Characteristic impedance
   (c) Standing wave ratio.

*3. Discuss the wave propagation in:
   (i) a lossy dielectric,
   (ii) a conductor.
   Derive relevant equations.

*4 Assuming that there is no accumulated free charge, (i) write Maxwell's equations in conductors, and (ii) write plane wave solutions to obtain an expression for the skin depth.

5. Derive the basic equations for electromagnetic waves in free space in terms of ***E*** and ***H***.

Or,

Write Maxwell's equations for vacuum, and derive the wave equation for the electric and magnetic fields in vacuum.

(a) What is a wave? Deduce the equation for the propagation of plane electromagnetic waves in free space.

*(b) What are the properties of a uniform plane wave? Show that the electric and magnetic field in the uniform plane wave will not have any component in the direction of the wave propagation.

Or,

Show that a uniform plane wave is a TEM wave.

Or,

Show that for a plane electromagnetic wave in free space, the unit vector in the direction of propagation, the electric field vector and the magnetic field vector are mutually perpendicular.

(c) Define and draw a neat diagram of an electromagnetic plane wave.

*(d) Show that the plane electromagnetic waves in free space travel with the velocity c.

*(e) Prove that the ratio of the (real) amplitudes of electric and magnetic fields (plane waves) is equal to the velocity of light in appropriate units.

*(f) A plane polarised wave is travelling along z-axis. Show graphically the variation of $\boldsymbol{E}$ and $\boldsymbol{H}$ with z.

***6.** (a) A plane electromagnetic wave is incident normally on a metal of electrical conductivity σ. Show that the electromagnetic wave is damped inside the conductor and find the skin depth.

Or,

Prove that in an imperfect dielectric medium of weak electrical conductivity σ, the electromagnetic wave is damped. Find expressions for the refractive index and the extinction coefficient in such a medium.

(b) Explain 'Skin effect' in a conductor. Show that for a good conductor the skin depth is $\lambda_m/2\pi$, where λ_m is the wavelength in the medium.

Or,

A plane electromagnetic wave is incident on a conductor of conductivity σ. Derive an expression for the 'skin depth'. What is its significance?

Or,

What is the skin effect? Define skin depth. Show that in case of semi-infinite solid conductor, the skin depth δ is given by $\delta = \sqrt{\dfrac{2}{\omega\mu\sigma}}$, where ω, μ and σ have usual significance.

(c) The term 'good conductor' and 'poor conductor' depend on frequency. Justify the statement.

(d) Introduce the complex index of refraction and explain the significance of its real and imaginary parts.

***7.** What is 'intrinsic impedance'? Derive an expression for it. A plane polarised wave is travelling along z-axis. Show that $E_y/H_z = 377\ \Omega$.

Or,

Show that for a lossy dielectric, the intrinsic impedance is given by, $\eta = \sqrt{\dfrac{j\omega\mu}{\sigma + j\omega\varepsilon}}$.

8. Establish the boundary conditions that must be satisfied by the field vectors when an electromagnetic wave is incident at the interface between two dielectric media.

9. Obtain the Poynting theorem for the conservation of energy in electromagnetic fields and discuss the physical meaning of each term in the resulting equation. Also, bring the Poynting theorem in complex form.

Or,

What is Poynting vector? Find the expression of Poynting vector. What is the physical interpretation of this vector?

10. (a) A current flows down a resistive wire of length l and radius r, subjected to a potential V. Calculate the energy per unit time delivered to the wire using Poynting vector and show that it is equivalent to Joule's heat loss.

(b) Considering a linearly polarised plane harmonic electromagnetic wave propagating in vacuum, show that the Poynting vector is given by the product of the phase velocity and the electromagnetic energy density.

Or,

* Show that the ratio of Poynting's vector to energy density is $\leq 3 \times 10^8$ m/s.

(c) For a plane monochromatic electromagnetic wave propagating in a homogeneous dielectric, show that the time-averaged electric and magnetic energy densities are equal.

11. Find the mean value over a period for the product of the real parts of two complex periodic quantities A and B. Hence show that the average intensity of the energy flow in a harmonic electromagnetic field [when $\vec{E}$ and $\vec{H}$ vary as $\exp(-j\omega t)$] can be represented (in SI units) by $S = \frac{1}{2}\mathrm{Re}(\vec{E} \times \vec{H}^*)$, where $\vec{H}^*$ is the complex conjugate of $\vec{H}$.

Or,

Show that average energy density in a harmonic electromagnetic field is

$$\langle u \rangle = \frac{1}{4}\mathrm{Re}[\vec{E} \cdot \vec{D}^* + \vec{H} \cdot \vec{B}^*]$$

where $\vec{D}^*$ and $\vec{B}^*$ are complex conjugates of $\vec{D}$ and $\vec{B}$.

***12.** Show that in a conductor, the electric and the magnetic fields are not in phase and that the energy is not shared equally between the electric and the magnetic fields.

Or,

Show that for an electromagnetic wave propagating in a conducting medium, the density of magnetic energy is greater than that of electric energy.

Or,

Show that in a conductor, the magnetic field lags the electric field in time but leads the electric field in position.

Or,

Show that the magnetic field lags the electric field in time by a phase angle in an electromagnetic field propagating in a conductor. Obtain an expression for the phase angle.

13. (a) Write Ampere's law of magnetomotive force. What is the deficiency of this law? How is it corrected by introducing the concept of displacement current?

(b) Define electromagnetic energy density and Poynting vector $\vec{S}$. What is the dimension of $\vec{S}$?

(c) Starting from Maxwell's equations show that in a nonconducting medium, $\nabla^2 \vec{E} - \mu\varepsilon \frac{\partial^2 \vec{E}}{\partial t^2}$ where μ and ε are the permeability and the permittivity of the medium.

14. (a) (i) Write the two equations of Maxwell which can be obtained from Ampere's law and Faraday's law of electromagnetic induction. (ii) Using these equations show that

$$\mathrm{div}(\vec{E} \times \vec{H}) = -\frac{1}{2}\frac{\partial}{\partial t}(\vec{E} \cdot \vec{D} + \vec{B} \cdot \vec{H}) - \vec{J} \cdot \vec{E}$$

in a linear medium, with $\vec{B} = \mu\vec{H}$ and $\vec{D} = \varepsilon\vec{E}$. Give physical interpretations of the terms on the two sides of the above equation.

(b) The wave equation in a homogeneous linear medium with zero charge density and conductivity σ is given by

$$\nabla^2 \vec{E} - \mu\varepsilon \frac{\partial^2 \vec{E}}{\partial t^2} - \sigma\mu \frac{\partial \vec{E}}{\partial t} = 0$$

(i) Obtain the solution for a plane wave with wavefront parallel to the *y-z* plane. (ii) Explain skin depth.

(c) Deduce an expression for the velocity of propagation of a plane electromagnetic wave propagating through a medium of dielectric constant ε and relative permeability μ.

15. Obtain the differential equations for the propagation of an electromagnetic wave in a conducting medium.

16. Write Maxwell's equations in a non-material medium and discuss the physical law implied by each of them. How would these equations be modified if magnetic monopoles existed?

17. Write Maxwell's equations in a non-conducting medium. Obtain plane wave solutions for $\vec{E}$ and $\vec{B}$. What is the relationship between the refractive index and the propagation constant?

18. Write short notes on:

(a) Poynting's theorem.

(b) Waves in conducting media.

(c) Skin effect.

19. Express Poynting's theorem in the form

$$\frac{\partial U}{\partial t} + \nabla \cdot \vec{S} = 0$$

where $\vec{S} = \frac{1}{\mu_0}(\vec{E} \times \vec{B})$ and U is the total energy density. Comparing this with the equation of continuity, give an interpretation of $\vec{S}$.

20. Prove that the average Poynting vector is given by $P_{avg} = \frac{1}{2}(H_m)^2 R_e(\eta)$. Find the value of P_{avg} for free space, dielectric medium and conducting medium.

21. Define polarisation of electromagnetic waves. What do you mean by linearly polarised, circularly polarised, and elliptically polarised waves? Give the expression for the $\vec{E}$ field in each case.

22. Explain what you understand by perpendicular polarisation and parallel polarisation.

A uniform plane electromagnetic wave is incident at an angle θ_1 at the surface of discontinuity between two homogeneous isotropic dielectrics with permittivity ε_1 and ε_2, ε_2 being the permittivity of the dielectric into which the wave gets refracted at an angle θ_2. If E_i, E_r and E_t are the electric intensities respectively of the incident, reflected and transmitted waves, show that

(a) Reflection co-efficient for parallel polarisation is given by $\frac{E_r}{E_i} = \frac{\tan(\theta_1 - \theta_2)}{\tan(\theta_1 + \theta_2)}$.

(b) Reflection co-efficient for perpendicular polarisation is given by $\frac{E_r}{E_i} = \frac{\sin(\theta_2 - \theta_1)}{\sin(\theta_2 + \theta_1)}$.

Or,

(a) A plane polarised electromagnetic wave is incident on an interface of two dielectric media. Find the relations between the angles of incidence, reflection and refraction. Also, find the reflection and transmission coefficients.

(b) Assuming the electric vector to lie in the plane of incidence, calculate the reflection coefficient. Hence, prove Brewster's law.

(c) Deduce the laws of refraction for plane waves at the boundary of two dielectrics from electromagnetic theory.

(d) What is Brewster's angle?

(e) For refraction of an electromagnetic wave from a denser to a rarer medium, explain the term 'critical angle of incidence'. What happens when the angle of incidence exceeds the critical value? How do the reflectances for *s* and *p* polarisations vary with the angle of incidence?

(f) Define and distinguish between Brewster's angle and critical angle with reference to an electromagnetic wave incident on a separating surface between two perfect dielectrics. Show that critical angle is normally greater than Brewster's angle.

(g) If an electromagnetic wave is incident on the surface of separation of two media, is it possible to have only a reflected wave, or only a refracted wave? Give reasons for your answer.

23. What is "radiation pressure"? Show that for a plane wave falling on an absorbing surface, the radiation pressure is the mean energy density of the electromagnetic radiation.

Multiple Choice Questions

1. Poynting vector has the unit of:
(a) Watt. (b) Watt/metre. (c) Watt/m^2. (d) None of the above.

2. The direction of propagation of electromagnetic waves is given by the direction of
(a) Vector **E**. (b) Vector **H**. (c) Vector (**E** × **H**). (d) None of the above.

3. The quantity $\frac{1}{\sqrt{\varepsilon_0 \mu_0}}$ in SI units has the
(a) value 330 m/s (b) value 1.73×10^4 (c) dimensions LT^{-1} (d) None of the above.

4. A plane electromagnetic wave in free space is specified by the electric field $\hat{a}_x$ $[20 \cos(\omega t - \beta z) + 5 \cos(\omega t + \beta z)]$ V/m. The associated magnetic field is

(a) $\frac{\hat{a}_y}{120\pi}[20 \cos(\omega t - \beta z) + 5 \cos(\omega t + \beta z)]$ A/m

(b) $\frac{\hat{a}_y}{120\pi}[20 \cos(\omega t - \beta z) - 5 \cos(\omega t + \beta z)]$ A/m

(c) $\frac{\hat{a}_x}{120\pi}[20 \cos(\omega t - \beta z) + 5 \cos(\omega t + \beta z)]$ A/m

(d) $\frac{\hat{a}_x}{120\pi}[20 \cos(\omega t - \beta z) - 5 \cos(\omega t + \beta z)]$ A/m

5. The velocity of the plane wave $\sin^2(\omega t - \beta x)$ is
(a) $\frac{2\omega}{\beta}$ (b) $\frac{\omega}{2\beta}$ (c) $\frac{\omega^2}{\beta^2}$ (d) $\frac{\omega}{\beta}$

6. When the load impedance is equal to the characteristic impedance of the transmission lines, then the reflection coefficient and standing wave ratio are, respectively
(a) 0 and 0 (b) 1 and 0 (c) 0 and 1 (d) 1 and 1

7. If the frequency of a plane electromagnetic wave increases four times, the depth of penetration, when the wave is incident normally on a good conductor will
(a) increase by factor of two. (b) decrease by a factor of four.
(c) remain same. (d) decrease by a factor of two.

8. The intrinsic impedance of a good conducting medium is given by (symbols have the usual meaning)

(a) $\sqrt{\dfrac{\mu\omega}{\sigma}}\angle -45°$ (b) $\sqrt{\dfrac{\omega\sigma}{\mu}}\angle 45°$ (c) $\sqrt{\dfrac{\mu\omega}{\sigma}}\angle 45°$ (d) $\sqrt{\mu\sigma\omega}\angle 0°$

9. A plane monochromatic electromagnetic wave travels in a perfect conducting medium, which is charge-free and external current free. Then,

(a) ***E*** field lags ***B*** field by $\pi/4$. (b) ***E*** field leads ***B*** field by $\pi/4$.
(c) ***E*** and ***B*** fields are co-phasal. (d) ***E*** and ***B*** fields differ in phase by $\pi/2$.

10. Which of the following properties pertain to a circularly polarised wave having no component of electric field in the *x*-direction but having components E_y and E_z?

1. E_y and E_z are equal in magnitude.
2. Direction of resultant electric vector varies with time.
3. Direction of resultant magnetic vector varies with time.
4. E_y and E_z have a time phase difference of 90°.

Select the correct answer from the codes given below.

(a) 2 and 4. (b) 1 and 3. (c) 1, 3 and 4. (d) 1 and 2.

11. A uniform plane wave travelling in a perfect dielectric is incident normally on the surface of a perfect conductor. Then

(a) The wave is transmitted into the conductor without attenuation.
(b) 50% of the incident wave is transmitted and 50% is reflected.
(c) A standing wave is set up in the conducting medium.
(d) A standing wave is set up in the dielectric.

12. The electric field of a uniform plane wave is given by

$$\boldsymbol{E} = 10 \cos (3\pi \times 10^8 t - \pi z)\boldsymbol{a}_x$$

Match List I with List II pertaining to the above wave and select the correct answer using the codes given below the lists:

List I (Parameters)	**List II (Values in MKS units)**
A. Phase velocity	1. 2
B. Wavelength	2. 3.14
C. Frequency	3. 377
D. Phase constant	4. 1.5×10^8
	5. 3.0×10^8

Codes:

	A	B	C	D
(a)	5	4	3	2
(b)	3	4	2	1
(c)	4	3	1	2
(d)	5	1	4	2

13. List II gives the mathematical expression for the variables given in List I. Match List I with List II and select the correct answer using the codes given below the lists:

List I	**List II**
A. Intrinsic impedance	1. $\dfrac{1}{\sqrt{\mu\varepsilon}}$
B. Velocity of wave propagation	2. $\sqrt{\dfrac{\mu}{\varepsilon}}$
C. Skin depth	3. $\dfrac{1}{\sqrt{\pi f \mu\sigma}}$
D. Attenuation constant	4. $\dfrac{1}{f\sqrt{\mu\omega}}$
	5. $\sqrt{\dfrac{\omega\mu\sigma}{2}}$

Codes:

	A	B	C	D
(a)	1	2	3	4
(b)	2	1	4	5
(c)	2	1	3	5
(d)	1	2	5	3

14. The ratio of velocity of propagation of EM waves in an overhead transmission line and in a cable with a dielectric of permittivity 4, is

(a) 0.25 (b) 0.5 (c) 2.0 (d) 4.0

15. Match List I with List II and select the correct answer using the codes given below the lists.

List I (Medium)	**List II** (Expression for intrinsic impedance for plane wave propagation)
A. Loss-less dielectric	1. $\sqrt{\dfrac{j\omega\mu}{\sigma + j\varpi\varepsilon}}$
B. Good conductor	2. $\sqrt{\mu/\varepsilon}\left(1 + j\dfrac{\sigma}{2\omega\varepsilon}\right)$
C. Poor conductor	3. $\sqrt{\mu/\varepsilon}$
D. Lossy	4. $(1 + j)\sqrt{\dfrac{\omega\mu}{2\sigma}}$

Codes:

	A	B	C	D
(a)	4	3	2	1
(b)	4	3	1	2
(c)	3	4	2	1
(d)	3	4	1	2

16. The intrinsic impedance of free space is

(a) 377Ω (b) $\sqrt{\mu_0 \varepsilon_0}$ (c) $j\sqrt{\frac{\varepsilon_0}{\mu_0}}$ (d) $\sqrt{\frac{\varepsilon_0}{\mu_0}}$

17. In the source free wave equation

$$\nabla^2 \vec{E} - \mu_0 \varepsilon_0 \mu \varepsilon \frac{\partial^2 \vec{E}}{\partial t^2} - \mu_0 \mu \sigma \frac{\partial \vec{E}}{\partial t} = 0$$

The term responsible for attenuation of the wave is

(a) $\mu_0 \mu \sigma \frac{\partial \vec{E}}{\partial t}$ (b) $\mu_0 \varepsilon_0 \mu \varepsilon \frac{\partial^2 \vec{E}}{\partial t^2}$

(c) $\nabla^2 \vec{E}$ (d) $\mu_0 \mu \sigma \frac{\partial \vec{E}}{\partial t}$ and $\mu_0 \varepsilon_0 \mu \varepsilon \frac{\partial^2 \vec{E}}{\partial t^2}$

18. Three media are characterised by

1. $\varepsilon_r = 8, \mu_r = 2, \sigma = 0$
2. $\varepsilon_r = 1, \mu_r = 9, \sigma = 0$
3. $\varepsilon_r = 4, \mu_r = 4, \sigma = 0$

ε_r is relative permittivity, μ_r is relative permeability and σ is conductivity.
The values of the intrinsic impedances of the media 1, 2 and 3, respectively are

(a) 188Ω, 377Ω and 1131Ω **(b) 377Ω, 1131Ω and 188Ω**
(c) 188Ω, 1131Ω and 377Ω **(d) 1131Ω, 188Ω and 377Ω**

19. For a perfect conductor, the field strength at a distance equal to the skin depth is X % of the field strength at its surface. The value of 'X %' is

(a) Zero (b) 50% (c) 36% (d) 26%

20. Consider the following statements:
The characteristic impedance of a transmission line can increase with the increase in

1. resistance per unit length
2. conductance per unit length
3. capacitance per unit length
4. inductance per unit length

Which of these statements are correct?

(a) 1 and 2 (b) 2 and 3 (c) 1 and 4 (d) 3 and 4

21. The equation $\vec{E} = \hat{a}_x \sin(\omega t - \beta z) + \hat{a}_y \sin(\omega t - \beta z)$ represents

(a) a left circularly polarised wave (b) a right circularly polarised wave
(c) a linearly polarised wave (d) an elliptically polarised wave

22. A monochromatic plane electromagnetic wave travels in vacuum in the positive x direction (x, y, z system of coordinates). The electric and magnetic fields can be expressed as

(a) $\vec{E}(x,t) = E_0 \cos(kx - \omega t)\vec{a}_y \quad \vec{H}(x,t) = H_0 \cos(kx - \omega t)\vec{a}_z$

(b) $\vec{E}(x,t) = E_0 \cos(kx - \omega t)\vec{a}_y \quad \vec{H}(x,t) = H_0 \cos\left(kx - \omega t - \frac{\pi}{2}\right)\vec{a}_z$

(c) $\vec{E}(x,t) = E_0 \cos(kx - \omega t)\vec{a}_y \quad \vec{H}(x,t) = -H_0 \cos(kx - \omega t)\vec{a}_z$

(d) $\vec{E}(x,t) = E_0 \cos(kx - \omega t)\vec{a}_y \quad \vec{H}(x,t) = -H_0 \cos\left(kx - \omega t - \frac{\pi}{2}\right)\vec{a}_x$

23. The incident wave on a lossless line carries an average power of 1.0 W. The load end reflection coefficient is 1/3. The average power absorbed by the load is

(a) 1/3 W (b) 2/3 W (c) 4/9 W (d) 8/9 W

24. Which one of the following is not true for waves in general?

(a) It may be a function of time only
(b) It may be sinusoidal or cosinusoidal
(c) It must be a funcion of time and space
(d) For practical reasons, it must be finite in extent.

25. What is the major factor for determining whether a medium is free space, lossless dielectric, lossy dielectric or good conductor?

(a) attenuation constant (b) constitutive parameters (σ, ε, μ)
(c) loss tangent (d) reflection coefficient

26. In a certain medium, $\vec{E} = 10\cos(10^8 t - 3y)\hat{a}_x$ V/m. The medium is:

(a) free space (b) perfect dielectric
(c) lossless dielectric (d) perfect conductor

Answers

1. (c)	**2.** (c)	**3.** (c)	**4.** (b)	**5.** (d)	**6.** (c)	**7.** (d)	**8.** (c)
9. (d)	**10.** (c)	**11.** (d)	**12.** (d)	**13.** (c)	**14.** (c)	**15.** (c)	**16.** (a)
17. (d)	**18.** (c)	**19.** (c)	**20.** (c)	**21.** (c)	**22.** (a)	**23.** (c)	**24.** (a)
25. (c)	**26.** (c)						

6

TRANSMISSION LINES

Learning Objectives

This chapter deals with the following topics:

- *Fundamentals of transmission lines*
- *Mathematics of transmission lines*
- *Input impedances in different types of transmission lines*
- *Introduction of Smith chart*
- *Concept of load matching*

6.1 INTRODUCTION

A transmission line is a device used for transmission of electromagnetic energy guided by two conductors in a dielectric medium. Transmission lines may consist of a set of conductors, dielectrics or combination thereof.

6.2 TYPES OF TRANSMISSION LINES

Depending upon the construction, transmission lines are of various types as given below.

1. One-wire lines (single conductor over a conducting ground plane),
2. Two-wire parallel lines,
3. Twisted lines,
4. Coaxial lines,
5. Parallel plate or Planar lines,
6. Microstrip lines, and
7. Optical fibres.

Figure 6.1 shows the configurations of these lines.

6.3 TRANSMISSION LINE MODES

Transmission line mode is the distinct pattern of electric and magnetic field induced on a transmission line under source excitation.

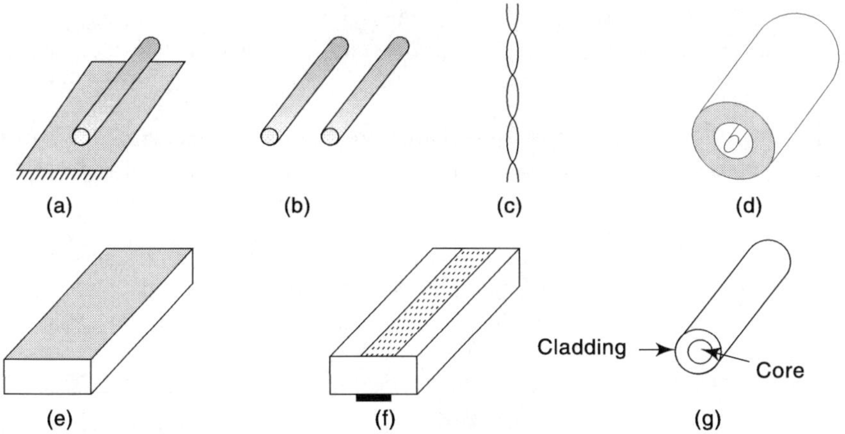

Fig. 6.1 *Configurations of typical transmission lines (a) one-wire lines, (b) two-wire parallel lines, (c) twisted lines, (d) coaxial lines, (e) planar lines, (f) microstrip lines, (g) optical fibre lines*

The field components in the direction of wave propagation are defined as *longitudinal* components while those perpendicular to the direction of propagation are defined as *transverse* components.

Assuming the transmission line is oriented with its axis along the z-axis (direction of wave propagation), transmission line modes may be classified as follows.

1. ***Transverse electromagnetic (TEM) mode:*** In this mode, the electric and magnetic fields are transverse to the direction of wave propagation with no longitudinal components [$E_z = H_z = 0$]. *TEM* modes cannot exist on single conductor guiding structures. *TEM* modes are sometimes called *transmission line modes* since they are the dominant modes on transmission lines. Plane waves can also be classified as *TEM* modes.
2. ***Quasi-TEM mode*:** This is the mode which approximates a true *TEM* mode for sufficiently low frequencies.

$$\lim_{f \to 0} E_z = \lim_{f \to 0} H_z = 0$$

3. ***Waveguide mode:*** In this mode, either E_z, H_z or both are non-zero. Waveguide modes propagate only above certain cutoff frequencies.

6.4 TRANSMISSION LINE PARAMETERS

A transmission line can be characterised by four parameters:

1. Resistance,
2. Inductance,
3. Capacitance, and
4. Conductance.

1. Resistance The resistance of a transmission line is uniformly distributed throughout the entire length of the transmission line. The value of the resistance depends upon the cross-sectional area of the line. This resistance is generally expressed as resistance per unit length (Ω/m).

2. Inductance When the conductors of a transmission line carry some currents, the magnetic flux linkages per unit current give the inductance of the line. This inductance is also not a lumped parameter, but is uniformly distributed along the entire length of the line. This is also expressed as Henry/m.

3. Capacitance The capacitance exists between the parallel conductors of a transmission line separated by an insulating medium. This capacitance is also not a lumped parameter, but is uniformly distributed along the entire length of the line. This is also expressed as Farad/m.

4. Conductance Because of the lossy nature of the insulating medium, some amount of leakage current (displacement current) flows through it and this gives rise to the conductance of a transmission line. This conductance is uniformly distributed along the entire length of the length. This is also expressed as mho/m.

These four distributed parameters are constant for a particular transmission line and are known as *primary line constants* of a transmission line. The electrical design and performance of the transmission line depend on these line constants.

Apart from these primary line constants, there are few other constants related to a transmission line. These include the characteristic impedance (Z_0), the propagation constant (γ), attenuation constant (α) and phase constant (β). These constants are known as the *secondary line constants*. This may be noted that the secondary constants are fixed when frequency is fixed, but change with the change of frequency.

6.5 TRANSMISSION LINE EQUATIONS (TELEGRAPHER'S EQUATIONS)

The *telegrapher's equations* (or just *telegraph equations*) are a pair of linear differential equations which describe the voltage and current on an electrical transmission line with distance and time.

Transmission lines are typically electrically long (several wavelengths) such that we cannot accurately describe the voltages and currents along the transmission line using a simple *lumped-element* equivalent circuit. We must use a *distributed-element* equivalent circuit which describes each short segment of the transmission line by a lumped element equivalent circuit.

The equivalent circuit of a short segment Δz of the two-wire transmission line may be represented by simple lumped-element equivalent circuit as shown in Fig. 6.2.

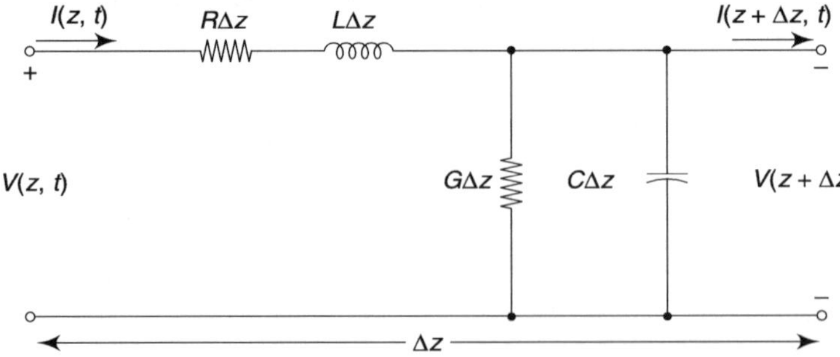

Fig. 6.2 *Equivalent circuit of a short segment of a two-wire transmission line*

Here,

R = series resistance per unit length (S/m) of the transmission line conductors,

L = series inductance per unit length (H/m) of the transmission line conductors (internal plus external inductance),

G = shunt conductance per unit length (S/m) of the media between the transmission line conductors (insulator leakage current), and
C = shunt capacitance per unit length (F/m) of the transmission line conductors.

By KVL for the equivalent circuit, we get

$$V(z, t) - R\Delta z I(z, t) - L\Delta z \frac{\partial}{\partial t} I(z, t) = V(z + \Delta z, t)$$

or
$$-RI(z, t) - L\frac{\partial}{\partial t} I(z, t) = \frac{V(z + \Delta z, t) - V(z, t)}{\Delta z} \quad \textbf{(6.1)}$$

By KCL for the equivalent circuit, we get

$$I(z, t) - G\Delta z V(z + \Delta z, t) - C\Delta z \frac{\partial}{\partial t} V(z + \Delta z, t) = I(z + \Delta z, t)$$

or
$$-GV(z + \Delta z, t) - C\frac{\partial}{\partial t} V(z + \Delta z, t) = \frac{I(z + \Delta z, t) - I(z, t)}{\Delta z} \quad \textbf{(6.2)}$$

Taking limit as $\Delta z \to 0$, the terms on the right hand side of the equations Eq. (6.1) and Eq. (6.2) become partial derivatives with respect to z as given.

$$-RI(z, t) - L\frac{\partial}{\partial t} I(z, t) = \frac{\partial}{\partial z} V(z, t) \quad \textbf{(6.3)}$$

$$-GV(z, t) - C\frac{\partial}{\partial t} V(z, t) = \frac{\partial}{\partial z} I(z, t) \quad \textbf{(6.4)}$$

For time-harmonic signals, the instantaneous voltage and current may be defined in terms of phasors such that

$$V(z, t) = \text{Re}\{V_S(z)e^{j\omega t}\}$$
$$I(z, t) = \text{Re}\{I_S(z)e^{j\omega t}\}$$

The derivatives of the voltage and current with respect to time yield $j\omega$ times the respective phasor. Therefore, Eq. (6.3) and Eq. (6.4) give

$$-(R + j\omega L)I_S(z) = \frac{dV_S(z)}{dz} \quad \textbf{(6.5)}$$

$$-(G + j\omega C)\, V_S(z) = \frac{dI_S(z)}{dz} \quad \textbf{(6.6)}$$

Now, taking derivative of Eq. (6.5) and using Eq. (6.6), we get

$$\frac{d^2V_S(z)}{dz^2} = -(R + j\omega L)\frac{dI_S(z)}{dz} = (R + j\omega L)(G + j\omega C)\, V_S(z) = \gamma^2 V_S(z)$$

or
$$\boxed{\frac{d^2V_S(z)}{dz^2} - \gamma^2 V_S(z) = 0} \quad \textbf{(6.7)}$$

where $\gamma = \alpha + j\beta = \sqrt{(R + j\omega L)(G + j\omega C)}$ is the complex propagation constant.

Similarly, taking derivative of Eq. (6.6) and using Eq. (6.5), we get

$$\frac{d^2I_S(z)}{dz^2} = -(G + j\omega C)\frac{dV_S(z)}{dz} = (R + j\omega L)(G + j\omega C)I_S(z) = \gamma^2 I_S(z)$$

or
$$\boxed{\frac{d^2 I_S(z)}{dz^2} - \gamma^2 I_S(z) = 0} \tag{6.8}$$

These two equations are known as *telegrapher's equations*.

The real part of the propagation constant (α) is the *attenuation constant* while the imaginary part (β) is the *phase constant*. The general equations for α and β in terms of the per-unit-length transmission line parameters are

$$\alpha = \frac{1}{\sqrt{2}}\sqrt{RG - \omega^2 LC + \sqrt{(R^2 + \omega^2 L^2)(G^2 + \omega^2 C^2)}} \tag{6.9a}$$

$$\beta = \frac{1}{\sqrt{2}}\sqrt{-RG - \omega^2 LC + \sqrt{(R^2 + \omega^2 L^2)(G^2 + \omega^2 C^2)}} \tag{6.9b}$$

The general solutions to the voltage and current wave equations, Eq. (6.7) and Eq. (6.8) are

$$V_s(z) = \underbrace{V_{S0}^{+} e^{-\gamma z}}_{+z \text{ directed wave}} + \underbrace{V_{S0}^{-} e^{\gamma z}}_{-z \text{ directed wave}}$$

and

$$I_s(z) = \underbrace{I_{S0}^{+} e^{-\gamma z}}_{+z \text{ directed wave}} + \underbrace{I_{S0}^{-} e^{\gamma z}}_{-z \text{ directed wave}}$$

where, $V_{S0}^{+}, V_{S0}^{-}, I_{S0}^{+}, I_{S0}^{-}$ are the wave amplitudes and are complex constants (phasors) which can be defined as

$$V_{S0}^{+} = |V_{S0}^{+}|\, e^{j\varphi_{v+}} \qquad V_{S0}^{-} = |V_{S0}^{-}|\, e^{j\varphi_{v-}}$$

$$I_{S0}^{+} = |I_{S0}^{+}|\, e^{j\varphi_{i+}} \qquad I_{S0}^{-} = |I_{S0}^{-}|\, e^{j\varphi_{i-}}$$

Thus, the instantaneous voltage and current as a function of position along the transmission line are given as

$$V(z,t) = \text{Re}\{V_S(z) e^{j\omega t}\} = \text{Re}\{|V_{S0}^{+}|\, e^{j\varphi_{v+}} e^{-\alpha z} e^{-j\beta z} e^{j\omega t} + |V_{S0}^{-}| e^{j\varphi_{v-}} e^{\alpha z} e^{j\beta z} e^{j\omega t}\}$$

$$\boxed{V(z,t) = |V_{S0}^{+}|\, e^{-\alpha z} \cos(\omega t - \beta z + \phi_{v+}) + |V_{S0}^{-}|\, e^{\alpha z} \cos(\omega t + \beta z + \phi_{v-})} \tag{6.10a}$$

and

$$I(z,t) = \text{Re}\{I_S(z) e^{j\omega t}\} = \text{Re}\{|I_{S0}^{+}|\, e^{j\phi_{i+}} e^{-\alpha z} e^{-j\beta z} e^{j\omega t} + |I_{S0}^{-}|\, e^{j\phi_{i-}} e^{\alpha z} e^{j\beta z} e^{j\omega t}\}$$

$$\boxed{I(z,t) = |I_{S0}^{+}|\, e^{-\alpha z} \cos(\omega t - \beta z + \phi_{i+}) + |I_{S0}^{-}|\, e^{\alpha z} \cos(\omega t + \beta z + \phi_{i-})} \tag{6.10b}$$

Given the transmission line propagation constant, the wavelength and velocity of propagation are found using the same equations as for unbounded waves.

$$\lambda = \frac{2\pi}{\beta} \qquad u = \frac{\omega}{\beta} = f\lambda$$

6.6 CHARACTERISTIC IMPEDANCE OF TRANSMISSION LINE

Characteristic impedance of a transmission line is defined as the ratio of positively travelling voltage wave to current wave at any point on the line.

The voltage and current wave equations are written as

$$V_S(z) = V_{S0}^+ e^{-\gamma z} + V_{S0}^- e^{\gamma z}$$

and

$$I_S(z) = I_{S0}^+ e^{-\gamma z} + I_{S0}^- e^{\gamma z}$$

Substituting these two equations in Eq. (6.5) and Eq. (6.6), respectively, we get

$$\frac{dV_S(z)}{dz} = -(R + j\omega L) I_S(z)$$

or,

$$-\gamma V_{S0}^+ e^{-\gamma z} + \gamma V_{S0}^- e^{\gamma z} = -(R + j\omega L)(I_{S0}^+ e^{-\gamma z} + I_{S0}^- e^{\gamma z}) \tag{6.11}$$

and,

$$\frac{dI_S(z)}{dz} = -(G + j\omega C) V_S(z)$$

or,

$$-\gamma I_{S0}^+ e^{-\gamma z} + \gamma I_{S0}^- e^{\gamma z} = -(G + j\omega C)(V_{S0}^+ e^{-\gamma z} + V_{S0}^- e^{\gamma z}) \tag{6.12}$$

Equating the coefficients of $e^{-\gamma z}$ and $e^{\gamma z}$, we get,

$$\gamma V_{S0}^+ = (R + j\omega L) I_{S0}^+ \qquad \gamma I_{S0}^+ = (G + j\omega C) V_{S0}^+$$

$$-\gamma V_{S0}^- = (R + j\omega L) I_{S0}^- \qquad -\gamma I_{S0}^- = (G + j\omega C) V_{S0}^-$$

Hence, the characteristic impedance of the line is obtained as

$$Z_0 = \frac{V_{S0}^+}{I_{S0}^+} = \frac{R + j\omega L}{\gamma} = \frac{\gamma}{G + j\omega C} = \sqrt{\frac{R + j\omega L}{G + j\omega C}}$$

and,

$$Z_0 = -\frac{V_{S0}^-}{I_{S0}^-} = -\frac{R + j\omega L}{-\gamma} = -\frac{-\gamma}{G + j\omega C} = \sqrt{\frac{R + j\omega L}{G + j\omega C}}$$

$\therefore$

$$\boxed{Z_0 = \frac{V_{S0}^+}{I_{S0}^+} = -\frac{V_{S0}^-}{I_{S0}^-} = \sqrt{\frac{R + j\omega L}{G + j\omega C}}} \tag{6.13}$$

In general, transmission line characteristic impedance is a complex quantity and can be defined by

$$Z_0 = R_0 + jX_0$$

where R_0 is the resistive component of Z_0
X_0 is the reactive component of Z_0

The voltage and current wave equations can be written in terms of the voltage coefficients and the characteristic impedance (rather than the voltage and current coefficients) using the relationships

$$I_{S0}^+ = \frac{V_{S0}^+}{Z_0} \qquad I_{S0}^- = -\frac{V_{S0}^-}{Z_0}$$

Then, the voltage and current equations become as follows.

$$V_S(z) = V_{S0}^+ e^{-\gamma z} + V_{S0}^- e^{\gamma z} \tag{6.14a}$$

$$I_S(z) = \frac{1}{Z_0}(V_{S0}^+ e^{-\gamma z} - V_{S0}^- e^{\gamma z}) \quad \textbf{(6.14b)}$$

These equations have unknown coefficients for the forward and reverse voltage waves only since the characteristic impedance of the transmission line is typically known.

In the succeeding sections, we will find out the characteristic impedance of the line for two extreme cases, one for a lossless line and the other for a distortionless line.

***Example 6.1** A 300 m long line has the following constants: $R = 4.5$ kΩ, $L = 0.15$ mH, $G = 60$ mho, $C = 12$ nF, operated at frequency 6 MHz. Find the propagation constant, characteristic impedance and velocity of propagation.

Solution Here, the line constant per unit length is obtained as follows.

$$R = \frac{4.5 \times 10^3}{300} = 15\ \Omega/\text{m}$$
$$L = \frac{0.15 \times 10^{-3}}{300} = 5 \times 10^{-7}\ \text{H/m}$$
$$C = \frac{12 \times 10^{-12}}{300} = 4 \times 10^{-11}\ \text{F/m}$$
$$G = \frac{60 \times 10^{-3}}{300} = 0.2 \times 10^{-3}\ \text{mho/m}$$

$$\omega = 2\pi f = 2\pi \times 6 \times 10^6 = 12\pi \times 10^6\ \text{rad/m}$$

So, the propagation constant is given as

$$\begin{aligned}\gamma &= \sqrt{(R + j\omega L)(G + j\omega C)} \\ &= \sqrt{(15 + j12\pi \times 10^6 \times 5 \times 10^{-7})(0.2 \times 10^{-3} + j12\pi \times 10^6 \times 4 \times 10^{-11})} \\ &= 0.19\angle 66.95° = (0.074 + j0.175)\ \text{per m}\end{aligned}$$

The characteristic impedance is given as

$$Z = \sqrt{\frac{R + j\omega L}{G + j\omega C}} = \sqrt{\frac{15 + j12\pi \times 10^6 \times 5 \times 10^{-7}}{0.2 \times 10^{-3} + j12\pi \times 10^6 \times 4 \times 10^{-11}}} = 126.71\angle -15.485°\ \Omega$$

Velocity of propagation is

$$v = \frac{1}{\sqrt{LC}} = \frac{1}{\sqrt{5 \times 10^{-7} \times 4 \times 10^{-11}}} = 2.236 \times 10^8\ \text{m/s}$$

Example 6.2 A lossy cable with $R = 2.5$ Ω/m, $L = 10$ μH/m, $C = 10$ pF/m and $G = 0$ operates at $f = 1$GHz. Find the attenuation constant of the line.

Solution Here, $R = 2.5$ Ω/m, $L = 10$ μH/m, $C = 10$ pF/m, $G = 0$, $f = 1$ GHz

The propagation constant is given as

$$\begin{aligned}\gamma &= \alpha + j\beta = \sqrt{(R + j\omega L)(G + j\omega C)}\\ &= \sqrt{(2.5 + j2\pi\times 1\times 10^{6}\times 10\times 10^{-6})\times (j2\pi\times 1\times 10^{6}\times 10\times 10^{-12})}\\ &= 0.00125 + j0.0628\end{aligned}$$

Hence, the attenuation constant of the line is

$$\alpha = 0.00125\ (\mathrm{m}^{-1})$$

6.6.1 Lossless Transmission Line

A transmission line is said to be lossless if

— the conductors of the line are perfect, i.e., the conductors have infinite conductivity and zero resistance ($\sigma = \infty$, $R = 0$), and

— the dielectric medium between the conductors is ideal, i.e., the medium has zero conductivity and infinite resistance ($\sigma = 0$, $G = 0$).

The equivalent circuit for a segment Δz of a lossless transmission line reduces to the circuit as shown in Fig. 6.3.

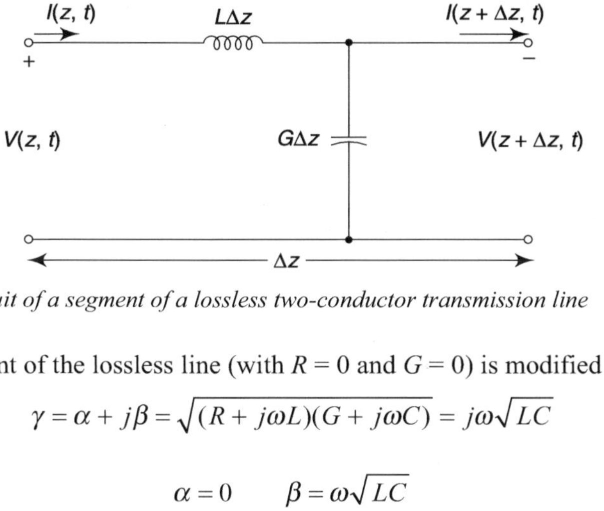

Fig. 6.3 *Equivalent circuit of a segment of a lossless two-conductor transmission line*

The propagation constant of the lossless line (with $R = 0$ and $G = 0$) is modified as

$$\gamma = \alpha + j\beta = \sqrt{(R + j\omega L)(G + j\omega C)} = j\omega\sqrt{LC}$$

$$\therefore \qquad \alpha = 0 \qquad \beta = \omega\sqrt{LC}$$

Thus, the characteristic impedance of the lossless transmission line is purely real and given by

$$Z_0 = \sqrt{\frac{R + j\omega L}{G + j\omega C}} = \sqrt{\frac{L}{C}}$$

The voltage and current equations for the lossless transmission line are given as

$$V(z) = V_0^{+}e^{-j\beta z} + V_0^{-}e^{j\beta z}$$

$$I(z) = \frac{1}{Z_0}(V_0^{+}e^{-j\beta z} - V_0^{-}e^{j\beta z})$$

The velocity of propagation on the lossless line is

$$u = \frac{\omega}{\beta} = \frac{1}{\sqrt{LC}}$$

***Example 6.3** A transmission line with air as dielectric has the characteristic impedance of 60 Ω and phase constant of 4 rad/m at 500 MHz. Calculate the inductance per metre and the capacitance per metre of the line.

Solution Here, $Z_0 = 60\ \Omega$; $\beta = 4$ rad/m; $f = 500$ MHz
An air line is regarded as a lossless line as $\sigma = 0$. Hence,

$$R = 0 = G \qquad \text{and} \qquad \alpha = 0$$

$\therefore$
$$Z_0 = R_0 = \sqrt{\frac{L}{C}}$$

$\therefore$
$$\gamma = \beta = \omega\sqrt{LC}$$

$\therefore$
$$\frac{Z_0}{\beta} = \frac{\sqrt{L/C}}{\omega\sqrt{LC}} = \frac{1}{\omega C}$$

So, the capacitance is given as

$$C = \frac{\beta}{\omega Z_0} = \frac{4}{2\pi \times 500 \times 10^6 \times 60} = 21.2\ \text{pF/m}$$

The inductance is given as

$$L = Z_0^2 C = (70)^2 \times 68.2 \times 10^{-12} = 334.2\ \text{mH/m}$$

***Example 6.4** In a lossless transmission line, the velocity of propagation is 2.5×10^8 m/s. Capacitance of the line is 30 pF/m, find:

(a) inductance per metre of the line.
(b) phase constant at 100 MHz.
(c) characteristic impedance of the line.

Solution Here, $v = 2.5 \times 10^8$ m/s; $C = 30 \times 10^{-12}$ F; $f = 100$ MHz

(a) We have, $v = \dfrac{1}{\sqrt{LC}} \Rightarrow L = \dfrac{1}{Cv^2} = \dfrac{1}{30 \times 10^{-12}(2.5 \times 10^8)^2} = 533\ \text{nH/m}$

(b) We have, $v = \dfrac{\omega}{\beta} \Rightarrow \beta = \dfrac{2\pi f}{v} = \dfrac{2\pi \times 100 \times 100^6}{2.5 \times 10^8} = 2.51\ \text{rad/m}$

(c) Characteristic impedance of the line is given as

$$Z_0 = \sqrt{\frac{L}{C}} = \sqrt{\frac{533 \times 10^{-9}}{30 \times 10^{-12}}} = 133.29\Omega$$

6.6.2 Distortionless Transmission Line

Waveform Distortion When the received signal is not identical with the transmitted signal, the signal is said to be distorted. There are two types of waveform distortion:

1. Frequency Distortion, and
2. Phase Distortion.

1. Frequency Distortion During the transmission of a signal on a transmission line, it gets attenuated through attenuation constant α. It is known that α is a function of frequency.

$$\alpha = \frac{1}{\sqrt{2}}\sqrt{RG - \omega^2 LC + \sqrt{(R^2 + \omega^2 L^2)(G^2 + \omega^2 C^2)}}$$

Hence, different frequencies transmitted along the line will then be attenuated to different extent. The received waveform will not be identical with the input waveform. This is known as *frequency distortion.*

2. Phase Distortion The phase constant is given as

$$\alpha = \frac{1}{\sqrt{2}}\sqrt{-RG - \omega^2 LC + \sqrt{(R^2 + \omega^2 L^2)(G^2 + \omega^2 C^2)}}$$

and velocity is given as

$$u = \frac{\omega}{\beta}$$

Since the phase constant of the line, β is a complicated function of frequency, all the frequencies applied to the transmission line will not have the same time of transmission; some frequencies being delayed more than the other. The received waveform will not be identical with the input waveform. This is known as delay or *phase distortion.*

A distortionless line has no frequency and phase distortions. For a distortionless line, α (and velocity u) should not be a function of frequency and β should be a direct function of frequency.

A transmission line can be made distortionless (linear phase constant) by designing the line such that the per unit length parameters satisfy the relation given as

$$\frac{R}{L} = \frac{G}{C}$$

NOTE

Derivation of Heaviside's condition for Distortionless Line

We have, $\gamma = \alpha + j\beta = \sqrt{(R + j\omega L)(G + j\omega C)}$

$\therefore$ $\gamma^2 = (\alpha + j\beta)^2 = (R + j\omega L)(G + j\omega C)$

or, $\alpha^2 - \beta^2 + j2\alpha\beta = RG - \omega^2 LC + j\omega(CR + LG)$

Equating the real and imaginary parts,

$$\alpha^2 - \beta^2 = RG - \omega^2 LC$$

$$2\alpha\beta = \omega(CR + LG)$$

Solving for α and β,

$$\alpha = \sqrt{\frac{RG - \omega^2 LC + \sqrt{(RG - \omega^2 LC)^2 + \omega^2 (CR + LG)^2}}{2}}$$

$$\beta = \sqrt{\frac{\omega^2 LC - RG + \sqrt{(RG - \omega^2 LC)^2 + \omega^2 (CR + LG)^2}}{2}}$$

Now α will be independent of Ω if the term under the second radical be reduced to $(RG + \omega^2 LC)$.

$\therefore \quad (RG - \omega^2 LC)^2 + \omega^2 (CR + LG)^2 = (RG + \omega^2 LC)^2$

or $\quad \omega^2 R^2 C^2 - 2\omega^2 RCLG + \omega^2 L^2 G^2 = 0$

or $\quad \omega^2 (RC - LG)^2 = 0$

or $\quad (RC - LG) = 0$

or $\quad RC = LG$

$\therefore \quad \frac{R}{L} = \frac{G}{C}$

Similarly, β will be a direct function of ω if the term under second radical be reduced to $(RG + \omega^2 LC)$. This again gives the same result.
Hence, the condition of a distortionless line is

$$\boxed{\frac{R}{L} = \frac{G}{C}}$$

The propagation constant of a distortionless line is given as

$$\gamma = \alpha + j\beta = \sqrt{(R + j\omega L)(G + j\omega C)} = \sqrt{R\left(1 + j\omega\frac{L}{R}\right)G\left(1 + j\omega\frac{C}{G}\right)}$$

$$= \sqrt{RG\left(1 + j\omega\frac{L}{R}\right)^2} \qquad \left\{\because \frac{L}{R} = \frac{C}{G}\right\}$$

$$= \sqrt{RG}\left(1 + j\omega\frac{L}{R}\right) = \sqrt{RG} + j\omega\sqrt{RG}\frac{L}{R}$$

$$= \sqrt{RG} + j\omega L\sqrt{\frac{G}{R}} = \sqrt{RG} + j\omega L\sqrt{\frac{C}{L}} = \sqrt{RG} + j\omega\sqrt{LC}$$

$$\therefore \qquad \alpha = \sqrt{RG} \qquad \beta = \omega\sqrt{LC}$$

Although the shape of the signal is not distorted, the signal will suffer attenuation as the wave propagates along the line since the *distortionless* line is a *lossy* transmission line.

The characteristic impedance of the distortionless transmission line is given by

$$Z_0 = \sqrt{\frac{R}{G}} = \sqrt{\frac{L}{C}}$$

The velocity of propagation on the distortionless line is

$$u = \frac{\omega}{\beta} = \frac{1}{\sqrt{LC}}$$

For most practical transmission lines, we find that $RC > GL$. In order to satisfy the distortionless line requirement, series loading coils are typically placed periodically along the line to increase L.

Example 6.5 A distortionless line has $Z_0 = 60\ \Omega$, $\alpha = 20$ mNp/m, $v = 0.6c$ where c is the speed of light in a vacuum. Find R, L, G, C and λ at 100 MHz.

Solution For distortionless line,

$$\frac{R}{L} = \frac{G}{C}$$

$$\therefore \quad Z_0 = \sqrt{\frac{L}{C}}, \quad \alpha = \sqrt{RG} = R\sqrt{\frac{C}{L}} = \frac{R}{Z_0}$$

$$\therefore \quad R = \alpha Z_0 = 20 \times 10^{-3} \times 60 = 1.2\ \Omega/\text{m}$$

Also,

$$v = \frac{1}{\sqrt{LC}}, \quad \therefore \quad \frac{v}{Z_0} = \frac{1}{\sqrt{LC}} \times \sqrt{\frac{C}{L}} = \frac{1}{L}$$

$$\therefore \quad L = \frac{Z_0}{v} = \frac{60}{0.6 \times 3 \times 10^8} = 333\ \text{nH/m}$$

$$\therefore \quad G = \frac{\alpha^2}{R} = \frac{(20 \times 10^{-3})^2}{1.2} = 333\ \mu\text{mho/m}$$

Also,

$$vZ_0 = \frac{1}{\sqrt{LC}} \times \sqrt{\frac{L}{C}} = \frac{1}{C}$$

$$\therefore \quad C = \frac{1}{vZ_0} = \frac{1}{0.6 \times 3 \times 10^8 \times 60} = 92.59\ \text{pF/m}$$

The wavelength is, $\lambda = \dfrac{v}{f} = \dfrac{0.6 \times 3 \times 10^8}{100 \times 10^6} = 1.8$ m

6.7 INPUT IMPEDANCE OF TRANSMISSION LINE

Input impedance is the line impedance seen at the beginning of a transmission line. The input impedance at any point on the transmission line is given by the ratio of voltage to current at that point.

We find the input impedance for the following three types of transmission lines:

1. Finite Lossy Transmission Line,
2. Finite Lossless High-frequency Transmission Line, and
3. Infinite (Lossy and Lossless) Transmission Line.

6.7.1 Finite Lossy Transmission Line

We have the general voltage and current equations given by

$$V_S(z) = V_{S0}^+ e^{-\gamma z} + V_{S0}^- e^{\gamma z}$$

and

$$I_S(z) = I_{S0}^+ e^{-\gamma z} + I_{S0}^- e^{\gamma z}$$

From the definition of characteristic impedance, $Z_0 = \dfrac{V_{S0}^+}{I_{S0}^+} = -\dfrac{V_{S0}^-}{I_{S0}^-}$, we get

$$I_S(z) = \frac{1}{Z_0}(V_{S0}^+ e^{-\gamma z} - V_{S0}^- e^{\gamma z})$$

At the sending end, $z = 0$, so the voltage and current equations reduce to

$$V_S(z) = V_{S0}^+ + V_{S0}^-$$

$$I_S(z) = \frac{1}{Z_0}(V_{S0}^+ - V_{S0}^-)$$

Solving for the voltage coefficients,

$$V_{S0}^+ = \frac{V_S + Z_0 I_S}{2} \qquad V_{S0}^- = \frac{V_S - Z_0 I_S}{2}$$

So, the voltage and current at any distance z from the sending end is

$$V_S(z) = \frac{1}{2}V_S[e^{\gamma z} + e^{-\gamma z}] - \frac{1}{2}Z_0 I_S[e^{\gamma z} - e^{-\gamma z}]$$

and,

$$I_S(z) = \frac{1}{2}I_S[e^{\gamma z} + e^{-\gamma z}] - \frac{1}{2}\frac{V_S}{Z_0}[e^{\gamma z} - e^{-\gamma z}]$$

But, $$\left(\frac{e^{\gamma z} + e^{-\gamma z}}{2}\right) = \cosh\gamma z \quad \text{and} \quad \left(\frac{e^{\gamma z} - e^{-\gamma z}}{2}\right) = \sinh\gamma z$$

∴ $$V_S(z) = V_S\cosh\gamma z - Z_0 I_S\sinh\gamma z$$

$$I_S(z) = I_S\cosh\gamma z - \frac{V_S}{Z_0}\sinh\gamma z$$

At the receiving end, $z = l$ (l is the length of the line), the voltage and current are given as

$$V_R = V_S\cosh\gamma l - Z_0 I_S\sinh\gamma l$$

$$I_R = I_S\cosh\gamma l - \frac{V_S}{Z_0}\sinh\gamma l$$

Now, if the receiving end is terminated with an impedance Z_R, then we have

$$V_R = Z_R I_R$$

or, $$(V_S\cosh\gamma l - Z_0 I_S\sinh\gamma l) = Z_R\left(I_S\cosh\gamma l - \frac{V_S}{Z_0}\sinh\gamma l\right)$$

or, $$V_S\left(\cosh\gamma l + \frac{Z_R}{Z_0}\sinh\gamma l\right) = I_S(Z_R\cosh\gamma l + Z_0\sinh\gamma l)$$

Hence, the input impedance is given as

$$Z_i = \frac{V_S}{I_S} = \frac{Z_R\cosh\gamma l + Z_0\sinh\gamma l}{\cosh\gamma l + \frac{Z_R}{Z_0}\sinh\gamma l} = Z_0\left(\frac{Z_R\cosh\gamma l + Z_0\sinh\gamma l}{Z_0\cosh\gamma l + Z_R\sinh\gamma l}\right) = Z_0\left(\frac{Z_R + Z_0\tanh\gamma l}{Z_0 + Z_R\tanh\gamma l}\right)$$

$$Z_i = Z_0\left(\frac{Z_R\cosh\gamma l + Z_0\sinh\gamma l}{Z_0\cosh\gamma l + Z_R\sinh\gamma l}\right) = Z_0\left(\frac{Z_R + Z_0\tanh\gamma l}{Z_0 + Z_R\tanh\gamma l}\right)$$

Three cases may appear.

Open-circuited Line In this case, the receiving end is kept opened. So, the receiving end current is zero.

$$\therefore \qquad I_R = \left(I_S\cosh\gamma l - \frac{V_S}{Z_0}\sinh\gamma l\right) = 0$$

Hence, the input impedance is

$$Z_i = \frac{V_S}{I_S} = Z_{OC} = Z_0\left(\frac{\cosh\gamma l}{\sinh\gamma l}\right) = Z_0\coth\gamma l$$

$$\boxed{Z_{OC} = Z_0\coth\gamma l}$$

Short-circuited Line In this case, the receiving end terminals are short-circuited by metallic strap. So, the receiving end voltage is zero.

$$\therefore \qquad V_R = (V_S\cosh\gamma l - Z_0 I_S\sinh\gamma l) = 0$$

Hence, the input impedance is

$$Z_i = \frac{V_S}{I_S} = Z_{SC} = Z_0\left(\frac{\sinh\gamma l}{\cosh\gamma l}\right) = Z_0\tanh\gamma l$$

$$\boxed{Z_{SC} = Z_0\tanh\gamma l}$$

NOTE

$$Z_0 = \sqrt{Z_{OC}\times Z_{SC}}$$

Matched Line In this case, $Z_R = Z_0$
Hence, the input impedance is

$$\boxed{Z_i = Z_0}$$

In this case, the whole electromagnetic wave is transmitted without reflection. The incident power is fully absorbed by the load.

6.7.2 Finite Lossless High-frequency Transmission Line

For lossless transmission line, $R = 0$ and $G = 0$.

$$\therefore \qquad \alpha = 0 \quad \text{and} \quad \beta = \omega\sqrt{LC}$$

$$\therefore \qquad \gamma = \sqrt{(R + j\omega L)(G + j\omega C)} = \alpha + j\beta = j\omega\sqrt{LC}$$

So, the general voltage and current equations reduce to

$$V_S(z) = V_{S0}^{+} e^{-j\beta z} + V_{S0}^{-} e^{j\beta z}$$

and

$$I_S(z) = \frac{1}{R_0}(V_{S0}^{+} e^{-j\beta z} - V_{S0}^{-} e^{j\beta z})$$

where, R_0 is the characteristic impedance for a lossless line $= \sqrt{\frac{L}{C}}$.

At the sending end, $z = 0$, so the voltage and current equations reduce to

$$V_S = V_{S0}^{+} + V_{S0}^{-}$$

$$I_S = \frac{1}{R_0}(V_{S0}^{+} - V_{S0}^{-})$$

Solving for the voltage coefficients,

$$V_{S0}^{+} = \frac{V_S + R_0 I_S}{2} \qquad V_{S0}^{-} = \frac{V_S - R_0 I_S}{2}$$

So, the voltage and current at any distance z from the sending end is

$$V_S(z) = \frac{1}{2} V_S [e^{j\beta z} + e^{-j\beta z}] - \frac{1}{2} R_0 I_S [e^{j\beta z} - e^{-j\beta z}]$$

and

$$I_S(z) = \frac{1}{2} I_S [e^{j\beta z} + e^{-j\beta z}] - \frac{1}{2}\frac{V_S}{R_0}[e^{j\beta z} - e^{-j\beta z}]$$

But $$\left(\frac{e^{j\beta z} + e^{-j\beta z}}{2}\right) = \cos\beta z \quad \text{and} \quad \left(\frac{e^{j\beta z} - e^{-j\beta z}}{2}\right) = \sin\beta z$$

$\therefore$ $$V_S(z) = V_S \cos\beta z - jR_0 I_S \sin\beta z$$

$$I_S(z) = I_S \cos\beta z - j\frac{V_S}{R_0}\sin\beta z$$

At the receiving end, $z = l$ (l is the length of the line), the voltage and current are given as

$$V_R = V_S \cos\beta l - jR_0 I_S \sin\beta l$$

$$I_R = I_S \cos\beta l - \frac{V_S}{R_0}\sin\beta l$$

Now, if the receiving end is terminated with an impedance Z_R, then we have

$$V_R = Z_R I_R$$

or $$(V_S \cos\beta l - jR_0 I_S \sin\beta l) = Z_R\left(I_S \cos\beta l - j\frac{V_S}{R_0}\sin\beta l\right)$$

or $$V_S\left(\cos\beta l + j\frac{Z_R}{R_0}\sin\beta l\right) = I_S(Z_R \cos\beta l + jR_0 \sin\beta l)$$

Hence, the input impedance is given as

$$Z_i = \frac{V_S}{I_S} = \frac{Z_R \cos \beta l + jR_0 \sin \beta l}{\cos \beta l + j\dfrac{Z_R}{R_0} \sin \beta l} = R_0\left(\frac{Z_R \cos \beta l + jR_0 \sin \beta l}{R_0 \cos \beta l + jZ_R \sin \beta l}\right) = R_0\left(\frac{Z_R + jR_0 \tan \beta l}{R_0 + jZ_R \tan \beta l}\right)$$

$$\boxed{Z_i = R_0\left(\frac{Z_R \cos \beta l + jR_0 \sin \beta l}{R_0 \cos \beta l + jZ_R \sin \beta l}\right) = R_0\left(\frac{Z_R + jR_0 \tan \beta l}{R_0 + jZ_R \tan \beta l}\right)}$$

Three cases may appear.

Open-circuited Line In this case, the receiving end is kept opened. So, the receiving end current is zero.

$\therefore$
$$I_R = \left(I_S \cos \beta l - j\frac{V_S}{R_0} \sin \beta l\right) = 0$$

Hence, the input impedance is

$$Z_i = \frac{V_S}{I_S} = Z_{OC} = -jR_0\left(\frac{\cos \beta l}{\sin \beta l}\right) = -jR_0 \cot \beta l$$

$$\boxed{Z_{OC} = -jR_0 \cot \beta l}$$

Short-circuited Line In this case, the receiving end terminals are short-circuited by metallic strap. So, the receiving end voltage is zero.

$\therefore$
$$V_R = (V_S \cos \beta l - jR_0 I_S \sin \beta l) = 0$$

Hence, the input impedance is

$$Z_i = \frac{V_S}{I_S} = Z_{SC} = jR_0\left(\frac{\sin \beta l}{\cos \beta l}\right) = jR_0 \tan \beta l$$

$$\boxed{Z_{SC} = jR_0 \tan \beta l}$$

NOTE

$$R_0 = \sqrt{Z_{OC} \times Z_{SC}}$$

Matched Line In this case, $Z_R = R_0$
Hence, the input impedance is

$$\boxed{Z_i = R_0}$$

Example 6.6 A transmission line is lossless and is 30 m long. It is terminated in a load impedance of $Z_L = (30 + j20)\Omega$ at a frequency of 10 MHz. The inductance and capacitance of the line are $L = 100$ nH/m, $C = 20$ pF/m. Find the input impedance of the line at the source end and at the mid-point of the line.

Solution Here, $Z_L = (30 + j20)\Omega$, $L = 100$ nH/m, $C = 20$ pF/m, $f = 10$ MHz, $l = 30$ m
The characteristic impedance of the line is

$$Z_0 = \sqrt{\frac{L}{C}} = \sqrt{\frac{100 \times 10^{-9}}{20 \times 10^{-12}}} = 70.71\,\Omega$$

The phase constant of the line is

$$\beta = \omega\sqrt{LC} = 2\pi f\sqrt{LC} = 2\pi \times 10 \times 10^6 \times \sqrt{100 \times 10^{-9} \times 20 \times 10^{-12}} = 0.08886 \text{ rad/m}$$

So, the input impedance at the source end is given as

$$Z_i = Z_0\left[\frac{Z_L + jZ_0 \tan\beta l}{Z_0 + jZ_L \tan\beta l}\right] = 70.1 \times \frac{30 + j20 + j70.1\tan[0.08886 \times 30]}{70.1 + j(30 + j20)\tan[0.08886 \times 30]}$$
$$= (48.14 + j52.35)\,\Omega$$

So, the input impedance at the mid-point ($z = 15$ m) is given as

$$Z_{\text{in}}(z = 15) = Z_0 \left.\frac{Z_L + jZ_0\tan[\beta(l - z)]}{Z_0 + jZ_L\tan[\beta(l - z)]}\right|_{z=15} = 70.1 \times \frac{30 + j20 + j70.1\tan[0.08886 \times 15]}{70.1 + j(30 + j20)\tan[0.08886 \times 15]}$$
$$= (30.41 + j21.49)\,\Omega$$

6.7.3 Infinite (Lossy and Lossless) Transmission Line

The general voltage equation is

$$V_S(z) = V_{S0}^{+} e^{-\gamma z} + V_{S0}^{-} e^{\gamma z}$$

At the sending end, $z = 0$ and hence the voltage equation becomes

$$V_S = V_{S0}^{+} + V_{S0}^{-}$$

At infinity, $z = \infty$, $V_s(z)$ must be zero.

$$\therefore \quad V_{S0}^{+} e^{-\gamma\infty} + V_{S0}^{-} e^{\gamma\infty} = 0$$

$$\Rightarrow \quad V_{S0}^{-} = 0$$

$$\therefore \quad V_S = V_{S0}^{+}$$

Therefore, voltage for infinite transmission line is

$$V_S(z) = V_S e^{-\gamma z}$$

Now,

$$\frac{dV_S(z)}{dz} = -\gamma V_S e^{-\gamma z}$$

From the differential equation of transmission line

$$I_S(z) = -\left(\frac{1}{R + j\omega L}\right)\frac{dV_S(z)}{dz} = -\frac{-\gamma V_S e^{-\gamma z}}{R + j\omega L} = \frac{\gamma V_S e^{-\gamma z}}{R + j\omega L}$$
$$= \frac{V_S}{Z_0} e^{-\gamma z} \qquad \{\because \gamma = \sqrt{(R + j\omega L)(G + j\omega C)}\}$$

So, the current for infinite transmission line is

$$\boxed{I_S(z) = \frac{V_S}{Z_0} e^{-\gamma z}}$$

At the sending end, $z = 0$.

$$\therefore \qquad I_S = \frac{V_S}{Z_0}$$

Therefore, the input impedance of an infinite transmission line is

$$\therefore \qquad \boxed{Z_i = \frac{V_S}{I_S} = Z_0}$$

Thus, it is observed that *for infinite transmission line, the input impedance is equal to the characteristic impedance of the line.*

$$\therefore \qquad Z_i = \sqrt{\frac{R + j\omega L}{G + j\omega C}}, \quad \text{for lossy infinite line}$$

$$= \sqrt{\frac{L}{C}}, \quad \text{for lossless infinite line}$$

6.7.4 Input Impedances in Some Special Cases

We will consider the input impedances of a lossless transmission line for some important load impedances and line lengths.

Case (1): When $l = \frac{\lambda}{2}$ If the length of the transmission line is exactly half wavelength $\left(l = \frac{\lambda}{2}\right)$, we have

$$\beta l = \frac{2\pi}{\lambda} \times \frac{\lambda}{2} = \pi$$

$$\therefore \qquad \cos \beta l = \cos \pi = -1 \quad \text{and} \quad \sin \beta l = \sin \pi = 0$$

$$\therefore \qquad Z_i = R_0 \left(\frac{Z_R \cos \beta l + jR_0 \sin \beta l}{R_0 \cos \beta l + jZ_R \sin \beta l} \right) = R_0 \left(\frac{-Z_R + j0}{-R_0 + j0} \right) = Z_R$$

$$\therefore \qquad \boxed{Z_i = Z_R}$$

Thus, if the transmission line is precisely one-half wavelength long, the input impedance is equal to the load impedance, regardless of Z_0 and β.

Case (2): When $l = \frac{\lambda}{4}$ If the length of the transmission line is exactly one-quarter wavelength $\left(l = \frac{\lambda}{4}\right)$, we have

$$\beta l = \frac{2\pi}{\lambda} \times \frac{\lambda}{4} = \frac{\pi}{2}$$

$$\therefore \qquad \cos \beta l = \cos \frac{\pi}{2} = 0 \quad \text{and} \quad \sin \beta l = \sin \frac{\pi}{2} = 1$$

$$\therefore \qquad Z_i = R_0 \left(\frac{Z_R \cos \beta l + jR_0 \sin \beta l}{R_0 \cos \beta l + jZ_R \sin \beta l} \right) = R_0 \left(\frac{Z_R \times 0 + jR_0 \times 1}{R_0 \times 0 + jZ_R \times 1} \right) = \frac{R_0^2}{Z_R}$$

$$\therefore \qquad \boxed{Z_i = \frac{R_0^2}{Z_R}}$$

Thus, if the transmission line is precisely one-quarter wavelength long, the input impedance is inversely proportional to the load impedance.

NOTE

If $Z_R = 0$, i.e., the receiving end is short-circuited, the input impedance is infinity ($Z_i = \infty$); thus, a quarter wave transmission line transforms a short-circuit into an open-circuit and vice-versa.

Case (3): When $Z_R = Z_0$ The line is said to be a matched line and the input impedance is $Z_i = Z_0 = R_0$.

Case (4): When $Z_R = jX_R$ (Purely Reactive Load) The input impedance is given as

$$\therefore \qquad Z_i = R_0\left(\frac{Z_R \cos\beta l + jR_0 \sin\beta l}{R_0 \cos\beta l + jZ_R \sin\beta l}\right)$$

$$= R_0\left(\frac{jX_R \cos\beta l + jR_0 \sin\beta l}{R_0 \cos\beta l + j \cdot jX_R \sin\beta l}\right)$$

$$= jR_0\left(\frac{X_R \cos\beta l + R_0 \sin\beta l}{R_0 \cos\beta l - X_R \sin\beta l}\right)$$

Thus, if the load is purely reactive, then the input impedance will also be purely reactive, regardless of the length of the line.

NOTE

The opposite is not true. Even if the load is purely resistive, the input impedance will be complex.

Case (5): When $l << \lambda$ If the transmission line is electrically long, i.e., its wavelength l is very small compared to signal wavelength λ, we have

$$\beta l = \frac{2\pi}{\lambda} \times l = 2\pi \times \frac{l}{\lambda} \approx 0$$

$$\therefore \qquad \cos\beta l \approx 1 \quad \text{and} \quad \sin\beta l \approx 0$$

$$\therefore \qquad Z_i = R_0\left(\frac{Z_R \cos\beta l + jR_0 \sin\beta l}{R_0 \cos\beta l + jZ_R \sin\beta l}\right) = R_0\left(\frac{Z_R + j0}{R_0 + j0}\right) = Z_R$$

$$\therefore \qquad \boxed{Z_i = Z_R}$$

Thus, if the transmission line length is much smaller than a wavelength, the input impedance will always be equal to the load impedance, Z_R. Such conditions of the line are frequently considered in circuit theory where the line is said to be a *lumped circuit*.

6.8 REFLECTION COEFFICIENTS OF TRANSMISSION LINE

Reflection Coefficient of a transmission line is the ratio of the reflected voltage (or current) to the incident voltage (or current), when a transmission line is terminated in an impedance (Z_R) not equal to the characteristic impedance (Z_0) of the line.

The general transmission line equations for the voltage and current as a function of position along the line are

$$V_S(z) = V_{S0}^{+} e^{-\gamma z} + V_{S0}^{-} e^{\gamma z}$$

and

$$I_S(z) = \frac{1}{Z_0}(V_{S0}^{+} e^{-\gamma z} - V_{S0}^{-} e^{\gamma z})$$

The voltage and current at the load ($z = l$) are given as

$$V_S(l) = V_{S0}^{+} e^{-\gamma l} + V_{S0}^{-} e^{\gamma l} = V_R$$

and

$$I_S(z) = \frac{1}{Z_0}(V_{S0}^{+} e^{-\gamma l} - V_{S0}^{-} e^{\gamma l}) = I_R$$

Solving these two equations, we get the voltage coefficients in terms of the load voltage and load current as follows.

$$V_{S0}^{+} = \left(\frac{V_R + Z_0 I_R}{2}\right) e^{\gamma l}$$

$$V_{S0}^{-} = \left(\frac{V_R - Z_0 I_R}{2}\right) e^{-\gamma l}$$

The *voltage reflection coefficient* as a function of position along the line [$\Gamma(z)$] is defined as the ratio of the reflected wave voltage to the transmitted wave voltage.

$$\therefore \qquad \Gamma(z) = \frac{V_{S0}^{-} e^{\gamma z}}{V_{S0}^{+} e^{-\gamma z}} = \frac{V_{S0}^{-}}{V_{S0}^{+}} e^{2\gamma z}$$

The *current reflection coefficient* at any point on the line is the negative of the voltage reflection coefficient at that point.

Inserting the expressions for the voltage coefficients in terms of the load voltage and current, we have

$$\Gamma(z) = \frac{\left(\frac{V_R - Z_0 I_R}{2}\right) e^{-\gamma l}}{\left(\frac{V_R + Z_0 I_R}{2}\right) e^{\gamma l}} e^{2\gamma z} = \frac{\frac{V_R}{I_R} - Z_0}{\frac{V_R}{I_R} + Z_0} e^{2\gamma (z-l)} = \frac{Z_R - Z_0}{Z_R + Z_0} e^{2\gamma (z-l)}$$

The reflection coefficient at the load ($z = l$) is

$$\boxed{\Gamma(l) = \Gamma_L = \frac{Z_R - Z_0}{Z_R + Z_0}}$$

Hence, the reflection coefficient as a function of position can be written as

$$\boxed{\Gamma(z) = \frac{Z_R - Z_0}{Z_R + Z_0} e^{2\gamma(z-l)} = \Gamma_L e^{2\gamma(z-l)}}$$

This is observed that the reflection coefficient is zero ($\Gamma_L = 0$) when $Z_R = Z_0$, i.e., when the transmission line is *perfectly matched*. This is the ideal case where no reflections occur from the load and all the energy associated with the forward travelling wave is delivered to the load.

If $Z_R \neq Z_0$, a *mismatch* exists and reflected waves are present on the transmission line. Just as plane waves reflected from a dielectric interface produce standing waves in the region containing the incident and reflected waves, guided waves on a transmission line reflected from the load produce standing waves on the transmission line (the sum of forward and reverse travelling waves).

The transmission line voltage and current equations can be written in terms of the reflection coefficients as follows.

$$V_S(z) = V_{S0}^+ e^{-\gamma z} + V_{S0}^- e^{\gamma z} = V_{S0}^+ e^{-\gamma z}\left(1 + \frac{V_{S0}^-}{V_{S0}^+} e^{2\gamma z}\right) = V_{S0}^+ e^{-\gamma z}[1 + \Gamma(z)]$$

$$I_S(z) = \frac{1}{Z_0}(V_{S0}^+ e^{-\gamma z} - V_{S0}^- e^{\gamma z}) = \frac{V_{S0}^+}{Z_0} e^{-\gamma z}\left(1 - \frac{V_{S0}^-}{V_{S0}^+} e^{2\gamma z}\right) = \frac{V_{S0}^+}{Z_0} e^{-\gamma z}[1 - \Gamma(z)]$$

$$\boxed{V_S(z) = V_{S0}^+ e^{-\gamma z}[1 + \Gamma(z)]}$$

$$\boxed{I_S(z) = \frac{V_{S0}^+}{Z_0} e^{-\gamma z}[1 - \Gamma(z)]}$$

Example 6.7 A lossless line is terminated with a load impedance of $(20 - j10)\Omega$. Find the phase constant and the reflection coefficient of the line of length 50 m. The characteristic impedance of the line is 70 Ω and the wavelength is 0.5 m.

Solution Here, $Z_L = (20 - j10)\Omega$, $Z_0 = 70\ \Omega$, $\lambda = 0.5$ m, $l = 50$ m

The phase constant is given as

$$\beta = \frac{2\pi}{\lambda} = \frac{2\pi}{0.5} = 12.566 \text{ rad/m}$$

The reflection coefficient of the line is

$$\Gamma_L = \frac{Z_L - Z_0}{Z_L + Z_0} = \frac{20 - j10 - 70}{20 - j10 + 70} = -(0.537 + j0.171)$$

Location of Voltage Maxima and Minima from Receiving End From Section 6.7.2, we have for a lossless line, the voltage and current given as

$$V_S(z) = V_{S0}^+ e^{-j\beta z} + V_{S0}^- e^{j\beta z}$$

$$I_S(z) = \frac{1}{R_0}(V_{S0}^+ e^{-j\beta z} - V_{S0}^- e^{j\beta z})$$

If we let x be the distance from the receiving end, the above two equations can be expressed in terms of x simply by putting $z = -x$.

$$V_S(x) = V_{S0}^+ e^{j\beta x} + V_{S0}^- e^{-j\beta x}$$

$$I_S(z) = \frac{1}{R_0}(V_{S0}^+ e^{j\beta x} - V_{S0}^- e^{-j\beta x})$$

Now, reflection coefficient is given as

$$\Gamma = |\Gamma|\, e^{j\theta} = \frac{V_{S0}^- e^{-j\beta x}}{V_{S0}^+ e^{j\beta x}} = \left(\frac{V_{S0}^-}{V_{S0}^+}\right) e^{-2j\beta x}$$

At receiving end, $x = 0$ and $\Gamma = \Gamma_L = |\Gamma|\, e^{j\theta} = \left(\frac{V_{S0}^-}{V_{S0}^+}\right)$

$\therefore$ $$V_{S0}^- = V_{S0}^+ \,|\Gamma|\, e^{j\theta}$$

Replacing this value, we get

$$V_S(x) = V_{S0}^+[e^{j\beta x} + |\Gamma|\, e^{j\theta} e^{-j\beta x}] = V_{S0}^+ e^{j\beta x}[1 + |\Gamma|\, e^{j\theta} e^{-2j\beta x}] = V_{S0}^+ e^{j\beta x}[1 + |\Gamma|\, e^{-j(2\beta x - \theta)}]$$

For voltage to be maximum, two components must be in phase, so that

$$\boxed{(2\beta x_{\max} - \theta) = 2n\pi}$$

where, $n = 0, 1, 2, 3, \ldots$

The first voltage maximum nearest to the load end will occur when $n = 0$, i.e., $2\beta x_{\max} = \theta$. The magnitude of the maximum voltage is given as

$$\boxed{V_{S\max} = V_{S0}^+[1 + |\Gamma|]}$$

For voltage to be minimum, two components must be in phase opposition, so that

$$\boxed{(2\beta x_{\min} - \theta) = (2n + 1)\pi}$$

where, $n = 0, 1, 2, 3, \ldots$

The first voltage minimum nearest to the load end will occur when $n = 0$, i.e., $(2\beta x_{\min} - \theta) = \pi$. The magnitude of the minimum voltage is given as

$$\boxed{V_{S\min} = V_{S0}^+[1 - |\Gamma|]}$$

NOTE

If we take the ratio of maximum and minimum voltages, we get

$$\frac{V_{S\max}}{V_{S\min}} = \frac{1 + |\Gamma|}{1 - |\Gamma|}$$

This expression is known as the voltage standing wave ratio (VSWR) as obtained in the next section.

6.9 STANDING WAVES AND STANDING WAVE RATIO (*S*) OF TRANSMISSION LINE

It has been discussed in Section 6.8 that if the receiving end of a transmission line is not perfectly matched, there will be reflection of the voltage and current. As a consequence of reflection, a *standing wave* may be visualised as an interference between the incident signal E_i at a given frequency, travelling in the forward direction, and the reflected signal E_r, at the same frequency, travelling in the reverse direction. At the load, the relationship between the amplitudes of E_r and E_i and the phase angle between them are uniquely determined by the load impedance. The phase angle between E_r and E_i, however, will vary along the line as a function of the distance from the load. Since the wave oscillates in amplitude but never moves laterally, it is called a *standing wave*.

In Fig. 6.4, the incident, reflected and the standing waves can be seen. The dashed lines are the E_r and the E_i, while the non-dashed one represents the standing wave.

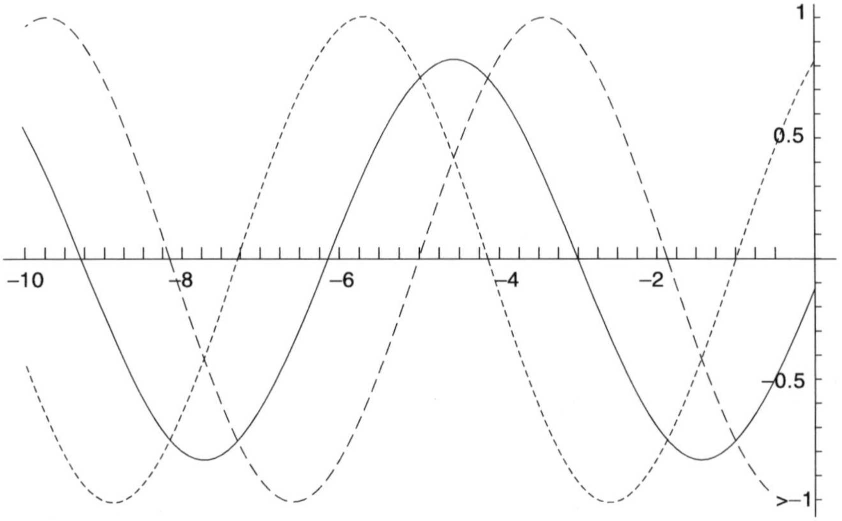

Fig. 6.4 *Standing waves in transmission line*

The following points may be noted:

1. At a position 180° and multiple of that from the load $\left(n\frac{\lambda}{2}\right)$, the voltage and current must have the same values they do at the load.
2. At a position 90° and odd multiple of that from the load $\left(n\frac{\lambda}{4}\right)$, the voltage and current must be inverted: if the voltage is lowest and the current is highest at the load, then at 90° from the load the voltage reaches its highest value and the current reaches its lowest value at the same point.

NOTE

No standing waves will be developed along a matched line. The voltage along the line is constant, so the matched line is also said to be a flat line.

Standing wave ratio is defined as the ratio of the maximum voltage (or current) to the minimum voltage (or current) of a line having standing waves.

$$\therefore \qquad \boxed{s = \frac{V_{\max}}{V_{\min}} = \frac{I_{\max}}{I_{\min}} = \frac{1+|\Gamma_L|}{1-|\Gamma_L|}}$$

In the case where the load contains no reactance, the *SWR* is equal to the ratio between the load resistance *R* and the characteristic impedance of the line.

Example 6.8 The transmission line of characteristic impedance of 50 Ω is terminated with a load of $(100 + j100)\Omega$. Find the reflection coefficient and the standing wave ratio (SWR).

Solution Here, $Z_0 = 50\ \Omega$; $Z_L = (100 + j100)$

Reflection coefficient is given as

$$\Gamma_L = \frac{Z_L - Z_0}{Z_L + Z_0} = \frac{100 + j100 - 50}{100 + j100 + 50} = 0.62\angle 29.77°$$

Standing wave ratio is given as

$$s = \frac{1 + |\Gamma_L|}{1 - |\Gamma_L|} = \frac{1 + 0.62}{1 - 0.62} = 4.263$$

***Example 6.9** A lossless line has a characteristic impedance of 50 ohm and is terminated in a load resistance of 75 ohm. If the length of the line is $\lambda/2$, determine (i) input impedance, (ii) reflection coefficient and (iii) VSWR. What will be the value of reflection coefficient, if the load impedance is 50 ohm?

Solution Here, $Z_0 = 50\ \Omega$, $Z_L = 75\ \Omega$, $l = \dfrac{\lambda}{2}$

(i) The input impedance for a $\lambda/2$ length line is given as (See Section 6.7.4),

$$Z_{\text{in}} = Z_L = 75\ \Omega$$

(ii) The reflection coefficient of the line is,

$$\Gamma_L = \frac{Z_L - Z_0}{Z_L + Z_0} = \frac{75 - 50}{75 + 50} = 0.2$$

(iii) VSWR is given as,

$$s = \frac{1 + |\Gamma_L|}{1 - |\Gamma_L|} = \frac{1 + 0.2}{1 - 0.8} = 1.5$$

If the load impedance is 50 ohm, the load is perfectly matched. Hence in this case, the reflection coefficient will be zero.

***Example 6.10** A line having Z_0 = 300 ohm, has a VSWR value of 4.48. The first voltage minimum occurs at a distance of 0.06 m from the receiving end. If the operating frequency is 200 MHz, determine the value of load impedance.

Solution Here, $Z_0 = 300\ \Omega$, VSWR = 4.48, $f = 200$ MHz, $x_{\text{min}} = 0.06$ m

$$\text{Wavelength, } \lambda = \frac{v}{f} = \frac{3 \times 10^8}{200 \times 10^6} = 1.5 \text{ m}$$

For the first minimum voltage, the condition is

$$2\beta x_{\text{min}} - \theta = \pi$$

$$\Rightarrow \qquad 2\frac{2\pi}{1.5} \times 0.06 - \theta = \pi$$

$$\Rightarrow \qquad \theta = -0.84\pi = -151.2°$$

Now, the reflection coefficient is given as

$$\Gamma = \frac{VSWR - 1}{VSWTR + 1} = \frac{4.48 - 1}{4.48 + 1} = \frac{3.48}{5.48} = 0.635$$

Also,

$$|\Gamma| e^{j\theta} = \frac{Z_R - Z_0}{Z_R + Z_0}$$

$$\Rightarrow \qquad 0.635(\cos 151.2° - j \sin 151.2°) = \frac{Z_R - 300}{Z_R + 300}$$

$$\Rightarrow \qquad Z_R = (71.2 - j73.1) = 102\ \Omega$$

So, the value of the load impedance is 102 Ω.

***Example 6.11** A lossless line having $Z_0 = 75$ ohm, has a VSWR value of 3.0. The first voltage minimum occurs at a distance 20 cm from the receiving end. If the operating frequency is 150 MHz, determine the value of load impedance.

Solution Here, $Z_0 = 75\ \Omega$, $VSWR = 3.0$, $f = 150$ MHz, $x_{min} = 0.2$ m

Wavelength, $\lambda = \frac{v}{f} = \frac{3 \times 10^8}{150 \times 10^6} = 2$ m

For the first minimum voltage, the condition is

$$2\beta x_{min} - \theta = \pi$$

$$\Rightarrow \qquad 2\frac{2\pi}{2} \times 0.2 - \theta = \pi$$

$$\Rightarrow \qquad \theta = -0.6\pi = -108°$$

Now, the reflection coefficient is given as

$$\Gamma = \frac{VSWR - 1}{VSWTR + 1} = \frac{3 - 1}{3 + 1} = \frac{2}{4} = 0.5$$

Also,

$$|\Gamma| e^{j\theta} = \frac{Z_R - Z_0}{Z_R + Z_0}$$

$$\Rightarrow \qquad 0.5(\cos 108° - j \sin 108°) = \frac{Z_R - 75}{Z_R + 75}$$

$$\Rightarrow \qquad |Z_R| = 98\ \Omega$$

So, the value of the load impedance is 98 Ω.

***Example 6.12** A lossless line in air having a characteristic impedance of 300 ohm is terminated by an unknown impedance. The first voltage minimum is located at 15 cm from the load. The standing wave ratio is 3.3. Calculate the frequency and terminated impedance.

Solution Here, $Z_0 = 300\ \Omega$, $VSWR = 3.3$, $x_{min} = 0.15$ m

Now, the reflection coefficient is given as

$$\Gamma = \frac{VSWR - 1}{VSWTR + 1} = \frac{3.3 - 1}{3.3 + 1} = \frac{2.3}{4.3} = 0.5348$$

Also,

$$\Gamma_L = \frac{Z_R - Z_0}{Z_R + Z_0}$$

$$\Rightarrow \quad 0.5348 = \frac{Z_R - 75}{Z_R + 75}$$

$$\Rightarrow \quad Z_R = 990\ \Omega$$

So, the value of the load impedance is 990 Ω.

Now, the first voltage minimum occurs at $x_{min} = 0.15$ m. By the condition

$$2\beta x_{min} - \theta = \pi$$

Since, the reflection coefficient is not having any phase angle, i.e. $\theta = 0$, we have

$$2\beta x_{min} = \pi$$

$$\Rightarrow \quad 2 \times \frac{2\pi}{\lambda} \times 0.15 = \pi$$

$$\Rightarrow \quad \lambda = 0.6 \text{ m}$$

So, the frequency is given as $f = \dfrac{v}{\lambda} = \dfrac{3 \times 10^8}{0.6} = 500$ MHz

6.10 INPUT IMPEDANCE AS A FUNCTION OF POSITION ALONG A TRANSMISSION LINE

The input impedance at any point on the transmission line is given by the ratio of voltage to current at that point.

Inserting the expressions for the phasor voltage and current $[V_S(z)]$ and $[I_S(z)]$ from the original form of the transmission line equations gives

$$Z_{in}(z) = \frac{V_S(z)}{I_S(z)} = Z_0 \frac{V_{S0}^{+} e^{-\gamma z} + V_{S0}^{-} e^{\gamma z}}{V_{S0}^{+} e^{-\gamma z} - V_{S0}^{-} e^{\gamma z}}$$

Inserting the expressions of voltage coefficients in terms of the load voltage and load current, we have

$$V_{S0}^{+} = \left(\frac{V_R + Z_0 I_R}{2}\right) e^{\gamma l} = \frac{I_R}{2}(Z_R + Z_0)e^{\gamma l}$$

$$V_{S0}^{-} = \left(\frac{V_R - Z_0 I_R}{2}\right) e^{-\gamma l} = \frac{I_R}{2}(Z_R - Z_0)e^{-\gamma l}$$

$$\therefore \quad Z_{in}(z) = Z_0 \frac{V_{S0}^{+} e^{-\gamma z} + V_{S0}^{-} e^{\gamma z}}{V_{S0}^{+} e^{-\gamma z} - V_{S0}^{-} e^{\gamma z}} = Z_0 \frac{e^{\gamma(l-z)}(Z_R + Z_0) + e^{-\gamma(l-z)}(Z_R - Z_0)}{e^{\gamma(l-z)}(Z_R + Z_0) - e^{-\gamma(l-z)}(Z_R - Z_0)}$$

$$= Z_0 \frac{Z_R[e^{\gamma(l-z)} + e^{-\gamma(l-z)}] + Z_0[e^{\gamma(l-z)} - e^{-\gamma(l-z)}]}{Z_R[e^{\gamma(l-z)} - e^{-\gamma(l-z)}] + Z_0[e^{\gamma(l-z)} + e^{-\gamma(l-z)}]}$$

Now, we know that $\sinh \gamma l = \dfrac{e^{\gamma l} - e^{-\gamma l}}{2}$ and $\cosh \gamma l = \dfrac{e^{\gamma l} + e^{-\gamma l}}{2}$

$$\therefore \qquad Z_{\text{in}}(z) = Z_0 \frac{Z_R[e^{\gamma(l-z)} + e^{-\gamma(l-z)}] + Z_0[e^{\gamma(l-z)} - e^{-\gamma(l-z)}]}{Z_R[e^{\gamma(l-z)} - e^{-\gamma(l-z)}] + Z_0[e^{\gamma(l-z)} + e^{-\gamma(l-z)}]}$$

$$= Z_0 \frac{Z_R \cosh[\gamma(l-z)] + Z_0 \sinh[\gamma(l-z)]}{Z_R \sinh[\gamma(l-z)] + Z_0 \cosh[\gamma(l-z)]}$$

$$= Z_0 \frac{Z_R + Z_0 \tanh[\gamma(l-z)]}{Z_0 + Z_R \tanh[\gamma(l-z)]}$$

Thus, the input impedance at any point along a general transmission line is given as

$$\boxed{Z_{\text{in}}(z) = Z_0 \frac{Z_R + Z_0 \tanh[\gamma(l-z)]}{Z_0 + Z_R \tanh[\gamma(l-z)]}}$$

For a lossless line, Z_0 is purely real, i.e. $Z_0 = R_0$, $\gamma = j\beta$. The hyperbolic tangent function reduces to

$$\tanh[\gamma(l-z)] = \tanh[j\beta(l-z)] = j\tan[\beta(l-z)]$$

So, the input impedance at any point along a lossless transmission line becomes

$$\boxed{Z_{\text{in}}(z) = R_0 \frac{Z_R + jR_0 \tan[\beta(l-z)]}{R_0 + jZ_R \tan[\beta(l-z)]}}$$

Two special cases are considered:

1. Open-circuited Lossless Line Here,

$$Z_R \to \infty \quad \text{and} \quad \Gamma_L = 1.$$

Input impedance becomes

$$\lim_{Z_R\to\infty} Z_{\text{in}}(z) = \lim_{Z_R\to\infty} \left\{ R_0 \frac{Z_R + jR_0 \tan[\beta(l-z)]}{R_0 + jZ_R \tan[\beta(l-z)]} \right\} = \lim_{Z_R\to\infty} \left\{ R_0 \frac{1 + j\dfrac{R_0}{Z_R}\tan[\beta(l-z)]}{\dfrac{R_0}{Z_R} + j\tan[\beta(l-z)]} \right\}$$

$$= \left\{ R_0 \frac{1}{j\tan[\beta(l-z)]} \right\} = -jR_0 \cot[\beta(l-z)] = Z_{OC}(z)$$

2. Short-circuited Lossless Line Here,

$$Z_R \to 0 \quad \text{and} \quad \Gamma_L = -1.$$

Input impedance becomes

$$\lim_{Z_R\to 0} Z_{\text{in}}(z) = \lim_{Z_R\to 0} \left\{ R_0 \frac{Z_R + jR_0 \tan[\beta(l-z)]}{R_0 + jZ_R \tan[\beta(l-z)]} \right\} = R_0 \frac{jR_0 \tan[\beta(l-z)]}{R_0} = jR_0 \tan[\beta(l-z)]$$
$$= Z_{SC}(z)$$

The variations of these impedances with the length of the line are shown in Fig. 6.5.

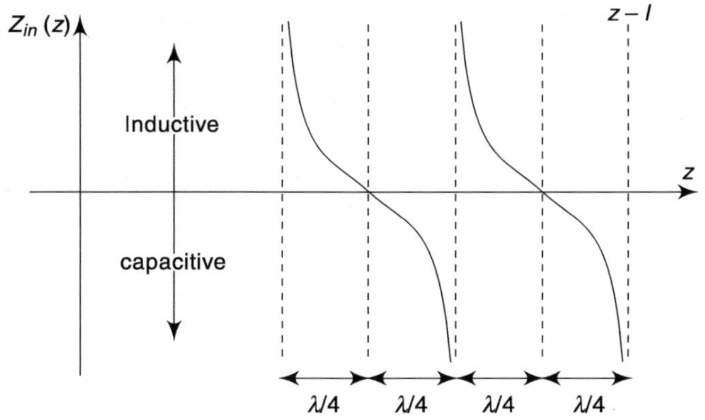

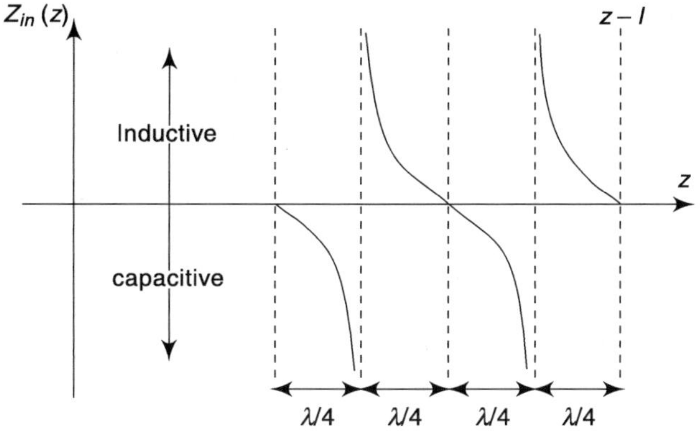

Fig. 6.5 *Input impedance of a lossless transmission line (a) when open (b) when short*

The impedance characteristics of a short-circuited or open-circuited transmission line are related to the positions of the voltage and current nulls along the transmission line. On a lossless transmission line, the magnitude of the voltage and current are given by

$$|V_S(z)| = |V_{S0}^{+} e^{-j\beta z}[1 + \Gamma(z)]| = |V_{S0}^{+}|\,|1 + \Gamma(z)|$$

$$|I_S(z)| = \left|\frac{V_{S0}^{+}}{R_0} e^{-j\beta z}[1 - \Gamma(z)]\right| = \frac{|V_{S0}^{+}|}{R_0}|1 - \Gamma(z)|$$

The equations for the voltage and current magnitude follow the *crank diagram* form and the minimum and maximum values for voltage and current are

$$|V_S(z)|_{\max} = |V_{S0}^{+}|\,(1 + |\Gamma(z)|) \qquad |V_S(z)|_{\min} = \left|V_{S0}^{+}\right|(1 - |\Gamma(z)|)$$

$$|I_S(z)|_{\max} = \frac{|V_{S0}^{+}|}{R_0}(1 + |\Gamma(z)|) \qquad |I_S(z)|_{\min} = \frac{|V_{S0}^{+}|}{R_0}(1 - |\Gamma(z)|)$$

For a lossless line, the magnitude of the reflection coefficient is constant along the entire line and thus, equal to the magnitude of Γ at the load.

$$|\Gamma(z)| = |\Gamma_L e^{j2\beta(z-l)}| = |\Gamma_L|$$

The *standing wave ratio*(s) on the lossless line is defined as the ratio of maximum to minimum voltage magnitudes (or maximum to minimum current magnitudes).

$$\therefore \qquad s = \frac{|V_S(z)|_{\max}}{|V_S(z)|_{\min}} = \frac{|I_S(z)|_{\max}}{|I_S(z)|_{\min}} = \frac{1+|\Gamma_L|}{1-|\Gamma_L|}$$

The standing wave ratio on a lossless transmission line ranges between 1 and ∞.

We now apply the previous equations to the special cases of open-circuited and short-circuited lossless transmission lines to determine the positions of the voltage and current nulls.

Voltage and Current Nulls in Open-Circuited Lossless Transmission Line In this case,

$$\Gamma_L = 1$$

To determine the voltage null, we have

$$|V_S(z)| = 0 \quad \text{when} \quad 1 + \Gamma_L e^{j2\beta(z-l)} = 0$$

or, $$e^{j2\beta(z-l)} = -1$$

or, $$2\beta(z-l) = n\pi \qquad \{n \text{ odd only}\}$$

or, $$2 \times \frac{2\pi}{\lambda}(z-l) = n\pi$$

or, $$(z-l) = n\frac{\lambda}{4} \qquad \{n \text{ odd only}\}$$

$$\therefore \qquad (z-l) = n\frac{\lambda}{4} \qquad \{n \text{ odd only}\}$$

To determine the current null, we have

$$|I_S(z)| = 0 \quad \text{when} \quad 1 - \Gamma_L e^{j2\beta(z-l)} = 0$$

or, $$e^{j2\beta(z-l)} = 1$$

or, $$2\beta(z-l) = n\pi \qquad \{n \text{ even only}\}$$

or, $$2 \times \frac{2\pi}{\lambda}(z-l) = n\pi$$

or, $$(z-l) = n\frac{\lambda}{4} \qquad \{n \text{ even only}\}$$

$$\therefore \qquad (z-l) = n\frac{\lambda}{4} \qquad \{n \text{ even only}\}$$

Thus, the nulls are:

Voltage Nulls ($Z_{in} = 0$): At $\frac{\lambda}{4}$ from the load and every $\frac{\lambda}{2}$ from that point.

Current Nulls ($Z_{in} = \infty$): At the load and every $\frac{\lambda}{2}$ from that point.

Voltage and Current Nulls in Short-Circuited Lossless Transmission Line In this case,

$$\Gamma_L = 1$$

To determine the voltage null, we have

$$|V_S(z)| = 0 \quad \text{when} \quad 1 + \Gamma_L e^{j2\beta(z-l)} = 0$$

or, $$e^{j2\beta(z-l)} = 1$$

or, $$2\beta(z-l) = n\pi \qquad \{n \text{ even only}\}$$

or, $$2 \times \frac{2\pi}{\lambda}(z-l) = n\pi$$

or, $$(z-l) = n\frac{\lambda}{4} \qquad \{n \text{ even only}\}$$

$\therefore$ $$(z-l) = n\frac{\lambda}{4} \qquad \{n \text{ even only}\}$$

To determine the current null, we have

$$|I_S(z)| = 0 \quad \text{when} \quad 1 - \Gamma_L e^{j2\beta(z-l)} = 0$$

or, $$e^{j2\beta(z-l)} = -1$$

or, $$2\beta(z-l) = n\pi \qquad \{n \text{ odd only}\}$$

or, $$2 \times \frac{2\pi}{\lambda}(z-l) = n\pi$$

or, $$(z-l) = n\frac{\lambda}{4} \qquad \{n \text{ odd only}\}$$

$\therefore$ $$(z-l) = n\frac{\lambda}{4} \qquad \{n \text{ odd only}\}$$

Thus, the nulls are:

Voltage Nulls ($Z_{in} = 0$): At the load and every $\frac{\lambda}{2}$ from that point.

Current Nulls ($Z_{in} = \infty$): At $\frac{\lambda}{4}$ from the load and every $\frac{\lambda}{2}$ from that point.

From the equations for the maximum and minimum transmission line voltage and current, we also find

$$\frac{|V_S(z)|_{\max}}{|I_S(z)|_{\max}} = \frac{|V_S(z)|_{\min}}{|I_S(z)|_{\min}} = Z_0$$

It will be shown that the voltage maximum occurs at the same location as the current minimum on a lossless transmission line and vice versa. Using the definition of the standing wave ratio, the maximum and minimum impedance values along the lossless transmission line may be written as

$$|Z_{\text{in}}(z)|_{\max} = \frac{|V_S(z)|_{\max}}{|I_S(z)|_{\min}} = s\frac{|V_S(z)|_{\min}}{|I_S(z)|_{\min}} = sZ_0 = Z_0\left(\frac{1+|\Gamma_L|}{1-|\Gamma_L|}\right)$$

$$|Z_{\text{in}}(z)|_{\min} = \frac{|V_S(z)|_{\min}}{|I_S(z)|_{\max}} = \frac{1}{s} \times \frac{|V_S(z)|_{\min}}{|I_S(z)|_{\min}} = \frac{Z_0}{s} = Z_0\left(\frac{1-|\Gamma_L|}{1+|\Gamma_L|}\right)$$

Thus, the impedance along the lossless transmission line must lie within the range of $\frac{Z_0}{s}$ to sZ_0.

Example 6.13 A source ($V_{sg} = 100\ \angle 0°\ V$, $Z_s = R_s = 50\ \Omega$, $f = 100$ MHz) is connected to a lossless transmission line ($L = 0.25\ \mu$H/m, $C = 100$ pF/m, $l = 10$ m). For loads of $Z_L = R_L = 0, 25, 50,$

100 and ∞ Ω, determine (a) the reflection coefficient at the load (b) the standing wave ratio (c) the input impedance at the transmission line input terminals

Solution Here,

$$V_{sg} = 100\angle 0°\ V,\ Z_s = R_s = 50\ \Omega,\ f = 100\ \text{MHz},$$
$$L = 0.25\ \mu\text{H/m},\quad C = 100\ \text{pF/m},\quad l = 10\ \text{m}$$

Characteristic impedance is, $Z_0 = \sqrt{\dfrac{L}{C}} = \sqrt{\dfrac{0.25\times 10^{-6}}{100\times 10^{-12}}} = 50\ \Omega$

$$\therefore \qquad \beta = \omega\sqrt{LC} = 2\pi \times 100 \times 10^6 \times \sqrt{0.25\times 10^{-6}\times 100\times 10^{-12}} = \pi = \frac{2\pi}{\lambda}$$

$$\therefore \qquad \lambda = 2\ m,\quad l = 5\lambda \quad \text{and} \quad \beta l = 10\,\pi$$

Reflection coefficient is given as

$$\Gamma_L = \frac{Z_L - Z_0}{Z_L + Z_0} = \frac{R_L - 50}{R_L + 50}$$

Standing wave ratio is given as

$$s = \frac{1+|\Gamma_L|}{1-|\Gamma_L|} = \frac{R_L}{50}$$

The input impedance at $z = 0$ is given as

$$Z_{in}(z=0) = R_0 \left.\frac{Z_L + jR_0 \tan[\beta(l-z)]}{R_0 + jZ_L \tan[\beta(l-z)]}\right|_{z=0} = R_0 \frac{R_L + jR_0 \tan \beta l}{R_0 + jR_L \tan \beta l} = R_0 \frac{R_L + jR_0 \tan 10\pi}{R_0 + jR_L \tan 10\pi} = R_L$$

So, for different values of the load, the values of the reflection coefficient, standing wave ratio and the input impedance is given in Table 6.1

Table 6.1 *Different values of the load*

R_L (ohm)	(a) Γ_L	(b) s	(c) Z_{in} (ohm)
0	–1	0	0
25	–0.333	0.5	25
50	0	1	50
100	+0.333	2	100
∞	+1	∞	∞

Example 6.14 An open-wire transmission line with characteristic impedance of 600 Ω is terminated by a load Z_L = 900 Ω. Find the reflection coefficient, transmission coefficient and the standing wave ratio (*SWR*).

Solution Here, Z_0 = 600 Ω; Z_L = 900 Ω

Reflection coefficient is given as

$$\Gamma_L = \frac{Z_L - Z_0}{Z_L + Z_0} = \frac{900-600}{900+600} = 0.2$$

Transmission coefficient is given as

$$\tau_L = 1 + \Gamma_L = 1 + 0.2 = 1.2$$

Standing wave ratio is given as

$$s = \frac{1+|\Gamma_L|}{1-|\Gamma_L|} = \frac{1+0.2}{1-0.2} = 1.5$$

***Example 6.15** A transmission line of characteristic impedance of $Z_0 = 50\ \Omega$ is terminated by a load $R_L = Z_L = 100\ \Omega$. Find *VSWR*, Z_{min} and Z_{max}.

Solution Here, $Z_0 = 50\ \Omega$; $R_L = Z_L = 100\ \Omega$
Reflection coefficient is given as

$$\Gamma_L = \frac{Z_L - Z_0}{Z_L + Z_0} = \frac{100-50}{100+50} = 0.33$$

Standing wave ratio is given as

$$s = \frac{1+|\Gamma_L|}{1-|\Gamma_L|} = \frac{1+0.32}{1-0.2} = 1.5$$

At the point of voltage maximum, we know that the current is minimum.

$$\therefore \qquad Z_{max} = Z_0 \frac{1+|\Gamma_L|}{1-|\Gamma_L|} = Z_0 s = 50 \times 2 = 100\ \Omega$$

At the point of voltage minimum, the current is maximum.

$$\therefore \qquad Z_{min} = Z_0 \frac{1-|\Gamma_L|}{1+|\Gamma_L|} = \frac{Z_0}{s} = \frac{50}{2} = 25\ \Omega$$

Example 6.16 A lossless transmission line with a characteristic impedance of 75 Ω is terminated by a load impedance of 120 Ω. If the magnitude of the incident wave is 10 V, calculate the minimum and maximum values of the voltages on the line.

Solution Here, $Z_0 = 75\ \Omega$, $Z_R = 120\ \Omega$, $V_0 = 10$ V
Reflection coefficient is given as

$$\Gamma_L = \frac{Z_R - Z_0}{Z_R + Z_0} = \frac{120-75}{120+75} = 0.243$$

So, the minimum and maximum values of the voltages on the line are given as

$$V_{min} = V_0[1 - \Gamma_L] = 10[1 - 0.243] = 7.567 \text{ V}$$

$$V_{max} = V_0[1 + \Gamma_L] = 10[1 + 0.243] = 12.432 \text{ V}$$

***Example 6.17** A transmission line with characteristic impedance of 300 ohm is terminated in a purely resistive load. It is found by measurement that the minimum line voltage upon it is 5 mV and maximum 7.5 mV. What is the value of load impedance?

Solution Here, $Z_0 = 75\ \Omega$, $V_{min} = 5$ mV, $V_{max} = 7.5$ mV

We know that the minimum and maximum values of the voltages on the line are given as

$$V_{\min} = V_0[1 - \Gamma_L]$$
$$V_{\max} = V_0[1 + \Gamma_L]$$

$$\therefore \quad \frac{V_{\min}}{V_{\max}} = \frac{1-\Gamma_L}{1+\Gamma_L} = \frac{1 - \dfrac{Z_R - Z_0}{Z_R + Z_0}}{1 + \dfrac{Z_R - Z_0}{Z_R + Z_0}} = \frac{Z_0}{Z_R}$$

$$\Rightarrow \quad \frac{5}{7.5} = \frac{300}{Z_R}$$

$$\Rightarrow \quad Z_R = \frac{300 \times 7.5}{5} = 450\ \Omega$$

So, the value of the load impedance is 450 Ω.

6.11 LOSSES IN TRANSMISSION LINES

Transmission line losses are of three types:

1. Copper losses,
2. Dielectric losses, and
3. Radiation or induction losses.

1. Copper Losses These losses occur because of the following reasons:

(a) I^2R Loss This is the I^2R Loss that occurs in a transmission line whenever current flows through the conductors. With copper braid, which has a resistance higher than solid tubing, this power loss is higher.

(b) Skin Effect Loss This is the copper loss that occurs to *skin effect* associated with alternating currents.

(c) Losses Due to Crystallisation of Conductors This type of copper loss occurs due to ageing of the transmission line, due to repeated bending of the line, producing different cracks on the conductor. This effect is known as *crystallisation of the conductors*. The effect is more when the line is subjected to high temperature, high winds, moisture, etc.

(d) Corona Losses Corona loss, caused by the ionisation of air molecules near the transmission line conductors, is the other major type of copper loss in transmission lines. Copper losses can be minimised and conductivity increased in a transmission line by plating the line with silver. Since silver is a better conductor than copper, most of the current will flow through the silver layer. The tubing then serves primarily as a mechanical support. Copper losses can also be minimised by terminating the line properly so that no standing waves are generated.

2. Dielectric Losses Dielectric losses result from the heating effect on the dielectric material between the conductors due to the distortions of the electron orbits when a potential difference is applied between the conductors.

The atomic structure of rubber is more difficult to distort than the structure of some other dielectric materials. The atoms of materials, such as polyethylene, distort easily. Therefore, polyethylene is often used as a dielectric because less power is consumed when its electron orbits are distorted.

3. Radiation or Induction Losses Radiation and induction losses are similar in that both are caused by the fields surrounding the conductors.

Induction losses occur when the electromagnetic field about a conductor cuts through any nearby metallic object and a current is induced in that object, resulting loss of power.

Radiation losses occur because some magnetic lines of force about a conductor do not return to the conductor when the cycle alternates. These lines of force are projected into space as radiation and these result in power losses.

6.12 SMITH CHART

The Smith chart is a useful graphical tool used to calculate the reflection coefficient and impedance at various points on a (lossless) transmission line system.

The Smith chart is actually a polar plot of the complex reflection coefficient $\Gamma(z)$ [ratio of the reflected wave voltage to the forward wave voltage] overlaid with the corresponding impedance $Z(z)$ [ratio of overall voltage to overall current].

Construction of the Smith Chart The construction of the Smith chart is based on the relation

$$\Gamma_L = \frac{Z_R - Z_0}{Z_R + Z_0} = |\Gamma_L| \angle\theta_\Gamma = \Gamma_{Lr} + j\Gamma_{Li} \quad \textbf{(6.15)}$$

where, Γ_L is the reflection coefficient at the load ($z = l$) and Γ_{Lr} and Γ_{Li} are the real and imaginary parts of Γ_L, respectively.

The corresponding standing wave ratio is

$$SWR,\ s = \frac{1 + |\Gamma_L|}{1 - |\Gamma_L|} \quad \textbf{(6.16)}$$

The magnitude of Γ_L is constant on any circle in the complex plane so that the standing wave ratio (s) is also constant on the same circle (see Fig. 6.6).

Once the position of Γ_L is located on the Smith chart, the location of the reflection coefficient as a function of position [$\Gamma(z)$] is determined using the reflection coefficient formula.

$$\Gamma(z) = \Gamma_L e^{j2\beta(z-l)} = |\Gamma_L| e^{j\theta_\Gamma} e^{j2\beta(z-l)} = |\Gamma_L| e^{j(\theta_\Gamma + \theta_z)} \quad \textbf{(6.17)}$$

This equation shows that to locate $\Gamma(z)$, we start at Γ_L and rotate through an angle of $\theta_z = 2\beta(z - l)$ on the constant SWR circle. With the load located at $z = l$, moving from the receiving end towards the sending end ($z < l$) defines a negative angle θ_z (clockwise rotation on the constant SWR circle).

Note that if $\theta_z = -2\pi$, we rotate back to the same point. The distance travelled along the transmission line is then obtained as

$$\theta_z = 2\beta(z - l) = -2\pi$$

$$\Rightarrow \quad 2 \times \frac{2\pi}{\lambda}(z - l) = -2\pi$$

$$\Rightarrow \quad l - z = \frac{\lambda}{2}$$

Thus, *one complete rotation around the Smith chart (360°) is equal to one half wavelength.*

Also, in the Smith chart, *clockwise rotation represents movement towards the sending end, whereas anti-clockwise rotation represents movement towards the receiving end.*

Instead of constructing separate Smith charts for transmission lines with different characteristic impedances, only one Smith chart is constructed within a circle of unit radius so that the magnitude of the complex-valued reflection coefficient ranges from 0 to 1 for any value of the load impedance ($|\Gamma_L| \leq 1$) as shown in Fig. 6.6. Note that as the reflection coefficient is constant on any circle in the complex plane, the standing wave ratio is also constant on the same circle ($s = 1$).

This is achieved by using a normalised chart in which all impedances are normalised with respect to the characteristic impedance Z_0 of the line. For example, the normalised load impedance is given as

$$z_R = \frac{Z_R}{Z_0} = r_R + jx_R \qquad \textbf{(6.18)}$$

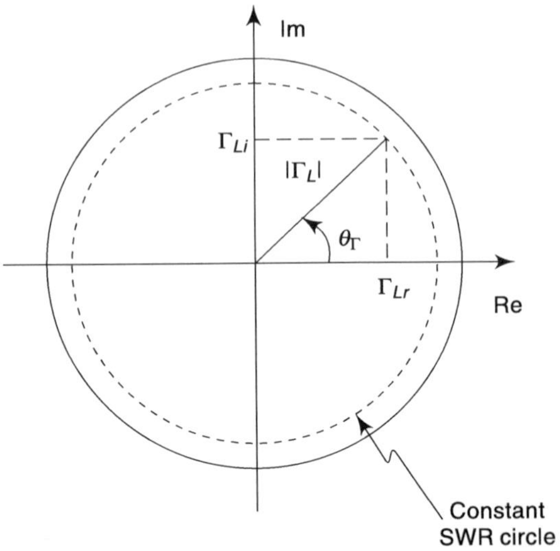

Fig. 6.6 *Unit circle to construct Smith chart*

So, the reflection coefficient is given as

$$\Gamma_L = \frac{Z_R - Z_0}{Z_R + Z_0} = \frac{Z_R/Z_0 - 1}{Z_R/Z_0 + 1} = \frac{z_R - 1}{z_R + 1} = \Gamma_{Lr} + j\Gamma_{Li} \qquad \textbf{(6.19)}$$

or,

$$z_R = r_R + jx_R = \frac{(1 + \Gamma_{Lr}) + j\Gamma_{Li}}{(1 - \Gamma_{Lr}) - j\Gamma_{Li}} \qquad \textbf{(6.20)}$$

Equating the real and imaginary components and simplifying, we get

$$r_R = \frac{1 - \Gamma_{Lr}^2 - \Gamma_{Li}^2}{(1 - \Gamma_{Lr})^2 + \Gamma_{Li}^2} \qquad \textbf{(6.21a)}$$

$$x_R = \frac{2\Gamma_{Li}}{(1 - \Gamma_{Lr})^2 + \Gamma_{Li}^2} \qquad \textbf{(6.21b)}$$

Rearranging the terms in Eq. (6.21*a*) and (6.21*b*), we have

$$\left[\Gamma_{Lr} - \frac{r_R}{1 + r_R}\right]^2 + \Gamma_{Li}^2 = \left[\frac{1}{1 + r_R}\right]^2 \qquad \textbf{(6.22a)}$$

and

$$[\Gamma_{Lr} - 1]^2 + \left[\Gamma_{Li} - \frac{1}{x_R}\right]^2 = \left[\frac{1}{x_R}\right]^2 \qquad \textbf{(6.22b)}$$

These two equations are the equations of circles.

Equations (6.22*a*) represents the *resistance circle* or *r-circle* with

Centre at: $\left(\dfrac{r_R}{1 + r_R}, 0\right)$

Radius $= \dfrac{1}{1 + r_R}$

Equation (6.22*b*) represents the *reactance circle* or *x-circle* with

Center at: $\left(1, \dfrac{1}{x_R}\right)$

Radius $= \dfrac{1}{x_R}$

As the normalised resistance r_R varies from 0 to ∞, we obtain a family of circles completely contained inside the domain of the reflection coefficient $|\Gamma_L| \leq 1$. Similarly, as the normalised reactance x_R varies from –∞ to ∞, we obtain a family of circles completely contained inside the domain of the reflection coefficient $|\Gamma_L| \leq 1$. Figure 6.7 shows typical *r-circles* and *x-circles*.

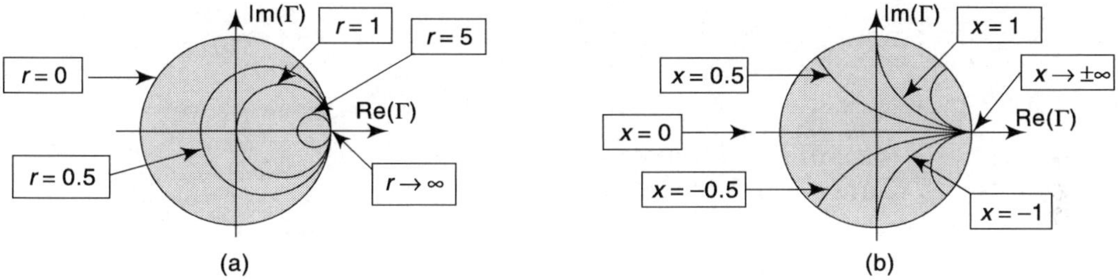

Fig. 6.7 *Typical r and x circles (a) r-circles (b) x-circles*

Applications of Smith Chart Smith chart can be used for the following purposes:

1. To find the normalised admittance from normalised impedance and vice-versa.
2. To find the parameters of mismatched transmission lines.
3. To find the *VSWR* for a given load impedance.
4. To find the reflection coefficient.
5. To find the input impedance of a transmission line.
6. To locate a voltage maximum on a transmission line.
7. To design stubs for impedance matching.

Determination of Normalised Impedance from Reflection Coefficient using Smith Chart The reflection coefficient as a function of position $\Gamma(z)$ along the transmission line can be related to the impedance as a function of position $Z(z)$. The general impedance at any point along the length of the transmission line is defined by the ratio of the phasor voltage to the phasor current.

$$V_S(z) = V_{S0}^{+} e^{-j\beta z}[1 + \Gamma(z)]$$

$$I_S(z) = \frac{V_{S0}^{+}}{Z_0} e^{-j\beta z}[1 - \Gamma(z)]$$

$$Z(z) = \frac{V_S(z)}{I_S(z)} = Z_0 \frac{[1 + \Gamma(z)]}{[1 - \Gamma(z)]}$$

The normalised value of the impedance $z_n(z)$ is

$$z_n(z) = \frac{Z(z)}{Z_0} = \frac{[1 + \Gamma(z)]}{[1 - \Gamma(z)]} = r(z) + jx(z) \quad \textbf{(6.23)}$$

Note that Eq. (6.16) is simply the above equation [Eq. (6.23)], evaluated at $z = l$.

Thus, as we move from point to point along the transmission line plotting the complex reflection coefficient (rotating around the constant SWR circle), we are also plotting the corresponding impedance.

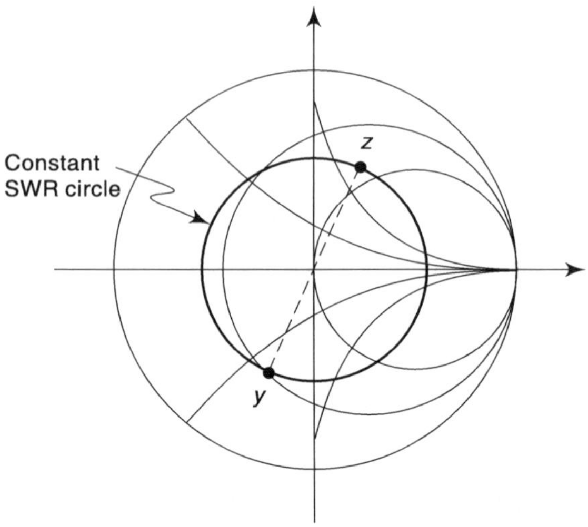

Fig. 6.8 *Determination of admittance from impedance in the Smith chart*

Determination of Admittance from Impedance using Smith Chart Once a normalised impedance is located on the Smith chart for a particular point on the transmission line, the normalised admittance at that point is found by rotating 180° from the impedance point on the constant reflection coefficient circle as shown in Fig. 6.8.

Determination of Maxima and Minima for Voltages and Currents using Smith Chart The locations of maxima and minima for voltages and currents along the transmission line can be located using the Smith chart given that these values correspond to specific impedance characteristics, e.g.,

Voltage maximum, Current minimum corresponds to Impedance maximum
Voltage minimum, Current maximum corresponds to Impedance minimum

These are shown in Fig. 6.9.

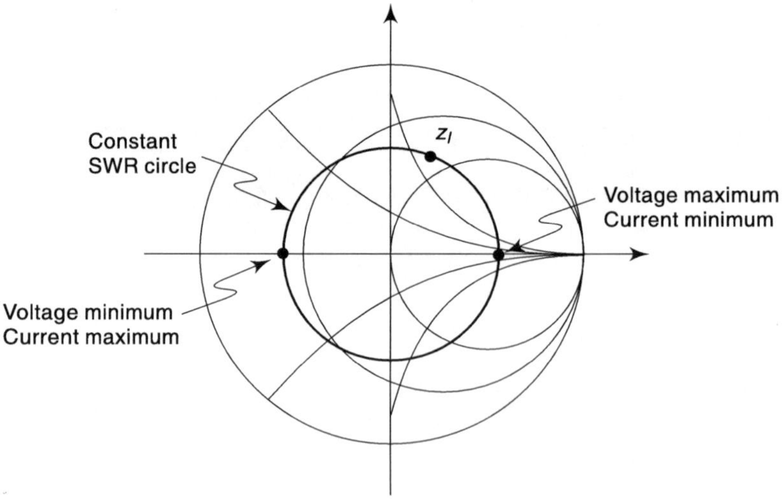

Fig. 6.9 *Determination of maxima and minima for voltages and currents using Smith chart*

The complete Smith chart is shown in Fig. 6.10.

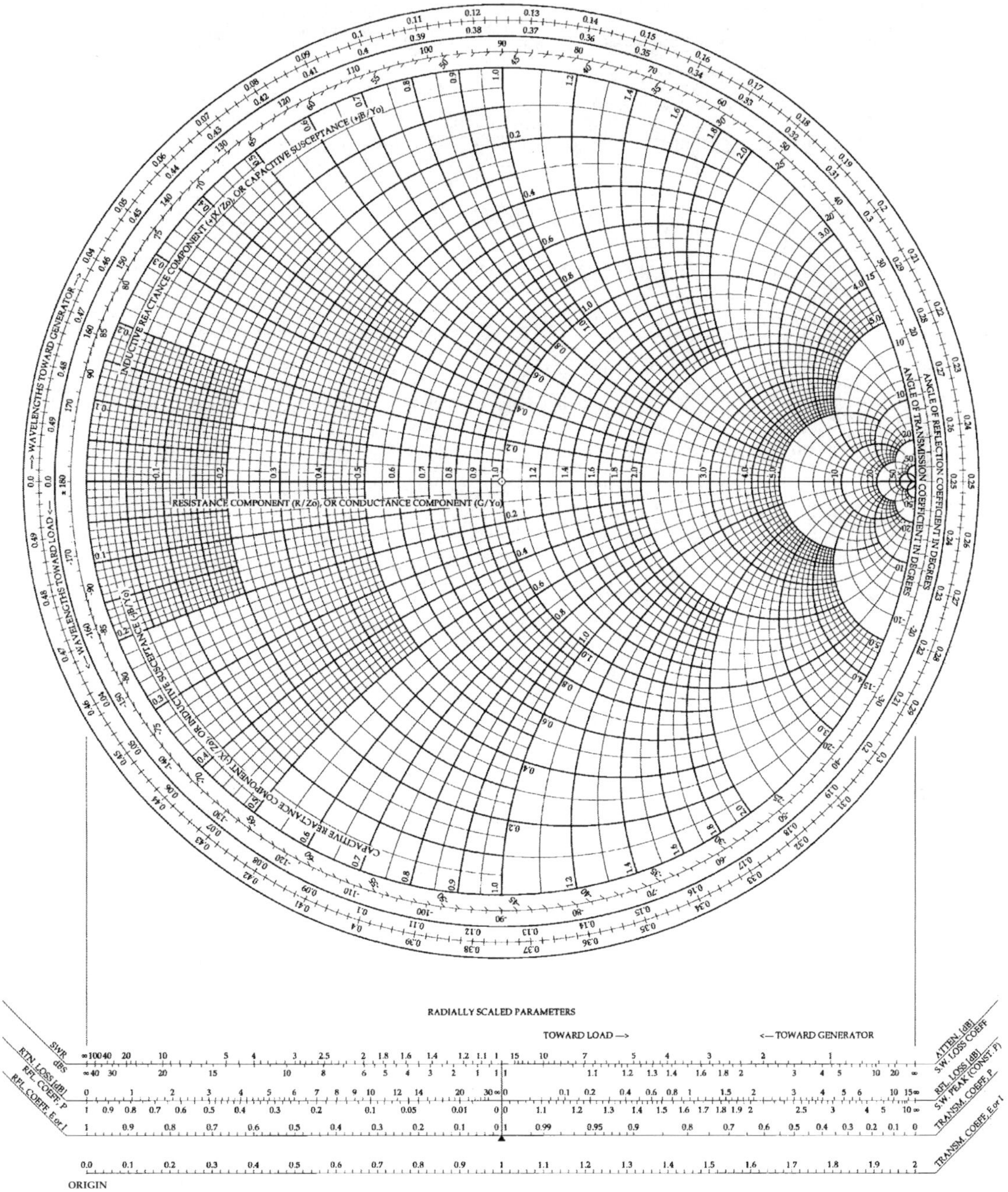

Fig. 6.10 *Complete Smith chart*

Example 6.18 Find the input impedance of a 75Ω lossless transmission line of length 0.1λ when the receiving end is (a) open-circuited, and (b) short-circuited. Use Smith chart. Compare the results with the theoretical results.

Solution Here, $Z_0 = 75\ \Omega$; $l = 0.1\lambda$

1. When receiving end is open-circuited: In this case, load impedance is, $z_L = \infty + j\infty$.

Smith Chart Method:

(a) We start at the point of $(\infty + j\infty)$ on Smith chart which is at the right of the rim of the chart. This corresponds to point P.

(b) We move clockwise from this point P through the perimeter of the chart by 0.1λ. This corresponds to point Q where we get

$$r = 0, \quad x = -1.38$$

Therefore, the normalised input impedance of the line is

$$z_{\text{in}} = (0 - j1.38)$$

(c) So, the actual input impedance is given as

$$Z_{\text{in}} = Z_0 \times z_{\text{in}} = 75 \times (0 - j1.38) = -j103.5\ \Omega$$

This is shown in Fig. 6.11(*a*).

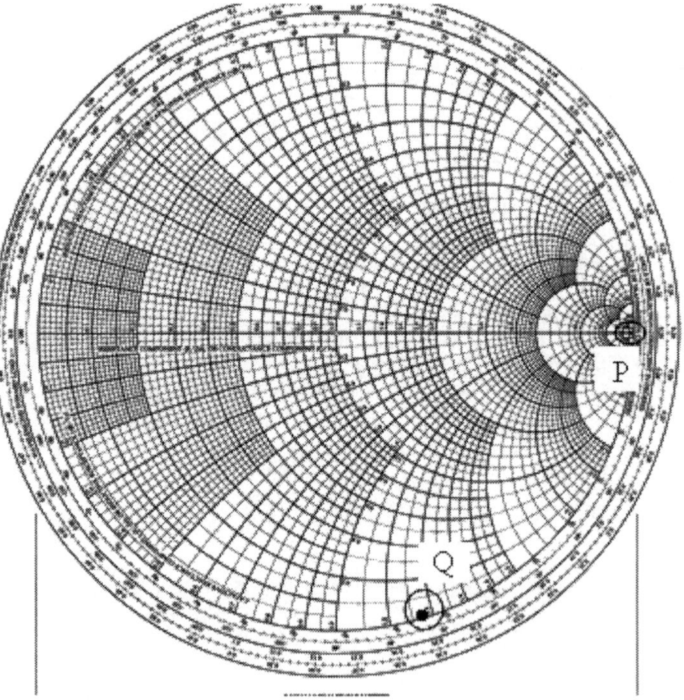

Fig. 6.11(a) *Smith chart to find input impedance of open-circuited line of Example 6.18*

Analytical Method:

For lossless open-circuited line, the input impedance is given as,

$$Z_{\text{in}} = Z_{OC} = -jR_0 \cot \beta l$$

For this line, the input impedance is

$$Z_{in} = -jR_0 \cot \beta l = -j75 \cot \beta l = -j75 \cot\left(\frac{2\pi}{\lambda} \times 0.1\lambda\right) = -j75 \cot (0.2\pi)$$

$$= -j75 \times 1.3764 = -j103.23\ \Omega$$

2. When the receiving end is short-circuited: In this case, load impedance is $z_L = 0 + j0$.

Smith Chart Method:

(a) We start at the point of $(0 + j0)$ on Smith chart, which is at the left of the rim of the chart. This corresponds to point P.

(b) We move clockwise from this point P through the perimeter of the chart by 0.1λ. This corresponds to point Q where we get

$$r = 0, \quad x = 0.73$$

Therefore, the normalised input impedance of the line is

$$z_{in} = (0 + j0.73)$$

(c) So, the actual input impedance is given as

$$Z_{in} = Z_0 \times z_{in} = 75 \times (0 + j0.73) = j54.75\ \Omega$$

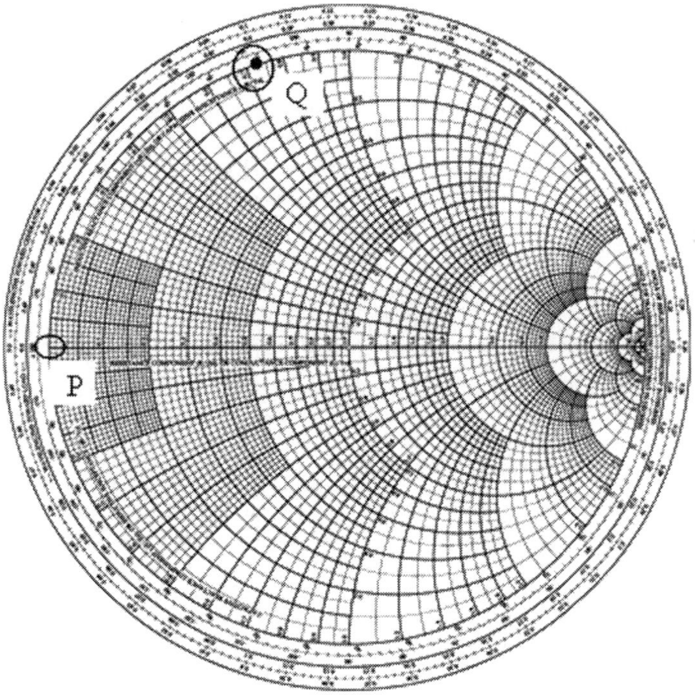

Fig. 6.11(b) *Smith chart to find input impedance of shorted line of Example 6.18.*

Analytical Method:

For lossless short-circuited line, the input impedance is given as

$$Z_{in} = Z_{SC} = jR_0 \tan \beta l$$

For this line, the input impedance is

$$Z_{\text{in}} = jR_0 \tan \beta l = j75 \tan \beta l = j75 \tan\left(\frac{2\pi}{\lambda} \times 0.1\lambda\right) = j75 \tan (0.2\pi)$$
$$= j75 \times 0.7265 = j54.5\ \Omega$$

***Example 6.19** A 75 Ω lossless transmission line of length 1.25λ is terminated by a load impedance of 120 Ω. The line is energised by a source of 100 V (rms) with an internal impedance of 50 Ω. Determine (a) the input impedance of the transmission line and (b) the magnitude of the load voltage. Use Smith chart.

Solution Here, $Z_0 = 75\ \Omega$; $l = 1.25\lambda$, $Z_L = R_L = 120\ \Omega$

In order to find the input impedance, the following steps are involved.

1. We find the normalised load impedance as

$$z_L = \frac{Z_L}{Z_0} = \frac{120}{75} = (1.6 + j0)\ \Omega$$

2. We start from the point corresponding to $(1.6 + j0)$ (i.e., point P) and move 1.25λ (2.5 revolutions) clockwise (towards generator) on the s-circle (i.e., point Q) to find the normalised input impedance as,

$$z_{\text{in}} = 0.625$$

3. So, the actual input impedance is given as

$$Z_{\text{in}} = Z_0 \times z_{\text{in}} = 75 \times 0.625 = 46.9\ \Omega$$

∴ $$Z_{\text{in}} = (46.9 + j0)\ \Omega$$

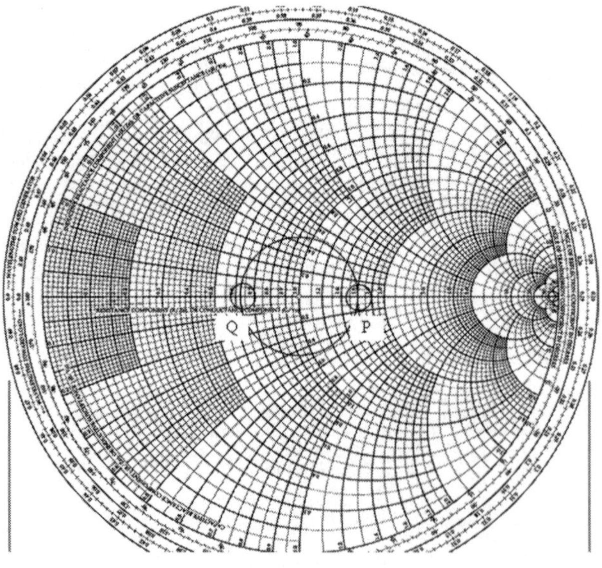

Fig. 6.12 *Smith chart to find the input impedance of Example 6.19*

In order to find the load voltage, we have the equivalent circuit as shown in Fig. 6.13.

$\therefore$ $$V_{s0} = 100\angle 0° \frac{46.9}{50 + 46.9} = 48.4\angle 0° \text{ (V)}$$

Also, $|V_{sL}| = s|V_s 0|$

Reflection coefficient,

$$\Gamma_L = \frac{Z_L - Z_0}{Z_L + Z_0} = \frac{120 - 75}{120 + 75} = 0.231$$

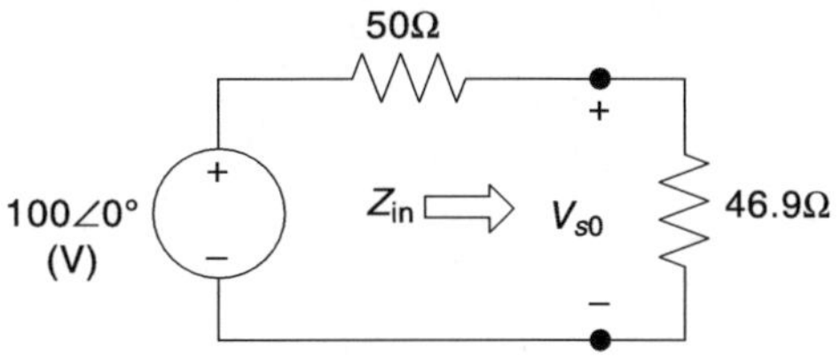

Fig. 6.13 *Equivalent circuit of Example 6.19*

Standing wave ratio, $s = \frac{1 + |\Gamma_L|}{1 - |\Gamma_L|} = \frac{1 + 0.231}{1 - 0.231} = 1.6$

$\therefore$ $$|V_{sL}| = s\,|V_{s0}| = 1.6 \times 48.4 = 77.4 \text{ (V)}$$

Example 6.20 Determine the input impedance of the transmission line of electrical length 28° with terminated load of $\frac{Z_L}{R_0} = 2.6 + j1$. Use Smith chart.

Solution Here, $\frac{Z_L}{R_0} = (2.6 + j1),\ l = \frac{28}{360}\lambda = 0.078\lambda$

In order to find the input impedance, the following steps are involved.

1. We find the normalised load impedance as

$$z_L = \frac{Z_L}{R_0} = (2.6 + j1)$$

2. We start from the point corresponding to $(2.6 + j1)$ (i.e. point P) and extend the line so that it cuts the rim of the Smith chart at point A.

3. We move 0.078λ clockwise (towards generator) on the rim to locate the point B.

4. Then we draw a line from the centre of the Smith chart (point O) to point B. This line intersects the constant-s circle at point Q.

5. The point B gives $r = 1.58$ and $x = -1.3$, so that, the normalised input impedance is given as

$$z_{in} = (1.5 - j1.3)$$

The Smith chart to find input impedance is given in Fig. 6.14.

***Example 6.21** A transmission line has standing wave ratio $s = 2.5$ and voltage minima exists at $0.15\ \lambda$ from the load. Find the load and input impedance for a line of 0.35λ length. Use Smith chart.

Solution Here, $s = 2.5$, V_{min} at 0.15λ from the load, length of the line, $l = 0.35\lambda$.

In order to find the load impedance, the following steps are involved.

1. We draw a constant-s circle with centre at the centre of the Smith chart (point O) that cuts the real axis of the Smith chart at point 2.5 (point P).

2. From the property of Smith chart, the voltage minimum occurs at a point on the real axis where the constant-s circle cuts the real axis (point Q). This point corresponding to the point of voltage minimum is at 0.4 $\left(\text{i.e. } \frac{1}{s} = \frac{1}{2.5} = 0.4\right)$. We extend this line so that it cuts the rim of the Smith chart at point A.

3. To locate the load point, we move in the lower half of the Smith chart in the anti-clockwise direction by a distance of 0.15λ. This point is point B. We join this point with the centre O. This line cuts the constant-s circle at point L which corresponds to $r = 0.88$, $x = j0.9$. This gives the load impedance as

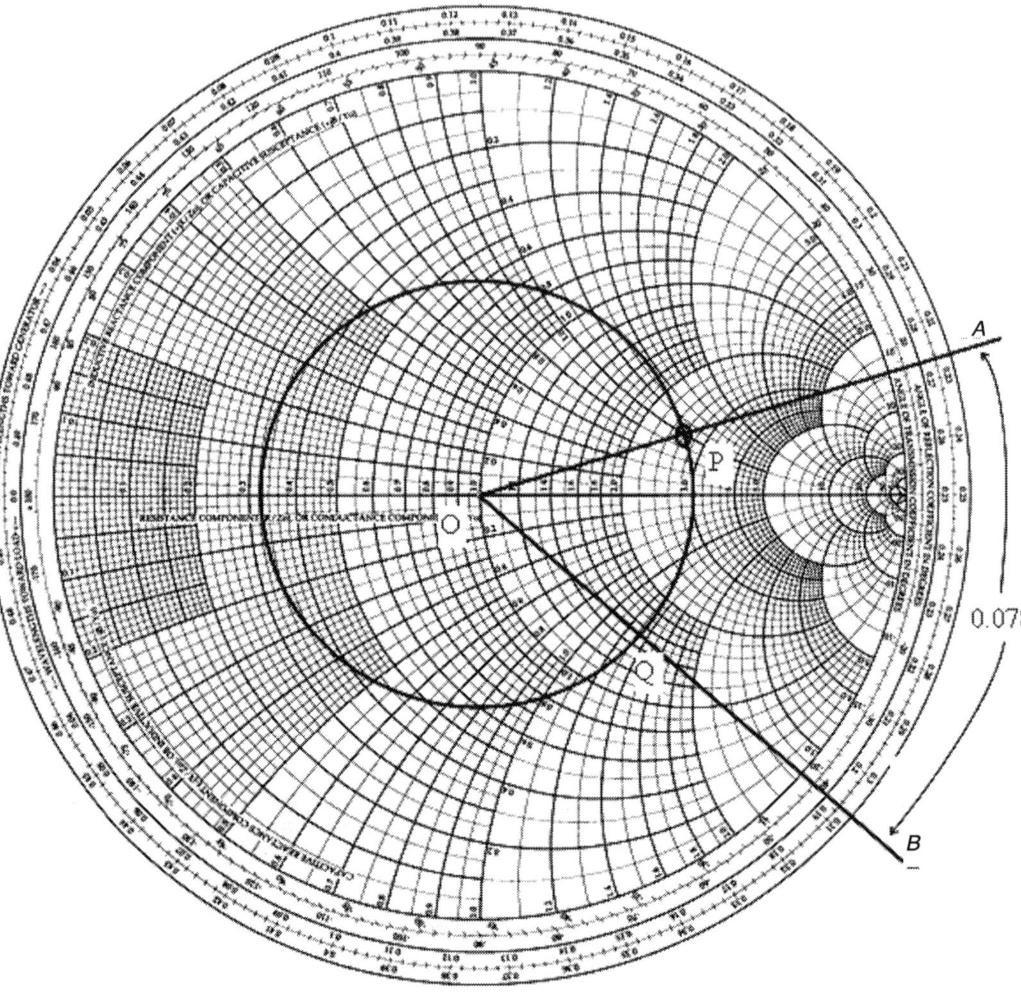

Fig. 6.14 *Smith chart to find the input impedance of Example 6.20*

$$z_L = \frac{Z_L}{Z_0} = (0.88 + j0.9)$$

In order to find the input impedance, we follow the following steps after step 2:

4. To locate the source point, we move from point *B* in the clockwise direction in the upper half of the Smith chart by a distance of 0.35λ. This point is point *C*. We join this point with the centre *O*. This line cuts the constant-*s* circle at point *S* which corresponds to $r = 1.7, x = j1$. This gives the input impedance as,

$$z_S = \frac{Z_S}{Z_0} = (1.7 + j1)$$

The Smith chart to find load and input impedance is given in Fig. 6.15.

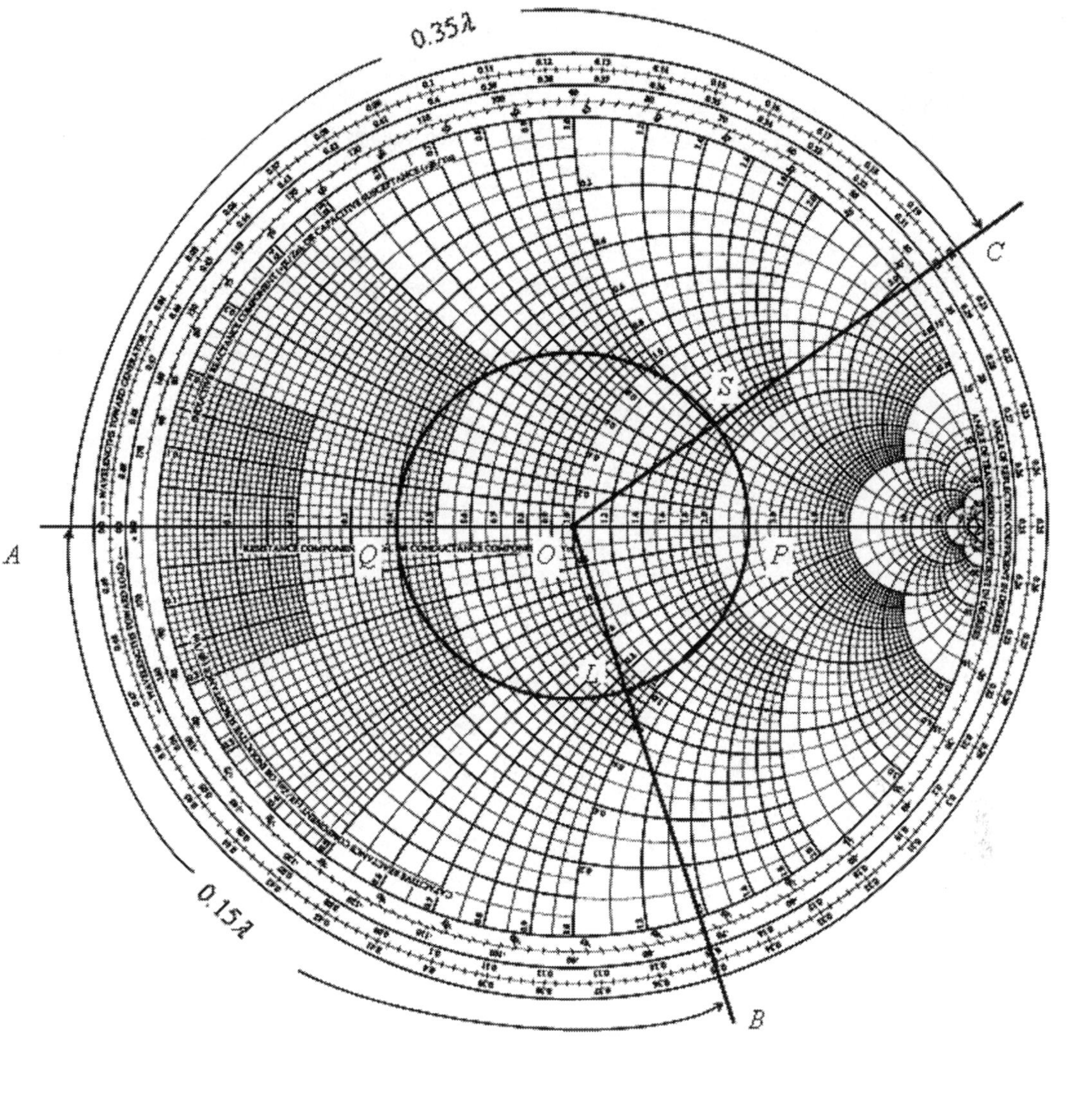

Fig. 6.15 *Smith chart to find the load and input impedance of Example 6.21*

***Example 6.22** A transmission line of length 0.4λ has a characteristic impedance of 100 Ω and is terminated in a load impedance of $(200 + j180)$ Ω. Find:

1. reflection coefficient,
2. standing wave ratio, and
3. input impedance of the line.

Use Smith chart. Compare the results with the theoretical results.

Solution Here, $Z_0 = 100\ \Omega$; $l = 0.4\lambda$, $Z_L = (200 + j180)\ \Omega$

Normalised impedance is, $z_L = \dfrac{(200 + j180)}{100} = (2 + j1.8)\ \Omega$

(1) To find reflection coefficient:

(a) We locate the point P in the Smith chart, which is the intersection of the circles $r = 2$ and $x = 1.8$.

(b) We draw a circle with radius equal to OP and centre at O. This OP is the magnitude of the reflection coefficient which is measured to be 0.591.

$$\therefore \qquad |\Gamma_L| = 0.591$$

(c) To find the phase angle of the reflection coefficient, we extend the line OP so that it cuts the $r = 0$ circle at point Q. This point is measured to be 0.207 in wavelength towards the generator. This point gives the phase angle of the reflection coefficient as,

$$\therefore \qquad \theta_\Gamma = (0.250 - 0.207) \times 720° = 31°$$

$$\therefore \qquad \Gamma_L = 0.591\angle 31°$$

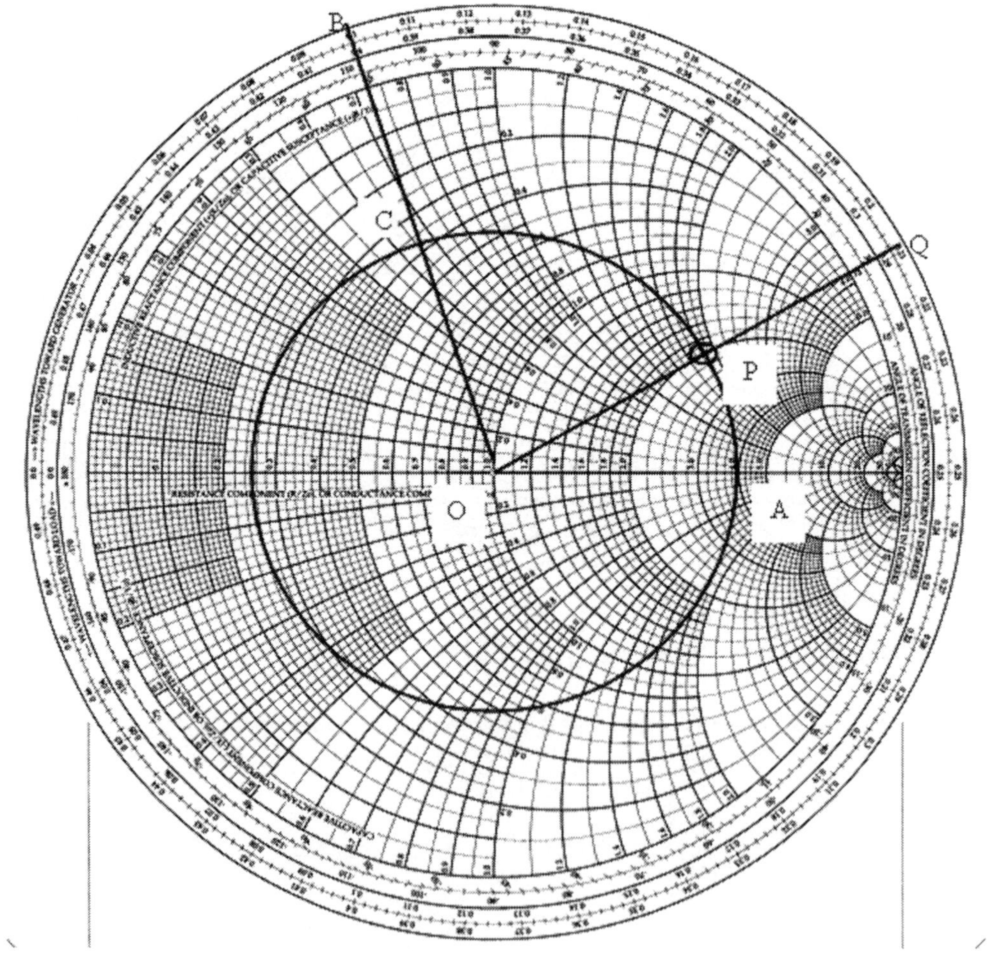

Fig. 6.16 *Use of Smith chart for Example 6.22*

(2) To find SWR:

The circle cuts the centre line of the chart at point A; OA gives the standing wave ratio. It is measured to be 4.

$$\therefore \qquad s = 4$$

(3) To find input impedance:

In order to find the input impedance, the following steps are involved.

(a) We start from the point corresponding to $(2 + j1.8)$ (i.e., point P) and move 0.4 wavelengths clockwise (towards generator) on the s-circle (i.e., point B). This line OB cuts the circle at point C. This OC gives the input impedance as

$$z_{\text{in}} = (0.4 + j0.72)$$

(b) So, the actual input impedance is given as

$$Z_{\text{in}} = Z_0 \times z_{\text{in}} = 100 \times (0.4 + j0.72) = (40 + j72)\ \Omega$$

This use of Smith Chart is shown in Fig. 6.16.

Example 6.23 A transmission line has a characteristic impedance of 300 Ω and is terminated in a load $Z_L = 150 + j150\ \Omega$. Find the following using Smith chart.

1. VSWR,
2. reflection coefficient,
3. input impedance at distance 0.1λ from the load,
4. input admittance from 0.1λ from load,
5. position of first voltage minimum and maximum from the load.

Solution Here, $Z_L = 150 + j150\ \Omega$, $Z_0 = 300\ \Omega$

Normalised load impedance is, $z_L = \dfrac{Z_L}{Z_0} = \dfrac{150 + j150}{300} = 0.5 + j0.5$

The load point is located at the intersection of $r = 0.5$ and $x = 0.5$ circles. This point is point A.

(1) To find VSWR:

We draw a circle with centre at origin (point O) and radius equal to OA. This circle cuts the real axis at 2.6. So, the VSWR is given as

$$s = 2.6$$

(2) To find reflection coefficient:

(a) This magnitude of OA as measured in the scale shown at the bottom of Smith chart is the magnitude of the reflection coefficient, which is measured to be 0.42.

$$\therefore \qquad |\Gamma_L| = 0.42$$

(b) To find the phase angle of the reflection coefficient, we extend the line OA so that it cuts the $r = 0$ circle at point P. This point is measured to be 0.088 in wavelength towards the generator. This point gives the phase angle of the reflection coefficient as

$$\therefore \qquad \theta_\Gamma = (0.250 - 0.088) \times 720° = 116.64°$$

$$\therefore \qquad \Gamma_L = 0.42\angle 116.64°$$

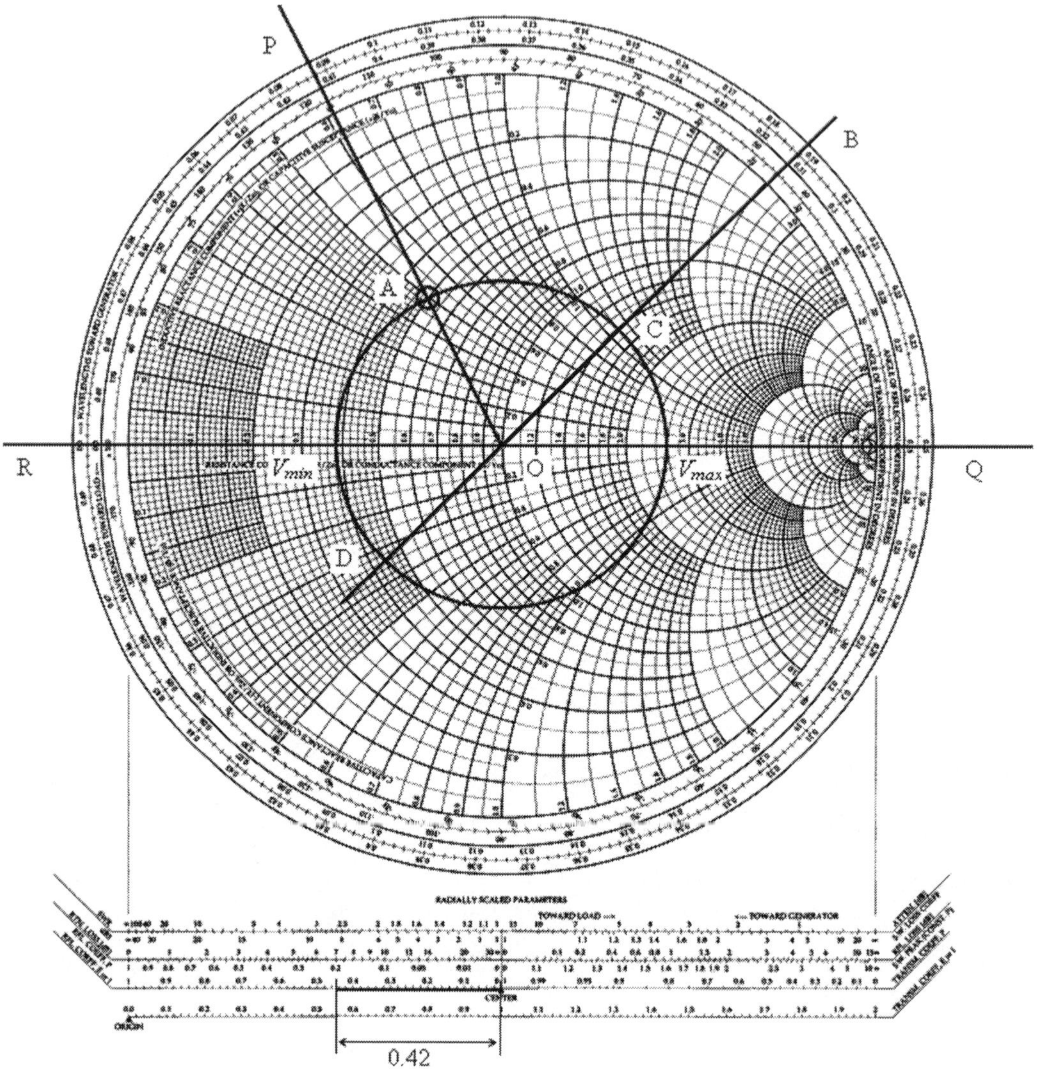

Fig. 6.17 *Use of Smith Chart for Example 6.23*

(3) To find input impedance:

(a) We start from the point A and move 0.1 wavelengths clockwise (towards generator) on the s-circle (i.e., point B). This line OB cuts the circle at point C which gives the input impedance as

$$z_{in} = (1.4 + j1.1)$$

(b) So, the actual input impedance is given as

$$Z_{in} = Z_0 \times z_{in} = 300 \times (1.4 + j1.1) = (420 + j330)\ \Omega$$

$$\therefore \quad Z_{in} = (420 + j330)\ \Omega$$

(4) To find input admittance:

(a) We extend the line OC so that it cuts the constant-s circle at point D. This point corresponds to $g = 0.44$ and $b = -0.34$ circles. So, the input admittance is given as

$$y_{in} = 0.44 - j0.34$$

(b) So, the actual input admittance is given as

$$Y_{in} = Y_0 \times y_{in} = \frac{(0.44 - j0.34)}{300} = (1.47 - j1.13) \times 10^{-3}$$

$$\therefore \quad Y_{in} = (1.47 - j1.13) \times 10^{-3} \text{ mho}$$

(5) To find position of first voltage minimum and maximum from the load:

The voltage minimum occurs to the left of the real axis at 0.39 while the voltage maximum occurs to the right of the real axis at $s = 2.6$.

(a) To locate voltage maximum from load:
We move through the rim of the chart from load point A to voltage maximum point (point Q). The distance between these two points is $(0.25 - 0.088)\lambda = 0.162\lambda$.
So, the first voltage maximum occurs at a distance 0.162λ from the load.

(b) To locate voltage minimum from load:
We move through the rim of the chart from load point A to voltage minimum point (point R). The distance between these two points is $(0.25 + 0.162)\lambda = 0.412\lambda$.
So, the first voltage minimum occurs at a distance 0.412λ. from the load. The Smith chart for this use is given in Fig. 6.17.

6.13 LOAD MATCHING TECHNIQUES IN A TRANSMISSION LINE

We discuss two techniques used for load matching.

6.13.1 Quarter Wave Transformer

A transmission line is said to be *perfectly matched* if the load impedance is exactly equal to the characteristic impedance of the line. In a matched transmission line, the power is transferred outward from the source until it reaches the load, where it is completely absorbed. Thus, the reflection coefficient for a matched line is zero ($|\Gamma| = 0$). However, if the characteristic impedance of the line does not match with the load impedance, a mismatch occurs. *Quarter wave transformer* is used as a matching technique to match the impedances between the transmission line and the load in order to eliminate the reflections on the feeder transmission line.

For matching, a quarter wavelength $\left(\frac{\lambda}{4}\right)$ section of different transmission line (characteristic impedance $= Z_0'$) is inserted between the original transmission line and the load. This quarter wave long section of the transmission line is called a *quarter wave transformer* as it is used for impedance matching like an ordinary transformer. This is demonstrated in Fig. 6.18(a) and Fig. 5.18(b).

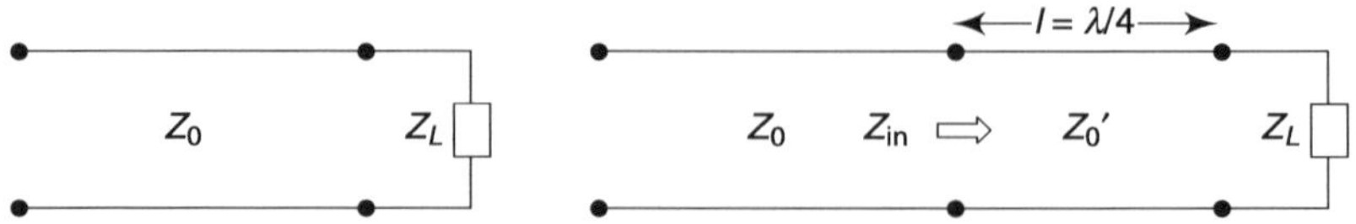

Fig. 6.18 *(a) If $Z_0 \neq Z_L$: mismatched line (b) Matching by quarter wave transformer*

The input impedance looking into the quarter wave transformer is given as

$$Z_{\text{in}}(z) = Z_0' \frac{Z_L + jZ_0' \tan \beta l}{Z_0' + jZ_L \tan \beta l}$$

Now,
$$\beta l = \frac{2\pi}{\lambda} \frac{\lambda}{4} = \frac{\pi}{2} \qquad \therefore \quad \tan \beta l = \infty$$

$\therefore$
$$Z_{\text{in}}(z) = Z_0' \frac{Z_0'}{Z_L} = \frac{(Z_0')^2}{Z_L}.$$

For the matched line, we require

$$Z_{\text{in}}(z) = Z_0 = \frac{(Z_0')^2}{Z_L} \quad \Rightarrow \quad Z_0' = \sqrt{Z_0 Z_L}$$

Hence, the required characteristic impedance of the quarter wave transformer is given as

$$\boxed{Z_0' = \sqrt{Z_0 Z_L}}$$

Example 6.24 Design a quarter wave transformer to match a load of 200 Ω to a source of 500 Ω. Operating frequency is 200 MHz.

Solution Here, $Z_L = 200\ \Omega,\ Z_{\text{in}} = 500\ \Omega$

For a quarter wave transformer, the input or source impedance is given as

$$Z_{\text{in}} = \frac{(Z_0')^2}{Z_L} \quad \Rightarrow \quad Z_0' = \sqrt{Z_{\text{in}} Z_L} = \sqrt{500 \times 200} = 316.22\ \Omega$$

So, the characteristic impedance of the line is 316.22 Ω.
Also, length of the quarter wave transformer is

$$l = \frac{\lambda}{4} = \frac{v}{4f} = \frac{3 \times 10^8}{4 \times 200 \times 10^6} = 0.375\ \text{m}$$

Example 6.25 Determine the length and impedance of a quarter wave transformer that will match a load of 150 Ω to a line of 75 Ω at a frequency of 12 GHz.

Solution Here, $Z_L = 150\ \Omega, \quad Z_0' = 75\ \Omega$

For a quarter wave transformer, the input or source impedance is given as

$$Z_{\text{in}} = \frac{(75)^2}{150} = 37.5\ \Omega$$

So, the input impedance of the quarter wave transformer is 37.5 Ω.

Also, length of the quarter wave transformer is

$$l = \frac{\lambda}{4} = \frac{v}{4f} = \frac{3 \times 10^8}{4 \times 12 \times 10^9} = 0.0625 \text{ m} = 6.25 \text{ cm}$$

6.13.2 Stub Tuner

The limitation of a quarter-wave transformer is that it can perform matching only a resistive load R_L to a transmission line of characteristic impedance Z_o when $R_L \neq Z_0$; but it cannot match a complex load impedance.

The *stub tuner* is used as a transmission line matching technique to match a complex load. If a point can be located on the transmission line where the real part of the input admittance is equal to the characteristic admittance $\left(Y_0 = \frac{1}{Z_0}\right)$ of the line ($Y_{in} = Y_0 \pm jB$), the susceptance B can be eliminated by adding the proper reactive component in parallel at this point. Theoretically, we could add inductors or capacitors (lumped elements) in parallel with the transmission line. However, these lumped elements usually are too lossy at the radio frequency range.

Rather than using lumped elements, we can use a short-circuited or open-circuited segment of transmission line to achieve any required reactance. Since parallel components are generally used, the use of admittances (as opposed to impedances) simplifies the mathematics.

Single Stub Tuner A single stub tuner is an open or shorted section of transmission line of length l connected in parallel at some distance d from the load (Fig. 6.19).

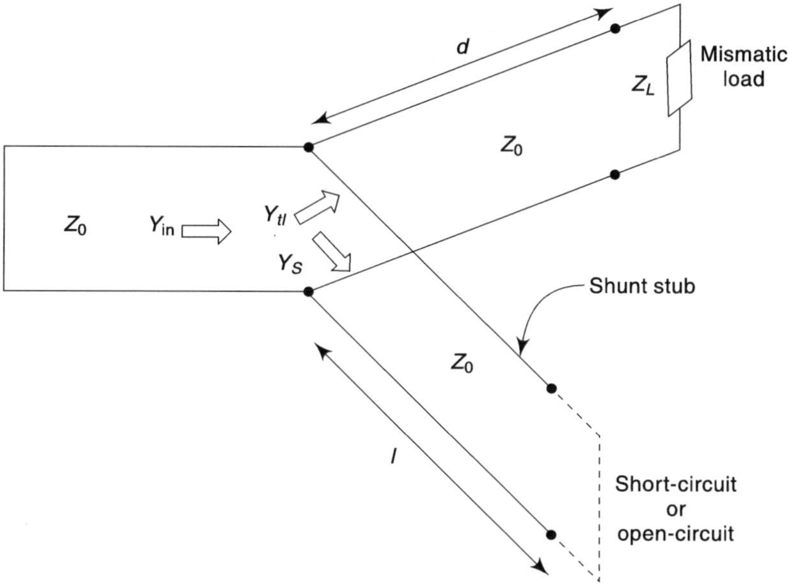

Fig. 6.19 *Single stub tuner*

In the figure,

l—length of the shunt stub,

d—distance of the stub connection from the load,
Y_S—input admittance of the stub,
Y_{tl}—input admittance of the terminated transmission line segment of length d,
Y_{in}—input admittance of the stub in parallel with the transmission line segment,

The admittance of the terminated transmission line section is

$$Y_{tl} = Y_0 + jB$$

The admittance of the stub (short-circuit or open-circuit) is

$$Y_S = -jB$$

Hence, the admittance of the short-circuited stub is

$$Y_{SC} = \frac{1}{Z_{SC}} = \frac{1}{jZ_0 \tan \beta l} = -jY_0 \cot \beta l$$

Hence, the admittance of the open-circuited stub is

$$Y_{OC} = \frac{1}{Z_{OC}} = \frac{1}{-jZ_0 \cot \beta l} = jY_0 \tan \beta l$$

Hence, overall input admittance is

$$Y_{in} = Y_{tl} + Y_S = Y_0$$

where, $Y_0 = \frac{1}{Z_0}$ and Z_0 is the characteristic impedance of the transmission line.

All admittances can be represented in normalised forms, by dividing by Y_0, so that we get,

$$y_{tl} = 1 + jb \qquad y_S = -jb \qquad y_{in} = 1$$

This is seen that the normalised conductance of the transmission line segment admittance is unity ($y_{tl} = g + jb$ and $g = 1$).

Short-circuited stub tuners are most commonly used because a shorted segment of transmission line radiates less than an open-circuited section. The stub tuner matching technique also works for tuners in series with the transmission line. However, series tuners are more difficult to connect since the transmission line conductors must be physically separated in order to make the series connection.

Design of Single Stub using Smith Chart The main objective for designing a single stub is to find out the distance of the stub from the length and the length of the stub. The design process of a single stub tuner consists of the following steps:

1. Given the load impedance Z_L and the characteristic impedance Z_0 of the transmission line, we find the normalised load impedance as,

$$z_L = \frac{Z_L}{Z_0} = r_L + jx_L$$

2. The point of intersection of the r_L-circle and x_L-circle is marked on the Smith chart.
3. A circle is drawn with the radius equal to the distance of this point from the centre of the chart.
4. Since the stub is connected in parallel with the main line, it is more convenient to deal with admittances. In order to locate the normalised admittance, a straight line is drawn from the point of intersection (obtained in step 2) through the centre of the chart to the other half of the chart. The point where this straight line intersects the circle drawn given the normalised admittance ($y_L = g + jb$).

5. The point where the circle drawn cuts the centre straight line of the chart to the right side of the centre gives the value of VSWR.
6. The points where the drawn circle intersects the $r = 1$ circle give the normalised admittance $(1 \pm jb)$. The stub to be designed for $\pm jb$ component will be placed at these points.
7. The distance travelled from the points $(y_L = g + jb)$ to $(1 \pm jb)$ round the circumference of the chart gives the distance of the stub from the load.
8. In order to find the length of the stub, we move clockwise round the perimeter of the chart and find the point at which the susceptance tunes out the $\pm jb$ susceptance of the line. For example, if the line admittance is $(1 + jb)$, the required susceptance is $-jb$. The distance in wavelengths from $(\infty, j\infty)$ of the chart to the new point (susceptance $-jb$) gives the length of the stub.

***Example 6.26** Design a stub to match a transmission line with a load impedance of $Z_L = (450 - j600)\ \Omega$. The characteristic impedance of the line is 300 Ω.

Solution Here, $Z_L = (450 - j600)\ \Omega$, $Z_0 = 300\ \Omega$

$\therefore$ Normalised load impedance is, $z_L = \dfrac{Z_L}{Z_0} = \left(\dfrac{450 - j600}{300}\right) = (1.5 - j2.0)$

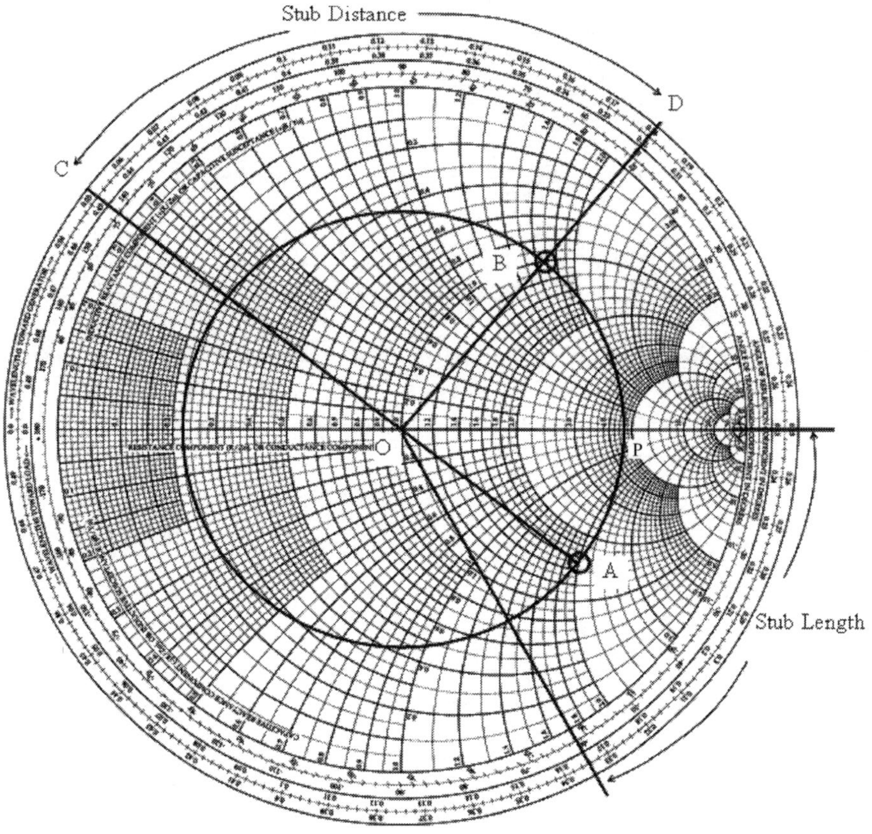

Fig. 6.20 *Use of Smith chart for Example 6.26*

Steps of Design:

1. The point of intersection of (r = 1.5)-circle and (x = –2.0)-circle is located on the Smith chart. This point is point A.
2. With the centre as the centre of Smith chart, point O, a circle is drawn with radius equal to OA.
3. The drawn circle cuts the r = 1 circle at point B which is measured to be (1 = j1.7).
4. The distance between points C and D on the rim of the chart gives the distance of the stub from the load. This is measured to be

$$CD = (0.181 - 0.053)\lambda = 0.128\lambda$$

5. As the load has a susceptance of +j1.7, the stub is required to provide a susceptance of –j1.7. Therefore, a point is marked by moving clockwise on the lower half of the chart where x = –1.7. This point is point E. The distance of this point from the short-circuit admittance point is the length of the stub. This is measured as
   Stub length = (0.3342 – 0.25) λ = 0.0842λ
   Hence, the specifications of the designed stub are:

$$\text{Stub Distance} = 0.128\lambda$$
$$\text{Stub Length} = 0.0842\lambda$$

   This can be seen in Fig. 6.20.

Example 6.27 Design a single stub match for a load of 150 + j255 ohm for a 75 ohm line at 500 MHz using Smith chart.

Solution Here, $Z_L = (150 + j255)\ \Omega$, $Z_0 = 75\ \Omega$

$$\therefore \text{ Normalised load impedance is, } z_L = \frac{Z_L}{Z_0} = \left(\frac{150 + j255}{75}\right) = (2 + j3.4)$$

Steps of Design:

1. The point of intersection of (r = 2)-circle and (x = 3.4)-circle is located on the Smith chart. This point is point A.
2. With the centre as the centre of Smith chart, point O, a circle is drawn with radius equal to OA.
3. The drawn circle cuts the r = 1 circle at point B, which is measured to be (1 + j2.5).
4. The distance between points C and D on the rim of the chart gives the distance of the stub from the load. This is measured to be

$$CD = 0.196 + (0.5 - 0.465)\lambda = 0.229\lambda$$

5. As the load has a susceptance of +j2.5, the stub is required to provide a susceptance of –j2.5. Therefore, a point is marked by moving clockwise on the lower half of the chart where x = –2.5. This point is point E. The distance of this point from the short-circuit admittance point is the length of the stub. This is measured as

$$\text{Stub length} = (0.31 - 0.25)\lambda = 0.06\lambda$$

   Hence, the specifications of the designed stub are:

$$\text{Stub Distance} = 0.229\lambda$$
$$\text{Stub Length} = 0.06\lambda$$

   Since the frequency is 500 MHz, the wavelength is, $\lambda = \dfrac{v}{f} = \dfrac{3 \times 10^8}{500 \times 10^6} = 0.6\text{ m}$

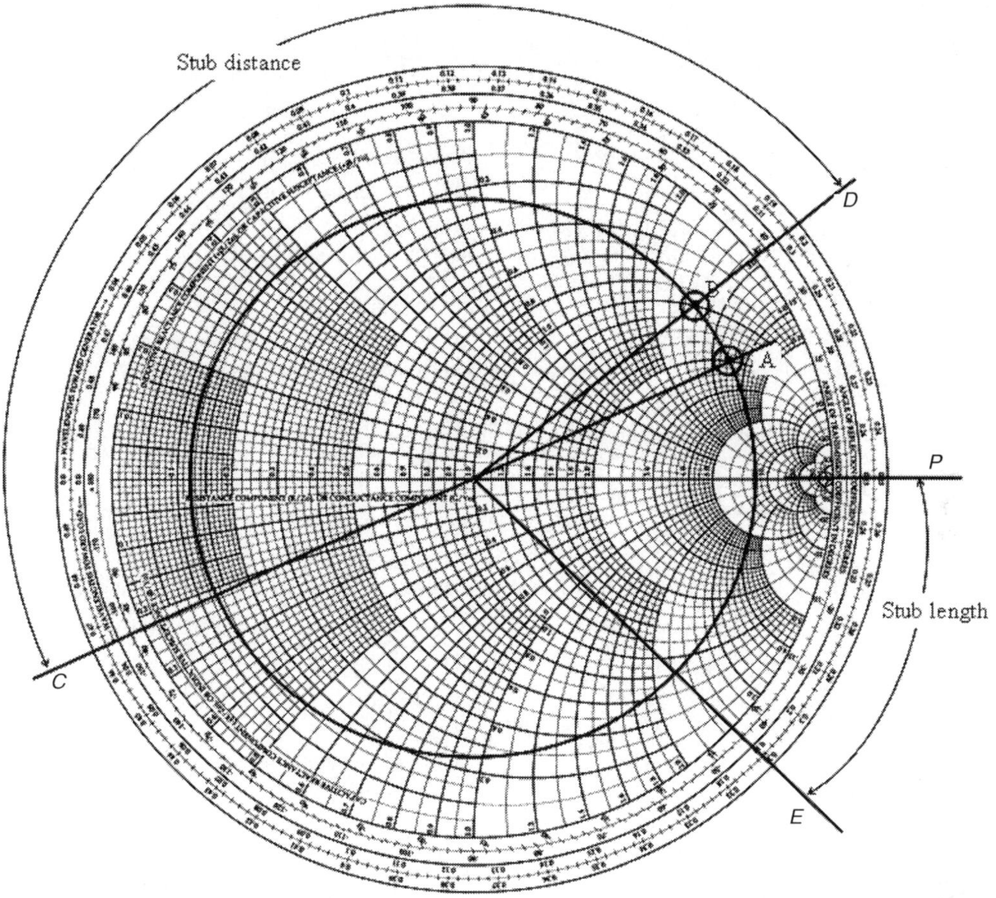

Fig. 6.21 *Use of Smith chart for Example 6.27*

Hence, the specifications of the designed stub are:
Stub Distance = $0.229\lambda = 0.229 \times 0.6 = 0.1374$ m = 13.74 cm
Stub Length = $0.06\lambda = 0.06 \times 0.6 = 0.036$ m = 3.6 cm
This is shown in Fig. 6.21.

***Example 6.28** A 75 Ω lossless line is to be matched to a $(100 - j80)$ Ω load with a shorted stub. Calculate the stub length, its distance from and the necessary stub admittance.

Solution Here, $Z_L = (100 - j80)\ \Omega$, $Z_0 = 75\ \Omega$

$\therefore$ Normalised load impedance is, $z_L = \dfrac{Z_L}{Z_0} = \left(\dfrac{100 - j80}{75}\right) = (1.33 - j1.067)$

Steps of Design:

1. The point of intersection of ($r = 1.33$)-circle and ($x = -1.067$)-circle is located on the Smith chart. This point is point A.

2. With the centre as the centre of Smith chart, point O, a circle is drawn with radius equal to OA.
3. The drawn circle cuts the $r = 1$ circle at point B, which is measured to be $(1 + j0.95)$.
4. The distance between points C and D on the rim of the chart gives the distance of the stub from the load. This is measured to be

$$CD = (0.161 - 0.071)\lambda = 0.09\lambda$$

5. As the load has a susceptance of $+j0.95$, the stub is required to provide a susceptance of $-j0.95$. Therefore, a point is marked by moving clockwise on the lower half of the chart where $x = -0.95$. This point is point E. The distance of this point from the short-circuit admittance point is the length of the stub. This is measured as

$$\text{Stub length} = (0.38 - 0.25)\lambda = 0.13\lambda$$

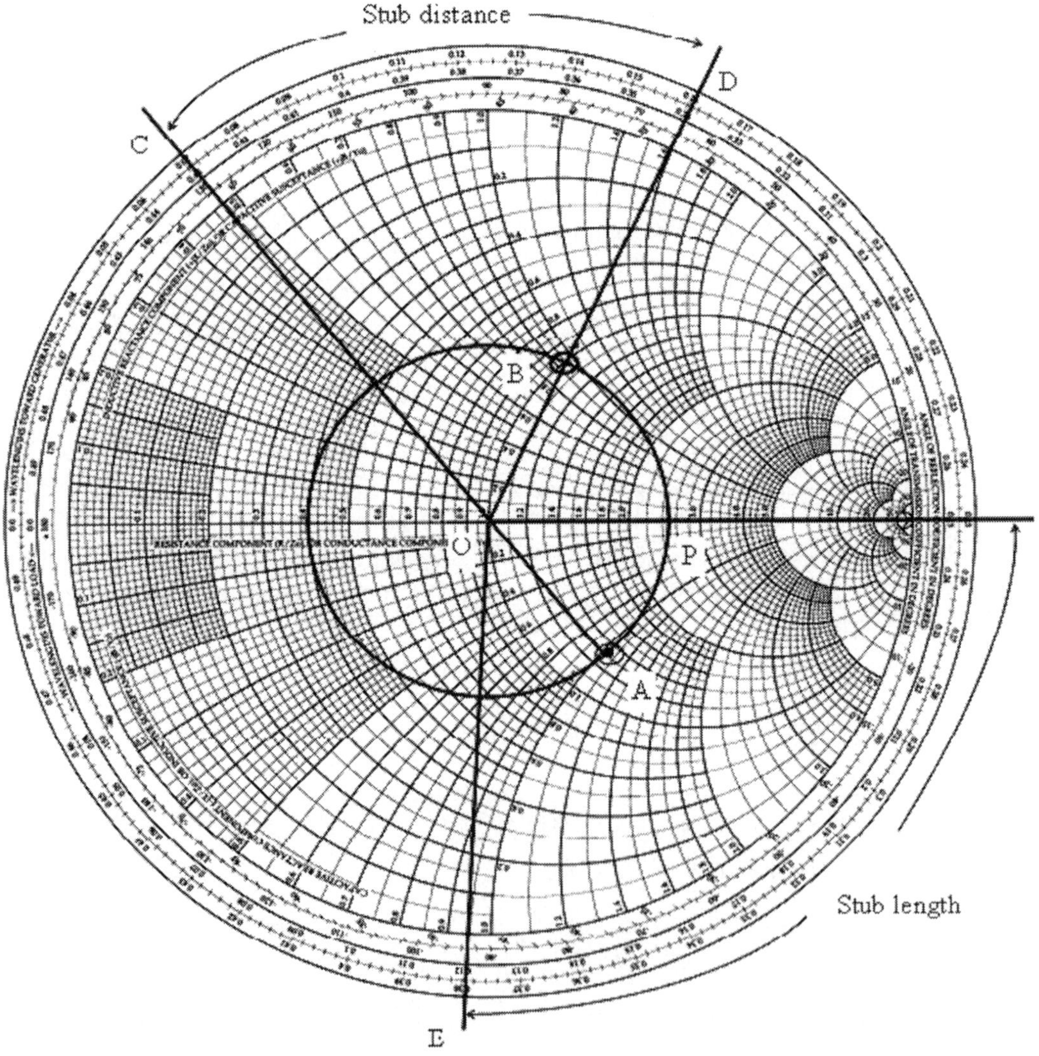

Fig. 6.22 *Use of Smith chart for Example 6.28*

Hence, the specifications of the designed stub are:

Stub Distance = 0.09λ

Stub Length = 0.13λ

This can be seen in Fig. 6.22.

***Example 6.29** An R.F. transmission line with a characteristic impedance of $300\angle 0°$ Ω is terminated in an impedance $100\angle -45°$ Ω. The load is to be matched to the transmission line by using a short-circuited stub. With the help of Smith chart, determine the length of the stub and the distance from the load.

Solution Here, $Z_L = 100\angle -45° = (70.71 - j70.71)\ \Omega$, $Z_0 = 300\ \Omega$

$\therefore$ Normalised load impedance is, $z_L = \dfrac{Z_L}{Z_0} = \left(\dfrac{70.71 - j70.71}{300}\right) = (0.2357 - j2357)$

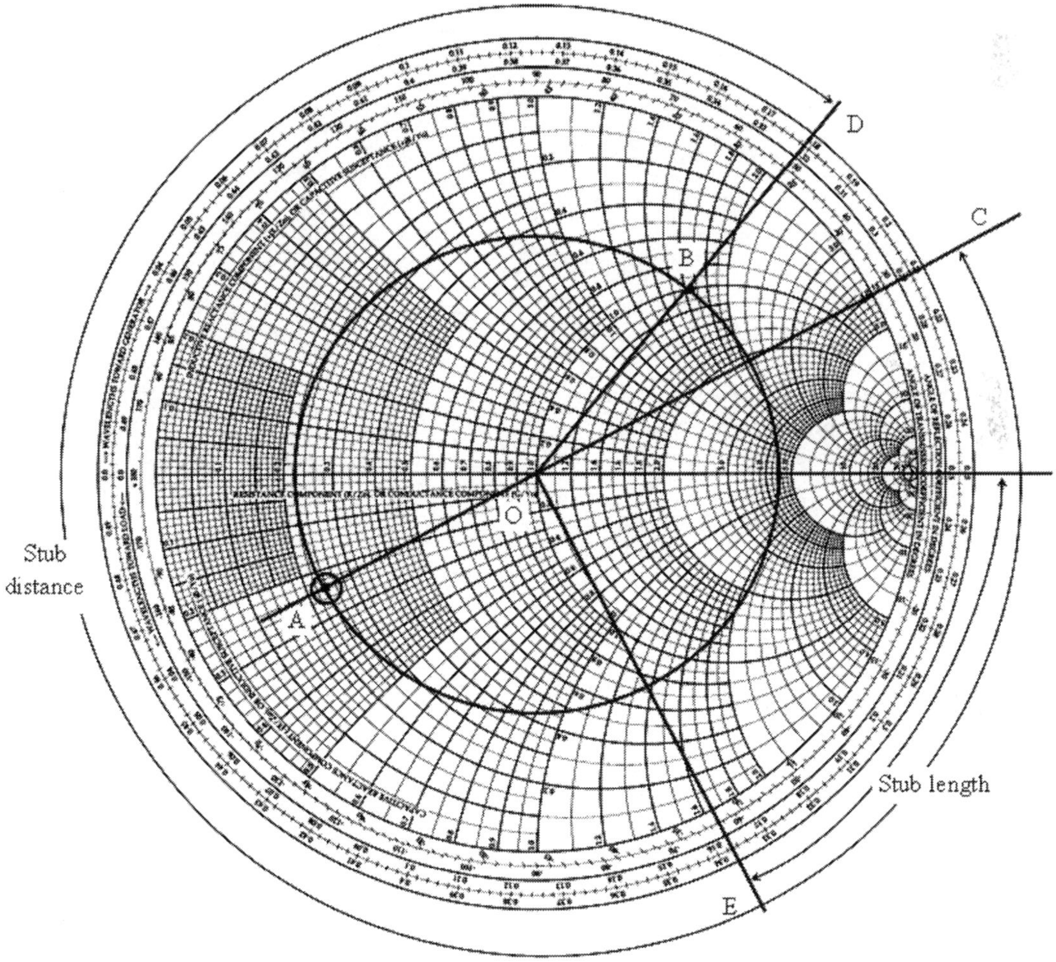

Fig. 6.23 *Use of Smith chart for Example 6.29*

Steps of Design:

1. The point of intersection of (r = 0.2357)-circle and (x = –0.2357)-circle is located on the Smith chart. This point is point A.
2. With the centre as the centre of Smith chart, point O, a circle is drawn with radius equal to OA.
3. The drawn circle cuts the r = 1 circle at point B, which is measured to be $(1 + j1.65)$.
4. The distance between points C and D on the rim of the chart gives the distance of the stub from the load. This is measured to be

$$CD = 0.25\lambda + 0.18\lambda + (0.25 - 0.21)\lambda = 0.47\lambda$$

5. As the load has a susceptance of $+j1.65$, the stub is required to provide a susceptance of $-j1.65$. Therefore, a point is marked by moving clockwise on the lower half of the chart where $x = -1.65$. This point is point E. The distance of this point from the short-circuit admittance point is the length of the stub. This is measured as

$$\text{Stub length} = (0.338 - 0.25)\lambda = 0.088\lambda$$

Hence, the specifications of the designed stub are:

Stub Distance = 0.47λ
Stub Length = 0.088λ

This use of the Smith chart is given in Fig. 6.23.

Double Stub Tuner In single stub matching, the stub is placed on the line at a specified point. Its location varies with the load impedance z_L and frequency. This creates some difficulties in changing the location of the stub with variation of load and/or frequency. In such cases, double stub matching is used.

Impedance matching is achieved by inserting two stubs at specified locations along transmission line as shown in Fig. 6.24.

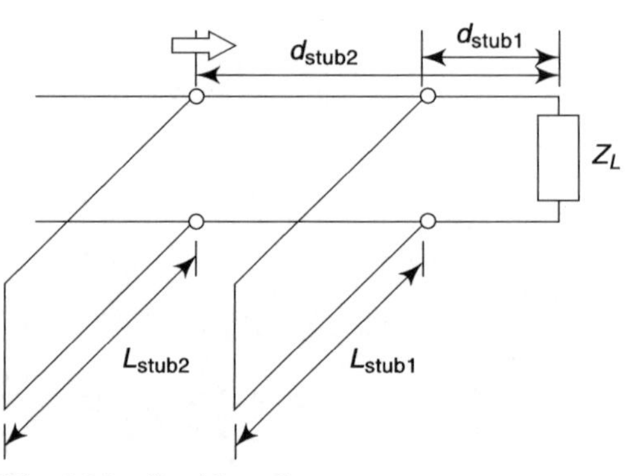

Fig. 6.24 *Double stub tuner*

In the double stub configuration, the distance between the stubs is fixed, such as $\frac{\lambda}{4}, \frac{\lambda}{8}, \frac{3\lambda}{8}, \frac{\lambda}{16}, \frac{3\lambda}{16}$ and so on and the lengths of the stubs are adjusted to match the load. In this way, if the load impedance is changed, one simply has to replace the stubs with another set of different length, without changing their locations.

There are two design parameters for double stub matching:

–The length of the first stub line L_{stub1}
–The length of the second stub line L_{stub2}

The length of the first stub is selected so that the admittance at the location of the second stub (before the second stub is inserted) has the real part equal to the characteristic admittance of the line. The length of the second stub is selected to eliminate the imaginary part of the admittance at the location of insertion.

At the location where the second stub is to be inserted, the possible normalised admittances required for matching are found on the circle of unitary conductance on the Smith chart. This circle is called *unitary conductance circle.*

At the location of the first stub, the allowed normalised admittances are found on an *auxiliary circle* which is obtained by rotating the *unitary conductance circle* counterclockwise, by an angle given as

$$\theta_{\text{aux}} = \frac{4\pi}{\lambda}(d_{\text{stub2}} - d_{\text{stub1}})$$

where, d_{stub1} and d_{stub2} are the distances of the first and second stub from the load, respectively. This is shown in Fig. 6.25 (a) and (b).

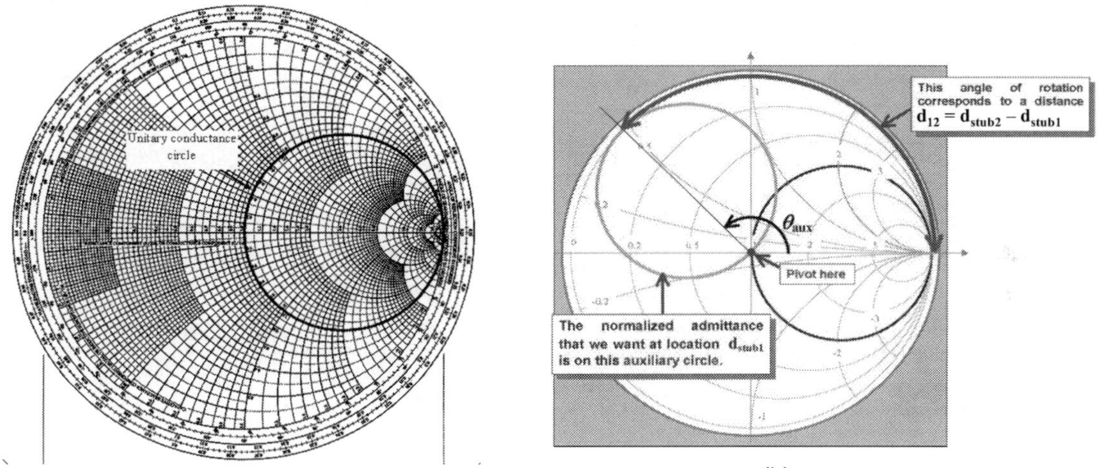

(a) (b)

Fig. 6.25 *(a) Unitary conductance circle (b) Auxiliary circle*

The positions of auxiliary circle will change depending upon the distance between the studs. This is shown in Fig. 6.26.

The main drawback of double stub tuning is that a certain range of load admittances cannot be matched once the stub locations are fixed.

Design of Double Stub Tuner

The main objective for designing a double stub is to find out the lengths of the stubs when the load impedance and the distance between the stubs are given. The design process of a double stub tuner consists of the following steps:

1. Given the load impedance Z_L and the characteristic impedance Z_0 of the transmission line, we find the normalised load impedance as

$$z_L = \frac{Z_L}{Z_0} = r_L + jx_L$$

2. The point of intersection of the r_L-circle and x_L-circle is marked on the Smith chart.
3. A circle is drawn with the radius equal to the distance of this point from the centre of the chart. This circle is actually the circle of constant magnitude of the reflection coefficient $|\Gamma|$ for the given load.
4. The normalised load admittance ($y_L = g_L + jb_L$) is found by rotating –180° on the constant $|\Gamma|$ circle, from the load impedance point.

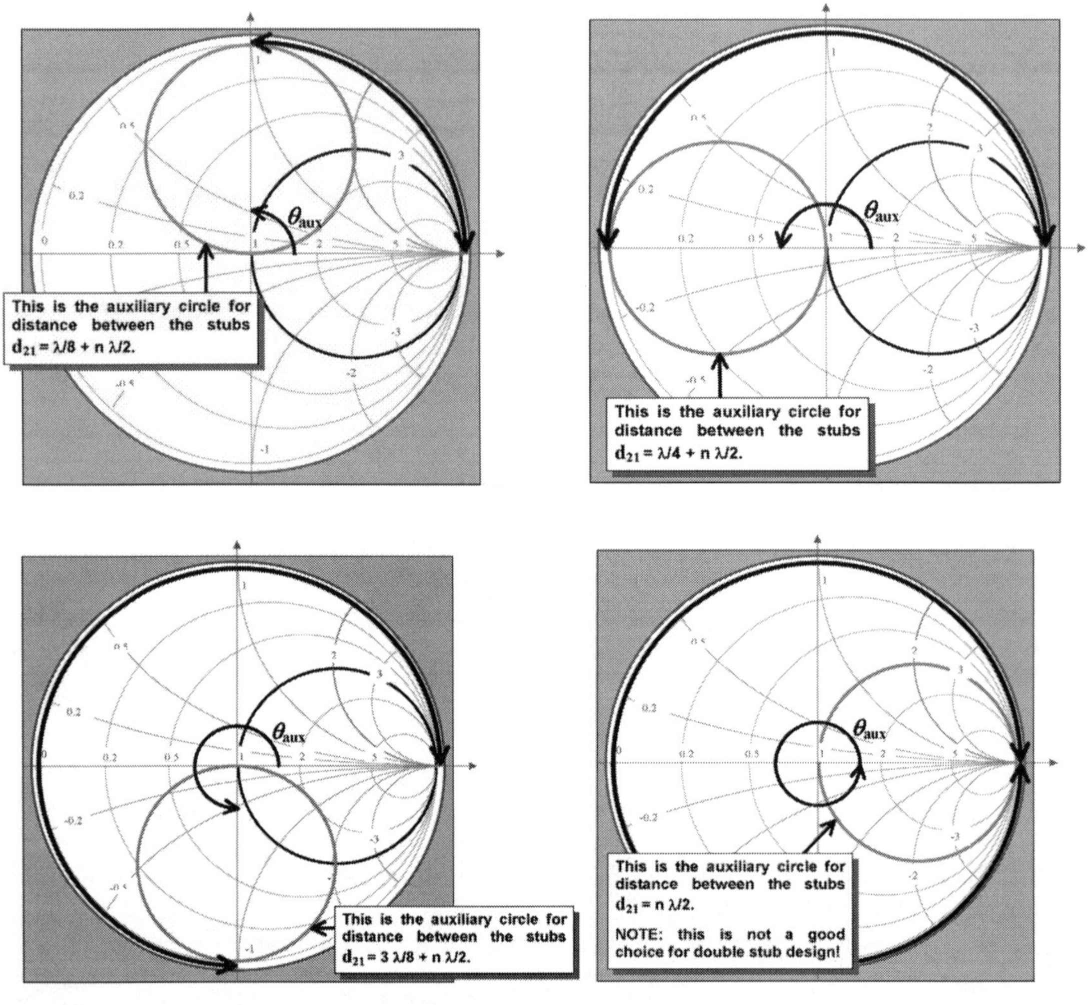

Fig. 6.26 *Position of auxiliary circle for different distances between stubs*

5. The normalised admittance at location d_{stub1} is found by moving clockwise on the constant $|\Gamma|$ circle.

NOTE

If the distance of the first stub from the load is not mentioned, it is assumed that the first stub is located at the load, i.e., $d_{stub1} = 0$.

6. The auxiliary circle is drawn considering the distance between the stubs.
7. The admittance of the first stub is added so that the normalised admittance point on the Smith chart reaches the auxiliary circle (two possible solutions). The admittance point will move on the corresponding conductance circle, since the stub does not alter the real part of the admittance.

8. The normalised admittance obtained on the auxiliary circle is mapped to the location of the second stub d_{stub2}. The point must be on the unitary conductance circle.
9. Finally, the admittance of the second stub is added so that the total parallel admittance equals the characteristic admittance of the line to achieve exact matching condition. The design steps of double stub turner are shown in Fig. 6.27.

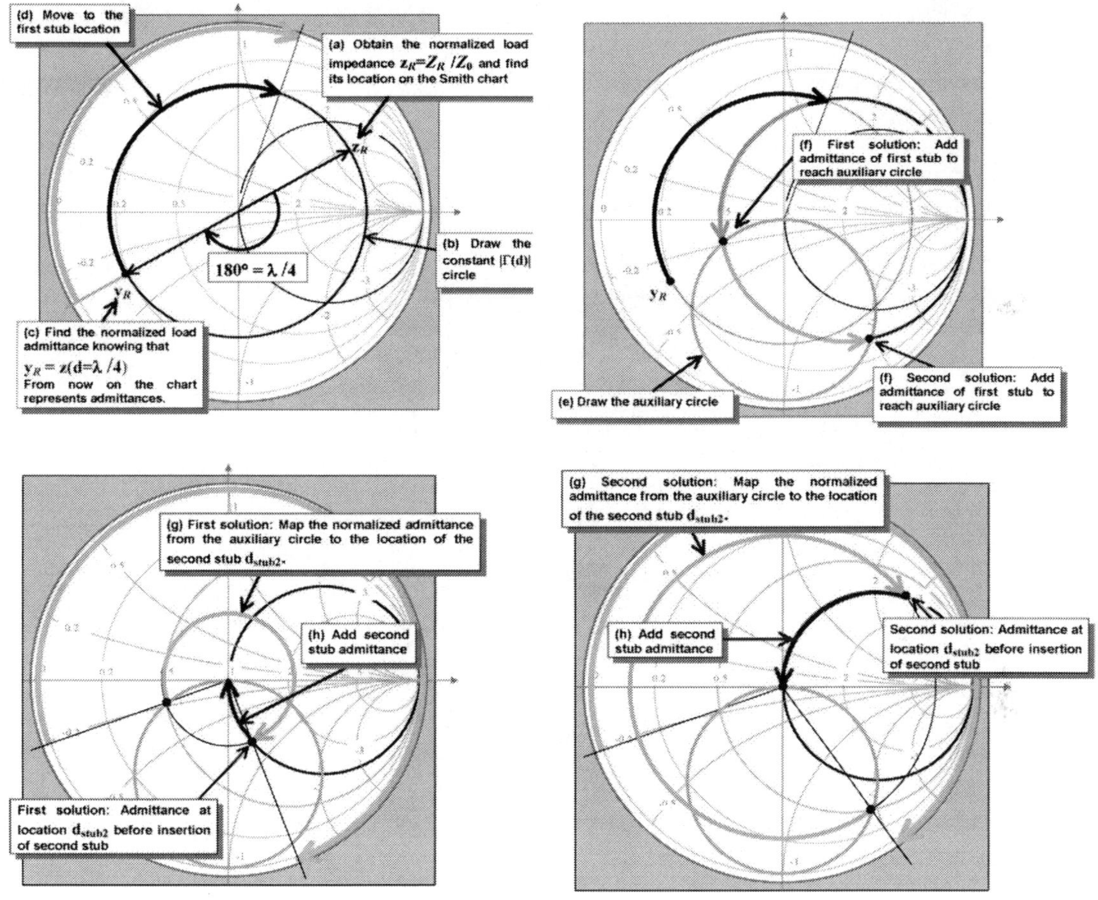

Fig. 6.27 *Design steps of double stub tuner*

***Example 6.30** For a load of $\frac{Z_R}{Z_0} = 0.8 + j1.2 = 0.8 + j1.2$, design a double stub tuner marking the distance between the stubs $\frac{3\lambda}{8}$. Specify the stub length and distance from the load to the first stub. The stubs are short-circuited. Use Smith chart.

Solution The normalised admittance is given as

$$\frac{Y_R}{Y_0} = \frac{\frac{1}{Z_R}}{\frac{1}{Z_0}} = \frac{Z_0}{Z_R} = \frac{1}{0.8 + j1.2} = (0.4 - j0.6)$$

Design Steps:

1. We locate the point of load admittance on the chart, say point A. The *first stub is located at this point.*
2. Since the distance between the stubs is $\frac{3\lambda}{8}$ or 0.375λ, the *auxiliary circle* is drawn by moving a distance of 0.375λ counter-clockwise (towards the load) as shown in Smith chart.
3. This auxiliary circle intersects the constant conductance circle (with conductance 0.4) at point B. This point gives the normalised admittance with first stub. This normalised admittance as read from the chart is

$$\frac{Y_1}{G_0} = (0.4 - j0.2)$$

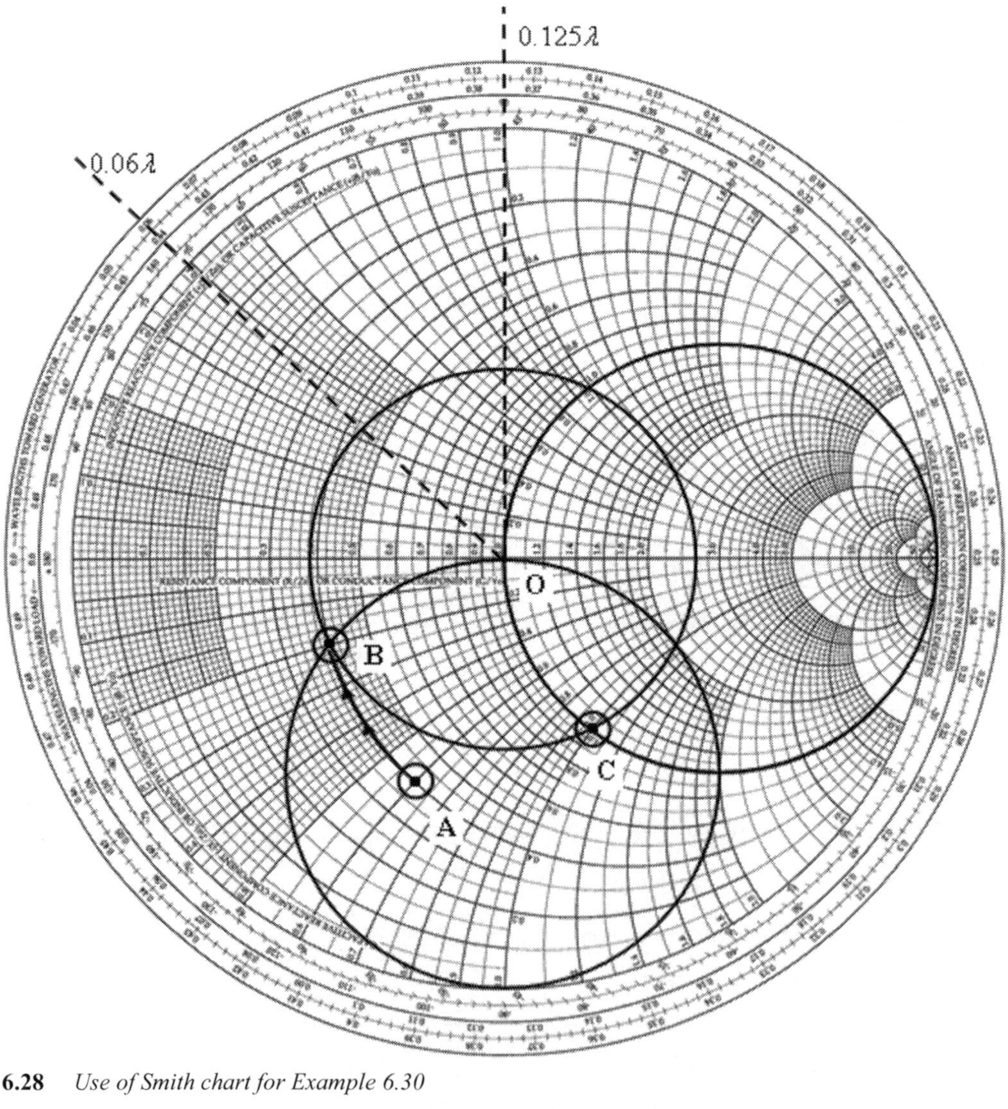

Fig. 6.28 *Use of Smith chart for Example 6.30*

4. The first stub must contribute to a susceptance of $j0.6 - j0.2 = j0.4$. The +0.4 constant susceptance circle intersects the rim of the chart at 0.06λ.

   So, the length of the first stub as measured from the short circuit admittance point (extreme left hand point on the chart) is given as

$$L_{\text{stub1}} = (0.5 - 0.25)\lambda + 0.06\lambda = 0.31\lambda$$

5. Now, we move from point B a distance of 0.375λ clockwise along the rim of the chart (or we move from point B along a circle with constant radius OB). This location cuts the unitary conductance circle at point C. Thus, we have moved from the location of stub 1 (point A) to the location of stub 2 (point C). The *second stub is located at point C.*
6. The normalised admittance at point C without the second stub as measured from the chart is $(1 - j1)$. For perfect matching, the second stub must eliminate the imaginary component of this admittance, i.e., the second stub must provide a susceptance of $+j1$. This $[+j1]$ circle cuts the rim of the chart at 0.125λ.

   So, the length of the second stub as measured from the short-circuit admittance point (extreme left hand point on the chart) is given as

$$L_{\text{stub2}} = (0.5 - 0.25)\lambda + 0.125\lambda = 0.375\lambda$$

   Hence, the lengths of the two stubs are as given:
   Length of first stub located at load, $L_{\text{stub1}} = 0.31\lambda$
   Length of second stub, $L_{\text{stub2}} = 0.375\lambda$
   This use of the Smith chart is shown in Fig. 6.28.

***Example 6.31** A 50 ohm line feeds an inductive load $Z_L = 35 + j35$ ohm. Design a double stub tuner to match this load to the line (make use of a Smith chart).

Solution The normalised load admittance is given as

$$\frac{Y_L}{G_0} = \frac{\frac{1}{Z_L}}{\frac{1}{Z_0}} = \frac{Z_0}{Z_L} = \frac{50}{35 + j35} = \frac{1}{0.7 + j0.7} = (0.7142 - j0.7142)$$

Design Steps:

1. Since the distance between the first stub and the load is not given, *we assume that the first stub is located at the load.* We locate the point of load admittance on the chart, say point A.
2. Also, as the *distance between the stubs* is not mentioned, we assume it to be $\frac{\lambda}{4}$. Therefore, the auxiliary circle is drawn as shown in Smith chart.
3. This auxiliary circle intersects the constant conductance circle (with conductance 0.7142) at point B. Hence, the *first stub is located at point B* where the normalised admittance as read from the chart is

$$\frac{Y_1}{G_0} = (0.7142 - j0.49)$$

4. The first stub must contribute to a susceptance of $(j0.7142 - j0.49) = j0.224$. The +0.224 constant susceptance circle intersects the rim of the chart at 0.036λ.

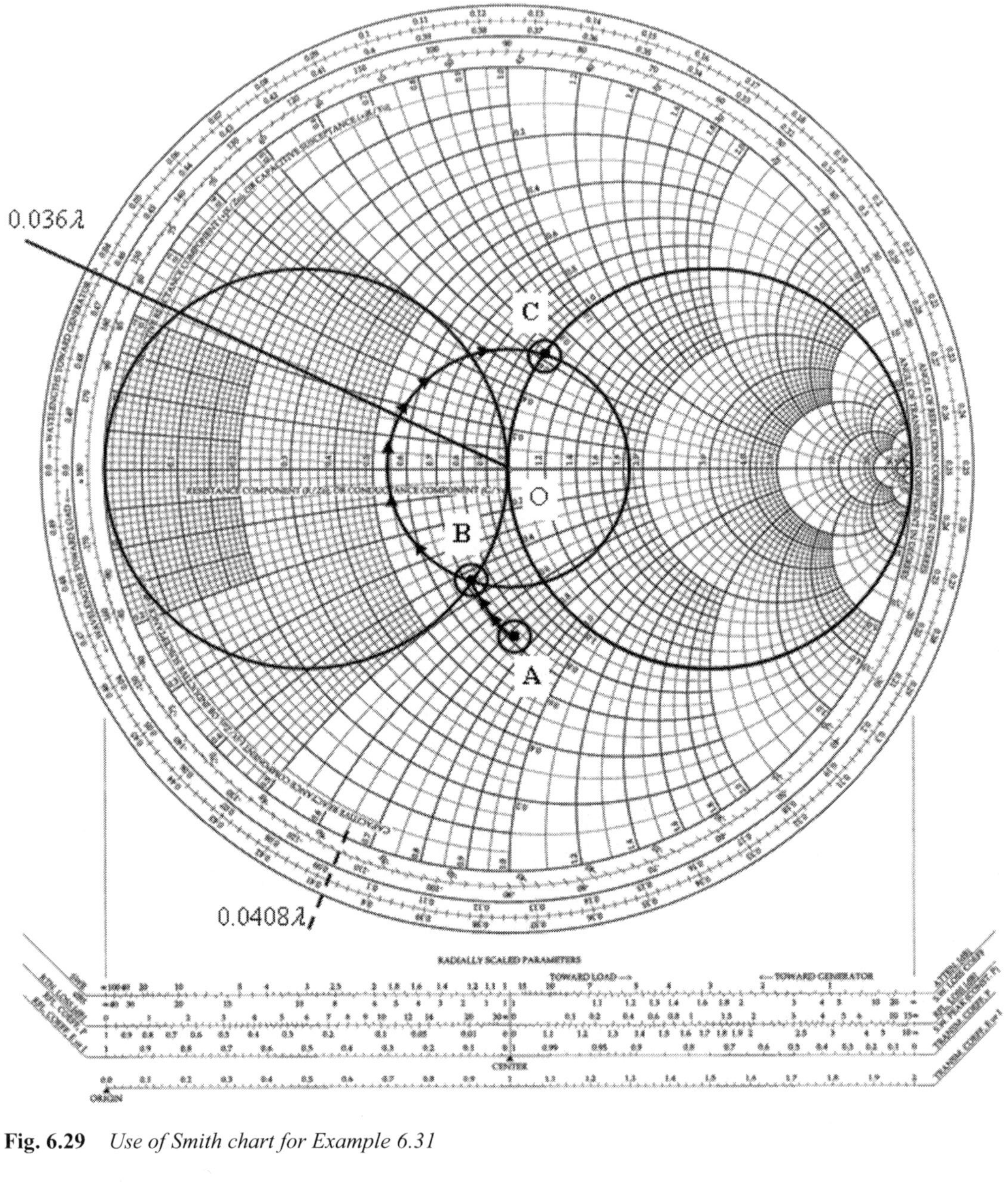

Fig. 6.29 *Use of Smith chart for Example 6.31*

So, the length of the first stub as measured from the short circuit admittance point (extreme left hand point on the chart) is given as

$$L_{\text{stub1}} = (0.5 - 0.25)\lambda + 0.036\lambda = 0.286\lambda$$

5. Now, we draw a circle with centre at O and radius equal to OA. This circle cuts the unitary conductance circle at point C. The *second stub is located at point C.*

6. The portion of the line between the stubs changes the admittance at point A to point C. The normalised admittance at point C without the second stub as measured from the chart is $(1 + j0.65)$. For perfect matching, the second stub must eliminate the imaginary component of this admittance, i.e., the second stub must provide a susceptance of $-j0.65$. This $[-j0.65]$ circle cuts the rim of the chart at 0.408λ.

   So, the length of the second stub as measured from the short-circuit admittance point (extreme left hand point on the chart) is given as

$$L_{\text{stub2}} = (0.408 - 0.25)\lambda = 0.158\lambda$$

Hence, the complete specifications of the double stub are as given:

Length of first stub located at load, $L_{\text{stub1}} = 0.286\lambda$

Length of second stub, $L_{\text{stub2}} = 0.158\lambda$

Distance between stubs $= \dfrac{\lambda}{4}$

This use of the Smith chart is shown in Fig. 6.29.

Example 6.32 A load of $\dfrac{Z_L}{Z_0} = (0.2 + j0.3)$ is located at the end of a transmission line. At a distance of 0.11λ from the load, an adjustable stub is placed. Another 0.175λ distance from the first stub, a second stub is placed. Using Smith chart, determine the lengths of the two stubs.

Solution The normalised load admittance is given as

$$\frac{Y_L}{G_0} = \frac{\dfrac{1}{Z_L}}{\dfrac{1}{Z_0}} = \frac{Z_0}{Z_L} = \frac{1}{0.2 + j0.3} = (1.54 - j2.31)$$

Design Steps:

1. We locate the point of load admittance on the chart, say point A.
2. Since the first stub is located at a distance of 0.11λ from the load, we move clockwise a distance 0.11λ along the rim of the chart from point A on a circle of constant radius OA and come to point B. The *first stub is located at point B.* The normalised admittance of this point (point without stub) as measured from chart is $(0.25 - j0.6)$.
3. Also, the distance between the stubs is 0.175λ. Therefore, the *auxiliary circle* is drawn by moving a distance of 0.175λ counter-clockwise (towards the load) as shown in the Smith chart.
4. This auxiliary circle intersects the constant conductance circle (with conductance 0.25) at point C. This point gives the normalised admittance with first stub. This normalised admittance as read from the chart is

$$\frac{Y_1}{G_0} = (0.25 + j0.05)$$

5. The first stub must contribute to a susceptance of $j0.05 - (-j0.6) = (j0.65)$. The +0.65 constant susceptance circle intersects the rim of the chart at 0.09λ.

   So, the length of the first stub as measured from the short-circuit admittance point (extreme left hand point on the chart) is given as

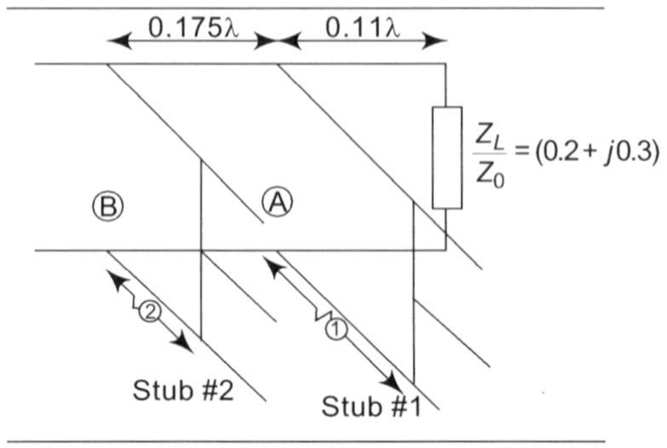

(a)

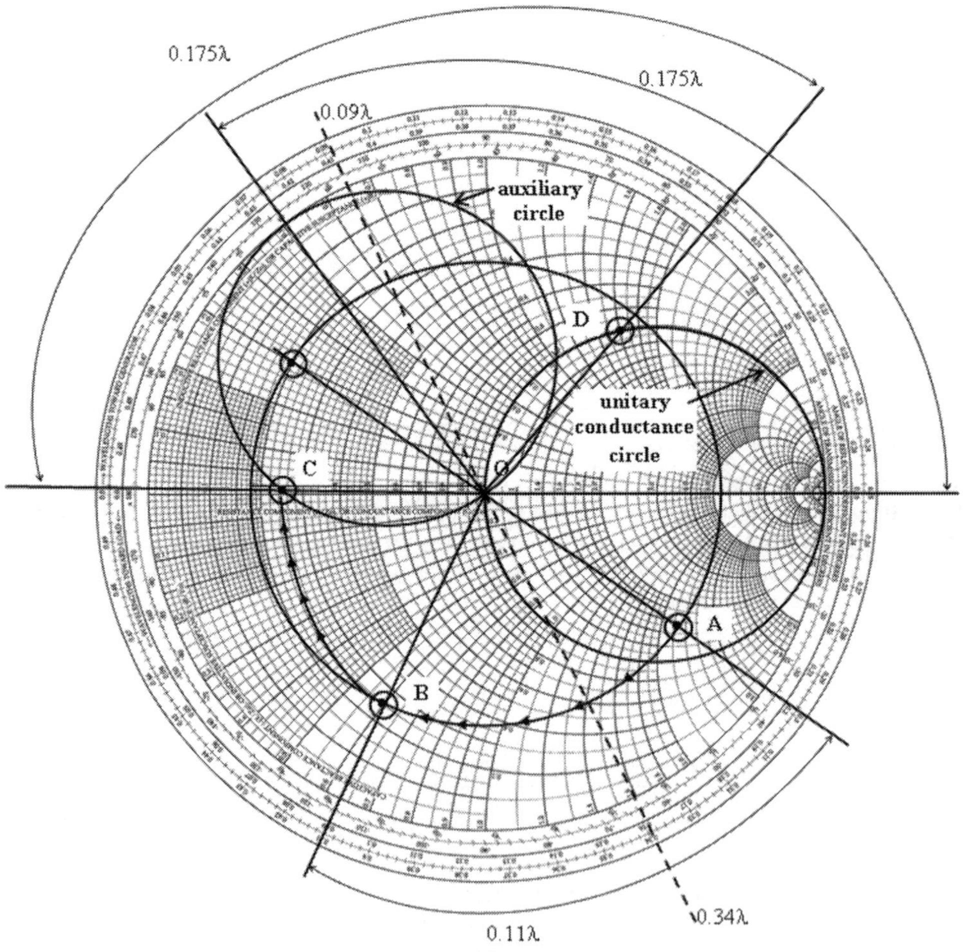

(b)

Fig. 6.30 *(a) Location of stubs (b) Use of Smith chart*

$$L_{\text{stub1}} = 0.25\lambda + 0.09\lambda = 0.34\lambda$$

6. Now, we move a distance of 0.175λ clockwise along the rim of the chart. This location cuts the unitary conductance circle at point D. Thus, we have to move from the location of stub 1 (point B) to the location of stub 2 (point D). The *second stub is located at point D.*

7. The normalised admittance at point D without the second stub as measured from the chart is $(1 + j1.6)$. For perfect matching, the second stub must eliminate the imaginary component of this admittance, i.e., the second stub must provide a susceptance of $-j1.6$. This $[-j1.6]$ circle cuts the rim of the chart at 0.34λ.

So, the length of the second stub as measured from the short-circuit admittance point (extreme left hand point on the chart) is given as

$$L_{\text{stub2}} = (0.34 - 0.25)\lambda = 0.09\lambda$$

Hence, the lengths of the two stubs are as given:

Length of first stub located at load, $L_{\text{stub1}} = 0.34\lambda$
Length of second stub, $L_{\text{stub2}} = 0.09\lambda$

The location of stubs and use of the Smith chart are shown in Fig. 6.30(a) and Fig. 6.30(b).

Summary

- A *transmission line* is a device used for transmission of electromagnetic energy guided by two conductors in a dielectric medium.
- *Transmission line mode* is the distinct pattern of electric and magnetic field induced on a transmission line under source excitation.

  Three types of transmission line modes are:

  1. Transverse electromagnetic (TEM) Mode,
  2. Quasi-TEM Mode, and
  3. Waveguide Mode

- A transmission line can be characterised by four distributed parameters: resistance (in Ω/m), inductance (in H/m), capacitance (in F/m), and conductance (in S/m). These four distributed parameters are constant for a particular transmission line and are known as *primary line constants* of a transmission line.
- Apart from these primary line constants, there are a few other constants related to a transmission line. These include the characteristic impedance (Z_0), the propagation constant (γ), attenuation constant (α) and phase constant (β). These constants are known as the *secondary line constants.*
- Transmission line voltage and current equations, known as *Telegrapher's equations*, are given as

$$\frac{d^2V_S(z)}{dz^2} - \gamma^2 V_S(z) = 0 \quad \text{and} \quad \frac{d^2I_S(z)}{dz^2} - \gamma^2 I_S(z) = 0$$

where, $\gamma = \alpha + j\beta = \sqrt{(R + j\omega L)(G + j\omega C)}$ is the complex propagation constant.

- *Characteristic impedance* of a transmission line is defined as the ratio of positively travelling voltage wave to current wave at any point on the line. It is given as

$$Z_0 = \frac{V_{S0}^+}{I_{S0}^+} = -\frac{V_{S0}^-}{I_{S0}^-} = \sqrt{\frac{R + j\omega L}{G + j\omega C}}$$

- A transmission line is said to be *lossless*, if
  - the conductors of the line are perfect, i.e., the conductors have infinite conductivity and zero resistance ($\sigma = \infty$, $R = 0$), and
  - the dielectric medium between the conductors is ideal, i.e., the medium has zero conductivity and infinite resistance ($\sigma = 0$, $G = 0$).

  For a lossless line, $\alpha = 0, \beta = \omega\sqrt{LC}, Z_0 = \sqrt{\frac{L}{C}}$.

- A transmission line is said to be *distortionless* if it has no frequency and phase distortions.

  The condition for a line to be distortionless is written as $\frac{R}{L} = \frac{G}{C}$.

  For a distortionless line, $\alpha = \sqrt{RG}, \beta = \omega\sqrt{LC}, Z_0 = \sqrt{\frac{R}{G}} = \sqrt{\frac{L}{C}}$.

- The *input impedance* at any point on the transmission line is given by the ratio of voltage to current at that point.

  Input impedance for different types of transmission lines are given as follows.

  For finite lossy transmission line:

$$Z_i = Z_0\left(\frac{Z_R \cosh \gamma l + Z_0 \sinh \gamma l}{Z_0 \cosh \gamma l + Z_R \sinh \gamma l}\right) = Z_0\left(\frac{Z_R + Z_0 \tanh \gamma l}{Z_0 + Z_R \tanh \gamma l}\right)$$

  For finite lossless high-frequency transmission line:

$$Z_i = R_0\left(\frac{Z_R \cos \beta l + jR_0 \sin \beta l}{R_0 \cos \beta l + jZ_R \sin \beta l}\right) = R_0\left(\frac{Z_R + jR_0 \tan \beta l}{R_0 + jZ_R \tan \beta l}\right)$$

  For infinite (lossy and lossless) transmission line:

$$Z_i = \frac{V_S}{I_S} = Z_0$$

- *Reflection Coefficient* of a transmission line is the ratio of the reflected voltage (or current) to the incident voltage (or current), when a transmission line is terminated in an impedance (Z_R) not equal to the characteristic impedance (Z_0) of the line. The reflection coefficient as a function of position can be written as

$$\Gamma(z) = \frac{Z_R - Z_0}{Z_R + Z_0} e^{2\gamma(z-l)} = \Gamma_L e^{2\gamma(z-l)}$$

- The reflection coefficient at the load ($z = l$) is

$$\Gamma(l) = \Gamma_L = \frac{Z_R - Z_0}{Z_R + Z_0}$$

- *Standing wave ratio* in a transmission line is defined as the ratio of the maximum voltage (or current) to the minimum voltage (or current) of a line having standing waves.

$$s = \frac{V_{\text{max}}}{V_{\text{min}}} = \frac{I_{\text{max}}}{I_{\text{min}}} = \frac{1+|\Gamma_L|}{1-|\Gamma_L|}$$

- The input impedance as a function of position at any point along a general transmission line is given as

$$Z_{\text{in}}(z) = Z_0 \frac{Z_R + Z_0 \tanh[\gamma(l-z)]}{Z_0 + Z_R \tanh[\gamma(l-z)]}$$

- Smith chart is a useful graphical tool used to calculate the reflection coefficient and impedance at various points on a (lossless) transmission line system.

  Smith chart can be used for different purposes, such as, to find the normalised admittance from normalised impedance and vice-versa, to find the parameters of mismatched transmission lines, to find the VSWR for a given load impedance, to find the reflection coefficient, to find the input impedance of a transmission line, to locate a voltage maximum on a transmission line, and to design stubs for impedance matching.

Important Formulae

Telegrapher's equation	$\frac{d^2V_S(z)}{dz^2} - \gamma^2 V_S(z) = 0$ and $\frac{d^2I_S(z)}{dz^2} - \gamma^2 I_S(z) = 0$
Propagation constant (γ)	$\gamma = \alpha + j\beta = \sqrt{(R+j\omega L)(G+j\omega C)}$
Attenuation constant (α)	$\alpha = \frac{1}{\sqrt{2}}\sqrt{RG - \omega^2 LC + \sqrt{(R^2+\omega^2L^2)(G^2+\omega^2C^2)}}$
Phase constant (β)	$\beta = \frac{1}{\sqrt{2}}\sqrt{-RG - \omega^2 LC + \sqrt{(R^2+\omega^2L^2)(G^2+\omega^2C^2)}}$
Characteristics impedance (Z_0) (lossy line)	$Z_0 = \sqrt{\frac{R+j\omega L}{G+j\omega C}}$
Characteristics impedance (Z_0) (lossless line)	$Z_0 = \sqrt{\frac{L}{C}}$
Condition for distortionless line	$\frac{R}{L} = \frac{G}{C}$
Input impedance (finite lossy line)	$Z_i = Z_0\left(\frac{Z_L + Z_0 \tanh \gamma l}{Z_0 + Z_L \tanh \gamma l}\right)$
Input impedance (finite lossless line)	$Z_i = R_0\left(\frac{Z_L + jR_0 \tan \beta l}{R_0 + jZ_L \tan \beta l}\right)$

Input impedance (infinite line)	$Z_i = \sqrt{\dfrac{R + j\omega L}{G + j\omega C}}$, for lossy infinite line $= \sqrt{\dfrac{L}{C}}$, for lossless infinite line
Reflection coefficient	$\Gamma_L = \dfrac{Z_L - Z_0}{Z_L + Z_0}$
Standing wave ratio	$s = \dfrac{1+\vert\Gamma_L\vert}{1-\vert\Gamma_L\vert}$

Exercises

[NOTE: * marked problems are important university problems]

- ***Easy***

*1. An open-wire transmission line has the following constants:

$$R = 5\ \Omega/\text{m},\quad L = 5.2 \times 10^{-8}\ \text{H/m},\quad C = 2.13 \times 10^{-13}\ \text{F/m},\quad G = 6.2 \times 10^{-3}\ \text{mho/m}$$

frequency = 4 GHz

Find the propagation constant, characteristic impedance and velocity of propagation.

[$\gamma = 103.37\angle 65.23°$; Z = 12.64 $\angle 24.49°$; 3×10^8 m/s]

*2. A transmission line with air as dielectric has Z_0 = 50 Ω and a phase constant of 3 rad/m at 10 MHz. Find the inductance and capacitance of the line. [2.39 μH/m, 0.955 nF/m]

3. A transmission line of characteristic impedance 50 Ω is terminated by resistor of 100 Ω. What will be the VSWR in the line? Calculate impedances at the voltage minimum and maximum positions. [2, 150 Ω, 16.67 Ω]

- ***Medium***

4. A 60 Ω lossless line has a maximum impedance Z_{in} = (180 + j0) Ω at a distance of λ/24 from the load. If the line is 0.3λ, determine (a) standing wave ratio, s (b) load impedance, Z_L and (c) the input impedance of the transmission line.

[3; (117.2 + j78.1) Ω; (20 + j2.8) Ω]

*5. A lossless transmission line having a characteristic impedance of 75 ohm is terminated in an unknown impedance Z_L. The VSWR is 3.0. The nearest minimum from the load is found to be at 20 cm. Calculate Z_L if the frequency is 150 MHz. [98.24 Ω]

- ***Hard***

6. A 2 m long lossless transmission line has an impedance of 300 Ω. The velocity of propagation is 2.5×10^8 m/s. The load has an impedance of 300 Ω with sending end voltage being 60 V at 100 MHz. Find:

(a) The phase constant
(b) The load voltage
(c) The load current

(d) The load reflection coefficient
(e) Standing wave ratio.

7. Design a short-circuited shunt stub tuner to match a load of $Z_L = (60 - j40)\Omega$ to a transmission line with characteristic impedance of 50 Ω.

   [stub length = 0.147λ, stub distance = 0.076λ]

*8. A transmission line of 100 m long is terminated in load of $(100 - j200)\Omega$. Determine the line impedance at 25 m from the load end at a frequency of 10 MHz. Assume line impedance $Z_0 = 100$ Ω. Determine the input impedance and admittance using Smith chart.

   [$(44 + j120)\Omega$; $(0.28 - j0.75)$ mho]

9. A 50 Ω transmission line is terminated in an impedance of $Z_L = (35 - j47.5)\Omega$. Find the position and length of the short-circuited stub to match it.

   [stub length = 0.059λ, stub distance = 0.111λ]

*10. A load $(50 - j100)\Omega$ is connected across a 50 Ω line. Design a short-circuited stub to provide matching between the two at a signal frequency of 30 MHz using Smith chart.

   [stub length = 0.072λ, stub distance = 0.126λ]

*11. For a load of Z_R = (50 – j50) ohm connected to a 50 ohm lossless line at 400 MHz, design a double stub tuner making the distance between the stubs $\frac{3\lambda}{8}$. Use Smith chart.

   [stub lengths: 4.5 cm and 4.3 cm]

12. A 90 ohm line feeds a load $Z_L = 270 + j0$ ohm at 600 MHz. Design a quarter wave double stub tuner to match this load to the line (make use of a Smith chart).

   [stub lengths: 9 cm and 4.95 cm]

Review Questions

[NOTE: * marked questions are important university questions.]

1. (a) What are the transmission line parameters?
   (b) Mention the different modes of transmission line.
2. What are *Telegrapher's equations*? Deduce the equations.

*3. (a) Explain what do you understand by the term 'Line parameters' in the context of a transmission line.
   (b) Draw the equivalent circuit of a transmission line and hence write the transmission line equations for an elemental section of a transmission line.

*4. Develop the analogy between the uniform plane E.M. Waves and the electric transmission line.

*5. (a) What is characteristic impedance of a transmission line? Derive its expression. What will be the characteristic impedance if the line is: (i) lossless line? (ii) distortionless line?
   (b) What is a distortionless line? How to achieve distortionless condition on the line? Derive the necessary conditions (Heaviside condition).

6. Discuss the field analysis to determine the line parameters R, L, G and C for a transmission line.
7. Starting from the equivalent circuit diagram, derive expressions for characteristic impedance and input impedance of a transmission line.
8. Deduce an expression for the input impedance of a finite lossy transmission line in terms of the line constants. What will be the expression if the transmission line is:
   (a) an open-ended line?
   (b) a short-ended line?
   (c) an infinite line?

9. Deduce an expression for the input impedance of an infinite lossy transmission line in terms of the line constants. What will be the input impedance if the line is a lossless one?

10. (a) Discuss the different losses associated with high-frequency transmission lines.

(b) Derive expressions for the input impedance of a lossless line of length l when the load terminals are (i) short-circuited, and (ii) open-circuited.

11. Bring out the differences between lossless and low-loss transmission lines.

12. Prove that the input impedance of a lossless radio frequency line of length l of characteristic impedance Z_0 and terminated in an impedance Z_L is given by,

$$Z_{in} = Z_0 \frac{Z_L + jZ_0 \tan \beta l}{Z_0 + jZ_L \tan \beta l}$$

where, β is the phase constant of the line.

13. (a) Derive the expression of input impedance, Z_{in} of a lossless transmission line in terms of relevant parameters, when the line is terminated in a load impedance, Z_L.

(b) Give a neat sketch of variation of Z_{in} as a function of electrical length of the line, when the line is terminated in a

(i) short circuit and

(ii) open circuit.

Discuss the significance of the plots.

***14.** Show that for a lossless transmission line the impedance of a line repeats over every $\frac{\lambda}{2}$ distance.

***15.** Show that a short-circuited lossless transmission line can offer reactances of any value by simply changing the length of the line.

***16.** Show that an open-circuited transmission line with length less than $\frac{\lambda}{4}$ is capacitive.

17. Sketch the input impedance offered by short-circuited and open-circuited transmission lines. Derive the expressions used.

18. (a) Mention the conditions under which a travelling wave and a standing wave form in a transmission line.

(b) Define, in relation to travelling waves in a transmission line, the followings:

(i) Reflection co-efficient,

(ii) Transmission co-efficient, and

(iii) Standing wave ratio (SWR).

19. What is Smith chart? Explain the characteristics of Smith chart.

20. Explain the formation of standing wave pattern on transmission line. Deduce the relation between the reflection coefficient and VSWR.

21. Derive an expression for reflection co-efficient of a transmission line. Prove there is no reflection if line is terminated in its characteristic impedance.

22. Discuss the principle of any one method for matching a transmission line with characteristic impedance z_0 to a load z_L.

23. Explain what you understand by the term 'quarter-wave transformer'. Write one application of such a transformer.

24. Show how a uniform transmission line can be used as an impedance transformer and explain how this property can be made use of in impedance matching.

25. Explain the principle of single stub-matching deriving expressions for stub length and stub location.

Multiple Choice Questions

1. In a transmission line, electric energy is transported by:
   (a) the flowing electrons (b) the flowing electrons and holes
   (c) the associated electric and magnetic field (d) none of the above.
2. In a transmission line, the distance between adjacent maxima and minima of a standing wave is
   (a) $\lambda/8$ (b) $\lambda/4$ (c) $\lambda/2$ (d) λ.
3. A transmission line is called a distortionless line, when
   (a) $\frac{R}{L}=\frac{G}{C}$ (b) $\frac{R}{G}=\frac{C}{L}$ (c) $RG=\frac{L}{C}$ (d) $\frac{R}{G}=LC$
4. A transmission line is said to be distortionless if:
   (a) $\frac{R}{G}=\frac{C}{L}$ (b) $\frac{R}{G}=\frac{L}{C}$ (c) $RG=LC$ (d) $R=0$.
5. A transmission line has R, L, G and C distributed parameters per unit length of the line. γ is the propagation constant of the lines. Which expression gives the characteristic impedance of the line?
   (a) $\frac{\gamma}{R+j\omega L}$ (b) $\frac{R+j\omega L}{\gamma}$ (c) $\frac{G+j\omega C}{\gamma}$ (d) $\sqrt{\frac{G+j\omega C}{R+j\omega L}}$
6. At low frequencies, the characteristic impedance of a transmission line is given as,
   (a) $\sqrt{\frac{R}{G}}$ (b) $\sqrt{\frac{G}{R}}$ (c) $\sqrt{\frac{L}{C}}$ (d) $\sqrt{\frac{C}{L}}$
7. At high frequencies, the characteristic impedance of a transmission line is given as,
   (a) $\sqrt{\frac{R}{G}}$ (b) $\sqrt{\frac{G}{R}}$ (c) $\sqrt{\frac{L}{C}}$ (d) $\sqrt{\frac{C}{L}}$
8. Consider the following statements:
   The characteristic impedance of a transmission line can increase with the increase in
   1. resistance per unit length
   2. conductance per unit length
   3. capacitance per unit length
   4. inductance per unit length

   Which of these statements are correct?
   (a) 1 and 2 (b) 2 and 3 (c) 1 and 4 (d) 3 and 4
9. SWR of a transmission line is measured by
   (a) reflectometer (b) voltmeter (c) ammeter (d) power meter
10. When the load impedance is equal to the characteristic impedance of the transmission lines, then the reflection coefficient and standing wave ratio are respectively
    (a) 0 and 0 (b) 1 and 0 (c) 0 and 1 (d) 1 and 1.
11. A transmission line of length $\frac{\lambda}{4}$ shorted at the far end behaves like
    (a) series resonant circuit (b) parallel resonant circuit
    (c) pure inductor (d) pure capacitor
12. The SWR on a lossless transmission line of characteristic impedance 100 ohm is 3. The line is terminated by
    (a) a resistance of 300 ohm (b) a reactance of $j300$ ohm.
    (c) a resistance of 100/3 ohm (d) a reactance of $j100/3$ ohm.

13. A transmission line is terminated by a pure capacitor. The VSWR in the line is
    (a) 1 **(b)** infinity
    (c) 0 **(d)** depends on the value of capacitor.
14. If the reflection coefficient of a transmission line is $(0.5 + j0.5)$ for a given load, VSWR will be
    (a) 1 **(b)** ∞ **(c)** 2 **(d)** $-\infty$.

Answers

1. (c)	**2.** (b)	**3.** (a)	**4.** (b)	**5.** (b)	**6.** (a)	**7.** (c)	**8.** (c)
9. (a)	**10.** (c)	**11.** (a)	**12.** (a)	**13.** (b)	**14.** (b)		

7 WAVEGUIDES

Learning Objectives

This chapter deals with the following topics:

- *Concepts of waveguides*
- *Analysis of three types of waveguides:*
  - *Parallel-plane waveguide,*
  - *Rectangular waveguide, and*
  - *Circular waveguide*
- *Power transmission,losses and attenuation in different waveguides*

7.1 INTRODUCTION

A *waveguide* is a hollow conducting pipe, of uniform cross-section, used to transport high frequency electromagnetic waves (generally, in the microwave band) from one point to another.

Classification of Waveguides Waveguides can be generally classified as

1. ***Metal waveguides*** These waveguides normally take the form of an enclosed conducting metal pipe. The waves propagating inside the metal waveguide may be characterised by reflections from the conducting walls.
2. ***Dielectric waveguides*** These waveguides consist of dielectrics only and employ reflections from dielectric interfaces to propagate the electromagnetic wave along the waveguide.

Advantages of waveguide over conventional transmission lines

1. Waveguides are simple and rigid. Uniform cross-section of a guide can be obtained much easily compared to uniform spacing between the conductors.
2. There are no radiation losses as the field is confined within the guide.
3. There is no dielectric loss due to absence of any inner conductor.
4. Ohmic power losses are also reduced in waveguides as compared to conventional transmission lines due to greater current carrying over waveguide walls and absence of an inner conductor with less diameter and higher current density.

Modes of Wave Propagation When an electromagnetic wave propagates through a hollow tube, only one of the fields, either electric or magnetic field, will actually be transverse (perpendicular)

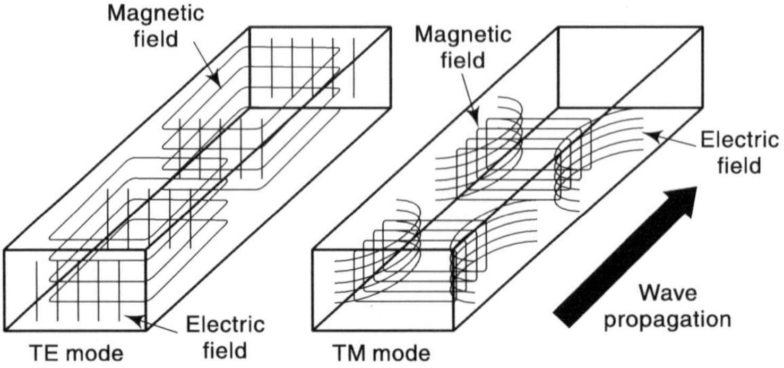

Fig. 7.1 *Waveguides TE and TM modes*

to the direction of wave propagation. The other field will 'loop' longitudinally to the direction of propagation, but still be perpendicular to the other field. Whichever field remains transverse to the direction of travel determines whether the wave propagates in *TE* mode (Transverse Electric) or *TM* (Transverse Magnetic) mode as shown in Fig. 7.1. It is observed that the electric flux lines appear with beginning and end points, whereas, the magnetic flux lines appear as continuous loops.

7.2 PARALLEL PLANES WAVEGUIDES BETWEEN TWO INFINITE PARALLEL CONDUCTING PLANES

Definition A *parallel-plane wave guide* is a waveguide formed by two infinite parallel perfectly conducting planes.

Derivation of Field Equations for Parallel-Plane Waveguide We consider two parallel perfectly conducting infinite planes separated by a distance *d*, as shown in Fig. 7.2.

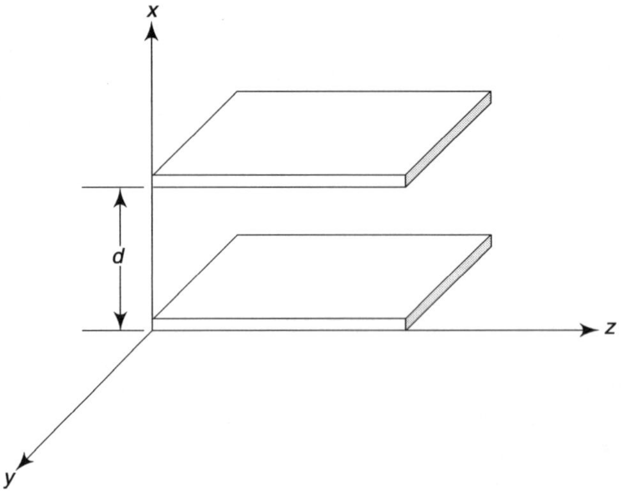

Fig. 7.2 *Wave between two infinite parallel conducting planes*

In order to determine the electromagnetic field configurations between the planes, Maxwell's equations are solved subject to the following two boundary conditions:

Boundary conditions:

1. Since the planes are assumed to be perfectly conducting, the tangential components of the electric field must be zero.

   $\therefore \qquad E_t = 0$

2. Since the planes are assumed to be perfectly conducting, the normal components of the magnetic field must be zero.

   $\therefore \qquad H_n = 0$

By Maxwell's equations for harmonically varying fields,

$$\nabla \times \bar{E} = -\mu \frac{\partial \bar{H}}{\partial t} = -j\omega\mu\bar{H} \tag{7.1}$$

and

$$\nabla \times \bar{H} = (\sigma + j\omega\varepsilon)\bar{E} \tag{7.2}$$

By wave equations,

$$\nabla^2 \bar{E} = \gamma^2 \bar{E} \tag{7.3}$$

and

$$\nabla^2 \bar{H} = \gamma^2 \bar{H} \tag{7.4}$$

where, $\gamma^2 = (j\omega\sigma\mu - \omega^2\mu\varepsilon)$

For non-conducting region between the planes ($\sigma = 0$), these equations reduce to,

$$\nabla \times \bar{E} = -j\omega\mu\bar{H} \tag{7.5}$$

$$\nabla \times \bar{H} = j\omega\varepsilon\bar{E} \tag{7.6}$$

$$\nabla^2 \bar{E} = -\omega^2\mu\varepsilon\bar{E} \tag{7.7}$$

$$\nabla^2 \bar{H} = -\omega^2\mu\varepsilon\vec{H} \tag{7.8}$$

From Eq. (7.5),

$$\begin{vmatrix} \hat{a}_x & \hat{a}_y & \hat{a}_z \\ \dfrac{\partial}{\partial x} & \dfrac{\partial}{\partial y} & \dfrac{\partial}{\partial z} \\ E_x & E_y & E_z \end{vmatrix} = -j\omega\mu(H_x\hat{a}_x + H_y\hat{a}_y + H_z\hat{a}_z)$$

Equating both sides,

$$\left(\frac{\partial E_z}{\partial y} - \frac{\partial E_y}{\partial z}\right) = -j\omega\mu H_x \tag{7.9a}$$

$$\left(\frac{\partial E_x}{\partial z}-\frac{\partial E_z}{\partial x}\right)=-j\omega\mu H_y \tag{7.9b}$$

$$\left(\frac{\partial E_y}{\partial x}-\frac{\partial E_x}{\partial y}\right)=-j\omega\mu H_z \tag{7.9c}$$

From Eq. (7.6) in the same way, we get

$$\left(\frac{\partial H_z}{\partial y}-\frac{\partial H_y}{\partial z}\right)=j\omega\varepsilon E_x \tag{7.10a}$$

$$\left(\frac{\partial H_x}{\partial z}-\frac{\partial H_z}{\partial x}\right)=j\omega\varepsilon E_y \tag{7.10b}$$

$$\left(\frac{\partial H_y}{\partial x}-\frac{\partial H_x}{\partial y}\right)=j\omega\varepsilon E_z \tag{7.10c}$$

Similarly, from Eqs. (7.7) and (7.8), we get

$$\frac{\partial^2 \bar{E}}{\partial x^2}+\frac{\partial^2 \bar{E}}{\partial y^2}+\frac{\partial^2 \bar{E}}{\partial z^2}=-\omega^2\mu\varepsilon\bar{E} \tag{7.11a}$$

and

$$\frac{\partial^2 \bar{H}}{\partial x^2}+\frac{\partial^2 \bar{H}}{\partial y^2}+\frac{\partial^2 \bar{H}}{\partial z^2}=-\omega^2\mu\varepsilon\bar{H} \tag{7.11b}$$

We assume that the wave is propagating in the positive z-direction so that the variations of the field components *w. r. t.* z is expressed as $e^{-\gamma z}$, where, γ is a complex propagation constant ($\gamma = \alpha + j\beta$).

Therefore, the field equations can be expressed as

$$\begin{aligned} E_x &= E_{x0}e^{-\gamma z}\\ E_y &= E_{y0}e^{-\gamma z}\\ E_z &= E_{z0}e^{-\gamma z}\\ H_x &= H_{x0}e^{-\gamma z}\\ H_y &= H_{y0}e^{-\gamma z}\\ H_z &= H_{z0}e^{-\gamma z} \end{aligned} \tag{7.12}$$

$$\therefore \qquad \frac{\partial E_y}{\partial z}=-\gamma E_y \quad \text{and} \quad \frac{\partial E_x}{\partial z}=-\gamma E_x$$

$$\frac{\partial H_x}{\partial z}=-\gamma H_z \quad \text{and} \quad \frac{\partial H_y}{\partial z}=-\gamma H_y \tag{7.13}$$

$$\therefore \qquad \frac{\partial^2 E_y}{\partial z^2}=\gamma^2 E_y \quad \text{and} \quad \frac{\partial^2 E_x}{\partial z^2}=\gamma^2 E_x, \ldots \text{etc.}$$

Also, as the planes are infinitely extended in the *y*-direction, there is no boundary condition in this direction and the fields in that direction can be assumed to be uniform or constant. This means that all derivatives *w. r. t. y* are zero. However, the planes are finite in the *x*-direction and thus, there are certain boundary conditions in that direction.

$$\therefore \qquad \frac{\partial E_x}{\partial y} = \frac{\partial E_z}{\partial y} = \frac{\partial H_x}{\partial y} = \frac{\partial H_z}{\partial y} = 0 \tag{7.14}$$

and

$$\frac{\partial^2 \overline{E}}{\partial y^2} = 0 \quad \frac{\partial^2 \overline{H}}{\partial y^2} = 0$$

Hence, from Eq. (7.9),

$$\gamma E_y = -j\omega\mu H_x \tag{7.15a}$$

$$\gamma E_x + \frac{\partial E_z}{\partial x} = j\omega\mu H_y \tag{7.15b}$$

$$\frac{\partial E_y}{\partial x} = -j\omega\mu H_z \tag{7.15c}$$

From Eq. (7.10),

$$\gamma H_y = j\omega\varepsilon E_x \tag{7.16a}$$

$$-\gamma H_x - \frac{\partial H_z}{\partial x} = j\omega\varepsilon E_y \tag{7.16b}$$

$$\frac{\partial H_y}{\partial x} = j\omega\varepsilon E_z \tag{7.16c}$$

From Eq. (7.11),

$$\frac{\partial^2 \overline{E}}{\partial x^2} + \gamma^2 \overline{E} = -\omega^2 \mu\varepsilon \overline{E} \tag{7.17a}$$

$$\frac{\partial^2 \overline{H}}{\partial x^2} + \gamma^2 \overline{H} = -\omega^2 \mu\varepsilon \overline{H} \tag{7.17b}$$

From Eq. (7.16*a*), $H_y = \dfrac{j\omega\varepsilon}{\gamma} E_x$

Putting this in Eq. (7.15*b*),

$$\gamma E_x + \frac{\partial E_z}{\partial x} = j\omega\mu \frac{j\omega\varepsilon}{\gamma} E_x$$

or

$$E_x \left(\gamma + \frac{\omega^2 \mu\varepsilon}{\gamma} \right) = -\frac{\partial E_z}{\partial x}$$

or

$$E_x = -\left(\frac{\gamma}{\gamma^2 + \omega^2 \mu\varepsilon} \right) \frac{\partial E_z}{\partial x} = -\frac{\gamma}{h^2} \frac{\partial E_z}{\partial x}$$

where $h^2 = (\gamma^2 + \omega^2\mu\varepsilon)$.

$$\therefore \qquad \boxed{E_x = -\frac{\gamma}{h^2}\frac{\partial E_z}{\partial x}} \tag{7.18}$$

Similarly, from Eq. (7.15*a*) and Eq. (7.16*b*), we get

$$-\gamma H_x - \frac{\partial H_z}{\partial x} = j\omega\varepsilon E_y = j\omega\varepsilon\left(-\frac{j\omega\mu}{\gamma}\right)H_x$$

or
$$H_x\left(\gamma + \frac{\omega^2\mu\varepsilon}{\gamma}\right) = -\frac{\partial H_z}{\partial x}$$

$$\therefore \qquad \boxed{H_x = -\frac{\gamma}{h^2}\frac{\partial H_z}{\partial x}} \tag{7.19}$$

Similarly, from Eqs. (7.16*b*) and (7.19), we get

$$-\gamma\left(-\frac{\gamma}{h^2}\right)\frac{\partial H_z}{\partial x} - \frac{\partial H_z}{\partial x} = j\omega\varepsilon E_y$$

or
$$\left(\frac{\gamma^2 - h^2}{h^2}\right)\frac{\partial H_z}{\partial x} = j\omega\varepsilon E_y$$

or
$$E_y = \frac{1}{j\omega\varepsilon}\left(\frac{\gamma^2 - h^2}{h^2}\right)\frac{\partial H_z}{\partial x} = \frac{-\omega^2\mu\varepsilon}{j\omega\varepsilon h^2}\frac{\partial H_z}{\partial x} \qquad \{\because h^2 = \gamma^2 + \omega^2\mu\varepsilon\}$$

or
$$E_y = \frac{j\omega\varepsilon}{h^2}\frac{\partial H_z}{\partial x}$$

$$\therefore \qquad \boxed{E_y = \frac{j\omega\varepsilon}{h^2}\frac{\partial H_z}{\partial x}} \tag{7.20}$$

Finally, from Eqs (7.15*b*) and Eq. (7.18), we get

$$\gamma\left(-\frac{\gamma}{h^2}\right)\frac{\partial E_z}{\partial x} + \frac{\partial E_z}{\partial x} = j\omega\mu H_y$$

or
$$\left(\frac{h^2 - \gamma^2}{h^2}\right)\frac{\partial E_z}{\partial x} = j\omega\mu H_y$$

or
$$H_y = \frac{1}{j\omega\mu}\left(\frac{h^2 - \gamma^2}{h^2}\right)\frac{\partial E_z}{\partial x} = \frac{\omega^2\mu\varepsilon}{j\omega\mu h^2}\frac{\partial E_z}{\partial x} = -\frac{j\omega\varepsilon}{h^2}\frac{\partial E_z}{\partial x}$$

$$\therefore \qquad \boxed{H_y = -\frac{j\omega\varepsilon}{h^2}\frac{\partial E_z}{\partial x}} \tag{7.21}$$

Equations (7.18) to (7.21) express the components of the electric and magnetic fields $\bar{E}$ and $\bar{H}$ in terms of E_z and H_z, i.e., the component of the fields in the direction of wave propagation.

It is observed that there must be a z-component of the field either $\bar{E}$ or $\bar{H}$; otherwise, all the components of $\bar{E}$ and $\bar{H}$ would be zero and there would be no fields in the region considered. In general, both E_z and H_z are present.

Depending upon the presence of the components of $\bar{E}$ and $\bar{H}$, the waves within the parallel planes are classified into three categories as

1. Transverse Electric Waves or *TE*-Waves or *H*-Waves,
2. Transverse Magnetic Waves or *TM*-Waves or *E*-Waves, and
3. Transverse Electromagnetic Waves or *TEM*-Waves

7.2.1 Field Equations for Transverse Electric (*TE*) Mode in Parallel-Plane Waveguide

In this case, there is a component of $\bar{H}$ in the direction of wave propagation, but no component of $\bar{E}$ is present in that direction. In other words, the electric field $\bar{E}$ lies in a plane transverse or perpendicular to the direction of wave propagation and the component of the magnetic field lies along the direction of wave propagation.

This implies that

$$E_z = 0 \quad \text{and} \quad H_z \neq 0$$

Substituting these conditions in Eqs. (7.18) and (7.21), we get

$$E_x = -\frac{\gamma}{h^2}\frac{\partial E_z}{\partial x} = 0$$

and

$$H_y = -\frac{j\omega\varepsilon}{h^2}\frac{\partial E_z}{\partial x} = 0$$

Thus, we get $E_x = 0$, $H_y = 0$; but in general, other components E_y, H_x, and H_z will be present.
From Eq. (7.17), the wave equation can be written in its component form as

$$\frac{\partial^2 E_x}{\partial x^2} + \gamma^2 E_x = -\omega^2 \mu\varepsilon E_x \tag{7.22a}$$

$$\frac{\partial^2 E_y}{\partial x^2} + \gamma^2 E_y = -\omega^2 \mu\varepsilon E_y \tag{7.22b}$$

$$\frac{\partial^2 E_z}{\partial x^2} + \gamma^2 E_z = -\omega^2 \mu\varepsilon E_z \tag{7.22c}$$

Similarly,

$$\frac{\partial^2 H_x}{\partial x^2} + \gamma^2 H_x = -\omega^2 \mu\varepsilon H_x \tag{7.23a}$$

$$\frac{\partial^2 H_y}{\partial x^2}+\gamma^2 H_y=-\omega^2\mu\varepsilon H_y \tag{7.23b}$$

$$\frac{\partial^2 H_z}{\partial x^2}+\gamma^2 H_z=-\omega^2\mu\varepsilon H_z \tag{7.23c}$$

From Eq. (7.22*b*), we get

$$\frac{\partial^2 E_y}{\partial x^2}=-(\gamma^2+\omega^2\mu\varepsilon)E_y=-h^2 E_y \tag{7.24}$$

Now, since $E_y = E_{y0}e^{-\gamma z}$, $\therefore \quad \frac{\partial E_y}{\partial x}=\frac{\partial E_{y0}}{\partial x}e^{-\gamma z}, \quad \therefore \quad \frac{\partial^2 E_y}{\partial x^2}=\frac{\partial^2 E_{y0}}{\partial x^2}e^{-\gamma z}$

From Eq. (7.24),

$$\frac{\partial^2 E_{y0}}{\partial x^2}e^{-\gamma z}=-h^2 E_{y0}e^{-\gamma z}$$

or

$$\frac{\partial^2 E_{y0}}{\partial x^2}=-h^2 E_{y0} \tag{7.25}$$

Solution of Eq. (7.25) yields

$$E_{y0}=A_1 \sin hx + A_2 \cos hx$$

Incorporating the z variation, we have

$$E_y=(A_1 \sin hx + A_2 \cos hx)e^{-\gamma z} \tag{7.26}$$

where A_1 and A_2 are arbitrary constants.

Applying the boundary conditions for the parallel plane guide, that the tangential component of $\bar{E}$ is zero at the surface of the conductor for all values of z and time, we get

$$E_y = 0 \text{ at } \quad x=0$$

$$E_y = 0 \text{ at } \quad x=d$$

$\therefore \quad 0=(A_1 \sin h0 + A_2 \cos h0)e^{-\gamma z}$

$\Rightarrow \quad A_2 = 0$

$$\therefore \quad E_y = A_1 \sin hxe^{-\gamma z}$$

Applying the second boundary condition,

$$\therefore \quad A_1 \sin hde^{-\gamma z}=0$$

$$\Rightarrow \quad \sin hd = 0$$

$$\Rightarrow \quad hd = n\pi \qquad n=\pm1, \pm2, \pm3,\ldots$$

$$\therefore \quad \boxed{h=\frac{n\pi}{d}} \tag{7.27}$$

Since for $n = 0$, all the field components become zero and there is no propagation of wave, we exclude zero value, i.e., $n \neq 0$.

$$\therefore \quad \boxed{E_y = A_1 \sin\left(\frac{n\pi}{d}x\right)e^{-\gamma z}} \quad (7.28)$$

The other components H_x and H_z are obtained as follows.
From Eqs. (7.19) and (7.20),

$$H_x = -\frac{\gamma}{h^2}\frac{\partial H_z}{\partial x} \quad \text{and} \quad E_y = \frac{j\omega\varepsilon}{h^2}\frac{\partial H_z}{\partial x}$$

Combining these two equations, we get

$$H_x = -\frac{\gamma}{j\omega\mu}E_y = -\frac{\gamma}{j\omega\mu}A_1 \sin\left(\frac{n\pi}{d}x\right)e^{-\gamma z}$$

$$\therefore \quad \boxed{H_x = -\frac{\gamma}{j\omega\mu}A_1 \sin\left(\frac{n\pi}{d}x\right)e^{-\gamma z}} \quad (7.29)$$

Also,

$$\frac{\partial E_y}{\partial x} = A_1 \frac{n\pi}{d}\cos\left(\frac{n\pi}{d}x\right)e^{-\gamma z}$$

From Eq. (7.15*c*),

$$H_z = -\frac{1}{j\omega\mu}\frac{\partial E_y}{\partial x} = -\frac{n\pi}{j\omega\mu d}A_1 \cos\left(\frac{n\pi}{d}x\right)e^{-\gamma z}$$

$$\therefore \quad \boxed{H_z = -\frac{n\pi}{j\omega\mu d}A_1 \cos\left(\frac{n\pi}{d}x\right)e^{-\gamma z}} \quad (7.30)$$

Therefore, we can summarise the components of the electric and magnetic fields for *TE* waves as follows.

$$\boxed{\begin{aligned} E_x &= 0 \\ E_y &= A_1 \sin\left(\frac{n\pi}{d}x\right)e^{-\gamma z} \\ E_z &= 0 \\ H_x &= -\frac{\gamma}{j\omega\mu}A_1 \sin\left(\frac{n\pi}{d}x\right)e^{-\gamma z} \\ H_y &= 0 \\ H_z &= -\frac{n\pi}{j\omega\mu d}A_1 \cos\left(\frac{n\pi}{d}x\right)e^{-\gamma z} \end{aligned}} \quad (7.31)$$

NOTE

The value of n specifies a particular field configuration or mode. The wave associated with integer n is designated as TE_n mode or TE_n wave. The smallest value of n is unity ($n = 1$) and the lowest order mode is TE_1.

In Eq. (7.31), the propagation constant, $\gamma = (\alpha + j\beta)$ is a complex quantity. For the wave propagation between two perfectly conducting planes, γ is either purely real or purely imaginary, depending upon the frequencies of the wave. For a real value of γ, i.e., $\gamma = \alpha$, or $\beta = 0$, there is no phase shift, but the wave is attenuated and there is no wave propagation. However, in the range of frequencies in which the wave propagation will occur and there will be no attenuation, the propagation constant γ will be purely imaginary, i.e., $\gamma = j\beta$ and $\alpha = 0$. Under this condition, the field equations of Eq. (7.31) is written as

$$\boxed{\begin{aligned} E_x &= 0 \\ E_y &= A_1 \sin\left(\frac{n\pi}{d}x\right)e^{-j\beta z} \\ E_z &= 0 \\ H_x &= -\frac{\beta}{\omega\mu}A_1 \sin\left(\frac{n\pi}{d}x\right)e^{-j\beta z} \\ H_y &= 0 \\ H_z &= -\frac{n\pi}{j\omega\mu d}A_1 \cos\left(\frac{n\pi}{d}x\right)e^{-j\beta z} \end{aligned}} \tag{7.32}$$

A sketch of these field distributions at some particular instant of time is shown in Fig. 7.3 for TE_{10} mode.

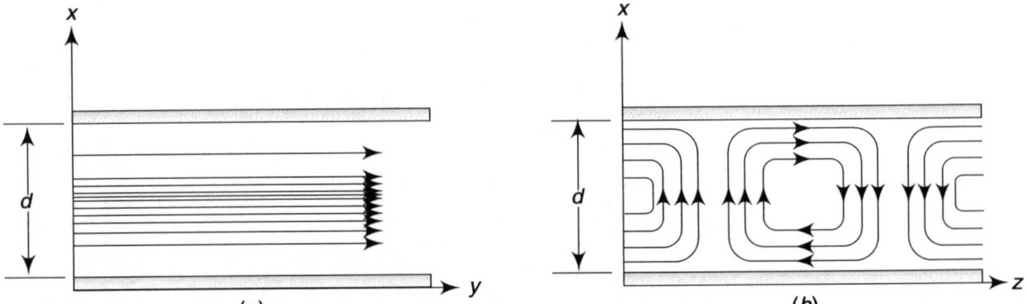

Fig. 7.3 *(a) Electric and (b) magnetic field distributions between parallel planes for TE_1 mode*

7.2.2 Field Equations for Transverse Magnetic (*TM*) Mode in Parallel-Plane Waveguide

In this case, there is a component of $\bar{E}$ in the direction of wave propagation, but no component of $\bar{H}$ is present in that direction. In other words, the magnetic field $\bar{H}$ lies in a plane transverse or perpendicular to the direction of wave propagation and the component of the electric field $\bar{E}$ lies along the direction of wave propagation.

This implies that

$$H_z = 0 \quad \text{and} \quad E_z \neq 0$$

Substituting these conditions in Eqs. (7.19) and (7.20), we get

$$H_x = -\frac{\gamma}{h^2}\frac{\partial H_z}{\partial x} = 0$$

and

$$E_y = \frac{j\omega\varepsilon}{h^2}\frac{\partial H_z}{\partial x} = 0$$

Thus, we get, $H_x = 0$, $E_y = 0$; but in general, other components E_x, E_z, and H_y will be present. From Eq. (7.17), the wave equation can be written in its component form as

$$\frac{\partial^2 E_x}{\partial x^2} + \gamma^2 E_x = -\omega^2 \mu\varepsilon E_x \tag{7.22a}$$

$$\frac{\partial^2 E_y}{\partial x^2} + \gamma^2 E_y = -\omega^2 \mu\varepsilon E_y \tag{7.22b}$$

$$\frac{\partial^2 E_z}{\partial x^2} + \gamma^2 E_z = -\omega^2 \mu\varepsilon E_z \tag{7.22c}$$

Similarly,

$$\frac{\partial^2 H_x}{\partial x^2} + \gamma^2 H_x = -\omega^2 \mu\varepsilon H_x \tag{7.23a}$$

$$\frac{\partial^2 H_y}{\partial x^2} + \gamma^2 H_y = -\omega^2 \mu\varepsilon H_y \tag{7.23b}$$

$$\frac{\partial^2 H_z}{\partial x^2} + \gamma^2 H_z = -\omega^2 \mu\varepsilon H_z \tag{7.23c}$$

From Eq. (7.23*b*), we get

$$\frac{\partial^2 H_y}{\partial x^2} = -(\gamma^2 + \omega^2\mu\varepsilon)H_y = -h^2 H_y \tag{7.33}$$

In a similar way as for *TE* waves, the solution of this equation can be written as

$$H_y = (A_3 \sin hx + A_4 \cos hx)e^{-\gamma z} \tag{7.34}$$

where A_3 and A_4 are arbitrary constants.

In this case, the boundary conditions cannot be applied directly to H_y to evaluate the constants A_3 and A_4 because, in general, the tangential component of $\bar{H}$ is not zero at the surface of the conductor. However, the expression for E_z can be obtained in terms of H_y and then the boundary conditions can be applied to E_z.

Using Eq. (7.16*c*),

$$E_z = \frac{1}{j\omega\varepsilon}\frac{\partial H_y}{\partial x} = \frac{h}{j\omega\varepsilon}(A_3 \cos hx - A_4 \sin hx)e^{-\gamma z}$$

Applying the boundary conditions,
$E_z = 0$ at $x = 0$

$$\therefore \qquad 0 = \frac{h}{j\omega\varepsilon}(A_3)e^{-\gamma z} \Rightarrow A_3 = 0$$

$E_y = 0$ at $x = d$

$$\therefore \quad \frac{h}{j\omega\varepsilon}(-A_4 \sin hd)e^{-\gamma z} = 0$$

$$\Rightarrow \quad \sin hd = 0$$

$$\Rightarrow \quad hd = n\pi$$

$$\Rightarrow \quad h = \frac{n\pi}{d}$$

where n is an integer.

$$\therefore \quad \boxed{E_z = -\frac{h}{j\omega\varepsilon} A_4 \sin\left(\frac{n\pi}{d}x\right)e^{-\gamma z}} \quad \textbf{(7.35)}$$

The other components E_x and H_y are obtained as follows.
From Eq. (7.16*c*),

$$\frac{\partial H_y}{\partial x} = j\omega\varepsilon E_z$$

$$\therefore \quad H_y = j\omega\varepsilon \int E_z dx = j\omega\varepsilon \int -\frac{h}{j\omega\varepsilon} A_4 \sin\left(\frac{n\pi}{d}x\right)e^{-\gamma z}dx = hA_4 \frac{d}{n\pi}\cos\left(\frac{n\pi}{d}x\right)e^{-\gamma z}$$

$$= A_4 \cos\left(\frac{n\pi}{d}x\right)e^{-\gamma z} \quad \left\{\because h = \frac{n\pi}{d}\right\}$$

$$\therefore \quad \boxed{H_y = A_4 \cos\left(\frac{n\pi}{d}x\right)e^{-\gamma z}} \quad \textbf{(7.36)}$$

From Eq. (7.16*a*),

$$E_x = \frac{\gamma}{j\omega\varepsilon} H_y = \frac{\gamma}{j\omega\varepsilon} A_4 \cos\left(\frac{n\pi}{d}x\right)e^{-\gamma z}$$

$$\therefore \quad \boxed{E_x = \frac{\gamma}{j\omega\varepsilon} A_4 \cos\left(\frac{n\pi}{d}x\right)e^{-\gamma z}} \quad \textbf{(7.37)}$$

Therefore, we can summarise the components of the electric and magnetic fields for *TM* waves as follows.

$$\boxed{\begin{aligned} &E_x = \frac{\beta A_4}{\omega\varepsilon}\cos\left(\frac{n\pi}{d}x\right)e^{-\gamma z} \\ &E_y = 0 \\ &E_z = -\frac{hA_4}{j\omega\varepsilon}\sin\left(\frac{n\pi}{d}x\right)e^{-\gamma z} \\ &H_x = 0 \\ &H_y = A_4 \cos\left(\frac{n\pi}{d}x\right)e^{-\gamma z} \\ &H_z = 0 \end{aligned}} \quad \textbf{(7.38)}$$

NOTE

As in TE waves, the value of n specifies a particular field configuration or mode. The wave associated with integer n is designated as TM_n mode or TM_n wave. Unlike TW waves, there is a possibility of n being zero, as for n = 0, some field components (E_x and H_y) still exist. Hence, the smallest value of n is zero (n = 0) and the lowest order mode is TM_0.

Similar to *TE* waves, for *TM* waves too, in the range of frequencies in which the wave propagation occurs, the propagation constant is purely imaginary, i.e., $\gamma = j\beta$ and thus, the field equations of Eq. (7.38) can be written as

$$\boxed{\begin{aligned} E_x &= \frac{\gamma A_4}{j\omega\varepsilon}\cos\left(\frac{n\pi}{d}x\right)e^{-j\beta z} \\ E_y &= 0 \\ E_z &= -\frac{hA_4}{j\omega\varepsilon}\sin\left(\frac{n\pi}{d}x\right)e^{-j\beta z} \\ H_x &= 0 \\ H_y &= A_4\cos\left(\frac{n\pi}{d}x\right)e^{-j\beta z} \\ H_z &= 0 \end{aligned}} \qquad (7.39)$$

A sketch of these field distributions at some particular instant of time is shown in Fig. 7.4 for TM_1 mode.

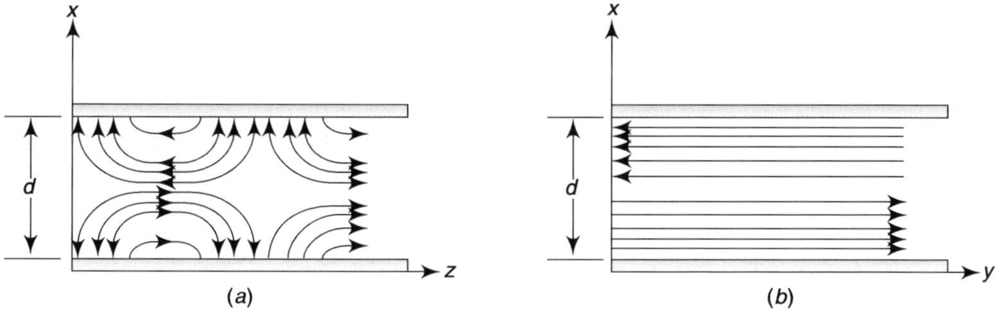

Fig. 7.4 *(a) Electric and (b) magnetic field distributions between parallel planes for TM_1 mode*

7.2.3 Characteristics of *TE* and *TM* Waves in Parallel-Plane Waveguide

From the summary of the different field components for *TE* and *TM* waves, it is seen that for each of the components of $\bar{E}$ or $\bar{H}$, there is a *sinusoidal or cosinusoidal standing-wave distribution* across the guide in the *x*-direction. This means that each of these components varies in magnitude, but not in phase, in the *x*-direction. In the *y*-direction, by assumption, there is no variation of either magnitude or phase of any of the field components.

Thus, any *xy*-plane is an *equiphase plane* for each of the field components. These equiphase surfaces propagate along the guide in the *z*-direction.

We find the following quantities for the *TE* and *TM* mode waves for parallel plane waveguide.

1. Propagation Constant (γ),
2. Cut-off Frequency (f_c) and Cut-off Wavelength (λ_c),
3. Guide Wavelength (λ),
4. Velocities of Wave Propagation, and
5. Wave Impedance (η).

1. Propagation Constant (γ):

We have, $$h^2 = \gamma^2 + \omega^2 \mu\varepsilon = \left(\frac{n\pi}{d}\right)^2$$

$$\therefore \quad \boxed{\gamma = \sqrt{\left(\frac{n\pi}{d}\right)^2 - \omega^2 \mu\varepsilon}} \tag{7.40}$$

The value of $\gamma = (\alpha + j\beta)$ depends on the value of the frequency of the wave and the medium through which the wave is propagating.

2. Cut-off Frequency (f_c) and Cut-off Wavelength (λ_c):

From Eq. (7.31), three conditions can be derived.

(a) *When the term under radical is zero:*

$$\therefore \quad \left(\frac{n\pi}{d}\right)^2 - \omega^2 \mu\varepsilon \quad \text{at any frequency } \omega = \omega_c$$

or $$\omega_c = \frac{n\pi}{d}\frac{1}{\sqrt{\mu\varepsilon}}$$

$$\therefore \quad \boxed{f_c = \frac{n}{2d}\frac{1}{\sqrt{\mu\varepsilon}}} \tag{7.41}$$

This frequency is known as the *cut-off frequency* or *critical frequency*.

At this frequency, the wave starts travelling along the guide or the attenuation condition changes. This is observed that for each value of n, there is a corresponding cut-off frequency.

(b) *When the frequency is below the cut-off frequency, $f < f_c$:*

The propagation constant γ is real, i.e., $\gamma = \alpha$, $\beta = 0$.

The field amplitudes will decrease very rapidly with distance z according to the exponential decay $e^{-\gamma z}$ even though the phase angle will remain constant.

(c) *When the frequency is above the cut-off frequency, $f > f_c$:*

Here, the propagation constant will be imaginary, i.e., $\gamma = j\beta$, $\alpha = 0$. This implies that the wave propagates without any attenuation. However, in a practical situation, owing to the finite conductivity of the planes, some attenuation is present at frequencies above the cut-off frequency.

Now, from Eqs. (7.40) and (7.41),

$$\gamma = \sqrt{\left(\frac{n\pi}{d}\right)^2 - \omega^2\mu\varepsilon} = \frac{n\pi}{d}\sqrt{1 - \frac{\omega^2\mu\varepsilon d^2}{n^2\pi^2}} = \frac{n\pi}{d}\sqrt{1 - \frac{(2\pi f)^2\mu\varepsilon d^2}{n^2\pi^2}} = \frac{n\pi}{d}\sqrt{1 - \frac{f^2}{\frac{n^2}{(2d)^2}\frac{1}{\mu\varepsilon}}}$$

$$= \frac{n\pi}{d}\sqrt{1 - \frac{f^2}{f_c^2}}$$

$$\therefore \qquad \boxed{\gamma = \frac{n\pi}{d}\sqrt{1 - \frac{f^2}{f_c^2}}} \qquad (7.42)$$

For $f > f_c$, $\gamma = j\beta$ and using Eq. (7.42), we have

$$\therefore \qquad \boxed{\beta = \sqrt{\omega^2 \mu\varepsilon - \left(\frac{n\pi}{d}\right)^2} = \frac{n\pi}{d}\sqrt{\frac{f^2}{f_c^2} - 1}} \qquad (7.43)$$

For $f >> f_c$,
$$\beta = \frac{n\pi}{d}\frac{f}{f_c} = \frac{n\pi}{d}\frac{\omega}{\omega_c} = \frac{n\pi}{d}\frac{\omega}{\frac{n\pi}{d}\frac{1}{\sqrt{\mu\varepsilon}}} = \omega\sqrt{\mu\varepsilon}$$

$$\therefore \qquad \boxed{\beta = \omega\sqrt{\mu\varepsilon}} \qquad (7.44)$$

Thus, from Eqs. (7.41) to (7.44), it is seen that the phase constant varies from zero to $\omega\sqrt{\mu\varepsilon}$ as the frequency approaches infinity.

Also, corresponding to the critical frequency (f_c), there is a *critical wavelength* or *cut-off wavelength* (λ_c) defined as that wavelength above which the wave will not be propagated through the region between the parallel planes. It is given as

$$\lambda_c = \frac{v_p}{f_c} = \frac{1/\sqrt{\mu\varepsilon}}{\left(\frac{n}{2d}\right)1/\sqrt{\mu\varepsilon}} = \frac{2d}{n}$$

where $v_p = \frac{1}{\sqrt{\mu\varepsilon}}$ is the phase velocity in an unbounded medium

$$\therefore \qquad \boxed{\lambda_c = \frac{2d}{n}} \qquad (7.45)$$

Equation (7.45) gives the relation between the plane separation distance and the critical wavelength $\left(d = \frac{n\lambda_c}{2}\right)$. Thus, critical wavelength is determined by the spacing between the planes as that at which the distance between the planes is exactly n times half wavelength.

Waves having wavelengths greater than the critical wavelength are attenuated and the waves with smaller wavelengths are propagated without losses.

The largest value of the critical wavelength is

$$\lambda_{max} = 2d \qquad \text{with } n = 1$$

This means that the largest free-space wavelength a signal may have and still be capable of propagating through parallel plane guide is just less than twice the plane separation. When $n = 1$, the signal is said to be propagated in the *dominant mode*, which yields the longest cut-off wavelength.

3. Guide Wavelength (λ_g):

This is the distance required to produce a phase sift of 360° or 2π radian within the guide.

$$\therefore \qquad \lambda_g = \frac{2\pi}{\beta} \text{ (in metre)} \qquad (7.46)$$

From Eq. (7.43),

$$\lambda_g = \frac{2\pi}{\sqrt{\omega^2\mu\varepsilon - \left(\frac{n\pi}{d}\right)^2}} = \frac{2\pi}{\omega\sqrt{\mu\varepsilon}\sqrt{1 - \frac{\left(\frac{n\pi}{d}\right)^2}{\omega^2\mu\varepsilon}}} = \frac{(2\pi/\omega)\frac{1}{\sqrt{\mu\varepsilon}}}{\sqrt{1 - \frac{\left(\frac{n\pi}{d}\right)^2\frac{1}{\mu\varepsilon}}{\omega^2}}}$$

$$= \frac{v_p/f}{\sqrt{1 - \left(\frac{n}{2d}\frac{1}{f\sqrt{\mu\varepsilon}}\right)^2}} \qquad \{\because \omega = 2\pi f\}$$

$$= \frac{\lambda}{\sqrt{1 - \left(\frac{n\lambda}{2d}\right)^2}} \qquad \left\{\because v_p = \frac{1}{\sqrt{\mu\varepsilon}} \text{ and } v_p = f\lambda\right\}$$

$$= \frac{\lambda}{\sqrt{1 - \frac{\omega_c^2}{\omega^2}}} = \frac{\lambda}{\sqrt{1 - \frac{f_c^2}{f^2}}} = \frac{\lambda}{\sqrt{1 - \frac{\lambda^2}{\lambda_c^2}}} \qquad \left\{\because v_p = f\lambda = \text{constant}, \quad \therefore \quad f \propto \frac{1}{\lambda}\right\}$$

$$\therefore \qquad \boxed{\lambda_g = \frac{\lambda}{\sqrt{1 - \left(\frac{n\lambda}{2d}\right)^2}} = \frac{\lambda}{\sqrt{1 - \frac{\omega_c^2}{\omega^2}}} = \frac{\lambda}{\sqrt{1 - \frac{f_c^2}{f^2}}} = \frac{\lambda}{\sqrt{1 - \frac{\lambda^2}{\lambda_c^2}}}} \qquad \textbf{(7.47)}$$

From Eq. (7.47), we can also show that

$$\boxed{\frac{1}{\lambda^2} = \frac{1}{\lambda_g^2} + \frac{1}{\lambda_c^2}} \qquad \textbf{(7.48)}$$

From Eq. (7.47), the equation of the critical wavelength can also be derived. As per the definition of critical wavelength as that wavelength which is unable to propagate in the waveguide, we derive that value of λ from Eq. (7.47) for which λ_g becomes infinite.

$$\therefore \qquad 1 - \left(\frac{n\lambda_c}{2d}\right)^2 = 0 \quad \Rightarrow \quad \lambda_c = \frac{2d}{n}$$

This is identical with the result obtained in Eq. (7.45).

4. Velocities of Wave Propagation:

We define two types of velocities:

(a) Phase Velocity, v_p and

(b) Group Velocity, v_g

(a) Phase Velocity, v_p: The *phase velocity* is defined as the velocity at which a point of constant phase moves. It is actually the velocity with which the wave travels within the guide.

For parallel-plane waveguide, it is given as

$$v_p = \frac{\omega}{\beta} = \frac{\omega}{\sqrt{\omega^2\mu\varepsilon - \left(\frac{n\pi}{d}\right)^2}} = \frac{1/\sqrt{\mu\varepsilon}}{\sqrt{1 - \frac{f_c^2}{f^2}}} = \frac{v_0}{\sqrt{1 - \frac{f_c^2}{f^2}}} = \frac{v_0}{\sqrt{1 - \frac{\lambda^2}{\lambda_c^2}}}$$

where, $v_0 = \frac{1}{\sqrt{\mu\varepsilon}}$ is the velocity of the wave in an unbounded medium.

$$\therefore \quad \boxed{v_p = \frac{v_0}{\sqrt{1-\frac{f_c^2}{f^2}}} = \frac{v_0}{\sqrt{1-\frac{\lambda^2}{\lambda_c^2}}}} \tag{7.49}$$

(b) Group Velocity, v_g: The *group velocity* (or *energy velocity*) is the velocity with which the energy is transported down the length of the waveguide.
For parallel-plane waveguide, it is given as

$$v_g = \frac{d\omega}{d\beta} = \frac{v_0^2}{v_p} \tag{7.50}$$

Now, $$\beta = \frac{n\pi}{d}\sqrt{\frac{f^2}{f_c^2}-1}$$

$$\therefore \quad \frac{d\beta}{d\omega} = \frac{1}{2\pi}\frac{d}{df}\left\{\frac{n\pi}{d}\sqrt{\frac{f^2}{f_c^2}-1}\right\} = \frac{1}{2\pi}\frac{n\pi}{d}\frac{1}{2}\left(\frac{f^2}{f_c^2}-1\right)^{-1/2} \times \frac{2f}{f_c^2} = \frac{n}{2df_c\sqrt{1-\frac{f_c^2}{f^2}}}$$

Substituting this value in Eq. (7.50),

$$v_g = \frac{d\omega}{d\beta} = \frac{1}{d\beta/d\omega} = \frac{1}{\frac{n}{2df_c\sqrt{1-\frac{f_c^2}{f}}}} = \frac{2df_c}{n}\sqrt{1-\frac{f_c^{2^2}}{f}} = v_0\sqrt{1-\frac{f_c^{2^2}}{f}} \quad \left\{\because f_c = \frac{n}{2d}v_0\right\}$$

$$\boxed{v_g = \frac{2df_c}{n}\sqrt{1-\frac{f_c^2}{f^2}} = v_0\sqrt{1-\frac{f_c^2}{f^2}}} \tag{7.51}$$

NOTE

From Eqs. (7.49) and (7.51), it is seen that

$$v_0 = \sqrt{v_p v_g}$$

i.e., v_0 is the geometric mean of v_p and v_g.

5. Wave Impedance (η):
From Eqs. (7.28) and (7.29), the intrinsic impedance of *TE* or *TM* wave is obtained as

$$\eta_{TE} = \eta_{TM} = \left|\frac{E_y}{H_x}\right| = \frac{j\omega\mu}{\gamma} = \frac{j\omega\mu}{\sqrt{\left(\frac{n\pi}{d}\right)^2 - \omega^2\mu\varepsilon}} \qquad \text{[Substituting the value of } \gamma \text{ from Eq. (7.40)]}$$

$$= \frac{\sqrt{\mu/\varepsilon}}{\sqrt{1-\frac{f_c^2}{f^2}}} = \frac{\eta_0}{\sqrt{1-\frac{f_c^2}{f^2}}}$$

$$\therefore \quad \boxed{\eta = \frac{\sqrt{\mu/\varepsilon}}{\sqrt{1-\frac{f_c^2}{f^2}}} = \frac{\eta_0}{\sqrt{1-\frac{f_c^2}{f^2}}}} \tag{7.52}$$

where $\eta_0 = \sqrt{\mu_0/\varepsilon_0}$ is the intrinsic impedance of free space.

7.2.4 Transverse Electromagnetic (*TEM*) Mode

A *transverse electromagnetic* (*TEM*) wave is a special type of wave in which there are no components of electric and magnetic field in the direction of wave propagation. This type of wave is entirely transverse.

We have seen in earlier section that the lowest order mode was

$n = 1$, i.e., TE_1 mode for *TE* wave

$n = 0$, i.e., TM_0 mode for *TM* wave

This special case with $n = 0$ in *TM* wave is known as *transverse electromagnetic wave* or *TEM* wave.

Although this is a special case of guided wave propagation, this is extremely important one, because, this is the most common type of wave propagation along all ordinary two-conductor transmission lines when operating in their customary (low-frequency) manner. For this reason, this wave is also known as *the principal waves*.

For $n = 0$, the field components are obtained from the results of *TM* wave as

$$\begin{aligned} E_x &= \frac{\gamma A_4}{j\omega\varepsilon} e^{-\gamma z} \\ E_y &= 0 \\ E_z &= 0 \\ H_x &= 0 \\ H_y &= A_4 e^{-\gamma z} \\ H_z &= 0 \end{aligned} \tag{7.53}$$

Equation (7.53) also reveals that for *TEM* waves between the parallel planes, the fields are not only entirely transverse but they are constant in amplitude across a cross section normal to the direction of wave propagation and their ratio is also constant.

Characteristics of *TEM* Waves:

We determine the following for *TEM* waves:

1. Propagation constant (γ),
2. Intrinsic impedance (η),
3. Cut-off wavelength (λ_c), and
4. Velocity of wave propagation.

1. Propagation constant (γ):

When $n = 0$, the propagation constant for *TEM* wave is given as

$$\gamma = \sqrt{\left(\frac{n\pi}{d}\right)^2 - \omega^2 \mu\varepsilon} = \sqrt{0 - \omega^2 \mu\varepsilon} = j\omega\sqrt{\mu\varepsilon} = j\omega\beta$$

$$\therefore \qquad \boxed{\gamma = j\omega\sqrt{\mu\varepsilon} = j\omega\beta} \qquad \textbf{(7.54)}$$

where $\beta = \sqrt{\mu\varepsilon}$, i.e., the value of β corresponds to that of free space.

Equation (7.54) implies that the attenuation constant for *TEM* wave is zero, $\alpha = 0$ and the *TEM* wave propagates without attenuation between the two perfectly conducting planes for all frequencies above zero.

Substituting the value of γ from Eq. (7.54) in Eq. (7.53), we have

$$E_x = \sqrt{\frac{\mu}{\varepsilon}} A_4 e^{-j\beta z} \qquad \textbf{(7.55a)}$$

$$H_y = A_4 e^{-j\beta z} \qquad \textbf{(7.55b)}$$

2. Intrinsic impedance (η):

This is denoted by η_{TEM} and is given as

$$\eta_{TEM} = \left|\frac{E_x}{H_y}\right| = \frac{\sqrt{\frac{\mu}{\varepsilon}} A_4 e^{-j\beta z}}{A_4 e^{-j\beta z}} = \sqrt{\frac{\mu}{\varepsilon}}$$

$$\boxed{\eta_{TEM} = \sqrt{\frac{\mu}{\varepsilon}} \; (\Omega)} \qquad \textbf{(7.56a)}$$

This is identical with the intrinsic impedance of uniform plane wave propagating through a perfect dielectric medium. For free space, this is given as

$$\boxed{\eta_{TEM} = \sqrt{\frac{\mu_0}{\varepsilon_0}} = 120\pi = 377\ \Omega} \qquad \textbf{(7.56b)}$$

3. Cut-off frequency (f_c) and cut-off wavelength (λ_c):

From Eq. (7.37) for $n = 0$, the cut-off frequency is

$$f_c = 0 \quad \text{or} \quad \lambda_c = \infty$$

This implies that for transverse electromagnetic waves, all frequencies down to zero can propagate through parallel plane guide.

4. Velocity of wave propagation:

From Eq. (7.45) with zero cut-off frequency, the velocity of wave propagation for *TEM* wave is given as

$$v_p = \frac{1}{\sqrt{\mu\varepsilon}} = v_0 = 3 \times 10^8 \text{ m/s}$$

This means that unlike *TE* and *TM* wave, the velocity of *TEM* wave is independent of frequency and has the free space value. However, this is true only when the planes are perfectly conducting and the space between them is a vacuum. The effect of finite conductivity for the conducting planes is to reduce the velocity by a small amount $\left[v = \dfrac{v_0}{\left(1 + \dfrac{\sigma^2}{8\omega^2\varepsilon^2}\right)}\right]$ as explained in Section 5.7 in Chapter 5.

A sketch of the field distributions for *TEM* waves at some particular instant of time is shown in Fig. 7.5.

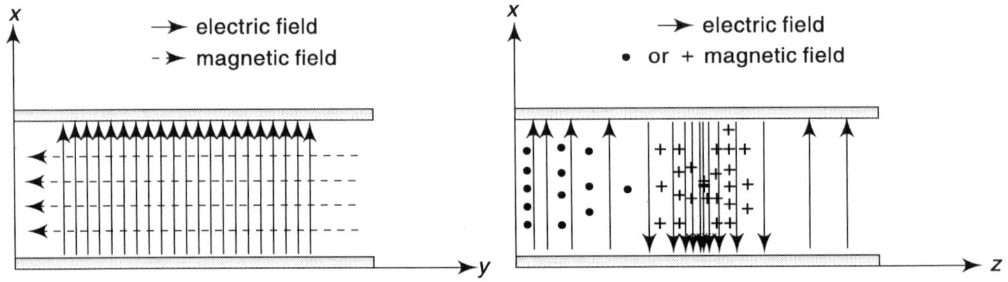

Fig. 7.5 *Electric and magnetic fields between parallel planes for TEM waves*

7.2.5 Manner of Wave Travel for *TE* and *TM* Modes

The velocity of propagation for a *TEM* wave (plane wave or transmission line wave) is referred to as the *phase velocity*. The phase velocity of a *TEM* wave is equal to the velocity of energy transport. The phase velocity of a *TEM* wave travelling in a lossless medium is given by

$$v_0 = \sqrt{\frac{1}{\mu\varepsilon}}$$

The phase velocity of *TE* or *TM* mode in a waveguide is defined in the same manner as that of a *TEM* wave. However, the waveguide phase velocity is not equal to the velocity of energy transport along the waveguide. The velocity at which energy is transported down the length of the waveguide is defined as the *group velocity*.

The differences between the waveguide phase velocity and group velocity can be illustrated using the field equations of the *TE* or *TM* rectangular waveguide modes. It can be shown that the field components of general *TE* and *TM* waveguide modes can be written as sums and differences of *TEM* waves.

We consider the field equations of a general *TE* wave in parallel plane guide.

$$E_y = A_1 \sin\left(\frac{n\pi}{d}x\right)e^{-j\beta z}e^{j\omega t}$$

$$H_x = -\frac{\beta}{\omega\mu}A_1 \sin\left(\frac{n\pi}{d}x\right)e^{-j\beta z}e^{j\omega t}$$

$$H_z = -\frac{n\pi}{j\omega\mu d}A_1 \cos\left(\frac{n\pi}{d}x\right)e^{-j\beta z}e^{j\omega t}$$

where, the term $e^{j\omega t}$ has been included to take into account the time variations.

Taking the real term of the first equation, we get

$$E_y = A_1 \sin\left(\frac{n\pi}{d}x\right)e^{-j\beta z}\cos\omega t$$

or $$E_y = \frac{1}{2}A_1 e^{-j\beta z}\left[\sin\left(\omega t + \frac{n\pi}{d}x\right) + \sin\left(\omega t - \frac{n\pi}{d}x\right)\right] \quad \textbf{(7.57)}$$

Similarly, taking the real terms of the other two equations, we get

$$H_x = -\frac{\beta}{\omega\mu}A_1 \sin\left(\frac{n\pi}{d}x\right)e^{-j\beta z}\cos\omega t$$

$$H_z = -\frac{n\pi}{\omega\mu d}A_1 \cos\left(\frac{n\pi}{d}x\right)e^{-j\beta z}\sin\omega t$$

Using the trigonometric formula,

$$\sin A \cos B = \frac{1}{2}[\sin(A+B) + \sin(A-B)]$$

$$\cos A \sin B = \frac{1}{2}[\sin(A+B) - \sin(A-B)]$$

The magnetic field components can be written as

$$H_x = -\frac{\beta}{2\omega\mu}A_1 e^{-j\beta z}\left[\sin\left(\omega t + \frac{n\pi}{d}x\right) - \sin\left(\omega t - \frac{n\pi}{d}x\right)\right] \quad \textbf{(7.58a)}$$

$$H_z = -\frac{n\pi}{2\omega\mu d}A_1 e^{-j\beta z}\left[\sin\left(\omega t + \frac{n\pi}{d}x\right) + \sin\left(\omega t - \frac{n\pi}{d}x\right)\right] \quad \textbf{(7.58b)}$$

If the directions of the field components are indicated by unit vectors $\hat{a}_x$ in the x direction and $\hat{a}_z$ in the z direction, the total magnetic field can be written as

$$\begin{aligned}\vec{H} &= H_x\hat{a}_x + H_z\hat{a}_z \\ &= -\frac{\beta}{2\omega\mu}A_1 e^{-j\beta z}\left[\sin\left(\omega t + \frac{n\pi}{d}x\right) - \sin\left(\omega t - \frac{n\pi}{d}x\right)\right]\hat{a}_x \\ &\quad - \frac{n\pi}{2\omega\mu d}A_1 e^{-j\beta z}\left[\sin\left(\omega t + \frac{n\pi}{d}x\right) + \sin\left(\omega t - \frac{n\pi}{d}x\right)\right]\hat{a}_z \\ &= -A_1 e^{-j\beta z}\frac{1}{2\omega\mu}\left[\left(\beta\hat{a}_x + \frac{n\pi}{d}\hat{a}_z\right)\sin\left(\omega t + \frac{n\pi}{d}x\right) - \left(\beta\hat{a}_x - \frac{n\pi}{d}\hat{a}_z\right)\sin\left(\omega t - \frac{n\pi}{d}x\right)\right]\end{aligned} \quad \textbf{(7.59)}$$

The two terms in the *TE* field equation above represent *TEM* waves moving in the directions shown in Fig. 7.6(*a*).

Equations Eq. (7.57) and Eq. (7.59) show that the wave propagation is the resultant of two *TEM* waves at an angle in the region between the planes. Thus, the *TE* wave in the rectangular waveguide can be represented as the superposition of two *TEM* waves reflecting from the upper and lower waveguide walls as they travel down the waveguide as represented in Fig. 7.6(b).

From Fig. (7.6), we have, $\tan\theta = \dfrac{\beta}{n\pi/d} = \dfrac{\beta d}{n\pi}$

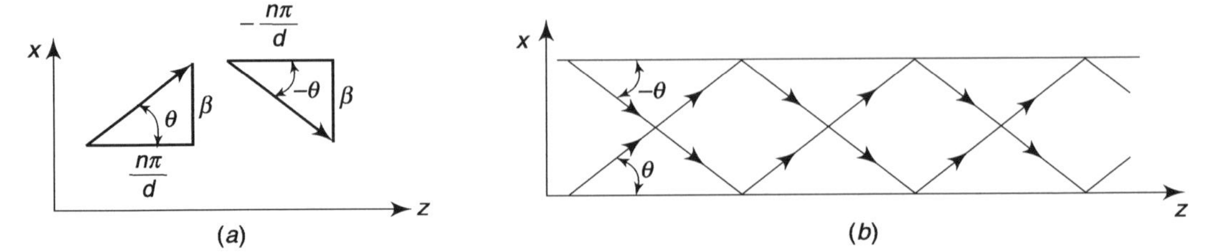

Fig. 7.6 *(a) Field components contributing to wave propagation and (b) Path of waves between parallel planes*

Replacing the value of β from Eq. (7.43) we get, $\tan\theta = \frac{d}{n\pi} \times \frac{n\pi}{d}\sqrt{\frac{f^2}{f_c^2} - 1} = \sqrt{\frac{f^2}{f_c^2} - 1}$

$$\therefore \qquad \boxed{\tan\theta = \sqrt{\frac{f^2}{f_c^2} - 1}} \qquad \textbf{(7.60)}$$

Although the above discussion pertains to *TE* mode, in general, it can be generalized to both *TE* and *TM* modes.

As the frequency is reduced and approach the critical frequency, tan θ and hence θ approaches zero. At $\theta = 0$, there is no transmission in the z direction since $\beta = 0$ and the waves simply bounce back and forth between the upper and lower planes. For frequencies much larger than the critical frequency, the angle of incidence θ becomes large and the wave propagates between the guiding planes by succession of glancing reflections.

For the general TE_{mn} of TM_{mn} waves, the phase velocity of the *TEM* component is given by

$$v_p = \frac{\beta}{\omega} = \frac{v_0}{\sqrt{1 - \frac{f_c^2}{f^2}}}$$

The waveguide phase velocity represents the speed at which points of constant phase of the component *TEM* waves travel down the waveguide.

The waveguide phase velocity is larger than the *TEM* wave phase velocity given that the square root in the denominator of the waveguide phase velocity equation is less than unity. The relationship between the waveguide phase velocity, waveguide group velocity, and the *TEM* component wave velocity is shown in Fig. 7.7.

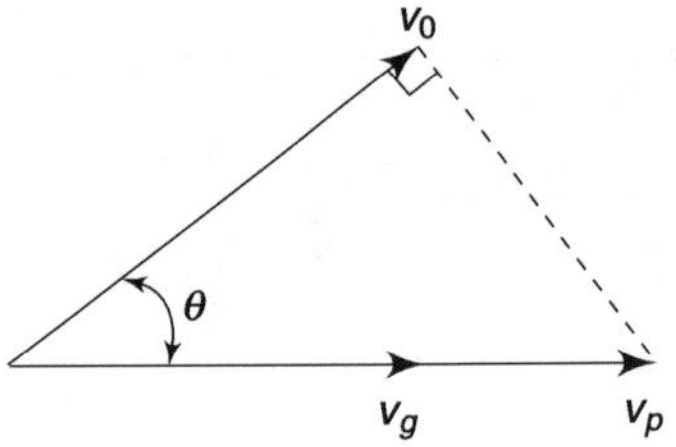

Fig. 7.7 *Relationship between phase and group velocity*

$$v_0 = v_p \cos\theta$$

$$\therefore \qquad \cos\theta = \frac{v_0}{v_p} = \sqrt{1 - \frac{f_c^2}{f^2}}$$

Also,

$$v_g = v_0 \cos\theta = v_0\sqrt{1 - \frac{f_c^2}{f^2}}$$

$$\therefore \qquad v_p v_g = v_0^2$$

The waveguide group velocity (the velocity of energy transport) is always smaller than the *TEM* wave phase velocity given the square root term in the numerator of the group velocity equation.

Example 7.1 Determine the maximum number of half cycles of electric field with which it may propagate in a waveguide with wall separation of 5 cm at a frequency of 10 GHz. Calculate the guide wavelength for this mode of propagation.

Solution Here, we have to find the largest value of n for which the cut-off wavelength is greater than the free space cut-off wavelength.

Free space wavelength, $\lambda_0 = \dfrac{c}{f} = \dfrac{3 \times 10^8}{10 \times 10^9} = 0.03 \text{ m} = 3 \text{ cm}$

For $n = 1$, cut-off wavelength, $\lambda_{c1} = \dfrac{2a}{n} = \dfrac{2 \times 5}{1} = 10 \text{ cm} > \lambda_0$

For $n = 2$, cut-off wavelength, $\lambda_{c2} = \dfrac{2a}{n} = \dfrac{2 \times 5}{2} = 5 \text{ cm} > \lambda_0$

For $n = 3$, cut-off wavelength, $\lambda_{c3} = \dfrac{2a}{n} = \dfrac{2 \times 5}{3} = 3.33 \text{ cm} > \lambda_0$

For $n = 4$, cut-off wavelength, $\lambda_{c4} = \dfrac{2a}{n} = \dfrac{2 \times 5}{4} = 2.5 \text{ cm} < \lambda_0$

Hence, the maximum value of n is, $n = 3$
Guide wavelength for $n = 3$, is

$$\lambda_{g3} = \frac{\lambda_0}{\sqrt{1 - (\lambda_0 / \lambda_{c3})^2}} = \frac{3}{\sqrt{1 - (3/3.33)^2}} = 6.88 \text{ cm}$$

Example 7.2 A parallel plane waveguide consists of two shells of good conductors separated by 10 cm. Find the propagation constant at frequencies 100 MHz and 10 GHz when operated in TE_1 mode. Does the wave propagate in each case?

Solution Here, $a = 10$ cm $= 0.1$ m, For TE_1 mode, $n = 1$
Hence, the propagation constant is given as

$$\gamma = \sqrt{\left(\frac{n\pi}{a}\right)^2 - \omega^2 \mu_0 \varepsilon_0} = \sqrt{\left(\frac{\pi}{a}\right)^2 - \left(\frac{\omega}{3 \times 10^8}\right)^2}$$

Since, $$\frac{1}{\sqrt{\mu_0 \varepsilon_0}} = c = 3 \times 10^8 \text{ m/s}$$

At frequency $f = 100$ MHz, $\gamma = \sqrt{\left(\dfrac{\pi}{a}\right)^2 - \left(\dfrac{2\pi \times 100 \times 10^6}{3 \times 10^8}\right)^2} = 31.35$ nepper per metre

At frequency $f = 10$ GHz, $\gamma = \sqrt{\left(\dfrac{\pi}{a}\right)^2 - \left(\dfrac{2\pi \times 10 \times 10^9}{3 \times 10^8}\right)^2} = j209.37$ radian per metre

At frequency $f = 100$ MHz, the propagation constant is real, i.e., phase shift is zero ($\alpha = 31.35$, $\beta = 0$). So, wave propagation does not take place at this frequency. However, at frequency $f = 10$ GHz, the propagation constant is imaginary, i.e., ($\alpha = 0$, $\beta = 209.37$). So, wave propagation takes place at this frequency.

Example 7.3 For a parallel plane waveguide, infinite in extent and a spacing of 20 cm between the plates, calculate the

(a) phase velocity for *TEM* mode at all wavelengths
(b) cut-off wavelength for dominant *TE* mode
(c) phase velocity for dominant mode at 80% of cut-off wavelength
(d) reflection angle for the above frequency of operation.

Solution Here, $d = 20$ cm

Phase velocity for *TEM* mode is given as, $v_0 = \dfrac{1}{\sqrt{\mu\varepsilon}} = \dfrac{1}{\sqrt{\mu_0\varepsilon_0}} = 3\times 10^8$ m/s

Cut-off wavelength for dominant mode is $\lambda_c = 2d = 40$ cm
Phase velocity for dominant mode at 80% of cut-off wavelength is given as

$$v_p = \frac{v_0}{\sqrt{1-\dfrac{f_c^2}{f^2}}} = \frac{v_0}{\sqrt{1-\dfrac{\lambda^2}{\lambda_c^2}}} = \frac{v_0}{\sqrt{1-0.8^2}} = \frac{3\times 10^8}{0.6} = 5\times 10^8 \text{ m/s}$$

Reflection angle is given as, $\tan\theta = \sqrt{\left(\dfrac{f}{f_c}\right)^2 - 1} = \sqrt{\left(\dfrac{\lambda_c}{\lambda}\right)^2 - 1} = \sqrt{\left(\dfrac{1}{0.8}\right)^2 - 1} = \dfrac{3}{4}$

$$\therefore \qquad \theta = 36.87°$$

***Example 7.4** A pair of perfectly conducting planes separated by 8 cm in air guides 5 GHz wave in TM_{10} mode. Find

(a) Cut-off frequency
(b) Characteristic impedance
(c) Phase constant
(d) Attenuation constant
(e) Phase velocity
(f) Group velocity
(g) Guide wavelength.

Solution Here, $d = 8$ cm, $f = 5$ GHz

(a) The cut-off frequency, $f_c = \dfrac{n}{2d\sqrt{\mu_0\varepsilon_0}} = \dfrac{1\times 3\times 10^8}{2\times 0.08} = 1.875$ GHz

(b) Characteristic impedance, $\eta = \dfrac{\eta_0}{\sqrt{1-\left(\dfrac{f_c}{f}\right)^2}} = \dfrac{120\pi}{\sqrt{1-\left(\dfrac{1.875}{5}\right)^2}} = 349.48\ \Omega$

(c) Phase constant, $\beta = \frac{n\pi}{d}\sqrt{\left(\frac{f}{f_c}\right)^2 - 1} = \frac{1 \times \pi}{0.08}\sqrt{\left(\frac{5}{1.875}\right)^2 - 1} = 97.08 \text{ rad/m}$

(d) Since the guide is operating above the cut-off frequency, the attenuation constant is zero, i.e., $\alpha = 0$.

(e) Phase velocity, $v_p = \frac{v_0}{\sqrt{1 - \left(\frac{f_c}{f}\right)^2}} = \frac{3 \times 10^8}{\sqrt{1 - \left(\frac{1.875}{5}\right)^2}} = 3.236 \times 10^8 \text{ m/s}$

(f) Group velocity, $v_g = v_0\sqrt{1 - \left(\frac{f_c}{f}\right)^2} = 3 \times 10^8 \times \sqrt{1 - \left(\frac{1.875}{5}\right)^2} = 2.781 \times 10^8 \text{ m/s}$

(g) Guide wavelength, $\lambda_g = \frac{2\pi}{\beta} = \frac{2\pi}{97.08} = 0.0647 \text{ m} = 6.47 \text{ cm}$

***Example 7.5** A parallel plane waveguide consists of two parallel perfectly conducting infinite planes separated by 10 cm. Determine the propagating TE_n modes for an electromagnetic wave of 5000 MHz assuming free space between the planes. For the propagating modes, find the following:

(a) cut-off frequency,
(b) guide wavelength,
(c) phase and group velocities,
(d) Intrinsic impedance.

Solution Here, $a = 10$ cm

Hence, the cut-off wavelength is $\lambda_c = \frac{2a}{n} = \frac{20}{n}$ cm

Also, $\lambda_0 = \frac{c}{f} = \frac{3 \times 10^8}{5 \times 10^9} \text{ m} = 0.06 \text{ m} = 6 \text{ cm}$

For $n = 1$, $\lambda_{c1} = \frac{20}{1} \text{ cm} = 20 \text{ cm} > \lambda_0$

For $n = 2$, $\lambda_{c2} = \frac{20}{2} \text{ cm} = 10 \text{ cm} > \lambda_0$

For $n = 3$, $\lambda_{c3} = \frac{20}{3} \text{ cm} = 6.67 \text{ cm} > \lambda_0$

For $n = 4$, $\lambda_{c4} = \frac{20}{4} \text{ cm} = 5 \text{ cm} < \lambda_0$

Therefore, the propagating modes are: TE_1, TE_2 and TE_3.

For TE_1 Mode:

Cut-off frequency, $f_{c1} = \frac{c}{\lambda_{c1}} = \frac{3 \times 10^8}{0.2} = 1.5 \text{ GHz}$

Guide wavelength, $\lambda_{g1} = \frac{\lambda_0}{\sqrt{1 - (\lambda_0/\lambda_{c1})^2}} = \frac{6}{\sqrt{1 - (6/20)^2}} = 6.29 \text{ cm}$

Phase velocity, $v_{p1} = \dfrac{c}{\sqrt{1-(\lambda_0/\lambda_{c1})^2}} = \dfrac{3\times 10^8}{\sqrt{1-(6/20)^2}} = 3.15\times 10^8$ m/s

Group velocity, $v_{g1} = \dfrac{c^2}{v_{p1}} = \dfrac{(3\times 10^8)^2}{3.15\times 10^8} = 2.86\times 10^8$ m/s

Intrinsic impedance, $\eta_1 = \dfrac{\eta_0}{\sqrt{1-(\lambda_0/\lambda_{c1})^2}} = \dfrac{377}{\sqrt{1-(6/20)^2}} = 395.19\ \Omega$

For TE_2 Mode:

Cut-off frequency, $f_{c2} = \dfrac{c}{\lambda_{c2}} = \dfrac{3\times 10^8}{0.1} = 3$ GHz

Guide wavelength, $\lambda_{g2} = \dfrac{\lambda_0}{\sqrt{1-(\lambda_0/\lambda_{c2})^2}} = \dfrac{6}{\sqrt{1-(6/10)^2}} = 7.5$ cm

Phase velocity, $v_{p2} = \dfrac{c}{\sqrt{1-(\lambda_0/\lambda_{c2})^2}} = \dfrac{3\times 10^8}{\sqrt{1-(6/10)^2}} = 3.75\times 10^8$ m/s

Group velocity, $v_{g2} = \dfrac{c^2}{v_{p2}} = \dfrac{(3\times 10^8)^2}{3.75\times 10^8} = 2.4\times 10^8$ m/s

Intrinsic impedance, $\eta_2 = \dfrac{\eta_0}{\sqrt{1-(\lambda_0/\lambda_{c2})^2}} = \dfrac{377}{\sqrt{1-(6/10)^2}} = 471.24\ \Omega$

For TE_3 Mode:

Cut-off frequency, $f_{c3} = \dfrac{c}{\lambda_{c3}} = \dfrac{3\times 10^8}{0.0667} = 4.5$ GHz

Guide wavelength, $\lambda_{g3} = \dfrac{\lambda_0}{\sqrt{1-(\lambda_0/\lambda_{c3})^2}} = \dfrac{6}{\sqrt{1-(6/6.67)^2}} = 13.76$ cm

Phase velocity, $v_{p3} = \dfrac{c}{\sqrt{1-(\lambda_0/\lambda_{c3})^2}} = \dfrac{3\times 10^8}{\sqrt{1-(6/6.67)^2}} = 6.88\times 10^8$ m/s

Group velocity, $v_{g3} = \dfrac{c^2}{v_{p3}} = \dfrac{(3\times 10^8)^2}{6.88\times 10^8} = 1.31\times 10^8$ m/s

Intrinsic impedance, $\eta_3 = \dfrac{\eta_0}{\sqrt{1-(\lambda_0/\lambda_{c3})^2}} = \dfrac{377}{\sqrt{1-(6/6.67)^2}} = 864.88\ \Omega$

7.3 RECTANGULAR WAVEGUIDES

Definition A *rectangular waveguide* is a hollow conducting device with four sides closed and two sides open.

In such a waveguide, the electric field varies with distance having a maximum at the centre. Magnetic field lines curve round and pass through the guide, tangential to the walls.

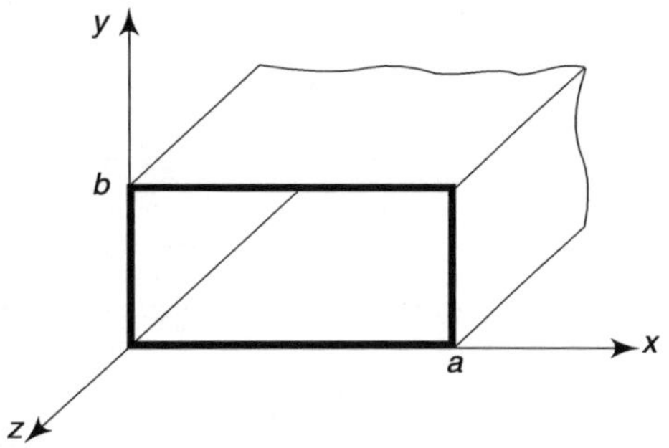

Fig. 7.8 *Rectangular waveguides*

Derivation of Field Equations for Rectangular Waveguide In order to determine the electromagnetic field configurations within a rectangular waveguide, Maxwell's equations are solved subject to the appropriate boundary conditions at the walls of the guide. Again, assuming perfectly conducting guide walls, these boundary conditions are as follows.

Boundary Conditions

1. The tangential components of the electric field must be zero.

$$\therefore \quad E_t = 0$$

2. The normal components of the magnetic field must be zero.

$$\therefore \quad H_n = 0$$

For the rectangular waveguide shown in Fig. 7.8, the boundary conditions are:

$$E_x = E_z = 0 \quad \text{at} \quad y = 0 \text{ and } y = b$$
$$E_y = E_z = 0 \quad \text{at} \quad x = 0 \text{ and } x = a$$

Assuming that the field variations in the z-direction may be expressed as $e^{-\gamma z}$, where γ is a complex propagation constant ($\gamma = \alpha + j\beta$), from Eqs. (7.9) and (7.10), we have Maxwell's equations for rectangular waveguides given as

$$\left(\frac{\partial E_z}{\partial y} + \gamma E_y\right) = -j\omega\mu H_x \tag{7.61a}$$

$$\left(\gamma E_x + \frac{\partial E_z}{\partial x}\right) = j\omega\mu H_y \tag{7.61b}$$

$$\left(\frac{\partial E_y}{\partial x} - \frac{\partial E_x}{\partial y}\right) = -j\omega\mu H_z \tag{7.61c}$$

and

$$\left(\frac{\partial H_z}{\partial y} + \gamma H_y\right) = j\omega\varepsilon E_x \tag{7.62a}$$

$$\left(\gamma H_x + \frac{\partial H_z}{\partial x}\right) = -j\omega\varepsilon E_y \tag{7.62b}$$

$$\left(\frac{\partial H_y}{\partial x} - \frac{\partial H_x}{\partial y}\right) = j\omega\varepsilon E_z \tag{7.62c}$$

Similarly, from wave equations

$$\nabla^2 \bar{E} + \gamma^2 \bar{E} = 0 \tag{7.63a}$$

$$\nabla^2 \bar{H} + \gamma^2 \bar{H} = 0 \tag{7.63b}$$

where $\gamma^2 = (\sigma = j\omega\varepsilon) j\omega\mu = -\omega^2\mu\varepsilon$ for non-conducting medium.

From Eq. (7.62*a*), $H_y = \frac{1}{\gamma}\left(j\omega\varepsilon E_x - \frac{\partial H_z}{\partial y}\right)$

Putting this in Eq. (7.61*b*),

$$\left(\gamma E_x + \frac{\partial E_z}{\partial x}\right) = j\omega\mu H_y = \frac{j\omega\mu}{\gamma}\left(j\omega\varepsilon E_x - \frac{\partial H_z}{\partial y}\right)$$

or $$E_x\left(\frac{\gamma^2 + \omega^2\mu\varepsilon}{\gamma}\right) = -\frac{j\omega\mu}{\gamma}\frac{\partial H_z}{\partial y} - \frac{\partial E_z}{\partial x}$$

or $$E_x\left(\frac{h^2}{\gamma}\right) = -\frac{j\omega\mu}{\gamma}\frac{\partial H_z}{\partial y} - \frac{\partial E_z}{\partial x}$$

where $h^2 = (\gamma^2 + \omega^2\,\mu\varepsilon)$.

$$E_x = -\frac{j\omega\mu}{h^2}\frac{\partial H_z}{\partial y} - \frac{\gamma}{h^2}\frac{\partial E_z}{\partial x} = -\frac{\gamma}{h^2}\frac{\partial E_z}{\partial x} - j\frac{\omega\mu}{h^2}\frac{\partial H_z}{\partial y}$$

∴ $$\boxed{E_x = -\frac{\gamma}{h^2}\frac{\partial E_z}{\partial x} - j\frac{\omega\mu}{h^2}\frac{\partial H_z}{\partial y}} \tag{7.64}$$

Similarly, from Eq. (7.61*a*) and Eq. (7.62*b*), we get

$$\left(\frac{\partial E_z}{\partial y} + \gamma E_y\right) = -j\omega\mu H_x = -\frac{j\omega\mu}{\gamma}\left(-j\omega\varepsilon E_y - \frac{\partial H_z}{\partial x}\right)$$

or $$E_y\left(\frac{\gamma^2 + \omega^2\mu\varepsilon}{\gamma}\right) = -\frac{\partial E_z}{\partial y} + j\omega\mu\frac{\partial H_z}{\partial x}$$

or $$E_y\left(\frac{h^2}{\gamma}\right) = -\frac{\partial E_z}{\partial y} + j\omega\mu\frac{\partial H_z}{\partial x}$$

∴ $$\boxed{E_y = -\frac{\gamma}{h^2}\frac{\partial E_z}{\partial y} + j\frac{\omega\mu}{h^2}\frac{\partial H_z}{\partial x}} \tag{7.65}$$

Putting this value in Eq. (7.62*b*),

$$H_x = \frac{1}{\gamma}\left(-\frac{\partial H_z}{\partial x} - j\omega\varepsilon E_y\right) = \frac{1}{\gamma}\left[-\frac{\partial H_z}{\partial x} - j\omega\varepsilon\left(-\frac{\gamma}{h^2}\frac{\partial E_z}{\partial y} + j\frac{\omega\mu}{h^2}\frac{\partial H_z}{\partial x}\right)\right]$$

$$= \frac{1}{\gamma}\left[-\frac{\partial H_z}{\partial x}\left(1 - \frac{\omega^2\mu\varepsilon}{h^2}\right) + j\frac{\omega\varepsilon\gamma}{h^2}\frac{\partial E_z}{\partial y}\right]$$

$$= \frac{1}{\gamma}\left[-\frac{\partial H_z}{\partial x}\left(\frac{h^2 - \omega^2\mu\varepsilon}{h^2}\right) + j\frac{\omega\varepsilon\gamma}{h^2}\frac{\partial E_z}{\partial y}\right]$$

$$= \frac{1}{\gamma}\left[-\frac{\partial H_z}{\partial x}\left(\frac{\gamma^2}{h^2}\right) + j\frac{\omega\varepsilon\gamma}{h^2}\frac{\partial E_z}{\partial y}\right]$$

$$= -\frac{\gamma}{h^2}\frac{\partial H_z}{\partial x} + j\frac{\omega\varepsilon}{h^2}\frac{\partial E_z}{\partial y}$$

$\therefore$

$$\boxed{H_x = -\frac{\gamma}{h^2}\frac{\partial H_z}{\partial x} + j\frac{\omega\varepsilon}{h^2}\frac{\partial E_z}{\partial y}} \qquad \textbf{(7.66)}$$

Finally, from Eq. (7.62*a*),

$$H_y = \frac{1}{\gamma}\left(j\omega\varepsilon E_x - \frac{\partial H_z}{\partial y}\right) = \frac{1}{\gamma}\left[j\omega\varepsilon\left(-\frac{\gamma}{h^2}\frac{\partial E_z}{\partial x} - j\frac{\omega\mu}{h^2}\frac{\partial H_z}{\partial y}\right) - \frac{\partial H_z}{\partial y}\right]$$

$$= \frac{1}{\gamma}\left[-\frac{\partial H_z}{\partial y}\left(1 - \frac{\omega^2\mu\varepsilon}{h^2}\right) - j\frac{\omega\varepsilon\gamma}{h^2}\frac{\partial E_z}{\partial x}\right]$$

$$= \frac{1}{\gamma}\left[-\frac{\partial H_z}{\partial y}\left(\frac{h^2 - \omega^2\mu\varepsilon}{h^2}\right) - j\frac{\omega\varepsilon\gamma}{h^2}\frac{\partial E_z}{\partial x}\right]$$

$$= \frac{1}{\gamma}\left[-\frac{\partial H_z}{\partial y}\left(\frac{\gamma^2}{h^2}\right) - j\frac{\omega\varepsilon\gamma}{h^2}\frac{\partial E_z}{\partial x}\right]$$

$$= -\frac{\gamma}{h^2}\frac{\partial H_z}{\partial y} - j\frac{\omega\varepsilon}{h^2}\frac{\partial E_z}{\partial x}$$

$\therefore$

$$\boxed{H_y = -\frac{\gamma}{h^2}\frac{\partial H_z}{\partial y} - j\frac{\omega\varepsilon}{h^2}\frac{\partial E_z}{\partial x}} \qquad \textbf{(7.67)}$$

Equations (7.64) to (7.67) express the components of the electric and magnetic fields $\overline{E}$ and $\overline{H}$ in terms of E_z and H_z, i.e., the component of the fields in the direction of wave propagation.

It is observed that there must be a *z*-component of the field either $\overline{E}$ or $\overline{H}$; otherwise, all the components of $\overline{E}$ and $\overline{H}$ would be zero and there would be no fields within the waveguide.

Also from Eq. (7.63), if we let

$$\overline{E} = (E_x, E_y, E_z) \quad \text{and} \quad \overline{H} = (H_x, H_y, H_z)$$

then each of Eq. (7.63*a*) and Eq. (7.63*b*) will consist of three scalar Helmholtz equations. This implies that we have to solve six scalar equations to obtain the fields $\overline{E}$ and $\overline{H}$. For the *z*-component Eq. (7.63*a*) is written as

$$\frac{\partial^2 E_z}{\partial x^2} + \frac{\partial^2 E_z}{\partial y^2} + \frac{\partial^2 E_z}{\partial z^2} = \omega^2\mu\varepsilon E_z \qquad \textbf{(7.68)}$$

This partial differential equation can be solved by using separation of variables method. Let

$$E_z(x, y, z) = X(x)Y(y)Z(z) \tag{7.69}$$

where $X(x)$, $Y(y)$, $Z(z)$ are functions of x, y and z respectively. Substituting this in Eq. (7.68), we get

$$YZ\frac{\partial^2 X}{\partial x^2} + XZ\frac{\partial^2 Y}{\partial y^2} + XY\frac{\partial^2 Z}{\partial z^2} = \omega^2\mu\varepsilon\, XYZ$$

or

$$\frac{1}{X}\frac{\partial^2 X}{\partial x^2} + \frac{1}{Y}\frac{\partial^2 Y}{\partial y^2} + \frac{1}{Z}\frac{\partial^2 Z}{\partial z^2} = \omega^2\mu\varepsilon \tag{7.70}$$

Since the solution of the three terms in Eq. (7.70) is constant, each term should be constant.

$$\therefore \quad \frac{1}{X}\frac{\partial^2 X}{\partial x^2} = -K_x^2, \quad \frac{1}{Y}\frac{\partial^2 Y}{\partial y^2} = -K_y^2, \quad \frac{1}{Z}\frac{\partial^2 Z}{\partial z^2} = -K_z^2 \tag{7.71}$$

$$\therefore \quad \frac{\partial^2 X}{\partial x^2} + K_x^2 X = 0 \tag{7.72a}$$

$$\frac{\partial^2 Y}{\partial y^2} + K_y^2 Y = 0 \tag{7.72b}$$

$$\frac{\partial^2 Z}{\partial z^2} + K_z^2 Z = 0 \tag{7.72c}$$

Hence, from Eq. (7.70) and Eq. (7.71),

$$(K_x^2 + K_y^2 + K_z^2) = \omega^2\mu\varepsilon \tag{7.73}$$

When the wave is propagating in the positive z-direction, the z-variation can be expressed as $e^{-\gamma z}$, so that

$$\frac{\partial^2 Z}{\partial z^2} = \gamma^2 Z$$

and hence,

$$K_z^2 = -\gamma^2$$

From Eq. (7.73),

$$K_x^2 + K_y^2 = \omega^2\mu\varepsilon + \gamma^2 = h^2$$

$$\therefore \quad \boxed{h^2 = \gamma^2 + \omega^2\mu\varepsilon = K_x^2 + K_y^2} \tag{7.74}$$

So, Eq. (7.72) can be written as

$$\therefore \quad \frac{\partial^2 X}{\partial x^2} + K_x^2 X = 0 \tag{7.75a}$$

$$\frac{\partial^2 Y}{\partial y^2} + K_y^2 Y = 0 \tag{7.75b}$$

$$\frac{\partial^2 Z}{\partial z^2} - \gamma^2 Z = 0 \quad \textbf{(7.75c)}$$

Solutions of these equations can be written as

$$X = C_1 \cos K_x x + C_2 \sin K_x x \quad \textbf{(7.76a)}$$

$$Y = C_3 \cos K_y y + C_4 \sin K_y y \quad \textbf{(7.76b)}$$

$$Z = C_5 e^{\gamma z} + C_6 e^{-\gamma z} \quad \textbf{(7.76c)}$$

where C_1, C_2, C_3, C_4, C_5, C_6 are arbitrary constants.

Again, as the wave is propagating in the positive z-direction, for the wave to be finite at infinity, we must have

$$C_5 = 0$$

Thus, the electric field can be written as

$$E_z(x, y, z) = (C_1 \cos K_x x + C_2 \sin K_x x)(C_3 \cos K_y y + C_4 \sin K_y y) C_6 e^{-\gamma z}$$

$$\therefore \quad E_z(x, y, z) = (A_1 \cos K_x x + A_2 \sin K_x x)(A_3 \cos K_y y + A_4 \sin K_y y) e^{-\gamma z} \quad \textbf{(7.77)}$$

where $A_1 = C_1C_6$, $A_2 = C_2C_6$, $A_3 = C_3C_6$, $A_4 = C_4C_6$.

Following similar steps, we can find the magnetic field as

$$H_z(x, y, z) = (B_1 \cos K_x x + B_2 \sin K_x x)(B_3 \cos K_y y + B_4 \sin K_y y) e^{-\gamma z} \quad \textbf{(7.78)}$$

Therefore, Eqs. (7.77) and (7.78) in conjunction with Eqs. (7.64) to (7.67) can be used to obtain the field configurations. It is seen from these equations that similar to a parallel plane guide, there are different field configurations or *modes*. Four different modes are:

1. Transverse Electric Waves or *TE*-Waves or *H*-Waves ($E_z = 0$, $H_z \neq 0$),
2. Transverse Magnetic Waves or *TM*-Waves or *E*-Waves ($E_z \neq 0$, $H_z = 0$),
3. Transverse Electromagnetic Waves or *TEM*-Waves ($E_z = 0$, $H_z = 0$), and
4. Hybrid Waves or *HE*-Waves ($E_z \neq 0$, $H_z \neq 0$).

7.3.1 Field Equations for Transverse Electric (*TE*) Mode in Rectangular Waveguide

In this mode, the electric field is entirely transverse (perpendicular) to the direction of wave propagation, i.e., $E_z = 0$. The other components, E_x, E_y, H_x, H_y and H_z are determined using Eqs. (7.64) to (7.67), Eq. (7.78) and the boundary conditions.

The boundary conditions for this case are:

$$E_x = 0 \quad \text{at} \quad y = 0 \quad \textbf{(7.79a)}$$

$$E_x = 0 \quad \text{at} \quad y = b \quad \textbf{(7.79b)}$$

$$E_y = 0 \quad \text{at} \quad x = 0 \quad \textbf{(7.79c)}$$

$$E_y = 0 \quad \text{at} \quad x = a \quad \textbf{(7.79d)}$$

Hence, from Eqs. (7.64) to (7.67),

$$\frac{\partial H_z}{\partial y} = 0 \quad \text{at} \quad y = 0 \tag{7.80a}$$

$$\frac{\partial H_z}{\partial y} = 0 \quad \text{at} \quad y = b \tag{7.80b}$$

$$\frac{\partial H_z}{\partial x} = 0 \quad \text{at} \quad x = 0 \tag{7.80c}$$

$$\frac{\partial H_z}{\partial x} = 0 \quad \text{at} \quad x = a \tag{7.80d}$$

Applying the boundary conditions of Eq. (7.80*a*) and Eq. (7.80*c*) to Eq. (7.78), we require that $B_2 = 0 = B_4$ and hence the magnetic field can be written as

$$H_z(x, y, z) = (B_1 \cos K_x x)(B_3 \cos K_y y)e^{-\gamma z} = H_0 \cos K_x x \cos K_y y e^{-\gamma z} \tag{7.81}$$

where $H_0 = B_1 B_3$.
Also, applying boundary conditions of Eqs. (7.80*b*) and (7.80*d*) in Eq. (7.81), we get

$$\sin K_y b = 0 \quad \text{and} \quad \sin K_x a = 0$$

This implies that

$$K_x a = m\pi, \quad m = 0, 1, 2, 3, \ldots$$

and

$$K_y b = n\pi, \quad n = 0, 1, 2, 3, \ldots$$

or

$$\boxed{K_x = \frac{m\pi}{a}, \quad K_y = \frac{n\pi}{b}} \tag{7.82}$$

Substituting this in Eq. (7.81),

$$\boxed{H_z(x, y, z) = H_0 \cos\left(\frac{m\pi}{a}x\right) \cos\left(\frac{n\pi}{b}y\right)e^{-\gamma z}} \tag{7.83}$$

The other field components are obtained from Eqs. (7.64) to (7.67) as follows.
From Eq. (7.64),

$$E_x = -\frac{\gamma}{h^2}\frac{\partial E_z}{\partial x} - j\frac{\omega\mu}{h^2}\frac{\partial H_z}{\partial y} = j\frac{\omega\mu}{h^2}\left(\frac{n\pi}{b}\right)H_0 \cos\left(\frac{m\pi}{a}x\right) \sin\left(\frac{n\pi}{b}y\right)e^{-\gamma z}$$

$$\therefore \quad \boxed{E_x = j\frac{\omega\mu}{h^2}\left(\frac{n\pi}{b}\right)H_0 \cos\left(\frac{m\pi}{a}x\right) \sin\left(\frac{n\pi}{b}y\right)e^{-\gamma z}} \tag{7.84a}$$

From Eq. (7.65),

$$E_y = -\frac{\gamma}{h^2}\frac{\partial E_z}{\partial y} + j\frac{\omega\mu}{h^2}\frac{\partial H_z}{\partial x} = -j\frac{\omega\mu}{h^2}\left(\frac{m\pi}{a}\right)H_0 \sin\left(\frac{m\pi}{a}x\right) \cos\left(\frac{n\pi}{b}y\right)e^{-\gamma z}$$

$$\therefore \quad \boxed{E_y = -j\frac{\omega\mu}{h^2}\left(\frac{m\pi}{a}\right)H_0 \sin\left(\frac{m\pi}{a}x\right) \cos\left(\frac{n\pi}{b}y\right)e^{-\gamma z}} \tag{7.84b}$$

From Eq. (7.66),

$$H_x = -\frac{\gamma}{h^2}\frac{\partial H_z}{\partial x} + j\frac{\omega\varepsilon}{h^2}\frac{\partial E_z}{\partial y} = -\frac{\gamma}{h^2}\left(\frac{m\pi}{a}\right)H_0 \sin\left(\frac{m\pi}{a}x\right)\cos\left(\frac{n\pi}{b}y\right)e^{-\gamma z}$$

$$\therefore \quad \boxed{H_x = -\frac{\gamma}{h^2}\left(\frac{m\pi}{a}\right)H_0 \sin\left(\frac{m\pi}{a}x\right)\cos\left(\frac{n\pi}{b}y\right)e^{-\gamma z}} \tag{7.84c}$$

From Eq. (7.67),

$$H_y = -\frac{\gamma}{h^2}\frac{\partial H_z}{\partial y} - j\frac{\omega\varepsilon}{h^2}\frac{\partial E_z}{\partial x} = \frac{\gamma}{h^2}\left(\frac{n\pi}{b}\right)H_0 \cos\left(\frac{m\pi}{a}x\right)\sin\left(\frac{n\pi}{b}y\right)e^{-\gamma z}$$

$$\therefore \quad \boxed{H_y = \frac{\gamma}{h^2}\left(\frac{n\pi}{b}\right)H_0 \cos\left(\frac{m\pi}{a}x\right)\sin\left(\frac{n\pi}{b}y\right)e^{-\gamma z}} \tag{7.84d}$$

where

$$\boxed{h^2 = K_x^2 + K_y^2 = \left(\frac{m\pi}{a}\right)^2 + \left(\frac{n\pi}{b}\right)^2} \tag{7.85}$$

NOTE

Here, m and n are integers where m denotes the number of half cycle variations of the fields in the x-direction and n denotes the number of half cycle variations of the fields in the y-direction. It is understood from Eqs. (7.83) and (7.84) that there will be different field configurations or modes depending upon the values of m and n, known as TE_{mn} modes. For TE mode, both m and n cannot be zero as this will vanish all the field components. Therefore, the lowest value of (m, n) may be either (0, 1) or (1, 0) and subsequently the lowest order mode may be either TE_{01} or TE_{10}.

Some field configurations for *TE* modes are shown in Fig. 7.9. The solid lines represent the electric field lines whereas the dotted lines represent the magnetic field lines.

7.3.2 Field Equations for Transverse Magnetic (*TM*) Mode in Rectangular Waveguide

In this mode, the magnetic field is entirely transverse (perpendicular) to the direction of wave propagation, i.e., $H_z = 0$. The other components, E_x, E_y, E_z, H_x and H_y are determined using Eqs. (7.64) to (7.67), Eq. (7.77) and the boundary conditions.

The boundary conditions for this case are given by Eq. (7.79).

Applying the boundary conditions of Eqs. (7.79*a*) and (7.79*c*) to Eq. (7.77), we require that $A_1 = 0 = A_3$ and hence the electric field can be written as

$$E_z(x, y, z) = (A_2 \sin K_x x)(A_4 \sin K_y y)e^{-\gamma z} = E_0 \sin K_x x \sin K_y y e^{-\gamma z} \tag{7.86}$$

where $E_0 = A_2 A_4$.

Also, applying boundary conditions of Eqs. (7.79*b*) and (7.79*d*) in Eq. (7.86), we get

$$\sin K_x a = 0 \quad \text{and} \quad \sin K_y b = 0$$

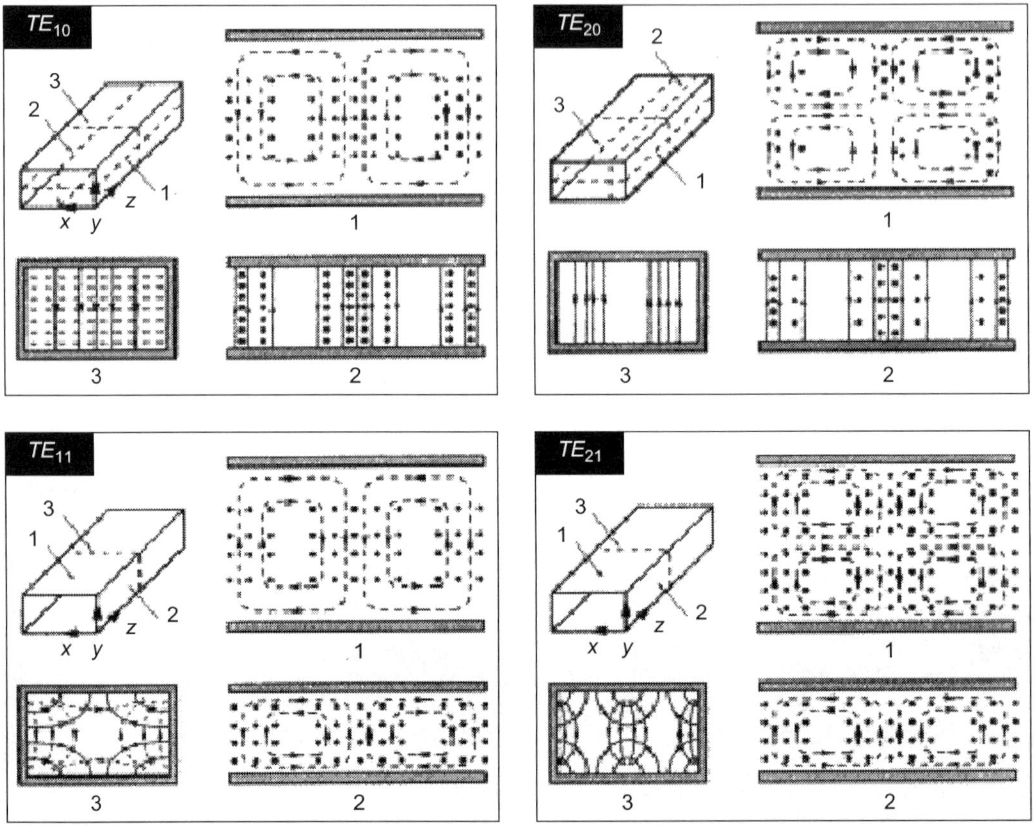

Fig. 7.9 *Field configurations for TE mode waves*

This implies that

$$K_x a = m\pi, \quad m = 1, 2, 3, \ldots$$

and

$$K_y b = n\pi, \quad n = 1, 2, 3, \ldots$$

This follows the same result as obtained for *TE* waves, i.e.,

$$\boxed{K_x = \frac{m\pi}{a}, \quad K_y = \frac{n\pi}{b}} \tag{7.87}$$

Substituting this in Eq. (7.86),

$$\boxed{E_z(x, y, z) = E_0 \sin\left(\frac{m\pi}{a}x\right)\sin\left(\frac{n\pi}{b}y\right)e^{-\gamma z}} \tag{7.88}$$

The other field components are obtained from Eqs. (7.64) to (7.67) as follows.

From Eq. (7.64),

$$E_x = -\frac{\gamma}{h^2}\frac{\partial E_z}{\partial x} - j\frac{\omega\mu}{h^2}\frac{\partial H_z}{\partial y} = -\frac{\gamma}{h^2}\left(\frac{m\pi}{a}\right)E_0 \cos\left(\frac{m\pi}{a}x\right)\sin\left(\frac{n\pi}{b}y\right)e^{-\gamma z}$$

$$\therefore \qquad \boxed{E_x = -\frac{\gamma}{h^2}\left(\frac{m\pi}{a}\right)E_0 \cos\left(\frac{m\pi}{a}x\right)\sin\left(\frac{n\pi}{b}y\right)e^{-\gamma z}} \qquad \textbf{(7.89a)}$$

From Eq. (7.65),

$$E_y = -\frac{\gamma}{h^2}\frac{\partial E_z}{\partial y} + j\frac{\omega\mu}{h^2}\frac{\partial H_z}{\partial x} = -j\frac{\gamma}{h^2}\left(\frac{n\pi}{b}\right)E_0 \sin\left(\frac{m\pi}{a}x\right)\cos\left(\frac{n\pi}{b}y\right)e^{-\gamma z}$$

$$\therefore \qquad \boxed{E_y = -j\frac{\gamma}{h^2}\left(\frac{n\pi}{b}\right)E_0 \sin\left(\frac{m\pi}{a}x\right)\cos\left(\frac{n\pi}{b}y\right)e^{-\gamma z}} \qquad \textbf{(7.89b)}$$

From Eq. (7.66),

$$H_x = -\frac{\gamma}{h^2}\frac{\partial H_z}{\partial x} + j\frac{\omega\varepsilon}{h^2}\frac{\partial E_z}{\partial y} = j\frac{\omega\varepsilon}{h^2}\left(\frac{n\pi}{b}\right)E_0 \sin\left(\frac{m\pi}{a}x\right)\cos\left(\frac{n\pi}{b}y\right)e^{-\gamma z}$$

$$\therefore \qquad \boxed{H_x = j\frac{\omega\varepsilon}{h^2}\left(\frac{n\pi}{b}\right)E_0 \sin\left(\frac{m\pi}{a}x\right)\cos\left(\frac{n\pi}{b}y\right)e^{-\gamma z}} \qquad \textbf{(7.89c)}$$

From Eq. (7.67),

$$H_y = -\frac{\gamma}{h^2}\frac{\partial H_z}{\partial y} - j\frac{\omega\varepsilon}{h^2}\frac{\partial E_z}{\partial x} = -j\frac{\omega\varepsilon}{h^2}\left(\frac{m\pi}{a}\right)E_0 \cos\left(\frac{m\pi}{a}x\right)\sin\left(\frac{n\pi}{b}y\right)e^{-\gamma z}$$

$$\therefore \qquad \boxed{H_y = -j\frac{\omega\varepsilon}{h^2}\left(\frac{m\pi}{a}\right)E_0 \cos\left(\frac{m\pi}{a}x\right)\sin\left(\frac{n\pi}{b}y\right)e^{-\gamma z}} \qquad \textbf{(7.89d)}$$

NOTE

Here too, m and n are integers where m denotes the number of half cycle variations of the fields in the x-direction and n denotes the number of half cycle variations of the fields in the y-direction. It is understood from Eqs. (7.88) and (7.89) that there will be different field configurations or modes depending upon the values of m and n, known as TM_{mn} modes. For TM mode, neither m nor n can be zero. This is because of the fact that the field expressions are identically zero if either m or n is zero. Therefore, the lowest mode for rectangular waveguide TM mode is TM_{11}.

Some field configurations for *TM* modes are shown in Fig. 7.10. Here too, the solid lines represent the electric field lines whereas the dotted lines represent the magnetic field lines.

7.3.3 Properties of *TE* and *TM* Waves in Rectangular Waveguide

We find the following quantities for *TE* and *TM* mode waves in rectangular waveguides:

1. Propagation constant (γ)
2. Cut-off frequency (f_c)
3. Cut-off wavelength (λ_c)
4. Phase constant (β)
5. Phase velocity (v_p)

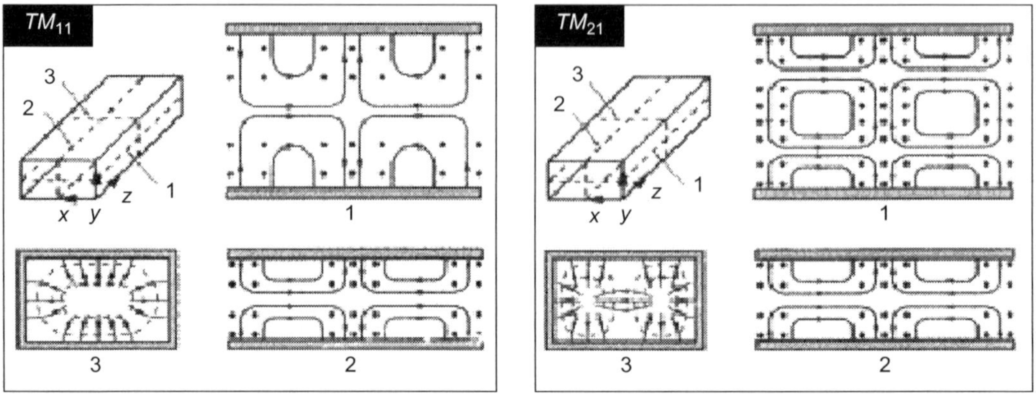

Fig. 7.10 *Field configurations for TM mode waves*

6. Guide Wavelength (λ_g)
7. Intrinsic wave impedance (η)

1. Propagation constant (γ): From Eqs. (7.74) and (7.82),

$$\gamma = \sqrt{h^2 - \omega^2\mu\varepsilon} = \sqrt{K_x^2 + K_y^2 - \omega^2\mu\varepsilon} = \sqrt{\left(\frac{m\pi}{a}\right)^2 + \left(\frac{n\pi}{b}\right)^2 - \omega^2\mu\varepsilon}$$

$$\therefore \quad \boxed{\gamma = \sqrt{\left(\frac{m\pi}{a}\right)^2 + \left(\frac{n\pi}{b}\right)^2 - \omega^2\mu\varepsilon}} \qquad \textbf{(7.90)}$$

Three cases may appear:

Case 1: When $\omega^2\mu\varepsilon = \left(\frac{m\pi}{a}\right)^2 + \left(\frac{n\pi}{b}\right)^2$ (Cut-off Mode)

There will be no wave propagation under this condition. This is called *cut-off region*. The corresponding frequency is called the *cut-off frequency* and corresponding wavelength is the *cut-off wavelength*.
In this case, $\gamma = 0$, i.e., $\alpha = \beta = 0$.

Case 2: When $\omega^2\mu\varepsilon < \left(\frac{m\pi}{a}\right)^2 + \left(\frac{n\pi}{b}\right)^2$ (Evanescent Mode)

The wave will be attenuated and there will be no wave propagation. This is called *evanescent mode*.
In this case, $\gamma = \alpha$ and $\beta = 0$.

Case 3: When $\omega^2\mu\varepsilon > \left(\frac{m\pi}{a}\right)^2 + \left(\frac{n\pi}{b}\right)^2$ (Propagation Mode)

Only under this condition, there will be wave propagation. Therefore, this mode is called the *propagation mode*.
In this case, $\gamma = j\beta$ and $\alpha = 0$.
From the three cases, it is understood that there will be no wave propagation in a rectangular waveguide for low frequencies. As the frequency is increased, at frequency equal to the cut-off frequency, the wave propagation takes place and for all frequencies above the cut-off frequency the wave propagates without any attenuation. Thus, *a rectangular waveguide behaves as a high pass filter*.

2. Cut-off frequency (f_c): The cut-off frequency of a waveguide is that operating frequency below which the wave is attenuated and above which the wave is propagated through the guide. This is given as

$$\omega_c = \frac{1}{\sqrt{\mu\varepsilon}}\sqrt{\left(\frac{m\pi}{a}\right)^2 + \left(\frac{n\pi}{b}\right)^2} = v_0\sqrt{\left(\frac{m\pi}{a}\right)^2 + \left(\frac{n\pi}{b}\right)^2}$$

where $v_0 = \frac{1}{\sqrt{\mu\varepsilon}}$ is the phase velocity of uniform plane waves in a lossless dielectric medium inside the waveguide.

$$\therefore \qquad f_c = \frac{\omega_c}{2\pi} = \frac{1}{2\pi\sqrt{\mu\varepsilon}}\sqrt{\left(\frac{m\pi}{a}\right)^2 + \left(\frac{n\pi}{b}\right)^2} = \frac{v_0}{2}\sqrt{\left(\frac{m}{a}\right)^2 + \left(\frac{n}{b}\right)^2}$$

$$\therefore \qquad \boxed{f_c = \frac{v_0}{2}\sqrt{\left(\frac{m}{a}\right)^2 + \left(\frac{n}{b}\right)^2}} \qquad \textbf{(7.91)}$$

From Eq. (7.91), it is seen that the wavelength is the smallest for TE_{10} mode of all TE mode waves and for TM_{11} modes of all TM mode waves.

For TE_{10} mode, $f_c = \frac{v_0}{2a}$

For TM_{11} mode, $f_c = \frac{v_0}{2ab}\sqrt{a^2 + b^2}$

3. Cut-off wavelength (λ_c): Corresponding to every cut-off frequency, there will be a *cut-off wavelength* given as

$$\boxed{\lambda_c = \frac{v_0}{f_c} = \frac{2}{\sqrt{\left(\frac{m}{a}\right)^2 + \left(\frac{n}{b}\right)^2}}} \qquad \textbf{(7.92)}$$

From Eq. (7.92), it is seen that the wavelength is the longest for TE_{10} mode of all TE mode waves and for TM_{11} modes of all TM mode waves.

For TE_{10} mode, $\lambda_c = 2a$

For TM_{11} mode, $\lambda_c = \frac{2ab}{\sqrt{a^2 + b^2}}$

4. Phase constant (β): From Eq. (7.90), the phase constant in the propagation mode may be written as

$$\boxed{\beta = \sqrt{\omega^2\mu\varepsilon - \left(\frac{m\pi}{a}\right)^2 + \left(\frac{n\pi}{b}\right)^2}} \qquad \textbf{(7.93)}$$

In terms of the cut-off frequency, it can be written as

$$\beta = \omega\sqrt{\mu\varepsilon}\sqrt{1 - \frac{1}{\omega^2\mu\varepsilon}\left[\left(\frac{m\pi}{a}\right)^2 + \left(\frac{n\pi}{b}\right)^2\right]} = \omega\sqrt{\mu\varepsilon}\sqrt{1 - \frac{\omega_c^2}{\omega^2}} = \beta_0\sqrt{1 - \frac{\omega_c^2}{\omega^2}} = \beta_0\sqrt{1 - \frac{f_c^2}{f^2}}$$

where $\beta_0 = \omega\sqrt{\mu\varepsilon}$ is the phase constant of uniform plane wave in a lossless medium.

$$\therefore \qquad \boxed{\beta = \beta_0\sqrt{1-\frac{f_c^2}{f^2}}} \tag{7.94}$$

5. Phase velocity (v_p): The phase velocity of the wave propagation is given by

$$\boxed{v_p = \frac{\omega}{\beta} = \frac{\omega}{\sqrt{\omega^2\mu\varepsilon - \left(\frac{m\pi}{a}\right)^2 + \left(\frac{n\pi}{b}\right)^2}} = \frac{v_0}{\sqrt{1-\frac{f_c^2}{f^2}}}} \tag{7.95}$$

$v_0 = \dfrac{1}{\sqrt{\mu\varepsilon}}$ is the phase velocity of uniform plane waves in a lossless dielectric medium inside the waveguide.

Equation (7.95) indicates that the phase velocity of wave propagation in the rectangular waveguide is greater than the phase velocity of uniform plane wave. As the frequency is increased above the cut-off frequency, the phase velocity decreases from an infinitely large value and approaches v_0.

6. Guide wavelength (λ_g): The wavelength in the rectangular waveguide is given as

$$\boxed{\lambda_g = \frac{2\pi}{\beta} = \frac{2\pi}{\sqrt{\omega^2\mu\varepsilon - \left(\frac{m\pi}{a}\right)^2 + \left(\frac{n\pi}{b}\right)^2}} = \frac{\lambda_0}{\sqrt{1-\frac{f_c^2}{f^2}}}} \tag{7.96}$$

where $\lambda_0 = \dfrac{2\pi}{\omega\sqrt{\mu\varepsilon}} = \dfrac{2\pi}{\beta_0}$ is the wavelength of the uniform plane wave in the lossless dielectric medium inside the guide.

7. Intrinsic wave impedance (η): The intrinsic wave impedance will be different for *TE* and *TM* modes.

Intrinsic wave impedance for *TE* Modes: For *TE* waves, from Eq. (7.84), it is given as

$$\eta_{TE} = \frac{E_x}{H_y} = -\frac{E_y}{H_x} = \frac{\omega\mu}{\beta} = \sqrt{\frac{\mu}{\varepsilon}}\frac{1}{\sqrt{1-\frac{f_c^2}{f^2}}} = \frac{\eta_0}{\sqrt{1-\frac{f_c^2}{f^2}}}$$

where $\eta_0 = \sqrt{\dfrac{\mu}{\varepsilon}}$ is the intrinsic impedance of uniform plane wave in a lossless dielectric medium.

$$\therefore \qquad \boxed{\eta_{TE} = \frac{\eta_0}{\sqrt{1-\frac{f_c^2}{f^2}}}} \tag{7.97}$$

Intrinsic wave impedance for *TM* Modes: For *TM* waves, from Eq. (7.89), it is given as

$$\eta_{TM} = \frac{E_x}{H_y} = -\frac{E_y}{H_x} = \frac{\beta}{\omega\varepsilon} = \sqrt{\frac{\mu}{\varepsilon}}\sqrt{1-\frac{f_c^2}{f^2}} = \eta_0\sqrt{1-\frac{f_c^2}{f^2}}$$

where $\eta_0 = \sqrt{\frac{\mu}{\varepsilon}}$ is the intrinsic impedance of uniform plane wave in a lossless dielectric medium.

$$\therefore \quad \boxed{\eta_{TM} = \eta_0 \sqrt{1 - \frac{f_c^2}{f^2}}} \tag{7.98}$$

From Eqs. (7.97) and (7.98), it is observed that the intrinsic wave impedance is purely resistive both for *TE* and *TM* waves. The variation of these impedances with frequency is shown in Fig. 7.11. Also,

$$\eta_{TE}\,\eta_{TM} = \eta_0^2 \tag{7.99}$$

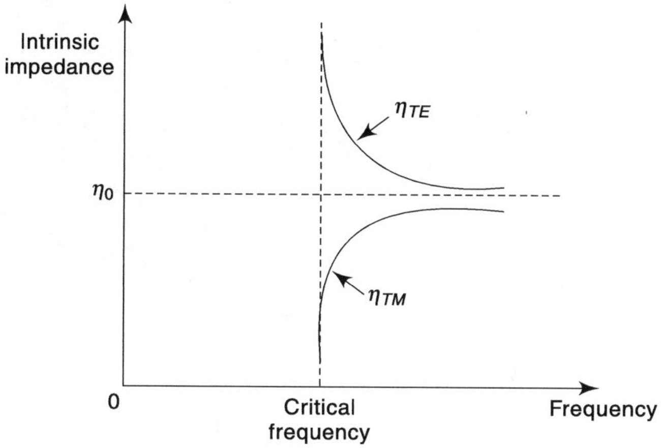

Fig. 7.11 *Variation of intrinsic wave impedance both for TE and TM modes*

Dominant Mode *Dominant mode* is that mode which has the lowest cut-off frequency or highest cut-off wavelength.

For *TE* modes, the lowest order mode can be either TE_{10} or TE_{01} depending upon the dimensions of the guide, i.e., the values of *a* and *b*.

For TE_{10} mode, the cut-off frequency is obtained [from Eq. (7.91)] as

$$f_{cTE_{10}} = \frac{v_0}{2a}$$

For TE_{01} mode, the cut-off frequency is obtained [from Eq. (7.91)] as

$$f_{cTE_{01}} = \frac{v_0}{2b}$$

Therefore, if $a > b$, then TE_{10} mode is the dominant mode and if $b > a$, then TE_{01} is the dominant mode. This may be noted that for the lowest order TM_{11} mode, the cut-off frequency is

$$f_{cTM_{11}} = \frac{v_0\sqrt{a^2 + b^2}}{2ab}$$

which is greater than the cut-off frequency of TE_{10} mode. Therefore, TM_{11} mode cannot be considered as the dominant mode.

Degenerate Modes Two or more modes having the same cut-off frequency are known as *degenerate modes*.

For example, TE_{mn} and TM_{mn} modes are always degenerate.

7.3.4 Field Equations for Transverse Electromagnetic (*TEM*) Mode in Rectangular Waveguide

For *TEM* mode, both the electric and magnetic fields are transverse to the direction of wave propagation. This implies that $E_z = 0 = H_z$ and hence from Eq. (7.64) to Eq. (7.67), we see that all field components will be zero. Thus, we conclude that *a TEM wave cannot exist in a rectangular waveguide.*

In Chapter 5, Section 5.6, we have derived different relations for uniform plane wave propagation through a lossless dielectric medium. This may be noted that those relations are valid for *TEM* wave, i.e., $\alpha = 0,\ \beta = \omega\sqrt{\mu\varepsilon}$, which implies that for *TEM* waves, $h = 0$.

Impossibility of Transverse Electromagnetic (TEM) Mode in Single-Conductor Waveguides This can be easily shown that a *TEM* wave, for which there is no axial component of either $\bar{E}$ or $\bar{H}$, cannot propagate within a single-conductor waveguide.

In order to prove this, we initially assume that a *TEM* wave exists within a hollow waveguide of any shape. Then the lines of magnetic field $\bar{H}$ lie entirely in the transverse plane. Also, in a non-conducting medium,

$$\nabla \cdot \bar{H} = 0$$

which requires that the lines of $\bar{H}$ be closed loops. Therefore, if a *TEM* wave exists inside a wave, the lines of $\bar{H}$ must be closed loops in a plane perpendicular to the axis.

However, by modified Ampere's circuital law, the magnetomotive force around each of these closed loops must be equal to the axial current (conduction or displacement) through the loop $\left(\oint_L \vec{H} \cdot d\vec{l} = I_{\text{enc}}\right)$.

In the case of a guide with an inner conductor, such as a coaxial transmission line, this axial current is the conduction current in the inner conductor. However, for a hollow waveguide having no inner conductor, this axial current is the displacement current. But, an axial displacement current requires an axial component of $\bar{E}$, which is not present in a *TEM* wave.

Hence, the *TEM wave cannot exist in a single-conductor waveguide.*

7.3.5 Hybrid (*HE*) Mode

In this mode, neither $\bar{E}$ nor $\bar{H}$ is transverse to the direction of wave propagation.

Example 7.6 A rectangular waveguide has a broad wall dimension of 2.29 cm and is fed by 10 GHz carrier from a coaxial cable to generate TE_{10} wave propagation. Find its guide wavelength, phase and group velocities.

Solution Here, $a = 2.29$ cm, $f = 10$ GHz, For TE_{10} mode, $m = 1$, $n = 0$

Free space wavelength, $\lambda_0 = \dfrac{v_0}{f} = \dfrac{3\times 10^8}{10\times 10^9} = 0.03 \text{ m} = 3 \text{ cm}$

Cut-off frequency is

$$f_c = \frac{v_0}{2}\sqrt{\left(\frac{m}{a}\right)^2 + \left(\frac{n}{b}\right)^2} = \frac{3\times 10^8}{2}\sqrt{\left(\frac{1}{2.29\times 10^{-2}}\right)^2} = 6.55\times 10^9 \text{ Hz} = 6.55 \text{ GHz}$$

Guide wavelength is given as

$$\lambda_g = \frac{\lambda_0}{\sqrt{1-\frac{f_c^2}{f^2}}} = \frac{0.03}{\sqrt{1-\left(\frac{6.55}{10}\right)^2}} = 0.0397 \text{ m} = 3.97 \text{ cm}$$

Phase velocity is given as

$$v_p = \frac{v_0}{\sqrt{1-\frac{f_c^2}{f^2}}} = \frac{3\times 10^8}{\sqrt{1-\left(\frac{6.55}{10}\right)^2}} = 3.97\times 10^8 \text{ m/s}$$

The group velocity is obtained as

$$v_p v_g = v_0^2$$

$$\therefore \quad v_g = \frac{v_0^2}{v_p} = \frac{(3\times 10^8)^2}{3.97\times 10^8} = 2.267\times 10^8 \text{ m/s}$$

Example 7.7 If the dimensions of a rectangular waveguide are 3.5 cm × 2.0 cm, and the frequency of operation is 10 GHz, determine all the possible *TE* and *TM* modes that can be propagated in this waveguide.

Solution Here, $a = 3.5$ cm $= 0.035$ m, $b = 2.0$ cm $= 0.02$ m, $f = 10$ GHz

Wavelength of the supply is $\lambda = \frac{v_0}{f} = \frac{3\times 10^8}{10\times 10^9} = 0.03$ m

The cut-off wavelength for any TE_{mn} or TM_{mn} wave is given as

$$\lambda_c = \frac{2}{\sqrt{\left(\frac{m}{a}\right)^2 + \left(\frac{n}{b}\right)^2}}$$

For a particular mode to propagate, the cut-off wavelength must be greater than the supply wavelength 0.03 m.

We calculate the cut-off wavelength for different values of m and n.

For TE_{10} mode, $m = 1$, $n = 0$

$$\lambda_c = \frac{2}{\sqrt{\left(\frac{1}{0.035}\right)^2 + 0}} = 0.07 \text{ m}$$

For TE_{01} mode, $m = 0$, $n = 1$

$$\lambda_c = \frac{2}{\sqrt{0 + \left(\frac{1}{0.02}\right)^2}} = 0.04 \text{ m}$$

For TE_{11} and TM_{11} modes, $m = 1$, $n = 1$

$$\lambda_c = \frac{2}{\sqrt{\left(\frac{1}{0.035}\right)^2 + \left(\frac{1}{0.02}\right)^2}} = 0.0347 \text{ m}$$

For TE_{20} mode, $m = 2$, $n = 0$

$$\lambda_c = \frac{2}{\sqrt{\left(\frac{2}{0.035}\right)^2 + 0}} = 0.035 \text{ m}$$

For TE_{02} mode, $m = 0$, $n = 2$

$$\lambda_c = \frac{2}{\sqrt{0 + \left(\frac{2}{0.02}\right)^2}} = 0.02 \text{ m} < \lambda$$

For TE_{21} mode, $m = 2$, $n = 1$

$$\lambda_c = \frac{2}{\sqrt{\left(\frac{2}{0.035}\right)^2 + \left(\frac{1}{0.02}\right)^2}} = 0.0263 \text{ m} < \lambda$$

For TE_{12} mode, $m = 1$, $n = 2$

$$\lambda_c = \frac{2}{\sqrt{\left(\frac{1}{0.035}\right)^2 + \left(\frac{2}{0.02}\right)^2}} = 0.0192 \text{ m} < \lambda$$

Hence, the possible *TE*/*TM* modes which can propagate through the guide are:

TE_{10}, TE_{01}, TE_{11} and TM_{11} and TE_{20} modes

***Example 7.8** In an air-filled rectangular waveguide with $a = 2.286$ cm and $b = 1.016$ cm, the y-component of the *TE* mode is given by

$$E_y = \sin\left(\frac{2\pi}{a}x\right)\cos\left(\frac{3\pi}{b}y\right)\sin\,(10\pi \times 10^{10}t - \beta z) \text{ V/m}$$

Find: (a) the operating mode, (b) the propagation constant γ and (c) the intrinsic impedance η.

Solution From the expression of E_y it is seen that, $m = 2$, $n = 3$, i.e., the guide is operating at TE_{23} mode.

Cut-off frequency is

$$f_c = \frac{v_0}{2}\sqrt{\left(\frac{m}{a}\right)^2 + \left(\frac{n}{b}\right)^2} = \frac{3\times 10^8}{2}\sqrt{\left(\frac{2}{2.286\times 10^{-2}}\right)^2 + \left(\frac{3}{1.016\times 10^{-2}}\right)^2}$$

$$= 46.19\times 10^9 \text{ Hz} = 46.19 \text{ GHz}$$

$$\therefore \qquad \beta = \omega\sqrt{\mu\varepsilon}\sqrt{1-\frac{f_c^2}{f^2}} = \frac{10\pi\times 10^{10}}{3\times 10^8}\sqrt{1-\left(\frac{46.19}{10\pi\times 10^{10}/2\pi}\right)^2} = 400.7 \text{ radian/s}$$

Hence, the propagation constant is, $\gamma = j\beta = j400.7/\text{m}$
Intrinsic impedance is

$$\eta_{TE_{23}} = \frac{\eta_0}{\sqrt{1-\frac{f_c^2}{f^2}}} = \frac{120\pi}{\sqrt{1-\left(\frac{46.19}{10\pi\times 10^{10}/2\pi}\right)^2}} = 985.2\ \Omega$$

***Example 7.9** An air-filled hollow rectangular conducting waveguide has a cross section of 8 × 10 cm. How many *TE* modes will this waveguide transmit at frequencies below 4 GHz? How are these waves designated and what are the cut-off frequencies?

Solution Here, $a = 10$ cm $= 0.1$ m, $b = 8$ cm $= 0.08$ m, $f = 4$ GHz
For a general *TE* mode, the cut-off frequency is given as

$$f_c = \frac{v_0}{2}\sqrt{\left(\frac{m}{a}\right)^2 + \left(\frac{n}{b}\right)^2} = \frac{3\times 10^8}{2\times 10^{-2}}\sqrt{\left(\frac{m}{10}\right)^2 + \left(\frac{n}{8}\right)^2} = 15\sqrt{0.01m^2 + 0.0156n^2} \text{ GHz}$$

For a particular mode to propagate, the cut-off frequency must be less than 4 GHz.
We calculate the cut-off frequency for different values of m and n.
For TE_{10} mode, $m = 1, n = 0$

$$f_c = 15\sqrt{0.01} \text{ GHz} = 1.5 \text{ GHz}$$

For TE_{01} mode, $m = 0, n = 1$

$$f_c = 15\sqrt{0.0156} \text{ GHz} = 1.875 \text{ GHz}$$

For TE_{11} modes, $m = 1, n = 1$

$$f_c = 15\sqrt{0.01 + 0.0156} \text{ GHz} = 2.4 \text{ GHz}$$

For TE_{20} mode, $m = 2, n = 0$

$$f_c = 15\sqrt{0.01\times 4} \text{ GHz} = 3 \text{ GHz}$$

For TE_{02} mode, $m = 0, n = 2$,

$$f_c = 15\sqrt{0.0156\times 4} \text{ GHz} = 3.75 \text{ GHz}$$

For TE_{21} mode, $m = 2, n = 1$,

$$f_c = 15\sqrt{0.01 \times 4 + 0.0156} \text{ GHz} = 3.54 \text{ GHz}$$

For TE_{12} mode, $m = 1, n = 2$,

$$f_c = 15\sqrt{0.01 + 0.0156 \times 4} \text{ GHz} = 4.04 \text{ GHz}$$

For TE_{31} mode, $m = 3, n = 1$,

$$f_c = 15\sqrt{0.01 \times 9 + 0.0156} \; GHz = 4.875 \text{ GHz}$$

Hence, the possible TE modes which can propagate through the guide with respective cut-off frequencies are given in Table 7.1.

Table 7.1 *TE modes*

Mode	Cut-off frequency (GHz)
TE_{10}	1.5
TE_{01}	1.875
TE_{11}	2.4
TE_{20}	3
TE_{02}	3.75
TE_{21}	3.54

Example 7.10 A rectangular waveguide with dimensions $a = 2$ cm and $b = 1$ cm filled with deionized water ($\mu_r = 1$, $\varepsilon_r = 81$) operates at 3 GHz. Determine all the propagating modes and the corresponding cut-off frequencies.

Solution For a general TE or TM mode, the cut-off frequency is given as

$$f_c = \frac{v_0}{2}\sqrt{\left(\frac{m}{a}\right)^2 + \left(\frac{n}{b}\right)^2} = \frac{3 \times 10^8}{2\sqrt{81}}\sqrt{\left(\frac{m}{0.02}\right)^2 + \left(\frac{n}{0.01}\right)^2} = 1.667\sqrt{0.25m^2 + n^2} \text{ GHz}$$

The cut-off frequencies as calculated from the above equation for different values of m and n are given in the table below.

Cut-off Frequency (in GHz) for *TM* modes				Cut-off Frequency (in GHz) for *TE* modes				
	$n = 1$	$n = 2$	$\cdots$		$n = 0$	$n = 1$	$n = 2$	
$m = 1$	1.863	3.436	$\cdots$	$m = 0$	$\times$	1.667	3.333	$\cdots$
$m = 2$	2.357	3.727	$\cdots$	$m = 1$	0.833	1.863	3.436	$\cdots$
$m = 3$	3.005	$\ddots$		$m = 2$	1.667	2.357	3.727	$\cdots$
	$\vdots$			$m = 3$	2.500	3.005	$\ddots$	
				$m = 4$	3.333	$\vdots$		
					$\vdots$			

Since the operating frequency is 3 GHz, all the propagating modes and their cut-off frequencies are given in the table below.

Mode	Cut-off Frequency (GHz)
TE_{10}	0.833
TE_{01}, TE_{20}	1.667
TE_{11}, TM_{11}	1.863
TE_{21}, TM_{21}	2.357
TE_{30}	2.500

***Example 7.11** A tunnel is modelled as an air-filled metallic rectangular waveguide with dimensions $a = 8$ m and $b = 16$ m. Determine whether the tunnel will pass: (a) a 1.5 MHz AM broadcast signal, (b) a 120 MHz FM broadcast signal.

Solution Here, $a = 8$ m and $b = 16$ m
Since, $b > a$, TE_{01} mode will have the lowest cut-off frequency.
The cut-off frequency is given as

$$f_c = \frac{v_0}{2}\sqrt{\frac{m^2}{a^2} + \frac{n^2}{b^2}} = \frac{3 \times 10^8}{2}\sqrt{0 + \frac{1^2}{16^2}} = 9.375 \text{ MHz}$$

(a) For operating frequency, $f = 1.5$ MHz, $f < f_c$ and so, the tunnel will not pass the signal.
(b) For operating frequency, $f = 120$ MHz, $f > f_c$ and so, the tunnel will pass the signal.

Example 7.12 Calculate the cut-off frequency for the dominant mode in a rectangular waveguide of dimensions 4 cm × 2 cm.

Solution Here, $a = 4$ cm, $b = 2$ cm, as $a > b$, TE_{10} mode is the dominant mode.

$$\therefore \qquad m = 1, n = 0$$

The cut-off frequency is given as

$$f_c = \frac{v_0}{2}\sqrt{\frac{m^2}{a^2} + \frac{n^2}{b^2}} = \frac{v_0}{2a} = \frac{3 \times 10^8}{2 \times 4 \times 10^{-2}} = 3.75 \text{ GHz}$$

***Example 7.13** Calculate the phase velocity and group velocity for the dominant mode in a rectangular waveguide (1D: 2.3 cm × 1.0 cm) at 10 GHz.

Solution Here, $f = 10$ GHz, $a = 2.3$ cm, $b = 1.0$ cm
Since $a > b$, TE_{10} mode is the dominant mode.

$$\therefore \qquad m = 1, n = 0$$

The cut-off frequency is given as

$$f_c = \frac{v_0}{2}\sqrt{\frac{m^2}{a^2} + \frac{n^2}{b^2}} = \frac{v_0}{2a} = \frac{3 \times 10^8}{2 \times 2.3 \times 10^{-2}} = 6.52 \text{ GHz}$$

The phase velocity is given as

$$v_p = \frac{v_0}{\sqrt{1-\dfrac{f_c^2}{f^2}}} = \frac{3\times10^8}{\sqrt{1-\left(\dfrac{6.52}{10}\right)^2}} = 3.957\times10^8 \text{ m/s}$$

The group velocity is given as

$$v_g = \frac{v_0^2}{v_p} = \frac{(3\times10^8)^2}{3.957\times10^8} = 2.274\times10^8 \text{ m/s}$$

Example 7.14 Design a rectangular waveguide to carry only the TE_{10} mode at a frequency of 5000 MHz.

Solution Here, $f = 5000$ MHz, $m = 1$, $n = 0$.
The dimensions of the waveguide should be such that the operating frequency is equal to 5000 MHz.
The cut-off frequency is given as

$$f_c = \frac{v_0}{2}\sqrt{\frac{m^2}{a^2}+\frac{n^2}{b^2}} = \frac{v_0}{2a} = \frac{3\times10^8}{2a} = 5000\times10^6$$

$\therefore$ $$a = 3 \text{ cm}$$

For standard waveguide, the aspect ratio is generally 2 : 1, so that we can write

$$b = \frac{a}{2} = 1.5 \text{ cm}$$

Hence, the dimensions of the waveguide are 3 cm × 1.5 cm.

***Example 7.15** A rectangular waveguide with dimensions 2.42 cm × 1.12 cm supporting TE_{10} mode at 6 GHz is filled with a dielectric of relative permittivity ε_r. What are the limits on ε_r if only the dominant mode propagates?

Solution Here, $a = 2.42$ cm, $b = 1.12$ cm, $m = 1$, $n = 0$, $f = 6$ GHz
If only dominant mode is to propagate, the cut-off frequency should be equal to the operating frequency.

$\therefore$ $$f = f_c = \frac{v_0}{2a}$$

$\Rightarrow$ $$6\times10^9 = \frac{1}{\sqrt{\mu_0\varepsilon_0\varepsilon_r}}\times\frac{1}{2\times2.42\times10^{-2}} = \frac{3\times10^8}{\sqrt{\varepsilon_r}}\times\frac{1}{2\times2.42\times10^{-2}} = \frac{6.198\times10^9}{\sqrt{\varepsilon_r}}$$

$\Rightarrow$ $$\varepsilon_r = \left(\frac{6.198}{6}\right)^2 = 1.0672$$

Also, the cut-off frequency can be given as

$$f = f_c = \frac{v_0}{2b}$$

$$\Rightarrow \qquad 6\times 10^9 = \frac{1}{\sqrt{\mu_0\varepsilon_0\varepsilon_r}}\times\frac{1}{2\times 1.12\times 10^{-2}} = \frac{3\times 10^8}{\sqrt{\varepsilon_r}}\times\frac{1}{2\times 1.12\times 10^{-2}} = \frac{13.39\times 10^9}{\sqrt{\varepsilon_r}}$$

$$\Rightarrow \qquad \varepsilon_r = \left(\frac{13.39}{6}\right)^2 = 4.9824$$

So, the limits are: $\quad 1.0672 < \varepsilon_r < 4.9824$

***Example 7.16** Design a rectangular waveguide which at 10 GHz, will operate in TE_{10} mode with 25% safety factor ($f \geq 1.25\, f_c$) when the interior of guide is filled with air. It is required that the mode with the next higher cut-off will operate at 25% below its cut-off frequency.

Solution Since the operating frequeny is 10 GHz, the cut-off frequency is given as

$$f_c = \frac{10}{1.25} = 8 \text{ GHz}$$

For dominant mode, the cut-off frequency is given as

$$f_c = \frac{v_0}{2}\sqrt{\left(\frac{m}{a}\right)^2 + \left(\frac{n}{b}\right)^2} = \frac{v_0}{2a}$$

or
$$8\times 10^9 = \frac{3\times 10^8}{2a}$$

∴
$$a = 0.01875\, m = 1.875 \text{ cm}$$

The next higher mode will be either TE_{01} or TE_{11}.
For TE_{01} mode, the operating frequency is 25% below the cut-off frequency. Hence, we can write

$$f_c = \frac{f}{0.75} = \frac{v_0}{2b}$$

or
$$\frac{10\times 10^9}{0.75} = \frac{3\times 10^8}{2b}$$

∴
$$b = 0.01125 \text{ m} = 1.125 \text{ cm}$$

For TE_{11} mode, the operating frequency is 25% below the cut-off frequency. Hence, we can write

$$f_c = \frac{f}{0.75} = \frac{v_0}{2}\sqrt{\left(\frac{1}{0.01875}\right)^2 + \frac{1}{b^2}}$$

or
$$\frac{10\times 10^9}{0.75} = \frac{3\times 10^8}{2}\sqrt{\left(\frac{1}{0.01875}\right)^2 + \frac{1}{b^2}}$$

∴
$$b = 0.0140625 \text{ m} = 1.40625 \text{ cm}$$

Hence, the dimensions of the guide may be 1.875 cm × 1.125 cm or 1.875 cm × 1.406 cm.

Example 7.17 An air-filled rectangular waveguide has cross-sectional dimensions $x = 8$ cm and $y = 4$ cm. Find the cut-off frequencies for the following modes: TE_{10}, TE_{20}, TE_{11} and the ratio of the guide velocity, v_p to the velocity in free space for each of these modes if $f = \frac{3}{2} f_c$.

Solution Here, $x = 8$ cm, $y = 4$ cm, $f = \frac{3}{2} f_c$

The cut-off frequency is given as

$$f_c = \frac{v_0}{2}\sqrt{\frac{m^2}{a^2}+\frac{n^2}{b^2}} = \frac{3\times 10^8}{2\times 10^{-2}}\sqrt{\frac{m^2}{8^2}+\frac{n^2}{4^2}} = 15\sqrt{0.0156m^2+0.0625n^2}\ \text{GHz}$$

For TE_{10} mode, $f_{c_{10}} = 15\sqrt{0.0156+0}\ \text{GHz} = 1.875\ \text{GHz}$

For TE_{20} mode, $f_{c_{20}} = 15\sqrt{4\times 0.0156+0}\ \text{GHz} = 3.75\ \text{GHz}$

For TE_{11} mode, $f_{c_{11}} = 15\sqrt{0.0156+0.0625}\ \text{GHz} = 4.192\ \text{GHz}$

The ratio of the guide velocity to the velocity in free space is given as

$$\frac{v_p}{v_0} = \frac{1}{\sqrt{1-\left(\frac{f_c}{f}\right)^2}} = \frac{1}{\sqrt{1-\left(\frac{f_c}{1.5 f_c}\right)^2}} = 1.34$$

Example 7.18 The wavelength measured in an air-filled rectangular waveguide, 20 cm × 5.0 cm in cross-section, is 12 cm. Calculate the frequency of the wave. Assume TE_{01} mode and $c = 3 \times 10^8$ m/s.

Solution Here, $a = 20$ cm, $b = 5.0$ cm, $\lambda = 12$ cm, $m = 0$, $n = 1$.

For a wavelength of 12 cm, the frequency is, $f = \frac{c}{\lambda} = \frac{3\times 10^8}{12\times 10^{-2}} = 2.5$ GHz

Also, for TE_{01} mode, the cut-off frequency is

$$f_c = \frac{v_0}{2}\sqrt{\frac{m^2}{a^2}+\frac{n^2}{b^2}} = \frac{v_0}{2b} = \frac{3\times 10^8}{2\times 5\times 10^{-2}} = 3\ \text{GHz}$$

Therefore, the operating frequency is 3 GHz or mode for the wave propagation to occur.

***Example 7.19** A TE_{11} mode of 10 GHz is propagated in an air filled rectangular waveguide. The magnetic field in the z-direction is given by $H_z = H_0 \cos\left(\frac{\pi x}{\sqrt{6}}\right)\cos\left(\frac{\pi y}{\sqrt{6}}\right)$ A/m. The phase constant $\beta = 1.0475$ rad/m, the quantities x and y are expressed in cm and $a = b = \sqrt{6}$. Determine the following (a) cut-off frequency, (b) phase velocity, and (c) guide wavelength.

Solution Here, $f = 10$ GHz, $\beta = 1.0475$ rad/m, $a = b = \sqrt{6}$ cm, $m = n = 1$,

(a) The cut-off frequency is

$$f_c = \frac{v_0}{2}\sqrt{\left(\frac{m}{a}\right)^2 + \left(\frac{n}{b}\right)^2} = \frac{3\times10^8}{2\times10^{-2}}\sqrt{\left(\frac{1}{\sqrt{6}}\right)^2 + \left(\frac{1}{\sqrt{6}}\right)^2} = 8.66 \text{ GHz}$$

(b) Phase velocity, $v_p = \dfrac{v_0}{\sqrt{1-\dfrac{f_c^2}{f^2}}} = \dfrac{3\times10^8}{\sqrt{1-\left(\dfrac{8.66}{10}\right)^2}} = 6\times10^8$ m/s

(c) Guide wavelength is

$$\lambda_g = \frac{\lambda_0}{\sqrt{1-\dfrac{f_c^2}{f^2}}} = \frac{2\pi/\omega\sqrt{\mu\varepsilon}}{\sqrt{1-\dfrac{f_c^2}{f^2}}} = \frac{1/f\sqrt{\mu\varepsilon}}{\sqrt{1-\dfrac{f_c^2}{f^2}}} = \frac{v_0/f}{\sqrt{1-\dfrac{f_c^2}{f^2}}} = \frac{3\times10^8}{10\times10^9\sqrt{1-\left(\dfrac{8.66}{10}\right)^2}} = 0.06 \text{ m} = 6 \text{ cm}$$

Example 7.20 An air-filled rectangular waveguide has cross-sectional dimensions $a = 6$ cm and $b = 3$ cm. Given that

$$E_z = 5\sin\left(\frac{2\pi}{a}x\right)\sin\left(\frac{3\pi}{b}y\right)\cos\left(10^{12}t - \beta z\right) \text{ V/m}$$

Calculate the intrinsic impedance of this mode.

Solution Since, $E_z = \neq 0$, it is understood that the wave is operating in TM_{23} mode, for which $m = 2$, $n = 3$.
The cut-off frequency is

$$f_c = \frac{v_0}{2}\sqrt{\frac{m^2}{a^2}+\frac{n^2}{b^2}} = \frac{3\times10^8}{2\times10^{-2}}\sqrt{\frac{2^2}{6^2}+\frac{3^2}{3^2}} = 15.81 \text{ GHz}$$

Operating frequency is, $f = \dfrac{\omega}{2\pi} = \dfrac{10^{12}}{2\pi} = 159.15$ GHz
Hence, the wave intrinsic impedance is given as

$$\eta_{TM_{23}} = \eta_0\sqrt{1-\left(\frac{f_c}{f}\right)^2} = 120\pi\times\sqrt{1-\left(\frac{15.81}{159.15}\right)^2} = 373.27\ \Omega$$

Example 7.21 A rectangular waveguide measures 3 × 4.5 cm internally and has a 10 GHz signal propagated in it. Calculate the cut-off wavelength, the guide wavelength and the characteristic impedance for the TE_{10} mode.

Solution Here, $a = 4.5$ cm, $b = 3$ cm, $f = 10$ GHz, $m = 1$, $n = 0$.
The cut-off wavelength is given as

$$\lambda_c = \frac{2}{\sqrt{\left(\dfrac{m}{a}\right)^2+\left(\dfrac{n}{b}\right)^2}} = \frac{2a}{m} = \frac{2\times4.5}{1} = 9 \text{ cm}$$

Then wavelength of the impressed signal is, $\lambda = \frac{c}{f} = \frac{3\times10^8}{10\times10^9} = 0.03 \text{ m} = 3 \text{ cm}$
The guide wavelength is given as

$$\lambda_g = \frac{\lambda}{\sqrt{1-\left(\frac{\lambda}{\lambda_c}\right)^2}} = \frac{3}{\sqrt{1-\left(\frac{3}{9}\right)^2}} = \frac{9}{\sqrt{8}} = 3.18 \text{ cm}$$

The characteristic impedance is given as

$$\eta_{TE_{10}} = \frac{\eta_0}{\sqrt{1-\left(\frac{\lambda}{\lambda_c}\right)^2}} = \frac{120\pi}{\sqrt{1-\left(\frac{3}{9}\right)^2}} = \frac{360\pi}{\sqrt{8}} = 399.86\ \Omega$$

Example 7.22 Calculate the cut-off frequencies for the TE_{01}, TE_{11} and TM_{12} modes in a rectangular metal waveguide of dimensions 2 cm × 1 cm.

Solution Here, $a = 2$ cm, $b = 1$ cm.

The cut-off frequency is given as $f_c = \frac{v_0}{2}\sqrt{\left(\frac{m}{a}\right)^2 + \left(\frac{n}{b}\right)^2}$
For TE_{01} mode, $m = 0, n = 1$,

$$f_{c01} = \frac{v_0}{2}\sqrt{\left(\frac{n}{b}\right)^2} = \frac{v_0}{2b} = \frac{3\times10^8}{2\times1\times10^{-2}} = 15 \text{ GHz}$$

For TE_{11} mode, $m = 1, n = 1$,

$$f_{c11} = \frac{v_0}{2}\sqrt{\left(\frac{m}{a}\right)^2 + \left(\frac{n}{b}\right)^2} = \frac{v_0}{2\times10^{-2}}\sqrt{\left(\frac{1}{2}\right)^2 + \left(\frac{1}{1}\right)^2} = \frac{3\times10^8\times\sqrt{5}}{2\times1\times10^{-2}\times2} = 16.77 \text{ GHz}$$

For TE_{01} mode, $m = 0, n = 1$,

$$f_{c01} = \frac{v_0}{2}\sqrt{\left(\frac{n}{b}\right)^2} = \frac{v_0}{2b} = \frac{3\times10^8}{2\times1\times10^{-2}} = 15 \text{ GHz}$$

Example 7.23 In an air-filled rectangular waveguide, the cut-off frequency of a TE_{10} mode is 5 GHz, whereas that of TE_{01} mode is 2 GHz. Calculate

(a) The dimension of the guide
(b) The cut-off frequencies of the next three higher TE modes
(c) The cut-off frequency for TE_{11} mode if the guide is filled with a lossless material having ε_r = 2.225 and $\mu_r = 1$.

Solution Here, $f_{cTE_{10}} = 5$ GHz, $f_{cTE_{01}} = 2$ GHz

(a) Since the cut-off frequency is lowest in TE_{01} mode, this is the dominant mode.This implies that for the rectangular waveguide, $b > a$.
Now, for TE_{01} mode, the cut-off frequency is given as

$$f_{cTE_{01}} = \frac{v_0}{2b} \quad \Rightarrow \quad b = \frac{v_0}{2f_{cTE_{01}}} = \frac{3\times 10^8}{2\times 2\times 10^9} = 0.075 \text{ m} = 7.5 \text{ cm}$$

Also, for TE_{10} mode, the cut-off frequency is given as

$$f_{cTE_{10}} = \frac{v_0}{2a} \quad \Rightarrow \quad a = \frac{v_0}{2f_{cTE_{10}}} = \frac{3\times 10^8}{2\times 5\times 10^9} = 0.03 \text{ m} = 3 \text{ cm}$$

So, the dimension of the rectangular waveguide is 3 cm × 7.5 cm.

(b) The cut-off frequency for TE_{mn} mode is given as

$$f_c = \frac{v_0}{2}\sqrt{\left(\frac{m}{a}\right)^2 + \left(\frac{n}{b}\right)^2} = \frac{v_0}{2}\sqrt{\left(\frac{m}{a}\right)^2 + \left(\frac{n}{b}\right)^2}$$

$$f_c = \frac{3\times 10^8}{2}\sqrt{\left(\frac{m}{0.03}\right)^2 + \left(\frac{n}{0.075}\right)^2} = 15\sqrt{0.111m^2 + 0.01778n^2} \text{ GHz}$$

The next higher order modes will be TE_{02}, TE_{11}, and TE_{03}.
The corresponding cut-off frequencies are given as follows.

$$f_{c_{02}} = 15\sqrt{0.01778\times 4} = 4 \text{ GHz}$$
$$f_{c_{10}} = 15\sqrt{0.111\times 1} = 5 \text{ GHz}$$
$$f_{c_{11}} = 15\sqrt{0.111 + 0.01778} = 5.385 \text{ GHz}$$
$$f_{c_{03}} = 15\sqrt{0.01778\times 9} = 6 \text{ GHz}$$

(c) If the guide is filled with a lossless material having $\varepsilon_r = 2.25$ and $\mu_r = 1$, the cut-off frequency for TE_{11} mode is

$$f_c = \frac{v}{2}\sqrt{\left(\frac{m}{a}\right)^2 + \left(\frac{n}{b}\right)^2} = \frac{v_0}{2\sqrt{\varepsilon_r}}\sqrt{\left(\frac{m}{a}\right)^2 + \left(\frac{n}{b}\right)^2}$$

$$f_{c_{11}} = \frac{15}{\sqrt{\varepsilon_r}}\sqrt{0.111 + 0.01778} \text{ GHz} = \frac{15}{\sqrt{2.25}}\sqrt{0.111 + 0.01778} \text{ GHz} = 3.59 \text{ GHz}$$

Example 7.24 The larger dimension of the cross section of a rectangular waveguide is 2 cm. Find the cut-off frequency and wavelength for the dominant *TE* mode.

Solution Let, $a = 2$ cm ($a > b$); then the dominant mode is TE_{10} mode. Hence, $m = 1$, $n = 0$. For this dominant mode, the cut-off frequency is given as

$$f_{c_{10}} = \frac{v_0}{2}\sqrt{\left(\frac{m}{a}\right)^2 + \left(\frac{n}{b}\right)^2} = \frac{v_0}{2a} = \frac{3\times 10^8}{2\times 2\times 10^{-2}} = 7.5 \text{ GHz}$$

The wavelength of this dominant mode is

$$\lambda_{c_{10}} = \frac{2}{\sqrt{\left(\frac{m}{a}\right)^2 + \left(\frac{n}{b}\right)^2}} = 2a = 4 \text{ cm}$$

Example 7.25 When the dominant H mode is propagated in an air-filled rectangular waveguide, the guide wavelength for a frequency of 9 GHz is 4.0 cm. Calculate the breadth of the guide. Deduce the formula used.

Solution Here, for dominant H mode, $m = 1$, $n = 0$.
Given that, $f = 9$ GHz, $\lambda_g = 4$ cm

The wavelength of the impressed signal is, $\lambda = \dfrac{c}{f} = \dfrac{3\times10^8}{9\times10^9} = 0.033 \text{ m} = 3.33 \text{ cm}$

We have the relation,

$$\lambda_g = \frac{\lambda}{\sqrt{1-\left(\frac{\lambda}{\lambda_c}\right)^2}}$$

or
$$4 = \frac{3.33}{\sqrt{1-\left(\frac{3.33}{\lambda_c}\right)^2}}$$

or
$$1-\left(\frac{3.33}{\lambda_c}\right)^2 = 0.694$$

or
$$\left(\frac{3.33}{\lambda_c}\right)^2 = 0.3055$$

or
$$\lambda_c = 6.03 \text{ cm}$$

Also, for dominant *H* mode, the cut-off wavelength is given as

$$\lambda_c = \frac{2}{\sqrt{\left(\frac{m}{a}\right)^2+\left(\frac{n}{b}\right)^2}} = \frac{2a}{m} = \frac{2a}{1} = 6.03 \text{ cm}$$

∴
$$a = 3.015 \text{ cm}$$

Hence, the breadth of the waveguide is given as $a = 3.015$ cm.

Example 7.26 A rectangular waveguide carries an electromagnetic wave having a frequency of 4000 MHz. A standing wave indicator shows that the wavelength of the wave in the guide is 11.4 cm. What is the cut-off wavelength of the waveguide and the velocity at which energy is propagated along the guide?

Solution Here, $f = 4000$ MHz, $\lambda_g = 11.4$ cm

Wavelength of the impressed signal is $\lambda = \dfrac{c}{f} = \dfrac{3\times10^8}{4000\times10^6} = 0.075 \text{ m} = 7.5 \text{ cm}$

We have the relation,

$$\lambda_g = \frac{\lambda}{\sqrt{1-\left(\frac{\lambda}{\lambda_c}\right)^2}}$$

or $$11.4 = \frac{7.5}{\sqrt{1-\left(\frac{7.5}{\lambda_c}\right)^2}}$$

or $$1-\left(\frac{7.5}{\lambda_c}\right)^2 = 0.433$$

or $$\left(\frac{7.5}{\lambda_c}\right)^2 = 0.567$$

or $$\lambda_c = 9.96 \text{ cm}$$

So, the cut-off wavelength is 9.96 cm.

The velocity with which the energy travels within the guide is the group velocity.

The group velocity is given as

$$v_g = v_0\sqrt{1-\frac{f_c^2}{f^2}} = v_0\sqrt{1-\frac{\lambda^2}{\lambda_c^2}} = 3\times 10^8\sqrt{1-\left(\frac{7.5}{9.96}\right)^2} = 1.298\times 10^8 \text{ m/s}$$

***Example 7.27** The cross section of a rectangular waveguide is 20 cm × 5 cm. If it is filled with air, find the first six lowest-order modes which will propagate through the waveguide and their cut-off frequencies.

Solution Here, $a = 20$ cm, $b = 5$ cm.

The cut-off frequency is given as

$$f_c = \frac{1}{2\sqrt{\mu_0\varepsilon_0}}\sqrt{\left(\frac{m}{a}\right)^2+\left(\frac{n}{b}\right)^2} = \frac{c}{2}\sqrt{\left(\frac{m}{a}\right)^2+\left(\frac{n}{b}\right)^2}$$

$$f_c = \frac{3\times 10^8}{2}\sqrt{\left(\frac{m}{0.2}\right)^2+\left(\frac{n}{0.05}\right)^2} = 15\sqrt{0.0025m^2+0.04n^2} \text{ GHz}$$

We know that for TE_{mnp} modes, any two indices have to be non-zero; whereas for TM_{mnp} modes, all the indices have to be non-zero. Since, $a > b$, the lowest order mode will be TE_{10}.

The next higher order modes will be TE_{20}, TE_{30}, TE_{40}, TE_{01}, TE_{11} and TM_{11}.

The corresponding cut-off frequencies are given as follows.

$$f_{c_{10}} = 15\sqrt{0.0025} = 0.75 \text{ GHz}$$

$$f_{c_{20}} = 15\sqrt{0.0025\times 4} = 1.5 \text{ GHz}$$

$$f_{c_{30}} = 15\sqrt{0.0025 \times 9} = 2.25 \text{ GHz}$$
$$f_{c_{40}} = 15\sqrt{0.0025 \times 16} = 3 \text{ GHz}$$
$$f_{c_{01}} = 15\sqrt{0 + 0.04} = 3 \text{ GHz}$$
$$f_{c_{11}} = 15\sqrt{0.0025 + 0.04} = 3.092 \text{ GHz}$$

Hence, the first six lowest order modes with their cut-off frequencies are given in the table below.

Mode	Resonant Frequency (GHz)
TE_{10}	0.75
TE_{20}	1.5
TE_{30}	2.25
TE_{40}	3
TE_{01}	3
TE_{11} & TM_{11}	3.092

***Example 7.28** A 6 GHz signal is to be propagated in the dominant mode in a rectangular waveguide. If its group velocity is 90% of the free space velocity of light, what must be the breadth of the waveguide?

Solution Free space wavelength is, $\lambda_0 = \dfrac{v_0}{f} = \dfrac{3 \times 10^8}{6 \times 10^9} = 0.05 \text{ m}$

From Eqs. (7.125) and (7.126), we know that

$$\frac{v_g}{v_0} = \frac{\lambda_0}{\lambda_g} = 0.9 \text{ (given)}$$

$\therefore$
$$\lambda_g = \frac{\lambda_0}{0.9} = \frac{0.05}{0.9} = 0.055 \text{ m}$$

Hence, the wavelength within the guide is 5.55 cm.
The cut-off wavelength is given as

$$\frac{1}{\lambda_c^2} = \left(\frac{1}{\lambda_0^2} - \frac{1}{\lambda_g^2}\right) = \left(\frac{1}{0.05^2} - \frac{1}{0.055^2}\right) = 76$$

$\Rightarrow$
$$\lambda_c = 0.1147 \text{ m}$$

So, the breadth of the guide is given as $a = \dfrac{\lambda_c}{2} = \dfrac{0.1147}{2} = 0.05735 \text{ m}$

$\therefore$
$$a = 5.735 \text{ cm}$$

7.4 CYLINDRICAL OR CIRCULAR WAVEGUIDES

Definition *Cylindrical* or *circular waveguides* are those that maintain a uniform circular cross-section along their length.

Derivation of Field Equations for Circular Waveguide The method of solution of the electromagnetic field equations for circular waveguides is similar to that for rectangular waveguides. However, in order to simplify the application of the boundary conditions that the tangential component of the electric field be zero, we convert all field equations in cylindrical coordinate systems. Cylindrical or circular waveguide is shown in Fig. 7.12.

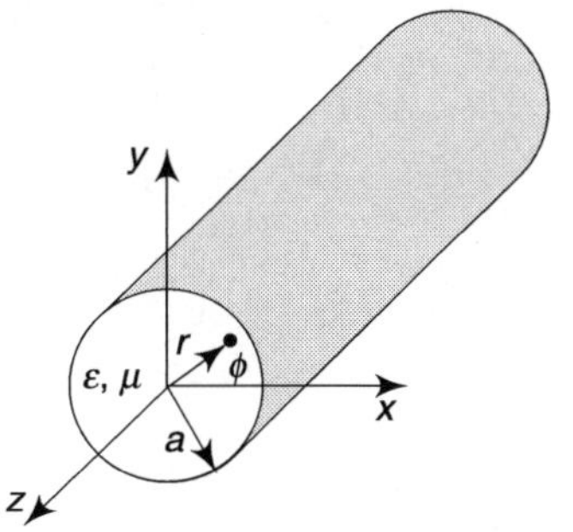

Fig. 7.12 *Cylindrical or circular waveguide*

By Maxwell's equations for time harmonic fields in non-conducting medium ($\sigma = 0$), we have

$$\nabla \times \overline{E} = -j\omega\mu\overline{H} \tag{7.100}$$

$$\nabla \times \overline{H} = j\omega\varepsilon\overline{E} \tag{7.101}$$

By wave equations,

$$\nabla^2\overline{E} + \gamma^2\overline{E} = 0 \tag{7.63a}$$

$$\nabla^2\overline{H} + \gamma^2\overline{H} = 0 \tag{7.63b}$$

where $\gamma^2 = (\sigma + j\omega\varepsilon)j\omega\mu = -\omega^2\mu\varepsilon$ for non-conducting medium.

Expanding Eq. (7.100) in cylindrical coordinate system,

$$\left(\frac{1}{r}\right)\begin{vmatrix} \hat{a}_r & r\hat{a}_\phi & \hat{a}_z \\ \dfrac{\partial}{\partial r} & \dfrac{\partial}{\partial \phi} & \dfrac{\partial}{\partial z} \\ E_r & rE_\phi & E_z \end{vmatrix} = -j\omega\mu\overline{H} = -j\omega\mu(H_r\hat{a}_r + H_\phi\hat{a}_\phi + H_z\hat{a}_z)$$

Equating both sides,

$$\frac{1}{r}\left[\frac{\partial E_z}{\partial \phi} - \frac{\partial}{\partial z}(rE_\phi)\right] = -j\omega\mu H_r \tag{7.102a}$$

$$\frac{\partial E_r}{\partial z} - \frac{\partial E_z}{\partial r} = -j\omega\mu H_\phi \tag{7.102b}$$

$$\frac{1}{r}\left[\frac{\partial}{\partial r}(rE_\phi) - \frac{\partial E_r}{\partial \phi}\right] = -j\omega\mu H_z \tag{7.102c}$$

Expanding Eq. (7.101) in cylindrical coordinate system,

$$\left(\frac{1}{r}\right)\begin{vmatrix} \hat{a}_r & r\hat{a}_\phi & \hat{a}_z \\ \frac{\partial}{\partial r} & \frac{\partial}{\partial \phi} & \frac{\partial}{\partial z}H \\ H_r & rH_\phi & E_z \end{vmatrix} = -j\omega\varepsilon\bar{E} = -j\omega\varepsilon(E_r\hat{a}_r + E_\phi\hat{a}_\phi + E_z\hat{a}_z)$$

Equating both sides,

$$\frac{1}{r}\left[\frac{\partial H_z}{\partial \phi} - \frac{\partial}{\partial z}(rH_\phi)\right] = j\omega\varepsilon E_r \qquad \textbf{(7.103a)}$$

$$\frac{\partial H_r}{\partial z} - \frac{\partial H_z}{\partial r} = j\omega\varepsilon E_\phi \qquad \textbf{(7.103b)}$$

$$\frac{1}{r}\left[\frac{\partial}{\partial r}(rH_\phi) - \frac{\partial H_r}{\partial \phi}\right] = j\omega\varepsilon E_z \qquad \textbf{(7.103c)}$$

If we assume that the wave is propagating in the positive z-direction with the variation expressed as $e^{-\gamma z}$, then we have

$$\frac{\partial E_r}{\partial z} = -\gamma E_r, \quad \frac{\partial E_\phi}{\partial z} = -\gamma E_\phi, \quad \frac{\partial E_z}{\partial z} = -\gamma E_z$$
$$\frac{\partial H_r}{\partial z} = -\gamma H_r, \quad \frac{\partial H_\phi}{\partial z} = -\gamma H_\phi, \quad \frac{\partial H_z}{\partial z} = -\gamma H_z$$

Substituting these in Eq. (7.102), we get

$$\frac{1}{r}\left[\frac{\partial E_z}{\partial \phi} - \frac{\partial}{\partial z}(rE_\phi)\right] = -j\omega\mu H_r$$

or

$$\frac{1}{r}\frac{\partial E_z}{\partial \phi} - \frac{1}{r}r\frac{\partial E_\phi}{\partial z} = -j\omega\mu H_r$$

or

$$\frac{1}{r}\frac{\partial E_z}{\partial \phi} + \gamma E_\phi = -j\omega\mu H_r$$

∴

$$\boxed{\frac{1}{r}\frac{\partial E_z}{\partial \phi} + \gamma E_\phi = -j\omega\mu H_r} \qquad \textbf{(7.104a)}$$

Similarly,

$$\frac{\partial E_r}{\partial z} - \frac{\partial E_z}{\partial r} = -j\omega\mu H_\phi$$

or

$$-\gamma E_r - \frac{\partial E_z}{\partial r} = -j\omega\mu H_\phi$$

∴

$$\boxed{\gamma E_r + \frac{\partial E_z}{\partial r} = j\omega\mu H_\phi} \qquad \textbf{(7.104b)}$$

and

$$\boxed{\frac{1}{r}\left[\frac{\partial}{\partial r}(rE_\phi) - \frac{\partial E_r}{\partial \phi}\right] = -j\omega\mu H_z} \quad \textbf{(7.104c)}$$

Similarly, from Eq. (7.103) we will have the following equations:

$$\boxed{\frac{1}{r}\frac{\partial H_z}{\partial \phi} + \gamma H_\phi = j\omega\varepsilon E_r} \quad \textbf{(7.105a)}$$

$$\boxed{-\gamma H_r - \frac{\partial H_z}{\partial r} = j\omega\varepsilon E_\phi} \quad \textbf{(7.105b)}$$

$$\boxed{\frac{1}{r}\left[\frac{\partial}{\partial r}(rH_\phi) - \frac{\partial H_r}{\partial \phi}\right] = j\omega\varepsilon E_z} \quad \textbf{(7.105c)}$$

From Eq. (7.104) and Eq. (7.105), the components E_r, E_ϕ, H_r and H_ϕ can expressed in terms of the components E_z and H_z as follows.
Combining Eq. (7.104*b*) and (7.105*a*),

$$\frac{1}{r}\frac{\partial H_z}{\partial \phi} + \gamma H_\phi = j\omega\varepsilon E_r$$

or
$$\frac{1}{r}\frac{\partial H_z}{\partial \phi} + \frac{\gamma}{j\omega\mu}\left(\gamma E_r + \frac{\partial E_z}{\partial r}\right) = j\omega\varepsilon E_r$$

or
$$E_r\left(\frac{\gamma^2}{j\omega\mu} - j\omega\varepsilon\right) = -\frac{\gamma}{j\omega\mu}\frac{\partial E_z}{\partial r} - \frac{1}{r}\frac{\partial H_z}{\partial \phi}$$

or
$$E_r\left(\frac{\gamma^2 + \omega^2\mu\varepsilon}{j\omega\mu}\right) = -\frac{\gamma}{j\omega\mu}\frac{\partial E_z}{\partial r} - \frac{1}{r}\frac{\partial H_z}{\partial \phi}$$

or
$$E_r\frac{h^2}{j\omega\mu} = -\frac{\gamma}{j\omega\mu}\frac{\partial E_z}{\partial r} - \frac{1}{r}\frac{\partial H_z}{\partial \phi}$$

where
$$h^2 = \gamma^2 + \omega^2\mu\varepsilon$$

∴
$$\boxed{E_r = -\frac{1}{h^2}\left(\gamma\frac{\partial E_z}{\partial r} + j\frac{\omega\mu}{r}\frac{\partial H_z}{\partial \phi}\right)} \quad \textbf{(7.106a)}$$

Combining Eqs. (7.104*a*) and (7.105*b*),

$$-\gamma H_r - \frac{\partial H_z}{\partial r} = j\omega\varepsilon E_\phi$$

or
$$-\frac{\gamma}{-j\omega\mu}\left(\frac{1}{r}\frac{\partial E_z}{\partial \phi} + \gamma E_\phi\right) - \frac{\partial H_z}{\partial r} = j\omega\varepsilon E_\phi$$

or
$$E_\phi\left(\frac{\gamma^2}{j\omega\mu} - j\omega\varepsilon\right) = -\frac{\gamma}{j\omega\mu}\frac{1}{r}\frac{\partial E_z}{\partial \phi} + \frac{\partial H_z}{\partial r}$$

or
$$E_\phi\left(\frac{h^2}{j\omega\mu}\right)=-\frac{\gamma}{j\omega\mu}\frac{1}{r}\frac{\partial E_z}{\partial \phi}+\frac{\partial H_z}{\partial r}$$

$\therefore$
$$\boxed{E_\phi=\frac{1}{h^2}\left(-\frac{\gamma}{r}\frac{\partial E_z}{\partial \phi}+j\omega\mu\frac{\partial H_z}{\partial r}\right)} \qquad \textbf{(7.106b)}$$

Combining Eqs. (7.104*a*) and (7.105*b*),
$$\frac{1}{r}\frac{\partial E_z}{\partial \phi}+\gamma E_\phi=-j\omega\mu H_r$$

or
$$\frac{1}{r}\frac{\partial E_z}{\partial \phi}+\frac{\gamma}{j\omega\varepsilon}\left(-\gamma H_r-\frac{\partial H_z}{\partial r}\right)=-j\omega\mu H_r$$

or
$$H_r\left(-\frac{\gamma^2}{j\omega\varepsilon}+j\omega\mu\right)=\frac{\gamma}{j\omega\varepsilon}\frac{\partial H_z}{\partial r}-\frac{1}{r}\frac{\partial E_z}{\partial \phi}$$

or
$$-H_r\left(\frac{h^2}{j\omega\varepsilon}\right)=\frac{\gamma}{j\omega\varepsilon}\frac{\partial H_z}{\partial r}-\frac{1}{r}\frac{\partial E_z}{\partial \phi}$$

$\therefore$
$$\boxed{H_r=-\frac{1}{h^2}\left(\gamma\frac{\partial H_z}{\partial r}-j\frac{\omega\varepsilon}{r}\frac{\partial E_z}{\partial \phi}\right)} \qquad \textbf{(7.106c)}$$

Combining Eq. (7.104*b*) and (7.105*a*),
$$\frac{1}{r}\frac{\partial H_z}{\partial \phi}+\gamma H_\phi=j\omega\varepsilon E_r=\frac{j\omega\varepsilon}{\gamma}\left(j\omega\mu H_\phi-\frac{\partial E_z}{\partial r}\right)$$

or
$$H_\phi\left(\gamma+\frac{\omega^2\mu\varepsilon}{\gamma}\right)=-\frac{j\omega\varepsilon}{\gamma}\frac{\partial E_z}{\partial r}-\frac{1}{r}\frac{\partial H_z}{\partial \phi}$$

or
$$H_\phi\frac{h^2}{\gamma}=-\frac{j\omega\varepsilon}{\gamma}\frac{\partial E_z}{\partial r}-\frac{1}{r}\frac{\partial H_z}{\partial \phi}$$

$\therefore$
$$\boxed{H_\phi=-\frac{1}{h^2}\left(\frac{\gamma}{r}\frac{\partial H_z}{\partial \phi}+j\omega\varepsilon\frac{\partial E_z}{\partial r}\right)} \qquad \textbf{(7.106d)}$$

Now from wave equations of Eq. (7.63),
$$\nabla^2\bar{E}+\omega^2\mu\varepsilon\bar{E}=0 \qquad \textbf{(7.63a)}$$
$$\nabla^2\bar{H}+\omega^2\mu\varepsilon\bar{H}=0 \qquad \textbf{(7.63b)}$$

Expanding the Laplacian of the electric field $\bar{E}$ in cylindrical coordinates, we have
$$\frac{1}{r}\frac{\partial}{\partial r}\left(r\frac{\partial \bar{E}}{\partial r}\right)+\frac{1}{r^2}\frac{\partial^2 \bar{E}}{\partial \phi^2}+\frac{\partial^2 \bar{E}}{\partial z^2}+\omega^2\mu\varepsilon\bar{E}=0$$

or
$$\frac{1}{r}\frac{\partial^2 \overline{E}}{\partial r^2}+\frac{1}{r^2}\frac{\partial^2 \overline{E}}{\partial \phi^2}+\frac{\partial^2 \overline{E}}{\partial z^2}+\frac{1}{r}\frac{\partial \overline{E}}{\partial r}+\omega^2\mu\varepsilon\overline{E}=0$$

Since the wave is propagating in the positive z-direction, we have

$$\frac{\partial^2 \overline{E}}{\partial z^2}=\gamma^2\overline{E}$$

Hence, the wave equation for E_z component in cylindrical co-ordinates is reduced to

$$\frac{1}{r}\frac{\partial^2 E_z}{\partial r^2}+\frac{1}{r^2}\frac{\partial^2 E_z}{\partial \phi^2}+\gamma^2 E_z+\frac{1}{r}\frac{\partial E_z}{\partial r}+\omega^2\mu\varepsilon E_z=0 \quad \textbf{(7.107)}$$

Proceeding in the similar way as for rectangular waveguides, let the solution of Eq. (7.107) be

$$E_z=P(r)Q(\phi)e^{-\gamma z}=E_{z0}e^{-\gamma z} \quad \textbf{(7.108)}$$

where $P(r)$ is a function of r alone and $Q(\phi)$ is a function of ϕ alone. Substituting this in Eq. (7.107),

$$Q\frac{\partial^2 P}{\partial r^2}+\frac{P}{r^2}\frac{\partial^2 Q}{\partial \phi^2}+\gamma^2 PQ+\frac{Q}{r}\frac{\partial P}{\partial r}+\omega^2\mu\varepsilon PQ=0$$

or
$$\frac{1}{P}\frac{\partial^2 P}{\partial r^2}+\frac{1}{Qr^2}\frac{\partial^2 Q}{\partial \phi^2}+\gamma^2+\frac{1}{Pr}\frac{\partial P}{\partial r}+\omega^2\mu\varepsilon=0$$

or
$$\frac{1}{P}\frac{\partial^2 P}{\partial r^2}+\frac{1}{Qr^2}\frac{\partial^2 Q}{\partial \phi^2}+\frac{1}{Pr}\frac{\partial P}{\partial r}+h^2=0 \quad \textbf{(7.109)}$$

As have been done for Eqs. (7.70), Eq. (7.109) too can be split into two ordinary differential equations written as

$$\frac{d^2 Q}{d\phi^2}=-n^2 Q \quad \textbf{(7.110a)}$$

$$\frac{d^2 P}{dr^2}+\frac{1}{r}\frac{dP}{dr}+\left(h^2-\frac{n^2}{r^2}\right)P=0 \quad \textbf{(7.110b)}$$

where n is a constant.
Solution of Eq. (7.110*a*) is given as

$$Q=(A_n\cos n\phi+B_n\sin n\phi) \quad \textbf{(7.111)}$$

Reducing Eq. (7.110*b*) to the form of Bessel's equation, we have

$$\frac{d^2 P}{d(rh)^2}+\frac{1}{rh}\frac{dP}{d(rh)}+\left[1-\frac{n^2}{(rh)^2}\right]P=0 \quad \textbf{(7.112)}$$

Solution of Eq. (7.112) is given as

$$P(rh)=J_n(rh) \quad \textbf{(7.113)}$$

where $J_n(rh)$ is Bessel's function of first kind of order n.

Substituting the solutions of Eq. (7.111) and Eq. (7.113) in Eq. (7.108)

$$\boxed{E_z = J_n(rh)(A_n \cos n\phi + B_n \sin n\phi)e^{-\gamma z}} \quad \textbf{(7.114a)}$$

Similarly, the solution for the magnetic field can be written as

$$\boxed{H_z = J_n(rh)(C_n \cos n\phi + D_n \sin n\phi)e^{-\gamma z}} \quad \textbf{(7.114b)}$$

For *TE* and *TM* waves, the solution for other components of $\bar{E}$ and $\bar{H}$ are obtained by substituting Eq. (7.114) in Eq. (7.106).

As for the rectangular waveguides, the solutions for the circular waveguides are also divided into two modes:

1. Transverse Electric (*TE*) Mode and
2. Transverse Magnetic (*TM*) Mode

It must be noted here that the relative amplitudes of the arbitrary constants A_n, B_n and C_n and D_n determine the orientation of the electric and magnetic fields inside the guide for *TM* and *TE* waves, respectively. It is known that for any value of n, the $\phi = 0$ axis can always be oriented to make either A_n and C_n or B_n and D_n equal to zero. Here, we assume that $\phi = 0$ axis is so oriented that

$$B_n = 0 \quad \text{and} \quad D_n = 0$$

Thus, the field equations of Eq. (7.114) can be written as

$$\boxed{E_z = J_n(rh)A_n \cos n\phi e^{-\gamma z}} \quad \textbf{(7.115a)}$$

$$\boxed{H_z = J_n(rh)C_n \cos n\phi e^{-\gamma z}} \quad \textbf{(7.115b)}$$

We will now determine the different characteristics of circular waveguides for *TE* and *TM* modes.

7.4.1 Field Equations for Transverse Electric (*TE*) Mode ($E_z = 0, H_z \neq 0$) in Circular Waveguide

In this case, E_z component is identically zero and H_z is given by Eq. (7.115*b*). The other field components for *TE* waves can be found by inserting Eq. (7.115*b*) in Eq. (7.106).

$$H_z = J_n(rh)C_n \cos n\phi e^{-\gamma z}$$

$$\therefore \quad E_r = -\frac{1}{h^2}\left(\gamma\frac{\partial E_z}{\partial r} + j\frac{\omega\mu}{r}\frac{\partial H_z}{\partial \phi}\right) = -\frac{1}{h^2}\left(0 - j\frac{\omega\mu}{r}J_n(rh)C_n n \sin n\phi e^{-\gamma z}\right)$$

$$= j\frac{n\omega\mu C_n}{rh^2}J_n(rh)\sin n\phi e^{-\gamma z}$$

$$\therefore \quad E_\phi = \frac{1}{h^2}\left(-\frac{\gamma}{r}\frac{\partial E_z}{\partial \phi} + j\omega\mu\frac{\partial H_z}{\partial r}\right) = \frac{1}{h^2}[0 + j\omega\mu h J_n'(rh)C_n \cos n\phi e^{-\gamma z}]$$

$$= j\frac{\omega\mu C_n}{h}J_n'(rh)\cos n\phi e^{-\gamma z}$$

where $J_n'(rh) = \frac{\partial}{dr}[J_n(rh)]$

$$\therefore \qquad H_r = -\frac{1}{h^2}\left(\gamma\frac{\partial H_z}{\partial r} - j\frac{\omega\varepsilon}{r}\frac{\partial E_z}{\partial\phi}\right) = -\frac{1}{h^2}[\gamma h J_n'(rh)C_n \cos n\phi e^{-\gamma z} - 0]$$

$$= -\frac{\gamma C_n}{h}J_n'(rh)\cos n\phi e^{-\gamma z}$$

$$\therefore \qquad H_\phi = -\frac{1}{h^2}\left(\frac{\gamma}{r}\frac{\partial H_z}{\partial\phi} + j\omega\varepsilon\frac{\partial E_z}{\partial r}\right) = -\frac{1}{h^2}\left[-\frac{\gamma}{r}J_n(rh)C_n n\sin n\phi e^{-\gamma z} + 0\right]$$

$$= \frac{n\gamma C_n}{rh^2}J_n(rh)\sin n\phi e^{-\gamma z}$$

Thus, the expressions for different field components for *TE* waves are written as

$$\boxed{\begin{aligned} H_z &= J_n(rh)C_n \cos n\phi e^{-\gamma z} \\ H_r &= -\frac{\gamma C_n}{h}J_n'(rh)\cos n\phi e^{-\gamma z} \\ H_\phi &= \frac{n\gamma C_n}{rh^2}J_n(rh)\sin n\phi e^{-\gamma z} \\ E_r &= j\frac{n\omega\mu C_n}{rh^2}J_n(rh)\sin n\phi e^{-\gamma z} = j\frac{\omega\mu}{\gamma}H_\phi \\ E_\phi &= j\frac{\omega\mu C_n}{h}J_n'(rh)\cos n\phi e^{-\gamma z} = -j\frac{\omega\mu}{\gamma}H_r \end{aligned}} \qquad \textbf{(7.116)}$$

The boundary condition to be satisfied for *TE* waves is that

$$E_\phi = 0 \quad \text{at} \quad r = a$$

where a is the radius of the circular waveguide (see Fig. 7.9).
Therefore, from Eq. (7.116), we have

$$J_n'(ah) = 0 \qquad \textbf{(7.117)}$$

Equation (7.117) shows that for a particular value of n, there is infinite number of roots of the equation and accordingly, there is infinite number of possible *TE* waves. However, as for rectangular waveguides, in the propagation range of frequencies, for circular guide too, the value of h ($h^2 = \gamma^2 + \omega^2\mu\varepsilon$) is very small ($h^2 << \omega^2\mu\varepsilon$). This implies that the first few roots of Eq. (7.117) will be of practical interest. The first few roots of Eq. (7.117) are given in Table 7.2.

Table 7.2 *Roots of* $J_n'(ah) = 0$

m \ n	0	1	2	3
1	3.83	1.84	3.05	4.20
2	7.02	5.33	6.70	8.01
3	10.17	8.53	9.97	11.34

The waves corresponding to different values of h_{nm} are obtained from Table 7.1 and are referred as TE_{01}, TE_{11}, TE_{02}, TE_{12}, and so on.

NOTE

Here, the first subscript refers to the value of n and the second refers to the roots in their order of magnitude.

7.4.2 Field Equations for Transverse Magnetic (*TM*) Mode ($E_z \neq 0, H_z = 0$) in Circular Waveguide

In this case, H_z component is identically zero and E_z is given by Eq. (7.115*a*). The other field components for *TM* waves can be found by inserting Eq. (7.115*a*) in Eq. (7.106).

$$E_z = J_n(rh) A_n \cos n\phi\, e^{-\gamma z}$$

$$\therefore \quad E_r = -\frac{1}{h^2}\left(\gamma \frac{\partial E_z}{\partial r} + j\frac{\omega\mu}{r}\frac{\partial H_z}{\partial \phi}\right) = -\frac{1}{h^2}[\gamma h J_n'(rh) A_n \cos n\phi e^{-\gamma z} + 0]$$

$$= -\frac{\gamma A_n}{h} J_n'(rh) \cos n\phi e^{-\gamma z}$$

where $J_n'(rh) = \frac{\partial}{dr}[J_n(rh)]$

$$\therefore \quad E_\phi = \frac{1}{h^2}\left(-\frac{\gamma}{r}\frac{\partial E_z}{\partial \phi} + j\omega\mu \frac{\partial H_z}{\partial r}\right) = \frac{1}{h^2}\left[\frac{\gamma}{r} J_n(rh) n A_n \sin n\phi e^{-\gamma z} + 0\right]$$

$$= \frac{n\gamma A_n}{rh^2} J_n(rh) \sin n\phi e^{-\gamma z}$$

$$\therefore \quad H_r = -\frac{1}{h^2}\left(\gamma \frac{\partial H_z}{\partial r} - j\frac{\omega\varepsilon}{r}\frac{\partial E_z}{\partial \phi}\right) = -\frac{1}{h^2}\left[0 + j\frac{\omega\varepsilon}{r} J_n(rh) n A_n \sin n\phi\, e^{-\gamma z}\right]$$

$$= -j\frac{n\omega\varepsilon A_n}{rh^2} J_n(rh) \sin n\phi e^{-\gamma z}$$

$$\therefore \quad H_\phi = -\frac{1}{h^2}\left(\frac{\gamma}{r}\frac{\partial H_z}{\partial \phi} + j\omega\varepsilon \frac{\partial E_z}{\partial r}\right) = -\frac{1}{h^2}[0 + j\omega\varepsilon h J_n'(rh) A_n \cos n\phi e^{-\gamma z}]$$

$$= -j\frac{\omega\varepsilon A_n}{h} J_n'(rh) \cos n\phi e^{-\gamma z}$$

Thus, the expressions for different field components for *TM* waves are written as

$$
\begin{aligned}
E_z &= J_n(rh)A_n \cos n\phi\, e^{-\gamma z} \\
H_r &= -j\frac{n\omega\varepsilon A_n}{rh^2} J_n(rh) \sin n\phi e^{-\gamma z} \\
H_\phi &= -j\frac{\omega\varepsilon A_n}{h} J_n'(rh) \cos n\phi e^{-\gamma z} \\
E_r &= -\frac{\gamma A_n}{h} J_n'(rh) \cos n\phi e^{-\gamma z} = \frac{\gamma}{j\omega\varepsilon} H_\phi \\
E_\phi &= \frac{n\gamma A_n}{rh^2} J_n(rh) \sin n\phi e^{-\gamma z} = -\frac{\gamma}{j\omega\varepsilon} H_r
\end{aligned} \quad \textbf{(7.118)}
$$

The boundary condition to be satisfied for *TM* waves is that

$$E_z = 0 \quad \text{at} \quad r = a$$

where a is the radius of the circular waveguide (see Fig. 7.11).
Therefore, from Eq. (7.118), we have

$$J_n(ah) = 0 \quad \textbf{(7.119)}$$

The first few roots of Eq. (7.119) are given in Table 7.3.

Table 7.3 *Roots of $J_n(ah) = 0$*

m \ n	0	1	2	3
1	2.40	3.83	6.38	7.59
2	5.52	7.02	9.76	11.06
3	8.63	10.17	13.01	14.37

The waves corresponding to different values of h_{nm} are obtained from Table 7.2 and are referred as TM_{01}, TM_{11}, TM_{02}, TM_{12}, and so on.

NOTE

As there is no roots of $(ah)_{00}$, TE_{00} or TM_{00} waves do not exist.

7.4.3 Properties of *TE* and *TM* Waves in Circular Waveguide

We find the following quantities for *TE* and *TM* mode waves in circular waveguides:

1. Propagation constant (γ),
2. Phase constant (β),
3. Cut-off frequency (f_c),
4. Cut-off wavelength (λ_c),
5. Phase velocity (v_p),
6. Group Velocity (v_g),
7. Guide wavelength (λ_g), and
8. Intrinsic wave impedance (η).

1. Propagation constant (γ): It is given as

$$\boxed{\gamma_{nm} = \sqrt{h_{nm}^2 - \omega^2 \mu \varepsilon}} \quad \textbf{(7.120)}$$

where, the value of h_{nm} is obtained from Tables 7.1 and 7.2 for *TE* and *TM* waves, respectively. Here too, wave propagation will occur only when $h_{nm}^2 < \omega^2 \mu\varepsilon$ in which case the propagation is entirely imaginary.

2. Phase constant (β): From Eq. (7.120), the phase constant in the propagation mode is obtained as

$$\gamma_{nm} = j\beta_{nm} = \sqrt{h_{nm}^2 - \omega^2 \mu\varepsilon}$$

or

$$\beta_{nm} = \sqrt{\omega^2 \mu\varepsilon - h_{nm}^2}$$

$$\boxed{\beta_{nm} = \sqrt{\omega^2 \mu\varepsilon - h_{nm}^2}} \quad \textbf{(7.121)}$$

3. Cut-off frequency (f_c): The cut-off frequency below which the wave is attenuated and above which the wave is propagated through the guide is obtained as

$$\gamma_{nm} = \sqrt{h_{nm}^2 - \omega^2 \mu\varepsilon} = 0$$

or

$$\omega_c = \frac{h_{nm}}{\sqrt{\mu\varepsilon}}$$

or

$$f_c = \frac{h_{nm}}{2\pi\sqrt{\mu\varepsilon}} = \frac{v_0}{2\pi} h_{nm}$$

where $v_0 = \dfrac{1}{\sqrt{\mu\varepsilon}}$ is the phase velocity of uniform plane wave in a lossless dielectric medium.

$\therefore$

$$\boxed{f_c = \frac{h_{nm}}{2\pi\sqrt{\mu\varepsilon}} = \frac{v_0}{2\pi} h_{nm}} \quad \textbf{(7.122)}$$

From Eq. (7.122) and from Tables 7.1 and 7.2, we see that the lowest cut-off frequency is with TE_{11} mode, the next higher modes being TM_{01}, TE_{21}, TE_{01}.

4. Cut-off wavelength (λ_c): Corresponding to every cut-off frequency, there will be a *cut-off wavelength* given as

$$\boxed{\lambda_c = \frac{v_0}{f_c} = \frac{2\pi}{h_{nm}}} \quad \textbf{(7.123)}$$

where, $v_0 = \dfrac{1}{\sqrt{\mu\varepsilon}}$ is the phase velocity of uniform plane wave in a lossless dielectric medium inside the guide.

5. Phase velocity (v_p): The phase velocity of the wave propagation is given by

$$\boxed{v_p = \frac{\omega}{\beta_{nm}} = \frac{\omega}{\sqrt{\omega^2\mu\varepsilon - h_{nm}^2}} = \frac{v_0}{\sqrt{1 - \dfrac{f_c^2}{f^2}}}} \quad \textbf{(7.124)}$$

$v_0 = \dfrac{1}{\sqrt{\mu\varepsilon}}$ is the phase velocity of uniform plane waves in a lossless dielectric medium inside the waveguide.

Equation (7.124) indicates that the phase velocity of wave propagation in the circular waveguide is greater than the phase velocity of uniform plane wave.

6. Group Velocity (v_g): From the relation that $v_0 = \sqrt{v_p v_g}$, the group velocity is obtained as

$$v_g = \frac{v_0^2}{v_p} = \frac{v_0^2}{\dfrac{v_0}{\sqrt{1-\dfrac{f_c^2}{f^2}}}} = v_0\sqrt{1-\frac{f_c^2}{f^2}}$$

$$\boxed{v_g = v_0\sqrt{1-\frac{f_c^2}{f^2}}} \qquad \textbf{(7.125)}$$

This shows that the group velocity in the guide is less than that in the free space.

7. Guide wavelength (λ_g): The wavelength in the circular waveguide is given as

$$\boxed{\lambda_g = \frac{2\pi}{\beta_{nm}} = \frac{2\pi}{\sqrt{\omega^2\mu\varepsilon - h_{nm}^2}} = \frac{\lambda_0}{\sqrt{1-\dfrac{f_c^2}{f^2}}} = \frac{\lambda_0}{\sqrt{1-\dfrac{\lambda^2}{\lambda_c^2}}}} \qquad \textbf{(7.126)}$$

where $\lambda_0 = \dfrac{2\pi}{\omega\sqrt{\mu\varepsilon}} = \dfrac{2\pi}{\beta_0}$ is the wavelength of the uniform plane wave in the lossless dielectric medium inside the guide.

8. Intrinsic wave impedance (η): As for rectangular waveguides, for circular waveguides too, the intrinsic wave impedance will be different for *TE* and *TM* modes.

(a) Intrinsic wave impedance for *TE* Modes in circular waveguides: For *TE* waves, from Eq. (7.116), it is given as

$$\eta_{TE} = \frac{E_r}{H_\phi} = -\frac{E_\phi}{H_r} = \frac{\omega\mu}{\beta} = \sqrt{\frac{\mu}{\varepsilon}}\frac{1}{\sqrt{1-\dfrac{f_c^2}{f^2}}} = \frac{\eta_0}{\sqrt{1-\dfrac{f_c^2}{f^2}}}$$

where $\eta_0 = \sqrt{\dfrac{\mu}{\varepsilon}}$ is the intrinsic impedance of uniform plane wave in a lossless dielectric medium.

$$\therefore \qquad \boxed{\eta_{TE} = \frac{\eta_0}{\sqrt{1-\dfrac{f_c^2}{f^2}}}} \qquad \textbf{(7.127)}$$

(b) Intrinsic wave impedance for *TM* modes in circular waveguides: For *TM* waves, from Eq. (7.118), it is given as

$$\eta_{TM} = \frac{E_r}{H_\phi} = -\frac{E_\phi}{H_r} = \frac{\beta}{\omega\varepsilon} = \sqrt{\frac{\mu}{\varepsilon}}\sqrt{1-\frac{f_c^2}{f^2}} = \eta_0\sqrt{1-\frac{f_c^2}{f^2}}$$

where $\eta_0 = \sqrt{\frac{\mu}{\varepsilon}}$ is the intrinsic impedance of uniform plane wave in a lossless dielectric medium.

$$\therefore \qquad \boxed{\eta_{TM} = \eta_0\sqrt{1-\frac{f_c^2}{f^2}}} \qquad \textbf{(7.128)}$$

NOTE

In this section in circular waveguides, h_{nm} should be replaced by h'_{nm} for all TE waves. For example for TE waves, the cut-off frequency is $f_c = \frac{v_0}{2\pi}h'_{nm}$.

Example 7.29 A circular waveguide has an internal radius of 2.5 cm. Calculate the cut-off wavelength, the guide wavelength and the wave impedance of the guide when operated at a frequency of 8 GHz and propagating in TE_{11} mode.

Solution Here, $n = 1$, $m = 1$, $f = 8$ GHz, $a = 2.5$ cm $= 0.025$ m

$$\therefore \qquad \lambda_0 = \frac{2\pi}{\omega\sqrt{\mu\varepsilon}} = \frac{3\times10^8}{8\times10^9} = 0.0375 \text{ m}$$

From Table 7.1, for TE_{11}, $h'_{nm} = 1.84$

Cut-off wavelength, $\lambda_c = \frac{2\pi a}{h'_{nm}a} = \frac{2\pi\times2.5}{1.84} = 8.54$ cm

Guide wavelength, $\lambda_g = \frac{\lambda_0}{\sqrt{1-\frac{\lambda^2}{\lambda_c^2}}} = \frac{0.0375}{\sqrt{1-\left(\frac{0.0375}{0.0854}\right)^2}} = 0.0417 \text{ m} = 4.17 \text{ cm}$

Wave impedance, $\eta_{TE} = \frac{\eta_0}{\sqrt{1-\frac{\lambda^2}{\lambda_c^2}}} = \frac{120\pi}{\sqrt{1-\left(\frac{0.0375}{0.0854}\right)^2}} = 419.64\ \Omega$

Example 7.30 A circular waveguide has an internal diameter of 5 cm. Calculate the cut-off frequency in it for the following modes: (a) TE_{11}, (b) TM_{01}.

Solution Here, $a = 2.5$ cm $= 0.025$ m

From Table 7.1, for TE_{11}, $h'_{nm} = 1.84$

Cut-off frequency is given as

$$f_c = \frac{h'_{nm}}{2\pi\sqrt{\mu\varepsilon}} = \frac{v_0}{2\pi}h'_{nm} = \frac{v_0}{2\pi a}h'_{nm}a = \frac{3\times10^8}{2\pi\times0.025}\times1.84 = 3.5 \text{ GHz}$$

From Table 7.2, for TM_{01}, $h_{nm} = 2.4$

Cut-off frequency is given as

$$f_c = \frac{h_{nm}}{2\pi\sqrt{\mu\varepsilon}} = \frac{v_0}{2\pi}h_{nm} = \frac{v_0}{2\pi a}h_{nm}a = \frac{3\times 10^8}{2\pi \times 0.025}\times 2.4 = 4.58 \text{ GHz}$$

Example 7.31 A lossless air-filled circular waveguide of inside diameter 3 cm is operated at 15 GHz. For the TM_{11} mode, find the cut-off frequency, the guide wavelength and wave impedance.

Solution Here, $a = \frac{3}{2} = 1.5$ cm, $f = 10$ GHz

From Table 7.2, for TM_{11} mode, the cut-off frequency is given as

$$f_c = \frac{h_{nm}}{2\pi\sqrt{\mu\varepsilon}} = \frac{v_0}{2\pi}h_{nm} = \frac{v_0}{2\pi a}ah_{nm} = \frac{3\times 10^8}{2\pi \times 1.5\times 10^{-2}}\times 3.83 = 12.19 \text{ GHz}$$

The guide wavelength is given as

$$\lambda_g = \frac{\lambda_0}{\sqrt{1-\frac{f_c^2}{f^2}}} = \frac{v_0/f}{\sqrt{1-\frac{f_c^2}{f^2}}} = \frac{3\times 10^8/15\times 10^9}{\sqrt{1-\left(\frac{12.19}{15}\right)^2}} = 5.89 \text{ cm}$$

The wave impedance is given as

$$\eta_{TM} = \eta_0\sqrt{1-\frac{f_c^2}{f^2}} = 120\pi\sqrt{1-\left(\frac{12.19}{15}\right)^2} = 219.64\ \Omega$$

Example 7.32 Find the radius and guide wavelength in an air-filled circular waveguide for the dominant mode at $f = 30\ GHz = 1.5 f_c$. Will TM_{11} mode propagate under these conditions?

Solution Here, $f = 30\ GHz = 1.3 f_c$

$$\therefore \qquad f_c = \frac{30}{1.5} = 20 \text{ GHz}$$

For circular waveguide, the dominant mode is TE_{11} mode.
The corresponding cut-off frequency is given as

$$f_c = \frac{v_0}{2\pi}h_{nm} = \frac{v_0}{2\pi a}ah_{nm}$$

$$\therefore \qquad a = \frac{v_0}{2\pi f_c}ah_{nm} = \frac{3\times 10^8}{2\pi\times 20\times 10^9}\times 1.84 = 0.44 \text{ cm}$$

The guide wavelength is given as

$$\lambda_g = \frac{\lambda_0}{\sqrt{1-\frac{f_c^2}{f^2}}} = \frac{v_0/f}{\sqrt{1-\frac{f_c^2}{f^2}}} = \frac{3\times 10^8/30\times 10^9}{\sqrt{1-\left(\frac{20}{30}\right)^2}} = 1.34 \text{ cm}$$

For this radius, for TM_{11} mode, the cut-off frequency is

$$f_c = \frac{v_0}{2\pi} h_{nm} = \frac{v_0}{2\pi a} ah_{nm} = \frac{3\times 10^8}{2\pi \times 0.44 \times 10^{-2}} \times 3.83 = 41.63 \text{ GHz}$$

Since this is greater than 20 GHz, TM_{11} mode will not propagate.

Example 7.33 Calculate the radius and guide wavelength for TM_{11} mode at $f = 30$ GHz $= 1.5 f_c$ in an air-filled circular waveguide.

Solution Here, $f = 30$ GHz $= 1.3 f_c$

$$\therefore \qquad f_c = \frac{30}{1.5} = 20 \text{ GHz}$$

From Table 7.2, for TM_{11} mode, the cut-off frequency is given as

$$f_c = \frac{h_{nm}}{2\pi\sqrt{\mu\varepsilon}} = \frac{v_0}{2\pi} h_{nm} = \frac{v_0}{2\pi a} ah_{nm}$$

$$\therefore \qquad a = \frac{v_0}{2\pi f_c} ah_{nm} = \frac{3\times 10^8}{2\pi \times 20 \times 10^9} \times 3.83 = 0.914 \text{ cm}$$

The guide wavelength is given as

$$\lambda_g = \frac{\lambda_0}{\sqrt{1 - \frac{f_c^2}{f^2}}} = \frac{v_0/f}{\sqrt{1-\frac{f_c^2}{f^2}}} = \frac{(3\times 10^8)/(30\times 10^9)}{\sqrt{1-\left(\frac{20}{30}\right)^2}} = 1.34 \text{ cm}$$

***Example 7.34** Evaluate the ratio of the area of a circular waveguide to that of a rectangular one, if both are to have the same cut-off frequency for dominant mode.

Solution Let r be radius of the circular waveguide and a be the larger dimension of the rectangular waveguide.

From Tables 7.1 and 7.2, we understand that the dominant mode in circular waveguide is TE_{11} mode. For this mode, the cut-off frequency is given as

$$f_c = \frac{v_0}{2\pi} h_{nm} = \frac{v_0}{2\pi r} rh_{nm} = \frac{v_0}{2\pi r} \times 1.84$$

For TE_{10} mode rectangular waveguide, the cut-off frequency is given as

$$f_c = \frac{v_0}{2}\sqrt{\left(\frac{m}{a}\right)^2 + \left(\frac{n}{b}\right)^2} = \frac{v_0}{2a}$$

Since both will have the same cut-off frequency, we get

$$\frac{v_0}{2\pi r} \times 1.84 = \frac{v_0}{2a}$$

or

$$\frac{r}{a} = 0.586$$

The area of the circular waveguide is, $A_c = \pi r^2$
For standard rectangular waveguide $a : b = 2 : 1$, so that the area of the rectangular waveguide is $A_r = a \times \frac{a}{2} = \frac{a^2}{2}$.

$$\therefore \quad \frac{A_c}{A_r} = \frac{\pi r^2}{a^2/2} = 2\pi\left(\frac{r}{a}\right)^2 = 2\pi \times (0.586)^2 = 2.155$$

***Example 7.35** Design an air-filled circular waveguide yielding a frequency separation of 1 GHz between the cut-off frequencies of the dominant mode and the next highest mode.

Solution The dominant mode in circular waveguide is TE_{11} mode and the next highest mode is TM_{01} mode. The respective cut-off frequencies are given as

$$f_{c11}^{TE} = \frac{v_0}{2\pi a} ah'_{11} = \frac{v_0}{2\pi a} \times 1.84$$

$$f_{c01}^{TM} = \frac{v_0}{2\pi a} ah_{01} = \frac{v_0}{2\pi a} \times 2.4$$

For a difference of 1 GHz between these two frequencies, we can write

$$f_{c01}^{TM} - f_{c11}^{TE} = 1 \times 10^9 = \frac{v_0}{2\pi a}(2.4 - 1.84) = \frac{v_0}{2\pi a} \times 0.556$$

$$\therefore \quad a = \frac{v_0}{2\pi \times 1 \times 10^9} \times 0.556 = \frac{3 \times 10^8}{2\pi \times 1 \times 10^9} \times 0.556 = 0.0267 \text{ m} = 2.67 \text{ cm}$$

The corresponding cut-off frequencies of this waveguide are

$$f_{c11}^{TE} = \frac{v_0}{2\pi \times 0.0267} \times 1.84 = 3.286 \text{ GHz}$$

$$f_{c01}^{TM} = \frac{v_0}{2\pi \times 0.0267} \times 2.4 = 4.286 \text{ GHz}$$

7.5 POWER TRANSMISSION IN WAVEGUIDES

The power transmitted along any waveguide as well as the transmission losses can be determined with the help of the field equations obtained in the preceeding sections and using the concept of Poynting vector as explained in Chapter 5, Section 5.10.

From Eq. (5.40), the total power carried by the fields along the guide direction (here z-direction) is obtained by integrating the z-component of the Poynting vector over the cross-sectional area of the guide.

$$\therefore \quad P_T = \int_S \vec{p}_{\text{ave}} \cdot d\vec{S} = \int_S \frac{1}{2} \text{Re}[\boldsymbol{E}_S \times \boldsymbol{H}_S^*] \cdot d\vec{S} \qquad \textbf{(7.129)}$$

7.5.1 Power Transmission in Rectangular Waveguides

The power transmitted in rectangular waveguide is obtained from Eq. (7.129) as

$$\therefore \quad P_T = \int_S \frac{1}{2}\text{Re}[\boldsymbol{E}_S \times \boldsymbol{H}_S^*]\cdot d\vec{S} = \int_S \frac{1}{2}\text{Re}[E_x H_y - E_y H_x]\hat{a}_z \cdot dxdy\hat{a}_z = \int_S \frac{1}{2\eta}(|E_x|^2 + |E_y|^2)dxdy$$

where η is the wave impedance of the guide.
For *TE* waves, the transmitted power is given as

$$P_{T_{TE}} = \frac{\sqrt{1-\frac{f_c^2}{f^2}}}{2\eta_0}\int_{x=0}^{a}\int_{y=0}^{b}(|E_x|^2 + |E_y|^2)dydx \tag{7.130}$$

where $\eta_0 = \sqrt{\frac{\mu}{\varepsilon}}$ is the intrinsic impedance of uniform plane wave.

For *TM* waves, the transmitted power is given as

$$P_{T_{TM}} = \frac{1}{2\eta_0\sqrt{1-\frac{f_c^2}{f^2}}}\int_{x=0}^{a}\int_{y=0}^{b}(|E_x|^2 + |E_y|^2)dydx \tag{7.131}$$

7.5.2 Power Transmission in Circular Waveguides

The power transmitted in circular waveguide is obtained from Eq. (7.129) as

$$\therefore \quad P_T = \int_S \frac{1}{2}\text{Re}[\boldsymbol{E}_S \times \boldsymbol{H}_S^*]\cdot d\vec{S} = \int_S \frac{1}{2}\text{Re}[E_r H_\phi - E_\phi H_r]\hat{a}_z \cdot rdrd\phi\hat{a}_z = \int_S \frac{1}{2\eta}(|E_r|^2 + |E_\phi|^2)rdrd\phi$$

where η is the wave impedance of the guide.
For *TE* waves, the transmitted power is given as

$$P_{T_{TE}} = \frac{\sqrt{1-\frac{f_c^2}{f^2}}}{2\eta_0}\int_{\phi=0}^{2\pi}\int_{r=0}^{a}(|E_r|^2 + |E_\phi|^2)rdrd\phi \tag{7.132}$$

where $\eta_0 = \sqrt{\frac{\mu}{\varepsilon}}$ is the intrinsic impedance of uniform plane wave.

For *TM* waves, the transmitted power is given as

$$P_{T_{TM}} = \frac{1}{2\eta_0\sqrt{1-\frac{f_c^2}{f^2}}}\int_{\phi=0}^{2\pi}\int_{r=0}^{a}(|E_r|^2 + |E_\phi|^2)rdrd\phi \tag{7.133}$$

Example 7.36 For TE_{10} mode rectangular waveguide, calculate the time-averaged power transmitted along the guide. Also, calculate the energy density of the fields and determine the velocity by which electromagnetic energy flows through the guide. Establish that this velocity is equal to the group velocity.

Solution For TE_{10} mode the non-zero field components are:

$$H_z = H_0 \cos\left(\frac{\pi}{a}x\right)$$

$$H_x = -\frac{j\beta}{h^2}\left(\frac{\pi}{a}\right)H_0 \sin\left(\frac{\pi}{a}x\right)$$

$$E_y = -j\frac{\omega\mu}{h^2}\left(\frac{\pi}{a}\right)H_0 \sin\left(\frac{\pi}{a}x\right)$$

The Poynting vector is obtained as

$$p_z = \frac{\sqrt{1-\frac{f_c^2}{f^2}}}{2\eta_0}|E|^2 = \frac{\sqrt{1-\frac{f_c^2}{f^2}}}{2\eta_0}|E_y|^2 = \frac{\sqrt{1-\frac{f_c^2}{f^2}}}{2\eta_0}\frac{\omega^2\mu^2}{h^4}\left(\frac{\pi}{a}\right)^2 H_0^2 \sin^2\left(\frac{\pi}{a}x\right)$$

The time-average power transmitted along the guide is obtained by integrating p_z over the cross-sectional area of the guide.

$$\therefore \qquad P_T = \int_0^a\int_0^b p_z dxdy = \frac{\sqrt{1-\frac{f_c^2}{f^2}}}{2\eta_0}\frac{\omega^2\mu^2}{h^4}\left(\frac{\pi}{a}\right)^2 H_0^2 \int_0^a\int_0^b \sin^2\left(\frac{\pi}{a}x\right)dxdy$$

$$= \frac{\sqrt{1-\frac{f_c^2}{f^2}}}{4\eta_0}\frac{\omega^2\mu^2}{h^4}\left(\frac{\pi}{a}\right)^2 H_0^2 \int_0^a\int_0^b \left[1-\cos\left(\frac{2\pi}{a}x\right)\right]dxdy$$

$$= \frac{\sqrt{1-\frac{f_c^2}{f^2}}}{4\eta_0}\frac{\omega^2\mu^2}{h^4}\left(\frac{\pi}{a}\right)^2 H_0^2 ab$$

For TE_{10} mode, $$h^2 = \left(\frac{m\pi}{a}\right)^2 + \left(\frac{n\pi}{b}\right)^2 = \left(\frac{\pi}{a}\right)^2$$

$$\therefore \qquad \boxed{P_T = \frac{1}{4\eta_0}\frac{\omega^2\mu^2}{h^2}H_0^2 ab\sqrt{1-\frac{f_c^2}{f^2}}}$$

We now calculate the distribution of electromagnetic energy along the guide, as measured by the time-averaged energy density. The energy densities of the electric and magnetic fields are given as

$$w_e = \frac{1}{2}\text{Re}\left(\frac{1}{2}\varepsilon\vec{E}\cdot\vec{E}^*\right) = \frac{1}{4}\varepsilon|E_y|^2 = \frac{1}{4}\varepsilon\frac{\omega^2\mu^2}{h^4}\left(\frac{\pi}{a}\right)^2 H_0^2 \sin^2\left(\frac{\pi}{a}x\right)$$

$$w_m = \frac{1}{2}\text{Re}\left(\frac{1}{2}\mu\vec{H}\cdot\vec{H}^*\right) = \frac{1}{4}\mu(|H_x|^2+|H_z|^2) = \frac{1}{4}\mu H_0^2\left[\frac{\beta^2}{h^4}\left(\frac{\pi}{a}\right)^2 \sin^2\left(\frac{\pi}{a}x\right) + \cos^2\left(\frac{\pi}{a}x\right)\right]$$

Energy distribution per unit length is obtained by integrating the above two quantities over the cross-sectional area of the guide.

$$\therefore \quad W_e = \int_0^a\int_0^b w_e dxdy = \frac{1}{4}\varepsilon\frac{\omega^2\mu^2}{h^4}\left(\frac{\pi}{a}\right)^2 H_0^2 \int_0^a\int_0^b \sin^2\left(\frac{\pi}{a}x\right)dxdy = \frac{1}{8}\varepsilon\frac{\omega^2\mu^2}{h^4}\left(\frac{\pi}{a}\right)^2 H_0^2 ab$$

$$= \frac{1}{8}\varepsilon\frac{\omega^2\mu^2}{h^2} H_0^2 ab$$

$$\therefore \quad W_m = \int_0^a\int_0^b w_m dxdy = \frac{1}{4}\mu H_0^2 \int_0^a\int_0^b \left[\frac{\beta^2}{h^4}\left(\frac{\pi}{a}\right)^2 \sin^2\left(\frac{\pi}{a}x\right) + \cos^2\left(\frac{\pi}{a}x\right)\right]dxdy$$

$$= \frac{1}{8}\mu H_0^2 ab\left[1 + \frac{\beta^2}{h^4}\left(\frac{\pi}{a}\right)^2\right]$$

Although these two expressions look different, they are actually the same as derived below.

For TE_{10} mode,
$$h^2 = \left(\frac{m\pi}{a}\right)^2 + \left(\frac{n\pi}{b}\right)^2 = \left(\frac{\pi}{a}\right)^2$$

$$\beta^2 = \omega^2\mu\varepsilon - h^2 = \omega^2\mu\varepsilon - \left(\frac{\pi}{a}\right)^2$$

$$\therefore \quad W_m = \frac{1}{8}\mu H_0^2 ab\left[1 + \frac{\beta^2}{h^4}\left(\frac{\pi}{a}\right)^2\right] = \frac{1}{8}\mu H_0^2 ab\left[1 + \frac{\beta^2}{h^2}\right] = \frac{1}{8}\mu H_0^2 ab\left[\frac{h^2 + \beta^2}{h^2}\right]$$

$$= \frac{1}{8}\mu H_0^2 ab\left(\frac{\omega^2\mu\varepsilon}{h^2}\right) = \frac{1}{8}\varepsilon\frac{\omega^2\mu^2}{h^2} H_0^2 ab = W_e$$

Hence, the total energy density per unit length is obtained as

$$W = W_e + W_m = 2W_e = \frac{1}{4}\varepsilon\frac{\omega^2\mu^2}{h^2} H_0^2 ab$$

$$\boxed{W = \frac{1}{4}\varepsilon\frac{\omega^2\mu^2}{h^2} H_0^2 ab}$$

The velocity by which electromagnetic energy flows through the guide is given by the ratio of the time-averaged power transmitted to the total energy density per unit length.

$$\therefore \quad v_{en} = \frac{P_T}{W} = \frac{\dfrac{1}{4\eta_0}\dfrac{\omega^2\mu^2}{h^2} H_0^2 ab\sqrt{1 - \dfrac{f_c^2}{f^2}}}{\dfrac{1}{4}\varepsilon\dfrac{\omega^2\mu^2}{h^2} H_0^2 ab} = \frac{1}{\eta_0\varepsilon}\sqrt{1 - \frac{f_c^2}{f^2}} = \sqrt{\frac{\varepsilon}{\mu}}\frac{1}{\varepsilon}\sqrt{1 - \frac{f_c^2}{f^2}}$$

$$= \frac{1}{\sqrt{\mu\varepsilon}}\sqrt{1 - \frac{f_c^2}{f^2}} = v_0\sqrt{1 - \frac{f_c^2}{f^2}}$$

$$\therefore \qquad \boxed{v_{en} = v_0\sqrt{1-\frac{f_c^2}{f^2}}}$$

By Eq. (7.51), this is the group velocity of the wave, i.e., the velocity with the energy is transported.

Example 7.37 A lossless air-filled rectangular waveguide has dimensions $a = 7.214$ cm and $b = 3.404$ cm. For the dominant mode propagation at 3 GHz, find the average power transmitted if the excitation level of the $\vec{E}$ field is 10 kV/m.

Solution Here, $a = 7.214$ cm, $b = 3.404$ cm, $f = 3$ GHz, $E_0 = 10$ kV/m
For the dominant mode, i.e., TE_{11} mode, the cut-off frequency is given as

$$f_c = \frac{v_0}{2}\sqrt{\left(\frac{m}{a}\right)^2+\left(\frac{n}{b}\right)^2} = \frac{v_0}{2a} = \frac{3\times10^8}{2\times7.214\times10^{-2}} = 2.08 \text{ GHz}$$

Hence, from Example 7.36, the average power transmitted is given as

$$P_{av} = \frac{E_0^2}{4\eta_0}ab\sqrt{1-\left(\frac{f_c}{f}\right)^2} = \frac{(10\times10^3)^2}{4\times120\pi}\times7.214\times3.404\times10^{-4}\sqrt{1-\left(\frac{2.08}{3}\right)^2}$$
$$= 115.92 \text{ Watt}$$

Example 7.38 A lossless air-filled rectangular waveguide has dimensions $a = 7.214$ cm and $b = 3.404$ cm. For the dominant mode propagation at 2.6 GHz, the guide transports 200 W of average power. Find the level of excitation of the $\vec{E}$ field.

Solution Here, $a = 7.214$ cm, $b = 3.404$ cm, $f = 3$ GHz, $E_0 = 10$ kV/m
For the dominant mode, i.e., TE_{11} mode, the cut-off frequency is given as

$$f_c = \frac{v_0}{2}\sqrt{\left(\frac{m}{a}\right)^2+\left(\frac{n}{b}\right)^2} = \frac{v_0}{2a} = \frac{3\times10^8}{2\times7.214\times10^{-2}} = 2.08 \text{ GHz}$$

Hence, from Example 7.36, the average power transmitted is given as

$$P_{av} = \frac{E_0^2}{4\eta_0}ab\sqrt{1-\left(\frac{f_c}{f}\right)^2}$$

$$\therefore \qquad E_0 = \sqrt{\frac{P_{av}\times4\eta_0}{ab\sqrt{1-\left(\frac{f_c}{f}\right)^2}}} = \sqrt{\frac{200\times4\times120\pi}{7.214\times3.404\times10^{-4}\sqrt{1-\left(\frac{2.08}{2.6}\right)^2}}} = 143.03 \text{ V/cm}$$

Example 7.39 For TE_{01} mode,

$$E_x = j\frac{\omega\mu\pi}{bh^2}H_0\sin\left(\frac{\pi}{b}y\right)e^{-\gamma z}, \quad E_y = 0$$

Find the average Poynting vector and the total average power transmitted across the guide.

Solution The average Poynting vector is given as

$$\vec{p}_{\text{ave}} = \frac{1}{2}\text{Re}[\boldsymbol{E}_S \times \boldsymbol{H}_S^*] = \frac{1}{2\eta}(|E_x|^2 + |E_y|^2)\,\hat{a}_z$$

where η is the wave impedance of the guide.

In this problem, $E_x = j\dfrac{\omega\mu\pi}{bh^2} H_0 \sin\left(\dfrac{\pi}{b}y\right)e^{-\gamma z}, \quad E_y = 0$

$$\therefore \qquad \vec{p}_{\text{ave}} = \frac{1}{2\eta}(|E_x|^2)\,\hat{a}_z = \frac{\omega^2\mu^2\pi^2}{2\eta b^2 h^4} H_0^2 \sin^2\left(\frac{\pi}{b}y\right)\hat{a}_z$$

Total average power transmitted across the guide is given as

$$\therefore \qquad P_T = \int_S \frac{1}{2}|E_x|^2\,\hat{a}_z \cdot dxdy\hat{a}_z = \int_S \frac{\omega^2\mu^2\pi^2}{2\eta b^2 h^4} H_0^2 \sin^2\left(\frac{\pi}{b}y\right)dxdy$$

$$= \frac{\omega^2\mu^2\pi^2}{2\eta b^2 h^4} H_0^2 \int_{y=0}^{b}\int_{x=0}^{a} \sin^2\left(\frac{\pi}{b}y\right)dxdy$$

$$= \frac{\omega^2\mu^2\pi^2}{4\eta b^2 h^4} H_0^2 \int_{y=0}^{b}\int_{x=0}^{a} \left\{1 - \cos\left(\frac{2\pi}{b}y\right)\right\}dxdy$$

$$= \frac{\omega^2\mu^2\pi^2}{4\eta b^2 h^4} H_0^2 ab$$

Here, $$h^2 = \left(\frac{m\pi}{a}\right)^2 + \left(\frac{n\pi}{b}\right)^2 = \frac{\pi^2}{b^2}$$

$$\therefore \qquad P_T = \frac{\omega^2\mu^2\pi^2}{4\eta b^2\left(\dfrac{\pi^2}{b^2}\right)^2} H_0^2 ab = \frac{\omega^2\mu^2 ab^3}{4\eta\pi^2} H_0^2$$

Example 7.40 In a lossless circular waveguide of radius 1 cm with dielectric of permittivity $\varepsilon_r = 2.1$, the transmitted power in the dominant mode at 15 GHz is 2 W. Find the level of excitation for the magnetic field.

Solution Here, $a = 1$ cm, $f = 15$ GHz, $P = 2$ Watt, $\varepsilon_r = 2.1$
The cut-off frequency for the dominant mode, i.e., for TE_{11} mode is

$$f_c = \frac{v_0}{2\pi}h_{nm} = \frac{v_0}{2\pi a}ah_{nm} = \frac{3\times 10^8}{2\pi \times 1 \times 10^{-2}} \times 1.84 = 8.785 \text{ GHz}$$

The average power transmitted for this mode is given as

$$P_{\text{av}} = \frac{\eta_0}{4}|H_0|^2\,\pi a^2 \left(\frac{f}{f_c}\right)^2 \sqrt{1 - \left(\frac{f_c}{f}\right)^2}\left[\left\{\frac{(ah_{11})^2 - 1}{(ah_{11})^2}\right\}\{J_n'(ah_{11})\}^2\right]$$

From the knowledge of Bessel's function, the value of the term within third bracket comes to be 0.239, as follows.

$$\left\{\frac{(ah_{11})^2-1}{(ah_{11})^2}\right\}\{J_n'(ah_{11})\}^2=\frac{(1.84)^2-1}{(1.84)^2}\times(0.58)^2=0.239$$

So, the average power transmitted is written as

$$P_{\text{av}}=\frac{\eta_0}{4}|H_0|^2\pi a^2\left(\frac{f}{f_c}\right)^2\sqrt{1-\left(\frac{f_c}{f}\right)^2}\times 0.239$$

$$\therefore \qquad H_0=\sqrt{\frac{4\times P_{\text{av}}}{\eta_0\times\pi a^2\times\left(\frac{f}{f_c}\right)^2\sqrt{1-\left(\frac{f_c}{f}\right)^2}\times 0.239}}=13.16\text{ A/m}$$

7.6 POWER LOSSES AND ATTENUATION IN WAVEGUIDES

In earlier discussions, we assumed a lossless waveguide ($\sigma = 0$, $\sigma_c = \infty$) for which the attenuation constant $\alpha = 0$ and the propagation constant $\gamma = j\beta$. However, in general, the guide walls are not perfectly conducting and the dielectric is lossy ($\sigma \neq 0$, $\sigma_c \neq \infty$) and there is a continuous loss of power as a wave propagates through the waveguide.
Consequently, for waveguides with conducting walls, the power (transmission) losses are primarily due to the following two causes:

1. Dielectric losses in the medium filling the space between the conductors in which the fields propagate, and
2. Ohmic losses or conduction losses due to conducting walls.

From Chapter 5, Section 5.11.5, the power flow in the guide is according to the equation,

$$P_T = P_0 e^{-2\alpha z} \tag{7.134}$$

From the law of conservation of energy, the rate of decrease of P_T must be equal to the time average power loss, i.e.,

$$P_L=-\frac{dP_T}{dz}=2\alpha P_0 e^{-2\alpha z}=2\alpha P_T$$

$$\therefore \qquad \alpha=\frac{1}{2}\left(\frac{P_L}{P_T}\right) \tag{7.135}$$

The attenuations in the wave due to the two types of losses are characterised by the attenuation constants α_c and α_d, respectively, so that the total attenuation constant of the waveguide is represented as

$$\alpha = \alpha_c + \alpha_d \tag{7.136}$$

1. Dielectric losses: This occurs in the medium filling the space between the conductors in which the fields propagate. The attenuation in the wave due to this loss is characterised by attenuation constant α_d.

Determination of α_d: For lossy dielectric, we replace the permittivity ε with the complex permittivity ε_c given as

$$\varepsilon_c = \left(\varepsilon - j\frac{\sigma}{\omega}\right) = (\varepsilon' - j\varepsilon'')$$

Hence, the propagation constant for the lossy dielectric is given as

$$\gamma = \alpha_d + j\beta_d = \sqrt{\left(\frac{m\pi}{a}\right)^2 + \left(\frac{n\pi}{b}\right)^2 - \omega^2\mu\varepsilon_c} = \sqrt{\left(\frac{m\pi}{a}\right)^2 + \left(\frac{n\pi}{b}\right)^2 - \omega^2\mu\varepsilon + j\omega\mu\sigma}$$

Squaring on both sides,

$$\alpha_d^2 - \beta_d^2 + 2j\alpha_d\beta_d = \left(\frac{m\pi}{a}\right)^2 + \left(\frac{n\pi}{b}\right)^2 - \omega^2\mu\varepsilon + j\omega\mu\sigma$$

Equating the real and imaginary parts,

$$\alpha_d^2 - \beta_d^2 = \left(\frac{m\pi}{a}\right)^2 + \left(\frac{n\pi}{b}\right)^2 - \omega^2\mu\varepsilon \qquad \textbf{(7.137a)}$$

$$2\alpha_d\beta_d = \omega\mu\sigma \quad \text{or,} \quad \alpha_d = \frac{\omega\mu\sigma}{2\beta_d} \qquad \textbf{(7.137b)}$$

For weakly conducting dielectrics, $\alpha_d^2 << \beta_d^2$ and we may make the approximation:

$$\alpha_d^2 - \beta_d^2 \approx -\beta_d^2 = \left(\frac{m\pi}{a}\right)^2 + \left(\frac{n\pi}{b}\right)^2 - \omega^2\mu\varepsilon$$

$$\therefore \qquad \beta_d = \sqrt{\omega^2\mu\varepsilon - \left(\frac{m\pi}{a}\right)^2 + \left(\frac{n\pi}{b}\right)^2} = \omega\sqrt{\mu\varepsilon}\sqrt{1 - \frac{f_c^2}{f^2}}$$

This is identical with Eq. (7.94). Substituting this in Eq. (7.137*b*), we have

$$\alpha_d = \frac{\omega\mu\sigma}{2\beta_d} = \frac{\omega\mu\sigma}{2\omega\sqrt{\mu\varepsilon}\sqrt{1 - \frac{f_c^2}{f^2}}} = \frac{\sqrt{\frac{\mu}{\varepsilon}}\,\sigma}{2\sqrt{1 - \frac{f_c^2}{f^2}}} = \frac{\sigma\eta_0}{2\sqrt{1 - \frac{f_c^2}{f^2}}}$$

where, $\eta_0 = \sqrt{\frac{\mu}{\varepsilon}}$ is the intrinsic impedance of uniform plane waves.

$$\boxed{\alpha_d = \frac{\sigma\eta_0}{2\sqrt{1 - \frac{f_c^2}{f^2}}}} \qquad \textbf{(7.138)}$$

2. Ohmic losses or conduction losses: This occurs due to the conductivity of the guide walls. The attenuation in the wave due to this loss is characterised by attenuation constant α_c.

Determination of α_c: The conduction losses are more complicated to calculate. In practise, the following approximate procedure is adequate.

(a) First, the fields are determined on the assumption that the conductors are perfect.

(b) Second, the magnetic fields on the conductor surfaces are determined and the corresponding induced surface currents are calculated by $\vec{J}_S = \hat{a}_n \times \vec{H}$, where $\hat{a}_n$ is the outward normal to the conductor.

Third, the ohmic losses per unit conductor area are calculated as

$$\frac{dP_L}{dA} = \frac{1}{2} R_S \, |\vec{J}_S|^2 \tag{7.139}$$

where $R_S = \sqrt{\frac{\omega\mu}{2\sigma}} = \eta\sqrt{\frac{\omega\varepsilon}{2\sigma}} = \frac{1}{2}\delta\omega\mu$ is the surface resistance of the conductor and $\delta = \sqrt{\frac{2}{\omega\mu\sigma}}$ is the skin depth.

Integrating Eq. (7.139) around the periphery of the conductor, we obtain the total power loss per unit z-length due to that conductor.

$$P_L = \oint_S \frac{1}{2} R_S \, |\vec{J}_S|^2 \, dl \tag{7.140}$$

(c) Fourth, total power transmitted through the guide is obtained from Eq. (7.129).

(d) Finally, the attenuation constant is calculated using Eq. (7.135) as

$$\boxed{\alpha_c = \frac{1}{2}\left(\frac{P_L}{P_T}\right) = \frac{1}{2}\frac{\oint_S \frac{1}{2} R_S \, |\vec{J}_S|^2 \, dl}{\int_S \frac{1}{2}\text{Re}[\boldsymbol{E}_S \times \boldsymbol{H}_S^*] \cdot d\vec{S}}} \tag{7.141}$$

The attenuation constants for different modes of propagation in different waveguides are given in Table 7.4.

The variation of the attenuation constant with frequency for *TE*, *TM* and *TEM* modes are shown in Fig. 7.13.

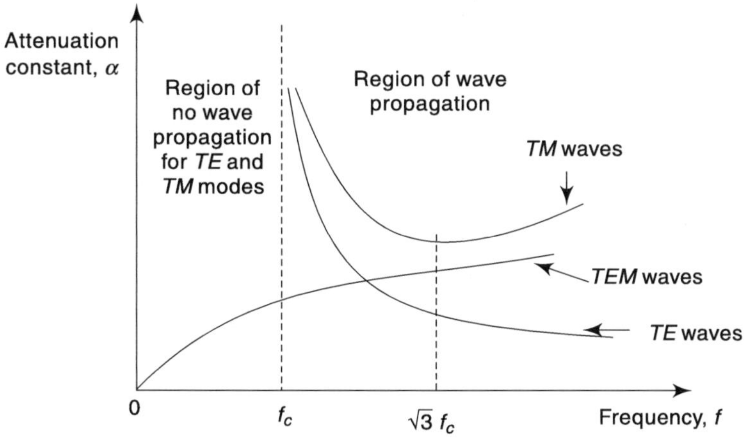

Fig. 7.13 *Variation of attenuation constant (α) with frequency*

In a particular case, we find out attenuation constant for TE_{10} wave in rectangular waveguide as follows.

Table 7.4 *Attenuation Constants for Different Modes of Wave Propagation in Different Waveguides*

Sl. No.	Type of Waveguides	Attenuation Constant		
		TE Mode	*TM* Mode	*TEM* Mode
1	Parallel Plane Waveguide	$\alpha_{TE} = \dfrac{2n^2\pi^2}{\omega\mu d^3\sqrt{\omega^2\mu\varepsilon - \left(\dfrac{n\pi}{d}\right)^2}}\sqrt{\dfrac{\omega\mu}{2\sigma}} = \sqrt{\dfrac{2n\pi}{\sigma\eta_0 d^3}}\dfrac{\left(\dfrac{f_c}{f}\right)^{3/2}}{\sqrt{1-\dfrac{f_c^2}{f^2}}}$	$\alpha_{TM} = \dfrac{2\omega\varepsilon}{d\sqrt{\omega^2\mu\varepsilon - \left(\dfrac{n\pi}{d}\right)^2}}\sqrt{\dfrac{\omega\mu}{2\sigma}} = \dfrac{2}{\eta_0}\sqrt{\dfrac{\pi f\mu/\sigma}{1-\left(\dfrac{f_c}{f}\right)^2}}$	$\alpha_{\text{TEM}} = \dfrac{1}{\eta d}\sqrt{\dfrac{\omega\mu}{2\sigma}}$
2	Rectangular Waveguide	$\alpha_c = \dfrac{2R_S}{b\eta_0\sqrt{1-\dfrac{f_c^2}{f^2}}}\left[\left(1-\dfrac{f_c^2}{f^2}\right)\dfrac{\dfrac{b}{a}\left(\dfrac{b}{a}m^2+n^2\right)}{\left(\dfrac{b^2}{a^2}m^2+n^2\right)}+\left(\dfrac{f_c^2}{f^2}\right)\left(1+\dfrac{b}{a}\right)\right]$	$\alpha_c = \dfrac{2R_S}{ab\eta_0\sqrt{1-\dfrac{f_c^2}{f^2}}}\left(\dfrac{m^2b^3+n^2a^3}{m^2b^2+n^2a^2}\right)$	—
3	Circular Waveguide	$\alpha_{cnm}^{TE} = \dfrac{R_S}{a\eta_0\sqrt{1-\dfrac{f_c^2}{f^2}}}\left[\left(\dfrac{f_c}{f}\right)^2+\dfrac{m^2}{(ha'_{nm})^2-m^2}\right]$	$\alpha_{cnm}^{TM} = \dfrac{R_S}{a\eta_0}\dfrac{1}{\sqrt{1-\dfrac{f_c^2}{f^2}}}$	—

Calculation of Attenuation Constant for TE_{10} Waves in Rectangular Waveguide In this case, only the components E_y, H_x and H_z exist. Also, $h = \left(\frac{\pi}{a}\right)$.

Hence, the time-average power transmitted through the guide is

$$P_{T_{TE10}} = \frac{\sqrt{1-\frac{f_c^2}{f^2}}}{2\eta_0} \int_{x=0}^{a} \int_{y=0}^{b} (|E_y|^2) dydx$$

$$= \frac{\sqrt{1-\frac{f_c^2}{f^2}}}{2\eta_0} \frac{\omega^2 \mu^2}{h^4} \left(\frac{\pi}{a}\right)^2 H_0^2 \int_{x=0}^{a} \int_{y=0}^{b} \left(\sin^2\left(\frac{\pi}{a}x\right) \cos^2(0) \right) dydx$$

$$= \frac{\sqrt{1-\frac{f_c^2}{f^2}}}{4\eta_0} \frac{\omega^2 \mu^2}{\left(\frac{\pi}{a}\right)^2} H_0^2 \int_0^a \left\{1 + \cos\left(\frac{2\pi}{a}x\right)\right\} dx \int_0^b dy$$

$$= \frac{\sqrt{1-\frac{f_c^2}{f^2}}}{4\eta_0} \frac{\omega^2 \mu^2}{\left(\frac{\pi}{a}\right)^2} H_0^2 \left\{x - \frac{a}{2\pi} \sin\left(\frac{2\pi}{a}x\right)\right\}_0^a b$$

$$= \frac{\sqrt{1-\frac{f_c^2}{f^2}}}{4\eta_0} \frac{\omega^2 \mu^2}{\pi^2} H_0^2 a^2 ab$$

$$= \frac{\sqrt{1-\frac{f_c^2}{f^2}}}{4\eta_0} \frac{\omega^2 \mu^2}{\pi^2} H_0^2 a^3 b \tag{7.142}$$

Similarly, total power loss for TE_{10} waves is obtained as follows.

$$P_{L_{TE10}}\Big|_{y=0} = \int_{x=0}^{a} \frac{1}{2} R_S |\vec{J}_S|^2 dl \Bigg|_{y=0} = \frac{1}{2} R_S \int_{x=0}^{a} (H_x^2 + H_z^2) dx \Bigg|_{y=0}$$

$$= \frac{1}{2} R_S \int_0^a \left[\frac{\gamma^2}{h^4} \left(\frac{\pi}{a}\right)^2 H_0^2 \sin^2\left(\frac{\pi}{a}x\right) + H_0^2 \cos^2\left(\frac{\pi}{a}x\right) \right] dx \Bigg|_{y=0}$$

$$= \frac{1}{2} R_S \int_0^a \left[\frac{\beta^2}{h^4} \left(\frac{\pi}{a}\right)^2 H_0^2 \sin^2\left(\frac{\pi}{a}x\right) + H_0^2 \cos^2\left(\frac{\pi}{a}x\right) \right] dx$$

$$= \frac{1}{2} R_S H_0^2 \int_0^a \left[\frac{\beta^2}{h^4} \left(\frac{\pi}{a}\right)^2 \sin^2\left(\frac{\pi}{a}x\right) + \cos^2\left(\frac{\pi}{a}x\right) \right] dx$$

$$= \frac{1}{4} R_S H_0^2 \int_0^a \left[\beta^2 \left(\frac{a}{\pi}\right)^2 \left\{1 - \cos\left(\frac{2\pi}{a}x\right)\right\} + \left\{1 + \cos\left(\frac{2\pi}{a}x\right)\right\} \right] dx$$

$$= \frac{1}{4} R_S H_0^2 \left[\beta^2 \left(\frac{a}{\pi}\right)^2 \left\{ x + \frac{a}{2\pi} \sin\left(\frac{2\pi}{a} x\right) \right\}_0^a + \left\{ x - \frac{a}{2\pi} \sin\left(\frac{2\pi}{a} x\right) \right\}_0^a \right]$$

$$= \frac{1}{4} R_S H_0^2 \left[\beta^2 \left(\frac{a^3}{\pi^2}\right) + a \right]$$

$$= \frac{1}{4} R_S H_0^2 a \left[1 + \beta^2 \left(\frac{a^2}{\pi^2}\right) \right]$$

$$\therefore \qquad P_L\big|_{y=0} = \frac{1}{4} R_S H_0^2 a \left[1 + \beta^2 \left(\frac{a^2}{\pi^2}\right) \right] \qquad \textbf{(7.143a)}$$

Similarly,

$$P_{L_{TE10}}\Big|_{x=0} = \int_{y=0}^{b} \frac{1}{2} R_S |\vec{J}_S|^2 \, dl \Bigg|_{x=0} = \frac{1}{2} R_S \int_{y=0}^{b} (H_z^2) dy \Bigg|_{x=0}$$

$$= \frac{1}{2} R_S \int_0^b \left[H_0^2 \cos^2\left(\frac{\pi}{a} x\right) \right] dy \Bigg|_{x=0}$$

$$= \frac{1}{2} R_S H_0^2 \int_0^b dy$$

$$= \frac{1}{2} R_S H_0^2 b$$

$$\therefore \qquad P_L\big|_{x=0} = \frac{1}{2} R_S H_0^2 b \qquad \textbf{(7.143b)}$$

Hence, total power loss is obtained as

$$P_{L_{TE}} = 2(P_L|_{y=0} + P_L|_{x=0})$$

$$= 2 \times \left\{ \frac{1}{4} R_S H_0^2 a \left[1 + \beta^2 \left(\frac{a^2}{\pi^2}\right) \right] + \frac{1}{2} R_S H_0^2 b \right\}$$

$$= R_S H_0^2 \left[b + \frac{a}{2} \left(1 + \frac{\beta^2 a^2}{\pi^2} \right) \right] \qquad \textbf{(7.144)}$$

From Eqs. (7.142) and (7.144), we get the attenuation constant for TE_{10} wave as

$$\alpha_{C_{TE10}} = \frac{1}{2} \left(\frac{P_{L_{TE10}}}{P_{T_{TE10}}} \right) = \frac{1}{2} \frac{R_S H_0^2 \left[b + \frac{a}{2} \left(1 + \frac{\beta^2 a^2}{\pi^2} \right) \right]}{\frac{\sqrt{1 - \frac{f_c^2}{f^2}}}{4\eta_0} \frac{\omega^2 \mu^2}{\pi^2} H_0^2 a^3 b} = \frac{2 R_S \eta_0 \pi^2}{\omega^2 \mu^2 a^3 b \sqrt{1 - \frac{f_c^2}{f^2}}} \left[b + \frac{a}{2} \left(1 + \frac{\beta^2 a^2}{\pi^2} \right) \right]$$

After some simplifications, we get the attenuation constant for TE_{10} wave as

$$\alpha_{C_{TE10}} = \frac{R_S}{\eta_0 b} \frac{\left(1 + \frac{2b}{a}\frac{f_c^2}{f^2}\right)}{\sqrt{1 - \frac{f_c^2}{f^2}}} \quad \textbf{(7.145)}$$

Similarly, we can find out the attenuation constant for TM_{11} wave by putting $m = n = 1$ in the equation of attenuation constant for TM mode of rectangular waveguide as,

$$\alpha_c = \frac{2R_S}{ab\eta_0} \frac{1}{\sqrt{1 - \frac{f_c^2}{f^2}}} \left(\frac{a^3 + b^3}{a^2 + b^2}\right) \quad \text{(attenuation constant for } TM_{11} \text{ wave)} \quad \textbf{(7.146)}$$

***Example 7.41** A brass waveguide ($\sigma_c = 1.1 \times 10^7$ mho/m) of dimensions $a = 4.2$ cm, $b = 1.5$ cm is filled with Teflon ($\varepsilon_r = 2.6$, $\sigma = 10^{-15}$ mho/m). The operating frequency is 9 GHz. For TE_{10} mode:

(a) Calculate α_d and α_c.

(b) What is the loss in decibels in the guide if it is 40 cm long?

Solution Since the guide is filled with Teflon ($\varepsilon_r = 2.6$, $\sigma = 10^{-15}$ mho/m) where conductivity is very small, the intrinsic impedance of the medium is given as

$$\eta_0 = \sqrt{\frac{\mu}{\varepsilon}} = \sqrt{\frac{\mu_0}{\varepsilon_0 \varepsilon_r}} = \frac{120\pi}{\sqrt{\varepsilon_r}} = \frac{120\pi}{\sqrt{2.6}} = 233.8\ \Omega$$

For TE_{10} mode, $m = 1$, $n = 0$. Hence, the cut-off frequency is

$$f_c = \frac{v_0}{2}\sqrt{\left(\frac{m}{a}\right)^2 + \left(\frac{n}{b}\right)^2} = \frac{v_0}{2a} = \frac{1}{2a}\frac{1}{\sqrt{\mu\varepsilon}} = \frac{1}{2a}\frac{1}{\sqrt{\mu_0\varepsilon_0\varepsilon_r}} = \frac{c}{2a\sqrt{\varepsilon_r}}$$

$$= \frac{3 \times 10^8}{2 \times 0.042 \times \sqrt{2.6}}$$

$$= 2.21\ \text{GHz}$$

At 9 GHz operating frequency, the attenuation constant due to dielectric loss is

$$\alpha_d = \frac{\sigma\eta_0}{2\sqrt{1 - \frac{f_c^2}{f^2}}} = \frac{10^{-15} \times 233.8}{2\sqrt{1 - \left(\frac{2.21}{9}\right)^2}} = 1.206 \times 10^{-13}\ \text{Np/m}$$

For TE_{10} mode, the attenuation constant due to ohmic loss is obtained from Eq. (7.145) as

$$\alpha_{C_{TE10}} = \frac{R_S}{\eta_0 b} \frac{\left(1 + \frac{2b}{a}\frac{f_c^2}{f^2}\right)}{\sqrt{1 - \frac{f_c^2}{f^2}}}$$

Here, $R_S = \sqrt{\dfrac{\pi f \mu}{\sigma_c}} = \sqrt{\dfrac{\pi \times 9 \times 10^9 \times 4\pi \times 10^{-7}}{1.1 \times 10^7}} = 5.683 \times 10^{-2}\ \Omega$

$$\therefore \qquad \alpha_{C_{TE10}} = \frac{R_S}{\eta_0 b} \frac{\left(1 + \frac{2b}{a}\frac{f_c^2}{f^2}\right)}{\sqrt{1 - \frac{f_c^2}{f^2}}} = \frac{5.683 \times 10^{-2}}{233.8 \times 0.015} \times \frac{\left[1 + \frac{2 \times 0.015}{0.042}\left(\frac{2.21}{9}\right)^2\right]}{\sqrt{1 - \left(\frac{2.21}{9}\right)^2}}$$

$$= 1.744 \times 10^{-2} \text{ Np/m}$$

Since, $\alpha_d << \alpha_c$, the total attenuation constant is

$$\therefore \qquad \alpha_{_{TE10}} = \alpha_{C_{TE10}} = 1.744 \times 10^{-2} \text{ Np/m}$$

For a length of 40 cm, the power loss is given as, $P_L = P_0 e^{2\alpha z}$

To convert it into decibels, we have

$$P_L = 10 \log_{10}\left(\frac{P}{P_0}\right) = \frac{10}{2.3026}\ln\left(\frac{P}{P_0}\right) = \frac{10}{2.3026}\ln\left(\frac{P_0 e^{2\alpha z}}{P_0}\right) = \frac{20}{2.3026}(\alpha z) = 8.686(\alpha z)$$

Hence, the power loss is

$$P_L = 8.686(\alpha z) = 8.686 \times 1.744 \times 10^{-2} \times 0.4 = 0.0606\ dB$$

Example 7.42 Find α_d for TE_1 mode propagating through an air-filled parallel plane guide of dimension $d = 2.5$ cm operating at 10 GHz. Given: $\sigma = 5.8 \times 10^7$ mho/m and $\sigma_d = 10^{-3}$ mho/m.

Solution Here, $n = 1$, $d = 2.5$ cm, $f = 10$ GHz, $\sigma = 5.8 \times 10^7$ mho/m, $\sigma_d = 10^{-3}$ mho/m

The cut-off frequency is given as $f_c = \dfrac{n}{2d\sqrt{\mu_0 \varepsilon_0}} = \dfrac{nc}{2d} = \dfrac{1 \times 3 \times 10^8}{2 \times 2.5 \times 10^{-2}} = 6$ GHz

$$\therefore \qquad \alpha_d = \frac{\sigma_d \eta_0}{2\sqrt{1 - \frac{f_c^2}{f^2}}} = \frac{10^{-3} \times 120\pi}{2\sqrt{1 - \left(\frac{6}{10}\right)^2}} = 0.2356 \text{ Np/m}$$

Example 7.43 If a rectangular waveguide is made of copper with $\sigma = 5.8 \times 10^7$ mho/m, find the attenuation constant in neper/metre for TE_{10} mode at $f = 1.5 f_c$. Waveguide dimensions are $a = 8$ cm and $b = 4$ cm.

Solution Here, $m = 1$, $n = 0$, $a = 8$ cm, $b = 4$ cm,
So, the cut-off frequency is given as

$$f_c = \frac{v_0}{2}\sqrt{\left(\frac{m}{a}\right)^2 + \left(\frac{n}{b}\right)^2} = \frac{v_0}{2a} = \frac{3 \times 10^8}{2 \times 0.08} = 1.875 \text{ GHz}$$

So, the operating frequency is

$$f = 1.5 f_c = 1.5 \times 1.875 = 2.8125 \text{ GHz}$$

The propagation constant is given as

$$\gamma = \sqrt{\left(\frac{m\pi}{a}\right)^2 + \left(\frac{n\pi}{b}\right)^2 - \omega^2 \mu\varepsilon} = \sqrt{\left(\frac{\pi}{a}\right)^2 - \frac{\omega^2}{c^2}}$$

$$= \sqrt{\left(\frac{\pi}{8\times 10^{-2}}\right)^2 - \frac{(2\pi \times 2.8125 \times 10^9)^2}{(3\times 10^8)^2}}$$

$$= j43.9$$

Since the propagation constant is purely imaginary, the attenuation constant is zero.

***Exampe 7.44** An air-filled rectangular waveguide has brass walls ($\mu = \mu_0$, $\sigma = 16$ MS/m) with $a = 2.286$ cm and $b = 1.016$ cm. Find the attenuation constant in dB/m due to the wall losses when the dominant mode is propagating at 9.6 GHz.

Solution Here, $\mu = \mu_0$, $\sigma = 16$ MS/m, $a = 2.286$ cm and $b = 1.016$ cm, $f = 9.6$ GHz
For the dominant mode, i.e., TE_{10} mode, the cut-off frequency is given as

$$f_c = \frac{v_0}{2}\sqrt{\left(\frac{m}{a}\right)^2 + \left(\frac{n}{b}\right)^2} = \frac{v_0}{2a} = \frac{3\times 10^8}{2\times 2.286 \times 10^{-2}} = 6.56 \text{ GHz}$$

The attenuation constant due to the wall losses is given as

$$\alpha_{C_{TE10}} = \frac{R_S}{\eta_0 b}\frac{\left(1 + \frac{2b}{a}\frac{f_c^2}{f^2}\right)}{\sqrt{1 - \frac{f_c^2}{f^2}}} = \frac{\sqrt{\pi f \mu/\sigma}}{\eta_0 b}\frac{\left(1 + \frac{2b}{a}\frac{f_c^2}{f^2}\right)}{\sqrt{1 - \frac{f_c^2}{f^2}}}$$

$$= \frac{\sqrt{\pi \times 9.6 \times 10^9 \times 4\pi \times 10^{-7}}}{\sqrt{16\times 10^6} \times 120\pi \times 1.016 \times 10^{-2}} \frac{\left[1 + \frac{2\times 1.016}{2.286}\left(\frac{6.56}{9.6}\right)^2\right]}{\sqrt{1 - \left(\frac{6.56}{9.6}\right)^2}}$$

$$= 0.0246 \text{ Np/m}$$

The attenuation in dB/m is given as

$$\alpha_{C_{TE10}} \text{(in dB)} = 0.0246 \text{ (in Np)} \times 8.686 = 0.214 \text{ dB/m}$$

***Exampe 7.45** A waveguide with dimensions $a = 2.286$ cm and $b = 1.016$ cm has perfectly conducting walls and is filled with lossy dielectric ($\sigma_d = 367.5$ μS/m, $\varepsilon_r = 2.1$, $\mu_r = 1$). Find the attenuation constant in dB/m for the dominant mode of propagation at a frequency of 9 GHz.

Solution Here, $a = 2.286$ cm and $b = 1.016$ cm, $\sigma_d = 367.5$ μS/m, $\varepsilon_r = 2.1$, $\mu_r = 1$, $f = 9$ GHz
For the dominant mode, i.e., TE_{10} mode, the cut-off frequency is given as

$$f_c = \frac{v_0}{2}\sqrt{\left(\frac{m}{a}\right)^2 + \left(\frac{n}{b}\right)^2} = \frac{v_0}{2a} = \frac{3\times 10^8/\sqrt{2.1}}{2\times 2.286\times 10^{-2}} = 4.53 \text{ GHz}$$

The attenuation constant is given as

$$\alpha_d = \frac{\sigma_d \eta_0}{2\sqrt{1-\frac{f_c^2}{f^2}}} = \frac{367.5\times 10^{-6}\times 120\pi/\sqrt{2.1}}{2\sqrt{1-\left(\frac{4.53}{9}\right)^2}} = 0.0553 \text{ Np/m}$$

The attenuation in *dB/m* is given as

$$\alpha_{C_{TE10}} \text{ (in dB)} = 0.0553 \text{ (in Np)} \times 8.686 = 0.48 \text{ dB/m}$$

Exampe 7.46 An air-filled rectangular waveguide of cross section 5 cm × 2 cm is operating in the TE_{10} mode at a frequency of 4 GHz. Determine:

(i) the group velocity
(ii) the guide wavelength
(iii) the attenuation to be expected at a frequency which is 0.95 times the cut-off frequency (assuming the guide walls to be made of perfect conductors)

Solution Here, $a = 5$ cm, $b = 2$ cm, $f = 4$ GHz, $m = 1$, $n = 0$
The cut-off frequency is given as

$$f_c = \frac{v_0}{2}\sqrt{\left(\frac{m}{a}\right)^2 + \left(\frac{n}{b}\right)^2} = \frac{v_0}{2a} = \frac{3\times 10^8}{2\times 5\times 10^{-2}} = 3 \text{ GHz}$$

The group velocity is given as

$$v_g = v_0\sqrt{1-\frac{f_c^2}{f^2}} = 3\times 10^8\sqrt{1-\left(\frac{3}{4}\right)^2} = 1.98\times 10^8 \text{ m/s}$$

The guide wavelength is given as

$$\lambda_g = \frac{\lambda_0}{\sqrt{1-\frac{f_c^2}{f^2}}} = \frac{v_0/f}{\sqrt{1-\frac{f_c^2}{f^2}}} = \frac{(3\times 10^8)/(4\times 10^9)}{\sqrt{1-\left(\frac{3}{4}\right)^2}} = 0.1134 \text{ m} = 11.34 \text{ cm}$$

For attenuation, the operating frequency is

$$f = 0.95 f_c = 0.95\times 3 = 2.85 \text{ GHz}$$

The propagation constant is given as

$$\gamma = \sqrt{\left(\frac{m\pi}{a}\right)^2 + \left(\frac{n\pi}{b}\right)^2 - \omega^2\mu\varepsilon} = \sqrt{\left(\frac{\pi}{a}\right)^2 - \frac{\omega^2}{c^2}}$$
$$= \sqrt{\left(\frac{\pi}{5\times 10^{-2}}\right)^2 - \frac{(2\pi\times 2.85\times 10^9)^2}{(3\times 10^8)^2}}$$
$$= 19.62$$

Since, the propagation is purely real, the attenuation constant is 19.62 Neper/m

Example 7.47 An air-filled circular waveguide with radius 5 mm operates in the TM_{01} mode at a frequency $f = 1.3 f_c$. Find the attenuation constant in *dB/m* due to wall losses in a short section of copper ($\sigma = 5.8 \times 10^7$ mho/m).

Solution Here, $a = 5$ mm, $\sigma = 5.8 \times 10^7$ mho/m

$$f = 1.3 f_c \quad \Rightarrow \quad \frac{f_c}{f} = \frac{1}{1.3}$$

For TM_{01} mode, the cut-off frequency is given as

$$f_c = \frac{v_0}{2\pi} h_{nm} = \frac{v_0}{2\pi a} a h_{nm} = \frac{3 \times 10^8}{2\pi \times 5 \times 10^{-3}} \times 2.4 = 22.92 \text{ GHz}$$

$\therefore$ $$f = 1.3 f_c = 1.3 \times 22.92 = 29.794 \text{ GHz}$$

The attenuation constant is given as

$$\alpha_{TM_{01}} = \frac{R_S}{a\eta_0} \frac{1}{\sqrt{1 - \left(\frac{f_c}{f}\right)^2}} = \frac{\sqrt{\pi f \mu / \sigma}}{a\eta_0} \frac{1}{\sqrt{1 - \left(\frac{f_c}{f}\right)^2}}$$

$$= \frac{\sqrt{\pi \times 29.794 \times 10^9 \times 4\pi \times 10^{-7}}}{\sqrt{5.8 \times 10^7} \times 5 \times 10^{-3} \times 120\pi} \frac{1}{\sqrt{1 - \left(\frac{1}{1.3}\right)^2}}$$

$$= 0.0374 \text{ Np/m}$$
$$= 0.0374 \times 8.686$$
$$= 0.325 \text{ dB/m}$$

Example 7.48 Show that the attenuation constant for TM_1 wave in parallel plane waveguide is a minimum at a frequency which is $\sqrt{3}$ times the cut-off frequency.

Solution For *TM* mode, the attenuation constant is given as

$$\alpha_{TM} = \frac{2}{\eta_0} \sqrt{\frac{\pi f \mu / \sigma}{1 - \left(\frac{f_c}{f}\right)^2}}$$

For the attenuation constant to be minimum

$$\frac{d\alpha}{df} = 0 \quad \Rightarrow \quad \frac{d}{df}\left[\frac{2}{\eta_0} \sqrt{\frac{\pi f \mu / \sigma}{1 - \left(\frac{f_c}{f}\right)^2}}\right] = 0 \quad \Rightarrow \quad \frac{d}{df}\left[\frac{\sqrt{f}}{\sqrt{1 - \left(\frac{f_c}{f}\right)^2}}\right] = 0$$

$$\Rightarrow \quad \frac{\sqrt{1-\left(\frac{f_c}{f}\right)^2}\left(\frac{1}{2}\frac{1}{\sqrt{f}}\right)-\sqrt{f}\,\frac{1}{2}\times\frac{1}{\sqrt{1-\left(\frac{f_c}{f}\right)^2}}\times\left(2\frac{f_c^2}{f^3}\right)}{1-\left(\frac{f_c}{f}\right)^2}=0 \quad \Rightarrow \quad \frac{1}{2}\left[1-\left(\frac{f_c}{f}\right)^2\right]=\left(\frac{f_c}{f}\right)^2$$

$$\Rightarrow \quad \frac{3}{2}\left(\frac{f_c}{f}\right)^2=\frac{1}{2}$$

$$\therefore \qquad \boxed{f=\sqrt{3}f_c}$$

Example 7.49 **(a)** Show that for a rectangular waveguide, the dominant mode exhibits a minimum attenuation due to conductor loss at a certain frequency. Find that frequency in terms of the cut-off frequency of the mode.
(b) If $a = 3b$, find the frequency for minimum attenuation.
(c) For a square waveguide, show that attenuation α_c is minimum for TE_{10} mode when $f = 2.962\, f_c$.

Solution

(a) For a TE_{10} mode rectangular wave, the attenuation constant is given as

$$\alpha_{C_{TE10}}=\frac{R_S}{\eta_0 b}\frac{\left(1+\frac{2b}{a}\frac{f_c^2}{f^2}\right)}{\sqrt{1-\frac{f_c^2}{f^2}}}$$

Substituting the value of the surface resistance $R_S=\sqrt{\frac{\pi f\mu}{\sigma}}$, we have

$$\alpha_{C_{TE10}}=\frac{\sqrt{\frac{\pi f\mu}{\sigma}}}{\eta_0 b}\frac{\left(1+\frac{2b}{a}\frac{f_c^2}{f^2}\right)}{\sqrt{1-\frac{f_c^2}{f^2}}}=\frac{\sqrt{\frac{\pi\mu}{\sigma}}}{\eta_0 b}\frac{\left(\sqrt{f}+\frac{2b}{a}\frac{f_c^2}{f^{3/2}}\right)}{\sqrt{1-\frac{f_c^2}{f^2}}}=K\frac{\left(\sqrt{f}+\frac{2b}{a}\frac{f_c^2}{f^{3/2}}\right)}{\sqrt{1-\frac{f_c^2}{f^2}}}$$

where $K=\frac{\sqrt{\frac{\pi\mu}{\sigma}}}{\eta_0 b}$ is constant.

For the attenuation constant to be minimum,

$$\frac{d\alpha_C}{df}=0$$

or $$\frac{d}{df}\left[\frac{\left(\sqrt{f}+\frac{2b}{a}\frac{f_c^2}{f^{3/2}}\right)}{\sqrt{1-\frac{f_c^2}{f^2}}}\right]=0$$

or $$\frac{\sqrt{1-\left(\frac{f_c}{f}\right)^2}\times\left[\frac{1}{2\sqrt{f}}-\frac{2b}{a}\times\frac{3}{2}\frac{f_c^2}{f^{5/2}}\right]-\left(\sqrt{f}+\frac{2b}{a}\frac{f_c^2}{f^{3/2}}\right)\times\frac{1}{2}\frac{1}{\sqrt{1-\left(\frac{f_c}{f}\right)^2}}\times\left(2\frac{f_c^2}{f^3}\right)}{1-\left(\frac{f_c}{f}\right)^2}=0$$

or $$\left(1-\frac{f_c^2}{f^2}\right)\left(\frac{1}{2\sqrt{f}}-\frac{3b}{a}\frac{f_c^2}{f^{5/2}}\right)-\left(\sqrt{f}+\frac{2b}{a}\frac{f_c^2}{f^{3/2}}\right)\frac{f_c^2}{f^3}=0$$

or $$\frac{1}{2\sqrt{f}}-\frac{1}{2}\frac{f_c^2}{f^{5/2}}-\frac{3b}{a}\frac{f_c^2}{f^{5/2}}+\frac{3b}{a}\frac{f_c^4}{f^{9/2}}-\frac{f_c^2}{f^{5/2}}-\frac{2b}{a}\frac{f_c^4}{f^{9/2}}=0$$

or $$\frac{1}{2}-3\left(\frac{1}{2}+\frac{b}{a}\right)\frac{f_c^2}{f^2}+\frac{b}{a}\frac{f_c^4}{f^4}=0$$

$$af^4-3(a+2b)f_c^2f^2+2bf_c^4=0$$

Solving this equation and taking only the positive root, we get

$$f^2=\frac{3(a+2b)f_c^2+\sqrt{9(a+2b)^2f_c^4-8abf_c^4}}{2a}$$

$$=\left[\frac{3(a+2b)}{2a}+\sqrt{\left\{\frac{3(a+2b)}{2a}\right\}^2-\frac{2b}{a}}\right]f_c^2$$

∴ $$\boxed{f=\left[\frac{3(a+2b)}{2a}+\sqrt{\left\{\frac{3(a+2b)}{2a}\right\}^2-\frac{2b}{a}}\right]^{\frac{1}{2}}f_c}$$

This is the frequency corresponding to the minimum attenuation due to conductor loss.

(b) If $a = 3b$, we have

$$f=\left[\frac{3(3b+2b)}{2\times 3b}+\sqrt{\left\{\frac{3(3b+2b)}{2\times 3b}\right\}^2-\frac{2b}{3b}}\right]^{\frac{1}{2}}f_c=f_c\sqrt{\frac{5}{2}+\sqrt{\frac{25}{4}-\frac{2}{3}}}=2.205f_c$$

∴ $$\boxed{f=2.205f_c}$$

(c) For square waveguide, $a = b$, so that we get

$$f = \left[\frac{3(a+2a)}{2a} + \sqrt{\left\{\frac{3(a+2a)}{2a}\right\}^2 - \frac{2a}{a}}\right]^{\frac{1}{2}} f_c = f_c\sqrt{\frac{9}{2} + \sqrt{\frac{81}{4} - 2}} = 2.962 f_c$$

$\therefore$ $\boxed{f = 2.962 f_c}$

7.7 CAVITY RESONATOR OR RESONANT CAVITIES

A *cavity resonator* or *resonant cavity* is a device where the inside space is completely enclosed by conducting walls that can contain oscillating electromagnetic fields and possess resonant properties.

Waveguide resonators are used in place of the lumped element *RLC* circuit to provide a tuned circuit at high frequencies. For this reason, at high frequencies, the ordinary *LC* resonant circuits are replaced by electromagnetic cavity resonators.

7.7.1 Rectangular Cavity Resonator

The rectangular waveguide resonator is basically a section of rectangular waveguide which is enclosed on both ends by conducting walls to form an enclosed conducting box.

Figure 7.14 shows a rectangular cavity the same cross-sectional dimensions as the rectangular waveguide (a, b) and z-length equal to c produced by replacing the sending and receiving ends of a rectangular waveguide by metallic walls.

A forward-moving wave will bounce back and forth from these walls, thus resulting in a standing-wave pattern along the z-direction.

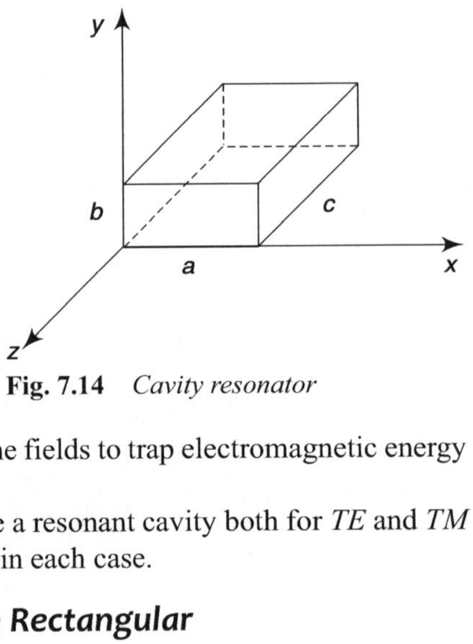

Fig. 7.14 *Cavity resonator*

In a cavity resonator, only those standing wave patterns will exist which satisfy the boundary conditions at each of the six walls. These boundary conditions on the cavity walls force the fields to trap electromagnetic energy only at certain quantized resonant frequencies.

Here, we will explain the different field configurations inside a resonant cavity both for *TE* and *TM* modes and determine the resonant frequency and quality factor in each case.

Field Equations for Transverse Electric (TE) Mode in Rectangular Cavity Resonator

In this case, $E_z = 0$ and we let,

$$H_z(x, y, z) = X(x)Y(y)Z(z) \quad \textbf{(7.147)}$$

Following the same procedure to solve Eq. (7.63) by separation of variables method as in Section 7.3, we get

$$X(x) = A_1 \cos K_x x + A_2 \sin K_x x \quad \textbf{(7.148a)}$$

$$Y(y) = A_3 \cos K_y y + A_4 \sin K_y y \quad \textbf{(7.148b)}$$

$$Z(z) = A_5 \cos K_z z + A_6 \sin K_z z \quad \textbf{(7.148c)}$$

where $$K^2 = K_x^2 + K_y^2 + K_z^2 = \omega^2 \mu\varepsilon \quad \textbf{(7.148d)}$$

$$\therefore \quad H_z(x, y, z) = (A_1 \cos K_x x + A_2 \sin K_x x)(A_3 \cos K_y y + A_4 \sin K_y y)(A_5 \cos K_z z + A_6 \sin K_z z) \tag{7.149}$$

The boundary conditions for this case are:

$$E_z = 0 \quad \text{at} \quad x = 0, a \tag{7.150a}$$

$$E_z = 0 \quad \text{at} \quad y = 0, b \tag{7.150b}$$

$$E_y = 0, E_x = 0 \quad \text{at} \quad z = 0, c \tag{7.150c}$$

Hence, from Eq. (7.64) to Eq. (7.67), the boundary conditions become,

$$\frac{\partial H_z}{\partial x} = 0 \quad \text{at} \quad x = 0, a \tag{7.151a}$$

$$\frac{\partial H_z}{\partial y} = 0 \quad \text{at} \quad y = 0, b \tag{7.151b}$$

$$H_z = 0 \quad \text{at} \quad z = 0, c \tag{7.151c}$$

Applying the boundary conditions of Eqs. (7.151*a*) and (7.151*b*) to Eq. (7.149), we require that $A_2 = 0 = A_4$ and

$$K_x = \frac{m\pi}{a}, \quad K_y = \frac{n\pi}{b}$$

where, $m = 0, 1, 2, 3, \ldots$ and $n = 0, 1, 2, 3, \ldots$
Applying the boundary condition of Eq. (7.151*c*), we require that, $A_5 = 0$ and

$$\sin K_z c = 0$$

$$\Rightarrow \quad K_z c = p\pi,$$

$p = 1, 2, 3, \ldots$

or

$$K_z = \frac{p\pi}{c}$$

Hence,

$$\boxed{K_x = \frac{m\pi}{a}, \quad K_y = \frac{n\pi}{b}, \quad K_z = \frac{p\pi}{c}} \tag{7.152}$$

Substituting this in Eq. (7.149),

$$\boxed{H_z(x, y, z) = H_0 \cos\left(\frac{m\pi}{a}x\right) \cos\left(\frac{n\pi}{b}y\right) \sin\left(\frac{p\pi}{c}z\right)} \tag{7.153}$$

where, $H_0 = A_1 A_3 A_6$.
The other field components are obtained from Eqs. (7.153) and (7.64) to Eq. (7.67) as follows.

$$\boxed{E_x = -j\frac{\omega\mu}{h^2}\frac{\partial H_z}{\partial y} = j\frac{\omega\mu}{h^2}\left(\frac{n\pi}{b}\right) H_0 \cos\left(\frac{m\pi}{a}x\right) \sin\left(\frac{n\pi}{b}y\right) \sin\left(\frac{p\pi}{c}z\right)} \tag{7.154a}$$

$$E_y = j\frac{\omega\mu}{h^2}\frac{\partial H_z}{\partial x} = -j\frac{\omega\mu}{h^2}\left(\frac{m\pi}{a}\right)H_0 \sin\left(\frac{m\pi}{a}x\right)\cos\left(\frac{n\pi}{b}y\right)\sin\left(\frac{p\pi}{c}z\right) \quad \textbf{(7.154b)}$$

$$H_x = \frac{1}{h^2}\frac{\partial^2 H_z}{\partial x \partial z} = -\frac{1}{h^2}\left(\frac{m\pi}{a}\right)\left(\frac{p\pi}{c}\right)H_0 \sin\left(\frac{m\pi}{a}x\right)\cos\left(\frac{n\pi}{b}y\right)\cos\left(\frac{p\pi}{c}z\right) \quad \textbf{(7.154c)}$$

$$H_y = \frac{1}{h^2}\frac{\partial^2 H_z}{\partial y \partial z} = -\frac{1}{h^2}\left(\frac{n\pi}{b}\right)\left(\frac{p\pi}{c}\right)H_0 \cos\left(\frac{m\pi}{a}x\right)\sin\left(\frac{n\pi}{b}y\right)\cos\left(\frac{p\pi}{c}z\right) \quad \textbf{(7.154d)}$$

NOTE

The integers m and n cannot be zero at the same time as in that case, the field components will be zero. The mode that has the lowest resonant frequency for a given cavity size (a, b, c) is known as the ***dominant mode****. For example, for a resonant cavity with a > b < c, TE_{101} mode is the dominant mode whereas for a cavity with a < b < c, TE_{011} mode is the dominant mode.*

Field Equations for Transverse Magnetic (TM) Mode in Rectangular Cavity Resonator

In this case, $H_z = 0$ and we let,

$$E_z(x, y, z) = X(x)Y(y)Z(z) \quad \textbf{(7.155)}$$

Following the same procedure to solve the Eq. (7.59) by separation of variables method as in Section 7.3, we get,

$$X(x) = B_1 \cos K_x x + B_2 \sin K_x x \quad \textbf{(7.156a)}$$

$$Y(y) = B_3 \cos K_y y + B_4 \sin K_y y \quad \textbf{(7.156b)}$$

$$Z(z) = B_5 \cos K_z z + B_6 \sin K_z z \quad \textbf{(7.156c)}$$

where $$K^2 = K_x^2 + K_y^2 + K_z^2 = \omega^2 \mu\varepsilon \quad \textbf{(7.156d)}$$

$\therefore$ $$E_z(x, y, z) = (B_1 \cos K_x x + B_2 \sin K_x x)(B_3 \cos K_y y + B_4 \sin K_y y) \times (B_5 \cos K_z z + B_6 \sin K_z z) \quad \textbf{(7.157)}$$

The boundary conditions for this case are:

$$E_z = 0 \quad \text{at} \quad x = 0, a \quad \textbf{(7.158a)}$$

$$E_z = 0 \quad \text{at} \quad y = 0, b \quad \textbf{(7.158b)}$$

$$E_y = 0, E_x = 0 \quad \text{at} \quad z = 0, c \quad \textbf{(7.158c)}$$

Applying the boundary conditions of Eqs. (7.158*a*) and (7.158*b*) to Eq. (7.157), we require that $B_1 = 0 = B_3$ and

$$K_x = \frac{m\pi}{a}, \quad K_y = \frac{n\pi}{b}$$

where, $m = 1, 2, 3, \ldots$ and $n = 1, 2, 3, \ldots$

In order to apply the boundary condition of Eq. (7.158*c*), we get by combining Eq. (7.61*b*) and Eq. (7.62*a*) that

$$j\omega\varepsilon E_x = \frac{\partial H_z}{\partial y} + \frac{1}{j\omega\mu}\left(\frac{\partial^2 E_x}{\partial z^2} - \frac{\partial^2 E_z}{\partial x \partial z}\right)$$

and with $H_z = 0$, this reduces to

$$j\omega\varepsilon E_x = \frac{1}{j\omega\mu}\left(\frac{\partial^2 E_x}{\partial z^2} - \frac{\partial^2 E_z}{\partial x \partial z}\right) \tag{7.159a}$$

Similarly, combining Eq. (7.61*a*) and Eq. (7.62*b*) with $H_z = 0$, we get

$$j\omega\varepsilon E_y = -\frac{1}{j\omega\mu}\left(\frac{\partial^2 E_x}{\partial y \partial z} - \frac{\partial^2 E_z}{\partial z^2}\right) \tag{7.159b}$$

Now, applying the boundary condition of Eq. (7.158*c*), from Eqs. (7.159*a*) and (7.159*b*), we require that $B_6 = 0$ and

$$\sin K_z c = 0$$

$\Rightarrow$

$$K_z c = p\pi,$$

$p = 0, 1, 2, 3, \ldots$

or

$$K_c = \frac{p\pi}{c}$$

Hence,

$$\boxed{K_x = \frac{m\pi}{a}, \quad K_y = \frac{n\pi}{b}, \quad K_z = \frac{p\pi}{c}} \tag{7.160}$$

Substituting this in Eq. (7.157),

$$\boxed{E_z(x, y, z) = E_0 \sin\left(\frac{m\pi}{a}x\right)\sin\left(\frac{n\pi}{b}y\right)\cos\left(\frac{p\pi}{c}z\right)} \tag{7.161}$$

where $E_0 = B_2 B_4 B_5$.

The other field components are obtained from Eqs. (7.161) and (7.61) and Eq. (7.62) as follows.

$$\boxed{E_x = \frac{1}{h^2}\frac{\partial^2 E_z}{\partial x \partial z} = -\frac{1}{h^2}\left(\frac{m\pi}{a}\right)\left(\frac{p\pi}{c}\right)E_0 \cos\left(\frac{m\pi}{a}x\right)\sin\left(\frac{n\pi}{b}y\right)\sin\left(\frac{p\pi}{c}z\right)} \tag{7.162a}$$

$$\boxed{E_y = -\frac{1}{h^2}\frac{\partial^2 E_z}{\partial y \partial z} = \frac{1}{h^2}\left(\frac{n\pi}{b}\right)\left(\frac{p\pi}{c}\right)E_0 \sin\left(\frac{m\pi}{a}x\right)\cos\left(\frac{n\pi}{b}y\right)\sin\left(\frac{p\pi}{c}z\right)} \tag{7.162b}$$

$$\boxed{H_x = j\frac{\omega\varepsilon}{h^2}\frac{\partial E_z}{\partial y} = j\frac{\omega\varepsilon}{h^2}\left(\frac{n\pi}{b}\right)E_0 \sin\left(\frac{m\pi}{a}x\right)\cos\left(\frac{n\pi}{b}y\right)\cos\left(\frac{p\pi}{c}z\right)} \tag{7.162c}$$

$$\boxed{H_y = -j\frac{\omega\varepsilon}{h^2}\frac{\partial E_z}{\partial x} = -j\frac{\omega\varepsilon}{h^2}\left(\frac{m\pi}{a}\right)E_0 \cos\left(\frac{m\pi}{a}x\right)\sin\left(\frac{n\pi}{b}y\right)\cos\left(\frac{p\pi}{c}z\right)} \tag{7.162d}$$

NOTE

Here in TM mode, both m and n can simultaneously be zero because even in that case the field components are not zero. Here also, the dominant mode depends on the cavity size. For a cavity with a > b > c, the dominant mode is TM_{110}.

Determination of Resonant Frequency for TE and TM Modes in Rectangular Cavity Resonator

The expression for the resonant frequency will be same both for *TE* and *TM* mode waves in cavity resonator, except some restrictions in the values of *m* and *n*. This obtained as follows.

The phase constant is obtained from Eq. (7.148*d*) and Eq. (7.152), as

$$\beta = K = \sqrt{K_x^2 + K_y^2 + K_z^2} = \sqrt{\left(\frac{m\pi}{a}\right)^2 + \left(\frac{n\pi}{b}\right)^2 + \left(\frac{p\pi}{c}\right)^2}$$

Since, $\beta^2 = \omega^2 \mu\varepsilon$, we have the resonant frequency as

$$\omega_r = \frac{\beta}{\sqrt{\mu\varepsilon}} = \frac{1}{\sqrt{\mu\varepsilon}}\sqrt{\left(\frac{m\pi}{a}\right)^2 + \left(\frac{n\pi}{b}\right)^2 + \left(\frac{p\pi}{c}\right)^2}$$

or

$$\boxed{f_r = \frac{1}{2\sqrt{\mu\varepsilon}}\sqrt{\left(\frac{m}{a}\right)^2 + \left(\frac{n}{b}\right)^2 + \left(\frac{p}{c}\right)^2} = \frac{v_0}{2}\sqrt{\left(\frac{m}{a}\right)^2 + \left(\frac{n}{b}\right)^2 + \left(\frac{p}{c}\right)^2}} \quad \textbf{(7.163)}$$

where $v_0 = \dfrac{1}{\sqrt{\mu\varepsilon}}$ is the phase velocity of uniform plane wave in free space.

The corresponding resonant wavelength is given as

$$\boxed{\lambda_r = \frac{v_0}{f_r} = \frac{2}{\sqrt{\left(\frac{m}{a}\right)^2 + \left(\frac{n}{b}\right)^2 + \left(\frac{p}{c}\right)^2}}} \quad \textbf{(7.164)}$$

The lowest order modes in a rectangular cavity are the TM_{110}, TE_{101}, and TE_{011} modes. Which of these modes is the dominant mode depends on the relative dimensions of the resonator.

Determination of Quality Factor for TE and TM Modes in Rectangular Cavity Resonator In an ideal case, a resonant cavity will have walls of infinite conductivity. However, in a practical cavity resonator, the walls have finite conductivity resulting in power loss. This loss is measured in terms of the quality factor of the resonator.

The quality factor of a cavity resonator is defined as

$$Q = 2\pi \frac{\text{Time average stored energy}}{\text{Energy loss per cycle of oscillation}} = 2\pi \frac{W}{P_L T} = \omega \frac{W}{P_L}$$

where $T = \dfrac{1}{f} = \dfrac{2\pi}{\omega}$ is the time period of oscillation

W—is the total time average energy stored in the electric and magnetic fields within the cavity
P_L—is the time average power loss in the cavity due to the wall ohmic losses (plus other losses, such as dielectric losses, if present)

The ratio $\Delta\omega = \frac{P_L}{W}$ is usually identified as the 3-dB bandwidth of the resonator centered at frequency ω. Therefore, the quality factor may be written as

$$\boxed{Q = 2\pi \frac{\text{Time average stored energy}}{\text{Energy loss per cycle of oscillation}} = \omega \frac{W}{P_L} = \frac{\omega}{\Delta\omega}} \quad \textbf{(7.165)}$$

In a particular case, for TE_{m0p} mode, the quality factor is given as

$$Q_{TE_{m0p}} = \frac{\omega\mu}{2R_S}\left[\frac{abc\left[\left(\frac{m\pi}{a}\right)^2 + \left(\frac{p\pi}{c}\right)^2\right]}{ac\left\{\left(\frac{m\pi}{a}\right)^2 + \left(\frac{p\pi}{c}\right)^2\right\} + 2bc\left\{\left(\frac{m\pi}{a}\right)^2 + \left(\frac{p\pi}{c}\right)^2\right\}}\right]$$

$$= \frac{\omega\mu}{2R_S}\left[\frac{\frac{m^2}{a^2} + \frac{p^2}{c^2}}{\frac{m^2}{a^2}\left(\frac{2}{a} + \frac{1}{b}\right) + \frac{p^2}{c^2}\left(\frac{2}{c} + \frac{1}{b}\right)}\right] = \frac{1}{\delta}\left[\frac{\frac{m^2}{a^2} + \frac{p^2}{c^2}}{\frac{m^2}{a^2}\left(\frac{2}{a} + \frac{1}{b}\right) + \frac{p^2}{c^2}\left(\frac{2}{c} + \frac{1}{b}\right)}\right]$$

$$\therefore \quad \boxed{Q_{TE_{m0p}} = \frac{1}{\delta}\left[\frac{\frac{m^2}{a^2} + \frac{p^2}{c^2}}{\frac{m^2}{a^2}\left(\frac{2}{a} + \frac{1}{b}\right) + \frac{p^2}{c^2}\left(\frac{2}{c} + \frac{1}{b}\right)}\right]} \quad \textbf{(7.166a)}$$

where $\delta = \frac{2R_S}{\omega\mu}$ is the skin depth.

For TE_{101} mode ($m = 1, n = 0, p = 1$) of cavity resonator, the quality factor is given as

$$\boxed{Q_{TE_{101}} = \frac{1}{\delta}\left[\frac{abc(a^2 + c^2)}{2b(a^3 + c^3) + ac(a^2 + c^2)}\right]} \quad \textbf{(7.166b)}$$

Example 7.50 A rectangular cavity resonator has dimensions $a = 3$ cm, $b = 6$ cm, $c = 9$ cm. If it is filled with polyethylene ($\varepsilon = 2.5\varepsilon_0$), find the resonant frequencies of the first five lowest-order modes.

Solution The resonant frequency is given as

$$f_r = \frac{1}{2\sqrt{\mu\varepsilon}}\sqrt{\left(\frac{m}{a}\right)^2 + \left(\frac{n}{b}\right)^2 + \left(\frac{p}{c}\right)^2} = \frac{v_0}{2}\sqrt{\left(\frac{m}{a}\right)^2 + \left(\frac{n}{b}\right)^2 + \left(\frac{p}{c}\right)^2}$$

Here, $$v_0 = \frac{1}{\sqrt{\mu\varepsilon}} = \frac{1}{\sqrt{\mu_0\varepsilon_0\varepsilon_r}} = \frac{c}{\sqrt{\varepsilon_r}} = \frac{3\times 10^8}{\sqrt{2.5}} = 1.89\times 10^8 \text{ m/s}$$

$$f_r = \frac{1.89 \times 10^8}{2}\sqrt{\left(\frac{m}{0.03}\right)^2 + \left(\frac{n}{0.06}\right)^2 + \left(\frac{p}{0.09}\right)^2}$$
$$= 9.5\sqrt{0.11m^2 + 0.0278n^2 + 0.0123p^2} \text{ GHz}$$

We know that for TE_{mnp} modes, any two indices have to be non-zero; whereas for TM_{mnp} modes, all the indices have to be non-zero. Since, $c > b > a$, the lowest order mode will be TE_{011}. The next higher order modes will be TE_{101}, TE_{110}, TE_{111} & TM_{111}, TE_{102}, TE_{022}, TE_{120}.
The corresponding resonant frequencies are given as follows.

$$f_{r_{011}} = 9.5\sqrt{0 + 0.027 + 0.0123} = 1.9 \text{ GHz}$$
$$f_{r_{101}} = 9.5\sqrt{0.11 + 0 + 0.0123} = 3.333 \text{ GHz}$$
$$f_{r_{110}} = 9.5\sqrt{0.11 + 0.027 + 0} = 3.535 \text{ GHz}$$
$$f_{r_{111}} = 9.5\sqrt{0.11 + 0.027 + 0.0123} = 3.689 \text{ GHz}$$
$$f_{r_{102}} = 9.5\sqrt{0.11 + 0 + 4 \times 0.0123} = 3.8 \text{ GHz}$$
$$f_{r_{022}} = 9.5\sqrt{0 + 4 \times 0.027 + 4 \times 0.0123} = 3.8 \text{ GHz}$$
$$f_{r_{120}} = 9.5\sqrt{0.11 + 4 \times 0.027 + 0} = 4.472 \text{ GHz}$$

Hence, the first five lowest order modes with their resonant frequencies are given in the table below.

Mode	Resonant Frequency (GHz)
TE_{011}	1.9
TE_{101}	3.333
TE_{110}	3.535
TE_{111} & TM_{111}	3.689
TE_{102} & TE_{022}	3.8

Example 7.51 A rectangular cavity resonator has dimensions $a = 3$ cm, $b = 2$ cm, $c = 4$ cm. Find the resonant frequencies of the first three lowest-order modes.

Solution The resonant frequency is given as

$$f_r = \frac{1}{2\sqrt{\mu\varepsilon}}\sqrt{\left(\frac{m}{a}\right)^2 + \left(\frac{n}{b}\right)^2 + \left(\frac{p}{c}\right)^2} = \frac{v_0}{2}\sqrt{\left(\frac{m}{a}\right)^2 + \left(\frac{n}{b}\right)^2 + \left(\frac{p}{c}\right)^2}$$

$$f_r = \frac{3 \times 10^8}{2}\sqrt{\left(\frac{m}{0.03}\right)^2 + \left(\frac{n}{0.02}\right)^2 + \left(\frac{p}{0.04}\right)^2}$$
$$= 15\sqrt{0.11\,m^2 + 0.25\,n^2 + 0.0625p^2} \text{ GHz}$$

We know that for TE_{mnp} modes, any two indices have to be non-zero; whereas for TM_{mnp} modes, all the indices have to be non-zero. Since, $c > a > b$, the lowest order mode will be TE_{101}. The next higher order modes will be TE_{011} & TE_{110}.

The corresponding resonant frequencies are given as follows.

$$f_{r_{101}} = 15\sqrt{0.11 + 0 + 0.0625} = 6.25 \text{ GHz}$$
$$f_{r_{011}} = 9.5\sqrt{0 + 0.25 + 0.0625} = 8.385 \text{ GHz}$$
$$f_{r_{110}} = 9.5\sqrt{0.11 + 0.25 + 0} = 9.014 \text{ GHz}$$

Example 7.52 An air-filled resonant cavity with dimensions $a = 10$ cm, $b = 4$ cm and $c = 5$ cm is made of copper ($\sigma = 5.8 \times 10^7$ mho/m). Find the resonant frequency and the quality factor for the dominant mode.

Solution Since $a > c > b$, the dominant mode is TE_{101} mode.
The resonant frequency for TE_{101} mode is given as

$$f_r = \frac{v_0}{2}\sqrt{\left(\frac{m}{a}\right)^2 + \left(\frac{n}{b}\right)^2 + \left(\frac{p}{c}\right)^2} = \frac{3 \times 10^8}{2 \times 10^{-2}}\sqrt{\left(\frac{1}{10}\right)^2 + \left(\frac{1}{5}\right)^2 + \left(\frac{0}{4}\right)^2} = 3.354 \text{ GHz}$$

The quality factor for the dominant mode is given as

$$Q_{TE_{101}} = \frac{1}{\delta}\left[\frac{abc(a^2 + c^2)}{2b(a^3 + c^3) + ac(a^2 + c^2)}\right] = \sqrt{\pi f \mu \sigma}\left[\frac{abc(a^2 + c^2)}{2b(a^3 + c^3) + ac(a^2 + c^2)}\right]$$

$$= \sqrt{\pi \times 3.354 \times 10^9 \times 4\pi \times 10^{-7} \times 5.8 \times 10^7}\left[\frac{10 \times 4 \times 5(100 + 25)}{8(1000 + 125) + 50(100 + 25)}\right] \times 10^{-2}$$

$$= 14,366.54$$

Example 7.53 A cubical cavity resonator made of copper ($\sigma = 5.8 \times 10^7$ mho/m) is to be operated at 15 GHz. Find the dimensions of the cavity, its quality factor and the bandwidth if it is operated in the dominant mode.

Solution The dominant mode in resonant cavity is TE_{101} mode. For this mode, for a cubical cavity, the resonant frequency is given as

$$f_r = \frac{v_0}{2}\sqrt{\left(\frac{m}{a}\right)^2 + \left(\frac{n}{b}\right)^2 + \left(\frac{p}{c}\right)^2} = \frac{3 \times 10^8}{2}\sqrt{\left(\frac{1}{a}\right)^2 + \left(\frac{0}{a}\right)^2 + \left(\frac{1}{a}\right)^2} = \frac{3 \times 10^8}{\sqrt{2}a}$$

This is given as 15 GHz.

$$\therefore \qquad 15 \times 10^9 = \frac{3 \times 10^8}{\sqrt{2}a}$$

$$\therefore \qquad a = \frac{3 \times 10^8}{\sqrt{2} \times 15 \times 10^9} = 1.414 \text{ cm}$$

Hence, the dimensions of the cavity are: 1.414 cm × 1.414 cm × 1.414 cm.
The quality factor for TE_{101} mode is given as

$$Q_{TE_{101}} = \frac{1}{\delta}\left[\frac{abc\left(a^2 + c^2\right)}{2b(a^3 + c^3) + ac(a^2 + c^2)}\right]$$

For cubical resonator,

$$Q_{TE_{101}} = \frac{1}{\delta}\left[\frac{a^3(a^2+a^2)}{2a(a^3+a^3)+a^2(a^2+a^2)}\right] = \frac{a}{3\delta}$$

where, skin depth, $\delta = \sqrt{\frac{1}{\pi f \mu \sigma}} = \sqrt{\frac{1}{\pi \times 15 \times 10^9 \times 4\pi \times 10^{-7} \times 5.8 \times 10^7}} = 5.396 \times 10^{-7}$ m

$$\therefore \qquad Q_{TE_{101}} = \frac{a}{3\delta} = \frac{0.01414}{3 \times 5.396 \times 10^{-7}} = 8736.4$$

The bandwidth of the resonator is

$$BW = \frac{f_r}{Q} = \frac{15 \times 10^9}{8736.4} = 1.717 \text{ MHz}$$

7.7.2 Circular Cavity Resonator

A circular cavity resonator is formed by shorting the both ends of a circular waveguide as shown in Fig. 7.15. This circular resonator is also used in microwave signal propagation. The tuning of the resonant frequency is done by the movable top wall.

For a cavity resonator with radius a and height d, we can calculate the resonant frequency as follows.

For a circular waveguide, the propagation constant is obtained fro Eq. (7.120) as

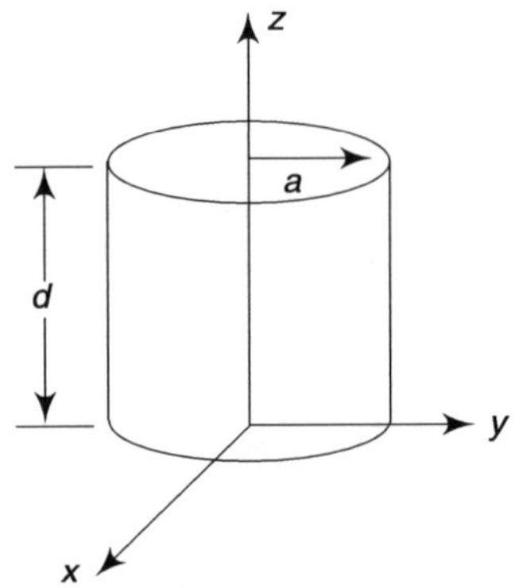

Fig. 7.15 *Circular cavity resonator*

$$\gamma_{nm} = \sqrt{h_{nm}^2 - \omega^2 \mu\varepsilon}$$

In the propagation condition, $\gamma_{nm} = j\beta_{nm}$. Replacing this

$$-\beta_{nm}^2 = h_{nm}^2 - \omega^2\mu\varepsilon$$

or $$\omega^2\mu\varepsilon = h_{nm}^2 + \beta_{nm}^2$$

Also, for a circular cavity resonator, the condition for propagation is that $\beta_{nm} = \left(\frac{p\pi}{d}\right)$ as explained for rectangular cavity resonator. Hence, we have

$$\omega^2\mu\varepsilon = h_{nm}^2 + \beta_{nm}^2 = h_{nm}^2 + \left(\frac{p\pi}{d}\right)^2$$

or $$\omega_r = \frac{1}{\sqrt{\mu\varepsilon}}\sqrt{h_{nm}^2 + \left(\frac{p\pi}{d}\right)^2}$$

$$\therefore \qquad \boxed{f_r = \frac{1}{2\pi\sqrt{\mu\varepsilon}}\sqrt{h_{nm}^2 + \left(\frac{p\pi}{d}\right)^2}} \qquad \textbf{(7.167)}$$

This is the expression for the resonant frequency for *TE* and *TM* waves except the fact that for *TE* waves, h_{nm} should be replaced by h'_{nm}.

7.8 DIELECTRIC SLAB WAVEGUIDES

A dielectric slab waveguide is a planar dielectric sheet or thin film of some thickness. A thin dielectric layer is deposited on another dielectric slab, called *substrate*. The dielectric constant of the deposited dielectric layer is greater than that of the substrate.
A schematic dielectric slab waveguide of thickness *d* is shown in Fig. 7.16.

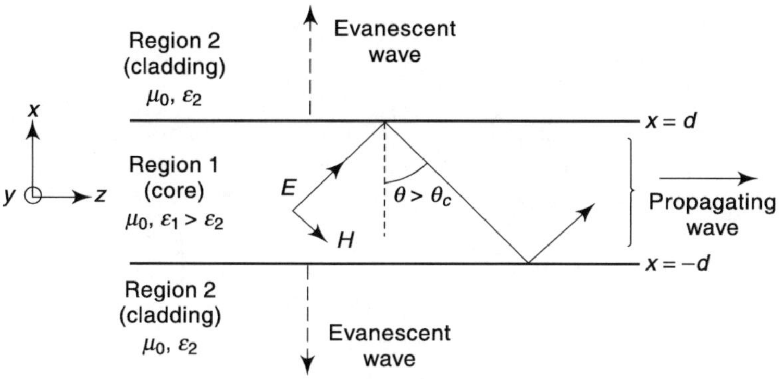

Fig. 7.16 *Dielectric slab waveguide of width 2d*

We consider a dielectric slab that is surrounded by another dielectric material that has a lower permittivity. A representative slab is shown in Fig. 7.16.

Salient Features

1. The waveguide thickness is $2d$ and the centre region (core) has a higher permittivity than the two outer regions (cladding) ($\varepsilon_1 > \varepsilon_2$).
2. The wave propagation in the z-direction is by *total internal reflection* from the top and bottom walls of the slab.
3. The propagating fields are confined primarily inside the slab; however, they also exist as evanescent waves outside it, decaying exponentially with distance from the slab.
4. Since the guide is infinitely extended in the y-direction, there are no boundary conditions of the fields along the y-direction, i.e., fields are independent of y.
5. Since the wave is propagating in the positive z-direction. Hence, the z-variation can be expressed as $e^{-j\beta z}$.

Therefore, we can write the electric and/or magnetic field as

$$E_z \text{ (or } H_Z) = X(x) e^{-j\beta z}$$

We assume the permittivity of the region inside the dielectric slab as ε_1 and outside the slab as ε_2 ($\varepsilon_1 > \varepsilon_2$).
Since the dielectric waveguide is intended to guide the wave, the fields in the cladding region should be evanescent or decaying in amplitude away from the slab. This guiding property requires the incident

angle to be more than the critical angle

$$\theta > \theta_C$$

or

$$\theta > \sin^{-1}\left(\sqrt{\frac{\varepsilon_2}{\varepsilon_1}}\right)$$

or

$$\theta > \sin^{-1}\left(\sqrt{\frac{\eta_2}{\eta_1}}\right)$$

where, η_1 and η_2 are the refractive indices of the two regions, respectively.
This requires the propagation constant to be in the range,

$$\eta_1 k_0 \sin\theta_c < \beta < \eta_1 k_0 \sin 90°$$

or

$$\eta_2 k_0 < \beta < \eta_1 k_0$$

We can use the concept of computing E_z or H_z as in the metallic waveguides and then applying Maxwell's equations we can obtain the remaining field components.

Points to remember:

1. *In case of the dielectric waveguide, due to the symmetry of the geometry, the fields will either be symmetric (even) or anti-symmetric (odd) about the y–z plane.*
2. *In order for the field to be guided by the high-permittivity dielectric slab, the fields outside the slab must be evanescent, i.e., they decay in the x direction. We will use these observations in the formulations that follow.*

We will now find the solutions for *TE* and *TM* modes.

7.8.1 Field Equations for Transverse Electric (*TE*) Mode in Dielectric Slab Waveguide

In this case, $E_z = 0$

The electric field for *TE* modes must satisfy the wave equation as given by

$$\nabla^2 E_y + \omega^2 \mu\varepsilon E_y = 0$$

or

$$\frac{\partial^2 E_y}{\partial x^2} + \frac{\partial^2 E_y}{\partial y^2} + \frac{\partial^2 E_y}{\partial z^2} + \omega^2 \mu\varepsilon E_y = 0$$

or

$$\frac{\partial^2 E_y}{\partial x^2} + 0 - \beta^2 E_y + \omega^2 \mu\varepsilon E_y = 0$$

$$\therefore \quad \frac{\partial^2 E_y}{\partial x^2} + (\omega^2 \mu\varepsilon - \beta^2) E_y = 0 \qquad \textbf{(7.168)}$$

Equation (7.168) is valid for all values of x in all regions. However, we must remember that the permittivity is different in the two regions.

Inside the dielectric slab: The solution of Eq. (7.168) inside the slab may have any of the two forms

$$E_y = A\cos k_{1x} x \quad \text{(even } TE \text{ modes)} \qquad \textbf{(7.169a)}$$

or

$$E_y = B \sin k_{1x}x \quad \text{(odd } TE \text{ modes)} \tag{7.169b}$$

where, k_{1x} is a real quantity.

Here, we have taken the sinusoidal (or cosinusoidal) forms rather than the exponential forms because we know that the fields within the slab will form standing waves.

Substituting Eq. (7.169) in Eq. (7.168), we have

$$k_{1x}^2 = \omega^2 \mu \varepsilon_1 - \beta^2 = k_1^2 - \beta^2 \tag{7.170}$$

Outside the dielectric slab: The fields outside the slab (cladding regions) are also of the basic form,

$$E_y = C \cos k_{2x}x \quad \text{(even } TE \text{ modes)} \tag{7.171a}$$

or

$$E_y = D \sin k_{2x}x \quad \text{(odd } TE \text{ modes)} \tag{7.171b}$$

where, k_{2x} is an imaginary quantity.
Or equivalently,

$$E_y = Ce^{-jk_{2x}x} \quad \text{for} \quad x \geq d \tag{7.171c}$$

or

$$E_y = De^{jk_{2x}x} \quad \text{for} \quad x \leq -d \tag{7.171d}$$

Substituting Eq. (7.171) in Eq. (7.168), we have

$$k_{2x}^2 = \omega^2 \mu \varepsilon_2 - \beta^2 = k_2^2 - \beta^2 \tag{7.172}$$

Here, in order to maintain guiding, the fields outside the slab (cladding region) must be evanescent or decay in amplitude with distance from the slab. This requirement caused the propagation constant to be in the range of ($\eta_2 k_0 < \beta$). Therefore, the propagation constant in the cladding regions must be imaginary.

Let $$k_{2x} = \pm j\alpha$$

where, $$\alpha = \pm\sqrt{\beta^2 - k_2^2}$$

The sign of k_{2x} is chosen such that the fields decay with distance away from the waveguide. This leads to

$$k_{2x} = -j\alpha \quad \text{for} \quad x \geq d \qquad \text{and} \qquad k_{2x} = j\alpha \quad \text{for} \quad x \leq -d$$

The resulting fields in the cladding regions are given by

$$E_y = Ce^{-\alpha x} \quad \text{and} \quad x \geq d \tag{7.173a}$$

or

$$E_y = De^{\alpha x} \quad \text{and} \quad x \leq -d \tag{7.173b}$$

Therefore, the electric field in the various regions is given by (let, $k_{1x} = k_x$),

Electric fields for odd *TE* modes

$$\begin{aligned}\vec{E} &= E_2 e^{-\alpha x - j\beta z}\,\hat{a}_y && \text{for} \quad x \geq d \\ &= E_1 \sin k_x x e^{-j\beta z}\hat{a}_y && \text{for} \quad |x| \leq d \\ &= -E_2 e^{\alpha x - j\beta z}\,\hat{a}_y && \text{for} \quad x \leq -d\end{aligned} \quad \textbf{(7.174a)}$$

Electric fields for even *TE* modes:

$$\begin{aligned}\vec{E} &= E_2 e^{-\alpha x - j\beta z}\,\hat{a}_y && \text{for} \quad x \geq d \\ &= E_1 \cos k_x x e^{-j\beta z}\hat{a}_y && \text{for} \quad |x| \leq d \\ &= E_2 e^{\alpha x - j\beta z}\,\hat{a}_y && \text{for} \quad x \leq -d\end{aligned} \quad \textbf{(7.174b)}$$

The magnetic fields can be computed by using Faraday's law as follows.

$$\vec{H} = -\frac{1}{j\omega\mu}(\nabla \times \vec{E}) = -\frac{1}{j\omega\mu}\begin{vmatrix} \hat{a}_x & \hat{a}_y & \hat{a}_z \\ \dfrac{\partial}{\partial x} & \dfrac{\partial}{\partial y} & \dfrac{\partial}{\partial z} \\ 0 & E_y & 0 \end{vmatrix}$$

Substituting the values of E_y from Eq. (7.173), we get the magnetic fields for different regions as given by:

Magnetic fields for odd *TE* modes:

$$\begin{aligned}\vec{H} &= \frac{E_2}{\omega\mu} e^{-\alpha x - j\beta z}(-\beta\hat{a}_x - j\alpha\hat{a}_z) && \text{for} \quad x \geq d \\ &= \frac{E_1}{\omega\mu}(-\beta \sin k_x x\hat{a}_x + jk_x \cos k_x x\hat{a}_z)e^{-j\beta z} && \text{for} \quad |x| \leq d \\ &= -\frac{E_2}{\omega\mu} e^{\alpha x - j\beta z}(-\beta\hat{a}_x + j\alpha\hat{a}_z) && \text{for} \quad x \leq -d\end{aligned} \quad \textbf{(7.174c)}$$

Magnetic fields for even *TE* modes:

$$\begin{aligned}\vec{H} &= \frac{E_2}{\omega\mu} e^{-\alpha x - j\beta z}(-\beta\hat{a}_x - j\alpha\hat{a}_z) && \text{for} \quad x \geq d \\ &= \frac{E_1}{\omega\mu}(-\beta \cos k_x x\hat{a}_x - jk_x \sin k_x x\hat{a}_z)e^{-j\beta z} && \text{for} \quad |x| \leq d \\ &= \frac{E_2}{\omega\mu} e^{\alpha x - j\beta z}(-\beta\hat{a}_x + j\alpha\hat{a}_z) && \text{for} \quad x \leq -d\end{aligned} \quad \textbf{(7.174d)}$$

In Eq. (7.174), there are four unknown quantities, E_1/E_2, k_x, α and β. We require four constraints for determining these unknowns.

From Eq. (7.170) and (7.172), replacing $k_{1x} = k_x$ and $k_{2x} = j\alpha$, we have

$$k_x^2 + \beta^2 = \omega^2\mu\varepsilon_1 = k_1^2 \quad \textbf{(7.175)}$$

$$-\alpha^2 + \beta^2 = \omega^2\mu\varepsilon_2 = k_2^2 \quad \textbf{(7.176)}$$

These two give two constraints. The other two constraints are obtained by applying the boundary conditions as follows. We will consider both the even and odd fields.

Solution for Odd *TE* Modes The tangential component of the electric fields must be continuous at the core-cladding interface. Therefore, at $x = d$,

$$E_1 \sin(k_x d) e^{-j\beta z} = E_2 e^{-\alpha d} e^{-j\beta z}$$

$$\therefore \qquad \sin(k_x d) E_1 = e^{-\alpha d} E_2 \tag{7.177a}$$

Note that applying the continuity at $x = -d$ results in an identical equation and thus does not help us. In fact, this stems from the symmetry of the problem and in reality we have already used this symmetry to break the problem into even and odd modes.

The other constraint equation is obtained by applying the continuity of the tangential component (i.e., z component) of the magnetic field at the boundary. At $x = d$,

$$jk_x \frac{E_1}{\omega\mu} \cos(k_x d) e^{-j\beta z} = -j\alpha \frac{E_2}{\omega\mu} e^{-\alpha d} e^{-j\beta z}$$

$$\therefore \qquad k_x \cos(k_x d) E_1 = -\alpha e^{-\alpha d} E_2 \tag{7.177b}$$

Again, we get the identical equation at $x = -d$. Combining Eq. (7.177*a*) and (7.177*b*), we get

$$\frac{k_x \cos(k_x d)}{\sin(k_x d)} = -\alpha$$

$$\therefore \qquad \boxed{\alpha d = -(k_x d)\cot(k_x d)} \tag{7.178}$$

Equations (7.175), (7.176) and (7.178) give the constraint equations solving which by some numerical method, we can get the different fields. Actually, these three equations can be combined as follows.

Subtracting Eq. (7.176) from (7.175),

$$k_x^2 + \alpha^2 = \omega^2\mu\varepsilon_1 - \omega^2\mu\varepsilon_2 = \omega^2\mu_0\varepsilon_0(\varepsilon_{r1} - \varepsilon_{r2}) = \omega^2\mu_0\varepsilon_0(\eta_1^2 - \eta_2^2)$$

$$(k_x d)^2 + (\alpha d)^2 = \omega^2\mu_0\varepsilon_0(\eta_1^2 - \eta_2^2)d^2 \tag{7.179}$$

Substituting the value of α from Eq. (7.178),

$$k_x^2 + k_x^2\cot^2(k_x d) = \omega^2\mu_0\varepsilon_0(\eta_1^2 - \eta_2^2)$$

or

$$1 + \cot^2(k_x d) = \frac{\omega^2\mu_0\varepsilon_0(\eta_1^2 - \eta_2^2)}{k_x^2}$$

or

$$\cot^2(k_x d) = \frac{\omega^2\mu_0\varepsilon_0(\eta_1^2 - \eta_2^2)}{k_x^2} - 1$$

$$\therefore \qquad \boxed{\cot(k_x d) = \sqrt{\frac{\omega^2\mu_0\varepsilon_0(\eta_1^2 - \eta_2^2)}{k_x^2} - 1}} \tag{7.180}$$

We can solve this simultaneous equation with the help of digital computer.

Solution for Even *TE* Modes The procedure is the same as for odd modes. The tangential component of the electric fields must be continuous at the core-cladding interface. Therefore, at $x = d$,

$$E_1 \cos(k_x d)e^{-j\beta z} = E_2 e^{-\alpha d} e^{-j\beta z}$$

$$\therefore \quad \cos(k_x d)E_1 = e^{-\alpha d} E_2 \tag{7.181a}$$

By applying the continuity of the tangential component (i.e., *z* component) of the magnetic field at the boundary, at $x = d$,

$$-jk_x \frac{E_1}{\omega\mu} \sin(k_x d)e^{-j\beta z} = -j\alpha \frac{E_2}{\omega\mu} e^{-\alpha d} e^{-j\beta z}$$

$$\therefore \quad k_x \sin(k_x d)E_1 = \alpha e^{-\alpha d} E_2 \tag{7.181b}$$

Combining Eq. (7.181*a*) and (7.181*b*), we get,

$$\frac{k_x \sin(k_x d)}{\cos(k_x d)} = \alpha$$

$$\therefore \quad \boxed{\alpha d = (k_x d)\tan(k_x d)} \tag{7.182}$$

We combine Eq. (7.175), (7.176) and (7.182) as follows.

Subtracting Eq. (7.176) from (7.175),

$$k_x^2 + \alpha^2 = \omega^2\mu\varepsilon_1 - \omega^2\mu\varepsilon_2 = \omega^2\mu_0\varepsilon_0(\varepsilon_{r1} - \varepsilon_{r2}) = \omega^2\mu_0\varepsilon_0(\eta_1^2 - \eta_2^2)$$

Substituting the value of α from Eq. (7.161),

$$k_x^2 + k_x^2 \tan^2(k_x d) = \omega^2\mu_0\varepsilon_0(\eta_1^2 - \eta_2^2)$$

or

$$1 + \tan^2(k_x d) = \frac{\omega^2\mu_0\varepsilon_0(\eta_1^2 - \eta_2^2)}{k_x^2}$$

or

$$\tan^2(k_x d) = \frac{\omega^2\mu_0\varepsilon_0(\eta_1^2 - \eta_2^2)}{k_x^2} - 1$$

$$\therefore \quad \boxed{\tan(k_x d) = \sqrt{\frac{\omega^2\mu_0\varepsilon_0(\eta_1^2 - \eta_2^2)}{k_x^2} - 1}} \tag{7.183}$$

We can solve this simultaneous equation with the help of digital computer.

Characteristics of *TE* Modes in Dielectric Slab Waveguide The following features for *TE* modes are observed in a dielectric slab waveguide.

1. For odd *TE* modes, solutions in the range $(m - 1)\pi/2 \le k_x d \le m\,\pi/2$, $m = 1, 3, 5, \ldots$ are known as odd TE_m modes. Similarly, for even *TE* modes, solutions in the range $(m - 1)\pi/2 \le k_x d \le m\pi/2$, $m = 2, 4, 6, \ldots$ are known as even TE_m modes.
2. Cut-off occurs when the mode is no longer guide, which occurs as soon as α becomes negative. So, we define the cut-off frequency as the frequency at which $\alpha = 0$.
   Using Eq. (7.178) and (7.182), at cut-off frequency,

$$\cot(k_x d) = 0 \quad \Rightarrow \quad k_x d = (m - 1)\frac{\pi}{2}, \quad m = 2, 4, 6, \ldots \text{ for odd } TE \text{ modes}$$

and,

$$\tan(k_x d) = 0 \quad \Rightarrow \quad k_x d = (m-1)\frac{\pi}{2}, \quad m = 1, 3, 5, \ldots \text{ for even } TE \text{ modes}$$

Also, using Eq. (7.179) with $\alpha = 0$, we have

$$\boxed{f_{c,m} = \frac{c(m-1)}{4d\sqrt{\mu_r \varepsilon_r - 1}}} \qquad \textbf{(7.184)}$$

This is the cut-off frequency both for odd and even *TE* modes.

This is seen that for $m = 0, f_{c,0} = 0$. This implies that the lowest order mode propagates at any frequency.

Also, at cut-off frequency, $\beta = k_2$ and $\beta^2 + k_x^2 = k_1^2$, the angle of incidence of the wave on the dielectric boundary can be expressed as,

$$\theta_i = \sin^{-1}\frac{\beta}{\sqrt{\beta^2 + k_x^2}} = \sin^{-1}\frac{k_2}{k_1} = \sin^{-1}\sqrt{\frac{\varepsilon_2}{\varepsilon_1}} = \theta_c$$

where, θ_c is the critical angle of incidence. So, cutoff occurs when the angle of incidence on the boundary is smaller than the critical angle.

This may also be noted that the cutoff condition of $\beta = k_2$ means that the propagation constant becomes that of the surrounding medium.

3. From Eq. (7.175), we have

$$\beta^2 = \omega^2 \mu \varepsilon_1 - k_x^2$$

From Eq. (7.179) at cut off,

$$k_x = \omega\sqrt{\mu(\varepsilon_1 - \varepsilon_2)}$$

From these two relations, we conclude that

$$\boxed{\omega\sqrt{\mu\varepsilon_2} \le \beta \le \omega\sqrt{\mu\varepsilon_1}} \qquad \textbf{(7.185)}$$

4. As the frequency is increased, $\alpha \to \infty$. This means that the field decays very rapidly outside the dielectric. The behaviour of the mode becomes like that of a parallel plane waveguide filled with a dielectric.
5. Here, k_x is frequency dependent, unlike in the rectangular waveguide.

7.8.2 Field Equations for Transverse Magnetic (*TM*) Mode in Dielectric Slab Waveguide

We follow the same procedure to find the fields in *TM* modes.

In this case, $H_z = 0$

The magnetic field for *TM* modes must satisfy the wave equation as given by

$$\nabla^2 H_y + \omega^2 \mu \varepsilon H_y = 0$$

or $$\frac{\partial^2 H_y}{\partial x^2}+\frac{\partial^2 H_y}{\partial y^2}+\frac{\partial^2 H_y}{\partial z^2}+\omega^2\mu\varepsilon H_y=0$$

or $$\frac{\partial^2 H_y}{\partial x^2}+0-\beta^2 H_y+\omega^2\mu\varepsilon H_y=0$$

∴ $$\frac{\partial^2 H_y}{\partial x^2}+(\omega^2\mu\varepsilon-\beta^2)H_y=0 \quad \textbf{(7.186)}$$

This Eq. (7.186) is valid for all values of x in all regions. However, we must remember that the permittivity is different in the two regions.

Inside the dielectric slab: The solution of Eq. (7.186) inside the slab may have any of the two forms,

$$H_y = A\cos k_{1x}x \quad \text{(even } TM \text{ modes)} \quad \textbf{(7.187a)}$$

or

$$H_y = B\sin k_{1x}x \quad \text{(odd } TM \text{ modes)} \quad \textbf{(7.187b)}$$

where, k_{1x} is a real quantity.

Here, we have taken the sinusoidal (or cosinusoidal) forms rather than the exponential forms because we know that the fields within the slab will form standing waves.

Substituting Eq. (7.187) in Eq. (7.186), we get the relation of Eq. (7.175) as of *TE* mode.

$$k_{1x}^2=\omega^2\mu\varepsilon_1-\beta^2=k_1^2-\beta^2 \quad \textbf{(7.188)}$$

Outside the dielectric slab: The fields outside the slab (cladding regions) are also of the basic form,

$$H_y = C\cos k_{2x}x \quad \text{(even } TM \text{ modes)} \quad \textbf{(7.189a)}$$

or

$$H_y = D\sin k_{2x}x \quad \text{(odd } TM \text{ modes)} \quad \textbf{(7.189b)}$$

where k_{2x} is an imaginary quantity.

Or equivalently,

$$H_y = Ce^{-jk_{2x}x} \quad \text{for} \quad x \ge d \quad \textbf{(7.189c)}$$

or

$$H_y = De^{jk_{2x}x} \quad \text{for} \quad x \le -d \quad \textbf{(7.189d)}$$

Substituting Eq. (7.189) in Eq. (7.186), we get the relation of Eq. (7.176) as

$$k_{2x}^2=\omega^2\mu\varepsilon_2-\beta^2=k_2^2-\beta^2 \quad \textbf{(7.190)}$$

Here, in order to maintain guiding, the fields outside the slab (cladding region) must be evanescent or decay in amplitude with distance from the slab. This requirement caused the propagation constant to be in the range of ($\eta_2\, k_0 < \beta$). Therefore, the propagation constant in the cladding regions must be imaginary.

Let $$k_{2x}=\pm j\alpha$$

where $$\alpha = \pm\sqrt{\beta^2 - k_2^2}$$

The sign of k_{2x} is chosen such that the fields decay with distance away from the waveguide. This leads to,

$$k_{2x} = -j\alpha \quad \text{for} \quad x \geq d \qquad \text{and} \qquad k_{2x} = j\alpha \quad \text{for} \quad x \leq -d$$

The resulting fields in the cladding regions are given by

$$H_y = Ce^{-\alpha x} \quad \text{for} \quad x \geq d \tag{7.191a}$$

or

$$H_y = De^{\alpha x} \quad \text{for} \quad x \leq -d \tag{7.191b}$$

Therefore, the magnetic field in the various regions is given by (let, $k_{1x} = k_x$),

Magnetic fields for odd *TM* modes:

$$\boxed{\begin{aligned} \vec{H} &= H_2 e^{-\alpha x - j\beta z}\,\hat{a}_y && \text{for} \quad x \geq d \\ &= H_1 \sin k_x x e^{-j\beta z}\hat{a}_y && \text{for} \quad |x| \leq d \\ &= -H_2 e^{\alpha x - j\beta z}\,\hat{a}_y && \text{for} \quad x \leq -d \end{aligned}} \tag{7.192a}$$

Magnetic fields for even *TM* modes:

$$\boxed{\begin{aligned} \vec{H} &= H_2 e^{-\alpha x - j\beta z}\,\hat{a}_y && \text{for} \quad x \geq d \\ &= H_1 \cos k_x x e^{-j\beta z}\hat{a}_y && \text{for} \quad |x| \leq d \\ &= H_2 e^{\alpha x - j\beta z}\,\hat{a}_y && \text{for} \quad x \leq -d \end{aligned}} \tag{7.192b}$$

The electric fields can be computed by using Ampere's law as follows.

$$\vec{E} = \frac{1}{j\omega\varepsilon}(\nabla \times \vec{H}) = \frac{1}{j\omega\varepsilon}\begin{vmatrix} \hat{a}_x & \hat{a}_y & \hat{a}_z \\ \dfrac{\partial}{\partial x} & \dfrac{\partial}{\partial y} & \dfrac{\partial}{\partial z} \\ 0 & H_y & 0 \end{vmatrix}$$

Substituting the values of H_y from Eq. (7.192), we get the electric fields for different regions as given by,

Electric fields for odd *TM* modes:

$$\boxed{\begin{aligned} \vec{E} &= \frac{H_2}{\omega\varepsilon_1} e^{-\alpha x - j\beta z}(\beta\hat{a}_x + j\alpha\hat{a}_z) && \text{for} \quad x \geq d \\ &= \frac{H_1}{\omega\varepsilon_2}(\beta \sin k_x x\hat{a}_x - jk_x \cos k_x x\hat{a}_z)e^{-j\beta z} && \text{for} \quad |x| \leq d \\ &= -\frac{H_2}{\omega\varepsilon_1} e^{\alpha x - j\beta z}(\beta\hat{a}_x - j\alpha\hat{a}_z) && \text{for} \quad x \leq -d \end{aligned}} \tag{7.193a}$$

Electric fields for even *TM* modes:

$$\begin{aligned}\vec{E} &= \frac{H_2}{\omega\varepsilon_1} e^{-\alpha x - j\beta z}(\beta\hat{a}_x + j\alpha\hat{a}_z) && \text{for} \quad x \ge d \\ &= \frac{H_1}{\omega\varepsilon_2}(\beta \cos k_x x\hat{a}_x + jk_x \sin k_x x\hat{a}_z)e^{-j\beta z} && \text{for} \quad |x| \le d \\ &= \frac{H_2}{\omega\varepsilon_1} e^{\alpha x - j\beta z}(\beta\hat{a}_x - j\alpha\hat{a}_z) && \text{for} \quad x \le -d\end{aligned} \tag{7.193b}$$

In Eq. (7.192) and Eq. (7.193), there are four unknown quantities, H_1/H_2, k_x, α and β. We require four constraints for determining these unknowns.

The two constraint equations remain the same as those for *TE* modes given as

$$k_x^2 + \beta^2 = \omega^2\mu\varepsilon_1 = k_1^2 \tag{7.175}$$

$$-\alpha^2 + \beta^2 = \omega^2\mu\varepsilon_2 = k_2^2 \tag{7.176}$$

The other two constraints are obtained by applying the boundary conditions as follows. We will consider both the even and odd fields.

Solution for Odd *TM* Modes The tangential component of the magnetic fields must be continuous at the core-cladding interface. Therefore, at $x = d$,

$$H_1 \sin(k_x d)e^{-j\beta z} = H_2 e^{-\alpha d} e^{-j\beta z}$$

$$\therefore \quad \sin(k_x d)H_1 = e^{-\alpha d} H_2 \tag{7.194a}$$

Note that applying the continuity at $x = -d$ results in an identical equation and thus does not help us. In fact, this stems from the symmetry of the problem and in reality, we have already used this symmetry to break the problem into even and odd modes.

The other constraint equation is obtained by applying the continuity of the tangential component (i.e., z component) of the electric field at the boundary. At $x = d$,

$$-jk_x \frac{H_1}{\omega\varepsilon_2}\cos(k_x d)e^{-j\beta z} = j\alpha\frac{H_2}{\omega\varepsilon_1}e^{-\alpha d}e^{-j\beta z}$$

$$\therefore \quad k_x \cos(k_x d)H_1 = -\alpha\frac{\varepsilon_2}{\varepsilon_1}e^{-\alpha d}H_2 \tag{7.194b}$$

Again, we get the identical equation at $x = -d$. Combining Eq. (7.194*a*) and (7.194*b*), we get,

$$\frac{k_x \cos(k_x d)}{\sin(k_x d)} = -\alpha\frac{\varepsilon_2}{\varepsilon_1}$$

$$\therefore \quad \alpha d = -\frac{\varepsilon_1}{\varepsilon_2}(k_x d)\cot(k_x d) \tag{7.195}$$

Equations (7.175), (7.176) and (7.195) give the constraint equations, solving which by some numerical method, we can get the different fields. Actually, these three equations can be combined as follows.

Subtracting Eq. (7.176) from (7.175),

$$k_x^2+\alpha^2=\omega^2\mu\varepsilon_1-\omega^2\mu\varepsilon_2=\omega^2\mu_0\varepsilon_0(\varepsilon_{r1}-\varepsilon_{r2})=\omega^2\mu_0\varepsilon_0(\eta_1^2-\eta_2^2)$$

$$(k_xd)^2+(\alpha d)^2=\omega^2\mu_0\varepsilon_0(\eta_1^2-\eta_2^2)d^2 \quad \textbf{(7.179)}$$

Substituting the value of α from Eq. (7.195),

$$k_x^2+k_x^2\left(\frac{\varepsilon_1}{\varepsilon_2}\right)^2\cot^2(k_xd)=\omega^2\mu_0\varepsilon_0(\eta_1^2-\eta_2^2)$$

or
$$1+\left(\frac{\varepsilon_1}{\varepsilon_2}\right)^2\cot^2(k_xd)=\frac{\omega^2\mu_0\varepsilon_0(\eta_1^2-\eta_2^2)}{k_x^2}$$

or
$$\cot^2(k_xd)=\left(\frac{\varepsilon_2}{\varepsilon_1}\right)^2\left[\frac{\omega^2\mu_0\varepsilon_0(\eta_1^2-\eta_2^2)}{k_x^2}-1\right]$$

∴
$$\boxed{\cot(k_xd)=\left(\frac{\varepsilon_2}{\varepsilon_1}\right)\sqrt{\frac{\omega^2\mu_0\varepsilon_0(\eta_1^2-\eta_2^2)}{k_x^2}-1}} \quad \textbf{(7.196)}$$

We can solve this simultaneous equation with the help of digital computer.

Solution for Even *TM* Modes The procedure is the same as for odd modes. The tangential component of the magnetic fields must be continuous at the core-cladding interface. Therefore, at $x=d$,

$$H_1\cos(k_xd)e^{-j\beta z}=H_2e^{-\alpha d}e^{-j\beta z}$$

∴
$$\cos(k_xd)H_1=e^{-\alpha d}H_2 \quad \textbf{(7.197a)}$$

By applying the continuity of the tangential component (i.e., z component) of the electric field at the boundary at $x=d$,

$$jk_x\frac{H_1}{\omega\varepsilon_2}\sin(k_xd)e^{-j\beta z}=j\alpha\frac{H_2}{\omega\varepsilon_1}e^{-\alpha d}e^{-j\beta z}$$

∴
$$k_x\sin(k_xd)H_1=\alpha\frac{\varepsilon_2}{\varepsilon_1}e^{-\alpha d}H_2 \quad \textbf{(7.197b)}$$

Combining Eq. (7.197*a*) and (7.197*b*), we get

$$\frac{k_x\sin(k_xd)}{\cos(k_xd)}=\alpha\left(\frac{\varepsilon_2}{\varepsilon_1}\right)$$

∴
$$\boxed{\alpha d=\left(\frac{\varepsilon_1}{\varepsilon_2}\right)(k_xd)\tan(k_xd)} \quad \textbf{(7.198)}$$

We combine Eq. (7.177), (7.176) and (7.198) as follows.
Subtracting Eq. (7.176) from (7.175),

$$k_x^2+\alpha^2=\omega^2\mu\varepsilon_1-\omega^2\mu\varepsilon_2=\omega^2\mu_0\varepsilon_0(\varepsilon_{r1}-\varepsilon_{r2})=\omega^2\mu_0\varepsilon_0(\eta_1^2-\eta_2^2)$$

Substituting the value of α from Eq. (7.198),

$$k_x^2 + k_x^2\left(\frac{\varepsilon_1}{\varepsilon_2}\right)^2 \tan^2(k_x d) = \omega^2\mu_0\varepsilon_0(\eta_1^2 - \eta_2^2)$$

or
$$1 + \left(\frac{\varepsilon_1}{\varepsilon_2}\right)^2 \tan^2(k_x d) = \frac{\omega^2\mu_0\varepsilon_0(\eta_1^2 - \eta_2^2)}{k_x^2}$$

or
$$\tan^2(k_x d) = \left(\frac{\varepsilon_2}{\varepsilon_1}\right)^2\left[\frac{\omega^2\mu_0\varepsilon_0(\eta_1^2 - \eta_2^2)}{k_x^2} - 1\right]$$

$\therefore$
$$\boxed{\tan(k_x d) = \left(\frac{\varepsilon_2}{\varepsilon_1}\right)\sqrt{\frac{\omega^2\mu_0\varepsilon_0(\eta_1^2 - \eta_2^2)}{k_x^2} - 1}} \tag{7.199}$$

We can solve this simultaneous equation with the help of digital computer.

***Example 7.54** How many modes exist in a dielectric waveguide that has the following parameters?
Index of refraction of the core = 1.6
Index of refraction of the cladding = 1.5
Wavelength = 1.0 μm
Waveguide core thickness, $2d = 4\ \mu$m
What is the waveguide thickness required for single mode operation?

Solution From Eq. (7.203), (7.204) and (7.205), we have the following equations:

$$\alpha d = k_x d \tan(k_x d)$$
$$\alpha d = -k_x d \cot(k_x d)$$

$$(k_x d)^2 + (\alpha d)^2 = \omega^2\mu_0\varepsilon_0(\eta_1^2 - \eta_2^2)d^2 = \left(\frac{2\pi}{\lambda}\right)^2(\eta_1^2 - \eta_2^2)d^2$$

$$\left\{\because c = f\lambda = \frac{\omega}{2\pi}\lambda \Rightarrow \frac{\omega}{c} = \frac{2\pi}{\lambda}\right\}$$

Now, if we put, $k_x d = x$ and $\alpha d = y$, we can write,

$$y = x\tan x$$
$$y = -x\cot x$$

and
$$x^2 + y^2 = \left(\frac{2\pi}{\lambda}d\right)^2(\eta_1^2 - \eta_2^2) = r^2$$

This is the equation of a circle for which the radius is

$$r = \left(\frac{2\pi}{\lambda}d\right)\sqrt{(\eta_1^2 - \eta_2^2)} = \left(\frac{2\pi}{1}\times 2\right)\sqrt{1.6^2 - 1.5^2} = 2.23\pi = 7\ \mu\text{m}$$

The equation $\quad x\tan x = 0$

when $\quad x = 0, \pi, 2\pi, 3\pi, \ldots, m\pi$

and $\quad x\tan x = \infty$

when $$x = \frac{\pi}{2}, \frac{3\pi}{2}, \frac{5\pi}{2}, \ldots, \left(\frac{\pi}{2} + m\pi\right)$$

Similarly, the equation $$-x \cot x = 0$$

when $$x = \frac{\pi}{2}, \frac{3\pi}{2}, \frac{5\pi}{2}, \ldots, \left(\frac{\pi}{2} + m\pi\right)$$

and $$-x \cot x = \infty$$

when $$x = 0, \pi, 2\pi, 3\pi, \ldots, m\pi$$

Also, at $x = 0$, $$-x \cot x = -1$$

Since, the radius of the circle is $r = 2.23\pi = 7\ \mu\text{m}$, there are three even possible modes ($(x = 0, \pi, 2\pi)$ and two odd possible modes $\left(x = \frac{\pi}{2}, \frac{3\pi}{2}\right)$

For single mode operation, we require that

$$r < \frac{\pi}{2}$$

or $$\left(\frac{2\pi}{\lambda} d\right)\sqrt{(\eta_1^2 - \eta_2^2)} < \frac{\pi}{2}$$

or $$d < \frac{\lambda}{4\sqrt{(\eta_1^2 - \eta_2^2)}}$$

or $$d < 0.449$$

Thus, for single mode operation, the waveguide thickness should be $2d = 2 \times 0.449 = 0.9\ \mu\text{m}$.

7.9 TRANSMISSION LINE ANALOGY FOR WAVEGUIDES

In this section, we will learn that there is an anlogy between the electric and magnetic field components of *TE* and *TM* waves in a waveguide and the voltages and currents of a transmission line. Based upon this analogy, one can easily draw an equivalent transmission line circuit of a waveguide to deal with many waveguide problems.

7.9.1. Analogy for *TE* waves

We have the two Maxwell's equations,

$$\nabla \times \vec{H} = j\omega\varepsilon\vec{E} \quad \textbf{(7.200a)}$$

$$\nabla \times \vec{E} = -j\omega\mu\vec{H} \quad \textbf{(7.200b)}$$

For *TE* waves, $E_z = 0$. Substituting this in Eq. (7.200*a*), we get

$$(\nabla \times \vec{H})_z = 0$$

This implies that the curl of the magnetic field in the *xy*-plane is zero and hence, we can represent H_x and H_y as a gradient of a scalar magnetic potential ϕ as

$$H_x = -\frac{\partial \phi}{\partial x} \quad \text{and} \quad H_y = -\frac{\partial \phi}{\partial y} \tag{7.201}$$

From Eq. (7.9*a*) and Eq. (7.10*a*) for *TE* waves,

$$\frac{\partial E_y}{\partial z} = j\omega\mu H_x \tag{7.202a}$$

$$\left(\frac{\partial H_z}{\partial y} - \frac{\partial H_y}{\partial z}\right) = j\omega\varepsilon E_x \tag{7.202b}$$

Analogy for voltage equations Using Eq. (7.20) and Eq. (7.23*a*), we get,

$$\frac{\partial}{\partial z}\left(\frac{j\omega\mu}{h^2}\frac{\partial H_z}{\partial x}\right) = -j\omega\mu\frac{\partial \phi}{\partial x}$$

or

$$\frac{\partial}{\partial z}\left(\frac{j\omega\mu}{h^2}H_z\right) = -j\omega\mu\phi \tag{7.203}$$

The term $\left(\frac{j\omega\varepsilon}{h^2}H_z\right)$ has the dimension of voltage and ϕ has the dimension of current so that Eq. (7.203) can be written as

$$\frac{\partial V}{\partial z} = -ZI \tag{7.204}$$

where

$$V = \frac{j\omega\mu}{h^2}H_z \quad I = \phi \quad \text{and} \quad Z = j\omega\mu$$

Equation (7.204) is analogous to the voltage equation of a transmission line.

Analogy for current equations: Using Eq. (7.64), Eq. (7.201) and Eq. (7.202*b*), we get,

$$\frac{\partial H_z}{\partial y} - \frac{\partial H_y}{\partial z} = j\omega\varepsilon E_x$$

or

$$\frac{\partial H_z}{\partial y} - \frac{\partial}{\partial z}\left(-\frac{\partial \phi}{\partial y}\right) = -j\omega\varepsilon\frac{j\omega\mu}{h^2}\frac{\partial H_z}{\partial y}$$

or

$$\frac{\partial \phi}{\partial z} = \left(\frac{\omega^2\mu\varepsilon}{h^2} - 1\right)H_z$$

or

$$\frac{\partial \phi}{\partial z} = -\left(\frac{h^2}{j\omega\mu} + j\omega\varepsilon\right)\left(\frac{j\omega\mu}{h^2}H_z\right) \tag{7.205}$$

As mentioned earlier, the term $\left(\frac{j\omega\mu}{h^2}H_z\right)$ has the dimension of voltage and ϕ has the dimension of current so that Eq. (7.205) can be written as

$$\frac{\partial I}{\partial z} = -YV \tag{7.206}$$

where $$V = \frac{j\omega\mu}{h^2} H_z \quad I = \phi \quad \text{and} \quad Y = j\omega\varepsilon + \frac{h^2}{j\omega\mu}$$

Equation (7.206) is analogous to the current equation of a transmission line.
Based upon the voltage and current equations of Eqs. (7.204) and (7.206), we draw transmission line equivalent circuit of a waveguide for *TE* waves as shown in Fig. 7.17.

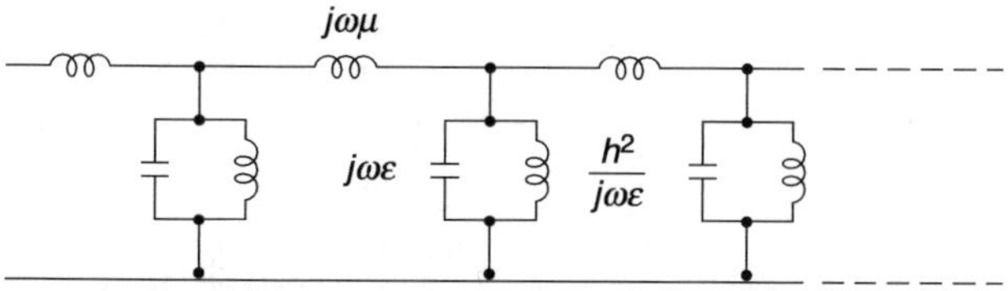

Fig. 7.17 *Transmission line equivalent circuit of a waveguide for TE waves*

7.9.2. Analogy for *TM* Waves

We have the two Maxwell's equations,

$$\nabla \times \vec{H} = j\omega\varepsilon \vec{E} \quad \textbf{(7.200a)}$$

$$\nabla \times \vec{E} = -\, j\omega\mu \vec{H} \quad \textbf{(7.200b)}$$

For *TME* waves, $H_z = 0$. Substituting this in Eq. (7.199*a*), we get

$$(\nabla \times \vec{E})_z = 0$$

This implies that the curl of the electric field in the *xy*-plane is zero and hence, we can represent E_x and E_y as a gradient of a scalar potential V as

$$E_x = -\frac{\partial V}{\partial x} \quad \text{and} \quad E_y = -\frac{\partial V}{\partial y} \quad \textbf{(7.207)}$$

From Eq. (7.9*b*) and Eq. (7.10*a*) for *TM* waves,

$$\left(\frac{\partial E_x}{\partial z} - \frac{\partial E_z}{\partial x}\right) = -\, j\omega\mu H_y \quad \textbf{(7.208a)}$$

$$\frac{\partial H_y}{\partial z} = -\, j\omega\varepsilon E_x \quad \textbf{(7.208b)}$$

Analogy for voltage equations: Using Eqs. (7.21) and (7.208*a*), we get

$$\left(\frac{\partial E_x}{\partial z} - \frac{\partial E_z}{\partial x}\right) = -\, j\omega\mu H_y = -\, j\omega\mu\left(-\frac{j\omega\varepsilon}{h^2}\frac{\partial E_z}{\partial x}\right)$$

or $$\frac{\partial E_x}{\partial z} - \frac{\partial E_z}{\partial x} = -\frac{\omega^2\mu\varepsilon}{h^2}\frac{\partial E_z}{\partial x}$$

or $$\frac{\partial}{\partial z}\left(-\frac{\partial V}{\partial x}\right) - \frac{\partial E_z}{\partial x} = -\frac{\omega^2\mu\varepsilon}{h^2}\frac{\partial E_z}{\partial x}$$

or
$$\frac{\partial V}{\partial z} = \left(\frac{\omega^2 \mu\varepsilon}{h^2} - 1\right) E_z$$

or
$$\frac{\partial V}{\partial z} = -\left(j\omega\mu + \frac{h^2}{j\omega\varepsilon}\right)\left(\frac{j\omega\varepsilon}{h^2} E_z\right) \quad \textbf{(7.209)}$$

The term $\left(\frac{j\omega\varepsilon}{h^2} E_z\right)$ has the dimension of current so that Eq. (7.209) can be written as

$$\frac{\partial V}{\partial z} = -Z I \quad \textbf{(7.210)}$$

where,
$$I = \frac{j\omega\varepsilon}{h^2} E_z \quad \text{and} \quad Z = j\omega\mu + \frac{h^2}{j\omega\varepsilon}$$

Equation (7.209) is analogous to the voltage equation of a transmission line.
Analogy for current equations: Using Eq. (7.21), (7.207) and Eq. (7.208*b*), we get

$$\frac{\partial H_y}{\partial z} = -\, j\omega\varepsilon E_x$$

or
$$\frac{\partial}{\partial z}\left(-\frac{j\omega\varepsilon}{h^2}\frac{\partial E_z}{\partial x}\right) = -\, j\omega\varepsilon\left(-\frac{\partial V}{\partial x}\right)$$

or
$$\frac{\partial}{\partial z}\left(-\frac{j\omega\varepsilon}{h^2} E_z\right) = -\, j\omega\varepsilon V \quad \textbf{(7.211)}$$

As mentioned earlier, the term $\left(\frac{j\omega\varepsilon}{h^2} E_z\right)$ has the dimension of current so that Eq. (7.211) can be written as

$$\frac{\partial I}{\partial z} = -YV \quad \textbf{(7.212)}$$

where
$$I = \frac{j\omega\varepsilon}{h^2} E_z \quad \text{and} \quad Y = j\omega\varepsilon$$

Equation (7.212) is analogous to the current equation of a transmission line.
Based upon the voltage and current equations of Eqs. (7.211) and (7.212), we draw transmission line equivalent circuit of a waveguide for *TM* waves as shown in Fig. 7.18.

It must be mentioned here that both Fig. 7.17 and Fig. 7.18 have the high-pass filter characteristics. For Fig. 7.17, the cut-off frequency occurs when shunt suceptance is zero. For Fig. 7.18, the cut-off frequency occurs when the series reactance is zero. In both the cases, the result is the same as follows.

$$h^2 = \omega_c^2 \mu\varepsilon$$

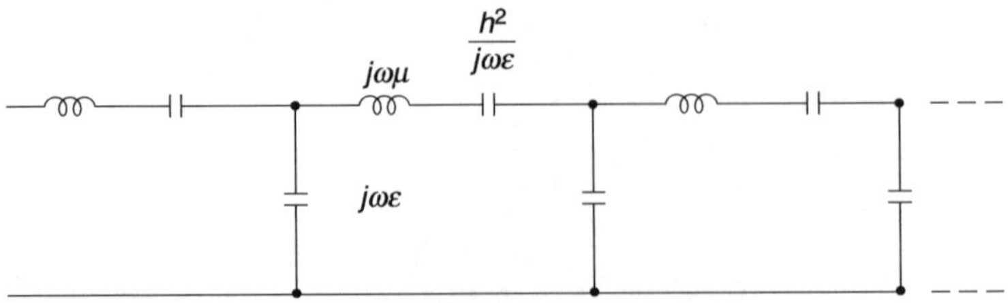

Fig. 7.18 *Transmission line equivalent circuit of a waveguide for TM waves*

This is the same result as obtained for *TE* or *TM* waves from the classical wave theory.

For Fig. 7.17 for *TE* wave, the characteristic impedance of the line is obtained as

$$Z_0(TE) = \sqrt{\frac{Z}{Y}} = \sqrt{\frac{j\omega\mu}{j\omega\varepsilon + \dfrac{h^2}{j\omega\mu}}} = \sqrt{\frac{\mu}{\varepsilon}}\sqrt{\frac{1}{1-(\omega_c/\omega)^2}} = \frac{\eta_0}{\sqrt{1-(\omega_c/\omega)^2}}$$

which is the same as the wave impedance of *TE* waves.

For Fig. 7.18 for *TM* wave, the characteristic impedance of the line is obtained as

$$Z_0(TM) = \sqrt{\frac{Z}{Y}} = \sqrt{\frac{j\omega\mu + \dfrac{h^2}{j\omega\varepsilon}}{j\omega\varepsilon}} = \sqrt{\frac{\mu}{\varepsilon}}\sqrt{1-(\omega_c/\omega)^2} = \eta_0\sqrt{1-(\omega_c/\omega)^2}$$

which is the same as the wave impedance of *TM* waves.

7.10 APPLICATIONS OF WAVEGUIDE

Waveguides are used to transfer electromagnetic power efficiently from one point in space to another. Some common guiding structures are shown in Fig. 7.19. These include the typical coaxial cable, the two-wire and microstrip transmission lines, hollow conducting waveguides, and optical fibres.

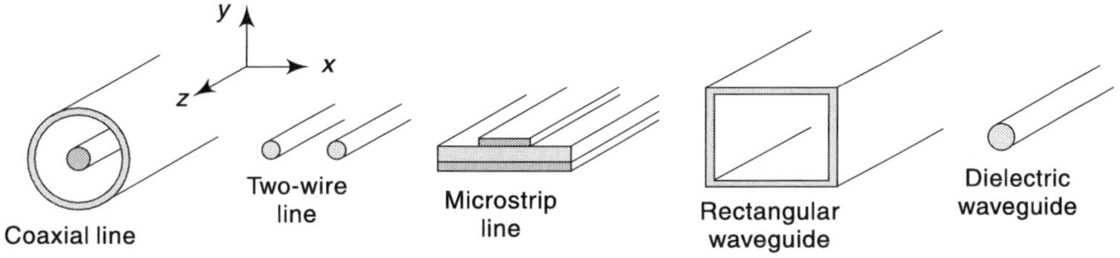

Fig. 7.19 *Typical Waveguide structures*

In practice, the choice of structure depends upon some factors, such as:

(a) the desired operating frequency band,
(b) the amount of power to be transferred, and
(c) the amount of transmission losses that can be tolerated.

Coaxial Cables These are widely used to connect *RF* components for frequencies below 3 GHz. Above that, the losses are too excessive. For example, the attenuation for different frequencies ranges for coaxial cable are as follows.

3 *dB* per 100 m at 100 MHz

10 *dB*/100 m at 1 GHz, and

50 *dB*/100 m at 10 GHz.

The power rating of coaxial cable is limited primarily because of the heating of the coaxial conductors and of the dielectric between the conductors. Typical power rating is of the order of 1 kW at 100 MHz, but only 200 W at 2 GHz. However, special short-length coaxial cables do exist that operate in the 40 GHz range.

Another issue in choice of waveguide structure is the single-mode operation of the line. At higher frequencies, in order to prevent higher modes from being propagated, the diameters of the coaxial conductors must be reduced, diminishing the amount of power that can be transmitted.

Two-Wire Lines These are not used at microwave frequencies because they are not shielded and can radiate. One typical use of two-wire line is for connecting indoor antennas to TV sets. Microstrip lines are used widely in microwave integrated circuits.

Rectangular Waveguides These are used routinely to transfer large amounts of microwave power at frequencies greater than 3 GHz. For example at 5 GHz, the transmitted power might be one megawatt and the attenuation only 4 dB/100 m.

Optical Fibres Optical fibres operate at optical and infrared frequencies, allowing a very wide bandwidth. Their losses are very low, typically, 0.2 dB/km. The transmitted power is of the order of milliwatt.

Summary

- A wave guide is a hollow conducting pipe, of uniform cross section, used to transport high frequency electromagnetic waves (generally, in the microwave band) from one point to another.
- Waveguides can be generally classified as either *metal waveguides* or *dielectric waveguides*.
- There are different mode of wave propagation through a waveguide, e.g., *TE* mode, *TM* mode, *TEM* mode and Hybrid mode.
  - — For *TE* mode, no component of electric field exists in the direction of wave propagation.
  - — For *TM* mode, no component of magnetic field exists in the direction of wave propagation.
  - — For *TEM* mode, no component of either electric or magnetic field exists in the direction of wave propagation.
  - — For hybrid mode, both electric and magnetic field exists in the direction of wave propagation.
- For parallel plane waveguide, consisting of two infinite parallel conducting planes, there exist three modes of wave propagation, e.g., *TE* mode, *TM* mode and *TEM* mode.
- Different parameters for parallel plane waveguide are given in the table below.

Parameters	*TE* and *TM* modes	*TEM* mode
Propagation constant	$\gamma = \sqrt{\left(\frac{n\pi}{d}\right)^2 - \omega^2\mu\varepsilon}$	$\gamma = j\omega\sqrt{\mu\varepsilon}$
Cut-off frequency	$f_c = \frac{n}{2d}\frac{1}{\sqrt{\mu\varepsilon}}$	$f_c = 0$
Cut-off wavelength	$\lambda_c = \frac{2d}{n}$	$\lambda_c = \infty$
Guide wavelength	$\lambda_g = \frac{\lambda}{\sqrt{1-\left(\frac{n\lambda}{2d}\right)^2}} = \frac{\lambda}{\sqrt{1-\frac{\lambda^2}{\lambda_c^2}}}$	$\lambda_g = \lambda$

Phase velocity	$v_p = \dfrac{\omega}{\sqrt{\omega^2\mu\varepsilon - \left(\dfrac{n\pi}{d}\right)^2}} = \dfrac{v_0}{\sqrt{1-\dfrac{f_c^2}{f^2}}}$	$v_p = \dfrac{1}{\sqrt{\mu\varepsilon}}$
Group velocity	$v_g = \dfrac{2df_c}{n}\sqrt{1-\dfrac{f_c^2}{f^2}} = v_0\sqrt{1-\dfrac{f_c^2}{f^2}}$	—
Wave impedance	$\eta = \dfrac{\eta_0}{\sqrt{1-\dfrac{f_c^2}{f^2}}}$; $\eta_0 = \sqrt{\dfrac{\mu_0}{\varepsilon_0}}$	$\eta_{TEM} = \sqrt{\dfrac{\mu}{\varepsilon}}$

- A rectangular waveguide is formed by placing four conducting planes.
- For rectangular waveguide, four modes of wave propagation, e.g., *TE* mode, *TM* mode, *TEM* mode and hybrid mode, exist. For TE_{mn} mode, $m = 0, 1, 2, \ldots; n = 1, 2, 3, \ldots$, For TM_{mn} mode, $m = 1, 2, 3, \ldots; n = 1, 2, 3, \ldots$
- Different parameters for rectangular waveguide are given in the table below.

Parameters	***TE* and *TM* modes**
Propagation constant	$\gamma = \sqrt{\left(\dfrac{m\pi}{a}\right)^2 + \left(\dfrac{n\pi}{b}\right)^2 - \omega^2\mu\varepsilon}$
Cut-off frequency	$f_c = \dfrac{v_0}{2}\sqrt{\left(\dfrac{m}{a}\right)^2 + \left(\dfrac{n}{b}\right)^2}$
Cut-off wavelength	$\lambda_c = \dfrac{2}{\sqrt{\left(\dfrac{m}{a}\right)^2 + \left(\dfrac{n}{b}\right)^2}}$
Guide wavelength	$\lambda_g = \dfrac{2\pi}{\sqrt{\omega^2\mu\varepsilon - \left(\dfrac{m\pi}{a}\right)^2 + \left(\dfrac{n\pi}{b}\right)^2}} = \dfrac{\lambda_0}{\sqrt{1-\dfrac{f_c^2}{f^2}}}$
Phase velocity	$v_p = \dfrac{\omega}{\sqrt{\omega^2\mu\varepsilon - \left(\dfrac{m\pi}{a}\right)^2 + \left(\dfrac{n\pi}{b}\right)^2}} = \dfrac{v_0}{\sqrt{1-\dfrac{f_c^2}{f^2}}}$
Group velocity	$v_g = \dfrac{2df_c}{n}\sqrt{1-\dfrac{f_c^2}{f^2}} = v_0\sqrt{1-\dfrac{f_c^2}{f^2}}$

Intrinsic wave impedance	$\eta_{TE} = \dfrac{\eta_0}{\sqrt{1-\dfrac{f_c^2}{f^2}}}$; $\eta_0\sqrt{1-\dfrac{f_c^2}{f^2}}$ where, $\eta_0 = \sqrt{\dfrac{\mu}{\varepsilon}}$.

- For *TEM* mode in rectangular waveguide, $\alpha = 0$, $\beta = \omega\sqrt{\mu\varepsilon}$, so that $\gamma = j\omega\sqrt{\mu\varepsilon}$.
- *TEM* wave cannot exist in a single-conductor waveguide.
- Cylindrical or circular waveguides are those that maintain a uniform circular cross-section along their length. Two modes of propagation, i.e., *TE* and *TM* exist in circular waveguides.
- Different parameters for circular waveguide are given in the table below.

Parameters	***TE* and *TM* modes**
Propagation constant	$\gamma_{nm} = \sqrt{h_{nm}^2 - \omega^2\mu\varepsilon}$ where, $h = \sqrt{\gamma^2 + \omega^2\mu\varepsilon}$
Phase constant	$\beta_{nm} = \sqrt{\omega^2\mu\varepsilon - h_{nm}^2}$
Cut-off frequency	$f_c = \dfrac{h_{nm}}{2\pi\sqrt{\mu\varepsilon}} = \dfrac{v_0}{2\pi}h_{nm}$
Cut-off wavelength	$\lambda_c = \dfrac{2\pi}{h_{nm}}$
Guide wavelength	$\lambda_g = \dfrac{2\pi}{\sqrt{\omega^2\mu\varepsilon - h_{nm}^2}} = \dfrac{\lambda_0}{\sqrt{1-\dfrac{f_c^2}{f^2}}}$
Phase velocity	$v_p = \dfrac{\omega}{\sqrt{\omega^2\mu\varepsilon - h_{nm}^2}} = \dfrac{v_0}{\sqrt{1-\dfrac{f_c^2}{f^2}}}$
Group velocity	$v_g = v_0\sqrt{1-\dfrac{f_c^2}{f^2}}$
Intrinsic wave impedance	$\eta_{TE} = \dfrac{\eta_0}{\sqrt{1-\dfrac{f_c^2}{f^2}}}$; $\eta_{TM} = \eta_0\sqrt{1-\dfrac{f_c^2}{f^2}}$ where, $\eta_0 = \sqrt{\dfrac{\mu}{\varepsilon}}$.

- In a waveguide, since the guide walls are not perfectly conducting and the dielectric is lossy ($\sigma \neq 0$, $\sigma_c \neq \infty$), there is a continuous loss of power as a wave propagates through the waveguide. These losses are mainly due to two reasons, represented by two attenuation constants:

$$\alpha_d = \frac{\sigma\eta_0}{2\sqrt{1-\frac{f_c^2}{f^2}}}, \text{ attenuation due to dielectric losses, and}$$

$$\alpha_c = \frac{1}{2}\left(\frac{P_L}{P_T}\right), \text{ attenuation due to conduction losses}$$

- The attenuation constants due to conduction losses in a parallel plane waveguide for three different modes are as

$$\alpha_{TE} = \sqrt{\frac{2n\pi}{\sigma\eta_0 d^3}}\frac{\left(\frac{f_c}{f}\right)^{3/2}}{\sqrt{1-\frac{f_c^2}{f^2}}}, \quad \alpha_{TM} = \frac{2\omega\varepsilon}{d\sqrt{\omega^2\mu\varepsilon-\left(\frac{n\pi}{d}\right)^2}}\sqrt{\frac{\omega\mu}{2\sigma}} \quad \text{and} \quad \alpha_{TEM} = \frac{\omega\varepsilon}{\beta d}\sqrt{\frac{\omega\mu}{2\sigma}}$$

- The attenuation constant due to conduction losses for TE_{mn} waves in a rectangular waveguide is given as

$$\alpha_c = \frac{2R_S}{b\eta_0\sqrt{1-\frac{f_c^2}{f^2}}}\left[\left(1-\frac{f_c^2}{f^2}\right)\frac{\frac{b}{a}\left(\frac{b}{a}m^2+n^2\right)}{\left(\frac{b^2}{a^2}m^2+n^2\right)}+\left(\frac{f_c^2}{f^2}\right)\left(1+\frac{b}{a}\right)\right]$$

- The attenuation constant due to conduction losses for TM_{mn} waves in a rectangular waveguide is given as

$$\alpha_c = \frac{2R_S}{ab\eta_0\sqrt{1-\frac{f_c^2}{f^2}}}\left(\frac{m^2b^3+n^2a^3}{m^2b^2+n^2a^2}\right)$$

- The attenuation constant due to conduction losses for TE_{nm} waves in a circular waveguide is given as

$$\alpha_{cnm}^{TE} = \frac{R_S}{a\eta_0\sqrt{1-\frac{f_c^2}{f^2}}}\left[\left(\frac{f_c}{f}\right)^2+\frac{m^2}{(ha'_{nm})^2-m^2}\right]$$

- The attenuation constant due to conduction losses for TM_{nm} waves in a circular waveguide is given as

$$\alpha_{cnm}^{TM} = \frac{R_S}{a\eta_0}\frac{1}{\sqrt{1-\frac{f_c^2}{f^2}}}$$

- A cavity resonator, usually used for energy storage at high frequencies, is one in which the electromagnetic waves exist in a hollow space inside the device.

- The rectangular waveguide resonator is basically a section of rectangular waveguide which is enclosed on both ends by conducting walls to form an enclosed conducting box.
- The resonant frequency in a rectangular cavity resonator for *TE* and *TM* mode waves is given as

$$f_r = \frac{v_0}{2}\sqrt{\left(\frac{m}{a}\right)^2 + \left(\frac{n}{b}\right)^2 + \left(\frac{p}{c}\right)^2}$$

- The quality factor of a cavity resonator is defined as

$$Q = 2\pi \frac{\text{Time average stored energy}}{\text{Energy loss per cycle of scillation}} = 2\pi \frac{W}{P_L T} = \omega \frac{W}{P_L}$$

- For the dominant mode in a rectangular cavity resonator (i.e., TE_{101} mode), the quality factor is given as

$$Q_{TE_{101}} = \frac{1}{\delta}\left[\frac{abc(a^2+c^2)}{2b(a^3+c^3)+ac(a^2+c^2)}\right]$$

- The resonant frequency in a circular cavity resonator for *TE* and *TM* mode waves is given as,

$$f_r = \frac{1}{2\pi\sqrt{\mu\varepsilon}}\sqrt{h_{nm}^2 + \left(\frac{p\pi}{d}\right)^2}$$

Exercises

[NOTE: * marked problems are important university problems]

- ***Easy***

1. A parallel plane waveguide with plate separation of 20 cm with TE_1 mode is excited at 1 GHz. Find the phase constant. [13.85 radian/m]

***2.** A pair of perfectly conducting planes is separated by 8 cm in air. For a frequency of 5000 MHz with TM_{10} mode excited, find cut-off frequency, phase shift, phase velocity and group velocity. [1.875 GHz, 97.08 radian/m, 3.236×10^8 m/s, 2.781×10^8 m/s]

3. Determine the separation between the planes of a parallel plane waveguide so that an electromagnetic wave of a frequency 10 GHz may propagate in a mode having three as the number of half cycles of the electric field ($n = 3$) and guide wavelength equal to 6.88 cm. [5 cm]

***4.** A frequency of 3 GHz is impressed on a hollow rectangular waveguide of dimensions 6 cm × 4 cm. Compute (a) cut-off wavelength, (b) guide wavelength, (c) phase constant, (d) phase velocity, (e) wave impedance for the TE_{10} mode of propagation. [12 cm, 18 cm, 35 radian/m, 5.4×10^8 m/s, 676.5 Ω]

5. A rectangular waveguide has the following dimensions $a = 2.54$ cm, $b = 1.27$ cm and waveguide thickness = 0.127 cm. Calculate the cut-off frequency for TE_{11} mode. [13.2 GHz]

***6.** An air-filled recatngualr waveguide with dimensions of $a = 8.5$ cm and $b = 4.3$ cm is fed by a 4 GHz carrier from co-axial cable. Determine the cut-off frequency, phase velocity and group velocity for TE_{11} mode. [3.909 GHz, 14.14×10^8 m/s, 0.6362×10^8 m/s]

- ***Medium***

7. The cut-off wavelengths of a rectangular waveguide are measured to be 8 cm and 4.8 cm for TE_{10} and TE_{11} mode respectively. Determine waveguide dimensions.

[$a = 4$ cm, $b = 3$ cm]

8. What do you mean by dominant mode? A rectangular waveguide has dimensions of 2.5 cm and 5 cm. Determine the phase constant, phase velocity and group velocity at a wavelength of 4.5 cm for the dominant mode.

[124.5 radian/s, 3.359×10^8 m/s, 2.679×10^8]

- ***Hard***

***9.** A waveguide has an internal breadth "a" of 3 cm and carries the dominant mode of a signal of unknown frequency. If the characteristic impedance is 500 ohm, what is the signal frequency?

[7.61 GHz]

10. A circular waveguide operated at 11 GHz has the internal diameter of 4.5 cm. For a TE_{01} mode propagation, calculate λ and λ_c. Given: $(ha)_{01} = 2.405$. [2.727 cm, 11.756 cm]

11. Given a circular waveguide of internal diameter 12 cm operating with an 8 GHz signal propagating TM_{22} mode. Calculate λ_0, λ_c, λ_g and η_g. Given: $(ha)_{22} = 8.42$.

[3.75 cm, 4.47 cm, 6.89 cm, 692.65 Ω]

12. A lossless rectangular waveguide with a dielectric medium of permittivity $\varepsilon_r = 1.8$ has dimensions $a = 7.214$ cm and $b = 3.404$ cm. For the dominant mode propagation at 2.6 GHz, the guide transports 200 W of average power. Find the level of excitation of the $\vec{E}$ field. [106.8 V/cm]

13. In a lossless air-filled circular waveguide with 1 cm radius, the transmitted power in the dominant mode at 15 GHz is 2 W. Find the level of excitation for the magnetic field. [0.11 A/m]

14. Find the inside diameter of a lossless air-filled circular waveguide so that a TE_{11} mode propagates at a frequency of 10 GHz, with the cut-off wavelength of the mode being 1.3 times the operating wavelength. [2.28 cm]

***15.** An aluminium waveguide ($a = 4.2$ cm, $b = 1.5$ cm, $\sigma_c = 3.5 \times 10^7$ mho/m) filled with Teflon ($\mu_r = 1$, $\varepsilon_r = 2.6$, $\sigma_d = 10^{-15}$ mho/m) operates at 4 GHz. Determine (a) α_c and α_d for the TE_{10} mode. (b) the waveguide loss in dB over a distance of 1.5 m.

[8.868×10^{-3} Np/m; 1.403×10^{-13} Np/m; 0.1154 dB]

16. A TE_{10} wave at 10 GHz propagates in a rectangular waveguide made of brass ($\sigma = 1.57 \times 10^7$ S/m) with inner dimensions $a = 1.5$ cm and $b = 0.6$ cm, filled with polyethylene ($\varepsilon_r = 2.25$, $\mu_r = 1$), loss tangent $= 4 \times 10^{-4}$. Determine (a) the phase constant, (b) the guide wavelength, (c) the phase velocity, (d) the wave impedance, (e) the attenuation constant due to loss in the dielectric, and (f) the attenuation constant due to loss in the guide walls.

[(*a*) 234 rad/m, (*b*) 0.027 m, (*c*) 2.68×10^8 m/s, (*d*) 337.4 Ω, (*e*) 0.084 Np/m or 0.73 dB/m, (*f*) 0.0526 Np/m or 0.457 dB/m]

17. Find the first five resonances of an air-filled rectangular cavity with dimensions of $a = 5$ cm, $b = 4$ cm and $c = 10$ cm ($c > a > b$).

$$\begin{bmatrix} f_{r_{101}} = 3.335\text{ GHz} & TE_{101} \\ f_{r_{011}} = 4.040\text{ GHz} & TE_{011} \\ f_{r_{102}} = 4.243\text{ GHz} & TE_{102} \\ f_{r_{110}} = 4.800\text{ GHz} & TM_{110} \\ f_{r_{111}} = 5.031\text{ GHz} & TE_{111}, TM_{111} \end{bmatrix}$$

18. An air-filled copper circular waveguide ($a = 5$ mm, $\sigma = 5.8 \times 10^7$ mho/m) is operated at 30 GHz. Determine the attenuation constant in dB/m for the TM_{01} mode. [0.3231 dB/m]

19. A resonant cavity with dimensions $a = 10$ cm, $b = 4$ cm and $c = 5$ cm is filled with a lossless dielectric ($\mu_r = 1$, $\varepsilon_r = 3$) and is made of copper ($\sigma = 5.8 \times 10^7$ mho/m). Find the resonant frequency and the quality factor for the dominant mode. [1.936 GHz, 10916.21]

20. A rectangular cavity resonator excited by TE_{101} mode, at 20 GHz, has the dimensions $a = 2$ cm, $b = 1$ cm. Calculate the length of the cavity. [0.809 cm]

21. A cubical cavity resonator made of copper ($\sigma = 5.8 \times 10^7$ mho/m) is to be operated at 10 GHz. Find the dimensions of the cavity, its quality factor and the bandwidth if it is operated in the dominant mode. [10693, 0.935 MHz]

Review Questions

[NOTE: * marked questions are important university questions.]

***1.** Explain, from fundamental principles, why a waveguide behaves as a high-pass filter.

2. Show that the waveguide acts as a high-pass filter.

3. Give the theory of propagation of microwaves between two parallel conducting planes considering the TE_n mode. Show, from inference conditions that the cut-off wavelength is given as $\lambda_c = \frac{2d}{n}$, where d is the separation between the planes.

4. **(a)** Derive an expression for the attenuation factor for the TM_1 wave between parallel conducting planes.

(b) Verify that the attenuation is a minimum at a frequency which is $\sqrt{3}$ times the cut-off frequency.

5. Discuss the propagation of *TE* and *TM* mode in a rectangular waveguide. Can *TEM* wave propagate in a rectangular waveguide? If not, why?

6. Explain wave impedance of a rectangular waveguide and derive the expression for the wave impedance of *TE*, *TM* and *TEM* waves.

7. Show that the wave impedance is real above a certain frequency and imaginary below that. What is the implication of the result?

8. A rectangular waveguide is propagating in the TE_{11} mode. Draw its field pattern. How do you extract energy from a wave propagating in this mode of propagation?

***9.** Show that a rectangular or circular metal waveguide cannot carry the *TEM* mode.

***10.** Show that a closed metal waveguide cannot support the *TEM* mode.

***11.** Explain why a *TEM* mode propagation is not possible in a single conductor waveguide.

12. Explain the terms "cut-off frequency" and "guide wavelength".

13. Explain why the guide wavelength in a rectangular waveguide is greater than the free-space wavelength.

14. What is dominant mode in a rectangular waveguide and why is it called so?

15. What is meant by the term "dominant mode" in a rectangular metal waveguide? Sketch the field distribution inside the waveguide in this mode.

16. What are dominant mode and degenerate modes in rectangular waveguide?

***17.** Show that at frequencies much higher than the cut-off frequency, the Q of a rectangular guide carrying the dominant TE_{10} wave approaches the value

$$Q \to b\alpha_m$$

where, $\alpha_m = \sqrt{\omega\mu_m\sigma_m/2}$ is the attenuation factor for a wave propagating in the metal of the guide walls. Assume $\mu_m = \mu_0$.

18. What are the respective advantages and disadvantages of waveguides of rectangular and circular cross section? What is meant by H_{01} mode of transmission in a rectangular waveguide? What factors influence the choice of the dimensions for a rectangular waveguide used to transmit an H_{01} wave?

***19.** Show the following relationship between guide wavelength and group velocity in an arbitrary air-filled waveguide: $v_g \lambda_g = c\lambda$, where $\lambda_g = 2\pi/\beta$ and λ is the free-space wavelength. Moreover, show that the λ and λ_g are related to the cut-off wavelength λ_c by:

$$\frac{1}{\lambda^2} = \frac{1}{\lambda_g^2} + \frac{1}{\lambda_c^2}$$

20. What is meant by cavity resonator? Derive the expression for the resonant frequency of a rectangular cavity resonator.

21. Derive the expression for electric and magnetic fields for the TE_{101} mode of rectangular waveguide cavity resonator. Draw the field lines.

22. Derive an expression for Q of a cavity resonator working in TE_{011} mode.

23. Find out the expression for the Q factor of a rectangular cavity resonator excited in TE_{011} mode.

24. What is meant by a dielectric slab waveguide? Derive field expressions both for TE and TM waves in a dielectric slab waveguide.

Multiple Choice Questions

1. The waveguide (a = 1.5 cm, b = 1 cm) is loaded with a dielectric (ε_r = 4). Which one of the following is correct?
The 8 GHz signal will
(a) pass through the waveguide
(b) not pass through the waveguide
(c) be absorbed in the guide
(d) none of the above.

2. For a rectangular waveguide of dimensions:
$a\sqrt{3}$ cm × a cm, the cut-off frequency for the TE_{10} mode is 2 GHz. What is the cut-off frequency for TM_{11} mode in the waveguide?
(a) 1 GHz
(b) 3.46 GHz
(c) 4 GHz
(d) 6 GHz

3. In an air-filled waveguide of dimensions a cm × b cm, at a given frequency, the longitudinal component of electric field of TM_{32} mode is of the form

$$E_z = 20 \sin (60\pi x) \sin (100\pi y)$$

Which form would E_z have for the lowest order TM mode?
(a) $E_z = 20 \sin (20\pi x)$
(b) $E_z = 20 \sin (20\pi y)$
(c) $E_z = 20 \sin (20\pi x) \sin (50\pi y)$
(d) $E_z = 20 \sin (20\pi x) \sin (100\pi y)$

4. The cut-off frequency of the dominant mode of a rectangular waveguide having aspect ratio more than 2 is 10 GHz. The inner broad wall dimension is given by
(a) 3 cm
(b) 2 cm
(c) 1.5 cm
(d) 2.5 cm.

5. The dominant mode in a circular waveguide is a
(a) TEM mode
(b) TM_{01} mode
(c) TE_{21} mode
(d) TE_{11} mode.

6. In a waveguide, the evanescent modes are said to occur if
(a) the propagation constant is real
(b) the propagation constant is imaginary
(c) Only the TEM waves propagate
(d) The signal has a constant frequency.

7. In a rectangular waveguide with broader dimension a and narrow dimension b, the dominant mode of microwave propagation would be
(a) TE_{10} (b) TM_{10} (c) TE_{01} (d) TM_{01}

8. If the height of a waveguide is halved, its cut-off wavelength will
(a) be halved (b) be doubled
(c) remain unchanged (d) be one-fourth of the previous value.

9. A rectangular waveguide measures 3×4.5 cm internally and has a 9 GHz signal propagated in it. The cut-off wavelength for TE_{10} mode is
(a) 5 cm (b) 10 cm (c) 15 cm (d) 9 cm

10. Which of the following is ***not*** possible in a circular waveguide?
(a) TE_{10} (b) TE_{01} (c) TE_{11} (d) TE_{12}

11. When a particular mode is excited in a waveguide, there appears an extra electric component in the direction of propagation. The resulting mode is
(a) transverse-electric (b) transverse-magnetic
(c) longitudinal (d) transverse-electromagnetic.

12. For a wave propagating in an air-filled rectangular waveguide
(a) guided wavelength is never less than the free space wavelength
(b) wave impedance is never less than the free space impedance
(c) *TEM* mode is possible if the dimensions of the waveguide are properly chosen
(d) propagation constant is always a real quantity.

13. A cylindrical cavity resonator has diameter of 24 mm and length 20 mm. The dominant mode and the lowest frequency band operated are as
(a) TE_{111} and X-band (b) TM_{111} and C-band
(c) TM_{011} and Ku-band (d) TM_{010} and X-band.

14. For a hollow waveguide, the axial current must necessarily be
(a) a combination of conduction and displacement currents
(b) conduction current only
(c) time-varying conduction current and displacement current
(d) displacement current only.

15. A cylindrical cavity resonates at 9 GHz in the TE_{111} mode. The bandwidth 3 dB is measured to be 2.4 MHz. The Q of the cavity at 9 GHz is
(a) $\dfrac{9000}{2.4\sqrt{3}}$ (b) $\dfrac{9000}{2.4\sqrt{2}}$ (c) $\dfrac{9000 \times 2.4\sqrt{3}}{\sqrt{2}}$ (d) $\dfrac{9000}{2.4}$

16. A cavity resonator can be represented by
(a) an *LC* circuit (b) an *LCR* circuit (c) a lossy inductor (d) a lossy capacitor.

17. The cut-off wavelength λ_c for TE_{20} mode for a standard rectangular waveguide is
(a) $\dfrac{2}{a}$ (b) $2a$ (c) a (d) $2a^2$

18. A cylindrical cavity operating in TE_{111} mode has a 3 dB bandwidth of 2.4 MHz and its quality factor is 400. Its resonant frequency would be
(a) 9.6 GHz (b) $\dfrac{9.6}{\sqrt{2}}$ GHz (c) $\dfrac{9.6}{\sqrt{3}}$ GHz (d) $\dfrac{9.6}{\sqrt{6}}$ GHz

19. Phase velocity v_p and the group velocity v_g in a waveguide (c is velocity of light) are related as

(a) $v_p v_g = c^2$ (b) $v_p + v_g = c$

(c) $\dfrac{v_p}{v_g} = \text{constant}$ (d) $v_p + v_g = \text{a constant}$

Answers

1. (a)	**2.** (c)	**3.** (c)	**4.** (c)	**5.** (d)	**6.** (a)	**7.** (a)	**8.** (c)
9. (d)	**10.** (a)	**11.** (b)	**12.** (a)	**13.** (a)	**14.** (d)	**15.** (d)	**16.** (b)
17. (c)	**18.** (a)	**19.** (a)					

SOME TYPICAL SHORT ANSWER TYPE QUESTIONS WITH ANSWERS

1. **State the assumptions made while defining Coulomb's law.**

   Or

   Explain the limitations of Coulomb's law.

Answer: Limitations (Assumptions) of Coulomb's Law:

1. The charges must be point charges; it is very difficult to apply Coulomb's law for charges of arbitrary shape.
2. The charges must be stationary.

2. **Show that the work done in moving a charge from one point to another in an electrostatic field is independent of the path between the points.**

Answer: We will consider a point charge Q, be moved from a point A to another point B in an electric field $\vec{E}$.

By Coulomb's law, the force on Q is, $\vec{F} = Q\vec{E}$

$\therefore$ Work done for a displacement of $d\vec{l}$ is $dW = -\vec{F} \cdot d\vec{l} = -Q\vec{E} \cdot d\vec{l}$

Hence, the total work done in moving the charge Q from A to B, i.e., the potential energy required is

$$W = -Q\int_A^B \vec{E} \cdot d\vec{l}$$

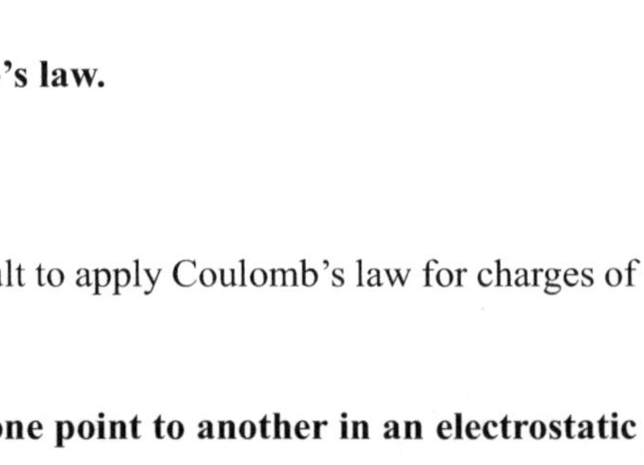

Fig. 1 *Potential difference due to a uniform electric field*

The potential energy per unit charge (W/Q) is known as the *potential difference* between the two points A and B, denoted by V_{AB}.

$$\therefore \qquad V_{AB} = \frac{W}{Q} = -\int_A^B \vec{E} \cdot d\vec{l}$$

Since $V = -\int_l \vec{E} \cdot d\vec{l}$ and $\vec{E}$ is in the radial direction, any contribution from a displacement in θ or ϕ direction is cancelled out by the dot product. Hence, $\vec{E} \cdot d\vec{l} = Edl \cos\theta = Edr$.

Thus, *the potential is independent of the path.*

For a closed path, $\oint_l \vec{E} \cdot d\vec{l} = 0$. Applying Stokes' theorem, $\oint_l \vec{E} \cdot d\vec{l} = \int_S (\nabla \times \vec{E}) \cdot d\vec{S} = 0$.

$$\therefore \qquad \nabla \times \vec{E} = 0$$

Thus, electrostatic field is conservative or irrotational.

3. (a) A conducting body is in the electric field of static charges. Explain why the net electric field at any point inside the conducting body will be zero.

(b) Use the result of (a) to show that

(i) the net volume charge density at any point inside the conductor is zero, and

(ii) the conductor is an equipotential body.

Or

For a conducting body in the electric field of static charges, explain what will be the

(a) net electric field inside the conductor, and

(b) volume charge density at any point inside the conductor.

Answer:

(a) When an electric field, $\vec{E}$ is applied to a conductor, the electrons will experience a force $(\vec{F} = -e\vec{E})$. As the electrons are not free in space, they will not be accelerated by the field; but they suffer constant collisions with atomic lattice and drift from one atom to another. This is called *drifting* of electrons.

After some time, the electrons will attain a constant average velocity, called *drift velocity* $(\vec{U})$ which is directly proportional to the applied field ($\vec{U} = -\mu\vec{E}$; μ is the mobility of electrons). The current associated with this drifting is known as the *drift current* or *conduction current* ($\vec{J} = \sigma\vec{E}$; σ is the conductivity of the conductor).

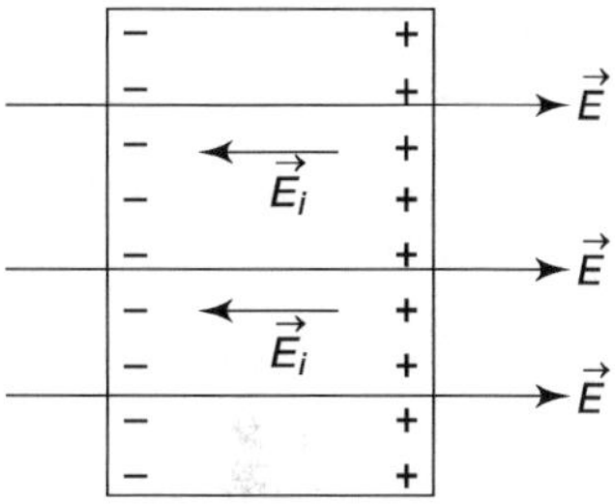

Fig. 2 *Isolated conductor under static electric field* ($\vec{E} = 0$, $\rho_v = 0$; $V_{12} = 0$) inside conductor

For a perfect conductor, the conductivity is infinite $(\sigma \to \infty)$. As the conduction current is $(\vec{J} = \sigma\vec{E})$, to maintain a finite current density $(\vec{J})$, the **electric field** $(\vec{E})$ **must be zero inside an isolated conductor.** All positive charges will move along the direction of $\vec{E}$ and negative charges will move in the opposite direction. Thus, all charges will accumulate on the surface and these induced surface charges will set up an internal induced field, $\vec{E}_i$; which cancels the externally applied field $\vec{E}$.

(b) (i) As field is zero inside a conductor $(\vec{E} = 0)$, by Gauss' law of electrostatics, the volume charge density at any point inside the conductor will also be zero $\left(\nabla \cdot \vec{E} = \dfrac{\rho}{\varepsilon};\ \because \vec{E} = 0 \Rightarrow \rho = 0\right)$.

(ii) As field is zero inside a conductor $(\vec{E} = 0)$, from the relation between field and potential, the potential inside a conductor is constant ($\vec{E} = -\nabla V$; $\because \vec{E} = 0 \Rightarrow V$ must be constant). Hence, a conductor is an equipotential body.

4. Derive Gauss' Law from Coulomb's Law.

Answer: Derivation of Gauss' Law from Coulomb's Law

Gauss' law can be derived from Coulomb's law, which states that the electric field due to a stationary point charge is

$$\vec{E} = \frac{Q}{4\pi\varepsilon r^2}\hat{a}_r$$

where

$\hat{a}_r$ is the radial unit vector,

r is the radius, $|\vec{r}|$,

q is the charge of the particle, which is assumed to be located at the origin.

Using the expression from Coulomb's law, we get the total field at $\vec{r}$ by using an integral to sum the field at $\vec{r}$ due to the infinitesimal charge at each other point $\vec{r}'$ in space, to give

$$\vec{E} = \frac{1}{4\pi\varepsilon}\int_v \frac{\rho(\vec{r}')(\vec{r}-\vec{r}')}{|\vec{r}-\vec{r}|^3}dv$$

where ρ is the charge density. If we take the divergence of both sides of this equation with respect to $\vec{r}$, and use the known theorem

$$\nabla\cdot\left(\frac{\vec{r}'}{|\vec{r}'|^3}\right) = 4\pi\delta(r')$$

where $\delta(r')$ is the Dirac delta function, the result is

$$\nabla\cdot\vec{E} = \frac{1}{\varepsilon}\int_v \rho(\vec{r}')\delta(\vec{r}-\vec{r}')dv$$

Using the shifting property of the Dirac delta function, we get

$$\nabla\cdot\vec{E} = \frac{\rho}{\varepsilon}$$

which is the differential form of Gauss' law.

5. **Derive Coulomb's law from Gauss' law.**

Answer: Derivation of Coulomb's Law from Gauss' Law

Gauss' law provides information only about the divergence of the electric field intensity, $\vec{E}$ and does not give any information about the curl $\vec{E}$. For this reason, Coulomb's law cannot be derived from Gauss' law alone.

However, we assume that the electric field from a stationary point charge has spherical symmetry. With this assumption, which is exactly true if the charge is stationary, and approximately true if the charge is in motion, Coulomb's law can be proved from Gauss' law. We consider a spherical surface of radius r, centered at a point charge Q. Applying Gauss' law in integral form, we have

$$\oint_S \vec{E}\cdot d\vec{S} = \frac{Q}{\varepsilon}$$

By the assumption of spherical symmetry, the integrand is a constant and can be taken out of the integral as

$$4\pi r^2 \hat{a}_r \vec{E} = \frac{Q}{\varepsilon}$$

where $\hat{a}_r$ is a unit vector directed radially away from the charge. Again by spherical symmetry, $\vec{E}$ is also in radially outward direction, and so we get

$$\vec{E} = \frac{Q}{4\pi\varepsilon r^2}\hat{a}_r$$

If another point charge q is placed on the surface, the force on that charge due to the charge Q is given as

$$\vec{F} = q\vec{E} = \frac{Qq}{4\pi\varepsilon r^2}\hat{a}_r$$

which is essentially equivalent to Coulomb's law.

Thus, the inverse-square law dependence of the electric field in Coulomb's law follows from Gauss' law.

6. Prove that electric field lines are perpendicular to the equipotential surfaces.

Answer: Since $\vec{E} = -\nabla V$, it can be shown that the direction of $\vec{E}$ is always perpendicular to the equipotential through the point. Below we give a proof.

Let the potential at a point $P(x, y, z)$ be $V(x, y, z)$. The difference in potential at a neighbouring point $P(x + dx, y + dy, z + dz)$ is given as

$$dV = V(x+dx, y+dy, z+dz) - V(x, y, z)$$
$$= \left[V(x, y, z) + \frac{\partial V}{\partial x}dx + \frac{\partial V}{\partial y}dy + \frac{\partial V}{\partial z}dz\right] - V(x, y, z)$$
$$= \frac{\partial V}{\partial x}dx + \frac{\partial V}{\partial y}dy + \frac{\partial V}{\partial z}dz$$

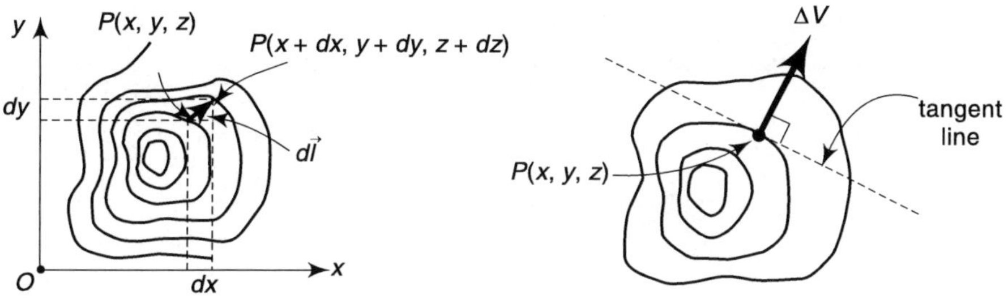

Fig. 3 *Change in V when moving from one equipotential curve to another*

With the displacement $P(x, y, z)$ vector given as, $\vec{dl} = dx\hat{a}_x + dy\hat{a}_y + dz\hat{a}_z$, we can rewrite dV as

$$dV = \frac{\partial V}{\partial x}dx + \frac{\partial V}{\partial y}dy + \frac{\partial V}{\partial z}dz = \left(\frac{\partial V}{\partial x}\hat{a}_x + \frac{\partial V}{\partial y}\hat{a}_y + \frac{\partial V}{\partial z}\hat{a}_z\right)\cdot(dx\hat{a}_x + dy\hat{a}_y + dz\hat{a}_z)$$
$$= (\nabla V)\cdot\vec{dl}$$
$$= -\vec{E}\cdot\vec{dl}$$

If the displacement $\vec{dl}$ is along the tangent to the equipotential curve through $P(x, y, z)$, then $dV = 0$ because V is constant everywhere on the curve. This implies that $\vec{E}$ is perpendicular to V along the equipotential curve.

7. Calculate the force between the plates of a capacitor.

Or

Show that the force experienced by the plate carrying the charge $+q$ of an isolated air-filled parallel-plate capacitor is $-\dfrac{q^2}{2\varepsilon_0 A}$, where A is the plate area.

Answer: Let d be the separation distance between the plates.

$\therefore$ Capacitance of the capacitor is $C = \dfrac{\varepsilon_0 A}{d}$

$\therefore$ Electrostatic energy is given as $W = \dfrac{1}{2}\dfrac{q^2}{C} = \dfrac{1}{2}\dfrac{q^2}{\dfrac{\varepsilon_0 A}{d}} = \dfrac{q^2 d}{2\varepsilon_0 A}$

$\therefore$ Force experienced by the plate is given as $F = -\dfrac{dW}{dd} = -\dfrac{d}{dd}\left(\dfrac{q^2 d}{2\varepsilon_0 A}\right) = -\dfrac{q^2}{2\varepsilon_0 A}$

$$\therefore \qquad \boxed{F = -\frac{q^2}{2\varepsilon_0 A}}$$

8. Explain how the current flowing through a capacitor differs from the normal conduction current.

Answer: The conduction current model does not characterise the capacitor current. The ideal capacitor is characterised by large, closely-spaced plates separated by a perfect insulator ($\sigma_C = 0$) so that no charge actually passes throughout the dielectric $[\vec{J}_C(t) = \sigma_C \vec{E}_C(t)]$. The capacitor current measured in the connecting wires of the capacitor is caused by the charging and discharging the capacitor plates. Let $Q(t)$ be the total capacitor charge on the positive plate.

Hence, the capacitor current, also termed as the *displacement current* is given as

$$i_C(t) = i_d(t) = \frac{dQ(t)}{dt} = C\frac{dv(t)}{dt} = \frac{\varepsilon A}{d}\frac{dv(t)}{dt} = \varepsilon A\frac{d}{dt}\left[\frac{v(t)}{d}\right] = \varepsilon A\frac{dE(t)}{dt} = A\frac{d}{dt}[\varepsilon E(t)] = A\frac{dD(t)}{dt}$$

So, the *displacement current density* is given as

$$\vec{J}_d(t) = \frac{d\vec{D}(t)}{dt}$$

As $\vec{D}$ may vary with space, the displacement current density is written as

$$\boxed{\vec{J}_d(t) = \frac{\partial \vec{D}(t)}{\partial t}}$$

9. Show that the displacement through a capacitor is equal to the conduction current if the supply voltage is $v = V_m \sin \omega t$.

Answer: Here the applied voltage is, $v = V_m \sin \omega t$

We consider a parallel plate capacitor of capacitance given as, $C = \dfrac{\varepsilon A}{d}$

where A is the area of the plate and d is the distance between the plates.

The conduction current is given as,

$$I_d = C\frac{dv}{dt} = \frac{\varepsilon A}{d}\frac{d}{dt}[V_m \sin \omega t] = V_m \omega \frac{\varepsilon A}{d}\cos \omega t \qquad \text{(i)}$$

However, as the medium between the plates is dielectric, there is no conduction current in a capacitor. A displacement current may be considered to flow which is given as

$$I_d = \varepsilon A\frac{dE}{dt} = \varepsilon A\frac{d}{dt}\left[\frac{v}{d}\right] = \frac{\varepsilon A}{d}\frac{d}{dt}[V_m \sin \omega t] = V_m \omega \frac{\varepsilon A}{d}\cos \omega t \qquad \text{(ii)}$$

From (*i*) and (*ii*), it is seen that the displacement in a capacitor is equal to the conduction current.

10. Show that the solution of Poisson's equation $V = \int \dfrac{\rho dv}{4\pi\varepsilon r}$ satisfies the Poisson's equation, $\nabla^2 V = -\dfrac{\rho}{\varepsilon}$.

Answer: We have the solution $\qquad V = \int \dfrac{\rho dv}{4\pi\varepsilon r}$

Taking Laplacian on both sides,

$$\nabla^2 V = \nabla^2 \int \frac{\rho dv}{4\pi\varepsilon r} = \int \frac{\rho}{4\pi\varepsilon} \nabla^2 \left(\frac{1}{r}\right) dv$$

With $r \neq 0$, $$\nabla^2 \left(\frac{1}{r}\right) = \frac{1}{r}\frac{d^2}{dr^2}\left(r \cdot \frac{1}{r}\right) = 0 \quad \text{(i)}$$

At $r = 0$, it is defined. However, we can find the value of $\int \nabla^2 \left(\frac{1}{r}\right) dv$ at $r = 0$, by the following procedure:

$$\int \nabla^2 \left(\frac{1}{r}\right) dv = \lim_{\alpha \to 0} \int \nabla^2 \left(\frac{1}{\sqrt{r^2+\alpha^2}}\right) dv$$

Now, $$\nabla^2 \left(\frac{1}{\sqrt{r^2+\alpha^2}}\right) = \frac{1}{r}\frac{d^2}{dr^2}\left(\frac{r}{\sqrt{r^2+\alpha^2}}\right) = -\frac{3\alpha^2}{(r^2+\alpha^2)^{5/2}}$$

$\therefore$ $$\lim_{\alpha \to 0} \int \nabla^2 \left(\frac{1}{\sqrt{r^2+\alpha^2}}\right) dv = \lim_{\alpha \to 0} \iiint -\frac{3\alpha^2 r^2}{(r^2+\alpha^2)^{5/2}} \sin\theta \, d\theta \, d\phi \, dr$$

$$= \lim_{\alpha \to 0} \left[-12\pi \int_0^\infty \frac{\alpha^2 r^2}{(r^2+\alpha^2)^{5/2}} dr\right] \quad \text{(ii)}$$

Now,

$$\int_0^\infty \frac{\alpha^2 r^2}{(r^2+\alpha^2)^{5/2}} dr = \int_0^{\pi/2} \frac{\alpha^2 \alpha^2 \tan^2\theta \alpha \sec^2\theta d\theta}{\alpha^5 (1 + an^2\theta)^{5/2}} \qquad \left\{\begin{array}{l}\text{Let, } r = \alpha \tan\theta \\ \therefore dr = \alpha \sec^2\theta d\theta\end{array}\right\}$$

$$= \int_0^{\pi/2} \frac{\tan^2\theta \sec^2\theta d\theta}{\sec^5\theta} = \int_0^{\pi/2} \frac{\sin^2\theta}{\cos^2\theta} \cos^3\theta d\theta = \int_0^{\pi/2} \sin^2\theta \cos\theta d\theta$$

$$= \int_0^1 p^2 dp \qquad \{\text{Let, } \sin\theta = p \quad \therefore \cos\theta d\theta = dp\}$$

$$= \left[\frac{p^3}{3}\right]_0^1 = \frac{1}{3}$$

From Eq (*ii*), $$\lim_{\alpha \to 0} \int \nabla^2 \left(\frac{1}{\sqrt{r^2+\alpha^2}}\right) dv = \lim_{\alpha \to 0} \left[-12\pi \int_0^\infty \frac{\alpha^2 r^2}{(r^2+\alpha^2)^{\frac{5}{2}}} dr\right] = -12\pi \times \frac{1}{3} = 4\pi$$

Hence, $$\nabla^2 V = \frac{\rho}{4\pi\varepsilon}(-4\pi) = -\frac{\rho}{\varepsilon}$$

Thus, it satisfies the Poisson's equation.

NOTE

Equations (i) and (ii) can be represented by a single relation,

$$\nabla^2 \left(\frac{1}{r}\right) = -4\pi\delta(r).$$

11. Differentiate between potential and EMF.

Answer: Differences between Potential (V) and EMF (ξ)

1. Potential field, i.e., electric field generated by static charges, is conservative; but emf-producing field is non-conservative.

$$\therefore \qquad \oint_l \vec{E}_C \cdot d\vec{l} = 0; \text{ but, } \oint_l \vec{E}_e \cdot d\vec{l} \neq 0 = \xi(emf)$$

2. Electric field produced by charges is not able to maintain a steady current; but emf-producing field can maintain a steady current.
3. Potential (V) is the negative of the line integral of the static field $\vec{E}_C$ while emf ξ is the line integral of $\vec{E}_e$. Thus, between two points a and b,

$$V_{ab} = (V_b - V_a) = -\int_a^b \vec{E}_C \cdot d\vec{l} \qquad \text{and} \qquad \xi_{ab} = \int_a^b \vec{E}_e \cdot d\vec{l}$$

Here, V_{ab} is independent of the path of integration between a and b, but ξ_{ab} is dependent on the path.

12. 'Any initial charge density in a conductor dissipates in a characteristic time $\tau = \varepsilon/\sigma$, where ε is the permittivity and σ is the electrical conductivity of the material'. Establish this statement and discuss how τ determines the quality of a conductor.

Or

Show that a charge placed anywhere in a conducting medium of conductivity σ and permittivity ε decays exponentially with a time constant ε/σ.

Answer: We consider a conductor carrying a surface current density $\vec{J}$, flowing perpendicular to the area $d\vec{S}$, having volume charge density ρ_v.

Thus, the total current coming out of the closed surface is

$$I = \oint_S \vec{J} \cdot d\vec{S}$$

Now, charges cannot be created or destroyed. Since, the current is simply the motion of charge, the total current flowing out of some volume must be equal to the rate of decrease of charge within the volume.

$$\therefore \qquad I = -\frac{dQ}{dt} = -\frac{d}{dt}\int_v \rho_v dv$$

$$\therefore \qquad \oint_S \vec{J} \cdot d\vec{S} = -\int_v \frac{\partial \rho_v}{\partial t} dv$$

This equation is known as *the integral form of the equation of continuity*.

From this equation, we get

$$\oint_S \vec{J} \cdot d\vec{S} = -\int_v \frac{\partial \rho_v}{\partial t} dv$$

$$\Rightarrow \qquad \int_v \nabla \cdot \vec{J} dv = -\int_v \frac{\partial \rho_v}{\partial t} dv$$

$$\therefore \qquad \nabla \cdot \vec{J} = -\frac{\partial \rho_v}{\partial t}$$

From the above equation, we have

$$-\frac{\partial \rho_v}{\partial t} = \nabla \cdot \vec{J} = \nabla \cdot (\sigma \vec{E}) = \sigma \nabla \cdot \vec{E} = \sigma \nabla \cdot \left(\frac{\vec{D}}{\varepsilon}\right) = \frac{\sigma}{\varepsilon} \nabla \cdot \vec{D} = \frac{\sigma}{\varepsilon} \rho_v$$

(Here, we have used, Ohm's law: $\vec{J} = \sigma \vec{E}$; Gauss' law: $\nabla \cdot \vec{D} = \rho_v$ and the constitutive relation: $\vec{D} = \varepsilon \vec{E}$)

$$\therefore \qquad \frac{\partial \rho_v}{\rho_v} = -\frac{\sigma}{\varepsilon} \partial t$$

Integrating $$\int_{\rho_0}^{\rho_v} \frac{\partial \rho_v}{\rho_v} = -\int_0^t \frac{\sigma}{\varepsilon} \partial t$$

where ρ_0 is the initial charge density at time $t = 0$.

$$\therefore \qquad \boxed{\rho_v = \rho_0 e^{-(\sigma/\varepsilon)t} = \rho_0 e^{-t/\tau}} \qquad \textbf{(3.29)}$$

This equation shows that whenever some charge is introduced at some interior point of a material, there is a decay of volume charge density ρ_v. This decay is associated with the movement of charge from the interior point to the surface of the material.

The time constant of the decay, called the *relaxation time* or *rearrangement time*, is given as

$$\boxed{\tau = \frac{\varepsilon}{\sigma}}$$

The relaxation time is inversely proportional to the conductivity of the medium. This means that the value of τ is *very small for good conductors* and *very large for a good dielectric.*

For example, for copper, $\sigma = 5.8 \times 10^7$ mho/m, and $\varepsilon = 1$, so that the value of relaxation time for copper is

$$\tau = \frac{\varepsilon}{\sigma} = \frac{\varepsilon_0 \varepsilon_r}{\sigma} = \frac{8.854 \times 10^{-12} \times 1}{5.8 \times 10^7} = 1.53 \times 10^{-19} \text{ seconds}$$

13. Establish the relation $\frac{G}{\sigma} = \frac{C}{\varepsilon}$, where G and C are respectively the conductance and capacitance between the two electrodes; σ and ε are respectively the conductivity and permeability of the intervening medium.

Or

Deduce the relation $G = C\frac{\sigma}{\varepsilon}$ and show that the total electrical conductance between any configuration of conductors embedded in a conducting medium of σ may be obtained by replacing permittivity ε by σ in the expression for the capacitance of the configuration where they are embedded in a dielectric of permittivity ε.

Answer:

By definition of capacitance, $$C = \frac{Q}{V} = \frac{\int_S \vec{Q}_S \cdot d\vec{S}}{\int_1^2 \vec{E} \cdot d\vec{l}}$$

But, at the surface of conductors, by boundary condition, $\varepsilon E_n = Q_s$

$$\therefore \qquad C = \frac{\varepsilon \int_S \vec{E}_n \cdot d\vec{S}}{\int_1^2 \vec{E} \cdot d\vec{l}} \qquad \text{(i)}$$

For the current flow, the conductance is

$$G = \frac{I}{V} = \frac{\int_S \vec{J}_n \cdot d\vec{S}}{\int_1^2 \vec{E} \cdot d\vec{l}} = \frac{\sigma \int_S \vec{E}_n \cdot d\vec{S}}{\int_1^2 \vec{E} \cdot d\vec{l}} \qquad \text{(ii)}$$

Combining (*i*) and (*ii*), we get $\qquad G = C \times \dfrac{\sigma}{\varepsilon}$

$$\therefore \qquad \boxed{\frac{G}{\sigma} = \frac{C}{\varepsilon}}$$

Thus, the conductance may be obtained at once by replacing ε by σ in the expression of the capacitance.

NOTE

Insulation resistance of the configuration is, $R = \dfrac{1}{G} = \dfrac{1}{C} \times \dfrac{\varepsilon}{\sigma}$.

14. A certain volume of dielectric has a polarisation $\vec{P}$ *C/m²*. Write the integral expression for the potential at any point due to this dielectric. Explain the different terms in this expression through a figure.

Answer: We consider a block of dielectric with polarisation $\vec{P}$ (dipole moment per unit volume).

We want to calculate the potential and field at an exterior point *P*.

By the relation of potential for a dipole,

$$dV = \frac{1}{4\pi\varepsilon_0} \frac{\vec{P} \cdot \hat{a}_r}{r^2} d\tau' \qquad \text{(i)}$$

for an elemental volume $d\tau'$ at (x', y', z') at the field point (x, y, z).

$$\therefore \qquad V = \frac{1}{4\pi\varepsilon_0} \int_v \frac{\vec{P} \cdot \hat{a}_r}{r^2} d\tau' \qquad \text{(ii)}$$

where, $r = (x - x')\hat{i} + (y - y')\hat{j} + (z - z')\hat{k} \qquad \therefore \quad \dfrac{\hat{a}_r}{r^2} = \nabla'\left(\dfrac{1}{r}\right)$

From Eq. (ii),

$$\therefore \qquad V = \frac{1}{4\pi\varepsilon_0} \int_v \vec{P} \cdot \nabla'\left(\frac{1}{r}\right) d\tau' \qquad \text{(iii)}$$

By the vector identity for a scalar S and a vector $\vec{V}$,

$$\nabla \cdot (S\vec{V}) \equiv S(\nabla \cdot \vec{V}) + \vec{V} \cdot \nabla S$$

Putting $S = \left(\frac{1}{r}\right)$, $\vec{V} = \vec{P}$, we get

$$\nabla \cdot \left(\frac{\vec{P}}{r}\right) = \frac{1}{r}(\nabla \cdot \vec{P}) + \vec{P} \cdot \nabla\left(\frac{1}{r}\right) \Rightarrow \vec{P} \cdot \nabla\left(\frac{1}{r}\right) = \nabla \cdot \left(\frac{\vec{P}}{r}\right) - \frac{1}{r}(\nabla \cdot \vec{P})$$

From Eq. (iii), we get

$$V = \frac{1}{4\pi\varepsilon_0}\int_v \left[\nabla' \cdot \left(\frac{\vec{P}}{r}\right) - \frac{1}{r}(\nabla' \cdot \vec{P})\right] d\tau'$$

$$= \frac{1}{4\pi\varepsilon_0}\int_S \frac{\vec{P} \cdot d\vec{S}}{r} - \frac{1}{4\pi\varepsilon_0}\int_v \frac{1}{r}(\nabla' \cdot \vec{P}) d\tau'$$

(By divergence theorem)

Both of the terms in the above equation have the form of potentials produced by the charge distribution; i.e.,

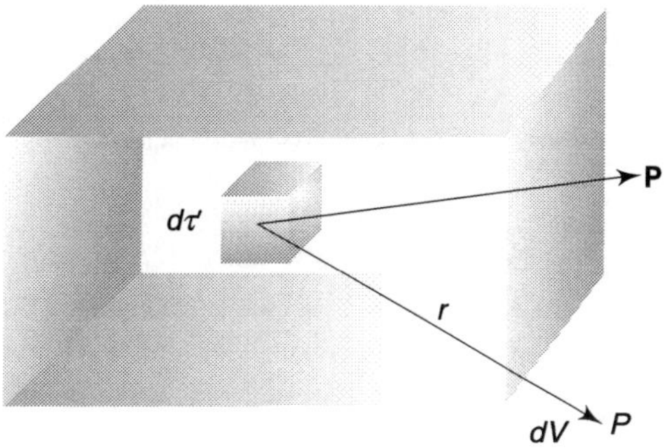

Fig. 4 *Block of dielectric with polarisation* $(\vec{P})$

- a *surface charge density*, $\sigma_b = \vec{P} \cdot \hat{n}$
- a *volume charge density*, $\rho_b = -\nabla'(\vec{P})$

Thus, the potential at any point due to this dielectric is given as

$$\boxed{V = \frac{1}{4\pi\varepsilon_0}\left[\int_S \frac{\sigma_b dS}{r} + \int_v \frac{\rho_b d\tau'}{r}\right]}$$

Significance of Terms:

Bound surface charge density σ_b: We consider a small volume inside the dielectric.

Let,

$\vec{E}$—electric field due to combined effect of an external field and the field due to the dipoles,

l—separation distance between positive and negative charges due to the influence of $\vec{E}$,

dS—elemental surface,

$d\tau$—elemental volume $(= \vec{l} \cdot d\vec{S})$, and

dQ—amount of charge crossing the elemental surface dS

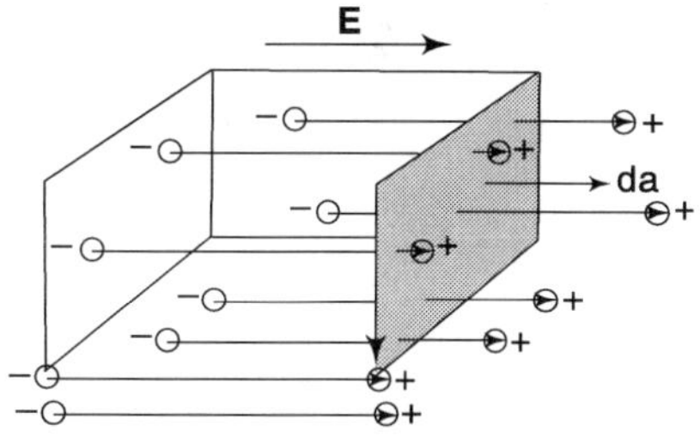

Fig. 5 *Illustration of bound charge densities*

$\therefore \quad dQ = NQ\vec{l} \cdot d\vec{S} \qquad Q\vec{l} = \vec{p} \qquad NQ\vec{l} = \vec{P}$

$\therefore \quad dQ = \vec{P} \cdot d\vec{S}$

If dS is on the surface of the material, this charge accumulates there in a layer of thickness $\vec{l} \cdot \hat{n}$ (which is small, of molecular dimensions) and the charge can be treated as a surface layer with density,

$$\boxed{\sigma_b = dQ/dS = \vec{P} \cdot \hat{n}}$$

Bound volume charge density ρ_b:

Total charge flowing out of the surface bounding volume τ, is the integral of this over the surface, i.e.,

$$Q = \int_S \vec{P} \cdot d\vec{S}$$

and net charge remaining within is $-Q$.

If the density of this remaining charge is ρ_b, then

$$\int_v \rho_b d\tau = -Q = -\int_S \vec{P} \cdot d\vec{S} = -\int_v (\nabla \cdot \vec{P}) d\tau$$

$$\therefore \qquad \boxed{\rho_b = -\nabla \cdot \vec{P}}$$

i.e., the bound charge density is numerically equal to minus the divergence of the polarisation vector.

15. Justify the statement 'Most of the electrical machines are working on electromagnetic principles rather than the electrostatic principles'.

Answer: We know that the electric force on a stationary charge Q kept in an electric field $\vec{E}$, given by Coulomb's law is

$$\vec{F}_e = Q\vec{E}$$

Since this force is in the same direction as the electric field, no torque is created.

When a stationary loop is kept in a time-varying magnetic field an emf induced given as

$$\mathcal{E}_S = \oint_C \vec{E} \cdot d\vec{l} = -\int_S \frac{\partial \vec{B}}{\partial t} \cdot d\vec{S}$$

This emf induced by the time-varying current (producing time-varying magnetic field) in a stationary loop is called *transformer emf*. This emf cannot produce any movement of the conductor.

But, when a charge is moving in the presence of a static magnetic field $\vec{B}$, the force on the charge is

$$\vec{F}_m = q\vec{v} \times \vec{B}$$

where q is the amount of charge, $\vec{v}$ is the velocity of movement.

Since this force acts in a direction perpendicular to both $\vec{v}$ and $\vec{B}$, it creates a torque on a closed loop. Thus, if a conducting loop is moving with a velocity $\vec{v}$, then a field is produced and an emf $(\vec{E}_m)$ is induced in the loop, known as *motional emf* or *flux-cutting emf*, given as

$$\vec{E}_m = \frac{\vec{F}_m}{q} = \vec{v} \times \vec{B}$$

This emf can produce the rotation of the conductor.

From the above discussion, we can conclude that most of the electrical machines are working on electromagnetic principles rather than the electrostatic principles.

16. Differentiate between magnetic scalar potential and magnetic vector potential.

Answer: Differences between magnetic scalar potential and magnetic vector potential

Magnetic Scalar Potential (V_m)	Magnetic Vector Potential ($\vec{A}$)
1. It is defined as, $\vec{H} = -\nabla V_m$	1. It is defined as, $\vec{B} = \nabla \times \vec{A}$
2. It is expressed in Ampere.	2. It is expressed in Weber per metre (Wb/m) or in Newton per Ampere (N/A) or in Volt-second per metre (V-s/m).
3. Magnetic scalar potential is defined only in regions where the current is zero $(\vec{J} = 0)$.	3. It is defined in regions with any finite value of current.
4. It satisfies Laplace's equation in source-free regions.	4. It satisfies Poisson's equation in a region with a source.
5. It is not a single-valued function.	5. It is always a single-valued function.
6. The concept of magnetic scalar potential is a useful tool in describing the magnetic field around a current source.	6. The concept of magnetic vector potential is extremely useful for studying radiation in transmission lines, wave guides, antennas, etc.

17. 'Ampere's law is bound to fail for non-steady currents'. Justify the statement. How did Maxwell remove this defect in Ampere's law?

Answer: Inconsistency in Ampere's law for time varying fields:

According to Ampere's circuital law in differential form, we have

$$\nabla \times \vec{H} = \vec{J} \qquad \text{(i)}$$

Taking divergence on both sides,

$$\nabla \cdot (\nabla \times \vec{H}) = \nabla \cdot \vec{J} = 0$$

(Since, the divergence of curl of any vector is zero.)

However, according to the continuity equation of current, we have

$$\nabla \cdot \vec{J} = -\frac{\partial \rho}{\partial t} \neq 0$$

Therefore, we see that Ampere's law is not applicable for time-varying fields. In order to make it compatible for time varying fields, we modify the law.

Modified Ampere's Law for Time Varying Fields

Maxwell removed this defect in Ampere's law by introducing an additional term in Eq. (i), known as *displacement current.*

$$\nabla \times \vec{H} = \vec{J} + \vec{J}_d \qquad \text{(ii)}$$

where $\vec{J}_d$ is to be determined and defined.

Taking divergence on both sides of Eq. (ii),

$$\nabla \cdot (\nabla \times \vec{H}) = \nabla \cdot \vec{J} + \nabla \cdot \vec{J}_d = 0$$

$$\therefore \qquad \nabla \cdot \vec{J}_d = -\nabla \cdot \vec{J} = \frac{\partial \rho}{\partial t} = \frac{\partial}{\partial t}(\nabla \cdot \vec{D}) = \nabla \cdot \frac{\partial \vec{D}}{\partial t}$$

$$\therefore \quad \boxed{\vec{J}_d = \frac{\partial \vec{D}}{\partial t}} \quad \text{(iii)}$$

This $\vec{J}_d$ is known as *displacement current density* and $\vec{J}$ is the *conduction current density* $(\vec{J} = \sigma \vec{E})$.

Now, substituting Eq. (iii) into Eq. (ii), we get

$$\boxed{\nabla \times \vec{H} = \vec{J} + \frac{\partial \vec{D}}{\partial t}}$$

This is the *modified Ampere's law*.

18. Can a magnetic field exist in a good conductor if it is static or time varying? Explain.

Or

Can a static magnetic field exist in a good conductor? Explain.

Answer: Yes, magnetic field can exist in a good conductor whether the field is static or time-varying. The conductivity of a good conductor is high. When the conductor carries some current, it produces some flux which can exist inside the conductor. Due to this flux, a magnetic field exists at any point inside a good conductor.

19. Starting from Maxwell's equations, establish Coulomb's law.

Answer: We will consider a spherical surface of radius r, centre at a point charge Q. Applying Maxwell equation (Gauss' law) in integral form, we have

$$\oint_S \vec{E} \cdot d\vec{S} = \frac{Q}{\varepsilon}$$

By the assumption of spherical symmetry, the integrand is a constant and can be taken out of the integral as

$$4\pi r^2 \hat{a}_r \vec{E} = \frac{Q}{\varepsilon}$$

where $\hat{a}_r$ is a unit vector pointing radially away from the charge. Again by spherical symmetry, $\vec{E}$ is also in radially outward direction, and so we get

$$\vec{E} = \frac{Q}{4\pi\varepsilon r^2}\hat{a}_r$$

If another point charge q is placed on the surface, the force on that charge due to the charge Q is given as,

$$\vec{F} = q\vec{E} = \frac{Qq}{4\pi\varepsilon r^2}\hat{a}_r$$

This is essentially equivalent to Coulomb's law.

20. Show that the equation continuity $\nabla \cdot \vec{J} + \frac{\partial \rho}{\partial t} = 0$ is contained in Maxwell's equation.

Answer: By Maxwell's equation (modified Ampere's law),

$$\nabla \times \vec{H} = \vec{J} + \frac{\partial \vec{D}}{\partial t}$$

Taking divergence on both sides,

$$\nabla \cdot (\nabla \times \vec{H}) = \nabla \cdot \left(\vec{J} + \frac{\partial \vec{D}}{\partial t} \right) = 0$$

∴ $$\nabla \cdot \vec{J} + \nabla \cdot \frac{\partial \vec{D}}{\partial t} = 0$$

or, $$\nabla \cdot \vec{J} + \frac{\partial}{\partial t}(\nabla \cdot \vec{D}) = 0$$

By Maxwell's equation, $$\vec{D} = \frac{\partial \rho}{\partial t}$$

∴ $$\boxed{\nabla \cdot \vec{J} + \frac{\partial \rho}{\partial t} = 0}$$

21. Starting from Maxwell's equation $\nabla \times \vec{E} = -\frac{\partial \vec{B}}{\partial t}$ and $\nabla \times \vec{H} = \vec{J} + \frac{\partial \vec{D}}{\partial t}$, show that $\nabla \cdot \vec{B} = 0$ and $\nabla \cdot \vec{D} = \rho$.

Answer: By Maxwell's equation

$$\nabla \times \vec{E} = -\frac{\partial \vec{B}}{\partial t}$$

Taking divergence on both sides, we get

$$\nabla \cdot (\nabla \times \vec{E}) = 0 = -\nabla \cdot \left(\frac{\partial \vec{B}}{\partial t} \right)$$

∴ $$\nabla \cdot \left(\frac{\partial \vec{B}}{\partial t} \right) = 0 \Rightarrow \frac{\partial}{\partial t}(\nabla \cdot \vec{B}) = 0 \Rightarrow \nabla \cdot \vec{B} = \text{constant}$$

As isolated magnetic monopole does not exist, we have

$$\boxed{\nabla \cdot \vec{B} = 0}$$

$$\nabla \times \vec{H} = \vec{J} + \frac{\partial \vec{D}}{\partial t}$$

Taking divergence on both sides, we get

$$\nabla \cdot (\nabla \times \vec{H}) = 0 = \nabla \cdot \left(\vec{J} + \frac{\partial \vec{D}}{\partial t} \right)$$

∴ $$\nabla \cdot \left(\vec{J} + \frac{\partial \vec{D}}{\partial t} \right) = 0 \Rightarrow \nabla \cdot \vec{J} = -\nabla \cdot \left(\frac{\partial \vec{D}}{\partial t} \right)$$

Again by continuity equation, we know

$$\nabla \cdot \vec{J} + \frac{\partial \rho}{\partial t} = 0$$

$$-\frac{\partial \rho}{\partial t} = -\nabla \cdot \left(\frac{\partial \vec{D}}{\partial t} \right) = -\frac{\partial}{\partial t}(\nabla \cdot \vec{D}) \Rightarrow \nabla \cdot \vec{D} = \rho$$

$$\boxed{\nabla \cdot \vec{D} = \rho}$$

22. Show that a uniform plane wave is a TEM wave.

Answer: An electromagnetic wave which has no electric or magnetic field components in the direction of propagation (all components of ***E*** and ***H*** are perpendicular to the direction of propagation) is called a *transverse electromagnetic* (*TEM*) *wave.*

A wave is said to be a *plane wave*, if

1. The electric field $\vec{E}$ and magnetic field $\vec{H}$ lie in a plane perpendicular to the direction of wave propagation.
2. The fields $\vec{E}$ and $\vec{H}$ are perpendicular to each other.
3. $\vec{E}$ and $\vec{H}$ are uniform in the plane perpendicular to the direction of propagation (i.e. $\vec{E}$ and $\vec{H}$ vary only in the direction of propagation).

Figure 6 shows the propagation of a uniform plane wave.

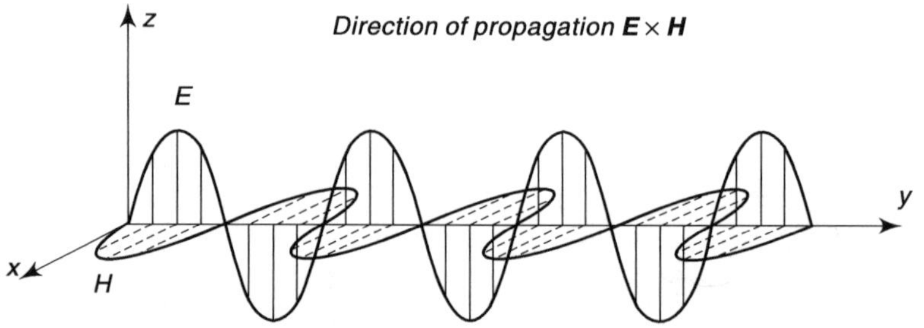

Fig. 6 *Propagation of uniform plane wave*

This uniform plane wave has only a z-component of electric field and an x-component of magnetic field which are both functions of only y.

Since a uniform plane wave has no electric or magnetic field components in the direction of propagation, it is always a *TEM* wave.

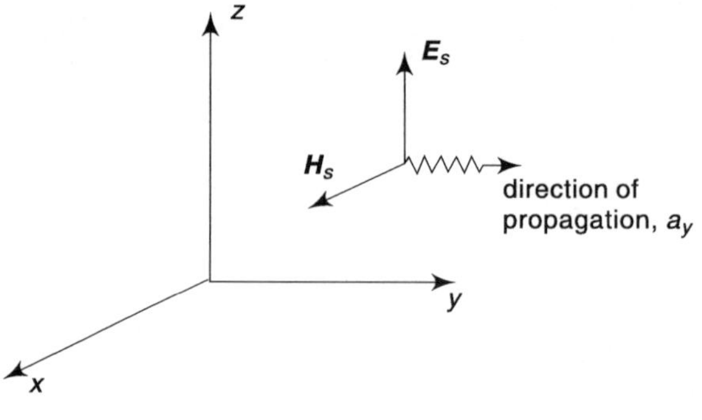

Fig. 7 *Uniform plane wave*

23. Show that the plane electromagnetic waves in free space travel with the velocity *c*.

Answer: The wave propagation in free space is obtained as follows. From three-dimensional wave equation,

$$\nabla^2 \boldsymbol{E}_S = \gamma^2 \boldsymbol{E}_S \qquad \text{and} \qquad \nabla^2 \boldsymbol{H}_S = \gamma^2 \boldsymbol{H}_S$$

where γ is known as the *propagation constant*, given as

$$\gamma = \sqrt{j\omega\mu(\sigma + j\omega\varepsilon)} = (\alpha + j\beta)$$

In free space, $\sigma = 0$, $\varepsilon = \varepsilon_0$, $\mu = \mu_0$, so that

$$\gamma = \sqrt{-\omega^2\mu_0\varepsilon_0} = j\beta$$

The wave equation becomes,

$$\nabla^2 \boldsymbol{E}_S = \gamma^2 \boldsymbol{E}_S = -\omega^2\mu_0\varepsilon_0 \boldsymbol{E}_S \quad \text{and} \quad \nabla^2 \boldsymbol{H}_S = \gamma^2 \boldsymbol{H}_S = -\omega^2\mu_0\varepsilon_0 \boldsymbol{H}_S$$

$\therefore$ Attenuation constant, $\alpha = 0$

$\therefore$ Phase constant, $\beta = \omega\sqrt{\mu_0\varepsilon_0} = \dfrac{\omega}{c}$

So, the velocity of wave propagation is given as

$$v = \frac{\omega}{\beta} = \frac{1}{\sqrt{\mu_0\varepsilon_0}} = c = 3\times 10^8 \text{ m/s, speed of light}$$

This shows that the electromagnetic wave travels with speed of light in the free space.

24. Prove that electric and magnetic fields in a uniform plane wave are perpendicular to each other.

Answer: We consider a uniform plane wave propagating in the z-direction. According to the definition of plane wave there will be no field component in the z-direction.

Now, we consider the dot product of two fields, given as

$$\vec{E}\cdot\vec{H} = E_xH_x + E_yH_y \qquad \{E_z = 0,\ H_z = 0\}$$

Also, by the definition of intrinsic impedance

$$\frac{E_x}{H_y} = \sqrt{\frac{\mu_0}{\varepsilon_0}} \quad \& \quad \frac{E_y}{H_x} = -\sqrt{\frac{\mu_0}{\varepsilon_0}}$$

$$\therefore \qquad \vec{E}\cdot\vec{H} = E_xH_x + E_yH_y = \sqrt{\frac{\mu_0}{\varepsilon_0}}H_yH_x - \sqrt{\frac{\mu_0}{\varepsilon_0}}H_yH_x = 0$$

Dot product of two vectors is zero only when the two vectors are perpendicular to each other.

This shows that the electric and magnetic fields in a uniform plane wave are perpendicular to each other.

25. Show that the ratio of Poynting's vector to energy density is $\leq 3\times 10^8$ m/s.

Answer: Average power can be calculated using Poynting vector as

$$\vec{p}_{\text{ave}} = \frac{1}{2}\text{Re}[\boldsymbol{E}_S \times \boldsymbol{H}_S^*]$$

Now, for a perfect dielectric medium

$\left|\dfrac{E_0}{H_0}\right| = \eta = \sqrt{\dfrac{\mu}{\varepsilon}}$ and if $\vec{E} = E_0\cos(\omega t - \beta x)\hat{a}_E$ and $\vec{H} = H_0\cos(\omega t - \beta x)\hat{a}_H$, then

$$\vec{p}_{\text{ave}} = \frac{1}{2}(E_0)\left(\frac{E_0}{\eta}\right)(\hat{a}_E \times \hat{a}_H) = \frac{1}{2}|E_0|^2 \frac{1}{\sqrt{\mu/\varepsilon}}\hat{a}_n = \frac{1}{2}|E_0|^2 \sqrt{\varepsilon/\mu}\,\hat{a}_n$$

where $\hat{a}_n$ is the unit vector in the direction of wave propagation, normal to both $\vec{E}$ and $\vec{H}$.

Now, the electromagnetic energy density 'u' is given by the sum of the electric energy density $\left[u_E = \frac{1}{2}\vec{E}\cdot\vec{D}\right]$ and the magnetic energy density $\left[u_M = \frac{1}{2}\vec{B}\cdot\vec{H}\right]$.

$$\therefore \quad u = \frac{1}{2}\vec{E}\cdot\vec{D} + \frac{1}{2}\vec{B}\cdot\vec{H} = \frac{1}{2}\varepsilon E^2 + \frac{1}{2}\mu H^2 \qquad (\because\ \vec{D} = \varepsilon\vec{E} \quad \vec{B} = \mu\vec{H})$$

where $\quad \vec{E} = E_0\cos(\omega t - \beta x)\hat{a}_E \quad$ and $\quad \vec{H} = H_0\cos(\omega t - \beta x)\hat{a}_H$

$$\therefore \quad u = \frac{1}{2}\varepsilon E_0^2\cos^2(\omega t - \beta x) + \frac{1}{2}\mu H_0^2\cos^2(\omega t - \beta x) = \frac{1}{2}[\varepsilon E_0^2 + \mu H_0^2]\cos^2(\omega t - \beta x) \quad \textbf{(5.49)}$$

So, the time-averaged energy density in the wave is

$$\langle U_d\rangle = \frac{1}{T}\int_0^T \frac{1}{2}[\varepsilon E_0^2 + \mu H_0^2]\cos^2(\omega t - \beta x)dt = \frac{1}{4}[\varepsilon E_0^2 + \mu H_0^2] \qquad \left(\because\ \langle\cos^2\omega t\rangle = \frac{1}{2}\right)$$

$$= \frac{1}{4}\left[\varepsilon E_0^2 + \mu\frac{E_0^2}{\eta^2}\right] \qquad \left(\because\ \eta = \sqrt{\frac{\mu}{\varepsilon}}\right)$$

$$= \frac{1}{4}\left[\varepsilon E_0^2 + \mu\frac{E_0^2}{\left(\sqrt{\frac{\mu}{\varepsilon}}\right)^2}\right] = \frac{1}{2}\varepsilon|E_0|^2$$

So, the average power is given as

$$\vec{p}_{\text{ave}} = \frac{1}{2}\sqrt{\varepsilon/\mu}\,|E_0|^2\,\hat{a}_n = \frac{1}{2}\varepsilon|E_0|^2\sqrt{\frac{1}{\mu\varepsilon}}\hat{a}_n = \langle U_d\rangle\sqrt{\frac{1}{\mu\varepsilon}}\hat{a}_n$$

$$\therefore \quad \boxed{\vec{p}_{\text{ave}} = \langle U_d\rangle\sqrt{\frac{1}{\mu\varepsilon}}\hat{a}_n = v\langle U_d\rangle\hat{a}_n}$$

where v is the velocity of the wave in the medium.

The above equation shows that the ratio of Poynting vector to energy density is always less than the velocity of light.

$$\frac{\vec{p}_{\text{ave}}}{\langle U_d\rangle} = v = \frac{c}{\sqrt{\mu_r\varepsilon_r}} \le 3\times10^8\text{ m/s}$$

26. Show that in a conductor, the energy is not shared equally between the electric and the magnetic fields.

Or

Show that for an electromagnetic wave propagating in a conducting medium, the density of magnetic energy is greater than that of electric energy.

Answer: From earlier answer, we get for a conductor, the time-averaged electric energy density given as

$$u_E = \frac{1}{2}\varepsilon\,|E_0|^2$$

and the time-averaged magnetic energy density given as

$$u_M = \frac{1}{2}\mu\,|H_0|^2 = \frac{1}{2}\mu\left|\frac{E_0}{\sqrt{\omega\mu/\sigma}}\right|^2 = \frac{1}{2}\frac{\sigma}{\omega}|E_0|^2$$

$\therefore$
$$\frac{u_E}{u_M} = \left(\frac{\omega\varepsilon}{\sigma}\right) << $$

This shows that for an electromagnetic wave propagating in a conducting medium, the density of magnetic energy is greater than that of electric energy, or in other words, the energy is not shared equally between the electric and the magnetic fields.

27. (a) Show that the magnetic field lags the electric field in time by a phase angle in an electromagnetic field propagating in a lousy dielectric medium.

(b) Show that in a conductor, the electric and the magnetic fields are not in phase.

Or

Show that in a conductor, the magnetic field lags the electric field in time but leads the electric field in position.

Or

Show that the magnetic field lags the electric field in time by a phase angle in an electromagnetic field propagating in a conductor. Obtain an expression for the phase angle.

Answer: We will consider the wave propagation along the y-direction, so that the electric field $\vec{E}$ has only z-component, E_z and the magnetic field has only x-component, H_x. Then the solution of wave equations gives

$$E_z(y,t) = E_0 e^{-\alpha y}\cos(\omega t - \beta y)\hat{a}_z \quad \text{and,} \quad H_x(y,t) = H_0 e^{-\alpha y}\cos(\omega t - \beta y)\hat{a}_x$$

where $H_0 = \dfrac{E_0}{\eta}$, η = intrinsic impedance of the medium

For a lousy dielectric medium

Here, we have the condition, $\dfrac{\sigma}{\omega\varepsilon} << 1$

Hence, the intrinsic impedance for the lousy dielectric is given as

$$\eta = \sqrt{\frac{j\omega\mu}{\sigma + j\omega\varepsilon}} = \sqrt{\frac{\mu}{\varepsilon}\left[\frac{1}{1+\dfrac{\sigma}{j\omega\varepsilon}}\right]} = \sqrt{\frac{\mu}{\varepsilon}}\left[1 + j\frac{\sigma}{2\omega\varepsilon}\right] = |\eta|\angle\theta_\eta$$

where
$$|\eta| = \frac{\sqrt{\mu/\varepsilon}}{\left[1+\left(\dfrac{\sigma}{\omega\varepsilon}\right)^2\right]^{1/4}} \qquad \theta_\eta = \frac{1}{2}\tan^{-1}\left(\frac{\sigma}{\omega\varepsilon}\right)$$

Hence, if the electric field is given as, $\vec{E} = E_0 e^{-\alpha y} \cos(\omega t - \beta y)\hat{a}_z$ and, then the magnetic field will be

$$\vec{H} = H_0 e^{-\alpha y} \cos(\omega t - \beta y)\hat{a}_x = \frac{E_0}{|\eta|} e^{-\alpha y} \cos(\omega t - \beta y - \theta_\eta)\hat{a}_x$$

This equation shows that the electric and magnetic fields in a lousy dielectric are not in phase; the magnetic field lags behind the electric field by some angle $\theta_\eta = \frac{1}{2}\tan^{-1}\left(\frac{\sigma}{\omega\varepsilon}\right)$.

For a conductor

In a conductor, $\sigma >> \omega\varepsilon$, i.e., $\frac{\sigma}{\omega\varepsilon} >>$, so that the intrinsic impedance is given as

$$\eta = \sqrt{\frac{j\omega\mu}{\sigma + j\omega\varepsilon}} = \sqrt{\frac{j\omega\mu}{\sigma\left(1 + j\frac{\omega\varepsilon}{\sigma}\right)}} = \sqrt{\frac{j\omega\mu}{\sigma}} = \sqrt{\frac{\omega\mu}{\sigma}}\angle 45^\circ$$

Thus, if the electric filed is given as $\vec{E} = E_0 e^{-\alpha y} \cos(\omega t - \beta y)\hat{a}_z$ and, then the magnetic field will be

$$\vec{H} = H_0 e^{-\alpha y} \cos(\omega t - \beta y)\hat{a}_x = \frac{E_0}{\sqrt{\frac{\omega\mu}{\sigma}}} e^{-\alpha y} \cos(\omega t - \beta y - 45^\circ)\hat{a}_x$$

Therefore, in a conductor the magnetic field lags behind the electric field by an angle of 45°.

28. Show that if a transmission line is precisely half wavelength long, the input impedance is equal to the load impedance.

Answer: If the length of a transmission line is exactly half wavelength $\left(l = \frac{\lambda}{2}\right)$, we have

$$\beta l = \frac{2\pi}{\lambda} \times \frac{\lambda}{2} = \pi$$

$$\therefore \qquad \cos\beta l = \cos\pi = -1 \quad \& \quad \sin\beta l = \sin\pi = 0$$

So, the input impedance is given as

$$\therefore \qquad Z_i = R_0\left(\frac{Z_R\cos\beta l + jR_0\sin\beta l}{R_0\cos\beta l + jZ_R\sin\beta l}\right) = R_0\left(\frac{-Z_R + j0}{-R_0 + j0}\right) = Z_R$$

$$\therefore \qquad \boxed{Z_i = Z_R}$$

Thus, if the transmission line is precisely one-half wavelength long, the input impedance is equal to the load impedance, regardless of Z_0 and β.

29. Show that for a lossless transmission line the impedance of a line repeats over every $\frac{\lambda}{2}$ distance.

Answer: The input impedance at any point on the transmission line is given by the ratio of voltage to current at that point.

Inserting the expressions for the phasor voltage and current $[(V_S(z)]$ and $[I_S(z)]$ from the original form of the transmission line equations gives

$$Z_{in}(z) = \frac{V_S(z)}{I_S(z)} = Z_0 \frac{V_{S0}^+ e^{-\gamma z} + V_{S0}^- e^{\gamma z}}{V_{S0}^+ e^{-\gamma z} - V_{S0}^- e^{\gamma z}}$$

Inserting the expressions of voltage coefficients in terms of the load voltage and load current, we have

$$V_{S0}^+ = \left(\frac{V_R + Z_0 I_R}{2}\right) e^{\gamma l} = \frac{I_R}{2}(Z_R + Z_0) e^{\gamma l}$$

$$V_{S0}^- = \left(\frac{V_R - Z_0 I_R}{2}\right) e^{-\gamma l} = \frac{I_R}{2}(Z_R - Z_0) e^{-\gamma l}$$

$$\therefore \quad Z_{in}(z) = Z_0 \frac{V_{S0}^+ e^{-\gamma z} + V_{S0}^- e^{\gamma z}}{V_{S0}^+ e^{-\gamma z} - V_{S0}^- e^{\gamma z}} = Z_0 \frac{e^{\gamma(l-z)}(Z_R + Z_0) + e^{-\gamma(l-z)}(Z_R - Z_0)}{e^{\gamma(l-z)}(Z_R + Z_0) - e^{-\gamma(l-z)}(Z_R - Z_0)}$$

$$= Z_0 \frac{Z_R[e^{\gamma(l-z)} + e^{-\gamma(l-z)}] + Z_0[e^{\gamma(l-z)} - e^{-\gamma(l-z)}]}{Z_R[e^{\gamma(l-z)} - e^{-\gamma(l-z)}] + Z_0[e^{\gamma(l-z)} + e^{-\gamma(l-z)}]}$$

Now, we know that $\sinh \gamma l = \dfrac{e^{\gamma l} - e^{-\gamma l}}{2}$ & $\cosh \gamma l = \dfrac{e^{\gamma l} + e^{-\gamma l}}{2}$

$$\therefore \quad Z_{in}(z) = Z_0 \frac{Z_R[e^{\gamma(l-z)} + e^{-\gamma(l-z)}] + Z_0[e^{\gamma(l-z)} - e^{-\gamma(l-z)}]}{Z_R[e^{\gamma(l-z)} - e^{-\gamma(l-z)}] + Z_0[e^{\gamma(l-z)} + e^{-\gamma(l-z)}]}$$

$$= Z_0 \frac{Z_R \cosh[\gamma(l-z)] + Z_0 \sinh[\gamma(l-z)]}{Z_R \sinh[\gamma(l-z)] + Z_0 \cosh[\gamma(l-z)]}$$

$$= Z_0 \frac{Z_R + Z_0 \tanh[\gamma(l-z)]}{Z_0 + Z_R \tanh[\gamma(l-z)]}$$

Thus, the input impedance at any point along a general transmission line is given as

$$\boxed{Z_{in}(z) = Z_0 \frac{Z_R + Z_0 \tanh[\gamma(l-z)]}{Z_0 + Z_R \tanh[\gamma(l-z)]}}$$

For a lossless line, Z_0 is purely real, i.e., $Z_0 = R_0$, $\gamma = j\beta$. The hyperbolic tangent function reduces to

$$\tanh[\gamma(l-z)] = \tanh[j\beta(l-z)] = j \tan[\beta(l-z)]$$

So, the input impedance at any point along a lossless transmission line becomes

$$\boxed{Z_{in}(z) = R_0 \frac{Z_R + jR_0 \tan[\beta(l-z)]}{R_0 + jZ_R \tan[\beta(l-z)]}}$$

Two special cases are considered:

1. **Open-circuited lossless line**

   Here, $Z_R \to \infty$ and $\Gamma_L = 1$.

Input impedance becomes

$$\lim_{Z_R \to \infty} Z_{in}(z) = \lim_{Z_R \to \infty} \left\{ R_0 \frac{Z_R + jR_0 \tan[\beta(l-z)]}{R_0 + jZ_R \tan[\beta(l-z)]} \right\} = \lim_{Z_R \to \infty} \left\{ R_0 \frac{1 + j\frac{R_0}{Z_R}\tan[\beta(l-z)]}{\frac{R_0}{Z_R} + j\tan[\beta(l-z)]} \right\}$$

$$= \left\{ R_0 \frac{1}{j\tan[\beta(l-z)]} \right\} = -jR_0 \cot[\beta(l-z)] = Z_{OC}(z)$$

2. **Short-circuited lossless line**

Here, $Z_R \to 0$ and $\Gamma_L = -1$.

Input impedance becomes

$$\lim_{Z_R \to 0} Z_{in}(z) = \lim_{Z_R \to 0} \left\{ R_0 \frac{Z_R + jR_0 \tan[\beta(l-z)]}{R_0 + jZ_R \tan[\beta(l-z)]} \right\} = R_0 \frac{jR_0 \tan[\beta(l-z)]}{R_0}$$

$$= jR_0 \tan[\beta(l-z)] = Z_{SC}(z)$$

The variations of these impedances with the length of the line are shown in Fig. 8.

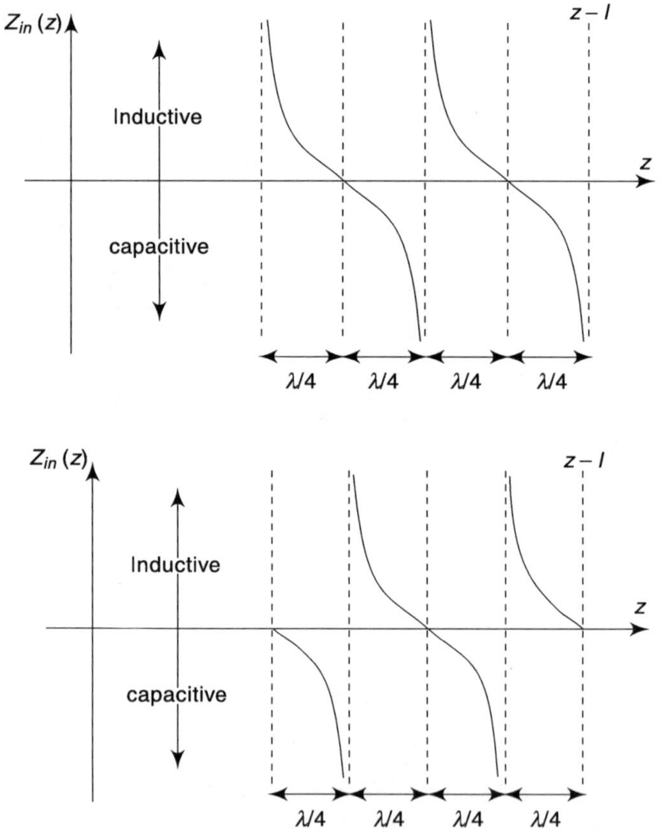

Fig. 8 *Input impedance of a lossless transmission line (a) when open and (b) when short*

Figure 8 reveals that the impedance of a lossless transmission line repeats over every $\lambda/2$ distance.

30. Show that a short-circuited lossless transmission line can offer reactances of any value by simply changing the length of the line.

Answer: For lossless transmission line, $R = 0$ and $G = 0$.

$$\therefore \quad \alpha = 0 \quad \& \quad \beta = \omega\sqrt{LC}$$

$$\therefore \quad \gamma = \sqrt{(R + j\omega L)(G + j\omega C)} = \alpha + j\beta = j\omega\sqrt{LC}$$

$$\therefore \quad \beta = \omega\sqrt{LC}$$

The input impedance is given as

$$Z_i = R_0\left(\frac{Z_R \cos \beta l + jR_0 \sin \beta l}{R_0 \cos \beta l + jZ_R \sin \beta l}\right) = R_0\left(\frac{Z_R + jR_0 \tan \beta l}{R_0 + jZ_R \tan \beta l}\right)$$

If the line is short-circuited at the receiving end, i.e., $Z_R = 0$, then the input impedance becomes

$$Z_i = Z_{SC} = jR_0\left(\frac{\sin \beta l}{\cos \beta l}\right) = jR_0 \tan \beta l$$

$$\boxed{Z_{SC} = jR_0 \tan \beta l}$$

This equation shows that a short-circuited lossless line can offer reactances of any value simply by changing the electrical length (βl) of the line.

31. Show that a lossless transmission line with length $\left(l << \frac{\lambda}{2\pi}\right)$ behaves

(a) like a capacitor when the receiving end is open-circuited, and

(b) like an inductor when receiving end is short-circuited.

Answer: For lossless transmission line, $R = 0$ and $G = 0$.

$$\therefore \quad \alpha = 0 \quad \text{and} \quad \beta = \omega\sqrt{LC}$$

$$\therefore \quad \gamma = \sqrt{(R + j\omega L)(G + j\omega C)} = \alpha + j\beta = j\omega\sqrt{LC}$$

$$\therefore \quad \beta = \omega\sqrt{LC}$$

The input impedance is given as

$$Z_i = R_0\left(\frac{Z_R \cos \beta l + jR_0 \sin \beta l}{R_0 \cos \beta l + jZ_R \sin \beta l}\right) = R_0\left(\frac{Z_R + jR_0 \tan \beta l}{R_0 + jZ_R \tan \beta l}\right)$$

(a) If the line is open-circuited at the receiving end, i.e., $Z_R = \infty$, then the input impedance becomes

$$Z_i = Z_{SC} = R_0\left(\frac{1}{j \tan \beta l}\right) = -jR_0 \cot \beta l$$

$$\boxed{Z_{OC} = -jR_0 \cot \beta l}$$

Since $l << \frac{\lambda}{2\pi} \Rightarrow \beta l << 1, \cot \beta l \approx \frac{1}{\tan \beta l} \approx \frac{1}{\beta l}$, and the input impedance becomes

$$Z_{OC} = -jR_0 \cot \beta l \approx -\frac{jR_0}{\beta l} = -j\sqrt{\frac{L}{C}} \times \frac{1}{\omega\sqrt{LC} \times l} = -\frac{j}{\omega Cl}$$

This equation shows that the line behaves like a capacitor.

(b) If the line is short-circuited at the receiving end, i.e., $Z_R = 0$, then the input impedance becomes

$$Z_i = Z_{SC} = jR_0\left(\frac{\sin \beta l}{\cos \beta l}\right) = jR_0 \tan \beta l$$

$$\boxed{Z_{SC} = jR_0 \tan \beta l}$$

Since $l << \frac{\lambda}{2\pi} \Rightarrow \beta l << 1, \tan \beta l \approx \beta l,$ and the input impedance becomes,

$$Z_{SC} = jR_0 \tan \beta l \approx jR_0 \beta l = j\sqrt{\frac{L}{C}} \times \omega\sqrt{LC} \times l = j\omega Ll$$

This equation shows that the line behaves like an inductor.

32. Show that a lossless $\frac{\lambda}{8}$ length line terminated as open circuit, behaves like a capacitor.

Answer: From earlier answer, we get the input impedance of a lossless line given as

$$Z_i = R_0\left(\frac{Z_R \cos \beta l + jR_0 \sin \beta l}{R_0 \cos \beta l + jZ_R \sin \beta l}\right) = R_0\left(\frac{Z_R + jR_0 \tan \beta l}{R_0 + jZ_R \tan \beta l}\right)$$

In this case, the receiving end is kept opened. So, the receiving end current is zero.

$$\therefore \quad I_R = \left(I_S \cos \beta l - j\frac{V_S}{R_0} \sin \beta l\right) = 0$$

Hence, the input impedance is

$$Z_i = \frac{V_S}{I_S} = Z_{OC} = -jR_0\left(\frac{\cos \beta l}{\sin \beta l}\right) = -jR_0 \cot \beta l$$

$$\boxed{Z_{OC} = -jR_0 \cot \beta l}$$

If the length of a transmission line is $l = \frac{\lambda}{8}$, we have

$$\beta l = \frac{2\pi}{\lambda} \times \frac{\lambda}{8} = \frac{\pi}{4}$$

$$\therefore \quad \cot \beta l = \cot \frac{\pi}{4} = 1$$

So, the input impedance is given as

$$\boxed{Z_{OC} = -jR_0}$$

This negative reactance shows that a lossless $\frac{\lambda}{8}$ length line terminated as open circuit, behaves like a capacitor.

33. Show that if a transmission line is precisely one-quarter wavelength long, the input impedance is inversely proportional to the load impedance.

Answer: If the length of a transmission line is exactly one-quarter wavelength $\left(l = \frac{\lambda}{4}\right)$, we have

$$\beta l = \frac{2\pi}{\lambda} \times \frac{\lambda}{4} = \frac{\pi}{2}$$

$$\therefore \qquad \cos \beta l = \cos \frac{\pi}{2} = 0 \qquad \& \qquad \sin \beta l = \sin \frac{\pi}{2} = 1$$

So, the input impedance is given as

$$\therefore \qquad Z_i = R_0 \left(\frac{Z_R \cos \beta l + jR_0 \sin \beta l}{R_0 \cos \beta l + jZ_R \sin \beta l} \right) = R_0 \left(\frac{Z_R \times 0 + jR_0 \times 1}{R_0 \times 0 + jZ_R \times 1} \right) = \frac{R_0^2}{Z_R}$$

$$\therefore \qquad \boxed{Z_i = \frac{R_0^2}{Z_R}}$$

Thus, if the transmission line is precisely one-quarter wavelength long, the input impedance is inversely proportional to the load impedance.

If $Z_R = 0$, i.e., the receiving end is short-circuited, the input impedance is infinity ($Z_i = \infty$); thus, a quarter-wave transmission line transforms a short-circuit into an open-circuit and vice versa.

34. Show that an open-circuited transmission line with length less than $\frac{\lambda}{4}$ is capacitive.

Answer: We know that the input impedance of a lossless line is given as

$$Z_i = R_0 \left(\frac{Z_R \cos \beta l + jR_0 \sin \beta l}{R_0 \cos \beta l + jZ_R \sin \beta l} \right) = R_0 \left(\frac{Z_R + jR_0 \tan \beta l}{R_0 + jZ_R \tan \beta l} \right)$$

If the receiving end is open circuited, $Z_R = \infty$ and the input impedance becomes

$$Z_{OC} = R_0 \left(\frac{\cos \beta l}{j \sin \beta l} \right) = -jR_0 \cot \beta l$$

$$\boxed{Z_{OC} = -jR_0 \cot \beta l}$$

If the length of a transmission line is $l < \frac{\lambda}{4}$, we have

$$\beta l < \frac{\pi}{2} \qquad \left\{ \because \beta l = \frac{2\pi}{\lambda} \times \frac{\lambda}{4} = \frac{\pi}{2} \right\}$$

Hence, cot βl is finite positive. So, the input impedance is always a negative reactance.

This negative reactance shows that an open-circuited lossless line with length less than $\frac{\lambda}{4}$ is capacitive.

35. Show that a short-circuited transmission line with length less than $\frac{\lambda}{4}$ is inductive.

Answer: We know that the input impedance of a lossless line is given as

$$Z_i = R_0\left(\frac{Z_R \cos\beta l + jR_0 \sin\beta l}{R_0 \cos\beta l + jZ_R \sin\beta l}\right) = R_0\left(\frac{Z_R + jR_0 \tan\beta l}{R_0 + jZ_R \tan\beta l}\right)$$

If the receiving end is short circuited, $Z_R = 0$ and the input impedance becomes

$$Z_{SC} = R_0\left(\frac{j\sin\beta l}{\cos\beta l}\right) = jR_0 \tan\beta l$$

$$\boxed{Z_{SC} = jR_0 \tan\beta l}$$

If the length of a transmission line is $l < \frac{\lambda}{4}$, we have

$$\beta l < \frac{\pi}{2} \qquad \left\{\because \beta l = \frac{2\pi}{\lambda} \times \frac{\lambda}{4} = \frac{\pi}{2}\right\}$$

Hence, tan βl is finite positive. So, the input impedance is always a positive reactance.

This positive reactance shows that a short-circuited lossless line with length less than $\frac{\lambda}{4}$ is inductive.

36. Show that if the load in a transmission line is purely reactive, then the input impedance will also be purely reactive, regardless of the length of the line.

Answer: When the load in a transmission line is purely reactive, i.e., $Z_L = jX_L$, the input impedance is given as

$$\therefore \qquad Z_i = R_0\left(\frac{Z_L \cos\beta l + jR_0 \sin\beta l}{R_0 \cos\beta l + jZ_L \sin\beta l}\right)$$

$$= R_0\left(\frac{jX_L \cos\beta l + jR_0 \sin\beta l}{R_0 \cos\beta l + j\cdot jX_L \sin\beta l}\right)$$

$$= jR_0\left(\frac{X_L \cos\beta l + R_0 \sin\beta l}{R_0 \cos\beta l - X_L \sin\beta l}\right)$$

Thus, if the load is purely reactive, then the input impedance will also be purely reactive, regardless of the length of the line.

37. In a waveguide, show that the wave impedance is real above a certain frequency and imaginary below that. What is the implication of the result?

Answer: The intrinsic impedance of *TE* or *TM* wave in parallel plane waveguide is obtained as

$$\eta_{TE} = \eta_{TM} = \left|\frac{E_y}{H_x}\right| = \frac{j\omega\mu}{\gamma}$$

Now, the propagation constant is given as

$$\gamma = \sqrt{\left(\frac{n\pi}{d}\right)^2 - \omega^2 \mu\varepsilon}$$

The cut-off frequency of the waveguide is given as

$$f_c = \frac{n}{2d}\frac{1}{\sqrt{\mu\varepsilon}}$$

When the frequency is below the cut-off frequency ($f < f_c$):
The propagation constant γ is real, i.e., $\gamma = \alpha, \quad \beta = 0$.

So, the intrinsic impedance is given as

$$\eta_{TE} = \eta_{TM} = \left|\frac{E_y}{H_x}\right| = \frac{j\omega\mu}{\gamma} = \frac{j\omega\mu}{\sqrt{\left(\frac{n\pi}{d}\right)^2 - \omega^2\mu\varepsilon}} = \frac{j\sqrt{\mu/\varepsilon}}{\sqrt{\frac{f_c^2}{f^2} - 1}} = j\frac{\eta_0}{\sqrt{\frac{f_c^2}{f^2} - 1}}$$

where $\eta_0 = \sqrt{\mu_0/\varepsilon_0}$ is the intrinsic impedance of free space.

Since $f < f_c$; $\frac{f_c}{f} > 1$ and hence, the intrinsic impedance is imaginary.

When the frequency is above the cut-off frequency ($f > f_c$):
The propagation constant will be imaginary, i.e., $\gamma = j \quad \beta, \alpha = 0$. This is given as

$$\gamma = \sqrt{\left(\frac{n\pi}{d}\right)^2 - \omega^2\mu\varepsilon} = \sqrt{\mu\varepsilon}\sqrt{\left(\frac{n\pi}{d\sqrt{\mu\varepsilon}}\right)^2 - \omega^2} = \omega\sqrt{\mu\varepsilon}\sqrt{\frac{\omega_c^2}{\omega^2} - 1} = \omega\sqrt{\mu\varepsilon}\sqrt{\frac{f_c^2}{f^2} - 1}$$

$$= j\omega\sqrt{\mu\varepsilon}\sqrt{1 - \frac{f_c^2}{f^2}}$$

So, the intrinsic impedance is given as

$$\eta_{TE} = \eta_{TM} = \left|\frac{E_y}{H_x}\right| = \frac{j\omega\mu}{\gamma} = \frac{j\omega\mu}{j\omega\sqrt{\mu\varepsilon}\sqrt{1 - \frac{f_c^2}{f^2}}} = \frac{\sqrt{\frac{\mu}{\varepsilon}}}{\sqrt{1 - \frac{f_c^2}{f^2}}} = \frac{\eta_0}{\sqrt{1 - \frac{f_c^2}{f^2}}}$$

Since $f > f_c$; $\frac{f_c}{f} < 1$ and hence, the intrinsic impedance is real.

This shows that the wave impedance is real above a certain frequency and imaginary below that.

NOTE

Similar types of equations can be derived in case of rectangular waveguide, where the propagation constant and cut-off frequency are given as

Propagation constant, $$\gamma = \sqrt{\left(\frac{m\pi}{a}\right)^2 + \left(\frac{n\pi}{b}\right)^2 - \omega^2\mu\varepsilon}$$

Cut-off frequency, $$f_c = \frac{v_0}{2}\sqrt{\left(\frac{m}{a}\right)^2 + \left(\frac{n}{b}\right)^2}$$

The implications of an imaginary value of wave impedance (i.e., reactance) below the cut-off frequency implies that no power is absorbed by the waveguide and no wave propagation takes place.

On the other hand, a real value of wave impedance (i.e., resistance) above the cut-off frequency implies that real power is absorbed by the waveguide and wave propagation takes place.

38. Evaluate the ratio of the area of a circular waveguide to that of a rectangular one if both are to have the same cut-off frequency for dominant mode.

Answer: Let r be radius of the circular waveguide and a be the larger dimension of the rectangular waveguide.

From Tables 7.1 and 7.2, we understand that the dominant mode in circular waveguide is TE_{11} mode. For this mode, the cut-of frequency is given as

$$f_c = \frac{v_0}{2\pi}h_{nm} = \frac{v_0}{2\pi r}rh_{nm} = \frac{v_0}{2\pi r}\times 1.84$$

For TE_{10} mode rectangular waveguide, the cut-off frequency is given as

$$f_c = \frac{v_0}{2}\sqrt{\left(\frac{m}{a}\right)^2 + \left(\frac{n}{b}\right)^2} = \frac{v_0}{2a}$$

Since both will have the same cut-off frequency, we get

$$\frac{v_0}{2\pi r}\times 1.84 = \frac{v_0}{2a}$$

or $$\frac{r}{a} = 0.586$$

The area of the circular waveguide is $A_c = \pi r^2$

For standard rectangular waveguide $a : b = 2 : 1$, so that the area of the rectangular waveguide is $A_r = a \times \frac{a}{2} = \frac{a^2}{2}$.

$\therefore$ $$\frac{A_c}{A_r} = \frac{\pi r^2}{a^2/2} = 2\pi\left(\frac{r}{a}\right)^2 = 2\pi \times (0.586)^2 = 2.155$$

39. Show that the attenuation constant for TM_1 wave in parallel plane waveguide is a minimum at a frequency which is $\sqrt{3}$ times the cut-off frequency.

Answer: For *TM* mode, the attenuation constant is given as

$$\alpha_{TM} = \frac{2}{\eta_0}\sqrt{\frac{\pi f \mu/\sigma}{1-\left(\frac{f_c}{f}\right)^2}}$$

For the attenuation constant to be minimum,

$$\frac{d\alpha}{df} = 0 \Rightarrow \frac{d}{df}\left[\frac{2}{\eta_0}\sqrt{\frac{\pi f \mu/\sigma}{1-\left(\frac{f_c}{f}\right)^2}}\right] = 0 \Rightarrow \frac{d}{df}\left[\frac{\sqrt{f}}{\sqrt{1-\left(\frac{f_c}{f}\right)^2}}\right] = 0$$

$$\Rightarrow \frac{\sqrt{1-\left(\frac{f_c}{f}\right)^2}\left(\frac{1}{2}\frac{1}{\sqrt{f}}\right) - \sqrt{f}\,\frac{1}{2}\times\frac{1}{\sqrt{1-\left(\frac{f_c}{f}\right)^2}}\times\left(2\frac{f_c^2}{f^3}\right)}{1-\left(\frac{f_c}{f}\right)^2} = 0 \Rightarrow \frac{1}{2}\left[1-\left(\frac{f_c}{f}\right)^2\right] = \left(\frac{f_c}{f}\right)^2$$

$$\Rightarrow \frac{3}{2}\left(\frac{f_c}{f}\right)^2 = \frac{1}{2}$$

$$\therefore \qquad \boxed{f = \sqrt{3}f_c}$$

40. (a) Show that for a rectangular waveguide, the dominant mode exhibits a minimum attenuation due to conductor loss at a certain frequency. Find that frequency in terms of the cut-off frequency of the mode.

(b) If $a = 3b$, find the frequency for minimum attenuation.

(c) For a square waveguide, show that attenuation α_c is minimum for TE_{10} mode when $f = 2.962\, f_c$.

Answer:

(a) For a TE_{10} mode rectangular wave, the attenuation constant is given as

$$\alpha_{C_{TE10}} = \frac{R_S}{\eta_0 b}\frac{\left(1+\frac{2b}{a}\frac{f_c^2}{f^2}\right)}{\sqrt{1-\frac{f_c^2}{f^2}}}$$

Substituting the value of the surface resistance $R_S = \sqrt{\frac{\pi f \mu}{\sigma}}$, we have

$$\alpha_{C_{TE10}} = \frac{\sqrt{\frac{\pi f \mu}{\sigma}}}{\eta_0 b}\frac{\left(1+\frac{2b}{a}\frac{f_c^2}{f^2}\right)}{\sqrt{1-\frac{f_c^2}{f^2}}} = \frac{\sqrt{\frac{\pi \mu}{\sigma}}}{\eta_0 b}\frac{\left(\sqrt{f}+\frac{2b}{a}\frac{f_c^2}{f^{\frac{3}{2}}}\right)}{\sqrt{1-\frac{f_c^2}{f^2}}} = K\frac{\left(\sqrt{f}+\frac{2b}{a}\frac{f_c^2}{f^{\frac{3}{2}}}\right)}{\sqrt{1-\frac{f_c^2}{f^2}}}$$

where $K = \dfrac{\sqrt{\dfrac{\pi\mu}{\sigma}}}{\eta_0 b}$ is constant.

For the attenuation constant to be minimum,

$$\frac{d\alpha_C}{df} = 0$$

$$\frac{d}{df}\left[\frac{\left(\sqrt{f} + \frac{2b}{a}\frac{f_c^2}{f^{\frac{3}{2}}}\right)}{\sqrt{1 - \frac{f_c^2}{f^2}}}\right] = 0$$

or $$\frac{\sqrt{1-\left(\frac{f_c}{f}\right)^2} \times \left[\frac{1}{2\sqrt{f}} - \frac{2b}{a} \times \frac{3}{2}\frac{f_c^2}{f^{5/2}}\right] - \left(\sqrt{f} + \frac{2b}{a}\frac{f_c^2}{f^{3/2}}\right) \times \frac{1}{2}\frac{1}{\sqrt{1-\left(\frac{f_c}{f}\right)^2}} \times \left(2\frac{f_c^2}{f^3}\right)}{1-\left(\frac{f_c}{f}\right)^2} = 0$$

or $$\left(1 - \frac{f_c^2}{f^2}\right)\left(\frac{1}{2\sqrt{f}} - \frac{3b}{a}\frac{f_c^2}{f^{5/2}}\right) - \left(\sqrt{f} + \frac{2b}{a}\frac{f_c^2}{f^{3/2}}\right)\frac{f_c^2}{f^3} = 0$$

or $$\frac{1}{2\sqrt{f}} - \frac{1}{2}\frac{f_c^2}{f^{5/2}} - \frac{3b}{a}\frac{f_c^2}{f^{5/2}} + \frac{3b}{a}\frac{f_c^4}{f^{9/2}} - \frac{f_c^2}{f^{5/2}} - \frac{2b}{a}\frac{f_c^4}{f^{9/2}} = 0$$

or $$\frac{1}{2} - 3\left(\frac{1}{2} + \frac{b}{a}\right)\frac{f_c^2}{f^2} + \frac{b}{a}\frac{f_c^4}{f^4} = 0$$

or $$af^4 - 3(a+2b)f_c^2 f^2 + 2bf_c^4 = 0$$

Solving this equation and taking only the positive root, we get

$$f^2 = \frac{3(a+2b)f_c^2 + \sqrt{9(a+2b)^2 f_c^4 - 8abf_c^4}}{2a}$$

$$= \left[\frac{3(a+2b)}{2a} + \sqrt{\left\{\frac{3(a+2b)}{2a}\right\}^2 - \frac{2b}{a}}\right] f_c^2$$

∴ $$\boxed{f = \left[\frac{3(a+2b)}{2a} + \sqrt{\left\{\frac{3(a+2b)}{2a}\right\}^2 - \frac{2b}{a}}\right]^{\frac{1}{2}} f_c}$$

This is the frequency corresponding to the minimum attenuation due to conductor loss.

(b) If $a = 3b$, we have

$$f = \left[\frac{3(3b + 2b)}{2 \times 3b} + \sqrt{\left\{\frac{3(3b + 2b)}{2 \times 3b}\right\}^2 - \frac{2b}{3b}}\right]^{\frac{1}{2}} f_c = f_c\sqrt{\frac{5}{2} + \sqrt{\frac{25}{4} - \frac{2}{3}}} = 2.205 f_c$$

$\therefore$ $\boxed{f = 2.205 f_c}$

(c) For square waveguide, $a = b$, so that we get

$$f = \left[\frac{3(a + 2a)}{2a} + \sqrt{\left\{\frac{3(a + 2a)}{2a}\right\}^2 - \frac{2a}{a}}\right]^{\frac{1}{2}} f_c = f_c\sqrt{\frac{9}{2} + \sqrt{\frac{81}{4} - 2}} = 2.962 f_c$$

$\therefore$ $\boxed{f = 2.962 f_c}$

BIBLIOGRAPHY

- Matthew N. O. Sadiku, *Principles of Electromagnetics*, Oxford University Press
- Bhag Singh Guru, Hüseyin R. Hiziroglu, *Electromagnetic Field Theory Fundamentals*, Cambridge University Press
- Nathan Ida, *Engineering Electromagnetics*, Springer Publishers
- Rajeev Bansal (edited by), *Fundamentals of Engineering Electromagnetics*, CRC Press
- Canstantine A. Balanis, *Advanced Engineering Electromagnetics*, Wiley India Pvt. Ltd.
- Canstantine A. Balanis, *Antenna Theory – Analysis and Design*, Wiley India Pvt. Ltd.
- Robert S. Elliott, *Antenna Theory and Design*, Wiley India Pvt. Ltd.
- Edward C. Jordan, Keith G. Balmain, *Electromagnetic Waves & Radiating Systems*, PHI Learning Pvt. Ltd.
- Ashutosh Pramanik, *Electromagnetism – Theory & Applications*, PHI Learning Pvt. Ltd.
- David J. Griffiths, *Introduction to Electrodynamics*, PHI Learning Pvt. Ltd.
- R. L. Yadava, *Antenna and Wave Propagation*, PHI Learning Pvt. Ltd.
- Saroj K. Dash, Smruti R. Khuntia, *Fundamentals of Electromagnetic Theory*, PHI Learning Pvt. Ltd.
- John D. Kraus, D. A. Fleisch, *Electromagnetics with Applications*, Tata McGraw Hill Education Pvt. Ltd.
- John D. Kraus, Ronald J. Marhefka, *Antennas for All Applications*, Tata McGraw Hill Education Pvt. Ltd.
- William H. Hayt, Jr., John A. Buck, *Engineering Electromagnetics*, Tata McGraw Hill Education Pvt. Ltd.
- Joseph A. Edminister, *Schaum's Outline of Theory and Problems of Electromagnetics*, Schaum's Outline Series, Tata McGraw Hill Education Pvt. Ltd.
- Clayton R. Paul, Keith W. Whites, Syed A. Nasar, *Introduction to Electromagnetic Fields*, Tata McGraw Hill Education Pvt. Ltd.
- R. K. Shevgaonkar, *Electromagnetic Waves*, Tata McGraw Hill Education Pvt. Ltd.
- Edward M. Purcell, *Electricity and Magnetism*, McGraw Hill Inc.
- David K. Cheng, *Field and Wave Electromagnetics*, Addison-Wesley Publishing Company
- B. B. Loud, *Electromagnetics*, New Age International Publishers
- S. Sivanagaraju, C. Srinivasa Rao, *Electromagnetic Fields*, New Age International Publishers
- Meena Kumari R., Subasri R., *Electromagnetic Fields*, New Age International Publishers
- Prantosh Chakraborty, *Electricity and Magnetism,* New Age International Publishers
- Rajeswari Chatterjee, *Antenna Theory and Practice*, New Age International Publishers
- G. S. N. Raju, *Electromagnetic Field Theory and Transmission Lines*, Pearson Education
- G. S. N. Raju, *Antenna and Wave Propagation*, Pearson Education
- N. N. Rao, *Elements of Engineering Electromagnetics*, Pearson Education

- A. K. Saxena, *Electromagnetic Theory and Applications*, Narosa Publishing House Pvt. Ltd.
- Liang Chi Shen, Jin Au Kong, Amalendu Patnaik, *Engineering Electromagnetics*, Cengage Learning India Pvt. Ltd.
- Deepak Sood, *Fields and Waves – A Fundamental Approach*, University Science Press
- K. D. Prasad, *Antenna and Wave Propagation*, Satya Prakashan
- K. K. Shah, *Introduction to Electromagnetics*, Dhanpat Rai & Co. Publishers
- D. Chattopadhyay, P. C. Rakshit, *Electricity and Magnetism*, New Central Book Agency
- Parul Dawar, *Concepts in Electromagnetic Field Theory*, S. K. Kataria & Sons
- R. Gowri, *Electromagnetic Fields & Waves*, S. K. Kataria & Sons
- Vipin Kumar, Rajesh Mishra, *Electromagnetic Field Theory*, S. K. Kataria & Sons
- P. V. Gupta, *Introductory Course in Electromagnetic Fields*, Dhanpat Rai Publishing Company (P) Ltd.
- S. P. Seth, *Elements of Electromagnetic Fields*, Dhanpat Rai Publishing Company (P) Ltd.
- K. A. Gangadhar, P. M. Ramanathan, *Field Theory*, Khanna Publishers
- D. V. Prasad, *Electromagnetic Fields Waves and Antennas*, Khanna Publishers
- G. Prasad, J. Prasad, J. Reddy Electromagnetic Waves and Transmission Lines, Scitech Publication India Ltd.
- K. A. Bakshi, A. V. Bakshi, *Electromagnetic Theory*, Technical Publications, Pune
- K. A. Bakshi, A. V. Bakshi, U. A. Bakshi, *Antenna and Wave Propagation*, Technical Publications, Pune
- K. A. Bakshi, A. V. Bakshi, *Transmission Lines and Waveguides*, Technical Publications, Pune

INDEX

Made in United States
Orlando, FL
04 November 2026

90209900R10409